Acid rain 2000

*Proceedings from the 6th International Conference on Acidic Deposition:
Looking back to the past and thinking of the future
Tsukuba, Japan, 10–16 December 2000*

Volume III / III

*Conference Statement
Plenary and Keynote Papers*

Edited by

KENICHI SATAKE, Editor-in-Chief *National Institute for Environmental Studies, Japan*
JUNKO SHINDO *National Institute for Agro-Environmental Sciences, Japan*
TAKEJIRO TAKAMATSU *National Institute for Environmental Studies, Japan*
TAKANORI NAKANO *Institute of Geoscience, University of Tsukuba, Japan*
SHIGERU AOKI *Tokyo National Research Institute of Cultural Properties, Japan*
TSUTOMU FUKUYAMA *National Institute for Environmental Studies, Japan*
SHIRO HATAKEYAMA *National Institute for Environmental Studies, Japan*
KAZUKAMASA IKUTA *National Research Institute of Aquaculture, Japan*
MUNETSUGU KAWASHIMA *Shiga University, Japan*
YOSHIHISA KOHNO *Central Research Institute of Electric Power Industry, Japan*
SATORU KOJIMA *Tokyo Woman's Christian University, Japan*
KENTARO MURANO *National Institute for Environmental Studies, Japan*
TOSHIICHI OKITA *Obirin University, Japan, ret.*
HIROSHI TAODA *Forestry and Forest Products Research Institute, Japan*
KINICHI TSUNODA *Gunma University, Japan*
MAKOTO TSURUMI *Hirosaki University, Japan*

Reprinted from *Water, Air, and Soil Pollution* 130: 1–4, 2001

SPRINGER SCIENCE+BUSINESS MEDIA, LLC

A C.I.P. Catalogue record for this book is available from the Library of Congress.

ISBN 978-94-010-3733-4   ISBN 978-94-007-0810-5 (eBook)
DOI 10.1007/978-94-007-0810-5

Printed on acid-free paper

All rights reserved
©2001 Springer Science+Business Media New York
Originally published by Kluwer Academic Publishers in 2001
No part of the material protected by this copyright notice may be reproduced
or utilized in any form or by any means, electronic or mechanical,
including photocopying, recording or by any information storage and
retrieval system, without written permission from the copyright owner.

PREFACE

Humankind cannot live independently in the biosphere of the earth. Air, water, soil, microorganisms, and many animals and plants on earth support human existence, and all are intimately interconnected in the irreplaceable ecosystem of the earth. Under these circumstances, environmental pollution and destruction of nature are likely to deleteriously affect not only human existence but also the future life cycle.

Humankind has developed science and technology as part of its civilization. Consequently, human beings have gained advantage in the biosphere by reducing competition from other living organisms and by developing new types of energy and resources. The growth in human population, spurred by rapid industrialization, has been exponential. In 2000, the human population exceeded 6 billion and it is still growing. However, this growth has not come without consequences. In the 20th Century, human activities such as industrialization, urbanization, and military actions have led to the extinction of many species. Human beings are consuming enormous amounts of nonrenewable resources and energy. Consequently, a great amount of material, some of it acidic, has been discharged into the environment. This has resulted in acid pollution of the atmosphere, lakes, and rivers, damaging forests, eroding cultural treasures such as sculptures and buildings, and even compromising human health.

Thus, extremely important issues concerning the global ecosystem have arisen. Acid pollution is a prominent global environmental problem since acid pollutants discharged into the air are carried from country to country by air currents, spreading the problem across national boundaries.

It is crucially important to solve these problems not only domestically but also internationally. All countries of the world are interconnected and the environmental problems of one country directly or indirectly relate to those of other countries. Governments, industrial sectors, institutes, and citizens need to collaborate with each other. At this Conference, scientists and specialists from all over the world have gathered to study acid pollution and to discuss how to prevent it and how to preserve the global ecosystem.

Against this backdrop, I believe it is very significant that we were able to hold this 6th International Conference on Acidic Deposition in Asia at the end of the 20th Century and just before the start of the 21st Century. As Chairman of the Organizing Committee, I am delighted that scientists and specialists from 35 countries chose to participate in this Conference. I sincerely hope that the outcome of this Conference will contribute to sustaining the global environment in the future. Finally, I would like to express my sincere appreciation to the members of the International Advisory Board for support and to the persons who worked to make this Conference a memorable one.

Jiro Kondo
Chairman of the Organizing Committee

TABLE OF CONTENTS

VOLUME III

Preface iii

Contents of Volume III v

EFFECTS ON TERRESTRIAL ECOSYSTEM

EUGENIJA KUPČINSKIENĖ
    Annual Variations of Needle Surface Characteristics of *Pinus Sylvestris*
    Growing Near the Emission Source 923-928

GALINA N. KOPTSIK, SERGUEI V. KOPTSIK and DAN AAMLID
    Pine Needle Chemistry Near a Large Point SO_2 Source in Northern Fennoscandia
    929-934

JESADA. LUANGJAME, BOONCHOOB BOONTAWEE and NIGORN. KLIANGPIBOOL
    Determination of Deposition and Leaves in Teak Plantations in Thailand 935-940

T. TAKAMATSU, H. SASE, J. TAKADA and R. MATSUSHITA
    Annual Changes in Some Physiological Properties of *Cryptomeria Japonica*
    Leaves from Kanto, Japan 941-946

ELINA J. OKSANEN
    Increasing Tropospheric Ozone Level Reduced Birch (*Betula Pendula*) Dry
    Mass within a Five Years Period 947-952

LUCY J. SHEPPARD, ALAN CROSSLEY, JUDITH PARRINGTON, FRANCIS J HARVEY and J. NEIL CAPE
    Effects of Simulated Acid Mist on a Sitka Spruce Forest Approaching Canopy
    Closure: Significance of Acidified Versus Non-Acidified Nitrogen Inputs 953-958

HIDEYUKI MATSUMURA
    Impacts of Ambient Ozone and/or Acid Mist on the Growth of 14 Tree Species:
    an Open-Top Chamber Study Conducted in Japan 959-964

TETSUSHI YONEKURA, YUKIKO DOKIYA, MOTOHIRO FUKAMI and TAKESHI IZUTA
    Effects of Ozone and/or Soil Water Stress on Growth and Photosynthesis of
    Fagus crenata Seedlings 965-970

TATSURO NAKAJI and TAKESHI IZUTA
    Effects of Ozone and/or Excess Soil Nitrogen on Growth, Needle Gas
    Exchange Rates and Rubisco Contents of *Pinus densiflora* Seedlings 971-976

Y. KOHNO, R. MATSUKI, S. NOMURA, K. MITSUNARI and M. NAKAO
    Neutralization of Acid Droplets on Plant Leaf Surfaces 977-982

JAKUB HRUSKA, PAVEL CUDLÍN and PAVEL KRÁM
    Relationship between Norway Spruce Status and Soil Water Base Cations/
    Aluminum Ratios in the Czech Republic 983-988

JAN MULDER, HELENE A. DE WIT, HELENA W.J. BOONEN and LARS R. BAKKEN
    Increased Levels of Aluminium in Forest Soils: Effects on the Stores of Soil
    Organic Carbon 989-994

HELENE A. DE WIT, JAN MULDER, PER H. NYGAARD and DAN AAMLID
    Testing the Aluminium Toxicity Hypothesis: a Field Manipulation
    Experiment in Mature Spruce Forest in Norway 995-1000

KAZUO SATO and TAKASHI WAKAMATSU
    Soil Solution Chemistry in Forests with Granite Bedrock in Japan 1001-1006

TAKESHI IZUTA, TAEKO YAMAOKA, TATSURO NAKAJI, TETSUSHI YONEKURA, MASAAKI YOKOYAMA, HIDEYUKI MATSUMURA, SACHIE ISHIDA, KENICHI YAZAKI, RYO FUNADA and TAKAYOSHI KOIKE
    Growth, Net Photosynthetic Rate, Nutrient Status and Secondary Xylem
    Anatomical Characteristics of *Fagus crenata* Seedlings Grown in Brown
    Forest Soil Acidified with H_2SO_4 Solution 1007-1012

RIE TOMIOKA and CHISATO TAKENAKA
  Differential Ability of the Root to Change Rhizosphere pH between
  Chamaecyparis Obtusa Sieb. (Hinoki) and *Quercus Serrata* Thumb.(Konara)
  under Aluminium Stress 1013-1018

MASAHIKO OHNO
  Sensitivity of a Japanese Earthworm (*Allolobophora japonica*) to Soil
  Acidity 1019-1024

SERGUEI V. KOPTSIK, NATALIYA BEREZINA and S. LIVANTSOVA
  Effects of Natural Soil Acidification on Biodiversity in Boreal Forest
  Ecosystems 1025-1030

LARS LUNDIN, MATS AASTRUP, LAGE BRINGMARK, SVEN BRÅKENHIELM,
  HANS HULTBERG, KJELL JOHANSSON, KARIN KINDBOM, HANS KVARNÄS
  and STEFAN LÖFGREN
  Impacts from Deposition on Swedish Forest Ecosystems Identified by
  Integrated Monitoring 1031-1036

CAROLE E.R. PITCAIRN, IAN D. LEITH, DAVID FOWLER, KEN J. HARGREAVES,
  MASOUD MOGHADDAM, VALERIE H. KENNEDY and LENNART GRANAT
  Foliar Nitrogen as an Indicator of Nitrogen Deposition and Critical Loads
  Exceedance on a European Scale 1037-1042

IAN D. LEITH, LUCY J. SHEPPARD, CAROLE E.R. PITCAIRN, J. NEIL CAPE, PAUL
  W. HILL, VALERIE H. KENNEDY, Y. SIM TANG, RON I. SMITH and
  DAVID FOWLER
  Comparison of the Effects of Wet N Deposition(NH_4Cl) and Dry N
  Deposition (NH_3) on UK Moorland Species 1043-1048

JOHAN BERGHOLM and HOOSHANG MAJDI
  Accumulation of Nutrients in above and below Ground Biomass in Response
  to Ammonium Sulphate Addition in a Norway Spruce Stand in Southwest
  Sweden 1049-1054

E. P. FARRELL, J. AHERNE, G. M. BOYLE and N. NUNAN
  Long-Term Monitoring of Atmospheric Deposition and the Implications
  of Ionic Inputs for the Sustainability of a Coniferous Forest Ecosystem 1055-1060

B. K. SITAULA, J. I. B. SITAULA, Å. AAKRA and L. R. BAKKEN
  Nitrification and Methane Oxidation in Forest Soil: Acid Deposition,
  Nitrogen Input and Plant Effects 1061-1066

J. R. HALL, B. REYNOLDS, T. SPARKS, A. COLGAN, I. THORNTON and
  S. P. MCGRATH
  The Relationship between Topsoil and Stream Sediment Heavy Metal
  Concentrations and Acidification 1067-1072

D. TANG, E. LYDERSEN, H. M. SEIP, V. ANGELL, O. EILERTSEN, T. LARSSEN,
  X. LIU, G. KONG, J. MULDER, A. SEMB, S. SOLBERG, K. TORSETH,
  R. D. VOGT, J. XIAO and D. ZHAO
  Integrated Monitoring Program on Acidification of Chinese Terrestrial
  Systems (IMPACTS) - a Chinese- Norwegian Cooperation Project 1073-1078

Y. G. XU, G. Y. ZHOU, Z. M. WU, T. S. LUO and Z. C. HE
  Chemical Composition of Precipitation,Throughfall and Soil Solutions
  at Two Forested Sites in Guangzhou, South China 1079-1084

YASUMI YAGASAKI, TAKASHI CHISHIMA, MASANORI OKAZAKI,
  DU-SIK JEON, JEONG-HWAN YOO and YOUNG -KULL KIM
  Acidification of Red Pine Forest Soil due to Acidic Deposition In Chunchon,
  Korea 1085-1090

HIROSHI OKOCHI and MANABU IGAWA
  Elevational Patterns of Acid Deposition into a Forest and Nitrogen Saturation
  on Mt. Oyama, Japan. 1091-1096

T. KAWAKAMI, H. HONOKI and H. YASUDA
  Acidification of a Small Stream on Kureha Hill Caused by Nitrate Leached

from a Forested Watershed 1097-1102

M. BABA, Y. SUZUKI, H. SASAKI, K. MATANO, T. SUGIURA and H. KOBAYASHI
Nitrogen Retention in Japanese Cedar Stands in Northern Honshu, with High
Nitrogen Deposition 1103-1108

K. MATANO, M. BABA, A. SHIBUYA, Y. SUZUKI, T. SUGIURA and
H. KOBAYASHI
Soil Solution Chemistry in Japanese Cedar Stands in Northern Honshu,
with High Nitrogen Deposition 1109-1114

TSUYOSHI YAMADA, SHUICHIRO YOSHINAGA, KAZUHITO MORISADA and
KEIZO HIRAI
Sulfate and Nitrate Loads on a Forest Ecosystem in Kochi in Southwest
of Japan 1115-1120

M. CRISTINA FORTI, ADILSON CARVALHO, ADOLPHO J. MELFI and
CELIA R. MONTES
Deposition Patterns of SO_4^{2-}, NO_3^- and H^+ in the Brazilian Territory 1121-1126

SANJAY KUMAR, R. DATTA, S. SINHA, T. KOJIMA, S. KATOH and M. MOHAN
Carbon Stock, Afforestation and Acidic Deposition: an Analysis of
Inter-relation with Reference to Arid Areas 1127-1132

MODELS FOR EVALUATING ECOLOGICATL EFFECTS

J.-P. HETTELINGH, M. POSCH and P.A.M. DE SMET
Multi-Effect Critical Loads Used in Multi-Pollutant Reduction Agreements
in Europe 1133-1138

M. POSCH, J.-P. HETTELINGH and P.A.M. DE SMET
Characterization of Critical Load Exceedances in Europe 1139-1144

SOON-UNG PARK and YOUNG-HEE LEE
Estimation of the Maximum Critical Load for Sulfur in South Korea 1145-1150

PETRA MAYERHOFER, JOSEPH ALCAMO, MAXIMILIAN POSCH and
JELLE G. VAN MINNEN
Regional Air Pollution and Climate Change in Europe: an Integrated
Assessment (AIR-CLIM) 1151-1156

JIMING HAO, XUEMEI YE, LEI DUAN and ZHONGPING ZHOU
Calculating Critical Loads of Sulfur Deposition for 100 Surface Waters in
China Using the Magic Model 1157-1162

C. J. CURTIS, R. HARRIMAN, M. HUGHES and M. KERNAN
Effects of Site Selection Strategy on Freshwater Critical Load Exceedances
in Wales 1163-1168

M. KERNAN, J. HALL, J. ULLYET and T. ALLOTT
Variation in Freshwater Critical Loads Across Two Upland Catchments in
the UK: Implications for Catchment Scale Management 1169-1174

JOHAN C.I. KUYLENSTIERNA, W. KEVIN HICKS, STEVE CINDERBY,
HARRY W. VALLACK and MAGNUZ ENGARDT
Variability in Mapping Acidification Risk Scenarios for Terrestrial
Ecosystems in Asian Countries 1175-1180

LEO SAARE, REET TALKOP and OTT ROOTS
Air Pollution Effects on Terrestrial Ecosystems in Estonia 1181-1186

FULU TAO and ZONGWEI FENG
Critical Loads of Acid Deposition for Ecosystems in South China
-- Derived by a New Method 1187-1192

MIKHAIL SEMENOV, VLANDIMIR BASHKIN and HARALD SVERDRUP
Critical Loads of Acidity for Forest Ecosystems of North Asia 1193-1198

LEI DUAN, SHAODONG XIE, ZHONGPING ZHOU, XUEMEI YE and JIMING HAO
Calculation and Mapping of Critical Loads for S, N and Acidity in China 1199-1204

JUNLING AN, LING ZHOU, MEIYUAN HUANG, HU LI, TSUNEHIKO OTOSHI and KAZUHIDE MATSUDA
A Literature Review of Uncertainties in Studies of Critical Loads for Acidic Deposition 1205-1210

KENTARO HAYASHI and MASANORI OKAZAKI
Acid Deposition and Critical Load Map of Tokyo 1211-1216

DANIEL KURZ, BEAT RIHM, MATTIAS ALVETEG and HARALD SVERDRUP
Steady-State and Dynamic Assessment of Forest Soil Acidification in Switzerland 1217-1222

BEAT RIHM and DANIEL KURZ
Deposition and Critical Loads of Nitrogen in Switzerland 1223-1228

K. R. BULL, J. R. HALL, J. COOPER, S. E. METCALFE, D. MORTON, J. ULLYETT, T. L. WARR and J. D, WHYATT
Assessing Potential Impacts on Biodiversity Using Critical Loads....................... 1229-1234

J. M. ULLYETT, J. R. HALL, M. HORNUNG and M. KERNAN
Mapping the Potential Sensitivity of Surface Waters to Acidification Using Measured Freshwater Critical Loads as an Indicator of Acid Sensitive Areas 1235-1240

GARETH J.P THORNTON
Calculating Weathering Rates of Stream Catchments in the English Lake District Using Critical Element Ratios, Mass-balance Budgets and the Magic Model..1241-1246

TAMON FUMOTO, JUNKO SHINDO, NORIKO OURA and HARALD SVERDRUP
Adapting the Profile Model to Calculate the Critical Loads for East Asian Soils by Including Volcanic Glass Weathering and Alternative Aluminum Solubility System 1247-1252

HIDESHI IKEDA and YOICHI MIYANAGA
Hydrogeochemical Conditions Affecting Acidification of Stream Water in Mountainous Watersheds.. 1253-1258

J. SHINDO, T. FUMOTO, N. OURA, T. NAKANO and T. TAKAMATSU
Estimation of Mineral Weathering Rates under Field Conditions Based on Base Cation Budget and Strontium Isotope Ratios 1259-1264

J. MILINDALEKHA, V. N. BASHKIN, and S. TOWPRAYOON
Calculation and Mapping of Sulfur Critical Loads for Terrestrial Ecosystems of Thailand 1265-1270

CHRISTINE ALEWELL
Predicting Reversibility of Acidification: The European Sulfur Story 1271-1276

SERGUEI V. KOPTSIK and GALINA N. KOPTSIK
Effects of Acid Deposition on Forest Soils in Northernmost Russia: Modelled and Field Data 1277-1282

HELEN J. FOSTER, MATTHEW J. LEES, HOWARD S. WHEATER, COLIN NEAL and BRIAN REYNOLDS
Dynamic Modelling of Spatially Variable Catchment Hydrochemistry for Critical Loads Assessment 1283-1288

WOJCIECH MILL
Integrated Modelling of Acidification Effects to Forest Ecosystems -Model Sonox..1289-1294

HU LI, AKIKAZU KAGA and KATSUHITO YAMAGUCHI
Prediction of Soil Acidification Using a Dynamic Model at a Bamboo Forest in Osaka Prefecture 1295-1300

PAVEL KRÁM, HJALMAR LAUDON, KEVIN BISHOP, LARS RAPP and JAKUB HRUSKA
Magic Modeling of Long-Term Lake Water and Soil Chemistry at Abborrträsket, Northern Sweden 1301-1306

ECOSYSTEM RECOVERY

D. T. MONTEITH, C. D. EVANS and S. PATRICK
Monitoring Acid Waters in the UK: 1988-1998 Trends 1307-1312

BJØRN WALSENG and LEIF R. KARLSEN
Planktonic and Littoral Microcrustaceans as Indices of Recovery in Limed
Lakes in SE Norway 1313-1318

BJØRN WALSENG and ANN KRISTIN.L. SCHARTAU
Crustacean Communities in Canada and Norway: Comparison of Species
along a pH Gradient 1319-1324

ANN KRISTIN L. SCHARTAU, BJØRN WALSENG and ED SNUCINS
Correlation between Microcrustaceans and Environmental Variables along
an Acidification Gradient in Sudbury, Canada 1325-1330

BJØRN WALSENG, ROY M. LANGÅKER, TOR E. BRANDRUD, PÅL BRETTUM,
ARNE FJELLHEIM, TRYGVE HESTHAGEN, ØYVIND KASTE,
BJØRN M. LARSEN and ELI-A. LINDSTRØM
The River Bjerkreim in SW Norway -Successful Chemical and Biological
Recovery after Liming 1331-1336

A. LYCHE-SOLHEIM, Ø. KASTE and E. DONALI
Can Phosphate Help Acidified Lakes to Recover? 1337-1342

S. SANDØY and R. M. LANGÅKER
Atlantic Salmon and Acidification in Southern Norway: A Disaster in
the 20TH Century, but a Hope for the Future? 1343-1348

FRODE KROGLUND, ØYVIND KASTE, BJØRN O. ROSSELAND and
TRYGVE POPPE
The Return of the Salmon 1349-1354

TRYGVE HESTHAGEN, TORBJØRN FORSETH, RANDI SAKSGÅRD,
HANS M. BERGER and BJØRN M. LARSEN
Recovery of Young Brown Trout in Some Acidified Streams in
Southwestern and Western Norway 1355-1360

TRYGVE HESTHAGEN, HANS M. BERGER, ANN KRISTIN LIEN SCHARTAU,
TERJE NØST, RANDI SAKSGÅRD and LEIDULF FLØYSTAD
Low Success Rate in Re-Establishing European Perch in Some Highly
Acidified Lakes in Southernmost Norway 1361-1366

M. RASK, H. PÖYSÄ, P. NUMMI and C. KARPPINEN
Recovery of the Perch (*Perca Fluviatilis*) in an Acidified Lake and Subsequent
Responses in Macroinvertebrates and the Goldeneye (*Bucephala Clangula*)
.... 1367-1372

K. NYBERG, J. VUORENMAA, M. RASK, J. MANNIO and J. RAITANIEMI
Patterns in Water Quality and Fish Status of Some Acidified Lakes
in Southern Finland during a Decade: Recovery Proceeding 1373-1378

ARNE FJELLHEIM and GUNNAR G. RADDUM
Acidification and Liming of River Vikedal, Western Norway. A 20 Year
Study of Responses in the Benthic Invertebrate Fauna 1379-1384

GODTFRED A. HALVORSEN, JOCELYNE H. HENEBERRY and ED SNUCINS
Sublittoral Chironomids as Indicators of Acidity (Diptera: Chironomidae)
.... 1385-1390

ARNE FJELLHEIM, ÅSMUND TYSSE and VILHELM BJERKNES
Reappearance of Highly Acid-Sensitive Invertebrates after Liming of
an Alpine Lake Ecosystem 1391-1396

ALUN S. GEE
A Strategic Appraisal of Options to Ameliorate Regional Acidification 1397-1402

ATLE HINDAR, MAXIMILIAN POSCH and ARNE HENRIKSEN

Effects of In-Lake Retention of Nitrogen on Critical Load Calculations 1403-1408

VILHELM BJERKNES and TORULV TJOMSLAND
Flow and pH Modelling to Study the Effects of Liming in Regulated, Acid
Salmon Rivers 1409-1414

K. BISHOP, H. LAUDON, J. HRUSKA, P. KRAM, S. KÖHLER and S. LÖFGREN
Does Acidification Policy Follow Research in Northern Sweden?
The Case of Natural Acidity during the 1990's 1415-1420

A. WILANDER
Effects of Reduced S Deposition on Large-Scale Transport of Sulphur in
Swedish Rivers... 1421-1426

JAAKKO MANNIO
Recovery Pattern from Acidification of Headwater Lakes in Finland.................... 1427-1432

BRIT LISA SKJELKVÅLE, KJETIL TØRSETH, WENCHE AAS and TOM ANDERSEN
Decrease in Acid Deposition - Recovery in Norwegian Waters 1433-1438

JOHN GUNN, ROD SEIN, BILL KELLER and PETER BECKETT
Liming of Acid and Metal Contaminated Catchments for the Improvement of
Drainage Water Quality 1439-1444

R. D. VOGT, H. M. SEIP, H. OREFELLEN, G. SKOTTE, C. IRGENS and J. TYSZKA
Trends in Soil Water Composition at a Heavily Polluted Site - Effects of
Decreased S-Deposition and Variations in Precipitation......................... 1445-1450

KAZUHIKO SAKAMOTO, YUGO ISOBE, XUHUI DONG and SHIDONG GAO
Simulated Acid Rain Leaching Characteristics of Acid Soil Amended with
Bio-Briquette Combustion Ash 1451-1456

EFFECTS ON MATERIALS AND CULTURAL PROPERITIES

JOHAN TIDBLAD, VLADIMIR KUCERA, ALEXANDRE A. MIKHAILOV,
JAN HENRIKSEN, KATERINA KREISLOVA, TIM YATES, BRUNO STÖCKLE
and MANFRED SCHREINER
UN ECE ICP Materials: Dose-Response Functions on Dry and Wet Acid
Deposition Effects after 8 Years of Exposure 1457-1462

M. KITASE, S. HATAKEYAMA, T. MIZOGUCHI and Y. MAEDA
Regional Characteristics of Copper Corrosion Components in East Asia
.... 1463-1468

JOHAN TIDBLAD, VLADIMIR KUCERA and ALEXANDRE A. MIKHAILOV
Mapping of Acid Deposition Effects and Calculation of Corrosion Costs on
Zinc in China 1469-1474

ELIN DAHLIN, PETER TORSSANDER, CARL -MAGNUS MÖRTH,
HELÉNE STRANDH, GÖRAN ÅBERG, JAN F. HENRIKSEN, ODD ANDA
and RUNO LÖFVENDAHL
Environmental Monitoring of Rock Carvings in Scandinavia................................ 1475-1480

TSUTOMU KANAZU, TAKURO MATSUMURA, TATSUO NISHIUCHI and
TAKESHI YAMAMOTO
Effect of Simulated Acid Rain on Deterioration of Concrete 1481-1486

Y. TSUJINO, Y. SATOH, N. KURAMOTO and Y. MAEDA
Effect of Acid Deposition on Urushi Lacquer in East Asia................. 1487-1492

ANALYTICAL METHODS AND MONITORING

KJETIL TØRSETH, WENCHE AAS and SVERRE SOLBERG
Trends in Airborne Sulphur and Nitrogen Compounds in Norway during
1985-1996 in Relation to Air Mass Origin 1493-1498

P. H. VIET, V. V. TUAN, P. M. HOAI, N. T.K. ANH and P. T. YEN
  Chemical Composition and Acidity of Precipitation: A Monitoring Program
  in North-Eastern Vietnam 1499-1504

MUNEHIRO WARASHINA, MASANOBU TANAKA, YOSHIO TSUJINO,
  TUGUO MIZOGUCHI, SIRO HATAKEYAMA and YASUAKI MAEDA
  Atmospheric Concentrations of Sulfur Dioxide and Nitrogen Dioxide in
  China and Korea Measured by Using the Improved Passive Sampling Method
  1505-1510

MOTONORI TAMAKI, TAKATOSHI HIRAKI, YOSHIHIRO NAKAGAWA,
  TOMIKI KOBAYASHI, MASAHIDE AIKAWA and MITSURU SHOGA
  Study on Sampling Method of Rainfall, Throughfall, and Stemflow to
  Monitor the Effect of Acid Deposition on Forest Ecosystem 1511-1516

MASAHIDE AIKAWA, TAKATOSHI HIRAKI, MITSURU SHOGA and
  MOTONORI TAMAKI
  Fog and Precipitation Chemistry at Mt. Rokko in Kobe, April 1997 - March
  1998 1517-1522

YASUSHI NARITA, KEI SATOH, KENICHI HAYASHI and SHIGERU TANAKA
  Development of Automatic Continuous Measurement System of Chemical
  Constituents in the Precipitation 1523-1528

ANNE -GUNN HJELLBREKKE and LEONOR TARRASON
  Mapping of Concentrations in Europe Combining Measurements and Acid
  Deposition Models 1529-1534

YUKIO KOMAI, SATOSHI UMEMOTO and TAKANOBU INOUE
  Influence of Acid Deposition on Inland Water Chemistry –A Case Study
  from Hyogo Prefecture, Japan – 1535-1540

C. D. EVANS, R. HARRIMAN, D. T. MONTEITH and A. JENKINS
  Assessing the Suitability of Acid Neutralising Capacity as a Measure of
  Long-term Trends in Acidic Waters Based on Two Parallel Datasets 1541-1546

RASA GIRGZDIENE and ALOYZAS GIRGZDYS
  Spatial and Temporal Variability in Ozone Concentration Level at Two
  Lithuanian Stations 1547-1552

AUDRONE MILUKAITE, ALDONA MIKELINSKIENE and
  BRONISLAVAS GIEDRAITIS
  Characteristics of SO_2, NO_2, Soot and Benzo(a)pyrene Concentration
  Variation on the Eastern Coast of the Baltic Sea 1553-1558

KOICHI WATANABE, YUTAKA ISHIZAKA, YUKIYA MINAMI and
  KOJI YOSHIDA
  Peroxide Concentrations in Fog Water at Mountainous Sites in Japan 1559-1564

OSAMU NAGAFUCHI, HITOSHI MUKAI and MINORU KOGA
  Black Acidic Rime Ice in the Remote Island of Yakushima, a World
  Natural Heritage Area 1565-1570

SITI ASIATI, TUTI BUDIWATI and LELY QODRITA AVIA
  Acid Deposition in Bandung, Indonesia 1571-1576

GÖRAN E. ÅBERG
  Tracing Pollution and its Sources with Isotopes 1577-1582

YORIKO YOKOO and TAKANORI NAKANO
  Sequential Leaching of Volcanic Soil to Determine Plant-Available Cations
  and the Provenance of Soil Minerals Using Sr Isotopes 1583-1588

KIN-ICHI TSUNODA, TOMONARI UMEMURA, KAZUMASA OHSHIMA,
  SHO-ICHI AIZAWA, ETSURO YOSHIMURA and KEN-ICHI SATAKE
  Determination and Speciation of Aluminum in Environmental Samples by
  Cation Exchange High-Performance Liquid Chromatography with High
  Resolution ICP-MS Detection 1589-1594

WENCHE AAS and ARNE SEMB
  Standardisation of Methods for Long-Term Monitoring 1595-1600

CHANG -JIN MA, MIKIO KASAHARA, SUSUMU TOHNO and TOMIHIRO KAMIYA
A New Approach for Characterization of Single Raindrops 1601-1606

LU -YEN CHEN, FU -TIEN JENG, YU -MEY HSU, SHIN -YUAN TSAI and UEI -RUEY PENG
Stability of Ionic Components in Precipitation Samples – a Case Study in Taipei 1607-1612

TSUNEHIKO OTOSHI, NORIO FUKUZAKI, HU LI, HIROSHI HOSHINO, HIROYUKI SASE, MASASHI SAITO and KATSUNORI SUZUKI
Quality Control and its Constraints during the Preparatory-Phase Activities of the Acid Deposition Monitoring Network in East Asia (EANET) 1613-1618

REGIONAL CASE STUDIES

N. KARVOSENOJA, P. HILLUKKALA, M. JOHANSSON and S. SYRI
Cost-Effective Abatement of Acidifying Emissions with Flue Gas Cleaning vs. Fuel Switching in Finland 1619-1624

DRIEJANA, D. W. RAPER, D. S. LEE, R. D. KINGDON and I. L. GEE
Wet Deposition to an Upland Area of England in1988 and 1999: Measurements and Modelling 1625-1630

DEXUAN WANG and WEI DENG
Atmospheric SO_2 Pollution and Acidity of Rain in Changchun China 1631-1634

T. LARSSEN, H. M. SEIP, G. R. CARMICHAEL and J. L. SCHNOOR
The Importance of Calcium Deposition in Assessing Impacts of Acid Deposition in China 1635-1640

MARTIN FORSIUS, SIRPA KLEEMOLA, JUSSI VUORENMAA and SANNA SYRI
Fluxes and Trends of Nitrogen and Sulphur Compounds at Integrated Monitoring Sites in Europe 1641-1648

YASUSHI NARITA, KEI SATOH, KEIICHI HAYASHI, TAKAMI IWASE, SHIGERU TANAKA, YUKIKO DOKIYA, MORIKAZU HOSOE and KAZUHIKO HAYASHI
Long Term Trend of Chemical Constituents in Tokyo Metropolitan Area in Japan 1649-1654

MASAKI ADACHI, KAZUHIKO HAYASHI and YUKIKO DOKIYA
Sea Ice Approach and Chemical Species in Precipitation at Abashiri, Japan 1655-1660

Y. MATSUURA, M. SANADA, M. TAKAHASHI, Y. SAKAI and N. TANAKA
Long-Term Monitoring Study on Rain, Throughfall, and Stemflow Chemistry in Evergreen Coniforous Forests in Hokkaido, Northern Japan 1661-1666

KAZUHIKO HAYASHI, YASUHITO IGARASHI, YUKITOMO TSUTSUMI and YUKIKO DOKIYA
Aerosol and Precipitation Chemistry During the Summer at the Summit of Mt. Fuji, Japan (3776 m a.s.l.) 1667-1672

NORIO FUKUZAKI, TSUYOSHI OHIZUMI and KAZUHIDE MATSUDA
Geographical and Temporal Variations of Chemical Constituents in Winter Precipitation Collected in the Areas along the Coast of the Sea of Japan 1673-1678

TSUYOSHI OHIZUMI, NAOKO TAKE, NOBORU MORIYAMA, OSAMU SUZUKI and MINORU KUSAKABE
Seasonal and Spatial Variations in the Chemical and Sulfur Isotopic Composition of Acid Deposition in Niigata Prefecture, Japan............................. 1679-1684

UMESH C. KULSHRESTHA, MONIKA J. KULSHRESTHA, R. SEKAR, M. VAIRAMANI, AJIT K. SARKAR and DANESH C. PARASHAR
Investigation of Alkaline Nature of Rain Water in India 1685-1690

AHTI LEPISTÖ and SANNA SYRI
Modeling Effects of Changing Deposition and Forestry on Nitrogen Fluxes
in a Northern River Basin 1691-1696

FRIDA EDBERG, HANS BORG and JAN-ERIK ÅSLUND
Episodic Events in Water Chemistry and Metals in Streams in Northern
Sweden during Spring Flood 1697-1702

N. L. ROSE, E. SHILLAND, T. BERG, K. HANSELMANN, R. HARRIMAN,
K. KOINIG, U. NICKUS, B. STEINER TRAD, E. STUCHLÍK, H. THIES and
M. VENTURA
Relationships between Acid Ions and Carbonaceous Fly-Ash Particles in
Deposition at European Mountain Lakes 1703-1708

H. FUSHIMI, T. KAWAMURA, H. IIDA, M. OCHIAI, T. NAKAJIMA and
Y. AZUMA
Internal Distribution of Acid Materials within Snow Crystals 1709-1714

CECILIA ANDRÉN, PAUL ANDERSSON and ELISABETH FRÖBERG
Temporal Variations of Aluminium Fractions in Streams in the Delsbo Area,
Central Sweden 1715-1720

H. HANSEN, T. LARSSEN, H. M. SEIP and R. D. VOGT
Trace Metals in Forest Soils at Four Sites in Southern China 1721-1726

BARBARA WALNA, JERZY SIEPAK and STANISLAW DRZYMALA
Soil Degradation in the Wielkopolski National Park (Poland) as an Effect of
Acid Rain Simulation 1727-1732

ZHAO DAWEI, T. LARSSEN, ZHANG DONGBAO, GAO SHIDONG, R. D. VOGT,
H. M. SEIP and O. J. LUND
Acid Deposition and Acidification of Soil and Water in the Tie Shan Ping
Area, Chongqing, China 1733-1738

M. JOHANSSON, R. SUUTARI, J. BAK, G. LÖVBLAD, M. POSCH, D. SIMPSON,
J.-P. TUOVINEN and K. TØRSETH
The Importance of Nitrogen Oxides for the Exceedance of Critical
Thresholds in the Nordic Countries 1739-1744

MAGNUS APPELBERG and TORBJÖRN SVENSON
Long-Term Ecological Effects of Liming - The ISELAW Programme 1745-1750

KERSTIN HOLMGREN
Biomass-Size Distribution of the Aquatic Community in Limed,
Circumneutral and Acidified Reference Lakes 1751-1756

H. BORG, J. EK and K. HOLM
Influence of Acidification and Liming on the Distribution of Trace Elements
in Surface Waters 1757-1762

FRIDA EDBERG, PAUL ANDERSSON, HANS BORG, CHRISTINA EKSTRÖM and
EINAR HÖRNSTRÖM
Reacidification Effects on Water Chemistry and Plankton in a Limed Lake
in Sweden 1763-1768

GUNNAR PERSSON and MAGNUS APPELBERG
Evidence of Lower Productivity in Long Term Limed Lakes as Compared
to Unlimed Lakes of Similar pH 1769-1774

HIDEHISA KAWAMURA, NOBUAKI MATSUOKA, SHINJI TAWAKI and
NORIYUKI MOMOSHIMA
Sulfur Isotope Variations in Atmospheric Sulfur Oxides, Particulate Matter
and Deposits Collected at Kyushu Island, Japan 1775-1780

JUNLING AN, MEIYUAN HUANG, ZIFA WANG, XINLING ZHANG, HIROMASA
UEDA and XINJIN CHENG
Numerical Regional Air Quality Forecast Tests Over the Mainland of China
.... 1781-1786

VALDO LIBLIK and MARGUS PENSA

Specifics and Temporal Changes in Air Pollution in Areas Affected by
Emissions from Oil Shale Industry, Estonia.. 1787-1792

TAKAHISA MAEDA, ZIFA WANG, MASAYASU HAYASHI and
MEIYUAN HUANG
Long-Range Transport of Sulfur from Northeast Asia to Chengshantu,
Shandong Peninsula: Measurement and Simulation.. 1793-1798

FRANK MURRAY, GORDON MCGRANAHAN and JOHAN C.I. KUYLENSTIERNA
Assessing Health Effects of Air Pollution in Developing Countries 1799-1804

PRABAL K. ROY and HIROSHI SAKUGAWA
Trends of Air Pollution and its Present Situation in Hiroshima Prefecture 1805-1810

SCIENCE & POLICY AND ENVIRONMENTAL EDUCATION

KEITH R. BULL
The Need for Education in Developing Acceptable Air Pollution Control
Strategies.. 1811-1816

EDUARD DAME, and MIKE HOLLAND
Cost-Benefit Analysis and the Development of Acidification Policy
in Europe.. 1817-1824

KENNETH E. WILKENING
Trans-Pacific Air Pollution: Scientific Evidence & Political Implications 1825-1830

SANNA. SYRI and NIKO KARVOSENOJA
Low-CO_2 Energy Pathways Versus Emission Control Policies
in Acidification Reduction 1831-1836

WAKANA TAKAHASHI and JUSEN ASUKA
The Politics of Regional Cooperation on Acid Rain Control in East Asia............ 1837-1842

HAIPING LAI, HIROYUKI KAWASHIMA, JUNKO SHINDO and KEIJI OHGA
Stages in the History of China's Acid Rain Control Strategy in the Light of
China-Japan Relations 1843-1848

YOSHIKAZU HASHIMOTO, YOSHIKA SEKINE, ZHI-MIN YANG and
KANJI YOSHIOKA
Profound Survival Program of Forests in Japan Islands –a 40 Year Strategy
for Environmental Conservation In Inland China – 1849-1854

TRENDS IN AIRBORNE SUPLHUR AND NITROGEN COMPOUNDS IN NORWAY DURING 1985-1996 IN RELATION TO AIR MASS ORIGIN

KJETIL TØRSETH, WENCHE AAS and SVERRE SOLBERG
Norwegian Institute for Air Research, P.O. Box 100, N-2027 Kjeller, Norway

Abstract. Major reductions in emissions of sulphur dioxide and nitrogen dioxide in Europe have significantly reduced the ambient concentrations of both sulphur dioxide, particulate sulphate and nitrogen dioxide, as well as of sulphate in precipitation at Norwegian monitoring sites. In this study, trends in ambient air concentrations were studied in relation to air mass origin by sector analysis. Associated trends in ambient concentrations were derived by non-parametric statistical methods and evaluated on the basis of emission figures within the various sectors. The observed trends correspond well with reported trends in emissions.

Keywords: trends, deposition, sulphur, nitrogen dioxide, monitoring

1. Introduction

Increased energy consumption and growth in population and standard of living within Europe has increased the emissions of sulphur and nitrogen compounds to the atmosphere during the last century. Correspondingly, ambient air concentrations have increased yielding enhanced deposition and even adverse effects in ecosystems (e.g. Drabløs and Tollan, 1980). Since the early 1970's, ambient concentrations have been monitored at regional sites in order to provide essential information for describing the transboundary advection of air pollutants. The monitoring data from this activity is combined with regional dispersion models, and with further inclusion of target deposition levels, legally binding and cost-effective emissions ceilings for Europe have been established in the framework of the Convention of Long-Range Transboundary Air Pollution (CLRTAP) (UN ECE, 1999). In 1999, a combined protocol to abate acidification, eutrophication and ground-level ozone set new emission ceilings of sulphur, nitrogen oxides, volatile organic compounds and ammonia to be reached by year 2010. When complied with, Europe's sulphur emissions will be reduced by at least 63%, its NO_x emissions by 41% and its ammonia emissions by 17% compared to their 1990 levels. Compared to 1980, the reduction in sulphur emission will be in the order of 75%. A major challenge for the future will be to verify the compliance with the protocols, and to see if the anticipated improvements in air quality occur.

In this study, available monitoring data of ambient sulphur dioxide (SO_2), particulate sulphate (SO_4) and of nitrogen dioxide (NO_2) at selected rural sites in Norway have been investigated in terms of trends in concentrations. In order to evaluate the effect of different emission reductions within various regions of Europe, trends are evaluated in relation to air mass origin and of emission trends in the corresponding regions.

2. Methods

Daily measurements of SO_2, SO_4 and NO_2 from the five regionally representative rural sites Birkenes, Skreådalen, Kårvatn, Tustervatn and Jergul were included in this work. The location of sites is shown in Fig. 1a while for a more detailed description of the monitoring programme and results, it is referred to Aas et al., (2000) and references therein. The sampling equipment for SO_2 and SO_4 in air consisted of a 3-stage filter pack with one Teflon (2 μm Zefluor) particle filter followed by a potassium hydroxide impregnated (KOH) Whatman 40 and finally an oxalic acid-impregnated Whatman 40 filter (EMEP, 1996). Filters were extracted in water and all analyses were by ion chromatography. Two methods for sampling NO_2 were applied; by a low volume air sampler, where initially NO_2 was absorbed by TGS absorbing solution, and later by NaI impregnated glass frits, both on a daily basis. The dates for the shifts in methodology were; Kårvatn 20/2.1992, Birkenes 1/1.1993, Jergul 4/5.1993, Tustervatn 1/6.1993 and Skreådalen 11/8.1994. For both methods, extracts were analysed for nitrite by spectrophotometry. The precision of the analysis based on regular analysis of control samples was in the order of 3-5% given as the relative standard deviation over one year of analysis.

To classify the origin of the air masses arriving at each measurement site, 96 h back trajectories were calculated every 6 h, using winds for the 0.925 sigma-level (approx. 700 m above ground) from the numerical weather prediction model at The Norwegian Meteorological Institute. A transport sector (see Fig. 1), between 1 and 5, was then allocated on a given day if more than 50% of the co-ordinate points between 150 and 1500 km from the receptor, on the 4 trajectories arriving that day, were within the sector. Otherwise the origin was classified as undetermined. For each sector, annual average concentrations were then calculated.

Time series analysis was performed using the Mann-Kendall test. To reduce the influence of inter-annual variations when estimating the relative changes, changes were estimated from polynomial functions fitted to the time series.

3. Results and discussion

Large reductions in concentration levels are seen both for SO_2 and SO_4, as well as for NO_2 (Table 1, Figures 1 and 2). During the examined period, trends in sulphur concentrations are generally monotonous, while for NO_x the reductions mainly have occurred from the late 1980's and with increasing levels in the previous years. The typical comparison between emission intensities and ambient concentrations is exemplified for the site Birkenes in figures 1b-1d. The concentration levels were significantly higher with transport from sectors 90-150° and 150-210°, corresponding well with the reported emission intensities within the corresponding regions. As illustrated, the trends are fairly similar for all air mass categories reflecting the similarity in relative change in emissions in the major emission areas (Table 1). However, the reductions tend to be relatively higher and in the order of 80-90 per cent in sectors 0-90° and 270-360°, which represents areas of low emission intensities. By 1985 concentration levels of SO_2 and SO_4 had already fallen significantly compared to the late 1970's (Aas et al., 2000). However, lack of historical wind fields limited this study to the period 1985-1996.

As also can bee seen from Table 1, the relative reductions in SO_2 has been significantly higher compared to SO_4. One likely explanation may be that the reductions in SO_2 emission close to the monitoring sites have been stronger compared to many more distant areas.

TABLE 1

Relative reductions (%) in ambient concentrations at selected rural sites in Norway (Fig 1a) in relation to air mass origin during 1985-1996 (*= data series for NO_2 at Kårvatn and Tustervatn started in 1988 and 1989 respectively).

SO_2	0-90°	90-150°	150-210°	210-270°	270-360°	Undeterm.
Birkenes	69	59	58	41	59	62
Skreådalen	90	78	58	63	83	63
Kårvatn	89	82	42	67	86	68
Tustervatn	83	83	70	81	94	80
Jergul	78	68	61	70	74	57

SO_4	0-90°	90-150°	150-210°	210-270°	270-360°	Undeterm.
Birkenes	48	40	39	9	42	37
Skreådalen	61	44	38	32	52	37
Kårvatn	58	49	36	46	47	24
Tustervatn	55	53	54	15	45	51
Jergul	61	48	57	58	60	52

NO_2	0-90°	90-150°	150-210°	210-270°	270-360°	Undeterm.
Birkenes	61	53	26	31	60	49
Skreådalen	49	37	20	29	47	32
Kårvatn*	60	50	41	45	50	49
Tustervatn*	66	67	41	52	62	42
Jergul	75	78	65	77	59	64

The trend in NO_2 concentrations is particularly of interest as the large reductions in sulphur emissions have increased the relative importance of nitrogen as an acidifying agent in Norwegian surface waters, and as being a precursor for surface ozone production. For the whole period, NO_2 concentrations are reduced by more than 20% in all sectors and typically by 40-60 per cent. It should be noted that the shift in the analytical method may have influenced the magnitude of the trend. However, trend analysis based on the running 90 percentile confirms the general decrease. If the trend is calculated using 1990 as initial year, even stronger downward trends are seen.

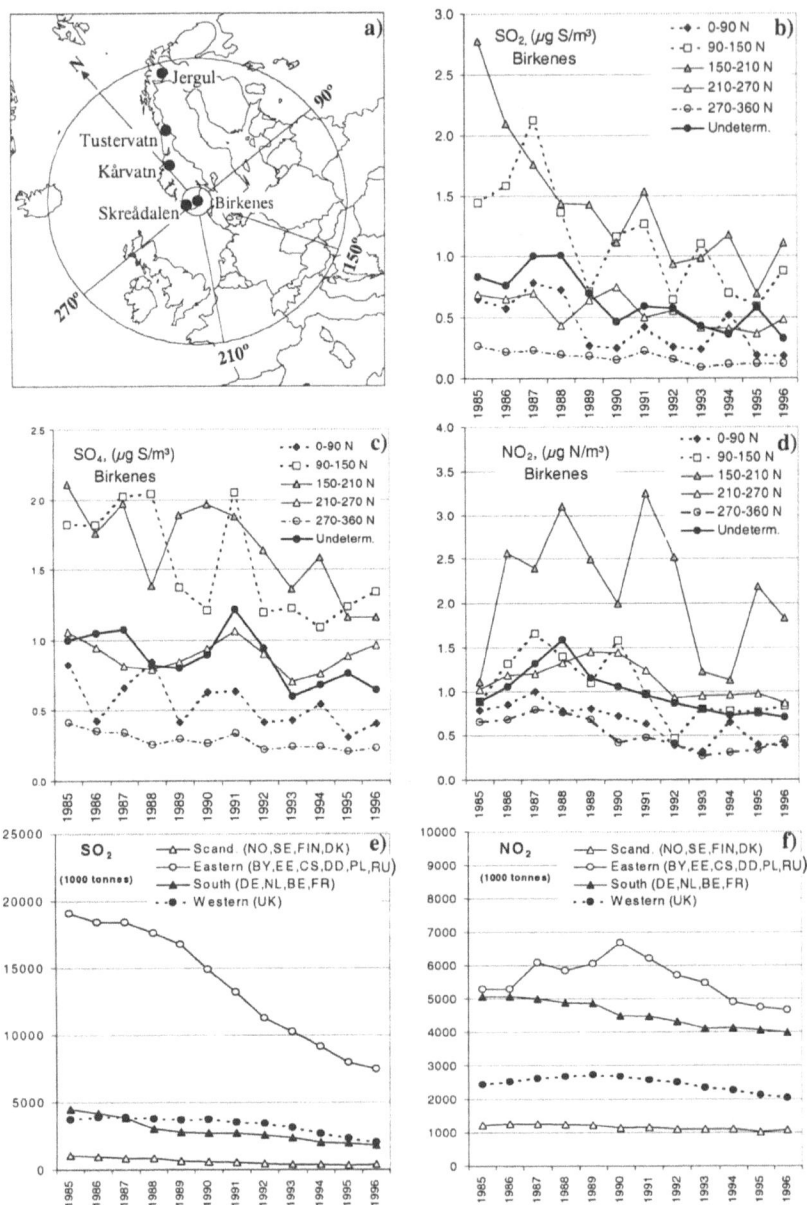

Figure 1. Location of sites and sector division (a), time series of SO_2, SO_4 and NO_2 by air mass origin at Birkenes (b-d) and official emission figures (e-f) (Tarrasòn and Schaug, 1999) during 1985-1996.

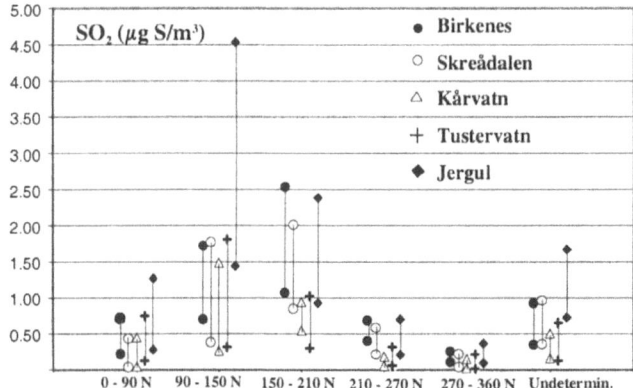
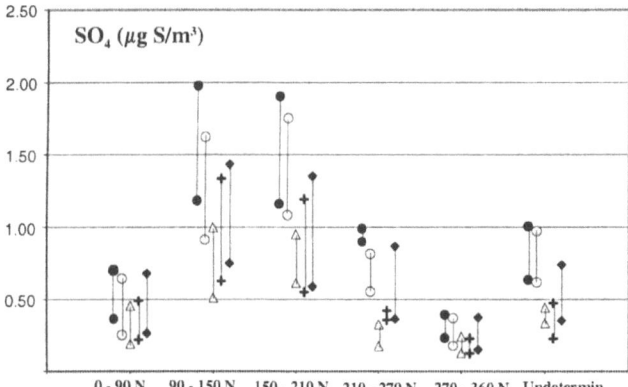
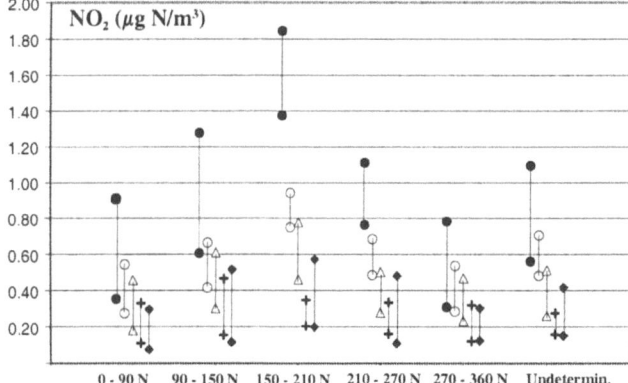

Figure 2. Change in ambient concentrations during 1985–96 estimated on the basis of polynomial functions fitted to the time series (changes in NO_2 at the sites Kårvatn and Tustervatn represents 1988-96 and 1989-96 respectively (upper point always represents initial value).

Variations in the advection patterns between years have a strong influence on the variation in annual mean concentration (i.e. the frequency of transport from major emission regions and particularly during winter). Even stronger variations are seen within the various sectors and particularly for sectors of low frequency (e.g. 90-150° and 150-210° typically representing 5 to 10 per cent each). On the other hand, years with a high frequency of transport from the sectors 210-270°, 270-0° and 0-90° during the winter season yield low annual concentrations. However, relatively high occurrence of transport from sectors 90-150° and 150-210° during 1993-1996 implies that reduction in emission intensities explains the reduction in ambient concentrations.

For NO_2 a clear spatial trend with decreasing levels with increasing latitude can be seen (Fig. 2). For SO_2 and SO_4 concentration levels also decrease with latitude in southern Norway, but tend to increase further north. This is due to a stronger influence of air transport from Eastern Europe. Particularly the site Jergul is influenced by large emission in the Nikel area, close to the Norwegian border in NW Russia, with transport from the sector 90-150°.

Acknowledgements

Thanks are due to Dr. Hilde Sandnes Lenschow and Dr. Leonor Tarrasòn at the Norwegian Meteorological Institute (EMEP MSC-W) for providing the windfield data and trajectory programme used in the sector analysis. The measurements of air chemistry were financed by the Norwegian Pollution Control Authority.

References

Aas, W., Tørseth, K. Solberg, S., Berg, T. and Manø, S.: 2000, Norwegian Institute for Air Research (NILU OR 23/2000), Kjeller. pp. 146.
Drabløs, D. and Tollan, A. (eds.): 1980, *Ecological impact of acid precipitation*. Proceedings of an international conference, Sandefjord, Norway, March 11-14, 1980. Oslo-Ås. pp 383.
EMEP: 1996, Norwegian Institute for Air Research (EMEP/CCC Report 1/95).
Tarrasòn, L. and Schaug, J.(eds.): 1999, Norwegian Meteorological Institute Research Report no. 83 (EMEP Report 1/99), Oslo. pp 246.
UN ECE: 1999, Document EB.AIR/1999/1, United Nations, Economic Commission for Europe, Geneva, pp 71.

CHEMICAL COMPOSITION AND ACIDITY OF PRECIPITATION: A MONITORING PROGRAM IN NORTHEASTERN VIETNAM

P. H. VIET[1*], V. V. TUAN[2], P. M. HOAI[1], N. T. K. ANH[2] and P. T. YEN[1]

[1] *Centre of Environmental Chemistry, Hanoi University of Science, 334-Nguyen Trai Street, Hanoi, Vietnam;* [2] *Institute of Hydrology and Meteorology, Lang Street, Hanoi, Vietnam.*
(*author for correspondence, e-mail: vietph@hn.vnn.vn)*

Abstract. Rainwater has been sampled weekly from five sites (nos. 1-5) in northeastern Vietnam in the period of May 1997 to Apr. 2000 (except Hoabinh site, from Aug. 1999 to Apr. 2000). Since Aug. 1999, weekly dry deposition samples including acidic gas and aerosol have been additionally collected at Hanoi (no. 4) and Hoabinh (no. 5) using filter pack system. In general, the pH in rainwater was frequently higher than 5.0. However, the trend of lower pH was observed during the winter and the beginning of autumn (from Nov. to Apr.). Interestingly, the highest frequency of the acidifying rainwater (32 %) and the lowest pH value (min. pH = 4.0) were observed in Hoabinh site. Acidic pH of rain water was also observed in Viettri (no. 3) and Hanoi (no. 4), indicating the local effects of human and industrial activities. Ca^{2+} and SO_4^{2-} were generally found as predominant in both rainwater and aerosol. SO_2 and NH_3 in Hanoi and Hoabinh were monitored out of corresponding environmental features.

Keywords: acidity, chemical composition, dry deposition, northeastern Vietnam, rainwater

1. Introduction

In different Asian countries (Hayami and Carmichael, 1997; Granat *et al.*, 1996; Ayers and Yeung, 1996), especially in Japan and China where acid precipitation are recognized (Yamaguchi *et al.*, 1991; Wang and Wang, 1996; Wang, 1995), acidity and/or chemical composition of atmospheric precipitation have been the subject of several research programs. In Vietnam, atmospheric precipitation has been monitoring based on the monitoring network setting up by the General Department of Hydrology and Meteorology. However, the acidity and chemical composition of wet and dry precipitation had been only simultaneously evaluated in short period in northern Vietnam (Viet *et al.*, 1998).

Although pH of event and monthly rainwater from previous study (Viet *et al.*, 1998) were generally basic or neutral, understanding the quality of rainwater and its spatial and temporal evolution of pH and chemical composition is essential. From May 1997 to Apr. 2000, we have continually collected weekly rainwater samples, instead of event ones, at those mentioned four sites. Additionally, from Aug. 1999 to Apr. 2000, rainwater, aerosol and gaseous components have been monitored at Hanoi and Hoabinh sites, which are included in a research program "Acid Deposition Monitoring Network in East Asia - EANET".

In this paper, we present the acidity and chemical composition of rainwater at five study sites (nos. 1-5) and of aerosol and acidic gas at two sites (nos. 4 and 5) in northeastern Vietnam (Fig. 1). The effects of environmental features to the acidity and the chemical composition will be also discussed when possible.

2. Materials and Methods

2.1. SAMPLING SITES

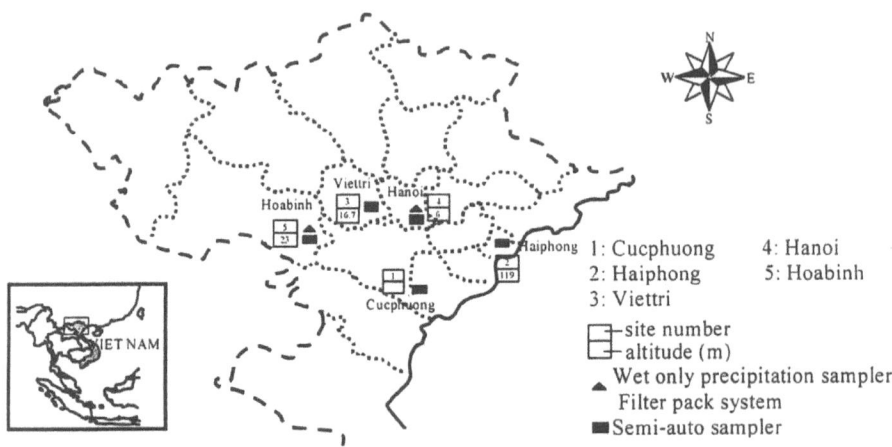

Figure 1. Sampling sites

Samples were collected on a weekly basis at five sampling sites, where characterized by typical tropical monsoon climate of the north and represented for various environmental features (Fig. 1). Cucphuong site (no. 1), Haiphong site (no. 2), Viettri site (no. 3) and Hanoi site (no. 4) are considered as representative for the clean atmospheric area, coastal area, industrial area and urban area, respectively. Detail characteristics of those sites were described elsewhere (Viet et al., 1998). Hoabinh site (no. 5) is located on the Da river basis and is near to Hoabinh hydropower station, the biggest power plant in Vietnam. Situated at the latitude of 20°49'N and the longitude of 105°20'E, Hoabinh is an important site which is representative for an ecological and rural area.

A semi-auto sampler has been equipped at site nos. 1-4 to collect weekly rainwater. The weekly rainwater was collected by mixing all event samples in a week. Details of that sampler, sampling procedures and sample preservation were described in Viet et al. (1998). Since Aug. 1999, a wet only collector and a filter pack system, including a pump, filter and flow rate meter, have been set up at site nos. 4 and 5 in order to collect rainwater, aerosols and acidic gas.

2.2. ANALYSIS

2.2.1. *Sample Preparation*
pH and conductivity measurement of rainwater as well as the use of quality control samples were followed by the method reported previously (Viet et al., 1998). Upon receipt in the laboratory, rainwater was filtered through 0.25 µm pore ID. filter. pH and conductivity were simultaneously measured within 24 hours. Collected gaseous and aerosol filters were extracted for 20 minutes by

deionized water or corresponding solutions in ultrasonic bath (Anonymous, 1997). The extracts were then filtered through 0.25 μm pore ID. Samples were then stored in precleaned polyethylene bottles in the dark at 4°C until analysis.

2.2.2. *Ion Measurements*
Prepared samples were analyzed for ion parameters by Ion Chromatography (Shimadzu LC-10A, Japan) equipped with a conductivity detector. Running program was followed by the method optimized by Viet *et al.* (1998). Briefly, anion concentrations (Cl^-, NO_3^- and SO_4^{2-}) were determined using column IC-A3 and the eluant of parahydroxybenzoic and tris (hydroxy) methyl-aminomethane. Cation concentrations (Na^+, NH_4^+ and K^+) were measured using IC-I1 and the eluant of 5 mM HNO_3. Meanwhile, Mg^{2+} and Ca^{2+} were measured using 4 mM tartaric acid/2 mM ethylendiamin as eluant and using the same IC-I1 column.

3. Results and Discussion

3.1. ACIDITY AND CHEMICAL COMPOSITION OF RAINWATER

3.1.1. *Ionic Concentrations*
Minimum, maximum and mean concentrations ($\mu eq\ l^{-1}$) of major ions, together with pH and conductivity ($\mu S\ cm^{-1}$) values of weekly rainwater are given in Table I. Ion balance between equivalent concentrations of the total cations and that of the total anions in rainwater from sites were examined. The ratios of 0.7 to 1.4 (Table I) indicated that a quantitative ion balance was achieved.

Depending up on the location, the level of an anion or cation in rainwater was different. In general, the major elements were SO_4^{2-}, Ca^{2+} and NH_4^+ (Table I). SO_4^{2-}, NO_3^- and Cl^- relatively dominated in industrial and urban areas (no. 3 and no. 4), explaining the high level of human activities such as domestic heating, automobile traffic, etc. and industries. The presence of the lowest concentrations of NH_4^+, SO_4^{2-} and NO_3^- at no. 1 is seem consistent with the clean atmospheric feature and can be due to the transportation of air masses from neighbor areas.

The major ion concentrations were generally measured relatively higher in Haiphong, Viettri and Hanoi comparing to Cucphuong and Hoabinh. This observation is probably consequence of two phenomena:
- Haiphong, Viettri and Hanoi sites are situated in the main cities in the north of Vietnam, where there is high level of human and industrial activities. Hanoi site is situated in Hanoi capital with the population of around 3 million. Although Viettri and Haiphong sites are situated in areas that have fewer inhabitants, these sites could be effected by many main industries located in these areas such as chemical factory, paper-sugar-wine plants, etc. in Viettri; and Haiphong cement plant, thermal power plant Thuongly, etc. in Haiphong. Whereas, Cucphuong and Hoabinh sites are situated in areas where human activities are less and industries are fewer.

- The rainfall amount at Haiphong, Viettri and Hanoi are less than that at Cucphuong and Hoabinh, thereby leading to less diluted pollutants. The average annual rainfall is recognized as 1786 mm, 1644 mm and 1661 mm at Haiphong, Viettri and Hanoi, respectively, compared to 2175 mm and 1948 mm at Cucphuong and Hoabinh, respectively.

TABLE I

Conductivity (μS cm^{-1}), pH and concentrations (μeq l^{-1}) of major ions in weekly rainwater collected from May 1997 to Apr. 2000 (except Hoabinh site, from Aug. 1999 to Apr. 2000)

	Cucphuong (n = 91)		Haiphong (n = 89)		Viettri (n = 75)		Hanoi (n = 83)		Hoabinh (n = 25)	
	Mean	Min-Max	Mean	Min-Max	Mean	Min-Max	Mean	Min-Max	Mean	Min-Max
Cl$^-$	13	<0.3 - 31	30	15 - 60	23	12 - 37	28	5 - 68	29	4 - 60
NO$_3^-$	19	7 - 40	18	5 - 31	26	13 - 43	25	4 - 76	16	9 - 35
SO$_4^{2-}$	30	14 - 60	41	20 - 75	54	27 - 86	46	11 - 97	41	4 - 92
Na$^+$	13	5 - 30	28	13 - 61	15	5 - 27	16	<0.3 - 73	9	1 - 50
NH$_4^+$	14	5 - 32	16	7 - 41	19	9 - 35	20	1 - 86	14	1 - 46
K$^+$	3	1 - 11	3	1 - 8	14	2 - 29	6	1 - 21	4	1 - 11
Mg^{2+}	2	<1 - 4	2	1 - 4	2	1 - 5	7	1 - 31	9	4 - 15
Ca^{2+}	20	9 - 64	30	10 - 59	35	10 - 55	40	16 - 92	25	2 - 86
H$^+$	2	0 - 6	1	0 - 6	3	0 - 50	4	0 - 79	13	0 - 100
pH	5.9	5.2 - 6.6	6.0	5.2 - 6.9	5.6	4.3 - 6.2	5.8	4.1 - 6.6	5.5	4.0 - 7.1
Cond.	9	5 - 14	12	7 - 18	14	7 - 33	13	4 - 41	15	6 - 53
Cation/Anion	0.7	0.9 - 1.2	0.7	0.9 - 1.3	0.7	0.8 - 1.3	0.7	1.0 - 1.4	0.8	1.0 - 1.4
Rainfall (mm)	37	0 - 91	35	0 - 80	34	0 - 76	35	0 - 78	38	0 - 102

3.1.2. pH of Rainwater

We have shown that pH of collected rainwater was frequently higher than 5.0, the value in pure rainwater (Charlson and Rodhe, 1982). However, acidic pH (pH< 5.0) was recognized in some samples collected during winter and beginning of autumn (from Nov. to Dec.); when the rainfall amount was low. Statistically, the frequency of acidic pH was 32 % (n = 8/25), 17 % (n = 14/83) and 11 % (n = 8/75) in Hoabinh, Hanoi and Viettri, respectively, whereas almost pH of collected rainwater were higher 5.0 in Cucphuong and Haiphong. It is recognized that the levels of NO$_3^-$, SO$_4^{2-}$, NH$_4^+$ and Ca^{2+} were higher in Hoabinh, Viettri and Hanoi comparing to those in Cucphuong and Haiphong. While most of the presence of Ca^{2+} can be attributed to the collection of anthropogenic origin such as from cement work, etc, the ratios NH$_4^+$/(SO$_4^{2-}$ + NO$_3^-$) could indicate the possible source of these ions and the considerable cause for the pH (Johnson et al., 1987). The ratios were much fluctuated in collected rainwater revealing the fluctuation of pH value. In general, this ratio was calculated much lower than 1 in rainwater which had low pH indicating that SO$_4^{2-}$ and NO$_3^-$ had not their origin only in the aerosols but also from the direct dissolution of oxides: SO$_2$, NO$_2$.

Surprisingly, the highest frequency of acidic pH (32 %) and the lowest pH of rainwater were observed in Hoabinh site (min. pH = 4.0) (Table I), an ecological and rural area. If high level of human activities and many industries are considered as a result of acidic pH in rainwater at Viettri site (11 %) and Hanoi site (17 %), long or intermediate range transport phenomenon should be accounted for explaining the obtained pH values at Hoabinh site. However, because this site have been set up for monitoring atmospheric deposition since Aug. 1999, the nine-month data in this study is not large enough to take in account for finding the sources of acidic rainwater. Further study and larger number of sampling points at the area should be undertaken.

Comparing to Hoabinh, higher mean pH was recognized in Hanoi (pH = 5.8) and Viettri (pH = 5.6), of which the environmental features of urban and industrial are seemly matched. These results are remarkable as Cl^-, NO_3^- and SO_4^{2-} concentrations were observed lower in Hoabinh site than others (Table I). According to Samara *et al.* (1992), this phenomenon means that the alkalinity, which is imported to the rainwater by NH_3 and fly ash components ($CaCO_3$, oxides, etc.) from anthropogenic origin such as traffic, cement work, etc., is more important in Viettri and Hanoi sites. Practically, the highest concentrations of NH_4^+ were found in Viettri and Hanoi.

TABLE II

Mean, minimum and maximum aerosol compositions (µg m^{-3}) and gaseous concentrations (10^{-3} ppm) of weekly samples collected in Hanoi and Hoabinh

	Hanoi (no. 4) (n = 37)		Hoabinh (no. 5) (n = 37)	
	Mean	Min - Max	Mean	Min - Max
Na^+	0.16	< 0.01 - 0.74	0.16	< 0.01 - 0.76
NH_4^+	0.54	< 0.01 - 2.66	0.50	< 0.01 - 1.69
K^+	0.44	< 0.03 - 1.08	0.08	< 0.03 - 0.16
Mg^{2+}	0.15	< 0.03 - 0.25	0.14	< 0.03 - 0.19
Ca^{2+}	0.94	< 0.01 - 2.88	0.88	< 0.01 - 2.19
Cl^-	0.24	< 0.01 - 1.11	0.27	< 0.01 - 1.54
NO_3^-	0.53	< 0.01 - 1.94	0.40	< 0.01 - 1.31
SO_4^{2-}	2.69	< 0.01 - 12.1	1.87	< 0.01 - 5.08
nss-SO_4^{2-}	2.65	< 0.01 - 11.87	1.83	< 0.01 - 4.89
NH_3	7.23	2.51 - 18.86	5.31	1.27 - 9.55
HCl	0.31	0.17 - 0.78	0.27	0.14 - 0.36
HNO_3	0.38	0.06 - 0.77	0.25	0.02 - 0.53
SO_2	0.91	0.39 - 2.27	1.22	0.63 - 2.28

3.2. AEROSOL COMPOSITION AND GASEOUS COMPONENTS

Aerosol composition from Aug. 1999 to Apr. 2000 (37 weekly samples) in Hanoi and Hoabinh is shown on Table II. It is recognized that the most dominant ions were SO_4^{2-}, Ca^{2+} and NH_4^+, of which 98 % is nss-SO_4^{2-} and account for one half of the total soluble aerosol composition. The higher level of Ca^{2+} and SO_4^{2-} in Hanoi (no. 4) compared to those in Hoabinh (no. 5) is probably accounted for the high level human activities and industries. This trend is significantly suitable with the levels of those anions observed in rainwater.

Also listed in Table II are the gaseous. It is revealed that NH_3 and SO_2 were predominant. Reversibly, SO_2 observed in Hoabinh (1.22 ppb), an ecological and rural area, was averagely relatively higher than in Hanoi (0.91 ppb), an urban area. Whereas NH_3 measured in Hoabinh (5.31 ppb) was averagely lower than in Hanoi (7.23 ppb). However, these results seem coincide with pH value of rainwater collected when pH is observed lower in Hoabinh than in Hanoi. Further study should be undertaken for explaining significantly this phenomenon.

4. Conclusions

Rainwater samples collected weekly at five areas in northeastern Vietnam from Mach 1997 to Apr. 2000 (except in Hoabinh site, from Aug. 1999) were analyzed to assess the acidity and chemical composition. Effects of environmental features to the acidity and the chemical composition were also briefly discussed when possible. In addition, aerosol composition and gaseous components were evaluated in Hanoi and in Hoabinh from Aug. 1999 to Apr. 2000.

We observed that generally SO_4^{2-}, Ca^{2+} and NH_4^+ were major elements in rainwater and aerosol. However, major ions in rainwater are relatively higher in Haiphong, Viettri and Hanoi comparing to Cucphuong and Hoabinh. The reasons are probably due to the high level of human and industrial activities at those sites.

Acidic pH was recognized in some rainwater samples. Especially the pH in rainwater was observed lowest at Hoabinh, an ecological and rural area. This phenomenon is possibly due to the transport of air mass from neighbor area, where the acidic components are highly produced. This suggestion was coincident when SO_2 concentration were found relatively higher in Hoabinh than those in Hanoi. Further study should be conducted to give more understanding on this phenomenon.

References

Anonymous: 1997, *Guidelines and Technical Manuals for Acid Deposition Monitoring Network in East Asia*, Environment Agency, Government of Japan.
Ayers, G. P. and Yeung, K. K.: 1996, *Atmospheric Environment* **30**, 1581.
Charlson, R. J., and Rodhe, H.: 1982, *Nature* **295**, 683.
Granat, L., Suksomsankh, K., Simachaya, S., Tabucanon, M. and Rodhe, H.: 1996, *Atmospheric Environment* **30**, 1589.
Hayami, H. and Carmichael, G. R.: 1997, *Atmospheric Environment* **31**, 3429.
Johnson, C. A., Sigg, L. and Zobrist, J.: 1987, *Atmospheric Environment* **21**, 2365.
Samara, C., Tsitouridou, R. and Balafoutis, C. H.: 1992, *Atmospheric Environment* **26B**, 359.
Viet, P. H., Hoai, P. M., Trung, N. X. and Nam, V. D.: 1998, *Composition and Acidity of Asian Precipitation (CAAP)*. Proceedings of the Forth Meeting of the IGAC/DEBITS Project. Chulalongkorn University, Bangkok, Thailand.
Wang, W.: 1995, *Water, Air and Soil Pollution* **85**, 2295.
Wang, W., and Wang, T.: 1996, *Atmospheric Environment* **30**, 4091.
Yamaguchi, K., Tatano, Tsutomo., Tanaka, F., Nakao, M., Gomyoda, M., and Hara, H.: 1991, *Atmospheric Environment* **25A**, 285.

ATMOSPHERIC CONCENTRATIONS OF SULFUR DIOXIDE AND NITROGEN DIOXIDE IN CHINA AND KOREA MEASURED BY USING THE IMPROVED PASSIVE SAMPLING METHOD

MUNEHIRO WARASHINA[1]*, MASANOBU TANAKA[1], YOSHIO TSUJINO[2], TUGUO MIZOGUCHI[3], SIRO HATAKEYAMA[4] and YASUAKI MAEDA[5]

[1] Osaka City Institute of Public Health and Environmental Sciences, 8-34 Tojo-cho, Tennouji-ku, Osaka 543-0026 Japan; [2] Osaka Prefecture Environmental Pollution Control Center, Nakamichi 1-3-62, Higashinari-ku, Osaka 537-0025, Japan; [3] Bukkyo University, 96 Murasakino-kitahananobo-cho, Kita-ku, Kyoto 603-8301 Japan; [4] National Institute for Environmental Studies, 16-2 Onogawa, Tsukuba, Ibaraki 305-0053 Japan; [5] Osaka Prefecture University, 1-1 Gakuen-cho, Sakai city, Osaka 599-8531 Japan.
(* author for correspondence, e-mail: wara@rinku.or.jp)

Abstract. The results of ambient sulfur dioxide (SO_2) and nitrogen dioxide (NO_2) concentrations measured in ten cities of China and Korea by the improved passive samplers are reported. The property of this sampler is the utilization for the long-term exposure to the high level of SO_2 and NO_2. In this method, the conversion coefficients from the analytical data to the ambient concentrations were obtained from the comparison with the direct concentrations through the automatic analyzers for SO_2 and for NO_2, respectively. The interesting monthly variations were observed in the ambient SO_2 and NO_2 concentrations measured by this passive sampler method, which seems to give important information to the formation of acid rain in these countries.

Keywords: China, Korea, NO_2, passive sampler, SO_2

1. Introduction

A passive sampler method is cheap and has the advantage of the small demand for the sampler setting area, and then it is easy to obtain data in many polluted places. Moreover, this method has the merit of the simultaneous analyses for some gases absorbed on the sampler. This method, however, has the disadvantages, i.e., the long-term exposure of the sampler to the ambient air generally gives the inaccuracy data easily: it is impossible to obtain the direct concentrations of air pollutants. Most of the inaccuracy data obtained with the sampler are caused by the change of the adsorption capacity in the sampler. The capacity is largely affected by variation factors of the ambient air such as the moisture, wind velocity and acidic gas concentrations.

At present, various passive samplers are on sale in Japan, especially for the analyses of ambient nitrogen oxides (NO and NO_2). These samplers have the advantage and disadvantage. For example, NO_2 sampler improved for reducing the effect of the wind velocity on the adsorption has too short diffusion length to be adapted to the long-term exposure of one month, although high accuracy can be obtained for a short term exposure, as the adsorption area toward the polluted gases is very small (Yanagisawa et al., 1980). Moreover, this sampler has the disadvantage of no recycling. Most of the recycled types of samplers for the analyses of ambient NO, NO_x and SO_2 are effected strongly by the wind velocity and the moisture (Hirano et al., 1981, 1991).

The Authors have reported the improved passive sampler for overcoming such problems; it has the longer diffusion length (the broader absorption area) and its air collection part is covered with a water-non-permeable filter (Warashina et al., 1994, 1996). This sampler obtains the following properties in comparison with those of others;

1) The longer diffusion length may make the adsorption of larger amounts of the ambient NO, NO_x and SO_2 possible. As a result, the long-term collection of air pollutants may be possible.

2) The water-non-permeable filter may reduce the large influence of moisture on the adsorption capacity in the long - term air collection.

In this paper, the following results on this sampler would be reported;

1) The conversion coefficients of the amounts of the adsorbed SO_2 and NO_2 to the passive sampler to the concentrations of the ambient SO_2 and NO_2 were determined from the comparison of these analytical data with the hourly concentrations measured by the conductometric analysis for SO_2 and Saltzman method for NO_2 in ambient air of Osaka City for a long time.

2) On the base of these conversion coefficients, the measurement results of ambient SO_2 and NO_2 concentrations in some cities of China and Korea through this improved passive sampler method were estimated.

2. Experimental

2.1. MATERIALS AND METHODS

The structure of the improved passive sampler is shown in Figure 1. The material of this sampler is composed of three parts made of polypropylene. A collection disc filter (19 mm ϕ) impregnated the reagent [70 μL of triethanolamine solution (10% in acetone) for NO_2 adsorption, 70 μL of 5% sodium carbonate aqueous solution for SO_2 adsorption] was placed at the bottom of polypropylene vessel. The collection disc filter was covered with the polyflon water-non-permeable filter and shielded with a cap with open part, through which the ambient air containing SO_2 and NO_2 was diffused. The distance between the collection disc and the cap was kept constant by inserting the spacer of 9 mm length.

2.2. SAMPLING METHODS AND ANALYTICAL PROCEDUREES

Six sets of three passive samplers were exposed to the ambient air to collect the acidic gases at the air monitoring station in Osaka City for the exposure period from one to six weeks between September 26 and November 7 in 1994. The filter in the exposed collection disc was taken out from each sampler and was extracted with the extraction reagents of 3 ~ 10 mL. The extraction reagents, 0.15%-H_2O_2 aqueous solution and distilled water were used for SO_2 and NO_2 respectively. The extraction was carried out under the ultrasonic irradiation for 20 min at 50 °C. Through these procedures, the adsorbed SO_2 and NO_2 were converted in the extraction reagents to sulfate (SO_4^{2-}) and nitrite (NO_2^-) anions respectively. The amounts of the adsorbed SO_2 and NO_2 were determined through the analyses for these anion concentrations by ion chromatography.

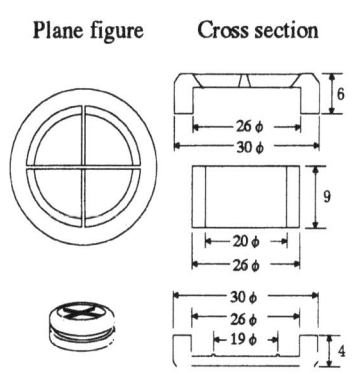

Figure 1. Diffusion passive sampler for NO_2 or SO_2 gas sampling.
Scales were described with the unit of mm.

The average conversion coefficients of the amounts of absorbed SO_2 and NO_2 on the collection disc filter to the ambient concentrations were obtained from the relationships between these amounts and the integrated values of hourly SO_2 and NO_2 concentrations measured by the automatic analyzers (the conductmetric method for SO_2 and Saltzman method for NO_2 respectively) in the ambient air of Osaka City. The improved passive samplers, whose conversion coefficients were determined, were placed at the several sites of Chongqing, Guiyang, Taiyuan, Shanghai, Shenyang, Beijing, Wuhan, Hong Kong in China and Taegu, Taejon in Korea for one month in two periods from July, 1994 to February, 1997 and from June, 1997 to February, 2000.

3. Results and Discussion

The relationships between the gas quantities collected by the passive samplers and the integrated concentration values of ambient SO_2 and NO_2 measured by the automatic analyzers for 1 ~ 6 weeks at the air monitoring station of Setsuyo in Osaka City are shown in Figure 2. A good correlation was obtained between the collected amounts and the cumulative values of ambient SO_2 and NO_2 concentrations for the periods of one and six weeks respectively as shown in Figure 2, which may suggest that the usage of the improved passive sampler for the long-term exposure to the ambient air having relatively high levels of acidic gases and moisture. On the base of the good correlations, the measurement of the ambient SO_2 and NO_2 concentrations by using the improved passive sampler was carried out in ten cities of China and Korea as described in experimental section.

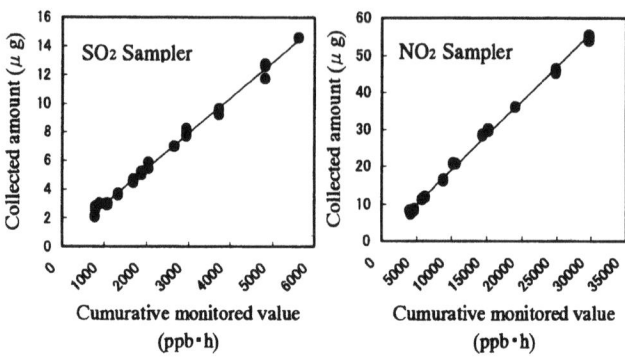

Figure 2. Relationship between collected amounts by passive samplers and monitored values by automatic analyzers for NO_2 and SO_2.

The measurement results of ambient SO_2 in Chinese cities were illustrated in Figure 3, which were classified into two categories, high SO_2 sites (Taiyuan, Guiyang, Beijing, Shenyang and Chongqing in Figure 3-a) and low SO_2 sites (Shanghai, Hongkong, Taegu, Taejon and Wuhan in Figure 3-b).

The following trends of the ambient SO_2 concentrations may be indicated from Figure 3-a. In Taiyuan, the high SO_2 concentrations above 200 ppb in the winter seasons and the lower values of about 80 ppb in the summer seasons were observed, respectively. The same seasonal variations were observed in the ambient SO_2 concentrations of Guiyang City. Moreover, the seasonal variations were found in the ambient SO_2 levels in Beijing City and Shenyang, which were in the range from 20 ppb in summer to 100 ppb in winter. On the other hand, such seasonal variations were not detected in the SO_2 concentrations in Chongqing City, which were high level near to 200 ppb in both seasons of 1995. The decrease of SO_2 levels of this city was observed in 1999 in spite of seasonal variation, which seems due to the effect of fuel change from coal to natural gas in some parts of combustion systems.

From Figure 3-b, the following suggestions may be proposed. The SO_2 levels in Shanghai and Taegu (Korea) were in the ranges of 20 ~ 50 ppb and 5 ~ 30 ppb respectively and the higher levels were detected in winter seasons between 1995 and 1996. Such seasonal variations, however, were not detected in the investigation of SO_2 concentrations in these cities between 1997 and 1999, which were in the lower levels. The SO_2 concentrations in Wuhan and Taejon were lower than those in Shanghai, e.g., in Taegu 10 ~ 20 ppb in winter and several ppb in summer. In Taegu and Taejon, the SO_2 levels detected between 1997 and 1999 were much lower than those between 1995 and 1996, which may suggest the effects of prevention countermeasures. In Hongkong, the around of 10 ppb were measured in both winter and summer, and the seasonal variation was not detected. The high SO_2 levels detected at many cities of China except Hongkong in winter seem due to the coal combustion, for the its consume is very low in Hongkong.

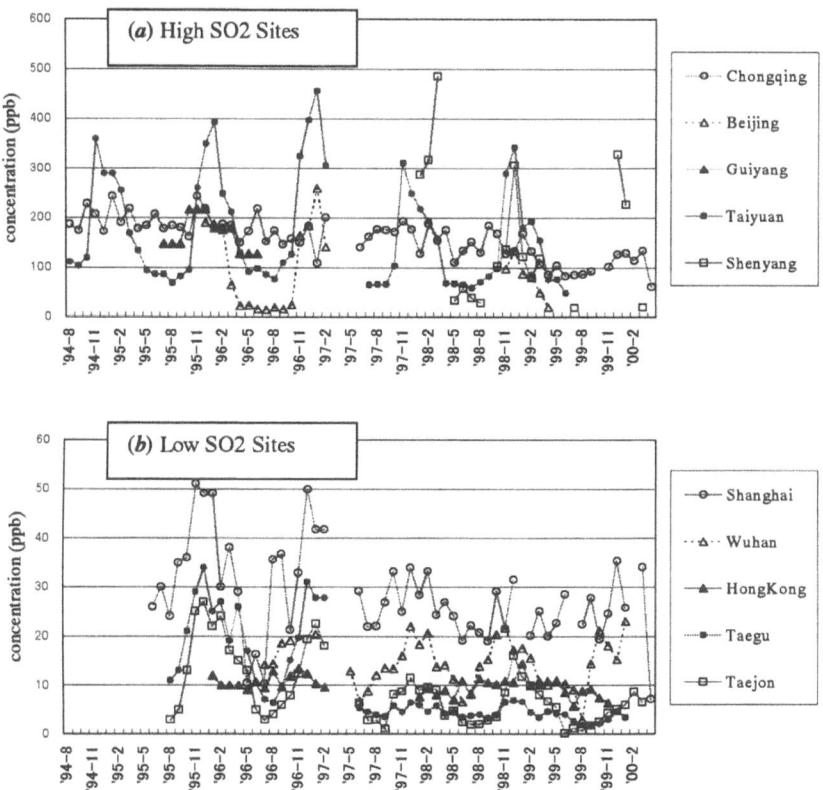

Figure 3. Monthly SO_2 concentrations in China and Korea.
a: Extremely high SO_2 was observed. b: High SO_2 was not observed.

The results of NO_2 concentrations measured by the improved passive sampler in these cities were shown in Figures 4-a and 4-b. The NO_2 levels detected at most of sites were in the range of 10~30 ppb, which were lower in comparison with the SO_2 levels described above. In Hongkong, Shanghai, Tegue and Taejon, a small seasonal variation of NO_2, i.e., the elevation of the level in autumn was observed. In Beijing, NO_2 concentration was low in winter, but was high in other seasons. In Chongqing, any seasonal variation was not detected.

In conclusion, the seasonal variations in the large range from 100 to several ppb were detected in the SO_2 and NO_2 concentrations measured by the improved passive sampler method. This method may be adapted to measure other air pollutants.

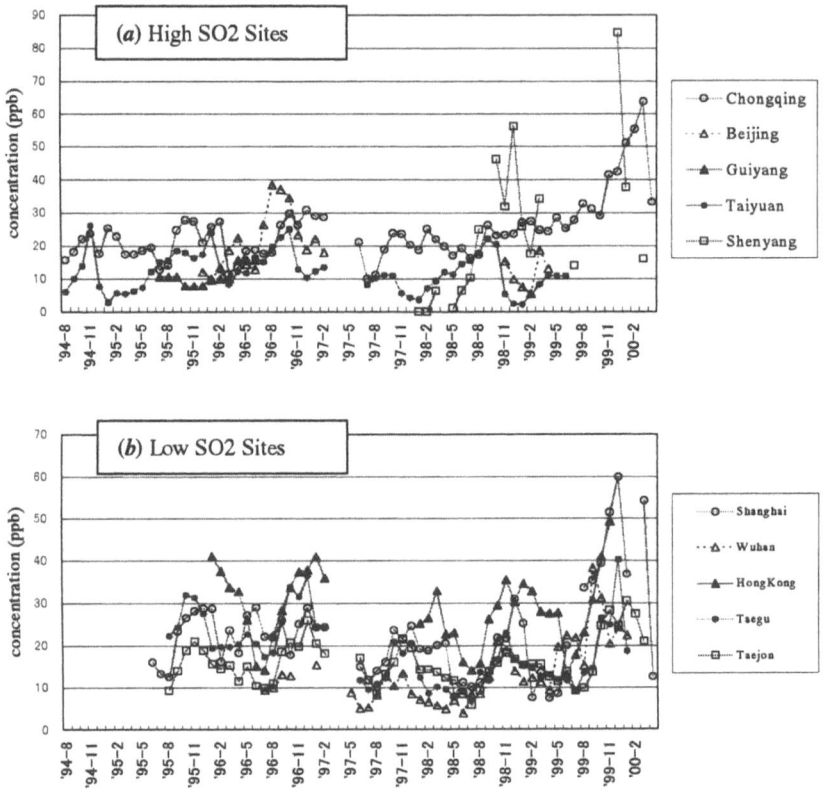

Figure 4. Monthly NO$_2$ concentrations in China and Korea.
a: Extremely high SO$_2$ was observed. b: High SO$_2$ was not observed.

References

Hirano, K., Maeda, H., Nakamura, M., Yoneyama, E.: 1981,'Relationship between personal exposure and NO, NO$_2$ gas concentration in ambient air', in *Annual Report of Yokohama Environmental Research Institute*, Yokohama Environmental Research Institute, Yokohama.

Hirano, K., Maeda, H., Matsuda, K.: 1991,'Diffusional Sampling Devices for Longer term Integrated Samples and the Simultaneous Determination NO, NO$_2$, SO$_2$ and Others in Ambient Air' in *Annual Report of Yokohama Environmental Research Institute*, Yokohama Environmental Research Institute, Yokohama.

Warashina, M., Nisiyama, Y., Tsujino, Y.: 1994, *Simple Measurement of Gaseous Pollutants Effective on Cultural Properties (2)*, Proceedings of the 35th Annual Meeting of the Japan Society of Air Pollution, Morioka City, Japan

Warashina, M.: 1996, Environmental Conservation Engineering, **25**, 10.

Yanagisawa, Y., Nishimura, H.: 1980, Journal of Japan Society of Air Pollution **15**, 316.

STUDY ON SAMPLING METHOD OF RAINFALL, THROUGHFALL, AND STEMFLOW TO MONITOR THE EFFECT OF ACID DEPOSITION ON FOREST ECOSYSTEM

MOTONORI TAMAKI[1*], TAKATOSHI HIRAKI[1], YOSHIHIRO NAKAGAWA[1], TOMIKI KOBAYASHI[1], MASAHIDE AIKAWA[1] and MITSURU SHOGA[2]

[1] *The Hyogo Prefectural Institute of Environmental Science, Yukihira-cho, Suma-ku, Kobe; Hyogo 654-0037, Japan;* [2]*Toyooka Health Center, Saiwai-cho, Toyooka, Hyogo 668-0025, Japan*
(*author for correspodence, e-mail: tamaki@pref.hyogo.jp)

Abstract. The use of samplers for rainfall, throughfall, and stemflow was studied in *Chamaecyparis obtusa* forest in Kobe to develop a suitable simplified collection method for long-term monitoring of the effect of acid deposition on the forest ecosystem. A filtrating bulk sampler, widely used in Japan due to its convenience, was modified for rainfall- and throughfall-sampling. The pH value, NH_4^+- and NO_3^-- concns. did not change within a two-week sampling period, and the collection efficiency of the modified type relative to the wet/dry sampler was 97% (mean). The three samplers (shampoo-hat, vinyl chloride tube, and gauze type) were used for stemflow sampling. Collection efficiency of the samplers was in the order of shampoo-hat> vinyl chloride tube> gauze and that of the gauze type varied significantly with rainfall condition.

Keywords: acid deposition, monitoring method, rainfall, stemflow, throughfall

1. Introduction

The first survey on rainwater in Japan was started in Tokyo and Kobe in 1935. Since then, nearly 4,000 reports on acid deposition have been published by ca. five hundred research groups in Japan. In the last decade, the trend of acid deposition over Japan has not changed, and the latest report by Japan Environment Agency, 1999, showed that the current mean pH values in wet deposition were 4.7-4.9. If acidification cannot be alleviated in the near future, the effects of acid deposition on Japan's terrestrial ecosystems will be greatly exacerbated. To evaluate of acid deposition on forest ecosystem, it is necessary to accumulate many kind of data. However, suitable sampling methods for throughfall and stemflow have not been established, comparing to those for rain- and fog water. In this paper, a simple method for monitoring the effect of acid deposition on a forest ecosystem is studied.

2. Experiment

2.1 FOREST STUDIED AND STUDY PERIOD

Water, Air, and Soil Pollution **130**: 1511–1516, 2001.
© 2001 *Kluwer Academic Publishers.*

The experiments were carried out at Kobe Municipal Arboretum (ca. 370 m above sea level) from Sept. 1991 to Mar. 1995, and routine measurement has been continued until the present. The arboretum has many kinds of trees on a 142 ha site, the largest in Japan, and is located in Kobe city, Hyogo prefecture, central Japan. The surrounding areas do not have major sources of air pollutants. The main forest used in the experiments was *Chamaecyparis obtusa*, and *Cryptomeria japonica-* and *Quercus serrata-* forest were also used for several experiments. All the forests were planted over fifty years ago.

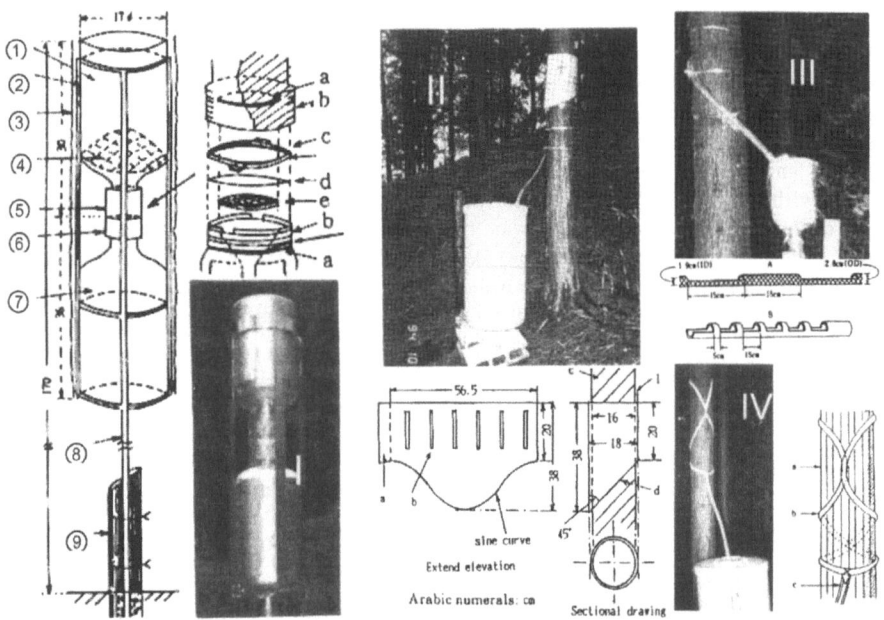

Figure 1. Rainfall- and throughfall- sampler (I: Modified filtrating sampler), (①polyethylene funnel, ②light shade: coating of silvery paint, ③light shade :black vinyl cover, ④nylon net, ⑤filter holder(filter:0.8 μ m, 47 mm ϕ, Nuclepore polycarbonate PC memb. or Millipore AAWP), ⑥air duct, ⑦polyethylene tank (100 L), ⑧stainless steel pole, ⑨stake, a:o-ring, b:screw, c:filter cover, d:filter, e:filter plate, f:air duct, Stemflow sampler (II:Shampoo-hat type) (a:overlap portion, b:prop, c:stem,d:bottom (polypropylene, fortified with silicone resin), Stemflow sampler (III:Vinyl chloride tube type) and Stemflow sampler (IV:Gauze type) (a:trunk, b:gauze,c:vinyl chloride tube)

2.2 Experiment method

2.2.1 *Rainfall And Throughfall*

A filtrating bulk sampler(Hara *et al.*, 1990), of the type widely used in Japan due to its convenience, was modified for rainfall- and throughfall-sampling. The main improvements were as follows. The sampler was made compact, fixed at 1.7 m above ground level using a stainless steel pole, and shaded from sunlight by silvery paint and a black vinyl cover (Fig.1-I). Two kinds of filter (0.8 μ m, 47mm ϕ Millipore AAWG and Nucleopore polycarbonate) were used for separation of insoluble components. Three samplers were set in different locations near *Cryptomeria japonica* forest.

The type of sampler used for rainfall sampling also was used for a throughfall experiment. Two samplers were set in *Chamaecyparis obtusa* forest, within 2 m apart. One was a conventional rainfall sampler and the other was slightly modified as follows: a conical nylon net was placed inside a polyethylene funnel, to prevent disturbance of rainwater flow by leaves. The rainwater amount and concns. of chemical components obtained by the two samplers were compared. Rainfall, throughfall, and stemflow were collected on an event basis as a general rule, and pH, electric conductivity, SO_4^{2-}, NO_3^-, Cl^-, NH_4^+, Ca^{2+}, Mg^{2+}, K^+, and Na^+ in the rainwater were measured within one week after sampling by the method reported previously (Japan Environment Agency,1997).

2.2.2 Stemflow
Three type samplers for stemflow were tested in the *Chamaecyparis obtusa* forest, and set on tree trunks at 1.5 m above ground level.
<u>Shampoo-hat type: Polypropylene-cylinder type (TypeA)</u> A trunk was surrounded with a polypropylene film cylinder at the height of 1.5 m as shown in Fig1-II. The cylinder was fixed with polypropylene props, and the bottom of the cylinder was made of polypropylene film fortified with a silicone resin. The sample was led through a vinyl chloride tube, to a 100 L polyethylene tank shaded from sunlight with a black vinyl cover.
<u>Vinyl chloride tube type (Japan Environment Agency type,Type B)</u> A vinyl chloride tube (i.d. 19mm×o.d. 26mm, strengthened type fortified with Tetron (Toray & Teijin)), which was cut off as shown in Fig1-III, was wound around the trunk twice in a spiral and fixed with string. The gap between the trunk and the tube was packed with a silicone resin.
<u>Gauze type (Sassa et al.,1991, Type C)</u> Two twisted strips of gauze were wound around the trunk in a spiral as shown in Fig.1-IV. The length of the gauze was slightly greater than the trunk circumference.

The three samplers were set on different trunks of *Chamaecyparis obtusa* having similar diameter at breast height and crown projection area. Using actual and artificial rainwater, the collection efficiency of the samplers was examined. In the experiment using artificial rainwater, one liter of the water, stemflow collected in advance, was poured from 10cm upper part of each samplers at a velocity of 100 mL/min using a nozzle with 0.8 mm ϕ.

3. Result and discussion

3.1 RAINFALL AND THROUGHFALL

A fltrating bulk sampler is inexpensive and can prevent the change of sample quality. The results of measurements showed that the mean rainfall amount collected by the modified filtrating bulk sampler was about 3% smaller than that collected by the wet/dry sampler in the forest, and that the pH value, and NH_4^+-, and NO_3^- concns. did not change in the two-week sampling period(average values for ten samples, NH_4^+:20.5→20.2, NO_3^-:39.7→39.6(μ equivalent /L), and pH:4.42 →4.44). The sampler has been used in two forests for long-term monitoring over more than eight years without problem.

The difference in the sample amount among three points within 330 m distance and 45 m height was within 8%. Choice of sampling point appears to pose little problem, if there is no significant obstruction by trees. Comparing the data obtained by an automatic acid rain analyzer (Kimoto AR107SNA), set 2 km away from the forest, rainfall amount was slightly lower and the pH value was slightly higher. These difference is, however, slight, so these results confirmed that the modified sampler is suitable for monitoring in a forest.

The effectiveness of a net for protection against fallen leaves was examined. The rainfall amount obtained by the sampler without the net was larger than that with the net, but the difference was less than 1%. The pH value obtained by the sampler with the net was slightly higher than that without a net, but the difference was smaller than 0.1 pH unit, and the corresponding result for concn. of chemical component was similar. These results showed that the collection efficiency was not reduced and the chemical properties of rainwater were not affected by the net. Accordingly, the use of a nylon net is recommended not only for throughfall sampling but also for rainfall sampling.

3.2 STEMFLOW SAMPLER

Three types of sampler were used for the stemflow study. Outline data was firstly obtained by using artificial rainwater. Then, the efficiency value obtained by using actual rainwater was corrected by the crown projection area. In Table I, the difference in collection amount for artificial rainwater among the samplers is shown by addition of one liter of rainwater to the trunk. The mean collection efficiency calculated by the collection amount(mL) /one liter(1,000 mL) of rainwater is 75%, 76 %and 34%, respectively. The efficiencies for types A and B were high, but that for type C was lower than 50%. Assuming type A to be the standard method, the relative efficiencies for types B and C were 101% and 45%, respectively. The difference in collection amount of actual rainwater is also shown in Table I. Assuming type A to be the standard method, the relative efficiencies for types B and C were 56% and 43%. After correction by the crown project area, the efficiencies were 91% and 59%. The efficiency of type C varied widely for each rainfall event. Unlike artificial rainwater, actual rainwater flowed down from the upper part of the trunk and additional throughfall may have occurred. This factor may have caused the difference in efficiency between the two experiments. These results indicated that monitoring can be done using the shampoo-hat type, which has a high collection efficiency, especially for heavy rainfall.

TABLE I

Effect of stemflow sampler on rainfall collection efficiency. Exp.A: using 1 L artificial rainwater, Exp.B: using actual rainwater. R.C.E.: Collection efficiency for each sampler/collection efficiency for sampler A. C.C.E.:Corrected collection efficiency(%), Corrected by each crown projection area (4.9, 3.0 and 3.6 m² for samplers A-C). Error is 1 σ.

Sampler	[Exp.A] (Artificial rainwater):Rainwater collected (mL)						Colected/Added (%) (Collection Efficiency)	R.C.E. (%)
	Experimental No.							
	I-1	I-2	I-3	I-4	II-1	II-2		
A	810	750	730	700	770	745	75±4	100
B	730	760	790	780	680	810	76±5	101±10
C	330	330	330	300	380	350	34±3	45±3
Sampler	[Exp.B] (Actual rainwater):Rainwater collected (mL)						R.C.E. (%)	C.C.E. (%)
	Experimental No.							
	1	2	3	4	5	6	7		
A	8870	6900	18300	14200	8000	7420	10300	100	100
B	5020	5550	9150	10500	2800	2510	6250	56±18	91
C	2930	2250	5390	7930	4750	4070	3760	43±13	59

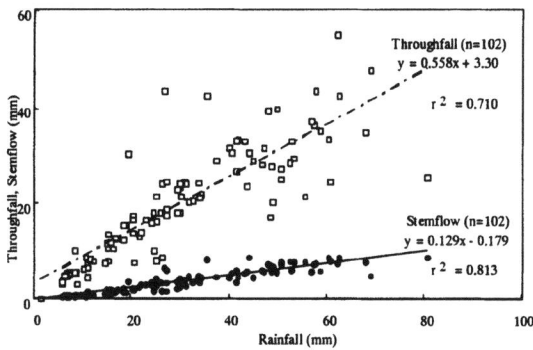

Figure 2. Relationship between throughfall □, stemflow ●, and rainfall. Kobe Municipal Arboretum, Apr. 1992-Mar. 2000, *Chamaecyparis obtusa* forest, monthly mean.

3.3 MEASUREMENT RESULTS

The rainfall, throughfall, and stemflow of *Chamaecyparis obtusa, Cryptomeria japonica,* and *Quercus serrata* were collected by a modified filtrating bulk sampler and a shampoo-hat type sampler. A tree studied, *Chamaecyparis obtusa* has 18.5 cm ϕ at breast height. The relationships between the collection amounts of rainfall, throughfall, and stemflow were expressed as linear equations, and in the case of Chamaecyparis obtusa forest(Fig.2), the equations for a period of eight years (Apr. 1992 to Mar. 2000) are as follows:

Throughfall amount (mm) = $0.558 \times$ Rainfall amount (mm) + 3.30,
Stemflow amount (mm) = $0.129 \times$ Rainfall amount (mm) - 0.179

The concns. of the main ions in the rainwater in *Chamaecyparis obtusa* forest were high. The analytical results (Apr. 1992-Mar. 1993) showed that the concns. of K^+ and Ca^{2+} in the throughfall were 13~19 times higher than those in the rainfall, and those in the stemflow were 30 times higher than those in the rainfall. The concns. of the main ions, except H^+, in the stemflow were higher than those in the rainfall. As the amount(mm) was 1~10% of the rainfall, the net loads of chemical components on the forest soil were in the order of rainfall> throughfall> stemflow.

4. Conclusion

Samplers for rainfall, throughfall, and stemflow were studied to develop a suitable method for long-term monitoring of the effect of acid deposition on a forest ecosystem. A filtrating bulk sampler was modified for rainfall and throughfall sampling. The pH value and the main components did not change in a two-week sampling period, and the sampler had high collection efficiency. Three types of samplers (shampoo-hat, vinyl chloride tube and gauze type)) were used for stemflow sampling. Collection efficiency of the samplers was in the order of shampoo-hat> vinyl chloride tube> gauze and that of gauze type varied significantly with rainfall conditions. These results indicated that monitoring can be done by the combination using a modified filtrated sampler which has simple design and shampoo-hat type sampler which has high collection efficiency and simple design.

References

Japan Environment Agency, 1997, *Guidelines and Technical Manuals for Acid Deposition Monitoring Network in East Asia*.

Hara,H.,Ito,E.,Katou,T.,Kitamura,Y., Komeiiji,T., Oohara,M., Okita,T., Sekiguchi,K., Taguchi,K., Tamaki,M., Yamanaka,Y., and Yoshimura, K.,1990, *Bull.Chem.Soc.Jpn.*, 3, 691.

Sassa,T., Gotoo,K., Hasegawa K. and Ikeda S.:1991, *Sinrin Ritti*,**32**,43 (In Japanese with English summary).

FOG AND PRECIPITATION CHEMISTRY AT MT. ROKKO IN KOBE, APRIL 1997-MARCH 1998

MASAHIDE AIKAWA[1*], TAKATOSHI HIRAKI[1], MITSURU SHOGA[2] and MOTONORI TAMAKI[1]

[1] *The Hyogo Prefectural Institute of Environmental Science, Yukihira-cho, Suma-ku, Kobe, Hyogo 655-0037, JAPAN;* [2] *Toyooka Health Center, Saiwai-cho, Toyooka, Hyogo 668-0025, JAPAN*
(*author for correspondence, e-mail: aikawa@pref.hyogo.jp)

Abstract. Fog water and precipitation were collected and analyzed to study fog and precipitation chemistry. The research was carried out through one year from April 1997 to March 1998 at Mt. Rokko in Kobe. Higher fog occurrence and larger volume of fog water were observed in summer, corresponding to the trend of seasonal variation in precipitation amount. The annual mean pH value of fog water (3.80) was lower by ca. one pH unit than that of precipitation (4.74). The concentration of chemical species in fog water was ca. 7 times that in precipitation. The highest anion and cation concentrations were SO_4^{2-} and NH_4^+ in fog water and Cl^- and Na^+ in precipitation, although the Cl^-/Na^+ equivalent ratio in both fog water and precipitation was almost the same value as that in sea water. It is considered that in the longest fog event, NH_4^+ and nss-SO_4^{2-} in fog water mainly scavenged as $(NH_4)_2SO_4$, mainly derived from $(NH_4)_2SO_4$ (aerosol) in the atmosphere, NH_3 was scavenged at the growing stage, and SO_2 was also scavenged after the mature stage. NO_3^- in this fog event was mainly absorbed as HNO_3.

Keywords: fog water, HNO_3, Kobe, Mt. Rokko, $(NH_4)_2SO_4$, precipitation

1. Introduction

Recently a substantial number of papers have been published on acid deposition, such as acid rain, acid fog and dry deposition, in Japan as well as in Europe and North America (Puxbaum *et al.*, 1998; Igawa *et al.*, 1998; Matsumoto and Okita, 1998; Fukuzaki *et al.*, 1999). Acid rain is the most well-known type of acid deposition, and has been researched since the 1970's, mainly in Europe (Linkens and Bormann, 1974; Penkett *et al.*, 1979; Durham *et al.*, 1981; Dillon *et al.*, 1988). Acid rain is partly responsible for forest decline, lake and soil acidification, etc. Thus, acid rain has been the subject of several research projects (NAPAP, 1985 ; Aneja V.P. *et al.*, 1992). In Japan, the Japan Environment Agency started a five-year precipitation chemistry monitoring program in the Kanto District in 1975. In the last two decades, in addition to acid rain, there has been increasing concern that acid fog may effect forest ecosystems to the same extent as acid rain (Fuzzi *et al.*, 1994; Kobayashi *et al.*, 1999; Vong *et al.*, 1991; Li and Aneja, 1992). However, there have been few year-long studies on both acid rain and acid fog, although there have been many studies on one or both for periods of less than one year. Therefore, in this research both fog and precipitation were collected for one year at the same sampling site. The purpose of this study is to examine the chemical characteristics of fog and precipitation and to discuss fog and precipitation chemistry.

2. Experimental

2.1. RESEARCH SITE

Research on fog and precipitation was carried out at Mt. Rokko in Kobe from April 1997 to March 1998. Mt. Rokko (931 m) is close to the urban area of Kobe, which is one of the biggest cities in Japan and is characterized by its intensive industrial development and high population density (over 1,460,000 people live within 550 km^2). The sampling was performed at the height of 800 m above sea level. The reason why Mt. Rokko was selected as the sampling site is its high frequency of fog occurrence in addition to the fact that forest damage has been observed.

Figure 1. Active string fog collector.

2.2. SAMPLING METHOD AND ANALYSIS

Fog sampling was performed by using an active string fog collector (Figure 1), which made it possible to collect 60 ml fog water samples in a separated by sampling bottle. Precipitation sampling was performed by using a wet only sampler.

The samples were immediately analyzed after collection for pH, electric conductivity (E.C.), and chemical species (Cl$^-$, NO$_3^-$, SO$_4^{2-}$, Na$^+$, NH$_4^+$, K$^+$, Mg^{2+}, and Ca^{2+}).

3. Results and Discussion

3.1. FOG OCCURRENCE AND PRECIPITATION

The precipitation amount and frequency of fog occurrence, number of hours of fog occurrence, and volume of fog water collected at Mt. Rokko are summarized in Table I. The precipitation at Mt. Rokko was higher in summer than in winter, showing a similar tendency to the volume of fog water, though frequency of fog occurrence and number of hours of fog occurrence were maximum in November.

TABLE I

Precipitation amount, frequency of fog occurrence, hours of fog occurrence, and volume of fog water at Mt. Rokko

	Precipitation amount (mm)	Fog		
		Frequency of occurrence (n)	Hours of occurrence (h)	Volume of fog water (ml)
Apr., 1997	1.1	5	32	749
May, 1997	152.6	6	64	1866
Jun., 1997	117.4	13	85	1874
Jul., 1997	332.3	19	226	7351
Aug., 1997	109.2	15	80	3032
Sep., 1997	217.8	10	99	2325
Oct., 1997	30.8	12	44	118
Nov., 1997	86.1	23	250	1397
Dec., 1997	56.2	2	-	774
Jan., 1998	122.7	2	-	132
Feb., 1998	62.3	5	36	197
Mar., 1998	200.0	8	104	2326
annual mean	124.0	10	102	1845
maximum	332.3	23	250	7351
minimum	1.1	2	32	118

3.2. pH AND ELECTRIC CONDUCTIVITY

Figure 2 presents the frequency distribution of pH value of fog water and precipitation. For fog pH value, which is shown for each fog event, values between 4.20 and 4.40 were observed with the highest frequency, which correspond to 16% for the whole fog pH data, and the annual mean, lowest and highest pH values were 3.80, 2.30 and 5.90, respectively. On the other hand, for precipitation pH value, which is shown for each sample collected every 2 weeks, the values between 4.60 and 4.80 occupied the largest proportion, which correspond to 30% for the whole precipitation pH data, and the annual mean, lowest and highest pH values were 4.74, 4.15 and 5.70, respectively. It can be seen that the pH value of fog water tended to be lower than that of

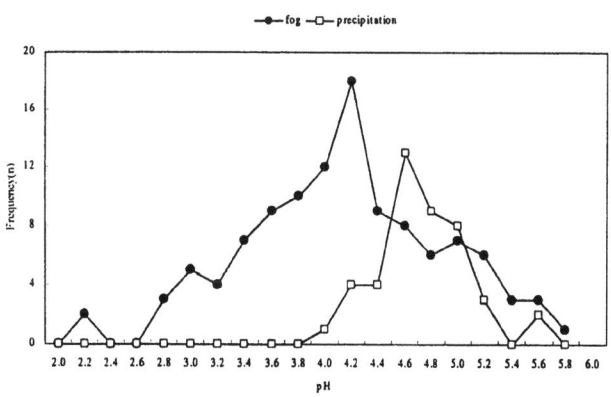

Figure 2. Frequency distribution of pH value of fog water and precipitation. Annual mean pH values of fog water and precipitation are 3.80 and 4.74.

precipitation.

The monthly mean E.C. values of fog water ranged from 49 μ S/cm to 343 μ S/cm, and the annual mean value was 115 μ S/cm. This annual mean value was larger than that of precipitation (16.8 μ S/cm). The E.C. values of fog water were larger than those of precipitation throughout the year.

3.3. CHEMICAL COMPOSITION OF FOG WATER AND PRECIPITATION

Figure 3 shows the annual mean chemical composition of fog water and precipitation. The total concentration of all chemical species in fog water (1240 μ eq./L) was ca. 7 times that in precipitation (170 μ eq./L). The Cl^-/Na^+ equivalent ratios were 1.25 and 1.13 for fog water and precipitation, respectively, which are almost equal to that of sea water (Cl^-/Na^+ equivalent ratio in sea water:1.17). Consequently, the Cl^- and Na^+ in fog water and precipitation are considered to mainly originate from the ocean. Among anions, SO_4^{2-} occupied the largest proportion in fog water and Cl^- in precipitation. Among cations, NH_4^+ occupied the largest proportion in fog water and Na^+ in precipitation. The Cl^- anion and Na^+ cation occupied the largest proportion in precipitation and the Cl^-/Na^+ equivalent ratio (1.13) is nearly the same as that for sea water (1.17), which indicates that the precipitation at Mt. Rokko was largely influenced by the ocean. Fog water was also affected by the ocean, since the Cl^-/Na^+ equivalent ratio (1.25) in fog water was also nearly equal to that of sea water. However, fog water was largely influenced by other substances, such as gases or particles in the atmosphere, which resulted in the ca. 7 times concentration in fog water compared with precipitation.

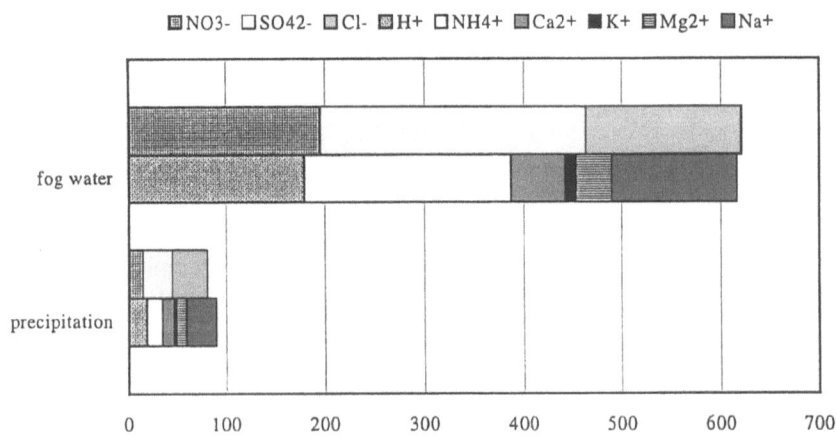

Figure 3. Annual mean chemical composition of fog water and precipitation.

Figure 4 shows the changes of pH value, H^+, NO_3^-, nss-SO_4^{2-}, NH_4^+, and nss-Ca^{2+} concentrations in a fog event that has 55 fog water fractions (one fraction corresponds to 60 ml of fog water). The event lasted from 5:57 July 26, 1997 to 16:29 July 29, 1997, and was the longest fog event in the study year. The concentrations of chemical species changed even within one fog event. It was found that these concentrations increased with the decrease of pH value. It suggests that pollutants such as SO_2, HNO_3 and $(NH_4)_2SO_4$ in the atmosphere mainly influence the change of pH. Among the above chemical species, NH_4^+ and nss-SO_4^{2-} showed good correlation. The correlation coefficient between NH_4^+ and nss-SO_4^{2-} was high, compared with other correlation coefficients, and the NH_4^+/nss-SO_4^{2-} equivalent ratio was near unity, which suggests that NH_4^+ and nss-SO_4^{2-} in fog water mainly derived from $(NH_4)_2SO_4$ (aerosol) in the atmosphere. Moreover, it was found that the NH_4^+/nss-SO_4^{2-} ratio was higher than unity at the growing stage, and approached a stable value (0.5 - 1.0) with time. At the mature stage and vanishing stage, the NH_4^+/nss-SO_4^{2-} ratio was relatively constant, but less than unity. An NH_4^+/nss-SO_4^{2-} ratio of unity suggests that nss-SO_4^{2-} is scavenged only as $(NH_4)_2SO_4$ (aerosol) in the atmosphere, and an NH_4^+/nss-SO_4^{2-} ratio higher than unity suggests other sources of NH_4^+; one possibility is NH_3 gas in the atmosphere. On the other hand, a ratio less than unity suggests other sources of nss-SO_4^{2-}; one possibility is SO_2 gas in the atmosphere. It is therefore considered that NH_4^+ and nss-SO_4^{2-} were mainly scavenged as $(NH_4)_2SO_4$ (aerosol) in the atmosphere, that NH_3 was scavenged at the growing stage, and that after the mature stage SO_2 was scavenged. NO_3^- showed good correlation with H^+, which suggests that the increase of NO_3^- was mainly due to the scavenging of HNO_3 gas in the atmosphere. The concentration of nss-Ca^{2+} was relatively constant, although a slight increase was observed at the growing and mature stages.

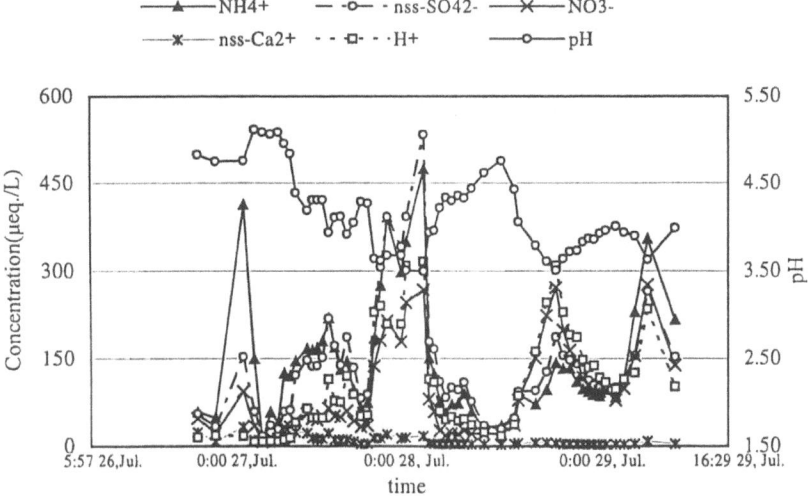

Figure 4. Changes of NO_3^-, nss-SO_4^{2-}, NH_4^+, and nss-Ca^{2+} concentrations in the longest fog event in the year, together with pH value. (5:57 July 26, 1997 to 16:29 July 29, 1997)

4. Conclusions

A study of fog and precipitation chemistry was performed at Mt. Rokko near Kobe city. Fog occurred with higher frequency in summer than in winter, showing a similar trend to precipitation amount. Low pH values were observed in fog water more frequently than in precipitation and the annual mean pH value in fog water was lower than that in precipitation. Both fog and precipitation were influenced by the ocean. However, the chemical component of fog was different from that of precipitation, being ca. 7 times larger. Even in one fog event, pH value and the concentration of chemical species fluctuated considerably. Nss-SO_4^{2-} was mainly scavenged as $(NH_4)_2SO_4$ (aerosol) in the atmosphere, and NO_3^- was absorbed as HNO_3.

References

Aneja,V.P., Robarge,W.P., Claiborn,C.S., Murthy,A., Kim,D.-S., Li,Z. and Cowling,E.B.: 1992, *Environmental Pollution* **75**, 89.
Dillon,P.J., Lusis,M., Reid,R. and Yap,D.: 1988, *Atmospheric Environment* **22**, 901.
Durham,J.L., Overton,J.H.,Jr. and Aneja,V.P.: 1981, *ibid.* **15**, 1059.
Fukuzaki,N., Oshio,T., Noguchi,I., Matsumoto,M., Morisaki,S., Oohara,M., Tamaki,M. and Hiraki,T.: 1999, *Chemosphere* **38**, 411.
Fuzzi,S., Facchini,M., Schell,D., Worbrock,W., Winkler,P., Arends,B.G., Kessel,M., Mols,J.J., Pahl,S., Schneider,T., Berner,A., Solly,I., Kruisz,C., Kalina,M., Fierlinger,H., Hallberg,A., Vitali,P., Santoli,L. and Tigli,G.: 1994, *Journal of Atmospheric Chemistry*. **19**, 87.
Igawa,M., Tsutsumi,Y., Mori,T. and Okochi,H.: 1998, *Environmental Science & Technology* **32**, 1566.
Kobayashi,T., Nakagawa,Y., Tamaki,M., Hiraki,T., Aikawa,M. and Shoga,M.: 1999, *Environmental Science* **12(4)**, 399. (in Japanese)
Li,Z. and Aneja,V.P.: 1992, *Atmospheric Environment* **26A**, 2001.
Linkens,E. and Bormann,F.H.: 1974, *Science* **184**, 1176.
Matsumoto,M. and Okita,T: 1998, *Atmospheric Environment* **32**, 1419.
NAPAP (1985) National acid precipitation assessment program, annual report, 1985. Interagency Task Force on Acid Precipitation, Washington, D.C.
Penkett,S.A., Jones,B.M.R., Brice,K.A. and Eggleton: 1979, *Atmospheric Environment* **13**, 123.
Puxbaum,H., Simeonov,V. and Kalina,M.F.: 1998, *ibid.* **32**, 193.
Vong,R.J., Sigmon,J.T. and Mueller,S.F.: 1991, *Environmental Science & Technology* **25**, 1014.

DEVELOPMENT OF AUTOMATIC CONTINUOUS MEASUREMENT SYSTEM OF CHEMICAL CONSTITUENTS IN THE PRECIPITATION

YASUSHI NARITA, KEI SATOH, KENICHI HAYASHI and
SHIGERU TANAKA*,

Keio University, Faculty of Science and Technology,3 -14-1 Hiyoshi, Kohoku -ku, Yokohama, 223-8522, Japan.
(* author for correspondence, e-mail: tanaka@applc.keio.ac.jp)

Abstract. Nowadays, acid rain is generally noticed as a global environmental problem. While acid rain has very much to do with the air pollutants, the relation between air pollution and chemical constituents in precipitation is not understood clearly yet. It is important to measure a variation of ion concentration in precipitation in short term for understanding the formation mechanism of acid rain. Therefore, an automatic continuous measurement system of chemical constituents in precipitation was developed and put into practical use in this study. The developed system was able to collect automatically every 1mm of precipitation and analyze major ions within 20 minutes.

Keywords: automatic continuous measurement system, chemical constituents in rain water, ion chromatography, major ions

1. Introduction

In order to establish and clarify the actual phenomenology and the formation mechanism of acid rain, it is important that the precipitation is collected not only as a sample representing the whole precipitation, but also as temporarily divided samples from the start of the precipitation and analysis of the concentration is made of its chemical constituents for each of these division samples (Coscio et al., 1982; Tanaka, et al., 1989; Bouchertall, 1990; Kronmiller et al., 1990). Today, it is possible to measure the acid gases (SO_2, HNO_3, HCl, and so forth) which cause the acid rain, and ammonia gas which acts as a neutralizing agent in the atmosphere in real time (Komazaki et al., 1999). Thus, temporarily divided sampling of precipitation has now become a requirement in order to elucidate the interrelation between the behavior of these chemical constituents that cause acid rain in the atmosphere and in the precipitation. But, since much labor and cost is required for divided sampling and analysis of precipitation, automation of the measuring method is indispensable in order to establish the long-term and wide area monitoring network using divided sampling. Thus, in the present study, a system has been developed that is capable of automatically sampling by 1mm by 1mm precipitation from the start of the precipitation, and measuring concentrations of the ionic constituents continuously. This system consists of a conventional ion chromatograph and precipitation sampler and is controlled by personal computer. Then, using this system, its practical applicability has been examined.

2. Automated Precipitation Divided Sampling System

2.1. OUTLINE OF THE AUTOMATED DIVIDED SAMPLING SYSTEM

A schematic diagram of automated precipitation divided sampling system is shown in the upper of Figure 1. A precipitation sampler (DRS-150W, manufactured by Kousin-Denki-Kogyo Co.) was used as an automated precipitation sampler. This sampler was installed on the rooftop of the 26th building in the Faculty of Science and Technology, Keio University, Hiyoshi, Yokohama. Only precipitation is collected automatically by means of a sampling funnel (18 cm in diameter) when the cover is opened by the signal from a sensor that detects rainfall. The collected precipitation sample is led through a Teflon tube into the building, to a filtration apparatus connected to a precipitation divider to remove insoluble constituent of the precipitation, where it is filtered in suction through a membrane filter (Millipore HAWP, pore size 0.45 μm) by a suction pump that is connected to the precipitation divider. Two solenoid valves (A, B) are provided at the top and the bottom of the precipitation divider such that the upper solenoid valve A is kept open and lower solenoid valve B is kept closed until the water level of precipitation reaches a specified height in the container. When the water level reaches the height corresponding to the precipitation amount of 1 mm (the liquid volume of 25.4 ml), an optical-fiber level sensor detects the precipitation, closes the solenoid valve A and opens the solenoid valve B, so that the precipitation flows into a sample changer to enable the accurate division of the precipitation in steps of 1 mm. The precipitation samples divided in step of 1 mm by the precipitation divider are successively taken into test tubes mounted on the sample changer (AS-1000, Hitachi Ltd.). Seventy-two test tubes can be arranged on the sample changer. The 72nd test tube is connected to a separate precipitation sampling container, so that the precipitation exceeding 71 mm can be sampled together collectively. The sampling time of each of the precipitation samples divided in steps of 1 mm is automatically recorded on a floppy disk using a personal computer. A proximity sensor is provided on the rooftop precipitation collector to detect the cover that is opened by the signal from the precipitation sensor detecting the rainfall. By taking the signal from this proximity sensor into a sequencer, it is possible to record the time of the beginning and the end of the rainfall likewise on the floppy disk.

2.2. THE ACCURACY OF THE AUTOMATED DIVIDED SAMPLING SYSTEM

Automated precipitation divided sampling in steps of 1 mm was actually performed using the present system as shown in Figure 1, at Hiyoshi, Yokohama. The variation of the amount of division samples in steps of 1 mm was small in any precipitation with the coefficient of variation not exceeding 2 %. Thus, it was confirmed that the present system used the optical fiber sensor allowed an accurate and automated divided sampling of precipitation.

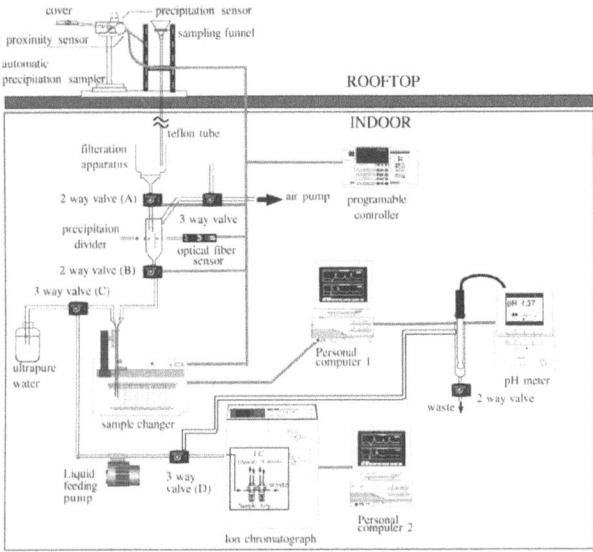

Figure 1. Schematic diagram of the automated precipitation divided sampling and the automated continuous measurement system for ionic constituents and pH of precipitation.

2.3. COMPARISON OF THE AMOUNT OF PRECIPITATION FROM AMeDAS DATA AND THAT FROM THE PRESENT SYSTEM

The sampling site where the present system was installed is adjacent to the observation site (Keio High School, Hiyoshi, Yokohama) of the amount of precipitation of the AMeDAS (Automated Meteorological Data Acquisition System) of the Meteorological Agency. In AMeDAS, precipitation is collected using a sampling funnel of 20 cm in inner diameter, and the tipping bucket precipitation gauge is used as the method for measuring the amount of precipitation.

The amount of precipitation obtained by the present system was compared with that from AMeDAS data, for 113 precipitations in the period from June, 1993 to September, 1996. It was confirmed that the amount of precipitation obtained with the present system was in good agreement with that from AMeDAS data over wide range of precipitation from a few mm to 100 mm (the slope of linear regression line: 0.982, correlation coefficient: 0.988).

3. Analysis of ionic constituents in the precipitation using by ion chromatography

In the present study, the concentration of the ionic constituents in the precipitation was analyzed using an ion chromatograph (IC-7000D, Yokokawa Analytical Systems Co.). The IC-7000D has two independent analysis lines, and is capable of analyzing simultaneously both anions (Cl$^-$, NO$_3^-$, SO$_4^{2-}$) and cations (Na$^+$, NH$_4^+$, K$^+$, Ca^{2+}, Mg^{2+}) within 15 minutes. The condition used for ion chromatographic analysis of the present study is shown in Table 1.

The analytical precision by the ion chromatograph was investigated by using standard solutions. It can be seen that this instrument has adequately high precision of analysis, since relative standard deviation is less than a few % for analysis of any of the ionic constituents. The detection limit of ionic constituents in precipitation using the ion chromatograph in the automated continuous measurement system is quite low (Na$^+$: 0.24; NH$_4^+$: 0.28; K$^+$: 0.34; Ca^{2+}: 0.81; Mg^{2+}: 0.81; Cl$^-$: 0.06; NO$_3^-$: 0.16; SO$_4^{2-}$: 0.17; unit: μeq./L), equal to or less than the lowest concentration of the ionic constituents in the precipitation.

TABLE I
Analytical condition of anions and cations in the precipitation sample by Ion chromatograph

Anions : Cl$^-$,NO$_3^-$,SO$_4^{2-}$	
Separation column	: ICS-A44(YAN)
Guard column	: ICS-A4G(YAN)
Eluent	: 4mM Na$_2$CO$_3$/4mM NaHCO$_3$ (1.0mL/min)
Suppressor	: HPS-SA1(YAN)
Scavenger	: 15mM H$_2$SO$_4$ (1.0mL/min)
Cations : Na$^+$,NH$_4^+$,K$^+$,Ca^{2+},Mg^{2+}	
Separation column	: ICS-C25(YAN)
Guard column	: ICS-C2G(YAN)
Eluent	: 4.0mM tartaric acid/0.8mM 2,6-pyridinedicarboxylic acid (1.0mL/min)

*YAN : Yokogawa Analytical Systems, co. ltd.

4. Automated divided sampling and continuous measurement system to measure ionic constituents and pH of precipitation

Figure 1 is a schematic diagram showing the construction of the system for automated divided sampling of precipitation and automated continuous measurement system for ionic constituents and pH of precipitation. The system consists of an automated precipitation sampling apparatus, a sample changer, an optical fiber liquid level sensor, an ion chromatograph (IC), and a pH meter in combination. By means of computer control, the system is capable of performing automated divided sampling of precipitation in steps of 1 mm of precipitation, and in addition, performing automated measurement of the concentration of ionic constituents and pH of the precipitation. The present system can perform automated continuous collection of precipitation samples up to 71 mm of precipitation (71 samples), and can measure the concentration of ionic constituents and pH of the sample simultaneously in about 20 minutes per sample. The system is controlled using two personal computers and a sequencer. The sequencer is used to control various solenoid valves and a pump.

The two personal computers have each entirely different function. One is for the management of the ion chromatograph, and is used for processing and recording of the result of ion chromatographic analysis. The program used for the data management is IC7000 WORK STATION prepared by Yokokawa Analytical Systems Co. The other is for the management of the sample changer, the pH meter and the sequencer.

Using the present system, the precision of analysis was examined for the automated continuous measurement of the concentration of ionic constituents and pH of the precipitation. Eight precipitation samples collected at Hiyoshi, Yokohama, during the period 1993~1994 were used. For each of the precipitation samples, the automated continuous measurement and the conventional manual measurement were performed in parallel. The measured values obtained by the two methods were in very good agreement and within a few percents of analytical accuracy both for each ionic constituent and for pH. It was confirmed that an accurate measurement was possible using the automated continuous measurement system.

5. Result of measurement of the concentration of ionic constituents and pH of the precipitation using the present system

During the period from June, 1993 to September, 1996, automated divided sampling of precipitation and automated continuous measurement of the concentration of its ionic constituents and pH were performed at Hiyoshi, Kouhoku-ku, Yokohama. Based on the measurement result of the precipitation, relation between the temporal variation of the concentration of ionic constituents and pH, and the meteorological condition was investigated. For example, Figure 2 shows a typical temporal variations of the concentrations of ionic constituents and pH of the precipitation on March 25, 1995. It can be seen from the temporal variations of ion concentrations that the ion

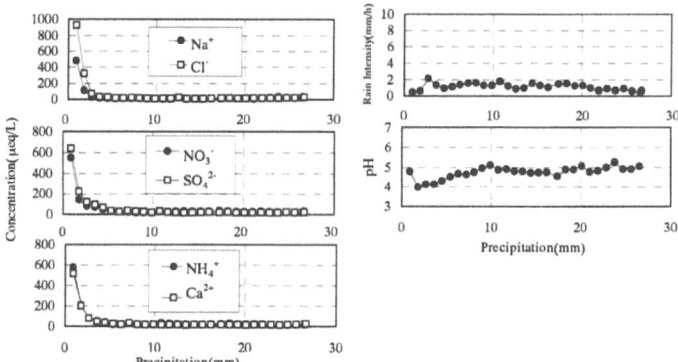

Figure 2 Temporal variations of ion concentrations and pH of the precipitation (Hiyoshi, Yokohama, March 25, 1995).

Concentrations are highest for any of the ionic constituents in the early period of the precipitation, and decline exponentially with the increase of the amount of precipitation (Ambe and Nishikawa, 1986; Durana et al., 1992) . In the early stage of precipitation (amount of precipitation 1, 2 mm), there are many substances in the atmosphere which supply these ionic constituents in the precipitation. These substances are taken into precipitation, thereby producing high concentrations of ionic constituents in the precipitation. As the precipitation continues and its amount increases, the amount of these substances in the atmosphere decreases substantially since much of these substances has been taken into the preceding precipitation. As a result, the concentrations of the chemical constituents in precipitation decrease rapidly. It is found that, in general, ion concentrations in precipitation become quite low after about 5 mm of early amount of precipitation, and remain to be low thereafter as shown in Figure 1.

6. Conclusion

In the present study, a system for automated divided sampling of precipitation and for automated measuring of chemical constituents of precipitation has been developed. As a result, automated continuous measurement of ionic constituents and pH in real time in steps of 1 mm of precipitation has now become possible. Automated divided sampling in steps of 1 mm of precipitation and automated continuous measurement of chemical constituents in precipitation was performed using the present system at Hiyoshi, Yokohama, in the period from 1993 to 1996. It has been found that the concentration of major ionic constituents in precipitation shows the highest value at the beginning of precipitation, and decreases with the increase of the amount of precipitation. From the result of measurement of temporal variations of the concentrations of ionic constituents and pH of precipitation, it has become evident that the chemical constituents of precipitation are greatly affected by meteorological conditions.

References

Ambe Y. and Nishikawa M.: 1986, *Atmospheric Envrionment* **20**, 10, 1931-1940.
Bouchertall, K. F.: 1990, *StaubReinhaltung der Luft* **50**, 157-159.
Coscio, M. R., Pratt, G. C. and Krupa, S.V.: 1982, *Atmospheric Environment* **16**, 8, 1939-1944.
Durana N., Casado H., Ezcurra A., Garcia C., Lacaux J. P. and Dinh P.V.: 1992, *Atmospheric Environment* **26A**, 13, 2437-2443.
Komazaki, Y., Hamada Y., Hashimoto, S., Fujita, T. and Tanaka, S.: 1999, *Analyst* **124**, 1151-1157.
Kronmiller, K. G., Ellenson, W. D., Baumgardner, R. E., Stevens, R. K. and Paur, R. J : 1990, *Atmospheric Environment* **24A**, 3, 525-536.
Tanaka, S., Yasue, K. and Hashimoto, Y.: 1989, *Proceeding of the 8th World Clean Congress 1989, Hague, Netherlands,* 669-674.

MAPPING OF CONCENTRATIONS IN EUROPE COMBINING MEASUREMENTS AND ACID DEPOSITION MODELS

ANNE-GUNN HJELLBREKKE[1] and LEONOR TARRASON[2]

[1]*EMEP/CCC, Norwegian Institute for Air Research, P.O Box 100, N-2027 Kjeller, Norway*
[2]*EMEP/MSC-W, Norwegian Meteorological Institute, P.O. Box 43 Blindern, N-0313 Oslo, Norway*

Abstract. Measurements of air and precipitation quality have been carried out within the EMEP programme under the Convention on Long Range Transboundary Air Pollution (LRTAP) since 1978. Approximately 100 rural sites are currently in operation. The Meteorological Synthesising Centre-West (MSC-W) operates two EMEP models estimating transboundary fluxes of air pollutants, a two-dimensional Lagrangian model and a three-dimensional Eulerian model. Traditionally kriging has been used to produce gridded concentration fields from observed data for comparison with modelled data. This paper describes a method for producing optimal fields based on both point measurements and. The difference between modelled and measured values in each measurement point is interpolated to give a smooth two-dimensional expression for the discrepancy between the two data sets. A combined map is derived by adjusting the modelled values with the interpolated difference weighted by the distance to the nearest measurement point. The method has been applied to sulphur and nitrogen measurements in air and precipitation from the EMEP network and modelled results in a 150x150 km grid from the EMEP Lagrangian model. The combined maps give improved regional concentration fields combining characteristics from both the measured and modelled data sets depending on the distance to the measurement points. Comparison with results from the higher resolution Eulerian model shows good agreement.

Keywords: mapping, acid deposition, deposition models, EMEP, kriging, interpolation

1. Introduction

Measurements of air quality in Europe have been collected by the Chemical Co-ordinating Centre of EMEP (CCC) since 1978 (Hjellbrekke, 1999). From the start, priority was given to measurements of sulphur dioxide (SO_2) and sulphate (SO_4^-) in air, and pH and sulphate in precipitation, gradually increasing to all main components in precipitation and nitrogen compounds in air. At present, approximately 100 rural and background sites are in operation.

Two acid deposition models are currently operated by the Meteorological Synthesizing Centre – West (MSC-W). The EMEP Lagrangian Acid Deposition Model is a receptor-oriented one layer trajectory model with a horizontal resolution of 150x150 km. Since the late 1980s the model has been employed at the EMEP/MSC-W to calculate concentrations and depositions of acidifying compounds in Europe, as well as transboundary fluxes and budget matrices (Barrett and Berge, 1996). However, the lack of vertical structure means that the model cannot parameterise atmospheric transport of pollutants above the mixing layer, and a three-dimensional Eulerian transport/deposition model has been developed since 1993.

In the past, a standard kriging technique (Journel and Huijbregts, 1978) has been used by the CCC to provide maps of measured concentration levels in Europe. However, the accuracy of the kriged fields depends on the location of measurement sites, and the uncertainty of the estimates increases with the distance to the sites and in areas with few sites. The EMEP Lagrangian model has better spatial coverage but a low resolution. This article describes a method for improving the concentration fields by combining both the measured and modelled data.

2. Method

Annual averages of EMEP measurements of sulphate and nitrate compounds from 1996 have been used together with annual averages calculated by the EMEP Lagrangian model from the same time period. In order to secure a consistent set of data for analysis, results from the 16th and 17th annual laboratory intercomparisons from the years 1995 and 1997 (Hanssen and Skjelmoen, 1996; 1997) have been used. Data from a laboratory have been excluded from the mapping if the average deviation from expected values obtained in the two intercomparisons is greater than 20 per cent or if the deviation from expected values in one of the intercomparisons is greater than 25 per cent. For a given component, the concentrations measured at a site may not be representative for the levels in the surrounding area, and are therefore excluded from the mapping.

Let $m(x_t, y_t)$, $t=1,...,n$ be the measurement value at each measurement site (x_t, y_t) and $mod(i,j)$, $i=1,...,i_{max}$, $j=1,...,j_{max}$ be the modelled values. For each measurement point (x_t, y_t), let the difference between the measured and modelled value in the corresponding grid cell (i_t, j_t) be

$$diff(x_t, y_t) = m(x_t, y_t) - mod(i_t, j_t).$$

The differences are interpolated using radial basis functions

$$f(x,y) = \sum_{k=1}^{n} c_k \varphi(d_k(x,y))$$

where $d_k(x,y)$ is the Euclidean distance to measurement point (x_k, y_k) and $\phi(r) = \sqrt(r)$. The parameters c_k are calculated by solving the linear system of equations obtained by letting $diff(x_t, y_t) = f(x_t, y_t)$, $t=1,...,n$. The function f is a two dimensional continuous function describing the difference at any point (x,y) within the modelled grid, and equals the difference between observed and modelled values at all measurement points.

The new optimal maps are produced by adjusting the model calculations with the interpolated difference. Close to measurements, we give larger weight to observed values whereas in regions with no observations modelled results are preferred. The actual region of influence of measured values depends on the type of component, in particular, on its characteristic transport distance, and is determined by the range of spatial covariance used in kriging.

The new concentration fields $N(x,y)$ are derived as follows:

$N(x,y) = f(x,y) + mod(x,y)$ $\qquad d \leq 1$
$N(x,y) = f(x,y)\, (D_{comp}-d)/(D_{comp}-1) + mod(x,y)$ $\qquad 1 < d \leq D_{comp}$
$N(x,y) = mod(x,y)$ $\qquad d > D_{comp}$

where d is the distance to the nearest measurement point and D_{comp} defines the region of influence for a given component. D_{comp} is set to 2 grid squares (300km) for SO_2, while for nitrate in precipitation D_{comp} is set to 5 grid squares. See (Tarrason et al, 1998) for details.

3. Results and discussion

Results for SO_2 in air are shown in Figure 1. Comparison of results from the Lagrangian model (Figure 1, upper left panel) with kriged measurement data (Figure 1, upper right panel) shows a similar geographical distribution but varying concentration levels. The interpolated difference of the measured and modelled values normalised by the sum of measured and modelled values (Figure 1, middle left panel) shows that the model overestimates the measurements in Central and Eastern Europe and underestimates in Spain and the British Isles. The overestimation may be related to the fact that SO_2 is a primary pollutant and that the EMEP network is designed to measure concentrations in background and rural areas. Because of the local character of most SO_2 sources (domestic heating, thermic power plants) we can expect to find sharp gradients in the SO_2 concentrations within relatively short distances. The EMEP Lagrangian model with one vertical layer and coarse horizontal resolution is not able to reproduce this and gives average values which overestimate the low concentrations measured at EMEP background sites located distant from sources. The combined results (Figure 1, middle right panel) show characteristics from both the modelled and observed values. The estimate reduces the model overestimation over Central Europe because of the use of EMEP background stations and traces the expected features around the Alps.

It is interesting to note that the combined fields to some extent reproduce transport features that the EMEP Lagrangian model is not able to simulate. The EMEP Eulerian model using a finer resolution (50x50 km) and 20 vertical layers is capable of reproducing the observed gradients in the Alps and Northern Scandinavia (Figure 1, lower panel). In source areas in Central Europe, Great Britain and Spain the Eulerian model manages to describe in greater detail the expected distribution of SO_2.

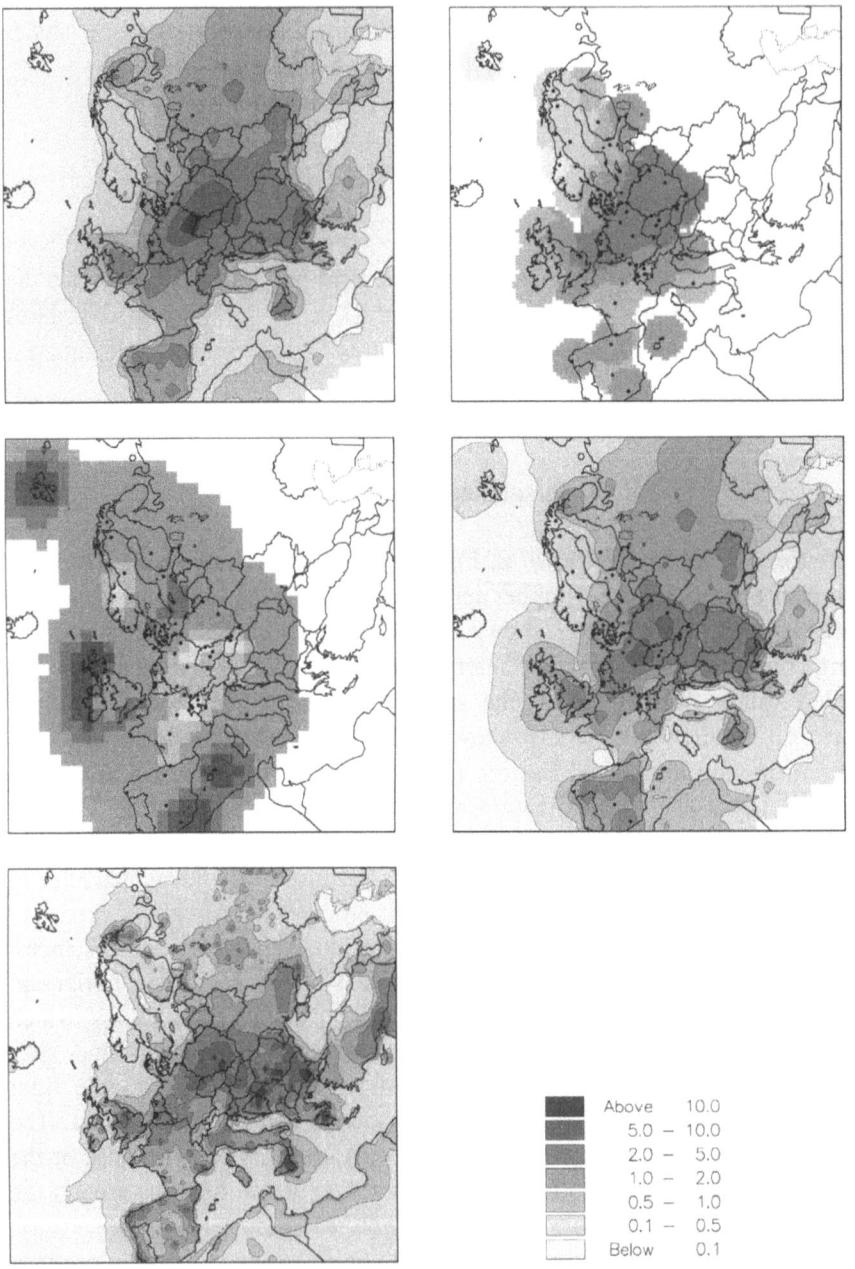

Figure 1. Analysis of yearly averaged SO$_2$ concentrations in air in 1996. Upper left panel: Results from the Lagrangian model. Upper right: Kriged measurement data. Middle left: Interpolated normalized differences. Light grey shows where the model overestimates the measurements, whereas dark grey shows model underestimation. Middle right: Combined estimate. Lower panel: Results from the Eulerian model, 1998. Black dots indicate the position of measurement sites.

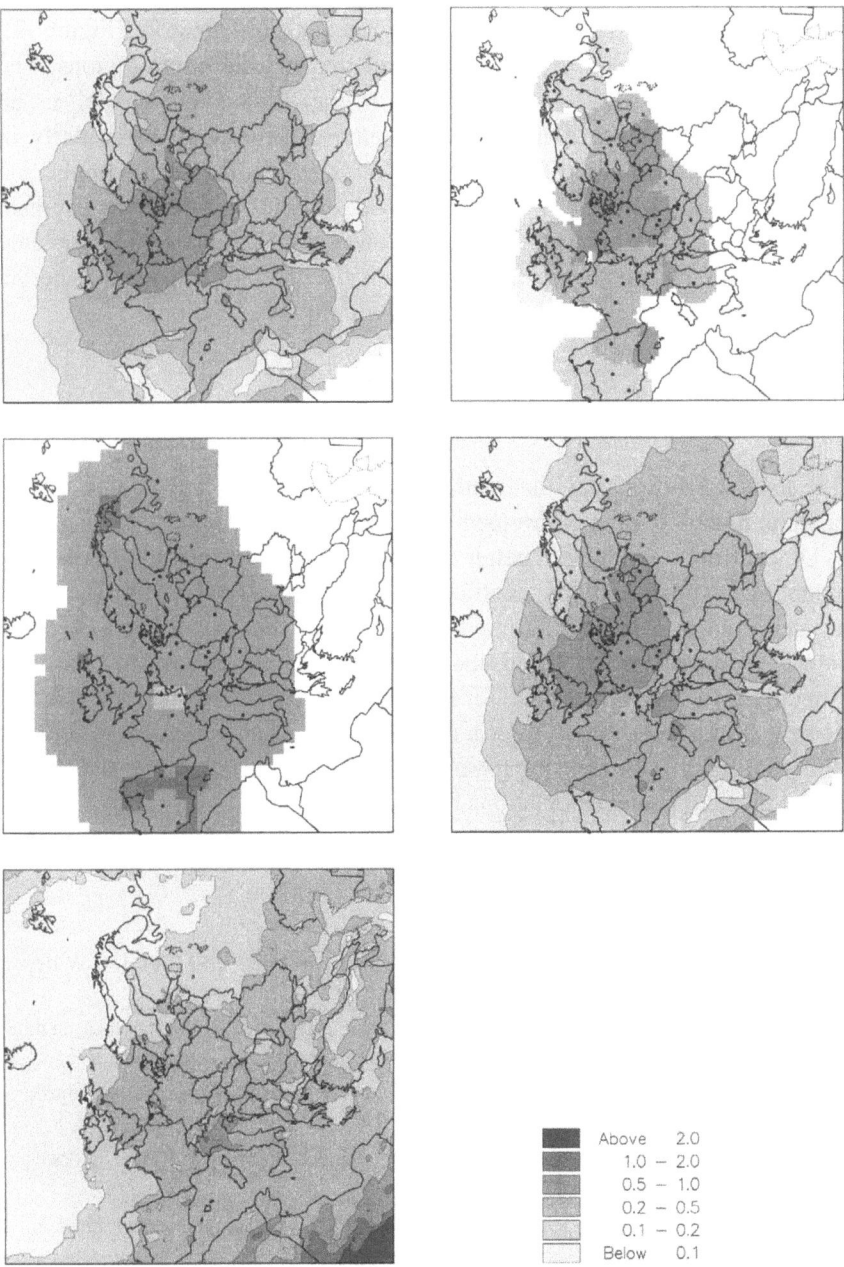

Figure 2. Analysis of yearly averaged NO_3^- concentrations in precipitation in 1996. Upper left panel: Results from the Lagrangian model. Upper right: Kriged measurement data. Middle left: Interpolated normalized differences. Light grey shows where the model overestimates the measurements, whereas dark grey shows model underestimation. Middle right: Combined estimate. Lower panel: Results from the Eulerian model, 1998. Black dots indicate the position of measurement sites.

Comparison of modelled with measured values for nitrate in precipitation shows good agreement, in terms of geographical distribution (Figure 2). However, the model tends to overestimate in the largest source regions. The lack of three-dimensional representation in the Lagrangian model is an important drawback when simulating wet removal processes. Particularly in connection with precipitation from frontal clouds, which may extend up to 2-5 km above the surface, the Lagrangian model will tend to underestimate concentrations in precipitation. Figure 2, lower panel depicts the estimated concentration of nitrate in precipitation from the three-dimensional Eulerian model in 1998, which are in good agreement with the observations.

4. Conclusions

The analysis shows that the method combining results from the EMEP Lagrangian model with measurement data improves the existing maps and yields concentration fields with better resolution and spatial coverage. For SO_2, the model overestimation in the central areas of Europe is reduced by the lower values measured at EMEP background sites. A significant part of the systematic differences between the modelled and observed concentrations are related to the limitations of the Lagrangian EMEP model to simulate the transport of pollution in complex terrain. Use of a finer horizontal and vertical resolution in the Eulerian EMEP model improves considerably the agreement between measurements and model calculations for SO_2.

References

Barrett, K. and Berge, E.: 1996, Transboundary Air Pollution in Europe. EMEP/MSC-W Report 1/96. The Norwegian Meteorological Institute, Oslo, Norway.

Hanssen, J.E. and Skjelmoen, J.E.: 1996, The fifteenth intercomparison of analytical methods within EMEP. Kjeller, Norwegian Institute for Air Research (EMEP/CCC Report 2/96).

Hanssen, J.E. and Skjelmoen, J.E.: 1997, The sixteenth intercomparison of analytical methods within EMEP. Kjeller, Norwegian Institute for Air Research (EMEP/CCC Report 2/97).

Hjellbrekke, A.-G.: 1999, Data Report 1997. Part 1: Annual summaries. Kjeller, Norwegian Institute for Air Research (EMEP/CCC Report 3/99).

Journel, A. G. And Huijbregts, C. G.: 1978, Mining Geostatistics. London, Academic Press.

Tarrasón, L., Semb, A., Hjellbrekke, A.-G., Tsyro, S., Schaug, J., Batnicki, J. and Solberg, S.: 1998, Geographical distribution of sulphur and nitrogen compounds in Europe derived both from modelled and observed concentrations. The Norwegian Meteorological Institute, Oslo, Norway (EMEP/MSC-W Note 4/98).

INFLUENCE OF ACID DEPOSITION ON INLAND WATER CHEMISTRY
-A CASE STUDY FROM HYOGO PREFECTURE, JAPAN-

YUKIO KOMAI[1]*, SATOSHI UMEMOTO[1] and TAKANOBU INOUE [2]

[1] Hyogo Prefectural Institute of Environmental Science, 3-1-27 Yukihiracho Suma Kobe, 654-0037 Japan; [2] Department. of Civil Engineering, Gifu University, 1-1 Yanagido Gifu, 501-1193 Japan

(*author for correspondence, e-mail:komai@pref.hyogo.jp)

Abstract. The Influence of acid deposition on stream and lake water chemistry was studied in a forested watershed of Hyogo prefecture, Japan. Monthly sampling of four streams, one artificial lake, and precipitation was carried out from 1995 to 2000. The pH of the monthly bulk precipitation and rainwater were ranged from 4.06 to 7.10. No trends were evident during the monitoring periods. The pH and alkalinity in the four streams, which flow into the artificial lake, ranged from 6.37 to 8.72 and 0.077 meqL^{-1} to 0.485 meqL^{-1}, respectively. The differences in the water quality of the four streams were related to the geology of each watershed. Lower pH and alkalinity were observed during high- discharge periods. On the other hand, the pH and alkalinity of the outflow from the lake ranged from 6.47 to 7.36 and 0.195 meqL^{-1} to 0.339 meqL^{-1}, respectively. No acidification of the aquatic environment was observed during the investigated periods. The results suggest that this forested ecosystem has the capacity to neutralize incoming acid deposition.

Keywords: forested watershed, water chemistry, monitoring, acid deposition

1. Introduction

East Asia emits one-third of the air pollutants being released worldwide (Acid Deposition Monitoring Network in East Asia, 2000), and acid rain is observed across a wide area of Japan, with a weighted average of pH4.8 (Hara, 1997). Global and regional ecosystems, including vegetation, soil, and the waters of streams and lakes are expected to face serious problems, as are buildings and other manmade structures. Therefore, it is important to monitor the influence of acid deposition on the environment.

The concentration and amount of acid deposition in Japan are monitored by the Japan Environment Agency (JEA) and by local governments. JEA has also surveyed the influence of acid deposition on several lakes and rivers every year since 1983. This monitoring has, however, been carried out independently for each environment, and the data for the inland aquatic environment regarding acid deposition influence is not extensive.

Thus, since 1995 we have been studying the influence of acid deposition on the inland aquatic environment, the water quality of four streams and an artificial lake in a forested watershed in the middle of Hyogo prefecture, Japan. In this paper, we show the results of that monitoring and evaluate the influence of acid deposition on the inland aquatic environment in Hyogo prefecture.

2. Methods

2.1. STUDY AREA

The study area is about 50 km from the Sea of Japan and the Seto Inland Sea (Fig. 1). Ginzan Lake is an artificial lake constructed in 1973, and it is a harmonic one. The lake has a watershed of 49 Km^2. The normal water depth is 47 m and the volume is 18,000,000 m^3. The calculated residence time is 0.15 year. Usually, the water flows out through a discharge pipe set at about 10 m above the bottom. The main stream of the Ichi River (M) and three streams (A, B, and C) flow into Ginzan Lake. The elevation is from 350 m to 800 m above sea level.

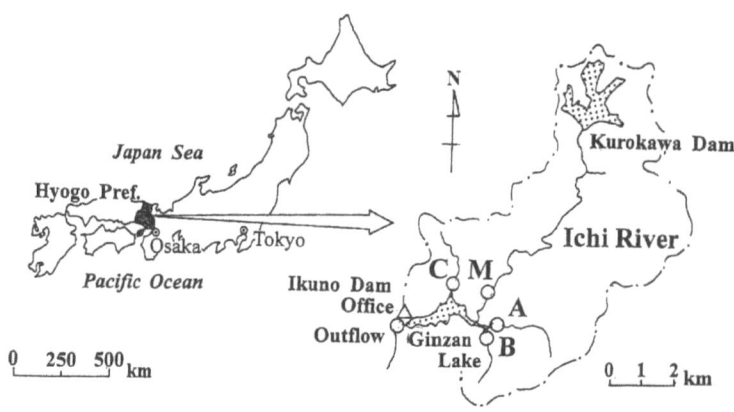

Figure 1. Map of sampling sites

The bedrock geology in the study area is a part of the Ikuno group, which is dominated by rhyolitic faces. The lithofaces in watershed B are andesitic lava and tuff breccia, which are relatively basic compared with lithofaces of the other watersheds. There is a large nonferrous-metal mine near watershed B, too (Komai et al., 1999). On the other hand, the lithofaces of watershed C are mainly rhyodacitic vitro-crystal welded tuff and lava, and rhyolitec sedimentary rock, both of which are relatively acid rock in these watersheds (Komai et al., 1999). The dominant soil is brown forest soil, which is common in Japan. The vegetation is mainly planted Japanese cedar, Japanese cypress, and red pine with deciduous stands. The precipitation was collected at the Ikuno dam office beside the dam. The average annual precipitation in the four years between August 1995 and July 1999 was 1740 mm. Due to the mild climate with little snow in the winter, most of the precipitation falls as rain. Ninety-two people live in watershed M, but there is no population in the watershed areas of the other three streams.

The main industry is forestry. There are rice paddy fields of 0.152 Km^2 and upland fields of 0.013 Km^2, too. However, we assume that human activities in

2.2. SAMPLING AND ANALYTICAL METHODS

Water samples have been taken at least once a month since July 1995 at the end point of each of the four streams and the outflow point of the discharge pipe from the lake. The bulk precipitation was collected once a week by a simple type collector (Kobayashi et al., 1993). A plastic can was used to collect the rainwater of a rain event every month, too.

Electrical conductivity (EC) and pH were measured by an EC meter and a pH meter, respectively. Alkalinity at pH 4.3 was measured by a titration method with 0.01 molL^{-1} sulfuric acid. The alkalinity at pH 4.3 was an average of 0.035 meqL^{-1} higher than that at pH 4.8, and the coefficient of variance for the difference between them was 12% in the water samples from the four streams. Major ionic species were analyzed by ion chromatograph.

3. Results and Discussion

3.1. CHANGE IN pH OF THE MONTHLY BULK PRECIPITATION AND RAINWATER OF A RAIN EVENT

Monthly precipitation and the pH of the monthly bulk precipitation and the rainwater of a rain event are shown in Figure 2. The precipitation from March to October is in general greater than in the rest of the year. The pH was in the range of 4.06 to 7.10. No specific tendency was observed during the period of the investigation, even though the pH of the rainwater of a rain event fluctuated at every rain event. The EC ranged from 0.5 mSm^{-1} to 7.1 mSm^{-1}.

The weighted average of pH in the bulk precipitation was 4.76. This was the

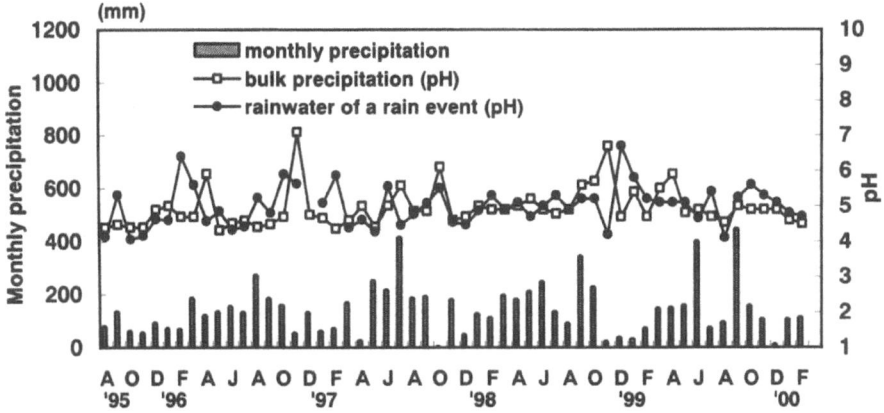

Figure 2. Change in monthly precipitation, and pH of bulk precipitation and rainwater of a rain event

average value of pH in Japan (Hara, 1997). There were no data on precipitation before 1995 in the Ikuno area. However, the acid deposition at Hikihara in a remote site, 20 km west of Ikuno, has been investigated since 1988. The annual weighted average of pH was 4.84 in 1988 (Hiraki et al., 1989). These results indicate that there has been acid deposition in the forested watershed at least since 1988.

3.2. WATER QUALITY OF STREAMS

The changes in pH and alkalinity of the water in the four streams are shown in Figure 3. These levels ranged of 6.37 to 8.72 and 0.077 meqL^{-1} to 0.485 meqL^{-1}, respectively. The average values of EC in the four streams were in the range of 4.05 mSm^{-1} to 7.56 mSm^{-1}. The pH fluctuated similarly in the water from the four streams, but the pH of stream M was sometimes more than 8.0. This may have been caused by the photosynthesis of attached algae. The change in the pH of stream B was similar to that of stream A and both were slightly higher than that of stream C.

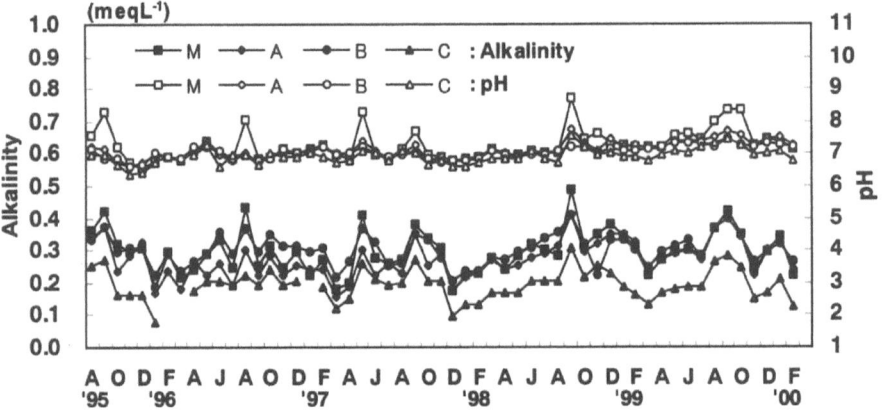

Figure 3. Change in pH and alkalinity of the water in four streams

The average concentrations of major ionic species of the four streams with lake water and of bulk precipitation are shown in Figure 4. SO_4^{2-}, Mg^{2+}, and Ca^{2+} were higher in stream B than in the other streams; this is not surprising, considering that the lithofaces in watershed B are relatively basic compared with those of the other watersheds and that there is a large nonferrous-metal mine near watershed B (Komai et al., 1999). On the other hand, the pH, alkalinity, and occurrence of these major ionic species were lowest in stream C among the four streams; the lithofaces of watershed C are relatively acid rock than those of the other watersheds. For example, the average alkalinity in stream C was 0.192 meqL^{-1} (0.163 meqL^{-1} at pH 4.8), and that of the other three streams ranged from 0.271 meqL^{-1} to 0.298 meqL^{-1}. The minimum value was 0.077 meqL^{-1} in February 1996, which became the concentration of

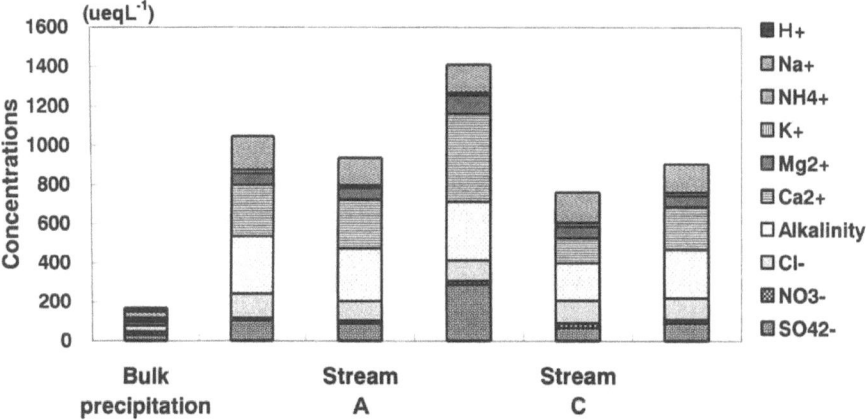

Figure 4. Average concentrations of major ionic species of the four streams with lake water and of bulk precipitation

acidification below 0.05 meqL^{-1} at pH 4.8. This is one example of an acid episode by snow melting.

During a storm event, decreases in pH and alkalinity were also observed according to the increasing discharge in stream A. However, they recovered rapidly to the original levels after the storm event. These changes were a temporary phenomenon, and the other stream water may behave similarly (Komai *et al.*, 2000). These results suggest that the differences in the water quality of the four streams were related to the geology of each watershed.

3.3. WATER QUALITY OF OUTFLOW

The water quality of the outflow from the lake is shown in Figure 5. The pH and alkalinity were ranged from 6.47 to 7.36 and 0.195 meqL^{-1} to 0.339 meqL^{-1}, respectively. The average value of EC was 4.78 mSm^{-1}. The fluctuation of

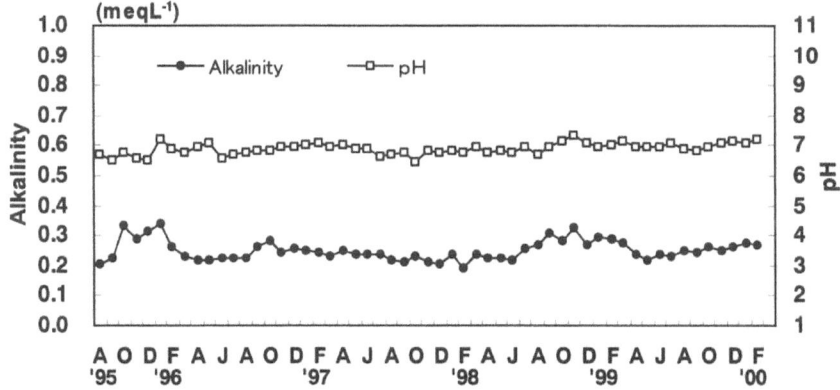

Figure 5. Change in pH and alkalinity of the outflow from the lake

outflow water quality was smaller compared with the fluctuations observed in stream water quality, and no trend was observed. These observations suggest that the lake water quality became uniform in a short time.

The water quality of the surface layer and outflow was investigated at the same time, once a month from August 1995 to July 1996. The pH of the surface layer fluctuated more than that of the outflow. This difference may be related to the photosynthesis of phytoplankton in the surface layer, which usually showed a higher pH than the outflow, especially from spring to autumn. The same tendency was also observed before 1995 and after 1996 (Himeji Civil Engineering Office of Hyogo prefecture and Association of Hyogo Environment Advance, 1999). The differences in alkalinity between surface and outflow were within approximately 10%. These results indicate that the lake water maintains a condition of non-acidification.

4. Conclusions

No acidification or trends in water chemistry in the lake and four streams of the forested watershed were observed during the monitoring periods 1995 to 2000. These results suggest that the forested ecosystem has the capacity to neutralize acid deposition even though the forested watershed has been subjected to acid deposition for more than a decade.

Acknowledgements

We would like to thank the Ikuno Dam Office for their cooperation on the rainfall sampling and precipitation data. This study was sponsored by Hyogo Science and Technology Association, Japan.

References

Acid Deposition Monitoring Network in East Asia:2000, Guidelines for Acid Deposition Monitoring in East Asia
Hara,H.:1997,*Chemical Society of Japan,No.11,*733(in Japanese).
Hiraki,T., Tamaki,M., and Torihashi,Y.:1989, *Report of Hyogo Prefectural Institute of Environmental Science,* No.21,15(in Japanese).
Himeji Civil Engineering Office of Hyogo prefecture and Association of Hyogo Environment Advance,:1999, *Report of water quality in Ikuno Dam,* Himeji Civil Engineering Office of Hyogo prefecture and Association of Hyogo Environment Advance.
Komai,Y. and Umemoto,S.:1999, *Research Report National Institute of Environmental Studies,Japan.,* No144,87(in Japanese).
Komai,Y., Umemoto,S., Inoue,T.:2000, *Proceeding of 4th International Conference on Diffuse Pollution,*239.
Kobayashi,T.and Nakagawa,Y.:1991, *Report of Hyogo Prefectural Institute of Environmental Science,* No.23,21(in Japanese).
Tamaki,M.,Shoga,M. and Hiraki, T.:1991 *Chemical Society of Japan,*No.6,930(in Japanese).

ASSESSING THE SUITABILITY OF ACID NEUTRALISING CAPACITY AS A MEASURE OF LONG-TERM TRENDS IN ACIDIC WATERS BASED ON TWO PARALLEL DATASETS

C.D. EVANS[1]*, R. HARRIMAN[2], D.T. MONTEITH[3] and A. JENKINS[1]

[1] *Centre for Ecology and Hydrology. Wallingford, Oxon., OX10 8BB, UK;* [2] *Freshwater Fisheries Laboratory, Faskally, Pitlochry, Perthshire, PH16 2LB, UK;* [3] *Environmental Change Research Centre, University College London, 26 Bedford Way, London, WC1H 0AP, UK.*
(* author for correspondence, email: cev@ceh.ac.uk)

Abstract. Acid Neutralising Capacity (ANC), calculated as the difference between base cations and acid anions, is widely used as a measure of freshwater acid status, and an indicator of biological conditions. Unlike pH and alkalinity, ANC is conservative with respect to CO_2 degassing and reactions with aluminium or organic species. However, since ANC is calculated as the residual of a large number of individual ion determinations, it is potentially sensitive even to relatively small analytical errors. For the Round Loch of Glenhead, SW Scotland, consistency of ANC estimation has been assessed based on a duplicate set of major ion analyses undertaken at different laboratories over an 11 year period. Results indicate that, while the two sets of individual ion determinations correspond well, correlation between calculated ANC values is poor. Consequently, estimated ANC trends exhibit severe discrepancies between datasets; one indicates substantial recovery, the other no apparent trend. These problems with ANC estimation are believed to be general to acidic waters and are of particular concern for long-term monitoring, where ANC changes may be small and difficult to detect (although nonetheless potentially biologically significant). In these situations, it is possible that a more stable measurement of ANC may be obtainable based on titration alkalinity, DOC and aluminium concentrations. Using this method, a small but highly consistent increase in ANC is observed over the study period, although much of this can be attributed to a shift from mineral to organic acidity, rather than an overall reduction in acidity.

Keywords: Acid Neutralising Capacity, alkalinity, analytical error, recovery, trend analysis

1. Introduction

Acid Neutralising Capacity (ANC) has been widely adopted as a standard measure of surface water acidity, notably as a chemical indicator of biological status for Critical Loads assessment; in the UK, ANC = 0 is taken as an effective threshold for ecosystem damage (CLAG, 1995). Unlike the other frequently-used measures of freshwater acidity, pH and alkalinity, ANC is conservative on mixing, being unaffected by CO_2 degassing or by reactions with aluminium (Al) or organic species (Neal *et al.*, 1999). However, whereas pH and alkalinity can be directly measured, ANC must be estimated indirectly, usually based on the charge balance. Charge balance ANC (CB ANC) is defined as the sum of base cations (ΣC_B) minus the sum of strong acid anions (ΣC_A) (Reuss *et al.*, 1986) which for most natural waters approximates to:

$$CB\ ANC = [Ca^{2+}] + [Mg^{2+}] + [Na^+] + [K^+] - [SO_4^{2-}] - [Cl^-] - [NO_3^-] \qquad (1)$$

(all units $\mu eq\ l^{-1}$). At least seven individual ion determinations are thus required to estimate ANC, with others such as NH_4^+ and F^- important for some systems.

Since ΣC_B and ΣC_A are relatively large in many acid waters (particularly those with high marine ion inputs) and the difference between them relatively small, a high level of analytical precision is required to accurately quantify ANC. This sensitivity to measurement error is particularly acute when examining temporal changes in ANC, often of only a few micro-equivalents per litre per year.

This study examines the sensitivity of CB ANC to analytical error by investigating temporal trends for a site in Scotland, where major ion analyses have been undertaken at two different laboratories over a long period. In addition to CB ANC, temporal trends are also assessed for ANC calculated from Gran titration alkalinity, inorganic Aluminium (Al_{inorg}) and DOC (Harriman and Taylor, 1999; Neal et al., 1999):

$$ANC = Alkalinity + (F \times DOC) - (3 \times Al_{inorg}) \qquad (2)$$

(with ANC and alkalinity in $\mu eq\ l^{-1}$, Al in $\mu mol\ l^{-1}$, and DOC in $mg\ C\ l^{-1}$). At the equivalence point, Al_{inorg} is assumed to be present entirely as Al^{3+}, while organic Al is assumed to be undissociated. F represents the change density ($\mu eq\ mg^{-1}$) of DOC at the equivalence point, estimated for Scotland as 4.5 for waters with pH 4.5 to 5.5, and 5 for waters with pH > 5.5 (Harriman and Taylor, 1999).

2. Study Area and Methods

The Round Loch of Glenhead is a small (12.5 ha) lake in Galloway, SW Scotland, with a ten-year mean pH of 4.9, ANC $-9\ \mu eq\ l^{-1}$, non-marine SO_4 47 $\mu eq\ l^{-1}$ and NO_3 7 $\mu eq\ l^{-1}$. The loch has an altitude of 295m, and a peaty moorland catchment. Non-marine S deposition is 37 $kg\ ha^{-1}\ yr^{-1}$, and total N deposition 44 $kg\ ha^{-1}\ yr^{-1}$ (Monteith and Evans, 2000). Declining emissions over the last decade have led to reductions in S deposition, but changes have been less than in other areas closer to emissions sources (Fowler and Smith, 2000), while N deposition has remained approximately constant. Round Loch has been sampled irregularly from 1979 to 1987, and monthly from 1988 to the present, by the Freshwater Fisheries Laboratory (FFL), Pitlochry. All analyses for this dataset were undertaken by FFL. Since 1988, Round Loch has also been part of the UK Acid Waters Monitoring Network (AWMN; Monteith and Evans, 2000). AWMN samples are collected quarterly, usually concurrent with those for FFL. Analysis of pH, alkalinity, NO_3, DOC and Al is carried out at FFL, but other determinands are analysed at the Centre for Ecology and Hydrology, Wallingford. Apart from NO_3, therefore, all the components of CB ANC have been analysed by both laboratories, each subject to rigorous QA/QC checks, for over a decade. These duplicate datasets provide an unusual and valuable opportunity to assess the consistency of CB ANC estimation.

A recent analysis of trends from 1988-1998 in the AWMN (Monteith and Evans, 2000) showed limited evidence for recovery at Round Loch. This was attributed to a combination of relatively small acid deposition reductions, and to confounding climatic factors, notably varying marine ion deposition. However, ANC was not included in this study, and here trends in CB ANC for both AWMN and FFL datasets are analysed for 1988-1999 using the same methodology. The procedure used is the non-parametric Seasonal Kendall Test (SKT; Hirsch and Slack, 1984). Data are grouped into seasonal blocks, and tested for monotonic trends using a ranking procedure. Results for each block are combined to give an overall trend test which is robust with respect to non-normality, missing data, seasonality and autocorrelation. Slopes are estimated as the median of all between-year differences within each seasonal block. SKT was also applied to H^+, Gran alkalinity and alkalinity-based ANC for the FFL dataset (all AWMN analyses for these determinands were undertaken by FFL). For multiple samples within one 'season', the mean was used. Finally, the two datasets were directly compared for the 43 samples analysed at both laboratories. For each major ion, and for CB ANC, measurements were compared using linear regression, with the intercept set to zero. This allowed analytical consistency, and sources of error in CB ANC, to be assessed.

3. Results

Overlaid time series for major ions analysed by the two laboratories are shown in Figure 1 (NO_3 is included for completeness, although all analyses for this anion were undertaken by FFL). Correspondence between the two datasets is generally good, particularly for Na and Cl. SO_4 and Mg also appear closely correlated, although for Ca the relationship between the two sets of analyses is clearly poorer. Both datasets generally show very similar patterns of temporal variation, regardless of differences in sampling frequency.

Despite this clear consistency between the two sets of major ion data, CB ANC values calculated from these measurements exhibit very marked discrepancies in terms of time series (Figure 2) and trend analyses (Table I). CB ANC calculated from FFL data (ANC_{FFL}) exhibits a highly significant increase, by an estimated 4 µeq l^{-1} yr^{-1}, whereas ANC calculated from AWMN data (ANC_{AWMN}) shows no significant trend. In general, temporal variations in ANC_{AWMN} and ANC_{FFL} appear very poorly related. Of the other acidity-related

TABLE I

Seasonal Kendall trend analysis of ANC and alkalinity estimates

Variable	Slope (µeq l^{-1} yr^{-1})	Significance
CB ANC (AWMN data)	+0.3	0.645
CB ANC (FFL data)	+4.0	0.007
Alkalinity	+0.3	0.232
Alkalinity-based ANC	+1.3	0.005
H^+	-0.3	0.023

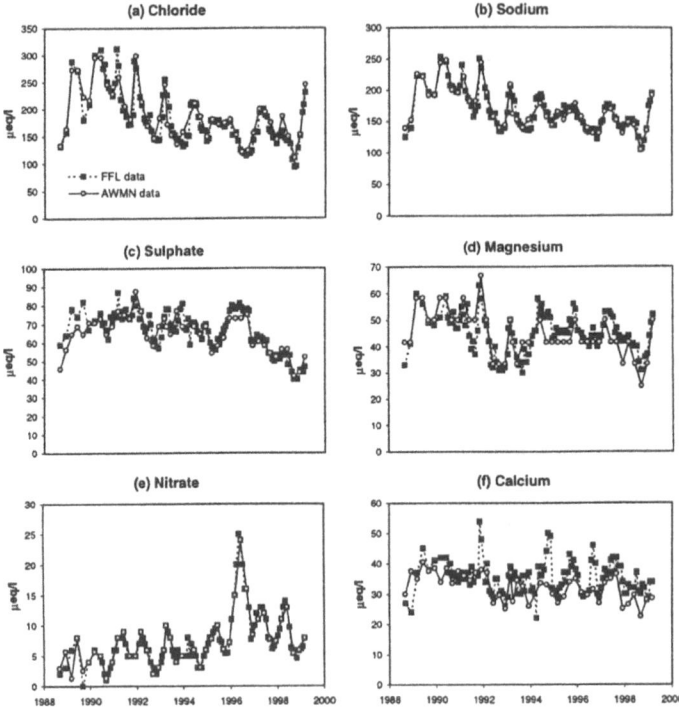

Figure 1. Overlaid major ion time series for the FFL and AWMN datasets.

variables, alkalinity shows a small, non-significant rise over the 11 years, but a highly significant rising trend is observed for alkalinity-based ANC, with an estimated slope of 1.3 µeq l^{-1} yr^{-1}. This suggests an actual ANC increase intermediate between those indicated by the two CB ANC estimates. A smaller but still significant decrease of 0.3 µeq l^{-1} yr^{-1} is also observed for H$^+$, again suggesting limited recovery.

Linear regression analysis of the subset of samples for which duplicate analyses were undertaken (Table II) confirms earlier observations, with good correlations between analyses for Na and Cl, and to a lesser extent for SO$_4$ and Mg. Ca and K are less well predicted, although for all ions, slope coefficients close to unity suggest little or no systematic bias between datasets. However, the correlation between estimates of CB ANC is extremely poor ($R^2 = 0.06$), in accordance with the inconsistent temporal patterns observed for ANC$_{AWMN}$ and ANC$_{FFL}$. An indication of the cause of these large discrepancies in CB ANC estimation is provided by the standard errors on individual ion determinations. These show that, in fact, the ions that are least well correlated between the two datasets, Ca and K, actually make a relatively minor contribution to errors in ANC. This reflects the small range of concentrations observed for these ions (although Figure 1f suggests that Ca errors are significant for some samples). Conversely, Na and Cl, despite having the strongest correlation between

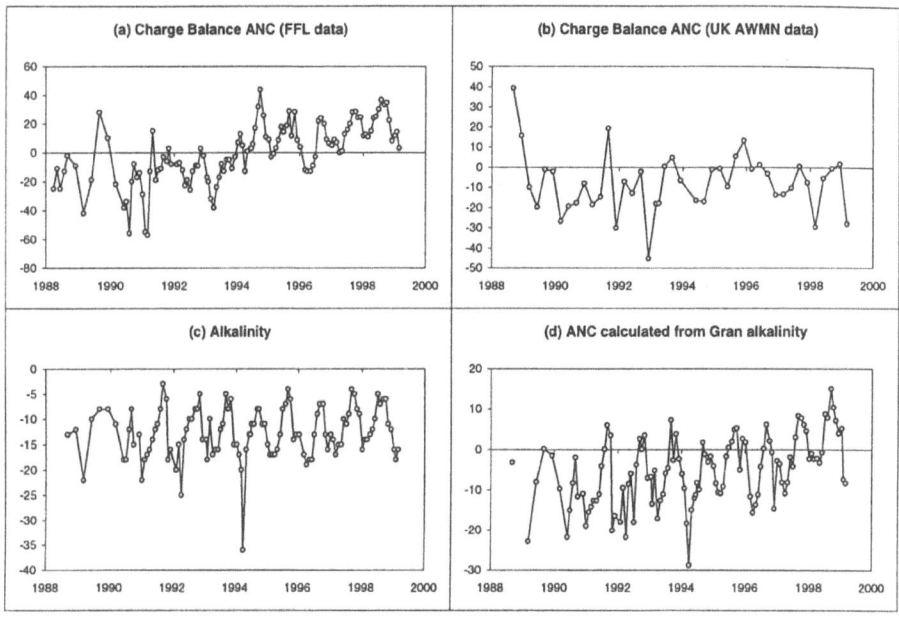

Figure 2. Time series for alkalinity, and different estimates of ANC, at Round Loch

datasets, also have the largest standard errors. These ions, which are present at high, and highly variable, concentrations at this near-coastal site, thus contribute most significantly to errors in ANC estimation.

TABLE II

Linear regression analysis of AWMN versus FFL measurements for samples analysed at both laboratories, and calculated ANC

Determinand	R^2	Slope Coefficient	Residual Mean Standard Error (μeq l^{-1})
Cl	0.90	1.01	16.5
SO$_4$	0.69	0.96	5.8
Na	0.93	1.00	8.9
Mg	0.63	0.99	5.4
Ca	0.19	0.90	4.5
K	0.35	0.99	1.7
CB ANC	0.06	0.20	14.4

4. Discussion and Conclusions

CB ANC represents the (often small, in acid waters) difference between two large values, ΣC_B and ΣC_A. These, in turn, are calculated from at least seven individual ion determinations. At Round Loch, an acidified lake typical of many in the UK and elsewhere, duplicated chemical analyses over the period 1988–1999 have clearly shown significant problems with this method of ANC estimation. The accumulation of relatively small errors on individual ion determinations, generally well within the range of expected analytical precision, generates a large cumulative error. As a result, two sets of ANC estimates based on duplicate analyses of the same sample set exhibit virtually no correlation, and inconsistent long-term trends.

An alternative estimation of ANC, based on titration alkalinity, indicates that the actual ANC change at this site has been a small but consistent increase of slightly over 1 $\mu eql^{-1}yr^{-1}$ since 1988. It is worth noting, however, that this is partly due to increases in DOC (Monteith and Evans, 2000), rather than to large reductions in overall acidity; the downward trend observed for H^+ is just 0.3 $\mu eq\ l^{-1}\ yr^{-1}$, and no significant trend was observed for alkalinity. Therefore, it appears that at Round Loch the last 11 years have in part witnessed a shift from mineral to organic acidity, rather than simply a reduction in acidity.

It is concluded, then, that CB ANC, although in many respects a convenient indicator of acid status, is subject to severe measurement error. In terms of spatial assessments, for example in regional freshwater critical loads assessments, these errors may be small relative to inter-site variations. However temporal changes in ANC at individual sites are typically much smaller, at least on the relatively short timescales available for most monitoring studies, and measurement errors thus attain greater significance. For assessments of acidification and recovery it is therefore suggested that, rather than using CB ANC, trends should be determined either for directly measured acidity indicators such as alkalinity or pH, or from a more stable estimate of ANC based on titration alkalinity, Al and DOC.

Acknowledgements

Funding for this work was provided by the UK Department of the Environment, Transport and the Regions (Contract no. EPG 1/3/92).

References

CLAG: 1995, *Critical Loads of Acid Deposition for United Kingdom Freshwaters*. Critical Loads Advisory Group summary report, Institute of Terrestrial Ecology, Bush, Edinburgh.

Fowler, D, and Smith, R.:2000, Spatial and temporal variability in the deposition of acidifying species in the UK between 1986 and 1997. In: Monteith, D.T., and Evans, C.D. (Eds). *The UK Acid Waters Monitoring Network: Ten Year Report*, ENSIS Publishing, London.

Harriman, R., and Taylor, E.M.: 1999, *Acid Neutralising Capacity and Alkalinity: Concepts and Measurement*. Report SR (99) 06F, Freshwater Fisheries Laboratory, Pitlochry.

Hirsch, R.M., and Slack, J.R.: 1984. *Water Resources Research* **20**, 727.

Monteith, D.T., and Evans, C.D. (Eds). *The UK Acid Waters Monitoring Network: 10 Year Report*. ENSIS Publishing, London.

Neal, N., Reynolds, B., and Robson, A.J.: 1999, *Science of the Total Environment* **243**, 233.

Reuss, J.O., Christophersen, N., and Seip, H.M.: 1986. Water, Air and Soil Pollution **30**, 909.

SPATIAL AND TEMPORAL VARIABILITY IN OZONE CONCENTRATION LEVEL AT TWO LITHUANIAN STATIONS

RASA GIRGZDIENE* and ALOYZAS GIRGZDYS

Institute of Physics, A. Gostauto 12, 2600 Vilnius, Lithuania
(* author for correspondence, e-mail:raseleg@ktl/mii/lt)

Abstract. The ozone concentration measurement results at two rural stations in Lithuania are presented. Ozone measurements have been carried out at the coastal site since 1980 and at the forested one since 1994. Both stations are far from the influence of the local pollution. The established ozone concentration trends at the coastal station were upward in cold and not so distinct in warm seasons during this monitoring time. The tendency even of a decreasing level in some summer months was established: in June and August. These trend results were very close to the results obtained at other European stations.

Comparing ozone levels and courses at both stations, significant differences were determined. Clear differences between monthly daytime and daily ozone concentration averages are observed during the warm season. They were three times larger at a forest site as compared with the results at a coastal one.

Keywords: monitoring, ozone, trend.

1. Introduction

Ozone is a natural component of the atmosphere. In the troposphere it plays an important role in the chemical processes. Anthropogenic ozone in the troposphere is produced through photochemical processes from nitrogen oxides and volatile organic compounds whose largest sources are motor vehicles and industry.

Some of the investigators (Staehelin *et al.*, 1994; Scheel *et al.*, 1999) observed a positive trend in ozone at various remote stations. On the other hand, a negative trend is observed, for example, at some stations of the Netherlands (Roemer and Bosschert, 1996). A more detailed analysis of ozone trend results each month was performed (Roemer, 2000; Scheel *et.al.*, 1999). Existing data, meanwhile, are not able to answer if there is the trend in ozone over Europe. Having in mind that ozone, one of the strongest photochemical oxidants, has negative effects on the forest and vegetation decline, human health, therefore, ozone concentration monitoring in different regions is necessary.

Surface ozone is routinely measured at four rural sites in Lithuania. However, in this paper data from two stations with the longest ozone data series are used.

The purpose of this paper is to examine the trend in the ozone concentration in Lithuania over different periods and to indicate the reason, which can have an influence on the trend estimation.

2. Sites and Methods

The surface ozone concentrations have been measured at two rural stations in Lithuania. The coastal station Preila (55°55' N and 21°00' E, 5 m a.s.l.) is located in the western part of Lithuania on the Curonian Spit, a narrow sandy strip separating the Baltic Sea and the Curonian Bay. There are no great sources of anthropogenic pollution of the atmosphere, soil or water close to the observation site. Data obtained at the station well reflect the air pollution level determined by the transported airmasses. These data are important especially taking into account a fact that 75% of airmasses entering Lithuania are from the southwest-northwest directions (Girgzdiene and Girgzdys, 1998).

The other rural station Rugsteliskes (55°26'N and 26°04'E) is located in the eastern forested part of Lithuania about 350 km from the seashore.

The surface ozone measurements have been carried out with electrochemical analysers in Preila since 1981 and in Rugsteliskes stations since 1994. The monitors were calibrated by the ozone generator periodically. Since 1992 the ozone concentration in Preila site has been measured with the commercial UV absorption ozone analyser 1003-RS. The instruments were calibrated against the reference standard UV photometer SRP11 at the Air Pollution Laboratory of the Institute of Applied Environmental Research of Stockholm University, Sweden. The UV photometer ML 9811 has been established in Rugsteliskes since 1996.

The ozone data for an analysis were selected according to some criteria. The ozone daytime values were calculated between 10-17 h and daily values between 01-24 h. Monthly means and monthly percentiles were based on data coverage of at least 75%. An analogous analysis of the ozone concentration data from both stations was carried out with the aim to reveal the differences and identities between the formation of daytime and daily ozone levels. The whole year was divided to two seasons – cold and warm. Six months period from 1 April to 30 September was comprehended as warm season. Roughly it is the growing season in our latitudes. This period is also distinguished for the intensive photochemical ozone production due to favourable weather conditions. Other six months (October – March) were attributed to cold season.

For the trend analysis of ozone concentrations, the statistical non-parametric seasonal Kendall test method was used (Gilbert, 1987), which is a test for a trend using relative magnitudes of the time-ordered data. The software for calculating the trend statistics was developed by Anders Grimvall, University of Linköping, Sweden.

3. Results

The established ozone concentration trend at the coastal station Preila in the period 1981-1999 was upward (Fig. 1).

Analysis of separate data during cold and warm seasons showed upward trends too (Fig. 2). The data series of the ozone monitoring at the Rugsteliskes station was too short for the trend estimation.

The daytime and daily ozone values were used in the trend analysis. Daytime ozone concentration values at the non-polluted ground level are less influenced by the local conditions than daily values and can better expose general tendencies of ozone level changes.

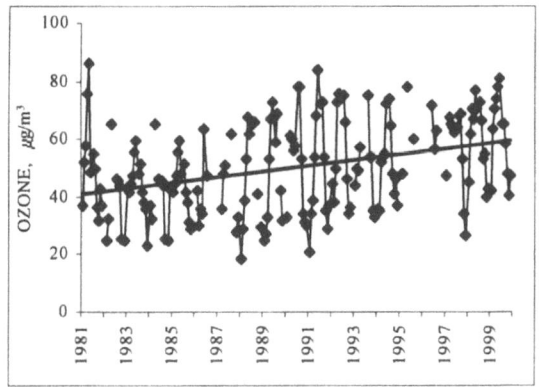

Figure 1. Variation of the monthly ozone concentrations and linear trend at the Preila station.

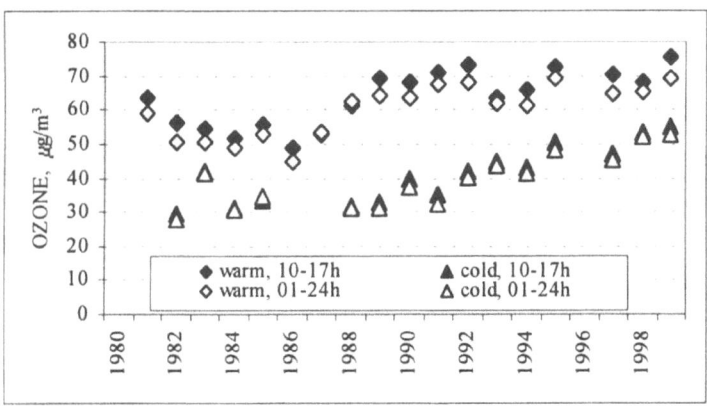

Figure. 2. Variation of the warm and cold periods ozone concentration means at the Preila station.

The detailed investigation of the data at the station Preila showed that after 1988 an upward tendency remained but at the enhanced ozone level as compared with the earlier established one. Analysing results of the two different seasons and periods it was observed that upward trends were more noticeable during the cold than the warm season over the last decade (1989-1999) (Fig. 2 and Table I). The average increase in mean values was larger during the cold season and it was smaller during the warm season over 1989-1999 as compared with that over the 1981-1999 period (Fig. 3). One of the possible explanations might be a determined change in the dominant air mass transport (Girgzdiene, 1998) over Lithuania. In this work the analysis of ozone level dependence on the origin of air masses back trajectories was performed. It was obtained that the average ozone level in Lithuania depends on from which and across which region they reached the country. The polluted and with high ozone concentration air masses originated in west Europe. If the air masses from west Europe predominated till 1988, so from 1989 "clean" air masses from the northern latitudes began to dominate. Owing to the economic recession, a local contribution to the air pollution level also changed after 1989.

TABLE I

Trend analysis based on annual average ozone concentrations of cold and warm seasons at the Preila station for two different time periods.

1981-1999				1989-1999			
Cold season		Warm season		Cold season		Warm season	
10-17 h	0-24 h	10-17 h	0-24 h	10-17 h	0-24 h	10-17 h	0-24 h
u***	u***	u***	u***	u***	u***	u	u

u = upward trend indicated, p<0.25,   u*** = p<0.01

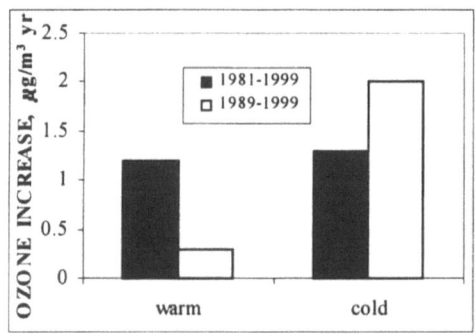

Figure 3. The average increases in the ozone daytime mean concentration ($\mu g/m^3$) per year at the Preila station during separate seasons.

The results of the monthly mean data evaluation during 1989-1999 show the increasing level in January – May and October – December. The tendencies of the decreasing ozone level were found in June and August, while no trend was observed in July. The analogous data analysis over a longer period 1981-1999 showed only upward tendencies. Similar tendencies were observed for high ozone values (98th percentiles). The results are shown in Table II. Resembling results were obtained by Roemer (2000) at some European stations.

The analysis of daily and daytime ozone seasonal courses was made with the aim to expose the differences and identities of the ozone level formation at two rural stations with different local factors. For example, ozone concentration means during the warm period at the station Preila were mostly higher than at Rugsteliskes one, but AOT40 exposure was higher at Rugsteliskes as compared with results at the station Preila (Table III).

TABLE II

Trend analysis based on the ozone monthly means and 98th percentiles at the Preila station.

1989-1999													
	10-17h												
	Jan	Feb	Mar	Apr	May	Jun	Jul	Aug	Sep	Oct	Nov	Dec
Mean	u**	u**	u***	u**	u*	d	–	d*	u	u*	u**	u
98th	u	u**	u**	u	u	d	–	d*	u	–	-	u
	01-24h												
Mean	u	u**	u***	u	u*	d	-	d	u*	u	u*	u
98th	u*	u**	u**	u*	u*	d	-	d***	u	u	–	u
1981-1999													
	10-17h												
Mean	u**	u***	u***	u*	u*	u**	u**	u**	u***	u***	u***	u**
98th	-	u*	u	u**	u	u	u	u**	u**	u**	u*	u***
	01-24h												
Mean	u**	u**	u***	u***	u*	u**	u**	u*	u**	u***	u***	u***
98th	u	u*	u***	u**	-	-	u*	-	u***	u***	u*	u**

u – upward trend indicated,   d- downward trend indicated,   – = no trend p>0.25
u – p<0.25   d – p<0.25
u* – p<0.1   d* – p<0.1
u** – p<0.05   d** – p<0.05
u*** – p<0.01   d*** – p<0.01

TABLE III

Ozone concentration means ($\mu g/m^3$) and AOT40 ($\mu g\ h/m^3$) at two stations during April – September in 1998 and 1999.

	1998		1999	
	Preila	Rugsteliskes	Preila	Rugsteliskes
Mean	66.9	57.7	70.3	61.7
AOT40	12960	15264	11850	21085

The comparison of the obtained results of daily and daytime ozone courses during 1994-1999 period is shown in Figure 4. As it was expected, the largest differences between the daytime and daily ozone concentration were observed during the warm season (April-September) at both stations. They did not exceed 6 $\mu g/m^3$ in summer and only 2 $\mu g/m^3$ in winter at the station Preila. However, this difference at the station Rugsteliskes was approximately three times larger than in Preila. It can be related to local topographical conditions of these stations. Preila is a coastal station. There, a formation of the temperature inversion is observed very seldom. Rugsteliskes is located at the forested site far from the sea influence and the night temperature inversions are very frequent. The ozone concentration decreases up to a few $\mu g/m^3$ during the inversion. The depletion of ozone by dry deposition to the ground is the main cause of the nocturnal depletion of ozone at unpolluted rural locations (Garland and Derwent, 1979).

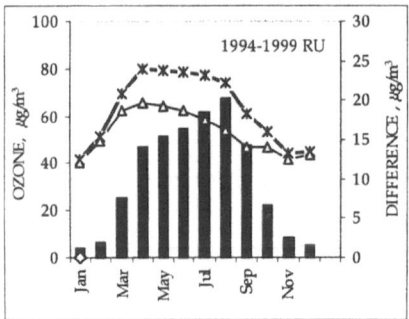 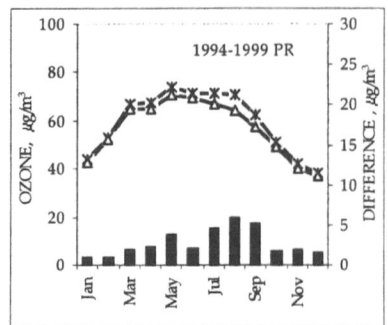

Figure 4. Average seasonal cycle of ozone at the Preila (PR) and Rugsteliskes (RU) stations: daytime (10-17 h) – stars and daily (01-24h) concentrations – opens triangles, the difference between ozone daytime and daily data – columns.

4. Conclusions

The upward trend in the ozone concentration of 0.93 $\mu g/m^3$ per year at the Preila station was established during 1981-1999.

The average increase in mean values went up from 1.3 up to 2.0 $\mu g/m^3$ per year during the cold season and it went down from 1.2 up to 0.3 $\mu g/m^3$ per year during the warm season over 1989-1999 as compared with the increase over the 1981-1999 period.

The largest differences between daytime and daily ozone concentrations were observed during the warm season (April-September) at both stations. They did not exceed 6 $\mu g/m^3$ in summer and only 2 $\mu g/m^3$ in winter at the station Preila. This difference at the station Rugsteliskes is approximately three times larger than in Preila.

References

Garland, J. A. and Derwent, R. G.: 1979, *Quart. J. R. Met. Soc.* **105**, 169.
Gilbert, R. O.; 1987, *Statistical Methods for Environmental Pollution Monitoring.* New York, Van Nostrand Reinhold Co.
Girgzdiene, R., Mikelinskiene, A. and Girgzdys, A.: 1998, *Environmental Physics* **20**, 13.
Roemer, M.: 2000, 'Trends of tropospheric ozone in Europe over the last 10 years', in P. M. Midgley, M. Reuther, M. Williams (Eds.), *Proceedings from EUROTRAC Symposium 2000*, Springer Verlag Berlin, Heidelberg (in print).
Roemer, M. and Bosschert, M.: 1996, *TNO-Publication P96/003*, Apeldoorn, The Netherlands.
Scheel, H. E., Sladkovic, R. and Kanter, H.-J.: 1999, 'Ozone variations at the Zugspitze (2962 m a.s.l.) during 1996-1997', in P. M. Borrell and P. Borrell (Eds.), *Proceedings of EUROTRAC Symposium '98*, WITPRESS, Southampeton, pp.264-268.
Staehelin, J., Thudium, J., Buehler, R., Volz-Thomas, A. and Graber, W.: 1994, *Atmospheric Environment*, **28**, 75.

CHARACTERISTICS OF SO_2, NO_2, SOOT AND BENZO(a)PYRENE CONCENTRATION VARIATION ON THE EASTERN COAST OF THE BALTIC SEA

AUDRONĖ MILUKAITĖ,[*] ALDONA MIKELINSKIENĖ, and BRONISLAVAS GIEDRAITIS

Institute of Physics, A.Goštauto 12, Vilnius 2600, Lithuania
(*author for correspondence, e-mail: amk@ktl.mii.lt)

Abstract. The investigation of SO_2, NO_2, soot and benzo(a)pyrene (BP) has been performed at the background station on the eastern coast of the Baltic Sea since 1980. A significant decreasing trend has been observed for SO_2 and NO_2, while soot and BP concentrations were changing insignificantly. The decreasing SO_2 and NO_2 high concentrations ($>10 \mu g \cdot m^{-3}$) have been determined in the air masses coming from the Western and Central Europe to Lithuania since 1990. The concentration of SO_2 in a range of $0-5 \mu g \cdot m^{-3}$ and the concentration of NO_2 in a range of $0-10 \mu g \cdot m^{-3}$ are characteric of the background atmospheric air.

Keywords: air mass trajectories, background site, concentration variation

1. Introduction

The acidification of environment mainly depends on the SO_2 and NO_2 emissions. Soot and BP are mainly accompanying SO_2 and NO_2 as they are also produced in the process of fuel burning. These pollutants are relatively stable in the atmosphere and may be transported with air masses for long distances (Bjorseth,1979; Duyzer and Fowler, 1994; Mylona, 1996; Milukaite,1995).

According to the 1992 EMEP (Environment monitoring evaluation program) data, the mean concentration of SO_2 mainly varied between 2 and 5 $\mu g\ m^{-3}$ and that of NO_2 between 5 and 10 $\mu g\ m^{-3}$ at the background sites of Europe (Duyzer and Fowler, 1994; Mylona, 1996). Furthermore, the data of SO_2 and NO_2 concentrations observed in various countries showed that they had a tendency to decrease in recent years (Hovmand, 1995; Zwozdiak *et al.*, 1995). The investigations of soot and benzo(a)pyrene are performed to a less extent. The concentration of BP ranged from 0.02 to 1.7 ng m^{-3} for the duration of 3 months in 1989-1990 (Lunden *et al.*,1994). High concentrations of BP differing between 8.7 and 41.6 ng m^{-3} were determined in upper Silesia in summer time (Bodzek *et al.*,1993). The annual average concentration of C_{el} was determined to be 1.3 $\mu g \cdot m^{-3}$ at 20 background sites in northern America, while monthly averaged concentration of soot was 0.77 $\mu g\ m^{-3}$ in November and 0.24 $\mu g\ m^{-3}$ in May 1999 in Iceland (Shah *et al.*, 1986; Jennings *et al.*, 1997). The lack of published data in Eastern Europe encouraged us to summarize data obtained on the eastern coast of the Baltic Sea since 1980 with the purpose to evaluate the level of some gaseous and particulate admixture concentrations as a consequence of different sources and different air masses passing Lithuania.

Water, Air, and Soil Pollution **130**: 1553–1558, 2001.
© 2001 *Kluwer Academic Publishers.*

2. Materials and Methods

Samples of gaseous and aerosols were collected at the Preila background station on the eastern coast of the Baltic Sea. The location of sampling site is presented

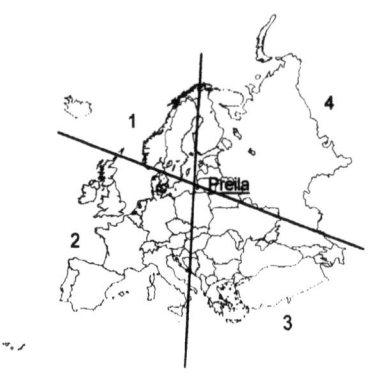

Figure 1. Site of sampling and sectors of trajectories

in Figure 1. The results of investigation at this site can be used for the evaluation of the regional air pollution background formation. The region of pollutants sources was determined from the back trajectories of the air masses transported to the background station. They were calculated from the maps of baric topography at a level of 850 mb for the preceding 48 hours. Trajectories of air masses were devided into 4 sectors and sector 0 characterizing air masses had been formed in the region of observation site.

SO_2 and NO_2 have been collected daily from 9^{00} a.m. to 9^{00} a.m. of the next day since 1980. SO_2 was adsorbed on the active layer of $NaHCO_3$ in the glass tubes with the air flow rate of 3 l min^{-1}. The concentration of sulphur dioxide was determined by the turbidimetric method. The sensitivity of determination was 0.2 µg m^{-3}. NO_2 was absorbed in the trietanolamine solution at the velocity of air drawing 0.5 l min^{-1} The concentration of nitrogen dioxide was determined by the colorimetric method. The sensitivity of the method was 0.3 µg m^{-3} (Giedraitis, 1986). Particulate pollutants (soot and BP) were collected daily until 1994 and twice a week after 1994. Soot was collected on the glass fibre filters (Whatman GF/B) at the velocity of drawing 0.5 m^3 h^{-1}. The concentration of soot (as $C_{el.}$) was determined by gas chromatography until 1990 and by light reflectance after 1990. The limit of determination was 0.04µg m^{-3} and 0.01µg m^{-3}, respectively (Armalis, 1986). Benzo(a)pyrene was collected with air flow rate of 1-2 m^3 h^{-1} on the fibre filters АФА-ХА-20 until 1995 and on the glass fibre filters since that time. BP was determined by the fluorescent method at the temperature of liquid nitrogen (77°K). The sensitivity of the method was 0.02 ng m^{-3} (Milukaitė, 1986).

3. Results and Discussion

Monthly mean concentrations of SO_2, NO_2, soot and BP are presented in Figure 2. The data of investigation showed that the monthly averaged concentration of these pollutants can vary in a wide range: 0.8-31.5 µg m^{-3}; 3.1-39.7 µg m^{-3}; 0.1-8.6 µg m^{-3} and 0.04-3.8 ng m^{-3}, respectively. Furthermore, each of pollutants has specific seasonal variations: SO_2, benzo(a)pyrene and soot being dependent on burning processes have clearly expressed seasonal variations, while NO_2 concentration

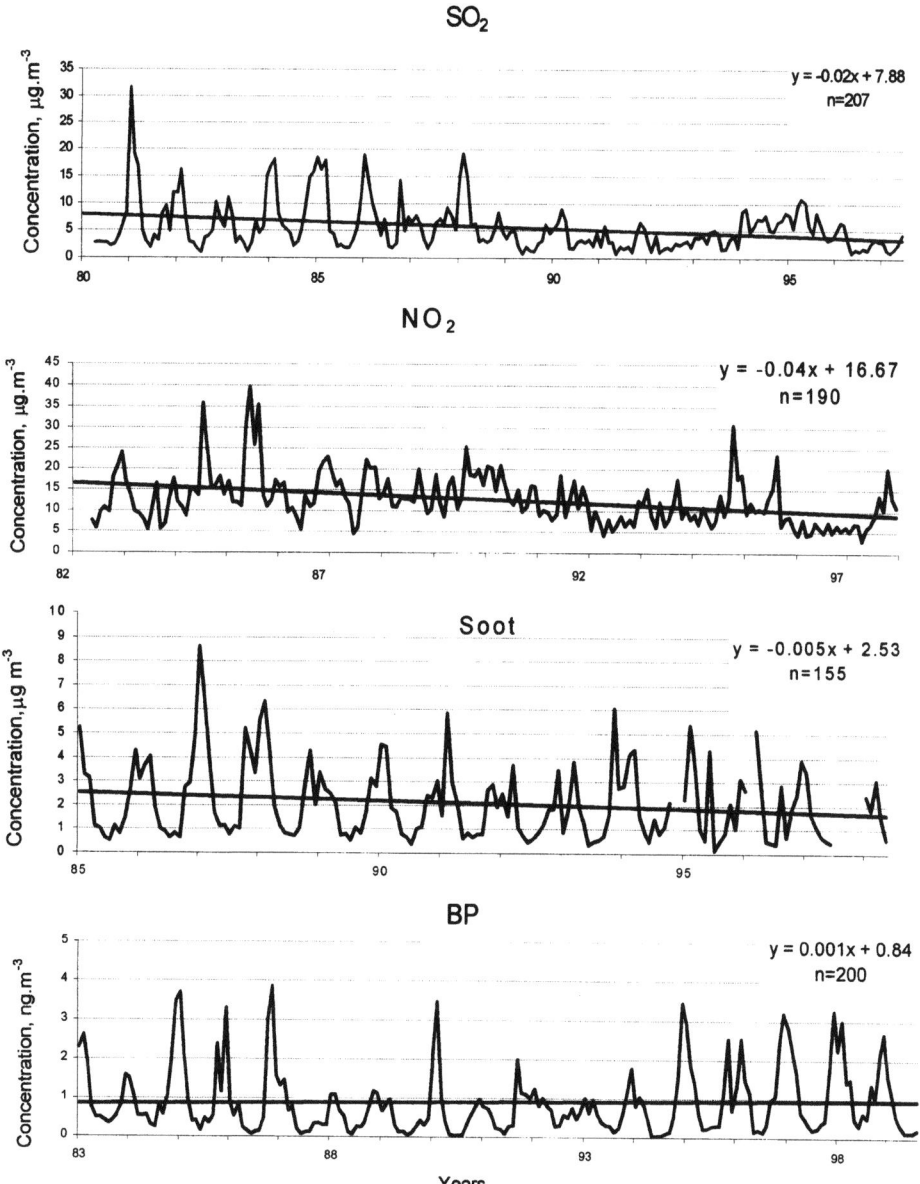

Figure 2. Monthly course and trend of pollutants at the background station.

in cold season of the year (October-April) and warm season of the year (May-September) is very similar. The best correlation is observed between SO_2, soot and BP concentration (r=0.56 and r=0.40) and the least correlation is observed between these pollutants and NO_2 (r=0.09–0.14). This correlation shows the differences in sources and fate of investigated pollutants in the atmospheric air. The concentrations of SO_2 and NO_2 have a decreasing trend with the monthly slope of 0.02 µg m^{-3} (P< 0.01) and of 0.04 µg·m^{-3} (P< 0.001), respectively, while soot and BP concentration changes are insignificant (P> 0.05) during the investigation.

For the estimation of pollutants trend, the data of concentrations were divided into 4 periods. Averaged data of warm and cold seasons for these periods are presented in Table I. As it is seen, the concentration of SO_2 decreases more

TABLE I
Average concentrations of pollutants in different periods during 1980 –1998

Periods	Cold season (October- April)				Warm season (May-September)			
	SO_2	NO_2	Soot	BP	SO_2	NO_2	Soot	BP
1980–1984	10.2±11.6	14.1±10.6	-	1.7±1.8	3.4±4.6	12.1±13.5	-	0.5±1.1
1985–1989	8.7±8.2	14.4±11.2	3.5±4.4	1.0±1.3	3.3±5.4	16.5±17.1	0.9±0.68	0.2±0.4
1990–1994	4.7±4.7	11.9±11.1	2.7±3.7	1.0±1.2	3.3±3.7	10.9±11.4	0.8±0.83	0.4±0.7
1995–1998	4.4±5.2	8.1±8.0	2.2±3.0	2.3±2.0	4.2±5.0	9.4±8.1	0.7±1.0	0.4±0.4

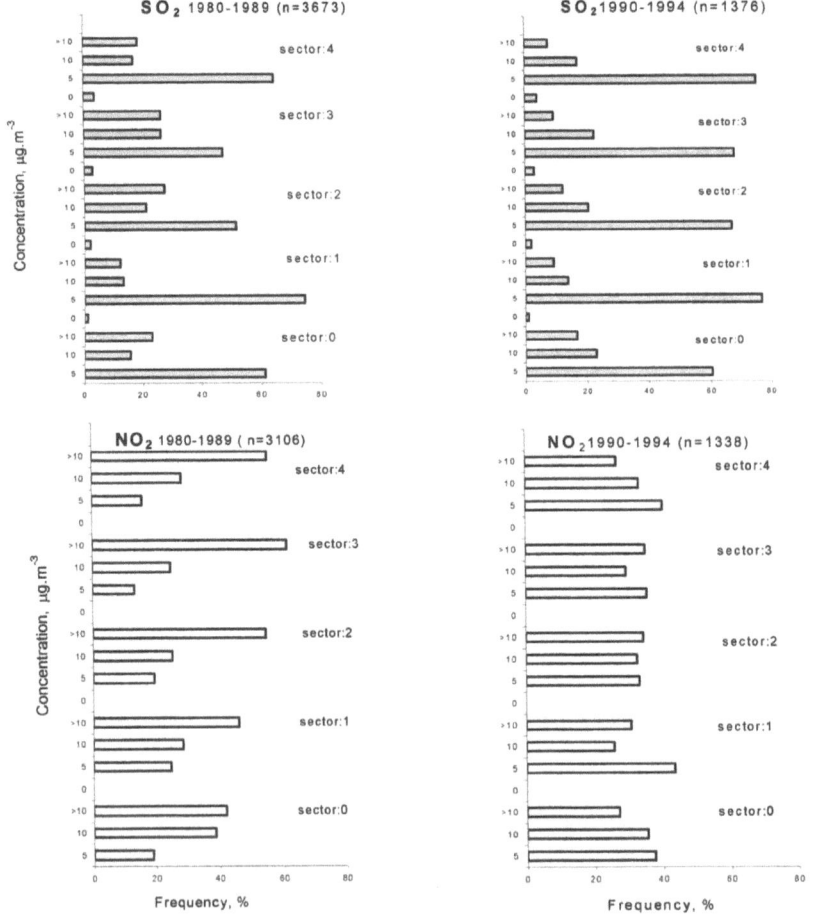

Figure 3. Frequency of SO_2 and NO_2 concentrations distribution in the periods 1980-1989 and 1990-1994 in the air masses of various directions.

than twice in cold season and the concentration of SO_2 is almost constant during warm season if data of two periods are compared:1980-1989 and 1990-1998. A decrease in the NO_2 concentration to 40 % is noticeable in cold season of the year and a less decrease in warm season of the year. An insignificant decrease in the soot concentration is observed in winter and relatively stable concentration in summer time since 1990. The lowest concentration of BP was determined during 1985-1989 in cold and warm season as well, while an increase in the BP concentration is evident during 1995-1998 in winter time.

Variation in the concentration of investigated pollutants shows that fuel and its burning process are improved significantlly after 1989 and that the sources of

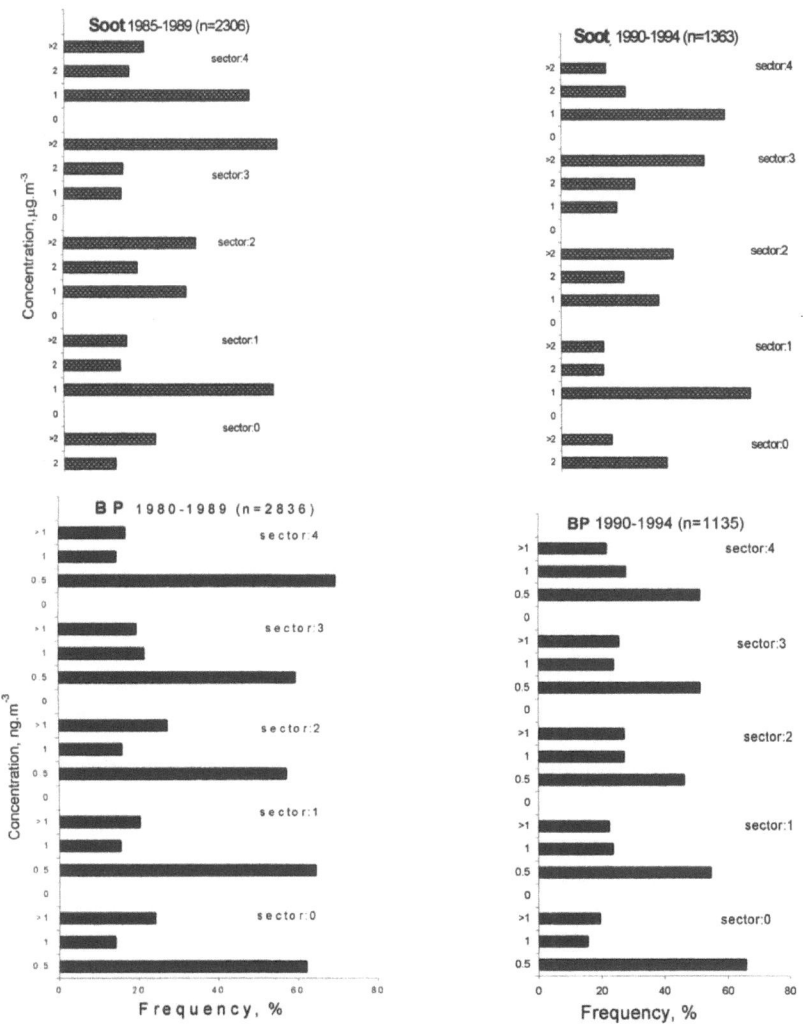

Figure 4. Frequency of soot and BP concentrations distribution in the periods 1980-1989 and 1990-1994 in the air masses of various directions.

pollutants in warm season are relatively stable, except NO_2 in 1980-1998.

The histograms of pollutants concentrations according to the air mass trajectories in 1980-1989 and 1990-1994 are presented in Fig.3 and Fig.4. As it is evident, the highest concentrations of SO_2 and NO_2 were coming to Lithuania from sectors 2 and 3, while the high frequency of low concentrations was from sector 1 in the period of 1980-1989. A significant decrease in SO_2 and NO_2 concentrations is evident from sectors 2 and 3 in the period of 1990-1994. The frequency of acidifying admixture concentrations higher than 10 µg m^{-3} decreased about twice in the last decade in these sectors. The frequency of high soot concentrations (>2 µg m^{-3}) decreased and the frequency of high BP concentrations (>1 ng m^{-3}) increased insignificantly in the air masses from sectors 2 and 3. The lower frequency of SO_2, NO_2, C_{el} and BP high concentrations in the local air masses (sector 0) may be the result of decreasing industry development in Lithuania after 1990.

Conclusions

A significant correlation has been observed between SO_2, soot and BP concentrations at the background site. Concentrations of the main acidifying admixtures as SO_2 and NO_2 during the period of 1980-1998 have been decreasing significantly at the Lithuanian background site since 1990. The concentration of soot was decreasing, while the concentration of benzo(a)pyrene was increasing insignificaly in the last decade. The variation in concentrations of SO_2, NO_2, soot and BP depends on the local pollution sources and air masses passing Lithuania.

References

Armalis S. and Nika A.: 1986, *Atmospheric Physics*, **11**, 155-159 (in Russian).
Bjorseth A., Lunde G., Lindskog A.: 1979, *Atmospheric Environment*, **13**, 45-53.
Duyzer J., Fowler D.: 1994, *Tellus*, **46** B, 353-372.
Giedraitis B.: 1986, *Protection of Atmosphere Against Pollution*, Vilnius, p.14-16 (in Russian).
Howmand M. F., Andersen H. V.:1995, *Water Air and Soil Pollution* **85**, 205-2210.
Jennings S.G., Geever M., McGovern F.M.:1997, *Atmospheric Environment*, **31**, 17 2795-2808.
Lunden E., Lindskog A., Mowrer J.: 1994, *Atmospheric Environment*, **28**, 3605-3615.
Milukaite A.: 1986, Unificated methods of the environment background monitoring. Moscow, p.p.46-53 (in Russian).
Milukaitė A., Juozefaitė V.: 1995, *Water, Air and Soil Pollution* **85**, 2003-2008.
Mylona S.,: 1996, *Tellus*, **48B**, 662-680.
Shah J., Johnson R., Heyerdahl E., Huntzicker J.: 1986, *J. Air Pollution Control Association*, **36**, 254-257.
Zwozdiak J., Zwozdiak A., Kmiec G., Kacperczyk K.: 1995, *Water Air and Soil Pollution*, **85**, 2009-2013.

PEROXIDE CONCENTRATIONS IN FOG WATER AT MOUNTAINOUS SITES IN JAPAN

KOICHI WATANABE[1*], YUTAKA ISHIZAKA[1], YUKIYA MINAMI[2] and KOJI YOSHIDA[3]

[1] *Institute for Hydrospheric-Atmospheric Sciences, Nagoya University, Nagoya, Japan;* [2] *Ishikawa Agricultural College, Ishikawa, Japan;* [3] *Japan Science and Technology Corporation, CREST, Graduate School of Bioagricultural Sciences, Nagoya University, Nagoya, Japan*
(* author for correspondence, e-mail: nabe@ihas.nagoya-u.ac.jp)

Abstract. Measurements of peroxide concentrations in fog water were conducted near the summit of Mt. Norikura (altitude, 2770m) in central Japan, and at the midslope of Mt. Oyama (altitude, 680m), southwest of the Kanto Plain. The concentrations of peroxide at Mt. Norikura, far from industrial regions, ranged from 3 to 120 μM during the summer and early autumn in 1993. The potential capacity for SO_2 oxidation appears to be very high near the summit of Mt. Norikura. Analysis of the chemical composition of three-stage size-fractionated fog water samples collected at Mt. Norikura showed that the concentrations of peroxide were apparently independent of droplet size, whereas the concentrations of chemical constituents mainly derived from secondary aerosols and the acidity were higher in smaller droplets. Peroxide concentrations in fog water were low (< 5 μM) at Mt. Oyama, located near heavy industrial areas, and lower than those in rain water sampled simultaneously (0.2-33 μM). Especially, peroxide was scarcely detected in strongly acidic fogs (< 0.2 μM). Peroxide might have been decomposed by SO_2 (S(IV)) oxidation in the aqueous-phase.

Key words: peroxide, mountainous sites, fog water, oxidation, size-fractionated

1. Introduction

Hydrogen peroxide (H_2O_2) is considered to be an important oxidant of SO_2 (S(IV)) in atmospheric liquid water. Hydrogen peroxide has been found to be highly soluble in cloud and fog droplets or aqueous aerosols (Yoshizumi et al., 1984; Watanabe et al., 1996). The reaction between H_2O_2 and S(IV) in water droplets is very rapid, and independent of the pH. This differs from the reaction between O_3 and S(IV), which does not occur under conditions below pH 5 (Martin and Damschen, 1981). Organic hydroperoxides (ROOH) are thought to play an important role in the oxidation of S(IV), as well as H_2O_2 (Lind et al., 1987). Therefore, the concentration of total peroxides (H_2O_2 + ROOH) in the aqueous phase is a useful indicator of oxidation capacity. Moreover, peroxide may contribute to vegetation damage as well as O_3 (Ennis et al., 1990; Hewitt and Terry, 1992). Peroxide concentration is mainly dependent on the meteorological conditions (UV radiation, temperature and humidity) and the concentrations of atmospheric pollutants (Sakugawa et al., 1990). Large seasonal variations in the concentrations of peroxide in the atmosphere have been observed (Sakugawa and Kaplan, 1989; Watanabe and Tanaka, 1995).

Peroxide concentrations in fog (or cloud) water have been measured at mountainous sites, especially in the United States and Europe (Romer et al.,

1985; Kelly et al., 1985; Chandler et al., 1988; Olszyna et al., 1988; Gallagher et al., 1991). Peroxide concentrations in fog and cloud water, recently reported in the literature, have ranged from below the level of detection to nearly 250 μM (Olszyna et al., 1988; Gunz and Hoffmann, 1990). Preliminary studies on peroxide in fog and cloud water in central parts of Japan have been reported (Watanabe et al. 1999, 2001). However, there is a shortage of such data on peroxide in East Asian countries.

In this paper, the concentrations of peroxide in fog and rain water near the summit of Mt. Norikura and at the midslope of Mt. Oyama in Japan are presented, and characteristics of peroxide concentrations at both mountainous sites are examined. Moreover, the chemical constituents of size-fractionated (three-stage) fog water samples collected at Mt. Norikura were investigated in an effort to elucidate the origin of the chemical components.

2. Methods

The observations were conducted at the Norikura Observatory of the Institute for Cosmic Ray Research, University of Tokyo, located near the summit of Mt. Norikura (altitude 2770m) in central Japan, and at the midslope of Mt. Oyama (altitude 680m), southwest of the Kanto Plain (Figure. 1). There are no heavy industrial areas around Mt. Norikura, whereas Mt. Oyama is located near heavy industrial areas. Due to severe winter/spring environmental conditions, observations at Mt. Norikura usually can be performed only in the summer and early autumn seasons.

Fog water collection was performed on the following days: 20-24 July and 12-14 and 16-17 September 1993 near the summit of Mt. Norikura, and 27-30 August and 2 September 1998 at the midslope of Mt. Oyama. Several fogs were associated with rainfall. The fogs were usually sampled with an active collector (FWG-400F, Usui Kogyo Kenkyusho. Inc.). Size-fractionated fog water (three-stage) samples were collected during July 1993 (Minami and Ishizaka, in preparation). The sampling and analysis of fog water at Mt. Norikura were also conducted from 1994 to 1996, and the results have been described in detail by Watanabe et al. (1999).

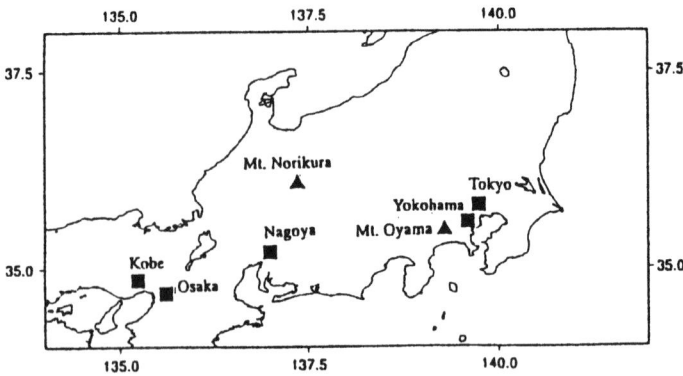

Figure1. Map of central Japan showing the location of Mt. Norikura and that of Mt. Oyama.

Peroxide was measued by the fluorometric method, using p-hydroxyphenyl acetic acid and peroxidase reagents (Lazrus et al., 1985), as described in detail by Watanabe et al. (1995). The peroxide measuring procedure was performed immediately after the sampling collections. Although organic hydroperoxides (ROOH) comprised a significant proportion of the total peroxides (H_2O_2 + ROOH) in the gas-phase near the summit of Mt. Norikura, ROOH rarely exceeded 10 % of the total peroxides in fog water (Watanabe et al., 1995; Watanabe et al., 1999). The pH and dissolved ionic species were measured by a pH meter and by an ion chromatograph, respectively (Watanabe et al., 1999).

3. Results and Discussion

Table I shows the ranges of pH and peroxide concentrations in fog and rain water near the summit of Mt. Norikura on 12-14 and 16-17 September 1993. The concentrations of peroxide in fog water were 5-90 μM, whereas those in rain water were much lower (1-7 μM). Chemical constituents are usually condensed in fog water. Tsuruta et al. (1991) also observed that the peroxide concentrations at Mt. Norikura were much higher in fog water than in rain water. These results seem to reflect characteristics of peroxide in fog and rain water at a high elevation site. However, higher concentrations of peroxide in rain water were observed at Mt. Oyama near industrial areas (discussed below).

Table II shows a summary of the chemistry of the three-stage size-fractionated fog water samples collected at Mt. Norikura during July 1993. The diameters of fog droplets trapped efficiently at each stage are 20 μm, 5.2 μm and 3.6 μm, respectively (Minami and Ishizaka, in preparation). High concentrations (> 100 μM) of peroxide were observed. Watanabe et al. (1999) also reported that the concentrations of peroxide (mainly H_2O_2) during the summer from 1994 to 1996 varied over the range of 3-180 μM, with the average concentration being 60 μM. Therefore, the potential capacity for SO_2 oxidation appears to be very high near the summit of Mt. Norikura.

TABLE I
Ranges of pH and peroxide concentrations in fog and rain water near the summit of Mt. Norikura on 12-17 September 1993.

	Date	pH	Peroxide (μM)
Fog	12-14 Sep 1993	3.9-4.8	11-45
Fog	16-17 Sep 1993	3.7-4.2	5-90
Rain	12-14 Sep 1993	4.9-5.2	1-4
Rain	16-17 Sep 1993	4.5-4.8	5-7

TABLE II
Summary of the chemistry of the three-stage size-fractionated fog water samples collected near the summit of Mt. Norikura on 20-24 July 1993.

Date	Time	Size	pH	SO_4^{2-} (μ eq/l)	Peroxide (μ M)
20 Jul 1993	11:30-13:30	Large	4.2	46	70
		Medium	4.1	51	70
		Small	4.1	55	68
	13:00-14:05	Large	4.2	53	107
		Medium	4.1	60	105
		Small	4.1	60	110
	17:00-19:00	Large	3.7	254	116
		Medium	3.5	296	117
		Small	3.3	303	118
	19:00-20:00	Large	3.7	207	5
		Medium	3.4	236	6
		Small	3.5	231	6
22 Jul 1993	22:30-22:55	Large	3.9	78	3
		Medium	3.7	151	3
		Small	3.2	493	3
	22:55-23:55	Large	4.0	59	3
		Medium	3.8	94	3
		Small	3.7	117	3
23 Jul 1993	1:00-2:00	Large	3.4	303	23
		Medium	3.2	564	24
		Small	3.1	638	20
24 Jul 1993	10:45-12:55	Large	4.4	16	12
		Medium	4.0	38	12
		Small	3.7	89	11

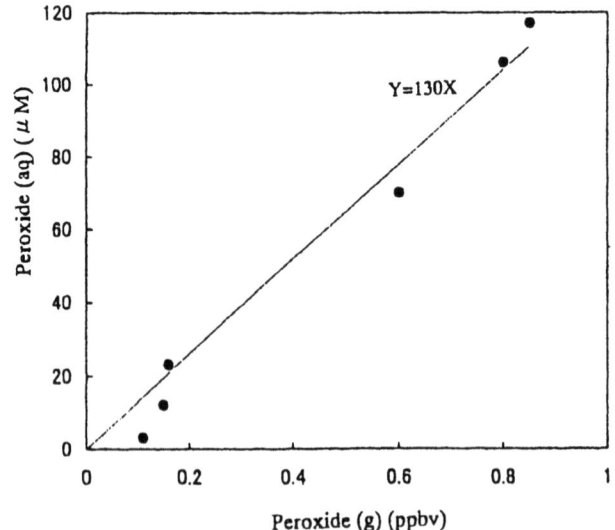

Figure 2. Relationship between the concentrations of peroxide in the gas-phase and those in the aqueous-phase near the summit of Mt. Norikura on 20-24 July 1993.

The concentrations of SO_4^{2-} and the acidity were found to be highly dependent on droplet size and were observed to decrease with droplet size (detailed results and discussion are now in preparation (Minami and Ishizaka, in preparation)). Similar results have been reported (Noone et al., 1992). According to Noone et al. (1992), solute concentrations were found to decrease with droplet size in a fog episode in the Po Valley. However, the concentrations of peroxide appeared to be independent of droplet size. Peroxide in fog water seems to be mainly derived from the gas-phase, and would be in equilibrium with the surrounding air.

Figure 2 shows the relationship between peroxide concentrations in the interstitial air (measured by the method of Watanabe et al., 1995) and those in the fog water. There is a linear relationship between the concentrations in the gas-phase and those in the aqueous-phase. The Henry's law constant for peroxide calculated from the slope of Figure 2 and the atmospheric pressure near the summit of Mt. Norikura (720hPa), was estimated to be $1.8 \times 10^5 Matm^{-1}$, which is comparable to the Henry's law constant for H_2O_2 (Yoshizumi et al., 1984; Lind and Kok, 1986), while the constant is dependent on temperature (Lind and Kok, 1986). Munger et al. (1989) reported that formic and acetic acids which exist in the gas-phase did not appear to be size-segregated. On the other hand, the chemical constituents mainly derived from secondary aerosols, such as SO_4^{2-} are diluted in the larger droplets.

A summary of the chemistry of the fog and rain water samples collected at the midslope of Mt. Oyama is shown in Table III. The fogs on 27-30 September were affected by a typhoon, therefore, sea-salt components were prevalent and the pH was relatively high. Peroxide concentrations in fog water were relatively low ($<5 \mu M$), and lower than those in rain water sampled simultaneously (0.2-33 μM). These results were quite different from those in the case of Mt. Norikura. Especially, peroxide was scarcely detected on 2 September 1998, when a strongly acidic fog was observed at Mt. Oyama (Table III). SO_2 concentrations at the midslope of Mt. Oyama were relatively high (< 5ppb), and SO_2 was detected even at the time of fog occurrence (about 0.5 ppb) on 2 September. The condition $[H_2O_2]<[SO_2]$ (*Oxidant Limitation*) may exist prior to the occurrence of fog. Peroxide might have been decomposed by SO_2 (S(IV)) oxidation in the fog droplets. On the other hand, the decomposition of peroxide is limited near the summit of Mt. Norikura because of the low SO_2 concentration (< 0.6 ppb) (Watanabe et al., 1995).

TABLE III
Summary of the chemistry of the fog and rain water sampled at the midslope of Mt. Oyama. Units of peroxide and ion concentrations are micromoles per liter and micro-equivalents per liter, respectively.

	Date	pH	Peroxide	Cl⁻	NO_3^-	SO_4^{2-}
Fog	27-30 Aug 1998	5.0-5.6	0.2-4	109-2290	6-466	28-690
Fog	2 Sep 1998	3.1-3.6	0-0.2	288-333	351-565	225-362
Rain	27-30 Aug 1998	4.5-5.3	0.2-33	9-180	2-24	3-46

Rain droplets are derived from upper altitudes where the concentrations of peroxide are high. According to Kleinman (1986), the peroxide concentration increases with altitude in the boundary layer, whereas SO_2 concentration decreases with altitude (there may be the condition $[H_2O_2]>[SO_2]$ in the higher atmosphere). Peroxide might have scarcely consumed in the precipitating cloud, and the rain droplets might have scarcely absorbed SO_2 during sampling the period. As a result, peroxide concentrations in the rain water were higher than those in the fog water (whereas major ions were diluted in the rain water) at the midslope of Mt. Oyama.

References

Chandler, A.S., Choularton, T.W., Dollard, G.J., Gay, M.J., Hill, T.A., Jones, A., Jones, B.M.R., Morse, A.P., Penkett, S.A. and Tyler, B.J.: 1988, *Atmospheric Environment* **22**, 683.
Ennis, C.A., Lazrus, A.L., Zimmerman, P.R. and Monson, R.K.: 1990, *Tellus* **42**, 183.
Gallagher, M.W., Choularton, T.W. Downer, R., Tyler, B.J., Stromberg, I.M., Mill, C.S., Penkett, S.A., Bandy, B., Dollard, G.J., Davies, T.J. and Jones, B.M.R.: 1991, *Atmospheric Environment* **25A**, 2029.
Gunz, D.W. and Hoffmann, M.R.: 1990, *Atmospheric Environment* **24A**, 1601.
Hewitt, C.N. and Terry, G.: 1992, *Environmental Science and Technology* **26**, 1891.
Kelly, T.J., Daum, P.H. and Schwartz, S.E.: 1985, *Journal of Geophysical Research* **90**, 7861.
Kleinman, L.I.: 1986, *Journal of Geophysical Research* **91**, 10889.
Lazrus, A.L., Kok, G.L., Gitlin, S.N., Lind, J.A. and Mclaren, S.: 1985, *Analytical Chemistry* **57**, 917.
Lind, J.A. and Kok, G.L.: 1986, *Journal of Geophysical Research* **91**, 7889.
Lind, J.A., Lazrus, A.L. and Kok, G.L.: 1987, *Journal of Geophysical Research* **92**, 4171.
Martin, L.R. and Damschen, D.E.: 1981, *Atmospheric Environment* **15**, 1615.
Munger, J.W., Collett, J.J., Daube, B. and Hoffmann, M.R.: 1989, *Atmospheric Environment* **23**, 2305.
Noone, K.J., Ogren J.A., Hallberg, A., Heintzenberg, J., Strom, J., Hansson, H.-C., Svenningsson I.B., Wiedensohler, A., Fuzzi, S., Facchini, M.C., Arends, B.G. and Berner, A.: 1992, *Tellus* **44B**, 489.
Olszyna, K.J., Meagher, J.F. and Bailey, E.M.: 1988, *Atmospheric Environment* **22**, 1699.
Romer, F.G., Viljeer, J.W., van den Beld, L., Slangewal, H.J., Veldkamp, A.A. and Reijnders, H.F.R.: 1985, *Atmospheric Environment* **19**, 1847.
Sakugawa, H. and Kaplan, I.R.: 1989, *Journal of Geophysical Research* **94**, 12957.
Sakugawa, H., Tsai, W., Kaplan, I.R. and Cohen, Y.: 1990, *Geophysical Research Letter* **17**, 93.
Tsuruta, H., Ohta, M. and Ishizaka, Y.: 1991, *Proceedings of the 32nd Annual Meeting of the Japan Society of Air Pollution* pp.168.
Watanabe, K. and Tanaka, H.: 1995, *Journal of the Meteorological Society of Japan* **73**, 839.
Watanabe, K., Ishizaka, Y. and Tanaka, H.: 1995, *Journal of the Meteorological Society of Japan* **73**, 1153.
Watanabe, K., Nagao, I. and Tanaka, H.: 1996, *Journal of the Meteorological Society of Japan* **74**, 393.
Watanabe, K., Ishizaka, Y. and Takenaka, C.: 1999, *Journal of the Meteorological Society of Japan* **77**, 997.
Watanabe, K., Ishizaka, Y. and Takenaka, C,: 2001, *Atmospheric Environment* **35**, 645.
Yoshizumi, K., Aoki, K., Nouchi, I, Kobayashi, T., Kamakura, S. and Tajima, M.: 1984, *Atmospheric Environment* **18**, 395.

BLACK ACIDIC RIME ICE IN THE REMOTE ISLAND OF YAKUSHIMA, A WORLD NATURAL HERITAGE AREA

OSAMU NAGAFUCHI[1*], HITOSHI MUKAI[2] and MINORU KOGA[3]

[1] *Fukuoka Institute of Health and Environmental Sciences, 39 Mukaizano, Dazaifu, 818-0135, Japan;*
[2] *National Institute for Environmental Studies, 16-2 Onogawa, Tsukuba, Ibaraki 305-0053, Japan;*
[3] *Prefectural University of Kumamoto, 3-1 Tsukide, Kumamoto, Kumamoto 862-0920, Japan*
(* *author for correspondence, e-mail: onaga@rb3.so-net.ne.jp*)

Abstract. In order to clarify the long-range transport of atmospheric pollutants in the East Asian regions, we have studied the components in rime ice on Yakushima Island, located in southwestern Japan near the China continent. The presence of a large number of particles has been found in the rime ice. Very small particles whose diameters are around 1 μm were identified as coal fly ash. The air mass at an altitude between 1500 m to 2000 m was probably long-range transported in association with a stable atmospheric layer in which the particles were efficiently scavenged by supercooled droplets. A back trajectory analysis also indicated the predominance of a north wind in the winter and in the other seasons as well. Such transport and deposition mechanisms may produce the greater pollutant deposition sometimes observed in mountain areas. Similar events may not be rare and could make an important contribution to the annual pollutants.

Keywords: rime ice, inorganic ash sphere, long-range transport, back trajectory, East Asia

1. Introduction

Rime ice deposits are common on high elevation trees and mountain top structures. Therefore, it is conceivable that the components of rime ice reflect the atmospheric environment at each elevation rather than those of snowfall at the same site, because rime ice occurs on structures which are exposed to a cloud of supercooled droplets. The droplets impact on structures and freeze causing the characteristic rime ice deposit. Rime ice occurs on Yakushima Island, located between the East China Sea and the Pacific Ocean, and is 800 km to the east of Shanghai and 1500km to the southeast of Beijing in China (Fig. 1). In addition, rime ice on the mountain sites of the Yakushima Island repeatedly forms and then falls off (Nagafuchi *et al.*, 1995). Namely, rime ice occurs due to the influence of cold air masses and it is separated from structures when the cold air masses leave. As mentioned above, Yakushima Island is located near China, which has the largest emission of acidic substances in East Asia that blows from the north or northwest during the winter season. Therefore, Yakushima Island will be directly exposed to the pollutants from the Asian Continents.

This study focused on the inorganic ash sphere (IAS) in the rime ice on mountain sites of Yakushima Island. We now report the results of the IAS in rime ice and discuss the long-range transport of the black substances.

2. Materials and Methods

Yakushima Island is not very big, but this island has seven mountains higher than 1800 m. Mt. Miyanoura is (1935 m) the highest mountain in the Kyushu regions (Fig. 1). This Island is known to have a vertical distribution of

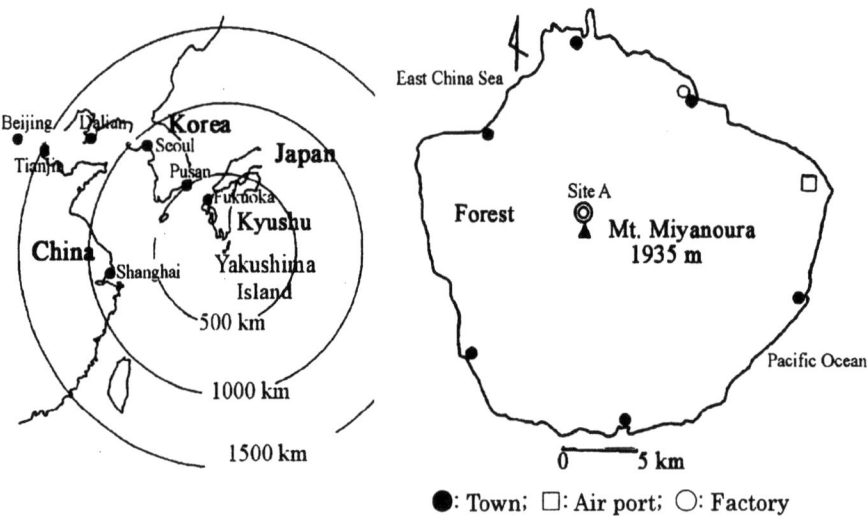

Figure 1. Location of sampling sites in East Asia and Yakushima Island

vegetation from sub-tropical forests to sub-frigid forests based on elevation. Because of the preservation of attractive primary forests, UNESCO assigned Yakushima Island as the world natural heritage area in Japan in 1993. We sampled snow and rime ice whose heights were typically higher than 1300 m, usually close to 2000 m above sea level on Yakushima Island. The collection site is isolated (Site A: 1900 m) and only accessible on foot. Samples were physically scraped from trees and then scooped into prerinsed polyethylene bottles and were frozen until analysis. This survey was carried out in Yakushima mountainous region on 10th –11th February 1998. Rime ice samples were left at room temperature and allowed to melt. They were then separated into soluble components and insoluble components using a membrane filter (Advantec, 47mmφ, pore size 0.45μm). For the insoluble components, the filtered matter was dried at room temperature and then carbon-gold evaporation was performed. It was analyzed using an Electron Microscope (JEOL, JEM-1200EX, equipped with ASID 10, KEVEX-7000J). Isentropic back-trajectories were calculated for the 700-, 800-, 850-, 900-, 950- and 1000- hP pressure levels, based on the program by Center for Global Environmental Research (CGER). The trajectories had a 3-day duration and ended at each elevation at 0000 and 1200 UT (0900 and 2100 LT) on the sampling day. Analyzed global wind data sets by the European Centre for Medium-Range Weather Forecasts (ECMWF) were used as input data.

3. Results and Discussion

Figure 2 shows a photograph of the suspended solid on filtered paper through which 200 ml of melted rime ice on February 11, 1998 was poured. These

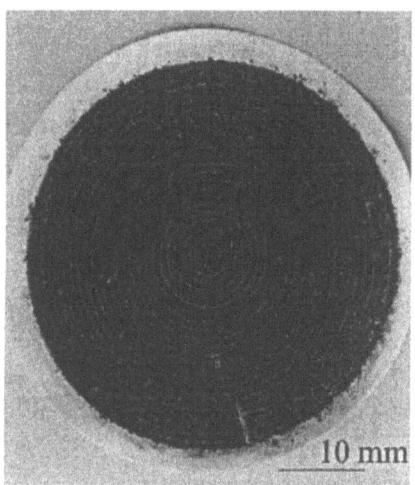

Figure 2. Photograph of a black substances on the filter

black substances were normally found on the filter paper. Figure 3 shows a scanning electron microscope photograph of the black filter paper. The IAS was generally 1-3 μm in diameter. Two types of spherical particles were identified using a scanning electron microscope (Fig. 4). The glassy spheres are called IAS and the porous particles are called spheroidal carboneous particles (SCP). The results obtained from the X-ray analysis of each particle using EDS are shown in Figure 5. The IAS is mainly composed of Si and Al, while SCP is of elemental carbon.

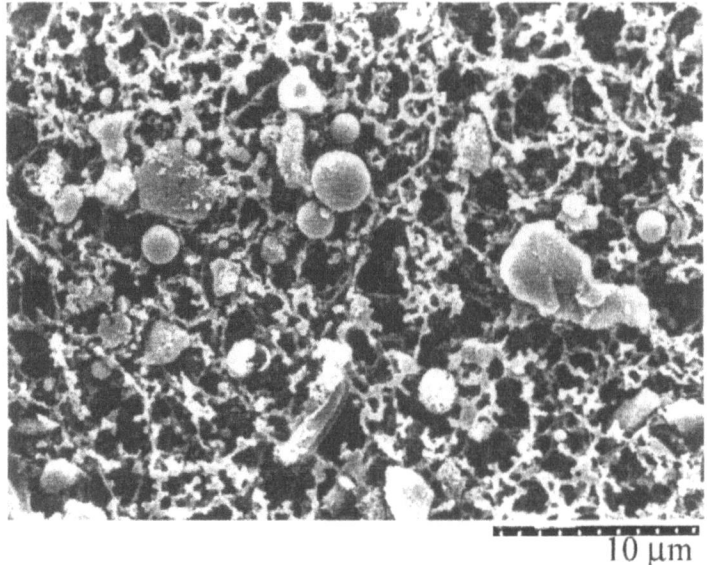

Figure 3. SEM photograph of IAS on the black filter.

However, we could not find a SCP in the rime ice. According to previous studies, over 80 % of the IAS is comprised of the particles formed from coal combustion (Watt and Thorne, 1965), but only 0.1 % from oil and much less from peat (Henry and Knapp, 1980). IAS particles generated in coal-fired

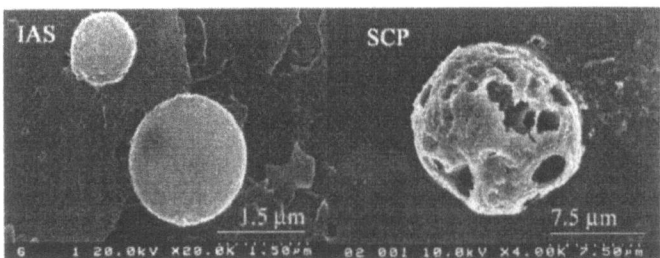

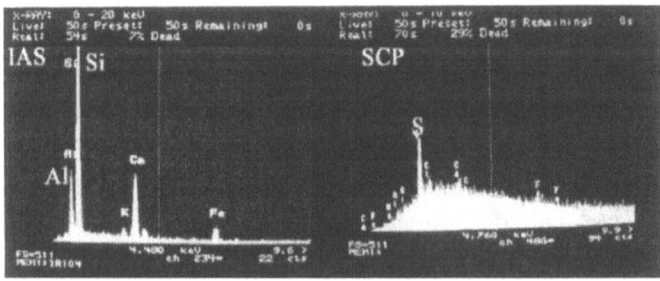

Figure 5. Result obtained from the X-ray analysis of each particle using EDS

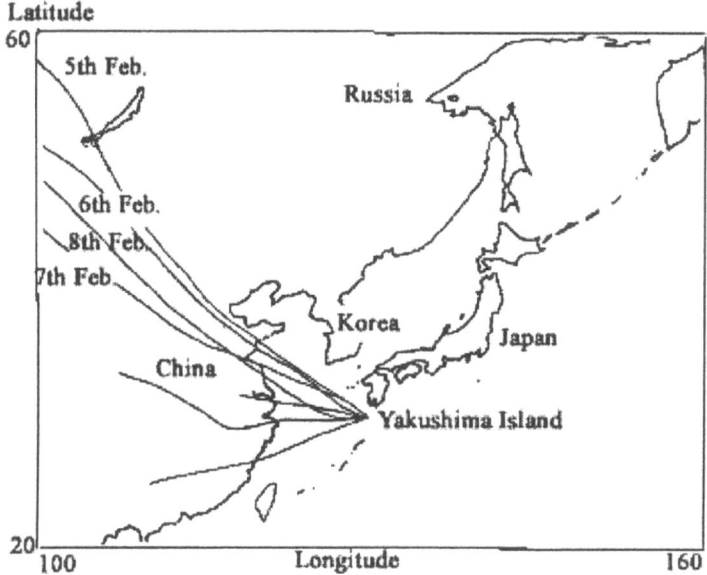

Figure 6. Air trajectory from on Yakushiama Island at an elevation of 1700 m (starting site corresponding Site A of Fig. 1). Each start at 0900LT (Japanese time) on 5th February 1998 to 11th February 1998.

power stations have diameters in the range 0.05 μm-10 μm (McElroy *et al.*, 1982), and most of the particles are deposited nearby. However, some particles travel for a long distance with air streams. The presence of these particles has been found in mid-ocean air samples (Folger, 1970; Parkin *et al.*, 1970), marine sediments in both coastal (Puffer, 1980) and deep-sea areas (Fredriksson and Martin, 1963), high latitude ice deposits, in both Greenland (Hodge *et al.*, 1963) and the Antarctic (Fredriksson and Martin, 1963; Hodge *et al.*, 1967), rime ice on a mountain top (Nagafuchi *et al.*, 1995), as well as estuarine (Allen, 1987) and lake sediments (Puffer, 1980; Nriagu and Bowser, 1969). Therefore, it has been determined that IAS can be used as a marker in the environment for long range transported atmospheric pollutants originating from fossil fuel combustion. Rime ice on mountain sites of Yakushima Island commonly adheres to trees and rocks on the northwestern slope of the ridge above 1300 m, especially above 1600 m. As shown in Figure 1, the northwestern area of the island has neither inhabitants nor any emissions of IAS. In addition, there is only the East China Sea on the northwest side, with the Asian Continent across the sea.

We used an air trajectory analysis to estimate the origin of the air mass when rime ice adheres. Figure 6 shows the back trajectory from Yakushima Island at 1800 m in February 1998. The air trajectory of Mt. Miyanoura in February 11th passed over the middle part of China. The location of the island

and this air trajectory clearly indicate that the origin of the IAS in the rime ice is not from our country.

4. Conclusion

Observations and analyses were made on rime ice and snow from the mountain site of Yakushima Island. A number of particles in the rime ice were found using electron microscopic observations. In addition, from the EDS analysis results, these particles were identified as IAS. The air trajectory from Mt. Miyanoura when rime ice adheres on February 11th passed over the middle part of China. These data clearly show that the IAS in the rime ice comes from the East Asian Continent.

Acknowledgments

The authors are very grateful to Mr. I. Ohta (The Japanese Alpine Club) and Mr. T. Ariga (Kyushu University) for assistance in collecting the samples. This work has been supported by the Toyota Foundation and the CREST (Core Research for Environmental Science and Technology, Japan).

References

Allen, J. R. L.: 1987, *Marine Pollution Bulletin* **18**, 13.

Folger, D. W.: 1970, *Deep-Sea Research* **17**, 337.

Fredriksson, K. and Martin, L. R.: 1963, *Geochimica et Cosmochimica Acta* **27**, 245.

Hodge, P. W., Wright, F. W. and Langway, C.C.: 1963, *J. Geophysical Research* **69**, 2919.

Hodge, P. W., Wright, F. W. and Langway, C.C.: 1967, *J. Geophysical Research* **72**, 1404.

Heenry, W. M. and Knapp, K. T.: 1980, *Environmental Science and Technology* **14**, 450.

MaElroy, M. W., Carr, R. C., Ensor, D. S and Markowski, G. R.: 1982, *Science* **215**, 13.

Nagafuchi, O., Suda, R., Mukai, H., Koga, M. and Kodama, Y.: 1995, *Water Air and Soil Pollution* **85**, 2351.

Nriagu, J. O. and Bowser, C. J.: 1969, *Water Research* **3**, 833.

Parkin, D. W., Phillips, D. R., Sullivan, R. A. and Johnson, L.: 1970, *J. Geophysical Research* **75**, 1782.

Puffer, J. H.: 1980, *J. Sedimentary Research* **50**, 247.

Watt, J. D. and Thorne, D. J.: 1965, *J. Applied Chemistry* **38**, 585.

ACID DEPOSITION IN BANDUNG, INDONESIA

SITI ASIATI*, TUTI BUDIWATI, LELY QODRITA AVIA

Atmospheric Research and Development Centre of National Institute Aeronautic and Space (LAPAN), Jl. Dr. Djundjunan 133, Bandung 40173, Indonesia
(* author for correspondence, e-mail: s_asiati@yahoo.com)

Abstract. LAPAN has measured rain acidity in Bandung, the location is Cipedes since 1985, with average pH in 1985 was 6.25. The pH condition 1985-1999 as follow: The monthly average of pH in period 1985-1992 was >5.6: in the middle of 1996-1997 it had big variation and than decrease until now. The monthly average of pH in 1997 until now was <5.6.The pH has decreasing trend, the reason was increasing fuel combustion for transportation and household because the area around the observation was change from rural to be transportation and settlement area.

The rain acidity comparison in Cipedes (rural site), Cicahuem (busy site), and Tanjungsari (remote site) hold in 1986-1987, the result was Tanjungsari the remote site had the lowest pH. It's suggested the reason was sulphur compound from Kamojang crater and air pollution from industrial area in south-east of Bandung were blown by the wind through this place.

The influence of air pollution to acid rain was studied by measurement NO_3^- and SO_4^{2-} in 5 places around Bandung, the results were: North of Bandung had the lowest NO_3^- concentration because the traffics were low: but had the highest SO_4^{2-} concentration; it's caused by emission of sulphur compounds from Tangkuban Perahu Montain. South of Bandung had the highest NO_3^- concentration because the traffics were crowded and a lot of industries around it. In general Bandung had SO_4^{2-} concentration higher than NO_3^- concentration, it's suggested due to the influence of sulphur compound from Tangkuban Perahu Montain.

The observation rain acidity in Ciater at Tangkuban Perahu Montain started in 1996, the result in period 1996-1998 as follow: The pH had decreasing trend, it's due to the traffic near this observation increase, so the air pollution around this area increase, it will influence the rain acidity. The maximum monthly average of ph was 6.78 and minimum was 4.63, the pH monthly average generally < 6. In El NINO year 1997, the monthly average of pH in April and December were > 6.5.

Keywords: nitric acid, pH, rain acidity, sulphuric acid

1. Introduction

Human activities can emit, SO_X, NO_X and it will be released to the atmosphere, primarily from combustion of fuel. SO_X naturally sources are from volcanoes, biological decay of organic matter and reduction of sulphate. NO_X naturally sources are from microbial activity in soil, lightning and oxidation of stratospheric N_2O (Manahan S.E., 1994). SO_X and NO_X are transformed into sulphuric and nitric acid by complex series of chemical reactions and removed from the atmosphere to the earth's surface by both wet and dry deposition processes. The acid deposition causes serious environmental damage to aquatic and terrestrial ecosystems, cultural properties and buildings.

In 1970's the economic growth, air pollution issue started in Indonesia. Due to urbanisation the population growth in big city like Bandung is increase. The result of economic growth and population growth are the increasing in the energy use, the impact is total amount of air pollutants which emitted into ambient air also increase, and serious air pollution problems also increase. To study the impact of air pollution in rainwater LAPAN in 1985 started monitoring the acid deposition. In this paper will discuss the result of monitoring acid deposition in Bandung and it's surrounding.

2. Materials and Methods

2.1. pH MEASUREMENT

The rainwater pH is measured by pH meter Orion model SA 720. This instrument has accuracy up to ± 0.002; its calibration is used the buffer solution with pH 4 and 7. The operation condition needs environmental temperature 25^0C, and sample preservation used refrigeration with temperature -4^0C.

2.2. NO_3^- AND SO_4^{2-} MEASUREMENT

The equipment for analysing nitrate and sulphate ion is Milton Roy UV/Visible spectrophotometer model spectronic 401; the operation temperature is 25^0C.

The nitrate ion is analysed by using reagents potassium sodium tartrate, sodium hydroxide and sodium salicylate. The NO_3 calibration curve is made from KNO_3 calibration solution, and it is measured by analysing equipment at 420 nm wavelengths.

The sulphate ion is analysed with turbidimeter method. The SO_4 calibration curve is used anhydrous sodium sulphate (Na_2SO_4) solution and it is measured by analysing equipment at 340 nm wavelengths.

3. Results and Discussion

3.1. MONITORING IN BANDUNG ($06^054'S$; $107^035'E$) and TANJUNGSARI ($6^0 54' 31''$ S; $107^0 47' 36''$ E)

The rain acidity (pH) has been measured at Cipedes (west of Bandung) since 1985 and the pH measurement still conducting until now. The result as follows: The average of pH measurement in 1985 was 6.25 and the monthly average of pH in 1985 – 1999 was in Figure 1 (Nurlaini et al., 1987; Budiwati et al. 1999).

Figure 1 showed that the rain acidity at Cipedes has increasing trend (pH decrease). In period 1985–1992 showed that the monthly average of pH > 5.6. In the middle 1996 until in the middle 1997 showed the monthly average of pH has big variation and than decrease < 5.6 until now. The changing of rain acidity is suggested as follow :

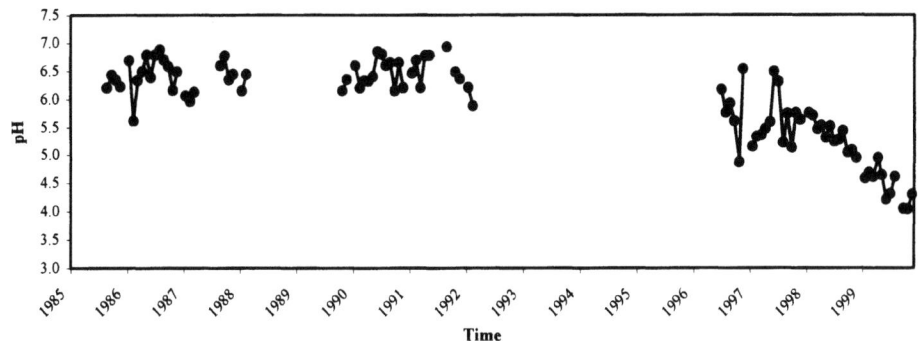

Figure 1. Monthly mean of pH in Bandung (Cipedes).

The fuel combustion for transportation and household produce SO_2 and NO_X; these gases enter to the atmosphere. SO_2 and NO_X in the atmosphere are converted to sulphuric and nitric acids respectively. These acids will removed by rain and may cause acidity on rain (Manahan S.E., 1994).

Cipedes was rural site until 1990, in the middle of 1990 the high way near the observation started to use and the settlement also improved in this area. After 1990 this place changed to be transportation and settlement area. This transportation and settlement area develop from time to time, as a result the fuel combustion in this area increase, and contribute to SO_2 and NO_X emission increase. The impact is sulphuric and nitric acids in rainwater will increase, its will influence the rain acidity, so that the pH decrease.

In 1986 -1987 the pH measurement also conducted at Cicaheum and Tanjungsari. It will be compared the pH in three place which have different conditions. Cicaheum is busy area, near bus station. Tanjungsari is remote site far from city and Cipedes is rural area. These three places can see in Figure 2. The monthly average of pH at the three places is in Figure 3.

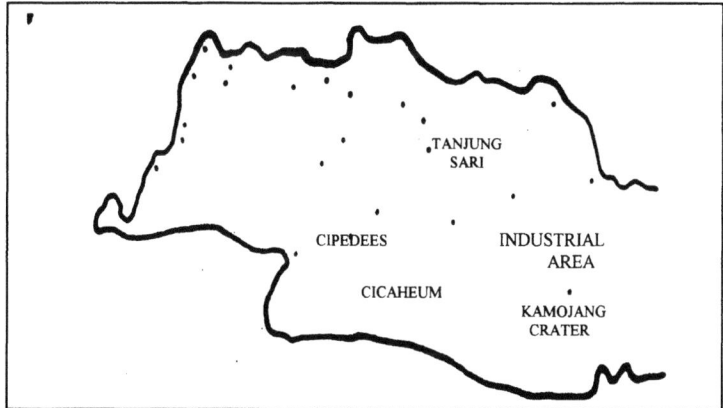

Figure 2. Area sorrounding Bandung

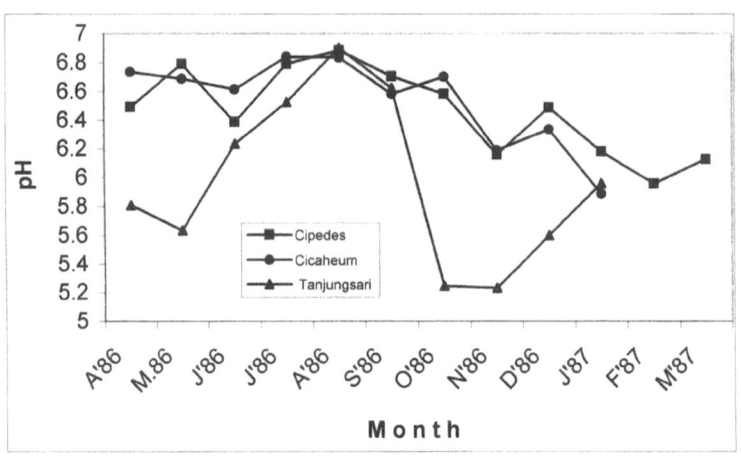

Figure 3. Monthly average of pH in Bandung Area.

The average pH since the measurement were: Cipedes 6.43, Cicaheum 6.55 and Tanjungsari 5.89. Figure 3 showed Tanjungsari had the lowest pH, it is remote site but had the lowest pH; the reasons were suggested as follow:

Figure 2 show at South of Tanjungsari there is Komojang crater. Some pollution consists of sulphur compound from Kamojang crater is blown by the wind through this place. The sulphur compound in the atmosphere is converted to be sulphuric acid, if there is rain the sulphuric acid will remove by rain. Figure 2, show at east of Bandung there is industrial area, the pollution from this industrial area and combustion of fuel for industrial energy are blown by wind through this place too. The sulphuric acid and another acid compounds from this industrial pollution will enhance the rain acidity in Tanjungsari.

3.2. NO_3^- AND SO_4^{2-} MEASUREMENT IN BANDUNG

To find out the influence of NO_3^- and SO_4^{2-} in acid rain Budiwati et al. (1990) were measured pH, NO_3^- and SO_4^{2-} at 5 places in Bandung with classification as follows: South as industrial area, North as settlement area and East, West and Centre of Bandung as urban area. The areas places are showed in Figure 4. The result is in Table I.

TABLE I
Rain acidity, sulphate, nitrate ion in Bandung measured in January–December 1990.

Location in Bandung	Average pH	Average SO_4^{2-} mg / l	Average NO_3^- mg / l
Centre (Suniaraja, KebonKelapa)	6.7	0.92	0.96
North (Dago)	6.48	1.08	0.49
South (Kopo, Teripang)	6.46	1.18	1.41
West (Cipedes)	6.48	1.31	0.77
East (Martadinata, Cicaheum)	6.55	1.03	0.77

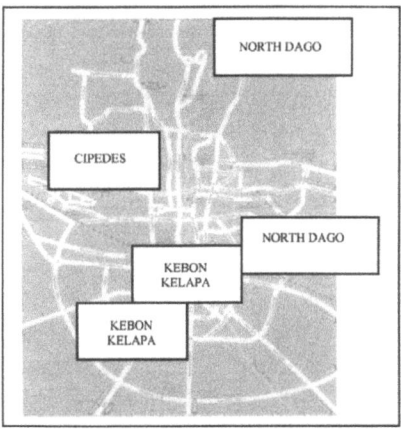

Figure 4. Location measurement in Bandung.

Centre and South of Bandung had high NO_3^- concentration because these places are urban areas with dense transportation so NO_X emission was high. In the atmosphere this gas converted to nitric acid, and will removed by rain, the result concentration NO_3^- in rainwater was high. West and East of Bandung had transportation lower than in Centre and South of Bandung so in the rainwater contains lower concentration NO_3^-. North of Bandung had NO_3^- concentration lower than the other because the traffic is lower. South of Bandung had the highest NO_3^- concentration because the traffic in this area very crowded and a lot industries in this place.

North of Bandung had lower traffic but the SO_4^{2-} concentration relative high, it's suggest due to the influence of sulphur compound from Tangkuban Perahu Mountain. In general Bandung had SO_4^{2-} concentration relative higher than NO_3^- concentration, it's suggested due to the influence of sulphur compound from Tangkuban Perahu Mountain.

3.3. MONITORING AT CIATER (6.43⁰ S; 107.41⁰E)

Ciater is in the north west of Bandung at Tangkuban Perahu Montain, the observation around 3 km from the road from Subang to Bandung. This road is the alternative road from Bandung to Jakarta, its can see in Figure 5. The observation was built in 1990; the air pollution was low that time. In 1996 - 1998 the rain acidity was

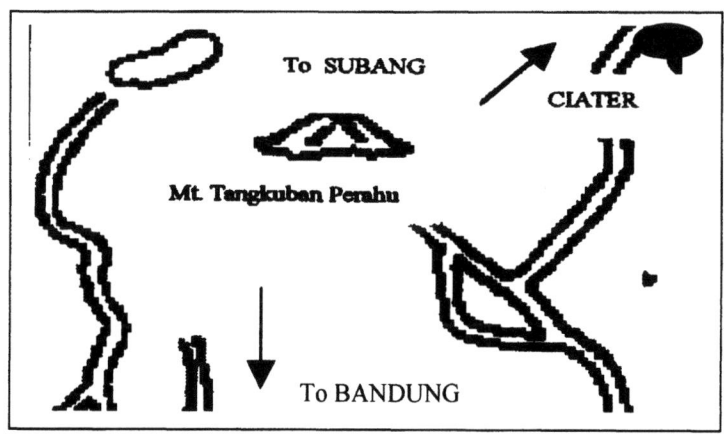

Figure 5. Area sorrounding Ciater

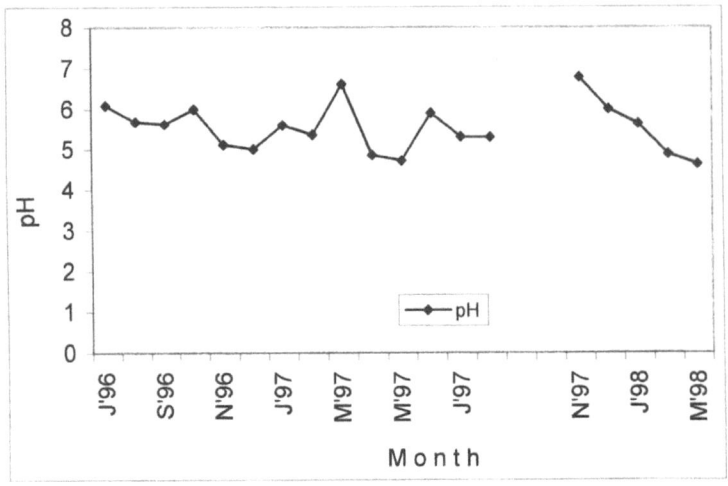

Figure 6. Monthly average of pH in Ciater.

measured in this place the result was in Figure 6.

Figure 6 showed the maximum monthly average of pH was 6.78 and the minimum monthly average of pH was 4.63 and the monthly average of pH generally < 6. In 1997 El Nino was happened and condition was dry. In April 1997 and December 1997, the pH > 6.5 it was suggested that in dry condition the rainfall was small so the acid compounds, which dissolved in rainwater was small, the result pH is high.

Figure 6 showed the pH had decreasing trend, it was suggested as follow:
National Bureau of Statistic data shows in the last five years automobile increase around 7% - 12% every year (Jaumil A. DS. and Firdausi Y., 1999). The traffics on the road from Subang to Bandung near this observation has increasing trend, especially in weekend or holiday because the road primarily from Bandung to Jakarta through Purwakarta often busy. The result is air pollution in this area increase; it will influence the rain acidity (decreasing pH trend) at Ciater observation.

Reference

Jaumil A. DS. And Firdaus Y., 1999, *Concentration Variation of SO2 Pollutant in Jakarta*, presented on National Workshop "Acid rain and Emissions Reduction in Asia Phase II" in Pusarpedal - Environmental Impact Management Agency, May 20 – 21, 1999.

Manahan S.E., 1994, *Environmental Chemistry*, Sixth Edition. Publish by Lewis Publishers, Boca Raton, Ann Arbor, London, and Tokyo. Page 330.

Nurlaini J., Yeti Priyati R., and Chunaeni L., 1987, *Rain Acidity Research 1986/1987*, Proceeding Program Research of Space Research and Development Centre of National Institute Aeronautic and Space (LAPAN) 1986/1987. Book I, ISSN 0216.4663, page 53-54.

Tuti Budiwati, Siti Asiati, and Nanang Efendi AR., 1991, *Chemical Composition of Rainwater in Bandung,* Proceeding Program Research of Space Research and Development Centre of National Institute Aeronautic and Space (LAPAN), Number: D-III/10-91, March, page 71-80.

TRACING POLLUTION AND ITS SOURCES WITH ISOTOPES

GÖRAN E. ÅBERG

Institute for Energy Technology, P.O.Box 40, NO-2027 Kjeller, Norway
(e-mail: gaa@ife.no)

Abstract. Naturally occurring isotope systems, such as strontium and lead, are very useful for characterizing sources of pollution and producing background information. The phase-out of lead additives to gasoline in Northern Europe has for example not phased out the lead contamination of the environment. The stable lead isotope method shows the contribution from small local sources to the lead contamination of the Oslo air during a period when the combustion of gasoline, the major source of lead contamination, has been decreasing. Wood combustion for domestic heating is one of these polluting sources in Norway. Strontium analyses show that roadsalt from de-icing of a nearby road is accumulated in the soil surrounding an about 4500 year old rock carving. At rainfall this salt is re-dissolved and drained over the rock carving. The impact from the road is confirmed by lead isotopes. Lead and strontium analysis of teeth show the contribution from industry and traffic on todays humans and the differences in nutrient intake during the Medieval era are compared with today. The isotope signatures of the Medieval teeth also show that a person living along the coast can be distinguished easily from a person living in the central parts of the country, and that the Medieval individual mainly lived on locally produce, while the contemporary person to a great degree lives on industrially manufactured food.

Keywords: pollution, sources, strontium isotopes, lead isotopes, Norway

1. Introduction

The Earth's biosphere has long been polluted, damaged and disturbed by human activities and one of the most important purposes of global environmental research today is to clarify the impact of human activities on the Earth in relation to our culture and future attitudes. A common characteristic of the above activities is the emission of lead (Pb) as part of the pollution. Pb mined and smelted for itself has contributed significantly to the total Pb production only after the Industrial Revolution when the Pb production rose from 100 000 t annually to 1 000 000 t 50 years ago, after the great expansion of motor vehicle traffic. Today about 4 000 000 t are produced annually worldwide (Nriagu, 1996).

Huge amounts of polluted waste material have been deposited through the years without sufficient security measures and control and also been added to the soil by leakage and outlet from industrial activity. These polluted areas are sources for a longtime diffusion of hazardous substances in soil and groundwater which may lead to a serious contamination of lakes and rivers being dangerous to human health.

A prerequisite, however, for taking decisions concerning future handling of waste is the availability of objective background material and an understanding of the relation between physical-chemical processes and waste cycle problems. To monitor the ongoing and ever increasing change of the environment there is a need for historical archives and appropriate methods enabling us to decipher them. Naturally occurring isotope systems have proven to be very useful in this sense. Of special interest is the interdisciplinary research where geoscientific and biological problems are integrated with the socio-economic aspects for society.

This presentation will show the application of the Pb isotope method to differentiate between the local contribution of Pb in Oslo and its portion from long-range transport of air pollution from sources a few thousand kilometers away; show how local pollution sources such as traffic may contribute to the chemical load on a 4500 year old Stone Age Rock Art Site at Ekeberg, in the town of Oslo; and show how stable Pb and Sr isotopes in teeth reveal the change of the natural and man-made sources of Pb and Sr intake during the course of history. An advantage of the above techniques is that the isotopic composition of an object will render a unique signature for each investigated object.

2. Lead emission and atmospheric transport

Analyses of the ^{206}Pb/^{207}Pb ratio on air filters from Oslo (Åberg et al.1999a), during the period from 1992 through 1996, show an interesting trend (Fig. 1). Although this ratio had been decreasing during 1992 and 1993, from 1.11 to 1.09, it started to increase continuously from 1.09 at the end of 1993 to over 1.15 at the beginning of 1998. The change of the ^{206}Pb/^{207}Pb ratio from the end of 1993 would indicate that although the impact of combustion of low-leaded gasoline on the air concentrations of this element after 1993 was still high, another source or sources have become more feasible.

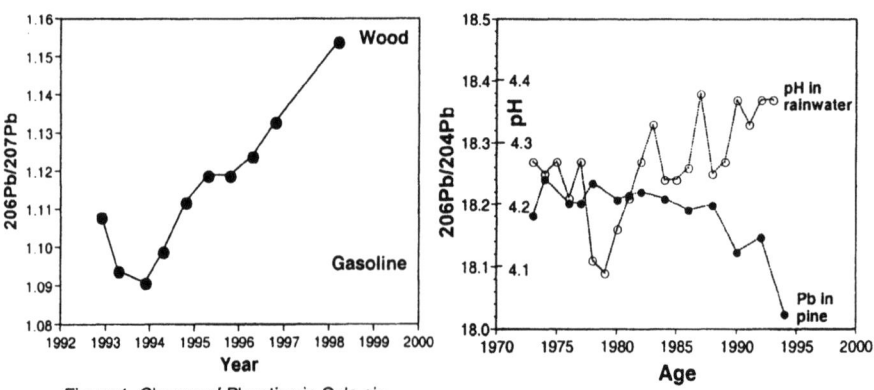

Figure 1. Change of Pb ratios in Oslo air from November 1992 to March 1998.

Figure 2. Change in Pb ratios in pine tree and pH in rainwater in southern Norway from 1973 to 1994.

Burning of wood can be an important source of Pb in certain regions in northern Europe and studies concluded that this factor may range from 1.0 to 5.0 g Pb/ tonne of wood burnt (Nriagu and Pacyna, 1988) depending on parameters related to the content of Pb in wood and combustion conditions. The Pb emission factors for wood combustion in fireplaces are comparable with the Pb emission factors for combustion of coal, ranging at least between 1 and 10 g Pb/ tonne coal burnt and combustion of residual oil, ranging from 2.0 and 6.0 g Pb/ tonne (Nriagu and Pacyna, 1988). Taking into account combustion of fuels in Oslo, both wood and residual oil are burnt in large amounts. An indication can be seen from the increasing importance of combustion of wood as a source of lead contamination in the Oslo area. Analyses of soot from domestic wood firing yielded $^{206}Pb/^{207}Pb$ ratios of about 1.16 and thus sustained this hypothesis.

Analyses of Pb concentration and $^{206}Pb/^{204}Pb$ ratio in pine tree-rings from the coast of southern Norway (1973 to 1994) show that the Pb concentration decreases from about 200 ppb to about 25 ppb. During the same time interval the $^{206}Pb/^{204}Pb$ ratio decreases from over 18.2 to 18.0 (Fig. 2). The decrease in Pb concentration in the environment during this time interval has been attributed to the decrease in use of Pb additives to gasoline. However, the $^{206}Pb/^{204}Pb$ ratio in Norwegian gasoline is quite low with a value of about 17 like on the continent. A decrease in the use of leaded gasoline would then hardly decrease the $^{206}Pb/^{204}Pb$ ratio but rather have the opposite effect of increasing the $^{206}Pb/^{204}Pb$.

Analyses of pH in rainwater from the same locality plotted together with $^{206}Pb/^{204}Pb$ ratios for tree-rings in the pine show that when pH decreases, the Pb ratio increases and vice versa (Fig. 2). This indicates that the $^{206}Pb/^{204}Pb$ ratios in the tree are strongly connected to the pH values in rainwater. Pollution from burning coal in England and Central Europe, with $^{206}Pb/^{204}Pb$ ratios of about 18.5, may after atmospheric transport to Norway be taken up by the trees. This lead with a higher isotopic ratio is transported together with the acid rain components. Lower pH caused by acid rain is thus reflected by the higher $^{206}Pb/^{204}Pb$ ratio of coal.

3. Pollution impact on a Stone Age Rock Art site in Oslo

The Ekeberg Rock Art Site is located in a rather steep slope close to and below a road with heavy traffic. It is covered by a shallow soil layer on the non-exposed part of the bedrock (Åberg et al. 1999 b). Analyses of precipitation and throughfall yielded $^{87}Sr/^{86}Sr$ ratios indicating rainwater as a major source but with a slight contribution of dust/salts/particles, seen from the Sr concentration, washed down from the canopy into the throughfall sample (Fig. 3).

The concentration of water leachable Sr is up to four times less for the soil sample uphill the road (30 ppm Sr) in comparison with the soil samples below the road (126 ppm). This increase in Sr concentration in the soil for the two samples at the carving is due to the use of roadsalt, of marine origin, for de-icing during winter time. The small increase in Sr ratio (0.710) above the marine Sr signature (0.709) is probably attributed to impact from the soil.

The $^{206}Pb/^{204}Pb$ ratios from acid leaching of the soil samples show the same trend as for the Sr ratios, with the highest ratio above the road, while analysis of the Pb concentration yielded values opposite to the Sr concentration. The uphill sample has a value of about 245 ppm Pb compared to the samples below the road with values about 95 ppm Pb (Fig. 4). The much lower concentration of Pb at the rock carving probably depends on a shallow soil layer (less than 15 cm), input of de-icing salt, and thus a stronger washing out of the elements while conditions uphill the road are characterized by deeper soil (up to 50 cm) and no addition of salt. The background concentration of lead in south-western Scandinavia varies about 20 ppm in comparison to the 245 ppm at Ekeberg.

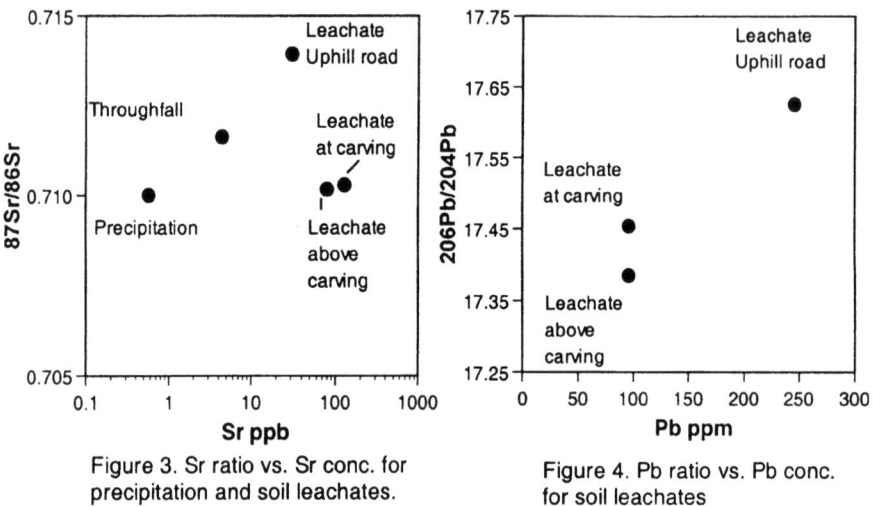

Figure 3. Sr ratio vs. Sr conc. for precipitation and soil leachates.

Figure 4. Pb ratio vs. Pb conc. for soil leachates

4. Pollution in Medieval and Contemporary teeth

Analyses of the $^{206}Pb/^{204}Pb$ ratio in the investigated teeth clearly define two groups, samples of Medieval teeth having ratios between 18.8 and 18.2, and contemporary teeth with ratios between 18.0 and 17.6 (Fig. 5). The only deviation from this trend is the contemporary Uvdal tooth which has similar $^{206}Pb/^{204}Pb$ ratio and Pb concentration as the Medieval one. Today's lower Pb ratios are more an artifact of Pb pollution from traffic and industrial emissions while the historic values are due to the ingestion of lead.

In Medieval time the Pb concentration could reach extremely high values as seen in the busy merchant port of Bergen where the living habits were different from those in the countryside (Åberg et al. 1998). Whilst people in Bergen had the possibility to consume imported food and beverages, the inland inhabitants lived on locally produce. Different kitchenware was also used, e.g. Pb glazed pottery and pewters in towns, and wooden cups for the inland farmers. Glazed pottery was probably one of the main sources for Pb poisoning of man during Late Medieval times, and especially if the glazed artifacts were used to contain acidic food or drink as acids may leach out Pb from the glazing.

The contemporary teeth have similar $^{87}Sr/^{86}Sr$ ratios which, however, may be an artifact of today's food industry, like dairy products, with the same origin for most products, and where the intake of locally produced food is of minor importance. Analyses of milk from the largest milk producer in Norway yielded a Sr value of 0.7086. This value of the milk most probably has its origin in the standard industrially produced livestock fodder distributed to the farmers. All dairy products such as milk, cream and cheese will thus have a similar Sr ratio. These products together with chicken, pork and meat make up a major proportion of the food consumed by the Norwegian people and will thus have a strong impact on the Sr signature in their teeth and bones which also can be seen from Figure 6.

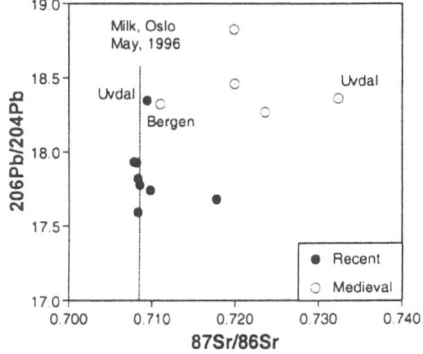

Figure 5. Pb ratio vs. Sr ratio for Norwegian teeth

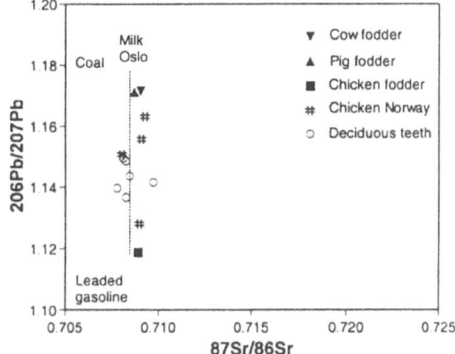

Figure 6. Pb and Sr data showing the relation between industrial livestock fodder, food products, and uptake in the human body.

When plotting the ^{206}Pb/^{204}Pb ratios for the teeth samples versus their ^{87}Sr/^{86}Sr ratios a complementary picture appears. The ^{206}Pb/^{204}Pb ratios for the teeth from Medieval Uvdal and Bergen are approximately the same (18.4), but when the ^{87}Sr/^{86}Sr ratios are compared there is a big difference between them (0.711 and 0.732, respectively). The Medieval Bergen tooth has a Sr ratio close to that of seawater (0.709), which is natural, considering the availability of seafood, while the Medieval Uvdal tooth has a Sr value strongly deviating from that of seawater. This latter value mirrors the Sr isotopic value of a geologically old area where the person had been eating the crop grown on the local substrate together with the dairy products from the grazing livestock.

5. Conclusions

The results of measurements presented in this paper show that the phase-out of Pb additives to gasoline in Northern Europe has not phased out the Pb contamination of the environment. Introduction of unleaded gasoline significantly reduced the Pb concentrations in the air over Scandinavia, but Pb is still measured due to certain amounts of Pb in unleaded gasoline (with no lead additives but with lead originating from crude oil), as well as contributions from other diffuse sources like industrial and domestic firing of oil and wood.

The use of a combination of Sr and Pb analyses at the Ekeberg Rock Art site in Oslo shows that it is possible to discriminate between different sources of pollution affecting the about 4500 year old rock carving which has a very sensitive setting being exposed to quite heavy traffic. During winter time de-icing salt from the road is continuously washed down and accumulated in the soil between the road and the carving. The soil then acts as a source for the salt which during influence of later precipitation is redissolved and drained over the carving. Strontium isotopic analyses of soil leachate above the rock carving link the high Sr concentration to the use of de-icing roadsalt.

The $^{206}Pb/^{204}Pb$ ratios of soil leachates show the earlier deposition and accumulation of coal lead, of local origin and atmospherically long-range transported, from Central Europe and England, and also from local firing of oil and wood for domestic heating and traffic lead. The high concentration of Pb in the soil implies that the overall deposition of lead is, or has been considerable in Oslo. The effects of the pollution in the Oslo area is also enhanced by the special atmospheric conditions prevailing due to the town´s topographic location.

The impact of Pb on man from modern industry and traffic is obvious compared with the intakes during the Medieval era. Moreover, Sr analyses of Medieval teeth show that a person living along the coast can easily be distinguished from a rural man. There is also the possiblity to present information on mobility between urban and rural people and to reveal historical changes of the magnitude of lead contamination of man due to changes and evolution of the local and international industry. The Sr signature shows that Medieval man lived on what was locally produced while contemporary man obviously lives to a great degree on industrially manufactured food.

Naturally occurring isotopic systems, like those for strontium and lead, are thus very useful for characterizing different sources, to define/verify the pathways of pollution transport from source to a receptor, and to produce a background information on the present and past levels of Pb in man in the environment and in various geographical regions.

The stable isotopic systems of Pb and Sr each give much valuable information, but the combination of the two techniques is a very powerful tool in environmental and archeological research. Still more information may be obatined by the combination with the light stable isotope systems of carbon, nitrogen and sulphur in problems concerning eutrophication, fertilization, mine drainage, weathering of stone in the cultural heritage etc.

References

Åberg, G., Fosse, G. and Stray, H.: 1998, *The Science of the Total Environment* **224**, 109.
Åberg, G., Pacyna, J.M., Stray, H. and Skjelkvåle, B-L.: 1999a, *Atmospheric Environment* **33**, 3335.
Åberg, G., Dahlin, E. and Stray, H.: 1999b, *Journal of Archaeological Science* **26**, 1483.
Nriagu J.O.: 1996, *Science* **272**, 223.
Nriagu J.O. and Pacyna J.M.: 1988, *Nature* **333**, 134.

SEQUENTIAL LEACHING OF VOLCANIC SOIL TO DETERMINE PLANT-AVAILABLE CATIONS AND THE PROVENANCE OF SOIL MINERALS USING Sr ISOTOPES

YORIKO YOKOO* and TAKANORI NAKANO

*Institute of Geoscience, University of Tsukuba, 1-1-1 Tennoudai, Ibaraki 305-8571, Japan
(* author for correspondence, e-mail: yokoo@ore.geo.tsukuba.ac.jp)*

Abstract. Sequential leaching experiments using H_2O, H_2O_2, NH_4Cl, and HCl were performed on surface soils on an andesite substrate at four sites in the Kawakami mountainous area, central Japan. The solutions extracted from the dehydrated soil by H_2O, H_2O_2, and NH_4Cl have relatively constant ratios with respect to Ca, Mg, and Sr, while they have variable $^{87}Sr/^{86}Sr$ ratios depending on the site. The elemental ratios and Sr isotopes in the extracted solution are different from those of the soil minerals but identical to those of the soil solution and the plants. Sr isotopic data indicate that the residues after extraction from fine-grained minerals by NH_4Cl and HCl are a mixture of acid-resistant minerals derived from bedrock and from arid areas in China. We suggest that there is a vital exchange of Sr and other cations between plants and the soil pool of exchangeable cations through the soil solution, while soil minerals, except chlorite, do not participate in the exchange reaction.

Keywords: Sr isotope, leaching, plant-available element, volcanic soil, Kosa

1. Introduction

Strontium isotopes have been used in recent years as tracers to improve our understanding of the circulation of elements in the soil-vegetation system and to assess acid rain impact on a forested ecosystem. This method is primarily based on the difference in the $^{87}Sr/^{86}Sr$ ratio between rainwater and bedrock, both of which are the ultimate sources of strontium in the vegetation and soil (*e.g.*, Åberg *et al.*, 1990; Miller *et al.*, 1993). However, previous studies have been conducted mainly on areas where the soil is underlain by granitic rocks or till deposits with wide $^{87}Sr/^{86}Sr$ variation. As a result, the contribution of the bedrock Sr to the soil (Graustein and Armstrong, 1983) and the exchange of cations between the plants and the associated soil fractions remain unclear.

In order to identify plant-available Sr and other elements in the soil, further extraction experiments are indispensable, particularly for soils developed on bedrock having a uniform $^{87}Sr/^{86}Sr$ ratio. Young volcanic rocks, generally having $^{87}Sr/^{86}Sr$ ratios within a restricted range irrespective of the constituent minerals, are widely distributed in Japan. We report here the results of an extraction experiment on volcanic soils in Japan that was performed using Sr isotopes in order to identify plant-available cations in the soil.

2. Samples and Experiment

Soil samples were collected at four sites along the slope of the Kawakami Experimental Basin of the University of Tsukuba, central Japan, on May 27, 1993. This mountainous basin of about 1500 to 1600 m in elevation is underlain by Meshimori

Water, Air, and Soil Pollution **130**: 1583–1588, 2001.
© 2001 *Kluwer Academic Publishers.*

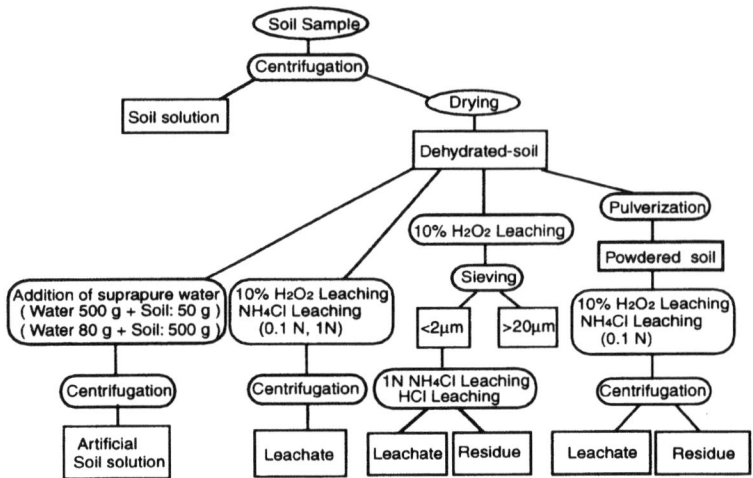

Figure 1. Flow chart of the soil leaching procedure.

-yama volcanic rocks of Late Pliocene age composed of andesite (Kawachi, 1977). The sampling area is covered with bamboo grass (*Sasa nipponica*) and Japanese oak (*Quercus mongolica Fisch*) at about 5-m intervals. The soil is classified as Kuroboku soil, which is characterized by a black A soil horizon. The four sites are located from the bottom (KC-1) to the top (KC-17) in the slope at about 15-m intervals. We collected about 3 kg of soil from the A horizon at a depth of about 10–20 cm at each site.

Analytical procedures used are shown in Figure 1. The soil solution corresponding to a pF value of 2.5–4.0 was collected by centrifugation. Then the dehydrated soil was dried in an oven at 100 °C for one night. Two hundred milligrams of the dried soil was reacted with 50 ml of 10% hydrogen peroxide solution (10% H_2O_2) at 70 °C to collect cations in organic matter or with 50 ml of ammonium chloride solution (0.1 N and 1 N NH_4Cl) at 100°C to collect exchangeable cations. After one day, the soil was centrifuged, and the extracted solution was used for the analysis. The same leaching experiment was carried out on a powdered sample, which was obtained by pulverizing 50 g dehydrated soil using a tungsten-carbide vessel. In order to reproduce the soil solution artificially, 500 g ultrapure water was also added to 50 g of the dried soil at room temperature, taking into account that this soil originally contained 5% to 9% H_2O; for reference, 80 g ultrapure water was also added to 500 g of the dried soil. The samples were preserved at room temperature from one day to two weeks. Then the sample was centrifuged to extract the water. About 100 mg of the residual soil was dissolved with 0.5 ml HNO_3, 0.5 ml $HClO_4$, and 1.0 ml HF in a Teflon vessel at 100°C for 12 hours and then dried.

Previous studies have demonstrated that aeolian particles from deserts in China, termed Kosa, are widely distributed in soils in Japan (Inoue and Naruse, 1987; Mizota *et al.*, 1992). Considering that the average grain size of the Kosa in Japan is 2 µm (Ishizaka, 1982), fine-grained minerals (FGM) below 2 µm in size and coarse-grained ones (CGM) above 20 mm in size were collected by sieving (Puri, 1949; Shimoda, 1985). The same leaching procedure mentioned above was carried out on the FGM fractions. Moreover, the FGM fraction was reacted with 15 ml of 20% hydrochloric acid (20% HCl) at 100°C for 45 minutes.

The Sr isotopic composition was determined using a Finnigan MAT 262RPQ multicollector mass spectrometer at the Institute of Geoscience in the University of Tsukuba. The $^{87}Sr/^{86}Sr$ ratio of the standard NIST-SRM987 throughout the analysis was 0.71024(±0.00001). The cation concentrations were determined using an inductively coupled plasma spectrometer (ICP-757V, Nippon Jarrell-Ash) at Chemical Analysis Center, University of Tsukuba.

3. Results and Discussion

3.1. ELEMENTAL COMPOSITION

The solutions extracted by H_2O, H_2O_2, and NH_4Cl from dehydrated raw soil had different concentrations of cations depending on the site, but the extracted compositions were in the order Ca>Mg, K>Na for all sites. The concentration of Ca in the extracted solution was highly variable depending on the site, but the variation in Na content was small. Sr in these extracted solutions (or leachates) and in the artificial soil solution positively correlated with Ca and Mg. As a result, these solutions all plotted in a distinct region on the ternary Ca-Mg-Sr diagram (Fig. 2). The solution extracted from FGM by NH_4Cl also plotted in the same region. Note that the concentration ratios of Ca, Mg, and Sr in the extracted solution were almost identical to those of the associated soil solution and they were close to those of the plants in the area. This compositional similarity indicates that the extracted solutions were representative of the exchangeable cations in the soil.

In contrast, the cation compositions of bulk soil were in the order K>Ca, Mg>Na, and they differed from those of the exchangeable soil fractions (Fig. 2). The high K content in the bulk soil indicates a dominant presence of illite. The chemical compositions of CGM were close to those of the bulk soil, while the FGM were enriched in Mg. Since chlorite, illite, vermiculite, and kaolinite were identified by X-ray diffraction analysis, the high Mg in FGM was attributable to the

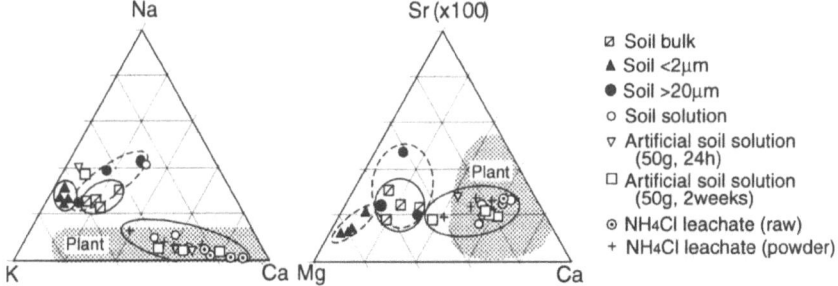

Figure 2. Ternary (A) Ca-K-Na and (B) Ca-Mg-Sr plots of soil and its extracted and residual fractions. Grey represents an area of plant.

TABLE I
$^{87}Sr/^{86}Sr$ of soilwaters, leachates and residues of A-horizon at for sites (KC-1, -7, -10, -17).

	KC-1	KC-7	KC-10	KC-17
Soilwater & Leachate				
Reaction with raw soil				
Soil solution (pF: 2.5–4.0)	0.70484	0.70578	0.70604	0.70680
Artificial soil solution (50 g, 1 day)	0.70496	0.70579	0.70616	0.70660
Artificial soil solution (50 g, 2 weeks)	0.70484	0.70579	0.70603	0.70677
Artificial soil solution (500g, 1 day)	0.70485	-	0.70605	-
Artificial soil solution (500g, 2 weeks)	0.70485	-	0.70603	0.70647
H_2O_2 leachate	0.70490	0.70580	0.70604	0.70658
NH_4Cl (0.1 N) leachate	0.70491	0.70578	0.70609	0.70654
NH_4Cl (1 N) leachate	0.70489	0.70577	0.70610	0.70658
NH_4Cl (1 N) leachate of FGM after H_2O_2 leaching	0.70495	0.70588	0.70609	0.70657
HCl leachate of FGM after H_2O_2 leaching	0.70558	-	0.70667	0.70702
Reaction with powdered soil				
H_2O_2 leachate	0.70524	0.70669	-	0.70718
NH_4Cl (0.1 N) leachate	0.70519	0.70658	0.70652	0.70717
Bulk & Residue				
Soil bulk	0.70641	0.70836	0.70778	0.70627
FGM after H_2O_2 leaching	0.71242	0.71315	0.71274	0.71000
CGM after H_2O_2 leaching	0.70462	0.70435	0.70434	0.70422
NH_4Cl (1 N)-residue of FGM after H_2O_2 leaching	0.71408	0.71416	0.71620	0.71258
HCl-residue of FGM after H_2O_2 leaching	0.71442	-	0.71466	0.71091
H_2O_2-residue of powdered soil	0.70649	0.70840	0.70798	0.70620
NH_4Cl (0.1 N)-residue of powdered soil	0.70674	-	0.70820	0.70638

presence of chlorite. The FGM and their exchangeable fractions are plotted in a region distinct from that of the bulk soil and CGM (Fig. 2), indicating a small amount of FGM in the soil.

3.2. Sr ISOTOPIC COMPOSITION

Nakano et al. (2001) reported that the $^{87}Sr/^{86}Sr$ ratios of plants change locally in the studied sites but are identical to those of the associated soil solution. The solutions extracted from the dehydrated raw soil by H_2O, H_2O_2, and NH_4Cl had $^{87}Sr/^{86}Sr$ ratios indistinguishable from the artificial soil solution (Table I, Fig. 3). The difference in the $^{87}Sr/^{86}Sr$ ratios between the 0.1 N and 1 N NH_4Cl leachates of the soil was negligible within an

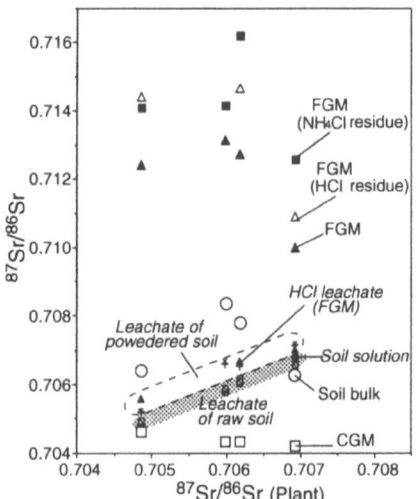

Figure 3. Plot of the $^{87}Sr/^{86}Sr$ ratio of bulk soil, its extracted and residual fractions, and plants.

analytical error of 0.00001. The NH_4Cl leachate, which extracted the exchangeable components in the soil, also had almost the same $^{87}Sr/^{86}Sr$ ratio as the H_2O_2 leachate, which dissolved organic materials (Shimoda, 1985). These data demonstrate the attainment of Sr isotopic equilibrium among plants, soil solution, organic material, and exchangeable sites in the soil, and a vital exchange of cations among them. Accordingly, it is likely that the relative uniformity of the cation compositions in the soil (Fig. 2) resulted from the interaction with plants rather than with the soil minerals.

In contrast, the dehydrated soil residues had different $^{87}Sr/^{86}Sr$ ratios from the associated soil-exchangeable components, indicating that Sr in the exchangeable pool did not interact with the soil minerals. CGM in the studied area had a relatively uniform $^{87}Sr/^{86}Sr$ ratio (0.7044±0.0002) close to that of the unaltered andesite bedrock (0.704), and they are considered to be largely composed of secondary minerals derived from bedrock (Nakano et al., 2001). The residues extracted from FGM by NH_4Cl and H_2O_2 had high $^{87}Sr/^{86}Sr$ ratios (0.7126-0.7162 and 0.7100-0.7132, respectively). The leachates and residues extracted from FGM by HCl after H_2O_2 showed variable $^{87}Sr/^{86}Sr$ ratios; the $^{87}Sr/^{86}Sr$ ratios of the HCl-extracted solution were low (0.7056-0.7070), while those of the HCl residues were high (0.7109-0.7147). This result shows that the soil minerals have different $^{87}Sr/^{86}Sr$ ratios. As the bedrock andesite is Late Pliocene in age, the variation of the $^{87}Sr/^{86}Sr$ ratios in the primary minerals due to the growth of ^{87}Sr is small (< 0.0001). Further, because the fresh andesite has uniformly low $^{87}Sr/^{86}Sr$ ratio (0.704), the Sr isotopic difference in the soil cannot be ascribed to heterogeneity of the bedrock-derived minerals.

Previous studies have shown that Kosa-derived minerals in Japan have high $^{87}Sr/^{86}Sr$ ratios, 0.716 in the snow of northern Japan and 0.724-0.726 in the soil of southern Japan (Mizota et al., 1992). The soil in the Central Loess Plateau (CLP) in China, considered to be the major provenance of Kosa, is mainly composed of quartz, feldspar, illite, kaolinite, and vermiculite. This mineral assemblage is similar to that of the NH_4Cl-extracted residue of the FGM. The extraction experiment shows that the acid-insoluble minerals in the CLP loess have high $^{87}Sr/^{86}Sr$ ratios (0.719±0.002), while weak-acid soluble minerals composed mainly of calcite have low $^{87}Sr/^{86}Sr$ ratios (0.7115±0.001) (Yokoo et al., 2000). Nakano and Tanaka (1997) reported that wet precipitation in the studied area contains Sr derived from water-soluble Kosa. It is therefore likely that the high $^{87}Sr/^{86}Sr$ ratios of the NH_4Cl- and HCl-extracted residues of FGM can be attributed to the presence of acid-resistant minerals in Kosa.

Chlorite is known to dissolve in HCl (Oinuma and Kobayashi, 1965), which was verified by X-ray diffraction of the H_2O_2 and HCl residues. The FGM extracted by HCl had slightly high $^{87}Sr/^{86}Sr$ ratios, but the ratios were close to the $^{87}Sr/^{86}Sr$ ratios of the exchangeable fractions. This result indicates that chlorite exchanged Sr and other cations with the soil solution and the pool of exchangeable cations in the soil. It is likely that the HCl residue of FGM was composed of a mixture of bedrock-derived secondary minerals (0.7044±0.0002) and acid-resistant Kosa minerals (0.719 ± 0.002).

3.3. POWDERED SOIL

The NH_4Cl- and H_2O_2-extracted residues of the powdered soil have $^{87}Sr/^{86}Sr$

ratios almost identical to those of the bulk soil. This finding is consistent with most Sr in the soil being contained in acid-resistant minerals. Although the cation compositions of the NH_4Cl leachate of powdered soil were almost the same as those of raw soil, there was a systematic difference in the $^{87}Sr/^{86}Sr$ ratio between the two; the leachate of powdered soil was enriched in radiogenic ^{87}Sr. Note that the $^{87}Sr/^{86}Sr$ ratios of the NH_4Cl- and H_2O_2-extracted solutions from the powdered soil were indistinguishable from those from the HCl extraction of FGM. The Sr isotopic difference in the NH_4Cl- and H_2O_2-extracted solutions between the raw soil and the pulverized soil indicates that the cations in most soil minerals were not exchanged with those in plants.

4. Conclusions

Leaching experiments using H_2O, H_2O_2, and NH_4Cl on an andesite soil with a uniform $^{87}Sr/^{86}Sr$ ratio in the Kawakami forested basin, central Japan, show that the extracted solution of the dehydrated raw soil had the same $^{87}Sr/^{86}Sr$ ratio as the associated soil solution, demonstrating an extensive exchange of Sr and other cations between the soil pool of exchangeable cations and vegetation. In contrast, the $^{87}Sr/^{86}Sr$ ratios of bulk soil and H_2O_2, NH_4Cl, and HCl residues were different from those of the exchangeable soil components, indicating that Sr and probably other cations in most soil minerals were not taken up by plants. The present extraction experiment demonstrated that (1) Sr isotopes have potential as a tool for determining plant-available Sr and identifying acid-resistant minerals, whether aeolian in origin or bedrock derived, and (2) the use of pulverized soil should be avoided because its leachate results in the extraction of elements that are not exchanged with cations in plants.

References

Åberg, G., Jacks, G., Wickman, T. and Hamilton, P.J.: 1990, *Catena* **17**, 1.
Graustein, W. C. and Armstrong, R. L.: 1983, *Science* **219**, 289.
Inoue, K. and Naruse, T.: 1987, *Soil Science and Plant Nutrition* **33**, 327.
Ishizaka, T.: 1982, *Időjárás* **86**, 249-253.
Kawachi, S.: 1977, *Geology of the Yatsugatake district*. Geological Survey of Japan (in Japanese with English abstract).
Miller, E. K., Blum, J. D. and Friedland, A. J.: 1993, *Nature* **362**, 438.
Mizota, C., Shimoyama, S., Kubota, M., Takemura, K., Izo, N. and Kobayashi, S.: 1992, *The Quaternary Research* **31**, 101. (in Japanese with English abstract).
Nakano, T. and Tanaka, T.: 1997, *Atmospheric Environment* **31**, 4237.
Nakano, T., Yokoo, Y., Yamanaka, M. : 2001, *Hydrological Processes* (in press).
Oinuma, K. and Kobayashi, K.: 1965, *Nendokagaku no Shinpo* **5**, 77 (in Japanese with English abstract).
Puri, A. N.: 1949, Soils: their physics and chemistry. pp. 550, Reinhold Publishing Co., New York.
Shimoda, S.: 1985, *A Guide to Study Clay Mineralogy*. Souzou-sha. (in Japanese).
Yokoo, Y., Nakano, T., Nishikawa, M. and Quan, H.: 2000, *Water, Air and Soil Pollution* (this volume).

DETERMINATION AND SPECIATION OF ALUMINUM IN ENVIRONMENTAL SAMPLES BY CATION EXCHANGE HIGH-PERFORMANCE LIQUID CHROMATOGRAPHY WITH HIGH RESOLUTION ICP-MS DETECTION

KIN-ICHI TSUNODA,[1,*] TOMONARI UMEMURA,[1] KAZUMASA OHSHIMA,[1] SHO-ICHI AIZAWA,[1] ETSURO YOSHIMURA[2] and KEN-ICHI SATAKE[3]

[1] *Department of Chemistry, Gunma University, Kiryu, Gunma, 376-8515, Japan;* [2] *Department of Applied Biological Chemistry, The University of Tokyo, Yayoi, Bunkyo-ku, Tokyo, 113-8657, Japan;* [3] *National Institute for Environmental Studies, Tsukuba, Ibaraki, 305-0053, Japan.*
(* *author for correspondence, e-mail:tsunoda@chem.gunma-u.ac.jp)*

Abstract. A cation-exchange high-performance liquid chromatography with high resolution inductively coupled plasma mass spectrometric detection (CE-HPLC/ICP-MS) was developed for the determination and the speciation of aluminum in environmental samples. Three types of aluminum species ($AlL_x^{<+2}$, AlL_x^{2+}, Al^{3+}) were separated from one another, and were determined with the present system. The comparison of the present system with an established CE-HPLC with fluorimetric detection using 5-sulfo-8-quinolinol (CE-HPLC/FL) was described. The present system showed better sensitivity for aluminum than CE-HPLC/FL. Moreover, the analytical results for soil extract and lake water samples obtained with both methods were in good agreement with each other.

Keywords: aluminum, determination, speciation, soil extract, lake water, high-performance liquid chromatography, inductively coupled plasma mass spectrometry

1. Introduction

Recently, much attention has been paid to the fact that acidic deposition into the environment, mainly as a product of all forms of combustion in human activities, has increased the geochemical mobility of aluminum in many aquatic and soil systems (Driscoll and Postek, 1996). Moreover, the differential toxicity of various Al species to aquatic and terrestrial organisms has been demonstrated (Driscoll et al., 1980). Thus, the speciation of Al in aqueous solutions has been recognized as an important subject in environmental analytical chemistry.

The methods for aluminum speciation so far reported can be classified into four categories: 1) kinetic or binding strength discrimination (Okura et al., 1962; Barnes, 1975; Lu et al., 1994); 2) separation by chromatography or electrophoresis (Jones et al., 1988; Gibson and Willett, 1991; Sutheimer and Cabaniss, 1995); 3) filtration (Shkinev et al., 1996); 4) non-invasive method (NMR) (Parker and Bertsch, 1993). Among them, high-performance liquid chromatographic (HPLC) techniques, particularly cation exchange (CE) HPLC, have been used as one of the most reliable methods for the purpose (Gibson and Willett, 1991; Sutheimer and Cabaniss, 1995). Tsunoda et al. (1997) also proposed a CE-HPLC method with fluorimetric detection using 5-sulfo-8-quinolinol, where the method of Jones et al. (1988) was substantially improved with regards to the elimination of the interference with concomitant ions, and it was applied to the aluminum speciation in soil extract samples. Although the method is quite sensitive and shows good reproducibilities in the measurements, it still lacks the sensitivity, enough to measure low aluminum levels normally found in soil extract samples. Moreover, as the method is based on the complex formation reaction between aluminum and 5-sulfo-8-quinolinol, very inert aluminum species can not be detected. Such problems can be avoided by the use of atomic spectrometric detection, where all the aluminum species are turned to atomic aluminum in high-temperature media.

In this paper, we describe the CE-HPLC system with high resolution inductively coupled plasma mass spectrometric detection (CE-HPLC/ICP-MS). The present system showed better sensitivity and selectivity for aluminum than the CE-HPLC/FL. The comparison of the present system with an established CE-HPLC with post-column fluorimetric detection using 5-sulfo-8-quinolinol (CE-HPLC/FL) was described. The analytical results for soil extract and lake water samples obtained with both methods were in good agreement with each other. Although Fairman et al. (1998) have already reported a similar system, several improvements have been made to avoid the problems that they encountered.

2. Materials and Methods

2.1. REAGENTS

All chemicals were of analytical grade from Wako, Japan. Water was first deionized and then distilled with a quartz distiller.

2.2. APPARATUS

A schematic diagram of the CE-HPLC/ICP-MS system is shown in Figure 1. Two Shimadzu LC-9A pumps for the eluent and a Rheodyne Model 9125 rotary valve injector (20 µl) were included in the CE-HPLC part. A Dionex Ionpac CG-2 (cation exchange column) was used for a separation column, while a Dionex Ionpac CS-2 (cation exchange column) was used for a guard column to remove the interference of aluminum contamination in eluents. For isocratic separation, 0.2 mol L^{-1} ammonium formate solution was used for the eluent. In gradient separation, on the other hand, the eluent A (0.2 mol L^{-1} ammonium formate solution) and the eluent B (distilled water) were sent with two pumps, respectively. The composition of the two eluents was changed linearly from A:B =1:4 at time t = 0 to A:B = 1:0 at t = 1 min, then, kept at constant. The total flow rate of the eluents was always kept at 0.4 ml min^{-1}. The column temperature was 40 C. For ICP-MS detection, a PLASMAX1 high-resolution ICP –MS from JEOL was used. When a high-resolution system is not used, some molecular ions, especially $^{13}C^{14}N^+$, often interfere with aluminum detection. Thus, it should be noted that the selectivity of the method is greatly improved by the use of the high-resolution system. For the interface of CE-HPLC and ICP-MS, an U-5000AT ultrasonic nebulizer from CETAC Technologies was used. The temperature of the tubing connecting the nebulizer and the plasma torch was kept at 200 C to prevent the condensation of water. The operating conditions of ICP-MS are as follows: RF power, 1.6 kW; plasma gas flow rate, 14 L min^{-1}; auxiliary gas flow rate, 0.8 L min^{-1}; nebulizer gas flow rate, 1.2 L min^{-1}; ion mass (m/z), ^{27}Al$^+$; sweep width, 1 amu; sweep time, 1 s (x 3); mass resolution, 3000, sampler, 1.0 mm Cu; skimmer, 1.2 mm Cu.

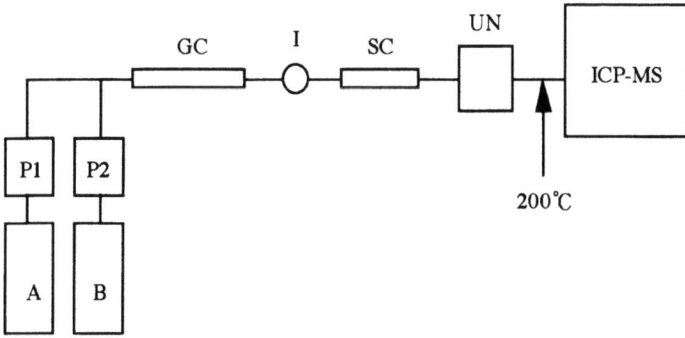

Figure 1. Schematic diagram of CE-HPLC/ICP-MS system.
A, eluent A; B, eluent B; P1 and P2, pumps; GC, guard column; I, injector valve; SC, separation column; UN, ultrasonic nebulizer.

2.3. PREPARATION OF SOIL EXTRACT AND LAKE WATER SAMPLES

A soil sample was collected in a Japanese cedar (*Cryptomeria japonica*) forest (brown forest soil, bedrock: granite) in Azuma, Gunma, Japan. The soil sample was dried in air at room temperature, then passed through a 2-mm sieve. Distilled water was added to the soil sample (1:1 in mass), and the mixture was shaken vigorously with a mechanical shaker for 10 min, then centrifuged at 5000g for 30 min. The supernatant was passed through a 0.45 μm membrane filter and applied to the analysis for aluminum. Aluminum extracted into distilled water was defined as water-soluble Al. A lake water sample was collected from Lake Inawashiro, Fukushima, Japan. The water of the lake (pH 5.0) is acidified by sulfuric acid of volcanic origin. The water sample was passed through a 0.45 μm membrane filter, and stored in a refrigerator until the measurements, although the water samples should usually be analyzed just after their sampling. This is due to that the sampling sites are far from our laboratory and the main purpose of this experiment is the evaluation of the CE-HPLC/ICP-MS method. As for soil extract samples, on the other hand, we have confirmed that the storage of the samples in a refrigerator gives almost no effect on the Al measurements. The samples were also analyzed by cation-exchange HPLC with post-column fluorometric detection. The instrumentation and procedure used for the measurements were described in a previous paper (Tsunoda et al., 1997). The detection limit of this method is ca. 0.2 µg L^{-1} for Al^{3+}.

2.4. EQUILIBRIUM CALCULATIONS

Estimates of Al species' distributions were made using MICROQL, a chemical equilibrium program for personal computers (Westall, 1986). Thermodynamic data used for calculation are the same as those of a previous paper (Tsunoda et al., 1997).

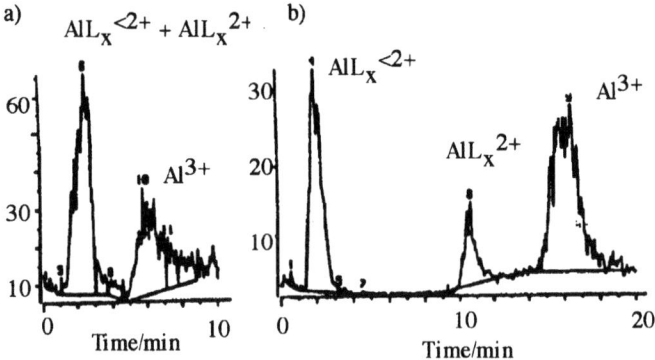

Figure 2. Chromatograms of soil extract sample for aluminum by CE-HPLC/ICP-MS. a) isocratic elution; b) gradient elution.

3. Results and Discussion

3.1. OPTIMIZATION OF MEASUREMENT CONDITIONS

Fairman et al. (1988) pointed out the problems of their system due to the high salt content of the mobile phase (0.08 mol L^{-1} K$_2$SO$_4$) which caused severe salt deposits in the injector tube to the ICP torch. In this study, we used 0.2 mol L^{-1} ammonium formate solution that is a volatile salt. In addition, the temperature of the tubing connecting the nebulizer and the plasma torch was kept at 200 C to prevent the condensation of water. Due to such changes, we did not encounter the problems of Fairman et al. Moreover, the gradient separation was applied to improve the separation between AlL$_x^{<2+}$ and AlL$_x^{2+}$. Figures 2a and 2b show the examples of the chromatograms of a soil extract sample for aluminum obtained by CE-HPLC/ICP-MS (a, isocratic elution; b, gradient elution). Although AlL$_x^{<+2}$ and AlL$_x^{2+}$ were not separated from each other with the isocratic elution, their complete separation was achieved with the gradient elution. Moreover, as the ion count level of the baseline was much lower in the gradient elution than in the isocratic elution, lower detection limit (ca. 4 nmol L^{-1} for each Al species) was achieved.

Aluminum contamination in the eluent produced a problem in the gradient elution. In the isocratic elution, the contamination level of aluminum in the eluent is always kept constant. In the gradient elution, aluminum in the eluent is first retained on the column, then was eluted together with Al^{3+} of the sample solution: it sometimes gave severe blank problem for the Al^{3+} determination. To prevent such problem, the eluent was purified with various methods such as Chelex-100 resin method or 8-quinolinol extraction. However, we could not always remove the contamination completely. In such cases, the guard column was introduced before the injector valve as shown in Figure 1. Although the complete separation could not be obtained between the Al^{3+} peak and the contamination peak, the contamination peak became much broader and the Al^{3+} peak could be distinguished from the peak.

3.2. COMPARISON BETWEEN CE-HPLC/ICP-MS AND EQUILIBRIUM CALCULATION.

The Al^{3+} standard solutions containing known amounts of fluoride ion, citrate, oxalate, respectively, were analyzed with the present system, and the results were compared with the estimated values by the equilibrium calculation. They gave comparable results with each other as the cases of the CE-HPLC with fluorometric detection (Tsunoda, et al., 1997). The case of fluoride was shown in Table 1 as an example. These results may suggest that the separation ability of this system for Al species should be similar to the other CE-HPLC methods (Sutheimer and Cabaniss, 1995, Tsunoda, et al., 1997) where Al^{3+}, $AlOH^{2+}$ and $Al(OH)_2^+$, $AlSO_4^+$, and Al complexes with monocarboxylic acids (e.g., formic and acetic acids) are found at the free Al^{3+} peak, on the other hand, Al complex with di-, and tricarboxylic acids (e.g., oxalic and citric acids) and fulvic acid at the $AlL_x^{<2+}$ peak (Tsunoda, et al., 1997).

TABLE I
Comparison of CE-HPLC/ICP-MS analysis and equilibrium calculation on aluminum standard solution containing fluoride ion[a]

Conc. of F⁻ (μmol L^{-1})	Method	Percentage of Aluminum Species		
		AlF_2^{1+}	AlF^{2+}	Al^{3+}
2.6	CE-HPLC/ICP-MS[b]	3	32	65
	Calculation	3	35	62
5.3	CE-HPLC/ICP-MS	11	57	32
	Calculation	7	57	32
10.5	CE-HPLC/ICP-MS	23	47	30
	Calculation	16	64	20

a) aluminum concentration, 3.7 μmol L^{-1} (pH, 3.0); b) gradient elution

3.3. ANALYSIS OF SOIL EXTRACT AND LAKE WATER

Table II summarizes the analytical results for water-soluble fraction of aluminum in soil extract and lake water samples by CE-HPLC/ICP-MS as well as by CE-HPLC/FL. The results of CE-HPLC/ICP-MS are in good agreement with those of CE-HPLC/FL. It means that the levels of inert aluminum species in the samples that could not be detected by CE-HPLC/FL were very low. The situation, however, may be dependent upon the types of samples (Tsunoda et al., 1997).

4. Conclusions

The present method was found to be very sensitive, and to provide consistent results with the equilibrium calculation and the CE-HPLC/FL method. Thus, it will be useful for the determination and the speciation of aluminum in environmental samples.

TABLE II
Analytical results of a soil extract sample[a] and a lake water sample[b] for aluminum by CE-HPLC/ICP-MS and CE-HPLC/FL.

Sample	Method	Total Al Conc. (μmol L^{-1})	Conc. of Al Species(μmol L^{-1})		
			AlL$_x^{<2+}$	AlL$_x^{2+}$	Al^{3+}
soil extract	ICP-MS (I)[c]	4.7		3.1	1.6
	ICP-MS (G)[d]	4.8	2.1	0.7	2.0
	FL(I)[e]	5	2	1	2
lake water	ICP-MS (I)	4.5		3.5	1.0
	ICP-MS (G)	3.2	1.2	1.0	1.0
	FL(I)	3	1	2	n.d.

a) Brown forest soil (bedrock: granite) in Gunma, Japan; b) Lake Inawashiro (pH 5.0) in Fukushima, Japan; c) CE-HPLC/ICP-MS with isocratic elution; d) CE-HPLC/ICP-MS with gradient elution; e) CE-HPLC/FL with isocratic elution.

Acknowledgements

The authors thank Dr. Akiko Takatsu for providing the lake water sample. This study was supported by the Global Environmental Research Program Budget (Acid Precipitation) of the Environment Agency of Japan.

References

Barnes, R.B. : 1975, *Chemical Geology* **15**, 177.
Driscoll, Jr., C. T., Baker, Jr., J. P., Bisogni, J. J. and Schofield, C. L.: 1980, *Nature* **284**, 161.
Driscoll, Jr., C.T. and Postek, K.M.: 1996, 'The chemistry of aluminum in surface water', in G. Sposito (ed.), *The environmental chemistry of aluminum* (2nd ed.), CRC Presss: Boca Raton, FL. pp. 363-418.
Fairman, B., Sanz-Medel, A., Jones, P, and Evans, E.H.: 1998, *Analyst* **123**, 699.
Gibson, J.A.E. and Willett, I.R.: 1991, *Communications in Soil Science and Plant Analysis* **22**, 1303.
Jones, P., Ebdon, L., and Williams, T. : 1988, *Analyst* **113**, 641.
Lu, Y., Chakrabarti, C.L., Back, M.H., Gregoire, D.C., and Schroeder, W.H. :1994, *Analytica Chimica Acta* **293**, 95.
Okura, T., K. Goto and Yotsuyanagi, T. : 1962, *Analytical Chemistry* **34**: 581.
Parker, D.R. and Bertsch, P.M. : 1993, *Environmental Science and Technology* **27**:2511.
Shkinev, V.M., Fedorova, O.M., Spivakov, B.Y. , Mattusch, J., Wennrich, R., and Lohse, M. : 1996, *Analytica Chimica Acta* **327**, 167.
Sutheimer, S.H. and Cabaniss, S.E. : 1995, *Analytical Chemistry* **67**, 2342.
Tsunoda, K., Yagasaki, T., Aizawa, S., Akaiwa H., and Satake, K.: 1997, *Analytical Sciences* **13**, 757.
Westall, J.C. : 1986, 'MICROQL - A chemical equilibrium program in BASIC Ver.2 for PCs', *Report 86-02*, Department of Chemistry, Oregon State University, Corvallis, Oregon.

STANDARDISATION OF METHODS FOR LONG-TERM MONITORING

WENCHE AAS[*] and ARNE SEMB

Norwegian Institute for Air Research, P.O. Box 100, N-2027 Kjeller, Norway
(* author of correspondence, email: wenche.aas@nilu.no)

Abstract. A monitoring programme should be designed for duration. This means that methods should not only be appropriate with respect to detection limits and accuracy, but they should also be as simple as possible and they should be documented in such a way that measurements will be comparable over many decades. In this connection, it is particularly important to understand that results are dependent on methods, instruments and procedures. Within the European monitoring network (EMEP) there are several different sampling procedures for the main air components, SO_2, NO_2, SO_4^{2-}, NO_3^- + HNO_3, and co-located experiments have therefore been initiated to quantify the difference between the measurements. Reference methods and reference instruments corresponding to the recommendation in the EMEP Manual have been run together with the usual measurements at EMEP sites in several countries. The results are generally satisfactory, especially in the case where identical methods are used. However, there are also some unacceptable differences, e.g. when comparing NO_2 and SO_2 monitors with the reference methods. The monitors do have a main advantage of providing easily accessible data with short time resolution; nevertheless, the accuracy at low concentrations is usually poor. The traditional reference methods need development and simplification in the direction of the more appealing automatic instruments.

Keywords: Monitoring, air pollution, sampling methods, EMEP manual

1. Introduction

The monitoring of precipitation chemistry originated as part of the scientific effort to understand the basis for plant growth. Even in this century, supply of plant nutrients and cycling of nitrogen and sulphur in the atmosphere was part of the motivation for air and precipitation monitoring. Monitoring programmes have evolved slowly, serving different purposes and interests. In the context of long-range transported air pollutants, the purpose is primarily to determine deposition of chemical compounds and exposure to pollutants, which are harmful to vegetation or to human health. It is necessary to carry out long-term monitoring to be able to determine historical trends and changes in deposition and exposure, following changes in emissions. These changes are strongly related to the Protocols in the Convention of the Long-Range Transportation of Air pollutants (LRTAP) in Europe, which means that monitoring of the changing airborne concentrations and the deposition is a method of testing the efficiency of the protocols. A co-operative programme for monitoring and evaluation of the long-range transmission of air pollutants in Europe (EMEP) was defined in 1985 to regularly provide governments and others under the LRTAP Convention with scientific information to support the development and further evaluation of the international protocols on emission reductions.

When models are used to derive the relationships between emissions and deposition or airborne concentrations, measurements are needed to provide information on processes and on the performance of the models. Monitoring the long-term changes in air and precipitation chemistry places the emphasis on long and consistent measurement time series. The time resolution can be low, but should allow for interpretations of the data in terms of different meteorological situations and even climatic changes. The EMEP program defines the components to be measured, as well as where and when these components should be measured, and it determines the requirements of the measurement methods, which are to be used for monitoring within EMEP and similar programmes. In the following, these requirements will be discussed in more detail using the experience gained during the last 20-30 years and results from field comparisons that has been carried out the last years (Schaug *et al.*, 1998; Aas *et al.*, 1999 and 2000) in Great Britain (Eskdalemuir GB2); Ireland (Valentia IE2); Portugal (Monte Velho PT4); France (Donon FR8); Germany (Shauinsland DE3); Poland (Diabla Gora PL5) and The Czech Republic (Košetice CZ3).

2. Sampling Procedures

2.1. CO-LOCATED SAMPLING

The field comparisons were carried out at the same site, where the national measurements were compared with a set of reference instruments, which correspond to the specifications in the EMEP Manual for Sampling and Chemical Analysis (EMEP, 1995). Samples from the reference instrumentation were shipped to the Norwegian Institute for Air Research (NILU) for chemical analyses. In order to make the comparison valid for a representative period, the measurements were distributed over a whole year; about 100 measurements were considered necessary and the sampling frequency were two days every week or one week every month.

2.2. INSTRUMENTATION

In the reference method, SO_2 is absorbed on a potassium hydroxide impregnated filter, which also will absorb other volatile acidic substances as HNO_3, and give *i.e.* solid potassium sulphite and nitrate. The recommended method is compared with the common method using absorption solutions of either hydrogen peroxide or tetrachloromercurate (TCM). Results from continuous SO_2 measurements, using UV-fluorescence, is also presented and discussed.

Several methods, both manual and continuous have also been used for the measurement of NO_2 in ambient air. In the manual reference method, nitrogen dioxide is absorbed on an alkaline sodium iodide impregnated glass sinter; iodide reduces NO_2 to nitrite (NO_2^-). Absorption solutions have also been used, such as the Saltzman method, a procedure based on a direct Griess reaction during sampling (ISO, 1985b). In urban air, the gas-phase chemiluminesence has been accepted as an ISO standard (ISO, 1985a). This automatic method is also being used at a few remote sites.

In the filterpack method, the first filter in the airstream is an aerosol filter for collecting the airborne particles containing i.e. sulphate and nitrate followed by alkaline and acid impregnated filters for collecting the gases. One disadvantage using the filterpack method is that separation of gaseous and particulate nitrogen compounds are not efficient. Sampling artefacts due to the volatile nature of ammonium nitrate, and possibly due to interaction with other atmospheric constituents make separation of these gases and particles by a simple aerosol filter unreliable. Therefore, the concentration of nitrates in air is given as the sum of the nitrate found on the aerosol filter and nitrate found on the alkaline impregnated filter, similar for the ammonium compounds.

3. Results and Discussion

3.1. SO_2 MEASUREMENTS

The co-located sampling measurements showed that H_2O_2 absorption tends to give too high SO_2 concentration compared with the reference method (Fig. 1 and Table 1). This has also been observed in earlier field comparisons performed in Vavihill (Semb *et al.*, 1991). The TCM absorption method, which was used at Schauinsland DE3, seems to suffer from negative interference by ozone or other oxidants, particularly during the summer season (Fig. 2). At Schauinsland, three different methods have been compared and the results were unsatisfactory for both the UV-fluorescence and TCM methods, but satisfactory for the recommended filterpack method (Table I). It should be pointed out that the monitor data are not used to provide official data from this station, only for supplementary purposes. Germany is now replacing their TCM measurements with the recommended method. Schauinsland is situated in a remote place in the mountain and maintenance of the equipment is therefore limited. Even though, the figure is illustrative for indicating how doubtful it is to use monitor data when the instrument is not frequently calibrated and maintained. Even when properly maintained, UV-fluorescence is not sufficiently precise at low SO_2 concentrations resulting in unsatisfactory annual average values at remote sites. Filterpack is the only available method that has been proven to give reliable and sensible results at the 0.1- 0.2 µg-S/m^3 concentration level for sulphur dioxide, and practical experience with the method now goes back more than 20 years. The main limitation of this method is the absorption capacity of the filter, which may cause problems at sulphur dioxide concentrations above 20-40 µg-S/m^3. Alternative methods include the use of aqueous hydrogen peroxide absorption method, followed by determination of sulphate. The detection limit for this method is limited by the blank value, or by the difficulty in determining a proper blank value for the absorption solution, which is partly evaporated during the sampling period.

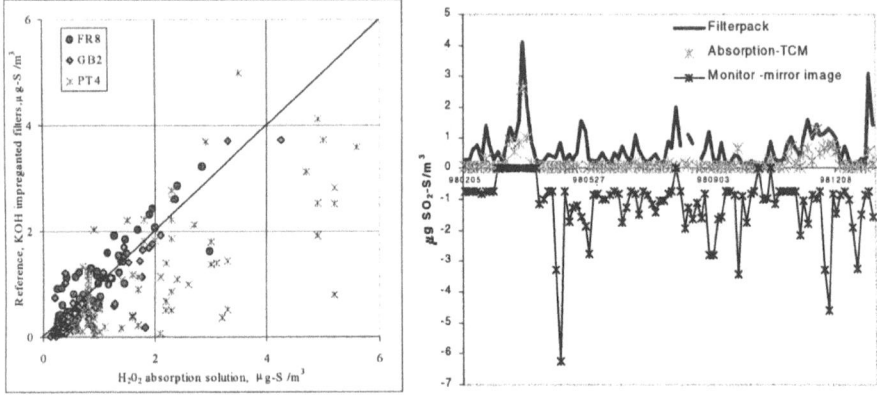

Figure 1. A comparison of SO_2 concentrations at three sites using H_2O_2 absorption solution and impregnated filters.

Figure 2. Comparisons of SO_2 concentrations at DE3 using three different methods

TABLE I

Average of the SO_2, SO_4^{2-}, NO_3^- + HNO_3 concentrations (μg-S(N)/ m^3), reference method in bold and number of samples in italic.

	GB2	IE2	PT4	FR8	DE3	PL5	CZ3
SO_2 filter	**0.62**	**0.59**	**1.79**	**0.72**	**0.54**	**1.39**	**1.57**
		0.57 (87)			*0.64 (93)*		*2.18 (80)*
abs. monitor	*0.86 (63)*		*2.96(101)*	*0.81 (94)*	*0.20 (93)*	*1.22 (95)*	
					1.15 (93)		
SO_4^{2-} filter	**0.63**	**0.85**	**1.56**	**0.82**	**0.61**	**1.02**	**1.09**
	0.64 (66)	*0.78 (87)*	*1.77(100)*	*0.62 (92)*	*0.66 (93)*	*1.24 (94)*	*1.24 (39)*
NO_2 sinter		**0.51**	**0.99**		**1.00**	**0.89**	**1.69**
							2.50 (75)
abs monitor		*0.65 (76)*			*1.13 (83)*	*0.94 (87)*	
			1.04(94)		*1.66 (83)*		
Σ nitrat filters	**0.44**				**0.46 (93)**	**0.54**	**0.66**
	0.38 (77)				*0.73 (93)*	*0.64 (92)*	*1.27 (78)*

3.2. NO_2 MEASUREMENTS

The NaI method is recommended at background stations with low concentrations of NO_2, and it is suitable when the analysis has to be performed in a laboratory far from the sampling site. Some chemiluminisence monitors can be as sensitive as the NaI method; however, the monitor is not specific because other reducible nitrogen compounds (*e.g.* HNO_3 and PAN) give a positive interference, which can be a serious problem at some sites. In difference from the reference method, chemiluminisence measures the sum of NO and NO_2. The Saltzmann method is

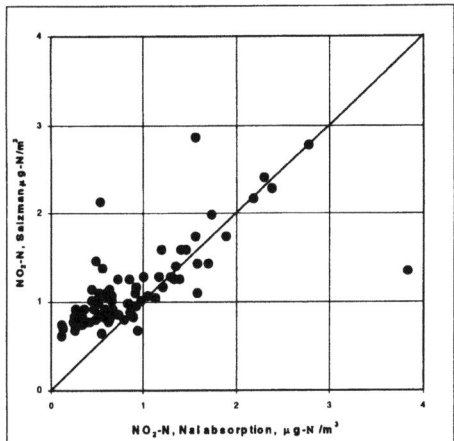

Figure 3. Comparison of NO₂ concentration using the NaI and Saltzman methods.

more selective than chemiluminescence, but the colour is developed during sampling, and the measurements have to be performed immediately after sampling due to instability. Saltzmann method is unsuitable if the exposed absorbing solution has to be transported to a chemical laboratory far from the sampling site, particularly if temperature and light exposure cannot be controlled. The field comparison at DE3 has also revealed that this method is not suitable at concentration levels below $1 \mu g N/m^3$ (Fig. 3).

3.3. SUM NITRATE (NO_3^- + HNO_3) MEASUREMENTS

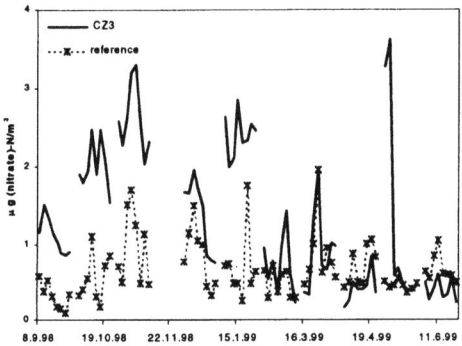

Figure 4. Comparison of sum nitrate at CZ3

In Table I there are large discrepancies between the co-located measurements at especially two sites, DE3 and CZ3. Statistical analysis showed that the correlation is reasonable, which is also seen in Figure 4, and that the biases are due to systematic errors. The most probable explanation is that there are problems when correcting for field blanks. At CZ3 the difference was more than 100% during the first 5 months, but a satisfactory difference of only 10% for the last 5 months.

4. Conclusions

The requirements with respect of accuracy of long-term deposition or average concentration measurements is determined by the use of these data in connection with effect assessments, but also for the comparison with model estimates of the same parameters. When used in conjunction with models, an accuracy requirement has just been defined in that the difference between the measurements and the model estimates should not exceed ±30%. This implies, taking into account the difficulties with the spatial resolution and site representativeness, that the absolute accuracy in the measurements should be better than ±10%, at least for long-term averages. Many of the measurement series satisfy this requirement, but as shown in Table I there are also some which fail, particularly when non-recommended methods are used.

Evaluation of trends particularly requires consistent data series, which call for documentation of methods, and the effects of changes in methods, equipment and procedures. This is a demanding task, since the more valuable data series represent more than 20 years of measurements, sometimes with changes of responsible laboratories and almost certainly with different persons involved. Ideally, methods should not be changed. However, equipment will have to be renewed and procedures are also liable to be changed. It is therefore important that the methods and sampling protocols are documented and defined in such a way that the effect of changes can be assessed and tested under representative conditions. Complicated instrumentation or calibration procedures should be avoided, if possible. In the selection of methods, the needs for precision, accuracy and consistency should be taken into account. At background sites it is particularly important that the detection limit is well below the ambient concentration levels, and that the equipment at the site is simple and robust, minimizing handling and operational errors. Simple sampling equipment, and detailed sampling procedures should be favoured over more sophisticated equipment, even if the latter could give more detailed or accurate results in the short term.

Acknowledgements

This work has been totally dependent on the cooperation from the participating laboratories and field personnel. Their work and willingness to contribute and share the results are highly appreciated.

References

Aas, W., Hjellbrekke, A.-G., Schaug, J.: 2000, 'Data quality 1998, quality assurance, and field comparisons'. Kjeller, Norwegian Institute for Air Research (EMEP/CCC 6/2000).

Aas, W., Hjellbrekke, A.-G., Semb, A., Schaug, J.: 1999, 'Data quality 1997, quality assurance, and field comparisons'. Kjeller, Norwegian Institute for Air Research (EMEP/CCC 6/99).

EMEP: 1996, 'Manual for sampling and chemical analysis.' Kjeller, Norwegian Institute for Air Research (EMEP/CCC 1/95).

International Organization for Standardization: 1985a, 'Ambient air - Determination of the mass concentration of nitrogen oxides-Chemiluminesence method.' Geneva (ISO. International Standard 7996:1985).

International Organization for Standardization: 1985b, 'Ambient air- Determination of the mass concentration of nitrogen dioxide - Modified Griess-Saltzman method.' Geneva (ISO. International Standard 6768:1985).

Schaug, J., Semb, A., Hjellbrekke, A.-G.: 1998, 'Data quality 1996, quality assurance, and field comparisons.' Kjeller, Norwegian Institute for Air Research (EMEP/CCC 6/98).

Semb, A., Andreasson, K., Hanssen, J.E., Lövblad, G. and Tykesson, A.: 1991, 'Vavihill, Field Intercomparison of Samplers for Sulphur Dioxide and Sulphate in Air.' Lillestrøm, Norwegian Institute of Air Research (EMEP/CCC Report 4/91).

A NEW APPROACH FOR CHARACTERIZATION OF SINGLE RAINDROPS

CHANG-JIN MA[1], MIKIO KASAHARA[1], SUSUMU TOHNO[1] and TOMIHIRO KAMIYA[2]

[1] Graduate School of Energy Science, Kyoto University Gokasho, Uji, Kyoto 611-0011, Japan;
[2] Advanced radiation technology center, Japan Atomic Energy Research Institute, Takasaki, Gunma 370-1292, Japan
(author for correspondence, e-mail: ma@uji.energy.kyoto-u.ac.jp)

Abstract. To determine the characteristics of single raindrops as a function of their size, the collodion film method was newly applied to the sampling of single raindrops. Sampling of single raindrops was performed at a height of 20 m above ground level of the Kyoto University building located in Uji, Japan during rain events from September to November 1999. It was possible to get successfully replicas of raindrop by collodion film method. And we tried to analyze the elemental components of the nuclei and pollutants that were incorporated into the developing raindrop. To the analysis of single raindrops, Particle Induced X-ray Emission (PIXE) was applied. Several elements including S in single raindrops were detected by PIXE. The concentrations of every element increased with decreasing drop diameter. Furthermore, to acquire more detailed information such as inner-structure and mixing state in single raindrops, micro-PIXE analysis was performed. The nucleation centers of single raindrops were successfully analyzed by micro-PIXE. Ca, S and Fe were apparently detected by scanning beam of $1\sim2$ μm diameter and about 70 pA beam current.

Keywords: collodion film, micro-PIXE, PIXE, single raindrop, replica, wet scavenging

1. Introduction

Chemical processes in the atmosphere as the washout of particles and gases, the damage of the biological system and the change of the radiation balance are not sufficiently described by the usual determination of elemental concentration in rainwater (Bächmann et al., 1993). Until recently, not much has been known about the chemical constitution of individual cloud and raindrops. Fortunately, however, recent technologies such as capillary zone electrophoresis (Bächmann et al., 1993) and micro-PIXE (Ma et al., 1999) have provided new methods by which chemical analysis of raindrops can be carried out.

The analysis of single raindrops is expected to give new and interesting information about anthropogenic air pollution and drop formation processes in the atmosphere.

In this study to determine the characteristics of single raindrops as a function of their size, the collodion film method was newly applied to the sampling of single raindrops. And we tried to analyze the elemental components of the nuclei and pollutants that were incorporated into the developing raindrop.

2. Materials and Methods

2.1. FORMATION OF SINGLE RAINDROP REPLICAS

Collodion solution was made by dissolving the nitrocellulose involving 11-12 % of nitrogen into the mixed solution of ether and alcohol. 200 $\mu\ell$ of collodion solution (3 %) was mounted onto the 47 mm diameter non-hole Nuclepore filter just before sampling. Fallen drizzle and small sized raindrops were allowed to settle on the surface of collodion film (130±10 μm) without bounce off. The replica formation process of single raindrop on the collodion film is illustrated diagrammatically in Figure 1.

2.2. SAMPLING OF SINGLE RAINDROP

Sampling of single raindrops using collodion film method was performed at a height of 20 m above ground level of the Kyoto University building located in Uji, Japan during rain events from September to November 1999. The surroundings of sampling site are residential and agricultural areas with no major point sources. During sampling period the temperature was around 11.5 - 19.6 ℃ and average relative humidity was 70.5 %. Sampling procedure consisted of placing a thin layer of collodion solution on non hole Nuclepore filter of 47 mm diameter laid on Petridish (ϕ 80 mm) that was installed on plate sampling stage (185 mm×185 mm). And then, the lid of Petridish was removed and raindrops were allowed to settle on the surface of collodion film. About 200 single raindrops were collected on 20 sheets of collodion film in four rain events.

2.3. CHEMICAL ANALYSIS OF SINGLE RINDRROP REPLICAS

For chemical analysis of raindrops collected on Collodion film, Particle Induced X-ray Emission (PIXE) analysis was applied. PIXE analysis was performed with a proton beam of 6 mm diameter and 2.0 MeV energy from a Tandem Cockcroft accelerator. Beam intensities from 10 to 60 nA were employed and the total does were about 20 μ C. X-ray with an energy up to 14.8 keV emitted from the target were detected by a Si (Li) detector which had a

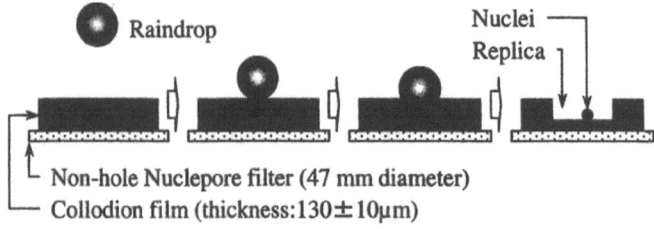

Figure 1. Replica formation of single raindrop on the collodion film.

resolution of 152 eV at 5.9 keV. The more detailed analytical procedures and experimental set-ups used for PIXE was described elsewhere (Kasahara et al., 1996).

To acquire more detailed information such as inner-structure and mixing state in the single raindrops, micro-PIXE analysis was performed using the facilities of the Takasaki Ion Accelerators for Advanced Radiation Application, Japan. Beam scanning, data acquisition, evaluation and the drawing of elemental maps are controlled by a computer on the basis of the system program. X-Y beam scanning control signals, which indicate the beam position, are also digitized at the same time. These data are addressed to the 3D matrices in the memory space, that consist of 1024 channels for the energy spectra and 128×128 pixels for corresponding the beam scan area. Micro-PIXE analysis was performed with a scanning 2.5 MeV H^+ micro beam accelerated by 3 MV single-end accelerator. Beam diameter and beam current were $1 - 2$ μm and > 100 pA, respectively. The more detailed analytical procedures and experimental set-ups used for micro-PIXE was described elsewhere (Ishii et al., 1996).

3. Results and Discussion

Application of collodion film method to the sampling of single rain droplets (diameter ≤ 2.1 mm) was very successful. But, unfortunately, this collodion film method did not allow to sampling of raindrops larger than 2.1 mm diameter because of breakup of large raindrops when they collide with collodion film surface. Separated clean raindrop replicas formed on the collodion film are shown in Photo 1. The circle that appear in the center of each replica, probably represent the nucleation center from which cloud condensation nucleation was initiated. Although the nucleation particle is not apparent, the diameter of nucleation center is about 20 ± 5 μm. Standard solution droplets collected on collodion film were tested from two kinds of aspect. One was the possibility of filter rupture by PIXE beam irradiation and the other was background concentration of collodion film. Consequently no apparent peak was found at the spectrum of collodion film blank. And collodion film kept the original shape without rupture. Furthermore we made certain of the high degree of analytical accuracy from the analysis of standard solution droplets collected on

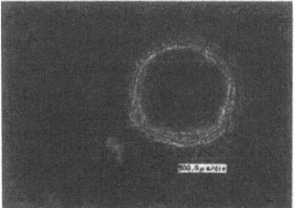 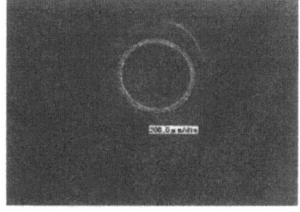

Photo 1. Raindrop replicas formed on the collodion film.

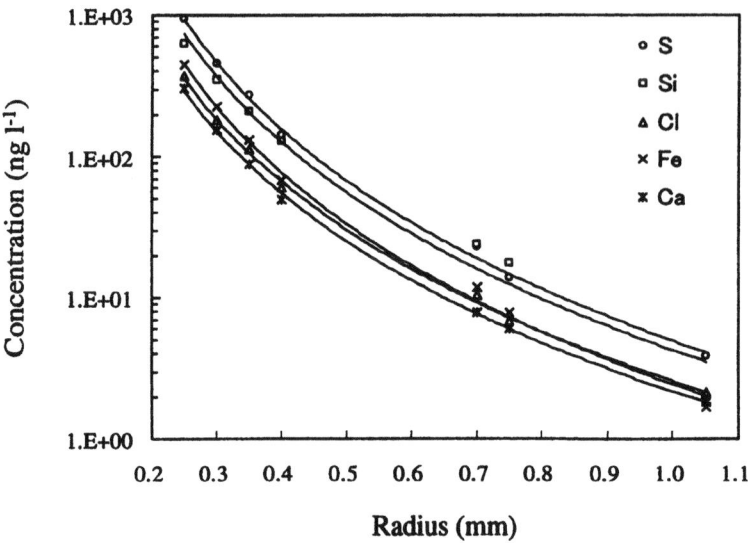

Figure 2. Variation of elemental concentration as a function of rain drop size.

collodion film. The size dependence of the elemental concentration of dissolved and suspended components in raindrops was obtained by PIXE analysis, and was illustrated in Figure 2. The concentrations of each data set in Figure 2 mean the average concentration of 10 – 15 raindrops. It shows a strong decrease of every elemental concentration with increasing droplet diameter.

Turner (1955) and Flossmann (1994) reported that raindrops within the cloud or near the cloud base have nearly identical concentrations regardless of drop size. On the other hand, it was found that dependence of concentration on drop diameter gave a continuous decrease in the concentration with increasing drop radius near and at the ground. It may, therefore, be conclude that dependence of elemental concentration on drop sizes measured near and at the ground is only due to below cloud influences. During rainfall from cloud base to ground, drops experience an increase in solute concentration due to evaporation and below cloud scavenging of particle and gases. In the size range of 200 to 2000 μm radius, the scavenging efficiency of particles and gases by drops decreases with increasing drop size (Hans *et al.*, 1998). In addition, evaporation of falling rain mainly dominates the smallest drops. In contrast to these, the field studies of Bächmann *et al.* (1993) showed that, during a precipitation event, the salt concentration across the raindrop spectrum evolves in time, and often develops the maximum concentration in drops of 200 to 300 μm radius. Cloud model studies of Tsias (1996) show that this maximum is an indirect result of collisional breakup of large raindrops. During such a breakup, small fragment drops are formed which have the same low concentration as their larger parent drops. Since many of these fragment drops evade further capture, they produce a concentration minimum in the range of 100 to 200 μm radius. The maximum of relative concentration thus occurs for slightly larger drops, which are not affected by this dilution mechanism. Unfortunately, this minimum concentration

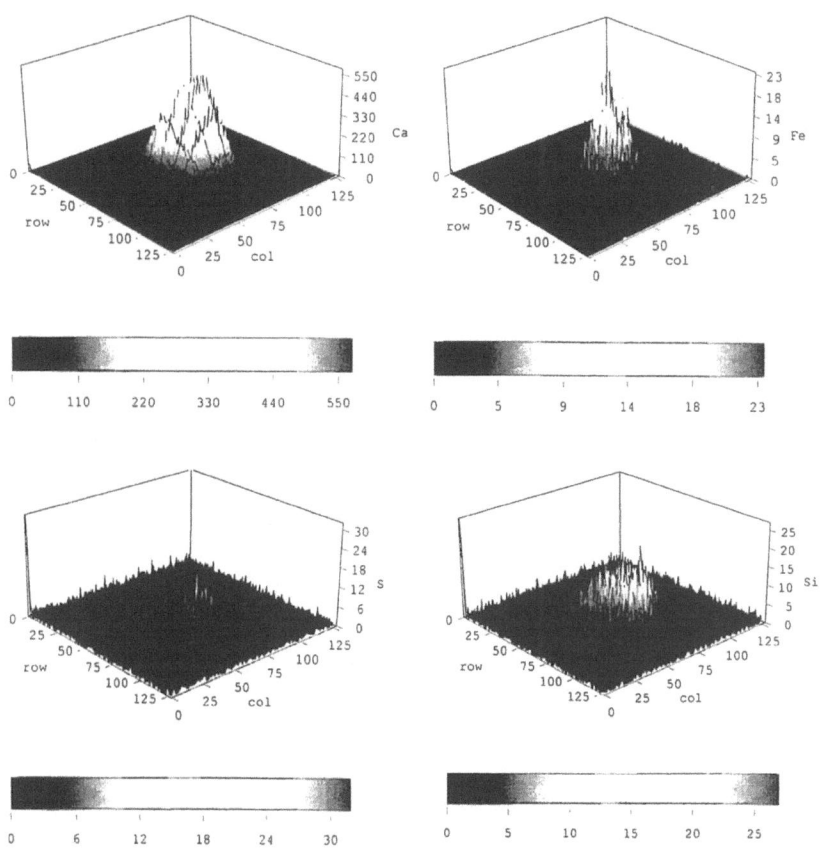

Figure 3. Elemental maps obtained by micro-PIXE at the nucleation center of raindrop.

in the range of 100 to 200 μm radius was not proved in this work because the minimum radius of raindrop collected by collodion film was 250 μm.

In order to investigate more detailed information such as mixing state of elements on the single raindrop, micro-PIXE analytical measurement was also applied. The circles appeared in the center of every raindrop replica can be considered to be nucleation centers from which cloud condensation nucleation was initiated, as mentioned earlier. Figure 3 shows the elemental maps obtained by micro-PIXE at the nucleation center of raindrop. Row and col are pixels corresponding beam scan area and the scale bar is the peak count of characteristic X-ray. Soil originated components such as Si, Ca and Fe were found to be the most abundant component in and/or on nucleation center of raindrop. S was found to be quite distributed. Kuroiwa (1961) observed the chemical composition of the residues from cloud and fog drops. His observations implied that particles derived from soil account for about 28 % of the residue left by evaporated cloud and fog drops. It is therefore, suggested that cloud condensation nucleation was mainly initiated from soil components in our sampling periods.

3. Conclusions

The size dependence of the elemental concentration of non-volatile material dissolved and suspended in rain droplets obtained by PIXE analysis shows a strong decrease of every elemental concentration with increasing droplet diameter. It can be suggested that smaller raindrops should have higher elemental concentration because they have lower velocities and consequently longer lifetimes than larger ones. And the reason of high concentration in small raindrops might be caused by the effect of evaporation, i.e. small raindrops show a much higher degree of evaporation than larger ones. It leads to an increase of the elemental concentration. Soil originated components such as Si, Ca and Fe were found to be the most abundant component in and/or on nucleation center of raindrop by micro-PIXE analysis. S was found to be widely distributed. It is, therefore, suggested that cloud condensation nucleation was mainly initiated with soil components in our sampling periods.

Acknowledgements

A part of this work has been done under the program of the Research for the Future (RFTF) of the Japan Society for the Promotion of Science (JSPS-RFTF97P01002) and Grant-in-Aid for Scientific Research (B) under Grant No. 09044161 from Ministry of Education, Science, Sports and Culture, Japan.

References

Bächmann, K., Haag, I., Prokop, Roder, T. A. and Wagner, P.: 1993, *J. Atmospheric Science* **24**, S421.

Ma, C. J., Kasahara, M., Tohno, S., Onishi, Y. and Whoang, K. C.: 1999, *The 3rd International Symposium on Bio-PIXE*, Kyoto, Japan, Nov. 16-19, 4O05.

Flossman, A. I.: 1994, *Atmospheric Research* **32**, 233.

Hans, R. P. and James, D. K.: 1998, *Atmos. and Ocean. Sci. Library*, **18**, 24.

Ishii, Y., Tanaka, R. and Isoya, A.: 1996, *Nucl. Instr. and Meth. in Phys. Res.* **B113**, 75.

Kasahara, M., Ogiwara, H. and Yamamoto, K.: 1996, *Nuc. Inst. and Methods in Phy.s Res.* **B118**, 400.

Kuroiwa, D.: 1961, *Tellus* **13**, 252.

Tsias, A.: 1996, Ph.D. Thesis, Dept. Atmos. Phys. Univ. of Mainz.

Turner J. S.: 1955, *Quart. JIR. Meteorological Society* **81**, 418.

STABILITY OF IONIC COMPONENTS IN PRECIPITATION SAMPLES – A CASE STUDY IN TAIPEI

LU-YEN CHEN*, FU-TIEN JENG, YU-MEY HSU, SHIH-YUAN TSAI, and UEI-RUEY PENG

National Taiwan University, 71 Chou-San Road, Taipei City, 106, Taiwan.
(* author for correspondence, e-mail: lychen@airlab.ntu.edu.tw)

Abstract. The changes in ionic contents were studied in acidic precipitation samples collected for precipitation events in Taipei, which is near the sea. The storage cases under investigation include filtration, refrigeration, and light. Thus the experimental design leads all precipitation samples collected in the same rain event stored under different conditions. They were then analyzed six times successively within two months to provide the information containing potential ionic composition change. The measured constituents are H^+, K^+, Na^+, Ca^{2+}, Mg^{2+}, NH_4^+, NO_3^-, SO_4^{2-}, and Cl^-. The comparison of measured ionic concentrations corresponding to different storage methods yield no significant difference. The increases of NO_3^- and decreases of NH_4^+ with time were observed to be of similar magnitude, while the variation of pH values is significant. The presented study indicated the important role played by sample storage in determining the ionic composition of precipitation samples.

Keywords: acid deposition, acid rain, precipitation, sample storage, wet deposition.

1. Introduction

The Taiwan Acid Precipitation Monitoring Network (TAPMON) collects daily rainwater samples. Though it was recommended and advantageous to analyze rainwater samples immediately after collection (Galloway and Likens, 1976), however, this is not easy to handle in common monitoring network both practically and economically. In TAPMON, samples are kept in containers during their transportation and storage, which possibly result in chemical changes.

There were reports of significant pH changes within a short time after sample collection (Hansen and Hidy, 1982). A 15% decrease in hydrogen ion activity, $a_{H+} = 10^{-pH}$, has been observed in Ontario between measurements of field pH and lab pH (Chau *et al.*, 1987). In Taiwan, the relationship between field pH and lab pH is characterized by the fitted line as: $[pH]_{lab} = 0.9496[pH]_{field} + 0.1736$, the interception hints a decrease of a_{H+} as 49% (Jeng, 1999a). Vesely (1990) surveyed H^+ ion change and proposed a corrected H^+ ion concentration (H_c^+) as an acidification indicator which is not very sensitive to storage.

Precipitation composition also caught research attention. Peden and Skowron (1978) analyzed the ionic concentration of event and weekly precipitation samples with different pre-treatment and storage conditions in Illinois, U.S. Ca^{2+}, Mg^{2+}, Na^+, K^+, H^+, NH_4^+, SO_4^{2-}, NO_3^-, and Cl^- were measured. Significant concentration variations of some species were observed. The counterparts of concentration change in sites near the sea are of interest because of its potential to affect the resolution of the long-term trends of wet deposition.

To resolve the chemical changes during storage for site near the sea, a sampling study was performed from October to December 1999 at the Taipei site of TAPMON. This site is 120°20' E and 25°02' N, and 17 meters above the sea level, only less than 50 km from the sea, and faces the prevailing winds coming from the north-east during the winter.

This paper compares the ionic concentrations of samples after different storage periods. This study is carried out following: (a) wet precipitation-only sampler was used; (b) all samples were analyzed in the same laboratory; (c) all sample handling procedures were identical.

2. Materials and Methods

2.1. FIELD OPERATIONS

Wet precipitation samples were collected at Taipei, an urban site in northern Taiwan. The samples were collected on a daily event basis in polyethylene buckets in a modified collector set to open only during precipitation periods. The collected rainwater samples were filtered utilizing filters with 0.4 μm pore to remove larger particles from precipitation.

All operators were trained in the sampling and pre-treatment operations based on the "Quality Assurance/Quality Control Program for Acidic Precipitation Monitoring"(Jeng, 1999b), which resembles the procedures of NADP/NTN (National Ambient Deposition Program/National Trend Network). Daily site visits were made at about 9:00 a.m. to check the equipment and remove any daily-collected samples. If no event occurred for one day, the sampling bucket was rinsed with de-ionized water and the bucket was then ready for new sampling. Several samples were selected for successive analyzing.

2.2. LABORATORY MEASUREMENTS

The following analysis order was taken to minimize chance of degradation of samples: pH, NH_4^+, SO_4^{2-}, NO_3^-, Cl^-, Ca^{2+}, Mg^{2+}, Na^+, and K^+ (Topol et al., 1987). The pH was measured electrometrically with Portable pH Meter WTW® pH 330. NH_4^+ and all anions were measured by ion chromatography with Dionex® DX-120, the separation columns adopted here is IONPAC® AS4A-SC for NH_4^+ and IONPAC® CS12A for anions, respectively. All metallic cations were analyzed by atomic absorption by Perkin Elmer® AA-5000.

All analytical instruments were calibrated once per day to catch the real-time machine performance. In addition, reagent blanks and spikes were run together with all analyzed samples in the analyzing days. A detail description of the quality assurance program is provided in another document (Jeng, 1999b).

2.3. EXPERIMENT DESIGN

Samples were analyzed in a successive six sequence to resolve the trends of the concentrations of ionic species. In addition to the temporal aspect, storage conditions were also investigated. Four different conditions, as Table I, were applied to precipitation samples collected on the same day from the same wet-only sampler. Filtration procedure was not applied to case 4. Samples were kept under room temperature (15 ~ 34°C) in case 2 and 4, others were refrigerated at 4°C. Only in case 1 the samples were kept from exterior light using brown storage bottle.

TABLE I
Pre-treatment and storage conditions of precipitation samples.

Case	Filtration	Refrigeration	Lighting
1	√	√	
2	√		√
3	√	√	√
4			√

3. Results and Discussion

The initial concentrations of all ionic species are shown in Figure 1. The dominating species are Na^+ and Cl^-.

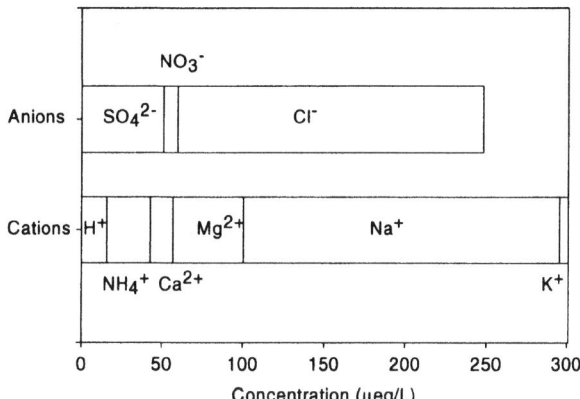

Figure 1. Initial ionic composition.

3.1. H^+

The H^+ concentration changes are illustrated in Figure 2. There was a decreasing trend of H^+ concentration. The greatest change appeared in Case 1, which is dominated by the analysis on Dec. 22. If the light was not avoided, the decrease of H^+ didn't differ quantitatively whether it was refrigerated or not. The difference between filtered and unfiltered samples is not obvious.

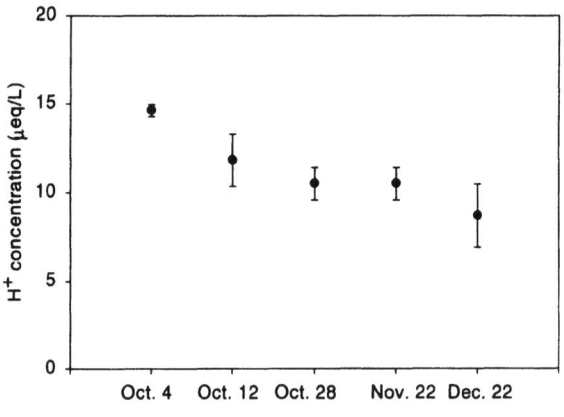

Figure 2. H^+ concentrations vs. time.

3.2. ANIONS

The mean anion concentration ratios (normalized by divided with initial concentration) for Case 1 is shown in Figure 3. The time series of Cl^- is a slightly increasing one except Dec. 22. This increase is thought to be the contribution from the air penetration into the sample. This site is near a municipal incinerator

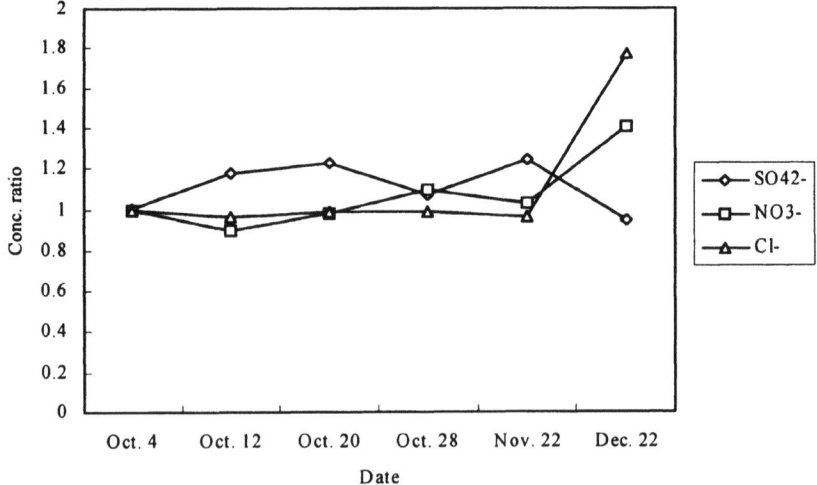

Figure 3. The mean concentration ratios for anions in case 1.

treating lots of PVC plastics. The contribution from sea salt is excluded because of the decreasing trend of Na^+. This differs from Peden and Skowron's work (1978), in which a significant change of Cl^- is observed.

The behavior of SO_4^{2-} differs from Cl^-. During the whole analysis period, the concentration ratio fell in the interval of 1 ~ 1.2. However, there was still a quantitative variation of SO_4^{2-} concentration observed. This result completes the lack of SO_4^{2-} time trend resolution in former work (Peden and Skowron 1978), and gives us the idea about potential SO_4^{2-} concentration change.

In this study, the concentration change of NO_3^- and NH_4^+ seem to compensate each other. The increase of NO_3^- is comparable with the decrease of NH_4^+, while the H^+ concentration decrease was about 25%.

3.3. CATIONS

Among four test cases, the behavior of cation concentration ratio is different from anions. The cases with/without filtration pre-treatment didn't differ so much. The cation concentration vs. time (Fig. 4) is presented. Though metallic cation's are more inert than anions, some significant changes were still observed for the behavior of K^+, Ca^{2+}, and Mg^{2+}. This is different from the one reported by Peden and Skowron (1978). For Na^+, the observed trend differs from their work and gave a relatively narrow interval of concentration ratio.

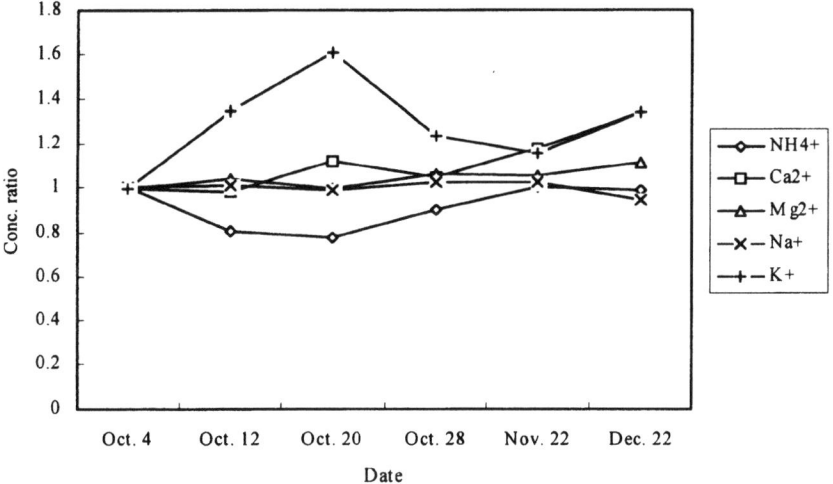

Figure 4. The mean concentration ratios for cations in case 1.

4. Conclusions

The results of this study indicate that the concentrations of H^+, K^+, Ca^{2+}, Mg^{2+}, NH_4^+, NO_3^-, and SO_4^{2-} change quantitatively during the analysis period for both filtered and unfiltered samples. The comparison of measured ionic concentrations corresponding to different storage methods yield some difference. The amount of NO_3^- increase and of NH_4^+ decrease with time was observed to be quite similar during an H^+ concentration increase of 25%.

References

Chau, W. H., Tang, A. J. S., Chung, D. H. S., and Reid N. W.: 1987, *Environ. sci. technol* **21**, 1219.

Galloway, J. N. and Likens, G. E.: 1976, *Water air soil pollut.* **6**, 241.

Hansen, D. A. and Hidy, G. M.: 1982, *Atmos. Environ.* **16**, 2107.

Peden, Mark E. and Skowron, Loretta M.: 1978, *Atmos. Environ.* **12**, 2343.

Jeng, Fu-Tien: 1999a, *Final Report of Survey Study on Acidic Deposition in Taiwan (in Chinese)*, Taiwan Environmental Protection Agency.

Jeng, Fu-Tien: 1999b, *Quality Assurance/Quality Control Program of Survey Study on Acidic Deposition in Taiwan(in Chinese)*, Taiwan Environmental Protection Agency.

Topol, L. E., Lev-On, M., and Pollack, A. K.: 1987, 'Comparison of Weekly and Daily Wet Deposition Sampling Results', in Russell W. Johnson, Glen E. Gordon, William Calkins, and A. Z. Elzerman (ed.), The Chemistry of Acid Rain – Sources and Atmospheric Processes, American Chemical Society, Washington, DC. pp. 229-241.

Vesely, Josef: 1990, *Atmos. Environ.* **24A**, 3085.

QUALITY CONTROL AND ITS CONSTRAINTS DURING THE PREPARATORY-PHASE ACTIVITIES OF THE ACID DEPOSITION MONITORING NETWORK IN EAST ASIA (EANET)

TSUNEHIKO OTOSHI[*], NORIO FUKUZAKI, HU LI, HIROSHI HOSHINO, HIROYUKI SASE, MASASHI SAITO and KATSUNORI SUZUKI

Acid Deposition and Oxidant Research Center, 1182, Sowa, Niigata, 950-2144 Japan
(* author for correspondence, e-mail: eanet@adorc.gr.jp)

Abstract. Participating countries of the Acid Deposition Monitoring Network in East Asia (EANET) launched the preparatory-phase activities from April 1998. For the recognition and improvement of the analytical precision and accuracy, the Interim Network Center (INC) carried out the inter-laboratory comparison on the analysis of artificial rainwater samples and soil samples. Relevant laboratories submitted their analytical data to the INC for the evaluation. Submitted data were summarized and evaluated in terms of precision and accuracy, and were compared with the Data Quality Objectives (DQOs) of EANET. These inter-laboratory comparisons made clear the present conditions of laboratories as well as the major constraints that should be solved in the future.

Keywords: inter-laboratory comparison, East Asia, monitoring network, rainwater, soil

1. Introduction

The First Intergovernmental Meeting on the Acid Deposition Monitoring Network in East Asia (EANET) held in March 1998 agreed to implement the preparatory-phase activities of the Network on an interim basis aimed at facilitating common understanding of the state of acid deposition in the region. Ten participating countries – China, Indonesia, Japan, Malaysia, Mongolia, Philippines, Republic of Korea, Russia, Thailand, and Viet Nam– have been carrying out monitoring on wet deposition, dry deposition, soil and vegetation, and inland aquatic environment, based on their own national monitoring plans. Though monitoring methodologies follow the Monitoring Guideline and the Technical Manuals of the Network, quality control of the data is one of the most important issues of the Network. In this context, the Quality Assurance/Quality Control (QA/QC) Programs during the preparatory-phase were developed.

The inter-laboratory comparison (round robin analysis survey of uniformly prepared artificial rainwater samples) was conducted among the analytical laboratories of EANET, as one of the QA/QC activities of the Network. The major purpose of the inter-laboratory comparison is to recognize the analytical precision and accuracy of the data in each participating laboratory, and give an opportunity to improve the quality of the analysis. Artificial rainwater samples that contain major ions were prepared and distributed by the Interim Network Center (INC). They are the first (1998) and the second (1999) trial of the inter-laboratory comparison in EANET, though there are inter-laboratory comparison projects with long history (WMO/GAW, 2000). Most of the participating laboratories submitted their analytical data, and obtained data for pH, EC, concentrations of SO_4^{2-}, NO_3^-, Cl^-, Na^+, K^+, Ca^{2+}, Mg^{2+}, and NH_4^+ were compared with prepared values and statistically treated. In the first trial, many outlying data caused by the incorrect calculations were submitted. In the second trail, remarkable improvement was seen in terms of numbers of outlying data. However, there are still constraints on analysis such as use of analytical tool that has poor sensitivity. This paper discusses the results obtained through these inter-laboratory comparisons, and pointed out constraints that should be solved in the future. Summary of the first attempt of inter- laboratory comparison on soil chemical analysis is also introduced.

2. Methods

2.1. INTER-LABORATORY COMPARISON ON ARTIFICIAL RAINWATER

INC prepared two artificial rainwater samples with different concentrations, and shipped them to all of participating laboratories (24 laboratories in 1998, and 23 laboratories in 1999) in EANET. Individual sample is contained in a clean 100ml polypropylene bottle, and the concentration of the sample is determined after diluted by de-ionized water exactly 100 times just before the analysis in each laboratory. Participating laboratories were expected to use analytical methods that are specified in the "Technical Manual for Wet Deposition Monitoring in East Asia" (The Second ISAG Meeting of EANET, 2000). Data on the concentrations of each constituent were submitted to INC as an average value of triple measurements of three aliquots collected from each sample bottle. The information on the analytical precision and accuracy on individual parameters can be obtained through the statistical treatment of submitted analytical data.

Submitted data are compared with prepared value in terms of the Data Quality Objectives (DQOs). DQOs on accuracy and precision of data obtained by the preparatory-phase activities of EANET were specified for every constituent as +/-15% in the "QA/QC program". The flag "E" is put on the data that exceed by a factor of 2 of the DQO (+/-15%~+/-30%), and the flag "X" is put on the data that exceed more than a factor of 2 of the DQO (<-30% or >30%).

Submitted data are also checked for ion balance (R_1) and conductivity agreement (R_2) as follows, where "C" is sum of cations (micro equivalent per liter), "A" is sum of anions (micro equivalent per liter), "Λ calc" is calculated electric conductivity, and "Λ meas" is measured electric conductivity.

$R_1 = 100 \times (C-A) / (C+A)$ (%)
$R_2 = 100 \times (\Lambda\,calc - \Lambda\,meas) / (\Lambda\,calc + \Lambda\,meas)$ (%)

Calculated R1 and R2 values are compared with acceptable levels of R1 and R2 specified in the "QA/QC program", and the flag "I" is put on the data sets that have poor ion balance, and the flag "C" is put on the data sets that have poor conductivity agreement.

Items of inter-laboratory comparison include of not only analytical results but also date of analysis, analytical methods and equipment used, years of experiences of the person in charge, and so on. Gathering above information, possible effects of these conditions to the data deviation are discussed.

2.2. INTER-LABORATORY COMPARISON ON SOIL CHEMICAL ANALYSIS

As one of the QA/QC activities, two kinds of soil samples (Acrisol and Gleysol) were distributed to participating laboratories that are in charge of chemical analysis of soil in EANET. The soil samples were collected in Japan which are recognized as typical high sensitive and low sensitive soil types in the region. Each sample is a fine soil powder contained in 500ml polyethylene bottle that are air-dried, sieved, mixed well, and sterilized using radio isotope. Each laboratory is expected to analyze pH, exchangeable base cations, and cation exchangeable capacity (CEC) as mandatory items, and other parameters as optional ones. Measurements of each parameter are carried out twice and then average values are determined.

3. Results and discussion

3.1. THE FIRST INTER-LABARATORY COMPARISON ON ARTIFICIAL RAINWATER ANALYSIS

Statistics that were calculated for each constituent of the artificial rainwater samples were : Average, Minimum (Min.), Maximum (Max.), Standard deviation (S.D.), and Number of data (N). For the calculation, outlying data that apart greater than a factor of 3 of S.D. from the Average were not included. As shown in Table I, average of submitted data were fairly well agreed with the prepared value/ concentration within a range of ±10%.

TABLE I
Summary of analytical results of the artificial rainwater samples (the 1st inter-laboratory comparison), (ADORC, 1999), (Reported data after outliers were removed)

Constituents	Prepared	Average	S.D.	N	Min.	Max.
[Sample No.1]						
pH	4.05	4.06	0.04	22	3.97	4.12
EC(mS/m)	7.94	7.27	0.71	24	5.16	7.86
SO_4^{2-}(μmol/L)	83.5	82.1	7.80	23	64.0	99.0
NO_3^-(μmol/L)	93.3	90.7	9.08	22	66.0	109.0
Cl^-(μmol/L)	129.0	126.9	9.02	22	110.0	149.0
Na^+(μmol/L)	95.8	93.2	7.87	21	71.6	107.0
K^+(μmol/L)	11.1	10.6	1.75	23	7.9	14.0
Ca^{2+}(μmol/L)	41.1	39.0	8.08	23	18.6	60.3
Mg^{2+}(μmol/L)	13.1	12.3	3.41	22	0.8	18.2
NH_4^+(μmol/L)	84.8	77.3	22.8	24	15.8	122.0
[Sample No.2]						
pH	4.51	4.53	0.06	22	4.41	4.69
EC(mS/m)	2.82	2.68	0.15	23	2.22	2.85
SO_4^{2-}(μmol/L)	29.1	28.1	3.34	22	19.3	36.0
NO_3^-(μmol/L)	36.1	35.6	3.78	23	27.0	43.6
Cl^-(μmol/L)	45.1	42.1	6.66	23	25.3	57.4
Na^+(μmol/L)	33.5	32.9	3.55	22	26.7	39.6
K^+(μmol/L)	7.4	7.0	1.56	23	3.3	10.1
Ca^{2+}(μmol/L)	14.3	14.5	2.93	23	7.5	20.7
Mg^{2+}(μmol/L)	4.6	4.5	1.20	22	0.4	6.2
NH_4^+(μmol/L)	29.5	27.2	8.70	24	3.2	43.2

(note) Prepared : Calculated vales from the amount of chemicals used for the preparation of samples.

Data on pH and EC varied less compared with other ionic constituents (Fig. 1). Measured data on pH were slightly higher than the prepared value. On the other hand, measured data on EC were slightly lower than the prepared value. Cause of this discrepancy is not clear by the obtained results. Analytical data of ionic constituents varied particularly for cations (Na^+, K^+, Ca^{2+}, Mg^{2+}, and NH_4^+). The cause of large deviation of analytical data for some cations (K^+, Ca^{2+}, and Mg^{2+}) was supposed to be the difficulty of analysis on lower concentration constituents.

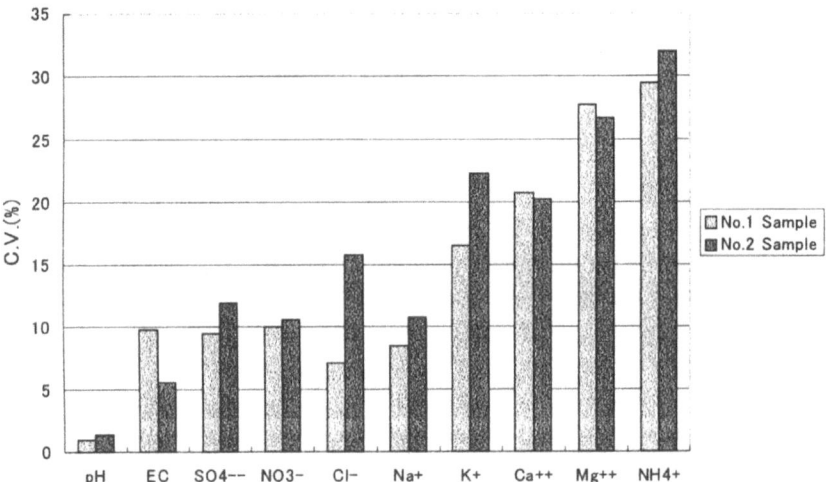

Figure 1. Coefficient of variation of each constituent in the 1st inter-laboratory comparison of rainwater (ADORC, 1999)

For the sample No.1 (higher concentration), about 78% of data were within DQOs, and about 12% of data were flagged by "E", and about 10% of data were flagged by "X" (TableII). Number of flags "E" for sample No.1 and sample No.2 (lower concentration) were almost similar. However, more data for sample No.2 were flagged "X" compared with data for sample No.1. It indicates the difficulty of the analysis of lower concentration constituents, particularly cations. For sample No.2, data within DQOs are about 75% in total.

Most of the participating laboratories employed recommended methods of EANET, particularly for measurements and analysis. Ion chromatography was a major analytical method adopted by the participating laboratories for chemical analysis of both

TABLE II

Number of flagged data for the Sample No.1 in the 1st inter-laboratory comparison of rainwater (ADORC, 1999) (*E : Value Exceeded the DQO by a factor of 2, *X : Value Exceeded the DQO more than a factor of 2)

Flag*	pH	EC	SO_4^{2-}	NO_3^-	Cl^-	Na^+	K^+	Ca^{2+}	Mg^{2+}	NH_4^+
E	0	2	4	3	1	1	10	1	1	4
X	0	1	1	2	2	3	1	4	6	4
Flagged (%)	0.0	13.0	21.7	21.7	13.0	17.4	47.8	21.7	30.4	34.8

anions and cations. Atomic absorption spectrometry for Na^+, K^+, Ca^{2+}, and Mg^{2+}, and spectrophotometry for NH_4^+ were also used for the determination of these cations. In general, much difference of data was not found among different analytical methods. However, some laboratories have to solve a problem of poor analytical sensitivity because of using traditional methods such as titration.

Measurement and analysis of rainwater samples were carried out by only one staff in 12 laboratories. In other laboratories, measurement was carried out by 2 to 4 staffs, and usually their responsibility were separated by the method used for analysis such as anions and cations. In more than half of laboratories, experiences of staff were less than 5 years. However, by information obtained through this project, clear evidence of data quality improvement was not found in terms of "years of experience of the staff".

3.2. THE SECOND INTER-LABORATORY COMPARISON ON ARTIFICIAL RAINWATER ANALYSIS

For the second inter-laboratory comparison, compared with the first one, similar but different concentration samples were distributed to the participating laboratories. Same as the first inter-laboratory comparison, analytical precision and concentration of constituents have a close relationship (Fig. 2). However, data within DQOs increased in the second inter-laboratory comparison, even concentrations of some constituents were lower than the first one. Data within DQOs at the first inter-laboratory comparison were between 75% to 80%, and around 90% of data were within DQOs in the second one (Table III).

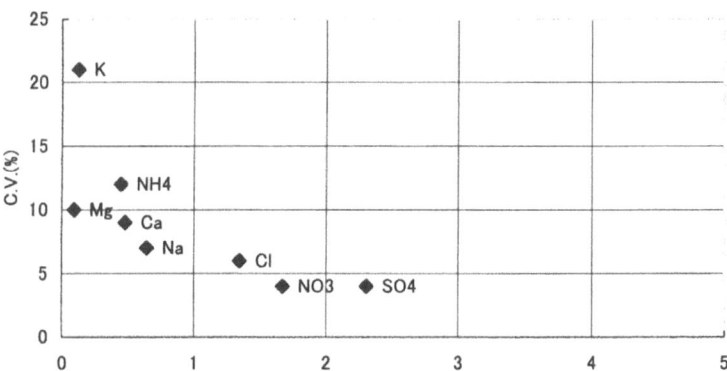

Figure 2. Relationship between concentration and the coefficients of variation (No.2 sample of the 2nd inter-laboratory comparison of rainwater)(ADORC, 2000)

TABLE III
Improvement of data quality between 1st and 2nd inter-laboratory comparison on rainwater

	Sample	Data within DQOs	Data with flag "X"	Data with flag "E"
1st inter-laboratory comparison	No.1	78%	10%	12%
	No.2	75%	13%	12%
2nd inter-laboratory comparison	No.1	95%	1%	4%
	No.2	89%	4%	7%

3.3. THE FIRST INTER-LABORATORY COMPARISON ON SOIL ANALYSIS

The first Inter-laboratory comparison on soil analysis was carried out in 1999. This kind of inter-laboratory comparison on soil analysis was the first attempt in East Asian region although similar program was carried out in Europe (EC-UN/ECE, 1997). Analytical results were submitted to INC from around ten laboratories. These data well represented different chemical composition of two soil samples. Average $pH(H_2O)$ for Acrisol was around 5, while average $pH(H_2O)$ for Gleysol was around 7, and amounts of exchangeable base cations and CEC were larger in Gleysol samples. In general, because of pretreatment of soil chemical analysis including extraction process, variation of data became wider than rainwater analysis.

4. Conclusions

In this paper, we have described the outline of the inter-laboratory comparison of chemical analysis on artificial rainwater and soil samples as part of the QA/QC activities of EANET. In the first attempt of inter-laboratory comparison, many data that had poor accuracy were submitted. Most of these problems were based on the incorrect calculation or incorrect conversion of unit, and were improved in the second inter-laboratory comparison. However, some laboratories are still using analytical methods that have poor detection limits. These points should be solved in the future to obtain reliable data on rainwater samples. We have just launched the inter-laboratory comparison on chemical analysis of soil, and evaluation of the results is still in progress. However, as seen in the example of artificial rainwater inter-laboratory comparison, accumulation of experiences is important particularly for the recognition and improvement of the data quality of the Network in preparatory-phase.

Acknowledgements

We would like to thank participating laboratories of the preparatory-phase activities of EANET, particularly for their contribution to inter-laboratory comparisons. We would also like to acknowledge the members of Interim Scientific Advisory Group for their useful comments.

References

Acid Deposition and Oxidant Research Center (ADORC): *Inter-laboratory Comparison Project 1998 (Round robin analysis survey) First Attempt* (1999).
Acid Deposition and Oxidant Research Center (ADORC): *Inter-laboratory Comparison Project 1999 (Round robin analysis survey) Second Attempt* (2000).
EC-UN/ECE, Interlaboratory variability. In: *Forest soil condition in Europe, Results of a large-scale soil survey*, Brussels, Geneva. EC-UN/ECE, p.201-239 (1997).
Technical Documents for Wet Deposition Monitoring in East Asia, adopted at: The Second Interim Scientific Advisory Group (ISAG) Meeting of EANET, March 2000.
WMO/GAW: *Report of the 22nd Intercomparison of WMO/GAW Precipitation Chemistry Laboratories*, WDCPC No.5, February 2000.

COST-EFFECTIVE ABATEMENT OF ACIDIFYING EMISSIONS WITH FLUE GAS CLEANING VS. FUEL SWITCHING IN FINLAND

N. KARVOSENOJA[1*], P. HILLUKKALA[2], M. JOHANSSON[1] and S. SYRI[1]

[1]*Finnish Environment Institute, P.O.Box 140, FIN-00251 Helsinki, Finland*
[2]*Tampere Power Utility, P.O.Box 175, FIN-33101 Tampere, Finland*
(*Author for correspondence, email: niko.karvosenoja@vyh.fi)

Abstract. Acidifying emissions from energy production and industry have decreased considerably during the last two decades in Finland. Especially the emissions of sulphur dioxide have dropped sharply with 85% in 1980-1998, although the energy use has increased 30% during the same period. The reduction has occurred through two mechanisms: by replacing the combustion of heavy fuel oil with cleaner energy carriers, and by direct emission reduction controls, *e.g.* flue gas desulphurization. In this study the Finnish cost curves for SO_2 and NO_x were first calculated to produce a consistent comprehensive view on further emission reduction costs and potentials. The data on technical and cost-related parameters were based on actual national experiences from power plants and industry. Most of the cost-efficient sulphur emission controls were already in use. For NO_x, a large share of further reduction potential still remained. Second, a case on the emission reductions and costs for fuel switching in a 205 MW_{th} peat power plant of Tampere Power Utility in Finland was studied. Fuel switching to natural gas was found less cost-efficient in SO_2 and NO_x emission reduction when compared to flue gas cleaning techniques. The findings provided new information on fuel switching as an alternative potential reduction measure, which is not considered in international assessments.

Keywords: air, cost curves, emissions, fuel switching, reduction costs

1. Introduction

Energy consumption in power production and industry has increased more than 30% in Finland during the 1980s and 1990s (Figure 1a). Nevertheless the acidifying emissions have decreased considerably during the same period; for sulphur dioxide 85% and nitrogen oxides 26% (Figure 1b and c). The sharp decrease in sulphur emissions has taken place by two mechanisms. During the 1980s the emissions have decreased mainly by measures which do not necessarily demand any direct investments, such as replacing the combustion of heavy fuel oil with cleaner energy carriers, *e.g.* natural gas and nuclear power. In the late 1980s and the 1990s the reduction has occurred mainly due to direct investments on emission control equipment, *e.g.* flue gas desulphurization.

In Finland the national emission reduction programmes, induced primarily by the international obligations (*e.g.* UN/ECE, 1979), have been prepared in extensive Committees with representatives from relevant Ministries, industry and NGOs (*e.g.* Acidification Committee, 1998). The reduction proposals have been designed based on cross-sectoral cost-effectiveness analysis, and they have been agreed upon in the Committees, which has enhanced the commitment of the groups affected by the stricter norms. The Committees have utilized estimates of reduction costs and average efficiencies provided by the power plant operators and industry in the design of the reduction programmes. National-scale integrated assessment models have been used in estimating the environmental impacts and additional reduction needs.

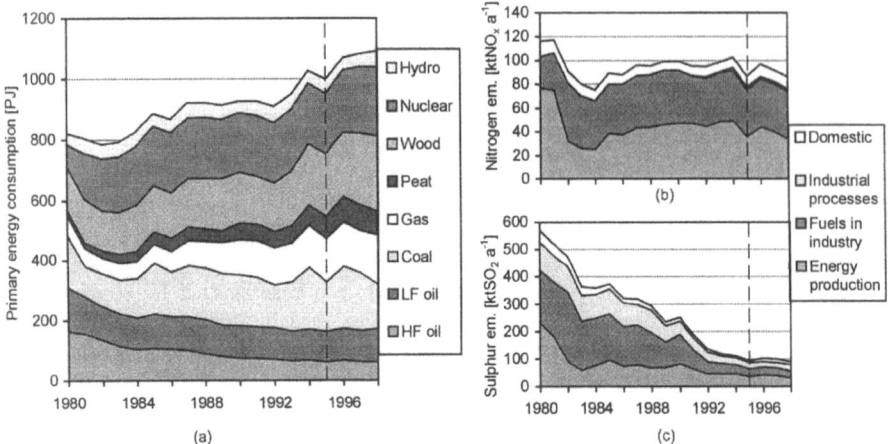

Figure 1. (a) Primary energy consumption in stationary sources by fuels, (b) NO$_x$ and (c) SO$_2$ emissions in stationary sources by emission sources in Finland in 1980 – 1998 (Statistics Finland, 1999; Statistics Finland at http://www.tilastokeskus.fi/tk/yr/yekoti.html)

The integrated assessment modelling of acidification in Finland has been carried out mainly at the Finnish Environment Institute (Johansson, 1999). The Finnish critical load integrated model (CLIM) includes the main elements from the international models: (i) emissions inventories and projections, (ii) deposition levels due to domestic sources and long-range transport, (iii) critical loads for forest soils and lakes, which describe the long-term tolerance of the ecosystems against pollutant loading. The calculations are carried out as scenario analyses, where the impacts due to emissions are assessed. An optimization mode is not available. The cost curves were included in the model system at a late stage (Johansson *et al.*, 2000). The effect of uncertainties in the cost curves to the integrated model uncertainty has not yet been assessed nationally (Johansson, in press).

This study assesses the measures to control emissions of sulphur and nitrogen oxides from power production and industry in Finland. The method is based on cost curves, which provide aggregated information on the control potential and related costs of individual control options for selected activity sector and fuel combinations. The method and data are presented shortly here and in more detail in Karvosenoja and Johansson (1999). The feasibility and costs of fuel switching, which is not considered in international assessments, are discussed based on a case study.

2. Methodology and the scope of the study

2.1 EMISSION AND COST CALCULATION

The emissions $em_{i,j,k,l}(t)$ in the critical load integrated model (CLIM) are calculated from activity data $a_{j,k}(t)$ and unabated emission factors $ef_{i,j,k}$. Several emission control technologies with removal efficiencies $\eta_{i,j,k,l}$ can be applied to each economic sector - fuel type combination (Table I) with defined penetration percentages $x_{i,j,k,l}(t)$. The emissions are:

$$em_{i,j,k,l}(t) = \sum_j \sum_k \sum_l (1 - \eta_{i,j,k,l}) \cdot x_{i,j,k,l}(t) \cdot a_{j,k}(t) \cdot ef_{i,j,k} \qquad (2.1)$$

where t = time step, i = pollutant, j = fuel, k = sector and l = control technology.

For each emission control technology the unit cost $uc_{i,j,k,l}$ is calculated from technical and cost parameters of control technologies. Total emission control costs $c_{i,j,k,l}(t)$ are calculated from reduced emission quantities and unit costs.

$$c_{i,j,k,l}(t) = \sum_j \sum_k \sum_l uc_{i,j,k,l} \cdot \eta_{i,j,k,l} \cdot x_{i,j,k,l}(t) \cdot a_{j,k}(t) \cdot ef_{i,j,k} \qquad (2.2)$$

In this study, the activity data for 1995 were determined using national statistics (Statistics Finland, 1999) (dashed line in Fig 1a). The unabated emission factors were based on national estimates (Acidification Committee, 1998; Statistics Finland, 1999; Boström *et al.*, 1992). The technical and cost parameters of control technologies were based on actual operation experiences of control technologies in use in Finnish power plants and industry during 1990s (Ministry of the Environment, 1998; 1995). The emission control investments were annualized using 20 years lifetime and 4% discount rate.

2.2 COST CURVES

A cost curve illustrates how to achieve a certain emission reduction with least cost, using the optimal cost-abatement combination. Technologies are ranked according to their costs for removing the last unit of emissions (marginal cost). A cost curve is usually compiled starting from a realized/predicted or a hypothetical "no-control" emission level of a chosen year. The composed curve is piece-wise linear, with individual segments determined by both the costs of applying the various technologies and the reduction potentials of the technologies in various sectors. In this study the Finnish cost curves were constructed for SO_2 and NO_x emission abatement in stationary sources in 1995.

TABLE I
(a) SO_2 and (b) NO_x control options for the power production and industrial sector applicable in the critical load integrated model (CLIM)

(a) SO_2 control options	(b) NO_x control options
Process emissions control	*Process emissions control*
Flue gas desulphurization (FGD)	*NO_x reduction in fuel combustion sectors*
• limestone injection in fluidized bed boilers	• combustion modifications (CM) in all the boilers and gas turbines
• wet FGD in boilers over 400 MW$_{th}$	
• spray dry scrubbers in boilers over 50 MW$_{th}$	• selective catalytic reduction (SCR) combined with CM in coal power plants
• NaOH-based scrubbers in black liquor recovery boilers	• selective non-catalytic reduction (SNCR) combined with CM in boilers over 50 MW$_{th}$
Low sulphur fuels	
• heavy fuel (HF) oil (1.0%S)	
• light fuel (LF) oil (0.045%S)	

2.3 CASE STUDY

The Naistenlahti 1 of Tampere Power Utility was started up in 1971 as a heavy fuel oil power plant to supply district heat and electricity. In 1982 it was converted to peat. In recent years, the existing boiler has become to the end of its economic lifetime, and modernization investments had to be made. In 1998 it was decided to replace the boiler with a new combined cycle gas turbine (CCGT) power plant, making use of the existing steam turbine and other equipment in the steam circuit. The conversion to natural gas will decrease sulphur and nitrogen emissions, and the use of CCGT increases the power-heat ratio from about 0.5 to nearly 1, thus cutting the emissions further.

In this study, two alternative ways to modernize Naistenlahti 1 power plant were studied: (1) conversion of the existing peat-fired boiler to natural gas, and (2) building a new CCGT power plant, *i.e.* the realized case. The differences in annual energy production costs and emissions of SO_2 and NO_x between the old plant and new alternatives indicate the cost-efficiencies of emission reductions in the two fuel switching cases. The first alternative would not require large investments, and the power and heat production capacities would remain in the same level as in the old plant, *i.e.* 62 MW_e for electricity and 120 MW_h for heat delivered to district heating system. The difference of annual fuel costs of using peat *vs.* natural gas indicates the cost difference of energy production.

In the realized case, the investment of the new CCGT plant was 51 M€. Both power and heat output were increased to 128 MW_e and 140 MW_h. In addition to the fuel costs, investment costs and income from the heat and electricity sales were included in the cost comparison. Current national values were used for prices of fuels (8.4 and 13 € MWh^{-1} for peat and gas) and heat (29 € MWh^{-1}) (Ministry of Trade and Industry, 2000). For the electricity the expected market price in 2003 (18 € MWh^{-1}) was used, instead of the current price (14 € MWh^{-1}, 12 month average) which is exceptionally low due to high availability of hydro power in the Nordic markets (pers. comm., K. Höysniemi, EL-EX Nord Pool Finland, 16 May 2000). The same discount rate and lifetime of investments as in the cost curve calculation were used. In the emission comparisons, the calculated annual emissions from the two fuel switching alternatives were compared with the cases, where the same amount of primary energy would have been combusted in the old plant. The same annual operating hours at full load 4 800 h a^{-1} as in the cost curve calculation for peat power plants were used.

3. Results

3.1 FINNISH COST CURVES FOR THE YEAR 1995

The Finnish national SO_2 and NO_x cost curves for stationary sources for 1995 were compiled taking into account the actual emission control situation (Figure 2). In both figures, the first curve corresponds to control measures which were already adopted to the energy production system in 1995, starting from the "no-control" situation and ending up with the actual emissions and control costs. The second curve represents additional abatement potential and costs.

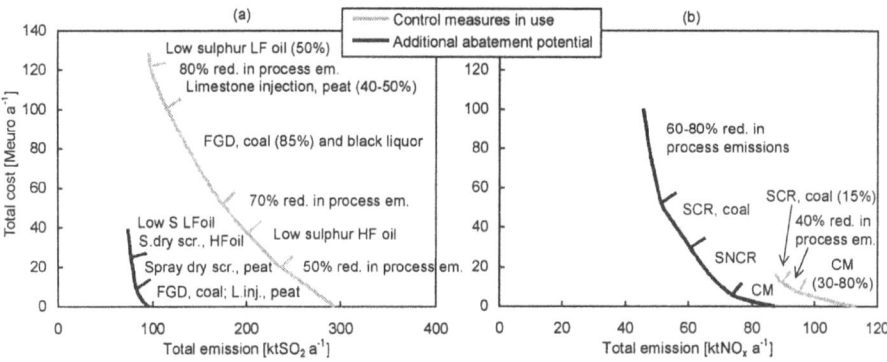

Figure 2. The cost curves for (a) SO_2 and (b) NO_x for stationary sources in Finland compiled for the actual emission control situation in 1995. Partial penetrations for control measures in use are given in brackets.

Almost all of the cost-efficient reduction methods for SO_2 emissions were already in use in 1995. Achieved emission reduction was 198 $ktSO_2$ a^{-1}. The corresponding reduction costs were 129 M€ a^{-1}, *i.e.* the average cost-efficiency was about 650 € tSO_2^{-1}. Further emission reduction potential was 20 $ktSO_2$ a^{-1}, with the cost of 40 M€ a^{-1}, *i.e.* the average of 2 000 € tSO_2^{-1}. For the NO_x emission reduction, the most cost-efficient methods were only partly implemented, and additional potential remained in all sectors. The emission reduction was 26 $ktNO_x$ a^{-1}, and the reduction costs 16 M€ a^{-1}, *i.e.* the average cost was 610 € tNO_x^{-1}. Further emission reduction potential was 42 $ktNO_x$ a^{-1}, and the emission reduction cost 100 M€ a^{-1}, the average of 2 400 € tNO_x^{-1}.

3.2 CASE STUDY

The direct conversion of peat-fired boiler to gas in Naistenlahti 1 would have decreased average emission factors for sulphur from 120 $mgSO_2$ MJ^{-1} to zero. Nitrogen oxides emissions would have remained at about the same level 150 $mgNO_x$ MJ^{-1}. This would have reduced annual sulphur emissions with 430 tSO_2 a^{-1} assuming 4 800 h a^{-1} operating hours at full load. The increase in annual total costs would have been 3.1 M€ a^{-1}. The emission factors of the new CCGT plant will be zero for sulphur and 60 $mgNO_x$ MJ^{-1} for nitrogen. This would decrease annual emissions 620 tSO_2 a^{-1} and 465 tNO_x a^{-1}, and increase costs 4.0 M€ a^{-1}.

The cost-efficiencies for emission reductions in the cases above were 7 200 € tSO_2^{-1} for direct conversion, and 6 500 € tSO_2^{-1} and 8 700 € tNO_x^{-1} for CCGT. The costs are high compared with the cost-efficiencies of flue gas cleaning on peat fired boilers estimated to be 2110 € tSO_2^{-1} for spray dry scrubbers, and 214 and 903 € tNO_x^{-1} for combustion modifications and SNCR, respectively (see Figure 2) (Karvosenoja and Johansson, 1999).

At national level, full application of fuel switching from peat to CCGT at industrial boilers and power plants sector would result in 14.7 $ktSO_2$ a^{-1} and 8.0 $ktNO_x$ a^{-1} less emissions and 108 M€ a^{-1} more costs. However, the applicability potential is restricted by gas distribution network and other factors.

4. Discussion and conclusions

Acidifying emissions from Finnish power plants and industry have decreased during the last two decades. Investments made on sulphur emission control during the late 1980s and the 1990s have led to an intense drop on emissions. The technical background work of the national emission reduction programmes, partly utilizing integrated assessment modelling approach, has substantially supported national planning of cost-effective emission reductions. Based on the cost curves constructed in this study, almost all of the cost-efficient reduction methods for sulphur emissions were already in use in 1995. Further reduction potential of acidifying emissions remained mainly for nitrogen oxides.

Fuel switching from peat to natural gas in the studied cases was less cost-efficient in SO_2 and NO_x emission reduction when compared to direct emission reduction controls. However, the fuel switching will also reduce the greenhouse gas emissions from 105 gCO_2 MJ_{th}^{-1} of peat to 56 gCO_2 MJ_{th}^{-1} of gas. It should be noted that, in reality, the old Naistenlahti 1 had to be modernized anyway, and the reduction of acidifying emissions was not the main reason for the investment decision.

The case study provided new information about fuel switching and its costs, which are not included in international integrated assessment models used in the negotiations under UN/ECE and EU. These results together with the national cost curves enhanced the composition and assessment of further national reductions of acidifying emissions in the Finnish integrated assessment modelling.

References

Acidification Committee: 1998, 'Report of the Acidification Committee', Ministry of the Environment, Helsinki, Finland, *The Finnish Environment* **219**. (In Finnish.)
Boström, S., Backman, R. and Hupa, M.: 1992, *Greenhouse gas emissions in Finland 1988 and 1990*, Insinööritoimisto Prosessikemia Ky, Turku, Finland.
Johansson, M.: 1999, 'Integrated models for the assessment of air pollution control requirements', Doctoral dissertation, *Monographs of the Boreal Environment Research* **13**.
Johansson, M.: in press, 'Variability in three Finnish integrated acidification models', *Water, Air, and Soil Pollution*.
Johansson, M., Ahonen, J., Amann, M., Bartnicki, J., Ekqvist, M., Forsius, M., Karvosenoja, N., Lindström, M., Posch, P., Suutari, R. and Syri, S.: 2000, 'Integrated environmental assessment modelling, Final report of the Finnish subproject, EU/LIFE project', Finnish Environment Institute, Helsinki, Finland, *The Finnish Environment* **387**.
Karvosenoja, N. and Johanson, M.: 1999, 'National cost curve analysis for SO_2 and NO_x emission control', Finnish Environment Institute, Helsinki, Finland, *The Finnish Environment* **362**.
Ministry of the Environment: 1998, 'Possibilities to reduce sulfur and nitrogen oxide emissions', Helsinki, Finland, *The Finnish Environment* **251**. (In Finnish with English summary.)
Ministry of the Environment: 1995, 'Reduction of nitrogen oxide emissions in energy production', Helsinki, Finland, *Report* **1/1995**. (In Finnish with English summary.)
Ministry of Trade and Industry: 2000, *Energy Review 1/2000*, Helsinki, Finland.
Statistics Finland: 1999, 'Energy Statistics 1998', Helsinki, Finland, *Energy* **1999:2**.
UN/ECE: 1979, *Convention on Long-Range Transboundary Air Pollution and the Protocols to the Convention*, United Nations, New York, Geneva.

WET DEPOSITION TO AN UPLAND AREA OF ENGLAND IN 1988 AND 1999: MEASUREMENTS AND MODELLING

DRIEJANA[1,3*], D. W. RAPER[1], D. S. LEE[2], R. D. KINGDON[2] and I. L. GEE[1]

[1] aric, *Dept. Environmental and Geographical Sciences, Manchester Metropolitan University, John Dalton Building, Chester St., Manchester M1 5GD, U.K*; [2] *Defence Evaluation and Research Agency, Propulsion and Performance Dept., Pyestock, Farnborough, Hants. GU14 OLS, U.K.;*
[3] *Dept. Environmental Engineering, Institute of Technology Bandung, Bandung 40132, Indonesia.*

(* author for correspondence, e-mail: r.driejana@mmu.ac.uk)

Abstract. Fine spatial-scale measurements of acidic deposition have been made over two time frames: 1987/1988 and 1999/2000. Over this period, dramatic reductions in precursor emissions have occurred. The two sets of data are compared in order to quantify deposition reductions. Orographic enhancement of wet deposition from the seeder-feeder mechanism is important in the UK at high-elevation sites such as these and the measurements were corrected for this process using a standard methodology. The data were also modelled using a simple long-term Lagrangian acid deposition model and comparisons made with both sets of observations at a resolution of 5 km _ 5 km. Future scenarios (2010) of emissions were modelled to examine the impact of UK and other European sources in order to determine optimal protection of ecosystems in this region.

Keywords: Wet deposition, seeder-feeder enhancement, modelling, emission reductions, UK.

1. Introduction

The Peak District, in the Northern England, is bordered by the major cities of Manchester to the west and Sheffield to the east. Consequently, it has a long history of acid deposition. In 1987/88, high deposition loadings of S and N were found and it was estimated that the buffering capacity was consumed or exceeded in 50% of the 20 km _ 30 km study area (Raper and Lee, 1996). Another one-year measurement campaign was undertaken in the same area in 1999/2000 in order to examine the impact of the emission reductions over the period. This paper discusses the impact of the past, current and future national and international precursor emission levels on this acid sensitive area. The measurement results from 1987/1988 are compared to those of 1999/2000 and future wet deposition loadings are modelled.

2. Site Location and Methodology

Ten measurement sites, operated on a weekly basis, were located in an area of approximately 25 km _ 30 km in May 1987/April 1988 and six were re-established in March 1999/February 2000. The site numbers in Fig. 1 correspond with the site details in Table I and II. The network uses bulk collectors identical to those used in the UK National Acid Deposition Network (Hall, 1986). The chemical species measured include H^+, Cl^-, NO_3^-, SO_4^{2-}, Ca^{2+}, Mg^{2+}, Na^+, K^+ and

NH_4^+. Details of sampling and analytical methodology have been given by Raper and Lee (1996) and Driejana *et al.* (2000a).

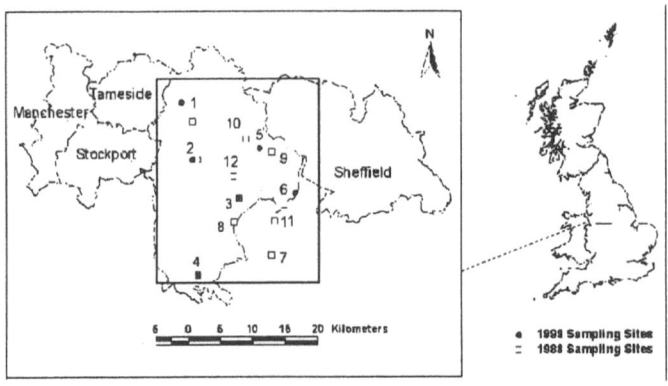

Figure 1. Location of the sampling sites and the two major conurbations

3. Annual Deposition in 1987/1988 and 1999/2000

Wet deposition from the 1987/1988 measurements (Table I) was calculated using rainfall collector volumes (Raper and Lee, 1996), while that for 1999/2000 (Table II) was calculated using rainfall data from the rain-gauge measurements. Rainfall amounts from bulk collectors are generally within 80-110 % of those collected by standard rain gauges (Raper, 1989). In comparing the two sets of measurements, two problems are apparent: firstly, year-to-year variability in rainfall amount can result in large differences in deposition. Secondly, in upland areas such as this the deposition measured at a particular location (which is generally at lower elevations for reasons of accessibility) does not represent the average deposition to a particular area because of the enhancement of wet deposition by the seeder-feeder process (Dore *et al.*, 1992). In order to overcome the year-to-year variations in rainfall amounts, the annual precipitation-weighted concentrations from the sites were multiplied by the long-term (1947–1970) 5 km rainfall. This allows a comparison of deposition for the different years that excludes the influence of rainfall amount. In order to account the seeder-feeder enhancement of wet deposition, the measurements were enhanced using the methodology proposed by Dore *et al.* (1992) and adopted by the United Kingdom Review Group on Acid Rain (RGAR, 1997). This involved separating the orographic and non-orographic components of rainfall (Dore *et al.*, 1992); for the orographic component the concentrations of ions in rain were multiplied by a factor of 2, a robust empirical factor based upon many studies in the U.K. (RGAR, 1997). This is hereafter referred to as enhanced measurements.

Comparison of 1987/1988 to 1999/2000 at the 4 sites which were located at the same position during both studies show that non sea-salt (nss) SO_4^{2-} annual precipitation-weighted mean concentrations (hereafter will be stated as concentrations) declined by 42–65% and NO_3^- by 22–42%. These reductions are of the same order as reductions in UK emissions over the same period of 56% and 31% for SO_2 and NO_x, respectively. Over the same period, European emissions of SO_2 declined by about 50% and by about 16% for NO_x.

However, deposition of S was reduced by 42%-52% between the study period. In addition, deposition of oxidised N was slightly reduced and at one site it even increased, with reductions between –8% to 15%. These differences may be partly explained by the higher rainfall amounts (approximately 30%) over the 1999/2000 measurement period. Enhanced deposition from both of the measurement periods were then calculated using the long-term rainfall in order to isolate only the effect of emission reduction. Using the long-term data, S deposition decreased by 53% to 62%, in line with the decrease in SO_2 emissions and in nss-SO_4^{-2} concentrations. However, the calculated oxidised N deposition decreased by 19% to 24%, which is less than the reduction in national emissions.

TABLE I

Annual Wet Deposition in 1987/1988 (kg ha^{-1} yr^{-1})

No.	Site Name	Alt. m	Conc. µeq l^{-1} SO_4^{2-}	Conc. µeq l^{-1} NO_3^-	Precip. mm*	S Dep.^{a} kg ha^{-1}	N Dep.^{a} kg ha^{-1}	S Dep.^{b} kg ha^{-1}	N Dep.^{b} kg ha^{-1}
1	Hurst	250	66.7	26.5	1069	11.4	4.0	20.6	7.8
2	Kinder	320	67.1	30.1	933	10.0	3.9	23.6	9.4
3	Oxlow House	435	87.1	38.2	934	13.0	5.0	16.4	6.4
4	Harpur Hill	380	84.3	35.8	881	11.9	4.4	12.2	4.4
7	Wardlow Hay Cop	350	80.6	37.9	896	11.5	4.7	7.8	3.2
8	Peak Forest	335	79.7	32.1	974	12.4	4.4	12.3	4.4
9	Derwent	270	69.2	33.3	988	10.9	4.6	14.0	5.6
10	Alport	270	63.9	30.6	1264	12.9	5.4	17.8	7.2
11	Hucklow	390	78.1	36.9	810	10.1	4.2	10.4	4.3
12	Edale	280	74.4	34.9	828	9.9	4.0	18.4	7.6

data taken from Raper and Lee (1996)
) mm rainfall from the network rainfall collectors; a) calculated using mm; b) calculated using long-term rainfall data

TABLE II

Annual Wet Deposition in 1999/2000 (kg ha^{-1} yr^{-1})

No.	Site Name	Alt. m	Conc. µeq l^{-1} SO_4^{2-}	Conc. µeq l^{-1} NO_3^-	Precip. mm*	S Dep.^{a} kg ha^{-1}	N Dep.^{a} kg ha^{-1}	S Dep.^{b} kg ha^{-1}	N Dep.^{b} kg ha^{-1}
1	Hurst	250	25.8	20.8	1334	5.5	3.9	9.0	6.3
2	Kinder	320	28.1	23.2	1297	5.8	4.2	10.2	7.4
3	Oxlow House	320	39.8	29.9	1109	7.1	4.6	7.7	5.1
4	Harpur Hill	380	29.3	20.9	1286	6.0	3.8	4.7	2.9
5	Lockerbrook	420	26.9	21.2	1359	5.9	4.0	6.3	3.9
6	Bamford	210	38.5	25.4	1062	6.6	3.8	5.7	3.3

) mm rainfall from the network rainfall collectors; a) calculated using mm; b) calculated using long-term rainfall data

4. The Potential Impact of Future Emission Reductions

A simple long-term Lagrangian acid rain model (Driejana et al., 2000b) was used to examine the effect of future emission reductions. The model uses the same long-term rainfall field to calculate annual average deposition. Emission data for the same years as the monitoring were not available, so emissions for 1992 (MOD92) were used and compared with the 1987/1988 enhanced measurement data (MEA88). The 1999/2000 measurement data (MEA99) were compared with modelled results, which used emission for 1996 (MOD96). Enhanced measured deposition values used for model evaluation were calculated using long-term rainfall. Model performance was first evaluated by comparing the deposition values in the grids where the measurement points were located (*Figs 2 to 5*) with the respective (grid-averaged) enhanced measured values.

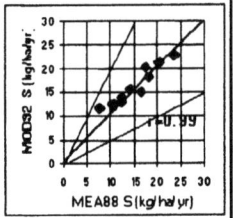

Figure 2. Enhanced measured 1988 (MEA88) vs modelled 1992 (MOD92) S deposition

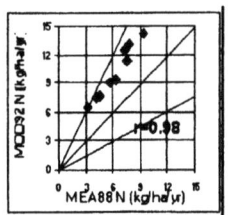

Figure 3. Enhanced measured 1988 (MEA88) vs modelled 1992 (MOD92) N deposition

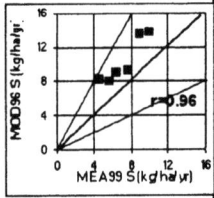

Figure 4. Enhanced measured 1999 (MEA99) vs modelled 1996 (MOD96) S deposition

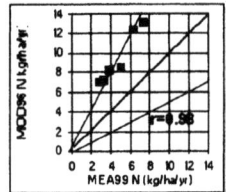

Figure 5. Enhanced measured 1999 (MEA99) vs modelled 1996 (MOD96) N deposition

Good correlations ($r > 0.95$) are shown between modelled and measured deposition, with modelled values fall within the factor of 2 lines. Total budgets in the study area are estimated from 5 km gridded deposition values. In term of S deposition budgets (Table III), MOD92 results of 1.17 ktonnes compared with the 1.04 ktonnes of MEA88. However, between 1988 and 1992 UK and European SO_2 emissions declined by ~9% and ~22%, respectively. The model estimate MOD92 was ~12% higher than MEA88 deposition.

The UK s SO_2 emissions in 2000 are predicted to be 33% lower those than in 1996, and those of European emissions are predicted to be at about the same level. MOD96 predicts S deposition of about 0.71 ktonnes, 58% higher than the MEA99 value of 0.45 ktonnes. With 1996 emission levels which were higher than those of 1999, the model over-predicts but it still less than a factor of two. Oxidised N deposition show a similar tendency to that of S deposition, MOD92 N deposition (0.69 ktonnes) is almost twice of that of MEA88 (0.40 ktonnes). A similar pattern was also shown when 1996 emission data were adjusted to the predicted 2000 emission levels.

TABLE III

Peak District measured and modelled total deposition budgets (ktonnes yr^{-1})

	MEA88	MEA99	MOD92	MOD96
S Deposition	1.04	0.45	1.17	0.71
N Deposition	0.40	0.30	0.69	0.64

Because measured and modelled values are relatively different in magnitude, spatial distributions were examined using normalised data (0 to 1 scale) rather than using the calculated deposition values. This comparison shows that despite the over prediction, the model predicted spatial distribution patterns that are in reasonably good agreement with the measurements, especially for N deposition. The 5-km modelled S deposition does not exactly reproduce the measurement field. Nevertheless, it still detects the high-deposition grids in the northern part of the study area and it shows variations in deposition loading better than the 20-km resolution. This validation exercise suggests that the model might be used for predicting deposition loadings with lower uncertainties than at 20 km resolution.

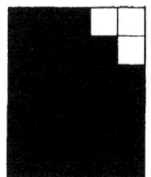
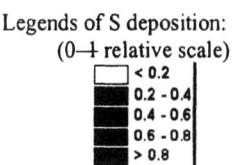

Figure 6. Normalised measured S deposition 1999

Figure 7. Normalised modelled S deposition 1996

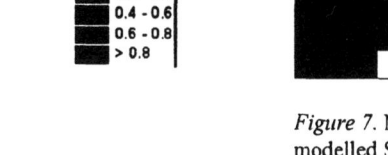
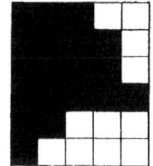

Figure 8. Normalised measured N deposition 1999

Figure 9. Normalised modelled N deposition 1996

The model was then used to predict deposition in 2010, using the current commitment (REF) and the most recent UNECE-LRTAP multi-effect protocol (the 1999 G teborg Protocol) scenarios. Under the REF and PRO scenarios, annual EMEP SO$_2$ emissions are reduced to 14,419 ktonnes and 13,975 ktonnes, respectively, and NO_x emissions to 14,567 ktonnes and 13,973 ktonnes. Annual UK emissions are expected to be 980 ktonnes and 625 ktonnes for SO_2 and 1,186 ktonnes and 1,181 ktonnes for NO_x for these respective scenarios. For the REF scenario, the model predicts reductions in S and N of 43% and 29%, respectively compared to 1996 budget levels. Using the PRO scenario, S and N deposition was calculated to be 55% and 29% less than the 1996 levels. If the proportions are applied to the current measurements, based on the REF scenario, annual S and N deposition is predicted to reduce to 0.25 ktonnes and 0.21 ktonnes in 2010. Under the PRO scenario, S deposition will be ~0.20 ktonnes, while N deposition is estimated to be the same as that from the REF scenario.

Raper and Lee (1996) described critical load exceedances for this area of the Peak District and predicted that in 2005 75% of the area was not subject to critical load exceedances (protected). The deposition levels predicted for 2005 are in fact of similar magnitude with the 1999 enhanced measured deposition. The scenario for 2010 is characterised by significant reductions of S and N depositions. It is therefore reasonable to expect that exceedances of critical loads will be confined to less that 25% of the study area.

5. Conclusions

Over the past decade, the nss-SO_4^{2-} and NO_3^- annual precipitation-weighted mean concentrations in rain and deposition of S and oxidized N have decreased substantially, in line with the decreases in precursor emissions. However, deposition is also dependent upon the year-to-year variation in precipitation.

Modelling showed that under the current agreement for emissions reductions for the year 2010 (REF scenario), S and N deposition will be reduced by 43% and 29%, respectively compared with 1996 levels. The latest protocol (PRO scenario) will result in greater reductions in S deposition compared with the REF scenario (55%), whilst reductions in oxidised N deposition will be similar. In terms of critical load exceedances, these reductions may increase the protected area to more than 75% of this acid-sensitive area the Peak District.

References

Dore, A. J., Choularton, T. W., Fowler, D.: 1992, *Atmospheric Environment* **26A**, 1375.
Driejana, Raper, D.W., Gee, I.L.: 2000a, *Spatial Variation and Source Attribution of the Chemical Composition in Acid Deposition in the Peak District, Northern England*. Proc. of the 93th Air & Waste Management Association s Annual Conference &Exhibition, Salt Lake City, UT.
Driejana, Lee, D. S., Kingdon, R. D., Raper, D. W. and Gee, I. L.: 2000b, *Water, Air and Soil Pollutions*, this volume.
Hall, D.J.; 1986, *The Precipitation Collector for Use in the Secondary National Acid Deposition Network*, Report LR 561 (AP)M, Warren Spring Laboratory, Hertfordshire.
Raper, D. W and Lee, D. S.: 1996, *Atmospheric Environment* **30**, 1193.
Raper, D.W.: 1989, Doctor of Philosophy Thesis, Manchester Metropolitan Polytechnic, Manchester.
Review Group in Acid Rain (RGAR): 1997, *Acid Deposition in the United Kingdom 1992-1994*, DETR. London.

ATMOSPHERIC SO₂ POLLUTION AND ACIDITY OF RAIN IN CHANGCHUN CHINA

DEXUAN WANG* and WEI DENG

Changchun Institute of Geography, the Chinese Academy of Sciences, Changchun 130021 China
(*author for correspondence, e-mail: wangdexuan@mail.ccig.ac.cn)

Abstract. It is mainly SO_2 that bring about acid rain in China. Changchun City, which is located in Northeast China, is a typical city that is polluted by SO_2 from coal combustion in winter. In winter, the daily mean concentration of atmospheric SO_2 is about 0.10mg/m^3 and about 5 times as high as in summer, and the daily highest concentration usually appears in daybreak and nightfall. The monitored lowest pH value of rainwater was 4.8 in spring and the range of pH value of rain /snow was 5.2-6.0 in winter, 4.8-5.8 in spring, 5.4-6.4 in summer, 5.6-6.4 in autumn, and the annual mean pH value of rainfalls was 5.8 (1999-2000). Because the alkaline aerosol from soil, meteorological conditions etc., is unfavorable to acid rain formation, even though high SO_2 emission intensity existed in winter, the acid rain did not appear obviously. The aerosol character, climate conditions in Northeast China are important factors for the acid rain formation, although SO_2 emission is the original cause.

Keywords: SO_2, acidity of rain, Changchun China

1. Introduction

Coal is a main energy resource and its combustion is a main cause that brings increase of atmospheric SO_2 emission in China. According to chemical analysis of precipitation, SO_4^{2-}/NO_3^- is 6, which indicates that the acid deposition is of sulfuric acid type in China. Acid rain is a comprehensive result of deposition of acidic matter and the acid buffering capacity of atmospheric particulate. Although SO_2 deposition in North China is not less than that in South China, soil is mainly alkaline in North China and acidic in South China, which causes acid rain to appear in South China.

As it is very cold in winter, coal consumption is increased for heating, which results in even more SO_2 emission in winter than in any of other season in Northeast China. Changchun is the capital city of Jilin province and a major city in Northeast China. Through studying atmospheric SO_2 and precipitation acidity in this city, one may have a general understanding of the conditions of SO_2 pollution and acid precipitation in cities of Northeast China.

2. Atmospheric SO₂ concentration in Changchun

Through out 1999, an auto-monitor, which is named KZL-SO₂ Auto-monitor made in China, was used to monitor the atmospheric SO_2 that was mainly caused by industry, power plant, and boilers for heating in Changchun. Monitoring was done once every five days, and every monitoring was done from 9 AM till 9 AM of the next day. As shown in Figure 1, the mean concentration of atmospheric SO_2 was about 0.110-0.129 mg/m^3 from January to March, 0.015-0.032 mg/ m^3 from April to October, 0.098-0.126 mg/m^3 from November to December, which indicates that atmospheric SO_2 pollution appears in winter. On the other hand, because the SO_2 emission source is mainly in urban area and SO_2 is emitted from low level sources, atmospheric SO_2 concentration in suburban area was about 0.03-0.05 mg/m^3 in winter, which indicates SO_2 pollution is a local phenomenon in northeastern China.

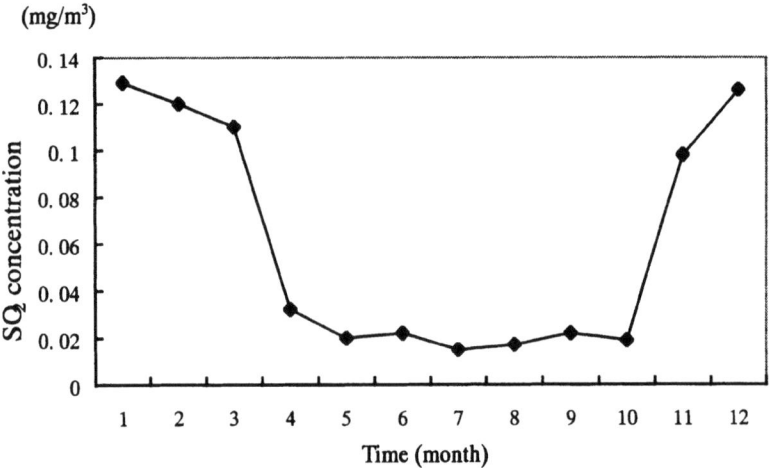

Figure 1. Annual change of mean atmospheric SO_2 concentration in Changchun in 1999.

3. Daily change of atmospheric SO_2 in winter

There are two high peaks of atmospheric SO_2 concentration that appeared at 5-8 AM and 5-9 PM when there was more SO_2 emission and radiation inversion obstructing SO_2 diffusion. The daily change of atmospheric SO_2 (mg/m^3) concentration in January 1999, is shown in Figure 2.

4. Precipitation acidity

From March, 1999 to February, 2000, the acidity of rain/snow was monitored using PHS-29 Acidity Instrument made in China. As shown in Figure 3, the lowest pH value of precipitation, 4.8, appeared in spring, the highest, 6.4, appeared in summer. The precipitation of pH<5.0, appeared only about 2% in spring, the precipitation of pH <5.6 appeared about 28% in winter and spring, precipitation of pH <5.6 did not appear in summer and autumn in Changchun. Although a heavy SO_2 emission resulted in pH value of precipitation which were lower in winter and early spring than in other season, acid snow (pH<5.0) did not appear. According to monitored data in some cities including Changchun, mean ion concentration in rainfalls in northern and southern China were reported (Wang, 1994). As shown in Table I, as the alkaline material of atmospheric particulate can neutralize acid material in northern China more strongly than in southern China, acid rain occurs mainly in southern China.

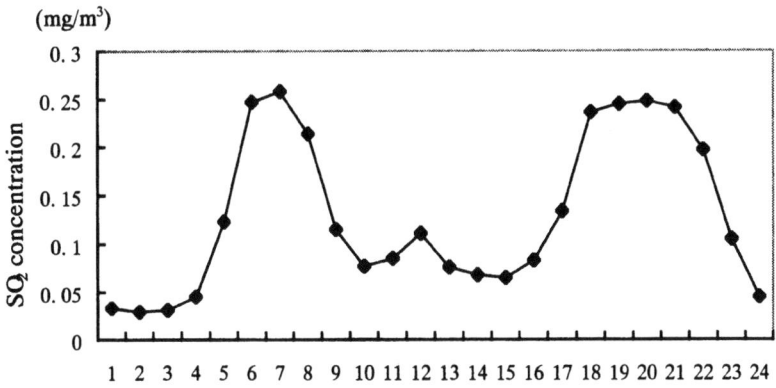

Figure 2. Daily change of mean atmospheric SO_2 concentration in January 1999, Changchun

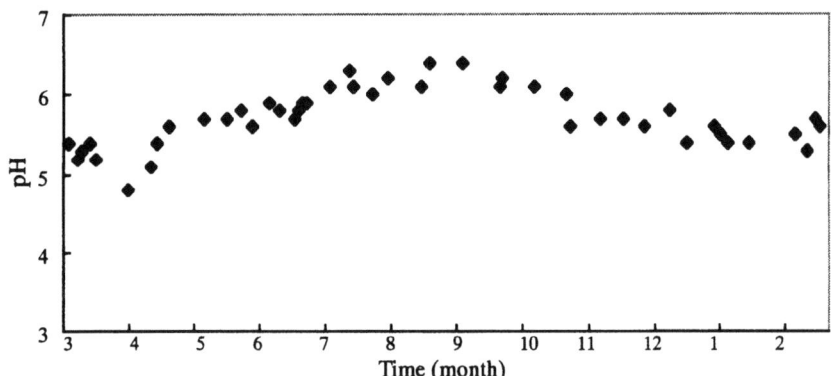

Figure 3. pH of rain (snow) during 1999-2000 in Changchun

TABLE I
Mean ion concentrations of rainwater in cities of northern and southern China (μ eq/L)

	SO_4^{2-}	NO_3^-	SO_4^{2-}/NO_3^-	NH_4^+	Ca^{2+}	Mg^{2+}
Northern	212.8	28.7	5.8	116.7	382.8	76.1
(Changchun)	156.5	21.2	7.38	61.3	256.5	51.2
Southern	125.1	20.0	6.0	92.5	90.2	16.1

5. Discussion

Because SO_2 emission is main precursor that result in acid rain formation, reducing SO_2 emission is main measures for controlling acid precipitation in China. Recently, to reduce SO_2 emission, many means, including space heating, were oprated in Changchun. During 1995-1999, the daily mean value of atmospheric SO_2 had been lowered from 0.14mg/m^3 to 0.10mg/ m^3 in winter in Changchun. On the other hand, neutralization of atmospheric particulate from soil is an important factor that affects acid precipitation. Changchun is located in alkaline Black Soil area. Although atmospheric SO_2 concentration in winter was five times as high as that in summer, acid precipitation did not appear in winter obviously because of the buffering action of atmospheric particulate in Changchun.

References

Wang, W.: 1994, *China Environmental Science* 14, 323.
Wang, D. *et al*: 1995, *Scientia Geographica Sinica* 15, 289.
Liu , B. *et al*: 1998, *China Environmental Science* 18, 1.
Wang, W., Zhang, W., Shi, Q. and Hong, S.: 1993, *China Environmental Science* 13, 401.

THE IMPORTANCE OF CALCIUM DEPOSITION IN ASSESSING IMPACTS OF ACID DEPOSITION IN CHINA

T. LARSSEN[1], H.M. SEIP[2], G.R. CARMICHAEL[3] and J.L. SCHNOOR[3]

[1] *Norwegian Institute for Water Research, P.O.Box 173 Kjelsas, 0411 Oslo, Norway. E-mail: thorjorn.larssen@niva.no;* [2] *Department of Chemistry, University of Oslo, P.O. Box 1033, Blindern, 0315 Oslo, Norway;* [3] *Center for Global and Regional Environmental Research, 204 IATL, The University of Iowa, Iowa City, IA 52242, USA.*

Abstract. The extensive use of coal as an energy carrier in China has led to high deposition of sulfur in a large part of the country. In the southern part of China large areas receive acid deposition, while in the northern part of the country the acidity of the emissions is neutralized by alkaline dust from the desert areas. In this paper we demonstrate the importance of knowing the sources and deposition patterns of base cations when assessing the effects of changes in sulfur emissions. Regional-scale data of both sulfur and calcium deposition from modeling and monitoring are combined in order to demonstrate how the acidity of deposition in China has changed historically and may change in the future. The importance of base cation deposition is also demonstrated using the dynamic acidification model MAGIC with input data from an intensive monitoring site outside Guiyang. It is not known what fraction of the deposited base cations is of natural origin and anthropogenic origin, respectively. The relative source strength varies greatly between regions. Future effects of emission changes are highly dependent on the relative reduction in sulfur and base cation emissions.

Keywords: acidification, alkaline dust, base cation deposition.

1. Introduction

In most studies of acid deposition in China, focus has been on precipitation pH or sulfur deposition flux. In this presentation we illustrate the importance of including base cation deposition in the discussion of acid deposition impacts in China. We illustrate the importance of increasing the knowledge on sources for base cation deposition in China in order to be able to assess the impacts of different future sulfur emission scenarios. We use both regional-scale mapping and a catchment-scale dynamic acidification model to illustrate the importance of base cation deposition scenarios.

2. Material and methods

We use the sulfur-to-calcium ratio in deposition when mapping historical and future changes in the deposition acidity in China. Sulfate is by far the dominant anion and calcium is generally the dominant cation in the deposition. Since calcium basically occurs as carbonate, oxide or sulfate, and the sulfur mostly occurs as SO_2, H_2SO_4 or $CaSO_4$, we can use the ratio S/Ca^{2+} (on equivalent basis) as a simple indicator of acidity. A ratio exceeding one indicates acid deposition.

The sulfur deposition for 1990 is taken from the RAINS-Asia model (Downing et al., 1997). The total calcium deposition was estimated by combining wet deposition monitoring data for 81 stations throughout the country and modeled dry deposition with source in the northern desert areas (Larssen and Carmichael, 2000; Chang et al., 1996).

We also present results from calculations with a dynamic soil acidification model (the MAGIC model; Cosby et al., 1985) to illustrate how the uncertainties in the base cation deposition scenarios influence the modeled soil acidification. The MAGIC model was applied to a data set from a research site outside Guiyang in the Guizhou province (Larssen et al., 1998). The model was calibrated in a Monte Carlo framework against observed soil water chemistry as described in Larssen et al. (2000).

For historical changes in sulfur deposition we used the data of Lefohn et al. (1999). In the simplified approach used here we divide the calcium deposition in an anthropogenic and a natural part and assign different scenarios to each of the two fractions. As emissions from different sources will change differently over time this is obviously a coarse simplification, but is still illustrative at the present state of knowledge. The fraction of the base cation deposition originating from anthropogenic sources is assumed to have developed historically similar to the sulfur deposition. The natural part of the base cations is assumed constant. Since the natural part includes contributions of dust from agricultural areas as well as desert areas, this is not necessarily true. For the purpose of this paper, two different scenarios for base cation deposition are used: In one scenario we assume 25 % of the base cations to be anthropogenic and 75% natural; in the other scenario we assume 75 % to be anthropogenic and 25% natural. The part of the calcium with origin in the northern desert areas is natural, and kept constant. The two scenarios for the anthropogenic:natural calcium ratio are applied for the fraction estimated from the monitoring network (cf. Larssen and Carmichael, 2000).

3. Results

With the methodology and assumptions used here, the S/Ca ratio in deposition started exceeding one on a regional-scale between 1950 and 1960. The ratio was below one in all grid cells in 1950. Figure 1 illustrates how the areas where S/Ca > 1 have developed historically for two different deposition scenarios. The figure illustrates how the historical S/Ca ratio in deposition, and hence the acidity, depends on what fraction of Ca^{2+} is anthropogenic. If a large fraction of the present calcium deposition has anthropogenic origin, a large area had S/Ca ratio > 1 already in 1960 while the increase from 1960 has been moderate (upper panel on Figure 1). If most calcium is of natural origin, the impacted area was small in 1960 since the Ca^{2+} deposition in this scenario increases much less than the S-deposition. The area receiving acid rain has been growing

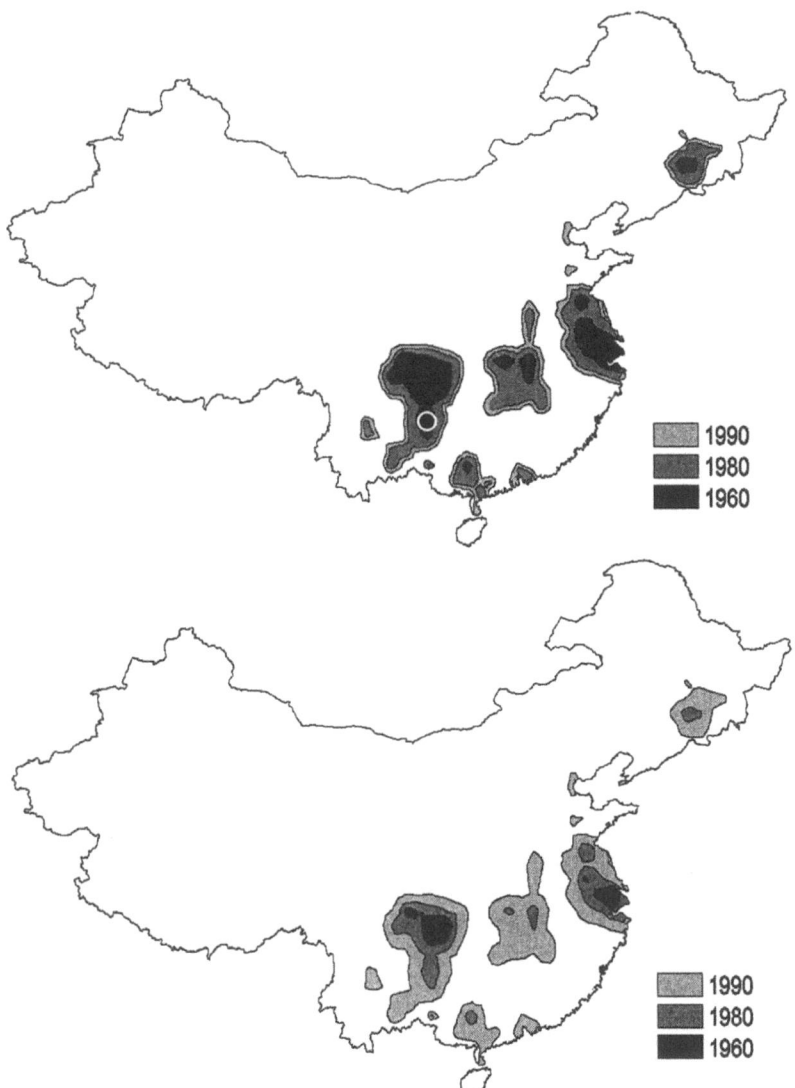

Figure 1. Maps showing the areas receiving deposition where S/Ca >1 in 1960, 1980 and 1990. In 1950 there was, according to our estimates, no area where S/Ca exceeded 1. The upper map shows the scenario assuming natural : anthropogenic calcium ratio 1:3. The lower map shows the scenario assuming natural : anthropogenic calcium ratio 3:1. The assumption of a large anthropogenic fraction results in a relatively small temporal change in the area receiving acid deposition in the period 1960-1990. The location of the sampling site near Guiyang is indicated with the white circle at the upper map.

considerably over the past few decades (lower panel on Figure 1). On the contrary, if Ca^{2+} deposition has been increasing parallel with the sulfur deposition, the potential acidity of the increased sulfur deposition has been counteracted by calcium: The anionic charge of sulfate in deposition is then

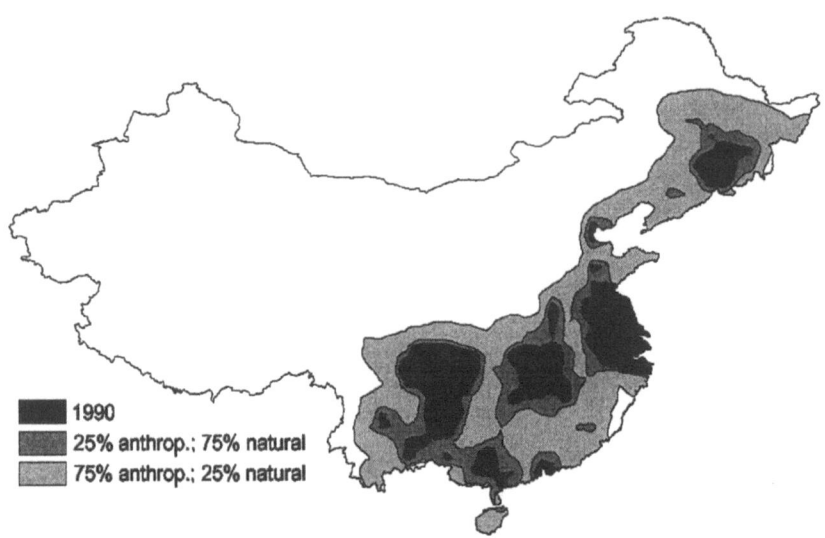

Figure 2. Maps showing future forecast in deposition S/Ca >1 assuming total removal of the anthropogenic fraction of Ca^{2+} and keeping S constant. Two scenarios, assuming respectively 1:3 and 3:1 ratio of anthropogenic : natural Ca^{2+} in 1990, are shown. Estimate for 1990 is also shown.

balanced largely by Ca^{2+} instead of H^+. In Figure 2 two forecast scenarios are compared to the 1990 S/Ca deposition ratio. In these forecasts the sulfur deposition is assumed constant at 1990 level while the anthropogenic Ca fraction is assumed removed. Hence we illustrate the situation if the anthropogenic alkaline dust emissions are removed without removing sulfur. Total removal of the anthropogenic alkaline dust without any reduction in sulfur emissions may be an overestimated worst case scenario, but faster reduction of base cations than acid gases is likely and has been observed in Europe and North America (Hedin *et al.*, 1994; Lee *et al.*, 1999). The area receiving acid deposition will increase more if a large calcium fraction is anthropogenic than if most of it is natural. The scenarios clearly illustrate the importance of increasing our knowledge regarding sources of base cations in deposition. As an illustrative case study we present results for two different calibrations of the MAGIC model. The historic base cation deposition is important when calibrating dynamic acidification models. The development of the base saturation in the upper soil horizon (15 cm) is shown in Figure 3 assuming anthropogenic : natural calcium ratios in 1990 of 1:3 (a and b) and 3:1 (c and d), respectively. With the higher anthropogenic fraction, the acidification (in terms of reduced base saturation) has been small, since increased input of Ca^{2+} has followed the increase in S deposition. Under the assumption of a high natural Ca^{2+} fraction, the modeled acidification from 1950 to 1995 is considerable.

Figure 3 also shows the effect of different forecast scenarios. The effect of reducing the anthropogenic base cation deposition fraction by 50% is compared to the situation if the input were kept constant at 1995 level; sulfate and nitrate

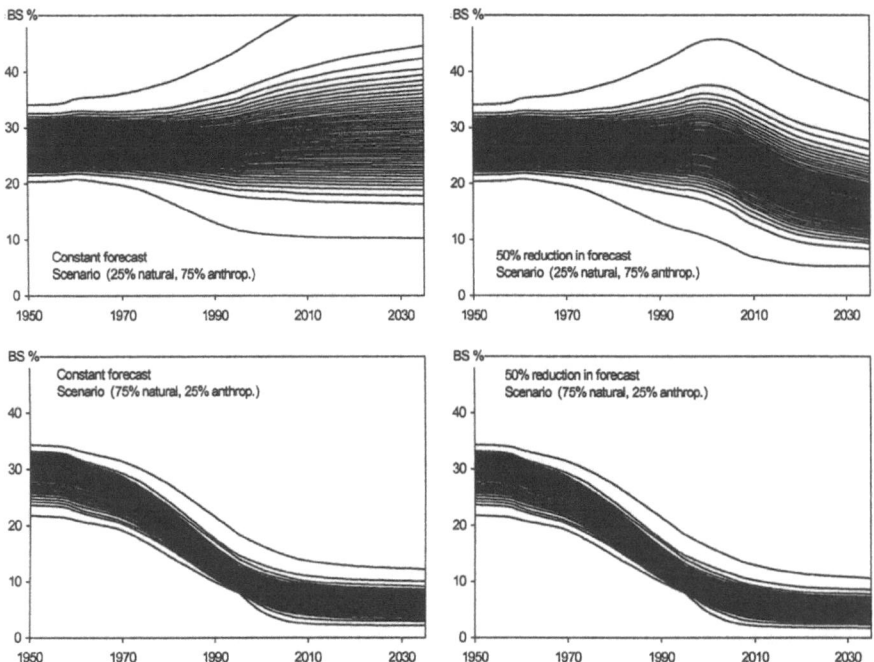

Figure 3. Changes in upper soil base saturation (BS) for the period 1950-2035 using MAGIC. The figures illustrate the importance of historic and future base cation deposition scenarios. In figures a) and b) the natural : anthropogenic calcium ratio is 1:3; in figures c) and d) the ratio is 3:1. For the two left hand figures, i.e. a) and c), the deposition in the forecasts is kept constant for all ions. For the two right hand figures, i.e. b) and d), the base cation deposition in the forecasts is reduced to 50% of 1995 level in one step (1995-1996), while the sulfate and nitrate depositions were kept constant. A Monte Carlo approach was used in the simulation; the figures show statistics for the accepted simulations. Each line represents percentiles of the results in 2% intervals (i.e. minimum, 2%-ile, 4%-ile, ..., 98%-ile and maximum). The hindcast scenarios are calibrated to fit within the measured base saturation and soil water chemistry 10- and 90-percentiles in 1995. For base saturation the 10- and 90-percentiles are 8% and 46%.

are kept constant at 1995 levels in all forecasts. Reduced base cation input gives faster reduction in soil base saturation.

4. Discussion

The uncertainties in the results presented here are obviously large. The estimate we have used for the total base cation deposition, has many sources of uncertainties, e.g. the representativeness of the monitoring stations; the quality of chemical analyses; and the complexity in modeling dust atmospheric dispersion. However, the results clearly illustrate the importance of including the base cations in the discussion when assessing impacts of emissions and acidification in China. In critical load mapping, the natural calcium (or base cation) fraction is an important input parameter with major impact on the result. More accurate calculations require further knowledge of the anthropogenic and natural sources of the deposition.

Our results also illustrate the importance of treating sulfur emissions and dust emissions simultaneously when setting emission standards for boilers and other combustion units. Reducing the dust emissions is cheaper and simpler than reducing the sulfur emissions. In addition, dust emission reductions will have a positive effect on local air quality. Our results show that such reductions, without also removing SO_2, will have negative impacts regarding acid rain and acidification. A faster reduction of base cation emission than sulfur emission will give increased acidification even in a situation where also sulfur is reduced. It is therefore highly recommended to include both sulfur and dust emissions when emission reduction measures are assessed.

5. Conclusion

The high concentration of base cations in air and precipitation in China must be taken into account when assessing the spread and future impacts of acid deposition in China. The size of the fraction of the present calcium deposition originating from anthropogenic sources is not known. Future response to continued acid deposition is dependent on the sources of the base cations in deposition. If anthropogenic calcium is removed from emissions faster than sulfur, the acidification situation in China may become worse in the future, even if the sulfur deposition is reduced. Efforts should be made to better understand the sources of calcium in deposition and the importance of the anthropogenically emitted dust should be taken into account together with sulfur when assessing emission reduction scenarios.

References

Chang, Y.-S., Arndt, R.L. and Carmichael, G.R.: 1996, *Atmospheric Environment* **30**, 2417.
Cosby B.J., Hornberger, G.M., Galloway, J.N. and Wright, R.F.: 1985, *Water Resources Research* **21**, 51.
Downing, R.J., Ramankutty, R. and Shah, J.J.: 1997, *Rains-Asia: an assessment model for acid deposition in Asia.* The World Bank, Washington DC.
Hedin, L.O., Granat, L., Likens, G.E., Buishand, T.A., Galloway, J.N., Butler, T.J. and Rodhe, H.: 1994, *Nature* **367**, 351.
Larssen, T. and Carmichael, G.: 2000, *Environmental Pollution* **110**, 89.
Larssen, T., Schnoor, J.L., Seip, H.M. and Zhao, D.: 2000, *Science of the Total Environment* **246**, 175.
Larssen, T., Xiong, J., Vogt, R.D., Seip, H.M., Liao, B. and Zhao, D.: 1998 *Water Air and Soil Pollution* **101**, 137.
Lee, D.S., Kingdon, R.D., Pacyna, J.M., Bouwman, A.F. and Tegen, I.: 1999, *Atmospheric Environment* **33**, 2241.
Lefohn, A.S., Husar, J.D. and Husar, R.B.: 1999, *Atmospheric Environment* **33**, 3435.

FLUXES AND TRENDS OF NITROGEN AND SULPHUR COMPOUNDS AT INTEGRATED MONITORING SITES IN EUROPE

MARTIN FORSIUS*, SIRPA KLEEMOLA, JUSSI VUORENMAA and SANNA SYRI

Finnish Environment Institute, P.O.Box 140, FIN-00251 Helsinki, Finland
(* author for correspondence, e-mail: martin.forsius@vyh.fi)

Abstract. The International Cooperative Programme on Integrated Monitoring (ICP IM) is part of the effects monitoring strategy of the UN/ECE Convention on Long-Range Transboundary Air Pollution. We calculated input-output budgets and trends of N and S compounds, base cations and hydrogen ions for 22 forested ICP IM catchments/plots across Europe. The site-specific trends were calculated for deposition and runoff water fluxes and concentrations using monthly data and non-parametric methods. The reduction in deposition of S and N compounds, caused by the new Gothenburg Protocol of the Convention, was estimated for the year 2010 using atmospheric transfer matrices and official emissions. Statistically significant downward trends of SO_4, NO_3 and NH_4 bulk deposition (fluxes or concentrations) were observed at 50% of the ICP IM sites. Implementation of the new UN/ECE emission reduction protocol will further decrease the deposition of S and N at the ICP IM sites in western and northwestern parts of Europe. Sites with higher N deposition and lower C/N-ratios clearly showed an increased risk of elevated N leaching. Decreasing SO_4 and base cation trends in output fluxes and/or concentrations of surface/soil water were commonly observed at the ICP IM sites. At several sites in Nordic countries decreasing NO_3 and H^+ trends (increasing pH) were also observed. These results partly confirm the effective implementation of emission reduction policy in Europe. However, clear responses were not observed at all sites, showing that recovery at many sensitive sites can be slow and that the response at individual sites may vary greatly.

Keywords: acidification, air pollution, forested catchments, deposition, emissions, surface waters

1. Introduction

In the past two decades, both national and international environmental regulations and agreements have led to widespread declines in the emissions of air pollutants in Europe and North America. In Europe the emissions of sulphur (S) and nitrogen (N) compounds have declined by 39 % (SO_2), 16 % (NO_2) and 18 % (NH_3) respectively, between the years 1988-1996 (EMEP/MSC-W, 1998). The protocols of the United Nations Economic Commission for Europe's Convention on Long-Range Transboundary Air Pollution (UN/ECE CLRTAP) and legislation of the European Union have been key international instruments causing this positive development. Clear signs of recovery in sensitive ecosystems have also been reported (*e.g.* Stoddard *et al.*, 1999; WGE, 1999).

Early in the discussions on the UN/ECE CLTRAP it was recognised that a good understanding of the harmful effects of air pollution was a prerequisite for reaching agreement on effective pollution control. The international multidisciplinary Integrated Monitoring programme (ICP IM) is one of the activities set up under the Convention to develop the necessary international co-operation in the assessment of pollutant effects. The key aim of the ICP IM is to quantify effects on the environment through monitoring, modelling and scientific review, using data from catchments/plots located in natural/seminatural areas (ICP IM Manual, 1998). The international Programme Centre is located at the Finnish Environment Institute. This paper presents fluxes and trends of S and N compounds, base cations and hydrogen ions from sites in the ICP IM network across Europe, in an attempt to assess the effects of emission reduction measures. We also estimate the reduction in deposition of S and N compounds at the sites, anticipated by the new 1999 Protocol to Abate Acidification, Eutrophication and Groundevel Ozone (Gothenburg Protocol) of the CLRTAP, by using atmospheric transfer matrices and official emissions for the target year 2010.

2. Materials and Methods

According to availability of internationally reported data in the ICP IM database, 22 sites were selected for the analysis (Fig. 1). A list of the National Focal Points contributing data is available at: www.vyh.fi/eng/intcoop/projects/icp_im/im.htm. The sites are covered mainly with coniferous forest. The trend assessment was performed mainly for the period 1988/89 - 1998. Time series with a minimum of five years monthly data were accepted for the statistical analyses. Trends were evaluated for non-marine (* denotes non-marine fraction) SO_4* and $(Ca+Mg)$*, H^+, NO_3 and NH_4 (except runoff). Deposition (bulk and through fall deposition) and output (runoff/soil water) fluxes were calculated from the quality and quantity of water using mean monthly values for water fluxes and chemical analyses (weighted means where available). Fluxes for the budget calculations were calculated as the average of the last three years with available data in order to reduce yearly variability. C/N-ratios (g/g) in the organic (Oh) soil layer were calculated for sites with available data. Methods for collection, storage and analysis of chemical samples are described in the programme manual (ICP IM Manual, 1998). The present paper is an extension and update of first trend assessment results of ICP IM sites, presented in Vuorenmaa *et al.* (2000).

The trend analyses were done with the DETECT software package (Cluis *et al.*, 1989). DETECT contains a suite of recognised non-parametric methods for trend analyses, and recommends the most appropriate method based on the presence and absence of statistical seasonality (defined by ANOVA testing differences in monthly means) and persistence (defines the presence and whether it is Markovian or first order). Monotonic trend was used as a default trend type. The following test were used: 1) Hirsch and Slack test for series with seasonality and persistence; 2) Seasonal Kendall test for series with seasonality only; 3) Kendall test for time series without seasonality and persistence, and 4) Spearman / Lettenmaier test for series with persistence only. By using monthly data (*i.e.* 12 seasons of one month each),

Figure 1. Location of ICP IM sites included in the calculations of fluxes and trends.

the programme requires a complete series of equidistant values. If a specific month was missing, the value was estimated using the mean of the month of the other years. By this interpolation method we tried to avoid creating persistence before it is studied.

The effect of the Gothenburg Protocol on S and N deposition at the sites was calculated with transfer matrices of the UN/ECE EMEP/MSC-W centre (150x 150 km grid), incorpo rated in the DAIQUIRI model (Syri et al., 1998), and reported official UN/ECE emissions (EMEP /MSC-W, 1998) for the reference year 1996 and the target year 2010. The change in deposition at the individual sites, caused by implementation of the protocol, was estimated as the difference in deposition between these two years (using average meteorology for the years 1985-96).

3. Results and Discussion

3.1. FLUXES

Ion mass budgets have proved to be useful for evaluating the importance of various biogeochemical processes that regulate the buffering processes in both terrestrial and aquatic portions of catchments (*e.g.* Likens *et al.,*1996). The results of the ICP IM site generally follow well-known patterns regarding ion fluxes: efficient retention of N compounds and hydrogen ions and release of base cations (due to weathering and ion exchange reactions) (Table I). For SO_4* both retention, apparent steady-state and release was observed. The results of the site CH01 indicate a clear geological S source with very high base cation and S output fluxes. High H^+ leaching was observed at the sites NO01, SE02 and SE04. At site DE01 the leaching results may have been influenced by an insect attack affecting tree health conditions, and at AT01 by changes in methodology for deposition monitoring during recent years.

According to Dise and Wright (1995), Dise *et al.* (1998) and Gundersen *et al.* (1998) empirical data from European forest ecosystems have indicated that three broad criteria are necessary (but not themselves sufficient) for sites to leach NO_3: 1) high fluxes of dissolved inorganic N in deposition (> ca. 65-70 meq m^{-2} a^{-1} = 9-10 kg N ha^{-1} a^{-1}), 2) low organic layer C/N-ratio (below about 25-30) and 3) low mineral soil pH (below about 4.3). The relationship between N deposition and N output flux (Fig. 2a), and C/N-ratio and N output flux (Fig. 2b) at the ICP IM sites seem to be generally

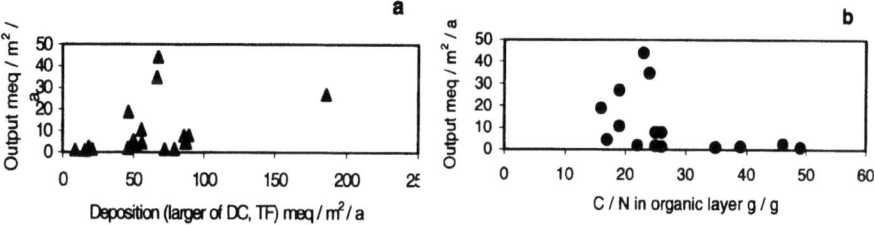

Figure 2. The relationship between N deposition and N output flux (2a), and C/N-ratio of the soil organic matter and N output flux (2b) at the ICP IM sites. The larger value of bulk deposition (DC) and throughfall deposition (TF) has been used for estimating N deposition.

consistent with these criteria. Sites with higher deposition and lower C/N-ratios clearly show higher risk of elevated N output fluxes. These results cannot be considered as an independent evaluation of the above criteria because data from some ICP IM sites have been used in their derivation. However, there is a great potential in using such relationships in conjugation with regional survey data for risk assessment and mapping of deposition effects on the large regional scale.

3.2. TRENDS

Statistically significant downward trends of SO_4^* bulk deposition (fluxes or concentrations) were observed at 11 of the 22 ICP IM sites (Table I). Significant downward trends for NO_3 and NH_4 bulk deposition were also observed at 11 sites. Significant decreasing H^+ trends were observed at 8 sites (fluxes or concentrations). These the decreases in SO_4^*. In addition, interaction between climate-induced changes and acidification processes h results seem to be consistent with the reported decreases in European SO_2, NO_2 and NH_3 emissions (EMEP/MSC-W, 1998).

A decreasing trend in concentrations and deposition of base cations, especially for calcium, has been observed in northern parts of Europe and USA over the last two or three decades, although the rate of decrease appears to have been slower in the more recent years (Hedin *et al.*, 1994). Such results have ecological significance, because the ecosystem effects of the deposition depend on the relative contribution of acidifying and neutralising compounds. Deposition of $(Ca+Mg)^*$ shows decreasing trends at IM sites in southern Fennoscandia, and sites NL01 and CZ01. Thus, the observed decrease in H^+ deposition is smaller than would be expected from the S and N trends alone.

Implementation of the new UN/ECE emission reduction protocol will further decrease the deposition of S and N at the ICP IM sites in western and northwestern parts of Europe (Table I). The decrease in SO_4 deposition (-36 % on the average at all the studied sites) is expected to be larger than for NO_3 (-24 % on the average). Changes in NH_4 deposition are expected to be rather small at all the ICP IM sites. At sites in more eastern and northeastern European regions (Baltic states, Finland, Belarus, Russia) the expected decrease of both SO_4 and NO_3 deposition will be smaller or non-existent. This is because the new UN/ECE protocol allows some growth in emissions from the present (reference year 1996) level in many eastern European countries. For western Europe the protocol implies a significant further decrease of sulphur and nitrogen oxides emissions.

The trend results regarding surface water fluxes/concentrations are less clear than those of deposition (Table I). This is to be expected, because the deposited compounds are involved in numerous complex (site-specific) processes in the ecosystems (*e.g.* Likens *et al.*, 1996) the end-result of which are not always evident. Decreasing SO_4^* and base cation trends in output fluxes and/or concentrations were commonly observed at the ICP IM sites (Table I). The decreasing base cation trends are a logical consequence of decreasing trends in deposition of both base cations and strong acid anions (*e.g.* Likens *et al.*, 1996). Statistically significant decreasing H^+ trends (increasing pH) is observed only at sites SE04, SE08, NO02, and LT01. Some sites (*e.g.* CH01, LT01) are well buffered and large changes in H^+ fluxes are not to be expected. The increasing trend in H^+ at site FI01 is apparently at least partly caused by observed increasing concentrations of organic carbon (and dissociated organic acids) in the surface water (observed TOC increase about 2 mg l^{-1} in ten years). The reason for the increase in total organic carbon (TOC) at this site is presently unknown. Decreasing NO_3 trends were observed at 5 sites in the Nordic countries. At several sites the time series are still too short (or sufficient data is lacking) for trend analysis. More work is also needed to analyse the reasons for the different response patterns at the sites.

Detailed studies at site SE04 (Gårdsjön) have indicated that recovery of surface water chemistry can proceed rapidly before soils begin to recover (Moldan, 1999). Likens *et al.* (1996) observed a large depletion of the base cation pool in the catchment soils at Hubbard Brook (USA), which was expected to retard the response of the ecosystem to the emission reductions. Dynamic model calculations at ICP IM sites have indicated that recovery of soil and water chemistry due to emission reductions often will be slow (Forsius *et al.*, 1998). Stoddard *et al.* (1999) observed lack of surface water recovery in several regions in North America which was attributed to strong regional declines in base cation concentrations exceeding the decreases in SO_4^*. In addition, interaction between climate-induced changes and acidification processes have been observed (Schindler *et al.*, 1996; Wright, 1998), further complicating evaluation of changes.

4. Conclusions

Statistically significant downward trends of SO_4^*, NO_3 and NH_4 deposition (fluxes or concentrations) were observed at 11 of the 22 ICP IM sites. Significant decreasing H^+ trends were observed at 8 sites. In some regions the observed decrease in H^+ deposition is smaller than would be expected from the S and N trends alone due to simultaneous decreases in base cation deposition. These results seem to follow the reported decreases in European emissions. Implementation of the new UN/ECE emission reduction protocol will further decrease the deposition of S and N at the ICP IM sites in western and northwestern parts of Europe. At sites in more eastern regions the decreases are anticipated to be smaller or non-existent. Changes in NH_4 deposition are expected to be rather small at all the ICP IM sites.

The relationships between N deposition and N output flux, and C/N-ratio and N output flux at the ICP IM sites were consistent with previous observations from European forested ecosystems. Sites with higher deposition and lower C/N-ratios clearly showed a higher risk of elevated N output fluxes. Such empirical relationships are useful for regional-scale mapping exercises and risk assessment. Accelerated N leaching in ecosystems may cause harmful effects regarding both eutrophication and acidification.

Decreasing SO_4^* and base cation trends in output fluxes and/or concentrations of surface/soil water were commonly observed at the ICP IM sites. At several sites in Nordic countries decreasing NO_3 and H^+ trends (increasing pH) were also observed. These results partly confirm the effective implementation of emission reduction policy in Europe. However, clear responses were not observed at all sites, showing that recovery at many sensitive sites can be slow and that the response at individual sites may vary greatly. Continued national and international research and monitoring efforts are needed to obtain scientific evidence on the recovery process to support future emission reduction policies.

References

Cluis, D., Langlois, C., van Coillie, R. and Laberge, C.:1989, *Environmental Monitoring and Assessment* **12**, 429.
Dise, N., and Wright, R. F.: 1995, *Forest Ecology and Management* **71**, 153.
Dise, N.B., Matzner, E. and Forsius, M.: 1998, *Environmental Pollution* **102**, 453
EMEP/MSC-W: 1998. Transboundary acidifying air pollution in Europe. *EMEP/MSC-W Status report* 1998. EMEP/MSC-W, Oslo, Norway.
Forsius, M., Alveteg, M., Jenkins, A., Johansson, M., Kleemola, S., Lükewille, A., Posch, M., Sverdrup, H. and Walse C.: 1998, *Water, Air and Soil Pollution* **105**, 21.
Gundersen, P., Callesen, I. and de Vries, W.: 1998, *Environmental Pollution* **102**: 403.
Hedin, L.O., Granat, L., Likens, G.E., Buishand, T.A., Galloway, J.N., Butler, T.J. and Rodhe, H.: 1994, *Nature* **367**, 351.
ICP IM Manual 1998. Finnish Environment Institute Helsinki. www-version: http://www.vyh.fi/eng/intcoop /projects/icp im/manual/index.htm
Likens, G.E., Driscoll, C.T. and Buso, D.C.: 1996, *Science* **272**, 244.
Moldan, F.: 1999, Doctor of Philosophy Disseration, *Acta Universitas Agriculturae Sueciae Silvestria* **117**. Swedish University of Agricultural Sciences, Umeå.
Schindler, D.W., Bayley, S.E., Parker, B.R., Beaty, K.G., Cruikshank, D.R., Fee, E.J., Schindler, E.U. and Stainton, M.P.: 1996, *Limnology and Oceanography* **41**, 1004.
Stoddard, J.L., Jeffries, D.S., Lükewille, A., Clair, T.A., Dillon, P.J., Driscoll C.T., Forsius, M., Johannessen, M., Kahl, J.S., Kellogg, J.H., Kemp, A., Mannio, J., Monteith, D.T., Murdoch, P.S., Patrick, S. Rebsdorf, A., Skjelkvåle, B.L, Stainton, M.P., Traaen, T., van Dam, H., Webster, K.E., Wieting, J. and Wilander, A.: 1999, *Nature* **401**, 575.
Syri S, Johansson M, and Kangas L.: 1998, *Atmospheric Environment* **32**, 409.
Vuorenmaa, J., Kleemola, S. and Forsius, M.: 2000. *Verh. Int. Verein. Limnol.* **27**, 384.
WGE 1999. *Trends in impacts of long-range transboundary air pollution.* Compiled by the International Cooperative programmes of the UN/ECE Working Group on Effects. Centre for Ecology and Hydrology, UK. ISBN 1 870 393 52X. 81 p.
Wright, R.F.: 1998, *Ecosystems* **1**, 216.

TABLE I

Fluxes and trends of SO_4^*, $(Ca+Mg)^*$, H^+, NO_3 and NH_4 at the ICP IM sites. DC = bulk deposition, TF = throughfall deposition, RW = runoff water, SW = soil water. Significance levels ($^*P<0.05$, $^{**}P<0.01$, $^{***}P<0.001$) and slopes for statistically significant trends (bold) are shown (n.s.t.= no significant trend; blank = insufficient data). Trend directions (+ or -) and the rate of change are expressed in slope values as meq m^{-2} a^{-1} (fluxes) and µeq l^{-1} a^{-1} (concentrations). The statistical methods and tests used are explained in section 2. The column '>% diff 2010' shows the estimated change in deposition (%) between years 1996 and 2010 according to the Gothenburg Protocol. Horisontal lines separate sites in Nordic countries, Baltic states+Belarus+Russia, and central+southern Europe.

Site	Variable	DC fl. meq/m²/a	TF fl. meq/m²/a	%diff 2010	RW/SW fl. meq/m²/a	DC fl. test	DC fl. slope	DC conc. test	DC conc. Slope	RW/SW fl. test	RW/SW fl. slope	RW/SW conc. test	RW/SW conc. slope	C/N g/g
FI01	SO4*	15.4	35.3	-13.9	33.4	4	-2.78	1	-3.06**	1	(n.s.t)	1	-1.30*	39
	(Ca+Mg)*	3.6	26.7		39.0	1	-0.52**	2	-0.35***	1	(n.s.t)	1	-1.20**	
	H+	14.8	11.4		10.4	4	-1.42	1	-1.41**	1	(n.s.t)	1	1.85**	
	NO3	12.0	6.0	-19.3	0.5	4	-0.86	1	-0.72*	1	-0.11*	1	-0.13*	
	NH4	8.7	4.2	-3.2	0.8	1	-1.36	1	-1.69*					
FI03	SO4*	13.9	16.8	-10.9	11.9	1	-2.00**	1	-1.98***	1	-1.10*	1	-1.87***	49
	(Ca+Mg)*	3.1	10.1		32.5	1	-0.23**	2	-0.16*	1	(n.s.t)	1	(n.s.t)	
	H+	15.2	14.5		0.3	4	-0.92	2	-1.29***	2	(n.s.t)	1	(n.s.t)	
	NO3	9.7	8.1	-11.9	0.5	4	-0.53	1	-0.69**	1	(n.s.t)	1	-0.08**	
	NH4	5.7	4.6	-11.5	0.4	3	-0.90***	2	-0.95***					
SE02	SO4*	47.5	117.2	-69.0	79.8	4	-4.32	4	-1.50	1	(n.s.t)	4	(n.s.t)	26
	(Ca+Mg)*	6.1	61.5		56.7	1	(n.s.t)	1	(n.s.t)	1	-3.07*	4	(n.s.t)	
	H+	40.0	12.3		22.7	2	-3.10*	3	-1.10**	1	(n.s.t)	1	(n.s.t)	
	NO3	42.7	31.5	-33.9	7.1	2	(n.s.t)	4	(n.s.t)	1	(n.s.t)	1	0.48**	
	NH4	43.0	15.0	-5.1	0.8	1	-3.19*	1	(n.s.t)					
SE04	SO4*	38.0	65.7	-44.0	118.8	4	-4.85	4	-4.23**	1	(n.s.t)	4	(n.s.t)	25
	(Ca+Mg)*	6.6	39.6		40.8	3	-0.49*	3	(n.s.t)	1	-2.96*	4	-4.48	
	H+	29.4	35.7		37.2	4	-7.13	4	-5.17	1	-1.96*	1	-1.12*	
	NO3	34.9	47.0	-31.2	0.6	1	-2.30*	2	-1.63*	1	-0.02**	3	-0.07**	
	NH4	32.1	26.4	-10.1	1.0	3	-1.56*	2	(n.s.t)					
SE08	SO4*	8.8		-27.2	16.9	1	-1.00*	1	(n.s.t)	2	-1.40**	4	-0.75	35
	(Ca+Mg)*	2.3			132.3	2	(n.s.t)	1	(n.s.t)	2	-15.4***	3	(n.s.t)	
	H+	11.1			0.3	3	-0.60*	3	-0.80**	3	-0.11*	3	(n.s.t)	
	NO3	5.8		-20.9	0.7	3	(n.s.t)	3	(n.s.t)	2	-0.11***	2	-0.17**	
	NH4	2.9		-4.4	0.4	4	-0.53	2	-0.58**					
NO01	SO4*	46.1	61.0	-43.1	82.0	3	-4.13***	1	-3.01**	1	(n.s.t)	4	-1.75	25
	(Ca+Mg)*	5.3	27.6		35.8	3	-0.53***	2	-0.67***	1	-2.99*	4	-1.83	
	H+	46.1	42.9		22.6	1	-4.68*	4	-2.38	1	(n.s.t)	1	(n.s.t)	
	NO3	47.0	36.0	-30.0	8.0	1	(n.s.t)	2	-1.45*	1	-0.94**	1	-0.75**	
	NH4	42.6	29.0	-9.6		3	-2.96*	3	-2.19**					
NO02	SO4*	7.3	8.3	-30.3	13.9	1	(n.s.t)	1	(n.s.t)	1	-0.71**	1	-0.29*	46
	(Ca+Mg)*	4.1	1.9		38.4	3	(n.s.t)	4	(n.s.t)	1	-0.97*	1	(n.s.t)	
	H+	9.5	9.4		1.4	2	-0.30*	2	-0.24*	1	-0.05*	1	-0.02*	
	NO3	6.6	5.5	-21.1	2.5	2	(n.s.t)	2	(n.s.t)	1	(n.s.t)	1	0.07***	
	NH4	11.7	6.2	-5.1		2	(n.s.t)	2	-0.14**					
DK01	SO4*	37.5		-48.1	27.9	4	(n.s.t)	4	(n.s.t)	1	(n.s.t)	4	(n.s.t)	26
	(Ca+Mg)*	4.7			0.1	3	(n.s.t)	3	(n.s.t)					
	H+	9.5			34.7	3	(n.s.t)	3	(n.s.t)	2	(n.s.t)	3	(n.s.t)	
	NO3	29.8		-31.5	0.7	3	(n.s.t)	4	(n.s.t)					
	NH4	42.2		-22.2	0.7	4	(n.s.t)	3	(n.s.t)					
BY02	SO4*	36.1		5.4		4	-2.92	2	-3.92***					
	(Ca+Mg)*	25.6				3	(n.s.t)	3	(n.s.t)					
	H+	10.7				1	(n.s.t)	1	(n.s.t)					
	NO3	21.3		-2.9		4	-1.14	4	-2.13					
	NH4	24.2		-13.1		4	-5.20	4	-8.75					
LT01	SO4*	21.9	42.5	-13.6	422.0	4	-5.03	4	-9.04					
	(Ca+Mg)*				1190.9									
	H+	3.9	6.8		0.0	4	-6.46	1	-10.3*	4	(n.s.t)	3	-0.01*	
	NO3	15.1	19.7	-8.3	5.3	1	(n.s.t)	1	(n.s.t)					
	NH4	23.2	30.3	21.4	0.5									

Site	Variable	DC ft. meq/m²/a	TF ft. meq/m²/a	%diff 2010	RW/SW ft. meq/m²/a	DC ft.test	DC ft.slope	DC conc.test	DC conc.slope	RW/SW ft.test	RW/SW ft. slope	RW/SW conc.test	RW/SW conc.slope	C/N g/g
LT02	SO4*	33.8	53.6	-28.3	165.0	3	(n.s.t)	3	-7.34***					
	(Ca+Mg)*				536.1									
	H⁺	3.8	3.8		0.0	4	-5.30	4	-12.0	3	-0.00***	3	-0.00*	
	NO₃	17.8	15.7	-14.0	2.4	3	(n.s.t)	3	(n.s.t)					
	NH₄	33.5	28.3	13.9	0.6	3	(n.s.t)	3	-6.53**					
LV01	SO4*	45.8	67.4	-29.9	107.5	3	(n.s.t)	4	(n.s.t)			3	-16.9**	
	(Ca+Mg)*	17.1	34.7		580.1							1	(n.s.t)	
	H⁺	16.3	18.9		0.0	1	(n.s.t)	1	(n.s.t)			4	-0.02	
	NO₃	34.8	31.7	-22.1	1.8	3	(n.s.t)	3	(n.s.t)			4	(n.s.t)	
	NH₄	53.0	33.3	36.3	2.8	3	(n.s.t)	4	(n.s.t)					
LV02	SO4*	34.3	37.5	-5.9	97.5	3	-6.34*	1	(n.s.t)	4	(n.s.t)	3	(n.s.t)	
	(Ca+Mg)*	18.0	26.2		587.8					4	(n.s.t)	2	(n.s.t)	
	H⁺	14.9	14.5		0.0	2	(n.s.t)	3	(n.s.t)	3	(n.s.t)	3	(n.s.t)	
	NO₃	20.5	19.7	-9.9	1.8	3	(n.s.t)	4	(n.s.t)	4	(n.s.t)	3	(n.s.t)	
	NH₄	35.6	20.7	34.5	2.8	3	-4.18**	4	(n.s.t)					
RU15	SO4*	50.5		-8.9	45.7	4	-13.4	4	-14.6	3	-18.3*	4	(n.s.t)	16
	(Ca+Mg)*	170.6	185.9		269.3	4	(n.s.t)	4	(n.s.t)	1	(n.s.t)	1	-30.6*	
	H⁺	51.5	28.0		0.1	1	(n.s.t)	1	(n.s.t)	3	(n.s.t)	3	0.01*	
	NO₃	10.4	9.0	0.4	8.5	4	-3.22	3	-2.51***	2	(n.s.t)	3	(n.s.t)	
	NH₄	31.1	36.1	3.5	10.3	4	(n.s.t)	1	(n.s.t)					
NL01	SO4*	36.8	74.9	-45.4		3	(n.s.t)	3	(n.s.t)					12
	(Ca+Mg)*	8.7	17.5			3	-2.50**	3	-3.02**					
	H⁺	18.0	9.6			4	-10.2	4	-15.7					
	NO₃	34.8	58.5	-33.8		3	(n.s.t)	3	-6.79*					
	NH₄	60.3	139.3	-16.0		2	(n.s.t)	4	(n.s.t)					
DE01	SO4*	33.3	53.6	-56.9	68.2	2	(n.s.t)	1	-4.23**	1	(n.s.t)	1	(n.s.t)	23
	(Ca+Mg)*	19.9	39.3		151.9	4	(n.s.t)	2	(n.s.t)	1	(n.s.t)	1	(n.s.t)	
	H⁺	24.4	28.2		0.6	3	(n.s.t)	4	-1.25	4	(n.s.t)	4	(n.s.t)	
	NO₃	34.4	23.9	-33.7	39.9	2	(n.s.t)	1	-2.93*	1	(n.s.t)	1	(n.s.t)	
	NH₄	33.1	12.1	-13.5	4.2	2	-2.94*	1	-5.15*					
GB02	SO4*	42.1		-59.4	84.4	3	(n.s.t)	2	(n.s.t)	1	(n.s.t)	3	(n.s.t)	24
	(Ca+Mg)*	14.2			75.3	1	0.78*	1	0.94**	1	(n.s.t)	4	(n.s.t)	
	H⁺	20.8			15.2	4	(n.s.t)	3	(n.s.t)	1	(n.s.t)	1	(n.s.t)	
	NO₃	28.5		-30.7	34.7	3	(n.s.t)	2	-0.77*	1	(n.s.t)	1	(n.s.t)	
	NH₄	38.1		-6.4		1	-1.20**	1	(n.s.t)					
CZ01	SO4*	42.6	114.8	-63.0	51.2	1	(n.s.t)	3	-4.39*	4	(n.s.t)	4	(n.s.t)	17
	(Ca+Mg)*	13.5	46.4		55.4	1	-1.07**	3	-4.84**	4	2.75	4	(n.s.t)	
	H⁺	18.2	30.0		0.0	2	-1.16**	3	-4.06**	3	(n.s.t)	3	(n.s.t)	
	NO₃	30.8	41.0	-30.8	4.5	2	-0.70*	2	(n.s.t)	4	(n.s.t)	4	-4.65	
	NH₄	38.6	45.5	3.2	0.1	1	(n.s.t)	2	(n.s.t)					
AT01	SO4	50.6	66.5	-50.1		2	(n.s.t)	4	(n.s.t)					
	(Ca+Mg)*													
	H⁺													
	NO₃	48.4	76.2	-32.8		3	(n.s.t)	2	(n.s.t)					
	NH₄	61.2	67.0	-15.9		4	(n.s.t)	4	(n.s.t)					
CH01	SO4*	55.3	69.9	-48.1	331.9	1	(n.s.t)	1	-1.44*	2	8.41*	1	(n.s.t)	19
	(Ca+Mg)*	26.2	48.9		4281.8	1	(n.s.t)	1	(n.s.t)	1	(n.s.t)	1	-23.6*	
	H⁺	29.7	32.3			3	(n.s.t)	1	(n.s.t)					
	NO₃	39.8	95.0	-37.8	23.3	1	(n.s.t)	1	-2.10*	1	(n.s.t)	1	(n.s.t)	
	NH₄	74.0	90.6	-6.9	3.7	1	(n.s.t)	1	(n.s.t)					
IT01	SO4*	25.6	30.1	-50.2	30.3	1	(n.s.t)	4	(n.s.t)	2	(n.s.t)	3	(n.s.t)	19
	(Ca+Mg)*	19.0	67.0		52.6	1	(n.s.t)	1	(n.s.t)	3	(n.s.t)	3	(n.s.t)	
	H⁺	6.1	5.9		2.0	3	(n.s.t)	3	(n.s.t)	3	(n.s.t)	3	(n.s.t)	
	NO₃	24.0	23.4	-39.7	9.9	1	(n.s.t)	2	-2.62*					
	NH₄	31.6	13.7	-2.1	0.7	1	(n.s.t)	1	(n.s.t)					
IT02	SO4*	20.8	25.2	-50.9	6.0	1	(n.s.t)	3	(n.s.t)	3	(n.s.t)	3	-8.90***	22
	(Ca+Mg)*	22.8	63.4		26.4	3	(n.s.t)	2	(n.s.t)	3	(n.s.t)	4	(n.s.t)	
	H⁺	5.3	2.0		0.4	3	(n.s.t)	3	(n.s.t)	3	(n.s.t)	3	(n.s.t)	
	NO₃	21.3	18.2	-40.4	1.6	1	(n.s.t)	2	(n.s.t)					
	NH₄	24.6	23.2	0.3	0.3	1	(n.s.t)	1	(n.s.t)					

LONG TERM TREND OF CHEMICAL CONSTITUENTS IN TOKYO METROPOLITAN AREA IN JAPAN

YASUSHI NARITA[1], KEI SATOH[1], KEIICHI HAYASHI[1],
TAMAMI IWASE[1], SHIGERU TANAKA[1*], YUKIKO DOKIYA[2],
MORIKAZU HOSOE[3] and KAZUHIKO HAYASHI[4]

[1]*Keio University, Faculty of Science and Technology, 3-14-1 Hiyoshi, Kohoku-ku, Yokohama, 223-8522, Japan;* [2]*Tokyo University of Agriculture and Technology, 3-5-8 Saiwai-cho,Fuchu, Tokyo, 183-8509, Japan;* [3]*National Defense Academy of Japan, Department of Earth and Ocean Sciences, Hashirimizu 1-10-20,Yokosuka, Kanagawa, 239-8686, Japan;* [4]*Meteorological College, 7-4-81, Asahi-cho, Kashiwa, Chiba, 277-0852, Japan*
(* author for correspondence, e-mail: tanaka@applc.keio.ac.jp)

Abstract. In recent years, acid rain has been a social problem all over the world. In Japan, it is also a big problem especially in the metropolitan area. Then, we have measured major ions such as H^+, Na^+, NH_4^+, K^+, Mg^{2+}, Ca^{2+}, Cl^-, NO_3^-, and SO_4^{2-} in precipitation and dry deposition samples which had been collected at 9 sampling sites at Hiyoshi, Mita, Kashiwa, Shiki, Fujisawa, Yokosuka, Mitaka, Hachiouji, and Ashikaga in Tokyo Metropolitan area for 10 years since 1990. The average pH of precipitation in their sites was 4.56 (n=1906). As the results of multiple regression analysis showed that pH of precipitation was determined by 5 ions such as NH_4^+, nssCa^{2+}(non sea salt calcium), nssCl$^-$(non sea salt chloride), NO_3^-, nssSO$_4^{2-}$(non sea salt sulfate) in the most of the sampling sites. Therefore, it is very important to investigate the behavior of these ions to understand the acidification of rain in Tokyo Metropolitan area. In this study, a long term trend of each ion concentration in precipitation and wet deposition was also investigated the base on the data we had observed at 7 sites for 10 years by the statistical method.

Keywords: precipitation, wet deposition, long term trend, chemical constituents, Tokyo Metropolitan area, network observation

1. Introduction

Destruction of ecosystem by acid rain water has been a serious social problem in recent years (Mohnen, 1988). This acid rain phenomenon occurs on account of air pollutant that is taken in by rain. In order to understand its actual state and mechanism, it is important to collect precipitation on a continuous and regular basis in wide areas and to observe pH and chemical components of the precipitation (Lynch et al., 1995; Ohkita, 1994). In Japan, wide-area research on acid rain has been implemented by the Environment Agency, to begin with, and municipalities.

In parallel with acid rain observation by the administration (Environment Agency Government of Japan, 1994), this study aims at carrying out acid rain monitoring in Tokyo Metropolitan area in Japan based upon network observations by educational organizations, mainly consisting of universities. These observations, which started in June 1990, have been continued to the present. Automatic precipitation samplers were installed at 9 educational to for measurements of pH and precipitation chemistry. Based upon enormous data on pH and precipitation chemistry collected through the 10-year network observations, and in order to understand the actual state of acid rain in Tokyo Metropolitan area, investigations were carried out with respect to the relationship between pH and concentrations of chemical constituents in precipitation chemistry with geographical factors and meteorological conditions the long term trend of pH and concentrations of chemical constituents in precipitation over 10 years in Tokyo Metropolitan area in Japan was also investigated.

2. Experiment

2.1. PRECIPITATION SAMPLING SITES

A rain-event precipitation has been collected at 9 sites in Tokyo Metropolitan area as shown in Fig. 1. Hiyoshi is located 10 km in the west of the Keihin industrial belt and about 10 km away also from the coastline (Tokyo Bay). Mita, which can be said a typical business area, is located in the central part of Tokyo and faces the coastline (Tokyo Bay). Fujisawa, Kashiwa, and Shiki are located on the outskirts of Tokyo. While Fujisawa is about 8 km away from the coastline (Sagami Bay), and Kashiwa is 20 km away from Tokyo Bay. Shiki, located comparatively on inland about 40 km away from the coastline, is a typical bedroom community in the suburb of Tokyo. Mitaka is in a residential area about 20 km away from the coastline. Ashikaga is located inland the most among the 9 sites and considerably (about 90 km) away from the Keihin industrial belt and Tokyo Bay. Yokosuka directly faces the coastline, and accordingly it is located nearest to the ocean among the 9 sites. Hachioji is a bedroom community located inland on the outskirts of Tokyo.

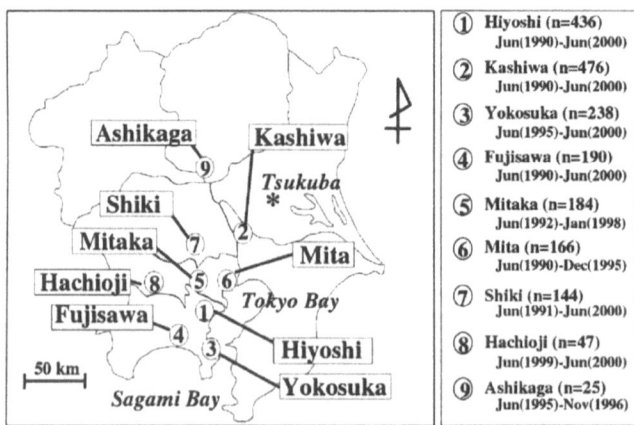

Figure 1. The precipitation sampling sites in Tokyo Metropolitan area.

2.2. COLLECTION AND ANALYSIS OF PRECIPITATION SAMPLES

The precipitation samplers manufactured by Koshin Denki Kogyo Co., Ltd.,

DRS-154W were used to collect the samples. These samplers can separately collect precipitation and dry deposition. The precipitation samplers were installed on the roof of building (10 m above the ground) at each sampling site. The devices are equipped with a rainfall sensor, and it is designed that the lid of the device can be automatically opened and shuts by rainfall. Regarding a continuous rainfall as one sample of precipitation, the sample was sucked and filtrated, immediately after its collection, with a membrane filter (Millipore type - HAWP, pore size: 0.45μm). The sample solution was then put in a polyethylene bottle. Precipitation samples that were collected at 8 sites (other than Hiyoshi) were delivered to the Faculty of Science and Technology, Keio University and were stored in a refrigerator (approx. 4℃). The pH of precipitation was measured using a pH meter (manufactured by Horiba Ltd. pH/ion meter F-24). The H^+ concentration in precipitation was determined by calculation out of pH values. The anions (Cl^-, NO_3^-, and SO_4^{2-}) and cations (Na^+, NH_4^+, K^+, Ca^{2+}, and Mg^{2+}) in precipitation were analyzed with an ion chromatograph (Yokokawa Analytical Systems Co., Ltd., IC7000D) (Yataki et al., 1994).

3. Results and Discussion

3.1. ANALYTICAL RESULT IN TOKYO METROPOLITAN AREA

Table I shows the results of measurement of chemical constituents concentrations in precipitation at respective sampling sites (Hiyoshi, Mita, Shiki, Fujisawa, Kashiwa, Mitaka, Ashikaga, Yokosuka, and Hachioji) in the Metropolitan area. Since concentrations in precipitation are largely affected by precipitation volume, the weighted mean value taking precipitation volume into consideration. It was found that rain within a range from pH 4.23 to 4.76 below pH 5.6, which was the defined value for acid rain, was fallen at all 9 sites and revealed acidification of precipitation all over Tokyo Metropolitan area.

3.2. REGIONAL FEATURES OF CONCENTRATIONS IN PRECIPITATION

It was recognized that Yokosuka and Fujisawa, which were located near the sea, had high concentrations of Na^+, Mg^{2+}, and Cl^- ions in precipitation originated from sea salt. Meanwhile, in Kashiwa and Hachioji, which were located comparatively in the suburbs and far from factories and the sea, it was found that almost all the chemical constituents concentrations were low. The precipitation at Mitaka had high concentrations of NH_4^+, Ca^{2+}, and NO_3^-, probably due to the influence from heavy traffic. At Ashikaga, which was located most inland, the concentrations of Na^+, Mg^{2+}, and Cl^- ions originated from sea salt were very low. On the contrary, NO_3^-, SO_4^{2-}, and NH_4^+ ions, which were generated from anthropogenic atmospheric pollutants generated in the center of Tokyo and long range transport showed high concentrations.

TABLE I
Measurement results of chemical constituents concentrations in precipitation
at 9 sampling sites in Tokyo Metropolitan area.

sampling sites	Number of Sample n	Total amount of precipitation (mm)		pH	H^+	Na^+	NH_4^+	K^+	Ca^{2+}	Mg^{2+}	Cl^-	NO_3^-	SO_4^{2-}	Cation	Anion	Balance
Hiyoshi[1]	436	13198	Av.	4.54	28.8	26.6	39.2	2.3	21.3	8.5	43.2	27.0	45.4	131.3	115.6	1.14
			Max.	5.97	219.3	461.5	587.1	49.5	647.4	104.7	616.7	349.9	598.4	1615.1	1354.4	
			Min.	3.66	1.1	0.3	5.8	0.1	0.9	0.1	6.3	2.2	6.2	25.9	18.4	
Kashiwa[2]	476	11468	Av.	4.62	23.8	27.7	38.2	2.9	23.9	9.5	40.5	28.7	46.3	130.5	115.5	1.13
			Max.	5.96	436.5	385.0	670.4	50.3	550.9	109.1	579.0	414.8	791.7	1522.4	1638.5	
			Min.	3.37	1.1	0.1	2.5	0.1	0.7	0.1	2.3	0.5	7.4	12.0	16.6	
Yokosuka[3]	238	6538	Av.	4.44	35.9	75.2	34.1	2.8	24.0	19.3	101.1	23.8	54.5	195.6	179.4	1.09
			Max.	5.98	302.0	981.8	266.6	24.6	262.4	216.0	1285.0	240.3	281.4	1467.9	1530.5	
			Min.	3.52	1.0	1.3	5.1	0.2	0.8	2.3	4.6	3.6	17.2	36.8	29.3	
Fujisawa[4]	190	10889	Av.	4.66	21.8	46.2	38.2	3.4	25.4	14.7	64.8	28.1	44.6	154.3	137.5	1.12
			Max.	5.98	186.2	1294.6	1736.4	105.1	1583.6	450.3	1454.2	1291.6	2132.8	5183.8	4878.6	
			Min.	3.73	1.0	2.2	4.3	0.1	2.1	0.3	11.0	2.5	6.4	23.0	20.8	
Mitaka[5]	184	6247	Av.	4.56	27.8	21.4	56.5	3.4	35.9	8.5	46.4	40.0	50.1	158.1	136.4	1.16
			Max.	5.97	537.0	309.5	1158.0	53.0	687.2	111.5	528.2	374.1	1044.0	1975.9	1572.2	
			Min.	3.27	1.1	1.5	12.0	0.1	1.5	1.1	9.5	5.7	12.1	36.1	36.2	
Mita[6]	166	7495	Av.	4.57	26.7	26.7	49.5	2.6	21.7	9.4	45.5	30.0	54.6	141.1	130.1	1.08
			Max.	5.91	190.5	420.3	698.9	29.3	775.8	126.0	542.5	341.7	836.4	1900.1	1530.4	
			Min.	3.72	1.5	0.9	4.3	0.1	0.8	0.9	6.8	2.6	7.1	20.7	20.0	
Shiki[7]	144	10905	Av.	4.54	29.1	21.5	45.2	3.9	36.8	8.9	42.8	42.8	47.1	150.0	132.6	1.13
			Max.	5.90	120.2	214.1	336.3	46.5	516.7	78.3	291.6	420.3	279.5	966.7	859.5	
			Min.	3.92	1.3	2.0	8.9	0.1	2.1	0.1	10.4	3.2	7.1	22.6	20.7	
Hachioji[8]	47	1203	Av.	4.63	23.2	16.9	35.1	3.2	12.6	6.6	23.5	29.6	29.6	102.2	82.7	1.24
			Max.	5.91	177.8	2.4	346.6	22.1	173.2	83.1	154.5	258.4	163.5	636.8	576.4	
			Min.	3.75	1.2	121.1	7.2	0.3	1.1	0.2	5.7	3.0	4.3	17.6	17.8	
Ashikaga[9]	25	353	Av.	4.23	59.3	13.4	67.3	4.1	33.5	6.2	34.1	73.6	76.2	188.0	183.8	1.02
			Max.	4.69	190.5	78.7	524.9	25.2	257.4	50.5	147.7	521.3	503.1	1088.1	1172.0	
			Min.	3.72	20.4	2.2	20.7	0.7	6.4	1.1	15.3	18.1	31.1	65.5	64.8	
Total	1906	68296		4.56	27.3	33.0	42.1	3.0	26.4	10.8	51.8	31.5	47.9	147.3	131.2	1.12

1) June 1990-June 2000; 2) June 1990 -June 2000; 3) June 1995 –June 2000; 4) June 1990 – June 2000; 5) June 1992-January 1998; 6) June 1990 – December 1995; 7) June 1991 - June 2000; 8) June 1999 - June2000; 9) June 1995 – November 1996; Cation: Total concentration of Cations; Anion: Total concentration of Anions; Balance: Cation / Anion

Thus, it is found that precipitation chemistry in Tokyo Metropolitan area are largely influenced by local geographical factors. And the results of multiple regression analysis showed that pH of precipitation was determined by 5 ions such as NH_4^+, $nssCa^{2+}$, $nssCl^-$ (non sea salt chloride), NO_3^-, $nssSO_4^{2-}$ (non sea salt sulfate) in the most of the sampling sites. It can be said that H^+ ion concentration in precipitation, namely pH, can be explained by the acidity potential defined by equation (1) as shown in Fig. 2.

[Acidity potential] = [nss-Cl^-] + [NO_3^-] + [nss-SO_4^{2-}] - [NH_4^+] - [nss-Ca^{2+}]   (1)

3.3. LONG TERM TREND OF CONCENTRATIONS IN PRECIPITATION

Precipitation pH and concentrations have been observed on a continuous basis for over 10 years since 1990 at 2 sites of Hiyoshi and Kashiwa. The upper part of Fig. 3 shows the long term trend of pH of precipitation over 10 years at Hiyoshi and Kashiwa. All data on pH of precipitation were plotted in Fig.3, and the moving averages of pH in precipitation for every 10 samples were also shown in Fig.3. Although Fig. 3 reveals a seasonal fluctuation in which pH decreases in the summer season (June, July, and August), pH of precipitation at

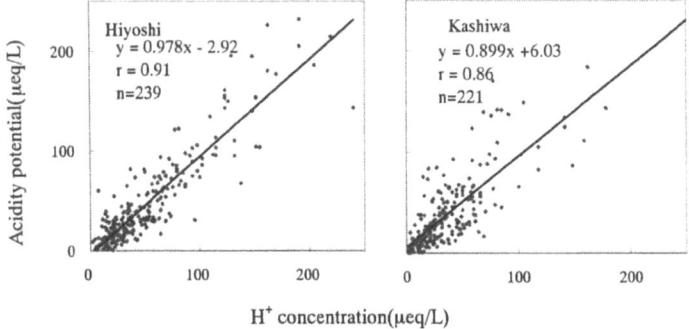

Figure 2. The relationship between H⁺ concentration and acidic potential (Jun. 1990- Jun.2000)

Hiyoshi and Kashiwa has been hovering at almost the same level for 10 years since 1990 when the observation started.

The lower part of Fig.3 shows the long term trend of SO_4^{2-} concentration in precipitation measured at Hiyoshi and Kashiwa during the period from June 1990 to June 2000. Assuming that the relationship between the sulfate concentration in precipitation (C, monthly weighted mean, µeq/L) and time (t, month) is given by the following equation (log (C) = a + bt). A correlation coefficient (r) was employed to test the significance of the primary regression equation that was calculated by the least square method.

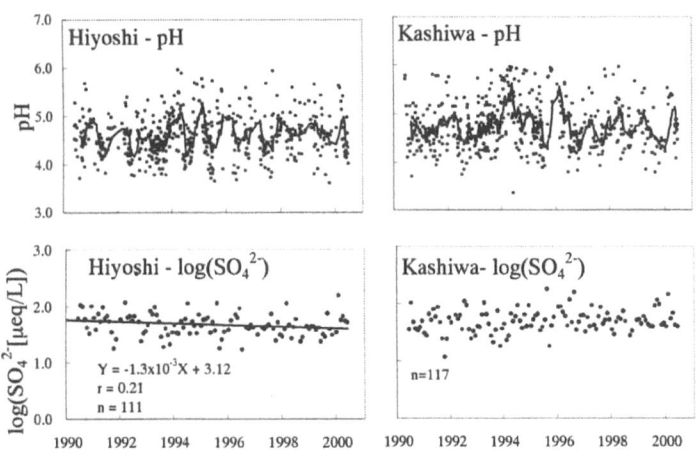

Figure 3. Long term trend of pH and sulfate concentration of precipitation at Hiyoshi and Kashiwa (June 1990-June 2000).

When the level of significance was assumed to be 5%, the slope of the primary regression equation was minus only at Hiyoshi, it was found that SO_4^{2-} concentration in precipitation had been on the decrease at an annual rate of 4.1% for the past 10 years. Likewise, the long term trend was investigated with respect to the NO_3^- and Ca^{2+} concentrations in precipitation measured at 6 sites,

Hiyoshi, Kashiwa, Yokosuka, Fujisawa, Shiki, and Mitaka. As regards the NO_3^- concentration in precipitation, a statistically significant result at the 5% level was obtained only at Kashiwa, the slope of linear line was plus, it was confirmed that the NO_3^- concentration in precipitation at Kashiwa had been on the increase at an annual rate of 6.4%. As to NH_4^+ ion, a significant result at the 5% level was obtained at Kashiwa and Mitaka where the NH_4^+ concentration in precipitation had been on the increase at annual rates of 7.6% and 8.9%, respectively. As to Ca^{2+} ion, furthermore, a significant result at the 5% level was obtained at Kashiwa, Fujisawa, Shiki, and Mitaka where Ca^{2+} concentration in precipitation had been on the increase at annual rates of 14.4%, 11.7%, 13.3% and 12.3%, respectively.

4. Conclusion

Precipitation was collected on a continuous basis for a period of 10 years from June 1990 to March 2000 at 9 sites in Tokyo Metropolitan area in Japan, and pH and chemical constituents concentrations in precipitation were measured. As regards concentrations of SO_4^{2-}, NO_3^-, H^+, NH_4^+, and Ca^{2+}, which were the main chemical constituents in precipitation at 7 sites in Tokyo Metropolitan area, their long term trend for 10 years was investigated. As a result, it was recognized that the SO_4^{2-} concentration was on the decrease in Hiyoshi, and the NO_3^- and NH_4^+ concentrations were on the increase in Kashiwa, and Ca^{2+} concentration was on the increase in Kashiwa, Fujisawa, and Shiki at the 5% level of significance. However, there was no outstanding increase or decrease in H^+ concentration, revealing the fact that the acidification of precipitation had not been progressing for the past 10 years in 1990s over Tokyo Metropolitan area.

Acknowledgements

The authors would like to thank the persons concerned for collecting precipitation at Keio Girl Senior High School, Keio Shiki Boys Senior High School, Keio Shonan Fujisawa Junior School and Senior High School, Yokogawa Analytical Systems, Co., Ltd., and Yokogawa Weathac Co., Ltd.

References

Environment Agency Government of Japan: 1994, *Technical Report for A Five-Year Study (Phase II Survey) of acid deposition*, Environment Agency Government of Japan (in Japanese).
Lynch, J.A., Grim J.W., Bowersox, V.C.:1995, *Atmospheric Environment* **29**, 11, 1231-1246.
Mohnen, V. A.: 1988, *Scientific American* **259**, 2, 14-22.
Ohkita T.:1994, *Kisho-kenkyu note* **23**, 41-60 (in Japanese).
Yataki, R., Umeda, M., Ohtsu, T., Onoguchi, A., Tanaka, S., Maruta, E., Tamura, S., Takagi, I., Dokiya, Y., Ono, M., Inomata, T. and Ohtoshi, T.: 1994, *Journal of Resources and Environment* **30**, 13, 238-244 (in Japanese).

SEA ICE APPROACH AND CHEMICAL SPECIES IN PRECIPITATION AT ABASHIRI, JAPAN

MASAKI ADACHI[1*], KAZUHIKO HAYASHI[2] and YUKIKO DOKIYA[3]

[1] *Sapporo Meteorological Observatory, North 2 West 18 Sapporo, Hokkaido 062-0002, Japan;*
[2] *Meteorological College, 7-4-81 Asahi-chou, Kashiwa, Chiba 277-0852, Japan;* [3] *Faculty of Agriculture, Tokyo Univ. of Agr.&Tech. 3-5-8 Saiwai-chou, Fuchu, Tokyo 183-8509, Japan*
(*author for correspondence, e-mail: m-adachi@met.kishou.go.jp)*

Abstract. Abashiri is a rural city on Hokkaido Island, Japan. It lies directly to the south of the Okhotsk Sea, which is the lowest latitude sea to freeze. We collected daily deposition samples over two periods: from Jan. 1997 to Mar. 1998, and from Nov. 1998 to Mar. 1999. The average concentrations of anthropogenic chemical species (NO_3^-, non-seasalt(nss)-SO_4^{2-}) were relatively low and those of seasalt species (Na^+, Cl^-, Mg^{2+}) were high in Japanese precipitation samples. During the period of study, we found that, when sea ice forms and approaches the coast, concentrations of seasalt species become lower, while almost no changes are found in the anthropogenic chemical species.

Keywords: daily deposition, sea ice, sea-salt and anthropogenic chemical species

1. Introduction

In Japan, various institutes have done many regional or nationwide studies on acidification. For example, the Japan Meteorological Agency (JMA) has one global station and two regional stations which monitor the chemical composition of precipitation and greenhouse gas concentrations (JMA, 2000), in accordance with the Global Atmosphere Watch (GAW) programme established by the World Meteorological Organization. The Japan Environmental Agency (JEA) has performed three studies on acidification nationwide (JEA, 1999). Ishii *et al.* (1993) carried out observations of snow cover in Hokkaido Island. However, there have only been a few studies on acidification of precipitation performed in the coastal area of the Okhotsk Sea.

The north and east sides of Hokkaido are two of the cleanest areas in Japan. Abashiri city is located in the eastern part of Hokkaido (44° N, 144° E), and faces the Okhotsk Sea (Fig.1). The population is approximately 40,000, and there is an agricultural zone surrounding the city. Annually, there is less snowfall in Abashiri than in the other coastal areas bordering the Japan Sea, due to the presence of high mountains in the west of Abashiri (see Fig.1). The amount of average annual precipitation is 815.3 mm (approximately 25% is snow).

In this paper, we discuss the chemical characteristics of precipitation in the Abashiri area, including its acidification and meteorological condition classification. Furthermore, we also discuss the change in concentrations due to the approach of sea ice to the coast.

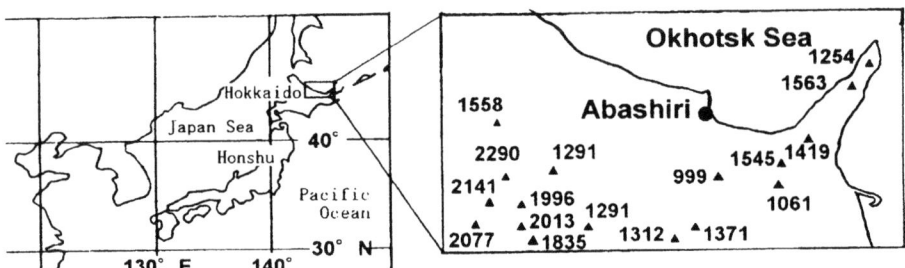

Figure 1. Map of Japan and Abashiri area (enlarged): The heights of mountains are shown in meters

2. Materials and Methods

In this study, daily total deposition samples were collected at Abashiri Meteorological Observatory over two periods: from Jan. 1997 to Mar. 1998, and from Nov. 1998 to Mar. 1999. Four polyethylene buckets (15cm height, 11.5cm diameter) were set out every morning in order to obtain deposition samples. When we had enough precipitation to analyze (usually approximately 40ml or more, which equals a precipitation depth of 1mm), samples were then weighed and filtered through a membrane filter (Millipore 0.45 μm pore size). Samples were then poured into polyethylene bottles and stored in a refrigerator. If we had no or insufficient precipitation, samples were not collected. We collected 134 samples in the first period, and 50 samples in the second period, which were then sent to the Meteorological College to be tested for determination of chemical species.

Electrical conductivity (EC) was measured with a conductivity meter (DKK, AOL-30). pH was measured with a pH meter (DKK IOL-50 with a 6157-W type electrode). There were 11 samples for which EC and pH could not be measured due to insufficient volume. The concentrations of major chemical species (Cl^-, NO_2^-, NO_3^-, SO_4^{2-}, NH_4^+, Na^+, K^+, Ca^{2+}, Mg^{2+}) were determined by ion chromatography (Yokogawa Electric Corporation, YEW-IC7000).

The mean concentrations are "precipitation amount weighted" averages. The nss-SO_4^{2-} was calculated on the assumption that all Na^+ originates from seasalt and the equivalent ratio of SO_4^{2-}/Na^+ (0.12) in the sea is conserved.

3. Results and Discussion

3.1. Concentrations and deposition amounts of chemical species

The results are shown in Table I, with the mean values of the earlier JEA study (Phase-II) at 29 sites from 1989 to 1993 (Hara et al., 1995) also shown for comparison. In this paper, these values are labeled as JEA.Mean. For purposes of comparison with the JEA.Mean, the 1997 mean of Abashiri samples was calculated, in addition to the mean of all samples.

The pH ranged from 4.01 to 6.60. There was no seasonal change. The annual mean pH value is equal to the JEA.Mean, but our mean concentrations of NO_3^- and nss-SO_4^{2-} were much lower than the JEA.Mean. NO_2^- was detected in only two samples (0.87 μ eq/l from thermal instability on 5/Aug/97 and 0.65 μ eq/l from an extratropical cyclone on 12/Aug/97). Our reports on the mean concentrations of seasalt species (Na^+, Cl^-, Mg^{2+}) are higher than those of the JEA.Mean. The sum of Na^+, Cl^-, Mg^{2+} and seasalt-SO_4^{2-} accounts for 75% of the total equivalent concentration, in contrast with 55% of the JEA.Mean. It should also be mentioned that the precipitation at Abashiri has a low anthropogenic pollution level and is influenced strongly by seasalt.

TABLE I

Weighted mean concentrations of chemical species in precipitation (μ eq/L)

	pH	Na^+	NH_4^+	K^+	Ca^{2+}	Mg^{2+}	Cl^-	NO_3^-	SO_4^{2-}	nssSO_4^{2-}
Abashiri (all)	4.9	125.9	20.3	7.6	19.7	31.4	161.5	10.1	37.0	21.9
25% percentile	4.7	25.2	12.7	1.8	13.3	8.2	35.8	7.3	23.8	14.2
75% percentile	5.5	178.7	35.2	7.0	37.9	43.0	238.9	16.8	61.7	36.0
Abashiri (1997)	4.8	103.9	21.6	4.1	17.3	26.5	128.6	10.4	35.0	22.5
JEA	4.8	49.1	18.3	3.1	16.0	13.6	63.5	14.1	44.4	38.6

The concentrations and depositions of some species showed seasonal changes. The concentrations of seasalt species were high in early winter. It is because of the strong north wind. The concentrations of NO_3^- and NH_4^+ were rather high in summer. During this study, heavy rain occurred in Aug. and Nov. of 1997. Consequently, the deposition amount of NO_3^- and nss-SO_4^{2-} increased during the summer, especially in Aug. The deposition amount of seasalt species had some peaks: Jan., Apr. and Nov. of 1997, and Nov. of 1998. It is known that the deposition of nss-SO_4^{2-} increases on the Japan Sea side of Honshu in winter (JEA, 1999). This trend was conspicuous when the winter monsoon pressure pattern intensified (Adachi et al., 1997). In our study, no similar trend was found at Abashiri.

3.2. Characteristic features of chemical species in the samples according to the meteorological conditions

We classified the samples into five groups according to meteorological conditions. The five groups are 1) cyclonic precipitation (type-C, 103 samples), 2) precipitation by trough or upper + vortex (type-V, 25 samples), 3) rain by thermal instability (type-I, 3 samples), 4) heavy rain caused by typhoon (type-T, 2 samples), and 5) snow with winter monsoon (type-M, 35 samples). We were unable to classify 16 of the samples because their meteorological conditions were not clear enough to define. The concentrations of each group are shown in Table II. The mean values of all samples are also shown for comparison.

Table II
The concentrations of chemical species in each meteorological condition (μ eq/L)
depth: the average of precipitation amount in millimeters. N/S ratio: $NO_3^-/nss\text{-}SO_4^{2-}$.

Type	depth	pH	Na^+	NH_4^+	K^+	Ca^{2+}	Mg^{2+}	Cl^-	NO_3^-	SO_4^{2-}	$nssSO_4^{2-}$	N/S ratio
C	6.8	4.9	91.8	17.7	8.3	16.7	23.4	119.0	9.8	31.2	20.2	0.5
V	2.0	4.9	165.2	40.3	6.1	35.4	43.2	206.9	12.2	48.7	28.9	0.4
I	17.8	4.6	6.3	22.9	0.6	6.5	0.8	6.4	11.8	32.7	31.9	0.4
T	13.3	5.3	132.1	12.1	3.4	12.6	30.5	164.9	5.7	22.8	7.0	0.8
M	3.5	5.1	367.6	20.4	9.8	36.2	87.0	465.9	9.1	64.8	20.8	0.4
all samples	5.4	4.9	125.9	20.3	7.6	19.7	31.4	161.5	10.1	37.0	21.9	0.5

The characteristic features of each group are summarized as follows: <u>Type-C</u>: More than half of the samples were classified as this type. The concentrations of seasalt species were comparatively low, while the concentrations of other ions were nearly equal to the mean concentrations of all samples at Abashiri. <u>Type-V</u>: The concentrations of chemical species were rather high, because the arithmetical mean of precipitation depth was the smallest among all the groups. <u>Type-I</u>: The concentrations of anthropogenic species were high, while the amounts of precipitation were also high. On the other hand, the concentrations of seasalt species were very low. Consequently, nss-SO_4^{2-} and NO_3^- accounted for about forty percent of the total equivalent concentration. Clouds grew inland and droplets captured pollutants including acidic sulfate and nitrate. <u>Type-T</u>: The equivalent ratio of $NO_3^-/nss\text{-}SO_4^{2-}$ was 0.8, different from other groups (about 0.4). This value suggests an influence of NO_3^- from Honshu Island, since typhoons made their way northward fast thorough Honshu. <u>Type-M</u>: The wind direction of this group was mainly NW-N (from the sea), so the concentrations of sea-salt species were very high. On the other hand, those of nss-SO_4^{2-} were almost equal to other groups. This suggests that winter monsoons did not transport SO_4^{2-} to Abashiri from the continent, presumably because of the high mountains in the middle of Hokkaido Island. Some different simulation (such as to use the back trajectory method) may be needed for more precise discussions.

3.3 The effect of the approach of sea ice

In the Okhotsk Sea, sea ice normally starts to form in the northern part of the sea around the first of November, it then spreads and maximizes around the end of February (Fig.2). At Abashiri, the normal first date of drift ice to the shore is Jan.28th.

The concentrations of NO_3^- did not change with the approach of sea ice (Fig.3.a), and the same is true of nss-SO_4^{2-}. On the other hand, the concentrations of seasalt species lessened after the approach of sea ice (Fig.3.b). This was due to a decrease in the supply of seasalt into the air, caused by the ice. In the case of seasalt, the change was significant at a level of 5 % in Z tests for each year.

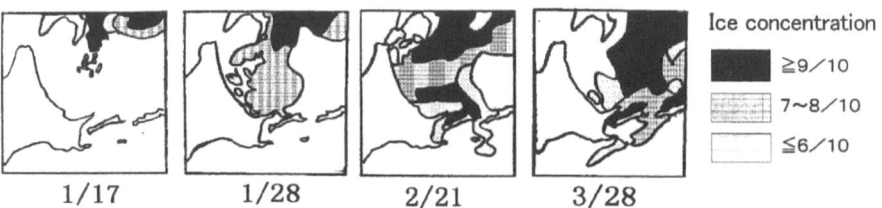

Figure 2. Distribution of sea ice in the southern Okhotsk Sea (1997)

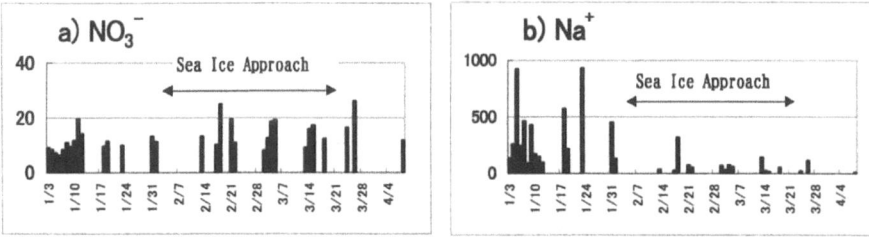

Figure 3. The changes of concentrations of chemical species in winter, 1997. a) NO_3^-, b) Na^+ (μ eq/L) In 1997, the "First date of drift ice to the shore" was Jan.31. After this date, the concentration of Na^+ decreased dramatically.

Conclusions

Daily total deposition samples were collected at Abashiri for precipitation events of approximately 1mm or greater. On average, the concentrations of anthropogenic chemical species were relatively low and those of seasalt species were relatively high.

After classifying the samples according to their meteorological conditions, some interesting features were found. The concentrations of anthropogenic species were especially high when it rained from thermal instability. The concentrations of seasalt species were very high when winter north monsoon winds blew from the sea.

When sea ice approaches the coast, the concentrations of seasalt species lessened, while the concentrations of anthropogenic species showed little change. In Abashiri area, this is the first report on the change of the concentrations of seasalt chemical species in the precipitation of Abashiri caused by sea ice approach, which may have implications for future research on the mechanism of in-cloud and below-cloud scavenging of chemical species.

Acknowledgment

The authors would like to express their gratitude to Mr. T. Aizawa and Miss C. Horikoshi for their help in chemical analysis, and Mr. K. Kimura, Mr. T. Tsuji and Miss A. Nakamura for their help in collecting daily samples. Thanks are also due to the anonymous reviewers who made this manuscript better.

References

Adachi, M., Yamamoto, O., Nabata, T., Dokiya, Y.: 1997, 'Chemical Species in Japanese Snow', *Proceedings of International Congress of Acid Snow and Rain 1997*, pp86-91

Hara, H., Kitamura M., Mori A., Noguchi I., Ohizumi T., Seto S., Takeuchi T. and Deguchi T.: 1995, *Water, Air and Soil Pollution* **85**, 2307

Ishii, Y., Akitaya, E., Nomura, M.: 1993, *Teionkagaku(Low Temperature Sciences)* A,**51**. Data Report. 9 (in Japanese)

Japan Environmental Agency: 1999, *The Summary of Phase-3 Survey on Acidification.* Japan Environmental Agency, Tokyo, Japan (in Japanese)

Japan Meteorological Agency: 2000, *Annual Report of Background Air Pollution Observation 1998.* Japan Meteorological Agency, Tokyo, Japan

World Meteorological Organization Global Atmosphere Watch: *Chemical Analysis of Precipitation for GAW: Laboratory Analytical Methods and Sample Collection Standards (TD-NO 550).* World Meteorological Organization, Geneva, Switzerland

LONG-TERM MONITORING STUDY ON RAIN, THROUGHFALL, AND STEMFLOW CHEMISTRY IN EVERGREEN CONIFOROUS FORESTS IN HOKKAIDO, NORTHERN JAPAN

Y. MATSUURA[1]*, M. SANADA[2], M. TAKAHASHI[3], Y. SAKAI[1], and N. TANAKA[1]

[1] *Hokkaido Research Center, Forestry and Forest Products Research Institute (FFPRI), Ministry of Agr., For., Fish., 7 Hitsujigaoka, Sapporo 062-8516, Japan;* [2] *Hokkaido Branch Office, Forest Development Technological Institute, 7 Hitsujigaoka, Sapporo 062-8516, Japan;* [3] *Forest Environment Division, FFPRI, Kukizaki, Ibaraki 305-8687, Japan*

(* author for correspondence, e-mail: orijoy@ffpri-hkd.affrc.go.jp)

Abstract. Long-term study on acid precipitation monitoring at suburban forests in Sapporo city showed that bulk precipitation pH were below 4.8 in recent years. Throughfall and stemflow chemistry for two main coniferous species (*Abies sachalinensis* and *Picea jezoensis*) showed different regime for pH and element deposition. The mean annual pH values of throughfall and stemflow in *Picea* stand were 1.0 to 1.3 units higher than that of rain collected outside the forest. In contrast, mean annual pH of throughfall and stemflow in *Abies* stand were 0.3 to 0.5 units higher than that of rain. Mean annual inorganic nitrogen input via throughfall and stemflow were estimated 0.41 ± 0.11 gN/m^2/yr in *Abies* stand, 0.44 ± 0.13 gN/m^2/yr in *Picea* stand. Cation input via throughfall, especially for K, in *Picea* stand was 1.4 times as large as that in *Abies* stand. Mean annual input of S in both stands was the same level. The possible effects on surface soil properties and nutrient cycling in northern evergreen conifers was discussed.

Key words: acid deposition, evergreen coniferous forest ecosystems, throughfall, stemflow

1. Introduction

Acid deposition effect on terrestrial ecosystems is one of the central environmental problems. Several researches focused on the effect of nitrogen deposition on forest growth, nitrogen retention in soil and critical load for forest soil buffering capacity in regional scale (Schulze, 1989; Aber *et al.* 1998; Nadelhoffer *et al.* 1999). From another point of view, acid deposition has an important effect on nutrient cycling and soil processes at the stand level. There are many studies on throughfall and stemflow chemistry and the long-term effect on soil properties in forest ecosystems (Parker, 1983; Falkengren-Grerup, 1989; Matschonat and Falkengren-Grerup, 2000; Nilsson *et al.*, 1998).

In this paper, we present inorganic nitrogen (inorganic-N; NH$_4$-N and NO$_3$-N), potassium (K) and Sulfur (S) inputs by throughfall and stemflow to the two northern evergreen coniferous stands and discuss the possible effects on forest ecosystem in northern Japan.

2. Materials and Methods

2.1. STUDY SITE

Acid rain monitoring study has been carried out in experimental forest stands of the Hokkaido Research Center, in the southeast of Sapporo city, located on about 43° 00'N-141° 24'E. The altitude of the experiment forest stands is about 150 m above sea level. Mean annual temperature of experiment station during these 25 years is 7.2°C. Mean annual precipitation is about 1000 mm, of which 20 to 30 % is precipitate as snow.

Experiment stands of evergreen coniferous species were planted in 1973. We have selected one *Abies sachalinensis* (Fr. Schmidt) Masters and one *Picea jezoensis* (Sieb. et Zucc.) Carriere stand for throughfall and stemflow sampling since 1990, except for snow sampling.

2.2. SAMPLING, CHEMICAL ANALYSIS AND NUTRIENT INPUT ESTIMATION

Three replicates of throughfall collect boxes and stemflow collect rubber bands were set in both stands. Bulk precipitation samples outside the forest were also collected by plastic funnel. From 1990 to 1996, we collected the samples at every precipitation event, and biweekly sampling from 1997. These routine sampling was conducted between snow-free period, normally from middle April to the beginning of December.

Samples taken into the laboratory were filtered. After filtering, we measured pH and stored in the refrigerator until sample processing for chemical analysis. Samples collected from 1992 were analyzed for cations and anions by ion chromatography method using IC-7000 (Yokogawa Analytical Systems, Ltd.).

We estimated nutrient input to the forest stands by multiplying dissolved nutrient concentrations in bulk precipitation, throughfall and stemflow samples and their amounts for each event during snow-free period, and sum up annual input from the year 1992 to 1999. Sulfur of non-sea salt origin (nss-S) was calculated by sodium (Na) concentrations in samples, assuming that Na origin was sea salt, and that weight ratio of $SO_4/Na = 0.2520$ (Hara, 1992).

3. Results

Mean annual pH of bulk precipitation was the lowest among precipitation, throughfall and stemflow (Fig. 1). Bulk precipitation pH changed below 4.8. Throughfall and stemflow pH in *Picea* stand showed 1.0 to 1.3 units higher than that of bulk precipitation. The pH of throughfall and stemflow in *Abies* stand showed 0.3 to 0.5 units higher than that of bulk precipitation. Throughfall pH in *Abies* stand was 0.1 to 0.3 units higher than that of stemflow pH, on the other hand, the rank of pH between throughfall and stemflow varied year to year in *Picea* stand (Fig. 1).

Mean annual inorganic-N input by bulk precipitation was estimated 0.25 ± 0.08 $gN/m^2/yr$. Total inorganic-N input by throughfall and stemflow in two evergreen conifers were estimated 0.41 ± 0.11 $gN/m^2/yr$ in *Abies* stand, $0.44 \pm$

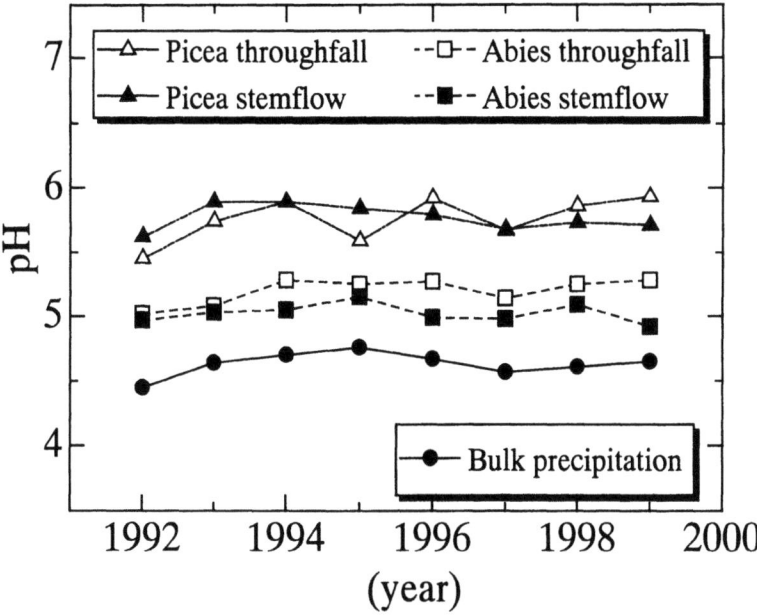

Figure 1. Mean annual pH values of bulk precipitation, throughfall and stemflow during snow-free period.

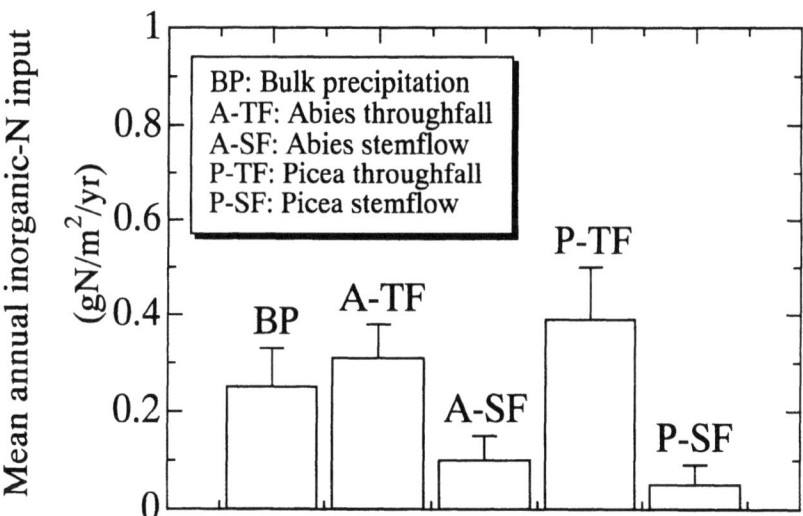

Figure 2. Mean annual inorganic-N input by bulk precipitation, throughfall and stemflow during snow-free period. Vertical bars represent 1 s.d.m. (n=8).

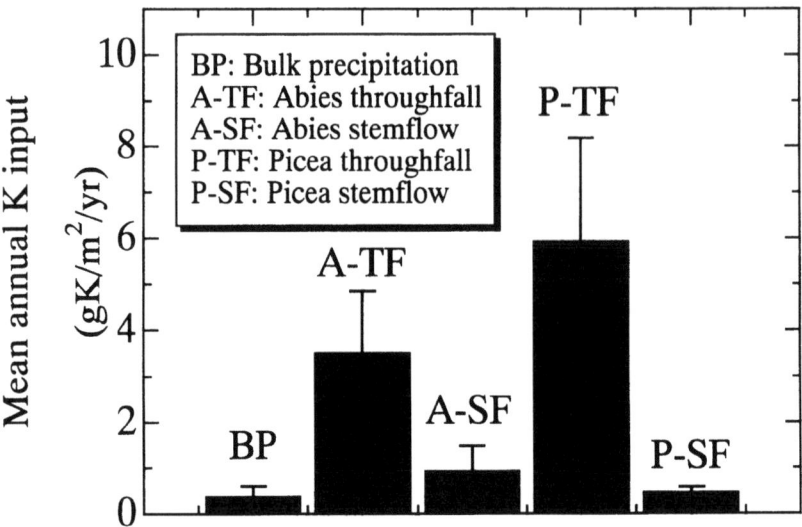

Figure 3. Mean annual K input by precipitation, throughfall and stemflow during snow-free period. Vertical bars represent 1 s.d.m. (n=8).

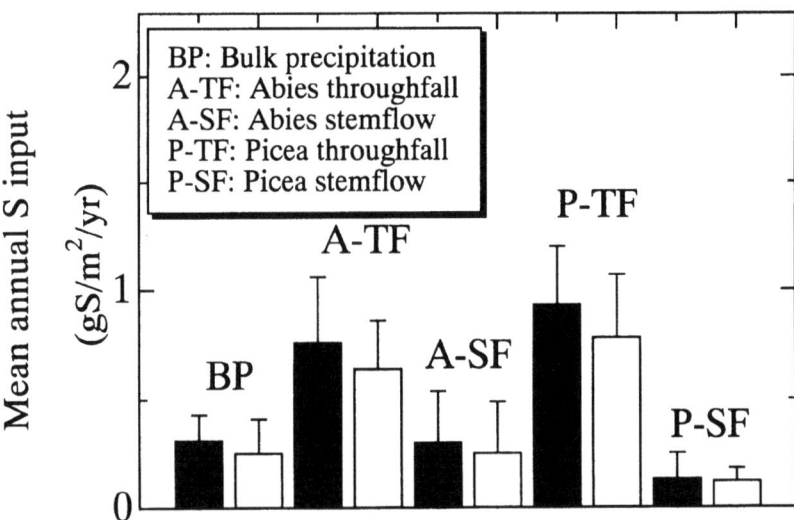

Figure 4. Mean annual input of SO_4-S and nss-SO_4-S by precipitation, throughfall and stemflow during snow-free period. Black bars and white bars represent total SO_4-S and nss-SO_4-S, respectively. Vertical bars represent 1 s.d.m. (n=8).

0.13 gN/m^2/yr in *Picea* stand. Throughfall mainly contributed to the annual inorganic-N input, with 76 % in *Abies* stand and 89 % in *Picea* stand (Fig. 2).

Mean annual K input by bulk precipitation was 0.36 ± 0.24 gK/m^2/yr. K input by throughfall and stemflow in *Abies* stand was 4.41 ± 1.36 gK/m^2/yr, and 6.37 ± 2.34 gK/m^2/yr in *Picea* stand. Throughfall was also main part of K input to the forest; with 73 % in *Abies* stand and 92 % in *Picea* stand by throughfall (Fig. 3).

Mean annual input of total S and nss-S input during snow-free period for 8 years was shown in Figure 4. The amount of nss-S in bulk precipitation was 0.25 ± 0.16 gS/m^2/yr, which was 81 % of total S input by bulk precipitation. Mean annual S input (the sum of S input by throughfall and stemflow) was almost the same in both *Abies* and *Picea* stands (1.06 ± 0.54 gS/m^2/yr), which was about three times larger than that of bulk precipitation. Except for *Picea* stemflow (P-SF in Fig. 4), nss-S composed of 83 to 84 % of total S input.

4. Discussion

The amount of deposition via throughfall and stemflow in forests depends on the location and distance from pollution sources. Annual input of inorganic-N and SO$_4$-S in a heavily damaged spruce ecosystem by air pollution was 2.534 gN/m^2/yr and 6.21 gS/m^2/yr (Ruzicka, 1995). Forests in urban area also have high level of nitrogen deposition. Matsuura (1992) reported that N input by bulk precipitation was 2.7 gN/m^2/yr in a *Cryptomeria japonica* stand located on northward of Tokyo area, Japan. Mean annual input of inorganic-N in this study was quite low level (0.25 gN/m^2/yr by bulk precipitation, 0.41 and 0.44 gN/m^2/yr by throughfall and stemflow), compared with those of heavily polluted or urban area. Inorganic-N input in forests located on remote from urban area was ranging from 0.4 to 1.3 gN/m^2/yr (Markewitz *et al*, 1998; Houle *et al*, 1999; Ollinger *et al*, 1993; Nilsson *et al*, 1998).

Since the amount of dissolved nutrients, such as K, controlled by internal cycling pathway, leaching supplies 60 to 90 % of K in throughfall (Parker, 1983). Because stemflow in forest stands concentrates at the stem base, with relatively high dissolved nutrient concentration, its chemical properties may affect topsoil condition near stem base. Stemflow in Swedish beech forests affected soil pH and base saturation (Falkengren-Grerup, 1989), and stemflow removal experiments showed recovery of base saturation (Matschonat and Falkengren-Grerup, 2000). As shown in Figure 1 and 3, throughfall and stemflow in *Abies* and *Picea* stands had different pH regime, and mean annual K input in *Picea* was higher than that of *Abies* stand. These chemical properties resulted in topsoil pH differences. Sanada *et al* (1991) reported that pH of 20 topsoil (0 to 5 cm) samples in *Picea* stands was 5.93 ± 0.15, and 5.54 ± 0.22 in *Abies* stand.

Estimated mean annual input of S and nss-S in this study were amount during snow-free period. The difference between S and nss-S input of bulk precipitation is roughly estimated seasalt origin S input. Though the difference in snow-free period is 0.06 gS/m^2/yr, this difference during whole year estimation is 0.16 gS/m^2/yr in 1999 (Matsuura et al., unpublished data). It means that seasalt origin S input concentrated during winter as snowfall. Sulfur in accumulated snow might affect soil solution chemistry during snowmelt in the spring.

Long-term monitoring study on nutrient input to the forest ecosystem in northern Japan suggested that input regime was not so high, despite in suburban area. The two evergreen conifers, *Abies* and *Picea*, had different regime of throughfall and stemflow chemistry. It might play a complementary role to create heterogeneous soil condition in subalpine natural forests, where these two species coexisted.

Acknowledgement

The authors would like to thank Etsuko Sanada and Reiko Takeuchi for their collaboration and field sampling. This study was partly funded by Acid Deposition Monitoring Study by Forestry Agency, Japan.

References

Aber, J. D., McDowell, W., Nadelhoffer, K. J. Magill, A., Berntson, G., Kamakea, M., McNulty, S., Currie, W., Rustad, L., Fernabdez, I.: 1998, *BioScience* **48**, 921.
Falkengren-Grerup, U.:1989, *Journal of Applied Ecology* **26**, 341.
Hara, H.: 1992,'8.Precipitation', in The Chemical Society of Japan (ed.), Chemistry of Terrestrial Water, Gakaki-Shuppan Center, Tokyo, pp. 69-78.
Houle, D., Ouimet, R., Paquin, R., Laflamme, J. -G.: 1999, *Canadian Journal of Forest Research* **29**, 1944.
Ollinger, S. V., Aber, J. D., Lovett, G. M., Millham, S. E., Lathrop, R. G., Ellis, J. M.: 1993, *Ecological Applications* **3**, 459.
Markewitz, D., Richter, D. D., Allen, H. L., Urrego, J. B.: 1998, *Soil. Sci. Soc. Am. J.* **62**, 1428.
Matschonat, G., Falkengren-Grerup, U.: 2000, *Scand. J. For. Res.* **15**, 39.
Matsuura, Y.: 1992, *Jpn. J. For. Environment* **34**, 20.
Nadelhoffer, K. J., Emmett, B. A., Gundersen, P., Kjonaas, O. J., Koopmans, C. J., Schleppi, P., Tietema, A., Wright, R. F.: 1999, *Nature* **398**, 145.
Nilsson, I., Berggren, D., Westling, O.: 1998, *Scand. J. For. Res.* **13**, 393.
Parker, G. G.: 1983, *Advances in Ecological Research* **13**, 57.
Ruzicka, S.: 1995, *Water, Air, and Soil Pollution* **83**, 205.
Sanada, M., Ohta, S., Ootomo, R., Sanada, E.: 1991, *Jpn. J. For. Environment* **33**, 8.
Schulze, E.-D.: 1989, *Science* **244**, 776.

AEROSOL AND PRECIPITATION CHEMISTRY DURING THE SUMMER AT THE SUMMIT OF MT. FUJI, JAPAN (3776m a.s.l.)

KAZUHIKO HAYASHI[1]*, YASUHITO IGARASHI[2], YUKITOMO TSUTSUMI[2#] and YUKIKO DOKIYA[3]

[1] *Meteorological College, 7-4-81 Asahi-cho, Kashiwa 277-0852, Japan;*
[2] *Meteorological Research Institute, 1-1 Nagamine, Tsukuba 305-0052, Japan*
(# *present address: Japan Meteorological Agency);*
[3] *Tokyo University of Agriculture and Technology, 3-5-8 Saiwai-cho, Fuchu 183-5809, Japan*
(* *author for correspondence, e-mail: hayashik@typhoon.mc-jma.ac.jp*)

Abstract. Aerosol and precipitation samples were obtained at the summit of Mt. Fuji, the highest peak (3776m a.s.l.) in Japan, in the summers of 1997, 1998, and 1999. The mountaintop might be affected by valley wind during the afternoon, but is located in the free troposphere during the morning. The temporal variations of chemical species in the aerosol and precipitation samples correspond with meteorological conditions. The SO_4^{2-} in the aerosol and precipitation exhibits high concentration with low temperature air mass, indicating the influence of long range transport from the Asian Continent. The contribution of the free troposphere to the chemical species obtained at the summit is estimated to be at least 30% during the summer season.

Key words: aerosol, free troposphere, long range transport, Mt. Fuji, precipitation, sulfate

1. Introduction

Mt. Fuji is an isolated dormant volcano and the highest elevation site (3776m a.s.l.) in Japan. At the summit there has been a weather station since 1932, where the Japan Meteorological Agency (JMA) staff have been continually in residence. The mountaintop is so small that most of the time it is located in the free troposphere (Tsutsumi *et al.*, 1994). It is possible that the atmospheric chemistry at the summit of Mt. Fuji represents that of the free atmosphere. Tsutsumi *et al.* (1994) have shown that the diurnal variation of ozone concentration is lower at the summit of Mt. Fuji than that of other high elevation sites (e.g., Mauna Loa Observatory) because the steep and small mountaintop prevents mountain and valley wind from generating there. At the summit of Mt. Fuji, the disturbance from lower air is so insignificant on both temporal and spatial scales that it is possible to investigate atmospheric chemistry and describe its long-term trends and long-range transport patterns around Japan.

In summer 1990, we obtained the first precipitation samples at the summit of Mt. Fuji (Dokiya *et al.*, 1993). So far we have samples of precipitation and aerosol from the summit: sampling were carried out for about 1 week every summer since 1997.

The main purpose of this study is to examine the temporal variations in aerosol and precipitation obtained at the summit of Mt. Fuji and to investigate their relationship to one another. The goal is to describe the long-range circulation patterns of natural and anthropogenic chemical species around Japan.

Water, Air, and Soil Pollution **130:** 1667–1672, 2001.
© 2001 *Kluwer Academic Publishers.*

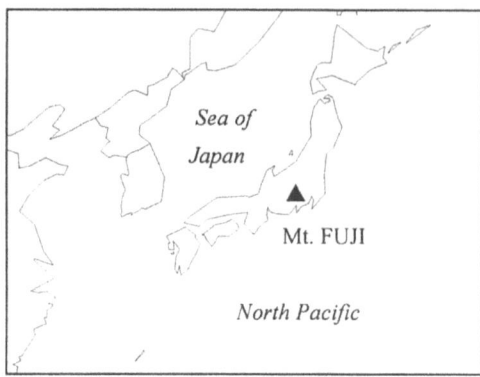

Figure 1. Geographic map showing the area around Japan. The triangle indicates Mt. Fuji (35.4N, 138.7E, 3776m).

2. Experimental Methods

All samples were collected at the Mt. Fuji Weather Station of the Japan Meteorological Agency (JMA), which is located at the highest point on the summit of Mt. Fuji (35.4N, 138.7E, Figure 1).

Aerosol samples were recovered on nuclepore membrane filters. The filter changing period usually lasted 4 hours. The air was drawn by a diaphragm pump through a long polyethylene tube (6mm inside-diameter, about 10m length), the inlet of which was fixed on the roof of the Weather Station. The volume of air sampled was measured by an in-line mass flow meter. The volume averaged $3.5m^3$ STP for all aerosol samples. The sample filters were mounted on plastic petri dishes and then wrapped in polyethylene film for refrigerated storage.

Precipitation samples were collected in plastic cylinder bottles (115mm diameter). A bottle was installed on the roof of the Weather Station and changed every morning (08:00 JST). The sampled bottles were stored in a cool dark place.

All samples from Mt. Fuji were transported to Meteorological College for analysis. Aerosol filters were extracted with 20ml Milli-Q water, during supersonic washing for approximately 10 minutes. The soluble components in both aerosol and precipitation were analyzed using pH glass electrode (DKK IOL-50), conductivity electrode (DKK AOL-30), and a Yokogawa IC7000 ion chromatograph. The chromatography conditions were as follows: for cations, ICS-C25 analytical column and 5mM tartaric acid eluent; for anions, ICS-A23 analytical column and 3mM Na_2CO_3 eluent.

3. Results

Major soluble components in the aerosol obtained at the summit of Mt. Fuji were sulfate and ammonium. SO_4^{2-} was fully neutralized by NH_4^+ in many of our samples. SO_2 gas concentration in the summit was too low (<1nmol/m³ in Aug. 10-15, 1997) to been oxidized to sulfate on the filter. On the other hand, NH_3 was the dominant gas in the summit (ca. 10nmol/m³ in Aug. 10-15, 1997). The NH_4^+ concentrations can be due to artifacts and/or post-sampling contamination (Silvente and Legrand, 1993). Other components, which originated in seasalt or soil, were rarely detected since the long drawn tube (about 10m length) was able to trap giant particles before filter-sampling. In July 1998, aerosol sampling was performed and some samples were collected every hour. A sampling period of 1 hour is too short to detect significant concentrations for the samples. Therefore sufficient continuous samples to describe inter-diurnal variations are not available.

Figures 2a and b show SO_4^{2-} concentrations in the aerosol in August 10-15, 1997 and July 6-11, 1999, respectively. In 1997, SO_4^{2-} concentrations in the aerosol gradually increased to 15 neq m⁻³ on August 15. In 1999, the concentrations, by contrast, rapidly decreased from over 20 neq m⁻³ on July 6.

Precipitation samples were often collected prior to passing of the pressure depressions (troughs) above Mt. Fuji. Figure 3 shows the ion concentrations in the four precipitation samples, whose precipitation amounts exceed 10mm. It is thought that the contribution of dry deposition and fog to the ion concentration of the samples is low. The total anion concentrations in the precipitation range from 9 to 25 μeq l⁻¹, whereas total cation concentration ranges from 10 to 37 μeq l⁻¹. The discrepancy between the total anion and cation appeared in some samples. This is due to their low total ion concentrations of 19 to 55μeq l⁻¹.

The precipitation collected during August 13-16, 1997 and July 11-13, 1998 had a similar cause; the dominant wind was westerly-northerly and accompanied a cold air mass from the Asian Continent. On the other hand, during July 15-18, 1998 and July 11, 1999 precipitation events, the developing cyclones passed eastward over the sea south of Japan. Hence at the summit of Mt. Fuji the prevailing wind was east to south, bringing a warm marine air.

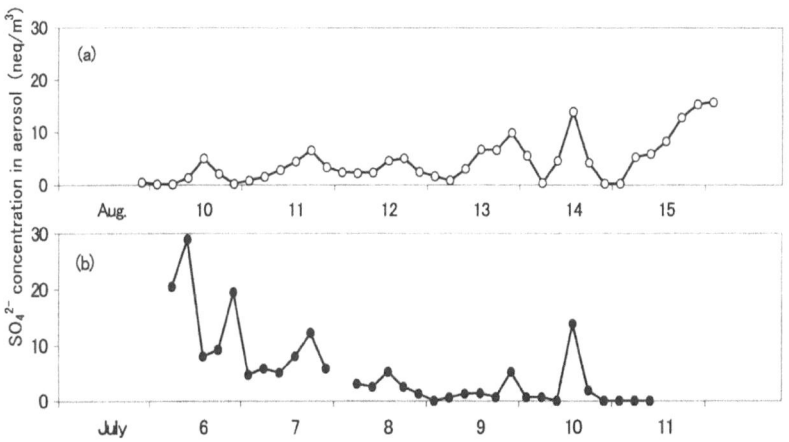

Figure 2. Concentrations of SO_4^{2-} in aerosol samples collected at the summit of Mt. Fuji: (a) August 10-15, 1997; (b) July 6-11, 1999.

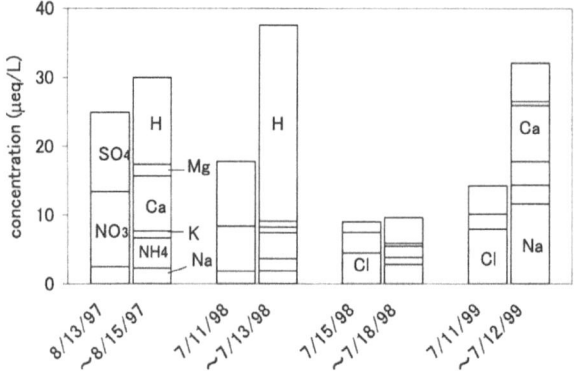

Figure 3. Concentrations of precipitation samples collected at the summit of Mt. Fuji

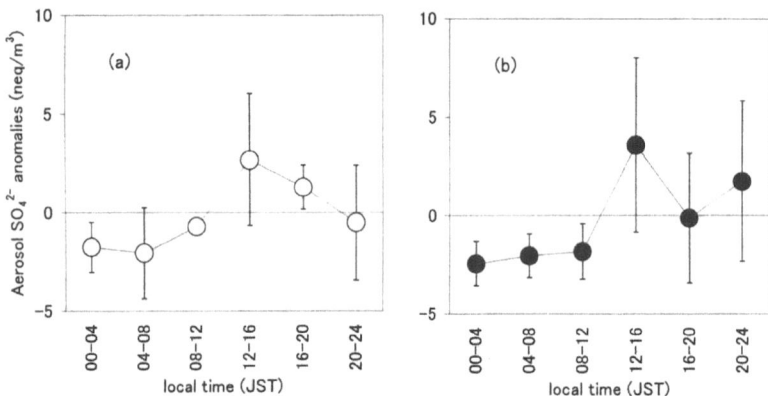

Figure 4. Averaged diurnal variations of the aerosol SO_4^{2-} anomalies from 24-hour running mean. Error bars indicate 1 standard deviations at each time: (a) August 10-15, 1997; (b) July 6-11, 1999.

4. Discussion

Figure 4a and b show the averaged diurnal variations of the aerosol SO_4^{2-} anomalies from 24-hour running mean values in August 10-15, 1997 and July 6-11, 1999, respectively. They both indicate peaks during 12:00-16:00 JST. The cause of the aerosol SO_4^{2-} peaks is thought to be that the lower local air containing pollutants is uplifted by the valley wind, which is generated by surface warming during the afternoon. On July 10, 1999, for example, because of direct solar radiation, the temperature at the summit rose during the daytime. In addition, the wind speed was very low (less than 1m s^{-1}). Therefore the air at the summit was easily be influenced by the valley wind. On that day, the aerosol SO_4^{2-} had a peak during 12:00-16:00 JST. Simultaneously CO concentration peaked and ozone decreased.

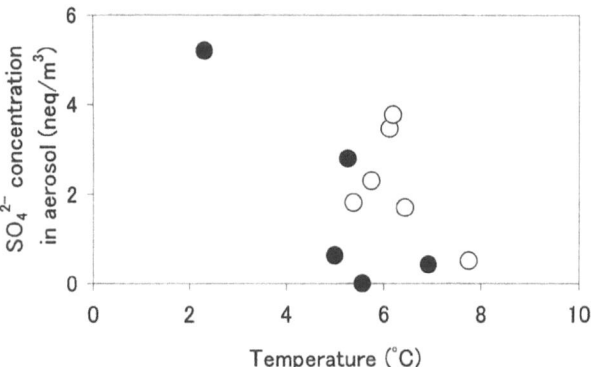

Figure 5. Relationship between the averaged aerosol SO_4^{2-} concentration and temperature at the summit of Mt. Fuji in the morning (00:00-12:00 JST). ○:August 10-15, 1997; and ●: July 6-11, 1999.

In contrast, in the morning (00:00-12:00 JST) the influence of the valley wind is so little that aerosol concentration is lowest. It is possible that the summit of Mt. Fuji is located in the free troposphere in the morning when the valley wind is very low. The averaged SO_4^{2-} concentrations in the morning and during the whole period are approximately 3 and 5 µeq l^{-1}, respectively. Therefore the contribution of the free troposphere to the SO_4^{2-} aerosol is estimated to be at least 30%. Figure 5 shows the relationship between the averaged aerosol SO_4^{2-} concentration and temperature in the morning (00:00-12:00 JST) at the summit of Mt. Fuji. This indicates that aerosol SO_4^{2-} concentration is inversely correlated with temperature. Therefore it is likely that the cold air mass contains high aerosol SO_4^{2-} concentration, whereas in the warm air mass SO_4^{2-} concentration is low. The cold air mass originated from the Continent in August 1997, and was subjected to the Okhotsk high in July 1999. It is not obvious why the cold air mass contains high SO_4^{2-} in the aerosol. But it is supposed that anthropogenic emissions from the Continent or biomass burning may be responsible.

Comparison between nss(non seasalt)-SO_4^{2-} concentration in the aerosol and precipitation obtained simultaneously at the summit is shown in Figure 6. They correlate positively. Therefore SO_4^{2-} concentration in the precipitation may also depend on the nature of the air mass, the same holding true for concentration in the aerosol.

5. Conclusions

The diurnal variation of the aerosol SO_4^{2-} concentration at the summit of Mt. Fuji is influenced by valley wind occurring during the afternoon. In the morning, however, the aerosol SO_4^{2-} concentration remains low level. Therefore, the mountaintop appears to be located in the free troposphere in the morning. In the free troposphere, the lower the temperature is, the higher the concentration of

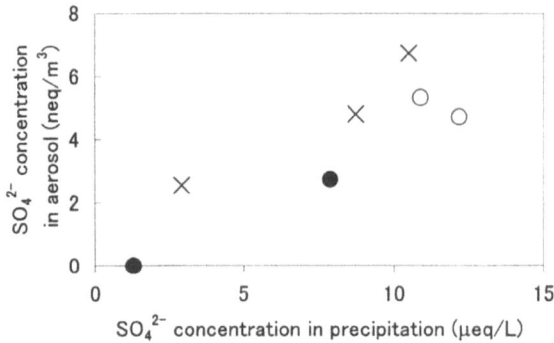

Figure 6. Relationship between nss-SO_4^{2-} concentrations in the aerosol and precipitation obtained simultaneously at the summit of Mt. Fuji. (○: August 1997; ×: July 1998; and ●: July 1999)

SO_4^{2-} in the aerosol. Therefore a cold air mass, which comes from the Continent, contains SO_4^{2-}-rich aerosol.

The nss-SO_4^{2-} concentration in the precipitation correlates with that in the aerosol. If nss-SO_4^{2-} in precipitation is derived from the scavenged aerosol, the contribution of the free troposphere can be at least 30% during the summer season. Year-round observation is necessary to estimate the transport of pollutants, such as SO_4^{2-}, precisely.

Acknowledgements

We thank the staff of the JMA Mt. Fuji Weather Station for their helpful cooperation. We also thank our colleagues, Dr. Kikuo Okada, Dr. Takayuki Tokieda, Dr. Yousuke Sawa, Mr. Hiroaki Naoe, Ms. Michiko Otsuka, Mr. Tatsuya Aizawa, Mr. Hiroki Shiozuru, Mr. Kentaro Murakami, and Mr. Tetsuo Yoshikawa for helping our observation.

References

Dokiya, Y., Tsuboi, K. and Maruta, E.: 1993, *Tenki (Bulletin Journal of Meteorological Society of Japan)* **40**, 539. (in Japanese)

Silvente, E. and Legrand, M.: 1993, *Geophysical Research Letters* **20**, 687.

Tsutsumi, Y., Zaizen, Y. and Makino, Y.: 1994, *Geophysical Research Letters* **21**, 1727.

GEOGRAPHICAL AND TEMPORAL VARIATIONS OF CHEMICAL CONSTITUENTS IN WINTER PRECIPITATION COLLECTED IN THE AREAS ALONG THE COAST OF THE SEA OF JAPAN

NORIO FUKUZAKI[1*], TSUYOSHI OHIZUMI[2] and KAZUHIDE MATSUDA[1]

[1] *Acid Deposition and Oxidant Research Center, 314-1 Sowa, Niigata, 950-2144 Japan*   [2] *Niigata Prefectural Research Laboratory for Health and Environment, 314-1 Sowa, Niigata, 950-2144 Japan*
(*author for correspondence, e-mail: fukuzaki@adorc.gr.jp)

Abstract. The geographical and temporal variations of chemical constituents in winter precipitation collected in the areas along the coast of the Sea of Japan (AASJ) were discussed by analyzing the data obtained in the 1st and 2nd National Acid Deposition Survey by Japan Environmental Laboratories Association. In western Tohoku (WT) and Hokuriku (HR) areas in AASJ, in spite of large amounts of precipitation in winter, concentrations of non sea salt (nss-) SO_4^{2-} are not as low as the other areas, and nss-Ca^{2+} in these areas is lower than the other areas. As a result, H^+ concentrations of precipitation in these areas are somewhat higher than other areas. From the temporal analysis of daily sampled data and back trajectory analysis of air mass, it was found that the concentrations of nss-SO_4^{2-}, NO_3^-, NH_4^+ and nss-Ca^{2+} are correlatively varied when air mass come from the Asian Continent, showing higher concentrations at the western sites in AASJ and depending on the meteorological conditions such as the direction of in flow air mass.

Key words: winter precipitation, concentration of constituents, geographical variations, temporal variations

1. Introduction

The Japanese Archipelago is located off the eastern edge of the Asian Continent in middle latitudes forming a boundary between the Sea of Japan and the Pacific, lying north to south over a distance corresponding to about 25 in latitude. In winter, prevailing monsoons from the Continent bring large amounts of precipitation as snow and/or rain to the areas along the coast of the Sea of Japan (AASJ). It is predicted that, through the prevailing monsoons, chemical constituents in winter precipitation in AASJ are affected by sea salt components (Tsunogai, 1975) and by the long range transport of continental soil dust which is observed as the Kosa (yellow sand) phenomenon in Japan (Fukuda and Tsunogai, 1975; Ichikuni, 1978) and by air pollutants released from the northeast Asian Continent (Ohizumi et al.,1994; Dokiya et al., 1995), as well as by air pollutants emitted in Japan (Chang et al.,1990). Due to the differences in the influences from the Asian Continent between the areas close to the continent and areas far away from it, the geographical and temporal variations in the chemical constituents in winter precipitation can be predicted.

The Environmental Laboratories Association composed of institutes for environmental research of local governments in Japan carried out nationwide surveys of acid deposition for two Phases using standardized methods. The surveys extended for two phases; Phase-1 was from April 1991 to March 1994, and Phase-2 was from 1996 to 1997 selectively done in winter and in the rainy season. We analyzed the data obtained from Pase-1 and Phase-2, respectively, to clarify the geographical and temporal variations of chemical constituents in winter precipitation in AASJ.

2. Methods

2.1. SAMPLING, SAMPLE ANALYSES AND QUALITY CONTROL

The distribution of the sampling sites in the Phase-1 and Phase-2 Survey is shown in Figure1. For geographical comparison, AASJ is divided into the following five sub-areas: western Tohoku (WT), Hokuriku (HR), eastern San-in (ES), western San-in (WS) and northern Kyushu (NK) in the Phase-1 Survey.

The monitoring period in Phase-1 was three years from April 1991 to March 1994. In this survey, precipitation was collected for every two weeks by a filtrating bulk sampler (JEA,1990; Hara et al.,1990). The winter survey in Phase-2 was carried out for two weeks from late January to early February in 1996 and 1997; the sampling was done on the 24-hour basis excluding the period from 9 AM on Friday to 9 AM on Monday using polyethylene buckets (Tosron Co., white type with a diameter 28.5 cm, height 38.6 cm).

Details of the sample handling and analytical methods have been described in the survey manual (JEA,1990). The accuracy of the analyses was appraised by the ion balance check and conductivity check (JEA,1990). The inter-laboratories precision was better than 5 percent for pH, conductivity, NH_4^+ and the anion concentrations, better than 8 percent for, Ca^{2+}, Mg^{2+}, and Na^+, and better than 15 percent for K^+ (Mimura et al., 1995).

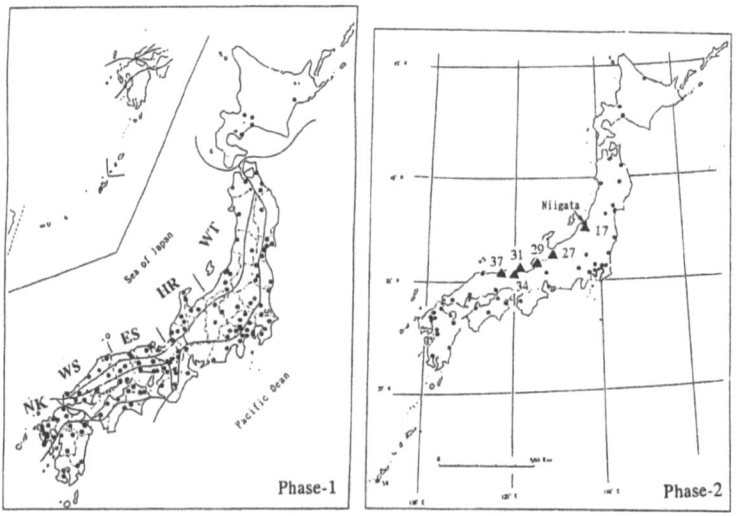

Figure 1. Sampling sites in Phase-1 and Phase-2 Survey

2.2. DATA ANALYSES AND BACK TRAJECTORY ANALYSIS

In the Phase-1 Survey, the depositions of each constituent collected in January and February were lumped as the winter depositions. The mean depositions divided by the mean precipitation depth at each site gave the precipitation-weighted mean concentrations. In the Phase-2 Survey, daily precipitation and concentration data were used without any treatment. The non sea salt (nss-) sulfate (nss-SO_4^{2-}) and nss-Ca^{2+} were calculated from the Na^+ concentration under the assumption that all of the Na^+ originated from sea salt.

In order to analyze the history of the air mass, isobaric back trajectories at the 850 hPa level were calculated using wind data from the Grid Point Value (GPV) of the Japan Meteorological Agency. The calculation was performed for a period between January 20 and 29, 1997 starting from Niigata Meteorological Observatory, which is located 20 km north of Niitsu.

3. Results and Discussion

3.1. GEOGRAPHICAL FEATURE

3.1.1. *Precipitation, Na^+ and H^+*

Precipitation, precipitation-weighed mean concentration of major constituents and pH are listed in Table I for the winter samples collected in each sub-area in AASJ. In the bottom row, the average over the whole country are given for comparison.

TABLE I
Concentrations of some major constituents, pH and indicator values in each area along the coast of the Sea of Japan (AASJ) and whole country in winter (Jan. and Feb.)

	Sites	H_2O^a (mm)	pH	Na^+	H^+	NH_4^+ (μeq/L)	nss-Ca^{2+}	NO_3^-	nss-SO_4^{2-}	$\dfrac{NO_3^-}{\text{nss-}SO_4^{2-}}$	$\dfrac{H^+}{PA^b}$
Western Tohoku (WT)	9	327	4.53	320	29.2	33.0	22.4	21.4	59.0	0.37	0.38
Hokuriku (HR)	10	421	4.50	270	31.7	30.0	15.6	22.5	55.8	0.40	0.42
Eastern San-in (ES)	5	356	4.55	268	28.3	31.2	14.4	19.3	57.1	0.34	0.37
Western San-in (WS)	4	299	4.63	294	23.7	37.3	29.7	34.5	67.4	0.51	0.23
Northern Kyushu1 (NK)	9	191	4.56	207	27.4	45.3	63.2	31.7	86.5	0.37	0.23
Whole Country	149	189	4.62	166	23.8	35.3	32.5	24.8	59.2	0.42	0.28

a)H_2O: Precipitation, b)PA: nss-SO_4^{2-}+NO_3^-.

As shown in Table 1, the mean winter precipitation depths in WT, HR, and ES are much more than the average over whole Japan. Sodium ions, which are referred to as sea salt components, in AASJ are also much higher than the Japan national average. The cold northwest winter monsoon, which blows from the eastern Asian Continent, takes up sea salt components as well as water vapor from the Sea of Japan. This is considered to be the cause of the high Na^+ concentration in the precipitation and heavy snowfall in AASJ (Tunogai, 1975).

The H^+ concentrations in each sub-area of AASJ except WS were higher than the average over the whole country. Daum (1984) used a fractional acidity (FA) defined by $H^+/(NO_3^- + \text{nss-}SO_4^{2-})$ as an indicator to show how much the original acid in precipitation remains unneutralized. The FA values are shown in the rightmost column of Table 1; FAs in WT, HR and ES far exceed the national average, indicating that much acid remains unneutralized in these sub-areas.

3.1.2. nss-SO_4^{2-}, NO_3^-, nss-Ca^{2+} and NH_4^+

The nss-SO_4^{2-} concentrations in WT, HR and ES are nearly equal to the national average in spite of the fact that the precipitation depths in those areas are almost twice as much as the national average precipitation. This means that SO_2 released from the northeast Asian Continent transported to AASJ being oxidized to SO_4^{2-} (Ohizumi at al., 1994; Dokiya et al., 1995). On the other hand, the NO_3^- concentrations in WS and NK were higher than other areas in AASJ and the national average.

A ratio NO_3^-/nss-SO_4^{2-} (=N/S, in equivalent) in each area is also shown in Table 1. The N/S in WS is higher than in WT, HR, ES and NK. The regional differences are conspicuous. In WS, the contribution of NO_3^- to the acidification of precipitation is relatively larger than that in other areas in AASJ. Its clear cause is not known yet. We need recent emission inventories of SO_2 and NOx in East Asian countries where motorization has progressed in recent years.

The concentrations of nss-Ca^{2+} in WT, HR, and ES are lower than the average over Japan, while that in NK is above the average and that in WS is slightly lower. This may be the cause of the remaining high concentrations of acid in WT, HR and ES in winter. The areal differences in AASJ are considered to be present because the calcium emissions from the ground's surface are suppressed by snow cover and/or rain in WT, HR and ES where much precipitation is observed and because continental dust which contains $CaCO_3$ (Ichikuni,1978;Tunogai et al.,1975) can more easily reach NK and WS, which are closer to the continent, than WT, HR and ES. On the other hand, there is not such large differences among NH_4^+ concentrations in AASJ, except NK.

3.2 TEMPORAL VARIATIONS

3.2.1. *The case in late January 1996*

The atmospheric pressure distribution in late January 1996 in the Phase-2 Survey was generally of the winter type. We compared the daily concentration variations from January 22 to February 2 (excluding the weekend, January 27 and 28) observed at several sites in AASJ. The precipitation depths at Niitsu (No.17 in Fig.1), Tateyama (No.27), Katsuyama (No.29), Mineyama (No.31), Toyooka (No.34) and Tottori(No.37) were 110 mm, 155 mm, 188 mm, 162 mm, 147 mm and 51mm, respectively. The precipitation depths in different sites were much the same except Tottori. The temporal concentration variations for nss-SO_4^{2-}, NO_3^-, NH_4^+ and nss-Ca^{2+} at these sites are shown in Figure 2.

The temporal variations of these parameters were clearly correlative, showing a higher concentration in western sites, Tottori > Toyooka $\geq$ Mineyama > Katsuyama > Niitsu $\leq$ Tateyama. It has been known that the concentration of nss-Ca^{2+} shows a remarkably high concentration in the Kosa phenomena (Ichikuni, 1978), however, it is noteworthy that not only nss-Ca^{2+} but NH$_4^+$, nss-SO$_4^{2-}$ and NO$_3^-$ increased when the Kosa phenomenon took place.

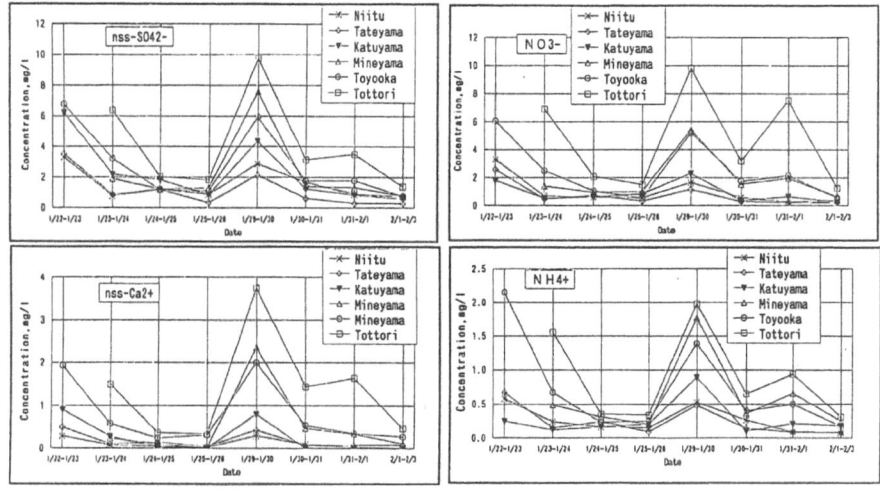

Figure 2. Temporal variation in January 1996

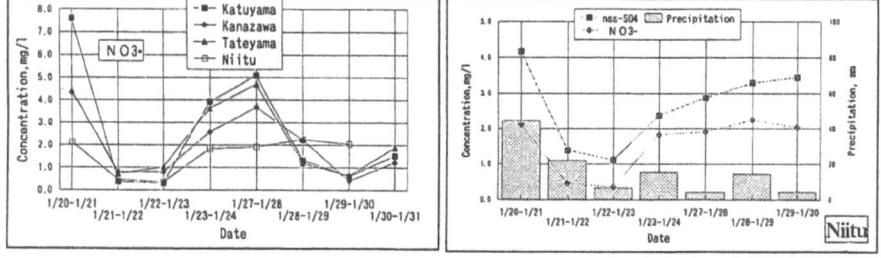

Figure 3. Temporal variation in January 1997

3.2.2. *The case in late January 1997*

The sites which exceeded 100 mm in total precipitation in the period (from 20 to 31, January 1997 excluding the 25 and 26) in AASJ were Niitsu (111 mm), Tateyama (127 mm), Kanazawa (136 mm) and Katsuyama (166 mm). The concentration of NO$_3^-$ at these sites showed a correlative variation pattern very similar to that observed in the preceding year, as shown in Figure 3. From these sites, we selected Niitsu to discuss the temporal variation of precipitation depth and concentrations because it was possible to analyze trajectory of the air mass from the site.

As shown in Figure 3, the concentrations of nss-SO$_4^{2-}$ and NO$_3^-$ are higher in the precipitation sampled on January 20-21 than those on 21-22 and 22-23, and they increased after latter sampling days. We compared these variations with the results of back trajectory analysis of the air mass. From the trajectory arrived at Niigata at 21:00 SJT, on 20 January, when the precipitation sampling is just intermediate, the flow direction of air mass bringing precipitation at Niitu was traced to back to the central China through southern part of the Korean Peninsula. It changed to near Vladivostok in Russia on the following 21, and further changed toward Korean Peninsula again on the 22. Comparing these situations with the concentration variation of nss-SO_4^{2-} and NO_3^-, it could be considered that the concentrations are higher when the air mass comes from central China through Korean Peninsula than from Russian area.

From these two limited case studies, it is considered that the concentrations of chemical constituents in winter precipitation in AASJ is correlatively changed when air mass come from the Asian Continent showing higher concentrations at the western sites, and meteorologically governed by inflow direction of the air mass, higher from the central China through southern part of Korean Peninsula than from Russian area.

4 Conclusion

In western Tohoku (WT) and Hokuriku (HR) areas in AASJ, in spite of large amounts of precipitation in winter, concentrations of nss-SO_4^{2-} are not as low as other areas, and nss-Ca^{2+} in these areas is lower than the other areas. As a result, H^+ concentrations of precipitation in these areas are somewhat higher than other areas. On the other hand, from the temporal analysis of daily sampled data and back trajectory analysis of air mass, it was found that the concentrations of nss-SO_4^{2-}, NO_3^-, NH_4^+ and nss-Ca^{2+} are correlatively varied when air mass come from the Asian Continent, showing higher concentrations at the western sites in AASJ and depending on the meteorological conditions such as the inflow direction of air mass.

Acknowledgements:

We are grateful for the effort of the staff of each institute in attendance at the National Acid Precipitation Survey of Environmental Laboratories Association, Japan.

References

Chang, Y.S., Ravishanker B.S., Kurita H. and Ueda, H.: 1990, *Atmos. Environ.* **24A**, 2035.
Daum, P.H., Kelly, T.J., Schwartz, S.E. and Newman, L.: 1984, *Atmos. Environ.* **18**, 2671.
Dokiyay, Y., Minakoshi, N., Hirooka, T., Yamashita, J., Ishikawa, S., Ohya, M. and Sugaya, J.: 1995, *J. Meteor. Soc. Jpn.* **73**, 873.
Fukuda, K. and Tsunogai, S.: 1975, *Tellus* **27**, 514.
Hara, H., Ito, E., Katou, T., Kitamura, Y., Komeiji, T., Oohara, M., Okita, T., Taguchi, K., Tamaki, M., Yamanaka Y. and Yoshimura K.: 1990, *Bull. Chem. Soc. Jpn.* **63**, 2691.
Ichikuni, M.: 1978, *J.Gephys.Res.* **83**, 6249.
JEA(Japan Environmental Egency), *Manuals for acid deposition survey(second ed.)*(in Japanese).
Mimura, H., Oshio T. and Kawamura Y., 1995, *Proceeding of the 36th annual meeting of the Japan Society of air pollution*, 524 (in Japanese).
Ohizumi, T., Fukuzaki, N. and Kusakabe, M.,: 1997, *Atmos. Environ.* **31**, 1339.
Tsunogai, S., Fukuda, K. and Nakaya, S.: 1975, *J. Meteorol. Soc. Japan*, **53**, 203.
Tsunogai, S., : 1975, *Tellus* **27**, 51.

SEASONAL AND SPATIAL VARIATIONS IN THE CHEMICAL AND SULFUR ISOTOPIC COMPOSITION OF ACID DEPOSITION IN NIIGATA PREFECTURE, JAPAN

TSUYOSHI OHIZUMI[1], NAOKO TAKE[1], NOBORU MORIYAMA[1], OSAMU SUZUKI[1] and MINORU KUSAKABE[2]

[1]*Niigata Prefectural Research Laboratory for Health and Environment, Sowa, Niigata 950-2144, Japan;* [2]*Institute of study for Earth's Interior, Okayama University, Misasa, Tottori 682-0193, Japan*

Abstract. The following measurements were carried out to clarify acid deposition in Niigata Prefecture, an area facing to the Sea of Japan. 1) Acid deposition fluxes and sulfur isotopic ratios of atmospheric deposition were measured at 9 sites in the prefecture in 1999. 2) Atmospheric deposition was collected daily at one site in 1997, to measure the sulfur isotopic variations of sulfate together with the air mass trajectory for each deposition. It became clear that: (a) The major component that acidifies atmospheric deposition is sulfuric acid. (b) Sulfate deposition increases in winter in the whole study area. (c) The sulfur isotopic ratios indicate that sulfur dioxide emitted from China affects the whole study area in winter. (d) Winter deposition of sulfate estimated to derive from coal combustion in China account for half of nss-sulfate deposition in average at 9 sites.

Keywords: sulfur isotopic ratio, sulfate deposition, coal combustion, seasonal variation, Niigata

1. Introduction

The acidification of precipitation is observed nationwide in Japan in recent years (Japan Environment Agency, 1999). The strongly acidic rains that give the acute damage to human beings and ecosystems has not been observed. However, there is a concern that such acid deposition may eventually make our environment acidic and exert an unfavorable effect on ecological environments of the areas on a long-term basis. To clarify acid deposition in Niigata Prefecture, an area facing to the Sea of Japan, in the present study we report the seasonal variation of concentration and deposition of sulfate in the whole study area on the basis of the monitoring at 9 sites evenly located in the area. We estimate the relative contribution of different sulfur sources to the sulfate deposition based on the sulfur isotopic variations of the deposition.

2. Methods

2.1. SAMPLE COLLECTION OF ATMOSPHERIC DEPOSITION AND ANALYTICAL METHOD

Samples of atmospheric deposition were collected using a bulk sampler every half a month at 9 sites that are evenly located in Niigata Prefecture from May 1999 to March 2000 (Fig. 1). In their sites, Niigata, Nagaoka and Joetsu were categorized as the plain sites, Aikawa as a coastal site and others as the mountainous sites. After the filtration through a membrane filter, pH (pH meter), electrical conductivity (electrical conductivity meter) and ionic concentrations (ion chromatography) were measured.

Atmospheric depositions were collected daily at Niigata from May 1997 to February 1998 to observe a short-term fluctuation of the sulfur isotopic ratios of the deposition.

Since a large amount of sample was required for sulfur isotopic analysis (>5mg $BaSO_4$), samples were combined in proportion to the precipitation amount in each site. In the 1997 observation at Niigata, a large amount of samples were collected to measure the daily sulfur isotopic variations.

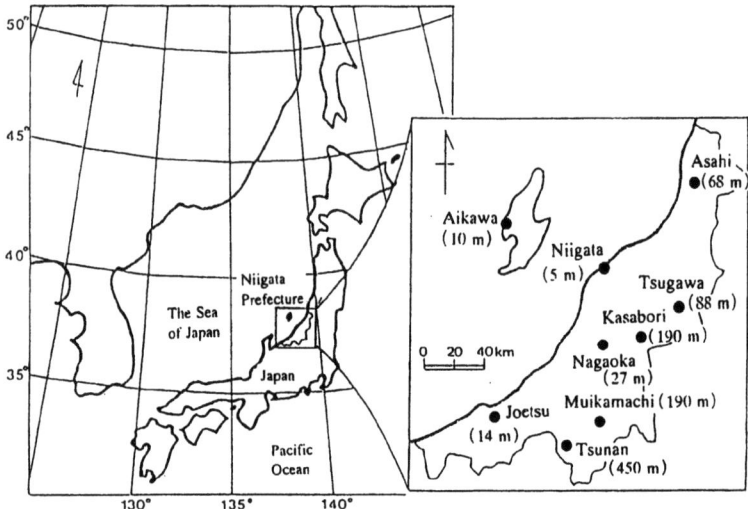

Figure 1. Map showing location of sampling sites and height of sampler from sea level

2.2. SULFUR ISOTOPIC ANALYSIS

Sulfate in the samples was precipitated as $BaSO_4$ which was then thermally decomposed to SO2 gas for sulfur isotopic analysis (Yanagisawa and Sakai, 1983). The SO_2 gas was then run on a stable isotope mass spectrometer (VG isogas SIRA10). Sulfur isotopic ratio was expressed in a conventional delta notation, $\delta^{34}S$, defined as

$$\delta^{34}S\ (\text{‰}) = ((^{34}S/^{32}S)_X/(^{34}S/^{32}S)_{CDT} - 1) \times 1000$$

where X and CDT stand for a sample and Canyon Diablo troilite (as standard substance), respectively. The overall accuracy of sulfur isotopic analysis was < 0.2 ‰.

3. Results and Discussion

3.1. MEAN ANNUAL COMPOSITION AND SEASONAL VARIATION OF ATMOSPHERIC DEPOSITION

Figure 2 shows the mean annual composition of the atmospheric deposition at Nagaoka averaged from April 1999 to March 2000. This area faces the Sea of Japan, and it is strongly affected by monsoon climate. Although there are differences in the mean annual concentration of seasalt components in the deposition between the sites depending on the distance from the coast, the differences are small for the concentration of non-seasalt components. The composition shown in Figure 2 is considered as the typical mean annual composition of the deposition in the area, because Nagaoka is located in the center of this area and the seasalt contribution there is also close to the average for the region. In the figure, the sum of cations of non-seasalt components is in balance with the sum of anions. From this fact, the acidity of the deposition seems to be determined by the degree of neutralization of sulfuric and nitric acids by ammonia gas and basic calcium aerosols. Since nss-sulfate concentration is higher than that of nitrate, sulfuric acid mainly causes the acidification of the deposition.

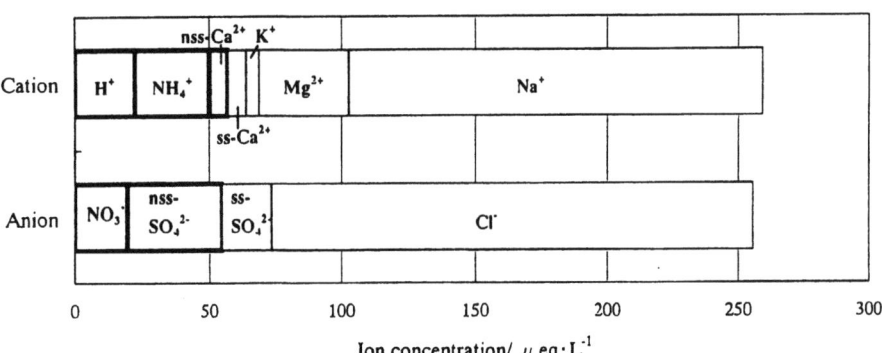

Figure 2. Mean chemical composition of atmospheric deposition at Nagaoka in 1999. Rainfall amount: 2381 mm•year-1, Mean pH: 4.65. The part surrounded in the bold line in the figure shows the components considered to derive other than sea salt.

Figure 3 shows the seasonal variation of the amount of precipitation, concentration and deposition of nss-sulfate. The nss-sulfate concentration increased in winter at 9 sites, and the amount of precipitation also increased in winter. It confirms that nss-sulfate deposition increases in winter over the whole study area.

3.2. SULFUR ISOTOPIC COMPOSITION OF ATMOSPHERIC DEPOSITION

Table I compares the sulfate deposition and $\delta^{34}S$ values of sulfate in at-mospheric deposition at 9 sites between summer and winter. The mean winter concentration

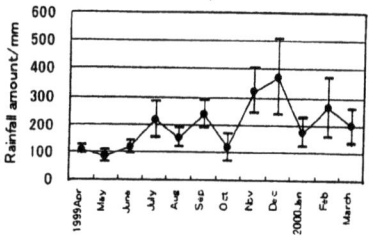
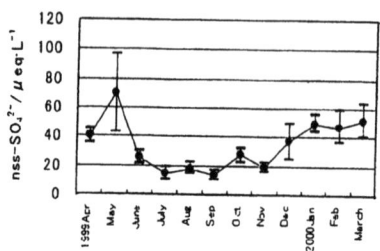
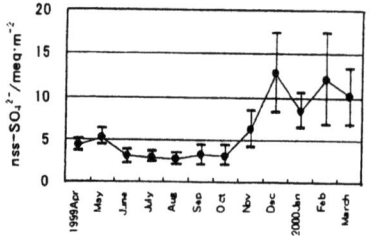

Figure 3. Seasonal variation of rainfall amount, nss-sulfate concentration and deposition at 9 sites. Mean values are plotted together with their standard deviation represented by vertical bars.

Table 1
Comparison with summer and winter in sulfate deposition at 9 sites in Niigata Prefecture.

Season	Site	Rainfall mm·day^{-1}	pH	SO_4^{2-} μ eq·L^{-1}	$nssSO_4^{2-}$	$\delta^{34}S$ ‰	$\delta^{34}Snss$	SO_4^{2-} μ eq·m^{-2}·day^{-1}	$ssSO_4^{2-}$	$_{LBV}SO_4^{2-}$	$ccSO_4^{2-}$
Summer	Asahi	7.3	5.08	16.7	14.9	4.3	2.3	123	13	110	
	Niigata	6.6	5.29	13.5	12.0	3.7	1.6	88.7	10	79	
	Tsugawa	8.0	5.06	11.0	10.2	3.4	1.9	88.3	7.0	81	
	Kasabori	7.8	4.85	17.0	15.7	3.2	1.8	132	10	120	
	Nagaoka	5.5	4.95	17.7	16.8	1.7	0.7	97.5	5.0	93	
	Joetsu	6.3	5.02	19.6	18.2	0.8	-0.6	123	8.4	120	
	Muikamachi	6.4	5.16	17.9	17.3	1.4	0.8	114	3.6	110	
	Tsunan	5.1	4.91	13.6	13.0	0.9	0.0	69.8	3.0	67	
	Aikawa	6.7	5.30	23.0	12.9	10.0	1.9	153	67	86	
	Mean	6.6	5.07	16.7	14.6	3.3	1.2	110	14.2	95.7	
	Std	1.0	0.16	3.6	2.7	2.8	1.0	26.0	20.2	19.1	
Winter	Asahi	12.1	4.80	67.4	30.8	13.3	5.0	817	440	140	230
	Niigata	8.0	4.64	69.9	31.1	12.6	3.0	559	310	180	70
	Tsugawa	14.8	4.68	42.3	28.4	9.6	4.3	627	210	200	220
	Kasabori	15.2	4.66	43.6	28.6	10.0	4.6	661	230	180	250
	Nagaoka	11.6	4.69	52.4	30.1	10.8	3.8	609	260	170	180
	Joetsu	9.6	4.93	49.7	26.9	10.4	2.0	476	220	160	95
	Muikamachi	11.8	4.92	31.8	21.3	8.7	3.0	375	120	160	95
	Tsunan	9.0	4.90	28.6	18.3	10.1	4.3	258	93	57	110
	Aikawa	6.4	4.95	212	33.6	17.9	5.1	1370	1200	67	150
	Mean	11.0	4.80	66.4	27.7	11.5	3.9	639	337	146	156
	Std	3.0	0.13	56.5	4.9	2.8	1.0	319	322	50.7	67.8

$ssSO_4^{2-}$: seasalt sulfate, $_{LBV}SO_4^{2-}$: Sulfate derived from local-biogenic-volcanic activity,
$ccSO_4^{2-}$: Sulfate derived from coal combustion in China
Sampling period: Summer; 1999/6/28-1999/9/27 (Asahi; 1999/6/30-1999/9/27), Winter; 1999/10/25-1999/12/27

of nss-sulfate is 2 times higher than the summer value. $\delta^{34}Snss$ value can be calculated by subtraction of the contribution of seasalt sulfate (20.3‰ , Sasaki,

1972). δ^{34}Snss values in summer range from -0.6 ‰ to 2.3 ‰. As the δ^{34}Snss values are larger in winter at all localities, the contribution of nss-sulfate sources varies with season in whole area in Niigata Prefecture, as already shown by in Nagaoka (Ohizumi *et al.*, 1997).

Figure 4 shows the variation with time in the δ^{34}Snss of the deposition and the results of air mass trajectory analysis. The δ^{34}Snss values in winter do not vary very much and range from 4 ‰ to 6 ‰. These isotopic values are close to those for the coal data for northern China (+6.6 ‰ in average, Maruyama *et al.*, 2000). From the air mass trajectory analysis, it is seen that many air masses brought the precipitation in winter pass the northern part of China. The trajectory of December 9th when a low δ^{34}Snss value of the precipitation was recorded, was different from the other winter trajectories. It is difficult to explain the rise in these d values in winter by increasing the contribution such as local anthropogenic activity, DMS from the ocean and volcanic gas, because local anthropogenic sulfur has low δ value (-2.7 ‰ in average, Ohizumi *et al.*, 1997) and the contribution of DMS (+15 ‰, McArdle and Liss, 1995) is not considered to increase in winter. And there is no active local volcano in this area. Thus a reasonable explanation for the isotopic variations is that the sulfur emitted by coal combustion in China is the source of much of the winter nss-sulfate

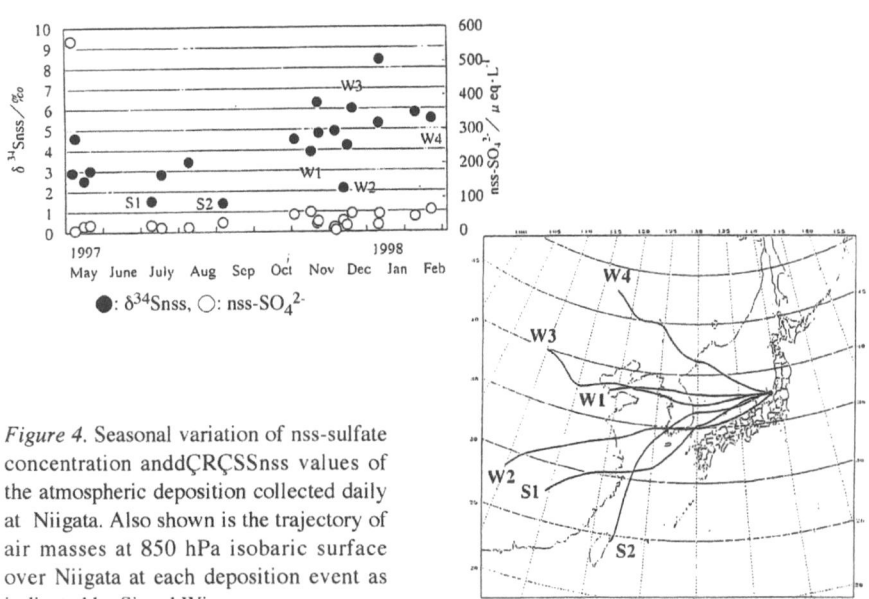

Figure 4. Seasonal variation of nss-sulfate concentration anddÇRÇSSnss values of the atmospheric deposition collected daily at Niigata. Also shown is the trajectory of air masses at 850 hPa isobaric surface over Niigata at each deposition event as indicated by Si and Wi.

deposition in this area of Japan.

The relative contributions of various sulfur sources for the sulfate in the atmospheric deposition were calculated by the method reported by Ohizumi *et al.* (1997). By solving the material balance equation which is constituted by δ

values and its fraction of sulfur sources to the bulk sulfur in atmospheric deposition, the contributions were calculated and the results were shown in Table I for 9 sites. The contribution of coal combustion in China ranges from 28 to 69 % with an average of 52 % for the nss-sulfate deposition in winter. Sulfate deposition derived from the coal combustion is high in Kasabori, but low in Niigata. Though the observation was carried out only for one year, the contribution of coal combustion was estimated to be high in winter sulfate deposition in the whole study area.

4. Conclusions

It has been shown from observations in the plain area within this study area that the effect of coal combustion in China is one of the causes of increased winter deposition flux of nss-sulfate (Ohizumi *et al.*, 1997). In this study, the similar observation was carried out to clarify acid deposition of the whole study area. Based on the chemical and sulfur isotopic analysis of atmospheric deposition with air mass trajectory analysis, the following points were clarified: 1) The deposition is acidified mainly by sulfuric acid. 2) The deposition flux of sulfuric acid increases in winter. 3) Coal combustion in China contributes to the increase of sulfur deposition flux in winter over the whole study area. 4) Winter deposition of sulfate estimated to derive from coal combustion in China account for 52 % of nss-sulfate deposition in average at 9 sites.

Acknowledgements

This work was carried out under the joint research program of Institute for Study of the Earth' Interior, Okayama University at Misasa. T. Nogi is thanked for her technical assistance.

Reference

Japan Environment Agency: 1999, Acid Deposition Survey Phase 3 Final Report.
Maruyama, T., Ohizumi, T., Taneoka, Y. Minami, N. Fukuzaki, N. Mukai, H. Murano, K. and Kusakabe, M.: 2000, Journal of the Chemical Society of Japan, *Chemistry and Industrial Chemistry* **2000**, 45.
McArdle, N. C. and Liss, P. S.: 1995, *Atmospheric Environment* **29**, 2553.
Ohizumi, T., Fukuzaki, N. and Kusakabe, M.: 1997, *Atmospheric Environment* **31**, 1339.
Sasaki, A.: 1972, Proc. 24th Int. Geological Congress, Section 10, pp.342.
Yanagisawa, F. and Sakai, H.: 1983, *Analytical Chemistry* **55**, 985.

INVESTIGATION OF ALKALINE NATURE OF RAIN WATER IN INDIA

UMESH C. KULSHRESTHA[1*], MONIKA J. KULSHRESTHA[1], R. SEKAR[1],
M. VAIRAMANI[1], AJIT K. SARKAR[2] and DANESH C. PARASHAR[2]

[1] *Indian Institute of Chemical Technology, Hyderabad 500007, India;*
[2] *National Physical Laboratory, New Delhi 110 01 India*
(* author for correspondence, e-mail: umesh@iict.ap.nic.in)

Abstract. Increased industrialization and urbanization lead to the atmospheric acidity which causes acid rain. However, in India, the nature of rain water has been observed to be alkaline. The reason for alkaline nature of rain water is found to be the buffering of acidity by soil-derived aerosols which are rich in Ca. Over the Indian Ocean where concentrations of soil dust are negligible, the acid rain has been observed to be a common phenomenon during INDOEX campaigns. In the Indian subcontinent, observations have indicated that rain becomes acidic when the buffering potential of rain water is weak. The weak buffering potential may be due to less interference of soil dust, acidic nature of soil or very high influence of industrial source.

Keywords: alkaline rain, soil-dust, precipitation, dustfall, acid rain, buffering, INDOEX.

1. Introduction

The acid rain phenomenon is an indicator of the dominance of species like H_2SO_4 and HNO_3 in the atmosphere. At deep continental sites, especially in tropical countries like India where loose soil is found in abundance in the atmosphere, the suspended dust particles also contribute to rain water composition (Kulshrestha *et al.*, 1996; Khemani, 1992). In most of the Indian studies reported so far, the pH of rain water has been found to be alkaline (pH > 5.6) due to the neutralization of the acidity by soil dust components like Ca and Mg (Khemani, 1992; Parashar *et al.*, 1996). In absence of dusty atmosphere, the pH of rain water has been reported acidic over Indian Ocean (Kulshrestha *et al.*, 1999) during four campaigns of INDOEX (Indian Ocean Experiment). In this study, an effort has been made to investigate the reason for the alkaline nature of rain water in Indian subcontinent. An analysis of data (especially SO_4 and Ca) has been performed with reference to soil derived particle interference.

2. Experimental

2.1. SAMPLING SITES

Rain water samples were collected at Delhi, Pune, Goa, Sinhagad, Agra, Drjiling, Haflong and Hyderabad which are continental sites as shown in Figure 1. Among these locations, Delhi, Pune, Goa and Sinhagad are the sites of Indo-Swedish network. Rain water samples were also collected over Indian ocean during 1st precampaign (SK cruie # 109), 2nd precampaign (SK cruise # 120),

FFP-98 (SK cruise # 133) and IFP-99 (SK cruise # 141) under INDOEX programme. The cruise tracks of these campaigns are shown in Figure 1. All these cruises covered a fairly remote marine atmosphere.

2.2. SAMPLE COLLECTION

At Delhi, Pune and Goa, the samples were collected with automatic wet-only samplers provided by International Meteorological Institute, Stockholm (Sweden) while at other sites, manual bottle and funnel collectors were used to collect the samples on event basis. During INDOEX, the samples were collected using wet-only collector. pH was measured in all samples immediately after the collection. Thymol was added to the samples as a biocide (Gillet, 1998).

2.3. ANALYSIS

The anions (Cl, NO_3, SO_4) were analyzed by Ion Chromatography or capillary electrophoresis while cations (Ca, Na, K, Mg) by Atomic Absorption Spectrophotometer. NH_4 was estimated by using capillary electrophoresis or Indo-phenol blue method. The details of analysis have been given elsewhere (Parashar et al., 1996).

3. Results and Discussion

3.1. DISTRIBUTION OF pH

Figure1 shows the pH values of rain water at different sites in India and over Indian ocean. At continental sites, the pH of rain water is alkaline (above 5.6) while over Indian ocean, the pH of rain water is acidic (below 5.6) in each event. At the continent, soil erosion and anthropogenic activities are the two main sources that play an important role in determining the nature and composition of rain water. The alkaline nature of rain water is due to the contribution of soil dust particles suspended in the atmosphere. Suspended soil particles are scavenged with rain water during below cloud scavenging process enhancing the pH of rain water (Parashar et al., 1996). Over the ocean, the non marine influence depends on the meteorology i.e. the transport of pollutants from continent to ocean. In the transport, the amount and type of pollutants depend on the particle size and chemical nature of the pollutants. Smaller sized particles and gaseous pollutants can travel long distance but coarse particles e.g. soil dust particles, can travel relatively short distance. Owing to these characteristics, under normal conditions, buffering by soil dust particles is not possible in far away places. Over the ocean, high nss-SO_4 and low nss-Ca concentrations have been observed due to which buffering becomes insufficient causing the rainfall to be acidic.

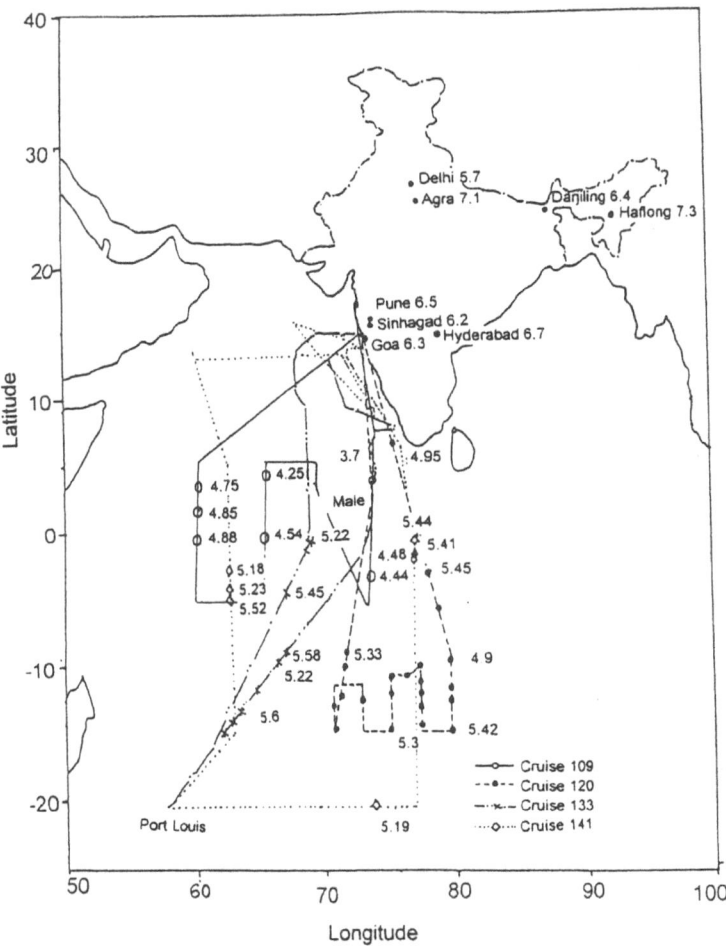

Figure 1. Distribution of pH of rain water over Indian Ocean and at continental sites in India

3.2. VARIATION OF pH WITH THE CONCENTRATIONS Ca AND SO_4

Table I gives the concentrations of Ca and SO_4 in rain water at various sites. It clearly shows that the sites in high alkaline dust area have high concentrations of Ca *i.e.*, Jodhpur, Agra, Srinagar, Allahabad and Haflong. At Jodhpur, the pH is highest indicating very high interference of suspended soil particles. The soil particles enhance the pH of rain water by contributing Ca possibly in the form of HCO_3. Similarly, SO_4 is also very high at this site which suggests that soil derived SO_4 does not lower the pH of rain water. The interference of soil derived SO_4 in rain water has been reported up to 50% depending on the site characteristics (Jain *et al.*, 2000). Jackes *et al* (1994) have also reported SO_4 in rain water due to resuspension of gypsyferous soil in India. These observations indicate that soil dust significantly alter the nature of rain water in India. At Delhi, some acid rain events have been observed when the ratios of $(Ca+Mg+NH_4)/(SO_4+NO_3)$ were around 0.5 (Kulshrestha et al., 1995a). All these events occurred after continuous rains of 2-3 days which washed off all the suspended matter in the atmosphere leading to the weak buffering potential. Acid rain events have been reported in India in the areas dominated by industrial sources (Khemani, 1993).

TABLE I

pH and concentration (µeq/l) of Ca and SO_4 at different sites in India

Site	pH	Ca	SO_4	Reference
Delhi	5.7	8	28	Parashar et al., 1996
Pune	6.5	34	16	-do-
Goa	6.3	42	27	-do-
Sinhagad	6.2	36	23	-do-
Agra	7.1	105	42	Kumar et al., 1993
Darjiling	6.4	61	22	Kulshrestha, 1998
Haflong	7.3	154	57	-do-
Hyderabad	6.7	43	37	New site
Indian Ocean	≤5.6	14	35	Kulshrestha et al., 1999
BAPMoN sites-				
Srinagar	7.0	205	23	Mukhopadyay et al., 1992
Jodhpur	8.3	311	39	-do-
Allahabad	7.1	143	18	-do-
Mohanbari	6.4	28	19	-do-
Nagpur	6.3	60	18	
Vishakhapatnam	6.5	70	38	-do-
Kodaikanal	6.1	28	12	-do-
Port Blair	6.1	31	18	-do-
Minicoy	6.5	60	11	-do-

Over the Indian Ocean, most of the nss SO_4 is anthropogenic (in the form of H_2SO_4) while at continental sites, the contribution of soil derived SO_4 is also significant. Over the ocean the SO_4 is secondary SO_4 which is responsible for acidic rainfall. It is possible that during the observation period of INDOEX, the prevailing NE trade winds transported SO_2 and anthropogenic SO_4 aerosols from the land to the ocean, which were subsequently scavenged by clouds. Over Indian Ocean, the nss Ca concentration was very low as compared to continental sites. On the other hand, nss SO_4 concentrations were observed to be higher due to which acidity is not fully neutralized and this may be the reason for acidic occurrence over Indian Ocean. The soil dust particles in the form of coarse mode cannot travel too long distance and hence buffering of acidity is incomplete over the ocean.

3.3. INFLUENCE OF SOIL-DERIVED PARTICLES ON CHEMICAL CHARACTERISTICS OF RAIN WATER

As discussed above, the pH is influenced by atmospheric concentrations of Ca. The Ca has been reported a major component of soil in India (Kulshrestha et al., 1995b).

At Delhi, Agra and Pune, the occurrence of Ca has been reported in coarse mode indicating its origin from natural source like soil (Kulshrestha *et al.*, 1998; Khemani, 1989).

Khemani (1992) observed that pH of the cloud base as 5.8 at Pune while at ground level, the observed pH of rain water was 7.3. He also noticed that the concentration of Ca in rain water was 40 times higher than that of cloud base, indicating that below cloud scavenging of suspended dust affect the rain water composition significantly.

In most parts of India, the pH of soil is very high (up to 10.6) except south west coast, some parts in Orissa and parts of north east states where the pH of soil itself is acidic. In these areas, the pH of rain water has been reported acidic (Khemani, 1993; Das *et al.*, 1998). The pH of the water extract of dustfall (24 hrs. exposure) ranged from 6.4 to 8.7 (Jain, 1999) which is in agreement with the pH of soil in this region. Ca and Mg are significantly scavenged by dry deposition process but at the same time, their wet deposition rates are also high in order to sufficiently buffer the acidity of rain water during below cloud scavenging (Parashar *et al.*, 1996). It is clear from the calculation of neutralization factor, the Ca had highest NF at all the sites showing that it is the major buffering agent for the acidity of rain water (Parashar *et al.*, 1996). In a comparison of wet-only and bulk deposition per day, Ca concentration was observed double in bulk sampler than in wet-only sampler showing its deposition in bulk sampler during 24 hrs exposure.

4. Conclusion

In Indian subcontinent, the dust particle influence is very high which is responsible for the high pH of rain water. Aerosols have high loadings of Ca which buffer the acidity caused due to the presence of H_2SO_4 and HNO_3. However, SO_4 is also contributed by soil being a component of soil in most of the parts of India. The soil SO_4 is not responsible for lowering the pH of rain water. In general, the nature of rain water is alkaline in India. However, acid rain may occur under the conditions- (a) if the site is located in industrial area, (b) after continuous rains which wash off soil-derived particles from the atmosphere leading to the weak buffering potential and (c) in the regions where soil is acidic.

Acknowledgements

Authors are grateful to Dr K V Raghavan, Director, IICT, Hyderabad and Dr A P Mitra, FRS, NPL, New Delhi for their encouragements. Benefit of discussions with Prof H Rodhe, IMI, Stockholm and Dr L Granat, IMI, Stockholm is gratefully acknowledged.

References

Das S.N., Thakur R S., Granat L. and Rodhe H.: 1998, *Prodeeding of 4th CAAP Workshop,* Bangkok Gillett R.W. : 1998, *Proceedings of 4th CAAP workshop,* Bangkok.

Jacks G., Sharma V.P., Torssander P. and Åberg G.: 1994, *Geochemical Journal,* **28,** 351.

Jain M. : 1999, *Multiphase measurements of atmospheric acidity in Delhi region. Ph D Thesis.* Dr B R Ambedkar University, Agra.

Jain M., Kulshrestha U.C., Sarkar A.K. and Parashar D.C.: 2000, *Atmospheric Environment,* **34,** 5129.

Khemani, L.T.: 1989, `Physical and chemical characteristics of atmospheric aerosols' in P.N. Cheremisinoff (ed), Air Pollution Control, **Vol 2,** *Encyclopedia of Environ. Control, Tech.*, Gulf Publ. Co. USA, pp 401-452.

Khemani L.T.: 1992, `Th e role of alkaline particulates on pH of rain water and implications for control of acid rain' in H.W. Ellasesser *(ed), Global 2000 revisited,* Pargon House, New York, pp 87-123.

Khemani, L.T.: 1993, *Indian J Radio and Space Physics,* **22,** 207.

Kulshrestha, U.C., Sarkar, A.K., Srivastava, S.S. and Parashar, D.C.: 1995a, *Water, Ai,r and Soil Pollution,* **85,** 2137.

Kulshrestha, U.C., Kumar, N., Saxena, A., Kumari, K.M. and Srivastava, S.S.: 1995b, *Energy Environment Monitor,* **11,** 177.

Kulshrestha, U.C., Sarkar, A.K., Srivastava, S.S. and Parashar, D.C.: 1996, *Atmos. Environ.,* **30,** 4149.

Kulshrestha U.C.: 1998, `Chemistry of atmospheric depositions in India' in S. Raha, P.K. Ray and B. Sinha (eds.), *Science at High Altitudes,* Allied Publishers, New Delhi, pp 187-194.

Kulshrestha, U.C., Jain, M., Sekar R; Vairamani M; Sarkar A.K.. and Parashar, D.C.: 2000. Current Science (in press).

Kulshrestha, U.C., Jain, M., Sarkar A.K., Kumar, A. and Parashar, D.C.: 1997, *Proceedings of IGAC Symposium on Atmospheric Chemistry,* Nagoya. **29,** 109.

Kulshrestha U.C., Jain M., Mandal T. K., Gupta P. K., Sarkar A. K. and Parashar D. C.: 1999, *Current Science,* **76,** 968.

Kumar N., Kulshrestha, U.C., Saxena, A., Kumari, K.M. and Srivastava, S.S.: 1993, *J. Geophys. Res.,* **98 D3,** 5135.

Mukhopadyay, B. Datar, S.V. and Srivastava, H.N.: 1992, *Mausam,* **43,** 249.

Parashar, D.C., Granat, L., Kulshrestha, U.C., Pillai, A.G., Naik, M.S., *et al.*: 1996, *A preliminary report on an Indo-Swedish Project on Atmospheric Chemistry. Report CM 90,* IMI, Stockholm University, Sweden.

Sehgal, J.: 1996, Pedology- Concepts and Applications, Kalyani Publishers, New Delhi.

MODELING EFFECTS OF CHANGING DEPOSITION AND FORESTRY ON NITROGEN FLUXES IN A NORTHERN RIVER BASIN

AHTI LEPISTÖ* and SANNA SYRI

Finnish Environment Institute, P.O.Box 140, FIN-00251 Helsinki, Finland
(* author for correspondence, e-mail: ahti.lepisto@vyh.fi)

Abstract. The application of a new, spatial nitrogen leaching/retention model N_EXRET to the Oulujoki river basin (22800 km^2) in Finland is discussed. The model utilizes remote sensing-based land use and forest classification and evaluated export coefficients obtained from detailed small catchment studies. The present and future N depositions were estimated with the regional deposition model DAIQUIRI. Based on source apportionment, N deposition, forestry and agriculture each contribute 16-17% of the total export, with pronounced variation between the different sub-basins. The effect of changing forestry and deposition on N fluxes is assessed by using N deposition scenarios based on recent international emission reduction agreements.

Keywords: N deposition, forestry, N fluxes, regional modeling, river basin

1. Introduction

Areal variability of nitrogen leaching may be explained by natural catchment characteristics, but may also be a response to climate, and to anthropogenic factors such as increased N deposition, forestry and agriculture activities, and different point sources. In order to create action plans for reducing nitrogen loadings to marine areas, it is important to know the contributions from the various sources and retention of N in the river basins with varying lake percentages.

Nitrogen oxides and ammonia act as fertilizers and cause eutrophication in freshwater and marine ecosystems. The relative importance of nitrogen deposition is likely to increase, due to the fact that international agreements to decrease the emissions have not up to now effectively addressed nitrogen emissions and resulting transboundary air pollution in Europe.

When modeling nutrient leaching in larger scales, it is extremely laborious, or even impossible, to obtain relevant manually collected data of forestry operations in large river basin scales. The use of satellite imagery for detecting forest changes has attracted increasing attention. The satellite image-based land cover and forest classification data has been developed for Finland (Vuorela, 1997), and offers new possibilities for GIS modeling purposes. The export coefficient modeling approach can nowadays be combined with land use classes based on remote sensing. The approach is simple and logical, and limited input requirements make the approach suitable for catchment management (Johnes, 1996).

In this paper, *first*, the application of the spatial, export coefficient -based N export and retention model N_EXRET (Lepistö *et al.*, 2001) to a large Oulujoki river basin is discussed, utilizing remote sensing -based land use and forest classification, and areal N deposition estimated by the DAIQUIRI model (Syri *et al.*, 1998). *Second*, the impact of two scenarios of N deposition change together with one scenario of forestry change are discussed.

2. Material and Methods

2.1. FACTORS CHARACTERIZING THE OULUJOKI RIVER BASIN

The Oulujoki river basin (Fig. 1) (22841 km^2), located in northern Finland, is dominated by mineral soil forests, together with wetlands and peatland forests (~30%), lakes (12%) and some arable land (3%). The central lake of the water system is Lake Oulujärvi (928 km^2), with two major waterways, Hyrynsalmi and Sotkamo discharging into it. The Oulujoki river discharges from Lake Oulujärvi into the Gulf of Bothnia of the Baltic Sea (Fig. 1). The average areal precipitation in the river basin during the same period is typically 650-700 mm, of which some 350 mm evaporates.

2.2. FOREST MANAGEMENT AND ITS CHANGE SCENARIOS

Clear-cut areas and plantation areas were obtained from the satellite image-based land cover and forest classification data (Vuorela, 1997). The TM images used for land use and forest classification were taken during the years 1989-1992. The 12 land use classes used in this study, including recent and elder cut areas in mineral and organic soil forests, plantations, pine- and spruce-dominated and open peatlands, peat harvesting areas, arable land and built-up areas, are described in more details by Lepistö *et al.*(2001). Two scenarios of forest management were used: one with the same cutting intensity as in the 1990s [scenario FOR_stable], and one with a 20% decrease in cut areas [scenario FOR_-20%] until 2010.

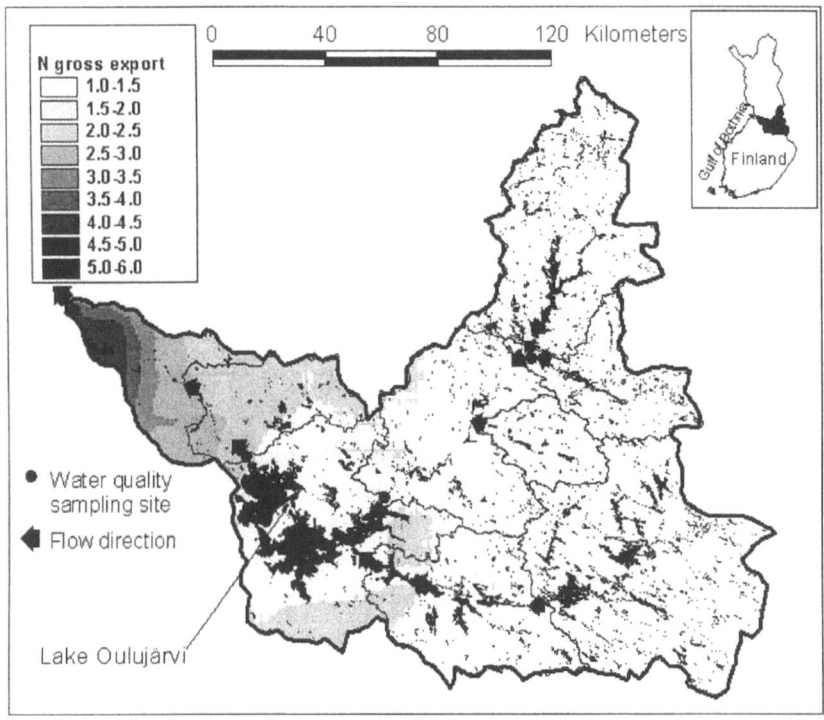

Figure 1. Modeled variability of gross N export (kg ha^{-1}a^{-1}) from soil into water in the Oulujoki river basin.

2.3. REGIONAL ATMOSPHERIC N DEPOSITION AND SCENARIOS

Inorganic N deposition (N_{dep}) was calculated using the regional deposition model DAIQUIRI (Syri *et al.*, 1998). N deposition from Finnish sources was calculated using a database of point-source emissions and agricultural emission data available at municipality level. Long-range atmospheric transport of nitrogen compounds was estimated using source-receptor matrices of the EMEP/MSC-W (EMEP/MSC-W, 1998). The expected future deposition level in 2010 was calculated from the emission reduction obligations recently agreed within the UN/ECE (UN/ECE,1999) [scenario UNECE]. The maximum possible reduction of nitrogen deposition in the near future was estimated using the Maximum Feasible Reductions emission scenario compiled by IIASA (Amann *et al.*, 1999) [scenario MFR]. Deposition straight into water surfaces was here taken into account. Of that N deposition into terrestrial parts of the basin, a very high percentage will retain in these northern, N-limited peatland and mineral soil forests.

2.4. THE MODEL N_EXRET

A spatial N export model N_EXRET (Lepistö *et al.*, 2001) was applied, with the total N load transported to a water body at any point along its length, calculated as the sum of the loss or export of nitrogen from each source in the river basin. The model is used by a raster-based GIS software. First, N export from soil of each grid to water (N_{ex_grid}) was estimated. The total terrestrial nitrogen export (N_{ex}) to a watercourse was calculated by summing the N exports of the grids (N_{ex_grid}) within the sub-basin n. Retention in wetlands and in lakes is assumed to be stable within a sub-basin and directly related to their areas.

Proportions of different land use classes within a grid were obtained from the above-mentioned satellite-based data base. The original 25x25m grids were aggregated into 1x1km grids. Export coefficients used were 'best available' values taken from the empirical studies assumed to be representative for the area. The representativeness of the empirical case study results was recently tested in a test catchment (Lepistö and Kenttämies, 1998). The average situation in the early 1990s was the time of focus.

3. Model Results and Discussion

3.1 GROSS N EXPORT AND ITS SOURCES

The estimated variability of N export from soil to water (N_{ex_grid}) in the Oulujoki river basin, as an average for the early 1990s, is shown in Figure 1. A total average of 5430 tonnes/year was estimated, with clear variability within the area when comparing natural leaching/forestry- dominated areas (1-2 kg N ha^{-1}a^{-1}) in the north-eastern parts of the basin with agricultural-dominated lowland areas (5-6 kg N ha^{-1}a^{-1}).

On average, forestry contributes 16% of the total export, with variation of 11-24% between the different sub-basins. Of the forestry contribution, about 80% was due to cuttings and 20% due to forest drainage. The contribution of forestry is highest to the north of Lake Oulujärvi, and close to the water divides. The contribution of atmospheric deposition was 16%, point sources 8 % and natural leaching 41%. On average, agriculture contributes 17% of the total export varying between 8% in the uppermost reach and 38% in the lowermost reach close to the sea.

3.2. IMPACT OF N DEPOSITION AND FORESTRY SCENARIOS

Contribution of N deposition and forestry to the gross N export and scenarios are discussed here. Other N sources, including agriculture, peat harvesting, point sources and built-up areas are discussed in Lepistö *et al.* (2001).

Table I shows the total modeled N export from the Oulujoki river basin to the sea, without retention modeling. According to UNECE and MFR scenarios, the atmospheric N deposition to the lakes would decrease clearly, 28-50%. This would affect to 4-8% decrease in the total N export to the sea areas. UNECE is based on the recent emission agreement within the UN/ECE, while MFR is the highest amount which could be reduced and can not be considered economically to be realistic.

According to the forestry scenario, a 20% decrease in cut areas would reduce N export from 644 tn a^{-1} to 547 tn a^{-1}, i.e. 15%. The total N export to the sea would reduce only 1.8% (6.1-4.3%) (Table I). This means that a relatively large decrease in cut areas would have a minor reduction in the total N export to the sea. This calls for more effective catchment-wide planning for forestry and increased use of buffer zones along the watercourses.

Atmospheric deposition, together with N point sources, have the highest inorganic fraction of the N sources, and contribute more to the *inorganic* N than to the total N export. The total measured N flux to the sea was 2850 tn a^{-1} in the 1990s, which includes about 530 tn a^{-1} of inorganic N (18%) the rest being organic.

TABLE I

N total export, source apportionment of export into N deposition and forestry (cut areas), and change due to deposition/forestry scenarios (sections 2.2 and 2.3) until 2010 in the Oulujoki river basin. Other N sources are not shown and no changes are assumed for them.

	Scenario	N tot. export	N dep.	N dep.	Cut areas	Cut areas	Decr.,N export
		tn a^{-1}	tn a^{-1}	%	tn a^{-1}	%	%
1990		5432	856	16	644	11.9	0
2010	UNECE,FOR_stable	5196	619	12	644	12.4	4.3
2010	MFR,FOR_stable	5005	429	8.6	644	12.9	7.9
2010	UNECE,FOR_-20%	5099	619	12	547	10.7	6.1
2010	MFR,FOR_-20%	4908	429	8.7	547	11.2	9.6

Retention in the river basin was on average 48% of the gross N export. On average 45% of the total N load was found to be retained in the southern half of Sweden (Arheimer and Brandt, 1998). Excess N is typical in these areas, if all the excess N will not be utilized along the - often P-limited - lakes of the long waterways in the river basin, it will retain or be transported to the sea areas contributing eutrophication.

The N_EXRET model can provide authorities with quantitative information about where in a river basin, and at which sources, attention should be focused so that remedial measures concerning the lakes and the sea areas are as effective as possible. In this river basin with high lake percentage, reduction of N deposition (UNECE scenario) would decrease total N export to the sea ~5%, and clearly more concerning inorganic N export.

The model will be further tested in river basins where land use and deposition patterns differ clearly from this northern basin.

References

Amann, M., Bertok, I., Gyarfas, F., Heyes, C., Klimont, Z., Makowski, M., Schöpp, W.and Syri, S.: 1999, Cost-effective control of acidification and ground-level ozone, *Technical Reports to the European Commission*, DG-XI, IIASA, Laxenburg, Austria.

Arheimer, B. and Brandt, M.: 1998, Modeling nitrogen transport and retention in the catchments of southern Sweden. *Ambio* **27**, 471.

EMEP/MSC-W: 1998, Transboundary air pollution in Europe. EMEP/MSC-W Status report 1998. EMEP/MSC-W, Oslo, Norway.

Johnes, P.: 1996, Evaluation and management of the impact of land use change on the nitrogen and phosphorus load delivered to surface waters: the export coefficient modelling approach. *Journal of Hydrology* **183**, 323.

Lepistö, A. and Kenttämies, K.: 1998, 'Towards the use of satellite-based forest change data in large-scale N leaching models - testing and modeling in a catchment scale', in Kajander J. (ed.) *Proc. XX Nordic Hydrological Conference*, Helsinki. NHP **44**, pp. 210-224.

Lepistö, A., Kenttämies, K. & Rekolainen, S.: 2001, Modeling combined effects of forestry, agriculture and deposition on nitrogen export in a northern river basin in Finland. *Ambio* (accepted).

Syri, S., Johansson, M. and Kangas, L.: 1998, Application of nitrogen transfer matrices for integrated assessment. *Atmospheric Environment* **32**, 409.

UN/ECE: 1999, Protocol to the 1979 Convention on long-range transboundary air pollution to abate acidification, eutrophication and ground-level ozone. UN/ECE Document EB/AIR/1999/1. United Nations, New York, Geneva.

Vuorela, A.: 1997, Satellite image based land cover and forest classification of Finland. Reports of the Finnish Geodetic Institute **97:2**, pp. 42-52.

EPISODIC EVENTS IN WATER CHEMISTRY AND METALS IN STREAMS IN NORTHERN SWEDEN DURING SPRING FLOOD

FRIDA EDBERG[1*], HANS BORG[1] and JAN-ERIK ÅSLUND[2]

[1] *Institute of Applied Environmental Research (ITM), Stockholm University, S-106 91 Stockholm, Sweden;* [2] *County Administration of Jämtland, S-831 86 Östersund, Sweden*
(*author for correspondence, e-mail: frida.edberg@itm.su.se)

Abstract. Thirteen streams in the province of Jämtland in northern Sweden were monitored during spring in 1995 (December 94 - July 95) to study changes in water chemistry and metal concentrations during snow melt. The brooks are not treated with lime, with one exception, and can be approximately divided into three groups according to watershed characteristics; A) > 65% above tree line, B) > 65 % wetland, C) > 55% forested. During peak flow, pH dropped 0.5-2.5 units and alkalinity generally to zero. The brooks above tree line were lowest in base cations and reached the lowest pH-values (4.4-4.6) during peak flow, while sulphate levels were about the same as in the forested watersheds. During peak flow, organic anions showed the highest increase in the wetland and forested catchments. Compared to base flow, Al, Zn, Pb and to some extent Mn was enriched during peak flow. The results also illustrate the difficulties in generalising the reasons for alkalinity losses during spring flood in this kind of streams. In some of the brooks, the use of either base cations or silica, when calculating dilution effects, gave deviating results concerning the relative contribution of strong acids in the snow pack.

Keywords: Acid episodes, brooks, metals, snow melt, Sweden, water chemistry

1. Introduction

Acidic episodes in brooks during snowmelt and rains have been well documented in Sweden (Bjärnborg, 1983; Jacks *et al.*, 1986; Jansson and Ivarsson, 1994; Borg *et al.*, 1995) as well as Canada (Molot *et al.*, 1989; Campbell *et al.*, 1992; Tranter *et al.*, 1994) and USA (Baker *et al.*, 1996; Heard *et al.*, 1997). The decreasing pH during these acidic episodes results in a change in aluminium (Al) speciation towards higher levels of toxic Al forms. This has been shown to cause stress and mortality in fish populations and other aquatic species (Henriksen *et al.*, 1984; Baker *et al.*, 1996; Heard *et al.*, 1997). The cause of the acidic episodes in these studies varies from anthropogenic sulphate from the snow pack combined with dilution of base cations (Jacks, 1986; Molot *et al.*, 1989; Borg *et al.*, 1995) to organic acids and naturally originated sulphate (Campbell, 1992; Jansson and Ivarsson, 1994).

This study was made during the spring of 1995 with the aim to study the acidic episodes during snowmelt in brooks in the province of Jämtland, northern Sweden, and the effects on water chemistry and metal concentrations. In most of the brooks in this study, changes in biological and water quality conditions have occurred over the last 30 years. The most profound changes are reduced densities of brown trout (*Salmo trutta L.*) (Olofsson *et al.*, 1995) and an increase in watercolour.

2. Material and methods

2.1 STUDY SITES

The studied brooks are situated in the province of Jämtland in the northwestern part of Sweden. Quartzite and porphyries dominate the pre-quartenary rocks in this area. Smaller fractions of greywackes and shales are frequent in some of the watersheds (see Table I, no 6 and 7). The quaternary deposits varies from boulder and gravel upon glaciofluvial sediments in high altitude areas to peat on till with fine sand in the watersheds dominated by wetland. The sulphur deposition in the area today (1999) is on average 2-3 kg ha^{-1} yr^{-1} in the lower parts of the watersheds and about 4 kg ha^{-1} yr^{-1} in the upper, high elevation parts. The sulphur deposition varies between different years, but since 1993 it has been reduced by approximately 50%.

Thirteen brooks were sampled weekly during the spring of 1995, from December 1994 to July 1995 and have been grouped by the type of dominating watershed (Table I). The brooks have never been limed, with one exception (see Table I), or subjected to forest or wetland management, with the exception for brooks 10,11,13, (see Table I) where a minor part of the forested watershed has been affected.

TABLE I
Watershed characteristics of the studied brooks. Groups: A) >65% of watershed above tree line, B) >65 % of watershed wetland, C) >55% of watershed forested.

Name	Watershed area km^2	Wetland	Lake	Forested	Above tree line	Farm land	Group
1. Arådalsbäcken	6.71	12.67%	0.15%	6.41%	80.77%	0.00%	A
2. Bastuån	46.86	7.59%	0.18%	25.26%	66.74%	0.22%	A
3. Kokdalsbäcken	11.89	1.79%	0.17%	3.87%	94.18%	0.00%	A
4. Stockån Skalet	9.07	11.80%	0.22%	2.87%	85.12%	0.55%	A
5. Tvärån, Gräftåvallen	16.45	0.90%	0.20%	3.28%	95.62%	0.00%	A
6. Blacklötumyrbäcken	12.64	71.89%	0.26%	27.85%	0.00%	0.00%	B
7. Hemmingsån	28.00	67.54%	0.71%	31.75%	0.00%	0.00%	B
8. Landsombods-Fuan	12.84	66.87%	2.11%	30.75%	0.00%	0.27%	B
9. Orrbäcken	4.45	73.71%	0.75%	21.97%	0.00%	3.57%	B
10. Brynnbäcken	22.92	17.82%	0.05%	75.44%	0.00%	6.69%	C
11. Lill-Trumman *	6.65	24.91%	0.00%	71.26%	3.11%	0.72%	C
12. Sandbäcken, Sånfjället	4.91	7.74%	0.00%	56.21%	36.25%	0.00%	C
13. Sandviksån	9.05	11.74%	0.08%	78.78%	9.13%	0.35%	C

*) Limed once in 1991

2.2 METHODS

All samples were analysed for pH, alkalinity and colour at the laboratory of the county Administration of Jämtland using Swedish and European standard methods. Major ions and total organic carbon (TOC) were analysed at ITM, Stockholm University, generally using Swedish standard methods, following general quality assurance and control routines of the laboratory accreditation. The measuring uncertainties are below 10 %. Metals were determined using ICP-AES and ICP-MS (SGAB, Luleå).

The discharge was calculated using a weir from River Hoån in the middle of the area. This is a representative and natural river not affected by waterpower, and the spring flood in River Hoån occurs at the same time as in the studied brooks. Base cations, [BC] (Na + K + Ca + Mg) and SO_4 was corrected for sea salt influence.

3. Results and Discussion

The spring of 1995 had a rapid snowmelt and drastic acidic episodes. During the spring flood the discharge at peak flow compared to base flow increased by a factor of 5-58 (mean 42). All the studied brooks showed a pH decrease at peak flow, ranging from 0.48 to 2.43 pH units, reaching pH values of 5.6-4.5. At peak flow, the alkalinity with only two exceptions reached zero, the levels of BC decreased and the concentrations of organic anions increased with two exceptions. The brooks had internal variation in waterflow and pH (Fig. 1) due to the different watershed characteristics (Table I). The densities of brown trout (*Salmo trutta L.*) were reduced markedly during the spring of 1995 in brook no 2 and 10 (see Table I), the only brooks with remaining fish populations (county Administration of Jämtland, unpublished data).

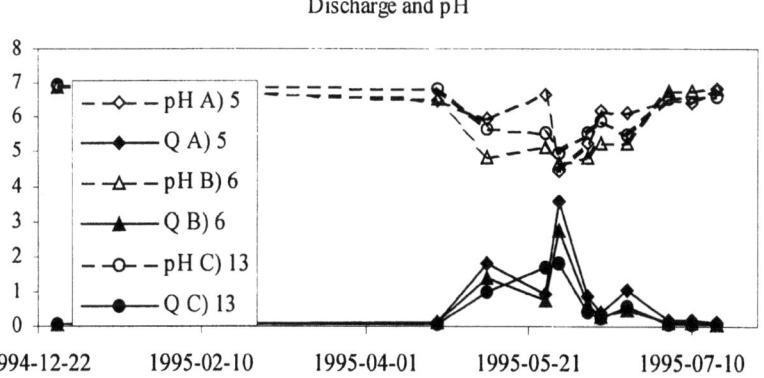

Figure 1. Discharge, Q (m^3 s^{-1}), and pH during the acidic episode for three brooks: A) 5-Tvärån, B) 6-Blacklötumyrbäcken, C)- 13 Sandviksån.

By grouping the brooks by dominating watershed type some different features of the groups could be seen. The group dominated by above tree line (bare mountain) (A) had the lowest levels of alkalinity during the study period as well as the lowest BC levels (Fig. 2). The levels of sulphate (SO_4) were slightly higher compared to the group of brooks dominated by wetland (B), but the same or lower than the group dominated by forested watershed (C). The brooks of group A also reached the lowest levels of pH during peak flow. The group dominated by wetland catchments (B) on the other hand had higher levels of BC than the group dominated by above tree line (A), but about the same as the group dominated by forested watersheds. All the group comparisons mentioned were significant at the <0.05 level according to variance analysis (one sided ANOVA).

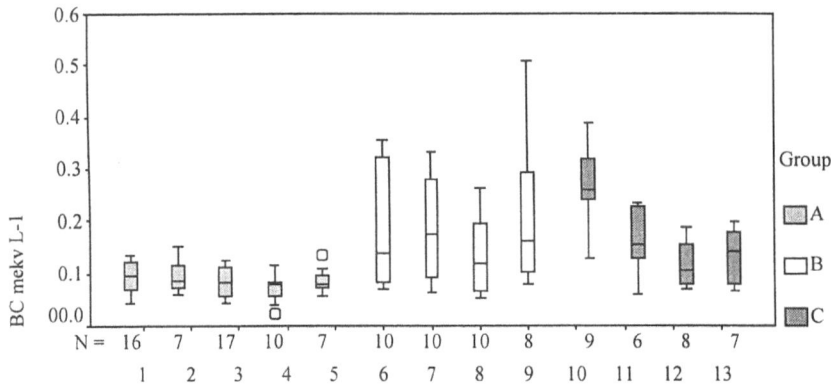

Figure 2. Base cations (BC) (mekv L^{-1}) in the different groups of brooks (median, 25-, 75- percentile, maximum and minimum values, N= no of observations). Group A), 1-Arådalsbäcken, 2-Bastuån, 3-Kokdalsbäcken, 4-Stockån, 5-Tvärån. Group B) 6-Blacklötumyrbäcken, 7-Hemmingsån, 8-Landsombudsfuan, 9-Orrbäcken. Group C) 10-Brynnbäcken, 11-Lill Trumman, 12-Sandbäcken, 13-Sandviksån. Groups according to dominating watershed, see Table I.

The discharge was used to determine the base flow and the peak flow in the brooks. Dilution indexes were calculated by dividing the concentrations at base flow with the concentrations at peak flow, and consequently a number >1 indicates dilution at peak flow, while <1 indicates an enrichment. The BC and SO_4 had dilution indexes >1, while colour, TOC, and H^+ had dilutions indexes <1. Since most metals increased during peak flow, this is presented in Figure 3 as enrichment factors, i.e. the opposite of dilution factors. Al, Zn, Pb, Cu and to some extent Mn showed enrichment factors ranging from 1 to 42, while Fe in about half of the cases was diluted (Fig. 3).

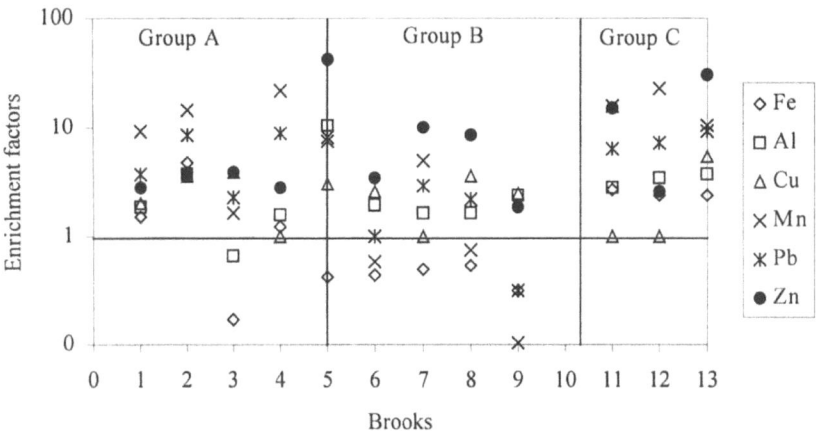

Figure 3. Enrichment factors for the metals (concentration at peak flow/concentration at base flow), logarithmic scale, in the streams. Group A), 1-Arådalsbäcken, 2-Bastuån, 3-Kokdalsbäcken, 4-Stockån, 5-Tvärån. Group B) 6-Blacklötumyrbäcken, 7-Hemmingsån, 8-Landsombudsfuan, 9-Orrbäcken. Group C) 11-Lill Trumman, 12-Sandbäcken, 13-Sandviksån. Groups according to dominating watershed, see Table I.

Silica (Si), considered to behave conservatively, i.e. only influenced by dilution, was diluted 4-13 times. This can be compared to SO_4, which was only diluted by a factor of 1-7, the dilution factor of Si being higher than SO_4 in all cases (Fig. 4). This indicates an anthropogenic contribution of SO_4 deposited in the snow pack, on the acid surges in the streams.

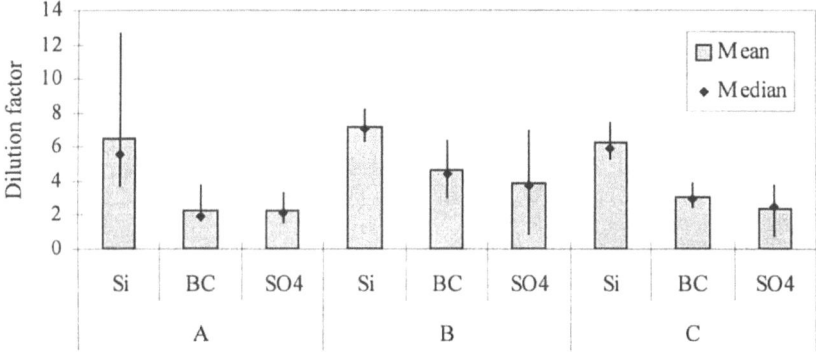

Figure 4. Dilution factors of the streams divided into tree groups according to dominating watershed, see Table I.

If we instead consider BC to behave conservatively, and use them to calculate dilution effects the results are more complicated to interpret, since the dilution factor of BC is lower compared to SO_4 in 50% of the cases. Compared to Si, BC might be more subjected to increased weathering during acid episodes in these watersheds, and thereby somewhat enriched during peak flow, contradicting the dilution effect.

4. Conclusions

The brooks in this study all had major acidic episodes during the spring flood of 1995. During these acidic episodes some metals were released and the concentrations of Al, Zn, Pb, Cu and to some extent also Mn were enriched. By dividing the brooks into three different groups according to dominating watershed characteristics some properties of the different groups could be seen. Streams above the tree line showed the lowest concentrations of BC (and alkalinity) and the most drastic decreases on pH. Enrichment of organic substances was most pronounced in the wetland and forested watersheds.

The use of dilution factors for Si or BC gave deviating results when calculating the causes of acidic episodes.

Acknowledgements

This study was financed by the Swedish Environmental protection agency. We also wish to thank the dedicated co-workers at ITM and the laboratory at the province of Jämtland for analysing all the water samples.

References

Baker, J.P., Van Sickle, J., Gagen, C.J., DeWalle, D.R., Sharpe, W.E., Carline, R.F., Baldigo, B.P., Murdoch, P.S., Bath, D.W., Kretser, W.A., Simonin, H.A. and Wigington, P.J.: 1996, *Ecological Applications* 6, 422.
Bjärnborg, B.; 1983, *Hydrobiologia* 101, 19.
Borg, H., Anderssson, P., Nyberg, P. and Olofsson, E.: 1995, *Water, Air and Soil Pollution* 85, 907.
Campbell, P.G.C., Hansen, H.J., Dubreuil, B. and Nelson, W.O.: 1992, *Canadian Journal of Fisheries and Aquatic Sciences* 49, 1938.
Heard, R.M., Sharpe, W.E., Carline, R.F. and Kimmel, W.G.: 1997, *Transactions of the American Fisheries Society* 126, 977.
Henriksen, A., Skogheim, O.K. and Rosseland, B.O.: 1984, *Vatten* 40, 255.
Jansson, M. and Ivarsson, H.: 1994, *Journal of Hydrology* 160, 71.
Jacks, G., Olofson, E. and Werme, G.: 1986, *Ambio* 15, 282.
Molot, L.A., Dillon, P.J. and LaZerte, B.D.: 1989, *Canadian Journal of Fisheries and Aquatic Sciences* 46, 1658.
Olofsson, E., Melin E. and Degerman E.: 1995, *Water, Air and Soil Pollution* 85, 419.
Tranter, M., Davies, T.D., Wigington, P.J. and Eshleman, K.N.: 1994, *Water, Air and Soil Pollution* 72, 19.

RELATIONSHIPS BETWEEN ACID IONS AND CARBONACEOUS FLY-ASH PARTICLES IN DEPOSITION AT EUROPEAN MOUNTAIN LAKES

N.L. ROSE[1,*], E. SHILLAND[1], T. BERG[2], K.HANSELMANN[3],
R. HARRIMAN[4], K.KOINIG[5], U. NICKUS[6], B. STEINER TRAD[7],
E. STUCHLÍK[8], H. THIES[5] and M. VENTURA[9].

[1] *Environmental Change Research Centre, University College London, 26 Bedford Way, London, WC1H 0AP U.K.;* [2] *Norwegian Institute for Air Research, P.O. Box 100, Instituttveien 18, N-2007 Kjeller, Norway;* [3] *Institute of Plant Biology / Microbiology, University of Zurich, Zollikerstr. 107, CH-8008 Zurich, Switzerland;* [4] *Freshwater Fisheries Laboratory, SOAFD, Faskally, Pitlochry, Perthshire, Scotland, UK;* [5] *Institut für Zoologie and Limnologie, Leopold Franzens Universität Innsbruck, Technikerstr. 25, A-6020 Innsbruck, Austria;* [6] *Institute of Meteorology and Geophysics, University of Innsbruck, Innrain 52, A-6020 Innsbruck, Austria;* [7] *Department of Chemistry, University of Bern, Freiestr. 3, CH-3012 Bern, Switzerland;* [8] *Department of Hydrobiology, Charles University, Vinicna 7, CZ 128 44 Prague 2, Czech Republic;* [9] *Departament d'Ecologia, Fac. Biologia, Universitat de Barcelona, Diagonal 645, Barcelona 08028, Spain.*
(* author for correspondence, e-mail: nrose@geog.ucl.ac.uk)

Abstract. The concentrations of major ions and spheroidal carbonaceous fly-ash particles (SCPs) in bulk deposition were determined in weekly samples from six European mountain lakes during 1997/98. SCPs are produced only from high temperature combustion of fossil-fuels and therefore provide an unambiguous indicator of atmospheric deposition from this source. Positive correlations were observed between SCPs and SO_4^{2-}, NO_3^- and NH_4^+ at all sites except for some determinands at Jorisee (Switzerland) and Starolesnienske (Slovakia). Correlations between SCPs and $SO_4^{2-} + NO_3^-$ were always more positive than for SCPs with 'total acid ions' ($SO_4^{2-} + NO_3^- + NH_4^+$). This is in agreement with the expectation that the contribution to NH_4^+ deposition made by fossil-fuels is negligible. Good positive correlations between SCPs and all acid anions were observed at Estany Redo (Pyrenees); lower but still positive correlations were observed for all acid ions with SCPs at Gossenköllesee (Austria), Lochnagar and Kårvatn (central Norway), whilst little trend in correlation was observed for Jorisee and Starolesnienske. It is suggested that this gradient reflects the influence of fossil-fuels on acid deposition in these areas. A high positive correlation was observed between SCP and Cl^- at Gossenkollesee possibly as a result of HCl from coal combustion.

Keywords: acid ions, atmospheric deposition, Europe, fly-ash particles, mountain lakes

1. Introduction

Spheroidal carbonaceous fly-ash particles are produced by the combustion of solid and liquid fossil-fuels at industrial temperatures. They are not produced by any natural processes and therefore have been used as unambiguous indicators of atmospheric deposition from sources such as the power generation and other industries (Renberg and Wik, 1985; Rose *et al.*, 1998). Historical trends in the atmospheric deposition of these contaminants to remote European mountain lakes were determined using their sediment records as natural archives and

Water, Air, and Soil Pollution **130:** 1703–1708, 2001.
© 2001 *Kluwer Academic Publishers.*

showed that spatial and temporal patterns of SCPs were in agreement with historical trends in known emission sources (Rose et al., 1999). However, as part of the EU funded MOLAR project (Wathne and Hansen, 1997), more detailed measurements of inputs to these remote lakes were required in order to more fully understand the movement and storage of pollutants within these sensitive systems. To this end, bulk deposition samples were collected for a period of c.15 months at a number of mountain lakes experiencing different regimes of pollutant deposition. Given that the main sources of a number of the measured parameters (e.g. acid anions, SCPs, certain trace metals) are expected to be the same, it might also be expected that depositional trends should be similar. This paper provides a preliminary assessment of these data in order that relationships between depositing pollutants at remote European mountain lakes might be identified.

2. Sites and Methods

Six sites were included in the sampling programme to provide broad geographical coverage of European mountain areas and a range of pollutant depositional regimes. Location information is provided in Table I.

TABLE I.
Site information.

Site	Region	Latitude	Longitude	Altitude (m.a.s.l.)
Kårvatn	Mid-Norway	62° 47' N	08° 53' E	210
Lochnagar	Scotland	56° 38' N	03° 13' W	785
Starolesnienske Pleso	Slovakian Tatra	49° 18' N	20° 23' E	2000
Gossenköllesee	Austrian Tyrol	47° 13' N	11° 01' E	2417
Jorisee	Swiss Alps	46° 47' N	09° 57' E	2500
Estany Redo	Pyrenees, Spain	42° 39' N	00° 46' E	2240

NILU (Norwegian Institute of Air Research) - type bulk deposition collectors were established at each site in late summer 1996 and, where weather conditions permitted, were sampled weekly in spring through autumn and fortnightly in winter for a period of 15 - 18 months. At Redo samples were collected monthly. Deposition volumes were recorded and linked to rainfall via automatic weather stations also located at each site. Analytical procedures for the measurement of major cations and anions along with details of inter-laboratory comparisons and analytical quality control and assessment are described elsewhere (MOLAR Water Chemistry Group, 1999). For SCP analysis, known volumes of sample were filtered through Whatman GF/C filters. These were digested using hydrofluoric acid to dissolve the filter followed by one or two washings in concentrated hydrochloric acid to dissolve the white precipitate formed by the HF treatment. Once clear, the resulting suspension was washed three times in distilled water.

Microscope slides were made and SCPs counted as described in Rose (1994). Data were transformed to fluxes using site specific rainfall data.

Principal components analysis (PCA) was used for initial data exploration. PCA was undertaken using CANOCO for Windows (ter Braak and Smilauer, 1997) and CALIBRATE (Juggins and ter Braak, 1997).

3. Results and Discussion

In terms of the relationships between acid ions (SO_4^{2-}, NO_3^-, NH_4^+) and SCPs the sites fall into three classes. First, Redo and Gossenköllesee where good positive correlations exist between the concentrations of SCPs and all acid ions (Table II). Second, Lochnagar, and Kårvatn where the relationships are still positive but low, and thirdly, Starolesnienske and Jorisee where the relationships show both positive and negative correlations.

TABLE II
Correlation coefficients between SCPs and acid ions at the study sites. N = number of samples.

Site	N	NO_3^-	SO_4^{2-}	NH_4^+
Estany Redo	19	0.74 (p = 0.09)	0.25	0.33
Gossenköllesee	25	0.32	0.17	0.32
Kårvatn	55	0.04	0.08	0.06
Lochnagar	32	0.13	0.19	0.28
Jorisee	8	0.30	-0.37	-0.02
Starolesnienske Pleso	20	-0.12	-0.20	0.04

Other studies have shown that Lochnagar and Starolesnienske are the most contaminated of these sites in terms of quantities of SCP deposited, whereas Kårvatn and Gossenköllesee are the least (Rose et al., in prep). The classification suggested above is therefore not one linked to the amount of contamination received but more likely the proportion of that contamination due to emissions from fossil-fuel combustion. If this is the case, then the more contaminated sites lower in the classification (e.g. Starolesnienske) are, by implication, receiving a higher proportion of deposited pollutants from industrial sources which do not directly utilise fossil-fuels for power.

Although an acid ion, positive correlations between NH_4^+ and SCPs are curious given the contrasting nature of their respective sources. SCPs are solely derived from high temperature fossil-fuel combustion whereas NH_4^+ is almost as exclusively from agricultural sources (RGAR, 1997). The reason for these positive correlations is uncertain, although it may be due to prevailing meteorological conditions at the sites bringing both atmospherically transported contaminants from the same source regions. However, at all sites the correlations between SCPs and $SO_4^{2-} + NO_3^-$ are always more positive than for SCPs with $SO_4^{2-} + NO_3^- + NH_4^+$ confirming that the sources of SCPs, SO_4^{2-} and NO_3^- are more closely related. The high positive correlations between SO_4^{2-} and NO_3^- with NH_4^+ may be due to the formation of the secondary aerosols $(NH_4)_2SO_4$ and $(NH_4)NO_3$ (Avila, 1996) and therefeore the positive correlations between SCPs and NH_4^+ may simply be an artefact of the positive correlations between SO_4^{2-} and NO_3^- with NH_4^+ and SCPs with SO_4^{2-} and NO_3^-.

At the sites closest to the coast e.g. Kårvatn and Lochnagar, PCA bi-plots of the major ion chemistry show an obvious marine influence with highly positive correlations between Na^+ and Cl^- ions (Fig. 1a & b). At the central European sites (Starolesnienske, Gossenköllesee, Jorisee) this is not observed but at Gossenköllesee there is good positive correlation between Cl^- and SCPs ($r = 0.38$; $p < 0.05$) (Fig. 1c). In contrast to the proposed meteorological link between SCPs and NH_4^+ proposed above, the SCP / Cl^- relationship at Gossenköllesee cannot be due to prevailing conditions bringing SCPs with marine Cl^-, or Na^+ would show a similar relationship and this is not the case. One major source of anthropogenic Cl^- is the combustion of coal emitting HCl (RGAR, 1997) and this may be the link between SCPs and Cl^- at the site. However, HCl emissions are thought only to have a significant influence close to the source (RGAR, 1990) and coal combustion sources are remote from Gossenköllesee. This relationship remains to be fully resolved.

Apart from acid ions, SCPs show high positive correlations with K^+ at Redo (0.70) and with Ca^{2+} and Mg^{2+} at Jorisee (both 0.55). For the former, K was found to be an important element in the characterisation of coal SCPs (Rose *et al.*, 1996) and this might provide an appropriate link. Although previous studies have suggested that Redo might be influenced by oil combustion (Rose *et al.*, 1999), large coal-fired power stations in the north-west of Spain (*e.g.* Puentes Garcia Rodriguez, Meirama) may have an influence. This is in agreement with bulk deposition data where samples were found to be more acidic when rainfall was derived from areas to the north-west of the region (Camarero and Catalan, 1993). For Ca^{2+} and Mg^{2+} at Jorisee, this may either be due to the inorganic component of fly-ash being enriched in these elements, or another source (*e.g.* cement production, gypsum plant) being located in the same region.

4. Conclusions

Despite the related industrial sources of the measured determinands, correlations between SCPs with acid ions were lower than expected. The reasons for this are possibly three-fold. First, although high temperature fossil-fuel combustion is the sole source of SCPs this is not the case for acid ions. Non-fossil-fuel sources of these ions will lower any relationship with SCP deposition and therefore the correlation between these pollutants might be seen as a crude measure of the relative influence of fossil-fuel combustion versus other industrial sources at a site. Second, it maybe that the sampling interval (weekly at best) is too coarse to observe short, episodic depositional events. The episodicity of pollutant deposition at these remote

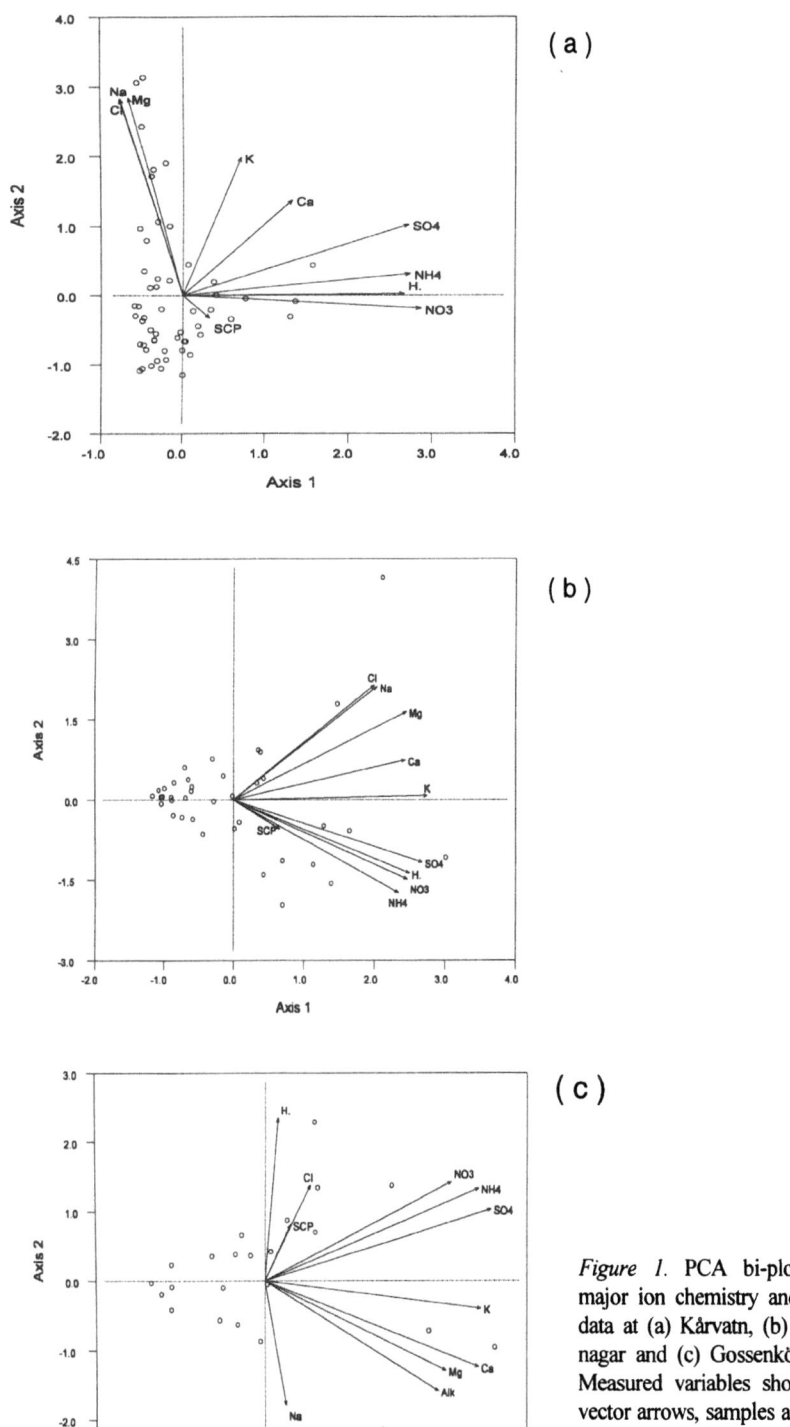

Figure 1. PCA bi-plots for major ion chemistry and SCP data at (a) Kårvatn, (b) Lochnagar and (c) Gossenköllesee. Measured variables shown as vector arrows, samples as open circles.

mountain lakes remains uncertain and finer interval sampling may be required to determine the true relationships of depositions from various sources. Third, sampling covered only a relatively short period. In mountain systems it is known that inter-annual variation in deposition can be considerable and depends on many factors. A longer dataset, such as that now developing at Lochnagar, may help interpret trends in depositing pollutants.

Acknowledgements

The authors would like to thank the EU for funding (MOLAR project: Contract No. (ENV4-CT95-0007) and our MOLAR partners and colleagues for their help with fieldwork and sample collection. NR would also like to thank Jo Porter for field sampling, Elaine Taylor at SOAFD Pitlochry, Scotland for handling the SCP samples and Martin Kernan (ECRC, University College London) for statistical advice.

References

Avila, A.: 1996, *Atmospheric Environment* **30**, 1363.
Camarero, L. and Catalan, J.: 1993, *Atmospheric Environment* **27A**, 83.
Juggins, S. and ter Braak, C. J. F.: 1997, *CALIBRATE Version 0.81*.
MOLAR Water Chemistry Group.: 1999, *Journal of Limnology* **58**, 88.
Renberg, I. and Wik, M.: 1985, *Ambio* **14**, 161.
Review Group on Acid Rain.: 1990, *Acid deposition in the United Kingdom 1986 - 1988*. Warren Spring Laboratory. 124pp.
Review Group on Acid Rain.: 1997, *Acid deposition in the United Kingdom 1992 - 1994*. AEA Technology plc. 176pp.
Rose, N. L.: 1994, *Journal of Paleolimnology* **11**, 201.
Rose, N. L., Juggins, S. and Watt, J.: 1996, *Proceedings of the Royal Society of London. Series A.* **452**, 881.
Rose, N. L., Alliksaar, T., Bowman, J. J., Boyle, J., Coles, B., Fott, J., Harlock, S., Juggins, S., Punning, J-M., St.Clair-Gribble, K., Vukic, J. and Watt, J.: 1998, *Water, Air, and Soil Pollution* **106**, 205.
Rose, N. L., Harlock, S. and Appleby, P.G.: 1999, *Water, Air and Soil Pollution* **113**, 1.
Rose, N. L., Shilland, E., Yang, H., Appleby, P .G., Berg, T., Camarero, L., Gabathuler, M., Hanselmann, K., Harriman, R., Koinig, K., Lien, L., Nickus, U., Steiner Trad, B., Stuchlík, E., Thies, H. and Ventura, M. (in prep.).
ter Braak, C. J. F. and Smilauer, P.: 1997, *CANOCO for Windows. Version 4.0*.
Wathne, B.W. and Hansen, H. (eds.): 1997, *MOLAR. Measuring and modelling the dynamic response of remote mountain lake ecosystems to environmental change: A programme of Mountain Lake Research. MOLAR Project Manual*. NIVA-report SNO 3710-97. 179 pp.

INTERNAL DISTRIBUTION OF ACID MATERIALS WITHIN SNOW CRYSTALS

H. FUSHIMI[1], T. KAWAMURA[2], H. IIDA[3], M. OCHIAI[4], T. NAKAJIMA[5] and Y. AZUMA[5]

[1] *School of Environmental Science, University of Shiga Prefecture, Hikone, Japan.* [2] *Institute of Low Temperature Science, Hokkaido University, Sapporo, Japan.* [3] *Tateyama Caldera Sabo Museum, Toyama Prefectural Government, Toyama, Japan.* [4] *Graduate School of Science, Tokyo Metropolitan University, Tokyo, Japan.* [5] *Lake Biwa Research Institute, Shiga Prefecture, Otsu, Japan*

Abstract. We found that river water is acidified not only in the first stage but also in the later stage of the snowmelt season in Japan, which differs from the so called acid shock occurring in the first stage of the snowmelt in the northern Europe. The acid shocks depend on the regional characteristics of the melt-refreeze processes forming the internal distribution of acid materials within snow crystals under the warm metamorphism. In a warm climatic region like the central Japan, there are possibilities to have the complicated distribution structure of acid materials within granular-snow crystals due to the repeated melt-refreeze processes even in midwinter. Consequently, the pH value of meltwater does not always increase as the snowmelt proceeds. Then, we showed the possibility by using the X-ray computed tomography that the domains with the acid materials exist in the inner parts of snow crystals. So, the acidification of the river water may occur even in the later stage of the snowmelt.

Key words: acidification, acid shocks, CT scanner, snow crystal, warm metamorphism

1. Introduction

Snow crystals in snow cover grow to be granular through the melt-refreeze process and the diameter of the granular snow may reach several millimeters. Along with a progress of the crystal growth caused by the melt-refreeze process, the acidic materials tend to be concentrated in the outer part of the granular snow (Suzuki, 1982; Bale et al., 1989; Suzuki, 2000). Consequently, the pH value of the meltwater at the first stage of melting becomes low, with acidic chemical constituents 2 to 5 times more than the average value of the snow cover (Johannessen and Henriksen, 1978), and the acid shock (acid flush) results. However the acid shocks have been occurring several times even in the later period of the snowmelt process and the pH value of the river water went down to 5.4 in the end of March, 1966, in Lake Biwa catchment area (Figure 1). The similar acid shocks have been also observed in March, 2000. This indicates that it is possible to estimate characteristic distribution structures of acidic materials within granular snow by observing the relative difference of the pH value of the meltwater between at the first stage and at the last stage of melting.

Fushimi (1994) concluded that actual conditions of the acid shock depend on the regional characteristics of the melt-refreeze processes forming the internal distribution of acidic materials within snow crystals. Therefore, it is important to clarify the changes in the pH values between the first stage and the last stage of the snow melt process by field observations, and the microscopic distribution patterns of the acidic materials within snow crystals by using the X-ray computed tomography.

2. Field observation

2.1. METHOD

Since the lowest pH value of the snow-meltwater was 3.8 in Lake Biwa catchment area, the acidification is progressing due to the increasing emission of acidic materials in East Asia. So, the field observations were carried out at Lake Biwa catchment area in Shiga Prefecture and at Mt. Tateyama area (2450m a.s.l.) in Toyama Prefecture in the central region of Honshu Island, Japan.

In order to predict acidification of meltwater from a snow cover, snow was directly sampled from each snow layer at a snow pit by using a sampling bottle with the volume of 100 ml and a weight of about 50 g. The samples were melted at room temperature of about 20 °C. The pH values were measured both at the first stage and at the last stage of the snowmelt process within a few hours after the sampling. In the first stage, the melting takes place mainly in the outer part of a snow grain and in the last stage, in the inner part.

2.2. RESULTS OF FIELD OBSERVATION

Figure 2 is the profile of the pH values of meltwater from granular snow in the upper layer to compacted snow in the lower layer with a snow depth of 600 cm at Mt. Tateyama area in April, 1999, when the upper layer of the snow cover was influenced by the snowmelt process and the loess sediments. The pH values ranged from 4.5 to 6.5 in the first stage and from 4.7 to 5.9 in the last stage of the snowmelt. The pH values of the meltwater at the last stage were lower in the upper layer, while the pH values of the meltwater at the last stage were higher in the lower layer.

Figure 3 shows a vertical profile of pH values of the meltwater, sampled from the snow cover of granular snow grains, with a snow depth of 100cm in the northern part of Lake Biwa catchment area in December, 1992. The pH values ranged from 4.0 to 4.7 in the first stage and from 3.8 to 4.3 in the last stage of the snowmelt. The pH values of the meltwater at the last stage were lower than that at the first stage, except for a few snow layers in the middle part of the snow cover which showed no considerable changes.

3. Analysis of X-ray computed tomography

3.1 METHOD

X-ray computed tomography (CT scanner) has led to remarkable achievements in the field of medical science for clinical diagnosis (Hounsield, 1973). The CT scanner has also been applied in a nondestructive manner to measure the distribution of voids in steady state two phase flows (Ikeda et al.,1983). Kawamura (1988 and 1990) has shown that the CT scanner is a useful tool for examining sea ice structure and for measuring core densities of three dimensional ice.

When X-ray was absorbed by samples, the transmitted intensity was detected by a fluorescence plate and a CCD camera in the CT scanner used in this study (SDD Inc.). The distributions of the intensity data were collected by successively rotating the sample through 360 degrees at intervals of 1 degree. The image of the object was then obtained as a map of an attenuation coefficient, designated as the CT value, of pixels for the sample of any desired section (slice). While the horizontal resolution was 100 μm at the selected field of view (FOV) of 50 mm, the vertical one was 300 μm, depending upon the slices. The CT values of both pure water and pure ice samples were measured as standards. If a material is composed of ice and air, the CT value of the pixels corresponds to the air volume. Impurities, e.g., H_2SO_4, have generally higher attenuation coefficients and therefore higher CT values than ice. The CT value of ice with only one kind of an impurity is related linearly with the impurity fraction. However, the CT value cannot determine the ratios of three- component-materials such as ice, air and impurities.

3.2. RESULTS OF COMPUTED TOMOGRAPHY ANALYSIS

For clarifying the microscopic distribution patterns of the acidic materials within snow crystals, granular snow layer of 480cm thickness were observed at Mt. Tateyama in 1994. Figure 4 shows the CT value distribution along the central line of the granular snow crystal whose higher CT values range from 220 to 240. Since the CT value of pure ice is 215, there are possibilities that the acidic materials with the higher CT values distributed from the outer part to the inner part of the snow crystal. Observations on granular snow cover were carried out in the northern part of the Lake Biwa catchment area in 1996. Figure 5 shows the microscopic distributions of the CT value across the center of the granular snow grain whose highest CT value is 240. Since the CT value of pure ice is 215, this also indicates that the acidic materials distribute even in the inner part of the granular snow crystal. So, the meltwater turns out to be more acidic in cases when the acidic materials concentrate in the central part of the granular snow.

4.Discussion and concluding remarks

In such a warm climate region like central Japan, there are many possibilities to form the granular snow with complicated distribution patterns of acidic materials

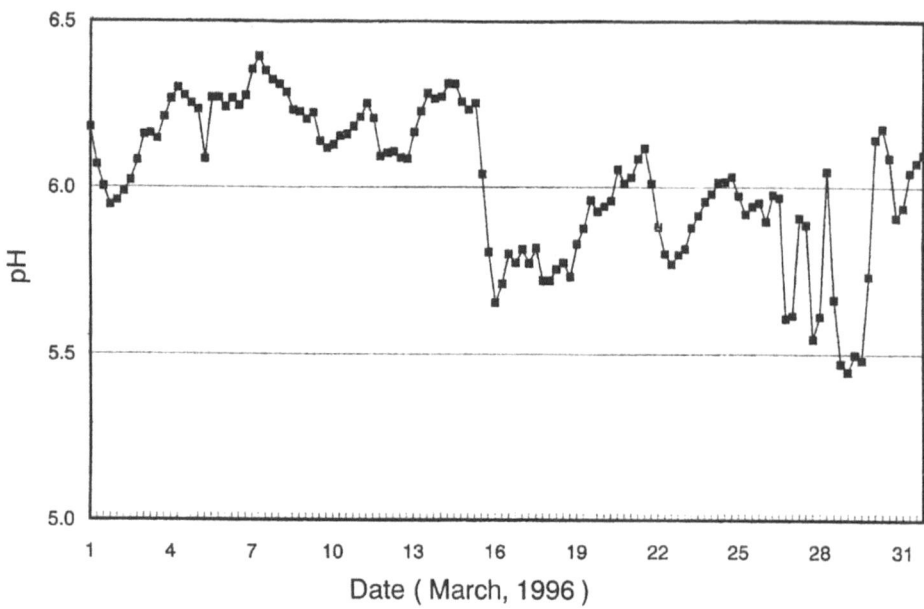

Fig. 1 Diurnal changes of the pH value of a river water in Lake Biwa catchment area.

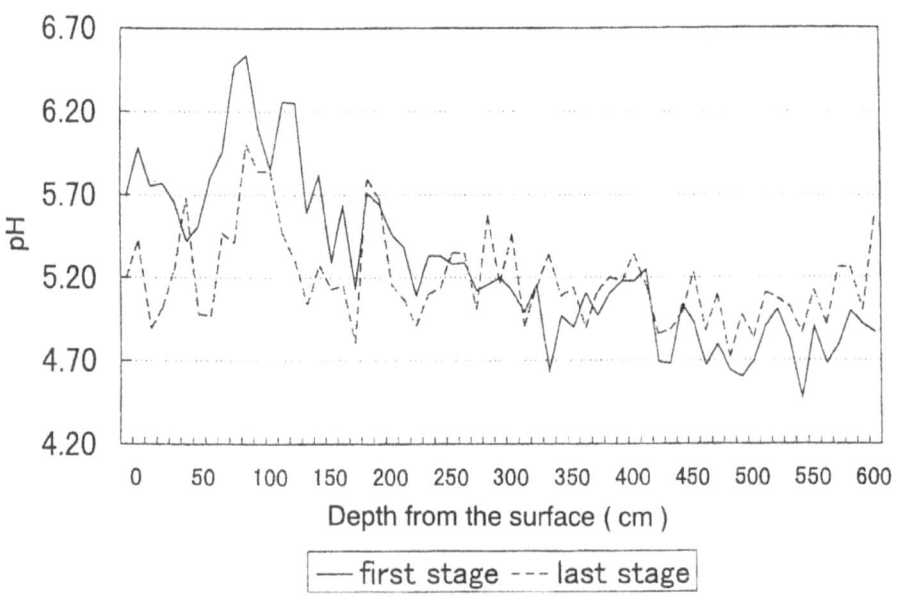

Fig. 2 pH profile of the snow layer in Mt. Tateyama area.

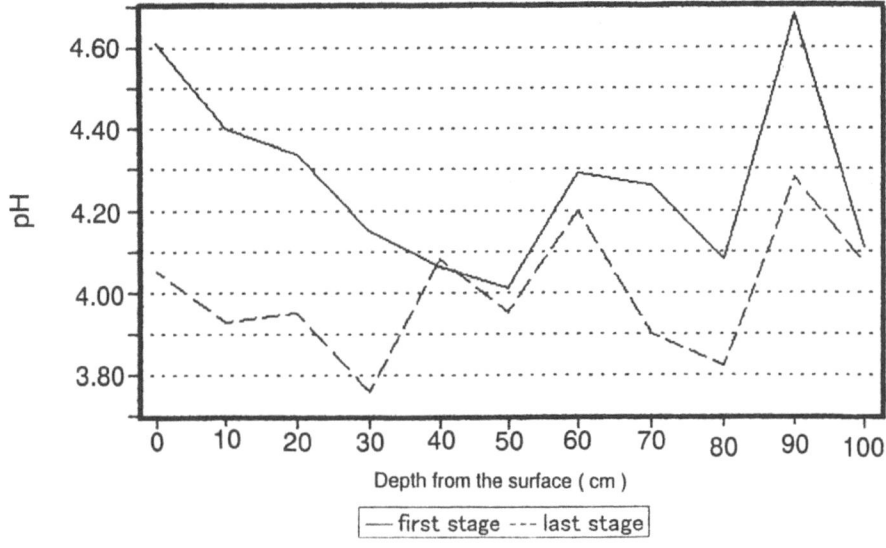

Fig. 3 pH profile of the snow layer in Lake Biwa catchment area.

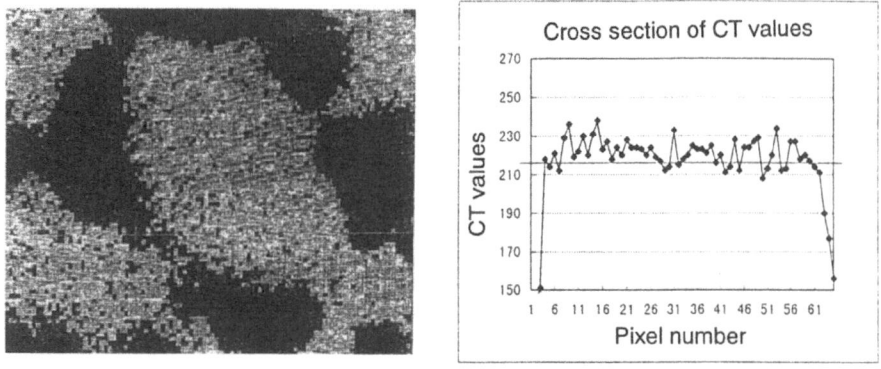

Fig. 4 CT images and CT value distribution of a granular snow in Mt. Tateyama area.

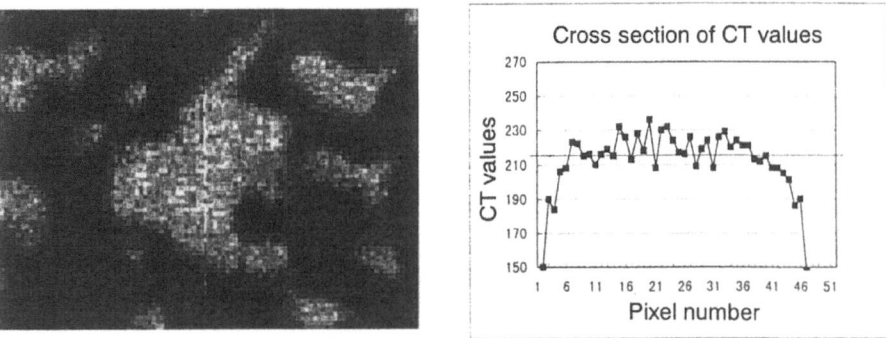

Fig. 5 CT images and CT value distribution of a granular snow in Lake Biwa catchment area.

within them since the melt-refreeze processes of snow cover layers take place repeatedly even in midwinter. Crystallographic thin-section studies indicate that the granular snow grain is composed of several single crystals, so it is considered that the formation of those granular snow grains makes a complicated distribution of acidic materials under warm metamorphism. The pH value of the meltwater does not always increase as the snowmelt progresses, but it does occasionally decrease in the last stage of the snowmelt. The meltwater turns out to be more acidic when the snowmelt process comes to the last stage and the inner parts of the snow crystals melt away, as the acidic materials concentrate even in the inner part of the snow crystals. Therefore, it is important to note that the acid shocks occur not only in the first stage but also even in the last stage of the snowmelt, which differs from the so-called acid shock phenomena reported from the northern part of Europe.

According to the analyses of granular snow by the X-ray computed tomography, the CT value distribution across the center of the granular snow sample showed the higher CT values ranging from 220 to 240. Since the CT value of pure ice is 215, there are impurities distributed from the outer part to the inner part of the granular snow crystal. According to results of artificial sample and chemical analyses, there is a possibility that the domains of the higher CT values correspond to those of acidic materials. Consequently, when the acidic materials mainly concentrate in the central part of snow crystals, the meltwater turns out to be more acidic and gives influences to the hydrological environments even in the last stage of the snowmelt.

References

Bale, R.C., Davis, R.E. and Stanley, D.A.: 1989, Ion elution through shallow homogeneous snow. *Water Resources Research*, **25**, 8, 1869-1877.
Fushimi, H.: 1994, Acid snow in Lake Biwa catchment area - microscopic distribution patterns of acidic materials -. *Seppyo*, **56**, 1, 19-29.
Hounsfield, G.N.: 1973, Computerized transverse axial scanning (tomography), part 1, Description of system, *Br. J. Radiol.*, **46**, 1016-1022.
Ikeda, T., Kotani, K., Maeda, Y. and Kohno, H. 1983: Preliminary study on application of X-ray CT scanner to measurement of void fractions, *J. Nucl. Sci. Technol.*, **20**, 1, 1-12.
Johannssen, M. and Henriksen, A.: 1978, Chemistry of snow meltwater changes in concentration during melting. *Water Resources Research*, **14**, 4, 615-619.
Kawamura, T. 1988 Observations of the internal structure of sea ice by X-ray-computed tomography, *J. Geophys. Res.*, **93**, C3, 2342-2350.
Kawamura, T.: 1990, Nondestructive, three-dimensional density measurements of ice core samples by X ray computed tomography, *J. Geophys. Res.*, **95**, B8, 12407-12412.
Suzuki, K.; 1982, Chemical changes of snow cover by melting. *Japanese Journal of Limnology*, **43**, 2, 102-112.
Suzuki, K.; 2000, Review on snow chemistry in Japan. *Journal of the Japanese Society of Snow and Ice*, **62**, 3, 185-196 (in Japanese).

TEMPORAL VARIATIONS OF ALUMINIUM FRACTIONS IN STREAMS IN THE DELSBO AREA, CENTRAL SWEDEN

CECILIA ANDRÉN[1*], PAUL ANDERSSON[2] and ELISABETH FRÖBERG[1]

[1] *Institute of Applied Environmental Research (ITM), Stockholm University, S-106 91 Stockholm, Sweden;* [2] *F:a SBV-analys, Östra Tolbo 4662, S-820 60 Delsbo, Sweden*
(* author for correspondence, e-mail: candren@itm.su.se)

Abstract. Stream waters were sampled weekly during spring and monthly during summer and autumn in 1998. The streams are more or less acidified, and some have been treated with lime. The aluminium fractions (total monomeric Al, organic monomeric Al, inorganic monomeric Al) were determined colourimetrically with pyrocatechol violet combined with cation exchange using Continuous Flow Analysis (Autoanalyzer I). The levels of inorganic monomeric aluminium varied substantially, between <3 to 271 µg/l. The levels were higher in untreated than in limed waters and twice as high in the most humic waters as in less humic waters. The importance of aluminium mobilisation from the catchments was obvious, with higher aluminium concentrations in surface runoff (unbalanced stream waters) compared to lake outlets (balanced and precipitated lake water). The highest mean levels were measured at spring, whereas the highest single peaks occurred during summer. Inorganic monomeric and total monomeric aluminium was best correlated to ion ratio and pH whereas acid soluble aluminium and organic monomeric aluminium was best correlated to TOC, water colour and iron.

Keywords: acidification, aluminium, liming, speciation, streams

1. Introduction

For more than three decades the interest for the role of aluminium in the deleterious effects of acidified waters has been large (*e.g.* Henriksen *et al.*, 1984; Rosseland *et al.*, 1990; Gensemer and Playle, 1999). Aluminium is leached from the soil by acid precipitation and elevated levels have been noted in Swedish surface waters as the acidification has proceeded (Dickson, 1975). Different methods have been developed to determine fractions of aluminium, namely separation of monomeric from polymeric aluminium and separation of inorganic from organic monomeric aluminium (Bloom and Erich, 1996). In our studies of acidified and/or limed water we have implemented and used an automated method using ion exchange and spectrometry with good results since 1991. There is still a need to gather more information about the behaviour of aluminium fractions especially at high flow episodes and their implications for the biological life.

The aim of this study was to examine the temporal variations of inorganic aluminium at high flow and to study the covariation of aluminium fractions with other factors in Swedish waters with high organic content. Initially the focus was at the spring flood but because of the extraordinary heavy summer rains in Sweden 1998, data from the whole year was collected, rendering two high flows.

2. Materials and Methods

2.1. STUDY AREA AND SAMPLING

The studied stream waters are located in central Sweden (Delsbo region), in boreal-forested mountains at altitudes ranging from 110 to 325 m (above sea level) and catchment areas covering from 0.5 to 17.5 km^2. The precipitation is sometimes still acidic when it reaches the waterways, because of differences in neutralising capacity in the ground material (sand or silt) and the amount/intensity of the runoff. Twelve streams were sampled weekly during spring for eleven weeks and out of these, eight was followed with six more monthly samplings covering the whole year. Data from eight additional streams in a parallell project was included with ten monthly samplings.

2.2. ANALYTICAL METHODS

The aluminium fractions was measured as total monomeric Al (Al_a) and organic monomeric Al (Al_o) by the colourimetric reaction of pyrocatechol violet described in the Swedish Standard (1992), combined with cation exchange (to distinguish inorganic monomeric aluminium (Al_i), Driscoll, 1984) using continuous flow analysis (Andrén, 1995) with an Autoanalyzer I system (Technicon). The fractionation of aluminium was performed at the field station in Delsbo within one day after sampling with small disturbance of the fractions during the short transportation and storage.

The samples were analysed at ITM for pH, conductivity (cond.), colour (col.), turbidity (turb.) alkalinity (alk.), sodium (Na), potassium (K), calcium (Ca), magnesium (Mg), sulphate (SO_4^{2-}), chloride (Cl), total organic carbon (TOC), total-nitrogen (Ntot), total-phosphorus (Ptot), sum nitrite-nitrate-nitrogen (NO23N), acid soluble aluminium (Al_r), iron (Fe), manganese (Mn) and zink (Zn) usually following Swedish Standard Methods. Our accredited laboratory followed general quality assurance and control (QA/QC) procedures giving measuring uncertainties (estimated as the relative standard deviation of the QC samples) below 10 %, except for low-level determinations of the aluminium fractions (<25 μ g/l), which have an uncertainity of 25%.

2.3. STATISTICAL METHODS

The analytical results of the stream waters were divided in subgroups according to treatment, stream type, colour (brown waters were defined to have a mean colour >100 mg Pt/l) and season (spring = March-May and summer/autumn = June-November). The significances of the differences in means between the groups were controlled by Student´s t-test (Table I). A correlation matrix (Table II) was constructed for the whole material on measured and calculated variables; sum of anions, base cations (BC = Ca + Na + K + Mg), acid neutralising capacity (ANC = (SO4 + Cl + NO23N)-BC), ANC divided with protons (ANC/H+) and ion ratio ((SO4 + Cl + NO23N + TOC/10)/BC in meqv/l). The ion ratio was constructed to illustrate the link between ionic load or pressure in the area and subsequent release of aluminium. One tenth of the organic carbon content is used (since not all is ionized) an amount largely eqivalent to the deficit in anions that usually is attributed to organic anions.

3. Results and Discussion

3.1. TEMPORAL VARIATIONS FOR ALUMINIUM FRACTIONS

Temporal curves for inorganic aluminium (Al_i) together with pH in five chosen locations are shown in Figure 1. The streams are of different character; in Havssvalgsbäcken (high altitude, small catchment area) the highest meanlevels

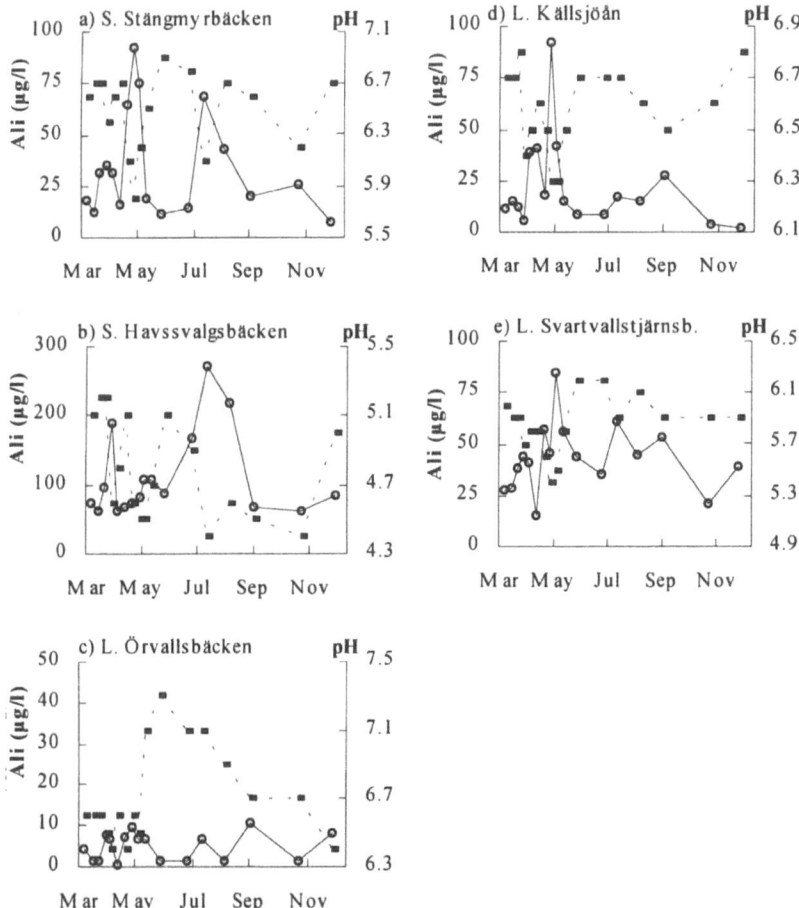

are found (pH 4.8, 100 μ g Al_i/l) whereas Stängmyrbäcken (low altitude, larger catchment area) has better values (pH 6.5, 35 μ g Al_i/l). Lake outlet Örvallssjön downstream stream Havssvalgsbäcken is limed to a high pH and low aluminium levels (pH 6.7, 5 μ g Al_i/l). The other two lake outlets are limed to more moderate pH with corresponding aluminium values, lake outlet Källsjöån (pH 6.6, 22 μ g Al_i/l) and lake outlet Svartvallstjärnbäcken (pH 5.9, 44 μ g Al_i/l). In the latter the whole catchment was limed and consequently, the aluminium levels are low even though the pH-levels are relatively low. Aluminium is appearently not released from the soil through ion exchange as readily here as in the other catchment areas.

The summer rains affect the streams more directly with high levels of inorganic aluminium in early summer (July) whereas a less pronounced peak occurs later in most lake outlets (August/September) when the precipitation has permeated downwards in the watershed. Generally there are higher and more fluctuating Al_i-levels (mobilized from the catchment) in the streams than in the lakes (where aluminium is complexed and precipitated). The aluminium levels are higher in the land-limed than in the wetland-limed and the lake-limed lake but the variations are moderate. The difference in aluminium content is an effect both of different liming strategy and lime dose and the resulting pH.

3.2. INFLUENCES BY pH AND TOC ON ALUMINIUM FRACTIONS

The results in Table I demonstrate the negative correlation of aluminium to pH (unlimed contra limed) and the retention of aluminium in lakes (streams contra lake outlets) and the positive correlation of aluminium to organic content (brown contra clear waters) previously described by Driscoll and Postek (1996). The difference between unlimed and limed waters was less valid for Al_r (not significant) and Al_o ($p<0.01$), which are humus-linked aluminium fractions than for the pH-linked fractions Al_a and Al_i. Even if it were not the case this year, significantly lower aluminium levels would probably have been registered during summer-autumn than in spring, in a year with "normal" seasonal precipitation patterns.

TABLE I
Mean values (μg/l), the differences are significant with p < 0.001 except for those in italics, marked with ns (non significant) or other probabilities noted.

Groups	Al_r	Al_a	Al_o	Al_i
Unlimed (n=105)	*258 ns*	179	*128 p<0.01*	52
Limed (n=155)	*241*	129	*107*	23
Stream (n=115)	282	183	137	48
Lake outlet (n=145)	221	122	72	24
Brown (n=174)	291	177	137	41
Clear (n=86)	161	92	70	22
Spring (n=164)	*239 ns*	*152 ns*	*114 ns*	*38 p<0.05*
Summer/autumn (n=96)	*264*	*144*	*118*	*29*

The levels of Al_a and Al_i seem to be more directly affected by surface runoff and leaching caused by strong acids manifested in lower pH (Driscoll and Schecher, 1990). The levels of Al_r and Al_o seem to be connected to processes of the ground-water runoff and leaching of organically bound aluminium in the soil together with TOC (ibid).

A similar pattern as above emerges in Table II where Al_r and Al_o groups together with high correlation to humus detected as TOC, col. and/or accompanied by Fe and nutrients. Al_a and Al_i is linked together by a good correlation to acidic variables as pH, alkalinity, and the ion ratio describing the ionic load (how acid anions are balanced by BC) and the subsequent release of aluminium by ion exchange in the soil. When correlations are run on the data divided in the two seasons the coefficients for Al_r, Al_a, Al_o and Al_i respectively are enhanced for pH (-0.56, -0.72, -0.60 and –0.75) and alk. (-0.38, -0.50, -0.38 and –0.56) in spring. In summer/autumn the coefficients are enhanced for TOC (0.82, 0.77, 0.83 and 0.46) and Fe (0.79, 0.73, 0.75 and 0.53).

TABLE II
Correlation coefficients for the aluminium fractions and other variables. The coefficients are significant with $p < 0.001$, except for those in italics.

Measured	pH	Alk.	TOC	Col.	Fe	Ntot
Al_r	-0.50	-0.30	0.79	0.71	0.71	0.61
Al_a	-0.71	-0.46	0.71	0.65	0.66	0.51
Al_o	-0.58	-0.34	0.78	0.70	0.69	0.57
Al_i	-0.74	-0.51	0.42	0.43	0.45	0.28
Calculated	Anions	BC	ANC	ANC/H+	ion ratio	
Al_r	-0.30	*-0.18*	*-0.18*	-0.22	0.58	
Al_a	-0.44	-0.35	-0.35	-0.31	0.77	
Al_o	-0.32	-0.21	-0.22	-0.26	0.64	
Al_i	-0.51	-0.47	-0.47	-0.30	0.79	
Aluminium	Al_r	Al_a	Al_o	Al_i		
Al_r	1					
Al_a	0.90	1				
Al_o	0.93	0.92	1			
Al_i	0.64	0.84	0.59	1		

4. Conclusion

This study confirms results in earlier studies of aluminium fractions and show that the use of ion ratio (an expression estimating the ionic load) may contribute to the understanding of aluminium species in streams and lakes. The meanlevels of inorganic aluminium are higher in spring but the highest single values occur in summer. The levels of acid-linked aluminium fractions Al_a and Al_i which have elevated levels in brown and stream waters (unbalanced without sedimentation reservoirs) are best correlated to ion ratio and pH. These fractions are clearly affected by liming and by processes in the lake. The organic-linked aluminium fractions Al_r and Al_o both exhibits higher levels in the same groups as above but without a significant effect of liming. They are best correlated to TOC, watercolour and Fe.

Acknowledgements

We thank the Swedish Environmental Protection Agency´s liming programme, which financed this work. We also like to thank all dedicated and hard-working co-workers at ITM´s laboratory for aquatic environmental chemistry.

References

Andrén, C.: 1995, *Water, Air, and Soil Pollution* **85**, 811.
Bloom, P.R. and Erich, M.S.: 1996, 'The Quantitation of Aqueous Aluminium' in G. Sposito (ed), *The Environmental Chemistry of Aluminium,* Boca Raton, Fla, Lewis Publ. pp 1-38.
Dickson, W.: 1975, *Report Institute Freshwater Research, Drottningholm* **54**, 8.
Driscoll, C.T.: 1984, *International Journal Environmental Analytical Chemistry* **16**, 267.
Driscoll, C.T. and Postek, K.M.: 1996, 'The chemistry of aluminium in surface waters' in G. Sposito (ed), *The Environmental Chemistry of Aluminium,* Boca Raton, Fla, Lewis Publ. pp 363-418.
Driscoll, C.T. and Schecher, W.D.: 1990, *Environmental Geochemistry and Health* **12**, 28.
Gensemer, R.W. and Playle, R.C.: 1999, *Critical Reviews in Environmental Science and Technology* **29**, 315.
Henriksen, A., Skogheim, O.K. and Rosseland, B.O.: 1984, *Vatten* **40**, 255.
Rosseland, B.O., Eldhuset, T.D. and Staurnes, M.: 1990, *Environmental Geochemistry and Health* **12**, 17.
Swedish Standard Methods, SS 0281-series and SS 028210.

TRACE METALS IN FOREST SOILS AT FOUR SITES IN SOUTHERN CHINA

H. HANSEN[1], T. LARSSEN[2*], H.M. SEIP[1] and R.D. VOGT[1]

[1] *Department of Chemistry, University of Oslo, P.O. Box 1033, Blindern, 0315 Oslo, Norway.*
[2] *Norwegian Institute for Water Research, P.O.Box 173 Kjelsas, 0411 Oslo, Norway.*

(*author for correspondence, e-mail: thorjorn.larssen@niva.no)

Abstract. Industrial development has increased fast in China the last decades. This has led to a range of environmental problems. Deposition of heavy metals to forest ecosystems via the atmosphere is one potential problem. In this paper we report results from a pilot study where the heavy metal levels in forest soils at four different sites have been measured. Three of the sites are located relatively close to the large cities Chongqing, Guiyang and Guangzhou; one site is located in a remote, mountainous area in Guizhou province. Total metal contents as well as fractions according to Tessier's scheme were determined. With a few exceptions, the metal concentrations can be characterized as low; i.e. in most cases within the range of what has been reported as typical background values in the literature. High content of arsenic (up to 100 ppm) was found in the samples from the site outside Guangzhou, most likely due to naturally high arsenic levels in the soil. Metals bound to organic matter and to iron- and manganese oxides were the dominant fractions. No clear differences in metal levels were found between topsoil and subsoil samples, indicating that the atmospheric deposition of heavy metals has been low.

Keywords: forest soils, heavy metals, China, atmospheric pollution.

1. Introduction

Environmental pollution has been increasing rapidly in China the last decades. High atmospheric emission of sulfur has been, and still is, of major concern (NEPA, 1997). Levels of heavy metals in Chinese cultivated soils have been studied, but little information exists on heavy metal contamination in forest soils. Contamination of agricultural soils is often related to fertilizer use and irrigation, while forest soils are normally contaminated from atmospheric deposition only. In Europe, coal combustion and other sources for atmospheric pollution are known to have contaminated forest soils in large areas (Schnoor et al., 1997). This has been of concern, especially in combination with acid deposition, as lowered solution pH increase the mobility of a range of heavy metals in soils.

In the present study we report data from a pilot study on heavy metal contents in Chinese acid forest soils. Both fractions and total amount of heavy metals in soils in the southeastern part of China are determined and the importance of atmospheric metal deposition is discussed.

Water, Air, and Soil Pollution **130**: 1721–1726, 2001.
© 2001 *Kluwer Academic Publishers.*

2. Soil sampling sites and analytical methods

The sampling sites selected for this study were forested areas assumed to receive pollution only from atmospheric deposition. The sites receive different atmospheric pollution loads as their location ranges from the suburbs of large metropolitan areas to a remote mountain region (Figure 1; Table I). The catchments were previously studied in an acidification assessment in China by Seip et al. (1999) and will be among the sites in a new acid rain study in China, IMPACTS (Tang et al., 2001). Soil chemical properties are given in Table II.

For the analysis of total content of heavy metals in the soil, samples from the top (O/A) horizon and the B-horizon were selected. Selected samples were also analyzed using a sequential extraction procedure.

Total amounts of heavy metals in the soils were extracted using hot nitric-acid for 18 hours (Berthelsen and Steinnes, 1995). Allthough this method may fail to digest all mineral material, and hence under estimate the true total metal concentration, we applied it in order to compare results with other studies using this method. The extracts were filtered and diluted prior to analysis. A sequential extraction was conducted to get information on how different metals were bound in the soil and hence potential for mobilization. The extraction scheme of Tessier et al. (1979) was followed, but an additional extractant, proposed by Novosamsky et al. (1993), was added to check for the weakest bound heavy metals (i.e. starting with a more dilute extractant, 0.01 M $CaCl_2$). The matrix matching method of Li et al. (1995) was applied.

Total concentrations of V, Cr, Ni, Co, Cu, Zn, As, Cd, Sn and Pb were analyzed using a semi-quantitative method on ICP-MS with one calibration standard for each element. The accuracy of the method is in the range 20%-60%. Lead was determined more accurately with flame-AAS; the precision was 5-10%. The extracts from the sequential procedure were analyzed for lead, copper and zinc using ICP.

3. Results and discussion

3.1. TOTAL CONCENTRATIONS

Total concentrations of heavy metals found for the Chinese soils are shown in Table III. For most metals the differences between the various sites are small. There is no clear gradient in metal concentrations with distance from cities along with sulfur deposition loads. One exception is the arsenic concentration, which is several times higher at the DingHuShan (DHS) site in the Guangzhou province than at the other sites. The Nickel concentration shows a tendency to be highest at the LiuChonGuang (LCG) and LeiGongShan (LGS) sites in the Guizhou province. As these are suburban and remote sites, respectively, it is likely that nickel concentrations are naturally derived. The results indicate that contamination from atmospherically transported heavy metals is low for soils in the investigated areas.

TABLE I
Soil sampling site locations and sulfur deposition. From Seip et al. (1999) and RAINS-Asia (Downing et al., 1997)

Site name	Code	Location	Sulfur deposition
DingHuShan	DHS	86 km W of Guangzhou	2-4 $gSm^{-2}yr^{-1}$
LiuChongGuan	LCG	10 km NE of Guiyang	8-10 $gSm^{-2}yr^{-1}$
LeiGongShan	LGS	140 km E/SE of Guiyang	1-2 $gSm^{-2}yr^{-1}$
TieShanPing	TSP	25 km NW of Chongqing	10-12 $gSm^{-2}yr^{-1}$

TABLE II
Soil quality parameters for the investigated soils at the different sites. Median values with ranges in parenthesis. CEC and BS at soil pH.

Site	Horizon	pH(H_2O)	CEC meq/kg	% Base saturation
DHS	O/A	3.7 (3,4-4,4)	107 (51-206)	10 (3-21)
	B	4.3 (4,0-5,1)	35 (7-68)	4.6 (2-11)
LCG	O/A	3.8 (3.1-4.1)	108 (64-223)	20 (5-60)
	B	4.0 (3.7-4.7)	82 (25-142)	7 (3-40)
LGS	A	4.1(3.8-4.9)	9(3-13)	38(24-100)
	B	4.3(4.1-4.8)	2.1(0.9-4.5)	23(9-81)
TSP	O/A	3.8 (3.5-4.0)	63 (55-90)	12 (7-17)
	B	4.0 (3.6-4.3)	37 (25-109)	8 (5-27)

Figure 1. Map with sampling locations.

TABLE III
Mean heavy metal concentrations (ppm) in soil horizons O/A and B for the four sites investigated.

Site	Horizon	V	Cr	Ni	Cu	Zn	As	Cd	Pb
LCG	O/A	38.5	30.3	9.5	22.4	39.6	7.4	0.5	44.5
	B	30.9	25.1	9.2	26.9	38.6	9.6	0.5	28.5
DHS	O/A	13.4	10.9	0.0	16.7	41.7	25.7	0.7	60.6
	B	21.4	21.6	0.2	13.4	12.8	57.7	0.5	34.0
LGS	O/A	14.4	14.6	6.8	22.5	51.7	9.8	0.7	42.6
	B	20.6	21.0	11.5	20.9	56.3	7.2	0.5	34.4
TSP	O/A	16.1	21.1	2.9	16.5	33.7	4.2	0.5	29.8
	B	20.4	20.3	4.2	13.0	33.1	4.1	0.7	31.9

TABLE IV
Heavy metal levels for some selected elements. Mean value and range (ppm). n.a.: data not available.

	China This study		Poland Andersen et al. 1994		Norway Steinnes et al. 1997		China Grade A Natl. standard[1]
	Mean	Range	Mean	Range	Mean	Range	Mean
Pb	38.3	19.0 – 97.4	56.4	1.9 – 236	40.0	8.5 - 111.3	35
Cd	0.59	0.47 – 1.01	0.34	0.01- 2.9	0.54	0.18 - 1.38	0.20
As	15.7	2.8 – 98.4	n.a.	n.a.	3.25	1.08 - 6.11	15
Zn	38.4	9.3 – 73.0	34.3	2.8 – 176	53.8	32.4 - 84.1	100
Ni	5.5	0 – 22.3	4.25	0.8 – 10	n.a.	n.a.	40
Cu	19.1	8.5 – 36.9	n.a.	n.a.	11.5	8.7 - 14.7	35

[1] The grade A standard corresponds to upper values in normal unpolluted soils (Chen et al., 1999)

TABLE V
Average values for the top and bottom horizons and detection limits. All values are in ppm dry soil. Fractions are denoted F1-5 in the table and can be interpreted as follows: F1: Bioavailable; F2: Exchangeable; F3: Bound to carbonates; F4: Bound to Fe/Mn oxohydroxides; F5: Bound to organic matter. < d.l. denotes below detection limit.

		F1	F2	F3	F4	F5	Det. limit
Cu	O/A	< d.l.	< d.l.	< d.l.	2.20	2.44	0.51
	B	< d.l.	< d.l.	< d.l.	3.17	2.39	
Pb	O/A	< d.l.	< d.l.	< d.l.	12.59	8.95	2.45
	B	< d.l.	< d.l.	< d.l.	15.53	6.88	
Zn	O/A	5.28	2.53	< d.l.	3.08	1.06	0.36
	B	< d.l.	< d.l.	< d.l.	1.57	0.42	

There are also no clear differences between the upper and lower soil horizons. Figure 2 shows percent difference in concentrations between the top and the bottom horizon. For most metals there is no clear pattern with depth; e.g. there is no indication that there is any heavy metal accumulation in the top horizon. This furthermore supports that the atmospheric deposition is low in the selected areas. Lead is an exception as there is a tendency towards higher lead concentrations in the upper horizon (Figure 2).

In Table IV the heavy metal concentrations for the Chinese sites are compared to data from Poland and Norway (Andersen et al., 1994; Steinnes et al., 1997), as well as the Chinese national standards (Chen et al., 1999). The Polish and Norwegian samples were selected, as they were analyzed with similar methods as our samples and the soils were from forest areas supposedly receiving only atmospherically transported pollutants. The concentration ranges reported for the different countries overlap and there are no general statistical differences between the countries, with the exception of arsenic, being considerably higher for the Chinese samples. However, the maximum concentrations reported for the Polish sites are higher than for the Norwegian and Chinese sites.

Compared to the National Grade A standard in China (Chen et al., 1999) all values for Cd in the Chinese samples are more than twice the standard (Table IV). The Pb values are generally below the National Grade A standard, except at one plot in DHS and one plot in LCG. The grade A standards correspond to upper values in natural unpolluted soils analysed after digestion with a mixture of acids, including HF (Chen et al., 1999). Hence our analysis, using nitric acid as the extractant is an under estimate of the total metal concentration.

3.2. SEQUENTIAL EXTRACTIONS

The fractions from the sequential extractions were analyzed for Cu, Pb and Zn.

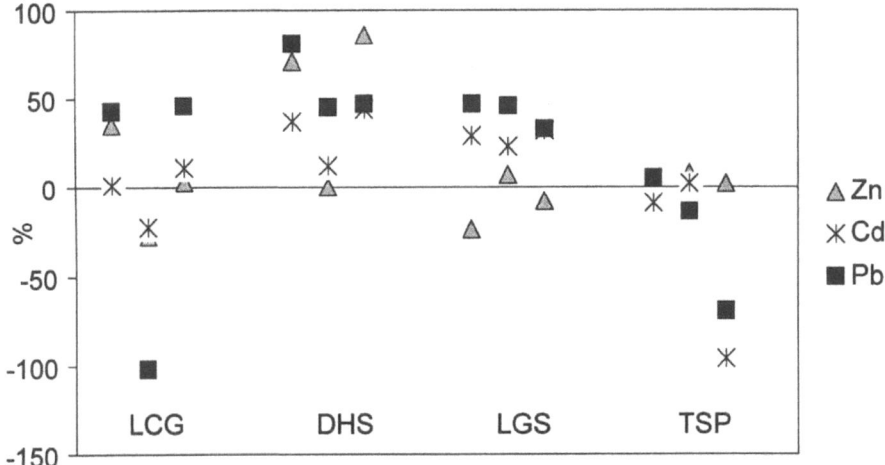

Figure 2. Percent difference between top and bottom horizons for concentrations of Zn, Cd, and Pb. There are generally no systematic differences between the top and bottom horizons. Lead is a possible exception, showing a slight tendency to have higher concentrations in the upper soil.

The results are presented in Table V. With the exception of Zn, the metals were mainly bound to the Mn/Fe oxohydroxide and organic fractions. On average, the largest Zn pool was the weakest bound (bioavailable). The bioavailable fractions of Cu and Pb were below the detection limit.

4. Conclusions

The heavy metal levels detected in the investigated Chinese forest soils were in general fairly low. The Cd concentrations were higher than the Chinese national grade A standard for all sites and at a concentration level similar to that found in Norwegian forest soils. There were neither clear pattern of higher concentration levels at sites located near large cities nor accumulation of metals in the upper soil horizon found, with lead as a possible exception.

References

Andersen, S., Ødegård, S., Vogt, R.D. and Seip, H.M.: 1994, *Ecological Engineering* **3**, 245.
Berthelsen, B. and Steinnes, E.: 1995, *Geoderma* **66**, 1.
Chen, H.M., Zheng, C.R., Tu, C. and Zhu, Y.G.: 1999, *Ambio* **28**, 130.
Downing, R.J., Ramankutty, R. and Shah, J.J.: 1997, *Rains-Asia: an assessment model for acid deposition in Asia.* The World Bank, Washington DC.
Li, X., Coles, B.J., Ramsey, M.H. and Thornton, I.: 1995, *Chemical Geology* **124**, 109.
NEPA, 1997. *The National Ninth Five-year Plan for Environmental Protection and the Long-term targets for the year 2010.* The National Environmental Protection Agency; Beijing.
Novozamsky, I., Lexmond, T.H.M. and Houba, V.J.G.: 1993, *International Journal of Environmental Analytical Chemistry* **51**, 47.
Schnoor, J.L., Galloway, J.N., and Moldan, B.: 1997, *Environmental Science and Technology* **31**, 412A.
Seip, H.M, Aagaard, P., Angell, V., Eilertsen, O., Larssen, T., Lydersen, E., Mulder, J., Muniz, I.P., Semb, A., Tang D., Vogt, R.D., Xiao J., Xiong J., Zhao D., and Zhou G.: 1999, Ambio **28**, 522.
Steinnes, E., Allen, R.O., Petersen, H.M., Rambæk, J.P. and Varskog, P.: 1997, *Science of the Total Environment* **205**, 255.
Tang, D., Seip, H.M., Angell, V., Eilertsen, O., Larrsen, T., Lydersen, E., Mulder, J., Semb, A., Torseth, K., Vogt, R.D., Xiao, J., Zhao, D. and Kong, G.: 2001. *Water, Air and Soil Pollution* This volume.
Tessier, A., Campbell, P.G.C. and Bisson, M.: 1979, *Analytical Chemistry* **51**, 844.

SOIL DEGRADATION IN THE WIELKOPOLSKI NATIONAL PARK (POLAND) AS AN EFFECT OF ACID RAIN SIMULATION

BARBARA WALNA[1]*, JERZY SIEPAK[2] and STANISŁAW DRZYMAŁA[3]

[1] *Adam Mickiewicz University, Jeziory Ecological Station, 62-050 Mosina, P.O. Box 40, Poland;*
[2] *Adam Mickiewicz University, Laboratory of Water and Soil Analysis, Drzymały 24, 60-613 Poznań;* [3] *Agricultural University of Poznań, Soil Science Department, Mazowiecka 42, 60-623 Poznań, Poland*
(* author for correspondence, e-mail: walna@amu.edu.pl.)

Abstract. The aim of this study was to evaluate qualitatively as well as quantitatively the response of typical soils of the Wielkopolski National Park to simulated acid rain. The experiments with simulated acid precipitation were performed on soil monoliths of unchanged structure. The artificial rain used for the simulations was diluted sulphuric acid of pH 2.0, 3.0 and for the control experiments distilled water was used. Changes in the soil pH and Ca/Al molar ratio of specified soil genetic horizons were described. A quantitative description of the leaching dynamics of calcium, magnesium, sodium, potassium and aluminium, dependent on the kind of sprinkling was given. The results proved that simulated acid rain washed out significant amounts of nutrients and released toxic amounts of aluminium. Humus and clay fraction colloids were noted to have a buffer effect.

Keywords: soil degradation, nutrients washing out, toxic aluminium, monoliths, the Wielkopolski National Park, Poland

1. Introduction

The Wielkopolski National Park (central Poland) is an area of unquestionable natural and cultural value in Poland which is in need of special protection. The considerable level of urbanisation of areas adjacent to the Park, and the accompanying excessive emission of gases and particulates, have disturbed its ecological equilibrium.

The acid rain research has been conducted since 1992 at the Adam Mickiewicz University Ecological Station (Walna and Siepak, 1999). A very low value of pH, even below 3.0, was observed in individual cases. The year 1994 was notable for the lowest average pH recorded, 3.92, as well as the lowest pH of an individual event, which was 2.94.

The chemical analyses of precipitation showed a considerable sulphate content confirming sulphur oxides as the dominant source of pollution.

The high acidity of precipitation recorded during the 8-year study must have a significant influence on plant and soil properties. The majority of forest soils in Poland are sandy, acid, and very poor in nutrients. This makes them less resistant to the degrading effect of acid rain. Acid rain undoubtedly induces the "mobility" of calcium and magnesium in soils (Lilkeville *et al.*, 1993; Rustad *et al.*, 1993; Matzner and Murach, 1995; Walna *et al.*, 1998). Soil acidification and deficiency of available Ca^{2+} and Mg^{2+} in the soil solution can affect nutrient uptake and root growth. This, in turn, contributes to forest decline (Schulze, 1989; Likens et al., 1996). Similar effects can be observed when studying the leaching of sodium and potassium (Walna et al., 2000).

Among the many detrimental effects of advancing soil acidification the release of aluminium whose high concentrations have a toxic effect on the element of the ecosystem (Mulder et al., 1987) seems particularly dangerous.

The aim of the study was to give a quantitative and qualitative evaluation of chemical changes in typical soils of the Wielkopolski National Park as a result of acid rain simulation.

2. Materials and Methods

The experiment was conducted on monoliths of naturally occurring soils under laboratory conditions. Soil cores with a diameter 15cm were sampled from the first 50cm of the soils. The acid rain simulation was conducted for 30 days by every day sprinkling the soil with an aqueous solution of sulphuric acid with a pH of 2.0 and 3.0 and distilled water with a pH of 5.6 (control sample). The total amount of simulated precipitation was approximately 400 mm, i.e. an amount equal to the average annual precipitation for 1994. The volume of the collected filtrate was measured each day and its composition, pH and electrical conductivity determined.

The kind of soils studied were: two types of sandy soil; no.1- previously used for agricultural purposes, no. 2 - sandy forest soil with high humus concentration and loamy forest soil (glacial till) – no. 3. Before sampling of the monolithic (core) samples an excavation was made and to determine basic physical and chemical properties (Table I).

TABLE I
Basic properties of the investigated soils (before treatment).
TEB– total exchangeable bases, CEC – cation exchange capacity, BS– base saturation

Soil	Depth and horizon	Bulk density	Ignit. loss	Soil particles Φ mm			pH	
				1–0.1	0.1–0.02	<0.02	H_2O	KCl
	(cm)	(Mg/m^3)	%		%			
1	0-12 A	1.51	1.84	73.6	22.4	4.0	4.75	3.96
	12-25 A	1.51	1.84	73.2	22.7	4.1	4.77	4.01
	25-50 C	1.60	0.51	62.7	31.0	6.3	5.22	4.64
2	0–5 A	0.99	8.32	80.1	11.4	2.5	3.90	3.28
	5-18 AE	n.d.	n.d.	81.7	9.8	3.4	4.24	3.56
	18-50 BC	1.53	1.17	75.7	9.7	5.9	4.65	4.15
3	0–10 A	1.46	3.58	53.1	13.7	31.2	4.80	3.96
	10-40 B	1.73	1.34	48.5	18.3	31.6	5.21	4.08
	40-50 C	n.d.	n.d.	41.9	22.1	33.5	6.25	5.06

The procedure of studies is schematically presented in Figure 1.

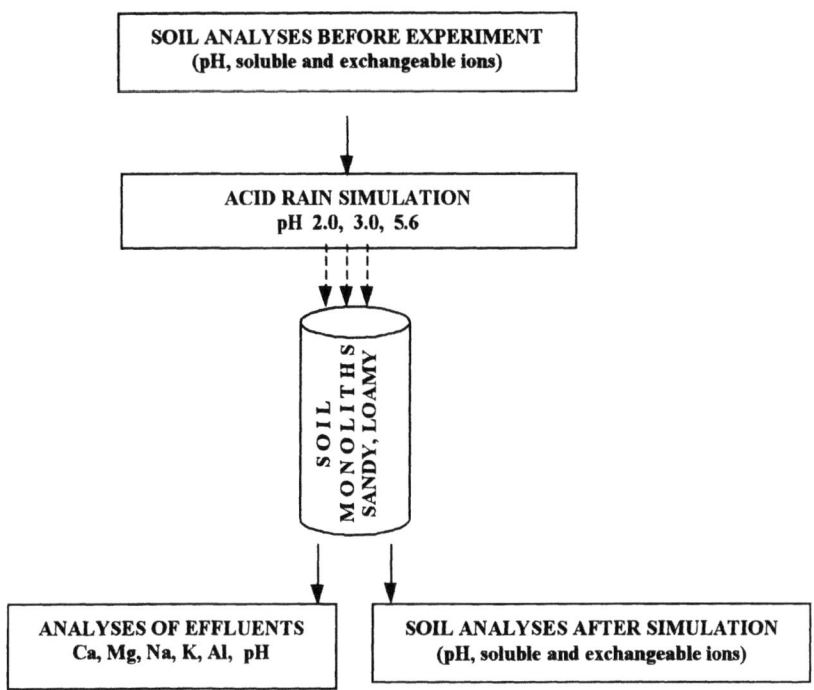

Figure 1. A general scheme of the experimental procedure applied.

Classical methods were used in the soil testing procedure. Determinations of calcium and magnesium were performed by means of atomic absorption, sodium and potassium by photometric technique, aluminium using the colorimetric method with eriochromocyanine. Following completion of this experiment, soil cores were subjected to analogous analysis. The values presented were obtained by repeating a given analysis for two monoliths.

3. Results

Concerning chemical degradation of the soils, the results of the study of the soils before and after acid rain simulation and the analyses of the chemical composition of the sprinkling eluates allow the following conclusions to be drawn.

The simulated extremely acid (pH 2.0) precipitation brings about a very significant drop in the soil pH, down to pH 3.0 in soil no. 1 (Figure 2) or even 2.68 in soil no. 2. The acid rain of pH 3.0, occurring sometimes in the study area in reality also, causes a noticeable lowering of the soil pH (Figure 2, line B), especially in sandy soils with a small admixture of humus.

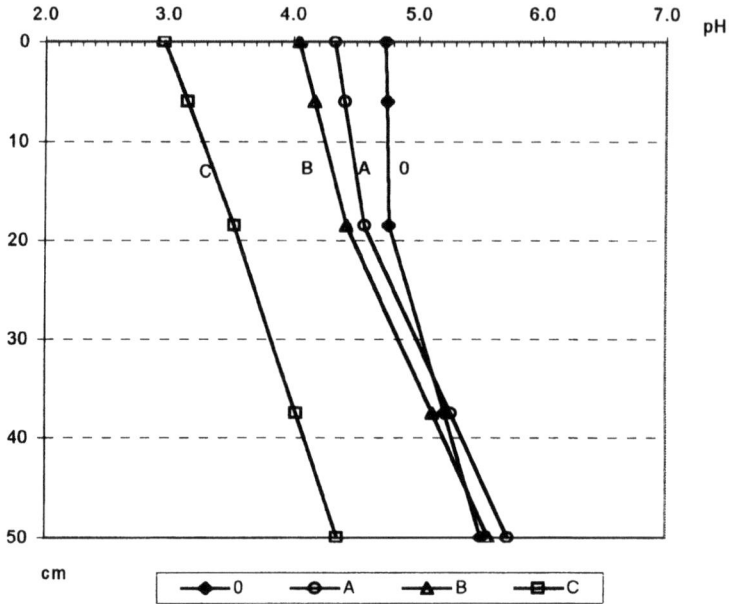

Figure 2. Changes of soil pH: A - after treatment with H_2O (pH 5.6), B - after treatment with acid solution of pH 3.0, C - after treatment with acid solution of pH 2.0, 0 - before treatment (natural).

The changes in the exchangeable complex are shown in Table II. The levels of exchangeable forms in the sandy soils (no.1 and no.2) are several times lower than in the loamy soil (no.3). Sandy soil, characterised by a very low cation exchange capacity, changes its base saturation most markedly in a 25-50cm deep layer in soil no.1 (from 18.7 to 8.2 %). The higher cation exchange capacity in soil no.2 (upper layer) is connected with its higher content of organic matter (Table I), but base saturation in those two sandy soils is similar. The loamy soil (no.3), with its about 30% content of soil particles < 0.02, shows big differences in base saturation after acid simulation. In the upper layer it declines from 34.5% to 18.5%. Detailed results of the contents of exchangeable and soluble forms of base cations and aluminium are reported in previous papers (Siepak et. al., 1999, Walna et al., 1998, Walna, et al., 2000).

One of the most conspicuous effects of the simulated acid rain on sandy soils was an increase in the amount of mobilised and washed out aluminium. The amount of the aluminium in sorption complex reached toxic levels (80-100 mg/kg) (Siepak et al., 1999), and the toxicity coefficient expressed by the Ca/Al ratio in soil solutions and eluates dropped below the critical value of 1.0 (Table III). Acid rain was observed to have a significant effect on the activation and leaching of base cations (Table IV), but also by changes in the order of cations in soil solutions (Walna et al., 1998; Walna et al., 2000). For instance, in the sandy soil the cation sequence in soil solution before sprinkling was Ca>K>Mg>Na>Al, and after sprinkling with solution of pH 2.0 was changed to Al>Ca>Mg>K>Na. This sequence was observed both in the upper soil layer exposed directly to the effects of acid rain as well as in the soil solution of the bottom layer. The amounts of washed-out components presented in Table IV change with the level of rain acidity and the type of soil.

TABLE II

Changes in exchangeable complex after treatment by simulated rain of pH 2.0.

Soil	Conditions	Depth cm	Base cations in complex cmol(+)/kg	Cation exchange capacity cmol(+)/kg	Base saturation %
1	Before treatment	0-12	0.14	2.64	5.3
		12-25	0.11	2.51	4.4
		25-50	0.23	1.23	18.7
	After pH 2.0	0-12	0.16	3.68	4.3
		12-25	0.13	3.05	4.3
		25-50	0.13	1.58	8.2
2	Before treatment	0-10	0.81	9.81	8.3
		10-18	0.25	4.45	5.6
		18-50	0.19	2.24	8.5
	After pH 2.0	0-10	0.46	10.31	4.5
		10-18	0.26	4.86	5.3
		18-50	0.24	3.37	7.1
3	Before treatment	0-10	2.06	5.96	34.6
		10-40	4.00	5.95	67.2
		40-50	6.34	7.29	87.0
	After pH 2.0	0-10	1.19	6.42	18.5
		10-40	2.83	5.28	53.6
		40-50	5.20	6.83	76.1

TABLE III

Soil solution Ca/Al molar ratio resulting from various sprinkling experiments.
A - after treatment with H_2O (pH 5.6), B - after treatment with acid solution of pH 3.0,
C - after treatment with acid solution of pH 2.0, 0 - before treatment (natural)

Depth cm	Ca/Al molar ratio			
	0	A	B	C
Soil no.1				
0–12	–	4.3	7.0	0.4
12–25	–	7.9	3.3	0.2
25–50	2.6	3.5	6.5	0.2
Soil no.2				
0–10	3.4	5.7	2.8	0.3
10–18	1.5	1.1	2.0	0.4
18–50	3.0	0.8	1.7	0.3
Soil no.3				
0–10	2.3	32.8	34.3	6.8
10–40	2.0	6.6	1.6	–
40–50	11.5	3.2	4.2	–

TABLE IV

Amounts of elements washed out at the depth of 50 cm from 1 m² of soil during simulation.
A - with H_2O (pH 5.6), B - with acid solution of pH 3.0, C - with acid solution of pH 2.0

Ions	Treatment	Amount (mg)	Treatment	Amount (mg)	Treatment	Amount (mg)
		Soil no.1		Soil no.2		Soil no.3
Al^{3+}	A	69	A	794	A	50
	B	375	B	1 019	B	<50
	C	44 438	C	31 238	C	3 238
Ca^{2+}	A	469	A	1 194	A	2 288
	B	1 369	B	1 475	B	4 613
	C	1 731	C	14 944	C	76 719
Mg^{2+}	A	81	A	650	A	1 094
	B	200	B	550	B	1 231
	C	1 438	C	3 863	C	12 488
K^+	A	556	A	531	A	688
	B	575	B	1 063	B	394
	C	1 463	C	3 038	C	1 919
Na^+	A	94	A	550	A	750
	B	150	B	463	B	675
	C	381	C	694	C	2 138

4. Conclusions

- The 8-year research on the reaction and composition of precipitation in the Wielkopolski National Park shows acid and very acid precipitation to be dominant, which poses a threat to the geoecosystem.
- The results show that a dramatic soil degradation is possible when the rain pH drops below 3.0. With the naturally acid forest soils the degrading effect of rain with a pH>3.0 is much smaller.
- Cores of natural soils can successfully be used in laboratory for studying the degrading effects of acid rain. Results obtained from such studies are comparable to the results found under field conditions.

References

Lilkeville, A., Bredemeier, M. and Ulrich, B.: 1993, *Agriculture, Ecosystem and Environment* **47**, 175.
Likens, G.E., Driscoll, C.T. and Buso, D.C.: 1996, *Science* **272**, 244.
Matzner, E., Murach, D.: 1995, *Water, Air and Soil Pollution* **85**, 63.
Mulder, J., van Grinsven, J. J. M. and van Breemen, N.: 1987, *Soil Sci. Soc. Am. J.* **51**, 1640.
Rustad, L.E., Fernandez, I.J., Fuller, R.D., David, M.B., Nodvin, S.C. and Halterman, W.A.: 1993, *Agriculture, Ecosystems and Environment* **47**, 117.
Schultze, E.D.: 1989, *Science*, **244**, 776.
Siepak, J., Walna, B. and Drzymała, S.: 1999, *Polish Journal of Environmental Studies* **8**, 55.
Walna, B., Drzymała, S. and Siepak, J.: 1998, *The Science of the Total Environment* **220**, 115.
Walna, B. and Siepak, J.: 1999, *The Science of the Total Environment* **239**, 173.
Walna, B., Drzymała, S. and Siepak, J.: 2000, *Water, Air and Soil Pollution* **121**, 31.

ACID DEPOSITION AND ACIDIFICATION OF SOIL AND WATER IN THE TIE SHAN PING AREA, CHONGQING, CHINA

ZHAO DAWEI[1], T. LARSSEN[2*], ZHANG DONGBAO[1], GAO SHIDONG[1], R.D. VOGT[3], H.M. SEIP[3] and O.J. LUND[3]

[1] *Chongqing Institute of Environmental Science and Monitoring, 37 Jia Ling VLG-1, Jiang Bei District, 630020, Chongqing, China;* [2] *Norwegian Institute for Water Research. P.O.Box 173 Kjelsas, 0411 Oslo, Norway.* [3] *Department of Chemistry, University of Oslo, P.O. Box 1033, Blindern, 0315 Oslo, Norway.*

(*author of correspondence, e-mail: thorjorn.larssen@niva.no)

Abstract. Chongqing is among the heaviest polluted cities in China. Combustion of coal with relatively high sulfur content causes high sulfur emission and deposition in the area. Effects on soils and waters of the acid deposition in the Chongqing area have been studied in the field at a forested site outside the city. Deposition chemistry and fluxes, soil and soil water chemistry as well as surface water chemistry are presented for the period 1996-1998. There are some stress symptoms at the forest in the area and severe forest damage has been reported at Nanshan, closer to Chongqing center. Monitoring of the acidification situation in the area must be followed closely as impacts may be expected if the deposition is not reduced in the future.

The deposition of sulfur, H^+ as well as calcium at the site is high. Wet deposition of sulfur is estimated to 4.7 - 5.7 g S m^{-2} yr^{-1} during the three years sampled; dry deposition is probably of similar size. Annual volume-weighted pH in bulk deposition was 4.0 - 4.2 and the calcium wet deposition flux was 2.6 - 3.6 g Ca^{2+} m^{-2}. There are considerable seasonal variations in the concentrations, related to the seasonal variations in precipitation amount (dry winter, wet summer). The soils at the site are acid with median base saturation of 12% and 8% in the topsoil and subsoil, respectively. In soil water, aluminum concentrations are typically in the range 3-8 mg L^{-1}. However, due to the high base cation deposition, the $Al/(Ca^{2+}+Mg^{2+})$ molar ratio is below unity in most samples, indicating little damage of forest due to aluminum in soil water.

Keywords: deposition fluxes, integrated monitoring, soil water, throughfall, China.

1. Introduction

Acid rain has become a serious environmental problem in southern China due to the use of coal with high sulfur content causing large emissions of SO_2. Chongqing City in southwestern China is among the areas receiving the highest sulfur deposition (e.g. Zhao et al., 1994).

In order to study the acid deposition fluxes and composition, and possible acidifying effects on soil, soil water and surface water, a small research catchment was equipped in the Tie Shan Ping (TSP) forest area about 25 kilometers northeast of downtown Chongqing.

Results from the first 3 years of monitoring of bulk and throughfall deposition, soil solution and surface water are presented in this paper. The catchment has recently been selected as an intensive integrated monitoring site in the IMPACTS project (Tang et al., 2001), and there will be broader and more intensified monitoring in the catchment in the future.

2. Site description and data material

The Tie Shan Ping (TSP) area (29°38'N, 106°41'E) is located approximately 25 km northeast from Chongqing center, at an elevation of about 450 meters. Large parts of the original deciduous forest were logged in 1958 and 1962 and later replaced with Masson pine (*Pinus massoniana*), which is now the predominant tree species. This is a typical forest history for large parts of China. There are stress symptoms and reduced tree vitality at the forest in the area. Insects attacks are at least partly responsible for the present damaged forest; wheter impacts of pollution, directly or indirectly, have contributed is not known at present. There is a relatively dense undervegetation of shrubs and ferns. The main soil type is locally called yellow mountain soil, which corresponds to Haplic Acrisol in the FAO classification system. The soils, developed from the red Jurassic sandstone and shale bedrock, are rather homogenous and relatively rich in finer particles.

Chongqing belongs to sub-tropical humid monsoon climate zone. Annual average temperature is around 18°C. Average temperature in January is 8°C and average highest temperature in summer is 27-29°C. There are little frost and snow and much fog all year round. Features are warm winter, hot summer, early spring and short autumn. The bulk of rain falls from April through October; the average annual total is about 1100 mm.

Equipment for sampling precipitation, throughfall and soil water has been installed in the catchment. Data presented here are for the sampling period 1995 - 1998. Samples were collected monthly from one open bulk precipitation sampler, 6 throughfall samplers, 14 ceramic cup lysimeters at 7 plots (two depths at each plot) and 3 surface water sampling locations. In addition soils were sampled at all lysimeter locations and analyzed for pH, cation exchange properties and other parameters. Water samples were analyzed for major anions (SO_4^{2-}, NO_3^-, Cl^-, F^-) and cations (Ca^{2+}, Mg^{2+}, K^+, Na^+, NH_4^+) by ion chromatography. Aluminum was fractionated into labile inorganic and organic complexes (Al_i and Al_o respectively) in accordance with the Barnes-Driscoll fractionation scheme (Barnes, 1975; Driscoll, 1984) and further speciated using the computer program ALCHEMI (Schecher and Driscoll, 1987). As a quality control/data screening measure, we rejected all samples having a discrepancy in the ion balance above 10%.

3. Results

3.1. DEPOSITION

The site experiences high airborne concentrations of sulfur dioxide and other pollutants. SO_2 concentration in air is only available for 1995. Average concentration was 76 µg m^{-3}. Assuming a deposition velocity of 0.5 - 0.7 cm s^{-1} gives dry deposition flux of SO_2 of 4.0 - 5.6 g S m^{-2} yr^{-1}. Preliminary studies indicate that deposition of particulate matter in addition may contribute with at least 1 g S m^{-2} yr^{-1}. The wet sulfate deposition calculated from bulk precipitation concentrations is 4.7 - 5.7 g S m^{-2} yr^{-1}.

The annual volume weighted pH in bulk precipitation was 3.8 - 4.2 and annual sulfate concentration was 200-400 µeq L^{-1} (Table I). Sulfate typically accounts for 80%-90% of the anionic charge.

The throughfall samples were more acid than the bulk precipitation with pH as low as 3.1 to 4.0. The low pH in the throughfall samples probably reflects high deposition to leaves and needles of SO_2, which is subsequently oxidized, dissolved in rainwater and collected as sulfuric acid. Ion concentrations were considerably higher in throughfall compared to the bulk precipitation.

3.2. SOIL WATER

Soil water chemistry is rather homogeneous with median soil water pH in the range 3.9 to 4.5 for all the lysimeters, except in a soil plot receiving water from an upwelling, where the median pH was 4.9. Due to the special features of this plot it will not be included in the further discussion.

The sulfate median concentrations for the different soil water sampling sites range from 0.5 to 1.8 meq L^{-1} (Figure 1). Sulfate is less dominant in soil water than in precipitation, but still accounts for typically 65 - 75 % of the sum of anionic charge.

Median concentrations of Al_i are in the range 2.0 - 8.9 mg L^{-1}, which must be characterized as high. Hydrated Al^{3+} is the dominant Al species in soil water. Fluoride- and sulfate complexes account for an additional 15-20 % each and hydroxo-complexes for only about 5%. The concentration of organically complexed Al is comparatively low, usually about 10% of the inorganic fraction.

TABLE I
Major ions in deposition in the Tie Shan Ping catchment. The given values are minimum and maximum of the annual volume weighted averages for 1996-1998. Unit: µeq L^{-1}.

	pH	NH_4^+	Ca^{2+}	Mg^{2+}	SO_4^{2-}	NO_3^-
Precipitation:	3.8 - 4.2	56 - 106	82 - 196	16 - 43	204 - 392	24 - 42
Throughfall:	3.1 - 4.0	119 - 332	281-852	68 - 262	584 - 2362	38 - 107

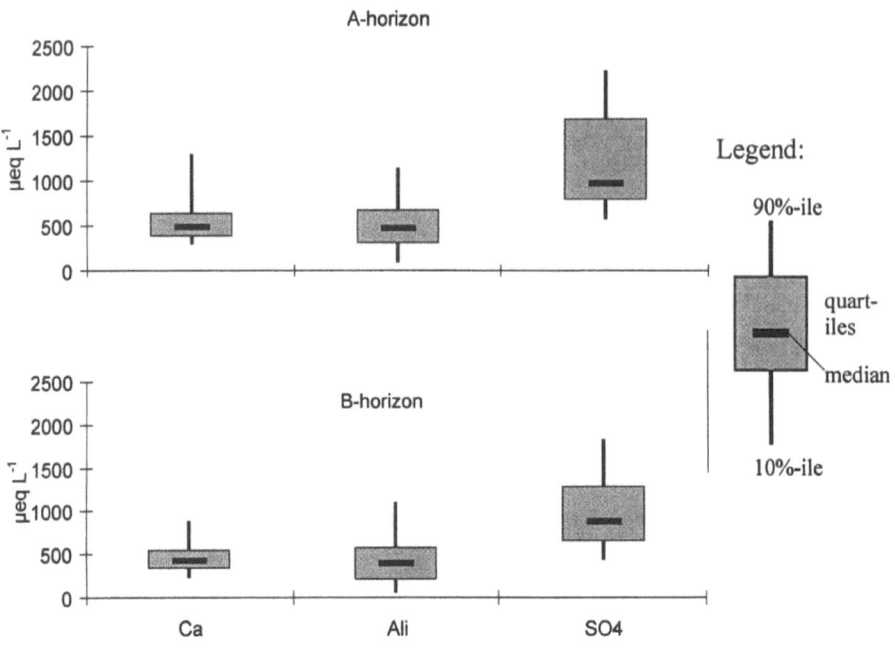

Figure 1. Soil water concentrations of Ca^{2+}, Al_i and SO_4^{2-}. Upper panel shows A horizon data, lower panel B horizon.

H^+ typically accounts for 3-7% of the total cationic charge in soil water; calcium accounts for 30-40% and aluminum 20-35%. The total base cation charge is 1.2 to 2.8 times larger than the sum of the acid cation charge $(H^+ + Al_i^{n+})$.

Quartiles for the molar ratio $Al_i/(Ca^{2+}+Mg^{2+})$ are 0.35 and 0.60 for all soil water samples. 17 of the 181 soil water samples in the data had $Al_i/(Ca^{2+}+Mg^{2+})$ between 1 and 2. High values for this ratio (e.g. > 1) are often regarded as a potential cause for forest damage (Cronan and Grigal, 1995).

3.3. SOIL

The soils at TSP have bulk densities increasing from 1.25 kg dm^{-3} in the A-horizon to 1.34 kg dm^{-3} in the B2-horizon. Soil porosity is about 50%. The generally high densities and low porosities are mainly due to a large fraction of fine silt and clay material. Water retention characteristics are similar for the A- and B-horizons and typical for clay-rich soils. Kaolinite, plagioclase and K-feldspar are important minerals in the clay fraction. The variability in soil properties between the different sampling locations is relatively low. The base saturation is low, with median values of 12% (range: 7-17) and 8% (range: 5-27) in the A- and B-horizons respectively.

Significant amounts of sulfate have been adsorbed in the soil; water extractable sulfate was in the range 1.7-2.7 meq kg^{-1} (21-34 g S m^{-2} considering the upper 60 cm of the soil). The capacity for further sulfate adsorption in the upper part (i.e. 60 cm) is low (Larssen, 1999).

3.4. SURFACE WATER

The water bodies investigated within TSP area, a pond and two small streams, have relatively low pH (quartiles: 4.7 and 5.3) and ANC (quartiles: -62 and 17 µeq L^{-1}). The concentrations of Al$_i$ are relatively high, although considerably lower than in soil water (quartiles: 170 and 570 µg L^{-1}). Sulfate is the dominant anion, but concentrations are again considerably lower than in soil water. The mobilization of base cations by cation exchange and weathering reactions in TSP is too low to prevent negative ANC-values in surface waters in the area.

Other waters sampled in the Chongqing region indicate that surface water generally have high concentrations of Ca^{2+} and SO_4^{2-}, high pH and high ANC (Lydersen et al., 1997; Xue and Schnoor, 1994). We may expect acid surface water to be neutralized due to mixing with other waters when entering larger water systems, although acidification may have some local effects.

4. Discussion

The concentration of inorganic aluminum in soil water must be characterized as high and possible toxic effects on plant roots may be expected. However, it is assumed that high concentrations of bivalent cations (i.e. Ca^{2+} and Mg^{2+}) counteract the toxic effect of Al. The $Al/(Ca^{2+}+Mg^{2+})$ molar ratio has been suggested used as an acidification indicator in Europe and North America (Cronan and Grigal, 1995). At TSP high concentration of calcium keeps the $Al/(Ca^{2+}+Mg^{2+})$ ratio moderately low. The high calcium concentration is primarily caused by high calcium deposition. The sources for the calcium deposition is not known, but if a significant part originates from combustion sources it is important to take this into consideration when setting emission reduction measures (Larssen and Carmichael, 2000).

The release of aluminum from the soil to the soil solution is strongly dependent on solution pH. However, the observed Al^{3+} concentrations in soil water are not in accordance with gibbsite ($Al(OH)_3$) equilibrium model, which is commonly used in acidification models. Instead there is the usually observed increasing undersaturation with decreasing pH. The relation between pH and aluminum in TSP soil water is very similar to that observed for a similarly studied catchment in Guiyang (Larssen et al., 1998).

There is a strong need for better knowledge of acidification and its effects on forest vegetation in China. Such studies will be carried out at TSP in the coming years as this catchment will be among the sites in the monitoring and research project IMPACTS (Integrated Monitoring Program on Acidification of Chinese Terrestrial Systems), where all aspects of acid deposition in Chinese forest ecosystems will be studied (Tang et al., 2001).

5. Conclusions

The Tie Shan Ping area receives a high load of acid deposition, estimated at 10-12 g S m^{-2} yr^{-1}. This leads to high concentration of aluminum in the soil solution. However, as the deposition of calcium also is high, severe toxic effects of aluminum on the forest is not yet expected. Intensive acid rain integrated monitoring and research will continue in the Tie Shan Ping catchment in the IMPACTS project.

References

Barnes, R.B.: 1975 *Chemical Geology* **15**, 177.
Cronan, C.S. and Grigal, D.F.: 1995 *Journal of Environmental Quality* **24**, 209.
Driscoll, C.T.: 1984 *International Journal of Environmental Analytical Chemistry* **16**, 267.
Larssen, T., 1999. *Acid deposition and acidification of soils and soil water in China.* Ph.D. thesis, University of Oslo.
Larssen, T., Xiong, J., Vogt, R.D., Seip, H.M., Liao, B. and Zhao, D.: 1998 *Water Air and Soil Pollution* **101**, 137.
Larssen, T. and Carmichael, G.R.: 2000 *Environmental Pollution* **101**, 89.
Lydersen, E., Angell, V., Eilertsen, O., Larssen, T., Mulder, J., Muniz, I.P., Seip, H.M., Semb, A. and Vogt, R.D.: 1997 *Planning an Integrateed Acidification Study and Survey on Acid Rain Impacts in China.* Final Report. NIVA report SNO 3719-97, Oslo.
Schecher, W.D. and Driscoll, C.T.: 1987 *Water Resources Research* **23**, 525.
Tang, D., Seip, H.M., Angell, V., Eilertsen, O., Larrsen, T., Lydersen, E., Mulder, J., Semb, A., Torseth, K., Vogt, R.D., Xiao, J., Zhao, D. and Kong, G.: 2001. *Water, Air and Soil Pollution.* This volume.
Xue, H.B., and Schnoor, J.L.: 1994 *Water Air and Soil Pollution* **75**, 61.
Zhao Dawei, Seip, H.M, Zhao Dianwu, and Zhang Dongbao.: 1994 *Water Air and Soil Pollution* **77**, 27.

THE IMPORTANCE OF NITROGEN OXIDES FOR THE EXCEEDANCE OF CRITICAL THRESHOLDS IN THE NORDIC COUNTRIES

M. JOHANSSON[1*], R. SUUTARI[1], J. BAK[2], G. LÖVBLAD[3], M. POSCH[4], D. SIMPSON[5], J.-P. TUOVINEN[6] and K. TØRSETH[7]

[1] *Finnish Environment Institute, P.O.Box 140, 00251 Helsinki, Finland;* [2] *National Environmental Research Institute, P.O.Box 314, Silkeborg 8600, Denmark;* [3] *IVL Swedish Environmental Research Institute, P.O.Box 47086, 40258 Gothenburg, Sweden;* [4] *RIVM/Coordination Center for Effects, P.O.Box 1, 3720 BA Bilthoven, The Netherlands;* [5] *EMEP/MSC-W, Norwegian Meteorological Institute, P.O.Box 43, Blindern, 0313 Norway;* [6] *Finnish Meteorological Institute, Sahaajankatu 20 E, 00810 Helsinki, Finland;* [7] *Norwegian Institute for Air Research, P.O.Box 100, 2027 Kjeller, Norway*
(* author for correspondence, e-mail: matti.johansson@vyh.fi)

Abstract. Impacts of air pollutants and especially acidification in ecosystems have been of serious concern in the Nordic countries since the 1970s. The current approach to assess several pollutants (sulfur and nitrogen oxides, ammonia, volatile organic compounds) and their effects (acidification, eutrophication and ground-level ozone) simultaneously is extremely complex. This study explored the relative role of nitrogen oxides in environmental impacts in the Nordic countries. The share of NO_x in the exceedances of critical loads, the long-term ecosystem protection targets, was found to be roughly 25% in acidification and 50% in eutrophication. The contribution of NO_x emissions to ground-level ozone formation was considered important, as NO_x is the limiting precursor in ozone formation in the Nordic countries. The comparison of observed and modeled accumulated ozone concentrations (AOT40) for the early 1990s shows noticeable differences in the Nordic area, partly due to the sensitivity of the AOT40 indicator to the 40 ppb threshold value.

Keywords: nitrogen oxides, integrated model, acidification, eutrophication, ozone

1. Introduction

International cooperation is required to combat the detrimental effects of long-range transboundary air pollutants. Extensive efforts have been made in Europe to utilize critical thresholds for acidification, eutrophication and the impacts of ground-level ozone on vegetation and human health to explore cost-efficient strategies to reduce emissions of sulfur and nitrogen oxides, ammonia and volatile organic compounds (VOCs). The negotiations within the Convention on Long-range Transboundary Air Pollution (CLRTAP) under the auspices of the United Nations Economic Commission for Europe (UN/ECE) resulted in the signing of the Protocol to Abate Acidification, Eutrophication and Ground-level Ozone in 1999 in Gothenburg, Sweden. The Commission of the European Union (EU) has put forward a proposal for a National Emission Ceilings Directive (NECD). Integrated assessment modeling has provided background information at international (*e.g.* Hordijk, 1995) and national (*e.g.* Johansson, 1999) levels.

The aim of this study was to assess the multi-pollutant/multi-effect air pollution problem from the viewpoint of nitrogen oxides in the Nordic countries (Denmark, Finland, Norway and Sweden). The study concentrated on acidification and eutrophication on aquatic and terrestrial ecosystems and ground-level ozone impacts on crops. The share of NO_x is presented by environmental effects, countries, years and emission scenarios.

2. Emission and Deposition Scenarios

The status for the emissions in the reference year 1990 is reported in EMEP/MSC-W (1998). Two scenarios for the target year 2010 were used: (i) the GP scenario based on the Gothenburg Protocol of the UN/ECE (UN/ECE, 1999), and (ii) the CP scenario, which includes national emission ceilings for the 15 member states of the European Union (Council of the European Union, 2000) and for other countries the emissions of the GP scenario. In general, emissions of SO_2 and also of NO_x will strongly decrease in Europe with the GP scenario, but less for NH_3. The reductions in the CP scenario are slightly bigger than in the GP scenario.

The deposition values in the 150 km×150 km grid were calculated using the average source-receptor relationships in 1985–1995 from the EMEP/MSC-W Lagrangian Acid Deposition Model (EMEP/MSC-W, 1998) (Figure 1). NO_x contributes to more than 50% of nitrogen deposition except in Denmark in 2010 with the GP scenario. The share in acidifying deposition is generally less than 40%.

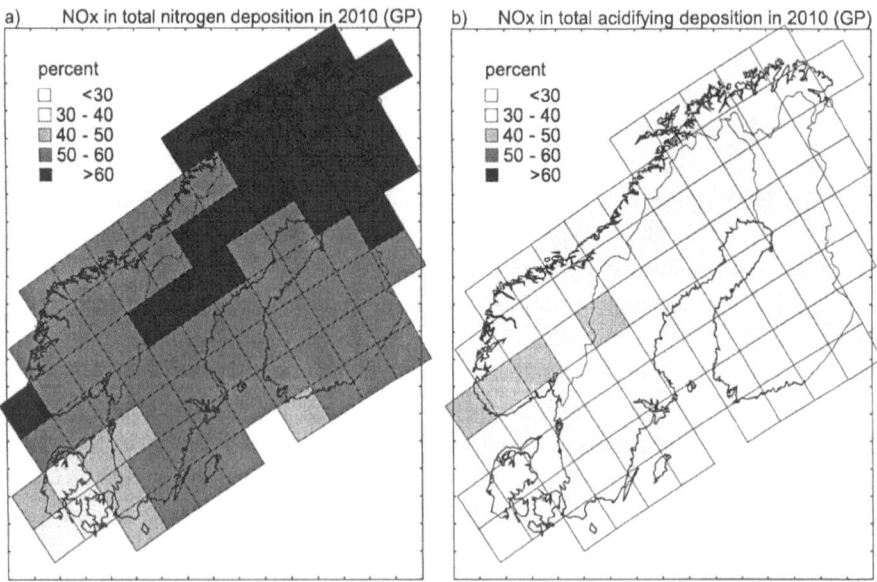

Figure 1. The share of nitrogen in % in a) total nitrogen and b) total acidifying deposition in 2010 with the GP scenario.

3. Critical Loads for Acidification and Eutrophication

A critical load is "a quantitative estimate of an exposure to one or more pollutants below which significant harmful effects on specified sensitive elements of the environment do not occur according to present knowledge" (Nilsson and Grennfelt, 1988). The methods to compute them are documented in UBA (1996), and data are reported by Posch *et al.* (1999).

The exceedances of critical loads for acidification and eutrophication were first studied with cumulative distribution functions, which are calculated by ranking ecosystem-specific exceedances of critical loads weighted with the corresponding ecosystem area (Figure 2). The relative ecosystem area with no exceedance increased from 82 to 94% for acidification and 70 to 88% for eutrophication from 1990 to 2010 with the GP scenario (Figure 2a). The highest absolute exceedances were reduced from 1700 to 500 and 1100 to 550 eq ha^{-1} a^{-1} for acidification and eutrophication, respectively. Figure 2b shows that the total ecosystem area exceeded would be reduced from 16 to 5.5 million hectares for acidification and from 15 to 6 million ha for eutrophication between 1990 and 2010 with the GP scenario. The share of SO$_2$ in acidification will decrease.

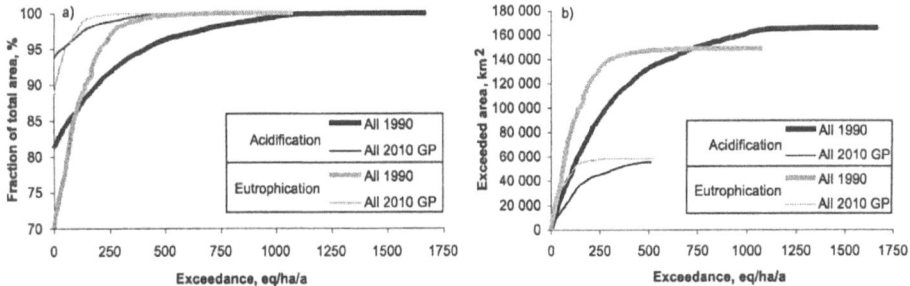

Figure 2. The cumulative distribution functions of the exceedances for acidification and eutrophication in 1990 and 2010 with the GP scenario in the Nordic countries. The exceeded area is shown a) relative to the total ecosystem area in % and b) in absolute ecosystem area in km^2.

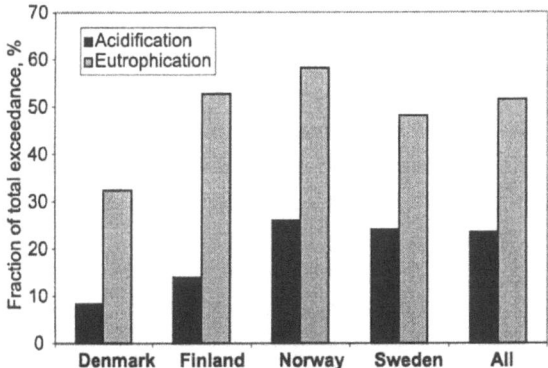

Figure 3. The proportion of nitrogen oxides in the exceedance (Ex) for acidification and eutrophication in the Nordic countries in 2010 with the GP scenario.

The share of nitrogen oxides in the exceedances of critical loads was examined using the critical load exceedance value, which is the difference between deposition and ecosystem critical load, in eq ha^{-1} a^{-1}. The proportion of nitrogen oxides in the ecosystem-specific exceedances was assumed to be the same as the relative fraction of NO$_x$ deposition in the modeled total acidifying deposition for acidification and in the total nitrogen deposition for eutrophication. The share of NO$_x$ in exceedances, averaged from area weighted ecosystem-specific critical load exceedances in a country, varied between 8 and 27% for acidification and between 32 and 58% for eutrophication at country level, and for all Nordic countries between 24 and 52%, respectively (Figure 3).

4. Ground-level Ozone

This study considered the ground-level ozone effects only on crops, for which the suggested threshold is an AOT40 of 3000 ppb·h (Kärenlampi and Skärby, 1996). AOT40 denotes the accumulated ozone concentration during daylight hours above a threshold of 40 ppb during growing season (May–July, for crops). The formation of ground-level ozone in the Nordic countries is largely determined by NO$_x$ emissions as the NO$_x$-VOC-ratio is relatively low, although in Denmark the contribution of VOCs becomes important (Simpson *et al.*, 1997). The emission reductions from 1990 to 2010 with the GP scenario are roughly 46% for both NO$_x$ and VOCs in the Nordic countries and 39% in Europe. The RAINS model, which includes a reduced-form description (Heyes *et al.*, 1996) of the EMEP ozone model, suggests consequent AOT40 reductions of 0–4 ppm·h in the Nordic area.

Ozone concentration measurements from 21 stations during 1992-96, provided by the EMEP Chemical Co-ordinating Centre at the Norwegian Institute for Air Research (EMEP-CCC/NILU), were used to calculate AOT40s. In general, the highest annual exposures occurred in 1992 and the lowest annual exposures in 1996, with considerable interannual variability. The AOT40s in 1992, which were calculated for each individual station and interpolated with ordinary kriging over the Nordic area, indicated exceedances in the southern and central parts of the Nordic countries (Figure 5a). The relative standard error in the kriged map was less than 30% in southern part of the Nordic countries, but more than 70% in the northernmost areas.

The modeled ozone concentration values from the EMEP ozone trajectory model (Simpson *et al.* 1997) were used to derive AOT40s for 1992 in the 150 km×150 km grid. AOT40 exceedances were found in Denmark and southern Sweden (Figure 5b), and they were much lower than the observed ones. Even though the number of observation stations is rather limited, the map shows the basic features on the differences between observed and modeled values.

The big differences between modeled and observed AOT40 are only partly a reflection of poor model performance. AOT40 values become very sensitive to even small biases in input data (either modeled or observed) for sites where ozone levels are often near the threshold value of 40 ppb (Simpson *et al.*, 1998; Tuovinen, 2000),

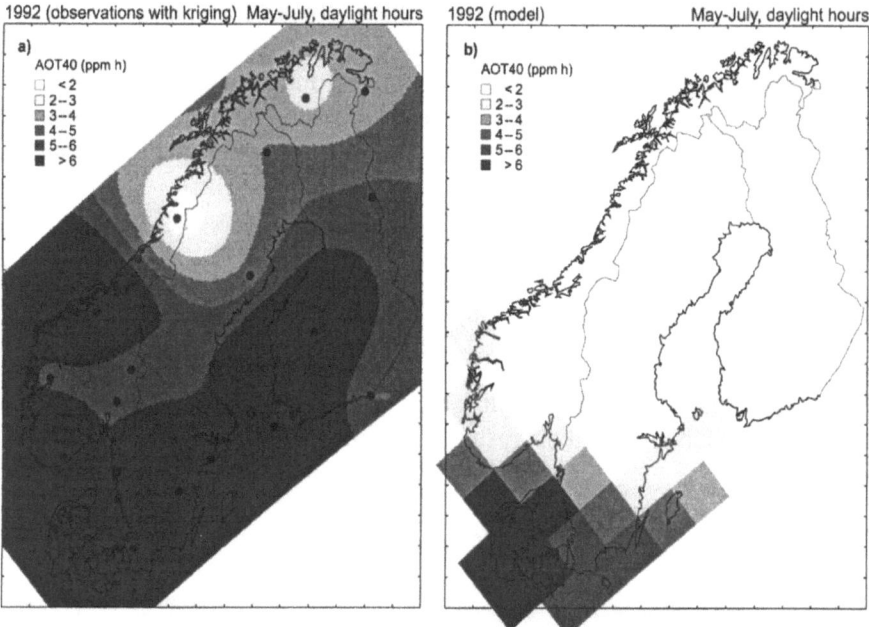

Figure 5. The yearly AOT40 for crops in 1992 calculated from a) observations at measurement stations (indicated with dots) and interpolated with kriging, and b) the EMEP ozone model results.

which is the case in the Nordic area. Tuovinen (2000) found that a 10% uncertainty in measured concentration data translates to 20–100% change in AOT40 in May–July for sites with high (Ispra, Italy) and low (Oulanka, Finland) ozone levels. Although the EMEP model performs reasonably well for most years (Simpson *et al.*, 1998), 1992 is an exceptional year with much worse model performance than average for all of the Nordic sites, which is apparently due to an episode of high free-tropospheric ozone at high latitudes not captured by the climatological boundary conditions.

5. Concluding Remarks

The importance of NO_x emissions will increase in environmental impacts due to transboundary air pollution in Europe in the future. The foreseen relative emission reductions are less than for sulfur. Ammonium, which has a shorter-range transport in the atmosphere than NO_x, will gain more relative importance locally. There is an increasing need for more information and knowledge on nitrogen-related processes in ecosystems and related potentially harmful effects.

The results from the calculations, which assumed the same share of NO_x for exceedance as in the deposition, suggested that NO_x contributes to the critical load exceedances with 25% for acidification and 50% for eutrophication in the Nordic countries. The results presented in this study should be interpreted cautiously especially for eutrophication, which still has considerable uncertainties in its critical load calculation.

The comparison of observed and modeled accumulated ozone concentrations for crops (AOT40) showed noticeable differences, although in the spatial assessment the kriging errors become significant in areas far from observation stations. The modeled values are used in integrated models providing scenarios for emission reduction negotiations, and should therefore be as well convergent with observations as possible.

Acknowledgments

The Nordic Council of Ministers is gratefully acknowledged for funding the project "NO_x emission reduction induced multi-pollutant multi-effect evaluation on Nordic ecosystems", which produced the results presented in this article.

References

Council of the European Union: 2000, 'Proposal for a directive of the European Parliament and of the Council on national emission ceilings for certain atmospheric pollutants', 9806/00, Brussels, Belgium.

EMEP/MSC-W: 1998, 'Transboundary air pollution in Europe. Part 1: Estimated dispersion of acidifying and eutrophying compounds and comparison with observations', EMEP/MSC-W Report 1/98, Norwegian Meteorological Institute, Oslo, Norway.

Heyes, C., Schöpp, W., Amann, M. and Unger, S.: 1996, 'A 'reduced-form' model to predict long-term ozone concentrations in Europe', Interim Report 96-12, IIASA, Laxenburg, Austria.

Hordijk, L.: 1995, *Water, Air and Soil Pollution* 85, 249–260.

Johansson, M.: 1999, Doctoral dissertation, *Monographs of the Boreal Environment Research* 13.

Kärenlampi, L. and Skärby, L. (eds.): 1996, 'Critical levels for ozone in Europe. Testing and finalizing the concepts', UN/ECE workshop report, University of Kuopio, Department of Ecology and Environmental Science, Kuopio, Finland.

Nilsson, J. and Grennfelt, P. (eds.): 1988. 'Critical loads for sulphur and nitrogen', Report from a workshop 19–24 March 1988, Skokloster, Sweden, Nordic Council of Ministers, Miljørapport 1988:15.

Posch, M., De Smet, P.A.M., Hettelingh, J.-P. and Downing, R.J. (eds.): 1999, 'CCE Status Report 1999', RIVM, Bilthoven, The Netherlands.

Simpson, D., Altenstedt, J. and Hjellbrekke, A.-G.: 1998, 'The Lagrangian oxidant model: status and multi-annual evaluation', in 'Transboundary photooxidant air pollution in Europe, Calculations of tropospheric ozone and comparison with observations', EMEP MSC-W Report 2/98, Norwegian Meteorological Institute, Oslo, Norway.

Simpson, D., Olendrzynski, K., Semb, A., Støren, E. and Unger, S.: 1997, 'Photochemical oxidant modelling in Europe: multi-annual modelling and source-receptor relationships', EMEP/MSC-W Report 3/97, Norwegian Meteorological Institute, Oslo, Norway.

Tuovinen, J.-P.: 2000, *Environmental Pollution*, 109, 361–372.

UBA: 1996, 'Manual on methodologies and criteria for mapping critical loads/levels and geographical areas where they are exceeded', Texte 71-96, Umweltbundesamt, Berlin, Germany.

UN/ECE: 1999, 'Protocol to Abate Acidification, Eutrophication and Ground-level Ozone', United Nations, Economic Commission for Europe, Geneva, Switzerland.

LONG-TERM ECOLOGICAL EFFECTS OF LIMING – THE ISELAW PROGRAMME

MAGNUS APPELBERG[1]* and TORBJÖRN SVENSON[2]

[1] *Institute of Freshwater Research, SE-178 93 Drottningholm, Sweden, and Dept. of environmental assessment, Swedish University of Agricultural Sciences, P.O.Box 7050, SE-750 07 Uppsala, Sweden;* [2] *Swedish Environmental Protection Agency, SE-106 48 Stockholm, Sweden (*author for correspondence, e-mail:magnus.appelberg@fiskeriverket.se)*

Abstract. The Swedish liming programme was initiated in 1977 to counteract the effects of anthropogenic acidification on aquatic ecosystems until the acid deposition has been reduced. Ecosystem development in limed waters has been followed since 1989 in a programme for integrated studies of the effects of liming acidified waters (ISELAW). The main objectives are to assess a) the long-term ecological effects of liming, b) to what extent ecosystems recover to a pre-acidification state, and c) to elucidate possible detrimental effects of lime treatment. The programme comprises monitoring of water chemistry, phyto- and zooplankton, vegetation, benthic invertebrates and fish in 13 limed and 5 non-limed lakes, and 12 limed and 10 non-limed streams. Paleolimnological studies are performed to reveal pre-acidification lake history. The results show that lime treatment detoxifies the water, although chemical and biological development varies among and within sites. In general the long-term changes are small compared to the initial changes associated with first treatment. Water chemical changes over time are reflected as reduced sulfur concentrations and increased nitrogen concentrations. Treated ecosystems seem not to recover fully to the situation before acidification, and due to re-colonization failure, several species are lacking in the limed waters.

Keywords: liming, acidification, long-term effects

1. Introduction

Measures to counteract the effects of anthropogenic acidification of aquatic ecosystems have been performed in Norway (Sandøy and Romundstad 1995), Finland (Kauppi *et al.*, 1990), Scotland (Howells and Dalziel, 1992), Wales (Ormerod *et al.*, 1990), USA and Canada (Schreiber and Villella, 1991, Olem, 1991). The probably most extensive programme is carried out in Sweden where more than 8,000 lakes and 12,000 km of running waters are limed repeatedly (Henrikson and Brodin, 1995; Svenson *et al.*, 1995). The major goals for the Swedish liming programme are to keep alkalinity above 0.05 meq l^{-1} and pH above 6.0 in order to protect existing flora and fauna and to let species re-colonize.

As long term studies (>10 years) of water chemical and biological development in limed lakes and streams are sparse (e.g. Wheaterley, 1988; Olem, 1991; Henrikson and Brodin, 1995), Swedish Environmental Protection Agency initiated "Integrated Studies of the Effects of Liming Acidified Waters" (ISELAW) in 1988. The three main objectives are a) to assess the long-term effects of liming acidified waters on lakes and streams, b) to assess if limed ecosystems recover to a pre-acidification situation, and c) to reveal possible detrimental effects in treated lakes and streams. The following paper presents an overview of the programme and a synthesis of the results obtained during the first 10 years of monitoring (1989-1998). Due to recent revisions of the stream programme, the results presented are mainly focused on lakes.

2. Material and methods

After revision in 1999 the programme consists of 13 limed lakes and 5 non-limed lakes and 12 limed and 10 non-limed streams, distributed over the country (Fig.1). As additional non-limed references 10 acid or circumneutral

Figure 1. Distribution of limed and non-limed lakes and streams included in the ISELAW and national environmental monitoring programme in Sweden.

lakes and 15 non-limed reference streams included in the national environmental monitoring programme, are used. Development of water chemical and biological variables are repeatedly followed in both lakes and streams. The lake programme includes water chemical variables, macrophytes, phyto- and zooplankton, benthic invertebrates and fish. The stream programme includes hydrology, water chemistry, benthic invertebrates and fish. Standard methods for environmental monitoring are mainly used according to Swedish Environmental Protection Agency. Sampling and analytical methods used have been described in Appelberg *et al.* (1995).

The first lime treatments were performed in 1974 and the average time passed since first liming is 20 years for lakes and 15 years for streams (Söderbäck 1997, Bergquist 2000). Liming methods comprise lake liming (or upstream lake liming), lime dosers in inlet streams and wetland liming. In several waters, methods have been changed over the years, and often several different methods are used in the same water system to optimize the effects of lime treatment. Frequency of treatment ranges from continuous lime additions using lime dozers (mainly in streams), to a 6 years interval using wetland and/or lake liming.

3. Results and discussion

3.1 LONG-TERM DEVELOPMENT AFTER LIMING

Although variation within and among objects is considerable, changes over time have in general been small compared to the initial changes after the first lime treatment. The water chemical goals (i.e. to keep minimum alkalinity>0.05 meq l^{-1} and minimum pH >6.0) have been achieved in all lakes except two (Table I) and in 11 out of 15 streams (Bergquist 2000).

TABLE I
Median and 5- and 95% percentiles of annual mean values for some water chemical variables in the 13 limed study lakes during the years 1989-1998. N=130.

	Alkalinity meq l^{-1}	pH mean	pH min.	Ca^{2+} meq l^{-1}	SO$_4$ meq l^{-1}	TOC mg l^{-1}	Total-N µg l^{-1}	Total-P µg l^{-1}
Median	0.140	6.76	6.48	0.242	0.138	9.02	402	10.2
Percentiles 5%	0.059	6,29	5.85	0.141	0.044	3.81	205	4.9
Percentiles 95%	0.251	7.22	7.02	0.397	0.207	14.97	614	19.8

Major significant trends in lake water chemistry over the ten years study period are decreased annual mean SO$_4$-concentrations (linear regression, p<0.05, Fig. 2 left), reduced minimum O$_2$-concentrations (p<0.01), and increased total nitrogen concentrations (p<0.01, Fig. 2 right). The largest reductions in SO$_4$ are found in lakes situated in southern Sweden, possibly reflecting a reduced sulfur deposition (Wilander, 1997). The increased nitrogen concentration could be an effect of both increased deposition as well as increased accumulation of nitrogen in the limed lakes (Persson et al., 1997). The limed streams did not show similar trends (Bergquist, 2000).

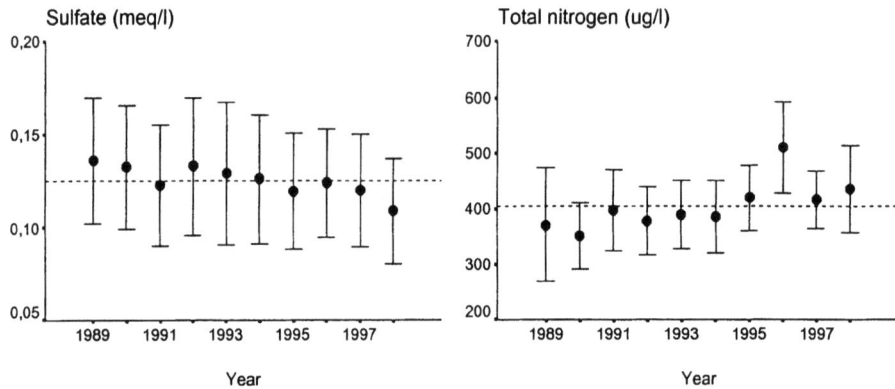

Figure 2. Mean and 95% C.L. for sulfate concentration (left) and total nitrogen concentration (right) in the 13 limed lakes 1989-1998. Dotted lines represents mean value for the study period.

In the limed lakes, changes in phyto-and zooplankton communities have been small (Hörnström and Ekström, 1997). Number of benthic invertebrate species in the profundal zone as well as number of species and abundance of invertebrates in the sub-littoral zone decreased during the years 1989-1998 (linear regression, $p<0.004$, $p<0.023$ and $p<0.083$ respectively). In contrast, number of benthic invertebrates species in the littoral zone increased significantly ($p<0.001$) over the same years. Fish biomass (measured as weight per unit effort) increased in some lakes and decreased in others during the ten years period. The reasons for this variability are possibly several, including expansion of colonizing populations, and biotic interactions within and between species (Appelberg, 1998).

3.2 DO LIMED ECOSYSTEMS RECOVER TO A PRE-ACIDIFICATION SITUATION?

Awaiting recently initiated paleolimnological studies and compilation of historical data records for reconstructing pre-acidification status of the limed lakes, post-liming situation is to be compared with as similar circumneutral references as possible to assess ecosystem recovery. pH and alkalinity in limed lakes resembled that of circumneutral reference lakes, although temporal and spatial variation was larger in the former (Söderbäck et al., 1997). Median metal-concentrations were low and stable in the limed lakes (Persson et al., 1997), and concentrations and temporal variations had decreased after liming in the limed streams. Despite this, toxic concentrations of labile aluminum (>20 µg l^{-1}, Kroglund et al., 1998) are still recorded in several of the limed streams (Bergquist, 2000). Phosphorus concentrations in the limed lakes were in general lower than in both circumneutral and acid reference lakes, which may result in low nitrogen utilization and accumulation of nitrogen in the limed lakes (Persson and Appelberg, this volume).

Biodiversity, measured as species richness, diversity and abundance at different trophic levels, was on average higher in the limed lakes compared to acidified lakes, but lower compared to circumneutral reference lakes (Söderbäck et al., 1997). Phyto- and zooplankton communities in limed lakes resembled those of circumneutral reference lakes, whereas sub-littoral benthic invertebrates showed significantly lower values for abundance, species number and diversity in the limed compared to reference lakes. Contradictory, benthic invertebrate fauna in both the littoral zone of limed lakes and that in streams has become more similar to that of circumneutral reference lakes and streams, partly as a result of re-colonization of gastropods (Lingdell and Engblom, 1997; Bergquist, 2000). Number and diversity of fish species was lower and more dominated by percids in the limed lakes compared to the circumneutral reference lakes (Beier et al., 1997,; Appelberg, 1998).

3.3 HAS LIMING CAUSED DETRIMENTAL EFFECTS IN THE LIMED LAKES?

Until now, no severe detrimental effects have been observed in the limed study lakes and streams. However, low total phosphorous concentrations, increasing nitrogen accumulation, impoverishment of the benthic invertebrate communities in profundal and sub-littoral zones, as well as an increasing dominance of percid fish species in the limed lakes, all indicate reduced productivity in treated lakes. It may be hypothesized that this development will affect the aquatic ecosystem in a long time perspective, resulting in reduced biodiversity. This is supported by the fact that biodiversity in the limed lakes still is lower compared to circumneutral reference lakes after 20 years of liming.

4. Conclusions

1. The long-term trends in studied limed lakes indicate that sulfur-concentrations decrease, possibly due to reduced deposition. A reduction in productivity in limed lakes was expressed both in abiotic and biotic variables.
2. The water chemical goals of the Swedish liming programme were accomplished in most lakes, but to a lesser degree in streams. In some cases, the liming methods used were not sufficient to cope with the temporal and spatial variation in acid run-off water, which may call for a more adaptable liming strategy.
3. No severe detrimental effects of long-term lime treatment have so far been observed in the studied lakes and streams. However, in a long time perspective reduced productivity may affect community composition and result in low biodiversity.

Acknowledgements

We thank the participants of the ISELAW-programme for providing basic data for this synthesis: Department of Environmental Assessment Swedish University of Agricultural Sciences; Institute of Applied Environmental Research Stockholm, University, and Limnodata HB. The ISELAW-programme is financially supported by the Swedish Environmental Protection Agency.

References

Appelberg, M., Lingdell, P.-E., and Andrén, C.: 1995, *Water, Air and Soil Pollution* **85**, 883.
Appelberg, M.: 1998, *Restoration Ecology* **6**, 343.
Baalsrud, K., Hindar, A., Johannessen, M. and Matzow, D.: 1985, *Liming of acidified waters. Liming project – final report*. Dep. Environ. Direct. Nat. Manag., Norway.
Beier, U., Reizenstein, M. and Andersson, P.: 1997, *Naturvårdsverket Rapport* **4816**, 46.
Bergquist, B., (ed).: 2000, *Naturvårdsverket Rapport* **5076**, 1.
Henrikson, L. and Brodin, Y. W. (eds): 1995, *Liming of acidified surface waters – A Swedish synthesis*. Springer-Verlag Berlin Heidelberg N.Y.
Howells, G. and Dalziel, T. R. K. (eds): 1991, *Restoring acidified waters: Loch Fleet 1984-1990*. Elsevier Applied Science, London.
Hörnström, E. and Ekström, C.: 1997, *Naturvårdsverket Rapport* **4816**, 25.
Johnson, R.K.:1997, *Naturvårdsverket Rapport* **4816**, 32.
Kauppi, P., Anttila, P. and Kenttämies, K.: 1990, *Acidification in Finland. Finnish Acidification Research Programme HAPRO 1985-1990*. Springer-Verlag.
Kroglund, F., Teien, H. C., Rosseland, B. O., Lucassen, E., Salbu, B. and Åtland, Å.: 1998, *NIVA Rapport* **3970**, 1.
Lingdell, P.-E. and Engblom, E.: 1997, *Naturvårdsverket Rapport* **4816**, 32.
Olem., H.: 1991, *Liming acidified surface waters*. Lewis Publishers, Chelsea, Michigan.
Ormerod, S.J., Weatherley, N.S. and Merett, W.J.: 1990, *Environmental Pollution* **64**, 67.
Persson, G., Edberg, F., Andrén, C. and Borg, H.: 1997, *Naturvårdsverket Rapport* **4816**, 10.
Persson, G. and Appelberg, M.: 2000., *Water, Air and Soil Pollution* (this volume).
Sandøy, S. and Romundstad, A. J.: 1995, *Water, Air and Soil Pollution* **85**, 997.
Schreiber, R. K. and Villella, R.: 1991, 'An overview of the U.S. fish and wildlife service mitigation research program'. In: Olem, H., Schreiber, R. K., Brocksen, R. W. and Porcella, D. B (eds). *International lake and watershed liming prcatices*. The Terrence Ins., Washington D.C., USA, pp 25-40.
Svenson, T., Dickson, W., Hellberg, J., Moberg, G., Munthe, N.: 1995, *Water, Air and Soil Pollution* **85**, 1003.
Söderbäck, B. (ed): 1997, *Naturvårdsverket Rapport* **4816**, 1.
Söderbäck, B., Appelberg, M. and Lingdell, P.-E.: 1997, *Naturvårdsverket Rapport* **4816**, 57.
Weatherley, N.S.: 1988, *Water, Air and Soil Pollution* **39**, 421.
Wilander, A.: 1997, *Naturvårdsverket Rapport* **4652**, 1.

BIOMASS-SIZE DISTRIBUTION OF THE AQUATIC COMMUNITY IN LIMED, CIRCUMNEUTRAL AND ACIDIFIED REFERENCE LAKES

KERSTIN HOLMGREN

Institute of Freshwater Research, SE-178 93 Drottningholm, Sweden
(e-mail: kerstin.holmgren@fiskeriverket.se)

Abstract. Semi-quantitative biomass-size distributions (BSD's) along a joint axis of individual size provided an integrated illustration of aquatic communities sampled at different taxonomic and trophic levels. The approach was applied within the Swedish ISELAW-programme (integrated studies of the effects of liming acidified waters) to test the general hypothesis that aquatic communities in limed lakes are not systematically different from communities in comparable non-limed circumneutral lakes. Input data included pelagic phytoplankton and zooplankton, sublittoral/profundal macroinvertebrates, and benthic fish, within twelve Swedish lakes (six limed, two acidic and four circumneutral reference lakes). The four compartments were sampled on different spatial scales, but each designed for between-lakes comparisons. There were no clear-cut differences in overall size distribution between the three categories of lakes. The mean BSD of limed lakes was indeed more similar to the mean BSD of circumneutral lakes than to that of two acidic lakes. Due to high variation within categories, however, acidification status alone can not be used for reliable prediction of BSD in a certain lake.

Keywords: biomass-size distribution, aquatic communities, phytoplankton, zooplankton, benthic invertebrates, benthic fish

1. Introduction

Many Swedish lakes have been severely affected by anthropogenic acidification (Brodin, 1995), and large-scale liming activities aim at restoring the aquatic communities of previously acidified lakes. Biological effects of liming have mainly been evaluated with a focus on re-colonisation, growth and recruitment of acid-sensitive species (*e.g.* Appelberg, 1995a; 1995b; 1998; Nyberg, 1995).

Body size and feeding ecology are informative descriptors of community function, irrespective of or in addition to taxonomic composition (Kerr, 1974; Sprules and Holtby, 1979). Theories of size distributions (Kerr, 1974; Platt and Denman, 1978) use size-dependent doubling time and metabolic rates to explain regularities in the size composition of oceanic, pelagic ecosystems (Sheldon *et al.*, 1972). Numerous studies have used biomass-size spectra for evaluation of general patterns in aquatic communities, including most organisms within the pelagic habitat (*e.g.* Sprules *et al.*, 1983; Ahrens and Peters, 1991; Gaedke, 1992), zoobenthos (Strayer, 1986; Hanson *et al.*, 1989; Rodríguez and Magnan, 1993; Rasmussen, 1993) or fish (Duplisea and Kerr, 1995; Macphearson and Gordoa, 1996; Holmgren, 1999). Size-classes with low or zero biomass found within a quantitative biomass-size distribution (BSD), may indicate insufficient sampling of some trophic groups, but the succession of domes may also be related to predator-prey interactions (Boudreau *et al.*, 1991).

The national monitoring programs in Swedish lakes include biomass estimates of phytoplankton, zooplankton, benthic invertebrates and benthic fish (Appelberg et al., 1995). Comparable sampling methods were used in twelve different lakes during 1997 and 1998. These data were used to screen the informative value of expressing relative distributions of biomass on a joint scale of individual size. They were also used to test the null hypothesis of the ISELAW-program (integrated studies of the effects of liming acidified waters); namely that aquatic communities in limed lakes are not systematically different from communities in non-limed circumneutral lakes, with similar ranges of environmental characteristics. Two acidic lakes were included for a preliminary comparison.

2. Material and methods

The study lakes (Table I) were part of either the ISELAW program (six limed lakes, first liming 1977-1984) or the national program for environmental monitoring (two acidic and four circumneutral lakes), both including long term chemical and biological sampling with Swedish standard procedures (Appelberg et al., 1995; Holmgren, 1999).

TABLE I
Lake characteristics, given as group means (SD) and individual values. Water transparency (Transp.), surface water temperature (Temp.), pH, conductivity (Cond.), and total phosphorus concentration (Total-P) represent means of samples taken in July and August in 1997 and 1998.

	Altitude (m)	Area (km2)	Maximum depth (m)	Transp. (m)	Temp. (°C)	pH	Cond. (mS/m)	Total-P µg/L
Limed lakes (n = 6)	135 (85)	1.00 (1.04)	25.8 (11.1)	4.6 (2.9)	19.6 (0.6)	7.0 (0.2)	5.6 (2.6)	7.4 (4.3)
Gyltigesjön	66	0.40	21.6	1.2	18.9	7.0	7.6	15.5
Stora Härsjön	89	2.77	47.0	6.9	19.9	7.4	9.8	4.0
Stensjön Åva	35	0.38	21.1	4.9	20.5	7.2	5.0	5.5
Lien	156	1.77	28.4	4.4	19.7	6.9	4.5	5.0
Västra Skälsjön	233	0.42	19.2	8.6	19.3	6.9	3.0	5.3
Källsjön	232	0.25	17.2	1.7	19.2	6.9	3.5	9.0
Neutral lakes (n = 4)	120 (83)	0.54 (0.56)	20.3 (13.7)	4.0 (1.5)	19.7 (1.1)	6.7 (0.3)	5.7 (2.7)	8.2 (2.4)
Stora Skärsjön	60	0.31	11.5	4.3	19.9	7.0	8.2	8.3
Allgjuttern	126	0.19	40.7	6.0	20.4	6.8	5.1	4.8
Fräcksjön	58	0.28	14.5	3.3	20.4	6.6	7.3	9.8
Remmarsjön	234	1.37	14.4	2.3	18.1	6.4	2.1	10.0
Acidic lakes (n = 2)	170 (69)	0.96 (1.11)	20.7 (16.0)	3.8 (1.4)	19.3 (0.8)	5.9 (0.0)	4.3 (1.7)	11.6 (4.4)
Rotehogstjärnen	121	0.17	9.4	2.8	19.9	5.9	5.5	14.8
Övre Skärsjön	219	1.74	32.0	4.7	18.8	5.9	3.1	8.5

Epilimnetic biomass of phytoplankton was provided for monthly samples, given as mm^3/m^3 per counted taxon and size-class (mean individual biovolume, IB = 0.1 - 87162 μm^3). Zooplankton biomass was similarly given for species and developmental stages ($IB = 5 \cdot 10^{-5} - 3 \cdot 10^{-1}$ mm^3), and original data from two depth strata were pooled to represent mean biomass (mm^3/m^3) in the water column. All plankton species were assumed to have a density of 1 μm^3 =10^{-12} g.

Mean individual masses of benthic invertebrate taxa, were derived by dividing total fresh mass (g) by total number. Biomass estimates (g/m^2) represent mean values for samples from the sublittoral and profundal zones. Benthic fish were sampled with multimesh Nordic gillnets. All individuals were measured, individual mass was estimated by using species-specific length-mass relationships, and biomass was expressed per effort used (g/m^2 net area).

Biotic compartments were sampled on different spatial scales, but biomass within each category could be partitioned into log$_2$-size-classes on a joint scale of fresh mass (g). For illustrations of between-lake differences, mean values for two years were taken from phytoplankton and zooplankton sampled in July and August, from benthic invertebrates sampled in October, and from benthic fish sampled in July or August. Biomass within size-classes was re-scaled by division with the mean total biomass (for all lakes) for each group of organisms, in order to reveal semi-quantitative BSD's for each lake.

Multivariate analysis of variance (MANOVA) was used to test for general differences between limed, circumneutral and acidic lakes. A first set of dependent variables consisted of log$_{10}$-transformed total biomass estimates, within each of four organism groups. Next, rescaled biomass within all (54) observed log$_2$-size-classes were log$_{10}$(x+1)-transformed and used as a second set of dependent variables.

3. Results and discussion

Semi-quantitative BSD's revealed a considerable variation between lakes (Fig. 1). In one acidic lake (L. Rotehogstjärnen), biomass of phytoplankton and benthic invertebrates was several times higher than in any other lake. The phytoplankton community was totally dominated by *Gonyostomum semen* (in log$_2$-mass class = -27), as reported from other acidified humic lakes (Stenson et al., 1993). High proportions of large zooplankton (copepods or predatory rotifers in log$_2$-mass class = -12) were observed in one lake of each category, i.e. in L. Stora Härsjön, L. Remmarsjön and L. Rotehogstjärnen.

Some of the present variance was obviously attributed to specific taxonomic groups. The shape of BSD's was, however, relatively invariant across other lakes differing in taxonomic structure (Rodríguez and Magnan, 1993), reflecting the importance of size-dependent processes. Predators are usually larger than their prey (Cohen et al., 1993), and larger species have on average lower production to biomass ratios (Banse and Mosher, 1980). Based on such relationships, use of semi-quantitative BSD's is a logical way to compress and integrate information from aquatic communities sampled at different taxonomic and trophic levels.

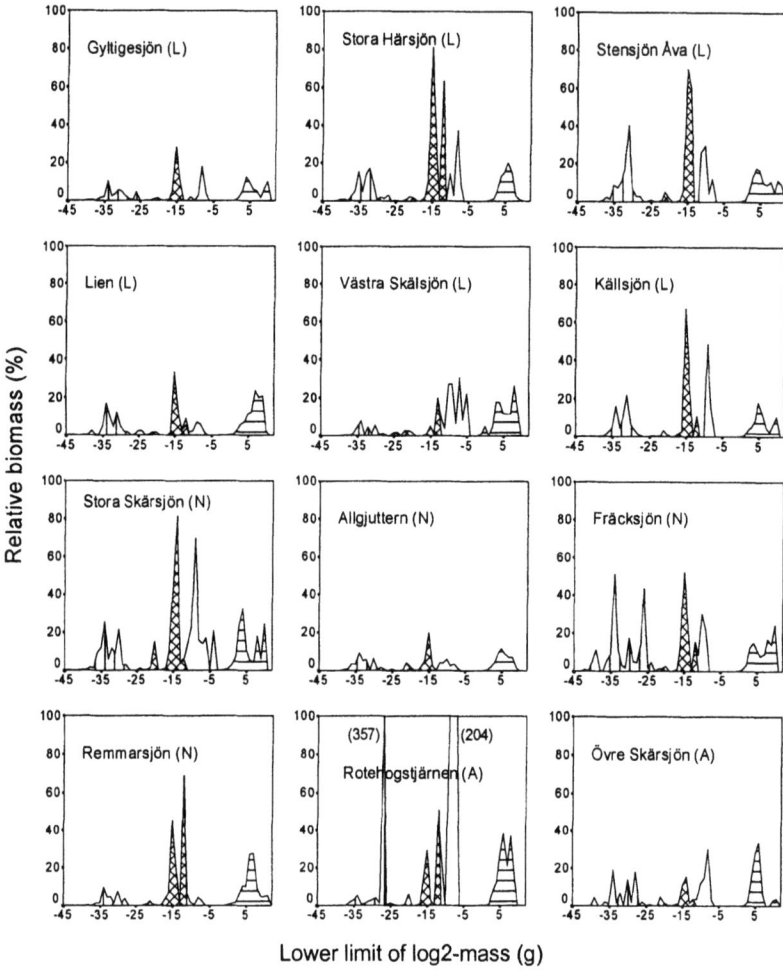

Figure 1. Relative biomass (%) in log_2-mass classes, within phytoplankton (vertically striped), zooplankton (crossed), benthic invertebrates (clear), and benthic fish (horizontally striped), respectively. Each lake is denoted by L = limed, N = circumneutral, or A = acidic. The highest peaks in Lake Rotehogstjärnen (values within parentheses) are truncated at 100%.

Mean biomass of limed and circumneutral lakes was more similar to each other than to means of two acidic lakes, both as mean total biomass within organism groups (Fig. 2) and as mean BSD (Fig. 3). However, neither of two MANOVA runs managed to detect any overall differences in biomass distributions between the three groups of lakes (Pillai's trace = 0.635 and P = 0.602 for organism groups, Pillai's trace = 1.684 and P = 0.484 for log_2-size-classes). The failure was partly expected due to a low sample size ($n = 2$) in the most variable group. In terms of species richness and diversity of plankton communities, limed lakes were also similar to circumneutral reference lakes (Appelberg *et al.*, 1995). Fish communities have also recovered after liming, but

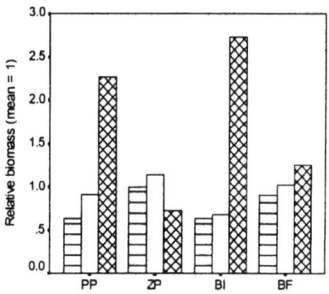
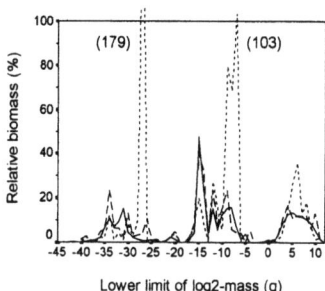

Figure 2. Relative biomass of phytoplankton (PP), zooplankton (ZP), benthic invertebrates (BI), and benthic fish (BF) in 6 limed (striped bars), 4 circumneutral (clear bars), and 2 acidic lakes (crossed bars). A mean of 1 is equal to 516 mm^3/m^3 (PP), 976 mm^3/m^3 (ZP), 2.65 g/m^2 (BI), and 16.7 g/m^2 gillnet area (BF).

Figure 3. Mean relative biomass distributions within 6 limed (solid line), 4 circumneutral (long dashed line), and 2 acidic lakes (short dashed line). The highest peaks in acidic lakes (values within parentheses) are truncated at 100%.

the species diversity 10-20 years after liming was still lower than in non-limed circumneutral lakes (Appelberg, 1998).

The BSD's of two acidic lakes in the present study were too dissimilar to make any general statements for communities in acidified lakes. It clearly indicates that acidification status alone can not be used for reliable prediction of BSD in a certain lake. Species richness and composition are influenced by additional factors, e.g. lake morphometry and productivity, that may also contribute to the variation in community size structure (Sprules *et al.*, 1983; Ahrens and Peters, 1991; Rasmussen, 1993; Cyr *et al.*, 1997a; 1997b).

4. Conclusions

To quantify the relative importance of small and large organisms within lakes, all biotic compartments have to be measured on the same spatial scale, but semi-quantitative BSD's can provide indications of functional differences between lakes. The preliminary result, *i.e.* similarity of mean BSD's of limed and comparable circumneutral lakes, is in accordance with the Swedish liming objective. A more powerful test would, however, require a higher number of lakes in each category and/or a less heterogeneous set of lakes in terms of other confounding characteristics of the lakes.

Acknowledgements

The ISELAW program and the national program for environmental monitoring were both financially supported by the Swedish Environmental Protection Agency (SEPA). Data were provided by SEPA's hosts of data, i.e. the Swedish

University of Agricultural Sciences, Institute of Environmental Assessment (water chemistry, phytoplankton, zooplankton, benthic invertebrates, www.ma.slu.se) and the National Board of Fisheries, Institute of Freshwater Research (fish, www.fiskeriverket.se). Further support for extended analyses were given by the EC Contract No. FAIR CT 96-1957.

References

Ahrens, M. A. and Peters, R. H.: 1991, *Canadian Journal of Fisheries and Aquatic Sciences* **48**, 1967.
Appelberg, M.: 1995a, 'Liming strategies and effects: the Lake Stora Härsjön case study', in L. Henrikson and Y.W. Brodin (eds.), *Liming of acidified surface waters,* Springer-Verlag, Berlin Heidelberg, pp. 339-351.
Appelberg, M.: 1995b, 'Liming strategies and effects: the Lake Gyslättasjön case study', in L. Henrikson and Y.W. Brodin (eds.), *Liming of acidified surface waters,* Springer-Verlag, Berlin Heidelberg, pp. 353.
Appelberg, M.: 1998, *Restoration Ecology* **6**, 343.
Appelberg, M., Lingdell, P. -E. and Andrén, C.: 1995, *Water, Air and Soil Pollution* **85**, 883.
Banse, K., and Mosher, S.: 1980, *Ecological Monograph* **50**, 355.
Boudreau, P. R., Dickie, L. M. and Kerr, S. R.: 1991, *Journal of Theoretical Biology* **152**, 329.
Brodin, Y. -W.: 1995, 'Acidification of lakes and water courses in a global perspective', in L. Henrikson, and Y.W. Brodin (eds.), *Liming of acidified surface waters,* Springer-Verlag, Berlin Heidelberg, pp. 45-62.
Cohen, J. E., S. L. Pimm, P. Yodzis and J. Saldaña.: 1993, *Ecology* 62: 67.
Cyr, H., Downing, J. A. and Peters, R. H.: 1997a, *Oikos* **79**, 333.
Cyr, H., Peters, R. H. and Downing, J. A.: 1997b, *Oikos* **80**, 139.
Duplisea, D. E. and Kerr, S. R.: 1995, *Journal of Theoretical Biology* **177**, 263.
Gaedke, U.: 1992, *Limnology and Oceanography* **37**, 1202.
Hanson, J. M., Prepas, E. E. and Mackay, W. C.: 1989, *Canadian Journal of Fisheries and Aquatic Sciences* **46**,1510.
Holmgren, K.: 1999, *Journal of Fish Biology* **55**, 535.
Kerr, S. R.: 1974, *Journal of the Fisheries Research Board of Canada* **31**, 1859.
Macphearson, E. and Gordoa, A.: 1996, *Marine Ecology Progress Series* **138**, 27.
Nyberg, P.: 1995, 'Liming strategies and effects: the Lake Västra Skälsjön case study', in L. Henrikson and Y. W. Brodin (eds.), *Liming of acidified surface waters*, Springer-Verlag, Berlin Heidelberg, pp.327-338.
Platt, T. and Denman, K.: 1978, *Rapports et Proces-verbaux des Réunions Conseil international pour L'Exploration de la Mer* **173**, 60.
Rasmussen, J. B.: 1993, *Canadian Journal of Fisheries and Aquatic Sciences* **50**, 2192.
Rodríguez, M. A. and Magnan, P.: 1993, *Canadian Journal of Fisheries and Aquatic Sciences* **50**, 800.
Sheldon, R. W., Prakash, A. and Sutcliffe, , W. R. Jr.: 1972, *Limnology and Oceanography* **17**, 327.
Sprules, W. G., Casselman, J. M. and Shuter, B. J.: 1983. *Canadian Journal of Fisheries and Aquatic Sciences* **40**, 1761.
Sprules, W. G. and L.B. Holtby, L. B.: 1979, *Journal of the Fisheries Research Board of Canada* **36**, 1354.
Stenson, J. A. E., Svensson, J.-E. and Cronberg, G.: 1993. *Ambio* **22**, 277.
Strayer, D.: 1986, *Oecologia (Berlin)* **69**, 513.

INFLUENCE OF ACIDIFICATION AND LIMING ON THE DISTRIBUTION OF TRACE ELEMENTS IN SURFACE WATERS

H. BORG*, J. EK, and K. HOLM

Institute of Applied Environmental Research (ITM), Stockholm University,
S-106 91 Stockholm, Sweden
(*author for correspondence, e-mail: hans.borg@itm.su.se)

Abstract. Metals in water have been monitored for up to 18 years in acidified regions of Sweden. The concentrations of metals (Al, Cd, Cu, Fe, Mn, Mo, Pb, Zn) were determined by AAS and ICP-MS, the dissolved fractions after separation by *in-situ* dialysis. Elements showing negative pH-correlation were primarily Al, Zn, Cd, Mn, and Pb, while Mo was positively correlated to pH, indicating a predominance of negatively charged ionic forms. Zn, Cd, and Mn occurred primarily in the dissolved fractions, especially at the lower pH levels. Fe, Al, Pb and Mn were further enriched in humic waters. During the study period, some of the sites were subject to lime treatment, which had a marked influence on most elements, causing the mean levels and the seasonal fluctuations to decrease. Treatment on the lake surface was less effective than wetland liming to reduce seasonal fluctuations, especially for metals mainly originating from the catchments, as Fe and Al.

Keywords: Acidification, metals, water, liming, Sweden

1. Introduction

The large-scale deposition of acidifying sulphur and nitrogen compounds has successively depleted the soils of base cations and buffer capacity. The decreased pH of the surficial ground water has also caused increased leaching of toxic metals such as labile inorganic Al, Zn, and Cd (Borg and Johansson 1989; Bergkvist 1987). As a consequence, elevated concentrations of metals were measured in lakes and streams in acidified areas (Dickson 1980; Borg 1983; Yan et al., 1990; Nelson and Campbell, 1991). The lower pH also promotes a dominance of dissolved, uncomplexed metal forms (Borg and Andersson 1984). These forms are potentially more bioavailable than particle bound forms, but a lower uptake of e.g. Cd in biota have also been found at low pH (Campbell and Stokes 1985; Yan et al., 1990; Lithner et al., 1995), probably caused by competition with H^+ ions (Campbell and Stokes 1985; Yan et al., 1990, Hare and Tessier 1998). As a remediation, a large-scale liming program was initiated in Sweden during the 70´s. Until now, about 7 500 lakes and 12 000 km watercourses have been treated (Svenson et al., 1995). This paper summarises the results of several studies performed in order to define the long-term change in distribution of trace metals during acidification and lime treatment of lakes and water systems.

2. Methods

Three different areas have been used; province of Kronoberg (Borg et al., 1989), province of Härjedalen (Borg et al., 1995), and province of Gävleborg (Andersson et al., 1989, Andrén et al., this volume). All studied lakes are small, oligotrophic

Water, Air, and Soil Pollution **130**: 1757–1762, 2001.
© 2001 *Kluwer Academic Publishers.*

forest lakes with catchments dominated by coniferous forests with some peatlands, and poorly weathered granitic bedrock. Most of the water systems were subject to lime treatment with doses following the recommendations from the Swedish Environmental Protection Agency. The lime was applied both directly on the lake surface, on selected wetlands, and by automatic lime dosers upstream.

Samples for water chemistry were generally taken 6-10 times annually. For determination of trace metals, the samples were collected directly into acid washed polypropylene bottles, kept in plastic bags. Filterable fractions were separated by pressure filtration through polycarbonate filters (0.4 µm). The truly dissolved metal fractions were collected by *in-situ* dialysis (pore-size 2 nm). Samples for trace metals were preserved with quarts distilled nitric acid.

Analysis of water chemistry was performed using Swedish standard methods. Trace metals were determined by replicate injection atomic absorption spectrometry with deuterium lamp or Zeeman (from 1985-) background correction. Since 1994, metals were directly determined by ICP-mass spectrometry (Sciex Elan 5000). All analyses followed quality control routines, defined in the accreditation of the laboratory. Sample handling and analysis of trace metals were performed in clean room laboratories.

2. Results and discussion

Elements showing increased concentrations at lower pH-levels were primarily Zn, Cd, Mn, Al, and Pb, in agreement with earlier lake surveys (Dickson 1980, Borg 1983) (Pb,Mn,Cd exemplified in Fig.1). The long-term acid induced leaching of metals from the soils may result in hazardous levels of e.g. Al, Zn and Cd in numerous lakes in acidified regions (Swedish EPA 1999). The dissolved metal fractions dominated strongly for Cd and Mn (Fig.1), as well as for Zn, and increased further in the more acidic waters, probably indicating an increased bioavailability. Pb and Fe occurred mostly in the particulate fraction, and similar to Al and Mn, they were further elevated in humic waters. Increased total concentrations of Pb and Mn were e.g. found in the most humic of the lakes in Fig.1 (pH 5.45). Fe and Pb occurred probably adsorbed to particulate and colloidal organic matter, as well as in dissolved metal-organic complexes.

In contrast to the other measured elements, molybdenum increased at higher pH (Fig 2). This is probably a result of the predominance of negatively charged molybdate ions. Positive correlations with pH for elements occurring as anions, such as Mo, V, As, and Se, have been reported in other studies (Lithner *et al.*, 2000).

A decrease of most metals after liming would be expected from a chemical point of view, and that was also what happened in the treated waters. Both annual mean values and the formerly large seasonal variations decreased after the treatment had started. Elements generally most affected by the liming were Cd and Zn (Table 1, Fig.3), but also metals of terrogenic origin, e.g. Al, and Fe, and particle bound elements such as Pb, decreased after treatment (Fig. 3-4, Table 1).

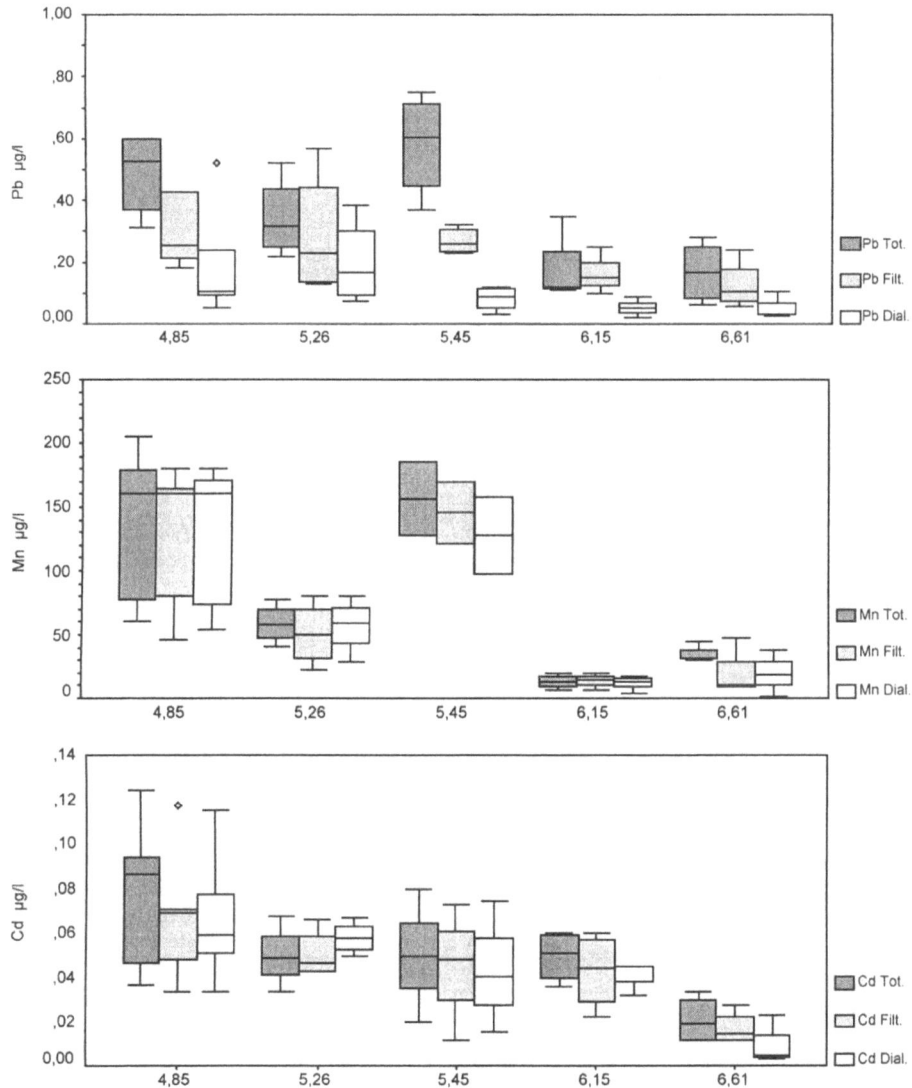

Figure 1. Total, filterable, and dialyzable metal fractions in five lakes of different 3 yr mean pH, in the prov. of Kronoberg, (25-, 50-, 75 percentiles, min.-, max. metal concentrations).

The higher pH-values obtained by liming probably favour the formation of Fe-oxide hydroxides, which leads to co-precipitation of other elements. The higher pH also contributes to an increased complex formation of metals with humic substances. These complexes are partly colloidal or particulate, and form aggregates with Fe oxide hydroxides, thus decreasing the dissolved, unbound metal fractions in the water column. Application of lime on the lake surface exclusively,

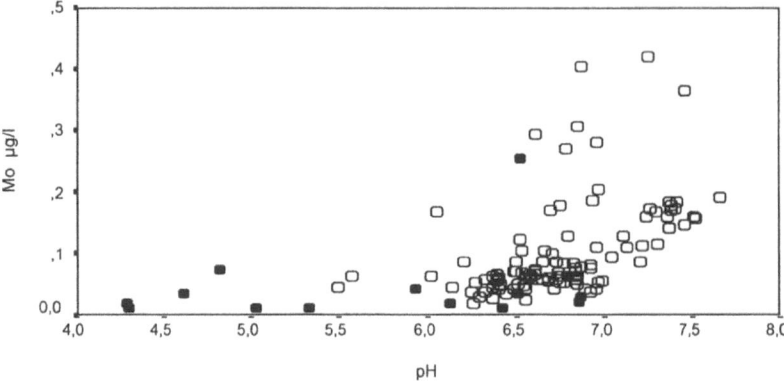

Figure 2. Molybdenum concentrations vs pH in eight streams in Lofsdalen, Sweden. Open symbols; lime treated, filled symbols; reference waters.

will not inhibit the release of metals from the acidified catchments. For example, the treatment of L. Gyslättasjön, that started with lake-liming in 1985, gave only a minor decrease in Al (Fig.3) and Fe. In 1988, a four times larger dose was applied and from 1991 and on, wetlands were also treated, causing a decreased leaching of these metals from the catchment. Similarly, a drastic decrease of Fe, Al, and Mn in mountain brooks was registered after liming on selected wetlands (areas with groundwater outflow) in several catchments in Lofsdalen, central Sweden (Borg et al 1995) exemplified in Fig. 4 (Table I).

TABLE I
Decrease of metal concentrations at lime treatments. Mean values, µg/l (standard deviation). Study period generally 5-15 years.

Study area		Before liming	After liming	% decrease	Sign
L.Gyslättasjön, prov of Kronoberg: lake liming/wetland liming	Al	204(60)	161(48)	21	p<0.001
	Cd	0.072(0.037)	0.021(0.014)	71	p<0.0001
	Fe	513(203)	386(208)	25	p<0.002
	Mn	135(90)	57(47)	58	p<0.0001
	Zn	12(3)	5(5)	58	p<0.0001
L. Örvallsjön catchment, prov of Hälsingland: lake liming/lime doser	Al	375(123)	254(79)	32	p<0.0001
	Cd	0.030(0.025)	0.017(0.009)	43	p<0.0001
	Fe	1111(517)	657(259)	41	p<0.0001
	Pb	0.48(0.21)	0.36(0.18)	25	p<0.0001
	Zn	5.6(2.4)	3.9(1.8)	30	p<0.0001
Eight catchments in Lofsdalen, prov of Härjedalen: wetland liming	Al	122(138)	70(49)	42	p<0.0001
	Cd	0.022(0.071)	0.010(0.017)	55	p<0.007
	Fe	2451(8497)	756(3123)	69	p<0.001
	Mn	221(1199)	80(184)	63	p<0.015
	Pb	0.35(0.63)	0.19(0.34)	46	p<0.001
	Zn	3.8(8.0)	2.2(2.1)	42	p<0.002

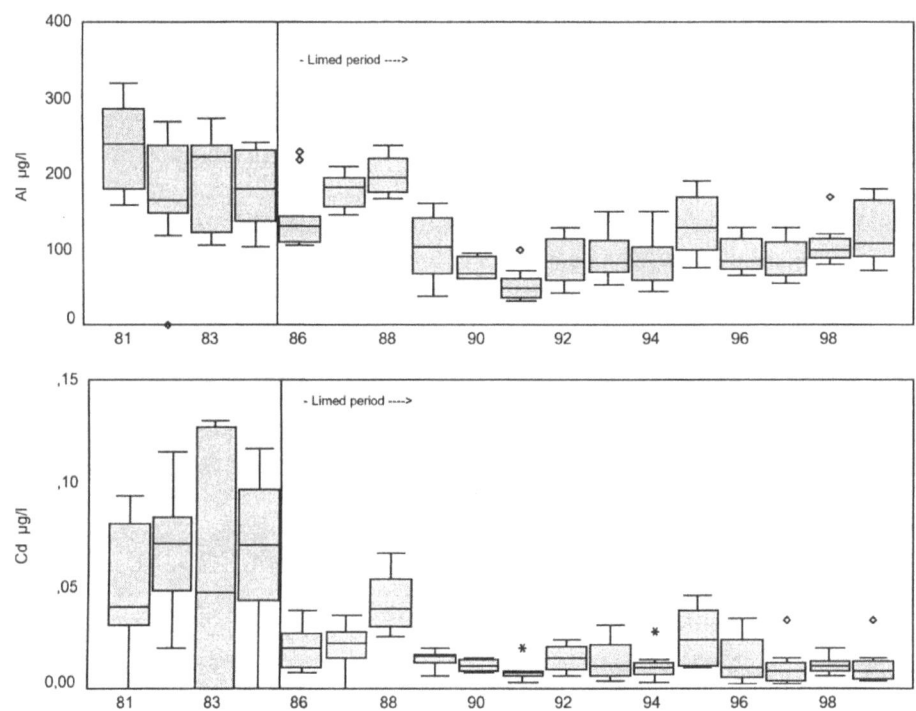

Figure 3. Metal concentrations before and after liming in L. Gyslättasjön, province of Kronoberg, Southern Sweden (annual 25-, 50-, and 75-percentiles, max and min., no sampling in 1985).

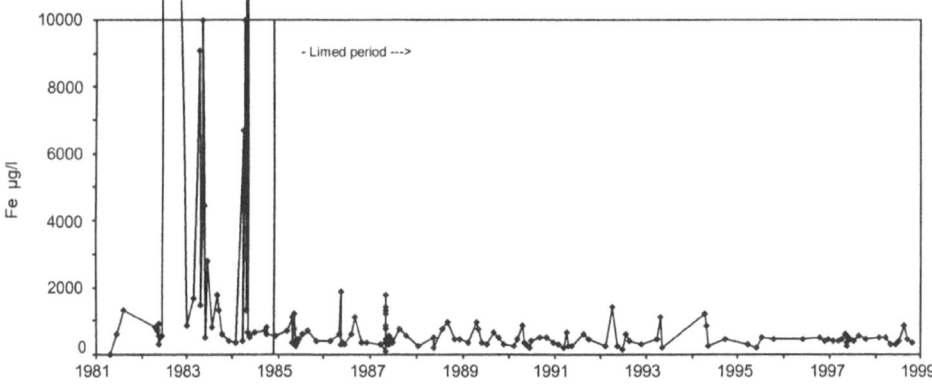

Figure 4. Concentration of iron before and after wetland liming of the River Djursvasslan catchment, Lofsdalen, Central Sweden (site 5004).

The decreased concentrations and dissolved portions of metals should provide an improved water quality for aquatic biota. If lime treatments manage to keep the pH-values neutral with low variations, toxicologically acceptable levels of metals may be reached, including Cd, an element occurring to a high degree in dissolved forms even at pH-values around seven.

4. Conclusions

As a secondary effect of acidic deposition, the concentration levels of many metals have been elevated in waters in the influenced regions. This is partly a result of the fact that the acidic conditions favoured a shift towards a larger portion of dissolved, metal forms, causing a decreased particle binding and sedimentation of metals. In contrast to the other measured elements, molybdenum decreased at lower pH, probably because of occurrence in anionic form.

Lime treatment was shown to markedly reduce the concentrations of the elements included in these investigations. A decrease of generally 30-70% was measured in different treated waters. Application of the lime also on wetlands effectively reduces the release of metals from the soils of the catchments, including elements of terrogenic origin, such as Fe, Al and Mn. The peak levels of metals registered during acid episodes in streams, e.g. during spring flood, was substantially reduced.

Acknowledgements

We thank Elisabeth Fröberg, for water chemical analyses, Paul Andersson and Erik Olofsson for planning and organization of liming operations and water sampling, and the Swedish EPA for financial support.

References

Andersson, P., Borg H., Olsson, B., Nilsson, Å., and Håkanson, L.: 1989, Bakrundstillstånd och genomförda åtgärder i PU-labs sjöar- Proj.kalkning-kvicksilver-cesium. Swedish EPA, report 3608 (in Swedish).
Andrén, C., Andersson, P., and Fröberg, E.: 2000, *Water, Air, and Soil Pollut.* (this volume).
Bergkvist, B.: 1987, *Water, Air, and Soil Pollut.* 33, 131.
Borg, H.: 1983, *Hydrobiologia*, **101**, 27.
Borg, H, and Andersson, P.: 1984, *Verh. Internat. Verein. Limnol.* 22, 725.
Borg, H., Andersson, P. and Johansson, K.: 1989, *Sci. Total Environ.* 87/88, 241.
Borg, H., and Johansson, K.: 1989, *Water, Air and Soil Pollut.* 47, 427.
Borg, H., Andersson, P., Nyberg, P. and Olofsson, E.: 1995, *Water, Air and Soil Pollut* 85, 907.
Campbell, P.G.C., and Stokes, P.M.: 1985, *Can. J. Fish. Aquat. Sci.* 42, 2034.
Hare, L. and Tessier, A.:1998, *Limnol. Oceanogr.* 43, 1850.
Lithner G., Holm K., and Borg H.: 1995, *Water Air and Soil Pollut.* **85**, 785.
Lithner, G., Borg, H., Ek, J., Fröberg, J., Holm, K., Johansson, A.M., Kärrhage, P., Rosen, G., and Söderström, M.: 2000, *Ambio* **29**, 217.
Nelson, W.O. and Campbell, P.G.C.:1991, *Environ. Pollut.* 71, 91.
Svenson, T., Dickson, W., Hellberg, J., Moberg, G. and Munthe, N.: 1995, *Water, Air and Soil Pollut.* 85, 1003.
Swedish Environmental Protection Agency: 1999, *Guidelines for Environmental Quality- Freshwater.* Report 4913 (in Swedish).
Yan, N.D., Mackie, G.L. and Dillon, P.J.: 1990, *Environ. Sci. Technol.* 24, 1367.

REACIDIFICATION EFFECTS ON WATER CHEMISTRY AND PLANKTON IN A LIMED LAKE IN SWEDEN

FRIDA EDBERG[1*], PAUL ANDERSSON[2], HANS BORG[1], CHRISTINA EKSTRÖM[3] and EINAR HÖRNSTRÖM[1]

[1] *Institute of Applied Environmental Research (ITM), Stockholm University, S-106 91 Stockholm, Sweden;* [2] *F:a SBV-analys, Ö. Tolbo 4662, S-820 60 Delsbo, Sweden;* [3] *Ekströms hydrobiologi, N.Mälarstrand 82, S-112 35 Stockholm, Sweden*
(*author for correspondence, e-mail: frida.edberg@itm.su.se)

Abstract. Water chemistry and plankton has been monitored in three Lakes in Tyresta National park SE of Stockholm since 1977. Liming operations started in Lake Långsjön and Lake Trehörningen in 1978 and were repeated every 3-5 years, while Lake Årsjön is an unlimed reference Lake. During 1991-1999, the annual pH median in Lake Långsjön and Lake Årsjön ranged between 6.6-7.1 and 5.2-5.8, respectively, and the composition of phyto- and zooplankton in these lakes did not change markedly. After a final treatment in 1991, the liming of Lake Trehörningen was terminated intentionally. As a result, pH decreased from an annual median 7.1 in 1991, to 6.1 in 1999 (5.8 in 1998). Total organic carbon (TOC) did not change markedly during this period, while the levels of calcium decreased. Metals, known to be influenced by acidification, especially cadmium, manganese and aluminium (Al), increased. The labile-inorganic forms of Al also reached higher levels, especially in 1998. Following the decreasing pH, the total number of phytoplankton taxa decreased by ca 40%. Among zooplankton, the cladocerans *Holopedium gibberum, Diaphanosoma brachyurum* and *Daphnia longispina*, common during the limed period, became rare.

Key words: acidification, liming, metals, phytoplankton, reacidification, zooplankton

1. Introduction

In Sweden the acidification has been successfully mitigated by large scale liming in more than 7 500 lakes and 11 000 km of running water since the 1980s (Svenson *et al.*, 1995). The deposition of acidifying sulphur has been reduced during the last decades (Ferm and Hultberg, 1998), which will cause reductions in the present Swedish liming programme. Terminated liming in several lakes, however, means a risk of reacidification in some lakes, with largely unknown effects.

Since only a few studies of reacidification effects in lakes have been made (Wright, 1985; Alenäs *et al.*, 1991; Keller *et al.*, 1992; Dickson *et al.*, 1995), and lake characteristics are highly varying, we have followed water chemistry and plankton in a small forest lake 8 years after terminated liming. An unlimed reference lake and a lake where the liming continued were also studied in the same lake system.

2. Material and methods

2.1 STUDY AREA AND SAMPLING

The lakes are situated in Tyresta National park, about 20 km south east of Stockholm. They are small forest lakes (Table I) with catchments of poorly weathered bedrock, dominated by gneissic-granite. The lakes have been monitored since 1977, one year before the liming started in the area. The treatments in Lake Trehörningen and Lake Långsjön were repeated with 4-5 years intervals, while Lake Årsjön was kept unlimed as a reference. In Lake Trehörningen the last liming was made in 1991 while liming continued in Lake Långsjön. In 1999 reacidification studies in these lakes were included in the national monitoring programme Integrated Studies of the Effects of Liming Acidified Waters (ISELAW) (Appelberg *et al.*, 1995).

TABLE I
Characteristics of Lakes Trehörningen, Långsjön and Årsjön.

Lakes	Trehörningen	Långsjön	Årsjön
Lake area (km^2)	0.04	0.10	0.16
Max depht (m)	4.4	8.0	8.4
Mean depth (m)	2.4	4.0	3.7
Turnover time (years)	0.42	0.8	1.4
Catchment (km^2)	0.75	1.7	1.4
% forested	88	93	94
% wetlands	7	5	5

Water sampling was made 4 or 8 times each year throughout the year and carried out just below surface (0.5-1.0 m). The sampling of both phyto- and zooplankton was made during May-September in the middle of the lakes, phytoplankton at a depth of 0.5-1.0 metres, and zooplankton with nets (mesh sizes 75 and 25 µm) hauled from near bottom to the surface. All samples were preserved with Lugol's solution.

2.2 ANALYTICAL METHODS

pH, colour (mg Pt L^{-1}), alkalinity, calcium (Ca) and total organic carbon (TOC) were determined with Swedish standard methods. Iron (Fe), manganese (Mn), copper (Cu), lead (Pb), cadmium (Cd) and zinc (Zn) were analysed by ICP-MS, total Al (acid soluble) was analysed by colourimetric reaction of pyrocatechol violet (SS028210, mod 1992) and the labile (inorganic) Al fraction according to Andrén (1995). All the analytical determinations were performed at our accredited laboratory by standardized and well-documented methods with continuous monitoring of the analytical quality. The measuring uncertainties (estimated as the relative standard deviation of continuously analysed quality control samples) are below 10 %, except for low-level determinations of the aluminium fractions (< 25 µg L^{-1}), which have an uncertainty of 25%.

50 ml of the phytoplankton samples was analysed under an inverted microscope, where all species were identified and the total number of taxa recorded. Phytoplankton volumes were determined according to Willén (1974). The zooplankton samples were analysed under stereomicroscope where taxon or species were identified.

Linear regression was performed on the data and an analysis of variance (one sided ANOVA) was made on the slope to determine if the concentrations increased or decreased significantly on the 95% level.

3. Results and Discussion

3.1 LIMING EFFECTS IN LAKE TREHÖRNINGEN AND LAKE LÅNGSJÖN

In 1977, before the liming started in the area, the annual median pH in the lakes ranged between 4.7-5.2, and the alkalinity was below detection limit. As a result of the treatment in Lakes Trehörningen and Långsjön in 1978, pH and alkalinity reached the recommended Swedish liming goals (pH>6.0 and alkalinity >0.1 mekv L^{-1}) and with few exceptions this has been maintained during the limed period. Metals known to be mobilized by acidification, e.g. Cd decreased as a result of the liming treatments (Andersson and Borg, 1988).

Before liming in 1978, only 12-29 phytoplankton taxa were recorded in Lakes Trehörningen and Långsjön, and 20-35 in the reference Lake Årsjön. The number of taxa in Lake Trehörningen gradually increased to 40-50 after the treatment and the development was similar in Lake Långsjön. Among zooplankton, the number of rotatorian species increased in all the lakes during the limed period 1978-1989. This change was probably caused by increased primary production, following increased transport of nutrients. The number of cladoceran species also increased which is a normal development some years after liming (Hörnström et al., 1993).

3.2 REACIDIFICATION EFFECTS IN LAKE TREHÖRNINGEN

As a result of the terminated liming in Lake Trehörningen, pH decreased significantly from annual median 7.1 in 1991 to 6.1 in 1999 (Fig. 1). Alkalinity and Ca concentrations also decreased significantly from annual median of 0.42 mekv L^{-1} in 1991 to 0.03 mekv L^{-1} in 1999, and 0.62 mg L^{-1} to 0.15 mg L^{-1}, respectively. The levels of TOC and colour remained relatively constant (Table II). The levels of total Al (Fig. 2) and Mn increased significantly (p<0.05), especially after 1995 when the annual median pH reached 6.0. Zn, Cd and labile Al (Fig. 2) reached their peak values during 1998 when also the lowest annual median pH was recorded. The levels of Cu and Pb did not change significantly during the period.

In Lake Långsjön and Lake Årsjön the metal concentrations showed no significant increase during 1991-1999.

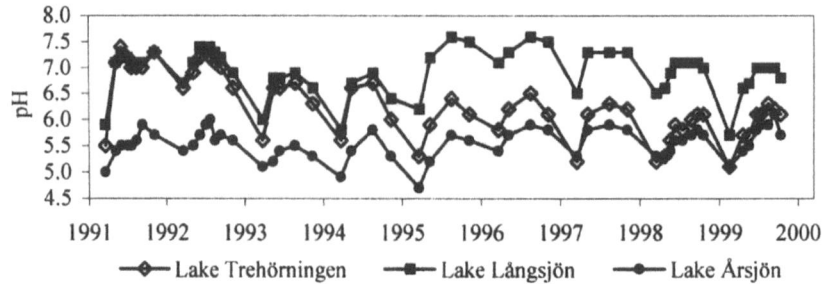

Figure 1. Variation of pH in Lakes Trehörningen, Långsjön and Årsjön, 1991-1999.

TABLE II
Annual medians of colour (mg Pt L^{-1}), TOC (mg L^{-1}), Mn, Fe, Cu, Zn, Cd, Pb, total Al (Al-tot) and labile Al (Al-labile) (µg L^{-1}) in the reacidified Lake Trehörningen.

	1991	1992	1993	1994	1995	1996	1997	1998	1999
Colour	140	85	70	100	70	65	75	70	90
TOC	18.5	16	14	15.5	15	14	15.5	14	14.5
Mn	43.5	18.5	23.3	47.6	34.1	31.9	56.5	48.5	69.6
Fe	425	260	185	374	244	230	345	344	522
Zn	8.7	7.4	10.8	12.8	12.8	12.8	11.1	13.9	11.6
Cd	0.02	0.038	0.031	0.038	0.042	0.042	0.049	0.07	0.058
Al-tot	205	170	210	300	300	220	310	355	355
Al-labile	.	.	8.3	35	58	33	29	71	43

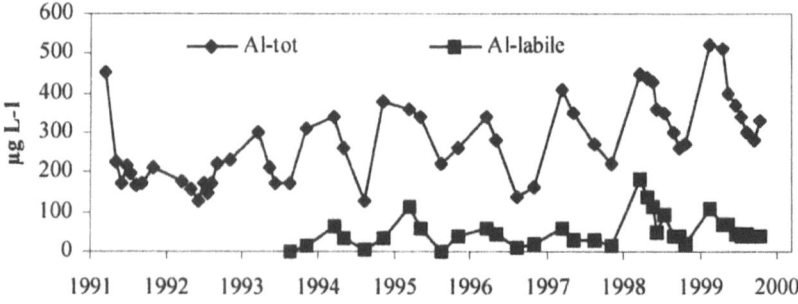

Figure 2. Concentrations of total Al (Al-tot) and labile Al (Al-labile) in µg L^{-1} in Lake Trehörningen, 1991-1999.

After the ceased liming activity in Lake Trehörningen, the number of phytoplankton taxa varied between 30 and 61, with a maxima in 1993-1994. In 1996-1998, during decreasing pH, only 30-36 taxa were recorded (Fig. 3). Several acid-sensitive species, e.g. *Scenedesmus spp* (Hörnström, 1999), were missing in Lake Trehörningen, while they still occurred in Lake Långsjön. In the acidic Lake Årsjön, the number of taxa was not markedly changed (Fig. 3).

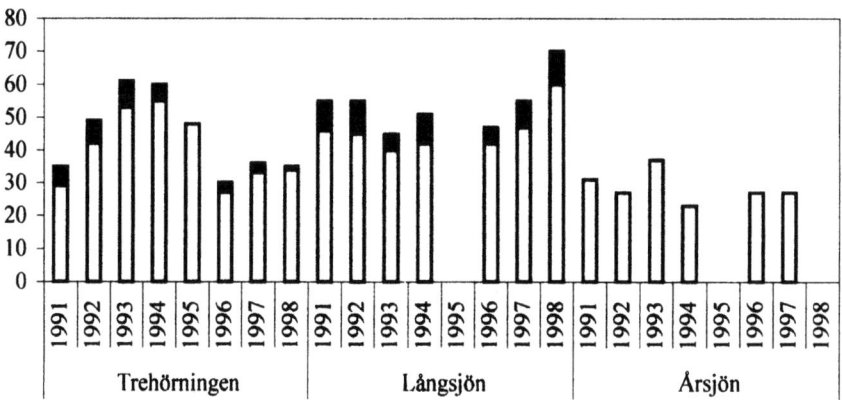

Figure 3. Total number of phytoplankton taxa in Lakes Trehörningen, Långsjön and Årsjön. Black areas represent acid sensitive taxa.

Among zooplankton, the abundance of rotatorians increased markedly in Lake Trehörningen, in contrast to that in the other lakes. The total number of Rotatoria species changed little, except for a significant decrease in Lake Årsjön.

The mean number of cladoceran species in Lake Trehörningen decreased significantly ($p<0,001$) from 1990 to 1999, in contrast to Lake Långsjön and Lake Årsjön (Fig. 4). In 1990-1991 *Daphnia longispina* and *Holopedium gibberum*, not present before liming, were dominating, together with *Diaphanosoma brachyurum* and *Ceriodaphnia quadrangula*. From 1993, there were no observations of *Ceriodaphnia* and the frequencies of *Holopedium, Daphnia* and *Diaphanosoma* strongly decreased, and *Bosmina* became rare. Especially *Bosmina* is not sensitive to low pH (Degerman *et al.*, 1995; Dickson *et al.*, 1995; Andersson and Hultberg, 1997), why their marked decrease in Lake Trehörningen must have other reasons than acidification, such as changed predatory conditions and food supply. In the competition of food, predation on cladocerans may favour the occurrence of rotifers (Nyberg, 1998).

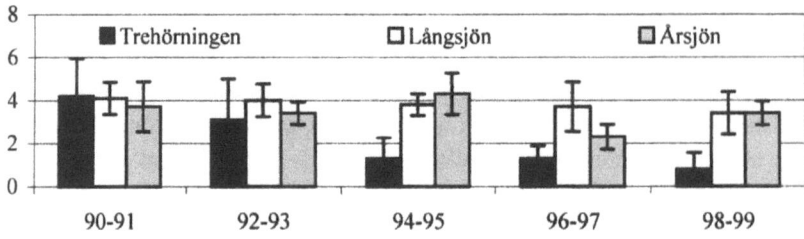

Figure 4. Mean number of cladoceran species in Lakes Trehörningen, Långsjön and Årsjön.

Conclusions

pH has decreased with approximately one unit during the 8 years following the termination of liming in Lake Trehörningen. As a result, the levels of Cd, Mn and Al (total Al and labile Al) has increased. The total number of phytoplankton taxa was reduced by the elimination of acid sensitive species, whereas the marked decrease of zooplankton cladocerans could not be explained by the increased acidity.

References

Alenäs, I., Andersson, B.I., Hultberg, H. and. Rosemarin, A.: 1991, *Water, Air, and Soil Pollution* **59**, 55.
Andersson, B. I. and Hultberg, H.: 1997, Swedish environmental research insitute, Report B1250.
Andersson, P. and Borg, H.: 1988, *Canadian Journal of Fisheries and Aquatic Sciences* **45**, 1154.
Andrén, C.: 1995, *Water, Air and Soil Pollution* **85**, 811.
Appelberg, M., Lingdell, P-E. and Andrén, C.: 1995, *Water, Air and Soil Pollution* **85**: 883
Degerman, E., Henrikson, L., Herrmann, J. and Nyberg, P.: 1995, 'The effects of liming on aquatic fauna', in Henrikson & Brodin (eds.), *Liming of acidified surface waters*, Springer - Verlag Berlin-Heidelberg. pp. 221- 266.
Dickson, W., Borg, H., Ekström, C., Hörnström, E. and Grönlund, T.: 1995, *Water, Air and Soil Pollution* **85**, 919.
Ferm, M. and Hultberg, H.: 1998, 'Atmospheric deposition to the Gårdsjön Research area', in H. Hultberg, and R.A. Skeffington, (eds.) *Experimental Reversal of Acid Rain Effects: The Gårdsjön Roof Project*, John Wiley & Sons Ltd. pp. 71-84.
Hörnström, E., Ekström, C., Fröberg, E., and Ek, J.: 1993, *Canadian Journal of Fisheries and Aquatic Sciences* **50**, 688.
Hörnström, E.: 1999. Doctor of philosophy Dissertation, University of Stockholm, Sweden.
Keller, W., Yan, N.D., Howell, T., Molot, L. and Taylor, W.D.: 1992, *Canadian Journal of Fisheries and Aquatic Sciences* **49** (suppl 1), 52
Nyberg, P.: 1998, *Water, Air and Soil Pollution* **101**, 257.
Svenson, T., Dickson, W., Hellberg, J., Moberg, G., Munthe, N.: 1995, *Water, Air and Soil Pollution* **85**, 1003.
Willén, E.: 1974. Methods of phytoplankton investigations. Nat. Swed. Environ. Protect. Board. PM 525, NLU Rep. 76, 45 p. (In Swedish with English summary).
Wright, R.F.: 1985, *Canadian Journal of Fisheries and Aquatic Sciences* **42**, 1103.

EVIDENCE OF LOWER PRODUCTIVITY IN LONG TERM LIMED LAKES AS COMPARED TO UNLIMED LAKES OF SIMILAR pH

GUNNAR PERSSON[1*] and MAGNUS APPELBERG[2]

[1]*Department of Environmental Assessment, Swedish University of Agricultural Sciences, P.O.Box 7050, SE-750 07 Uppsala, Sweden;* [2]*Institute of Freshwater Research, SE-178 93 Drottningholm, Sweden*
(* *author for correspondence, e-mail: Gunnar.Persson@ma.slu.se*)

Abstract. Ecosystem development in lime-treated waters in Sweden has been followed since 1989 in a programme for integrated studies of the effects of liming acidified waters (ISELAW). Observations after prolonged liming (>10 y) indicate a phosphorus depletion in the limed lakes which contrasts to the increased phosphorus supply often following within the initial years after lime treatment. After prolonged liming, the levels of total phosphorus are lower as compared to neutral reference lakes at identical TOC, and the phosphorus/TOC –ratio is consequently lower in limed lakes. Depletion of dissolved inorganic nitrogen during the summer is also lower in limed as compared to neutral reference lakes. Phytoplankton biomass and species number also lower in the limed lakes as compared to unlimed neutral references. Furthermore the bacterial number per unit TOC is lower in the long term limed lakes, possibly as a result of phosphorus limitation. As to the higher trophic levels, the benthic soft-bottom fauna of limed lakes (specifically the sublittoral fauna) is poorer in terms of species diversity and abundance. Also fish community composition indicates lower productivity in the limed lakes. Taken together there is thus evidence that the long term limed lakes have a lower trophic level than reference lakes.

Keywords: liming, productivity, lakes, oligotrophication, phosphorus, nitrogen

1. Introduction

It has been estimated that ca 14000 (ca 15%) Swedish lakes have a pH<6. Half of them are considered to suffer from anthropogenic acidification and are subject to liming activities in a large scale national programme started in 1976. The effects of Swedish liming were studied within a number of research programmes during the last two decades (review see: Henrikson and Brodin, 1995). These programmes generally confirmed that liming had the desired positive effects on water chemistry (pH>6.0, alkalinity >0.10 mekv/l) and biota. However, long term changes among the different communities and at an ecosystem level still need study. Such aspects are now the targets of a programme for integrated studies of the effects of liming in acidified waters (ISELAW) initiated by the Swedish Environmental Protection Agency in 1989 (Appelberg and Aldén, 1992; Appelberg and Svenson, 2000). The programme comprises 14 lakes limed for on average 18 years. The lakes are basically meso- to ultraoligotrophic and the productivity is generally regulated by phosphorus (additional Internet information: http://www.ma.slu.se).

There is a general impression that lakes undergoing acidification also suffer from oligotrophication (Grahn *et al.*, 1974; Jansson *et al.*, 1986; Appelberg *et al.*, 1993) although this is not generally accepted (Schindler, 1988; Olsson and Pettersson, 1993). Upon liming, nutrients accumulated in detritus, algae and macrophytes during the acid period may be set free and increase nutrient availability at least during a short time. Phosphorus is also introduced through commercial lime products (Broberg, 1988). A study of 63 Swedish lakes revealed that the phosphorus level increased over the short term in ca 50% of the lakes but decreased in the rest (Broberg, 1987). A similar result was seen for nitrogen.

None of these studies addressed long term changes, however, and we here approach this question based on the complete ecosystem effects after prolonged liming and a more far reaching equilibration between watershed nutrient input and lake status, and the successive changes of the biological communities.

2. Material and Methods

Regular chemical and biological monitoring of the limed lakes started in August 1989. Samplings for water chemistry were made monthly April– Oct. and in Feb./Mar. at a central station in each lake. Results from 0.5 and 5 m depth were used. Analyses of ionic balance, organic material and nutrients followed standard procedures described by Wilander (1997). Samples for quantitative and taxonomic evaluation of the phytoplankton communities were taken on the same occasions (except for the winter sample) as a composite sample from the surface down to ca 1 m above the thermocline. Samples were preserved, brought to the laboratory, enumerated and biovolumes calculated (Utermöhl, 1956; Willén, 1976)

Both water chemistry and phytoplankton biovolumes were assessed by identical methods over the 5-year period 1989–93 in i)17 circumneutral reference lakes and ii) in 7 acid lakes spread all over Sweden additional Internet information: http://www. ma.slu.se). The former lakes met the criteria of having mean pH 6.0–7.0 over the period and the latter of having mean pH< 5.5 with one extreme of 4.4. The data for these references were compared to the corresponding data from the limed lakes for the period 1990–93.

For phytoplankton taxonomy different time periods were used for the comparision (1997-98 for ISELAW- and 1992-93 for reference lakes) in order to be able to use results from one laboratory.

Enumerations of aquatic bacteria (Bergström 1999), were made during one year (1998) in the limed lakes, which had to be compared to data from other reference lakes than the above cited two groups of lakes.

3. Results and Discussion

3.1. ACIDITY AND ORGANIC MATTER IN THE LAKE GROUPS

Means of chemistry variables for the group of limed lakes, the group of circumneutral untreated lakes and acid lakes showed the expected differences in acidity variables (Table I). The limed lakes had significantly higher pH and alkalinity than the acid lakes (Kruskal-Wallis test). On the other hand, the mean pH of the limed lakes was almost similar to the neutral lakes and the alkalinity mean was insignificantly higher in the limed lakes as compared to the neutral lakes. These conditions offer opportunities to compare these two groups to find out if they are also similar in other respects. The comparisions of means for TOC, Secchi depth, and organic nitrogen (DON) showed small differences which were insignificant. When DON/TOC-quotients were calculated it was evident that the neutral lakes had an organic matter slightly richer in nitrogen than the other groups.

TABLE I

Means of selected chemical and biological variables for limed, acid and circumneutral reference lakes. Reference lakes were grouped in neutral (pH6.0–7.0; 17 lakes) and acid lakes (pH 4.6–5.5; 7 lakes). Quotients between ISELAW-means and means for the two groups of reference lakes are given to the right. Means April–Nov. 1990–93 in ISELAW-lakes and 1989–93 in acid and neutral references were used. Kruskal-Wallis rank test was used to test for differences between groups and significance denoted by ** ($p<0.05$) and * ($p<0.10$)

	ISELAW	Concentrations		Quotients	
		Neutral refs	Acid refs	ISELAW/neutral	ISELAW/acid
pH	6.70	6.53	5.23	1.03	1.25**
Alk.or Acid (mekv/l)	0.143	0.095	0.003	1.51	47.7**
TOC (mg/l)	8.04	6.82	8.41	1.18	0.96
Secchi depth (m)	4.9	4.3	4.2	1.14	1.17
Org-N (µg N/l)	281	306	319	0.92	0.88
Tot-P (µg P/l)	10.7	11.7	12.1	0.91	0.88
Tot-P/TOC	1.44	1.92	1.72	0.75**	0.84
DIN (µg N/l)	92	45	113	2.04**	0.81
DIN Feb-March(µg N/l)	135	119	188	1.14	0.72*
DIN Sept (µg N/l)	69	21	75	3.29*	0.92
N-dep (kg N/ha,y)	7.7	7.1	10	1.08	0.77*
Pot DIN (µg/l)	1924	2147	2820	0.90	0.68
Phytopl. species nr	78	86	50	0.91	1.56*
Phytopl.vol.(mm3/l)	0.21	0.29	0.16	0.72	1.28
Phytopl.-B/Tot-P	0.021	0.045	0.030	0.47*	0.70

3.2. DIFFERENCES IN TOTAL PHOSPHORUS

Table I indicates that the total phosphorus (Tot-P) concentration was slightly lower in the limed lakes as compared to the references. If, however, the P concentration per unit of TOC was compared between the lake groups this quotient was significantly lower in the group of limed lakes as compared to neutral references. The mean Tot-P concentration was therefore lower in this group in spite of a fairly high TOC concentration (Table I) which would normally coincide with higher phosphorus concentrations.

More adequate data for group comparison were obtained at a standard TOC level (TOC=8 mg C/l) from regression equations between TOC and tot-P concentration within each group. The phosphorus concentration was 1.2 µgP/l (or ca 12%) lower in the limed group as compared to the acid lakes whereas the difference as compared to the neutral lakes was 2 µg P/l or 20%.

Bergström (1999) expanded the dataset on Tot-P/TOC-quotients with data from another 14 Swedish reference lakes sampled during one year (mostly 1996). The total data (Fig. 1) show the typical aggregation of ISELAW-data along the lower periphery of the plot. Estimated data from the equations (at TOC=8mg C/l) give 6,4 and 12.8 µg P/l for the limed and reference lakes respectively. This is a 50% difference but there are overlaps between the different data sets.

Taken together these data indicate differences in the behaviour of TOC and Tot-P between the lake groups. Either there is a more efficient sink for phosphorus in limed lakes or there is a lower input of phosphorus relative to organic matter, perhaps due to phosphorus precipitation in acidic soils rich in dissolved aluminum (Jansson et al., 1986). A different behaviour of TOC is an alternate explanation.

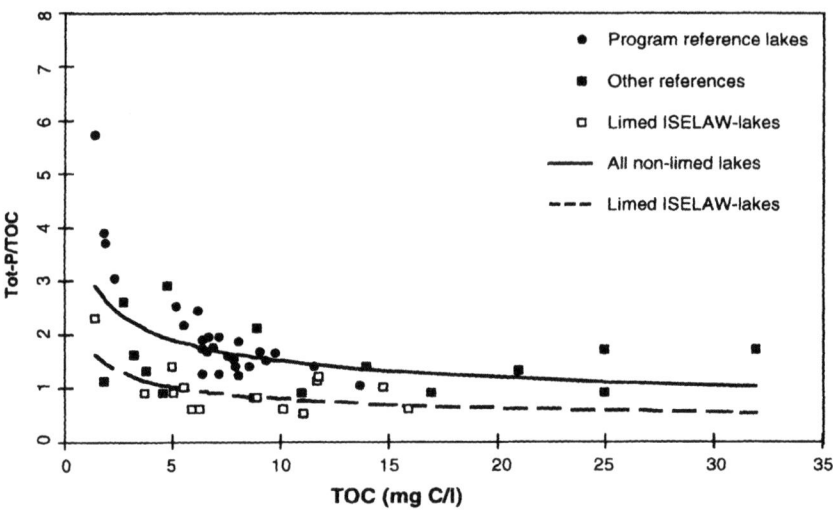

Figure 1. The seasonal mean Tot-P/TOC-quotients as a function of TOC. Equations for limed lakes Y= 1.83*TOC$^{-0.34}$ (r^2=0.34) and for all unlimed references Y= 3.25*TOC$^{-0.34}$ (r^2=0.37). From Bergström (1999).

3.3. DIFFERENCES IN INORGANIC NITROGEN CONCENTRATION

The seasonal mean of DIN was significantly lower in the neutral references as compared to the limed lakes (Table I). This was not primarily due to differences in deposition as shown by mean nitrogen deposition and potential DIN concentration (deposition divided by runoff) for the lake groups (Table I). The similarity between the limed lakes and the neutral references was high, whereas the nitrogen deposition was actually slightly higher for the acid references. Furthermore, the stored DIN in February/March, before the productive season started, was significantly higher in the acid references (as expected from the potential concentration differences) as compared to the other groups. The DIN-depletion during the productive season turned out to be higher in the acid (113 μg N/l) and neutral lakes (98 μg N/l) than in the limed lakes (66 μg N/l) which gave similar September levels in both limed and acid lakes but a significantly lower level in the neutral reference lakes.

The processes involved in the DIN-decline are not distinctly defined but assimilation and sedimentation followed by ammonification-nitrification-denitri-fication would be the main depletion factors, to some degree counteracted by ammonification.

Nitrification and denitrification are both considered to be negatively affected at pH-values slightly below 6 (Fenchel and Blackburn, 1979) or at somewhat lower levels (Gahnström et al., 1993) and would thus not affect differences between the limed lakes and the neutral reference lakes.

Phosphorus availability may also interfere with the DIN decline both by regulating the DIN assimilation by plants, algae and bacteria and by supplying organic matter for degradation by the denitrifying bacteria in the sediment as well as contributing to the close-to-anoxic conditions needed by these bacteria. Phosphorus may thus have an important but unproven role in these respects (Hellström, 1996). In this material the tentative phosphorus role can only be seen in the cases with excess DIN left in September. In other cases DIN deficiency may restrict the decline.) There were in total 11 such lakes with excess DIN (>30 µg N/l) at the end of summer out of which 6 were ISELAW-lakes, 3 were acid lakes and 2 were neutral. The DIN-decrease during the season in these lakes could be described as a linear function of Tot-P concentration and DOC (Y=27.8*TP-16.7*TOC-56.7; r^2=0.66) or phosphorus concentration alone (Y=14.9*TP-38.0; r^2=0.43).

From this it may appear that phosphorus regulates the primary productivity and thus the drop in DIN during summer and that the lower DIN consumption (and resulting autumn DIN-excess) particularly in the group of limed lakes may be due to the low phosphorus levels in that group.

3.4. DIFFERENCES IN PHYTOPLANKTON AND BACTERIA

Yearly means of phytoplankton biovolumes were also compared between the lake groups. All lakes with the species *Gonyostomum semen* were excluded (two limed lakes and four acid lakes) since the extremely high biomasses in such lakes otherwise would bias the comparison. The mean biovolumes ranked from high in the neutral lakes, intermediate in the limed lakes and low in the acid lakes but differences were not significant. The means of quotients between phytoplankton biovolumes and Tot-P ranked with the limed lakes lowest followed by limed and neutral lakes (Table I). The quotient was in fact significantly lower in the ISELAW lakes as compared to the neutral references.

Phytoplankton species number followed the same ranking between the groups as the biomass, with the acid group at the low end and the neutral lakes at the high end. In this case the species number was significantly lower in the acid lakes as compared to the others (Table I).

These phytoplankton data thus pinpoint the group of acid lakes to be significantly different but it is also evident that neutral references – not limed lakes – are at the high end of the compared groups

Bergström (1999), noted that bacterial numbers were lower in the humic (TOC >9 mg C/l) limed lakes as compared to her references. She further hypothesized that this was due to the lower Tot-P concentration at a given TOC-level and pointed out the implications for the food web in this kind of lake where bacteria transform allocthonous TOC into an important particulate food source in the food chain (Jansson *et al.*, 1999). In less humic waters (TOC< 9 mg C/l) bacterial numbers did not deviate much between limed and reference lakes which might indicate that the bacteria there were limited by carbon, not by phosphorus.

3.5. DIFFERENCES IN BENTHOS AND FISH

Johnson (1997) compared profundal and sublittoral benthos in ISELAW lakes to 8 neutral and 4 acid lakes (a subgroup of those cited above). He found the groups neutral, limed and acid lakes to rank in descending order as judged by mean species number, abundance, biomass and diversity both in the profundal and sublittoral zones. Differences between neutral and limed lakes were significant in the sublittoral zone for species number, abundance and diversity. Since these communities rely largely on the supply of food derived from primary production (and usually are not affected by acidity) these data indicate a lower trophic level for the limed than for the neutral lakes.

Fish assessments were carried out in similar lake groups as above using multi-mesh Nordic gill nets (Appelberg, 1998). Species occurrence, abundance, size and biomass, proportions of cyprinids, and proportion piscivores as well as diversity were assessed. The species number and diversity ranked highest in neutral lakes followed by limed lakes and acid lakes. In particular, fish community compositions indicate lower productivity in the limed lakes as compared to the neutral references. The limed lakes are dominated by percids in contrast to neutral reference lakes which to a larger extent are dominated by cyprinids Appelberg op. cit.).

In conclusion, there is thus multiple evidence that the long term limed lakes have a lower trophic level than reference lakes.

Acknowledgements

We thank our collegues within the ISELAW-programme for stimulating discussions and comments. The programme is financially supported by the Swedish Environmental Protection Agency.

References

Appelberg, M.: 1998, *Restoration Ecology* **6**, 343.
Appelberg, M., Lingdell, P.E. and Andrén, C.: 1999, *Water, Air, and Soil Pollution* **85**, 883.
Appelberg, M. and Svenson, T.: 2000, *Water, Air, and Soil Pollution* (THIS VOLUME)
Appelberg, M., Henrikson, B.I., Henrikson, L. and Svedäng, M.: 1993, *Ambio* **22**, 290.
Bergström, A-K.: 1999, Dept. Environmental Assessment, Uppsala, Sweden Report **1999, 11** (in Swedish).
Broberg, O.: 1987, *Hydrobiologia* **150**, 11.
Fenchel, T. and Blackburn,T.H.: 1979, *Bacteria and mineral cycling*. Academic Press, London.
Gahnström, G., Blomqvist, P. and Fleischer, S.: 1993, *Ambio* **22**, 318.
Hellström, T.: 1996, *Wat. Environ. Res.* **68**, 55.
Henrikson, L. and Brodin, Y.W.: 1995, (eds) *Liming of acid waters, a Swedish synthesis*. Springer-Verlag, Berlin.
Grahn, O., Hultberg, H. and Landner, L.: 1974, *Ambio* **3**, 93.
Jansson, M., Persson, G. and Broberg, O.: 1986, *Hydrobiologia* **139**, 81.
Jansson, M., Bergström, A.-K., Blomqvist, P., Isakson, A. And Jonsson, A.: 1999, *Arch Hydrobiol.* **144**, 409.
Johnson, R.J.: 1997 Swedish Environmental Protection Agency, Stockholm, Report **4816** (in Swedish).
Olsson, H. and Pettersson, A.: 1993, *Ambio* **22**,312.
Schindler, D.W.: 1988, *Science* **239**,149.
Wilander, A.: 1997, Swedish Environmental Protection Agency, Stockholm, Report **4652** (in Swedish).
Wilander, A., Andersson, P., Borg, H. and Broberg, O.: 1995, The effects of liming on water chemistry, in Henrikson, L. and Brodin, Y.W. (eds) *Liming of acid waters, a Swedish synthesis*. Springer-Verlag, Berlin.
Willén, E. 1976. *Br. Phycol. J.* **11**, 265.
Utermöhl, H. 1958.. *Mitt. Int. Verein. Theor. Angew. Limnol.* **9**, 1.

SULFUR ISOTOPE VARIATIONS IN ATMOSPHERIC SULFUR OXIDES, PARTICULATE MATTER AND DEPOSITS COLLECTED AT KYUSHU ISLAND, JAPAN

HIDEHISA KAWAMURA[1,2*], NOBUAKI MATSUOKA[1,3], SHINJI TAWAKI[4] and NORIYUKI MOMOSHIMA[5]

[1] *Kyushu Environmental Evaluation Association, Higashi-ku, Fukuoka 813-0004, Japan;*
[2] *Department of Chemistry and Physics of Condensed Matter, Graduate School of Science, Kyushu University, Higashi-ku, Fukuoka 812-0053, Japan;* [3] *Institute of Environmental Systems, Kyushu University, Higashi-ku, Fukuoka 812-0053, Japan;*
[4] *Research Laboratory, Kyushu Electric Power Co., Inc., Minami-ku, Fukuoka 815-0032, Japan;* [5] *Department of Environmental Science, Faculty of Science, Kumamoto University, Kurokami, Kumamoto 860-8555, Japan*
(* author for correspondence, e-mail: kawamura@keea.or.jp)

Abstract. Atmospheric sulfur oxides, particulate matter and deposits (wet and dry deposits) were collected from July 1998 to June 1999 at Kyushu Island, Japan. The isotopic composition of sulfur ($\delta^{34}S$) was measured to identify the source of sulfur in the samples. The monthly $\delta^{34}S$ values were always low in the order of the sulfur oxides, sulfate in particulate matter and deposits. The $\delta^{34}S$ values of the sulfur oxides ranged from -2.7 ‰ to -0.4 ‰ and were close to those of fossil fuels used in Japan. The $\delta^{34}S$ values of sulfate in the particulate matter and deposits correlated with seasalt contribution, so that the $\delta^{34}S_{nss}$ value was calculated for non-seasalt sulfate. The $\delta^{34}S_{nss}$ values of sulfate in the particulate matter and deposits trended higher in winter than summer, suggesting the possibility of isotopic fractionation during chemical transformation (SO_2 to SO_4^{2-}) and of contribution of sulfate derived from sulfur sources with higher $\delta^{34}S$ values.

Keywords: sulfur isotopic composition ($\delta^{34}S$), sulfur oxides, particulate matter, deposits, sulfur source

1. Introduction

Recently, air pollution originated from the consumption of large amounts of fossil fuels have caused greenhouse effects and acid rain, damaging ecosystems on a worldwide scale. It is recognized that acid rain is due to sulfur compounds, which are one of the main air pollutants. Therefore, it is important to investigate the movement of sulfur in the environment in order to evaluate the effects of acid rain.

Sulfur has four stable isotopes (^{32}S, ^{33}S, ^{34}S and ^{36}S). The isotopic composition of sulfur ($\delta^{34}S$) can be highly variable and highly source dependent. Therefore, the $\delta^{34}S$ value will give more useful information about the source of sulfur in the environment.

In this investigation, atmospheric sulfur oxides, particulate matter and deposits (dry and wet deposits) were collected at Kyushu Island, Japan. The $\delta^{34}S$ values were measured for the source identification of sulfur in these samples.

2. Experimental

2.1. SAMPLES

The sampling location is Fukuoka, which is a major city facing the sea and located in the north of Kyushu Island, Japan (Fig. 1). Atmospheric sulfur oxides were collected on cellulose filter paper (Toyo Roshi Kaisha, Ltd.; No. 526, area: 0.05 m^2) impregnated with 30% K_2CO_3 (Fukui, 1963), which was exposed to the open air for the period of one month in a louvered box. Simultaneously, atmospheric deposits were collected using a deposits collector (area: 0.07 m^2). Atmospheric particulate matter was collected on quartz filter paper (Toyo Roshi Kaisha, Ltd.; QR100) using a high-volume air sampler (SHIBATA; HVC-1000N, flow rate: 1 m^3/min) for the period of one week. These samples were collected from July 1998 to June 1999.

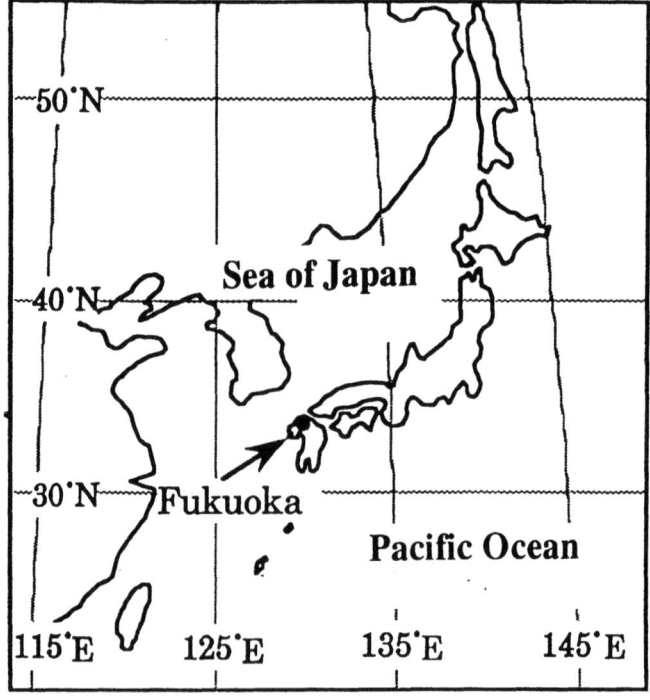

Figure 1. Sampling location.

2.2. PRETREATMENT AND MEASUREMENT

The sulfur oxides on the cellulose filter paper and sulfate in particulate matter on the quartz filter paper were eluted with distilled water. The sulfur was recovered as SO_4^{2-} in the solution. The pH of the solution was adjusted to between 2 and 3 with HCl, and $BaSO_4$ was precipitated by adding 5% $BaCl_2$ solution. The sulfate in the deposits was concentrated on a hotplate and prepared in the same manner as the precipitation. The BaSO$_4$ was filtrated using a membrane filter (Toyo Roshi Kaisha, Ltd.). After separation of the BaSO$_4$, direct conversion of BaSO$_4$ to SO$_2$ was performed by the V$_2$O$_5$ method (Yanagisawa and Sakai, 1983). The δ^{34}S value was determined using a mass spectrometer (VG Isogas; SIRA10), where δ^{34}S is defined as

$$\delta^{34}S(‰) = \{(^{34}S/^{32}S)_{sample}/(^{34}S/^{32}S)_{standard} - 1\} \times 1000$$

The δ^{34}S values were reported on the CDT (Canyon Diablo troilite as as standard substance) scale.

In addition to the δ^{34}S values, cation concentrations in the deposits, Na$^+$ and Mg^{2+}, were determined by atomic absorption spectrometry (Seiko; SAS7500) and anion concentrations, Cl$^-$ and SO$_4^{2-}$, by ion chromatography (DIONEX; 2000i).

3. Results and Discussion

3.1. SULFUR CONCENTRATION AND ISOTOPIC COMPOSITION

The ratios of Cl$^-$ and Mg^{2+} to Na$^+$ concentration in the deposits are consistent with those in seawater, suggesting these ions originated from seawater (Fig. 2). On the basis of Na$^+$ as a conservative tracer for seasalt, non-seasalt sulfate (nssSO$_4^{2-}$) was calculated. Concentrations of SO$_4^{2-}$ and nssSO$_4^{2-}$ in the deposits are shown in Figure 3. The concentrations of SO$_4^{2-}$ and nssSO$_4^{2-}$ were always higher in winter than summer throughout the study period.

The δ^{34}S values of sulfate in the deposits tend to be higher in winter than summer (Fig. 4) and correlate with seasalt contribution (f), where f is defined as

$$f = (Na^+)_{obs} \times (SO_4^{2-}/Na^+)_{sea} / (SO_4^{2-})_{obs} \times 100$$

obs: concentration in the deposits,
sea: concentration in seawater (Na$^+$; 10.56 g kg^{-1}, SO$_4^{2-}$; 2.65 g kg^{-1})
It is well known that the δ^{34}S value of sulfate in seawater is about 20 ‰. Therefore, the δ^{34}S value of non-seasalt sulfate (δ^{34}Snss) was calculated, where δ^{34}Snss is defined as (Rees et al., 1978)

$$\delta^{34}Snss = \{\delta^{34}S_{spl} - 20.3 \times (f/100)\} / \{1-(f/100)\}$$

spl: the δ^{34}S value of total sulfate in the deposits,
20.3: the δ^{34}S value of sulfate in seawater
The δ^{34}Snss values are shown in Figure 4. The δ^{34}Snss values ranged from 2.9 ‰ to 6.5 ‰ and showed a seasonal variation like the δ^{34}S values.

The δ^{34}S values of the sulfur oxides are also shown in Figure 4. The δ^{34}S values ranged from –0.4 ‰ to –2.7 ‰ with a mean of –1.5 ‰. Compared to the δ^{34}Snss values of sulfate in the deposits, the δ^{34}S values of the sulfur oxides

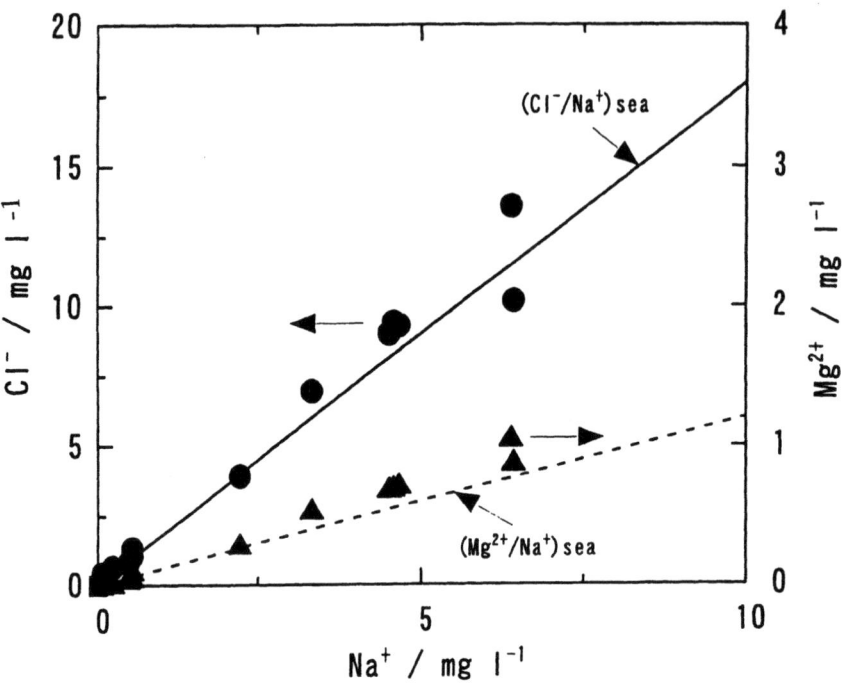

Figure 2. Relation of Cl⁻ and Mg^{2+} to Na^+ concentration in atmospheric deposits from July 1998 to June 1999.

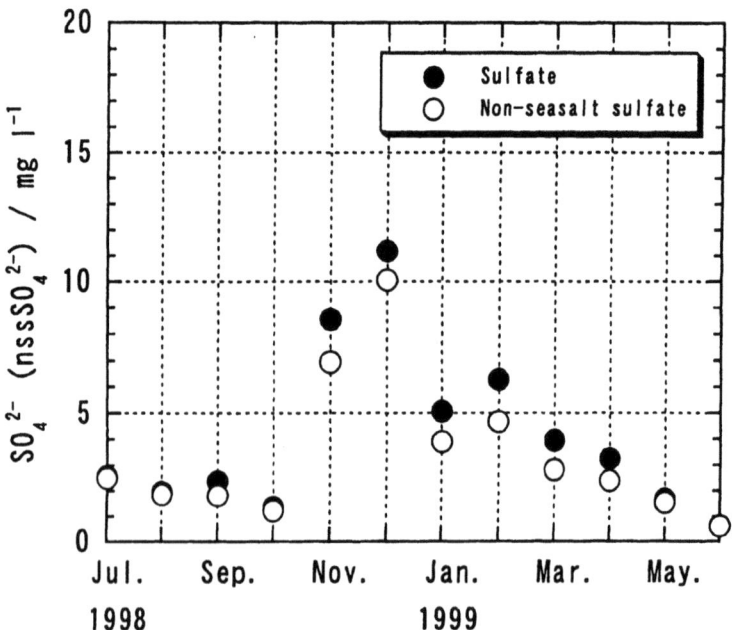

Figure 3. Seasonal variations in SO_4^{2-} and nss SO_4^{2-} concentrations in atmospheric deposits.

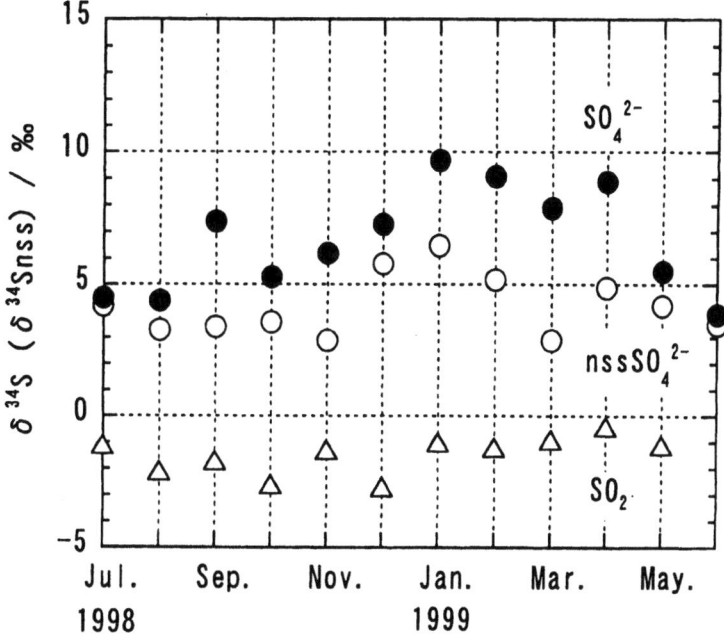

Figure 4. Seasonal variations in $\delta^{34}S$ and $\delta^{34}S_{nss}$ values of sulfate in atmospheric deposits. The $\delta^{34}S$ values of atmospheric sulfur oxides are also shown.

TABLE I
The $\delta^{34}S$ and $\delta^{34}S_{nss}$ values of sulfate in atmospheric particulate matter.

Sample #	Sampling dates	$\delta^{34}S$ (‰)	$\delta^{34}S_{nss}$ (‰)
1	1999/1/22 ~ 1999/2/1	4.6	3.5
2	1999/2/15 ~ 1999/2/12	5.5	4.4
3	1999/3/16 ~ 1999/3/23	5.9	3.4
4	1999/4/16 ~ 1999/4/23	2.1	1.2
5	1999/5/17 ~ 1999/5/25	4.8	3.5
6	1999/6/21 ~ 1999/6/29	1.9	0.4

were lower and the seasonal variation was very small. The $\delta^{34}S$ values of sulfate in the particulate matter are presented in Table I. The $\delta^{34}S_{nss}$ values was also calculated as the same manner for sulfate in the deposits and presented. The $\delta^{34}S_{nss}$ values of sulfate in the particulate matter showed a seasonal variation.

3.2. SOURCES OF SULFUR IN ATMOSPHERIC SULFUR OXIDES, PARTICULATE MATTER AND DEPOSITS

Sulfur compounds in the atmosphere can originate naturally (sea spray, volcanic, biogenic) or anthropogenically (combustion of fossil fuels). It is reported that the $\delta^{34}S$ values of fossil fuels used in Japan show less than 0 ‰ (Maruyama *et al.*, 2000). The $\delta^{34}S$ values of the sulfur oxides collected at Fukuoka were close to those of fossil fuels, suggesting that the combustion of fossil fuels is one of the sources releasing sulfur oxides to the atmosphere.

In Kyushu Island, there are many volcanoes, such as Mt. Sakurajima and Mt. Aso. It is estimated that the amount of sulfur originated from volcanic emanation is comparable to that from combustion of fossil fuels (Fujita *et al.*, 1992). However, the $\delta^{34}S$ values of volcanic sulfur as well as biogenic sulfur have not been investigated in detail. Further investigation into these sources is required to apply the $\delta^{34}S$ value to the source identification of sulfur.

It is very important to consider whether the isotopic composition is preserved during chemical transformations in the atmosphere (*e.g.* SO_2 to SO_4^{2-}). Nriagu and Coker (1978) observed isotopic fractionation associated with a washout of sulfur oxides.

Compared to the $\delta^{34}S$ values of the sulfur oxides, the $\delta^{34}Snss$ values of sulfate in the particulate matter and deposits were higher, suggesting the possibility of isotopic fractionation during chemical transformation. The $\delta^{34}Snss$ values also showed seasonal variation, although the $\delta^{34}S$ values of the sulfur oxides were constant. It was reported that sulfate in rainwater collected in winter was mainly derived from sources with higher $\delta^{34}S$ values, such as coal combustion in northern China (Kitamura *et al.*, 1993). The higher $\delta^{34}Snss$ values of sulfate in the deposits observed in winter at Fukuoka may reflect the $\delta^{34}S$ value of sulfur in coal used in mainland China.

Acknowledgments

We would like to thank Dr. M. Noto and Ms. Y. Wakiyama for determination of $\delta^{34}S$ and their useful advice, and Ms. M. Shibasaki for her abundant help in making this work possible.

References

Fujita, S., Tonooka, Y. and Ohta, K.: 1992, *Journal of Japan Society of Air Pollution* **27**, 336.
Fukui, S.: 1963, *Bunseki kagaku* **12**, 1005.
Kitamura, M., Sugiyama, M., Ohhashi, T. and Nakai, N.: 1993, *Chikyukagaku* **27**, 109.
Maruyama, T., Ohizumi, T., Taneoka, Y., Minami, N., Fukuzaki, N., Mukai, H., Murano, K. and Kusakabe, M.: 2000, *Journal of the Chemical Society of Japan* **2000**, 45.
Nriagu, J.O. and Coker, R.D.: 1978, *Nature* **274**, 883.
Rees, C.E., Jenkins, W.J. and Monster, J.: 1978, *Geochimica et Cosmochimica Acta* **42**, 377.
Yanagisawa, F. and Sakai, H.: 1983, *Analytical Chemistry* **55**, 985.

NUMERICAL REGIONAL AIR QUALITY FORECAST TESTS OVER THE MAINLAND OF CHINA

JUNLING AN[1, 2*], MEIYUAN HUANG[2], ZIFA WANG[2, 4], XINLING ZHANG[2]
HIROMASA UEDA[3] and XINJIN CHENG[2]

[1] *Acid Deposition and Oxidant Research Center (ADORC), 1182 Sowa, Niigata-shi, 950-2144, Japan;* [2] *LAPC, Institute of Atmospheric Physics, Chinese Academy of Sciences, Beijing 100029;* [3] *Disaster Prevention Research Institute, Kyoto University, Uji, Kyoto 611-0011, Japan;* [4] *Frontier Research System for Global Change - IGCR Yokohama, 3173-25 Showa-machi, Kanazawa-ku, Yokohama 236-0001, JAPAN*
(* author for correspondence, e-mail: anjl@lycos.com)

Abstract. The paper gives a general description of a numerical regional-scale air quality forecast model, with emphasis on solution schemes for all possible processes (emissions, transport, deposition, chemistry, and initial boundary conditions) considered in the Eulerian transport/deposition model. In order to improve forecast efficiency we first introduce a looking-up table method for treatment of NO_x and ozone chemical processes instead of the coupling method. Meteorological field is forecasted by the Eta model, which is driven by NCEP data. Several-week regional-scale air quality in China is forecasted using the numerical model system. Comparison and analysis indicate that the air quality levels of key cities over China and the time evolution of pollutants over most places of China can well be forecasted by the numerical model system. Further improvements in some important aspects are needed and presented.

Keywords: numerical forecast model, air pollution index, air quality, pollutants

1. Introduction

Economic expansion and rapid consumption of natural resources have led to a large increase in SO_2 and NO_x emissions (Wang and Wang, 1996; Fenger, 1999; Streets and Waldhoff, 2000). These emissions are causing severe environmental problems (Wang and Wang, 1996; An *et al.*,1999a; Tong and Wan, 1999). One of them is significant degradation of urban air quality in the mainland of China (Tong and Wan, 1999). *e.g.*, TSP (total suspended particles) is the major pollutant for almost all big cities in Northern parts of China. The yearly averaged occurrence frequency of TSP is between 70% and 100% (An *et al.*, 2000). For Beijng both NO_x and TSP are major pollutants (An *et al.*, 2000). The annual averaged NO_x occurrence frequency for Shanghai and Guangzhou reaches 95% (Tong and Wan, 1999). Gradually China government has realized the serious consequences of air pollution and taken measures to mitigate degradation of air quality (The controlled regions of acid rain and SO_2 pollution has been mapped (CEPA, 1998; Tong and Wan, 1999)). During the latest years China government has stressed the importance of air quality forecast (Tong and Wan, 1999). Under this conditions we have developed a numerical air quality forecast model (Numerical air quality forecast methods are still under the way (Scheffe and Morris, 1993; Peters *et al.*, 1995; Dennis *et al.*, 1996).) and done several-week forecast tests over the mainland of China. Currently major pollutants in the mainland of China are SO_2, NO_x, and TSP. Ozone will be considered as a major one in the near future. The reason is that photochemical smog has occurred in some big cities (Zhang *et al.*, 1998) and automobiles in most big cities are still increasing (Elliott *et al.*, 1997).

Considering forecast efficiency and limited computer resources, we first introduce a look-up table method for calculating NO_x and ozone levels instead of directly coupling gas-phase chemical processes. This can save CPU time approximately 85%. Some other processes are also simplified for forecast efficiency considerations and their improvements requires further understanding of the physical and chemical processes occurring in the atmosphere and advances in computing power and numerical methodologies. The paper has three parts. Part one generally describes a numerical regional air quality forecast model and several-week forecast tests are shown in part two. Conclusions and discussions are presented in the last part.

2. Numerical Regional Air Quality Forecast Model and Input Data

2.1. NUMERICAL REGIONAL AIR QUALITY NODULE (RAQM)

This Eulerian module, a 1°×1° model with 8 layers in the vertical direction (Bottom layer ~50m), covers most parts of the whole East Asia (75-146°E, 16-60°N). Transport, deposition, and chemical processes are all included in the module (An et al., 1999a). Dry deposition velocity of pollutants is chosen typical values from studies of Brost et al., 1988; Erismann et al., 1994; Lei, 1996. In-cloud and below-cloud scavenging of pollutants is calculated according to rain intensity, raindrop spectrum and pollutant concentrations (He and Huang, 1995; An et al., 1999a). Advection scheme uses an upstream finite difference (Huang et al., 1995; Wang et al., 1997). Vertical and horizontal eddy diffusivity takes different constants for different layers (An et al., 1999a). Integration of the chemistry rate equations at each grid point within a simulation region can consume as much as 98% of the total simulation CPU time (Peters et al., 1995). For air quality forecast the model run time must be much less than 8 hours for 24-hour forecasts (Lei et al., 1998). Computer resources in China are still limited now. Considering these limited factors we choose a look-up table method instead of coupling the gas-phase chemical module into a 3-D transport module and this can save CPU time 85%. The main idea of the looking-up table method is selection of dominant factors that influence conversion rates of major pollutants and calculation of gas-phase conversion rates under different combinations of the factors. A conversion rate table for major pollutants is formed and saved in a computer in advance. Looking up the table according to different conditions such as humidity and temperature while running the RAQFM model. For SO_x processes detailed description can be seen from Huang et al., 1995; He and Huang, 1996; An et al., 1999a. Compared with SO_x processes related chemical reactions for NO_x and ozone are extremely complicated. Zhu (1999) has succeeded in making a big table for NO_x and ozone conversion rates by selecting 6 dominant factors (zenith, height, temperature, cloud fraction, NO_x levels and initial ozone concentrations). For each factor the possible range of values occurring in the atmosphere is partitioned into different bins, e.g., initial NO_x levels from 0.005 to 20 ppb are divided into 65 bins. An et al. (1999b) demonstrates that the looking-up table method is applicable in the mainland of China according to current NO_x levels and VOC/NO_x ratios in the air over typical urban polluted areas. Liquid-phase chemical conversion rates are taken different constants also for forecast efficiency.

2.2. METEOROLOGICAL MODULE

China has varied topography. The Tibetan plateau and other Chinese steep mountains play an important role in forecasting weather system in China, even in the world (Yu, 1994). The varied topography can be well treaded by the LASG Eta-coordinate model (Yu, 1989; 1994), which is a regional weather forecasting model developed by the State Key Laboratory of Numerical Modelling for Atmospheric Sciences and Geophysical Fluid Dynamics (LASG), Institute of Atmospheric Physics (IAP), Chinese Academy of Sciences. The Eta model has succeeded in forecasting of typical weathers (An et al., 1999a), and is widely applied to different levels of meteorological departments in the mainland of China. Compared with MM4 or MM5 models the Eta model also consumes less CPU time.

2.3. INPUT DATA

Emissions are taken from Bai (1996) as a base case. For 1998 and 1999 years emissions are estimated from the work of Wang and Wang (1996) based on the base year emissions. Terrain and vegetation data are from the Data Center of IAP. Initial meteorological fields uses NCEP data taken from the net. All observed data are from the net (www.envir.online.sh.cn/airnews/).

2.4. PROCEDURE OF RUNNING THE NUMERICAL REGIONAL AIR QUALITY FORECAST MODEL

Run the RAQM module using the previous three-week NCEP data (day by day) as a meteorological field and save the final simulation result as an initial concentration field. Forecast next-day meteorological fields by running the Eta module with current NCEP data downloaded from the net and then forecast next-day concentrations of major pollutants by running the RAQM module, whose meteorological field is the output of the Eta module. Gas-phase chemical conversion rates of major pollutants in the RAQM module are obtained by the look-up table method according to specific meteorological conditions provided by the Eta module. The output of the RAQM module at the end of 24-hour simulations will be saved as a next-day initial concentration field.

3. Forecasting Tests

Air pollution index (API) can be calculated by,

$$I = \frac{I_1 - I_2}{C_1 - C_2}(C - C_2) + I_2$$

where I denotes API, I_1 and I_2 are the reference API corresponding to C_1 and C_2, which are the reference concentrations listed in Table I, respectively. C is the concentration of pollutants ($C_2 < C < C_1$). Say, I_1, I_2, C_1, C_2 should be 300, 200, 1.6mg/m^3, 0.25mg/m^3 (read from Table I) if an observed SO$_2$ level is 0.4mg/m^3. I will equal I_2 if C equals C_2. For ozone when $C > C_{III}$, which is a reference concentration issued by China Environmental Protection Agency, the

API is estimated by ($C_{III} = 0.20\, mg.m^{-3}$) $I = 100 \cdot C / C_m$

TABLE I Levels of API

API (I)	Pollutant concentration ($mg.m^{-3}$)		
	C (TSP)	C (SO$_2$)	C (NO$_x$)
500	1.000	2.620	0.940
400	0.875	2.100	0.750
300	0.625	1.600	0.565
200	0.500	0.250	0.150
100	0.300	0.150	0.100
50	0.120	0.050	0.050

Air quality is I, II, III, IV and IV when API is $0 \sim 50$, $51 \sim 100$, $101 \sim 200$, $201 \sim 300$, and larger than 300, respectively. For ozone air quality is I, II, and III as the hourly average concentration is 0.12, 0.16 and $0.20\, mg.m^{-3}$.

TABLE II
Comparison of modeled and observed weekly averaged air quality of some big cities in the mainland of China from Oct.23 to 29 of 1998

City	Observed API	Modeled API	Major pollutant	Air quality
Beijing	215	213	NO$_x$	IV
Shanghai	85	71	NO$_x$	II
Hefei	62	44	NO$_x$	II
Qingdao	77	62	SO$_2$	II
Wuhan	98	80	NOx	II
Changsha	103	75	SO$_2$	III (II)
Guiyang	141	129	SO$_2$	III
Chongqing	202	181	SO$_2$	IV (III)

From Oct.23 to 29 in 1998 strong wind and precipitation occurred in most areas of the mainland of China. Large amount of pollutants are removed under this conditions, so air quality for most big cities is fine whereas few cities, e.g., Beijing and Chongqing have very bad air quality due to unfavorably meteorological conditions (Fig. 1 and Table II). Table II also shows that modeled values for API are in good agreement with observed ones whereas modeled air quality for Changsha and Chongqing has some deviations.

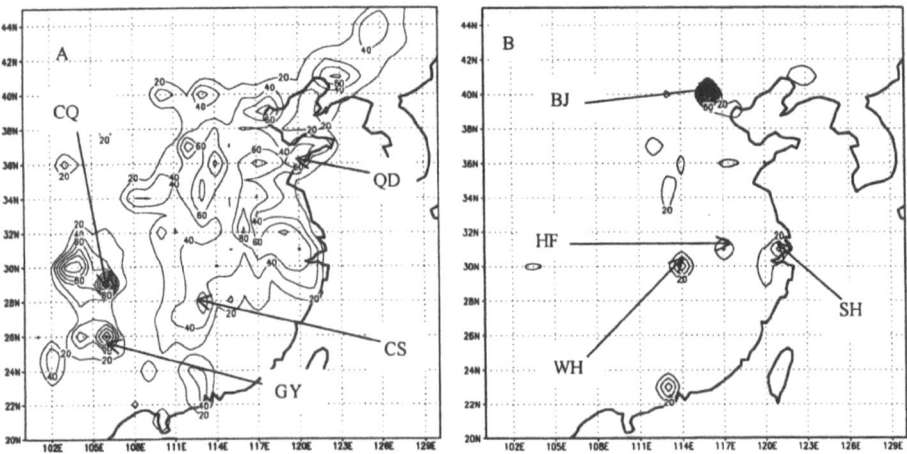

Figure 1. Distributions of forecasted weekly averaged API for SO$_2$ (A) and NO$_x$ (B) Over most parts of China from Oct.23-29, 1998 (BJ, SH, HF, WH, CS, GY, QD and CQ denotes Beijing, Shanghai, Hefei, Wuhan, Changsha, Guiyang, Qingdao, and Chongqing, respectively.)

Temporal development of modeled and observed API is shown in Figure 2, which indicates that modeled NO$_x$ levels follows the observed ones but changes in modeled SO$_2$ levels are smaller than those in observed ones.

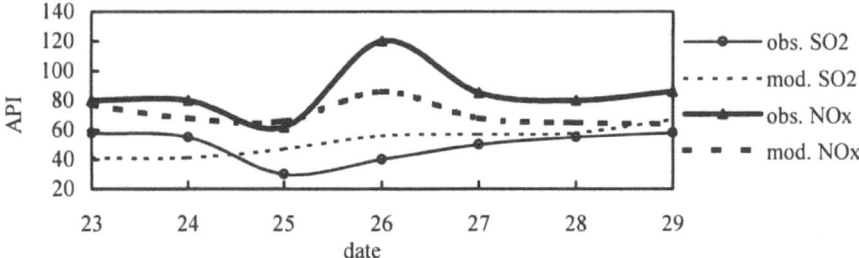

Figure 2. Comparison of observed and modeled daily averaged API for SO_2 and NO_x in Shanghai from Oct.23 to 29 of 1998

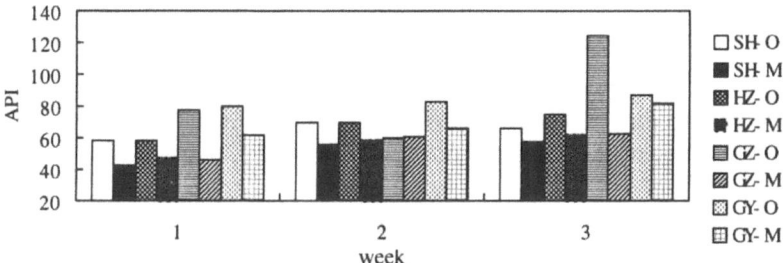

Figure 3. Comparison of observed and modeled weekly averaged API for Shanghai (SH), Hangzhou (HZ), Guangzhou (GZ), and Guiyang (GY) within 3 weeks in June of 1999 (1, 2, and 3 denotes first week, second week, and third week, respectively, O = observed; M = modeled). The major pollutant for GY is SO_2 and that for other three cities is NO_x.

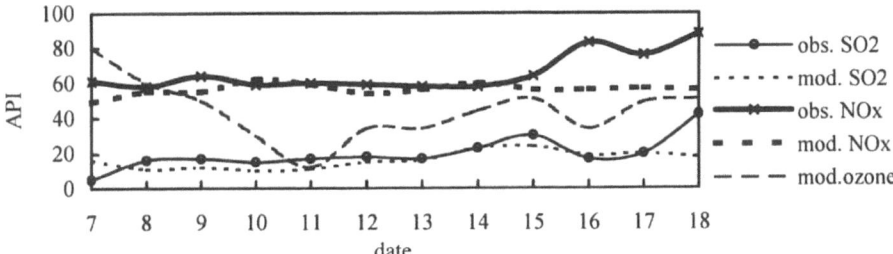

Figure 4. Comparison of observed and modeled daily averaged API for SO_2, NO_x and O_3 in Shanghai from Jun.7 to 18 of 1999

Generally precipitation often occurs in summers over the mainland of China. Atmospheric pollutants can be efficiently removed by precipitation. The air quality in most places is good. Modeled values for weekly averaged API (except Guangzhou in the third week) shows good agreement with observed ones (Fig. 3). The reason why modeled API for Guangzhou is much less than observed API needs further investigation. From June 7 to 18 in 1999 the fitness between modeled and observed values is good and ozone levels are generally higher than SO_2 levels (Fig. 4) and more attention should be focused in the future.

4. Conclusions and Discussions

A general description of a numerical regional air quality forecast model is given and treatment of physical and chemical processes with emphasis on the looking-up table method for NO_x and ozone processes in the Eulerian module are shown. Several-week forecast tests have been done. Results indicate that the API and the corresponding air quality of big cities over the mainland of China and time evolution of pollutants (SO_2, NO_x, and ozone) can be well forecasted by the numerical model. Further improvements include: (1) investigation of sources of SO_2 and NO_x, especially of TSP and VOC ; (2) monitoring data from different layers above the ground; (3) A good PBL (planetary boundary layer) model; (4) developing a new module which can both efficiently and accurately treat liquid-phase chemical processes occurring in the atmosphere.

Acknowledgements

The funding for this research was supported by Chinese National Key Program for Developing Basic Sciences (G1999032801) and the Key Project A of Chinese Academy of Sciences (Grant number is KZ951-A1-403-03-03).

References

An Junling, Wang Zifa, Huang Meiyuan, et al.: 1999a, *Climatic and Environmental Research* **4**, 244.
An Junling, Han Zhiwei, Wang Zifa, et al.: 1999b, *Chinese Journal of Atmospheric Sciences* **23**, 753.
An Junling, Zhang Renjian and Han Zhiwei: 2000, *Climatic and Environmental Research* **5**, 25.
Bai Naibin: 1996, 'CO_2, SO_2 and NO_x Emissions in the Mainlanf of China', in Zhou Xiuji (Ed.), *Ozone Changes in China and their Effects on Climate*, China Meteorological Press, Beijing, pp. 145-150.
Brost, R. A., Chatfield, R. B., Greenberg, J. P., et al.: 1988, *Tellus* **40B**, 385.
China Environmental Protection Agency (CEPA): 1998, *Environmental Protection* **3**, 7.
Dennis, R. L., Byun, D. W., Novak, J. H.: 1996, *Atmospheric Environment* **30**, 1925.
Elliot, S., Shen, M., Blake, D. R.: 1997, *Journal of Atmospheric Chemistry* **27**, 31.
Erisman, J. W., Pul, A. V. And Wyers, P.: 1994, *Atmospheric Environment* **28**, 2595.
Fenger Jes: 1999, *Atmospheric Environment* **33**, 4877.
He Dongyang and Huang Meiyuan: 1996, *Atmospheric Environment* **30**, 2449.
Huang Meiyuan, Wang Zifa, He Dongyang, et al.: 1995, *Water, Air, and Soil Pollution* **85**, 1921.
Lei Xiaoen: 1996, *Acta Meteorologica Sinica* **10**, 118.
Lei Xiaoen, Zhang Meigen, Han Zhiwei, et al.: 1998, *Numerical Forecast Models and Fundamentals on Air pollution*. China Meteorological Press, Beijing.
Lurmann, F. W., Lloyd A.C. and Atkinson, R.: 1986, *Journal of Geophysical Research* **91**, 10905.
Peters, L. K., Berkowitz, C. M., Carmichael, G. R., et al.: 1995, *Atmospheric Environment* **29**, 189.
Scheffe, R. D. And Morris R. E.: 1993, *Atmospheric Environment* **27B**, 23.
Streets, T. G. And Waldhoff, S. T.: 2000, *Atmospheric Environment* **34**, 363.
Tong Yanchao and Wan Bentai: 1999, *Climatic and Environmental Research* **4**, 275.
Wang, W. And Wang, T.: 1996, *Atmospheric Environment* **30**, 4091
Wang Zifa, Huang Meiyuan, He Dongyang, et al.: 1997, *Chinese Journal of Atmospheric Sciences* **21**, 366.
Yu Rucong: 1989, *Chinese Journal of Atmospheric Sciences* **13**, 139.
Yu Rucong: 1994, *Documentation of the LASG Regional Eta-coordinate Model*, State Key Laboratory of Numerical Modelling for Atmospheric Sciences and Geophysical Fluid Dynamics (LASG), Institute of Atmospheric Physics (IAP), Chinese Academy of Sciences.
Zhang Yuanhang, Shao Kesheng, Tang Xiaoyan, et al.: 1998, *Acta Scientiarum Naturalium Universitatis Pekinensis* **34**, 392.
Zhu Bin: 1999, *Numerical Simulation of Tropospheric Ozone and NO_x evolving Mechanism and Their distributions over Eastern Asia*. Master's thesis, Nanjing Institute of Meteorology.

SPECIFICS AND TEMPORAL CHANGES IN AIR POLLUTION IN AREAS AFFECTED BY EMISSIONS FROM OIL SHALE INDUSTRY, ESTONIA

VALDO LIBLIK* and MARGUS PENSA

North-East Estonian Department, Institute of Ecology, Tallinn University of Educational Sciences, Pargi St 15, 41537 Jõhvi, Estonia
(* author for correspondence, e-mail: valdo@ecoviro.johvi.ee)

Abstract. Atmospheric air pollution levels and long-term effects on the environment caused by simultaneous presence of SO_2 and oil shale alkaline fly ash during the last five decades (since 1950) were investigated. The annual critical value of SO_2 for forest (20 µg m^{-3}) was surpassed in 1% (~35 km^2) of the study area where the load was 30–40 µg m^{-3}. No effect of long-term SO_2 concentrations of up to 10–11 µg m^{-3} (0.5-h max up to 270 µg m^{-3}) and simultaneous fly ash loads of up to 95 µg m^{-3} (1000 µg m^{-3}) on the growth and needle longevity of *Pinus sylvestris* was established. The yearly deposition (average load up to 20–100 kg S ha^{-1}) was alkaline rather than acidic due to an elevated base cation deposition in 1960–1989. Since 1990, the proportion of SO_2 in the balance of components increased: about 70–85% of the total area was affected while the ratio of annual average concentrations of SO_2 to fly ash was over 1. The limit values of fly ash for *Sphagnum* mosses and conifers in the presence of SO_2 are recommended.

Keywords: air pollution, environmental effects, oil shale, solid particles, sulphur dioxide

1. Introduction

Contamination of atmospheric air in the north-eastern part of Estonia is regionally different and dependent on spatial-temporal distribution of fluxes from the combustion of oil shale in the power plants (PP). As a result, since 1950 the ecosystems in this region have been seriously affected by a simultaneous presence of large amounts of SO_2 (up to 200,000 t yr^{-1}) and alkaline oil shale fly ash (Liblik *et al.*, 1995; Liblik *et al.*, 1997[1]).

During the period of 1960–1988 the increased atmospheric input of ash caused important changes in bog plant cover (Karofeld, 1997). Since 1989 the atmospheric air situation has changed: in the balance of pollutants the proportion of SO_2 in comparison with fly ash has significantly increased (Liblik *et al.*, 1997[2]). In NE Estonia the critical pollution load of sulphur was established at the level of 33 kg S $ha^{-1}yr^{-1}$ (2080 eqv ha^{-1}) (Saare and Oja, 1997). Lower loads guarantee the preservation of terrestrial ecosystems with 95% probability. So far the limit values for SO_2 and fly ash in the air at their simultaneous presence have not been studied. In the vicinity of mines and metallurgic plants the average annual concentrations of dust at the level of 40–90 µg m^{-3} would initiate a degradation of vegetation (Krychkov, 1991). The annual limit concentration for dust, emitted from PP operating on coal, should not exceed 10–20 µg m^{-3} for lichen, 20–30 µg m^{-3} for conifers, 50–80 µg m^{-3} for deciduous trees, etc. The critical level for SO_2 depending on the type of vegetation is 10–30 µg m^{-3} (Leeuwen, 1997); for sensitive lichen and *Sphagnum* mosses a limit value of 3–9 µg m^{-3} was established (Krychkov, 1991).

The present paper deals with the specifics and temporal changes in air pollution in the area of oil shale industry in NE Estonia during more than 50 years and in the "new situation". The effects of long-term air concentration fields and pollutant precipitation loads on the environment under simultaneous discharge of SO_2 and alkaline oil shale fly ash are discussed.

2. Study Area and Methods

Figure 1. The study area: north-eastern part of Estonia. The power plants (+): 1– Kohtla-Järve, 2–Ahtme, 3–Estonian and 4– Baltic. ☐–Kurtna Landscape Reserve (LR).

The study area (~3360 km^2), the so-called Estonian Oil Shale Basin is located in the north-eastern part of Estonia between the Gulf of Finland and Lake Peipsi (Figure 1).

To study the dynamics of long-term temporal changes in air pollution, the method of Air Quality Complex Index (Liblik and Kundel, 1998[1], 1998[2]) was used. Using this index pollution zones were distinguished dependent on the average annual (C_a) and short-time (0.5-h) maximum (C_m) concentration of pollutants in the overground air layer. The concentrations were calculated applying a semi-empirical modelling method (OND-86, 1987; Liblik *et al.*, 1995), based on Gaussian theory, as well as pollutant's limit concentrations in terms of health and nature. In retrospective studies of the impact of air pollution on needle longevity of Scots pine (*Pinus sylvestris*) the Needle Trace Method (Jalkanen, 1995) was used.

3. Results and Discussion

3.1. Temporal Changes in Air Pollution

Figure 2 shows the general changes in the emission of sulphur dioxide and oil shale fly ash from NE Estonian power plants (Figure 1) since 1950. Three main periods can be distinguished: before (pre-1970), during (1970–1988 and 1989–1991) and after (post-1991) the period of peak combustion of oil shale and emissions (Punning et al., 1997). As it can be clearly seen, the role of SO_2 in the total amount of emissions was continuously increasing. The ratio of the emitted SO_2 amount to fly ash was within the range of 0.17–0.21 in the period of 1950–1969, 0.45–0.95 in 1970–1988 and 1.1–1.23 in the post-1988 period. This phenomenon was caused mainly by a decrease in fly ash emissions in 1976–1977 and 1987–1988 from the Ahtme and Kohtla-Järve PPs (after the installation of dust filters). In some regions near the town of Kohtla-Järve the ratio SO_2/fly ash was much higher (5.8–36.4 in 1990–1999).

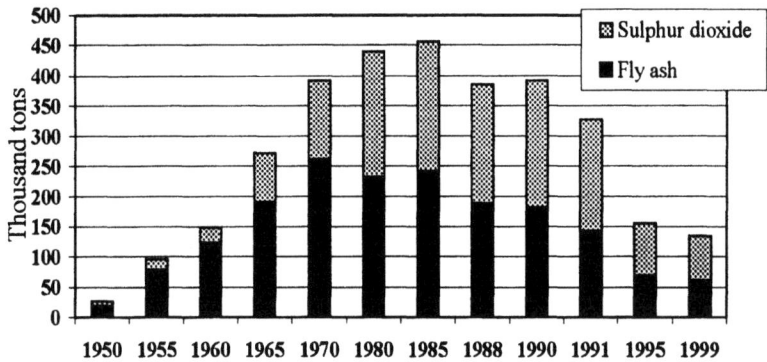

Figure 2. Emission of SO_2 and oil shale fly ash from power plants in 1950–1999.

Table I gives a characterisation of air pollution zones (Liblik and Kundel, 1998[1]) by calculated C_a and C_m values and Table II – corresponding approximate precipitation loads of S and Ca-ions for some regions.

The proportion of areas with serious fly ash pollution decreased notably during 40 years (Tables II and III). In 1960–1988 the total area of zones with very high, high and relatively high fly ash pollution covered more than 32% of the territory; in 1989–1991 only 6.8% and in the post-1991 period no such zones existed. At the same time the area of low and very low pollution level increased about 2.6 times – from 35% to 90.2%. Since 1992 SO_2 prevailed in about 70–80% of the area (zones 1 and 2) when compared with fly ash (zone 1).

Air pollution with SO_2 was relatively stable (Table III): the areas with low and moderate pollution levels were prevailing, making up 33.4–44.4% and 30.5–61.2% respectively, whereas the values of C_a remained within the limits 1–10 μg m^{-3} over most of the study area. Deposition loads of S in zones 1–2 varied between 2 and 120 kg ha^{-1}yr^{-1} (Table II).

3.2. LONG-TERM EFFECTS OF SO_2 AND FLY ASH POLLUTION FIELDS

Until 1988 the critical value of SO_2 for forest (20 μg m^{-3}) was surpassed only in 1% of the study area (Table III – cf zone 4, ~35 km^2). The sensitive lichens and

TABLE I
Characterisation of air pollution zones

Zone No	Air pollution level	Concentration of pollutants, µg m^{-3}				Effects on nature
		Sulphur dioxide		Oil shale fly ash		
		C_a	C_m	C_a	C_m	
0	Very low	≤2	≤20	≤2	≤20	No effects
1	Low	≤10	>20 ≤50	≤10	>20 ≤100	Slight effects on materials and natural objects
2	Moderate	≤20	>50 ≤500	≤16	>100 ≤300	Effects on materials and natural objects
3	Relatively high	≤20	>500	≤20	>300 <1000	Significant effects on plants (lichens, trees, etc)
4	High	>20 ≤50	>500	>20 ≤100	>300 <1500	Significant and strong effects on natural objects
5	Very high	>50 ≤500	>500	>100 ≤300	>1000	Strong effects on nature

TABLE II
Deposition loads (kg ha^{-1}yr^{-1}) by monitored (Kallaste et al., 1992; Estonian..., 1998) and calculated data and corresponding air pollution zones (in brackets) in some studied regions

Region	1960–1988		1994–1998	
	S-SO$_4^{2-}$	Ca^{2+}	S-SO$_4^{2-}$	Ca^{2+}
Kohtla-Järve town	45–200 (3, 4)	230–750 (4, 5)	15–50 (2)	10–35 (1)
Jõhvi–Ahtme–Kurtna LR	30–120 (2)	20–300 (3)	2–28 (1,2)	5–30 (1)
Narva town	20–100 (2)	100–400 (3, 4)	5–30 (2)	20–160 (2)
Central and southern part	3–41 (0, 1)	10–150 (1, 2)	2–10 (0, 1)	5–25 (0, 1)

TABLE III
The area of air pollution zones during different emission periods (the mean % from total area)

Period	Pollutant	Zone No					
		0	1	2	3	4	5
1960–1969 (pre-1970)	SO$_2$	13.2	44.4	41.6	0.8	0	0
	Fly ash	0.3	37.1	29.9	24.2	7.7	0.8
1970–1988 (peak)	SO$_2$	4.7	32.6	61.6	0.1	1.0	0
	Fly ash	0.4	34.6	32.6	25.0	6.7	0.7
1989–1991 (peak)	SO$_2$	0	45.8	54.2	0	0	0
	Fly ash	0.5	75.7	17.0	5.6	1.2	0
1992–1999 (post-1991)	SO$_2$	36.1	33.4	30.5	0	0	0
	Fly ash	25.1	65.1	9.8	0	0	0

mosses (*Sphagnum*) were affected with $C_a>3$ µg m^{-3} on a much larger territory: until 1991 about 30% and in 1992–1999 5–10%. Simultaneous presence of acidic and alkaline components has twofold effect on the local environment:

• During the period of very high, high and relatively high fly ash pollution level (1960–1988) the sulphur deposition load (max 100–200 kg ha^{-1}yr^{-1}) did not actually cause acidification, rather, the environment suffered alkalisation due to an increased base cation deposition of up to 750 kg ha^{-1}yr^{-1} (Table II). The pH of bog waters increased from the normal value of 2–3 to 5–6 and the precipitation was often alkaline with pH reaching 7.5–9.5 (Kallaste *et al.*, 1992).

• At the present time (post-1991), when the emission of SO$_2$ in respect to fly ash has increased, the presence of relatively small concentrations of fly ash (also low level of Ca^{2+} deposition) does not result in alkalisation and dangerous effects of SO$_2$ to plants are possible. During 1994–1997 the pH of snow and rainwater fell to 5.8–7.1 (Mandre *et al.*, 1996; Estonian..., 1998).

Table IV shows simultaneous effects of SO$_2$ and fly ash on nature on the ground of biomonitoring data (Mandre *et al.*, 1996; Karofeld, 1996) recorded in different pollution zones. The leading role of fly ash in the disappearing of *Sphagnum* is evident: therefore, the limit values of about $C_a \leq 15-20$ µg m^{-3} and $C_m = 100-150$ µg m^{-3} (Ca^{2+} deposition up to 80 kg ha^{-1}yr^{-1}) may be recommended. The tolerance of conifers to fly ash is higher: $C_a \leq 50-100$ µg m^{-3}, $C_m \leq 500-1000$ µg m^{-3}, Ca^{2+}>80 kg ha^{-1}yr^{-1}. In the case SO$_2$ concentrations fall below pollution level for zone 2, limit values may be lower than given above.

Study of conifers (Table IV) indicated a possible role of SO$_2$ in damages (declining needle and shoot length, etc) in zones 2–4 (by C_a values) after alkaline fly ash concentrations decreased noticeably. The retrospective investigations in Scots pine stands proved that during 1960–1998 the moderate pollution level of SO$_2$ (C_a=10–11 µg m^{-3}, $C_m \leq 270$ µg m^{-3}) with simultaneous concentrations

TABLE IV
Effects of different concentrations (µg m^{-3}) of SO$_2$ and fly ash in the air on the nature

SO$_2$			Fly ash			Effects
Zone	C_a *	C_m	Zone	C_a	C_m	
Sphagnum species (Kurtna Landscape Reserve)						
2	3–9	60–250	3–4	10–100	400–1500	Complete degradation
2	2–5	50–200	1–2	3–16	100–250	Recurrence is beginning
1	1–4	30–50	1	1–3	25–90	Recurrence
Conifers (Kohtla-Järve area)**						
4	20–45	500–800	4–5	100–300	2000–8000	Strong changes
2–4	10–30	100–320	1	1–5	20–60	Significant changes
2	<15	<200	4	50–100	500–1000	First signs of damages

* The background level of 2.5–5.6 µg m^{-3} must be taken into account (Liblik *et al.*, 1997^2)
** Organic compounds may be also affected

of fly ash of C_a=80–95 µg m^{-3} and $C_m \leq 1000$ µg m^{-3} had no effect on needle longevity and growth (radial and height increment) of trees.

4. Conclusion

In 1960–1989 alkaline fly ash was the prevalent component of air pollution in NE Estonia. Since 1990 the proportion of SO_2 has surpassed that of fly ash in the balance of these components. However, as both the total emission and pollution loads have also decreased, neither noticeable alkalisation nor acidification processes have occurred. Simultaneous presence of SO_2 and fly ash in the air may decrease their individual critical values for nature though not significantly. If the decrease in the emission of fly ash continues, the sensitive lichens and mosses may be affected by SO_2, and so may conifers in some more polluted regions.

Acknowledgements

This study was supported by the Estonian Science Foundation (Grant No 3776) and the Estonian Ministry of Education (Project No 0280340s98).

References

Estonian Environmental Monitoring 1996: 1998, Estonian Ministry of Environment.
Jalkanen, R.: 1995, *Needle Trace Method for Retrospective Needle Retention Studies on Scots pine*. Acta Universitas Ouluensis A264.
Leeuwen, F.X.R.: 1997, 'Update and Revision of WHO Air Quality Guidelines for Europe', in *Environmental Research Forum* Vols 7-8, Trans Tech Publ, Switzerland, pp. 22–31.
Kallaste, T., Roots, O., Saar, J., Saare, L.: 1992, *Environmental Report 3*. EDC, Helsinki, 62.
Karofeld, E: 1996, *Suo* 47 (4), 105.
Kryuchkov, V.: 1991, *Ecology* 3, 28 (Akademy of Sciences of SSSR, in Russian).
Liblik, V., Rätsep, A., Kundel, H.: 1995, *Water, Air and Soil Pollution* 85, 1903.
Liblik, V., Kundel, H., Rätsep, A.: 1997[1], Dynamics of Air Pollution with Oil Shale Fly Ash in North-East of Estonia. *Environmental Research Forum* Vols 7–8, Trans Tech Publ, Switzerland, pp. 641–647.
Liblik, V., Kundel, H., Rätsep, A.: 1997[2], 'Dynamics of Acid Deposition Precursors in the Areas of Oil Shale Industry in Estonia', in *Proceedings of Int Congress of Acid Snow and Rain, Niigata*, pp. 616–621.
Liblik, V., Kundel, H.: 1998[1], *Oil Shale* 15/1, 75.
Liblik, V., Kundel, H.: 1998[2], 'Estimation of Air Quality by Complex Index', in J. Breuste, H. Feldmann, O.Uhlmann (eds), *Urban Ecology*, Springer, pp. 114–120.
Mandre, M., Liblik,V., Rauk, J., Rätsep, A., Tuulmets, L.: 1996, *Oil Shale* 13/4, 309.
OND-86: 1987, *The Method for Calculation of Concentrations of Harmful Substances in the Atmospheric Air*. Leningrad (in Russian).
Punning, J.-M., Liblik, V., Alliksaar, T.: 1997, *Oil Shale* 14/3, 347.
Saare, L., Oja, T.: 1997, 'Estonia', in M. Posch *et al.* (eds), *Calculation and Mapping of Critical Thresholds in Europe*, Bilthoven, Netherlands, pp. 76–77.

LONG-RANGE TRANSPORT OF SULFUR FROM NORTHEAST ASIA TO CHENGSHANTU, SHANDONG PENINSULA: MEASUREMENT AND SIMULATION

TAKAHISA MAEDA[1], ZIFA WANG[2,3]*, MASAYASU HAYASHI[1], and MEIYUAN HUANG[3]

[1] National Institute for Resources and Environment, 16-3 Onogawa, Ibaraki 305-8569 Japan;
[2] Frontier Research System for Global Change, IGCR, Yokohama 236-0001, Kanagawa, Japan;
[3] LAPC, Institute of Atmospheric Physics, Chinese Academy of Sciences, Beijing, China
(*author for correspondence, e-mail: zifawang@frontier.esto.or.jp)

Abstract. Observations of air pollutants were conducted at Chengshantu in January 1996 to clarify the extent of trans-boundary pollution from the Asian continent. A nested air quality prediction modeling system (NAQPMS), which included parameters on emission, transport, diffusion, deposition and transformation of sulfur oxides, was performed to compile sulfur concentrations over the observation site. The model calculation reproduced the observed variations of sulfur oxides and sulfate well, although the model calculations could not reproduce the extensively low sulfate concentrations. Using nesting improves the ability of the modeling system to capture peak episodes seen in the observations. Calculation of the origins of deposited sulfur in each country with the NAQPMS shows that the long-range transport of sulfur is very serious. For Japan, 49% of sulfur deposition was from other countries during the campaign period.

Keywords: transboundary transport, East Asia, nested modeling, sulfur deposition.

1. Introduction

The rapid growing, highly industrial East Asia economy has been characterized by increasing emissions of sulfur dioxide and nitrogen oxides (Akimoto and Narita, 1994). These emissions are oxidized to sulfate and nitrate and result in acid rain, which has harmful impacts on health, the natural environment and the ecological system. Trans-boundary pollution is one of key issues in East Asia and has been discussed in numerous publications (Ichikawa and Fujita, 1995; Huang et. al., 1995; Xiao et. al., 1997; Muranto et. al., 2000; Chang et. al., 2000). However, estimates of the major source of deposited sulfur are quite different among these simulations. Some of these models used simplified chemical processes and meteorological fields taken directly from reanalysis data with low temporal and spatial resolution or from the off-line results of meteorological models.

Chengshantu (122°E, 37°N), located at the northeastern corner of the Shandong peninsula, is an ideal sampling site to monitor the transport of air pollutants among the surroundings countries, China, Korea and Japan. A continuous 20-day series of daily aerosol samples and 1-minute mean sulfur dioxide concentration in January 1996 were collected at this site. A nested air quality prediction modeling system (NAQPMS) has been evaluated and extensive detailed are provided by Wang et. al. (2001). Here we conducted simulations of the characteristics of gaseous SO_2 and sulfate aerosol and their transport in the atmosphere using the NAQPMS.

2. Observations

To clarify the extent of trans-boundary air pollution from the continental Asia to Japan, we conducted an extensive field survey at a remote site, Chengshantu. The sampling period was 20 days, covering the period of January 10-29, 1996. Continuous 1-minute mean SO_2 concentrations and 31 particle sample were collected on fiber filters using a high-volume sampler (GMWL-2000). The flow rate was $1.5m^3min^{-1}$ and the sampling time was normally 12 hours. Elemental compositions of particles were determined by instrumental neutron activation analysis (INAA) at the laboratory of the Central Electric Power Research Institute.

Figure 1 shows variations of observed SO_2 and non-sea-salt sulfate concentrations in the atmosphere. It can be seen that there was an obvious high peaks during January 20-21. The nss-SO_4^{2-} concentration increased to $12.5\mu g/m^3$, and SO_2 concentration was up to 60ppb. This demonstrates the long-range transport of sulfur from other polluted regions to the observed site. During the period, a northwesterly wind prevailed for three days in north China. Backward trajectory analysis shows that this high peak of sulfur dioxide and sulfate can be attributed to the emissions from north China and the base of the Shandong Peninsular (Figures not shown). Wang et. al. (1996) concluded the sulfate can be transported longer distance than sulfur dioxide. That's why there are much more obvious peaks of sulfate than those of SO_2.

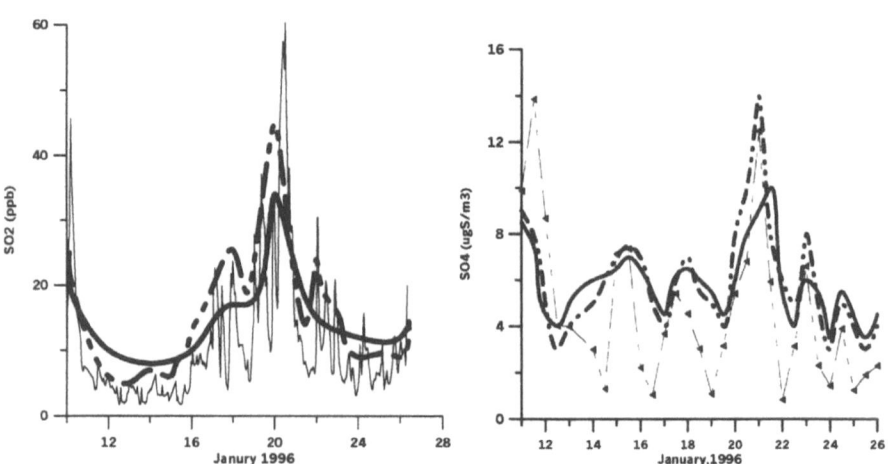

Figure 1. Observed surface SO_2 and sulfate concentration during January 11-28 1996 at Changshantu together with values simulated with NAQPMS. The dark line and dashed dark line represent the output from the coarse domain and nested domain, respectively.

3. Model analysis

3.1 DESCIPTION OF NAQPMS AND VALIDATION

A nested air quality prediction modeling system (NAQPMS) is a three-dimensional Eulerian model, which includes parameters on emission, transport, diffusion, chemistry and deposition. It employs a flexible horizontal grid resolution with multi-level nested grids with options for one-way and two-way nesting procedures. The NCAR/Penn State Fifth-Generation Mesoscale Model (MM5) is used as a meteorological driver. This study used a coarse domain and a nested domain with 27 km resolution. The gas-phase chemical reaction scheme used in the present work is a slightly modified version of the CBM-IV (Wang et. al., 1996). The modified mechanism includes 34 species and 86 photochemical, inorganic and organic reactions. The NAQPMS have been evaluated. Aqueous phase chemistry used the RADM-2 mechanism, and the effects include sub-grid scale vertical re-distribution, dissolution, dissociation, and wet deposition are also considered. The coarser domain uses emissions of SO_2, sulfate, NO, NO_2, NH_3, CO and VOC, which are derived from GEIA datasets and RAINS-ASIA emission datasets. The emissions are obtained from EPA of China. Details of model construction and performance are provided by Wang et. al.(2001).

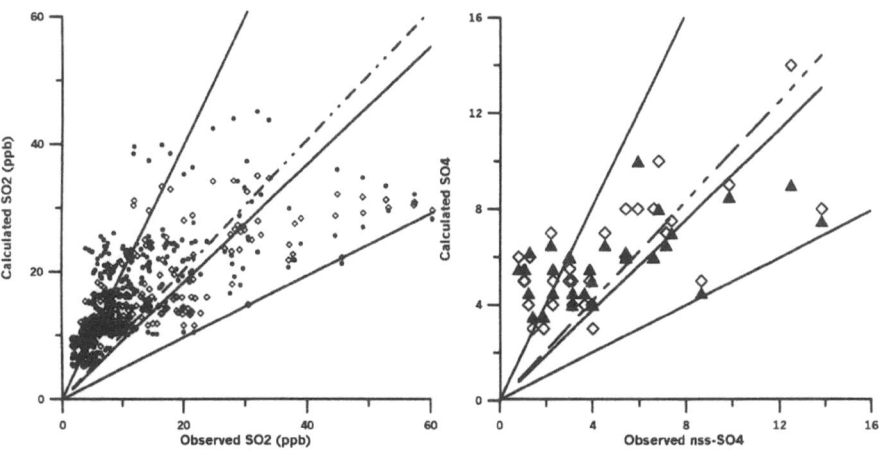

Figure 2. Scatterplot diagram for the calculated and observed (a) SO_2 and (b) SO_4^{2-} concentration at Chengshantu. The dark closed triangles and open squares represent the calculations from the coarse domain and nested domain, respectively.

Figure 1 also shows the model results for the observation site from the coarse domain and nested domain. A reasonable agreement is reached between the predicted SO_2 concentrations and observed values, especially for the nested domain. The model calculations can reproduce the variations of sulfur dioxide and sulfate well, although the model calculations could not reproduce the extensively low sulfate concentrations. It can be seen that using nesting in the numerical modeling system can improve the simulation results.

A scatter plot gives a clear visual picture of correlation between observed and predicted values. Thus, scatter diagrams for SO_2 and SO_4^{2-} concentration are shown in Figure 2 in which all the observed values at monitoring site are plotted. A reasonable correlation between observed and predicted values is obtained in these figures, although some points lay far from the straight line. Most of points are located in the confident area. The better predictive performance for the nested domain is obtained. The over-estimated values at the right side of each figure may be caused by the over transport due to the errors of the wind speed simulations.

3.2 DISTRIBUTION OF SO_2 AND SULFUR TRANSPORT

The horizontal winds and distributions of surface SO_2 concentrations are shown in Figure 3 and Figure 4 for the coarse domain and nested domain, respectively. On January 20, the prevailing wind over north China and the Shandong peninsular was northwesterly. From the simulation, extremely high concentration of SO_2 occurred in Sichuan province and Shanghai area. The weather showed a typical winter pattern and air pollutants over north China were transported southeastward to Chengshantu with the strong northwesterly winds. The results from the nested domain can reproduce the structure of wind fields and spatial distribution of sulfur dioxide more clearly. From the nested domain, there is a lesser extent over Bohai Sea and would be transported to the Shangdong Peninsular in the next few hours.

Estimates of the major sources of deposited sulfur are quite different in different model; for example, the foreign contributions to sulfur deposition in Japan vary from 5% to 65% (Ichikawa and Fujita, 1995; Huang et al., 1995; Chang et al., 2000). Here we also calculate the transport of sulfur among four countries in northeast Asia using the model system for the campaign period. To obtain the origin for each country, we estimate the contributions to sulfur deposition by assuming zero emissions in each country in turn, and the results are given in Table I. It can be seen that the origins of deposited sulfur in each country are quite different from each other, varying from 51% to 99%. The origins of total sulfur from China to Japan, South Korea and North Korea are 28.36%, 23.27% and 40.14%, respectively. For Japan, nearly 49% of deposited sulfur is from the continent. This suggests that the long-range transport of sulfur is very serious during the campaign period. Of course, the estimations of these values are different with various synoptic systems.

1797

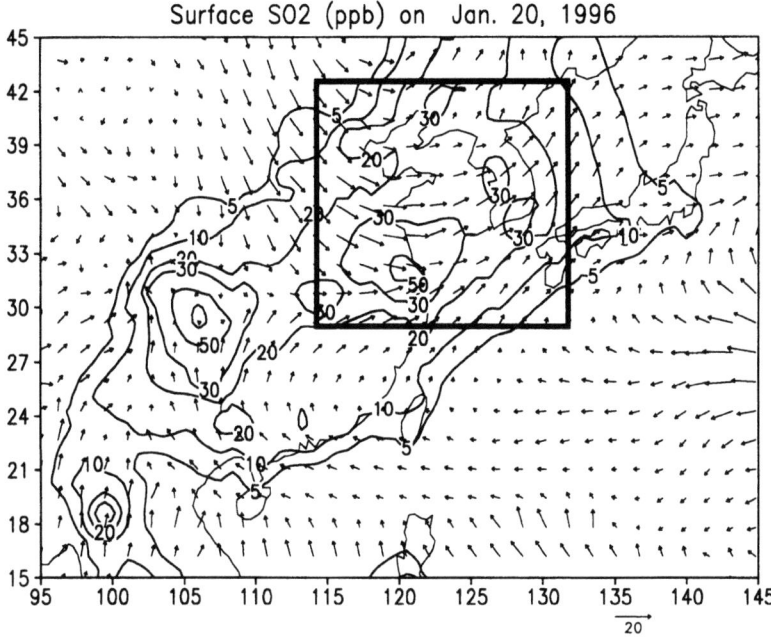

Figure 3. Simulated surface SO$_2$ distribution with NAQPMS on January 20 1996 from the coarse domain.

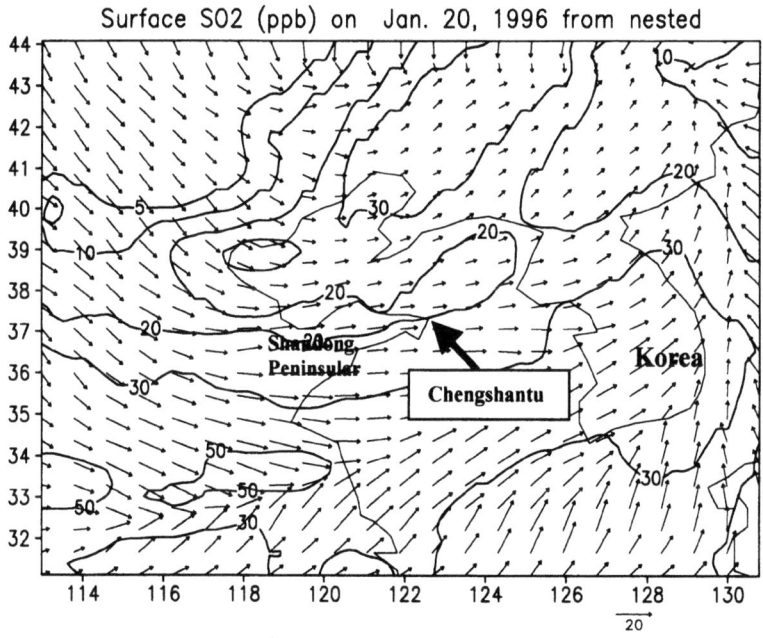

Figure 4. Simulated surface SO$_2$ distribution with NAQPMS on January 20 1996 from the nested domain.

TABLE I

Percentage contribution of the deposition of total sulfur to each country in East Asia from other countries during January 10-24, 1996

Receptor	Source			
	China	Japan	S. Korea	N. Korea
China	98.90	0.00	0.21	0.05
Japan	28.36	51.53	17.41	2.45
S. Korea	23.27	0.17	72.80	3.72
N. Korea	40.14	0.00	8.19	51.19

4. Conclusion

Ground-based observations performed in winter at Chengshantu, the northeast corner of Shangdong peninsular, showed that the influence of the continental outflow of air pollutants contributed to the observed high concentration of SO_2 and sulfate on January 20 1996 by the prevailing northwesterly winds. Results of a nested air quality prediction modeling system (NAQPMS) simulations indicted that nesting improves the ability of the modeling system to catch the peaks of pollutants. The calculated origins of deposited sulfur in each country with the NAQPMS show that the long-range transport of sulfur is a serious problem in winter. For Japan, 49% of sulfur deposition in Japan was from other countries during the campaign period. It is necessary to conduct multi-point observations of air pollutants in order to compare the performance of various trans-boundary models applied to East Asia through international co-operations.

Acknowledgements

The authors would like to thank anonymous referees for valuable critical reviews which helped to improve this manuscript.

References

Akimoto, H., and Narita, H.: 1994, *Atmos. Environ.* **28**, 213.
Chang, K.-H., et. al.: 2000, *Atmos. Environ.* **34**, 3281.
Huang, et al.: 1995, *Water, Air and Soil Pollution*, **85**, 1921.
Ichikawa, Y. and Fujita, S.: 1995, *Water, Air and Soil Pollution*, **85**, 1927.
Murano, K., et. al.: 2000, *Atmos. Environ.* **34**, 5139.
Wang, Z., et. al.: 1996, *Advances in Atmospheric Sciences* **13**, 399.
Wang, Z., Maeda, T., and Hayashi, T.: 2001, *Water, Air and Soil Pollution, (this issue).*
Xiao, H., et. al.:1996, *J. Geophys. Res.* **102**, 28589.

ASSESSING HEALTH EFFECTS OF AIR POLLUTION IN DEVELOPING COUNTRIES

FRANK MURRAY[1], GORDON MCGRANAHAN[2,4] and
JOHAN C.I. KUYLENSTIERNA[3*]

[1] *Environmental Science, Murdoch University, Perth, Western Australia;* [2] *Stockholm Environment Institute (SEI), Box 2142, S-103, 14 Stockholm, Sweden;* [3] *Stockholm Environment Institute at York (SEI-Y), Box 373, University of York, York, UK;* [4] *Current address: International Institute for Environment and Development, 3 Endsleigh Street, London WC1H 0DD, UK.*
(* *author for correspondence, e-mail: jck1@york.ac.uk*)

Abstract. Health effects of air pollution in Asia have been assessed as part of a programme on Regional Air Pollution in Developing Countries. The impacts of air pollution on health have been studied in North America and Europe for many decades, but research on effects on health in developing countries is less advanced. A key question is whether the dose-response models that are based on research conducted in developed countries can be applied to exposures to air pollution in developing countries. The study considered this issue and examined the factors that may lead to either increased sensitivity or increased human tolerance of air pollutants. It is suggested that although there are factors in developing countries that may increase or decrease human sensitivity to air pollution, overall, a similar range of sensitivity can be expected by individuals in these countries responding to the same effective dose as those in developed countries.

Keywords: air pollution, health effects, developing country, dose-response, risk assessments

1. Introduction

In public opinion, local impacts of air pollution on health are closely associated with industrialisation. Deaths resulted from air pollution in the 18th, 19th and the early part of the 20th centuries in London, UK; the Meuse Valley (Belgium); Donora (USA); New York City (USA); Osaka (Japan) and elsewhere. High air pollution levels in these cities resulted in excess deaths, including more than 4000 excess deaths in London from a stagnant atmosphere of fog, smoke and sulphur dioxide in five days in December 1952 (Brimblecombe, 1987).

With public pressure for less polluted air in cities, availability of relatively cheap clean fuels, strong economic growth and increasing incomes, governments in developed countries slowly introduced measures to improve ambient air quality in cities. In response to growing regional problems, more effective international action has been implemented. International guidelines on ambient air quality have been produced by organisations such as WHO (WHO, 1987; 2000), and international policies are being co-ordinated under conventions such as the Convention on Long-range Transboundary Air Pollution (UN/ECE, 1996).

Much of the world's population lives in areas where levels of ambient air pollution exceed World Health Organisation (WHO) guidelines. More than 1200 million people may be exposed to excessive levels of sulphur dioxide, more than 1400 million people may be exposed to excessive levels of suspended particulate matter. Although data from developing countries are incomplete, about 15-20 per cent of the population of Europe and North America could be exposed to excessive levels of nitrogen dioxide (UNEP, 1991). Many of the published studies on the effects on human health of air pollution relate to the effects of outdoor air pollution on residents of North America and Europe of Caucasian descent, usually of good nutritional status, living in uncrowded conditions, without physical stress or untreated chronic diseases. There are fewer studies on populations of other ethnic backgrounds, nutritional status, living conditions, stress, or history of chronic diseases, or of indoor air pollution. These factors may alter the dose-response relations derived for exposure to outdoor air pollutants (Table I).

To help to clarify these issues, the Stockholm Environment Institute is co-ordinating a programme on Regional Air Pollution in Developing Countries, funded by the Swedish International Development Co-operation Agency (Sida). A key question is whether the dose-response models that are based on research conducted in developed countries can be applied to exposures to air pollution in developing countries. The study considered this issue and examined the factors that may lead to either increased sensitivity or increased human tolerance of air pollutants.

The purpose of this study is to synthesise policy-relevant knowledge on air pollution and health, and thereby provide a firmer basis for improving public health in developing countries. The emphasis is on scientific knowledge and its use in air pollution management. The goal is to support locally driven processes of air pollution management, not to summarise current conditions in the region. Studies from Europe and North America, where the research is more advanced, were reviewed to provide insights into the relationships between some of the most critical air pollutants and health. Studies from a range of developing countries, where studies are less exhaustive but the results are probably more comparable, were also reviewed.

Although the database on indoor air quality is considerably weaker than the database on outdoor air quality, many premature deaths are caused by indoor air pollution (Smith, 1996). Indoor air pollution is of particular concern when smoky fuels are used for cooking and/or heating in poorly ventilated rooms. This tends to be more common in low-income rural areas, where fuelwood and biofuels are plentiful and people cannot easily afford cleaner fuels. Thus, economic development tends to be associated with declining indoor air pollution problems.

There is much to learn from the studies relating air pollution to health in North America and Europe. They include studies of the common air pollutants and their adverse effects on a wide range of health indicators, such as mortality rates, hospital admissions, and lung function. Only a few of the more advanced studies have been replicated in developing countries, so many health assessments in developing countries have used these Northern studies.

2. Development of Standards

Epidemiological studies in the late 1980s and 1990s based on time-series analyses have raised new concerns about health impacts of some of the most common air pollutants. The results of these studies have been remarkably consistent and have withstood critical examination (WHO, 2000). They demonstrate associations between air pollutants and health impacts at levels of pollution previously expected to be quite safe. Partly as a result, WHO has developed new air quality guidelines (WHO, 2000).

New insights into air pollution are also providing the basis for new tools for air pollution management. The recent assessments of WHO show that for particles and ozone there is no indication of any threshold of effect; that is, there are no safe levels of exposure, but risk of adverse health effects increases with exposure. Similar difficulties in identifying a threshold of effect at a population level apply to lead. This is important as no single guideline value can be recommended. It is a significant departure from the concept of a guideline value as a level of exposure at which the great majority of people, even in sensitive groups, would be unlikely to experience any adverse effects. Translation of this new form of guideline into an air quality standard is likely to be difficult.

There are a several issues to be considered when translating the models based on results of studies conducted in Europe or North America into national or state air quality standards in developing countries (WHO, 2000):

i. The chemical composition of the particles may differ from published studies
The mixture of particles in the communities studied in the development of the particulate response models was dominated by emissions from motor vehicles, power generation, and space heating by natural gas and light oil combustion. The mixtures of particles in communities in developing countries may be different. They may be dominated by different emissions sources with different chemical characteristics, and by wind-blown soil with different toxic properties from those in most published studies.

ii. The concentration range may differ from published studies
The response-concentration models for particulate matter are based on a linear model of response, within the range of particulate concentrations typically found in developed countries. There are no grounds for simple extrapolation of the concentration-exposure relationship to high levels of particulate pollution sometimes found in large cities in some developing countries. Several studies have shown that the slope of the regression line is reduced when the concentration of particulates is at high concentrations.

iii. The responsiveness of the population may differ from published studies
Many of the response-concentration models are based on responses of populations in developed countries that were mostly well nourished and with access to modern health services. In contrast, the populations exposed to higher concentrations of particles in less developed countries may have a lower level of quality of both nutrition and health care, and may have different age structure and other characteristics (see Table I). It is unclear whether the responsiveness of the populations in other parts of the World differs from those studies in North and South America and Europe.

The epidemiological evidence on ambient air pollution and health in developing

TABLE I
Summary of Effects of Factors on Response to Air Pollution (adapted from WHO/UNEP, 1993)

Outdoor Pollutant	Extremes of Temperature		Extremes of Humidity		High Altitude	High Natural Particle Levels	Endemic Disease	Inadequate Medical Care	Nutritional Deficiency and Debilitating disease	Genetic Factors	Local Habits	Life Style	Susceptible Groups
	High	Low	High	Low									
PM$_{10}$	+	(+)	+	+	+	+	?	+	+	+	+	+	a-f
O$_3$	+	0	(+)	0	(+)	?	(+)	+	+	?	+	+	a,c,d
SO$_2$	+	+	+	+	?	+	+	+	+	?	+	+	a,b,c,f
NO$_2$	(+)	+	?	(+)	(+)	(+)	+	+	+	?	+	+	a,b,c,e
CO	(+)	(+)	0	0	+	0	+	+	+	+	+	+	b,d,e,f
Pb	0	0	0	0	+	(+)	+	+	+	+	+	+	b,d,e

Response:
- \+   increase can exacerbate the effect
- (+)   increase probably enhances the effect
- 0   increase produces no effect
- (-)   increase probably mitigates the effect
- -   increase can mitigate the effect
- ?   insufficient evidence

Susceptible Groups:
- (a) Asthmatics
- (b) Elderly
- (c) COPD[1]
- (d) Myocardial pathology
- (e) Children, infants and/or foetus
- (f) Pulmonary fibrotic disease

[1] COPD is an acronym for Chronic Obstructive Pulmonary Disease

countries taking into account factors such as nutritional status and population structure, suggests that the adverse health effects may be even greater than those found in developed countries (Romieu and Hernandez, 1999). The data required for the more in-depth studies are not generally available, but the available evidence tends to confirm the view that residents of polluted cities in developing countries are at considerable risk. For example, appreciable risks were found in studies relating particulate concentrations and mortality in Sao Paulo, Santiago, Mexico City and Bangkok. Clearly, more research is needed in developing countries, but the indications are that the effects of air pollution are at least roughly comparable.

Models suggest that a $10\mu g/m^3$ increase in particulate matter with a diameter less than 10 µm (PM_{10}) is associated with an increase in daily mortality of about 0.74%, and an increase of 1-4% in emergency visits and hospital admissions (WHO, 2000). Studies in cities in developing countries reviewed by Romieu and Hernandez (1999), as part of this project, show that although direct extrapolation of health effects observed in populations living in urban areas of developed countries to populations living in developing countries is difficult, most available evidence suggests they experience similar or greater adverse effects of air pollution at the same ambient concentrations.

In large parts of developing countries, however, an exclusive focus on ambient air quality is potentially misleading. Indoor air pollution may be having a large impact on health owing to the use of biofuels, such as fuelwood, to cook (and sometimes heat) in enclosed spaces, especially in rural areas. Among the principal health risks are acute respiratory infection in children, and chronic obstructive lung disease and lung cancer in women. Despite its potentially great importance, most of the research on the health risks of indoor air pollution is recent, with sample sizes and study designs far less sophisticated than those used to study the effects of outdoor air concentrations. There is emerging evidence that indoor air pollution is associated with important health effects, but much research is needed to understand these effects (Smith, 1993).

Vehicles are a major contributor to ambient air pollution in many Asian cities, and can lead to high exposures among people situated in heavily trafficked areas. The vehicle fleets of Asia are not as large as in many developed countries. However, the popularity of highly polluting motorcycles and scooters, the age and poor maintenance of the vehicles, the high usage rates, and the continued use of leaded and poor quality fuels, lead to high emissions per vehicle for a number of health-damaging pollutants. However, there have been effective measures to reduce the vehicle pollution in a number of Asian cities and countries.

There is, of course, enormous variation in the air pollution problems different Asian cities face, and the extent to which their health implications have been studied. A review of recent findings in Hong Kong suggests that the relative risks to health from a number of pollutants are higher than in Western European cities, but that recent control measures have had an appreciable effect on air quality and health (Barron *et al.*, 1995).

Taken together, these studies raise serious concerns about the health hazards of air pollution in developing countries, and indicate that much can be done to reduce these concerns. There is, however, a wide range of actors involved, and co-ordination is of critical importance. There is little point in collecting information if it will not be used, and it is unfortunate when actions are taken on the basis of insufficient information. This is one of the justifications of developing air quality management systems at the local level, and involving all stakeholders in the process. National governments need to provide support for local initiatives, as well as an appropriate regulatory and policy framework. In addition, regional and international co-operation can also lead to important contributions.

Regional co-operation on transboundary air pollution could also be linked to air pollution and health concerns. Improved health is one of the most compelling reasons to prevent air pollution in developing countries.

3. Conclusions

The evidence of studies conducted in developing countries suggests that the results of studies on impacts of air pollution on health in developed countries may be generally applicable to developing countries. Care is needed to characterise the air pollution situation, and to avoid extrapolations beyond accepted concentration-response models. Clearly, further studies are required to assess the relationship between air quality and health parameters in the developing countries, and to characterise particular situations in developing countries that are not commonly found in the developed nations.

References

Barron, W.F. Liu, J. Lam, T.H. Wong, C.M. Peters, J. and Hedley, A.J.: 1995, Costs and benefits of air quality improvement in Hong Kong. *Contemporary Economic Policy* **9**, 105.

Brimblecombe, P.: 1987, *The Big Smoke*. Methuen, London.

Romieu I. and Hernandez, M.: 1999, 'Air pollution and health in developing countries: review of epidemiological evidence', in G. McGranahan and F. Murray (eds.), *Health and Air Pollution in Developing Countries*. Stockholm Environment Institute, York. pp. 43- 65.

Smith, K.R.: 1993, Fuel combustion, air pollution exposure and health: the situation in developing countries. *Annual Review of Energy and the Environment* **18**, 529.

Smith K.R.: 1996, 'Indoor air pollution in developing countries: growing evidence of its role in the global disease burden', in *Proceedings of Indoor Air '96. The 7th International Conference on Indoor Air Quality and Climate*. Institute of Public Health, Tokyo.

UN-ECE: 1996, *1979 Convention on Long-range Transboundary Air Pollution and its Protocols*. ECE/EB.AIR/50. United Nations Economic Commission for Europe, Geneva, Switzerland.

UNEP: 1991, *Urban Air Pollution*. United Nations Environment Programme, Nairobi.

WHO: 1987, *Air Quality Guidelines for Europe*. WHO Regional Office for Europe, Copenhagen.

WHO: 2000, *World Health Organization Guidelines for Air Quality*. WHO, Geneva.

WHO/UNEP: 1993, *Factors Affecting the Impact of Air Pollutants on Health*. WHO, Geneva.

TRENDS OF AIR POLLUTION AND ITS PRESENT SITUATION IN HIROSHIMA PREFECTURE

PRABAL K. ROY* and HIROSHI SAKUGAWA

*Faculty of Integrated Arts and Sciences, Hiroshima University,
1-7-1 Kagamiyama, Higashi-Hiroshima 739-8521, Japan
(* author for correspondence, e-mail: kanti@hiroshima-u.ac.jp)*

Abstract. The ambient concentration of SO_2, NOx and O_x in the atmosphere of Hiroshima, Fukuyama and Fuchu city which were monitored by the prefectural monitoring stations, are examined to give a picture of the typical air pollution at these sites. Results show that the yearly concentrations of SO_2 in these areas are significantly fall from 20 to 6 ppb during 1978–1996 when the NOx concentrations having no such significant change which varies from 40 to 30 ppb. The Photochemical Oxidant (Ox) increases annually at the rate of 0.3 ppb to 0.6 ppb in Hiroshima city only. To know the present situation of air pollution the Differential Optical Absorption Spectroscopy (DOAS) system is used in the city of Higashi Hiroshima. The daily average concentrations of SO_2, NO_2, O_3 and HONO measured during the period of August 1999 to March 2000 ranged from 1.4 ppb to 2.8 ppb, 13 ppb to 26.9 ppb, 21 ppb to 53.6 ppb and 1 ppb to 4.3 ppb respectively. The patterns of concentrations of NO_2 and O_3 measured by DOAS look similar to the seasonal patterns of NOx and Ox by the conventional system.

Keywords: air pollution, SO_2, NOx, Ox, DOAS

1. Introduction

Like the other big cities in Japan, Hiroshima and Fukuyama are expanding by its population with an increasing trends of 12% and 8% respectively from 1980 to 1995, while the total population is decreased by 5% in the small Fuchu city. During this period though the total number of factories is decreased in all the areas, the total sulfur dioxide (SO_2) discharge remains same in these areas. The changing trends of population and their activities also change the urban and industrial developments in these areas, which have sharp impacts on the local air qualities. To examine the air pollution conditions and to maintain a better living environment, the prefecture and the cities established the Air Pollution Monitoring Telemeter System in 1971. Ambient Air Pollution Monitoring Stations collected the hourly data of air pollution for 24 hours and send the information to the monitoring center through radio communications.

Due to the strong abatement policy of the Govt. the control of SO_2 has been remarkably improved in recent years but there is still a great concern for the nitrogen oxides (NOx), which is mainly for its role to ignite the ground level ozone (Ox), which ultimately effects human health and crops (Chameides et al., 1994). NOx, which is the combination of Nitric Oxide (NO) and Nitrogen dioxide (NO_2) have its anthropogenic sources from the fossil fuel combustion and biomass burning (Garnett, 1979; Glasson, 1981). NO and NO_2 serve as the catalysts in many atmospheric reaction chains as well as controlling the distribution of atmospheric ozone (Chameides and Walker, 1973). In this paper the trends of air pollution from 1978-1996 in the cities of Hiroshima, Fukuyama and Fuchu and its present situation in Higashi Hiroshima city is analyzed.

2. Materials and Methods

For this study the SO_2, NOx and Ox data are collected from the total 15 individual general monitoring stations among which 7 stations are in each city areas of Hiroshima and Fukuyama and 1 station in Fuchu city. Besides those, NO_2 data from 3 roadside monitoring stations in Hiroshima city and 1 in Fukuyama city are also considered separately only to compare the trends of NO_2 concentrations in those cities. The electroconductivity analyzer is used for measuring SO_2 from change in the conductivity of absorbent (hydrogen peroxide solution), whereas NO_2 is measured by the Salzman automatic colorimeter. Since nitric oxide (NO) does not react with the Saltzman reagent, after oxidizing it by passing through the sulfuric acid potassium permanganate solution the concentration of NO is determined. For oxidant (Ox) the neutral buffered automatic KI colorimeter is used at the monitoring stations (HPG, 1990).

The individual hourly pollutants are than summarized as for seasonal and yearly variations. The pollutants of SO_2, NO_2, O_3 and gaseous nitrous acid (HONO) are monitored in the Higashi Hiroshima city from August 1999 to March 2000 by using the Differential Optical Absorption Spectrometer (DOAS), which is installed, at the 5th floor at about 13 meter above the ground level of the main faculty building of the Hiroshima University Higashi Hiroshima campus. The university campus is 150 meter above the sea level. The retro reflector was mounted at the top of the opposite building at 260 meter far from the emitting telescope. The DOAS (Thermo Electron Co., Ltd.- Model 2000) system has the specifications by which the measurement can be possible within the optical path length from 50 to 2000 m, averaging time from 10 sec to 24 hours, normal wavelength from 200 to 650 nm and the scan range of 32 nm. For our study the wavelengths were fixed differently for individual elements, which were 262 nm for O_3, 300 nm for SO_2 and 355 nm for both NO_2 and HONO. The averaging time to determine each compound was taken as 5 min and was carried out in every 15 min. The concentration data as stored in the computer are later rearranged by the hourly values and analyzed. The meteorological data are taken from NIES and Hiroshima University Meteorological Data Acquisition System online service.

3. Results and Discussion

3.1. TRENDS OF SO_2, NOX AND OX CONCENTRATIONS

To clarify the trends of air pollution the annual variation of SO_2, NOx and Ox concentrations from 1978-96 for the cities of Hiroshima, Fukuyama and Fuchu are shown in Figure 1. The annual average concentration of sulfur dioxide (SO_2), the source of which is mainly from oil and gas processing, ore smelting and the burning of coal and heavy oil, is significantly decreased from 18 ppb to 6 ppb in Hiroshima city, 11 ppb to 8 ppb in Fukuyama city and 12 ppb to 7 ppb in Fuchu city. It is seen that even during the higher annual average of SO_2 of 1991 in Fukuyama and of 1992 in Hiroshima, the hourly values of maximum concentrations have never exceeded the 100 ppb (the NAAQS of Japan for SO_2).

In Hiroshima, the trends of Photochemical Oxidant (Ox) is increased by 0.6 ppb annually ($P < 0.05$) when its NOx concentration is decreased (Fig. 1). This may be the cause of the precursors change from the place of occurrence of higher concentrations (S. Wakamatsu et al., 1996) or the long range transport from other region of Japan and the Asian continent (Mauzerall et al., 2000). The concentrations of NO at the general monitoring stations and NO_2 at the automobile exhaust monitoring stations showing no trends (Fig. 2). Similar also the case of Ox in Fuchu (Fig. 1). But the NOx concentration (Fig. 1) in Fuchu has tended to be increased significantly by 0.4 ppb annually ($P < 0.05$). In Fukuyama, the concentration of NOx and Ox virtually having no trends during the study period (Fig. 1) though the concentrations of NO and NO_2 at the general monitoring stations are found significantly increasing ($P < 0.05$) (Fig. 2).

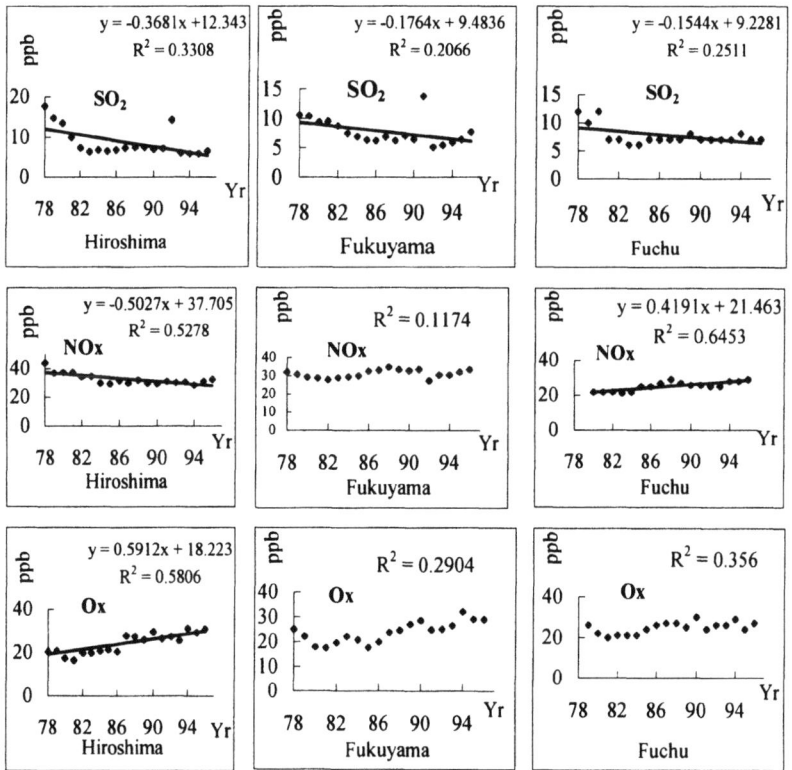

Figure 1. Annual average concentrations of SO_2, NOx and Ox in Hiroshima, Fukuyama and Fuchu city from 1978-96.

Though there is no automobiles emission data available, but during the period of 1988 to 1995, in Hiroshima city, the numbers of automobiles have increased 34% and jointly in Fukuyama and Fuchu City it is increased 36.5%. From the factory emission it is seen that from 1988 to 1995 the emission of NOx in Hiroshima is increased 47% (2657.32 ton), when at the same period it is increased 9% (18630 ton) jointly in Fukuyama and Fuchu (HPG, 1998). To clarify the characteristics of the annual trends of NOx in this region, the NOx emission data from the automobiles should be needed.

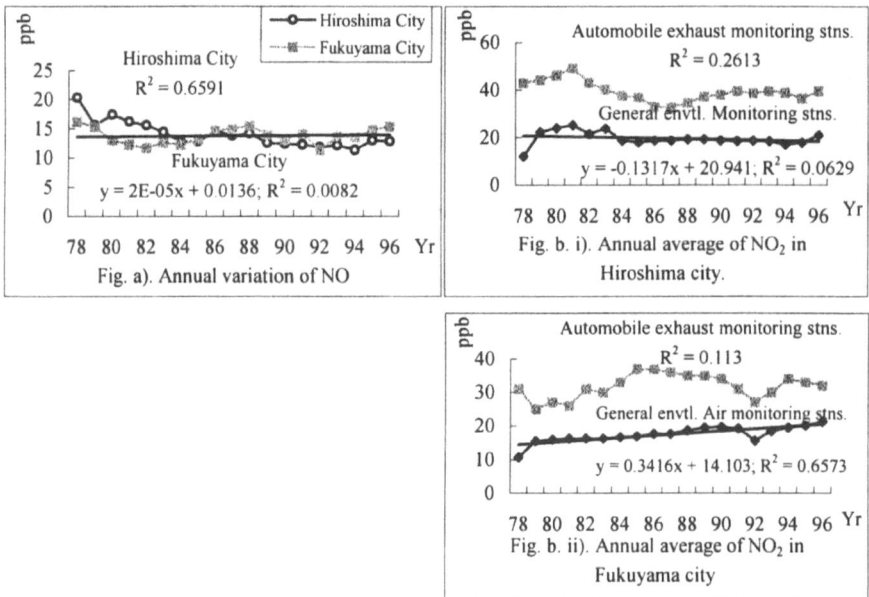

Figure 3 shows the annual variations of seasonal NOx and Ox concentrations in Hiroshima city. NOx is found having higher concentrations in winter and lower in summer when it has been gradually decreasing in all the seasons from 1980 to 1996 (Fig. 3 a). The higher concentration of NOx in winter is mainly due to the result of the local meteorology and the lower rate of photochemical oxidation (Sakugawa and Kaplan, 1989). Ox is found its higher concentration in spring and lowers in winter (Fig. 3 b) in Hiroshima city. The cause of this spring maximum

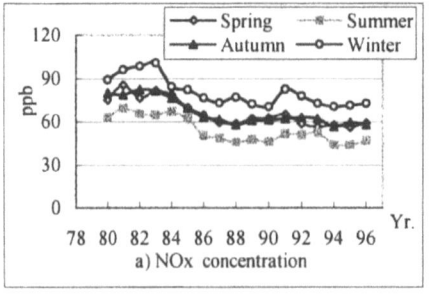
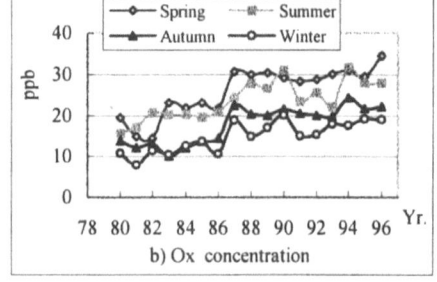

Figure 3. Annual variation of seasonal a) NOx and b) Ox concentrations in Hiroshima.

of ozone is a Northern Hemispheric phenomenon, which occurs widely across the mid-latitudes in the Northern Hemisphere (Monks, 2000).

Figure 4 shows the air pollution situation in Higashi Hiroshima city measured by DOAS from August 1999 to March 2000 when only 22 to 24 days per month data from August to October and 11 to 14 days per month from November to March data are considered for analysis. The concentration data could not be monitored in other days because of the reduction of lamp intensity generated from the focus of the telescope used or the occurrence of strong fog at early morning which can block the light if the path is long (Edner *et al.*, 1993). The characteristic feature of the air pollution condition in this place is that except in November and December the concentration of NO_2 was much lower than the O_3 concentration in this area. The mean of the daily average of O_3 ranges from the lowest 21 ppb in December to the highest 53.6 ppb in March, when the NO_2 has its lowest daily average concentration of 13 ppb in August and the highest 27 ppb

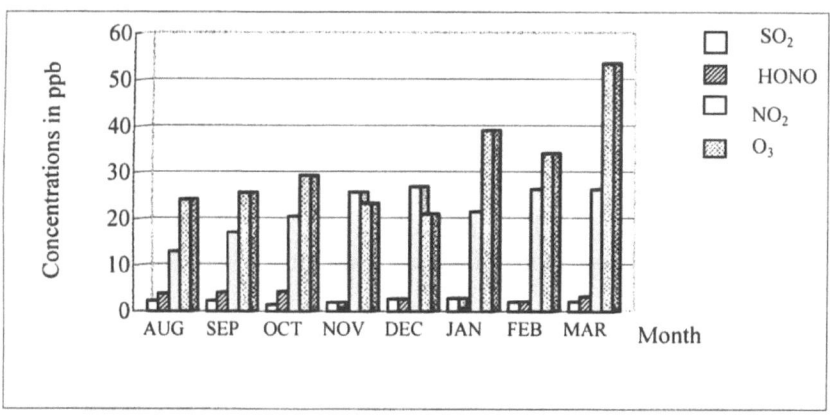

Figure 4. Mean of the daily average of O_3, SO_2, HONO and NO_2 in Higashi Hiroshima city from Aug.'99 to Mar. 2000.

in December respectively. During the entire study period, the daily average of SO_2 remains at the same level of 2.0 ppb. The gaseous nitrous acid (HONO), which is a well-known photolytical source of OH radicals, shows its daily average concentration of 3.6 ppb during the measuring period, when its lowest daily average measured was 1.0 ppb in November and highest was 4.3 ppb in October. From this analysis, SO_2 and HONO looks no trends when the concentration of NO_2 looks lower in summer and higher in winter and O_3 looks lower concentration in winter and higher in spring indicating their similarity to the seasonal patterns as in Figure 3 of traditional monitoring system data. During the 8 months study period by DOAS a calm (0.1 to 2 m/s) wind speed prevails in Higashi Hiroshima city. In our separate study we have also found that the SO_2, NO_2 and O_3 concentrations measured by DOAS having strong correlations with the data of the conventional methods (Hirakawa *et al.*, 2000).

4. Conclusions

The annual trends of SO_2 in all the cities of the study area and NOx only in Hiroshima are found decreasing during the study period. Photochemical oxidant (Ox) in Hiroshima and also the NOx in Fuchu are increasing when the average wind speed was around 2 m/s in these areas. The increasing trends of Ox in Hiroshima and its springtime higher concentrations are believed to be the causes of the precursors changes from the point of origin of other domestic sources or the causes of the long range transport from the region of east Asia are needed to be further studied to verify it. The concentration pattern of NOx and Ox in Fukuyama and Ox in Fuchu are not clear about their annual trends when the increasing trends of NOx in Fuchu are needed to be considered for our concern.

Among the DOAS data though SO_2 and HONO in Higashi Hiroshima city looks no trends of their concentrations but the patterns of concentrations of NO_2 and O_3 look similar to the seasonal patterns of NOx and Ox of the conventional system.

Acknowledgements

The authors are indebted to National Institute of Environmental Studies (NIES), Tsukuba for providing the air pollution data. Thanks also to the CREST of Japan Science and Technology (JST) for providing the DOAS. We gratefully acknowledge the support of the Japan Society for the Promotion of Science through a grant (P 99122).

References

Chameides, W., Kasabhatla, P., Yienger, J. and Levy, H.: 1994, *Science* **264**, 74.
Chameides, W.L. and J. C. G. Walker.: 1973, *Journal of Geophysical Research* **78**, 8751.
Edner, H., Ragnarson, P., Spannare, S. and Svanberg, S.: 1993, *Applied Optics* **32**, 327.
Garnett, A.: 1979, *Atmospheric Environment* **13**, 845.
Glasson, W.A.: 1981, *Journal of Air Pollution Control Association* **31**, 1169.
Hirakawa, T., Roy, P. K. and Sakugawa, H.: 2000, 'Measurements of Ozone, Nitrogen Oxides and Sulfur Dioxide by Differential Optical Absorption Spectroscopy (DOAS) System and Other Monitoring System in Higashi Hiroshima, Western Japan' in H. Akimoto and H. Sakugawa (eds.), Oxidants / Acidic Species and Forest Decline in East Asia , Japan Science and Technology Corporation. pp 197-200.
HPG: 1990, Hiroshima Prefectural Government (HPG) published report on 'Air Pollution in Hiroshima Prefecture', pp 9-20.
HPG: 1998, Hiroshima Prefectural Government (HPG) published report on 'Air Pollution in Hiroshima Prefecture', pp 1-127.
Mauzerall, L. D., Narita, D., Akimoto, H., Horowitz, L., Walters, S., Hauglustaine, D. A. and Brasseur, G., 2000, *Journal of Geophysical Research* **105**, 17895.
Monks, P. S.: 2000, Atmospheric Environment **34**, 3545.
Sakugawa, H. and Kaplan, R. I.: 1989, *Journal of Geophysical Research* **94**, 12957.
Wakamatsu, S., Ohara, T. and Uno, I.: 1996, *Atmospheric Environment* **30**, 715.

THE NEED FOR EDUCATION IN DEVELOPING ACCEPTABLE AIR POLLUTION CONTROL STRATEGIES

KEITH R. BULL

Environment and Human Settlements Division, United Nations Economic Commission for Europe, Palais des Nations, CH-1211 Geneva 10, Switzerland
(for correspondence, e-mail: keith.bull@unece.org)

Abstract. The acute effects of air pollution on human health and the environment are well understood and the arguments for measures to prevent local, gross pollution are strong. Governments and the public will accept the need for controls where effects are obvious. At the broader scale where effects may be more subtle, and where the costs of abatement are high, a convincing case is necessary before acceptable solutions are adopted. An education process is needed to provide the relevant facts in an understandable form. For major air pollutants, where international agreement for control of emissions is required, effect-based instruments have proved successful in Europe. These are designed to be cost effective by offering protection to the more sensitive areas by targeting the cheapest emission controls on the sources responsible for effects. This level of complexity has demanded improved education and communication for all those involved in the decision-making process. The principles and approaches that have provided success are discussed. Attention to these is needed in the future if more stringent and costlier measures are to be agreed.

Key words: abatement strategies, communication, education.

1. Introduction

Effective communication between scientists, negotiators, decision makers and the public at large is an essential part of the development of air pollution control strategies. Each group needs to understand the concepts, ideas and information behind the decision making process and appreciate the arguments used to arrive at a final strategy. If we understand the principles behind past successes they can be applied with effect in the future. If communication is poor, it may be difficult to involve all groups effectively in the process and there is a risk of failure with negotiations. The media, the public and groups with vested interests, such as industry, must also be well-informed if a strategy is to achieve general acceptance.

Air pollution is often included in student courses designed to educate and to provide information and examples of important environmental issues and how they may be resolved. But much less consideration has been given to the education of those actually involved in the development of control strategies themselves. As control measures become more stringent and costlier, there is increasing risk of losing general acceptance of policies, and hence increased need to educate all those involved to improve understanding, enhance communication and increase transparency of decision making. The risk of failure also continues to increase as the basis of defining measures becomes more technical and more complex; there is a strong emphasis on making use of science to define targets, but this may not always prove simple. The time has come to think carefully about what is required to ensure that the levels of knowledge and understanding are adequate, in particular what mechanisms are available and what general principles can be employed to ensure success. Problems need to be appreciated and dealt with at an early stage, if possible before negotiations begin.

This paper draws upon experience gained in the development of Protocols under the Convention on Long-range Transboundary Air Pollution (CLRTAP) (Nordberg, 1999). It considers the successes of this regional Convention and the role of education and communication. It describes important principles and mechanisms that might be considered for similar negotiations in the future. The paper is intended to educate as well as present the needs for communication and education.

2. Geographic scale effects

It should be recognized that there is an inherent geographic scale effect when considering education needs related to air pollution effects. There are two main issues causing this. The first is the closer involvement of the population at the local level, an effect not confined to air pollution. The second results from the increased technical complexity of the problem for large geographic areas. These two effects are inter-related, e.g. it is difficult to explain complex problems to many people.

At a local level there is usually an appreciation of acute, local air pollution problems. There is an awareness of the sources of pollution and of the evidence for effects, since the issue may have been the subject of publicity and action. There may also be an appreciation of the "costs", financial or otherwise, and a recognition that any control measures will be for the benefit of the local population. While the issue itself may be technically complex, involvement of the media is likely to result in a simple distillation of the facts to provide most with a grasp of the problem and its solution. The close links between the population and the local decision-maker may also be an important factor in ensuring effective communication and agreement.

At the national scale, the public and decision-maker alike are usually more removed from both sources and effects. Effects are less observable to most, and there may be a lack of awareness for the need to take action on emissions. Even so, publicity and communication by the national press may provide sufficient information and there may be general public concern without any first hand knowledge of effects. Because governments are concerned with the protection of their own environment, they are likely to initiate action where they believe there is sufficient need. This may be determined by real, or potential, public pressure, or a plan to avoid such pressures in the future. At this scale technical issues are more complex. Pollutant transport modelling is more complex, and effects are more difficult to quantify for the purposes of taking action.

At the international scale, the population and policy maker are usually even more removed from sources and effects. Environmental damage may seem very remote. Costs for decreasing emissions affecting other countries may seem very high with little apparent benefit for the country as a whole. There is likely to be a lack of appreciation and understanding of remote effects as there will be little attention from the press and little to attract the public" attention. The complexity of the issue is greater too. There are few models of long-range transport, and these operate at coarse resolution and have large uncertainties. While they provide good policy tools internationally, they may be difficult to relate to local pollution levels.

So at national and international scales especially a knowledge gap is likely. Since the technical complexity is also greater and the political negotiations more difficult, the need to provide clear, simple information is increased. The need for education and understanding becomes greater with larger areas.

3. Effects-based approaches to air pollution control

Until relatively recently, air pollution control measures were simply aimed at reducing emissions but with little understanding of the environmental benefits resulting. With severe and acute pollution problems these measures may be sufficient to make marked improvements, for example with the introduction of smokeless fuels in cities to decrease the incidence of smog. With chronic effects improvements are not so easy to link with any emission decreases.

In the last decade so called "effects-based" approaches for emission controls have evolved. This is partly a result of improved science providing quantitative targets that are linked to environmental goals, mainly using the critical loads approach (Bull, 1995). It is also a result of the large costs of advanced technology, or other measures such as fuel switching or transport policy, that may be necessary for decreasing pollutant emissions to very low levels.

Work under the CLRTAP has developed critical loads approaches and linked science and policy through integrated assessment (IA) modelling (Hordijk, 1995). The IA models bring together data on critical loads, the transport of pollutants and the costs of abatement, and calculate optimal solutions that achieve environmental goals at least cost. The critical loads data provide concise information on environmental targets or goals for environmental protection. The transport models indicate the movement of emissions from source countries to where the concentrations or deposition may affect sensitive receptors. The costs are considered sequentially, applying cheaper technologies first and moving to more expensive measures to achieve more ambitious environmental goals.

The modelling mechanism is simple to understand in principle, and its component parts are easy to understand in more detail (Amman *et al*, 1999). While the model optimization program and some of the associated models are quite complex, it is not necessary to understand these details, provided there is confidence in the experts applying the models and programs. To ensure that there is appreciation, understanding, and acceptance of the component parts of such an approach, it is essential to consider all participants in the decision-making progress (figure 1).

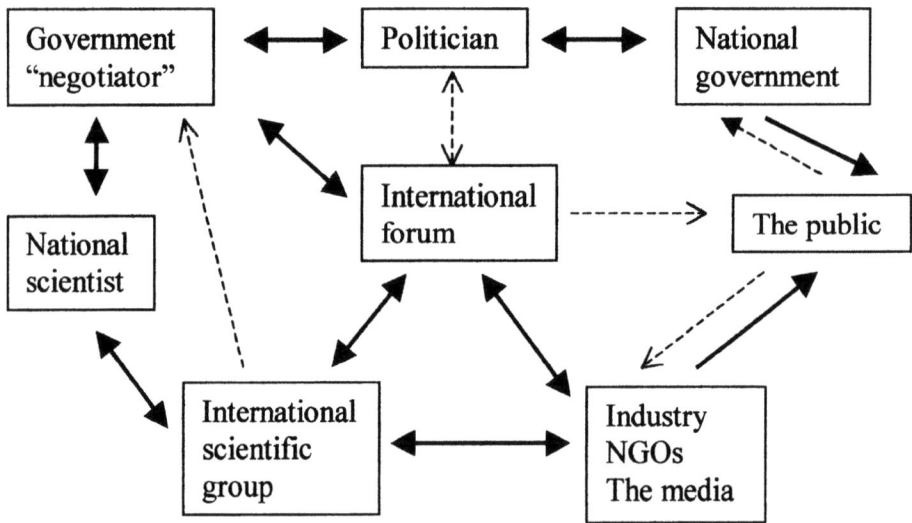

Figure 1. The participants and communication links for the decision-making process associated with the Convention on Long-range Transboundary Air Pollution (line thickness indicates strength of communication)

While the strength of these links can be debated, it is clear that there is often effective communication between some participants, while between others it can be poor. There is potential for additional links, for example between international scientific groups and the public, but these may be well-exploited. Improved communication and education efforts should focus on two areas:
- Further support strong links where concepts or issues are complex or difficult; or

- Address links that are weak or non-existant, e.g. between the international forum and the politician and public, whose understanding may be poor but who may have more influence on decision making in the long term, through policy development and elections.

4. Principles on which to base scientific education

There are a number of factors that have proved important for developing agreements under the CLRTAP. These may be summarized as general "principles" on which to build effective policy development. As part of *our* education we should recognize the importance of the principles and work with them to aid education, communication and policy development in the future.

A major asset in the recent achievements under CLRTAP has been the concept of critical loads. This illustrates several important principles of value:
- Simplicity;
- Scientific acceptability;
- Easy to understand;
- Flexibility, providing useful and variable targets; and,
- Transparency.

At an early stage critical loads caught the imagination of both policy-maker and scientist. It created a bridge for developing an effects-based policy, but the principles on which it was based were an important aid for education and acceptance of decisions.

Other important principles are those of sharing and cost/benefit. These can be considered together. Sharing both costs and benefits between parties adds considerably to the chances of successful agreement (Figure 2). If either is not shared, consensus is difficult. Similarly the overall benefits should be seen to outweigh the overall costs, i.e. the costs should be justified. If there is unequal sharing and poor return for money spent consensus is likely to be impossible. Sharing of costs and benefits with a high return for money spent will provide a good basis for negotiations, and make education easier.

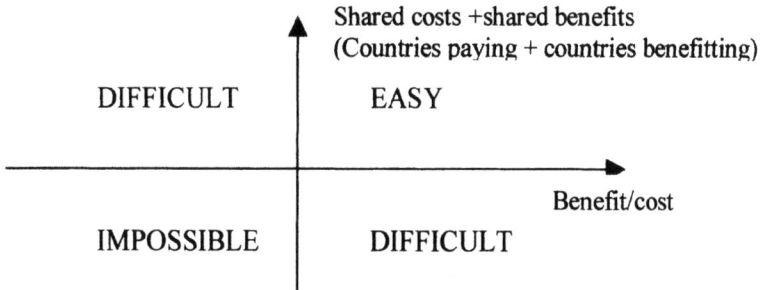

Figure 2. The relative difficulties of reaching consensus agreement when considering the benefits and costs and how these are shared between participating countries.

5. Presenting the case - understanding the important details

While the simple concept and principles above are important, it is also essential that communication of issues (i.e. education) fails because of poor or confusing presentation. The usual rules apply as for other scientific and technological topics but with even greater emphasis on simplicity and clarity. There is a major risk of attempting to be too scientific and confusing the basic facts with issues that make little difference to the overall policy decisions. For example, much has been said about the need to take account of uncertainties in critical loads, but this needs to be treated simply so that does not confuse the main issues.

In science, diagrams and graphs are used to explain relationships and their consequences. For developing effects-based policy, diagrams must be simple and self-explanatory. While diagrams may be useful, they may also create problems for those unfamiliar with scientific and technical issues.

There are a variety of communication tools that can be used. CLRTAP has employed many of these both in development, negotiating and adoption phases of its protocols. They include (with target audiences in parentheses): presentations at meetings (all participants, NGOs, industry), education seminars (all active participants), in-house documents (participants), brochures (public/policy-maker, others), scientific papers and books (scientists, media), conferences (scientists, media), the media (public, policy-maker), press conferences (the media).

Finally, it should be stressed that education is a two-way process. All active participants need to learn from one another, and all need to ensure that the public and the politician are properly informed. Good communication and education will help ensure agreement throughout the policy development process. Of course, where there are vested interests and possible financial gains, there will always be risks of deliberate misinterpretation. However, openness and a positive approach to providing information and education will greatly aid the confidence building and consensus needed for international agreement.

References

Amman, M.., Cofala, J., Heyes, C., Klimont, Z., Schopp, W.: 1999, *Pollution Atmosphérique, Special Edition 1999*, 41- 63.
Bull, K. R.: 1995, *Water, Air and Soil Pollution*, **85**, 201 - 212.
Hordijk, L.: 1995, *Water, Air and Soil Pollution*, **85**, 249 – 260.
Nordberg, L., 1999, *Pollution Atmosphérique, Special Edition 1999*, 5 - 7.

COST-BENEFIT ANALYSIS AND THE DEVELOPMENT OF ACIDIFICATION POLICY IN EUROPE

EDUARD DAME[1]* and MIKE HOLLAND[2]

[1] *Ministry of the Environment, Climate Change and Acidification Department, The Hague, The Netherlands;* [2] *Mike Holland, AEA Technology, United Kingdom*
(* *author for correspondence, eduard.dame@minvrom.nl*)

Abstract. Countries that will ratify the new *Protocol to abate acidification, eutrophication and ground-level ozone*, also known as the Göteborg-protocol, are committed to meet national emission ceilings for SO_2, NOx, VOCs and NH_3 in 2010. AEA Technology calculated impacts and monetised benefits for four scenarios used during the preparation of the new Protocol, each scenario compared with the situation 1990. The calculated benefits were compared with the costs calculated by the International Institute for Applied Systems Analysis (IIASA) using the RAINS model. The overall conclusion is that the benefits are likely to exceed the costs of implementing the scenarios considered in the study, by a factor of between two and three. Based on the principal set of assumptions followed in the study, it appeared that impacts on haity and materials were negligible. Impacts on ecosystems remained unquantified.

Keywords: cost-benefit analysis, integrated assessment modelling, Gothenburg protocol, national emission ceilings

1. Introduction

Measures to abate acidifying emissions have to be paid for. In general these costs will be paid by industry, the agricultural sector and consumers following the principle that the polluter pays. Unfortunately (for the environment) there exists no economic driving force for improving the environment. Polluters need to be obliged to account for the environmental, health and other damages that they cause by making the necessary investments. At the same time it is necessary to guard against the imposition of excessive costs for environmental protection that could endanger the international competitive position of a country or group of countries, leading to a decrease in the economic growth rate.

The necessary comparison between costs and benefits appears at first as a comparison between apples and oranges. However the use of methods developed by environmental economists allows monetisation of a range of environmental 'goods', which in turn permits direct comparison of the costs and benefits of pollution abatement. That said, it is necessary to add that the use of Cost-Benefit Analysis in the development of environmental regulation is relatively new. Results are prone to significant uncertainties, though research has gone some way to quantifying these and describing them in an open manner. The results indicate the benefits of control of emissions on a number of kinds of impact, allowing direct comparison with the costs of abatement, and consideration of the potential effects of uncertainty on the analysis. However, there is a

marked reluctance in some circles to quantify the adverse effects of air pollution, and an even greater reluctance to proceed beyond that point to monetisation, particularly for effects on health, ecological sustainability and cultural heritage. The reasons given for this are broadly as follows:
• Quantification of costs and benefits is subject to significant uncertainty, including inability to quantify some aspects. In view of this uncertainty it is thought by many that Cost-Benefit Analysis cannot provide definitive answers.
• The costs made and the benefits earned don't always follow the same bank account. In daily live practice additional economic mechanisms, such as prices of products, taxes, health insurance contributions, are needed to bring costs and benefits together. In general these mechanisms are however far from perfect to do so.
• Monetisation of effects on health, the natural environment and cultural heritage is often viewed as unethical as these goods are typically thought of as being 'not for sale'. In relation to ecology it is also viewed as a challenge to sustainability (under some, but not all definitions of the word).

For the first time in the field of European policy making on acidification, an extensive Cost-Benefit Analysis was used in the period 1997 to 1999 during the negotiations of the so called Gothenburg Protocol under the Convention on Longe-range Transboundary Air Pollution of the UN/ECE. In this paper the Cost-Benefit Analysis for the Gothenburg Protocol is examined to judge the importance of the use of the analysis and to learn for future occasions.

2. The Gothenburg Protocol and the Scenarios Used

The Protocol to Abate Acidification, Eutrophication and Ground-level Ozone has been signed in Gothenburg (Sweden) by 31 countries of the UN/ECE and will enter into force after ratification by at least 16 countries, which is expected during 2002. The Protocol sets national emission ceilings for 2010 for four pollutants: SO_2, NO_x, VOCs and ammonia. These ceilings were negotiated on the basis of scientific assessments of pollution effects and abatement options using the RAINS model developed by the International Institute for Applied Systems Analysis (Amann *et al.*, 1999). The RAINS model has modules for emission generation (with databases on current and future economic activities, energy consumption levels, fuel characteristics, etc.), for emission control options and costs, for atmospheric dispersion of pollutants and for environmental sensitivities (databases on critical loads). The time horizon of the model extends from the base year 1990 up to the year 2010.

During the negotiations analysis was carried out on a large number of scenarios. The following are the most relevant:
1. a reference (REF) scenario, in which emissions for 2010 are estimated taking into account current reduction plans and current Legislation. This represents the lowest possible ambition level.
2. a medium level ambition (MLA) scenario which provided the focus for the

final negotiations.

3. a Maximum Feasible Reduction (MFR) scenario, demonstrating how far emissions could be reduced by 2010 through the adoption of all technically feasible options. This scenario represents an upper bound ambition level.

The 1990 emissions were used as the baseline against which other scenarios were compared. A scenario taking the emissions adopted under the finalised Protocol (UN/ECE) is included as well.

3. Costs of the Gothenburg Protocol

Once the Protocol is fully implemented, Europe's sulphur emissions should be cut by at least 63%, its NO_x emissions by 41%, its VOC emissions by 40% and its ammonia emissions by 17% compared to 1990. The Protocol also sets tight limit values for specific emission sources (*e.g.* combustion plant, electricity production, dry cleaning, cars and lorries) and requires best available techniques to be used to keep emissions down. VOC emissions from such products as paints or aerosols will also have to be cut. Farmers will have to take specific measures to control ammonia emissions. Guidance documents adopted together with the Protocol provide a wide range of abatement techniques and economic instruments for the reduction of emissions in the relevant sectors, including transport. All the measures involved have specific costs. Table I gives the control costs for the abatement of the emissions of SO_2, NO_x, VOC and NH_3 for the 2010 scenarios calculated by the RAINS model (Amann *et al.*, 1999).

TABLE I
Estimated costs in Million Euro/year for the four pollutants and for each scenario.

Pollutant	Country	REF	From REF to UN/ECE	From REF to MLA	From REF to MFR
SO_2	EU-15	10813	285	935	4309
	other UN/ECE	3202	81	879	5552
	total Europe	14016	365	1814	9860
NOx and	EU-15	47358	951	2318	27414
VOC	other UN/ECE	5332	213	917	16875
		52590	1165	3235	57797
NH_3	EU-15	361	256	2450	11239
	other UN/ECE	0	967	991	10813
	otal Europe	361	1223	3442	22052
All	EU-15	58433	1492	5704	42961
pollutants	other UN/ECE	8534	1261	2787	46748
	total Europe	66967	2753	8490	89709

The control costs of NO_x and VOC are given jointly because control technologies in the transport sector simultaneously reduce the emissions of those two pollutants. It is notable that the costs for the EU-15 (the 15 current member states of the European Union) are estimated to be greater as a result of the EU-directive on

National Emissions Ceilings, which is likely to have more stringent ceilings for 2010 for most EU-countries. This directive is expected to come into force at the beginning of 2001.

There exists a widespread opinion that cost estimates are likely to be exaggerated as they fail to account for some relatively cheap measures that would reduce emissions (*e.g.*, fuel switching, energy efficiency), and technological advances. The RAINS model in its present state has a focus on end-of pipe technologies and also does not take account of the fact that technology becomes cheaper in time.

The Protocol increases the total costs above the REF-scenario for the EU by 2.6%, for the other UN/ECE countries by 14.8% and for the total of Europe by 4.1%. The difference between EU-15 and the other UN/ECE countries is a logical outcome of the optimisation by the RAINS model to find the most cost-effective solutions and reflects the lead in environmental policy by the EU-15 with respect to the rest of Europe.

Optimisation for the most cost-effective solution using integrated assessment modelling gives major financial benefits compared to alternative abatement strategies, such as a flat rate reduction over all countries. For the EU-directive on National Emission Ceilings the total costs for the EU-15 would be approximately 50% higher (European commission, 1999). Comparable cost savings apply to the Gothenburg Protocol.

4. Benefits of the Gothenburg Protocol

With the use of the RAINS model it has been estimated that once the Protocol is implemented, the area in Europe with excessive levels of acidification will shrink from 93 million hectares in 1990 to 15 million hectares. That with excessive levels of eutrophication will fall from 165 million hectares in 1990 to 108 million hectares. The number of days with excessive ozone levels will be halved. It is estimated that life-years lost as a result of the chronic effects of exposure to secondary particles linked to SO_2, NO_x and ammonia emissions will be about 2,300,000 lower in 2010 than in 1990, in addition to numerous other effects on health. The exposure of vegetation to excessive ozone levels will be 44% down on 1990.

The next question that rises is: what does this mean in financial terms? In the period of preparation for the Gothenburg Protocol AEA Technology (United Kingdom) had already quantified and monetised the impacts of numerous scenarios, for those effects of which sufficient knowledge exists to permit quantification. After the final negotiations the Ministry of the Environment of the Netherlands commissioned AEA Technology to complete the calculations by taking into account the final results of the negotiations, and to make a Cost-Benefit Analysis on the issue, including an extensive review of the uncertainties involved. The benefits in terms of reduced damage to health, materials and some aspects of managed ecosystems in agriculture and forestry, were quantified using the ALPHA (Atmospheric Long-range Pollution Health/environment Assessment) Model. This model and its inputs have been widely reviewed by many groups including the UN/ECE taskforces, Member States, Industry and NGO's. An overview of the analytical process is shown in Figure 1.

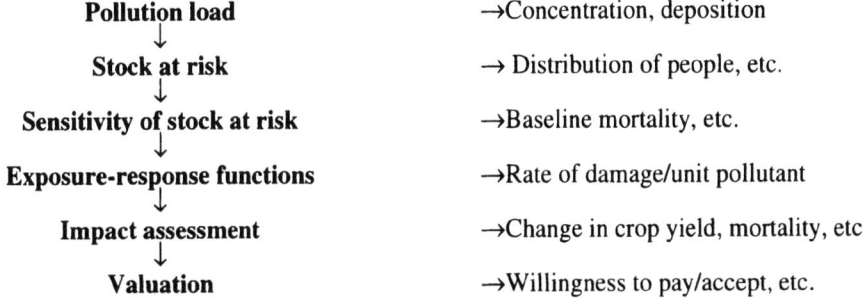

Figure 1. Pathway for the analysis of the benefits of pollution abatement

The benefits have been calculated only for those effects of which sufficient knowledge exist (Holland *et al.*, 1999). This is the case for most of the impacts on human health, the acid effects on utilitarian buildings, the effects on timber production and the effects on crops. Impacts on ecosystems remain unquantified beyond determination of the area in each country subject to exceedence of critical loads.

The health assessment considered both acute and chronic effects of the Protocol pollutants and associated secondary species (such as ozone and sulphate aerosols). Acute effects are those that arise from short-term exposures to air pollution. Exposure response functions for them are typically derived from observations of daily fluctuations in concentrations of air pollutants, compared with the daily variation in the incidents of death or disease. Chronic effects, in contrast, arise from long term exposures. Exposure-response functions in this case are derived by comparing (*e.g.*) annual concentrations of air pollution against age at death, or against the incidence of diseases such as bronchitis which themselves are log-term in nature. Drawing on the literature that is available it would seem likely that more complete account is taken of the health effects caused by NO_3 and SO_4-aerosols compared with those of O_3 in particular. Potential effects of chronic exposures to O_3 are not included in the analysis through a lack of data.

The benefits in terms of changing mortality rates are quantified based on the value of a statistical life (VOSL) as well on the value of a life year (VOLY). This latter method, derived from the VOSL method, takes into account the limited life expectancy of those who are most likely to be affected by air pollution. Not surprisingly, therefore, the method gives lower values. This abstract presents the results based on use of the VOLY values, as the choice of the lowest values makes the final conclusions of the cost-benefit analysis more robust in the event that benefits exceed costs. It is acknowledged that the methods used to derive the VOLY are preliminary. Further research in this area is already underway in a number of countries, though results are as yet unavailable.

Table II shows that the benefit (as avoided damage) caused by increased mortality is dominant amongst those effects that it has been possible to quantify. The benefits related to materials and crops and the damage related to the decrease in timber production, are relatively small compared with the calculated health benefits.

TABLE II

Damage in Million Euro/year for all the UN/ECE countries for each impact. The damage related to mortality is calculated on basis of VOLY.

Impact	1990	REF	UN/ECE	MLA	MFR
Increased morbidity	47000	30000	29000	26000	18000
· acute and chronic morbidity by NO_3- and SO_4 aerosols					
· acute and chronic morbidity by O_3 and SO_2					
Increased mortality	230000	140000	140000	120000	76000
· acute and chronic mortality by NO_3- and SO_4 aerosols					
· acute mortality by O_3 and SO_2					
Materials	1800	700	600	500	200
· SO_2/ acid effects on utilitarian buildings					
Crops	27000	20000	19000	18000	9000
· direct effects of SO_2 and O_3 on crop yield					
· indirect SO_2 and O_3 effects on livestock					
· N deposition fertiliser					
· acidification / liming					
Timber production	2200	1500	1400	1300	600
· O_3 effects on timber production					

5. Cost-Benefit Analysis

The overall conclusion is that the benefits quantified for the scenarios investigated here are likely to exceed the costs of implementation by a factor two or higher. The benefit:cost factor (b/c) will be considerably higher when using the VOSL method instead of the more conservative VOLY method. A number of effects, including those on ecosystems, remained largely outside this analysis because of limitations on the availability of data. These would add to the quantified benefits. The third factor that points in the direction of a factor higher than two is that the real costs incurred by 2010 will probably be lower than now predicted.

Table III gives the benefit:cost ratio b/c not only for all the European UN/ECE countries together, but also for the EU-15 collectively and for three specific countries: the Netherlands, Portugal and Poland. Actually, for all the countries this b/c ratio is available, but the importance of this ratio for each country separately is limited, because of the fact that reduction of emission in a country has not only benefits for that specific country, but also for a number of neighbouring countries. Conversely, the total benefits that occur in any specific country are partly due to the investment made by other countries and vice versa.

The country specific ratio averaged over Europe is positive, as it is for most individual countries (for example the Netherlands and Poland). For a restricted number of countries however the ratio is lower than 1. For instance Ireland and Portugal make a loss on the abatement of acidification, eutrophication and ground level ozone, mainly because of their position in the extreme west of Europe. Because of the prevailing wind directions both countries receive little benefit from neighbouring countries, but are nevertheless obliged to reduce their emissions.

TABLE III
Benefit:cost factor for four scenarios based on VOLY

	REF	UN/ECE	MLA	MFR
The Netherlands	2.5	2.5	2.2	2.2
Portugal	0.4	0.3	0.5	0.6
Poland	2.9	2.9	2.9	1.6
EU-15	1.3	1.4	1.4	1.3
non EU-countries	4.9	4.6	4.7	1.5

6. Considerations and Conclusions

During the negotiations for the Protocol the RAINS model appeared a useful instrument for acidification policy making based on integrated assessment modelling instead of flat rate reduction scenario's. There is however a school of thought that RAINS could even be more successful because its conservative nature (as explained in chapter 3) held countries back from making more significant commitments to abatement. In this regard it is interesting that a number of EU-countries have gone further than Gothenburg under the NEC-directive, presumably because they found abatement costs to be lower than predicted by RAINS. It must be noted also that while the model results strongly guided the negotiations, the commitments made by the Parties were significantly influenced by domestic considerations and - less decisive - the comparison of costs and benefits. The results of the Cost-Benefit Analysis have been treated as additional information and were particularly useful for building consensus within some member states and within some of the pan-European legislative bodies. For decision makers the reassuring outcome of the analysis was that there is no reason to reject any scenario (not even the most severe) for the reason that the benefits would less than the costs.

The Cost-Benefit Analysis did not take into account a number of effects that cannot yet be quantified. This was for instance the case for the benefits of ecosystems which might attribute considerably to the total amount of benefits. The Ministry of the Environment in the Netherlands has recently funded research to explore methods to quantify the benefits of terrestrial ecosystems in the Netherlands.

Much concern was expressed about the uncertainties of this analysis.

However, the simple fact that analysis of environmental benefits is subject to significant uncertainty is of little relevance, because policy makers routinely have to account for all kind of uncertainties when making decisions. An explicit assessment of uncertainties will assist policy maker take account of them. For that reason such an overview is part of the Cost-Benefit Analysis made for the Gothenburg Protocol (Holland *et al.*, 1999). What is more, quantification is necessary in order to go beyond an assessment of general indices of risk that simply show whether or not an effect is likely. This provides opportunity to assess the significance of effects. Having made an initial quantification it is of course possible to consider whether an estimate is likely (taking consideration of the uncertainties) to be larger or smaller than reality, and to investigate the consequences of possible errors using sensitivity analysis.

Finally, the use of monetisation simply seeks to establish an index that allows inspection of society's preference for the allocation of resources (*i.e.* money). This enables comparison of different environmental improvement options and prioritisation of the measures that are likely to be most effective. Without explicit use of monetisation comparisons will of course still be made. However, the valuation process is likely to be implicit and therefore less open to debate.

References

Amann, M., Bertok, I., Cofala, J., Gyarfas, F., Heyes, C., Klimont, Z., Makowski, M., Schopp, W. and Syri, S.: 1999, *Integrated Assessment Modelling for the Protocol to Abate Acidification, Eutrophication and Ground-level Ozone in Europe*. Air & Energy 132, Ministry of Housing, Spatial Planning and the Environment, Directorate Air and Energy, The Hague, The Netherlands

European Commission: 1999, *Proposal for a directive of the European Parliament and the Council on national emission ceilings for certain atmospheric pollutants*, COM(1999) 125 final

Holland, M.R., Forster, D. and King, K: 1999, *Cost-Benefit Analysis for the Protocol to Abate Acidification, Eutrophication and Ground-level Ozone in Europe*. Air & Energy 133, Ministry of Housing, Spatial Planning and the Environment, Directorate Air and Energy, The Hague, The Netherlands

TRANS-PACIFIC AIR POLLUTION: SCIENTIFIC EVIDENCE & POLITICAL IMPLICATIONS

KENNETH E. WILKENING

*International Studies Program, University of Northern British Columbia
Prince George, B.C., Canada V2N 4Z9
(author for correspondence, e-mail: kew@unbc.ca)*

Abstract. Long-range tropospheric transport of acidic and non-acidic contaminants into or across the Pacific Ocean from sources originating in Pacific Rim countries and beyond (trans-Pacific air pollution, for short) is an emerging international environmental issue. This paper provides a definition of trans-Pacific air pollution; summarizes some of the scientific evidence for what seems to be the dominant pollutant pathway in the Asia-Pacific region, transport on westerly winds from Asia to North America; discusses a recent conference on trans-Pacific air pollution; and analyzes some of the political implications of the problem. Evidence for trans-Pacific air pollution comes from three main sources: observational data, computer simulations, and research on concentrations of pollutants in various media. Trans-Pacific air pollution is of political consequence. There may be wide-ranging physical and chemical effects if the relatively pristine air of the Pacific troposphere is altered. Domestic air quality along the west coast of North America may be altered. Degradation of the Arctic environment may be accelerated. The issue reinforces the need for greater international cooperation on environmental issues in the Asia-Pacific region.

Keywords: Asia-Pacific air quality, East Asia, international environmental policy, long-range transport of air pollutants, Pacific Rim, trans-Pacific air pollution

1. Introduction

A long-range, large-scale air quality problem of potentially serious international concern is gaining attention in the Asia-Pacific region—tropospheric transport into and across the Pacific Ocean of "significant" quantities of atmospheric contaminants. The chemistry of huge swaths of the Pacific Ocean may represent some of the last vestiges of the natural processes extant prior to major human influences on the global atmosphere. Preliminary, and growing, scientific evidence, however, indicates that the chemical composition of the troposphere above the Pacific Ocean is being altered by anthropogenic emissions. This has major scientific and political implications.

The fact that atmospheric contaminants are transported long distances in the troposphere has been known for a long time. However, it wasn't until the late 1960s and early 1970s that regional-scale air quality problems involving significant transport of contaminants across national borders were identified. Since this time many regional-scale, international air quality problems have come to light associated with a wide variety of contaminants. In many cases the horizontal scale of the problems has increased over time.

This paper provides a definition of trans-Pacific air pollution, summarizes some of the scientific evidence related to the problem, discusses a recent conference on trans-Pacific air pollution, and analyzes some of the political implications of the problem.

2. Definition of Trans-Pacific Air Pollution

Trans-Pacific air pollution can be defined as long-range tropospheric transport of contaminants into or across the Pacific Ocean from emission sources in Pacific Rim countries and beyond. Such contaminants can alter the properties of the atmosphere above the Pacific Ocean, adversely affect Pacific Ocean marine and terrestrial ecosystems, and negatively impact human and wildlife health in the Pacific Rim. The types of pollutants involved include aerosols, gaseous species, and toxics (such as persistent organic pollutants (POPs), heavy metals, and radionuclides).

The problem potentially encompasses the entire region within and around the Pacific Ocean. As currently understood, there is a bifurcation at the equator into northern and southern hemispheric domains. Within the northern Pacific Rim, the aspect that has drawn the most attention to date is transport of pollutants on westerly winds from emission sources in East Asia. While currently drawing the lionshare of attention, it is by no means the only pathway of pollutant transport in the northern Pacific region. However, due to global atmospheric circulation patterns, the East Asia-North America axis is of primary interest because East Asia contains the greatest concentration of emission sources upwind of the Pacific Ocean.

3. Scientific Evidence for Trans-Pacific Air Pollution

While there is a modestly long history of investigating atmospheric transport of chemical species (primarily soil aerosols) into the Pacific region (Rex and Goldberg 1958; Tsunogai *et al.*, 1972; Duce *et al.*, 1980), detailed research on atmospheric pathways and deposition of contaminants is still in the formative stages. The first large-scale program was the Asia-Pacific Atmospheric Research Experiment (APARE) initiated in the early 1990s under the auspices of the International Global Atmospheric Chemistry Program (IGAC) to study the impact of emissions from the Asian continent on the atmospheric chemistry of the North Pacific off the coast of East Asia. Individual projects under the APARE umbrella included the Perturbation by the East Asian Continental Air Mass to the Pacific Oceanic Troposphere (PEA-CAMPOT) project begun in 1991 and funded by the Environment Agency of Japan (Hatakeyama *et al.*, 1995; Hatakeyama *et al.*, 1995), and the Pacific Exploratory Mission-West (PEM-West) A & B campaigns conducted in 1991 and 1994, respectively, and funded by the U.S. National Aeronautic and Space Administration (NASA) (Hoell *et al.*, 1996; Hoell *et al.*, 1997). Following these campaigns, and partly stimulated by them, evidence for trans-Pacific air pollution continues to accumulate. It comes from three main sources: observational and monitoring data, computer simulations, and research on concentrations in various media. Some of this evidence is discussed below. See also Wilkening *et al.* (2000).

3.1. OBSERVATIONAL DATA

The work of Dan Jaffe of the University of Washington-Bothell and his colleagues was the first to document transport of air pollutants from Asia to the west coast of North America (Jaffe *et al.*, 1999). Jaffe *et al.* announced their results at the December 1998 annual meeting of the American Geophysical Union.

In 1998 satellite remote sensing images of a massive April dust storm in western China detected Asian dust being transported to the west coast of North America (for an on-line view of the dust storm, visit http://capita.wustl.edu/Asia-FarEast/). This event captured the attention of numerous scientists across North America. Other observational data on various atmospheric species also suggest trans-Pacific transport of pollutants (Perry *et al.*, 1999; Bailey *et al.*, 2000).

3.2. COMPUTER SIMULATIONS

Numerical models are now capable of doing reasonably detailed analysis of atmospheric transport of trace species in the Pacific region. The models are generally of two types: Pacific-wide, regional-scale models and global tropospheric chemistry models applied to the Pacific region. Results using both types models demonstrate the possibility of cross-Pacific transport of pollutants. Some simulation results roughly correlate with monitoring data. In particular, the models of Berntsen and Uno correlate tolerably well with Jaffe *et al.*'s monitoring data (Jaffe, *et al.*, 1999).

Recent modeling study by Berntsen *et al.* (1999) and Jacob *et al.* (1999) suggest that "significant" quantities of pollutants arrive from Asia to North America, especially during spring, and that North American domestic air quality may be adversely affected.

3.3. CONCENTRATIONS IN MEDIA

Pollutant concentrations (especially POPs) measured in various media indirectly suggest trans-Pacific transport of pollutants. Numerous studies document high levels of POPs in the Arctic or near-Arctic regions. Recent research now demonstrates that the high mountains of the Canadian west also have surprisingly high levels of POPs (Blais *et al.*, 1998). The origin of these pollutants is at present unknown. Long-range atmospheric transport across the Pacific Ocean from Asia is one possible source. Toxic airborne pollutants from Asia may be a source of contamination in fish stocks and lake sediments in four British Columbia lakes (MacDonald *et al.*, 2000).

Anthony *et al.* (1999) associated elevated levels of certain organochlorine pesticides with low reproduction of nesting bald eagles on remote islands in the Aleutian

Archipelago in Alaska. Concentrations of organochlorine contaminants increased in eggs from east to west along the Aleutian Island chain suggesting that Asia may be a potential source of the pollutants. A recent study on orcas (killer whales) in the Pacific Northwest revealed very high PCB concentrations in some of the orca populations (Ross et al., 2000). Where the PCBs come from is not yet known. However, one possible source is long-range atmospheric and/or marine transport from Asia.

POPs are not the only chemical species transported across the Pacific that may be accumulating in ecosystems. Robert Edmonds of the University of Washington found what may be evidence for trans-Pacific transport during testing of nitrates and sulfates in the streams of the pristine Olympic National Forest in Washington (Robert Edmonds, 1998).

In conclusion, there is a confluence of scientific evidence documenting trans-Pacific transport of air pollutants and its potentially detrimental effects.

4. Conference

The evidence was reviewed at the First International Conference on Trans-Pacific Transport of Atmospheric Contaminants convened in Seattle, Washington, USA, 27-29 July 2000. The conference was sponsored by over a dozen government agencies in the United States and Canada, and was attended by about 100 scientists and government officials from both sides of the Pacific. The main objectives of the conference were to:
1) map the state of science on long-range transport of air pollutants into and across the Pacific Ocean, including integrating and evaluating findings from diverse research disciplines;
2) craft recommendations and establish a research agenda that will clarify uncertainties in our knowledge;

The two primary conference outputs were a consensus statement (http://www.epa.gov/oia/iepi/transpac.htm) and an article in *Science* (Wilkening et al., 2000). The consensus statement contains a description of the problem, and statement of research needs. The *Science* article contains a review of scientific evidence.

5. Political Implications

Trans-Pacific air pollution has significant, and as yet largely unrecognized, policy implications. Since the end of atmospheric testing of nuclear weapons, trans-Pacific air pollution is the first large-scale, region-specific environmental pollution issue to confront the countries surrounding the Pacific Ocean. The following are some areas in which accumulating scientific evidence is important to policymakers in the United States, Canada, Japan, China, and other countries of the Asia-Pacific region.
1) *Changing chemistry of the Pacific troposphere*: Trans-Pacific transport of

air pollutants could have a profound impact on the chemistry of the Pacific troposphere (Elliott *et al.*, 1997). It could change the oxidizing capacity of the atmosphere, and affect marine nutrient cycles.

2) *Impact on air quality standards*: Trans-Pacific air pollution could affect air quality along the west coast of North America (Jacob *et al.*, 1999). Import of pollutants may push some cities or regions (such as Class I wilderness regions in the U.S.) over their standards.

3) *Mix of technologies transferred to developing countries*: The trans-Pacific air pollution issue urges a rethinking of the current North American technology transfer programs to Asia. Perhaps greater support for a different blend of emission-reducing technologies is appropriate.

4) *Degradation of Arctic environment*: The Arctic Monitoring and Assessment Programme (AMAP)'s assessment report documents, among other things, the role of Eurasian air pollutants in degradation of the Arctic environment (Arctic Monitoring and Assessment Programme, 1998). Arctic ecosystems, scientific research, and policymaking are thus intimately tied to issues raised by trans-Pacific air pollution.

5) *Desertification*: Years ago researchers thought aerosol dust coming from Asia constituted a "natural background" of such material. However, a significant fraction may be the result of human-induced desertification. Policies toward desertification issues may need to be reexamined.

6) *International environmental cooperation in the Asia-Pacific*: The trans-Pacific air pollution issue will hopefully trigger expanded international environmental cooperation in the Asia-Pacific region.

6. Conclusion

The newly emerging issue of trans-Pacific air pollution once again demonstrates that the scale of human-induced environmental problems is expanding (Vitousek *et al.*, 1997). Action is urgently needed on both the scientific and political dimensions of the problem before it turns serious. In the author's opinion, an international Pacific Monitoring and Assessment Program (PMAP) and a Pacific Environmental Protection Strategy (PEPS) is required to ensure protection of the Pacific atmospheric and marine commons. Such a program and strategy is one large-scale link in the development of a global system of environmental governance.

Acknowledgements

I wish to thank the many individuals, too numerous to name, from many disciplines and institutions on both sides of the Pacific who have contributed to my understanding of the newly emerging trans-Pacific air pollution issue.

References

Arctic Monitoring and Assessment Programme (AMAP): 1998, *AMAP Assessment Report: Arctic Pollution Issues*, Oslo, Norway: AMAP.

Anthony, R. G., Miles, A. K., Estes, J. A. and Isaacs, F. B.: 1999, *Environmental Toxicology And Chemistry* **18**, 2054.

Bailey, R., Barrie, L. A., Halsall, C. J., Fellin, P. and Muir, D. C. G.: 2000, *Journal of Geophysical Research*, **105**, 11,805.

Berntsen, T. K., Karlsdóttir, S. and Jaffe, D. A.: 1999, *Geophysical Research Letters* **26**, 2171.

Blais, J. M., Schindler, D. W., Muir, D. C. G., Kimpe, L. E., Donald, D. B. and Rosenberg, B.: 1998, *Nature* **395**, 585.

Duce, R. A., Unni, C. K., Ray, B. J., Prospero, J. M. and Merrill, J. T.: 1980, *Science* **209**, 1522.

Edmonds, R. L., Blew, R. D., *et al.*: 1998, Vegetation Patterns, Hydrology, and Water Chemistry in Small Watersheds in the Hoh River Valley, Olympic National Park (US Department of Interior, National Park Service, Scientific Monograph NPSD/NRUSGS/NRSM-98/02).

Elliott, S., Blake, D. R., Duce, R. A., Lai, C. A., McCreary, I., McNair, L. A., Rowland, F. S., Russell, A. G., Streit, G. E. and Turco, R. P.: 1997, *Geophysical Research Letters* **24**, 2671.

Hatakeyama, S., Murano, K., Bandow, H., Mukai, H. and Akimoto, H.: 1995, *Terrestrial, Atmospheric and Oceanic Sciences* **6**, 403.

Hatakeyama, S., Murano, K., Bandow, H., Sakamaki, F., Yamato, M., Tanaka, S. and Akimoto, H.: 1995, *Journal of Geophysical Research* **100**, 23.

Hoell, J. M., Davis, D. D., Liu, S. C., Newell, R. E., Akimoto, H., McNeal, R. J. and Bendura, R. J.: 1997, *Journal of Geophysical Research* **102**, 28.

Hoell, J. M., Davis, D. D., Liu, S. C., Newell, R. E., Shipman, M., Akimoto, H., McNeal, R. J., Bendura, R. J. and Drewry, J. W.: 1996, *Journal of Geophysical Research* **101**, 1641.

Jacob, D. J., Logan, J. A. and Murti, P. P.: 1999, *Geophysical Research Letters* **26**, 2175.

Jaffe, D., Anderson, T., Covert, D., Kotchenruther, R., Trost, B., Danielson, J., Simpson, W., Berntsen, T. K., Karlsdottir, S., Blake, D. R., Harris, J. M., Carmichael, G. and Uno, I.: 1999, *Geophysical Research Letters* **26**, 711.

MacDonald, R. W., Shaw, D. P. and Gray, C. B. J.: 2000, Contaminants in lake sediments and fish, in Gray, C. B. J. and Tuouminen, T. (eds.), *Health of the Fraser River Aquatic Ecosystem: A synthesis of research conducted under the Fraser River Action Plan, Vancouver, BC*, Vancouver, BC: Environment Canada pp. 23.

Perry, K. D., Cahill, T. A., Schnell, R. C. and Harris, J. M.: 1999, *Journal of Geophysical Research* **104**, 18.

Rex, R. W. and Goldberg, E. D.: 1958, *Tellus* **10**, 153.

Ross, P. S., Ellis, G. M., Ikonomou, M. G., Barrett-Lennard, L. G. and Addison, R. F.: 2000, *Marine Pollution Bulletin* **40**, 504.

Tsunogai, S., Saito, O., Yamada, K. and Nakaya, S.: 1972, *Journal of Geophysical Research* **77**, 5283.

Vitousek, P. M., Mooney, H. A., Lubchenco, J. and Melillo, J. M.: 1997, *Science* **277**, 494.

Wilkening, K. E., Barrie, L. A., Engle, M.: 2000, *Science* 290, 65.

LOW-CO_2 ENERGY PATHWAYS VERSUS EMISSION CONTROL POLICIES IN ACIDIFICATION REDUCTION

SANNA SYRI[*] AND NIKO KARVOSENOJA

Finnish Environment Institute, P.O.Box 140, FIN-00251 Helsinki, Finland
([*] *Author for correspondence, email: sanna.syri@vyh.fi*)

Abstract. Integrated assessment models were applied to analyze the side-benefits of structural changes in the energy systems of Finland and the EU to reduce acidity critical load exceedances in Finland. The acidification reduction potential of structural measures, induced by the need to restrict carbon dioxide (CO_2) emissions, was compared with international agreements limiting acidifying emissions. The impacts of the UN/ECE Gothenburg protocol and the European Commission's proposal for a National Emission Ceilings (NEC) Directive were assessed together with domestic and bilateral policy scenarios. The study utilized point-source emission databases of Finland and the neighboring areas and a meso-scale deposition model together with the Europe-wide transfer matrices of the EMEP/MSC-W and European emission scenarios of IIASA. Further technical reduction measures in the adjacent large sulfur emission sources in Russia would reduce considerably deposition to nearby Finnish areas. Energy saving and shifts in the EU energy system towards less carbon-intensive fuels instead of further technical emission controls were found to have significant potential in limiting acidifying deposition. Implementation of the Kyoto protocol in the EU could reduce the Finnish areas at risk of acidification more than the Gothenburg protocol and the NEC Directive together.

Keywords: acidification, critical loads, Gothenburg protocol, National Emission Ceilings Directive, CO_2 emissions, Kyoto protocol

1. Introduction

Acidification of waters and forest soils caused by long-range transported air pollution has been a considerable environmental problem in Europe for decades. International work to combat acidification has been organized under the Convention on Long-range Transboundary Air Pollution (CLRTAP) established within the United Nations Economic Commission for Europe (UN/ECE) (UN/ECE, 1995). Research and monitoring of the environmental effects is organized under the Working Group on Effects under the Convention, with National Focal Centers (NFCs) in signatory countries responsible for calculating and mapping critical loads or levels for acidification, eutrophication and ground-level ozone. The Finnish NFC has actively participated in the development of national, bilateral and international emission reduction programs (*e.g.* Acidification Committee, 1998) by estimating the emission reduction needs for protecting ecosystems and by calculating regional reduction options to achieve certain environmental targets specified at a political level.

This paper presents scenario analysis on the side-impacts of structural measures restricting CO_2 emissions in the EU, required to fulfill the Kyoto Protocol of the United Nations Framework Convention on Climate Change (UN/FCCC, 1998), in reducing acidifying emissions and protecting ecosystems.

If realized, the Kyoto protocol will induce significant changes in the energy production and consumption structures of the EU countries. Its reflections on acidifying emissions are an interesting issue, as the cost-effective technical reduction potential of acidifying emissions is to a large extent exhausting in western Europe, and yet there are still considerable areas affected by excess deposition.

The plausible impacts of the Kyoto protocol were compared with the impacts of the UN/ECE Gothenburg protocol, the European Commission's proposal for National Emission Ceilings (NEC) Directive, and bilateral reduction scenarios with Russia. The effectiveness of alternative policies is illustrated using Finnish ecosystem areas at risk of acidification as measure.

2. Models used

The environmental impacts of alternative energy and emission control scenarios are analyzed with a linked model system comprising modules for air emissions, atmospheric transport and environmental impacts. Point-source emission databases of Finland and the neighboring areas, regional atmospheric deposition modules and the Finnish ecosystem critical load database are linked with European scale models (Syri *et al.*, 1999). The regional deposition modules have a resolution of about 14 km × 14 km, and they are used for estimating sulfur, nitrogen oxides and ammonia deposition from Finnish emission sources and sulfur deposition from sources in the Leningrad District, Karelia, Murmansk District and Estonia. European emission projections have been compiled by IIASA with the RAINS model (Amann *et al.*, 1999, Schöpp *et al.*, 1999), and long-range atmospheric transport of pollutants is estimated using the source-receptor matrices of the EMEP/MSC-W (EMEP/MSC-W, 1998).

3. Energy and emission projections and their acidifying impacts

3.1. TRENDS IN EMISSIONS AND ACIDITY CRITICAL LOAD EXCEEDANCES

Energy consumption has increased in Finland during the 1980's and the 1990's, yet the acidifying emissions have been controlled with increasingly stringent regulations, resulting in a sharp drop of sulfur emissions and some decline in nitrogen oxides emissions (Karvosenoja *et al.*, this Vol). As similar trends have occurred in other European countries as well, wide ecosystem recovery trends have already been observed (*e.g.* Stoddard *et al.*, 1999). The areas at risk of acidification in Finland have decreased from about 14% of the total ecosystem area in 1990 to 5.4% in 1995 (Fig. 1).

3.2. ALTERNATIVE ENERGY SCENARIOS AND ACIDIFYING EMISSIONS

The EU countries are foreseeing a continuation of strong economic growth, and their official energy projections are based mainly on an increased use of oil products and gas. The Kyoto protocol, however, demands reductions in the greenhouse gas emissions of the industrialized countries. Considerable structural changes in the energy sector are required to fulfill the protocol. The bulk of CO_2 emission reductions in the EU will probably be realized by switching fuels from oil and coal to natural gas and biomass and by energy-saving measures. As oil and coal are fuels with high sulfur contents, fuel switching reduces sulfur emissions, especially where flue gas desulfurization is not required by legislation. In Europe, installations using gas usually have lower NO_x limit values than those burning oil or coal, thus fuel switching will have some positive effects on NO_x emissions as well. As energy–saving measures reduce fuel consumption, they will also have beneficial impacts on acidifying emissions.

The two Finnish energy projections utilized in this study were compiled by the Ministry of Trade and Industry (1997, 2001). The 'Energy Market Scenario' (EMS), predicts an increase of 31% in the energy consumption during 1990-2010, which would mainly be met by expanded hard coal and gas use. With EMS, CO_2 emissions would increase by about 33% from 1990 to 2010. The impacts of energy policies fulfilling the Kyoto greenhouse gas emission targets were assessed using the 'KIO1' scenario used in the preparation of the Finnish national climate program (Ministry of Trade and Industry, 2001). The 'KIO1' scenario fulfills the EU burden sharing agreement for Finland, i.e. the stabilization of net greenhouse gas emissions at their 1990 level by strong shifting from coal to biomass and gas, by accelerated penetration of renewable energy technologies and by decreased total energy consumption.

The 'Baseline' energy projections of the EU countries used in this study predict some decrease of coal use, but large growth in transport fuels consumption and in gas consumption for power production. The 'Baseline' would result in an 8% increase of CO_2 emissions by 2010 (Syri *et al.*, accepted). At present, the EU countries are preparing their action plans to meet the Kyoto requirements, thus they were not available for this study. Instead, an illustrative 'Kyoto' energy scenario (Syri *et al.*, accepted) compiled with the PRIMES model at the National Technical University of Athens (Capros *et al.*, 1998) was utilized to demonstrate the impacts of structural measures in the energy system on acidifying emissions. The 'Kyoto' energy scenario used here assumes that no emission trading would take place. Thus it probably overestimates the CO_2 limitation measures to be taken within the EU-15. For the non-EU countries, the 'Official Energy Pathway', as reported to the UN/ECE, forecasts an 11% drop in CO_2 emissions (Amann *et al.*, 1999), thus it was also used in the 'Kyoto' estimate. Table I shows acidifying emissions with the alternative energy scenarios, assuming otherwise present legislation but without the control policies described in Section 3.3.

TABLE I

Emissions in 1990 and scenarios for 2010 with the alternative energy projections (kilotons year^{-1}).

	1990			2010 'Baseline'			2010 'Kyoto'		
	SO$_2$	NO$_x$	NH$_3$	SO$_2$	NO$_x$	NH$_3$	SO$_2$	NO$_x$	NH$_3$
Finland [1]	260	300	38	102	186	31	86	172	31
Estonia [2]	275	84	29	175	73	29	175	73	29
Russian Federation [1], [2]	4460	3485	1282	2344	2675	894	2344	2675	894
EU-15 [3]	16340	13230	3580	4680	6880	3150	3620	6050	3150
Non-EU Europe [2]	21600	10170	3980	9640	6980	3420	9640	6980	3420

[1] Syri et al., 1999. [2] Amann et al., 1999. [3] Syri et al., accepted.

3.3. POLICIES TO CONTROL ACIDIFICATION

In 1999, a UN/ECE protocol to abate emissions causing acidification, ozone formation and eutrophication was signed in Gothenburg, Sweden (UN/ECE, 1999). Acidifying emissions in 2010 with the protocol are shown as 'UN' in Table II. The European Commission has put forward a proposal for a National Emission Ceilings (NEC) directive, which aims at more ambitious environmental improvements than the Gothenburg protocol (European Commission, 1999). 'NEC' shows the emissions in 2010 assuming the implementation of NEC directive and the Gothenburg protocol. Finland has conducted bilateral negotiations with the Russian Federation on reducing acidifying emissions in the bordering areas. 'BIL' includes additionally the planned bilateral reductions.

TABLE II

Emissions in 2010 with the control policies considered (kilotons year^{-1}) (Syri et al., 1999, Amann et al., 1999, Syri et al., accepted).

	'UN'			'NEC'			'BIL'		
	SO$_2$	NO$_x$	NH$_3$	SO$_2$	NO$_x$	NH$_3$	SO$_2$	NO$_x$	NH$_3$
Finland	116	170	31	116	152	31	116	152	31
Estonia	175	73	29	175	73	29	175	73	29
Russian Federation	2352	2653	894	2352	2653	894	2202	2653	894
EU-15	4040	6650	3130	3640	5920	2830	3640	5920	2830
Non-EU Europe	9930	7330	3150	9930	7330	3150	9780	7330	3150

3.4. EFFECTIVENESS TO REDUCE ACIDIFICATION IN FINLAND

Figure 1 shows the percentage of ecosystem areas in Finland with critical load exceedances under the different scenarios. The bars 'Baseline' and 'Kyoto' show the difference made by the alternative energy structure (without the policies described in 3.3). Gray bars illustrate the impacts of the further emission control policy options (with the 'Baseline' energy scenario as background assumption). For reference, Figure 1 also shows the area at risk of acidification if the European legislation would not have been tightened from the situation in the mid-nineties (illustrated as scenario 'CRP –96'), after the signing of the Second Sulphur Protocol in 1994 (UN/ECE, 1995; Amann *et al.,* 1996).

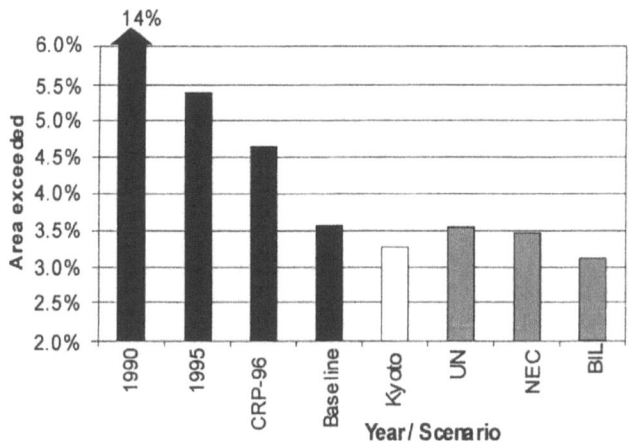

Figure 1. Share of ecosystem area at risk for acidification in Finland in 1990, 1995 and in 2010 with the scenarios studied.

As the majority of total deposition in Finland is imported, the expected decrease of long-range transported deposition will bring considerable benefits to Finnish ecosystems. Already with the legislation amendments in Europe after the mid-nineties (situation 'Baseline' in Fig. 1) the ecosystem protection will improve significantly. Structural measures in the EU reflecting the Kyoto protocol without further technical emission controls could protect the Finnish ecosystems more than the NEC Directive and the Gothenburg protocol together. For Estonia and Russia, the Gothenburg protocol allows growth in emissions from their present level. Further measures at the neighboring large emission sources in Russia would have significant beneficial effects for Finnish ecosystems.

4. Discussion and Conclusions

The potential of structural changes in the Finnish and EU energy systems to reduce acidifying deposition to Finland was assessed and compared with the impacts of international policies limiting acidifying emissions. The model results illustrate the efficacy of the different control scenarios for use in further policy development.

Shifts in the EU energy system towards less carbon-intensive fuels and energy saving could bring notable further reduction of acidifying emissions. The case with Finnish ecosystems does not fully display the environmental side-benefits of these measures, as Finland is strongly affected by deposition from outside EU, mainly Russia and Estonia. Yet the improvement in ecosystem protection level was found greater than that expected with the NEC Directive, which has been estimated to cost about 7500 million € annually with the 'Baseline' energy pathway, as the cheap technical emission reduction potential is exhausting in most EU countries. This demonstrates that policies to control air pollutants and greenhouse gas emissions should be designed jointly to ensure cost-efficacy of the measures.

References

Acidification Committee: 1998, 'Report of the Acidification Committee' (English summary), Ministry of the Environment, Helsinki, Finland, *The Finnish Environment* **219**.

Amann, M., Bertok, I., Cofala, J., Gyarfas, F., Heyes, C., Klimont, Z. and Schöpp, W.: 1996, Cost-effective control of acidification and ground-level ozone, *Second Interim Report to the European Commission*, DG-XI, IIASA, Laxenburg, Austria.

Amann, M., Bertok, I., Cofala, J., Gyarfas, F., Heyes, C., Klimont, Z., Makowski, M., Schöpp, W. and Syri, S.: 1999, Cost-effective control of acidification and ground-level ozone, *Technical Reports to the European Commission*, DG-XI, IIASA, Laxenburg, Austria.

Capros, P., Georgakopoulos, T. and Mantzos, L.: 1998, *Int. J. Environment and Pollution* **10**(3-4), pp.403-427.

EMEP/MSC-W: 1998, Transboundary air pollution in Europe. EMEP/MSC-W Status report 1998. EMEP/MSC-W, Oslo, Norway.

European Commission: 1999, COM(99)125 final, Brussels, Belgium.

Karvosenoja, N., Hillukkala, P., Johansson, M. and Syri, S.: (accepted) Cost-effective abatement of acidifying emissions with flue gas cleaning vs. fuel switching. *Water, Air and Soil Pollution*, this Volume.

Ministry of Trade and Industry: 1997, Energy economics 2025 - Scenario studies (English summary), Ministry of Trade and Industry, Publication 3/97, Helsinki, Finland.

Ministry of Trade and Industry: 2001. The needs of greenhouse gas emission reductions and reduction possibilities. Background report for the National climate program. Ministry of Trade and Industry, Helsinki, Finland.

Schöpp, W., Amann, M., Cofala, J., Heyes, C., Klimont, Z.; 1999, *Environmental Modelling & Software* **14**:1-9.

Stoddard, J., Jeffries, D., Lükewille, A. and 20 others: 1999, Nature **401**:575-578.

Syri, S., Johansson, M., Grönroos, J. and Ekqvist, M.: 1999, *Environmental Modeling & Assessment* **4**:103-113.

Syri, S., Amann, M, Capros, P., Mantzos, L., Cofala, J. and Klimont, Z.: (accepted) Low-CO_2 energy pathways supporting control strategies of acidification and ground-level ozone in Europe. *Energy Policy*.

UN/ECE: 1995, Strategies and policies for air pollution abatement – 1994 major review. UN/ECE Document EB.AIR/44, United Nations, New York, Geneva.

UN/ECE: 1999, Protocol to the 1979 Convention on long-range transboundary air pollution to abate acidification, eutrophication and ground-level ozone. UN/ECE Document EB/AIR/1999/1. United Nations, New York, Geneva.

UN/FCCC: 1998, the Kyoto Protocol to the United Nations Framework Convention on Climate Change. UN/FCCC Document FCCC/CP/1997/7/Add.1. United Nations, New York, Geneva.

THE POLITICS OF REGIONAL COOPERATION ON ACID RAIN CONTROL IN EAST ASIA

WAKANA TAKAHASHI[1*] and JUSEN ASUKA[2]

[1] *Institute for Global Environmental Strategies, 1560-39 Kamiyamaguchi, Hayama Kanagawa 240-0198, Japan;* [2] *The Center for North East Asian Studies, Tohoku University, 2-1-1 Katahira, Aoba-ku, Sendai, Miyagi 980-8577, Japan*

(*Author for correspondence, e-mail: togo@iges.or.jp)

Abstract. Several multilateral initiatives on acid rain control were advocated in the 1990s, and extended through multiple channels in East Asia. There is, however, little coordination between those initiatives. The geopolitical make-up of the region has also hampered further progress in regional cooperation on the issue.

In this paper we first examine the major features and weaknesses of existing regional environmental cooperative schemes. Second, we review collective initiatives on the acid rain issue, and examine how further progress has been hindered by the inherent inadequacies of the schemes. Finally, we consider whether and how regional cooperation on the acid rain issue may be promoted, and try to draw some implications from the European experience that might help enhance regional cooperation.

Keywords: acid rain, East Asia, Europe, politics, regional cooperation

1. Introduction

Economic development in East Asia has been accompanied by increased emissions of acidifying pollutants. This situation is likely to worsen in future because much of the energy driving industrialization comes from the combustion of coal with a high sulfur content.

Accordingly, countries in the region are paying more attention to, and strengthening national laws and regulations on, controlling air pollution. Although there was little collective regional action to address this problem, several initiatives were begun in the 1990s. Nevertheless, there is little coordination of those initiatives. Countries in the region have significant differences in their viewpoints and approaches to acid rain issue. It can be said that weaknesses and inadequacies of environmental cooperation scheme in this region has hindered further progress in regional cooperation on acid rain issues.

Here we wish to raise the following questions: 1) What are the major features and weaknesses of environmental cooperation scheme in the East Asian region? 2) How has the acid rain issue been addressed and identified at regional level in East Asia? Have such cooperative initiatives been influenced by weaknesses and inadequacies of the environmental cooperation scheme? 3) How could the situation be improved? What are implications from European experience of acid rain control to East Asia?

2. Regional Environmental Cooperation Scheme

Until the late 1980s East Asia lacked political, economic, or social cohesion. Except for certain bilateral initiatives, there was little cooperation on environmental issues. It was only during the first half of the 1990s that diverse multilateral environmental programs, forums, and bilateral agreements began to emerge in the region[1]. Some of the initiatives appeared to be comprehensive while some only focussed on a single issue. The initiatives were established through different channels, including environmental ministries, official diplomatic channels, officers of environmental agencies and ministries, NGOs, and academics with little coordination between the various channels. Consequently, some initiatives contain much material that is redundant.

Some multilateral initiatives target the subregion of Northeast Asia, while others are directed to the whole region of East Asia or, even more broadly, Asia and the Pacific. The status of participating states differs from one initiative to another, depending on the international membership of the host country/organization[2].

Since most of the regional environmental initiatives have no institutional structure or financial foundation, cooperation has made only slow progress. Indeed, some initiatives have stagnated in terms of institutional and financial development (Takahashi, 2000b).

The weaknesses of the environmental cooperation scheme in East Asia can be summarized as follows: 1) in the absence of regional organizations there is little coordination among multilateral initiatives; 2) because of this absence UN organizations have been active in facilitating and promoting regional cooperation[3]. There is, however, a limit to what UN organizations can achieve given that they have to cover the entire Asia-Pacific region (and the world) and their human and financial resources are limited; 3) national governments differ significantly in their viewpoints and approaches to environmental cooperation, mostly due to lack of economic and political homogeneity in the region[4]; 4) participation of civilians and

---

[1] These include the Acid Deposition Monitoring Network in East Asia (EANET), the Asia-Pacific Network for Global Change Research, the East Asian Seas Action Plan, the Environment Congress for Asia and Pacific, the North Asia-Pacific Environment Partnership, the Northeast Asian Conference on Environmental Cooperation, the North-East Asian Subregional Programme of Environmental Cooperation (NEASPEC), the Northwest Pacific Action Plan, and the Environmental plans on Tumen River Area Development Programme. For further information, see Takahashi (2000b).
[2] For example, North Korea does not attend most subregional programs, except those hosted by UN Organizations. Taiwan has no access to many initiatives because its position in international politics is uncertain although it is a member of APEC and ADB.
[3] UN Economic and Social Commission for Asia and the Pacific (UN/ESCAP) has striven to facilitate and promote communication between countries (UN/ESCAP, 1998). It has played the role of interim secretariat for the subregional environmental program, and has also endeavored to persuade the Asian Development Bank (ADB) to provide funding for implementing projects of the regional program. The UN Environment Programme (UNEP) has provided technical assistance for a number of projects developed under subregional initiatives. It has also initiated the Regional Seas Programme in both the Northwest Pacific and East Asian Seas. The UN Development Programme (UNDP) and the Global Environmental Facility (GEF) have also provided financial and technical assistance to subregional environmental initiatives in the developing world.
[4] For example, China has suffered significant environmental deterioration caused by industrial pollution and inland & marine water pollution, and believes that subregional cooperation should be

NGOs in multilateral environmental cooperation has thus far been limited[5].

3. Regional Cooperation on Acid Rain Issue

Regional cooperation on the acid rain issue in East Asia was initiated in the 1990s, when countries became aware that the region was facing threats from air pollution and acid rain. The need for regional cooperation to investigate and monitor acid deposition was repeatedly stressed at several regional and subregional intergovernmental and interagency conferences.

Accordingly, the establishment of the EANET was advocated by Environment Agency of Japan[6] with the aim of developing a common understanding of the region's "acid rain" problem. EANET conducted its preparatory-phase activities in 1998, and full operation is expected to start by the end of the year 2000.

Apart from monitoring acid deposition, an expert meeting on long-range air pollutants has been held annually since 1995. These meetings, organized by the National Institute for Environmental Research of South Korea, were attended by experts from South Korea, Japan, and China. Plans to carry out joint monitoring of, and modeling research on, long-range transboundary air pollutants were adopted. In 1999 UN/ESCAP hosted a parallel expert group meeting on the same issue in Japan, in which experts from nine East Asian countries participated.

International joint research on modeling long-range transport to assess the distribution of acid deposition was also initiated on a non-official channel. Although it is expected that long-range transport models, already developed by several institutes, will play a critical role in science and policy analysis, there are significant differences between those models as well as gaps in information about the processes involved, giving rise to differences in predicted outcomes. This has prompted research institutes and researchers from several countries to initiate a model inter-comparison study in 1998 (Ichikawa and Hayami, 1999)[7].

---

focused on those issues. China is quite sensitive to the use of the term "transboundary." It believes that developed countries in the subregion should offer substantial financial and technical support to developing countries. Japan, who has focused on bilateral, rather than multilateral, initiatives apparently believes that multilateral initiatives should not overlap with its existing bilateral and multilateral assistance projects but instead focus on monitoring the state of the environment and transboundary pollution. Japan appears wary of multilateral initiatives becoming another channel for development assistance, and suggests the principle of burden sharing (Valencia, 1998).
South Korea is keen to promote multilateral environmental cooperation in Northeast Asia. It apparently believes that multilateral initiatives should include both technical projects as preferred by China and monitoring-type environmental management projects as preferred by Japan (Valencia, 1998).
[5] One exception is wildlife conservation. The North East Asian Crane Site Network, launched in 1997, is promoted by an international NGO, the Wetlands International Asia Pacific and Wetlands International. A number of NGOs in the subregion as well as governmental organizations collaborated on the network. It has encouraged information exchange and the sharing of experience about wetlands sites important to cranes.
[6] China, Indonesia, Japan, South Korea, Malaysia, Mongolia, Philippines, Russia, Thailand and Vietnam attended.
[7] These include the International Institute for Applied System Analysis (IIASA) in Austria, the Central Research Institute of Electric Power Industry of Japan, the University of Iowa, and several research

The Atmosphere Action Network East Asia (AANEA) was established in 1994 to encourage regional cooperation of NGOs and to address the concerns of governments and citizens about atmospheric problems, including transboundary air pollution and climate change. The network was formed by seventeen NGOs from seven East Asian countries (China, Hong Kong, Japan, Korea, Mongolia, Russia, and Taiwan). The AANEA held a workshop in 1998 during which a project for the joint monitoring of air pollution in East Asia was proposed.

Thus, regional cooperation on the acidification issue has been growing, and this trend is likely to continue. That is not to say, however, that every state and relevant actors have actively participated in the process, nor that talks on such collaboration have gone smoothly or without controversy.

For example, not only EANET but also AANEA attempted to establish a network for monitoring acid deposition. Similar projects are being conducted in Northeast Asia. Thus, in keeping with the objectives of Northeast Asian Subregional Program on Environmental Cooperation (NEASPEC)[8], the National Institute for Environmental Research (NIER) in Korea is establishing a regional center for collecting data on acid deposition (UN/ESCAP, 2000). However, there is little coordination and communication between EANET, AANEA, and NEASPEC.

Securing financial sources is also a critical issue. Since the start of the Expert Meetings, Japan has paid all of the operating costs of EANET. This is also true for the preparatory phase of EANET. Japanese representatives want each participating country to carry some of the financial burden so that every country can have a real stake in the network when it becomes regulatively operational. However, most of the other countries call on Japan to shoulder the entire burden.

4. Conclusion: Implication from European experiences

The analysis of this paper showed that weaknesses and inadequacies of environmental cooperation scheme in East Asia have hindered the progress in regional cooperation on the acid rain issue. The question arises whether and how intra-regional cooperation may be promoted.

The European experience of fostering regional cooperation on long-range transboundary air pollution (LRTAP) control seems to offer good guidance for East Asia. The formation of this cooperative body can be traced back to 1972 when OECD launched an international monitoring program. In 1977 UN/ECE took over the responsibility, and the LRTAP Convention was agreed upon in 1979. Protocols on reducing sulfur emissions by 30% were adopted in 1984, and those on NO_x emissions (by the same percentage) in 1988. The Second Sulfur Protocol was

---

institutes and experts from South Korea, Taiwan, Japan, Sweden, and China.

[8] NEASPEC is a comprehensive and intergovernmental program on environmental cooperation in Northeast Asia. Air pollution issue is identified as one of the priority areas, and several fundamental projects on the issue have been identified and implemented.

agreed upon in 1992. In 1999 agreement was reached on a protocol to attenuate acidification, eutrophication, and ground-level ozone concentrations (UN/ECE, 1999).

East Asia apparently tries to follow in the footsteps of Europe in terms of gathering scientific facts. The current stage of EANET is roughly equivalent to that of the OECD program when it was initiated the mid 1970s. This region is also trying to develop a scientific consensus on monitoring emissions and estimating the transmission of air pollutants.

There are many differences between East Asia and Europe in terms of institutional arrangements and processes, however. East Asia apparently has more difficulty than Europe in reaching a consensus because it lacks economic, political, and social homogeneity (Takahashi, 2000a). The LRTAP Convention has been well maintained by strong links to EU policies and aid programs but no regional organizations, equivalent to the EC/EU, exist in East Asia. Moreover, the establishment of the LRTAP in Europe was promoted by the overall politics of improving East-West relations during the Cold War[9]. These regional characteristics of Europe also helped enhance the cooperation. It is perhaps premature to judge whether East Asia will follow the European footstep. However, it seems unlikely that legally binding treaty/protocols, similar to CLRTAP, will appear in East Asia, given the fundamental differences that exist between the two regions.

Nevertheless, there are lessons to be learned from the European experience. In Europe, EC/EU environment aid programs have been formally and informally linked to common environmental policies of the EU as well as to other regional and international policies such as the LRTAP Convention (CEC, 1996).

This is particularly the case for the Baltic Sea region (Levy, 1993). Although the subregion has no legally binding environmental policies equivalent to EC/EU directives, there are several multilateral cooperative frameworks. Each framework contains a set of clear objectives together with various environmental action plans and programs. Such decisions have been efficiently linked to multilateral and bilateral environmental aid programs of Scandinavian countries.

On the other hand, the collaborative activity in East Asia has been fragmented. Results are deemed to be too meager in relation to the amounts of environmental aid money expended. East Asia should learn from the European experience of forming some kind of organic link between individual initiatives and financial mechanisms,

---

[9] The 1960s saw a growing tendency toward détente between the U.S. and the USSR, culminating in the 1975 Conference on Security and Cooperation in Europe (CSCE) Final Act. In the talks, the environment turned out to be the least controversial topic, and the task of realizing cooperation in this area -- with implications for transboundary air pollution -- was assigned to the UN/ECE (Gehring, 1992). On the other hand, East-West relations influenced the talks leading up to the 1979 Convention in a different way. That is, most of the Western European economies opposed the suggestion by Scandinavian countries that agreements on pollutant emission reduction should be legally binding. It was the communist governments that supported the Scandinavian stance wholeheartedly, if for no other reason than to divide the Western camp (Björkbom, 2000). The communist governments also believed that environmental degradation was "the result of the capitalist form of organization of the domestic and world economy" and had "nothing to do with the socialist states" (Churchill et al., 1995).

between bilateral and multilateral aid programs, between donor agencies, and between regional politics and financial aid mechanisms. Such a linkage will help bring about an effective regional response to the issue of acid rain.

Acknowledgements

We thank Prof. Akio Morishima and Prof. Kazu Kato for the precious opportunity of study. We wish to acknowledge Mr. Katsunori Suzuki, Mr. Eisaku Toda, Mr. Masakazu Ichimura, Dr. Kentaro Murano, Prof. Yutaka Tonooka, Mr. Taishi Sugiyama, Dr. Yoichi Ichikawa, Dr. Rezaul Karim, Mr. Mahesh Pradhan, Mr. Lars Nordberg, Mr. Jurgen Henningsen, Mr. Lars Lindau, Mr. Lars Bjorkbom, and other policymakers and researchers for responding to interviews and providing valuable information. We are grateful to Prof. Hiroshi Matsushita, Prof. Makoto Iokibe, and Dr Benny K.G. Theng for valuable advice.

References

Bjorkbom, Lars (a former chairman of the LRTAP WG on strategies): 2000, interview with author, Sweden.
CEC: 1996, *Progress Report from the Commission on the Implementation of the European Community Programme of Policy Action in Relation to the Environment and Sustainable Development 'Towards Sustainability,'* COM (95) 624 final, Brussels.
Churchill, R. R., Kütting, G. and Warren, L. M.: 1995, 'The 1994 UN ECE Sulfur Protocol,' *Journal of Environmental Law*, 7 (2), 169.
Environment Agency of Japan (Interim Secretariat of EANET): 1998, *Acid Deposition Monitoring Network in East Asia, the First Intergovernmental Meeting, the First and the Second Meeting of the Working Group.*
Gehring, Thomas: 1994, *Dynamic International Regimes*, Peter Lang, Berlin.
Ichikawa, Yoichi and Hayami, Hiroshi 1999, Report on "Workshop on the Transport of Air pollutants in Asia," *Taiki Kankyo Gakkai –shi*, 34 (6), A53, in Japanese.
Levy, M. A.: 1993, "East-West Environmental Politics after 1989: The Case of Air Pollution," in Keohane, R. O., Nye, J. S. and Hoffman S.: 1993, *After the Cold War: International Institutions and State Strategies in Europe*, Harvard UP, Cambridge.
Takahashi, Wakana: 2000a, "Formation of an East Asian regime for acid rain control," *International Review for Environmental Strategies* 1, 97.
Takahashi, Wakana: 2000b, "Review and Future of Northeast Asian environmental cooperation: From an institutional viewpoint," a paper presented at the 9th Northeast Asian Conference on Environmental Cooperation, July 26-28, 2000, Ulaanbaatar, Mongolia.
UN/ECE: 1999, *Strategies and Policies for Air Pollution Abatement: Major Review Prepared under the Convention on Long-range Transboundary Air Pollution*, Geneva.
UN/ESCAP: 1998, "Selected Issues with Reference to the Work of the Committee on Environment and Natural Resources Development," E/ESCAP/ENRD/1.
UN/ESCAP, 2000, "Report of the Preparatory Meeting for the Sixth Meeting of Senior Officials on Environmental Cooperation in Northeast Asia, Seoul, 8-9, March 2000," ENR/SO/ECNA(6)/Rep.
Valencia, Mark (1998), "Ocean Management Regimes in the Sea of Japan," presented at ESENA workshop: energy-related marine issues in the Sea of Japan, Tokyo: 11-12 July.

STAGES IN THE HISTORY OF CHINA'S ACID RAIN CONTROL STRATEGY IN THE LIGHT OF CHINA-JAPAN RELATIONS

HAIPING LAI[1], HIROYUKI KAWASHIMA[1*], JUNKO SHINDO[2] and KEIJI OHGA[1]

[1] *Department of Global Agricultural Sciences, The University of Tokyo, 1-1-1 Yayoi, Bunkyo-Ku, Tokyo 113-8657, Japan;* [2] *National Institute of Agro-Environmental Sciences, 3-1-3 Kannondai, Tsukuba, Ibaraki 305-8604, Japan*
(*corresponding author; e-mail: akawashi@mail.ecc.u-tokyo.ac.jp)

Abstract. We analyze the history of acid rain in China over the last three decades in the light of the relationship between China and Japan. China has been aware of the problem of acid rain and has been conducting scientific research since the late 1970s. However, little effort has been put into practical countermeasures. In the early 1990s, acid rain was re-evaluated as an international environmental problem. Meanwhile, Japan became concerned about trans-boundary air pollution from China. What is important is that Japan's concern strongly influences China's acid rain policy. In the context of growing international concern for environmental problems, China has taken remarkable steps to control acid rain. In this respect, dealing with acid rain is an issue of national pride for China.

Keywords: acid rain, China, international cooperation, Japan, long-range transport, policy-making

1. Introduction

China has accomplished rapid economic development since the government took its 'open policy' decision in the late 1970s. The growing demand for energy has been met nearly entirely by the heavy use of fossil fuels, but the burning of high-sulfur- content coals has brought problems of acid rain and other air pollution. Since acid rain was discussed at the United Nations Conference on the Human Environment, held in Stockholm in 1972, a good deal of effort has been made to control acid rain in Europe and North America. The problem is now generally well under control in these regions, but in East Asia, particularly in China, it remains serious. Agenda 21, adopted at the United Nations Conference on Environment and Development in June 1992, states that experiences dealing with acid rain in Europe and North America are to be shared with other regions of the world and, in particular, points out the importance of regional cooperation in East Asia. As atmospheric pollutants can be transported for long distances, acid rain becomes a trans-boundary problem for surrounding countries. Both regional and international efforts are essential to prevent acid rain. In this paper, we discuss the policy toward acid rain in China from the late 1970s to the late 1990s, in the light of the relationship between China and Japan. This brings us further into a consideration of how increasing international environmental concern influences the acid rain issue in China, how Japan plays a role in this matter as the only economically well established country in Asia, and the nature of the driving forces for China's rapid policy change in the 1990s.

2. History of Acid Rain in China

China's environmental policy is essentially decided in accordance with the

Water, Air, and Soil Pollution **130**: 1843–1848, 2001.
© 2001 *Kluwer Academic Publishers.*

TABLE I

Acid rain in China in the light of the relationship between China and Japan

Period	Progress in China		Related activity in Japan	
1979–1985: Early awareness	1979	Pilot research on precipitation chemistry in Chongqing and Guiyang	1977	Japanese environmental delegation visited China and bilateral environmental cooperation began
	1982	Acid rain survey in south-west China	1983–1988	EAJ conducted the Acid Deposition Survey Phase 1 project
	1982	Issuing of the Standard of Air Quality and a pollutant concentration-based levying system		
	1982–1984	Nationwide survey on acid deposition		
	1984–1986	Nationwide survey on acid deposition		
1986–1990: National recognition	1986–1990	National Key Project on acid rain in the 7th five-year plan	1988–1993	EAJ conducted the Acid Deposition Survey Phase 2 project
	1987	Air Pollution Control Act adopted	1988	Bilateral agreement on establishing the China–Japan Friendship Center for Environmental Protection
	1990	Notifications on controlling the expansion of acid rain		
	1990	Participated in the RAINS-ASIA project	1990	Japan set up new monitoring sites on solitary islands in the Sea of Japan
1991–1995: Regional issue	1991–1995	National Key Project on acid rain in the 8th five-year plan	1992	Japan's initiative to establish a regional monitoring network: later EANET
	1992	Comprehensive policy on acid rain pollution in model cities	1992	Green Aid Plan launched
	1994	China's Agenda 21 adopted	1993–1997	EAJ conducted the Acid Deposition Survey Phase 3 project
	1995	Air Pollution Control Act amended	1993	First Expert Meeting for EANET held
	1995	Proposal for planning for acid deposition control zone and sulfur dioxide control zone	1994	Japan–China Environmental Cooperation Agreement signed
1996–2000: Rapid policy development	1996–2000	The Total-amount Control Plan and the Trans-century Green Project Plan in the 9th five-year plan	1996	Completion of the China–Japan Friendship Center for Environmental Protection
	1996	Standards of Air Quality and Integrated Emission Standard of Air Pollutants adopted. System for Controlling the Total Amount of Major Pollutants started.	1997	Japan announced the Initiatives for Sustainable Development toward the 21st Century (ISD), with EANET being the main item.
	1998	Planning for Acid Deposition Control Zone and Sulfur Dioxide Control Zone approved	1998	1st intergovernmental conference on EANET held. EANET preparatory phase started.
	1998	SO$_2$ emission levying system applied in the two control zones	1998	Joint communiqué on environmental cooperation for the 21st century
	1998	China declared official participant in EANET	2000	2nd Intergovernmental Conference on EANET held
	1999	National monitoring for the EANET preparatory phase started	2001	EANET started operating officially

National Economic and Social Development Plan on a five-year basis. The history of acid rain in China is summarized in Table I. We have characterized the progress of China's acid rain issue into four stages: early awareness, national recognition, regional issue, and rapid policy development. We examine each stage in the light of the China–Japan relationship.

2.1. 1979–1985: EARLY AWARENESS

This stage corresponded to the sixth five-year-plan period, during which the government was reforming the country's political and economic systems. Nationwide surveys on precipitation chemistry were carried out. As a result, widespread acidic pollution was observed and acid rain emerged as an issue in China (Yong, 1997). In particular, acid rain in southwest China received world wide attention. Japan had worked out an effective system for controlling sulfur dioxide emissions in the early1970s. The Environment Agency of Japan (EAJ) conducted the Acid Deposition Survey Phase 1 project from 1983 to 1988.

2.2. 1986–1990: NATIONAL RECOGNITION

This stage corresponded to the seventh five-year-plan period, during which the Chinese economy stabilized. The acid rain issue was adopted as a National Key Project. Systematic studies suggested that acid rain pollution was going from bad to worse in terms of the extent of land affected and the level of acidity, and thus in terms of the severity of damage to forests and crops (CEY, 1995, 1996). Southern China was the focal area. China's Air Pollution Control Act was adopted in September 1987, but sulfur pollution was still critical, because many sulfur emission facilities, such as power stations, were excluded from the subjects requiring control under the Act. In response to the urgent need for environmental remediation, in 1988 China and Japan agreed to establish the China–Japan Friendship Center for Environmental Protection. This symbolized the real beginning of environmental cooperation between China and Japan.

2.3. 1991–1995: REGIONAL ISSUE

This stage corresponded to the eighth five-year plan period, during which China's economy was booming and the concept of sustainable development was being advocated. Global environmental change attracted considerable attention, and various action plans were implemented in China. In 1992, Japan proposed the establishment of a regional network for monitoring acid deposition; this became known as the Acid Deposition Monitoring Network in East Asia (EANET). Japan hosted the first Expert Meeting in October 1993 – the first substantial action by the network. In March 1994 the Japan–China Environmental Cooperation Agreement was signed, and bilateral environmental activities were heavily promoted, particularly in the area of air pollution. It was acknowledged that acid rain was a regional problem. Meanwhile, Japan became anxious about trans-boundary air pollution coming from China. In 1990, aiming to explore long-range transport of sulfate from China, the EAJ set up sampling sites on six solitary islands in the Sea of Japan, in addition to the Acid Deposition Survey Phase 2 network (EAJ, 1995). Under these conditions, the Chinese government strengthened acid rain projects in its National Plan and began to conduct related policies in model area (CEY, 1995, 1996). Eastern China was the focal area. In response to growing public concern about the environment and increasing requests to make information about pollution openly available, official publications began to release data relating to acid rain.

2.4. 1995–2000: RAPID POLICY DEVELOPMENT

This stage corresponded to the ninth five-year-plan period, during which environmental protection was included in the plan for economic development. The Air Pollution Control Act 1987 was amended, and articles dealing with sulfur and acid rain pollution were revised in 1995. The new act prescribes provisions relating to the acid deposition control zone and the sulfur dioxide control zone. It has been a remarkable step in China's policy toward acid rain control. In 1996, sulfur dioxide was listed as one of the pollutants requiring control under the System for Controlling the Total Amount of Major Pollutants. Furthermore, international environmental cooperation on acid rain has developed actively. Japan announced its Initiatives for Sustainable Development toward the 21st Century (ISD) in 1997; one of the pillars of ISD was the establishment of EANET. The first intergovernmental meeting on EANET was held in March 1998, and the preparatory phase started the next month. Japan and China reached agreement for further effort on economic assistance and environmental cooperation in November 1998. At the same time, Japan achieved China's announcement of official participation in EANET. The governments issued a joint communiqué on environmental cooperation for the 21st century, and decided to actively promote the establishment of EANET. Other actions, such as the creation of the China–Japan Environmental Model City Plan (later Guiyang, Chongqing, and Dalian were selected), also gave high priority to acid rain control.

3. Discussion and Conclusions

We have analyzed the history of management of the acid rain issue in China over the last three decades. The history is well characterized by the implementation of national regulations and initiatives. It is obvious that the development of China's acid rain policy accelerated in the 1990s. Below, we consider the context of this acceleration in the light of the relationship between China and Japan.

3.1. GROWING ENVIRONMENTAL CONCERN: MEETING THE NEEDS OF THE TIMES

The age of global environmental concern began in the late 1980s and early 1990s, as the Cold War structure was collapsing and the North-South problem was coming to prominence in the world. Under this new paradigm, the acid rain issue in China was re-evaluated in the light of growing international environmental attention, and its priority as an environmental problem was enhanced. In the 1980s China became aware of the acid rain issue and began to deal with it. Since the 1990s, the environmental situation in China has become more critical. Although air pollution controls may restrict China's economic development temporarily or locally, the Chinese government has decided to take a strong policy stand (China's Agenda 21, 1994).

3.2. JAPAN'S ATTENTION TO LONG-RANGE TRANSPORT: A GAP AND ALSO A MOTIVE

Japan has paid more and more attention to the long-range transport of atmospheric pollutants. In recent years, a considerable number of researches have been conducted with the aim of exploring the long-range transport of acid rain between China and Japan. In Table II we list some of the results.

TABLE II
Research on long-range transport of acid rain in East Asia

Investiga-	Method	Result
Wu et al. (1998)	In situ observation or monitoring	Precipitation samples taken over the East China Sea in April 1994 under a north-east wind system were quite acidic. The results suggested pollution sources from the Korean Peninsula and Japan.
Hatakeyama et al. (1995)		High concentrations of SO_2 and sulfate aerosol were observed over the Sea of Japan in November 1992 when wind blew from the west or north. The results suggested pollution contributions from Korea and northern China.
EAJ (1995, 1997)		Concentrations of non-seasalt SO_4^{2-} at sites on the coast of the Sea of Japan were higher than in other areas, particularly in winter. The results suggested pollution contributions from China.
Huang et al. (1995, 1996)	Modeling estimation	China's contribution to Japan's total sulfur deposition was only **3.5%**. The external contribution to Japan's total sulfur was only **7%** (4% for SO_2 and 26% for sulfate deposition). A considerable amount of sulfur emission in East Asia is deposited over the sea.
Ichikawa and Fujita (1995)		China's contribution to Japan's total sulfur deposition was **25%** (50% for anthropogenic sulfur deposition).
Arndt et al. (1998)		External contribution to Japan's total sulfur deposition was about **10%** (17% for anthropogenic sulfur deposition). China's contribution to Japan's sulfur deposition had significant spatial and seasonal variations.

These examples illustrate the differences between the attitudes of research groups, revealing a divergent understanding of the trans-boundary issue. The scientific conclusions are open to further study (Ichikawa and Hayami, 1999), but "more important is the need for widespread recognition that science cannot answer policy questions" (Herrick and Jamieson, 1995). Nevertheless, the uncertainty of science enhances the political sensitivity between Beijing and Tokyo. When the first intergovernmental conference on EANET was held in Japan, China attended the meeting as an observer only. There seems to be a need for environmental negotiation in the East Asia region, as "Japan is keen to demonstrate that an increasing percentage of the acid rain falling on the country originates in the industrial districts of north China and Korea" (Triendl, 1998). Nevertheless, the debate on the long-range-transport issue has led the Chinese government to become aware of the impending crisis in acid rain control (CEY, 1998).

3.3. JAPAN'S ECONOMIC ASSISTANCE TO CHINA: A DRIVING FORCE FOR POLICY CHANGE

The Official Development Assistance (ODA) programs financially support Japan's Economic Assistance to China. The ODA programs for China began in 1979 and expanded rapidly over the last decade. The first environmental project supported by the ODA program was conducted in 1993. A special lower-interest loan has been given for environmental projects since 1995. The yen loan package of 1996–1998 involved a number of projects relating to acid rain and air pollution. The projects have been effective in helping to control China's acid rain. This is a reflection of Japan's concern with and enthusiasm for the acid rain issue. Because of increasing attention to environmental issues by Japan and pressure in negotiations, China has started investing in environmental programs. This is an important driving force for China's policy-making on acid rain control.

Overall, Japan has shown a great willingness to lead the region in environmental conservation and has played an active role in promoting a framework for preventing acid rain in East Asia. China has taken practical steps from an initial awareness of the problem to the formulation of a control strategy. China now recognizes that its acid rain problem is of regional importance, rather than being a simple domestic environmental issue.

Acknowledgments

This research was supported by the global environmental fund, Environment Agency of Japan.

References

Arndt, R. L., Carmichael, G. R. and Roorda, J. M.: 1998, Seasonal source-receptor relationships in Asia, *Atmospheric Environment* **32 (8)**, 1397.
China's Agenda 21 editorial committee: 1994, *China's Agenda 21*, China Environmental Science Publishing House, Beijing.
CEY (China Environmental Yearbook) editorial committee: 1995–1998, *China Environmental Yearbook*, China Environmental Yearbook Publishing House, Beijing.
EAJ (Environment Agency of Japan): 1995, *Final Report of the Acid Deposition Survey: Phase 2*, Environment Agency of Japan, Tokyo.
EAJ (Environment Agency of Japan): 1997, *An Interim Report of the Acid Deposition Survey: Phase 3*, Environment Agency of Japan, Tokyo.
Hatakeyama, S., Murano, K., Bandow, H., Mukai, H. and Akimoto, H.: 1995, High concentration of SO_2 observed over the Sea of Japan, *Terrestrial, Atmospheric and Ocean Sciences* **6**, 403.
Herrick, C. and Jamieson, D.: 1995, The social construction of acid rain, *Global Environmental Change* **5(2)**, 105.
Huang, M., Wang, Z., He, D., Xu, H. and Zhou, L.: 1995, Modeling studies on sulfur deposition and transport in East Asia, *Water Air and Soil Pollution* **85(4)**, 1921.
Huang, M., Wang, Z., He, D., Xu. H., Zhou, L. and Gao, H.: 1996, Studies on Cross-Boundary Transport of Sulfur in China and East Asia, *Climatic and Environmental Research* **1(1)**, 55.
Ichikawa, Y. and Fujita, S.: 1995, An analysis of wet deposition of sulfur using a trajectory model for East Asia, *Water Air and Soil Pollution* **85(4)**, 1927.
Ichikawa, Y. and Hayami, H.: 1999, Report on "Workshop on the Transport of Air Pollutants in Asia", *Journal of Japan Society for Atmospheric Environment* **34(6)**, A53.
Triendl, R.: 1998, Asian states take 'first step' on acid rain, *Nature* **392(6675)**, 426.
Wu, Y., Shen, Z. and Huang, M.: 1998, Chemical character of spring precipitation over the East Sea Region, *ACTA Scientiae Circumstantiae* **18(4)**, 362.
Yong, Y.: 1997, in Z. Chen (ed.), *Acid Deposition Research in China*, China Environmental Science Publishing House, Beijing, pp.1–10.

PROFOUND SURVIVAL PROGRAM OF FORESTS IN JAPAN ISLANDS A 40 YEAR STRATEGY FOR ENVIRONMENTAL CONSERVATION IN INLAND CHINA

YOSHIKAZU HASHIMOTO[1*], YOSHIKA SEKINE[2], ZHI-MIN YANG[3]
and KANJI YOSHIOKA[4]

[1]*Center for Area Studies, Keio University, Mita, Tokyo 108-8345 Japan;* [2] *Faculty of Science, Tokai University, Hiratsuka 259-1292 Japan;* [3] *China-Japan Environment Center, Chengdu, 610072 China ;* [4] *Keio Economic Observatory, Mita, Tokyo 108-8345 Japan.*
(* author for correspondence , e-mail:yosherg@aol.com)

Abstract. 15 years ago an interim report for an intense environmental program was compiled concerning the location of the islands of Japan at the meteorological down stream of the East Asian Countries. Parts of sulfur dioxide and other air pollutants, those supposed to cause acid deposition are emitted from the eastern parts of the Asian continent, especially in China. The air pollutants flow down to the east to spread over these islands. This acid deposition is projected to cause damage to forest resources of Japan in the future by increasing air pollutants emanating from the continent. A long term project by a research group at Keio University commenced in 1985 to identify ways of preventing this damage to the forests. The group formed the JACK Air Surveillance Network in China and South Korea in order to collect the first precise air pollution data in this region, as well as to identify a reliable partner for the project. On the completion of the JACK project, a highly cooperative group was formed between the researchers of Keio University and those in Chengdu, Sichuan province of China and has effectively worked since 1991. The goal of the project is the formation of an Inland Environmental Information Center in inland China to be accomplished by 2005. To launch a 10 year project by the center, a nation wide campaign is planned to raise the awareness of the population, specifically the lower socio-economic group on the effect of environmental issues. It is suggested that this education campaign take effect in inland China no later than 2025 for the preservation of the forests on the Islands of Japan.

Keywords: socio-technological experiment, risk control, inland China, forests on Japan Islands, a 40 year program, environmental information

1. Introduction

This is an intermediate report of a long term socio-technological experiment to protect the forests of Japanese islands from damage by future acid rain.

The Islands of Japan have a precipitation of 1000-2000 mm per annum, and are located in the 35 degrees NL temperature zone. Two thirds of the area is covered with rich forest. At present these forests only receive minimal damage from air pollution and acid rain from abroad. In the 1970's, there were some air pollution and acid rain episodes reported in Europe, specially damages to the forests. This news aroused world interest, even in Japan of air pollution. Although the situation in Japan differs in cross border pollution from that of the European countries as the Islands of Japan are distant from neighboring countries due to the boundaries of ocean and sea.

However, damage by air pollution and acid rain is still a factor for Japan due to the rising polluted air from the Asian continent—specifically China. At present there is a few scientific proofs of the damage to the forests of Japan by acid rain from the rest of Asia (Murano, 2000). However, wind transport pollutants from the continent to the Islands of Japan will be greatly increased in 10-20 years, due to the economic development of China. This development requires large amounts of energy that is supplied mainly by coal combustion. As Japan's Islands are geographically located down stream of this industrial air pollutant, damages to the forests and natural resources are logical. The forest is very important in the absorption of carbon dioxide and protecting the diversity of the ecosystem.

Although it seems necessary that China will either decrease or limit the consumption of fossil fuels, that economy requires much more consumption far beyond the present amount. The Chinese Government must adopt a policy to protect the environment, however, there is no intensive provision for environmental conservation as seen in the Chinese constitution. China has to feed a population of 1.2 billion in a great area of land, therefore, lowering the air pollution to an acceptable level to maintain the people's health and to avoid damaging the environment with pollutants is not easy even to the Chinese government.

Even so, the foreign countries cannot control China's environment. We can only cooperate with the Chinese people to develop the action necessary to conserve the natural environment. A long perspective of environment conservation has been suggested by giving environment education to the people of inland China. We received advice to visit the large cities along the coast of the Yellow Sea and the South China Sea. However, much after consideration, it was decided instead to target the cities of inland China at the beginning of the program. Cities in the coastal area will be restricted with pollution emission in the near future as they will be developed rapidly. This development will be carried out by foreign investment during the early stages and will be observed by an international forum on environmental conservation. This group cannot go into action as the research budget is too small to achieve any results, and due to the urgency of this development, this group is out of its depth.

2. Drafting a Plan

2.1. ENVIRONMENTAL CAPACITY TO PROCESS WASTES

A special feature of modern civilization is mass production and mass consumption. Benefits of the modern materialistic civilization are great, however, on the other hand, the flooding of chemicals and artificial goods produced by modern industry are causing environmental pollution. With the wastes by the mass production being a very real threat to society that breaks the balance between the amount of production and the natural power of processing wastes.

We can't predict when this time we will call X will arrive. We don't know when the balance of wastes from production and natural processing power will be broken. Unfortunately we do not have the answer. Even if someone estimated X by collecting a large amount of data and the ensuing knowledge we will not be satisfied. We have no reliable estimation.

However, we found needed to decide the time X for the purpose of drafting the environmental conservation plan. It was decided the time X hypothetically by arbitration will be between 2020 and 2030. It is not thought that the function of modern society or cities will be destroyed or stopped by this "crisis" point.

Environmental protection processes must be completed, therefore it was decided to make this a hypothetical time and thought that in order to achieve this goal time X was set for 2025.

2.2. A 40 YEAR STRATEGY AND A 30 YEAR OPERATION

Environmental treatment takes a long time, more than several years, in some cases it can take up to ten years. Therefore, it was decided that the time for one operation be 10 years. A rough schedule was planned for 40 years with an estimation of the following unit operations; The entire time necessary for the project is 40 years. This time is too long for one generation. The most successful component of the project would be considered in the middle part of the time frame. Time is also required to look at the program operation without the establishment of the Inland Environmental Information Center, and the many short term projects planned in the preliminary stages before the main program commences.

The project commenced in 1985, with completion due in 2025. However, there is 5 years remaining before the establishment of Inland Environmental Information Center, since it is to be set up no later than 2005 according to the plan.

3. Operation, Results and Discussion

3.1. ESTABLISHMENT OF AN AIR MONITORING NETWORK IN EAST ASIA

The establishment of an air monitoring network from Korean peninsula to Chinese continent was the first step. This network has a length of 3000 km east and west, with 6 stations, Seoul, Beijing, Baotou, Lanzhou, Urumuqi and Chengdu. It took 7 years to set up, starting in 1985 and with completion in 1991. This network is known as the JACK Network. The initials are those of the names of esch of the countries involved. This monitoring network collected valuable atmospheric data and ceased operation only one year after its commencement. If the network continued measuring, it would have been destroyed by the economic collapse experienced in Japan.

In a short period, the network collected precise and useful atmospheric data from 6 stations. These measurements were the first detailed observations of atmospheric air in this area. The measured data of airborne particulate matter for traces of 30 elements, and mainly analyzed by neutron activation method and 2 anions were measured by ion chromatography. The data can be directly compared with the data from National Air Surveillance Network (NASN), Japan (Hashimoto et al., 1994).

However, the most useful result from the network was to find a totally reliable partner for the program.

3.2. ESTABLISHMENT OF COOPERATIVE SYSTEM

The first approach for the establishment of a cooperative system for the operation was to have a joint project in an environmental measurement plan- this was carried out in the city of Chengdu.

To test the first joint experimental work an overall environmental investigation was planned. This environmental investigation covered a wide range of the research items, with normal measurements for atmospheric air plus indoor pollution and water quality. Additional research components were medical, economic and social inquiries and examinations.

The Government of Chengdu funded 400,000 Chinese yuens for the project. This funding for the research was a very large amount, beyond our expectations and hard to believe by those in Japan. The joint project of the researchers of Keio University group and those of the city of Chengdu was a total success (Yamada, 1995) and the city government presented prizes for both groups of Keio University and Chengdu. The success of this joint work gave momentum to the cooperation resulting in the project proceeding very smoothly.

3.3. DEFINITE TARGET OF THE PROGRAM AND RELATED ACTIVITY

When the stage began for the establishment of a center for environmental information, it was necessary to make the final vision and roles of those involved clear. The center with multi-functions covering a wide area, was thought to be the final aspect and fitted the purpose of relating information on environment conservation to the people.

Economic development and cooperation are important in this kind of activity, as obtaining the economics of the standard of living must precede public contact. Therefore the center is required at least some functions like the exchange of economic and industrial knowledge for foreign and Chinese people. This function can give social information to foreigners. In Inland China, contact between Chinese and foreigners and exchange between business people is still not enough to achieve better economic development.

The other working policy of the center is to give the knowledge on the environmental protection to the people while doing business activity. The center formed a commercial company (China-Japan Environment Center, Chengdu) and already began activity in such areas as importing and selling efficient burning devices for oil and gas imported from Japan. Higher efficiency of fuel burning makes the fuel consumption decreases, resulting in lower emission of complex organic substances from the unburned or imperfect burning of fuels. The company is acting an operation center in the business for Inland Environmental Information Center that will established in the future.

The first requirement in building the center, is a large amount of money. Therefore this plan requires the assistance of entrepreneurs of proper sizes. The group is planning to look for investment from Japanese entrepreneurs who are looking at establishing small and medium enterprises in inland China. The program group is now preparing a proposal for investment.

However, there is still a long way to go and the group needs to be more active in direct environmental activities and in the education of environmental conservation. As a trial, a drawing contest was developed by the operation center with the incentive of small prizes on the environment for school children in the city of Chengdu. This was accepted favorably by the children and more than 50,000 drawings were submitted. A similar trial was also developed for school children of a minority race near the city of Chengdu. Once again a large number (12,000) entries were collected of writing depicting nature. These entries were of excellent quality with some entries showing a much clearer reality than the surrounding environment. It is anticipated that this experience will hold fast in the minds of these children for a long time, resulting in ecologically aware adults in 10-20 years hence.

3.4. A SHORT HISTORY OF THE PROGRAM

The program is now in the middle stage, 15 years from the beginning. A short history of the program can be seen in Table I below.

TABLE I
A short chronological table of the program

Year	Items
1985	Generation of the idea of an environmental protection line along the Islands.
1986	The plan for JACK Network was developed.
1990	Preparation of the Joint research project in Chengdu.
1991	Completion of the JACK Network.
1992	Cessation of the JACK Network. Joint investigation in Chengdu.
1993	Completion of the joint investigation.
1994	Setting up of the operation center(China-Japan Environment Center, Chengdu).
1998	Introduction of a biobriquette machine and technique to Chengdu.
...	
(2005)	To set up Inland Environmental Information Center at Chengdu.

When the information center was completed as expected, the first activity is to support the industrial and commercial enterprises from foreign countries. The center does not offer financial assistance, but can give support by having an understanding of the Chinese community, by introducing foreign enterprises through the right channels of the local governments, teaching practical Chinese language to dispatched business people, and giving practical study and training of the social custom.

The center is to give other related services to foreign visitors, like the translation and interpretation of Chinese language into foreign language necessary for the inspection of inland China, including the local dialogue of minority races.

The program's goal is that the information center sends indispensable information on environmental conservation to the people of inland China. Although vitally important too is having scientific knowledge, social and cultural education on the environment. This kind of groundwork is useful and effective both for administrative bodies to execute the environmental protection and for nongovernmental people to cooperate with Chinese people in environmental conservation on a voluntary basis.

As is described the activity of the information center is to cover a wide area. The necessary preparation of the center is not only for building the center building, but also for the effective activities. More important is the function of the information center, therefore, the operation center has accumulated various individual activities necessary for the future information center and is to succeed them to the Inland Environmental Information Center.

On the other hand, it is not optimistic to think the operation center fulfills its expected function even if the center was set up on schedule and functioning properly. To date, the propelling force of the program has been based on strong personal relationships of mutual trust. These relationships of understanding assisted the business in an efficient and substantial way. However, this process of management based on human relation was successful only in cities in inland China where modern processes of business were not entirely completed. Such an old fashion form of management based on human relations could be the most reliable, successful and the safest way in a place, where a modern or western way of social and business system has not been established.

However, this style of management, founded on human relations can be, in some cases hindered by unexpected and sudden meetings of the people concerned. Therefore, the adoption of modern management of the operation center must be a necessary condition, without this method a new and extremely reliable partnership with strong faith will not be established between the groups of the university and the city of Chengdu. This is the destiny of this program, although the program has proceeded as planned on the basis of personal efforts over years.

4. Conclusion

This article is not entirely presented in a format seen in a scientific paper. This is due to the content being mainly of an empirical and observational nature and not purely scientific. It is but a social and technological study, the study of environmental affairs requires researchers to concentrate on a wide field of learning and investigation. The accumulation of empirical social knowledge in inland Chinese cities obtained from this program is to be published somewhere else for the aid of future practice of this kind of environmental program.

This program requires a very long time in the lives of the people concerned. Before we can observe and confirm proof of the justice of our attempt, the progress and part of the results were submitted in a short paper. The authors believe that this program will open a window to reveal a small picture in environmental conservation to give courage to the people working in this area.

Acknowledgement

The authors greatly appreciate to Prof. T. Yamada and Prof. T. Kojima, Keio University, for their leadership and kind assistance to perform this program. The authors are fully grateful to give research fund to the following person, funds and foundations (chronological order); the late Mr. Takashi Hashimoto, the Sumitomo Foundation, Takahashi Foundation, the Toyota Foundation, AEON Group Environment Foundation, the Asahi Glass Foundation, the Kajima Foundation, Ministry of Education, the Japan Foundation, Keio University Special Grant-in-Aid for Innovative Collaborative Research Projects and Japan Society for the Promotion of Science.

References

Hashimoto, Y. *et al.*: 1994, *Atmospheric Environment* **28**, 1437.
Murano, K.: 2000, (Lecture)"Expanded threat of acid precipitation by economic development in East Asia, Environment Research (Symposium) –A New Millennium, at Tokyo Forum, June 6.
Yamada, T. and Hashimoto, Y.: 1995,"Studies on Chinese Environment —A case study—" (in Japanese), pp. 207, Keiso Shobo, Tokyo.

If you have any concerns about our products,
you can contact us on
ProductSafety@springernature.com

In case Publisher is established outside the EU,
the EU authorized representative is:
**Springer Nature Customer Service Center GmbH
Europaplatz 3, 69115 Heidelberg, Germany**

Printed by Libri Plureos GmbH
in Hamburg, Germany

Acid rain 2000

*Proceedings from the 6th International Conference on Acidic Deposition:
Looking back to the past and thinking of the future
Tsukuba, Japan, 10–16 December 2000*

Volume III / III

*Conference Statement
Plenary and Keynote Papers*

Edited by

KENICHI SATAKE, Editor-in-Chief *National Institute for Environmental Studies, Japan*
JUNKO SHINDO *National Institute for Agro-Environmental Sciences, Japan*
TAKEJIRO TAKAMATSU *National Institute for Environmental Studies, Japan*
TAKANORI NAKANO *Institute of Geoscience, University of Tsukuba, Japan*
SHIGERU AOKI *Tokyo National Research Institute of Cultural Properties, Japan*
TSUTOMU FUKUYAMA *National Institute for Environmental Studies, Japan*
SHIRO HATAKEYAMA *National Institute for Environmental Studies, Japan*
KAZUKAMASA IKUTA *National Research Institute of Aquaculture, Japan*
MUNETSUGU KAWASHIMA *Shiga University, Japan*
YOSHIHISA KOHNO *Central Research Institute of Electric Power Industry, Japan*
SATORU KOJIMA *Tokyo Woman's Christian University, Japan*
KENTARO MURANO *National Institute for Environmental Studies, Japan*
TOSHIICHI OKITA *Obirin University, Japan, ret.*
HIROSHI TAODA *Forestry and Forest Products Research Institute, Japan*
KINICHI TSUNODA *Gunma University, Japan*
MAKOTO TSURUMI *Hirosaki University, Japan*

Reprinted from *Water, Air, and Soil Pollution* 130: 1–4, 2001

SPRINGER SCIENCE+BUSINESS MEDIA, LLC

A C.I.P. Catalogue record for this book is available from the Library of Congress.

ISBN 978-94-010-3733-4   ISBN 978-94-007-0810-5 (eBook)
DOI 10.1007/978-94-007-0810-5

Printed on acid-free paper

All rights reserved
©2001 Springer Science+Business Media New York
Originally published by Kluwer Academic Publishers in 2001
No part of the material protected by this copyright notice may be reproduced
or utilized in any form or by any means, electronic or mechanical,
including photocopying, recording or by any information storage and
retrieval system, without written permission from the copyright owner.

PREFACE

Humankind cannot live independently in the biosphere of the earth. Air, water, soil, microorganisms, and many animals and plants on earth support human existence, and all are intimately interconnected in the irreplaceable ecosystem of the earth. Under these circumstances, environmental pollution and destruction of nature are likely to deleteriously affect not only human existence but also the future life cycle.

Humankind has developed science and technology as part of its civilization. Consequently, human beings have gained advantage in the biosphere by reducing competition from other living organisms and by developing new types of energy and resources. The growth in human population, spurred by rapid industrialization, has been exponential. In 2000, the human population exceeded 6 billion and it is still growing. However, this growth has not come without consequences. In the 20th Century, human activities such as industrialization, urbanization, and military actions have led to the extinction of many species. Human beings are consuming enormous amounts of nonrenewable resources and energy. Consequently, a great amount of material, some of it acidic, has been discharged into the environment. This has resulted in acid pollution of the atmosphere, lakes, and rivers, damaging forests, eroding cultural treasures such as sculptures and buildings, and even compromising human health.

Thus, extremely important issues concerning the global ecosystem have arisen. Acid pollution is a prominent global environmental problem since acid pollutants discharged into the air are carried from country to country by air currents, spreading the problem across national boundaries.

It is crucially important to solve these problems not only domestically but also internationally. All countries of the world are interconnected and the environmental problems of one country directly or indirectly relate to those of other countries. Governments, industrial sectors, institutes, and citizens need to collaborate with each other. At this Conference, scientists and specialists from all over the world have gathered to study acid pollution and to discuss how to prevent it and how to preserve the global ecosystem.

Against this backdrop, I believe it is very significant that we were able to hold this 6th International Conference on Acidic Deposition in Asia at the end of the 20th Century and just before the start of the 21st Century. As Chairman of the Organizing Committee, I am delighted that scientists and specialists from 35 countries chose to participate in this Conference. I sincerely hope that the outcome of this Conference will contribute to sustaining the global environment in the future. Finally, I would like to express my sincere appreciation to the members of the International Advisory Board for support and to the persons who worked to make this Conference a memorable one.

Jiro Kondo
Chairman of the Organizing Committee

VOLUME III

Preface iii

Contents of Volume III v

EFFECTS ON TERRESTRIAL ECOSYSTEM

EUGENIJA KUPCINSKIENĖ
Annual Variations of Needle Surface Characteristics of *Pinus Sylvestris*
Growing Near the Emission Source 923-928

GALINA N. KOPTSIK, SERGUEI V. KOPTSIK and DAN AAMLID
Pine Needle Chemistry Near a Large Point SO_2 Source in Northern Fennoscandia
.... 929-934

JESADA. LUANGJAME, BOONCHOOB BOONTAWEE and NIGORN. KLIANGPIBOOL
Determination of Deposition and Leaves in Teak Plantations in Thailand 935-940

T. TAKAMATSU, H. SASE, J. TAKADA and R. MATSUSHITA
Annual Changes in Some Physiological Properties of *Cryptomeria Japonica*
Leaves from Kanto, Japan 941-946

ELINA J. OKSANEN
Increasing Tropospheric Ozone Level Reduced Birch (*Betula Pendula*) Dry
Mass within a Five Years Period 947-952

LUCY J. SHEPPARD, ALAN CROSSLEY, JUDITH PARRINGTON,
FRANCIS J HARVEY and J. NEIL CAPE
Effects of Simulated Acid Mist on a Sitka Spruce Forest Approaching Canopy
Closure: Significance of Acidified Versus Non-Acidified Nitrogen Inputs 953-958

HIDEYUKI MATSUMURA
Impacts of Ambient Ozone and/or Acid Mist on the Growth of 14 Tree Species:
an Open-Top Chamber Study Conducted in Japan 959-964

TETSUSHI YONEKURA, YUKIKO DOKIYA, MOTOHIRO FUKAMI and
TAKESHI IZUTA
Effects of Ozone and/or Soil Water Stress on Growth and Photosynthesis of
Fagus crenata Seedlings 965-970

TATSURO NAKAJI and TAKESHI IZUTA
Effects of Ozone and/or Excess Soil Nitrogen on Growth, Needle Gas
Exchange Rates and Rubisco Contents of *Pinus densiflora* Seedlings 971-976

Y. KOHNO, R. MATSUKI, S. NOMURA, K. MITSUNARI and M. NAKAO
Neutralization of Acid Droplets on Plant Leaf Surfaces 977-982

JAKUB HRUŠKA, PAVEL CUDLÍN and PAVEL KRÁM
Relationship between Norway Spruce Status and Soil Water Base Cations/
Aluminum Ratios in the Czech Republic 983-988

JAN MULDER, HELENE A. DE WIT, HELENA W.J. BOONEN and LARS R. BAKKEN
Increased Levels of Aluminium in Forest Soils: Effects on the Stores of Soil
Organic Carbon 989-994

HELENE A. DE WIT, JAN MULDER, PER H. NYGAARD and DAN AAMLID
Testing the Aluminium Toxicity Hypothesis: a Field Manipulation
Experiment in Mature Spruce Forest in Norway 995-1000

KAZUO SATO and TAKASHI WAKAMATSU
Soil Solution Chemistry in Forests with Granite Bedrock in Japan 1001-1006

TAKESHI IZUTA, TAEKO YAMAOKA, TATSURO NAKAJI, TETSUSHI
YONEKURA, MASAAKI YOKOYAMA, HIDEYUKI MATSUMURA, SACHIE
ISHIDA, KENICHI YAZAKI, RYO FUNADA and TAKAYOSHI KOIKE
Growth, Net Photosynthetic Rate, Nutrient Status and Secondary Xylem
Anatomical Characteristics of *Fagus crenata* Seedlings Grown in Brown
Forest Soil Acidified with H_2SO_4 Solution 1007-1012

RIE TOMIOKA and CHISATO TAKENAKA
Differential Ability of the Root to Change Rhizosphere pH between
Chamaecyparis Obtusa Sieb. (Hinoki) and *Quercus Serrata* Thumb.(Konara)
under Aluminium Stress 1013-1018

MASAHIKO OHNO
Sensitivity of a Japanese Earthworm (*Allolobophora japonica*) to Soil
Acidity 1019-1024

SERGUEI V. KOPTSIK, NATALIYA BEREZINA and S. LIVANTSOVA
Effects of Natural Soil Acidification on Biodiversity in Boreal Forest
Ecosystems.. 1025-1030

LARS LUNDIN, MATS AASTRUP, LAGE BRINGMARK, SVEN BRÅKENHIELM,
HANS HULTBERG, KJELL JOHANSSON, KARIN KINDBOM, HANS KVARNÄS
and STEFAN LÖFGREN
Impacts from Deposition on Swedish Forest Ecosystems Identified by
Integrated Monitoring 1031-1036

CAROLE E.R. PITCAIRN, IAN D. LEITH, DAVID FOWLER, KEN J. HARGREAVES,
MASOUD MOGHADDAM, VALERIE H. KENNEDY and LENNART GRANAT
Foliar Nitrogen as an Indicator of Nitrogen Deposition and Critical Loads
Exceedance on a European Scale.. 1037-1042

IAN D. LEITH, LUCY J. SHEPPARD, CAROLE E.R. PITCAIRN, J. NEIL CAPE, PAUL
W. HILL, VALERIE H. KENNEDY, Y. SIM TANG, RON I. SMITH and
DAVID FOWLER
Comparison of the Effects of Wet N Deposition(NH_4Cl) and Dry N
Deposition (NH_3) on UK Moorland Species 1043-1048

JOHAN BERGHOLM and HOOSHANG MAJDI
Accumulation of Nutrients in above and below Ground Biomass in Response
to Ammonium Sulphate Addition in a Norway Spruce Stand in Southwest
Sweden.. 1049-1054

E. P. FARRELL, J. AHERNE, G. M. BOYLE and N. NUNAN
Long-Term Monitoring of Atmospheric Deposition and the Implications
of Ionic Inputs for the Sustainability of a Coniferous Forest Ecosystem 1055-1060

B. K. SITAULA, J. I. B. SITAULA, Å. AAKRA and L. R. BAKKEN
Nitrification and Methane Oxidation in Forest Soil: Acid Deposition,
Nitrogen Input and Plant Effects 1061-1066

J. R. HALL, B. REYNOLDS, T. SPARKS, A. COLGAN, I. THORNTON and
S. P. MCGRATH
The Relationship between Topsoil and Stream Sediment Heavy Metal
Concentrations and Acidification 1067-1072

D. TANG, E. LYDERSEN, H. M. SEIP, V. ANGELL, O. EILERTSEN, T. LARSSEN,
X. LIU, G. KONG, J. MULDER, A. SEMB, S. SOLBERG, K. TORSETH,
R. D. VOGT, J. XIAO and D. ZHAO
Integrated Monitoring Program on Acidification of Chinese Terrestrial
Systems (IMPACTS) - a Chinese- Norwegian Cooperation Project 1073-1078

Y. G. XU, G. Y. ZHOU, Z. M. WU, T. S. LUO and Z. C. HE
Chemical Composition of Precipitation,Throughfall and Soil Solutions
at Two Forested Sites in Guangzhou, South China 1079-1084

YASUMI YAGASAKI, TAKASHI CHISHIMA, MASANORI OKAZAKI,
DU-SIK JEON, JEONG-HWAN YOO and YOUNG -KULL KIM
Acidification of Red Pine Forest Soil due to Acidic Deposition In Chunchon,
Korea 1085-1090

HIROSHI OKOCHI and MANABU IGAWA
Elevational Patterns of Acid Deposition into a Forest and Nitrogen Saturation
on Mt. Oyama, Japan. 1091-1096

T. KAWAKAMI, H. HONOKI and H. YASUDA
Acidification of a Small Stream on Kureha Hill Caused by Nitrate Leached

from a Forested Watershed 1097-1102

M. BABA, Y. SUZUKI, H. SASAKI, K. MATANO, T. SUGIURA and H. KOBAYASHI
Nitrogen Retention in Japanese Cedar Stands in Northern Honshu, with High
Nitrogen Deposition 1103-1108

K. MATANO, M. BABA, A. SHIBUYA, Y. SUZUKI, T. SUGIURA and
H. KOBAYASHI
Soil Solution Chemistry in Japanese Cedar Stands in Northern Honshu,
with High Nitrogen Deposition 1109-1114

TSUYOSHI YAMADA, SHUICHIRO YOSHINAGA, KAZUHITO MORISADA and
KEIZO HIRAI
Sulfate and Nitrate Loads on a Forest Ecosystem in Kochi in Southwest
of Japan 1115-1120

M. CRISTINA FORTI, ADILSON CARVALHO, ADOLPHO J. MELFI and
CELIA R. MONTES
Deposition Patterns of SO_4^{2-}, NO_3^- and H^+ in the Brazilian Territory 1121-1126

SANJAY KUMAR, R. DATTA, S. SINHA, T. KOJIMA, S. KATOH and M. MOHAN
Carbon Stock, Afforestation and Acidic Deposition: an Analysis of
Inter-relation with Reference to Arid Areas 1127-1132

MODELS FOR EVALUATING ECOLOGICATL EFFECTS

J.-P. HETTELINGH, M. POSCH and P.A.M. DE SMET
Multi-Effect Critical Loads Used in Multi-Pollutant Reduction Agreements
in Europe 1133-1138

M. POSCH, J.-P. HETTELINGH and P.A.M. DE SMET
Characterization of Critical Load Exceedances in Europe 1139-1144

SOON-UNG PARK and YOUNG-HEE LEE
Estimation of the Maximum Critical Load for Sulfur in South Korea 1145-1150

PETRA MAYERHOFER, JOSEPH ALCAMO, MAXIMILIAN POSCH and
JELLE G. VAN MINNEN
Regional Air Pollution and Climate Change in Europe: an Integrated
Assessment (AIR-CLIM) 1151-1156

JIMING HAO, XUEMEI YE, LEI DUAN and ZHONGPING ZHOU
Calculating Critical Loads of Sulfur Deposition for 100 Surface Waters in
China Using the Magic Model 1157-1162

C. J. CURTIS, R. HARRIMAN, M. HUGHES and M. KERNAN
Effects of Site Selection Strategy on Freshwater Critical Load Exceedances
in Wales 1163-1168

M. KERNAN, J. HALL, J. ULLYET and T. ALLOTT
Variation in Freshwater Critical Loads Across Two Upland Catchments in
the UK: Implications for Catchment Scale Management 1169-1174

JOHAN C.I. KUYLENSTIERNA, W. KEVIN HICKS, STEVE CINDERBY,
HARRY W. VALLACK and MAGNUZ ENGARDT
Variability in Mapping Acidification Risk Scenarios for Terrestrial
Ecosystems in Asian Countries 1175-1180

LEO SAARE, REET TALKOP and OTT ROOTS
Air Pollution Effects on Terrestrial Ecosystems in Estonia 1181-1186

FULU TAO and ZONGWEI FENG
Critical Loads of Acid Deposition for Ecosystems in South China
-- Derived by a New Method 1187-1192

MIKHAIL SEMENOV, VLANDIMIR BASHKIN and HARALD SVERDRUP
Critical Loads of Acidity for Forest Ecosystems of North Asia 1193-1198

LEI DUAN, SHAODONG XIE, ZHONGPING ZHOU, XUEMEI YE and JIMING HAO
Calculation and Mapping of Critical Loads for S, N and Acidity in China 1199-1204

JUNLING AN, LING ZHOU, MEIYUAN HUANG, HU LI, TSUNEHIKO OTOSHI and KAZUHIDE MATSUDA
A Literature Review of Uncertainties in Studies of Critical Loads for Acidic Deposition 1205-1210

KENTARO HAYASHI and MASANORI OKAZAKI
Acid Deposition and Critical Load Map of Tokyo 1211-1216

DANIEL KURZ, BEAT RIHM, MATTIAS ALVETEG and HARALD SVERDRUP
Steady-State and Dynamic Assessment of Forest Soil Acidification in Switzerland 1217-1222

BEAT RIHM and DANIEL KURZ
Deposition and Critical Loads of Nitrogen in Switzerland 1223-1228

K. R. BULL, J. R. HALL, J. COOPER, S. E. METCALFE, D. MORTON, J. ULLYETT, T. L. WARR and J. D, WHYATT
Assessing Potential Impacts on Biodiversity Using Critical Loads 1229-1234

J. M. ULLYETT, J. R. HALL, M. HORNUNG and M. KERNAN
Mapping the Potential Sensitivity of Surface Waters to Acidification Using Measured Freshwater Critical Loads as an Indicator of Acid Sensitive Areas 1235-1240

GARETH J.P THORNTON
Calculating Weathering Rates of Stream Catchments in the English Lake District Using Critical Element Ratios, Mass-balance Budgets and the Magic Model 1241-1246

TAMON FUMOTO, JUNKO SHINDO, NORIKO OURA and HARALD SVERDRUP
Adapting the Profile Model to Calculate the Critical Loads for East Asian Soils by Including Volcanic Glass Weathering and Alternative Aluminum Solubility System 1247-1252

HIDESHI IKEDA and YOICHI MIYANAGA
Hydrogeochemical Conditions Affecting Acidification of Stream Water in Mountainous Watersheds 1253-1258

J. SHINDO, T. FUMOTO, N. OURA, T. NAKANO and T. TAKAMATSU
Estimation of Mineral Weathering Rates under Field Conditions Based on Base Cation Budget and Strontium Isotope Ratios 1259-1264

J. MILINDALEKHA, V. N. BASHKIN, and S. TOWPRAYOON
Calculation and Mapping of Sulfur Critical Loads for Terrestrial Ecosystems of Thailand 1265-1270

CHRISTINE ALEWELL
Predicting Reversibility of Acidification: The European Sulfur Story 1271-1276

SERGUEI V. KOPTSIK and GALINA N. KOPTSIK
Effects of Acid Deposition on Forest Soils in Northernmost Russia: Modelled and Field Data 1277-1282

HELEN J. FOSTER, MATTHEW J. LEES, HOWARD S. WHEATER, COLIN NEAL and BRIAN REYNOLDS
Dynamic Modelling of Spatially Variable Catchment Hydrochemistry for Critical Loads Assessment 1283-1288

WOJCIECH MILL
Integrated Modelling of Acidification Effects to Forest Ecosystems -Model Sonox 1289-1294

HU LI, AKIKAZU KAGA and KATSUHITO YAMAGUCHI
Prediction of Soil Acidification Using a Dynamic Model at a Bamboo Forest in Osaka Prefecture 1295-1300

PAVEL KRÁM, HJALMAR LAUDON, KEVIN BISHOP, LARS RAPP and JAKUB HRUSKA
Magic Modeling of Long-Term Lake Water and Soil Chemistry at Abborrträsket, Northern Sweden 1301-1306

ECOSYSTEM RECOVERY

D. T. MONTEITH, C. D. EVANS and S. PATRICK
Monitoring Acid Waters in the UK: 1988-1998 Trends 1307-1312

BJØRN WALSENG and LEIF R. KARLSEN
Planktonic and Littoral Microcrustaceans as Indices of Recovery in Limed
Lakes in SE Norway 1313-1318

BJØRN WALSENG and ANN KRISTIN.L. SCHARTAU
Crustacean Communities in Canada and Norway: Comparison of Species
along a pH Gradient 1319-1324

ANN KRISTIN L. SCHARTAU, BJØRN WALSENG and ED SNUCINS
Correlation between Microcrustaceans and Environmental Variables along
an Acidification Gradient in Sudbury, Canada 1325-1330

BJØRN WALSENG, ROY M. LANGÅKER, TOR E. BRANDRUD, PÅL BRETTUM,
ARNE FJELLHEIM, TRYGVE HESTHAGEN, ØYVIND KASTE,
BJØRN M. LARSEN and ELI-A. LINDSTRØM
The River Bjerkreim in SW Norway -Successful Chemical and Biological
Recovery after Liming 1331-1336

A. LYCHE-SOLHEIM, Ø. KASTE and E. DONALI
Can Phosphate Help Acidified Lakes to Recover? 1337-1342

S. SANDØY and R. M. LANGÅKER
Atlantic Salmon and Acidification in Southern Norway: A Disaster in
the 20TH Century, but a Hope for the Future? 1343-1348

FRODE KROGLUND, ØYVIND KASTE, BJØRN O. ROSSELAND and
TRYGVE POPPE
The Return of the Salmon 1349-1354

TRYGVE HESTHAGEN, TORBJØRN FORSETH, RANDI SAKSGÅRD,
HANS M. BERGER and BJØRN M. LARSEN
Recovery of Young Brown Trout in Some Acidified Streams in
Southwestern and Western Norway 1355-1360

TRYGVE HESTHAGEN, HANS M. BERGER, ANN KRISTIN LIEN SCHARTAU,
TERJE NØST, RANDI SAKSGÅRD and LEIDULF FLØYSTAD
Low Success Rate in Re-Establishing European Perch in Some Highly
Acidified Lakes in Southernmost Norway 1361-1366

M. RASK, H. PÖYSÄ, P. NUMMI and C. KARPPINEN
Recovery of the Perch (*Perca Fluviatilis*) in an Acidified Lake and Subsequent
Responses in Macroinvertebrates and the Goldeneye (*Bucephala Clangula*)
.... 1367-1372

K. NYBERG, J. VUORENMAA, M. RASK, J. MANNIO and J. RAITANIEMI
Patterns in Water Quality and Fish Status of Some Acidified Lakes
in Southern Finland during a Decade: Recovery Proceeding 1373-1378

ARNE FJELLHEIM and GUNNAR G. RADDUM
Acidification and Liming of River Vikedal, Western Norway. A 20 Year
Study of Responses in the Benthic Invertebrate Fauna 1379-1384

GODTFRED A. HALVORSEN, JOCELYNE H. HENEBERRY and ED SNUCINS
Sublittoral Chironomids as Indicators of Acidity (Diptera: Chironomidae)
.... 1385-1390

ARNE FJELLHEIM, ÅSMUND TYSSE and VILHELM BJERKNES
Reappearance of Highly Acid-Sensitive Invertebrates after Liming of
an Alpine Lake Ecosystem 1391-1396

ALUN S. GEE
A Strategic Appraisal of Options to Ameliorate Regional Acidification 1397-1402

ATLE HINDAR, MAXIMILIAN POSCH and ARNE HENRIKSEN

Effects of In-Lake Retention of Nitrogen on Critical Load Calculations 1403-1408

VILHELM BJERKNES and TORULV TJOMSLAND
Flow and pH Modelling to Study the Effects of Liming in Regulated, Acid
Salmon Rivers 1409-1414

K. BISHOP, H. LAUDON, J. HRUSKA, P. KRAM, S. KÖHLER and S. LÖFGREN
Does Acidification Policy Follow Research in Northern Sweden?
The Case of Natural Acidity during the 1990's 1415-1420

A. WILANDER
Effects of Reduced S Deposition on Large-Scale Transport of Sulphur in
Swedish Rivers.. 1421-1426

JAAKKO MANNIO
Recovery Pattern from Acidification of Headwater Lakes in Finland.................... 1427-1432

BRIT LISA SKJELKVÅLE, KJETIL TØRSETH, WENCHE AAS and TOM ANDERSEN
Decrease in Acid Deposition - Recovery in Norwegian Waters 1433-1438

JOHN GUNN, ROD SEIN, BILL KELLER and PETER BECKETT
Liming of Acid and Metal Contaminated Catchments for the Improvement of
Drainage Water Quality 1439-1444

R. D. VOGT, H. M. SEIP, H. OREFELLEN, G. SKOTTE, C. IRGENS and J. TYSZKA
Trends in Soil Water Composition at a Heavily Polluted Site - Effects of
Decreased S-Deposition and Variations in Precipitation... 1445-1450

KAZUHIKO SAKAMOTO, YUGO ISOBE, XUHUI DONG and SHIDONG GAO
Simulated Acid Rain Leaching Characteristics of Acid Soil Amended with
Bio-Briquette Combustion Ash 1451-1456

EFFECTS ON MATERIALS AND CULTURAL PROPERITIES

JOHAN TIDBLAD, VLADIMIR KUCERA, ALEXANDRE A. MIKHAILOV,
JAN HENRIKSEN, KATERINA KREISLOVA, TIM YATES, BRUNO STÖCKLE
and MANFRED SCHREINER
UN ECE ICP Materials: Dose-Response Functions on Dry and Wet Acid
Deposition Effects after 8 Years of Exposure 1457-1462

M. KITASE, S. HATAKEYAMA, T. MIZOGUCHI and Y. MAEDA
Regional Characteristics of Copper Corrosion Components in East Asia
.... 1463-1468

JOHAN TIDBLAD, VLADIMIR KUCERA and ALEXANDRE A. MIKHAILOV
Mapping of Acid Deposition Effects and Calculation of Corrosion Costs on
Zinc in China 1469-1474

ELIN DAHLIN, PETER TORSSANDER, CARL -MAGNUS MÖRTH,
HELÉNE STRANDH, GÖRAN ÅBERG, JAN F. HENRIKSEN, ODD ANDA
and RUNO LÖFVENDAHL
Environmental Monitoring of Rock Carvings in Scandinavia................................. 1475-1480

TSUTOMU KANAZU, TAKURO MATSUMURA, TATSUO NISHIUCHI and
TAKESHI YAMAMOTO
Effect of Simulated Acid Rain on Deterioration of Concrete 1481-1486

Y. TSUJINO, Y.SATOH, N. KURAMOTO and Y. MAEDA
Effect of Acid Deposition on Urushi Lacquer in East Asia.................................... 1487-1492

ANALYTICAL METHODS AND MONITORING

KJETIL TØRSETH, WENCHE AAS and SVERRE SOLBERG
Trends in Airborne Sulphur and Nitrogen Compounds in Norway during
1985-1996 in Relation to Air Mass Origin 1493-1498

P. H. VIET, V. V. TUAN, P. M. HOAI, N. T.K. ANH and P. T. YEN
  Chemical Composition and Acidity of Precipitation: A Monitoring Program
  in North-Eastern Vietnam 1499-1504

MUNEHIRO WARASHINA, MASANOBU TANAKA, YOSHIO TSUJINO,
TUGUO MIZOGUCHI, SIRO HATAKEYAMA and YASUAKI MAEDA
  Atmospheric Concentrations of Sulfur Dioxide and Nitrogen Dioxide in
  China and Korea Measured by Using the Improved Passive Sampling Method
  1505-1510

MOTONORI TAMAKI, TAKATOSHI HIRAKI, YOSHIHIRO NAKAGAWA,
TOMIKI KOBAYASHI, MASAHIDE AIKAWA and MITSURU SHOGA
  Study on Sampling Method of Rainfall, Throughfall, and Stemflow to
  Monitor the Effect of Acid Deposition on Forest Ecosystem 1511-1516

MASAHIDE AIKAWA, TAKATOSHI HIRAKI, MITSURU SHOGA and
MOTONORI TAMAKI
  Fog and Precipitation Chemistry at Mt. Rokko in Kobe, April 1997 - March
  1998 1517-1522

YASUSHI NARITA, KEI SATOH, KENICHI HAYASHI and SHIGERU TANAKA
  Development of Automatic Continuous Measurement System of Chemical
  Constituents in the Precipitation 1523-1528

ANNE-GUNN HJELLBREKKE and LEONOR TARRASON
  Mapping of Concentrations in Europe Combining Measurements and Acid
  Deposition Models 1529-1534

YUKIO KOMAI, SATOSHI UMEMOTO and TAKANOBU INOUE
  Influence of Acid Deposition on Inland Water Chemistry –A Case Study
  from Hyogo Prefecture, Japan – 1535-1540

C. D. EVANS, R. HARRIMAN, D. T. MONTEITH and A. JENKINS
  Assessing the Suitability of Acid Neutralising Capacity as a Measure of
  Long-term Trends in Acidic Waters Based on Two Parallel Datasets 1541-1546

RASA GIRGZDIENE and ALOYZAS GIRGZDYS
  Spatial and Temporal Variability in Ozone Concentration Level at Two
  Lithuanian Stations 1547-1552

AUDRONE MILUKAITE, ALDONA MIKELINSKIENE and
BRONISLAVAS GIEDRAITIS
  Characteristics of SO_2, NO_2, Soot and Benzo(a)pyrene Concentration
  Variation on the Eastern Coast of the Baltic Sea... 1553-1558

KOICHI WATANABE, YUTAKA ISHIZAKA, YUKIYA MINAMI and
KOJI YOSHIDA
  Peroxide Concentrations in Fog Water at Mountainous Sites in Japan 1559-1564

OSAMU NAGAFUCHI, HITOSHI MUKAI and MINORU KOGA
  Black Acidic Rime Ice in the Remote Island of Yakushima, a World
  Natural Heritage Area 1565-1570

SITI ASIATI, TUTI BUDIWATI and LELY QODRITA AVIA
  Acid Deposition in Bandung, Indonesia 1571-1576

GÖRAN E. ÅBERG
  Tracing Pollution and its Sources with Isotopes 1577-1582

YORIKO YOKOO and TAKANORI NAKANO
  Sequential Leaching of Volcanic Soil to Determine Plant-Available Cations
  and the Provenance of Soil Minerals Using Sr Isotopes 1583-1588

KIN-ICHI TSUNODA, TOMONARI UMEMURA, KAZUMASA OHSHIMA,
SHO-ICHI AIZAWA, ETSURO YOSHIMURA and KEN-ICHI SATAKE
  Determination and Speciation of Aluminum in Environmental Samples by
  Cation Exchange High-Performance Liquid Chromatography with High
  Resolution ICP-MS Detection 1589-1594

WENCHE AAS and ARNE SEMB
  Standardisation of Methods for Long-Term Monitoring 1595-1600

CHANG-JIN MA, MIKIO KASAHARA, SUSUMU TOHNO and TOMIHIRO KAMIYA
A New Approach for Characterization of Single Raindrops 1601-1606

LU-YEN CHEN, FU-TIEN JENG, YU-MEY HSU, SHIN-YUAN TSAI and UEI-RUEY PENG
Stability of Ionic Components in Precipitation Samples – a Case Study in Taipei 1607-1612

TSUNEHIKO OTOSHI, NORIO FUKUZAKI, HU LI, HIROSHI HOSHINO, HIROYUKI SASE, MASASHI SAITO and KATSUNORI SUZUKI
Quality Control and its Constraints during the Preparatory-Phase Activities of the Acid Deposition Monitoring Network in East Asia (EANET) 1613-1618

REGIONAL CASE STUDIES

N. KARVOSENOJA, P. HILLUKKALA, M. JOHANSSON and S. SYRI
Cost-Effective Abatement of Acidifying Emissions with Flue Gas Cleaning vs. Fuel Switching in Finland 1619-1624

DRIEJANA, D. W. RAPER, D. S. LEE, R. D. KINGDON and I. L. GEE
Wet Deposition to an Upland Area of England in 1988 and 1999: Measurements and Modelling 1625-1630

DEXUAN WANG and WEI DENG
Atmospheric SO_2 Pollution and Acidity of Rain in Changchun China 1631-1634

T. LARSSEN, H. M. SEIP, G. R. CARMICHAEL and J. L. SCHNOOR
The Importance of Calcium Deposition in Assessing Impacts of Acid Deposition in China 1635-1640

MARTIN FORSIUS, SIRPA KLEEMOLA, JUSSI VUORENMAA and SANNA SYRI
Fluxes and Trends of Nitrogen and Sulphur Compounds at Integrated Monitoring Sites in Europe 1641-1648

YASUSHI NARITA, KEI SATOH, KEIICHI HAYASHI, TAKAMI IWASE, SHIGERU TANAKA, YUKIKO DOKIYA, MORIKAZU HOSOE and KAZUHIKO HAYASHI
Long Term Trend of Chemical Constituents in Tokyo Metropolitan Area in Japan 1649-1654

MASAKI ADACHI, KAZUHIKO HAYASHI and YUKIKO DOKIYA
Sea Ice Approach and Chemical Species in Precipitation at Abashiri, Japan
.... 1655-1660

Y. MATSUURA, M. SANADA, M. TAKAHASHI, Y. SAKAI and N. TANAKA
Long-Term Monitoring Study on Rain, Throughfall, and Stemflow Chemistry in Evergreen Coniforous Forests in Hokkaido, Northern Japan 1661-1666

KAZUHIKO HAYASHI, YASUHITO IGARASHI, YUKITOMO TSUTSUMI and YUKIKO DOKIYA
Aerosol and Precipitation Chemistry During the Summer at the Summit of Mt. Fuji, Japan (3776 m a.s.l.) 1667-1672

NORIO FUKUZAKI, TSUYOSHI OHIZUMI and KAZUHIDE MATSUDA
Geographical and Temporal Variations of Chemical Constituents in Winter Precipitation Collected in the Areas along the Coast of the Sea of Japan 1673-1678

TSUYOSHI OHIZUMI, NAOKO TAKE, NOBORU MORIYAMA, OSAMU SUZUKI and MINORU KUSAKABE
Seasonal and Spatial Variations in the Chemical and Sulfur Isotopic Composition of Acid Deposition in Niigata Prefecture, Japan............................. 1679-1684

UMESH C. KULSHRESTHA, MONIKA J. KULSHRESTHA, R. SEKAR, M. VAIRAMANI, AJIT K. SARKAR and DANESH C. PARASHAR
Investigation of Alkaline Nature of Rain Water in India 1685-1690

AHTI LEPISTÖ and SANNA SYRI
Modeling Effects of Changing Deposition and Forestry on Nitrogen Fluxes
in a Northern River Basin 1691-1696

FRIDA EDBERG, HANS BORG and JAN -ERIK ÅSLUND
Episodic Events in Water Chemistry and Metals in Streams in Northern
Sweden during Spring Flood 1697-1702

N. L. ROSE, E. SHILLAND, T. BERG, K. HANSELMANN, R. HARRIMAN,
K. KOINIG, U. NICKUS, B. STEINER TRAD, E. STUCHLÍK, H. THIES and
M. VENTURA
Relationships between Acid Ions and Carbonaceous Fly-Ash Particles in
Deposition at European Mountain Lakes 1703-1708

H. FUSHIMI, T. KAWAMURA, H. IIDA, M. OCHIAI, T. NAKAJIMA and
Y. AZUMA
Internal Distribution of Acid Materials within Snow Crystals 1709-1714

CECILIA ANDRÉN, PAUL ANDERSSON and ELISABETH FRÖBERG
Temporal Variations of Aluminium Fractions in Streams in the Delsbo Area,
Central Sweden 1715-1720

H. HANSEN, T. LARSSEN, H. M. SEIP and R. D. VOGT
Trace Metals in Forest Soils at Four Sites in Southern China 1721-1726

BARBARA WALNA, JERZY SIEPAK and STANISLAW DRZYMALA
Soil Degradation in the Wielkopolski National Park (Poland) as an Effect of
Acid Rain Simulation 1727-1732

ZHAO DAWEI, T. LARSSEN, ZHANG DONGBAO, GAO SHIDONG, R. D. VOGT,
H. M. SEIP and O. J. LUND
Acid Deposition and Acidification of Soil and Water in the Tie Shan Ping
Area, Chongqing, China 1733-1738

M. JOHANSSON, R. SUUTARI, J. BAK, G. LÖVBLAD, M. POSCH, D. SIMPSON,
J. -P. TUOVINEN and K. TØRSETH
The Importance of Nitrogen Oxides for the Exceedance of Critical
Thresholds in the Nordic Countries 1739-1744

MAGNUS APPELBERG and TORBJÖRN SVENSON
Long-Term Ecological Effects of Liming - The ISELAW Programme 1745-1750

KERSTIN HOLMGREN
Biomass-Size Distribution of the Aquatic Community in Limed,
Circumneutral and Acidified Reference Lakes 1751-1756

H. BORG, J. EK and K. HOLM
Influence of Acidification and Liming on the Distribution of Trace Elements
in Surface Waters 1757-1762

FRIDA EDBERG, PAUL ANDERSSON, HANS BORG, CHRISTINA EKSTRÖM and
EINAR HÖRNSTRÖM
Reacidification Effects on Water Chemistry and Plankton in a Limed Lake
in Sweden 1763-1768

GUNNAR PERSSON and MAGNUS APPELBERG
Evidence of Lower Productivity in Long Term Limed Lakes as Compared
to Unlimed Lakes of Similar pH 1769-1774

HIDEHISA KAWAMURA, NOBUAKI MATSUOKA, SHINJI TAWAKI and
NORIYUKI MOMOSHIMA
Sulfur Isotope Variations in Atmospheric Sulfur Oxides, Particulate Matter
and Deposits Collected at Kyushu Island, Japan 1775-1780

JUNLING AN, MEIYUAN HUANG, ZIFA WANG, XINLING ZHANG, HIROMASA
UEDA and XINJIN CHENG
Numerical Regional Air Quality Forecast Tests Over the Mainland of China
.... 1781-1786

VALDO LIBLIK and MARGUS PENSA

Specifics and Temporal Changes in Air Pollution in Areas Affected by
Emissions from Oil Shale Industry, Estonia... 1787-1792

TAKAHISA MAEDA, ZIFA WANG, MASAYASU HAYASHI and
MEIYUAN HUANG
Long-Range Transport of Sulfur from Northeast Asia to Chengshantu,
Shandong Peninsula: Measurement and Simulation.. 1793-1798

FRANK MURRAY, GORDON MCGRANAHAN and JOHAN C.I. KUYLENSTIERNA
Assessing Health Effects of Air Pollution in Developing Countries 1799-1804

PRABAL K. ROY and HIROSHI SAKUGAWA
Trends of Air Pollution and its Present Situation in Hiroshima Prefecture 1805-1810

SCIENCE & POLICY AND ENVIRONMENTAL EDUCATION

KEITH R. BULL
The Need for Education in Developing Acceptable Air Pollution Control
Strategies... 1811-1816

EDUARD DAME, and MIKE HOLLAND
Cost-Benefit Analysis and the Development of Acidification Policy
in Europe..1817-1824

KENNETH E. WILKENING
Trans-Pacific Air Pollution: Scientific Evidence & Political Implications 1825-1830

SANNA. SYRI and NIKO KARVOSENOJA
Low-CO_2 Energy Pathways Versus Emission Control Policies
in Acidification Reduction 1831-1836

WAKANA TAKAHASHI and JUSEN ASUKA
The Politics of Regional Cooperation on Acid Rain Control in East Asia............ 1837-1842

HAIPING LAI, HIROYUKI KAWASHIMA, JUNKO SHINDO and KEIJI OHGA
Stages in the History of China's Acid Rain Control Strategy in the Light of
China-Japan Relations 1843-1848

YOSHIKAZU HASHIMOTO, YOSHIKA SEKINE, ZHI-MIN YANG and
KANJI YOSHIOKA
Profound Survival Program of Forests in Japan Islands –a 40 Year Strategy
for Environmental Conservation In Inland China – 1849-1854

ANNUAL VARIATIONS OF NEEDLE SURFACE CHARACTERISTICS OF *PINUS SYLVESTRIS* GROWING NEAR THE EMISSION SOURCE

EUGENIJA KUPČINSKIENĖ

Department of Botany, Lithuanian University of Agriculture, Kaunas - Noreikiškės, LT - 4324, Lithuania
(author for correspondence, e-mail: likup@takas.lt)

Abstract. In present study, pollutant effects on needle surface characteristics of *Pinus sylvestris* in the area affected by a nitrogen fertilizer plant have been investigated over 1994-1997 year period. Near the factory, sites with 15-25-year-old trees on a 0.5-22 km interval were chosen. Mean monthly concentrations of NO_2 and NH_3 varied across the transect in the range of 1.8-8.8 $\mu g\ m^{-3}$ and 1.8 - 69.3 $\mu g\ m^{-3}$, respectively. NH_3 concentrations exceeded the critical level (>23 $\mu g\ m^{-3}$) only in the 0.5 km vicinity. Assesment of needle surface wettability by measuring contact angles (CA) of water droplets and surface quality by measuring stomatal area covered by structural wax (SW) revealed significant ($p<0.05$) needle age, site, and year of sampling related differences. Comparison of SW between sites showed reliably ($p<0.05$) higher surface wax erosion on one-year-old needles sampled in the area, where ammonia concentration exceeds critical level. Significant correlations between site SW on one-year-old needles and distance from the pollution source, NO_2 and NH_3 concentrations were detected ($r = 0.539$; $r = - 0.495$; $r = - 0.426$; $p<0.001$, respectively). Correlations between CA and factors mentioned were lower.

Keywords: nitrogen dioxide, ammonia, conifers, surface wettability, wax structure

1. Introduction

In seventies-eighties, a nitrogen fertilizer plant was one of the heaviest polluters in Lithuania. Improved technologies and lower production of the factory during the last decade have resulted in reduced amounts of emissions, although acidification of forest soils still continues (Kairiūkštis and Rudzikas, 1999).

The largest aerial interface between trees and surrounding atmosphere is formed by the plant cuticles. The use of the wax structure as an indicator has been mainly based on direct observations under scanning electron microscope or by using indirect methods such as assessment of wettability of the needles by measuring contact angles of water droplets (Turunen and Huttunen, 1990). The major objective in the survey was determining the concentrations of NH_3 and NO_2 on sites situated at different distances from the nitrogen fertilizer plant also evaluation of the surface quality of needles of *Pinus sylvestris* under the influence of pollution.

2. Materials and Methods

The area affected by a nitrogen fertilizer plant was investigated. In 1994-1997, total emissions from the factory comprised 4.9-6.8 kt yr^{-1}; ammonia, - 0.3-1.3 kt yr^{-1}; NO_x, - 0.12-0.50 kt yr^{-1} (the data provided by the Ministry of Environment of Lithuania). 15-25-year-old *Pinus sylvestris L.* stands were studied along a 22 km transect. Forest sites mainly on podzols in the direction of the prevailing north-east wind from the pollution source were selected. Opened areas near the woods were places for aerial NH_3 and NO_2 estimations.

Sampling of pine needles was done in September months of 1994-1997. Current-year (c) and one-year-old (c+1) needles were used. The needle samples were coated with silver and examined under JEOL-JXA-50A scanning electron microscope (12-15 kV accelerating voltage). Needle surface rating into classes according to the area covered by structural wax (SW) and estimation of wettability by measuring contact angles (CA) of water droplets, were done as it was described by Cape *et. al.*, (1988), Turunen and Huttunen (1996).

Monthly exposures of diffusion tubes (Gradko International Ltd, UK) were performed twice a year (in summer and late autumn; totally, 7 times for NO_2 and 4 times for NH_3) in the period of 1995-1999. NO_2 was determined at the Lithuanian University of Agriculture according to the method described by Ashenden and Bell (1989) and NH_3 analyses were performed by Gradko International Ltd, UK.

All the statistical analyses were performed by the General Linear Models and Regression Analysis in the SAS package. Error bars in the figures indicated an interval of 95% confidence.

3. Results

The concentration of NO_2 was found to vary (1.8-8.8 µg m^{-3}) across the transect and it was to some extent (1.4-3.4 times) greater at 0.5 km distance from the pollution source (Fig. 1). In the period 1995-1999 the concentration of NO_2

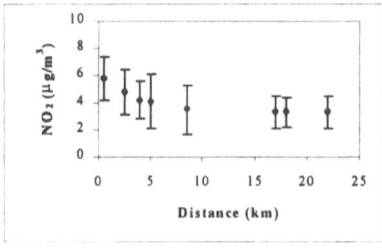 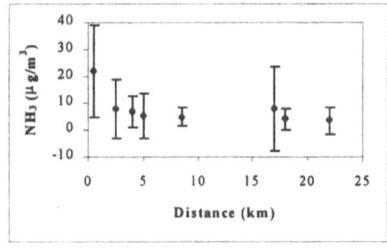

Figure 1. Mean monthly concentrations of NO_2 (1995-1998) and NH_3 (1998-1999) in *Pinus sylvestris* sampling sites located at different distance from the nitrogen fertiliser plant.

decreased from 5.3-8.8 µg m^{-3} to 1.9-4.6 µg m^{-3}. In winter months the amount of NO_2 was slightly higher than in summer. Across the transect the scale of variations in the concentrations of NH_3 was higher (1.8-45.2 µg m^{-3}), and the most contrasting concentrations differed 7.2-25.1 times. Near the factory, the highest concentration of NH_3 (69.3 µg m^{-3}) was found in the point located 200 m eastwards of pine sampling site. The amount of NH_3 was twice higher in summer months than in late autumn. In some cases in the furthest points, the concentration of ammonia was higher than on nearer located sites.

In 1994-1997, needle stomatal areas covered by structural wax varied from 42.4 to 69.3 % on c needles (data are not shown) and from 11.6 to 45.2 % on c+1 needles (Fig. 2). At the same time, contact angles of water droplets on c needles ranged between 55.7-82.5° (data are not shown) and 39.6-71.9° on c+1 needles (Fig. 3). Significant differences between the sites according to the SW or CA data, were found for one-year-old needles in all years of investigation and for current-year needles only in some years. Especially in 1995 and 1996, eroded wax occupied much larger areas on the surface of c+1 needles taken from the sites at a 0.5 km distance from the factory when compared with further (5-22 km) located sites. Like for c needles, c+1 needles from the highest pollution site had higher wettability, i.e. lower CA values, as compared with the most further located sites.

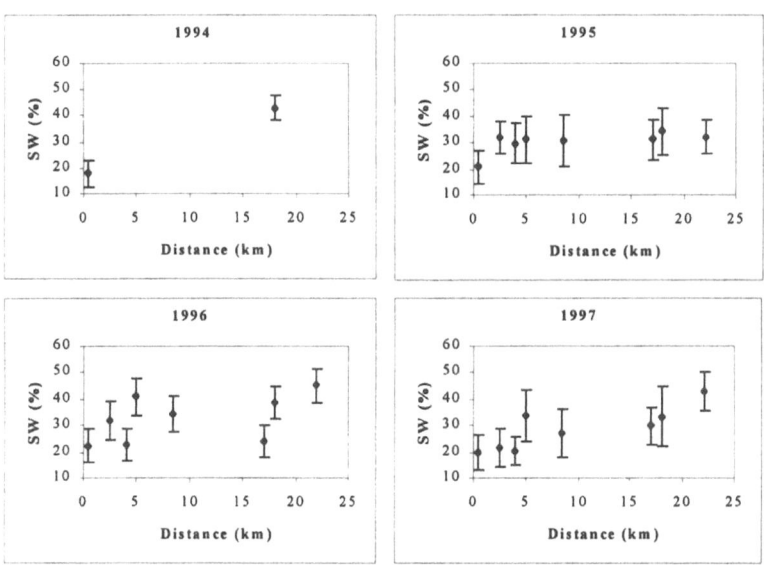

Figure 2. The mean values of area covered by structural wax (SW) on one-year-old needles of *Pinus sylvestris* growing at different distances from the nitrogen fertilizer plant (1994-1997).

Assessment of needle surface quality by measuring stomatal area covered by structural wax and contact angles of water droplets revealed significant ($p<0.05$) needle age, site and year of sampling related differences (Table I).

Structural surface wax and wettability of the needles showed a significant relationship to distance from the factory. Middle- and low-level significant relations were found between needle surface characteristics and concentrations of NH_3 and NO_2 (Table II).

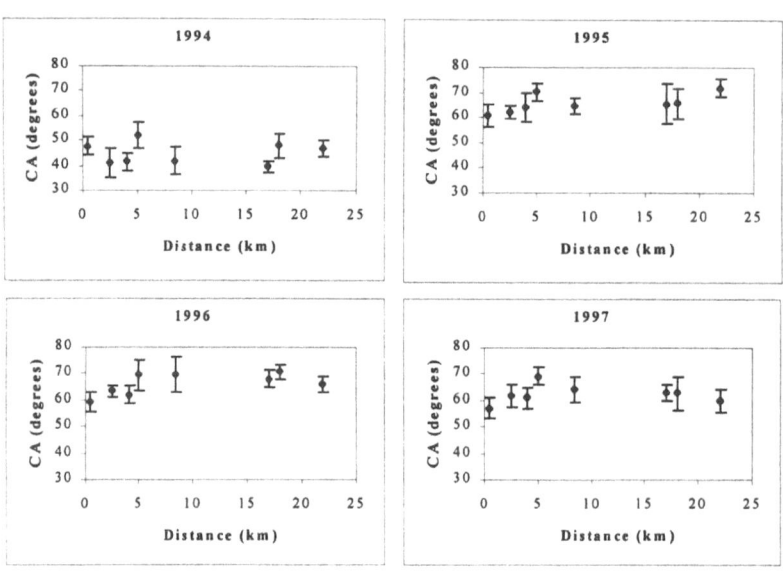

Figure 3. The mean values of contact angles (CA) of water droplets on one-year-old needles of *Pinus sylvestris* growing at different distances from the nitrogen fertilizer plant (1994-1997).

TABLE I

Main effects of needle age, site and year of sampling, and their interaction tested by ANOVA for P*inus sylvestris* needle area covered by structural wax (SW) and contact angles (CA) of water droplets

Variable	Needle age	Site	Year	Interaction			
	1	2	3	1*2	1*3	2*3	1*2*3
SW	xxx	xxx	xxx	n.s.	xxx	xxx	n.s.
CA	xxx	xxx	n.s.	xxx	x	xx	n.s.

Statistical significances: $^x p<0.05$; $^{xx} p<0.01$; $^{xxx} p<0.001$; n.s. = non-significant

The regression co-efficients of the curves describing relationship between SW, and NO_2, NH_3 concentrations were generally low: SW=52.61-5.23 * NO_2, r^2=0.27; standard error, SE=10.04; SW=35.99-0.73 * NH_3, r^2=0.22; SE=10.4.

In accordance with data obtained in 1996, the quantity of wax extracted from the surface of c needles ranged between 1.8 and 2.2 % dry weight (d. w.) and from c+1 needles - between 1.7 and 2.0 % d. w. No significant differences between sites were detected comparing the wax amount on the needles.

TABLE II
Site-specific Spearman's rank correlations between needle surface characteristics (area covered by structural wax - SW, contact angles - CA) and distance from the pollution source, concentrations of NH_3 and NO_2 in the air

Variable	CA		SW	
	c	c+1	c	c+1
Distance	0.262^{xxx}	0.301^{xxx}	0.226^{xx}	0.539^{xxx}
NO_2	-0.234^{xxx}	-0.296^{xxx}	-0.176^{x}	-0.495^{xxx}
NH_3	-0.210^{xxx}	-0.284^{xxx}	-0.165^{x}	-0.426^{xxx}

Statistical significances: ^{x} p<0.05; ^{xx}p<0.01; ^{xxx}p<0.001

4. Discussion

NO_2 concentrations (1.9-8.8 µg m^{-3}) detected near the nitrogen fertilizer plant were of low range when compared with the other parts in Lithuania (1.5-82.0 µg m^{-3}, Kairiūkštis and Rudzikas, 1999). The lowest concentrations of NO_2 (1.9-2.2 µg m^{-3}) and NH_3 (<1.0-1.8 µg m^{-3}) found in our study are the same like background concentrations in unpolluted areas of Europe (Fangmaier et al., 1994; Ashenden and Edge, 1995). The highest monthly mean concentrations of NH_3 (up to 45.2-69.3 µg m^{-3}) documented at a 0.5 km distance from the factory exceeded monthly critical level for ammonia (23 µg m^{-3}). It is in support to the opinion that NH_3 is rather readily subjected to dry deposition. Literature about gradual even distribution of NH_3 along transects is mainly based on the data of studies of livestock caused pollution (Fangmaier et al., 1994). The fact that in each exposure the peaks of NH_3 were detected also in further located points of the 0.5-22 km transect, received a support from the deposition studies (Kairiūkštis and Rudzikas, 1999) and could be explained by functioning of several various height (50-150 m) stacks emmiting ammonia.

Like in an earlier (1990-1993) performed study (Kupčinskienė et al., 1996) present data about surface quality of Scots pine needles demonstrate possibility of a damage to the needles under influence of pollutant mixtures with concentrations of ammonia exceeding the critical level. Ammonia-caused erosion of tubular wax of needles were shown by studies of *Pseudotsuga menziesii* in climate-controlled growth chambers, open-top chambers or stands polluted with NH_y (Van der Eerden et al., 1992). Low-level pollution can be overridden by other abiotic and biotic factors influencing surface quality. In agreement to other studies (Turrunen and Huttunen, 1996), revealed relations between the concentrations of air pollutants and *Pinus sylvestris* needle surface characteristics were not strong.

5. Conclusions

Gradients of NO_2 (1.8 - 8,8 µg m^{-3}) and NH_3 (1.8 - 69.3 µg m^{-3}) were determined along 22 km transect from the nitrogen fertilizer plant according to the prevailing wind direction. The comparison of the sites of the transect has revealed reliably larger areas of eroded surfaces of the one-year-old needles of trees at the 0.5 km distance than in the plots located further (2.5-22.0 km) from the polluter. Structure quality data obtained have shown to the possibility for a present level of pollution to induce detectable damage to the surface of needles of *Pinus sylvestris*. Contact angles on the needles taken from the pines growing at a 0.5 km vicinity of the factory only in some cases had reliably lowest values.

Acknowledgements

The investigation was supported by grants from CEU (Hungary), ERSS Contract, Project No. 91-5 (1994); J.S.C. ACHEMA, (Lithuania) Projects No 448F (1996), No 155F (1998), No 223F (1999). The author thanks Prof. Huttunen S., Prof. Ashenden, T.W., Dr. Cape, N., Dr. Janilionis V., Dr. Armolaitis K., Dr. Grigaliūnas V., Mr. Kliūčius, A., for advices and assistance in the research. The English of the manuscript was revised by Mr. Simaška P.

References

Ashenden, T. W. and Bell, S. A.: 1989, *Environmental Pollution* **58**, 179.
Ashenden, T. W. and Edge, C. P.: 1995, *Environmental Pollution* **87**, 11.
Cape, J. N., Paterson, I. S., Wellburn, A. R., Wolfenden, J., Mehlhorn, H., Freer-Smith, P. H. and Fink, S.: 1988, *The early diagnosis of forest decline*. HMSO, London.
Fangmeier, A., Hadwiger-Fangmeier, A.H., Van der Eerden, L. and Jäger, H.-J.: 1994, *Environmental Pollution* **86**, 43.
Kairiūkštis, L. and Rudzikas, Z. 1999, *Ecological Sustainability of Lithuania in a Historical Perspective*. Vilnius.
Kupčinskienė, E., Huges, S. and Matulionis, E.: 1996, *Landschaftsentwicklung und Umweltforschung* **104**, 101.
Van der Eerden, L.J.M., Lekkerkerk, L.J.A., Smeulders, S.M. and Jansen, A.E.: 1992, *Environmental Pollution* **76**, 1.
Turunen, M. and Huttunen, S.: 1990, *Environmental Quality* **19**, 35.
Turunen, M. and Huttunen, S.: 1996, *Environmental Pollution* **93**, 175.

PINE NEEDLE CHEMISTRY NEAR A LARGE POINT SO$_2$ SOURCE IN NORTHERN FENNOSCANDIA

GALINA N. KOPTSIK[1*], SERGUEI V. KOPTSIK[2] and DAN AAMLID[3]

[1] *Soil Science Faculty, Moscow State University, Moscow 119899, Russia;* [2] *Faculty of Physics, Moscow State University, Moscow 119899, Russia;* [3] *Norwegian Forest Research Institute, Høgskoleveien 12, N-1432 Ås, Norway*

(* *author for correspondence, e-mail: koptsik@soil.msu.ru)*

Abstract. Air pollution induced changes in pine needle chemistry were observed at sample sites in the surroundings of the Pechenganikel smelter. Close to the smelter, elevated concentrations of Ni, Cu and S were found (Ni: 0.7-1 mmol/kg, Cu: 0.4-0.5, and S 40-60 mmol/kg). Close to the pollution source, needles were enriched in Ni and Cu by needle age. Correlation and principal component analyses show that changes in the element composition of pine needles depended on air pollution and on natural factors as well. The contribution from air pollution increased with needle age. Besides direct input of pollutants from atmosphere, soil contamination and nutritional disturbance contributed significantly to the observed changes.

Keywords: pine needle chemistry, air pollution, sulphur, heavy metals, soils

1. Introduction

The area along the Norwegian-Russian border has been affected by air pollution from the nickel-processing industry in the Kola Peninsula, Russia, for several decades. The studied boreal forests are among the northernmost coniferous forests of the world. In addition to strong natural stress even minor loads of air pollutants may have severe effects on forest vitality. The high concentrations of sulphur dioxide have caused severe damage to trees and other vegetation due to direct injuries (Aamlid, 1992; Aamlid and Venn, 1993). Indirect impact, via soil and roots caused by soil acidification and release of aluminium or other harmful elements (*e.g.* heavy metals) or nutrient deficiency is also possible. Abrahamsen *et al.* (1994) concluded that acid deposition could result in nutrient deficiency in the long term at sensitive growth sites. In the study area, boreal forest ecosystems with acid soils and bedrocks (Koptsik *et al.*, 1999a), low weathering (Koptsik *et al.*, 1999b), and low nutrient status (Lukina and Nikonov, 1998) may be sensitive to acidification, leading to nutrient leakage followed by tree damage. Critical loads of acid deposition are exceeded on large areas of the Kola Peninsula (Koptsik and Koptsik, 1995). The aim of this study was to investigate pine needle chemistry and to compare it with soil chemistry near the Pechenganikel smelter, the largest SO$_2$ emitter in Northern Europe.

2. Materials and Methods

2.1. STUDY AREA

The study area was located to South Varanger, eastern Finnmark, Norway, and in the vicinity of the nickel smelters in the Kola Peninsula, Russia, close to the Norwegian border (Fig. 1). The area is situated at the polar tree line. Pine and birch forests are characterised by scarce tree stands and ground vegetation dominated by crowberry, bilberry, mosses and lichens. Thin sandy podzols with high acidity and low base cation content are typical for the area.

Figure 1. Plot locations.

2.2. SAMPLING AND CHEMICAL ANALYSIS

Scots pine (*Pinus sylvestris* L.) needles were collected for chemical analyses from forests at different distances and directions from the smelter, 21 sites in total (Fig. 1). The pine needles were taken from the top third of 10 dominant trees on each sample site and divided into current year needles and needles of the previous years. Concentration of mineral nutrients and some heavy metals (Al, Ca, Cu, Fe, K, Mg, Mn, Na, Ni, P, S, Zn) were determined by an ICPAES in acid digested samples and by XRF. Total nitrogen was determined as ammonium after Kjeldahl digestion. Using the mean S/N ratio of 0.028 for the species found in the background plots, the S concentration required to balance the N and the difference between the measured S and this "predicted" S was described as "excess" S. Correlation and principal component analyses (PCA) were used to assess interactions among plant and soil elements and their changes with needle ageing in the vicinity of the pollution source.

3. Results and Discussion

Concentrations of both S and heavy metals (HM) in the pine needles were notably higher in the vicinity of the smelter compared with background sites (Table I). Close to the smelter, the concentrations of Ni ranged from 0.7 to 1 mmol/kg and from 0.4 to 0.5 mmol/kg for Cu. At the remote sites (more than 30 km from the smelter), the concentrations of these metals were 0.15-0.20 mmol/kg for Ni and about 0.13 mmol/kg for Cu. Elevated concentrations of Ni and Cu in pine needles near smelters have been reported in other studies (Aamlid, 1992; Steinnes et al. 2000; Lukina and Nikonov, 1998; Rautio et al., 1998). The total sulphur concentrations in the needles of different age were in the range of 40-60 mmol/kg close to the smelter, and 26-38 mmol/kg at remote sites (Table I).

TABLE I.
Total element concentrations in the current year needles at different distances from the smelter

Area	Para	N	P	S	Ca	Mg	K	Al	Fe	Ni	Cu	Zn
	Meters				mmol/kg						μmol/kg	
Nikel, >30 km	Mean	1180	88.8	37.8	41.5	46.7	177	8.03	0.90	196	127	598
	Std	137	16.5	8.9	14.7	5.8	36	2.00	0.62	86	50	124
	Min	989	61.3	15.6	24.1	34.2	105	4.96	0.52	70	61	423
	max	1440	121.0	48.3	72.5	57.0	227	12.50	2.99	333	209	840
Nikel, 10-30 km	mean	1210	92.0	42.3	45.8	49.8	214	9.46	1.07	241	161	678
	std	87	12.1	5.5	13.7	6.6	38	4.63	0.65	92	40	155
	min	1100	73.9	31.2	31.1	37.1	162	4.37	0.57	137	117	467
	max	1330	111.0	52.8	75.0	58.5	269	18.50	2.56	414	240	926
Nikel, <10 km	mean	1210	88.0	48.4	45.2	47.9	244	6.30	1.50	333	181	634
	std	175	16.4	8.2	7.7	5.3	41	2.10	1.17	73	54	206
	min	956	58.2	32.0	29.6	38.9	164	3.41	0.72	204	93	369
	Max	1520	120.0	59.9	57.5	55.8	306	11.40	5.35	469	319	1010
8 km [1]	Mean	-	63.3	57.3	37.5	35.8	230	5.11	4.71	4650	2890	245
40 km [1]	Mean	-	53.9	39.7	78.1	39.9	207	11.60	1.04	629	351	566
81 km [1]	Mean	-	47.7	24.2	63.9	54.1	166	7.19	0.36	94	58	505
Apatity [2]	Mean	928	67.8	15.6	27.5	41.1	194	11.10	3.58	-	-	-
Lapland [3]	Mean	-	42.1	27.3	35.2	-	136	-	-	119	50	-

[1] Monchegorsk area, Lukina and Nikonov, 1998; [2] Ushakova, 1997; [3] Finland, Raitio, 1998.

Normal S concentration in Scots pine needles is 15-25 mmol/kg in background areas of the Kola Peninsula (Lukina and Nikonov, 1998; Ushakova, 1997), as well as in Norway (Aamlid, 1992), and in Finland (Manninen et al., 1996; Raitio

2000). According to our data elevated concentrations of sulphur occurred over the entire area studied. Strongly raised concentrations (>30 mmol/kg) were found at sites situated up to 30 km from the smelter. These sites are roughly located within the area defined by the isocurve of the annual mean atmospheric SO_2 concentration exceeding 10 $\mu g/m^3$ as given by Sivertsen et al. (1994). Similar patterns were observed by Raitio et al. (1995). Calculations based on the N concentrations of the needles and a S/N ratio of 0.028 gave mean values of 5-14 mmol/kg of "excess" S in the needles of polluted sites compared with remote sites. Manninen et al. (1996) suggest that a concentration of about 28 mmol/kg may be considered a critical level for foliar S in areas with low N supply. The S concentration in the current year needles exceeded this level nearly at all sites within 40 km from the smelter.

The concentration of K in pine needles, as well as the HM, increased towards the smelter. The correlation coefficients between the distances from the smelter and the K concentrations in the current (C), C+1 and C+2 year needles were -0.69, -0.62 and -0.59, respectively. Elevated concentrations of mineral nutrients of mobile elements may be connected with their intensive uptake and translocation from the older needles. The endurance of needle life decreased from 5-6 years in the background areas to 2-3 years close to the pollution source. Similar findings were observed by Lukina and Nikonov (1998) and Rautio et al. (1998) close to the Severonikel smelter. Besides, increased availability of K in soils due to its replacement by HM in the exchange sites (Koptsik et al., 1999) may contribute to foliar accumulation of K.

The concentrations of N, P, K, Mg, S decreased, while Ca, Al, Fe, Mn increased with needle age both in the polluted and remote plots. Close to the smelter concentrations of Ni and Cu increased with needle age. However, S did not follow this tendency.

A large number of natural and anthropogenic factors affect the foliar chemistry. We applied PCA to get insight into the structure of these factors. Foliar total elements were chosen as independent variables for PCA. The site scores in the first two principal components (PC) are shown in Figure 2a, b. The diagrams characterise similarity-dissimilarity of foliar chemistry as a whole for the sites studied. The element scores in the first two PCs are shown in Figure 2c, d with solid arrows and external variables (distance from the smelter and soil properties) - with dotted arrows; cosine of angle between any two arrows approximates correlation coefficient between the corresponding variables. The first two PCs counted for 45 and 20% of the total variation for needles of all years, 36 and 25% for current year needles, and 31 and 21% for C+3 – C+5 year needles.

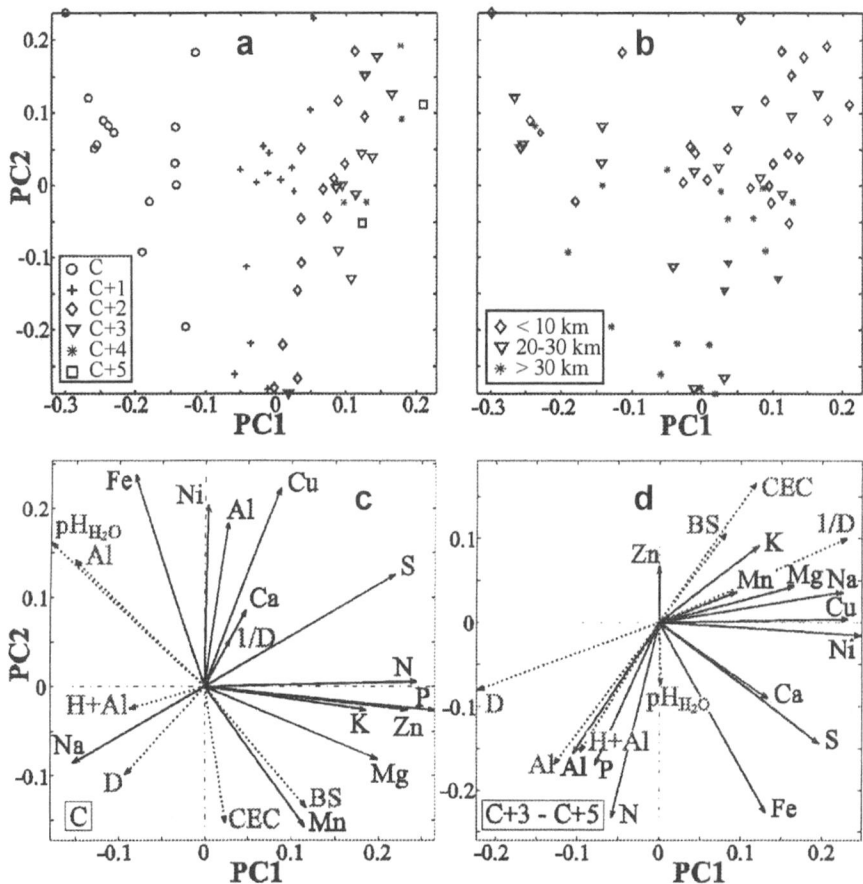

Figure 2. Ordination of sites (a, b) and loadings of foliar total element contents (c, d). External variables are shown with dotted arrows: D - distance, 1/D - inverse distance from the smelter, H and Al - soil exchangeable H and Al, BS - base saturation, CEC - cation exchange capacity.

The similarity-dissimilarity of sites pictorially show that natural changes with needle ageing play a decisive role in transformation of foliar chemistry – points along the first PC are neatly split with needle age (Fig. 2a). However, by significance the next factor is the airborne pollution: points along the second PC tend to change with distance from smelter (Fig. 2b).

Systematic changes of element composition with needle age are shown in ordination diagrams 2c, d, which represent the youngest and oldest age classes. For the current year needles, N, P, and K are the main contributors to the first PC; their approximated correlation coefficients are positive. By age positions of these nutrients change (negative correlation); their contribution to the first PC lowers while contribution to the second PC increases and becomes dominant by the third year. While HM determine the second PC for current year needles, their positions shift to the first PC with time, with their accumulation, and determine it for the C+3 – C+5 year needles. Diagrams clearly reflect positive correlation of foliar Mg and K with base saturation and their negative correlation with exchangeable Al in soils which in turn positively correlates with foliar Al.

4. Conclusions

The Scots pine needles close to the Pechenganikel smelter showed higher concentrations of both S and heavy metals compared to remote sites. The concentration of Ni and Cu in needles near the smelter was as high as 0.7-1 and 0.4-0.5 mmol/kg, respectively; these values being by an order of magnitude higher than those in background areas. Foliar total S concentrations were observed to be above the normal S level (15-24 mmol/kg) within over 30 km from the pollution source. At polluted sites, the pine needles contained 5-14 mmol/kg of "excess" S (determined from S/N ratio) comparing with background sites. Elevated levels of Fe and K in the needles have also been found near the pollution source. The concentrations of N, P, K, Mg, S decreased, while those of Ca, Al, Fe, Mn increased with needle age on both polluted and background plots. Close to the smelter the concentrations of Ni and Cu increased with needle age, while concentration of S did not show this tendency. Despite high natural variability, pine needle chemistry may indicate the status of forest ecosystems under stress.

References

Aamlid, D.: 1992, *Forest Pathological Research in Northern Forests with Special Reference to Abiotic Stress Factors*. Finnish Forest Research Instiute. Research Papers 451.
Aamlid, D. and Venn, K.: 1993, *Norwegian Journal of Agricultural Sciences* **7**, 71.
Abrahamsen, G., Stuanes, A.O. and Tveite, B. (eds.): 1994, Long-Term Experiments with Acid Rain in Norwegian Forest Ecosystems. *Ecological Studies* **104**, 342 pp.
Koptsik, G. and Koptsik, S.: 1995, *Water, Air and Soil Pollution* **85**, 2553-2558.
Koptsik, G., S. Koptsik, K. Venn, D. Aamlid, L. Strand and Zhuravleva, M.: 1999, *Eurasian Soil Science* **32** (7), 787.
Koptsik, G., Teveldal, S., Aamlid, D. and Venn, K.: 1999, *Applied Geochemistry* **14** (2), 173.
Lukina, N.V. and Nikonov, V.V.: 1998, *Nutrient Status of North Taiga Forests* (Natural Regularities and Pollution-Induced Changes). Apatity, KSC RAS. 316 pp. (In Russian).
Manninen, S., Huttunen, S., Ratio, P. and Perämäki, P.: 1996, *Environmental Pollution* **93** (1), 27.
Raitio, H., Tuovinen, J.-P. and Anttila, P.: 1995, *Water, Air and Soil Pollution* **85**, 1361.
Raitio, H.: 2000, 'Tree nutrient status', in E Mälkönen (ed.), *Forest condition in a changing Environment - The Finnish case*, Kluwer Academic Publishers, Dordrecht, pp. 93-102.
Rautio, P., Huttunen, S. and Lamppu, J.: 1998, *Water, Air and Soil Pollution* **102**, 389.
Sivertsen, B., Baklanov, A., Hagen, L.O. and Makarova, T.: 1994, *Air pollution in the Border Areas of Norway and Russia*. Summary report 1991-1993. NILU OR 56/94. 14 pp.
Steinnes, E., Lukina, N., Nikonov, V., Aamlid, D. & Røyset, O.: 2000. - *Environmental Monitoring and Assessment* **60** (1):71-88.
Ushakova, G.I.: 1997, *Biogeochemical migration of elements and soil formation in forests of the Kola Peninsula*. Apatity, KSC RAS. 150 pp. (In Russian).

DETERMINATION OF DEPOSITION AND LEAVES IN TEAK PLANTATIONS IN THAILAND

JESADA. LUANGJAME[1*], BOONCHOOB BOONTAWEE[1],
and NIGORN. KLIANGPIBOOL[1]

[1]*Royal Forest Department, Bangkok 10900, Thailand*
(* author for correspondence, e-mail: jesada@mozart.inet.co.th)

Abstract. This paper outlines the results to date of the continuing acidic deposition study from 1997 in three teak plantations at the Na Pralan, Klangdong, and Donglan villages of Thailand. The aim of this study was to examine the impact on teak plantations of acid deposition – the increasing flow of chemical compounds including CO_2, SO_2, NO_2, and NO_x into the atmosphere. The 1997-1999 results showed no symptoms of acidity of the precipitation in the teak plantations at the treated sites (Na Pralan and Klangdong). During this period, the pHs of stemflow and throughfall were still over 7; and the pHs of the rainfall were around 7, except at Klandong where it dropped to around 5 in 1999. The pH and EC values were higher at the polluted sites than at the control site; this may be attributed to contamination with lime dust from nearby industrial plants, including cement factories and quarrying mills.

Fresh leaves were contaminated with Sulphur to quite high levels at the Na Pralan site. This contamination seems to have affected the physiology, biomass and chemical content of the leaves. It might be due to pollution gases (SO_2 and NO_x) from the heavy trucks on the nearby Phaholyothin road. However, these gases could not be detected by a gas detector even though pollution seemed quite heavy. This study did not detect acid rain damage to the teak plantations.

Keywords: concentrations, stemflow, throughfall, rainfall, leaves, teak, Thailand

1. Introduction

Several processes in a forest plantation and in the soil can be substantially affected by large scale industrialisation. These effects can occur in at least two principal ways. Firstly, there appear compounds which have a direct effect on the metabolism of trees, and secondly, there arise other compounds which change the potential of the environmental factors. The changes due to these direct and indirect effects are then reflected in the growth characteristics and morphology of trees (Hari and Raunemaa, 1984).

The hypothesis was that acid deposition could change the chemical properties of precipitation, which passes through the teak canopy and follow the stems into the soil. Thus, the chemical input could affect the teak plantations. This change, could, ultimately, continue to change the environment. This paper provides results from a study in three teak plantations in Thailand for the period 1997-1999. The aim of the study was to examine the impact of acid rain and the air pollution due to industrialisation on teak plantations. In particular, acid deposition in teak plantations due to increasing flow of chemical compounds such as CO_2, SO_2, NO_2 NO_x, *etc.* into the atmosphere, was monitored over time.

2. Monitoring Sites

The study examined three teak plantations planted at a spacing of 4 x 4 m in Na Pralan, Klangdong, and Donglan villages in 1953, 1956, and 1968, respectively. These teak plantations are located in varying distances from industrial plants or factories, which are potential sources of acid rain and air pollution. The approximate distances from the factories (pollution sources) to the study plot were different. They are 0.5-5 km to the Na Pralan, 2-20 km to the Klangdong, and over 20 km to the Donglan (a control) study plots. In the first two study sites, we found some insect defoliators (*Hyblaea peura*) and skeletors (*Eutectona machaeralis*) in the dry seasons but not in the rainy season.

3. Materials and Methods

3.1. SAMPLING AND CHEMICAL ANALYSES OF RAIN WATER

At each studied site, rain water from the stemflow, throughfall and rainfall were collected and analysed. They were 4, 3 and 1 replications per site, respectively. The rainfall (precipitation) gauge was set in an open place within the study site. It was located more than 100 m from heavy traffic roads, and 50 m from tall trees and houses. At the teak plantations, stemflow was measured by setting a trap on dominant teak-tree trunks. A vinyl tube with a plastic container at one end was attached to the trunk to collect the stemflow water. The tubes were installed carefully so as not to injure the stem cambium, and no adhesives were used except silicone. Four tree trunks were set per plantation. The throughfall gauges were placed at a distance of 2 m from the tree trunks.

The pH and electrical conductivity (EC) were immediately measured every time of the rainfall at the site. If the pH and EC could not be determined at the site on the same day, the rainfall water would be brought to the laboratory, stored in the refrigerator and the pH and EC measured within 7 days. One litre of the rain sample determined the chemical properties. The stemflow, throughfall, and rainfall of the first rainfall of the month were analysed for pH, EC, Sodium, (Na), Potassium (K), Calcium (Ca), Magnesium (Mg), Nitrate (NO_3), Chloride (Cl), Ammonium (NH_4), and Sulphate (SO_4).

3.2. SAMPLING AND CHEMICAL ANALYSES OF LEAVES

Fresh leaves were examined for Nitrogen and Sulphur contents in order to evaluate the nutrient and pollution levels and dry deposition of Sulphur dioxide. (Studies of areal and seasonal variations of Sulphur content are necessary for evaluating air pollution.) Approximately 500 g of fresh leaf samples from the top crown or outer canopy of the plot were collected. To obtain a more representative sample of the area, leaves from more than one tree were mixed. Only mature leaves growing in sunny positions were selected. The leaf samples were then dried in the oven at 75-80°C for two days, weighed, and then ground into small (less than 0.5 mm) particles. About 100 g of the particles were placed in a plastic bag and sent away for analysis of Sulphur content (mg/100 g) by the LECO Sulphur analyser.

Fallen leaves were used to detect growth-reduction heavy metals such as Aluminium (Al), Copper (Cu), Zinc (Zn), and Lead (Pb). The contents (in mg/100 g) were determined based on the atomic absorption method. The fallen leaves were collected from leaf samples in 1 m x 1 m litter traps installed at random in a sample plot located within each plantation. The litter fall, collected twice a month, was separated into the leaves, stem, bark, flower, seed and other droppings. A similar analysis procedure as used for fresh leaves was done for the fallen leaves. The litter fall of teak was analysed to estimate the amount of litter fall biomass (kg/ha/yr).

4. Results and Discussions

4.1. RAIN WATER ANALYSIS

The annual average pH and EC of stemflow, throughfall and rainfall in 1997, 1998, and 1999 are shown in Tables I–III.

TABLE I
Average acidity (pH) and electrical conductivity (EC, mS/cm) of the rain water at the Na Pralan, Klangdong, and Donglan teak plantations in 1997.

Site	Stemflow		Throughfall		Rainfall	
	pH	EC	pH	EC	pH	EC
Na Pralan	7.2	<0.01	7.4	<0.01	7.3	<0.01
Klangdong	7.4	0.01	7.5	0.04	7.6	0.04
Donglan	7.2	0.12	7.0	0.04	6.6	0.04

TABLE II
Average acidity (pH) and electrical conductivity (EC, mS/cm) of the rain water at the Na Pralan, Klangdong, and Donglan teak plantations in 1998.

Site	Stemflow		Throughfall		Rainfall	
	pH	EC	pH	EC	pH	EC
Na Pralan	7.3	<0.01	7.2	<0.01	7.0	<0.01
Klangdong	7.1	0.11	7.1	0.06	6.6	0.16
Donglan	7.1	0.10	6.8	0.03	6.0	0.02

TABLE III
Average acidity (pH) and electrical conductivity (EC, mS/cm) of the rain water at the Na Pralan, Klangdong, and Donglan teak plantations in 1999.

Site	Stemflow		Throughfall		Rainfall	
	pH	EC	pH	EC	pH	EC
Na Pralan	7.4	<0.01	7.1	<0.01	6.7	<0.01
Klangdong	7.2	0.13	7.0	0.05	5.1	0.05
Donglan	7.1	0.15	6.9	0.06	5.9	0.66

The lowest pH of rainfall was found in the Klangdong plantation (5.1 in 1999), while in the Donglan plantation the pHs ranged between 5.9 to 6.6 in the period 1997 to 1999. For the pHs of stemflow and throughfall, they were the lowest (6.8-7.2) in the Donglan teak plantation. The pHs were more alkaline at Na Pralan (pH 7.2-7.4) than at Donglan (6.8-7.2). This suggests that the quarry mills near the Na Pralan site may have affected the teak plantations by polluting them with lime. However, there was no symptom of acidity. In other studies, the pHs of stemflow in China was about 3-4 (Haoubao et al., 1996), of rainfall in Malaysia was 5-5.5 (Philip and Rizal, 1996), and of rainfall in Indonesia was 5.2- 7.0 (Forest and Nature Conservation Research and Development Centre, 1996).

For the chemical properties of rain water at the monitoring sites in 1997, 1998 and 1999, almost every element or compound, especially, K, Ca, Mg, and SO_4, of the stemflow was at the highest level among throughfall and rainfall, e.g. Ca and SO_4 were 128 and 90 ml/l in the stemflow, 90 and 39 ml/l in the throughfall, and 73 and 39 ml/l in the rainfall at the Na Pralan teak plantation in 1997, respectively. Throughfall contains dry deposition on leaves and leached from the leaves, when stemflow contain dry deposition on and the leaches from the trunk and branches as well as a fraction of the throughfall.. These trends were similar to the Malaysian data (Philip and Rizal, 1996). As well, Suksawang et al. (1996) found that chemicals of the rain water were dependent on season, date and time of rainfall.

4.2. CHEMICAL PROPERTIES OF LEAVES

The results of chemical properties analysis of fresh and fallen leaves are shown in Tables IV, V, and VI, for 1997, 1998, and 1999, respectively. The Nitrogen and Sulphur contents were high at Na Pralan teak plantation. For Sulphur, It was assumed that it might come from the many heavy trucks passing by everyday near this plantation. However, when gas detectors were used to check the SO_2 and NO_x, it was found that the concentrations of these gases were very low. They may accumulate in the forest over time. The heavy metals (Zn, Cu, Al and Pb) were higher in Na Pralan than Klangdong and Donglan. Among these heavy metals, Al was at the highest level (71.50 mg/100 g dry weight) at the Na Pralan teak plantation in 1997. Yet in the previous year (1996) it was recorded as 33.65 mg/100 g dry weight (Luangjame et. al., 1996).

TABLE IV
Chemical analysis of teak leaves (mg/100gm dry weight) in December 1997.

Site	Fresh leaves		Fallen leaves			
	N	S	Zn	Cu	Al	Pb
Na Pralan	1562.50	74.09	3.65	1.80	71.50	1.55
Klangdong	1460.00	48.29	1.55	0.31	17.25	1.31
Donglan	1000.00	37.46	1.13	0.24	14.00	0.60

TABLE V
Chemical analysis of teak leaves (mg/100gm dry weight) in December 1998.

Site	Fresh leaves		Fallen leaves			
	N	S	Zn	Cu	Al	Pb
Na Pralan	1095.00	39.96	1.15	0.27	15.38	0.48
Klangdong	1097.50	43.29	0.65	0.17	7.18	0.57
Donglan	1047.50	34.97	1.68	0.33	9.50	0.77

TABLE VI
Chemical analysis of teak leaves (mg/100gm dry weight) in December 1999.

Site	Fresh leaves		Fallen leaves			
	N	S	Zn	Cu	Al	Pb
Na Pralan	1270.00	45.79	1.70	0.58	7.13	0.44
Klangdong	1515.00	39.13	0.95	0.58	6.94	0.36
Donglan	1017.00	21.65	1.40	0.63	6.98	0.29

4.3. LITTER FALL BIOMASS ESTIMATION

The results of the litter fall biomass estimation of the year 1997, 1998, and 1999 were found that the highest litter fall in 1997, 1998, and 1999 was at Na Pralan (12,719; 9,412; and 8,211 kg/ha/yr, respectively), followed by Klangdong (7,558; 6,095; and 6,602 kg/ha/yr, respectively) and Donglan (4,998; 3,071; and 5,130 kg/ha/yr, respectively). Donglan had the highest litter fall (8,198 kg/ha/yr) followed by Na Pralan (7,299 kg/ha/yr) and Klangdong (7,181 kg/ha/yr) in 1996 (Luangjame *et. al.*, 1996).

5. Conclusions and Recommendation

The 3-year results showed no symptoms of acidity of the precipitation in the teak plantations at the polluted sites (Na Pralan and Klangdong). There were high pH and EC values at these polluted sites than at the control site; this may be because these polluted sites were contaminated with lime dust (Ca) from the cement factories, quarrying mills, *etc*. However, Neal (1991) cited that the mild contamination of industrialisation might take time; this may suggest that acid deposition may not have happened yet in the Na Pralan and Klangdong sites.

Fresh leaves were contaminated with Sulphur to quite high levels at the Na Pralan site. This might be because of polluted elements from the heavy trucks that passed by on Phaholyothin road. However, these polluting gases could not be detected by the gas-detector even though pollution was quite heavy. In general, based on the determination of pH, EC and chemical elements in the teak plantations at the study sites, it appears that acid rain did not cause damage to the teak plantations. Moreover, lime dust may have neutralised the acid at the treated sites, but, it affected the teak due to its exposure.

Acknowledgements

We are very grateful to the Japan International Forestry Promotion and Cooperation Centre (JIFPRO) to initiate this research. We are also grateful to the Royal Forest Department for financial assistance throughout the study and the staff of the Watershed Research Sub-division who helped in chemical analysis. We are also indebted to Mr Viroj Krongkitsiri who kindly helped in collecting the data and Dr A Y Omule for reviewing this paper.

References

Luangjame, J., Boontawee, B. and Kliangpibool. N.: 1996, Monitoring of acidification in precipitation in Thailand. *In*: FORTROP'96 International Conference on Tropical Forestry in the 21st Century at the Kasetsart University in Bangkok, Thailand on 25-29 November 1996, Bangkok, Thailand.

Forest and Nature Conservation Research and Development Centre, Bogor, Indonesia.: 1996, Monitoring forest damage caused by acid rain and air pollution at Cikampek, West Java, Indonesia. In: the Meeting on Forest Damage Caused by Acid Rain and Air Pollution, Thailand on 14-18 July 1996. 20 p.

Hari, P. and Raunemaa, T.: 1984, The effect of changing environmental factors of forest growth. *In* E. Klimo and R. Sally (eds.), Air Pollution and Stability of Coniferous Forest Ecosystems. International Symposium on October 1-5, 1984 Institution of Forest Ecology, Faculty of Forestry, University of Agriculture Bruno, Czechoslovakia. pp 249-261

Houbao, F., M. Zhuang and Yichi. L.: 1996, Report on monitoring forest damage caused by acid rain and air pollution. Forestry Department, Fujian Forestry College, Nanping, China. *In*: the Meeting on Forest Damage Caused by Acid Rain and Air Pollution, Thailand on 14-18 July 1996. 39 p.

Neal, P.: 1991, Considering Conservation Acid Rain. Dryad Press Limited, London. 48 p.

Philip, E. and Rizal, M.K.M.: 1996, Report on monitoring forest damage caused by acid precipitation and air pollution. Forest Research Institute Malaysia, Kuala Lumpur, Malaysia. In: the Meeting on Forest Damage Caused by Acid Rain and Air Pollution, Thailand on 14-18 July 1996. 16 p.

Suksawang, S., Tangtham, N., Yavut, C., and Jirasuktaweekul. W.: 1996, Nutrients balance of mixed deciduous forest with bamboo watershed at Kanchanaburi. Royal Forest Department, Bangkok, Thailand. 50 p.

ANNUAL CHANGES IN SOME PHYSIOLOGICAL PROPERTIES OF *CRYPTOMERIA JAPONICA* LEAVES FROM KANTO, JAPAN

T. TAKAMATSU[1]*, H. SASE[2], J. TAKADA[3] and R. MATSUSHITA[3]

[1] *Soil Science Section, National Institute for Environmental Studies, 16-2 Onogawa, Tsukuba, Ibaraki 305-0053, Japan;* [2] *Acid Deposition and Oxidant Research Center, 314-1 Sowa, Niigata 950-2144, Japan;* [3] *Research Reactor Institute, Kyoto University, Sennan, Osaka 590-0494, Japan*
(* author for all correspondence, e-mail: takamatu@nies.go.jp)

Abstract. *C. japonica* leaves were sampled monthly in a heavily damaged area (*a*-I: Saitama), a slightly damaged area (*a*-II: plains in Ibaraki), and a healthy area (*a*-III: mountainous areas in Ibaraki) in Kanto. The leaves were analyzed for apparent cuticular transpiration rates, amounts of epicuticular wax, and contact angles. Sb in aerosols deposited on the leaves was also analyzed. The transpiration rates and the increase in transpiration with leaf aging were higher in *a*-I than in *a*-II and *a*-III. Erosion rates of the wax were higher in *a*-I and *a*-II than in *a*-III. The decrease of contact angles with aging tended to be higher in *a*-I and *a*-II than in a-III. Rates of aerosol-Sb deposition on the leaves were in the order *a*-I >> *a*-II > *a*-III. The transpiration rates correlated with the values obtained from a linear binomial function that included the amounts of wax and aerosol-Sb as variables (r = 0.855, $P < 0.01$). In *a*-I, the large quantity of aerosols on the leaves (and probably gaseous air pollutants) may have increased the transpiration rates by ca. 50% owing to erosion of the wax and stomatal malfunction, placing *C. japonica* under chronic water stress.

Keywords: contact angles, *Cryptomeria japonica*, cuticular transpiration, epicuticular wax, Sb

1. Introduction

In Japan, since the 1950s, Japanese cedar (*Cryptomeria japonica* D. Don) has been declining in the Kanto and Kansai districts. Several theories have been proposed to explain this phenomenon, including direct effects of gaseous air pollutants (Kohno *et al.*, 1994), soil acidification due to acidic air pollutants (Nashimoto, 1988), and deterioration of water condition due to urbanization (Sase *et al.*, 1998b). At present, the water-stress scenario is the likeliest, but its mechanism is still unclear, because of insufficient data on tree physiology.

Air pollutants, such as acidic and oxidizing gases and aerosols, cause thinning of leaf cuticles (Berlyn, *et al.*, 1993) and erosion and deterioration of epicuticular wax (Turunen and Huttunen, 1990; Percy *et al.*, 1993; Sase *et al.*, 1998b). Aerosols often intrude into stomata, resulting in stomatal malfunction (Crossley and Fowler, 1986; Sase *et al.*, 1998b). These changes increase water and nutrient loss from the leaf surface, and consequently cause water and nutrient stress to trees (Turunen and Huttunen, 1990; Sase *et al.*, 1998b).

We analyzed *C. japonica* leaves, sampled monthly from areas of decline in Kanto, for physiological properties related to water stress (transpiration rate, amount of epicuticular wax, and contact angles), and discuss our results in relation to air pollution.

Water, Air, and Soil Pollution **130:** 941–946, 2001.
© 2001 *Kluwer Academic Publishers.*

2. Materials and Methods

2.1. LEAF SAMPLING

Three areas in Kanto were selected on the basis of tree decline: *a*-I (Saitama; level of decline 2.0-4.0), *a*-II (plains in Ibaraki; 1.0-1.4), and *a*-III (mountainous areas in Ibaraki; 1.0). Decline was rated as 1, healthy; 2, defoliation > 30% without treetop dieback; 3, defoliation > 30% with some treetop dieback; 4, treetops showing total dieback (Nashimoto, 1988). Two *C. japonica* trees (> 30 years old) in each area, which were isolated or situated on the southern margin of the stands, were selected for sampling. Leaves and branches from south-facing sunny positions ca. 5 m in height were collected monthly (June 1997–March 1998) in a plastic bag containing damp tissue, and kept in the dark at 5°C until analysis (within 1 week; without abrasion) (Sase *et al.*, 1998a, b).

2.2. MEASUREMENT OF APPARENT CUTICULAR TRASPIRATION RATE

Zero-year and 1-y leaves (together with ca. 5-cm small stems) were cut off, rinsed with water, and then dried gently at room temperature. After the cut surface was covered with silicone grease, leaves (36 samples, ca. 10 g) were placed in a desiccator containing a fixed amount (150 g) of silica gel, kept at 20 ± 2°C in a dark room. Leaves were weighed at regular intervals (every 10 h initially) until the weight was constant. The rate of water loss was rapid during the first 10 h, due probably to stomatal transpiration, and then became nearly constant over the following several days. This constant phase is due to cuticular transpiration (Mengel *et al.*, 1989), although closure of some stomata may be incomplete (Crossley and Fowler, 1986). The selected range, in which the relative water content of leaves was higher than 60%, was used to calculate the rate (% water loss per h; initial water content = 100%) (number of determination: 9 for each sample) (Sase *et al.*, 1998b).

2.3. DETERMINATION OF EPICUTICULAR WAX

Five grams of 0-y or 1-y leaves (together with small stems) were washed in water for 1 min with an ultrasonic cleaner (Branson, 80W), and dried (< 50°C). The leaves were shaken for 15 s in 20 mL $CHCl_3$ to extract epicuticular wax. The extract was filtered with quartz wool and a filter paper (Toyo, No. 5C) to remove suspended aerosols. The $CHCl_3$ was then evaporated, and the wax that remained was determined gravimetrically (number of determination: 3 for each sample). The water content of leaves was determined from another aliquot of the sample, and the amount of wax was expressed as mg/g of dry leaves (DL) (Sase *et al.*, 1998a). Aerosols on quartz wool ("leaf-aerosols") were analyzed by neutron activation (section 2.5).

2.4. MEASUREMENT OF CONTACT ANGLE

Water (0.2 µl) was dropped from a microliter syringe on 0-y or 1-y needles (1-1.5 cm) held on a glass slide. Within 15 s, a side image of the droplet was observed through a small prismatic mirror and a microscope and recorded in a computer *via* a digital camera. On the image on screen, the basal diameter (BD) and height (H) of the droplet were measured, and the contact angle was calculated as $2\tan^{-1}(H/BD/2)(180/\pi)$ (Cape, 1983). Measurements were taken twice on each of 5 needles for each sample.

2.5. NEUTRON ACTIVATION ANALYSIS OF LEAF-AEROSOLS

Leaf-aerosols and a neutron flux monitor (30 µg Co) were irradiated together for 50 min in Pn-2 ($\phi_{th}= 2.75 \times 10^{13}$ n/cm^2/s) at Kyoto University Reactor. After cooling of the samples for 4-5 days, the γ-ray spectra of the samples were measured for 8000 s by a Ge(Li) diode detector coupled to a 4096-channel pulse-height analyzer, and then computer-analyzed for semi-long-lived nuclides. After a month, the γ-ray spectra were measured again for 100 000 s for long-lived nuclides (Koyama and Matsushita, 1980). Although 23 elements were analyzed, Sb is used here for discussion. Its detection limit and relative standard deviations of the analytical values were ca. 1 ng and < 5%, respectively.

2.6. STATISTICAL ANALYSIS

Statistical calculations—ANOVA and Fisher's PLSD—were done with StatView 4.5j (Abacus Concepts, Inc., Berkeley, CA, USA).

3. Results

3.1. LEAF PHYSIOLOGY

Annual changes of apparent cuticular transpiration rates, amounts of epicuticular wax, and contact angles are shown in Figures 1A–C.

Apparent cuticular transpiration rates increased with aging of the leaf (0-y: 0.44± 0.15%/h; 1-y: 0.60±0.17%/h; $P < 0.01$), and their rates of increase were higher in 0-y leaves (after August) than in 1-y leaves (after July) (0-y: 0.026%; 1-y: 0.00074%/h/month). In *a*-I, the transpiration rates (0-y: 0.56±0.18%[a]; 1-y: 0.77±0.16%/h[b] ; different italic letters: $P < 0.01$) and their rates of increase with aging (0-y: 0.041%; 1-y: 0.019%/h/month) were clearly higher than those in *a*-II (0-y: 0.37±0.06%/h[c], 0.015%/h/month; 1-y: 0.51±0.06%/h[a], 0.0042%/h/month) and in *a*-III (0-y: 0.39±0.10%/h[c], 0.023%/h/month; 1-y: 0.52±0.12%/h[a], 0.0011%/h/month).

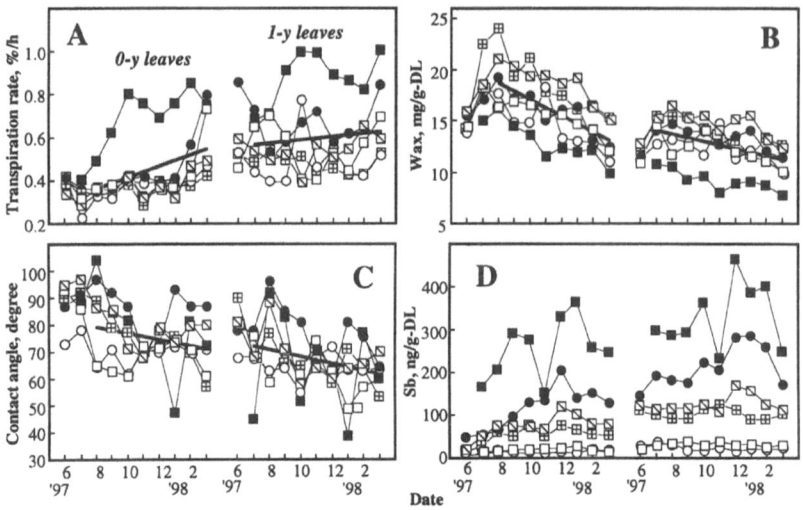

Figure 1. Annual changes in physiological properties of *C. japonica* leaves and amounts of aerosol-Sb on the leaves. (A) apparent cuticular transpiration rate, (B) amount of epicuticular wax, (C) contact angle, (D) amount of aerosol-Sb. (●, ■) *a*-I, (⊞,◨) *a*-II, (○,□) *a*-III. Thick lines: regression for all.

Amounts of epicuticular wax were lower in 1-y leaves (12.5±2.1 mg/g DL) than in 0-y leaves (16.1±2.9 mg/g DL; $P < 0.0.1$), and their rates of decrease with aging were higher in 0-y leaves (after August) (0.80 mg/g DL/month) than in 1-y leaves (after July) (0.36 mg/g DL/month). Although the cause is not clear, the amounts of wax tended to be low in *a*-I and high in *a*-II. Also, the apparent annual erosion of the wax (wax on 0-y leaves minus wax on 1-y leaves) and its rate of erosion (100 × the difference / wax on 0-y leaves) were 3.5±1.3 mg/g DL and 22.0±6.8%, in *a*-I and *a*-II. These were clearly higher ($P < 0.05$) than those in *a*-III (2.5±1.1 mg/g DL and 16.5±5.7%). The erosion rates of wax in 0-y leaves, from regression of the plots in Figure 1B, were also higher in *a*-I and *a*-II (0.85 mg/g DL/month) than in *a*-III (0.70 mg/g DL/month), although those in 1-y leaves were the same in all areas (0.36 mg/g DL/month).

Contact angles were clearly smaller ($P < 0.01$) on 1-y leaves (68°±11°) than on 0-y leaves (78°±11°), and thus leaf wettability increased with aging. Although the angles had no clear differences among areas, their rates of decrease with aging tended to be higher in *a*-I and *a*-II (0-y: 2.2°; 1-y: 1.5°/month) than in *a*-III (0-y: slightly increased; 1-y: 0.92°/month).

3.2. LEAF-AEROSOLS

Figure 1D shows annual changes in concentration of Sb on leaves, which is a typical anthropogenic element in aerosols (anthropogenic fraction > 90%), as an indication of aerosol accumulation on leaves (Takamatsu *et al.*, 2000). Aerosol deposition began just after leaf expansion (May–June) and reached the maximum in winter. After that, the amounts of leaf-aerosols were maintained at roughly constant levels depending on the degree of air pollution. Thus, amounts of aerosol-Sb were higher on 1-y leaves than on 0-y leaves. The order among areas was a-I (0-y: 182 ± 93^a; 1-y: 269 ± 86^b ng/g DL) > a-II (0-y: 63 ± 25^c; 1-y: 115 ± 21^d ng/g DL) > a-III (0-y: 16 ± 6^e; 1-y: 28 ± 7^e ng/g DL).

There were large differences in the annual profiles, especially of transpiration rates and amounts of aerosol-Sb between two trees in a-I. In our other survey (on March 1997), high transpiration rates (0-y: 1.08 ± 0.51; 1-y: 1.01 ± 0.24 %/h) and large amounts of aerosol-Sb (0-y: 330 ± 190; 1-y: 440 ± 230 ng/g DL) have been observed for 4 other trees in a-I, which were comparable to the values in plot-■. Therefore, one tree may have been typical for a-I, whereas the other tree (plot-●) may have been shielded to some extent from the polluted air by an adjacent forest, which was separated by a small open area.

4. Discussion

Apparent cuticular transpiration rates are a measure of uncontrollable water loss from leaves, and are thus responsible for water stress of trees. Although cuticular transpiration is controlled by cuticles (DeLucia and Berlyn, 1984), the epicuticular wax has the most effective shielding effect (total thickness of the *C. japonica* cuticles was almost constant in Kanto, i.e., 8.7 ± 0.7 μm for 0-y, 9.5 ± 0.7 μm for 1-y, n = 12; Takamatsu unpublished). This is supported by the facts that the transpiration rates usually correlate with the amount of wax (Jordan et al., 1984; Sase et al., 1998b) and increase significantly by removal of the wax (Jordan et al., 1984). Thus, damage to the wax increases transpiration. Also, stomatal malfunction (Turunen and Huttunen, 1990; Sase et al., 1998b; Takamatsu et al., 2001) increases apparent cuticular transpiration. Therefore, the transpiration rates may depend on amounts of epicuticular wax, leaf wettability (contact angles), and rates of stomatal malfunction. Although the latter parameter was not analyzed here, it is correlated approximately with the amount of aerosol-Sb on leaves (rate of stomatal malfunction [%] = 0.092 × amount of aerosol-Sb [ng/g DL] − 1.3, r = 0.939, $P < 0.01$; Takamatsu et al., 2001), and thus aerosol-Sb can be used as a substitute for the rate of stomatal malfunction.

Figure 2 shows the relationship between the transpiration rates and the values calculated from a linear binomial function (based on multiple regression analysis) that includes the amounts of wax and aerosol-Sb as variables. (Contact angles were excluded from variables because of the relatively large deviations in repeated and monthly analyses.) The transpiration rates and values show a fairly good correlation. In a-I (Saitama), the large amount of aerosols on the *C. japonica* leaves may have accelerated the erosion and deterioration of the epicuticular wax. Some of the aerosols, which had intruded into stomata (identified by SEM observation), may have caused the stomatal malfunction, resulting in a 50% increase in the transpiration rates. In Saitama, concentrations of gaseous air pollutants were also high in 1997 (eg. NO_x: a-I, $39±3^a$; a-II, $25±4^b$; a-III, $16±4^b$ ppb; Japan Environment Agency, 1998). Thus, they may have worked together (Turunen and Huttunen, 1990; Saxe, 1991). In Saitama, the air has been becoming drier since the 1950s (Sase et al., 1998b) and the groundwater lowered significantly from around 1950 to 1970 (Isozaki and Sugawara, 1986). Therefore, it is likely that the physiological deterioration of leaves owing to air pollution, working with the meteorological and hydrological changes, have placed C. japonica under chronic water stress, causing the decline.

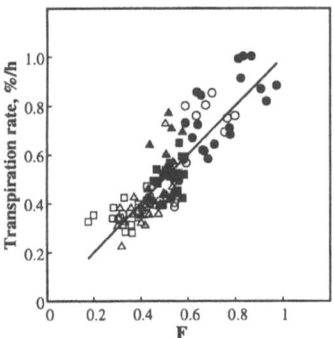

Figure 2. Relationship between apparent cuticular transpiration rates and the values calculated from a linear binomial function including amounts of epicuticular wax and aerosol-Sb on the leaves as variables. F = 0.861 − 0.0305a + 0.000829b, a: wax amounts (mg/g DL), b: amounts of aerosol-Sb (ng/g DL). Regression line: y = 0.000806 + 0.999x, r = 0.855 ($P < 0.01$). (●,○) a-I, (■, □) a-II, (▲,△) a-III. Open and closed symbols: 0-y and 1-y leaves, respectively.

References

Berlyn, G.P., Anoruo, A.O., Johnson, A.H., Vann, D.R., Strimbeck, G.R., Boyce, R.L. and Silver, W.L.: 1993, *J. Sustainable Forest*, **1**, 25.
Cape, J.N.: 1983, *New Phytol.*, **93**, 293.
Crossley, A. and Fowler, D.: 1986, *New Phytol.*, **103**, 207.
DeLucia, E.H. and Berlyn, G.P.: 1984, *Can J. Bot.*, **62**, 2423.
Isozaki, Y. and Sugawara, T.: 1986, in Res. Group Agric. Groundwater (ed.), Groundwater in Japan, Chikyusya, Tokyo, pp. 237-269. (in Japanese)
Japan Environment Agency: 1998, *Report 1997 on Measurement Results at General Air Monitoring Stations*, Tokyo, 751 p. (in Japanese)
Jordan, W.R., Shouse, P.J., Blum, A., Miller, F.R. and Monk, R.L.: 1984, *Crop Sci.*, **24**, 1168.
Kohno, Y., Nashimoto, M., Matsumura, H., Kobayashi, T. and Takahashi, A.: 1994, *Denchuken Review*, Central Res. Inst. Electric Power Ind., Abiko, No. **31**, 77. (in Japanese)
Koyama, M. and Matsushita, R.: 1980, *Bull. Inst. Chem. Res., Kyoto Univ.*, **58**, 235.
Mengel, K., Hogrebe, A.M.R. and Esch, A.: 1989, *Physiol. Plant.*, **75**, 201.
Nashimoto, M.: 1988, *Abiko Res. Rep.*, Central Res. Inst. Electric Power Ind., Abiko, No. **U87091**, 18 p. (in Japanese)
Percy, K.E., Jagels, R., Marden, S., McLaughlin, C.K. and Carlisle, J.: 1993, *Can. J. For. Res.*, **23**, 1472.
Sase, H., Takamatsu, T. and Yoshida, T.: 1998a, *Can J. For. Res.*, **28**, 87.
Sase, H., Takamatsu, T., Yoshida, T. and Inubushi, K.: 1998b, *Can. J. For. Res.*, **28**, 546.
Saxe, H.: 1991, *Adv. Bot. Res.*, **18**, 1.
Takamatsu, T., Sase, H. and Takada, J.: 2001, *Can. J. For. Res.*, **31**, 663.
Takamatsu, T., Takada, J., Matsushita, R. and Sase, H.: 2000, *Global Environ. Res.*, **4(1)**, 49.
Turunen, M. and Huttunen, S.: 1990, *J. Environ. Qual.*, **19**, 35.

INCREASING TROPOSPHERIC OZONE LEVEL REDUCED BIRCH (*BETULA PENDULA*) DRY MASS WITHIN A FIVE YEARS PERIOD

ELINA J. OKSANEN*

Department of Ecology and Environmental Science, University of Kuopio, POB 1627, Kuopio, Finland
(*author for correspondence, e-mail elina.oksanen@uku.fi)

Abstract. Soil-growing randomized mixtures of ten European silver birch (*Betula pendula* Roth) clones, showing different ozone sensitivity, were exposed to ambient air (control) or 1.4-1.7x ambient (elevated) ozone over five growing seasons using free-air fumigation (FACE) approach. During the last season, the juvenile trees were measured for growth, net assimilation rate and starch content. In elevated-ozone plants, significant effects were observed as 21-28% reduced new leaf development, 44.8% lower dry mass of leaves and 33.8% lower dry mass of roots, as well as 7.6% lower RGR of leaves and 27.8% lower RGR of roots, leading to 16% lower root/shoot ratio. In addition, net assimilation rate and starch content were slightly (8.9% and 14.3%) reduced in ozone-stressed plants. The results indicated cumulative ozone-induced growth reductions over five years. Ozone-stressed trees with declined root growth may become susceptible to other environmental stresses such as water and nutrient deficiency, and lose belowground competitiveness, which may affect tree survival.

Keywords: *Betula*, birch, clone, carry-over effects, FACE, long-term exposure, open-field experiment, ozone, sensitivity

1. Introduction

Tropospheric concentration of ozone is predicted to increase in proportion to the increasing world population at a rate of 0.2-2% per year, especially in the Northern Hemisphere (e.g. Keeling *et al.*, 1995). Increasing ozone concentrations have been estimated to be one of the major environmental risk factors to European forest ecosystems in the 21th century, besides increasing CO_2 and nitrogen deposition (Fuhrer *et al.* 1997; Matyssek and Innes, 1999). Although the short-term action of ozone on seedlings or young forest trees are well documented today (for reviews, see e.g. Skärby *et al.*, 1998; Matyssek and Innes, 1999), knowledge about long-term experiments using relevant chamberless exposure regimes and older trees are still far from complete. The ozone induced changes in plant internal resource allocation and disturbed assimilate transport resulting in altered root-shoot ratio, as reported in potted plants, are predicted to be extremely critical for structure and function of forest trees, biodiversity and long-term stand development and stability of European forests under chronic ozone exposure (Küppers, 1994; Skärby *et al.*, 1998; Matyssek and Innes, 1999).

In our large-scale open-field experiments in Finland, 33% of 46 tested birch (*Betula pendula* and *Betula pubescens*) clones showed obvious ozone sensitivity (Pääkkönen *et al.*, 1993, 1997ab). In these fast-growing, commercially used clones ozone effects appeared as reduced growth rate, reduced root/shoot ratio, accelerated leaf senescence (according to yellowing and ultrastructural changes of mesophyll cells), lower net photosynthesis, decreased Rubisco, starch and chlorophyll concentrations, and as visible and ultrastuctural leaf injuries (especially in chloroplast membranes). Visible foliar injuries and accelerated senescence were accompanied by induction of genes for stress proteins PAL and PR-10 (Pääkkönen *et al.*, 1998c), by cell wall modifications, as well as by accumulation of tannin and phenolic deposition in leaf mesophyll tissue (Pääkkönen *et al.*, 1998b), indicating activated ozone stress defence mechanisms. In the sensitive *B. pendula* clone 5 and tolerant clone 2, effects by mild drought stress and ozone were additive (Pääkkönen *et al.*, 1998ab). Significant interaction was also found after low soil nitrogen combined with elevated ozone exposure, leading to greater ozone damages in N-deficient plants (Pääkkönen and Holopainen, 1995).

In tolerant birch clone 2, several persistent acclimation and defence reactions have been detected in ozone-stressed plants on the biochemical (increased Rubisco- and K-concentration), morphological (increased number of emerging buds and thickened cell walls) and physiological (increased stomatal conductance and photosynthetic capacity) level. On the other hand, increased phenolic compounds and decreased growth indicated a change in carbon allocation, which was typical for the sensitive clone 5 under ozone stress (Oksanen and Saleem, 1999).

Recently, we also tested different genotypes of Japanese white birch (*B. platyphylla*), Alaskan white birch (*B. resinifera*), European white birch (*B. pendula*), and hybrids between them for ozone tolerance. In most genotypes a slightly elevated ozone concentrations over two growing seasons resulted in altered resource allocation leading to reduced root/shoot ratio. In fast-growing birch origins, stimulated stem height and foliage growth in the second growing season indicated compensatory growth, which was accompanied by accelerated senescence of mature leaves. Highest ozone sensitivity according to growth responses and leaf visible injuries was determined for seedlings of *B. platyphylla* and crossings of Finnish plus trees, which were bred for fast growth (Oksanen and Rousi, manuscript).

In this paper, new results from our open-field low-stress ozone exposure over five growing seasons are presented to achieve realistic information about actual ozone risks in soil-growing juvenile birch trees. The chamberless free-air fumigation technique allows exposure to elevated ozone levels without microclimatic bias, thereby providing ecologically relevant data for long-term responses of birch.

2. Materials and Methods

Saplings of three ozone-sensitive, four intermediate, and three ozone-tolerant *Betula pendula* Roth clones (determined earlier in Pääkkönen *et al.*, 1997a) were grown in soil under ambient air (serving as control) and elevated ozone according to 1.4-1.7 x ambient profile from May 19, 1994 until August 31, 1998. Twenty saplings per clone were randomly planted into the two control fields and two elevated-ozone fields (ten saplings per treatment). The ozone gas was continuously monitored, and pure oxygen was used for ozone generation. The ozone dispensing apparatus was operating 24 hours per day. The free-air ozone enrichment (FACE) system in Kuopio Botanical Garden (central Finland) has been technically described in Wulff *et al.* (1992). The cumulative ozone exposures, AOT00 (total cumulative exposure, ppm.h), AOT40 (cumulative exposure accumulated over threshold 40 ppb, ppm.h) and the 7-h (11.00-18.00) and 24-h mean concentrations (ppb) for each growing seasons are presented in Table I. The plants were fertilized with Superex-9, and irrigated when needed.

TABLE I

Ozone data from May 19, 1994 until August 31, 1998, calculated for each growing season. Values are mean concentrations for the two ambient-air (control) and the two elevated-ozone fields. The increase in total ozone exposure as compared to ambient air is given as a factor in parenthesis.

Year	Treatment	AOT00 (ppm.h)	AOT40 (ppm.h)	7-h mean (ppb)	24-h mean (ppb)
1994	Ambient	65.2	0.8	25.9	20.3
	Elevated	102.0 (1.56x)	18.3	41.7	32.6
1995	Ambient	62.7	0.5	22.5	22.4
	Elevated	108.9 (1.74x)	18.6	37.8	34.2
1996	Ambient	71.2	1.0	30.2	24.1
	Elevated	107.0 (1.50x)	17.3	45.8	36.3
1997	Ambient	67.6	0.8	27.5	22.4
	Elevated	107.1 (1.58x)	13.9	42.0	35.5
1998	Ambient	57.7	0.3	26.1	21.8
	Elevated	80.8 (1.39x)	6.9	33.9	30.4

In the last growing season, all the trees were measured for new leaf development (number of emerging buds and mean leaf size) between May 15 and June 5, 1998. At final harvest on August 31, 1998, the plants were determined for height growth, number of leaves, dry mass and relative growth rates (RGR) of leaves, stem wood and roots, net assimilation rate (NAR), and foliar starch concentrations. For RGRs [= ln(final dry mass) – ln(mean initial dry mass) /number of years], additional saplings were harvested before the start of the experiment for initial biomass determination for each clone. NAR = [2 (final dry mass –mean initial dry mass)]/ [(final foliage area + initial foliage area)(number of years)].

3. Results

At the beginnig of the fifth growing season (1998), elevated-ozone exposure significantly decreased the average size of new developing leaves by 21.3-

28.2%, whereas the number of emerging buds on May 15 was not significantly different between the treatments (Table II).

At final harvest on August 31, significantly decreased dry mass of leaves (44.8%) and roots (33.8%), as well as relative growth rates of shoot (7.6%) and roots (27.8%) were observed in ozone-exposed trees, leading to significantly increased shoot/root ratio (Table II). Stem height growth was unaffected by elevated-ozone, whereas number of leaves, net assimilation rate and starch concentration were reduced by 13.0%, 8.9% and 14.3%, respectively (not a significant difference) (Table II).

TABLE II

Effects of open-field ozone stress on growth and starch concentration of ten *Betula pendula* clones (data of ten trees pooled per treatment), determined during fifth growing season. RGR = relative, growth rate, NAR = net assimilation rate. Values are means ± SE. Analysis of variance, $p < 0.05$, n = 10.

Response	Ambient-ozone (control)	Elevated-ozone	Growth reduction, %
May - June 1998:			
Number of buds, 15 May	195.9 ± 24.2	211.9 ± 14.5	
Mean leaf size, cm^2			
18 May	4.01 ± 0.22	2.88 ± 0.14***	28.2
28 May	8.98 ± 0.38	7.07 ± 0.33***	21.3
5 June	13.84 ± 0.67	10.86 ± 0.47***	21.5
August 31, 1998:			
Height, cm	327.6 ± 15.5	338.6 ± 10.2	
Height growth in 1998, cm	46.9 ± 4.7	52.9 ± 7.1	
Number of leaves	702.6 ± 89.5	611.4 ± 79.0	13.0
Dry mass of leaves, g	145.0 ± 5.9	80.1 ± 8.1***	44.8
Dry mass of wood, g	197.8 ± 36.3	190.7 ± 14.7	3.6
Dry mass of shoot, g	342.8 ± 38.3	270.8 ± 17.1	21.0
Dry mass of roots, g	299.6 ± 34.1	198.4 ± 15.6*	33.8
Shoot/root ratio, g/g dwt	1.14 ± 0.12	1.36 ± 0.32*	
RGR of shoot	0.976 ± 0.03	0.902 ± 0.01*	7.6
RGR of roots	0.911 ± 0.05	0.658 ± 0.04**	27.8
NAR, g/cm^2 year	0.0236 ± 0.002	0.0215 ± 0.003	8.9
Starch concentration, mg/g DWT	31.08 ± 6.44	26.57 ± 5.01	14.3

4. Discussion and Conclusions

The results indicated deleterious impact of long-term ozone stress on both foliage and belowground growth in this selected clonal birch population of various genotypes. The wood production, however, was not significantly changed. Previously, also Maurer *et al.* (1997) and Matyssek *et al.* (1992, 1998) reported, that stem height growth was less affected by ozone in birch than other biomass parameters. In the present experiment, reduced foliage mass was accompanied by slightly lower net assimilation rates and starch content, suggesting impaired photosynthetic production under ozone stress. In addition, ozone-induced premature leaf shedding (as reported in our previous experiments) may have contributed to lower foliage mass. The great ozone impact on root growth can be explained by corresponding ozone-induced reduction in foliage mass, as well as likely increased demand for ozone-detoxification and repair processes, as reported in birch e.g. by Maurer *et al.* (1997) and Landolt *et al.* (1997), and impaired assimilate translocation from leaves to roots, as found by Matyssek *et al.* (1992) and Günthardt-Goerg *et al.* (1993). In birch, ozone-induced disturbance in translocation and phloem loading has been observed concomitant with stomatal closure, impaired photosynthetic capacity, end-product inhibition of photosynthesis, reduction in sucrose synthesis and shift towards glycolytic and anaplerotic metabolism (involved in detoxification and repair processes), progressive collapse of mesophyll cells and accumulation of starch along leaf veins (Matyssek *et al.* 1991; Günthardt-Goerg *et al.* 1993; Landolt *et al.* 1997; Einig *et al.* 1997).

In our previous two-season ozone experiment using the same, although potted clonal birch material, ozone-induced reduction in contents of Rubisco, chlorophyll, carotenoids, starch and nutrients, and decreased new shoot growth and net assimilation rate were determined as long persistent effects, accompanied with 7.2% and 9.6% reduction in shoot and root dry weights (respectively) at the end of experiment (Oksanen and Saleem, 1999). These carry-over ozone effects were remaining at least two seasons after the end of exposure. Thereby, according to the present results, ozone-induced growth reductions in birch seem to be cumulative over five years, although in contrast to significant root dry mass reduction, the shoot dry mass reduction was not significant.

To conclude, significantly changed shoot/root ratio at the expence of roots was demonstrated in this long-term ozone experiment in birch. The changed balance between root and shoot, and altered crown and root architecture may affect competitiveness of trees. Reduced root/shoot ratio in ozone-exposed plants may promote predisposition to summer drought (Mansfield, 1988; Davidson *et al.*, 1992), winter dessication (reviewed in Skärby *et al.*, 1998), shortage in nutrient supply and parasite attack (Langebartels *et al.*, 1997) through limited root development. It remains to be verified, that trees with a slightly changed shoot/root ratio will be a risk factor in future forest ecosystems, as suggested e.g. in reviws by Skärby *et al.* (1999) and Matyssek and Innes (1999).

Acknowledgements

This research was funded by Academy of Finland. The author acknowledges Timo Oksanen for ozone exposures and for providing ozone data.

References

Davidson, S.R., Ashmore, M.R. and Garretty, C.: 1992, *Forest Ecology Management* **51**, 187.
Einig, W., Lauxmann, U., Hauch, B., Hampp, R., Landolt, W., Maurer, S. and Matyssek, R.: 1997, *New Phytologist* **137**, 673.
Fuhrer, J., Skärby, L. and Ashmore, M.: 1997, *Environmental Pollution* **97**, 91.
Günthardt-Goerg, M.S., Matyssek, R., Scheidegger, C. and Keller, T.: 1993, *Trees* **7**, 104.
Keeling, C.D., Whort, T.P., Wahlen, M. and van der Plicht, J.: 1995, *Nature* **375**, 666.
Küppers, M.: 1994, 'Canopy Gaps: Competitive Light Interception and Economic Space Filling – a Matter of Whole-plant Allocation', in M.M. Caldwell and R.W. Percy (eds), *Exploitation of Environmental Heterogeneity by Plants – Ecophysiological Processes Above and Below-Ground*, Academic Press, San Diego, pp. 111-144.
Landolt, W., Günthardt-Goerg, M.S., Pfenninger, I., Einig, W., Hampp, R., Maurer, S. and Matyssek, R.: 1997, *New Phytologist* **137**, 389.
Langebartels, C., Ernst, D., Heller, W., Lütz, C., Payer, H.-D. and Sandermann, H.: 1997, 'Ozone Responses of Trees: Results from Controlled Chamber Exposures at the GSF Phytotron', in H. Sandermann, A.R. Wellburn and R.L. Heath (eds.), *Forest Decline and Ozone, a Comparison of Controlled Chamber and Field Experiments*, Ecological Studies **127**, Springer-Verlag Berlin, pp.163-200.
Mansfield, T.A.: 1988, 'Factors Determining Root:Shoot Partioning', in J.N. Cape and P. Mathy (eds.), *Scientific Basis of Forest Decline Symptomology*, Brussels, COST/CEC, pp. 171-180.
Matyssek, R., Günthardt-Goerg, M.S., Keller, T. and Schneidegger, C.: 1991, *Trees* **5**, 5.
Matyssek, R., Günthardt-Goerg, M.S., Saurer, M.S. and Keller, T.: 1992, *Trees* **6**, 69.
Matyssek, R., Günthardt-Goerg, M.S., Schmutz, P., Saurer, M., Landolt, W. and Bucher, J.B.: 1998, *Journal of Sustainable Forestry* **6**, 3.
Matyssek, R. and Innes, J.L.: 1999, *Water, Air and Soil Pollution* **116**, 199.
Maurer, S., and Matyssek, R.: 1997, *Trees* **12**, 11.
Pääkkönen, E. and Holopainen, T.: 1995, *New Phytologist* **129**, 595.
Pääkkönen, E., Paasisalo, S., Holopainen, T. and Kärenlampi, L.: 1993, *New Phytologist* **125**, 615.
Pääkkönen, E., Holopainen, T. and Kärenlampi, L.: 1997a, *Environmental Pollution* **95**, 37.
Pääkkönen, E., Holopainen, T., and Kärenlampi, L.: 1997b, *Environmental Pollution* **96**, 117.
Pääkkönen, E., Vahala, J., Pohjola, M., Holopainen, T. and Kärenlampi, L.: 1998a, *Plant,Cell and Environment* **21**, 671.
Pääkkönen, E., Günthardt-Goerg, M. and Holopainen, T.: 1998b, *Annals of Botany* **82**, 49.
Pääkkönen, E., Seppänen, S., Holopainen, T., Kärenlampi, S., Kärenlampi, L. and Kangasjärvi, J.: 1998c, *New Phytologist* **138**, 295.
Oksanen, E. and Saleem, A.: 1999, *Plant, Cell and Environment* **22**, 1401.
Skärby, L., Ro-Poulsen, H., Wellburn, F.A.M. and Sheppard, L.J.: 1998, *New Phytologist* **139**, 109.
Wulff, A., Hänninen, O., Tuomainen, A. and Kärenlampi, L.: 1992, *Annals of Botanici Fennici* **29**, 253.

EFFECTS OF SIMULATED ACID MIST ON A SITKA SPRUCE FOREST APPROACHING CANOPY CLOSURE: SIGNIFICANCE OF ACIDIFIED VERSUS NON-ACIDIFIED NITROGEN INPUTS

LUCY J. SHEPPARD[1*], ALAN CROSSLEY[1], JUDITH PARRINGTON[2], FRANCIS J. HARVEY[1] and J. NEIL CAPE[1]

[1] *Centre for Ecology and Hydrology, Edinburgh Research Station, Bush Estate, Penicuik, Midlothian EH26 0QB, UK;* [2] *Centre for Ecology and Hydrology, Merlewood Research Station, Grange over Sands, Cumbria LA11 6JU, UK*
(* author for correspondence, e-mail:ljs@ceh.ac.uk)

Abstract. Effects of enhanced N, S and NS Acid additions, up to approximately 100 kg N and S ha^{-1} y^{-1}, are described for a 15-year-old Sitka spruce forest growing on an acid peat in Scotland. Groups of 10 trees, replicated over 4 blocks, have been treated at canopy height on approximately 50 or 100 occasions with 2 mm precipitation equivalent, between April and November, since 1996. Relative stem volume increment (RSVI) has been consistently higher in the NS Acid treated trees compared with control or N treated trees since the second year of treatment, although no dose response was found. Litterfall was also considerably increased in the NS Acid treatments and showed a clear dose effect but was not correlated with RSVI. Base cation concentrations in soil waters, collected using zero tension lysimeters reflected the presence or absence of the spray treatments and showed a dose related increase in response to NS Acid inputs. Treatment with 2NS Acid increased NH_4-N and NO_3-N by an order of magnitude. Results after 4 years of treatment showed a positive growth and litterfall response to NS Acid inputs but no effect of N alone. Enhanced stemwood growth may be linked to the higher base cation and phosphate concentrations measured in the soil water in the presence of NS Acid and S treatments.

Keywords: Acid mist, nitrogen, sulphur, Sitka spruce, litter, soil water chemistry, stemwood.

1. Introduction

In Europe, nitrogen (N) has superseded sulphur (S) as the dominant anthropogenic component of both wet and dry deposition over the last decade (Fowler *et al.*, 1998; Goulding *et al.*, 1998). Consequences of this change in N:S balance for forest growth are still to be addressed. Open-top chamber studies which have exposed spruce seedlings to simulated cloud water have shown most detrimental effects when the ratio of S to N, and acidity were high (Cape *et al.*, 1991; Jacobson *et al.*, 1990). These observations suggest that a decline in S deposition in relation to N should improve conditions for spruce growth. Detrimental effects of acidic mist containing S and N on stemwood growth of Sitka spruce were confirmed in a field experiment (Crossley *et al.*, 1997). A monoclonal stand of Sitka spruce, growing on a base-rich soil, was sprayed at canopy height with acid mist at pH 2.5 (H_2SO_4 + NH_4NO_3 equimolar 1.6 mol m^{-3}), providing 48 kg N and 50 kg S ha^{-1} y^{-1} for 3 years. Stemwood growth was rapidly and consistently reduced by acid mist at inputs corresponding to N and S deposition to a high altitude site in the Scottish Borders (Crossley *et al.*, 1998).

However, assessment of the potential impacts of acid deposition on UK forestry required a more rigorous experiment using a wider genetic base and more typical soil type. A large-scale experiment, providing minimal disturbance of the microclimate (no physical enclosures) was established in 1995 within a commercial Sitka spruce plantation, including several provenances, growing on a deep peat. After 2 years of treatment with NH_4NO_3 (N), Na_2SO_4 (S), H_2SO_4 + NH_4NO_3 (single and double dose, NS Acid and 2NS Acid at pH 2.5), no spray and a pH 5 control, growth trends were the reverse of those seen on the base rich soil (Sheppard et al., 1999). Since 1998, additional below-ground measurements have been undertaken to explore the interactive effects of acidified S on tree response to N deposition. This paper reports some initial findings for soil water chemistry and litterfall in relation to tree growth.

2. Materials and Methods

2.1. EXPERIMENTAL DESIGN

The experiment covers 1.5 ha on a very acid (< pH 3.0, $CaCl_2$) drained basin peat in the Scottish Borders at 290 m. The trees, *Picea sitchensis* Bong. Carr, were planted in 1986 at 2 m spacing on inverted mounds formed from the drainage furrows. Each plot was replicated 4 times and contained 10 trees in 2 lines. When treatment commenced in April 1996, basal branches had begun to interlock and the 10 year-old trees had a mean height of 3.8m. The experimental design, site information and methodologies were described in Sheppard et al. (1999). Treatments were applied in rainfall, collected at the site, to the upper canopy as fine droplets (100 – 250 :m diam.) by full cone sprayers (2 per tree) supported by a framework of galvanized steel poles. The 6 treatments (see earlier) provided N at 48 kg ha^{-1} y^{-1} or S at 50 kg ha^{-1} y^{-1} or both, or twice this for the double dose where the treatment was applied twice as often. Ions were supplied at a concentration of 1.6 mol m^{-3}(NH_4^+, NO_3^-, SO_4^{2-}) and 3.2 mol m^{-3} (H^+, Na^+). Background N deposition at the site is approximately 8 kg N ha^{-1} (wet + dry).

2.2. MEASUREMENTS

Stem diameter was read, two or four times a month, from vernier girth bands placed between whorls 8 and 9 below the leader in 1995. Once a year, diameters at breast height and below the leader (1995) were also measured so that annual volume increments could be calculated (Sheppard et al., 1999). Stem volume increments are expressed in proportion to initial tree size. Soil water has been collected since May 1999 using zero-tension lysimeters. These collectors, 10 per plot, spiked with the biocide thymol were inserted into the side of the furrow 5-10 cm below the surface. Samples were bulked by plot in the field. Volume, pH and conductivity were measured before filtering with a Whatman glass fibre pre-filter and a 0.2 :m cellulose nitrate filter. Base cations (Ca^{2+}, Mg^{2+} & K^+) and Al^{3+} were measured by ICP-OES-Sequential Inductively Coupled Plasma Optical Emission (Jobin-Yvon JY38 Plus Spectrometer) with Meinhardt nebulisation procedures, NH_4^+ and PO_4^{3-} by colorimetry (Automated colorimetric system (Skalar SAN^{plus})) using indophenol blue and molybdenum blue respectively, Cl^-, NO_3^- and SO_4^{2-} by ion chromatography (Automated liquid chromatograph, Dionex, DX-100T) (Rowland 1996). Throughfall (Cape et al., this volume) and litter were collected using a system of open gutters, 0.3 m above the ground, running under each tree (1 m^2 collecting area). Litter was removed twice a year in April and August from 1998, dried and weighed.

3. Results

3.1. STEM VOLUME INCREMENT

TABLE I

Effects of treatment on annual relative stem volume increment (RSVI) (percent change) and litter fall (g m^{-2}) (April 1998 – April 2000) in Sitka spruce trees treated from 10 years old with a range of N and S inputs up to 100 kg N and S $ha^{-1}y^{-1}$.

		No Spray	N	S	Control	NS Acid	2NS Acid	P value	LSD
1996-97	%	54	52	56	53	55	54	0.62	6
1996-98	%	151	145	160	156	165	162	0.33	21
1996-99	%	209	189	209	191	222	223	0.29	59
1996-00	%	315	279	316	275	335	329	0.15	57
Litterfall	g	124	187	144	156	267	343	0.05	47

Relative stem volume increment (RSVI) values (Table I) for the consecutive treatment years indicate that the site is capable of supporting rapid tree growth. Initially all treatments increased their stem volumes by similar amounts and one year of treatment produced no treatment effects. However, in subsequent years large increases in stem volume have resulted from treatment with NS Acid and 2NS Acid, although not in proportion to the NS Acid dose. The addition of N alone, as NH_4NO_3, has had no effect on volume increment. Additions of S as Na_2SO_4 have increased stem increment in the last 2 seasons, likewise the no spray treatment.

3.2. LITTERFALL

In this relatively young stand litterfall was negligible when treatments began, but after 2 years litter began to accumulate on the floor of the 2NS Acid treatment. Accumulation of litter in the throughfall gutters continues to be significantly higher in the NS Acid treatments in a positive response to the NS Acid dose (Table I). While litterfall was in proportion to NS Acid dose, growth enhancement was equal in both acid treatments and not in proportion to dose. Treatment with N alone did not increase litterfall or stemwood growth. Relative stemwood volume increment was not related to litterfall ($r^2 = 0.067$).

3.3. SOIL WATER CHEMISTRY

TABLE II

Effects of treatment on the volume weighted mean (4) ion concentrations (mmol$_c$ l^{-1}) and the Base Cation : Al ratio (BC:Al) measured in soil waters collected at 5-10cm depth using zero tension lysimeters. The spraying period ran from May to mid-October 1999 (4 collections), spray was withheld from mid-October to mid-April 2000 (5 collections)

	Plus Spray									
	H^+	NH_4^+	NO_3^-	Al^{3+}	Ca^{2+}	Mg^{2+}	K^+	BC:Al	SO_4	PO_4
Control	0.054	0.007	0.009	0.01	0.09	0.04	0.009	14	0.10	0.003
NS Acid	0.151	0.026	0.031	0.03	0.17	0.09	0.022	9	0.43	0.007
2NS Acid	0.396	0.232	0.181	0.11	0.25	0.14	0.026	4	1.01	0.008
N	0.056	0.033	0.064	0.01	0.11	0.04	0.006	16	0.07	0.003
S	0.078	0.013	0.009	0.03	0.14	0.04	0.016	7	0.57	0.008
No spray	0.036	0.008	0.008	0.02	0.10	0.05	0.011	8	0.09	0.003

	Spray Withheld									
	H^+	NH_4^+	NO_3^-	Al^{3+}	Ca^{2+}	Mg^{2+}	K^+	BC:Al	SO_4	PO_4
Control	0.115	0.006	0.003	0.03	0.05	0.03	0.001	4	0.07	0.004
NS Acid	0.147	0.008	0.004	0.02	0.05	0.03	0.002	4	0.10	0.004
2NSAcid	0.236	0.021	0.017	0.03	0.05	0.03	0.002	3	0.16	0.006
N	0.137	0.012	0.006	0.02	0.07	0.04	0.001	6	0.09	0.004
S	0.078	0.006	0.003	0.02	0.04	0.02	0.002	3	0.12	0.006
No spray	0.078	0.007	0.003	0.02	0.06	0.04	0.001	5	0.07	0.005

Soil water collections for the 'with spray 'and 'spray withheld' periods (Table II) show that concentrations of NH_4^+, NO_3^-, the base cations (BC = Ca^{2+}, Mg^{2+} and K^+) and SO_4^{2-} are highly dependent on the presence of the spray treatment. Ion concentrations generally, increased markedly during the spray period. Both during and in the absence of treatment, concentrations in the 2NS Acid treatment, which supplied approximately 96 kg N and 100 kg S ha^{-1} y^{-1}, were different from all the others, usually much higher. Soil water acidity reflected the treatment proton dose. Concentrations of NH_4-N and NO_3-N were > 20-fold higher in the 2NS Acid treatment compared with the control. By comparison addition of N as NS Acid or N, at 50% of the N input of 2NS Acid, caused relatively small increases in NH_4-N and NO_3-N. Nitrate tended to be higher in the N treatment. Base cations were doubled / trebled in the NS Acid / 2NS Acid treatments and K and Ca were also increased in the S treatment. Al increased 10-fold in the 2NS Acid treatment. The BC:Al ratio was lowest over the winter period when spray was withheld. Changes in SO_4^{2-} generally reflected treatment inputs. PO_4-P concentrations were least affected by treatment, but were highest in treatments that provided S.

4. Discussion

In contrast to Crossley *et al.'s* (1997) experiment on a base rich soil, here, the addition of NS Acid to an acid peat has stimulated stemwood increment. These increases in stemwood growth appear to be driven by the addition of acidified S rather than the addition of N, since N alone failed to stimulate stemwood growth. Doubling the dose of S/N/acidity did not further increase stemwood growth, but almost doubled litterfall and had a profound effect on soil water chemistry substantially increasing the concentrations of the major ions. The effects of 2NS Acid on NH_4-N and NO_3-N concentrations greatly exceeded that predicted from a linear dose response, with the 20-fold enhancement suggesting the soil had become saturated with respect to N. Effects on base cations were more consistent with a linear dose response implying a process of active exchange between H^+ ions and BC. Treatment effects derived from zero tension lysimeters (which tend to respond to large water fluxes/rainfall) did not always reproduce the effects seen by leaching soil cores with the appropriate treatment solution in the laboratory (Sheppard *et al.*, 1999: Sheppard and Crossley, 2000). In those experiments the NS Acid treatments leached more BC than the 2NS Acid treatment, which was explained by the higher water inputs, increasing the mobile anion flux through the profile causing BC depletion in the 2NS Acid treatment. The *in situ* zero tension lysimeters show that even after 4 years of treatment soil water BC concentrations remain higher in the acid treatment. Enhanced stemwood growth in the Na_2SO_4 treatment also appears to be linked to enhanced BC concentrations in response to Na^+ inputs.

Despite the very large increase in NH_4-N and NO_3-N in the 2NS Acid treatment compared with NS Acid, growth was not relatively increased. This implies that tree growth on this acid peat is not N limited, but rather that a large proportion of the added N in the 2NS Acid treatment is surplus to demand (Skiba *et al.*, 1999). What then is stimulating growth in the acid treatments and why is the response not dose related? In addition to enhancing soil water BC concentrations the 2NS Acid treatment has very significantly increased soil water acidity and Al concentrations, reducing the BC:Al ratio to 4. In addition, the enhanced loss of litter suggests the double dose of acid could be exerting a negative impact on needle retention, possibly via negative effects on water retention by foliage. It is known that acid treatment can increase cell wall elasticity (Eamus *et al.*, 1989) and membrane permeability (Sheppard *et al.,* 1995). Overall, these observations suggest that the 2NS Acid treatment is having a wide-ranging impact on factors influencing stemwood increment, and the increased growth is the net effect of some negative and positive benefits. By contrast the single dose treatment may be providing smaller beneficial changes that are better matched to tree growth. The available data suggest the enhanced stemwood increment on this acid peat, like some mineral soils, is linked to higher base cation and phosphate concentrations in the soil solution that accompany the addition of S with H^+ and to a lesser extent Na^+.

Acknowledgements

The UK Department of the Environment, Transport and the Regions (contract EPG 1/3/94) and the Natural Environment Research Council are thanked for their financial support.

References

Cape, J.N., Leith, I.D., Fowler, D., Murray, M.B., Sheppard, L.J., Eamus, D. and Wilson, R.H.F.: 1991, *New Phytologist*, **118**, 119.

Crossley, A., Sheppard, L.J., Cape, J.N., Smith, R.I. and Harvey, F.J.: 1997, *Environmental. Pollution*. **96**, 185.

Crossley, A., Harvey, F., Cape, J.N., Guillevic, C., Binnie, J., Wilso, D.B. and Fowler, D.: 1998, Proc 1st Int. Conf. Fog and Fog Collection. Ed R. Schemenauer and H. Bridgman Vancouver, Canada, July 19-24 1998, 321.

Eamus, D., Leith, I.D. and Fowler, D.: 1989, *Tree Physiology* **5**, 387.

Fowler, D., Flechard, C., Skiba, U., Coyle, M. and Cape, J.N.: 1998, *New Phytologist* **139**, 11.

Goulding, K.W.T., Bailey, N.J., Bradbury, N.J., Hargreaves, P., Howe, M., Murphy, D.V., Poulton, P.R. and Willison, T.W.: 1998, *New Phytologist*. **139**, 49.

Jacobson, J.S., Bethard, T., Heller, L.I. and Lassoie, J.P.: 1990, *Physiologia. Plantarum*. **78**, 595.

Rowland, A.P.: 1996, Analytical guidelines for water samples. *In: United Kingdom Environmental Change Network: protocols for standard measurements at terrestrial sites*, ed. J.M. Sykes and A.M.J. Lane, HMSO London, 149.

Sheppard, L.J., Leith, I.D., Smith, C.M.S. and Kennedy, V.: 1995, *Water Air, and Soil Pollution* **84**, 34.

Sheppard, L.J., Crossley,A., Cape, J.N., Harvey, F.J., Parrington, J. and White, C.: 1999, *Phyton* **39**, 1.

Sheppard, L.J. and Crossley, A.: 2000, *Phyton* **40**, 169.

Skiba, U., Sheppard, L.J., Pitcairn, C.E.R., Van Dijk, S. and Rossall, M.J.: 1999, *Water Air, and Soil Pollution* **116**, 89.

IMPACTS OF AMBIENT OZONE AND/OR ACID MIST ON THE GROWTH OF 14 TREE SPECIES: AN OPEN-TOP CHAMBER STUDY CONDUCTED IN JAPAN

HIDEYUKI MATSUMURA

Biology Department, Abiko Res. Lab., Central Research Institute of Electric Power Industry, 1646 Abiko, Abiko City, Chiba 270-1194, Japan

Abstract. Young trees of 14 species were exposed to ambient ozone (O_3), (charcoal-filtered air [CF] or non-filtered air [NF]) and/or acid mist (pH 5 or 3: SO_4^{2-}, NO_3^-, Cl^- at equivalent 1:2:1 ratio) over three seasons (from June 1993 to November 1995) using tunnel-type open-top chambers at two sites (Abiko: 25 m a.s.l. and Akagi: 540 m a.s.l.) in Japan. Ambient, 12-hr (0600-1800) mean O_3 concentration for April-September during the period of experiment at Akagi (41 ppb) was 40% higher than that at Abiko (30 ppb). The NF- and CF-chambers had 90% and 30% of ambient O_3, respectively. Significant decreases in biomass in the NF treatments were observed in *Pinus densiflora*, *Larix kaempferi*, *Picea abies*, *Abies firma*, *Abies homolepis*, *Abies veitchii*, *Cryptomeria japonica*, *Populus maximowiczii*, *Betula platyphylla*, *Fagus crenata* and *Zelkova serrata* as compared with the CF treatment. These results indicate that the current ambient level of O_3 in Japan is high enough to have adverse effects on the growth of all tree species examined, except *Pinus thunbergii*, *Chamaecyparis obtusa* and *Quercus mongolica*. Increasing acidity of mist caused no growth decreases in all tree species examined. However, the growth decreases by O_3 were greater at pH 3 mist treatment than at pH 5 mist treatment on *Abies veitchii* and *Fagus*. This suggested that the O_3 effect on tree growth can be exacerbated by the deposition of acid mist, possibly associated with nitrate.

Keywords: acid mist, biomass, Japanese tree, open-top chamber, ozone

1. Introduction

A combination of the field survey and atmospheric monitoring suggested that the primary cause of decline of Japanese cedar in the Kanto plain was photochemical oxidant pollution, primarily ozone (O_3) (Nashimoto, 1993). Acid fog with pH at 3 or below has been reported in many mountains in the Kanto district as well as O_3 at 100 ppb or higher (Murano, 1993; Hatakeyama and Murano, 1996), suggesting the possibility of attribution of the declines of trees such as fir, birch, oak and beech to co-occurrence of these pollutants. However, limited information is available on the responses of Japanese trees to O_3 or acid mist (Izuta, 1998). Much less attention has been paid to interactive effects of both pollutants by multiple-season exposure under field conditions.

In the present study, the effects of ambient O_3 and/or acid mist on young trees of 14 native and introduced tree species were investigated over the three growing seasons using open-top chambers at two sites in the Kanto district. The objectives of this study were to (i) test the hypothesis that ambient O_3 may affect the growth of Japanese tree species; (ii) determine whether repeated exposure to acid mist of pH 3 might cause the alterations in tree growth; and (iii) examine the interactive effects of acid mist and ambient level of O_3 on tree growth.

2. Materials and Methods

The experiments were carried out at two sites in the Kanto district: Abiko Research Laboratory (Abiko city, Chiba) at an elevation of 25 m a.s.l., 30 km NE of Tokyo, and Akagi Testing Center (Seta county, Gunma) at an elevation of 540 m a.s.l., 100 km NNW of Tokyo on the south slope of Mt. Akagi.

The experimental plants consisted of fourteen tree species. Details of tree species and experimental period and site for each species are given in Table I. Trees were planted individually in pots filled with 2, 5, or 10 dm³ of a volcanic

TABLE I
Tree species examined in this study.

Scientific name	Common name	Age[1] (years)	Period of experiment[2]	Site
Conifers				
Pinus densiflora Sieb. et Zucc.	Japanese red pine	1-2[3]	(I), (II)	AB, AK
Pinus thunbergii Parl.	Japanese black pine	2	(I)	AB, AK
Larix kaempferi (Lamb.) Carr.	Japanese larch	2	(II)	AB, AK
Picea abies (L.) Karst.	Norway spruce	6	(I)	AB
Abies firma Sieb. et Zucc.	Japanese fir	6	(I)	AB
Abies homolepis Sieb. et Zucc.	Nikko fir	5	(I)	AB, AK
Abies veitchii Lind.	Veitch's silver fir	5	(II)	AB, AK
Chamaecyparis obtusa Endl.	Japanese cypress	2	(I)	AB, AK
Cryptomaria japonica D.Don	Japanese cedar	1-2[3]	(I), (II)	AB, AK
Deciduous broad-leaves				
Populus maximowiczii Henry	Poplar	4	(II)	AK
Betula platyphylla Sukatchev var. japonica	Japanese white birch	1	(I)[4], (II)	AB, AK[4]
Quercus mongolica Fisch. var. grosseserrata	Oak	4	(I)	AK
Fagus crenata Blume	Siebold's beech	1	(II)	AB, AK
Zelkova serrata Makino	Japanese zelkova	1	(II)	AB, AK

Seedlings examined for all the tree species with the exception of poplar (cutting). [1] at the start of experiment.
[2] (I): Jun. 1993 to Nov. 1995, (II): May 1994 to Nov. 1995. [3] 2-yr and 1-yr old seedlings were examined for period (I) and (II) respectively. [4] examined only at Abiko until Oct. 1994 in the experiment of period (I).

ash soil (andosol). 60 and 40 trees per species were transferred into each chamber in May 1993 and in April 1994 respectively. Each spring and summer all pots were fertilized at a rate of 80-80-80 kg ha^{-1} of N-P-K. Trees were supplied with tap water by drip irrigation to the soil for May-November, and with deionized water *via* mist generating systems for December-May.

Exposure was conducted using tunnel-type open-top chambers (OTC) (3 m wide x 15 m long x 2.8 m high), originally designed by Kobayashi *et al.* (1994). Four OTCs were employed at each experimental sites. OTCs were equipped with two fans on both sides, and covered by an outer framework to exclude ambient precipitation. Turbulent air was supplied to each OTC, sufficient to achieve one and half air changes per minute. During the experiment, increases in annual average of mean air temperature inside OTC as compared with outside was < 1 °C at both sites. Global radiation inside the chambers was 50-60 % of full sunlight. Detailed description of the OTC is provided in Matsumura and Kohno (1997).

Trees were exposed to all combinations of two levels of ambient O_3 and two pHs of simulated acid mist at both experimental sites. At both sites, two OTCs were supplied with charcoal-filtered air (CF) all day for every day, while the other two with non-filtered air (NF). The charcoal-filter consisted of activated-charcoal surrounded by dust filters. Each OTC with non-filtered air was equipped with dust filters. Ambient O_3 concentrations averaged for 12-hr mean between April and September during the period of experiment at Akagi was 40% higher than that at Abiko (Table II). The NF- and CF-chambers at both sites had 90% and 30% of ambient O_3 in daytime 12-hr respectively. The highest observations of daily 1-hr maximum concentration were higher at Abiko than at Akagi, but the seasonal means of that were higher at Akagi. Annual mean NO_2 concentration in AA at Abiko (15 ppb) was higher than that at Akagi (3 ppb). NO_2 could be scarcely removed by charcoal-filters. Mean SO_2 concentrations in the OTCs and AA at both sites were very low (< 2 ppb) over the whole period of experiment.

Simulated acid mist contained anions of sulfate, nitrate and chloride in an equivalent ratio of 1:2:1. It was volumetrically diluted with deionized water to prepare simulated mist designated at pH of 3.0 or 5.0 for the exposure. This anion composition was similar to that of the chemistries of highly acid fog in

TABLE II

24-hr seasonal mean, 12-hr (0600-1800) seasonal mean and daily 1-hr max. concentrations of O_3, and AOT40 in open-top chambers with charcoal-filtered air (CF) and non-filtered air (NF) and in ambient air (AA) at two sites (Abiko and Akagi) between April and September during the period of experiment (June 1993 to November 1995). Values of AOT40 were calculated for the hours of each day when global radiation exceeded 50 W m^{-2}. Values in CF and NF represent the means of two chambers per treatment.

Period		Abiko Concentration (ppb)					Akagi Concentration (ppb)				
		Seasonal mean			Highest		Seasonal mean			Highest	
		24-hr	12-hr	1-hr max.	1-hr max.	AOT40 (ppb•hr)	24-hr	12-hr	1-hr max.	1-hr max.	AOT40 (ppb•hr)
Jun.-Sep. 1993	CF	5	7	16	63	70	9	10	19	56	0
	NF	12	17	33	109	1,333	25	28	46	103	4,707
	AA	16	22	38	117	2,122	27	31	51	111	6,237
Apr.-Sep. 1994	CF	6	8	17	68	57	12	14	22	53	115
	NF	18	27	50	179	9,661	36	41	62	119	20,203
	AA	21	31	56	190	14,056	42	49	73	139	31,784
Apr.-Sep. 1995	CF	7	9	17	44	12	10	11	18	47	12
	NF	22	30	52	133	10,742	36	38	58	124	14,335
	AA	25	35	59	156	15,348	37	40	60	125	16,883
Total	CF	6	8	17	68	139	11	12	20	56	126
	NF	18	26	46	179	21,736	33	37	56	124	39,245
	AA	21	30	52	190	31,526	36	41	63	139	54,904

Mt. Akagi (Murano, 1993). pH 3 was selected because it closely mimicked the highly acidic fog observed in the mountainous areas in Japan (Murano, 1993). pH 5 of mist treatment was the control. Mist treatment was conducted using mist generating system equipped with the chamber. Acid mist was delivered to the trees through fogger nozzles (SFL-408, Sun Hope, Japan), producing droplets between 30 and 50 µm diameter. There are sixty-five nozzles per chamber at a height of 1.8 m. Deposition rate of mist averaged 0.25 mm min.$^{-1}$ (± 30% C.V.) within each chamber. Trees were exposed to acid mist three days per week from 20:00 for June-November every year at both experimental sites. The system was operated in one day three to four times in a cycle of three-minute on and fifty-seven off. Trees received an annual average of 230 mm of the acid mist.

Fourteen, fourteen and twenty-four trees per chamber, per tree species were harvested in the end of season in 1993, 1994 and 1995 respectively. Trees were separated into shoot (with foliage) and root and then oven-dried before determining the dry weights of component plant-parts. Statistical significance was tested by SPSS statistical package using analysis of variance at $P < 0.05$.

3. Results and discussion

Many studies on the effects of ambient O_3 on the juvenile trees of native forest species to North America or Europe by multiple-season exposure using open-top chambers under field conditions have demonstrated decreased biomass in sensitive tree species (e.g., Braun and Flückiger, 1995). In this study, at pH 5 mist treatment, significant decreases in biomass of shoot or total plant in NF O_3 treatments were observed in seven tree species (*Pinus densiflora, Larix kaempferi, Abies homolepis, Abies veitchii, Betula platyphylla, Fagus crenata* and *Zelkova serrata*) as compared with CF O_3 treatment during the period of experiment at both experimental sites (Table III). The decreases in total biomass in the NF-trees were observed in *Populus maximowiczii* as compared with the CF-trees at the site of Akagi. In general, the effects of O_3 on trees became more apparent on root biomass than on shoot biomass as a result of retention of carbon resources in the foliage and decreased carbohydrate allocation to roots (Grulke et al., 1998). This study also demonstrated that root biomass of three conifers (*Picea abies, Abies firma* and *Cryptomeria japonica*) were decreased in the NF treatments versus the CF treatments, though decreases in shoot or total biomass could not be seen (Table III). No significant decreases in the growth by the three seasons of O_3 treatment became apparent on three species (*Pinus thunbergii, Chamaecyparis obtusa* and *Quercus mongolica*). No visible foliar injury by O_3 treatment could not be found on all tree species examined. These results indicated that current ambient levels of O_3 in Japan are high enough to cause biomass loss of the tree species examined.

The seasonal mean O_3 concentrations were higher at Akagi than at Abiko (Table II). However, the percent reductions in biomass of *P. densiflora* and *Larix* were greater at Abiko than at Akagi among ten tree species examined at both sites (Table III), possibly as a result of effects of O_3 in combination with higher NO_2 concentration and/or higher air temperature at the urban site (Abiko).

TABLE III

Changes in biomass for 14 tree species after one, two or three seasons of exposure to O_3. Changes are expressed as percent difference of the mean value in non-filtered air treatment at pH 5 mist from that in charcoal-filtered air treatment (CF) at the same pH mist. Significance from CF denoted: * $P < 0.05$, ** $P < 0.01$, *** $P < 0.001$. n.e.: not examined at that site.

Tree species	No. of seasons	Year of harvest	Abiko Shoot	Root	Total	Akagi Shoot	Root	Total
Pinus densiflora	1	1993	-7	+14	0	-8	+5	-5
	2	1994	-6	-31**	-13	-4	-19	-7
	3	1995	-22***	-31***	-24***	-11	-22***	-12*
	1	1994	+14	-2	+9	+23	+15	+15
	2	1995	+8	-10*	+3	-16*	-9*	-15*
Pinus thunbergii	1	1993	+4	+21	+5	-11	+2	-9
	2	1994	0	-16	-3	-15	-12	-14
	3	1995	+9	-1	+7	-9	+3	-7
Larix kaempferi	1	1994	-8	-8	-8	-2	+3	0
	2	1995	-24*	-26*	-24*	-5	-21**	-8*
Picea abies	1	1993	-13	-6	-11	n.e.	n.e.	n.e.
	2	1994	-10	0	-8	n.e.	n.e.	n.e.
	3	1995	-10	-14*	-11	n.e.	n.e.	n.e.
Abies firma	1	1993	-15	-8	-13	n.e.	n.e.	n.e.
	2	1994	+2	-19*	-4	n.e.	n.e.	n.e.
	3	1995	-4	-17*	-7	n.e.	n.e.	n.e.
Abies homolepis	1	1993	-11	+9	-4	-8	+1	-5
	2	1994	-12	-18*	-14*	-18*	-25**	-20**
	3	1995	-1	-13*	-4	-8	-18**	-10*
Abies veitchii	1	1994	-12	-11	-12	+17	+10	+15
	2	1995	-9	-21**	-13*	-15**	-15*	-15**
Chamaecyparis obtusa	1	1993	+12	+12	+12	-2	+7	0
	2	1994	+4	0	+3	+11	+27	+14
	3	1995	+3	+2	+2	+13*	-3	+10*
Cryptomeria japonica	1	1993	+1	+2	+1	+7	+19	+9
	2	1994	-5	-16*	-8	-13	-29**	-16
	3	1995	+8*	-6*	+4	+4	-7*	+3
	1	1994	+15	+3	+12	+5	-11	+2
	2	1995	+1	-10*	-2	-9	-17**	-10*
Populus maximowiczii	1	1994	n.e.	n.e.	n.e.	-2	+1	-1
	2	1995	n.e.	n.e.	n.e.	-7	-19***	-10
Betula platyphylla	1	1993	+5	+7	+6	n.e.	n.e.	n.e.
	2	1994	-11	-13*	-12*	n.e.	n.e.	n.e.
	1	1994	+4	+4	+4	-6	-41**	-16
	2	1995	-13**	-16*	-14**	-21*	-22*	-21*
Quercus mongolica	1	1993	n.e.	n.e.	n.e.	+1	+5	+3
	2	1994	n.e.	n.e.	n.e.	+20	+3	+11
	3	1995	n.e.	n.e.	n.e.	-6	-1	-4
Fagus crenata	1	1994	-17	+21	-6	+8	-5	+2
	2	1995	-6	-15**	-10*	-8	-19*	-12*
Zelkova serrata	1	1994	-18**	-1	-12*	-23*	-8	-19*
	2	1995	-6*	-8	-6*	-15*	-21***	-17**

In the CF O_3, the effects of the mist treatment became more apparent on shoot or total biomass than on root biomass in all tree species (Table IV). Shoot or total biomass of all tree species were significantly increased in pH 3 mist as compared with pH 5 mist at either Abiko or Akagi, except three species (*Picea abies*, *A.homolepis* and *Quercus mongolica*). These stimulations by pH 3 mist were attributed to the nitrogen fertilization from acid mist (Temple, 1988). As a result, no significant decreases in the biomass and no visible foliar injuries due to acid mist could be seen on all tree species. It was reported that not only did the fogwater samples collected in some mountainous areas in the Kanto district average pH 3.5 or higher for the seasonal mean, but the fogwater with pH at 3 or below was a few % of the samples (Murano, 1993; Takahashi and Fujita, 1992). Therefore, the current acidity of fog observed in Japan could not be high enough to cause direct adverse effect on the tree growth.

TABLE IV

Changes in biomass for 14 tree species after one, two or three seasons of exposure to acid mist. Changes are expressed as percent difference of the mean value at pH 3 mist treatment in CF O_3 from that at pH 5 mist treatment in CF O_3. Significance from pH 5 denoted: * $P < 0.05$, ** $P < 0.01$, *** $P < 0.001$. n.e.: not examined at that site.

Tree species	No. of seasons	Year of harvest	Abiko			Akagi		
			Shoot	Root	Total	Shoot	Root	Total
Pinus densiflora	1	1993	-3	+5	-1	-2	-7	-3
	2	1994	+7	-10	+2	0	-2	0
	3	1995	+1	+6	+3	-4	-8	-5
	1	1994	+21	0	+15	+11	+1	+8
	2	1995	+22*	+2	+17*	+11	+15	+11
Pinus thunbergii	1	1993	+19*	+22	+20*	-8	-18	-10
	2	1994	+20	+1	+17	+5	+5	+5
	3	1995	+16	+14	+16	+1	+8	+3
Larix kaempferi	1	1994	+9	+21	-5	-1	+10	+2
	2	1995	+96***	+91***	+95***	+2	+3	+2
Picea abies	1	1993	-7	-13	-9	n.e.	n.e.	n.e.
	2	1994	-17	+6	-11	n.e.	n.e.	n.e.
	3	1995	+1	-3	0	n.e.	n.e.	n.e.
Abies firma	1	1993	-4	+5	-1	n.e.	n.e.	n.e.
	2	1994	0	-6	-2	n.e.	n.e.	n.e.
	3	1995	+13*	+24**	+16**	n.e.	n.e.	n.e.
Abies homolepis	1	1993	-10	+6	-5	-8	+5	-4
	2	1994	-13	-5	-10	-3	+8	0
	3	1995	+5	+3	+4	-10	-7	-9
Abies veitchii	1	1994	-11	+8	-6	+16	+23	+18**
	2	1995	+5	-11	+1	+20***	+7	+17**
Chamaecyparis obtusa	1	1993	+5	+14	+8	-8	+9	-5
	2	1994	+2	-10	-1	+19	+3	+16
	3	1995	+18***	+1	+14**	+19**	+4	+16**
Cryptomeria japonica	1	1993	+48***	+18	+39***	+2	+20*	+6
	2	1994	+16*	+22*	+18*	-1	-10	-2
	3	1995	+29***	+4	+23***	+13**	+14**	+13**
	1	1994	+14	-12	+7	+7	+4	+7
	2	1995	+31***	+11	+26***	-4	-15	-6
Populus maximowiczii	1	1994	n.e.	n.e.	n.e.	+18*	+1	+12
	2	1995	n.e.	n.e.	n.e.	+15*	-2	+11*
Betula platyphylla	1	1993	+15	+6	+11	n.e.	n.e.	n.e.
	2	1994	+21	+34**	+26*	n.e.	n.e.	n.e.
	1	1994	+9	+17	+11	-14	0	-10
	2	1995	+9	+14	+11*	+20*	+11	+18*
Quercus mongolica	1	1993	n.e.	n.e.	n.e.	+19	-5	+7
	2	1994	n.e.	n.e.	n.e.	+24	+3	+13
	3	1995	n.e.	n.e.	n.e.	+1	+1	+1
Fagus crenata	1	1994	-24	+20	-6	+21	+20	-20
	2	1995	+12*	-4	+5	+32***	+13	+25**
Zelkova serrata	1	1994	+50**	+24**	+40**	+26*	+32**	+28*
	2	1995	+32***	+31***	+32***	+28***	+9	+23***

Experimental evidence for significant interactive effects of O_3 and acid mist or fog on trees has been contradictory (Temple, 1988). In this study, significant interactions between O_3 and acid mist treatment could hardly be found on the biomass of all tree species with the exception of two tree species. Synergistic interactions were found on the biomass of *A.veitchii* and *Fagus* examined at Akagi for two seasons. The relative reductions in shoot or total biomass of the NF-trees to the CF-trees at the pH 3 mist treatment were greater than those at the pH 5 mist treatment (Fig. 1). Matsumura *et al.* (1998) reported that the O_3 effects on growth exacerbated by acid precipitation in the form of simulated acid rain with pH 3 may be due to the input of nitrate from the rain in the experiments investigating the combined effect of O_3 and simulated acid rain on the growth of four Japanese tree species. These results suggested that the O_3 effect on tree growth can be exacerbated by the deposition of acid mist, possibly associated with nitrate.

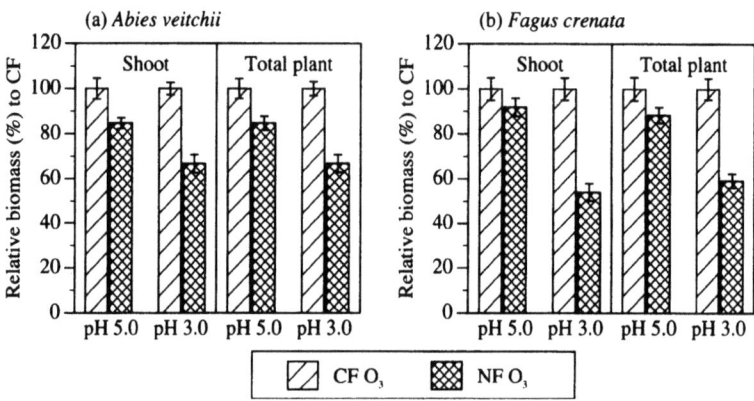

Figure 1. Responses of shoot and total-plant biomass of (a) *Abies veitchii* and (b) *Fagus crenata* harvested in November 1995 after two seasons of exposure to the combinations of O_3 (charcoal-filtered air [CF] or non-filtered air [NF]) and acid mist (pH 5 or pH 3) at Akagi. Growth responses were expressed as the relative value of the mean of 24 trees (± 1 SE) in the NF O_3 as compared with the CF O_3 at each mist pH.

References

Braun, S., Flückiger, W.: 1995, *New Phytologist* **1 2 9**, 33.
Grulke, N.E., Andersen, C.P., Fenn, M.E., Miller, P.R.: 1998, *Environmental Management* **1 0 3**, 63.
Hatakeyama, S., Murano, K.: 1996, *Journal of Japan Society for Atmospheric Environment* **3 1**, 106.
Izuta, T.: 1998, *Journal of Plant Research* **111**, 471.
Kobayashi, K., Okada, M., Nouchi, I.: 1994, *New Phytologist* **1 2 6**, 317.
Matsumura, H., Kohno, Y.: 1997, 'Effects of ozone and/or sulfur dioxide on tree species', in Y. Kohno (ed.), *Proceedings of CRIEPI International Seminar on Transport and Effects of Acidic Substances*, CRIEPI, Tokyo, pp. 190-205.
Matsumura, H., Kobayashi, K., Kohno, Y.: 1998, *Journal of Japan Society for Atmospheric Environment* **3 3**, 16.
Murano, K.: 1993, *Journal of Japan Society of Air Pollution* **2 8**, 185.
Nashimoto, M.: 1993, *Decline of Japanese cedars* (Cryptomeria japonica D.Don) *and secondary air pollutants.* CRIEPI Research report, U93017, CRIEPI, Tokyo.
Takahashi, A., Fujita, S.: 1992, *Contribution of nitrogen oxides to the acidification of rain - uptake and reaction of nitric acid/nitrate in the fog droplet -.* CRIEPI Research report, T91082, CRIEPI, Tokyo.
Temple, P.J.: 1988, *Environmetal and Experimental Botany* **2 8**, 323.

EFFECTS OF OZONE AND/OR SOIL WATER STRESS ON GROWTH AND PHOTOSYNTHESIS OF *FAGUS CRENATA* SEEDLINGS

TETSUSHI YONEKURA[1], YUKIKO DOKIYA[1], MOTOHIRO FUKAMI[2] and TAKESHI IZUTA[1]*

[1] *United Graduate School of Agricultural Science,*
Tokyo University of Agriculture and Technology, Fuchu, Tokyo 183-8509, Japan.
[2] *Faculty of Agriculture, Utsunomiya University, Utsunomiya, Tochigi 321-8505, Japan.*
(* *author for correspondence, e-mail: izuta@cc.tuat.ac.jp*)

Abstract. The effects of ozone (O_3) and soil water stress, singly and in combination, on the growth and photosynthesis of *Fagus crenata* seedlings were investigated. Four-year-old seedlings were exposed to charcoal-filtered air (< 5 nmol mol^{-1} O_3) or 60 nmol mol^{-1} O_3, 7 hours per day (11:00-18:00), for 156 days from 10 May to 11 October 1999 in naturally-lit growth chambers at 20/15 °C (6:00-18:00/18:00-6:00). During the same period, half of the seedlings in each gas treatment received 250 mL of water at the 3-day intervals (well-watered treatment), while the rest received 175 mL of water at the 3-day intervals (water-stressed treatment). The exposure of the seedlings to O_3 caused reductions in the leaf, stem, root and whole-plant dry weights. The net photosynthetic rate at 350 μmol mol^{-1} CO_2, the maximum net photosynthetic rate at saturated CO_2-concentration, carboxylation efficiency of photosynthesis and Rubisco content were significantly reduced by the exposure to O_3. The soil water stress induced reductions in the stem, bud and whole-plant dry weights, transpiration rate and leaf water potential during the midday. The additive effects of O_3 and soil water stress were observed on the dry matter production, leaf gas exchange rates and leaf water potential. As a result, the whole-plant dry weight of the seedlings exposed to both stresses was markedly reduced compared with that of the seedlings exposed to charcoal-filtered air and grown in the well-watered treatment.

Keywords: ozone, soil water stress, *Fagus crenata* Blume, dry weight growth, photosynthesis

1. Introduction

Tropospheric ozone (O_3) is recognized as a widespread phytotoxic air pollutant and atmospheric concentrations of this gas have been increasing year by year not only in Europe and North America, but also in Asia (Akimoto *et al.*, 1994; Stokwell *et al.*, 1997). In Europe and North America, O_3 is considered to be closely related to forest decline (Chappelka and Samuelson, 1998; Skärby *et al.*, 1998). In Japan, although forest decline and tree dieback can be observed in many mountainous areas, the causes and mechanisms of these phenomena are still not clarified. However, relatively high concentrations of O_3 above 100 nmol mol^{-1} are recently recorded especially in summer at several mountainous areas (Hatakeyama, 1999). Furthermore, relatively high concentrations of O_3 and drought stress usually co-occur as major phytotoxic factors in forest areas especially in the mid-summer. However, the combined effects of both stresses on growth and physiological functions of Japanese forest tree species are not clarified at the present time.

In the present study, we investigated the effects of O_3 and soil water stress, singly and in combination, on dry weight growth and photosynthesis of *Fagus crenata* seedlings. *Fagus crenata* was selected because this tree is one of the most representative broad-leaved deciduous tree species native to Japan, but showing severe decline at several mountainous areas (Maruta *et al.*, 1999).

2. Materials and Methods

On 28 April 1999, four-year-old seedlings of *Fagus crenata* Blume were transplanted into 5.3-L pots filled with brown forest soil. Seedlings were grown in four naturally-lit growth chambers (Koito Co. Ltd., Japan) from 28 April to 12 October 1999. In the growth chambers, air temperature, relative air humidity and atmospheric CO_2 concentration were maintained at $20.0\pm1.0/15.0\pm1.0$ °C (6:00-18:00/18:00-6:00), $70\pm5\%$ and 350 ± 10 µmol mol^{-1}, respectively. All the seedlings were fertilized with a liquid fertilizer (N:P:K = 5:10:5, Hyponex, USA) at the 3-week intervals.

The experiment was a factorial, split-plot in randomized blocks with O_3 concentration as the whole-plot treatment. The whole-plot treatment comprised two levels of O_3 replicated two times for a total of four chambers. The sub-plot treatment consisted of two levels of water supply to soil in each chamber. The 15 seedlings were randomly assigned to each O_3-water-chamber combination for a total of 120 seedlings. These treatments were conducted for 156 days from 10 May to 12 October 1999.

Charcoal-filtered (CF) air with O_3 at < 5 nmol mol^{-1} was daily introduced into the two growth chambers, whereas the air with O_3 at 60 ± 5 nmol mol^{-1} was daily dispensed into the other two growth chambers for 7 hours a day from 11:00 to 18:00. The CF air was daily introduced into the two chambers for O_3-exposure from 18:00 to 11:00. The O_3 concentration and exposure duration were determined based on the data obtained at Tanzawa Mountains in Kanagawa Prefecture where dieback of mature *Fagus crenata* is observed and relatively high concentrations of O_3 are detected from May to October with an average O_3 concentration of approximately 60 nmol mol^{-1} in the afternoon (Kanagawa Prefecture, unpublished data). The O_3 was generated from dried ambient air with an electrical discharge O_3 generator (MO-5A, Nippon Ozone Co., Japan), and then injected into the two growth chambers through a water trap to remove nitrogen by-products produced by the O_3 generator (Brown and Roberts, 1988). The concentrations of O_3 at the plant canopy height in the four growth chambers were continuously monitored at the 6-minute intervals with a UV absorption O_3 analyzer (Model 1100, Dylec Inc., Japan).

Half of the seedlings within each growth chamber were grown in the well-watered soil (WW). In the WW treatment, the seedlings received 250 mL of water per pot at the 3-day intervals during the growth period of 156 days, which corresponds to 1193 mm of water supply to the potted soil surface. The amount of irrigated water was determined based on the annual mean precipitation at many forest areas of Japanese deciduous broad-leaved tree species such as *Fagus crenata* (National Astronomical Observatory, 1998; Murai *et al.*, 1991). The remaining seedlings were grown in the water-stressed soil (WS). In the WS treatment, the seedlings received 175 mL of water per pot at the 3-day intervals during the same growth period. The average values of soil pF during the growth period were 1.80 ± 0.06 and 2.28 ± 0.19 in the WW and WS treatments, respectively.

On 12 October 1999, all the seedlings were harvested to determine the dry weights of plant organs. On 14 August 1999, leaf water potential was determined at 15:00 (Ψmid, midday) with a psychrometer (SC-10-A, Decagon Devices Inc., USA). The content of ribulose-1,5-bisphosphate carboxylase/oxygenase (Rubisco) in the leaves were measured on 14 August 1999. The leaf tissue (100 mg) was frozen in liquid nitrogen, homogenized with 1.5 mL extraction buffer containing 100 mM HEPES (pH 7.5),

5 mM EDTA, 1 mM PMSF, 2% PVPP (w/v), 0.7 % polyethylene glycol 20000 (w/v), 1% Tween-80 and 24 mM 2-mercaptoethanol, and then Rubisco content was measured according to the methods of Dann and Pell (1989).

On 26 August 1999, net photosynthetic rate (A), stomatal diffusive conductance to water vapor (Gs) and transpiration rate (E) were determined with a portable gas exchange system (LCA-4, ADC Co., U.K.). During the measurements of A, Gs and E, atmospheric CO_2 concentration, air temperature and relative air humidity in the leaf cuvette were maintained at 350±5 μmol mol^{-1}, 20.0±0.5 °C and 60±5%, respectively. Photosynthetic photon flux density (PPFD) was supplied from a cold lighting system (PICL-NEX twin, Nippon P. I. Co., Japan), and maintained at approximately 1200 μmol m^{-2} s^{-1} at the leaf surface. To obtain the intercellular CO_2 concentration (Ci)-response curve of net photosynthetic rate (A), the A was determined at atmospheric CO_2 concentrations in the leaf cuvette of 5, 50, 100, 200, 350, 700 and 1000 μmol mol^{-1}. The maximum net photosynthetic rate at saturated CO_2-concentration ($Amax$) and carboxylation efficiency (CE) of photosynthesis were regarded as the A at 1000 μmol mol^{-1} CO_2 and the initial slope of the linear portion of the A/Ci curve, respectively. To describe the light-response curve of net photosynthetic rate (A/light curve), the A was determined under the PPFD at the leaf surface of 0, 50, 100, 300, 500, 700, 1000 and 1200 μmol m^{-2} s^{-1}. The quantum yield (QY) of photosynthesis was regarded as the initial slope of the linear portion of the A/light curve.

The statistical analyses were performed with SPSS statistical package. In analysis of variance, O_3 effects (1 d.f.) were tested against whole-plot chamber variation (2 d.f.), whereas the main effects of water stress (1 d.f.) and O_3×water stress interactions (1 d.f.) were tested against the water stress×chamber/O_3 error term (2 d.f.) (Tjoelker and Luxmoore, 1991). An alternative analysis of water stress and O_3×water stress effects using a pooled error term (water stress×chamber/O_3+ experimental [subsampling] error term) did not change any inferences.

3. Results

Figure 1 shows the dry weights of *Fagus crenata* seedlings on 12 October 1999.

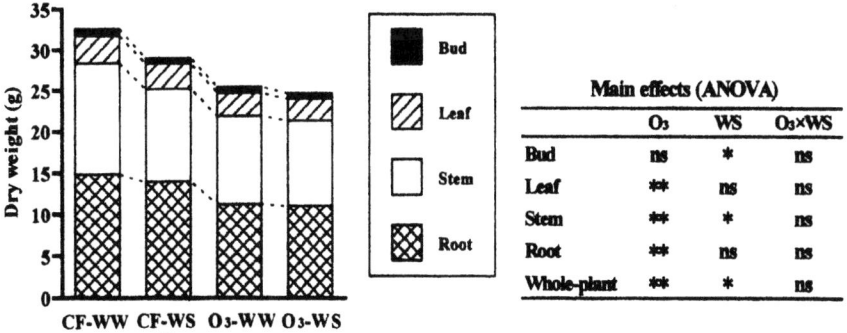

Figure 1. Effects of O_3 and/or soil water stress on dry weights of *Fagus crenata* on 12 October 1999. The seedlings were exposed to charcoal filtered air (CF) or 60 nmol mol^{-1} O_3 (O_3), and grown in well-watered (WW) or water-stressed (WS) soil. Each value is the mean of 14 determinations. ANOVA: * p<0.05, ** p<0.01, ns Not significant.

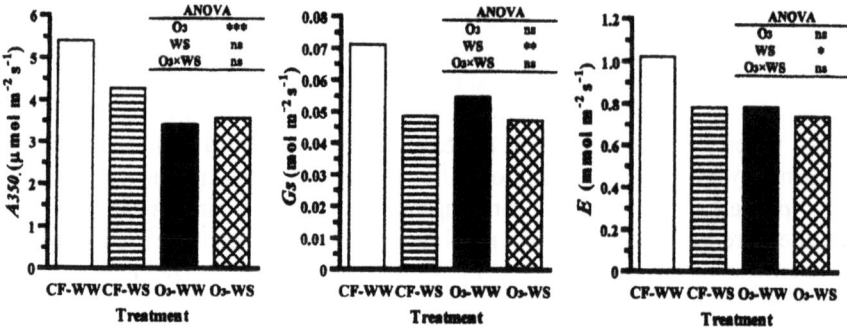

Figure 2. Effects of O_3 and/or soil water stress on net photosynthetic rate at 350 μmol mol^{-1} CO_2 (*A350*), stomatal diffusive conductance to water vapor (*Gs*) and transpiration rate (*E*) of *Fagus crenata* seedlings. Each value is the mean of 8 determinations. ANOVA: * $p<0.05$, ** $p<0.01$, *** $p<0.001$, ns Not significant.

The O_3 caused significant reductions in the leaf, stem, root and whole-plant dry weights. The soil water stress induced significant reductions in the bud, stem and whole-plant dry weights. However, there were no significant interactive effects of both stresses on the dry weights of the seedlings.

The photosynthetic and stomatal parameters at 350 μmol mol^{-1} CO_2 on 26 August 1999 are indicated in Figure 2. The net photosynthetic rate at 350 μmol mol^{-1} CO_2 (*A350*) was significantly reduced by exposure to O_3. However, O_3 did not caused significant reductions in the stomatal diffusive conductance to water vapor (*Gs*) and transpiration rate (*E*). The soil water stress induced significant reductions in the *Gs* and *E*. There were no significant interactive effects of both stresses on the *A350*, *Gs* and *E*.

As shown in Figure 3, O_3 caused significant reductions in the maximum net photosynthetic rate at saturated CO_2-concentration (*Amax*) and the carboxylation efficiency (*CE*), but not in the quantum yield (*QY*) on 26 August 1999. No significant effects of soil water stress and interactive effects of both stresses were found on the *Amax*, *CE* and *QY*.

Figure 4 illustrates Rubisco content and leaf water potential at midday (Ψmid) on 14 August 1999. The Rubisco content was significantly reduced by O_3, but was not

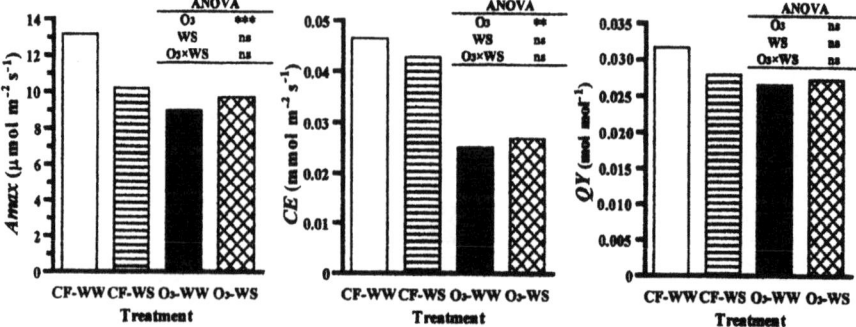

Figure 3. Effects of O_3 and/or soil water stress on maximum net photosynthetic rate at saturated CO_2-concentration (*Amax*), carboxylation efficiency (*CE*) and quantum yield (*QY*) of *Fagus crenata* seedlings. Each value is the mean of 8 determinations. ANOVA: ** $p<0.01$, *** $p<0.001$, ns Not significant.

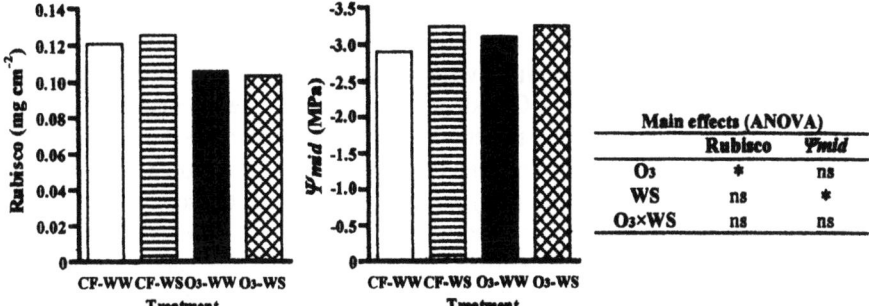

Figure 4. Effects of O_3 and/or soil water stress on Rubisco content and leaf water potential at 15:00 (Ψmid, midday) of *Fagus crenata* seedlings. Each value is the mean of 6 to 8 determinations. ANOVA: * $p<0.05$, ns Not significant.

significantly affected by soil water stress. Furthermore, there was no significant interactive effect of both stresses on the Rubisco content. The soil water stress caused a significant reduction in the Ψmid. However, there were no significant effect of O_3 and interactive effect of both stresses on the Ψmid.

4. Discussion

In the present study, the combined effects of O_3 and soil water stress were additive on photosynthetic parameters in August and dry weights in October of *Fagus crenata* seedlings (Figures 1, 2 and 3). Pearson and Mansfield (1993) and Karlsson *et al.* (1995) reported that drought stress caused a reduction in the stomatal diffusive conductance, and could reduce negative effects of O_3 on the growth of *Fagus sylvatica* and *Picea abies* seedlings. In their experiments, the seedlings were exposed to O_3 with water deficiency. Therefore, reduced negative effects of O_3 on the seedlings might be mainly due to the drought stress-induced stomatal closure, which reduces uptake of atmospheric O_3 into the leaves. In our experiment, negative effects of O_3 on photosynthetic parameters were observed from June, and then negative effects of soil water stress on photosynthesis appeared from August (data not shown). These results suggest that *Fagus crenata* seedlings were firstly suffered from negative effects of O_3 in May and June, and then were negatively affected by O_3 and water stress since mid-summer. As a result, the whole-plant dry weight of the seedlings exposed to both stresses was markedly less than that of the seedlings exposed to charcoal-filtered air and grown in the well-watered treatment on 12 October (Figure 1). Therefore, it is considered that additive effects of O_3 and soil water stress on trees are induced in the case of relatively longer period of exposure to O_3 with water stress, but stomatal closure protects trees against the detrimental effects of O_3 in the case of water deficiency.

As shown in Figure 1, the exposure to O_3 caused great reduction in the root dry weight of *Fagus crenata* seedlings. Similar results have been well documented in European and North American forest tree species, and the O_3-induced reduction in the root dry weight is considered to be mainly due to the inhibition of assimilate partitioning from the shoot to roots (Anderson et al., 1997). Net photosynthetic rate (A_{350}) of *Fagus crenata* seedlings was reduced, whereas stomatal diffusive conductance to water vapor (Gs) and transpiration rate (E) of the seedlings were not significantly affected by the exposure to O_3 (Figure 2), indicating that the O_3-induced reduction in the net photosynthetic rate was not due to the stomatal closure. Furthermore, the exposure of the seedlings to O_3 caused the reductions in the maximum net photosynthetic rate at saturated CO_2-concentration ($Amax$), carboxylation efficiency (CE) of photosynthesis and Rubisco content (Figure 3 and 4), suggesting that the O_3-induced reduction in the net photosynthetic rate of the seedlings was mainly due to the reduction in the regeneration rate of ribulose-1,5-bisphosphate (RuBP) and the quantity and/or activity of Rubisco (Sharkey et al., 1985; von Caemmerer and Farquhar, 1981). When the soil water stress was imposed, *Fagus crenata* seedlings showed reductions in the whole-plant dry weight, leaf water potential at midday (Ψmid), Gs and E (Figures 1, 2 and 4). These results suggest that soil water stress caused stomatal closure induced by the lack of leaf water content, and the growth of water-stressed seedlings was limited primarily by reduction in the uptake of atmospheric CO_2 from stomata (Beyers et al., 1992).

In conclusion, the combined effects of O_3 and soil water stress were additive on the photosynthesis and dry weight growth of *Fagus crenata* seedlings. This result suggests that *Fagus crenata* may be suffered from negative interactive effects of O_3 and drought stress in some mountainous areas of Japan.

References

Akimoto, H., Nakane, H. and Matsumoto, Y.: 1994, `The chemistry of oxidant generation: tropospheric ozone increase in Japan´ in Calvert, J. G. (ed.), *Impact of Global Change*, Blackwell Scientific Publications, Oxford. pp. 261-273.

Anderson, C. P., Wilson, R., Plocher, M. and Hogsett, W. E.: 1997, *Tree Physiology* 17, 805.

Beyers, J. L., Riechers, G. H. and Temple, P. J.: 1992, *New Phytologist* 122, 81

Brown, K. A. and Roberts, T. M.: 1988, *Environmental Pollution* 55, 55.

Chappelka, A. H. and Samuelson, L. J.: 1998, *New Phytologist* 139, 91.

Dann, M. S. and Pell, E. J.: 1989, *Plant Physiology* 91, 427.

Hatakeyama, S.: 1999, *Environmental Science* 12, 227. [In Japanese]

Karlsson, P. E., Medin, E-L., Wickström, H., Selldén, G., Wallin, G., Ottosson, S. and Skärby, L.: 1995, *Water, Air and Soil Pollution* 85, 1325.

Maruta, E., Shima, K., Horie, K., Aoki, M., Dokiya, Y., Izuta, T., Totsuka, T., Yokoi, Y. and Sakata, G.: 1999,*Environmental Science* 12, 241. [In Japanese]

Murai, H., Yamatani, K., Kataoka, Y. and Yui, M.: 1991, *Natural environment and its conservation on Buna (Fagus crenata) forest*, Soft science Inc., Tokyo. pp. 142-144. [In Japanese]

National Astronomical Observatory.: 1998, *RIKA NENPYO (Chronological Scientific Tables 1998)*, Maruzen Co. Ltd., Tokyo. pp. 210-213. [In Japanese]

Pearson, M. and Mansfield, T. A.: 1993, *New Phytologist* 123, 351.

Sharkey, T. D.: 1985, *Botanical Review* 51, 53.

Skärby, L., Ro-Poulsen, H., Wellburn, F. A. M. and Sheppard, L. J.: 1998, *New Phytologist* 139, 109.

Stokwell, W. R., Kramm, G. Schell, H-E., Mohnen, V. A., Seiler, W.: 1997, `Ozone formation, destruction and exposure in Europe and the United States´ in Sandermann, H. Jr., Wellburn, A. R. and Heath, R. L.(eds.), *Forest Decline and Ozone*, Springer, Berlin. pp. 1-38.

Tjoelker, M. G. and Luxmoore, R. J.: 1991, *New Phytologist* 119, 69.

von Caemmerer, S. and Farquhar, G. D.: 1981, *Planta* 153, 376.

EFFECTS OF OZONE AND/OR EXCESS SOIL NITROGEN ON GROWTH, NEEDLE GAS EXCHANGE RATES AND RUBISCO CONTENTS OF *Pinus densiflora* SEEDLINGS

TATSURO NAKAJI [1] and TAKESHI IZUTA [2*]

[1] *United Graduate School of Agricultural Science;* [2] *Faculty of Agriculture, Tokyo University of Agriculture and Technology, Fuchu, Tokyo 183-8509, Japan*

(*author for correspondence, e-mail: izuta@cc.tuat.ac.jp)

Abstract. The effects of ozone (O_3) and excess soil nitrogen (N), singly and in combination, on growth, needle gas exchange rates and ribulose-1,5-bisphosphate carboxylase/oxygenase (Rubisco) contents of *Pinus densiflora* seedlings were investigated. One-year-old seedlings were grown in 1.5-L pots filled with brown forest soil with 3 levels of N supply (0, 100 or 300 mg $N \cdot L^{-1}$ fresh soil volume). The seedlings were exposed to charcoal-filtered air or 60±5 $nL \cdot L^{-1}$ O_3 (8 hours a day) in naturally-lit phytotrons for 173 days from 22 May to 11 November.

The exposure to O_3 or high N supply to the soil caused a significant reduction in the dry weights of the seedlings. Although no significant interactive effects of O_3 and excess soil N were detected on the dry weight growth of the seedlings, the whole-plant dry weight of the O_3-exposed seedlings grown in the soil treated with 300 mg $N \cdot L^{-1}$ was greatly reduced compared with the control value. Ozone reduced net photosynthetic rate at 350 $\mu mol \cdot mol^{-1}$ CO_2 (A_{350}), carboxylation efficiency (*CE*) of photosynthesis and Rubisco content without a significant change in the gaseous phase diffusive conductance to CO_2 (gs) of the needles. The excess soil N reduced the A_{350}, *CE*, gs and Rubisco content of the needles. These results suggest that the reduction in the dry weight growth of *Pinus densiflora* seedlings induced by the exposure to O_3 and/or excess soil N was caused by reduction in the net photosynthetic rate mainly due to the decrease of Rubisco quantity in the chloroplasts.

Keywords: ozone, excess soil nitrogen, *Pinus densiflora* Sieb. et Zucc., growth, photosynthesis

1. Introduction

Ozone (O_3) is considered to be a major phytotoxic air pollutant and has been associated with forest decline observed in Europe and North America (Skärby *et al.*, 1998). In Japan, relatively high concentrations of O_3 above 100 $nL \cdot L^{-1}$ (ppb) are detected in several mountainous areas where forest decline or tree dieback have been observed (Totsuka *et al.*, 1997).

In general, soil nitrogen (N) is one of the most important limiting factors in forest ecosystems when N load to the soil is relatively low. However, elevated N deposition to forest ecosystems is considered to be one of the environmental stresses adversely affecting forest tree species (Nihlgård, 1985; Aber *et al.*, 1989). When N supply to forest ecosystems is much more than its demand, excess soil N may lead to a reduction in the growth and physiological functions of forest tree species as a result of soil acidification and/or imbalance of plant nutrient status (Skeffington and Wilson, 1988; Nilsson *et al.*, 1988). Therefore, there is the possibility that forest tree species will be adversely affected by elevated O_3 and excess soil N, singly and in combination, in the near future.

However, there is no information about the combined effects of O_3 and excess soil N on Japanese forest tree species.

In the present study, we investigated the effects of O_3 and excess soil N, singly and in combination, on dry weight growth, needle gas exchange rates and ribulose-1,5-bisphosphate carboxylase/oxygenase (Rubisco) content of *Pinus densiflora* seedlings. We used *Pinus densiflora* as the plant material because this tree is one of the most representative Japanese coniferous tree species, and forest decline and dieback of this tree are observed mainly in the western parts of Japan.

2. Materials and Methods

Nitrogen was added as NH_4NO_3 to brown forest soil collected from the A-horizon below coniferous tree stands (Kusaki, Gunma Prefecture, Japan) at 0, 100 or 300 mg $N \cdot L^{-1}$ fresh soil volume, which corresponds to 0, 135 or 405 kg $N \cdot ha^{-1}$, respectively. These soil N treatments were designated as N-0, N-100 or N-300, respectively. On 28 April, one-year-old seedlings of *Pinus densiflora* Sieb. et Zucc. were transplanted into 1.5-L pots filled with the soil of these three different N contents. All the seedlings were grown in a naturally-lit greenhouse for 24 days, and then were grown in four naturally-lit phytotrons (Koito Industry Co.) for 173 days from 22 May to 11 November. In the phytotrons, air temperature and relative air humidity were maintained at 25/18 °C (6:00-18:00/18:00-6:00) and 70±5%, respectively. During the experimental period, the potted soil was irrigated daily with deionized water. From 22 May to 11 November, the seedlings were exposed to charcoal-filtered (CF) air (< 5 $nL \cdot L^{-1}$ O_3) or 60±5 $nL \cdot L^{-1}$ O_3, 8 hours a day, from 9:00 to 17:00 in the phytotrons. Two replicated chambers were randomly assigned to each gas treatment. A mixture of charcoal-filtered air and that with O_3, which was generated with a silent electrical discharge O_3 generator (MO-5, Nihon Ozone Co.), was introduced into the phytotrons through a water trap to remove nitrogen by-products produced by the O_3 generator (Brown and Roberts, 1988). The concentrations of O_3 in the phytotrons were continuously monitored with an UV absorption O_3 analyzer (Model 1100, Dylec Inc.). On 11 November, all the seedlings were harvested for measuring the dry weights of plant organs.

On 8 September, measurements of gas exchange rates of current-year needles were made using an infrared gas analyzer system with a 12 cm^3 leaf chamber (LCA-4, ADC Co. Ltd.). During the measurements of needle gas exchange rates, air temperature, relative air humidity and photosynthetic photon flux density in the leaf chamber were maintained at 25.0±0.5 °C, 65±5% and 1500±50 $\mu mol \cdot m^{-2} \cdot s^{-1}$, respectively. Charcoal-filtered air was introduced into the leaf chamber at a rate of 273 $\mu mol \cdot s^{-1}$. Net photosynthetic rate at 350 $\mu mol \cdot mol^{-1}$ CO_2 (A_{350}) and gaseous phase diffusive conductance to CO_2 (gs) were determined on the basis of needle dry weight. The intercellular CO_2 concentration (Ci)-response curves of net photosynthetic rate (A) were generated by measuring the A at 8 different atmospheric CO_2 concentrations. The carboxylation efficiency (CE) of photosynthesis was determined as the initial slope of A/Ci curve.

Immediately after the measurements of needle gas exchange rates, current-year needles were collected to analyze the contents of total soluble protein (TSP), Rubisco and total needle N. The fresh needles (100 mg) were frozen in liquid nitrogen, and then were homogenized in 1 mL extraction buffer containing 100 mM HEPES (pH 7.5), 5 mM EDTA, 2% PVPP (w/v), 0.7% polyethylene glycol 20000 (w/v) and 24 mM 2-mercaptoethanol. Procedures were carried out at 4 °C. The homogenate was centrifuged at 9000 g for 30 seconds, and the supernatant was used in the assays of TSP and Rubisco contents. The content of TSP was determined according to the colorimetric method of Bradford (1976) using the Bio-Rad Protein Assay (Bio-Rad Laboratories Inc.). The supernatant was subjected to SDS-PAGE (Laemmli, 1970) for separation of subunit bands of Rubisco protein. The amount of Rubisco was calculated from the density of subunit bands scanned with a gel image analysis system (Densitograph AE-6920MF, Atto Co.). A calibration curve of protein content was made with bovine serum albumin. Needle N content was determined using a C/N analyzer (MT-500, Yanagimoto Co.). In the present study, we assumed that N content in the Rubisco is 16.67% of Rubisco protein content (Steer *et al*, 1968).

Soil solution was taken from the potted soil using a soil moisture sampler (Eijkelkamp Co.). The pH of soil solution was measured with a pH meter (M-12, Horiba Co.). The concentrations of anions (NO_2^- and NO_3^-) and cations (Ca, Mg, K, Na and Al) in the soil solution were determined with an ion chromatograph (IC200, Yokogawa Co.) and an atomic absorption spectro-photometer (AA-670, Shimadzu Co.), respectively. The concentration of NH_4^+ in the soil solution was determined by colorimetrical methods (Scheiner, 1976).

The statistical analyses were performed with SPSS® statistical package. The split-plot factorial experiment was designed as a random block design. Ozone levels were whole-plots (1 d.f.), and N treatments were sub-plots (2 d.f.). Because chamber to chamber (block) variation proved to be non-significant, the data were analyzed as a two-factor analysis of variance (ANOVA).

3. Results and Discussion

Table I shows the results of soil solution analysis at the transplantation of *Pinus densiflora* seedlings into the potted soil. The concentrations of nitrate and ammonium in the soil solution were increased by increasing the amount of N added as NH_4NO_3 to the soil. In all the treatments, nitrite was not detected in the soil solution. Because Al concentration in the soil solution

TABLE I

Chemical property of soil solution at the transplantation of *Pinus densiflora* seedlings.

Treatment	pH	Element concentration of soil solution (mM)							
		NO_3^-	NO_2^-	NH_4^+	Ca	Mg	K	Na	Al
N-0	4.6	5.89	nd	0.11	1.66	0.45	0.35	0.18	0.02
N-100	4.6	10.53	nd	5.22	1.99	0.53	0.44	0.17	0.03
N-300	4.7	21.59	nd	16.51	1.86	0.53	0.51	0.17	0.02

Each value is the mean of 4 determinations. nd = not detected.

was very low in all the treatments, detrimental effects of Al on the seedlings are considered to be small in this study.

As shown in Table II, the exposure to O_3 or high N supply to the soil caused a significant reduction in the whole-plant dry weight of *Pinus densiflora* seedlings at the end of growth period. Although the whole-plant dry weight of the O_3-exposed seedlings grown in the N-300 treatment was greatly reduced compared with that of the CF air-exposed seedlings grown in the N-0 treatment, there were no significant interactive effects of O_3 and soil N on the dry weights of the seedlings. This result indicates that the combined effects of O_3 and excess soil N was additive on the dry weight growth of the seedlings. Pell *et al.* (1995) reported that exposure to O_3 induced a pronounced growth reduction in *Populus tremuloides* seedlings when the seedlings were grown in the soil with optimal N supply. Conversely, Pääkkönen and Holopainen (1995) reported that relatively high N supply to the soil induced greater resistance to O_3 in *Betula pendula* seedlings. Therefore, combined effects of O_3 and soil N content on dry weight growth are considered to be quite different among the tree species.

The exposure of *Pinus densiflora* seedlings to O_3 caused a significant reduction in the dry weight of fine roots (< 1mm in diameter) and ratio of fine root dry weight to shoot dry weight (Table II). The O_3-induced reduction in the root dry weight was detected in several forest tree species such as *Pinus taeda* and *Pinus ponderosa*, and this reduction is considered to be mainly due to an inhibition of allocation of photosynthate from the shoot to the roots (Spence *et al.*, 1990; Andersen *et al.*, 1997). The dry weight of fine roots of *Pinus densiflora* seedlings was also reduced by increasing the soil N content (Table II). Furthermore, poor mycorrhizal fine roots were observed in the seedlings grown in the N-100 and N-300 treatments. Several researchers have reported that excess soil N induced a reduction in the root dry weight of coniferous seedlings (Seith *et al.*, 1996). One explanation for this reduction is that root growth may be restricted by a shortage of carbohydrate in the roots, which results from accelerated consumption of photosynthate in the shoot to assimilate excess N components (Chapin *et al.*, 1986; Wallenda *et al.*, 1996). This excess soil N-induced inhibition of fine root development is considered to

TABLE II

Effects of O_3 and/or soil N on the dry weights of *Pinus densiflora* seedlings (11 November).

Treatment	Dry weight (g)										Fine root/shoot ratio	
	Needle		Stem		Coarse root[a]		Fine root[a]		Whole-plant			
	CF	O_3	CF	O_3	CF	O_3	CF	O_3	CF	O_3	CF	O_3
N-0	3.64	3.10	1.39	1.34	0.98	1.01	1.63	1.26	7.63	6.72	0.33	0.28
N-100	2.67	2.20	1.19	1.23	1.03	0.86	0.81	0.64	5.70	4.92	0.21	0.19
N-300	2.42	2.12	1.28	1.06	0.89	0.78	0.60	0.47	5.19	4.42	0.16	0.15
ANOVA[b]												
O_3	***		ns		ns		***		***		*	
Soil N	***		***		ns		***		***		***	
O_3 × Soil N	ns		ns		ns		ns		ns		ns	

[a] Coarse root, ≥ 1 mm in diameter ; fine root, < 1 mm in diameter (including mycorrhizal root).

[b] Two-way ANOVA : * p < 0.05, *** p < 0.001, ns = not significant.

Each value is the mean of 8 determinations.

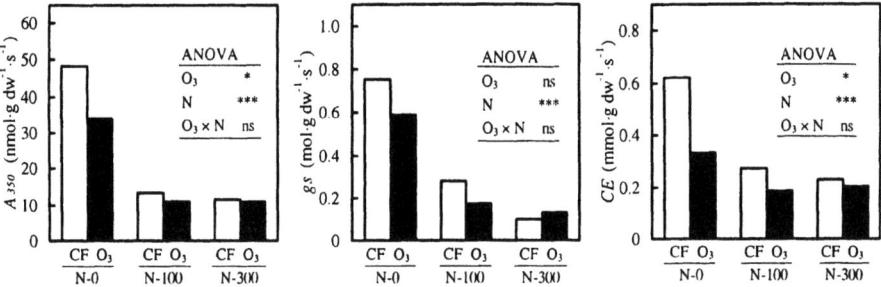

Figure 1. Effects of O_3 and/or soil N on net photosynthetic rate at 350 µmol·mol^{-1} (A_{350}), gaseous phase diffusive conductance to CO_2 (gs) and carboxylation efficiency (CE) of *Pinus densiflora* seedlings. Each value is the mean of 4 determinations.

be very important for tree health because absorption of nutrients and water in many tree species such as *Pinus densiflora* is closely associated with mycorrhizal fine roots.

As shown in Figure 1, although no significant effect of O_3 was detected on the gaseous phase diffusive conductance to CO_2 (gs), O_3 caused a reduction in the net photosynthetic rate at 350 µmol·mol^{-1} CO_2 (A_{350}) of *Pinus densiflora* seedlings. Furthermore, the carboxylation efficiency (CE) of photosynthesis was significantly reduced by the exposure to O_3. These results indicate that the O_3-induced reduction in the net photosynthetic rate was not mainly caused by stomatal closure, but by reduction in the activity and/or quantity of ribulose-1,5-bisphosphate carboxylase/oxygenase (Rubisco), which is a key enzyme involved in the initial capture of CO_2 in the chloroplasts (von Caemmerer *et al.*, 1981). Similar results were obtained in *Picea abies* and *Fagus crenata* seedlings exposed to ambient levels of O_3 (Wallin *et al.*, 1992; Izuta *et al.*, 1996). By comparison, high N loads to the soil caused a reduction in the A_{350}, gs and CE of *Pinus densiflora* seedlings (Figure 1). This result shows that excess soil N-induced reduction in the net photosynthetic rate was due to stomatal closure and reduction in the activity and/or quantity of Rubisco.

The O_3-induced reduction in the Rubisco content has been reported in several

TABLE III

Effects of O_3 and/or soil N on the N content, total soluble protein content, ribulose-1,5-bisphosphate carboxylase/oxygenase (Rubisco) content and Rubisco-N as a percentage of nitrogen content in the current-year needles of *Pinus densiflora* seedlings.

Treatment	Needle N		Total soluble protein				Rubisco			
	(mg·g dw^{-1})		(mg·g dw^{-1})		(g·g needle N^{-1})		(mg·g dw^{-1})		(N as % of needle N)	
	CF	O_3	CF	O_3	CF	O_3	CF	O_3	CF	O_3
N-0	32.5	32.0	22.8	18.4	0.71	0.58	10.8	8.9	5.5	4.5
N-100	32.6	33.8	13.1	13.1	0.40	0.38	7.2	6.5	3.8	3.2
N-300	39.9	39.6	13.0	12.5	0.33	0.31	6.4	6.1	2.9	2.5
ANOVA[a]										
O_3	ns		ns		ns		*		ns	
Soil N	*		***		***		*		***	
O_3 × Soil N	ns		ns		ns		ns		ns	

[a] Two-way ANOVA : * $p < 0.05$, ** $p < 0.01$, *** $p < 0.001$, ns = not significant.
Each value is the mean of 4 determinations.

tree species (Pell et al., 1994). As shown in Table III, there were no significant interactive effects of O_3 and soil N on the contents of needle N, total soluble protein (TSP) and Rubisco of *Pinus densiflora* seedlings. However, the exposure to O_3 reduced Rubisco content in the current-year needles of the seedlings. The needle N content, TSP content and Rubisco-N expressed as a percentage of needle N content were not reduced by the exposure to O_3 (Table III). These results suggest that O_3 reduced Rubisco content without a depression of N use efficiency in the protein synthesis of *Pinus densiflora* seedlings. Although needle N content was increased by increasing the N supply to the soil, excess soil N reduced the contents of TSP and Rubisco in the current-year needles of *Pinus densiflora* seedlings (Table III). The TSP content per needle N content and Rubisco-N expressed as a percentage of needle N content were also significantly reduced by increasing the N supply to the soil (Table III). In the present study, therefore, excess soil N may affect *Pinus densiflora* seedlings as a toxic substance rather than an element essential for plant growth.

In conclusion, the combined effects of O_3 and excess soil N were additive on the dry weight growth and photosynthesis of *Pinus densiflora* seedlings. The reduction in the dry weight growth of the seedlings induced by the exposure to O_3 and/or excess soil N was caused by reduction in the net photosynthetic rate mainly due to the decrease of Rubisco quantity in the chloroplasts.

References

Aber, J.D., Nadelhoffer, K.J., Steudler, P. and Melillo, J.M.: 1989, *Bioscience* **39**, 378.
Andersen, C.P., Wilson, R., Plocher, M. and Hogsett, W.E.: 1997, *Tree Physiol.* **17**, 805.
Arnon, D. I.: 1949, *Plant Physiol.* **24**, 1.
Bradford, M.: 1976, *Anal. Biochem.* **72**, 248.
Brown, K.A. and Roberts, T.M.: 1988, *Environ. Pollut.* **55**, 55.
Chapin III, F.S., Shaver, G.R. and Kedrowski, R.A.: 1986, *J. Ecology* **74**, 167.
Izuta, T., Umemoto, M., Horie, K., Aoki, M. and Totsuka, T.: 1996, *J. Jpn. Soc. Atmos. Environ.* **31**, 95.
Laemmli, U.K.: 1970, *Nature* **227**, 680.
Nihlgård, B.: 1985, *Ambio* **14**, 2.
Nilsson, S.I., Berdén, M. and Popvic, B.: 1988, *Environ. Pollut.* **54**, 233.
Pääkkönen, E. and Holopainen, T.: 1995, *New Phytol.* **129**, 595.
Pell, E.J., Eckardt, N.A. and Glick, R.E.: 1994, *Photosynth Res.* **39**, 453.
Pell, E.J., Sinn, J.P. and Vinten-Johansen, C.: 1995, *New Phytol.* **130**, 437.
Scheiner, D.: 1976, *Water Research.* **10**, 31.
Seith, B., George, E., Marschner, H., Wallenda, T., Schaeffer, C., Einig, W., Wingler, A. and Hampp, R.: 1996, *Plant and Soil* **184**, 291.
Skärby, L., Ro-Poulsen, H., Wellburn, F.A.M. and Sheppard, L.J.: 1998, *New Phytol.* **139**, 109.
Skeffington R.A. and Wilson, E.J.: 1988, *Environ. Pollut.* **54**, 159.
Spence, R.D., Rykiel, E.J. and Sharpe, P.J.H.: 1990, *Environ. Pollut.* **64**, 93.
Steer, M.W., Gunning, B.E.S., Graham, T.A. and Carr, D.J.: 1968, *Planta* **79**, 254.
Totsuka, T., Aoki M., Izuta, T., Horie, K. and Shima, K.: 1997, *Tanzawa Range Environmental Studies*, Kanagawa Prefecture, 93 (in Japanese).
von Caemmerer, S., Farquhar, G.D.: 1981, *Planta* **153**, 376.
Wallenda, T., Schaeffer, C., Einig, W., Wingler, A., Hampp, R., Seith, B., George, E. and Marschner, H.: 1996, *Plant and Soil* **186**, 361.
Wallin, G., Ottosson, S. and Selldén, G.: 1992, *New Phytol.* **121**, 395.

NEUTRALIZATION OF ACID DROPLETS ON PLANT LEAF SURFACES

Y. KOHNO[1*], R. MATSUKI[1], S. NOMURA[2], K. MITSUNARI[2], and M. NAKAO[2]

[1] *Biology Department, Central Research Institute of Electric Power Industry, 1646, Abiko, AbikoCity, Chiba 270-1194, Japan;* [2] *Horiba Ltd., Minami-Ku, Kyoto 601-8510, Japan*
(* author for correspondence: E-mail: kohno@criepi.denken.or.jp)

Abstract. Sulfuric acid mist exposure of bush bean leaves at a low rate of precipitation suggested that acid on the leaf surface was neutralized by cations leached from leaf tissues and that Ca-S compounds were accumulated on the leaf surface (Kohno, 1994). This report summarizes visual observations of the neutralization process of acid on leaf surfaces as determined by a pH-imaging microscope. Small droplets of sulfuric acid were placed on the adaxial leaf surface and allowed to air dry under laboratory conditions. Droplets (0.1µl) of sulfuric acid took about 7-8 minutes to dry. Leaf samples were cut at various times after the acid droplets dried. The adaxial leaf surface was placed on the pH-adjusted agar film layer on the pH-imaging sensor of the microscope. Hydrogen ions dispersed into the film layer and resulting pH distributions were visualized as pH distribution patterns. The size of the acidic area generated became smaller with time after the acid was added and allowed to dry. Results indicate that leaves could neutralize the surface acid probably by ion exchange with cations from their surface tissues and could recover from strong temporary acid stress imposed by acid rain or acid fog in a relatively short period of time. Our findings indicate that acidic precipitation at current acidity levels does not pose a direct threat to plants.

Keywords: acid precipitation, leaching, neutralization, pH-imaging analysis, plant response,

1. Introduction

Evaluation of plant leaf responses to acidic solutions is very important in discussing effects of acidic deposition, when plants are exposed to acidic rain or fog. Precipitation causes leaching of solutes from plant surfaces (Tukey, 1970). Loss of cations such as potassium, magnesium and calcium is especially marked. Their concentration generally increases with increasing acidity of the rain (Kobayashi & Matsumura, 1998). Cryo-scanning electron microscopy has revealed accumulation of Ca-S compounds after sulfuric acid mist exposure at a low rate of precipitation (Kohno, 1994). These results suggest that acids on the plant surface are neutralized. Even if dew or fog may dissolve substances deposited on the leaf surface, surface solutes with high acidity will be neutralized with cations supplied from the plant surface through ion-exchange. In order to confirm these phenomena, we applied a newly developed pH-imaging microscope (scanning chemical microscope) to visualize plant leaf surface response to acid solutions.

2. Materials and Methods

2.1. INSTRUMENT

A newly developed scanning chemical microscope (SCHEM-100, Horiba Ltd., Japan) was used for pH-imaging of residual acids on the plant leaf surface (Nakao *et al.*, 1998). This microscope utilizes a flat pH-imaging sensor, which can function as an array of multiple pH sensing spots. The pH dependent electric signal can be obtained when the backside of the sensor is illuminated by a light source with bias voltage applied between the backside and electrolyte solution placed on the sensor surface. Since only the illuminated area of the sensor can work as the pH sensing area, multiple point pH measurement can be conducted by scanning the illuminated spot. The light scanning mechanics and electric circuit for the sensor function are incorporated into the microscope. The resulting pH dependent electric signal at each measurement point is processed to the pH-image.

An agar film with 1mm thickness was formed on the surface of the pH-imaging sensor before every experiment. The 1.5% agar solution containing 0.1M of potassium chloride was adjusted by hydrochloric acid to pH 5.5 to 6.5 depending on the sample. A heated agar solution was poured onto the sensor surface and cooled to room temperature, in order to obtain the agar film layer.

2.2. PLANT SAMPLES

Mature leaves of Yoshino cherry (*Prunus* x *yedoensis* cv. Yedoensis) and Oshima cherry (*P. lannesiana* var. *speciosa*) grown in the campus of our institute were used in this experiment. Yoshino cherry is relatively sensitive to simulated acid rain, while Oshima cherry is relatively insensitive (Matsuki *et al*, 2000).

2.3. ACID TREATMENTS

Small droplets (0.05 to 10µl) of sulfuric, nitric or hydrochloric acid at pH 1.0 to 4.0 (100 to 0.1mN) were added with a micropipette to the adaxial leaf surfaces of freshly cut branches in the laboratory. The small leaf discs were marked to indicate the location of the acid droplets and cut out at various times after the droplets were air-dried. The adaxial leaf surface was thus contacting the top of the agar film layer formed on the pH-imaging sensor. The pH-imaging was conducted soon after the disc of the sample leaf was placed on the agar film layer. The pH distribution formed on the agar film was imaged by computer processing.

3. Results

3.1 TIME COURSE CHANGES IN LEAF SURFACE RESPONSE

A 0.05µl droplet of sulfuric acid at 100mN (pH 1.0) was placed on the leaf surface and air-dried. A 15 x 15mm disc including the area of the acid droplet was located, cut out and measured. The lower pH region of residual protons was confirmed by a pH-imaging microscope, soon after the droplet dried which was within a few minutes at laboratory conditions. Figure 1 shows the area of residual acid detected by pH-imaging produced in the agar film layer transferred from Yoshino cherry leaves. The leaf surface showed a very clear acidic area immediately following treatment with sulfuric acid. The background gel pH was adjusted to pH 6.5 and pH of the center part in the acidic area indicated pH 5.4. By 4 hours, the acidic area had disappeared.

A 0.1µl droplet of sulfuric acid at 100mN (pH 1.0) to 1mN (pH 3.0) was placed on the leaf surface and air-dried under laboratory conditions (TABLE I). Droplets took about 7 minutes to dry at laboratory conditions. The pH-image of the residual acid area was very clear at first; however, it became obscure with time and disappeared overnight. An acid area at pH 3.0 was not obtained after the droplet was dried.

TABLE I
Residual acid detection by pH-imaging microscope after the placement of sulfuric acid droplets with different pH on Yoshino cherry leaves.

Time (hour)	Droplet pH		
	1.0	2.0	3.0
0	+	+	-
2	+	+	-
4	+	+	
Overnight	-	-	

Droplet volume was 0.1µl. The pH-imaging analysis was conducted in the different discs at every observation. Agar gel film layer was adjusted at pH 5.5. +: Residual acid area as a pH-image was detected. - : Not detected.

3.2 EFFECT OF VOLUME AND pH OF SULFURIC ACID

The pH-imaging data were dependent on the amount of residual protons on the leaves (TABLE II). For a sulfuric acid droplet of 0.05µl at pH 2.0, an acidic area was not detected after 4 hours. As its volume at pH 2.0 increased from 0.05µl to 0.1µl, an acidic area was still observed after 4 hours, but disappeared overnight.

In contrast to pH 2.0, an acidic area was not detected in a 0.1µl droplet at pH

3.0. It is suggested that residual protons are required to generate a pH distribution. However, a droplet of 1μl or greater took about 30 minutes or more to evaporate. Even if the amount of given protons in the 1μl droplet at pH 3.0 was equivalent to 0.1μl at pH 2.0, an acidic area was not detected in a 1μl droplet at pH 3.0 after 2 hours. An acidic area was also not observed in a 10μl droplet at pH 4.0. These results suggest that acid on the leaf surface might be neutralized during the drying process.

TABLE II
Residual acid detection by pH-imaging microscope after placement of different volumes of sulfuric acid droplets on Yoshino cherry leaves.

Time (hour)	pH 2.0			pH 3.0			pH 4.0
	0.05μl	0.1μl	1μl	0.1μl	1μl	5μl	10μl
0	+	+	+	-	+/-	+	-
2	+/-	+	+	-	-		
4	-	+/-				+/-	
5						-	
7		-					

The pH-imaging analysis was conducted in a different disc at every observation. Agar gel film layer was adjusted at pH 5.5. +: Residual acid area as a pH-image was detected. - : Not detected.

3.3 DIFFERENTIAL RESPONSE IN CHERRY SPECIES

TABLE III shows results obtained in the two cherry leaves: Yoshino and Oshima cherry. Residual acid was imaged at 0, 1, 3, 6 hours and over night after the droplets dried. In Yoshino cherry leaves, the pH region disappeared in the case of 0.05μl after 3 hours. However, in leaves of Oshima cherry, pH-imaging could be observed at 0.05μl after 6 hours and at 0.1μl after remaining overnight. This result suggests that leaf surface responses to sulfuric acid might be affected by leaf surface properties of the plant.

3.4 EFFECT OF NITRIC AND HYDROCHLORIC ACID

Nitric and hydrochloric acid droplets were placed together with those of sulfuric acid. The 0.1μl droplets of nitric, sulfuric and hydrochloric acid at 10mN (pH 2.0) were placed respectively on the adaxial cherry leaf surface. Leaf discs were cut out after all acid droplets had dried. Residual acid area for sulfuric acid was observed in the center portion (Figure 2). Acidic areas due to nitric or hydrochloric acid were slightly detected, but were obscure in comparison to sulfuric acid even though the imaging was conducted just after the droplets had dried. This phenomenon was also confirmed on the chemically non-reactive plate that the two acids evaporated during a drying process at laboratory conditions, but sulfuric acid did not.

TABLE III
Residual acid detection by a pH-imaging microscope after placement of different volumes of sulfuric acid droplets at pH 2.0 on Yoshino and Oshima cherry leaves

Time (hour)	Yoshino cherry		Oshima cherry	
	0.05µl	0.1µl	0.05µl	0.1µl
0	+	+	+	+
1	+		+	
3	-	+	+	+
6			+	+
Overnight		-	-	+

The pH-imaging analysis was conducted in a different disc at every observation. Agar gel film layer was adjusted to pH 5.5. +: Residual acid area as a pH-image was detected. -: Not detected.

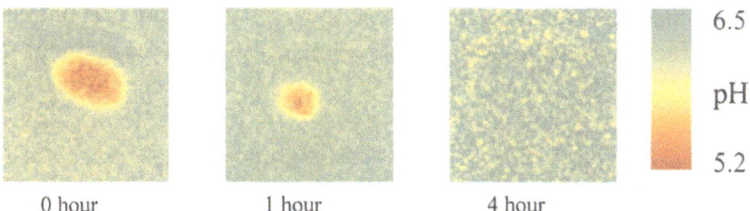

Figure 1. Residual acid detection by pH-imaging analysis after placement of a sulfuric acid droplet (0.05µl, 100mN) on Yoshino cherry leaves. Agar gel film layer was adjusted to pH 6.5.

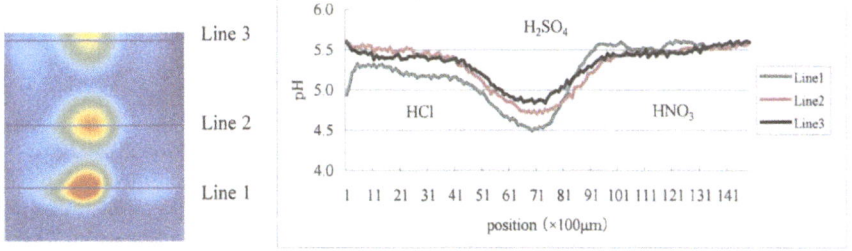

Figure 2. Comparative responses to sulfuric, hydrochloric, and nitric acid droplets placed on a Yoshino cherry leaf. The volume of each droplet was 0.1µl at pH 2.0. Three droplets for an acid were placed vertically in the figure (left). Profile of the area from left to right: HCl, H_2SO_4, and HNO_3.

4. Discussion

The neutralization of acid droplets on the plant leaf surface was visually confirmed as a change in pH-images, using a scanning chemical microscope. These results indicate that the leaves will neutralize the surface acid, probably by ion exchange with basic cations from the surface tissues. This suggests that leaves can recover from temporal strong acid stress in a relatively short term. We also confirmed differences in neutralization time between two species of cherry; leaves of acid tolerant Oshima cherry showed lower pH regions for a longer time than those of acid sensitive Yoshino cherry. Thus, Oshima cherry may have a mechanism of resistance such as a thick waxy layer or chemical properties to protect accelerated leaching of cations due to acids, while Yoshino cherry lacks such a mechanism of resistance.

The current experiments were conducted and evaluated under strongly acidic conditions. Generally, continuous exposures of such extremely low pH solutions to plants will cause direct effects such as visible injury (Kohno *et al*, 1994, 1998). However, current experiments suggest that dry deposition of sulfuric acid would be neutralized even if it is dissolved after the absorption of water on the plant surface. In contrast, hydrochloric and nitric acids appear to disappear as these solutions dry due to vaporization itself. These results indicate that sulfuric acid deposition may be more important in estimating direct effects of acidic deposition on plants than deposition of nitric and hydrochloric acids.

Acknowledgment

The authors would like to express great appreciation to Dr. Donald T. Krizek, Sustainable Agricultural Systems Laboratory, ANRI, ARS, USDA, BARC-West, Beltsville, MD 20705-2350, U.S.A. for his critical reading of this manuscript.

References

Kobayashi, T., and Matsumura, H.: 1998, *Central Research Institute of Electric Power Industry (CRIEPI) Report U97069*.
Kohno, Y.: 1994, *J. Japan Soc. Air. Pollut. 29(2): 71-79*.
Kohno, Y., Matsumura, H., and Kobayashi, T.: 1994, *J. Japan Soc. Air. Pollut. 29(4):206-219*.
Kohno, Y., Matsumura, H., and Kobayashi, T.: 1998, 'Differential sensitivity of trees to simulated acid rain or ozone in combination with sulfur dioxide', in V. Bashkin and S-U. Park(eds.), *Acid Deposition and Ecosystem Sensitivity in East Asia*, Nova Sciences Publishers: Commack, NY, U.S.A. pp.143-188.
Matsuki, R., Matsumura, M., Kobayashi, T., and Kohno, Y.: 2000, *Central Research Institute of Electric Power Industry (CRIEPI) Report U99027*.
Nakao, M., Nomura, S., Takamatsu, S., Tomita, K., Yoshinobu, T., and Iwasaki, H.: 1998, *T. IEE Japan 118-E(12):584-589*.
Tukey, H. B. Jr.: 1970, *Ann. Rev. Plant Physiol. 21:305-324*.

RELATIONSHIP BETWEEN NORWAY SPRUCE STATUS AND SOIL WATER BASE CATIONS/ALUMINUM RATIOS IN THE CZECH REPUBLIC

JAKUB HRUŠKA[1]*, PAVEL CUDLÍN[2] and PAVEL KRÁM[1,3]

[1] *Czech Geological Survey, Klárov 3, 118 21 Prague 1, Czech Republic;* [2] *Institute of Landscape Ecology, Academy of Sciences of the Czech Republic, Na Sádkach 7, 370 05 České Budějovice, Czech Republic;* [3] *Swedish University of Agricultural Sciences, B. 7050, 750 07 Uppsala, Sweden.*
(* author for correspondence, e-mail: hruska@cgu.cz)

Abstract. There is a concern that soil acidification by acidic deposition, along with the resulting depletion of the labile pool of nutrient cations (e.g. Ca, Mg) and enhanced leaching of Al from soil may contribute to forest dieback. The molar ratios of Ca/Al or (Ca+Mg+K)/Al in the soil solution have been widely used as a criterion for risk of tree damage due to acidification. Intensity and quality of the crown and branch structure transformation due to formation of secondary shoots in successive series is a very sensitive indicator of long-term tree damage, and the subsequent regenerative processes. Soil water chemistry and crown structure transformation of Norway spruce were observed at 16 forest plots within the Czech Republic with the following results: parameters, expressing degradation processes in the crown (defoliation of primary structure), regeneration processes (percentage of secondary shoots) or synthetic stages of crown structure transformation showed high correlation with soil water (Ca+Mg+K)/Al ratio in organic horizons. No relationships were found for mineral horizons. The correlations between soil water and crown status parameters were considerably stronger when using the (Ca+Mg+K)/Al ratio rather than the Ca/Al ratio.

Keywords: soil acidification, Norway spruce, base cations/aluminum ratio, Czech Republic, defoliation, forest decline, soil water

1. Introduction

The Czech Republic (CR) has a serious decline of Norway spruce (*Picea abies*) forests. About 1000 km^2 of Norway spruce stands died during the last three decades. Direct effect of SO_2 on the spruce canopy together with climatic stress was ascribed as a main factor in the Krušné hory (Ore Mts.; Kubelka et al., 1993). SO_2 emissions in the CR peaked in 1987 (2.2 million t yr^{-1}). A desulfurization program was enacted (1992-1998) and emissions declined to 0.4 million t yr^{-1} in 1998, which is 18% of 1987 emissions. This change did not improve forest health in the CR. For example annual average defoliation of Norway spruce stands in the Krkonoše Mts. (Figure 1) increased significantly during 1980's and 1990's (17% in 1980, 51% in 1990, 54% in 1995 and 64% in 1998; derived from Vacek, 1998). These observations did not support the hypothesis of direct SO_2 effect as a major cause of the forest decline in the CR. Therefore, soil acidification and related effects can be the major cause of the alarming situation. Depletion of nutrient cation soil pools , (e.g. Ca and Mg), and enhanced concentrations of potentially toxic Al in soil solution may contribute to the observed forest dieback. Molar ratios of Ca/Al and (Ca+Mg+K)/Al in the

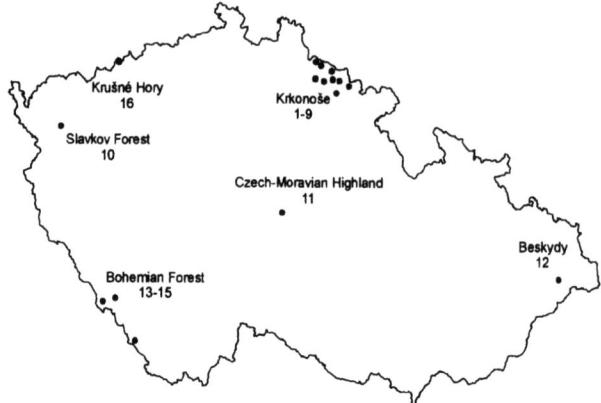

Figure 1. Sampling site locations within the Czech Republic.

soil solution has been widely used as a criterion for risk of tree damage (Sverdrup *et al.*, 1992; Cronan and Grigal, 1995). This concept was criticized because it is largely based on experiments with seedlings and not with mature trees (Högberg and Jensen, 1994; Løkke *et al.*, 1996; de Wit *et al.*, 1999).

The objective of our study was to evaluate forest status in relation to soil solution Ca/Al and (Ca+Mg+K)/Al molar ratios using regional evaluation of available data from spruce stands in the CR. Intensity and quality of the crown and branch structure transformation and formation of secondary shoots in successive series (Gruber, 1994; Cudlín *et al.*, 1999) were used in this study as a very sensitive indicator of long-term tree damage and of the subsequent regenerative processes. Such comparison may facilitate reconstruction of a forest stand's stress response history and predict their subsequent development (Cudlín *et al.*, 1999).

2. Site Description and Methods

Soil water chemistry and crown structure transformation of Norway spruce were observed at 16 forest plots situated on the major mountain ranges within the CR (Figure 1, Table I). One third of the CR is forested. Managed stands of Norway spruce cover 55% of the forested areas.

2.1. SOIL WATER

Suction lysimeters were used at 9 sites (plots 5-9, 12-15), zero-tension lysimeters were used at the other sites (plots 1-4, 10-11, 16). Total Al and base cations (BC: Ca, Mg, K) were analysed using atomic absorption spectroscopy (AAS) or inductively coupled plasma spectroscopy (ICP). The molar ratios of Ca/Al and

(Ca+Mg+K)/Al were calculated separately for organic soil solution (O horizons) and mineral soil solution (E or B1 horizons) (Table I).

2.2. CROWN STRUCTURE TRANSFORMATION

Twenty to thirty trees on plots were evaluated for total defoliation and other parameters according to the ICP-Forest Program (ICP Forest, 1994), including defoliation of the primary structure (Gruber, 1994), percentage of secondary shoots, and types of defoliation, using the modified approach of Lesinski and Landman (1985). These parameters were determined only for the middle (productive) part of the crown, i.e. exclusive of upper, juvenile part and bottom, shadowed part. Progressive stages (from 0 to 4) of crown structure transformation were derived from last two parameters (Cudlín *et al.,* 1999).

3. Results and Discussion

Soil solutions were sampled predominantly between 1994-1998. The lowest ratios for Ca/Al were observed for mineral soil water at plot 4 at Krkonoše (0.2), and only slightly higher Ca/Al was observed at plot 12 (Beskydy Mts.) and Krkonoše Mts. (plot 1: 0.3). Four plots showed Ca/Al ratio of 0.4. The highest ratio was found again for Krkonoše Mts. – (plot 2: 1.2) and an extremely high ratio was found for plot 11 (17.6), the site at the lowest altitude (440 m a.s.l.). Ca/Al ratios in organic soil solution were lowest for several plots in the Krkonoše Mts. (three sites 0.6, and one site 0.4). The highest one was observed at Želivka again (plot 11: 5.2).

BC/Al ratio for mineral soil water was lowest at plot 4, Krkonoše Mts. (0.4), plot 16, Krušné hory Mts. (0.7) and three other plots showed a ratio of 0.8. The highest BC/Al ratio was observed at plot 11 (10.7). BC/Al ratio for organic soil was lowest at plot 8 (Krkonoše) – 0.7 and the highest as expected was at the low altitude plot 11 (10.7).

The highest total defoliation 47 % (TD in Table I) was found for plot 7 in the Krkonoše. All TD higher than 40% were also observed in the Krkonoše. (plots 4,5,6,8,9). The lowest TD was observed in the Bohemian Forest (25% at plot 14) and again in the Krkonoše (plot 1: 28%). The range of TD was between 25%-47%, which is a generally narrow interval if we assume the large variability of altitudes, forest age, soils and soil water chemistry (Table I). Thus quantification of this parameter is very sensitive to potential inaccuracy created by an from observer's subjectivity.

Defoliation of primary structure (DPS) showed a much larger span of observed values (40%-94%).

TABLE I

Geographical and soil characteristics, soil water ratios (molar) for organic and mineral soils, and crown structure parameters for examined plots

No.	Site Name	Mountain Region	Altitude (m)	Sampling Horizon	Soil Type	Forest age (years)	Ca/Al organic	BC/Al organic	Ca/Al mineral	BC/Al mineral	TD (%)	DPS (%)	SH (%)	SCT
1.	Rýchory	Krkonoše	742	O, B1	Ferric Podzol	100	0.7		0.3	1	28	54	41	1.2
2.	Bílé Labe	Krkonoše	1100	O, B1	Humic Podzol	100	1.1	2	1.2	2.2	37	72	58	1.9
3.	Přední Žaly	Krkonoše	918	O, B1	Ferric Podzol	100	1.2	2.7	0.6	1.1	38	74	61	2.0
4.	Bílá Voda	Krkonoše	1009	O, B1	Humic Podzol	100	0.6	1.3	0.2	0.4	41	78	67	2.5
5.	Alžbětinka	Krkonoše	1192	O, E	Humic Podzol	200	1.3	2.2			44	78	68	2.4
6.	Modrý důl	Krkonoše	1237	O, E	Humic Podzol	140	0.6	1.3	0.5	0.8	42	73	64	2.3
7.	Pašerácký chodníček	Krkonoše	1317	O, E	Leptic Podzol	145	0.6	1.2	0.4	1.1	47	94	90	3.4
8.	Pudlava	Krkonoše	1140	O, E	Humic Podzol	102	0.4	0.7	0.4	0.8	43	77	69	2.4
9.	Sluneční údolí	Krkonoše	1241	O, E	Humic Podzol	154	0.7	1.4	0.9	2.2	40	82	75	2.8
10.	Lysina	Slavkov Forest	910	O, E, B1	Podzol	120	2.5	8.3	1.4	3.5	31	57	41	1.4
11.	Želivka	Czech-Morav. Highland	440	O, B1	Luvisol	95	5.2	10.7	17.6	35.9	32	40	23	1.0
12.	Staré Hamry	Beskydy	580	B1	Gleyic Cambisol	88			0.3	0.9	36	46	32	1.0
13.	Filipova Huť	Bohemian Forest	1160	B1	Humic Podzol	116			0.4	0.8	31	51	36	1.8
14.	Javoří pila	Bohemian Forest	1125	B1	Humic Podzol	124			0.7	1.3	25	50	37	1.7
15.	Trojmezí	Bohemian Forest	1245	B1	Humic Podzol	110			0.7	1.1	37	68	59	2.2
16.	Načetín	Krušné hory	770	O, B1	Dystric Cambisol	120	2.0	3.8	0.4	0.7	32	74	63	1.9

TD-total defoliation; DPS-defoliation of primary structure; SH-secondary shoots; SCT-stage of crown structure transformation; BC-base cations (Ca+Mg+K)

TABLE II
Spearman's correlation coefficients between soil solution molar ratios and crown structure transformation parameters. BC = Σ(Ca+Mg+K)

	Total defoliation	Defoliation of primary structure	Secondary shoots	Stage of crown transformation
Ca/Al organic soil	-0.60*	-0.51*	-0.66*	-0.67*
BC/Al organic soil	-0.81**	-0.65*	-0.80**	-0.81**
Ca/Al mineral soil	NS	NS	NS	NS
BC/Al mineral soil	NS	NS	NS	NS

*$P<0.05$; **$P<0.01$; NS-no statistically significant correlation

The synthetic stage of crown structure transformation (SCT) involved the percentage of secondary shoots and types of defoliation and was classified from 0-4. Trees with higher stage have more damage and simultaneously more regenerated assimilative organs.

Spearman's correlation coefficients were used to evaluate the relationship between soil water chemistry and tree status. No statistically significant correlations were found for mineral horizons (E and B1 horizons) for both of the examined ratios (Table II). Significant negative correlation ($P<0.05$) were observed between Ca/Al ratio in organic soil water and all examined parameters of crown structure status. The highest correlation was found for stage of crown transformation and percentage of secondary shoots (Table II).

Very significant negative correlations ($P<0.01$) were found for BC/Al ratio in organic soil water. Correlation coefficients were almost identical for total defoliation, percentage of secondary shoots and stage of crown transformation.

The root zone of Norway spruce was almost exclusively located in the organic horizons at the sites examined (except low elevation plot 11 Želivka). Very shallow root depth is probably responsible for the lack of correlation between mineral soil solution chemistry and forest status.

4. Conclusions

1. Total defoliation, defoliation of primary structure, percentage of secondary shoots, and degree of crown structure transformation showed a statistically significant negative relationship with molar ratios of Ca/Al ($P<0.05$) and (Ca+Mg+K)/Al ($P<0.01$) in organic horizon soil water.
2. No similar relationships were found for mineral horizons. This phenomenon was probably caused by only a limited amount of roots in the mineral soil horizons.
3. Based on this study, we could not easily set strict criteria for forest dieback due to decline of Ca/Al or BC/Al molar soil solution ratios. However, we were able to show that the response history of Norway spruce forest stands

was significantly affected in plots with decreasing values of these ratios in organic soil horizons.

Acknowledgements

This study was funded by the Grant Agency of the Czech Republic (grant No. 206/94/0832 and 205/99/1685) and the Czech Geological Survey. The authors would like to thank Robert Kaufman, Václav Lochman, Pavel Moravčík, Martin Novák, Ota Rauch, and Otakar Schwarz for supplying us with their unpublished soil water chemistry data. We appreciate the help of Vladimír Majer with statistical analyses and Kevin Bishop's comments on the language.

References

Cronan., C.S. and Grigal, D.F.: 1995, *Journal of Environmental Quality* **24**, 209.
Cudlín, P., Novotný, R. and Chmelíková, E.: 1999, *Phyton* **39**, 149.
De Wit, H., Mulder, J., Nygaard, P.H., Aamlid, D., Huse, M. and Kortnes, E.: 1999, 'Aluminium: the need for a re-evaluation of its solubility and toxicity in mature spruce stands', in *Critical Loads Copenhagen Conference Abstracts*, National Environmental Research Institute, Silkeborg, Denmark, 165-168.
Gruber, F.: 1994, 'Morphology of coniferous trees: Possible effects of soil acidification on the morphology of Norway spruce and silver fir', in D.L. Godbold and A. Hütterman (eds.), *Effects of acid rain on forest processes*, Wiley-Liss: New York, pp. 265-324.
Högberg, P. and Jensén, P.: 1994, *Water, Air, and Soil Pollution* **75**, 121.
ICP Forest: 1994, *Manual on methods and criteria for harmonized sampling, assessment, monitoring and analysis of the effects of air pollution on forests. UN ECE Convention on Long-Range Transboundary Air Pollution.* International Co-operative Programme on Assessment and Monitoring of Air Pollution Effects on Forest. Hamburg and Prague.
Kubelka, L., Karásek, A., Rybář, V., Badalík, V. and Slodičák, M.: 1993, *Forest regeneration in the heavily polluted NE, "Krušné hory" Mts.* Czech Ministry of Agriculture, Agrospoj: Prague.
Lesinski, J.A. and Landman, G.: 1985, 'Crown and branch malformation in conifers related to forest decline', in J.N. Cape and P. Mathy (eds.), *Scientific basis of forest decline symptomatology, Air Pollution Research Report 15,* Commission of European Communities: Brussels, pp. 95-105.
Løkke, H., Bak, J., Falkengren-Grerup, R.D., Finlay, H., Ilvesniemi, P.H., Nygaard, P.H. and Starr, M.: 1996, *Ambio* **25**, 365.
Sverdrup, H., Warfvinge, P. and Rosén, K.: 1992, *Water, Air, and Soil Pollution* **61**: 365.
Vacek, S.: 1998, *Dynamics of the forest ecosystem development in the Krkonoše National Park* (In Czech), Research Report, Forestry and Game Research Institute, Research Station Opočno, Czech Rep.

INCREASED LEVELS OF ALUMINIUM IN FOREST SOILS: EFFECTS ON THE STORES OF SOIL ORGANIC CARBON

JAN MULDER[1*], HELENE A. DE WIT[2], HELENA W.J. BOONEN[1]
and LARS R. BAKKEN[1]

[1]*Department of Soil and Water Sciences, Agricultural University of Norway, Box 5028, N-1432 Aas, Norway;* [2]*Norwegian Forest Research Institute, Hoegskoleveien 12, N-1432 Aas, Norway*
(** author for correspondence, e-mail:jan.mulder@ijvf.nly.no*)

Abstract. Acid atmospheric deposition results in increased levels of mobile aluminium (Al) in forest soils. Laboratory studies suggest that increased binding of Al to soil organic matter (SOM) in the forest floor results in decreased mobility of organic matter in soil water, viz. lower concentrations of dissolved organic carbon (DOC). Other laboratory studies indicate decreased decomposition rates of SOM as a result of Al binding. So far, little field evidence supporting these effects of Al on the lability of SOM have been reported. Here we present a field manipulation experiment in mature Norway spruce forest in Norway, where the content of Al in soil and soil water was increased. Increased Al in the forest floor caused a pronounced decrease in the leaching of DOC. Simultaneously, the decomposition rate of SOM decreased by 30% to 40%. This suggests that elevated Al in the forest floor stimulates accumulation of SOM. In a companion paper we present the effect of increased Al on forest vitality.

Keywords: Al chemistry, Al toxicity, acidification, forest floor, soil organic matter, SOM, DOC, decomposition rate

1. Introduction

Acid atmospheric deposition may result in decreased contents of calcium (Ca) and increased contents of aluminium (Al) in the organic horizon of forest soils (Lawrence et al., 1995). Aluminium binds strongly to organic matter and laboratory studies suggest that increased binding of Al causes a decrease in the mobility (De Wit et al., 1999) and the decomposition rate of soil organic matter (SOM; Boudot et al., 1989).

The mobility of SOM (measured as the concentration of DOC in soil water) is hypothesised to increase with decreasing hydrophobicity and increasing charge of organic macro-molecules. This hypothesis, in which hydrophobicity is an intrinsic property of the material and where charge depends on the binding of proton and metals, is embodied in the model WHAM (Tipping, 1994). WHAM was successfully used to describe equilibrium DOC concentrations in laboratory studies using soil suspensions (Lofts et al., 2000).

Negative effects of Al on the decomposition rate of SOM were reported for soils rich in Al (e.g. Klemmedson and Blaser, 1988; Boudot et al., 1986). A decrease in decomposition rate may be due to direct toxicity of Al to soil organisms, or to decreased "bioavailability" due to complexation of reactive groups with Al.

If the content of Al in the forest floor increases due to acidification, we expect reduced mobility and decomposition rate of SOM. In the long-term, this may lead to a considerable increase in the accumulation rate of SOM. However, so far there is little evidence that these mechanisms also operate at the field scale.

Here we present results of a 3-year manipulation study in the field, where concentrations of Al in soil and in soil water were increased. The study involved four different Al treatments, including a reference plot. We measured concentrations of DOC and other macro-solutes in O horizon drainage water as well as the decomposition rate of SOM in the forest floor of selected treatments. Effects of increased Al on forest vitality are presented in a companion paper (De Wit et al., this volume).

2. Materials and methods

2.1 FIELD MANIPULATION AND SAMPLING OF SOIL AND SOIL WATER

The field manipulation experiment was conducted in a 45-year old Norway spruce stand (*Picea abies* (L.) Karst.) in South-eastern Norway (11°06' E., 60°16' N). The site is 200 m above sea level and has an annual precipitation of 860 mm and a mean air temperature of 4°C. The growing season is from May to early October. The area is a homogeneous, flat plain with 60 m deep fluvioglacial sands, in which well drained Typic Udipsamments have developed. Soils have a thin (2-4 cm) O horizon, a 1 to 3 cm thick E horizon and a weakly developed B horizon. Base saturation of the O horizon is 65%, whereas base saturation in the mineral soil is below 8%. Calcium is the dominant exchangeable base cation in the O horizon.

Field manipulation of the content of Al in soil and soil water was initiated in 1996. Dilute $AlCl_3$ solutions were added to twelve adjacent forest plots (20m x 20m) by means of an irrigation system (De Wit et al., in press). Three plots were assigned to each of four treatments according to a randomised block design. The four treatments included Al additions at three different levels (i.e. treatment A-0, A-1 and A-2, representing increasing Al additions; Table I) and a deionised water addition treatment (Reference treatment). Table I summarizes the Al additions from 1996 to 1999.

During irrigation with dilute $AlCl_3$ free drainage water was sampled below the O horizon, using two zero-tension lysimeter plates (30cm x 30cm) per plot. Soil solution samples, pooled per plot, were stored at 4°C prior to filtration (0.45 µm, nitrocellulose filters) and analysis (pH, major elements (Ogner et al., 1999), and DOC (high temperature catalytic oxidation, Dohrmann DC-190, Rosemount Analytical, Inc)).

Soils were sampled at the end of the experiment (autumn 1999). For each plot, 16 samples of the forest floor, collected according to a grid pattern of 2.5 m x 2.5 m, were pooled. Pooled field-moist samples were stored at 4°C until further use. Before chemical analysis the samples were air-dried for a week at 28°C, sieved (2 mm diameter sieve) and homogenized. CEC was determined as the sum of cations extracted by 1 M NH_4NO_3 (base cations, Mn^{2+} and exchangeable acidity). Base cations, Al and Mn^{2+} in the extract were determined

TABLE I

Al added (as $AlCl_3$) in each treatment from 1996 to 1999 ('TOTAL'). Total Al was added as an initial dose at a high concentration of Al ('INITIAL') and subsequently as weekly 10 mm doses during the growing season at relatively low Al concentration ('WEEKLY'). Concentrations of added Al in the weekly additions were varied depending the expected dilution in the soil, which depends on soil water content. For further details reference is made to de Wit et al. (in press).

	Al added from 1996 to 1999			Al concentrations in weekly 10 mm addition	
	TOTAL	WEEKLY	INITIAL	Minimum	maximum
	1996-1999	1997-1999	1996-1997		
Treatment	(10^{-3} mol/m^2)			(10^{-6} mol L^{-1})	
C	0	0	0	0	0
A0	301	0	301	0	0
A1	418	117	301	148	370
A2	534	233	301	296	741

using ICP (Ogner et al., 1999). Exchangeable acidity was estimated by titration of the extract to pH 7 (Ogner et al., 1999). The content of soil organic carbon (SOC) was determined by dry combustion (LECO CHN-1000).

2.2 SOIL ORGANIC MATTER DECOMPOSITION

Fresh forest floor samples (pooled from 10 sub-samples) were taken in August 1999 at the reference and the A2 (highest Al addition) plots. Thirteen replicate samples, 1 g each, were brought to 50% of the water holding capacity and incubated at room temperature in the dark in glass bottles with rubber septum. The concentration of CO_2 was measured during 4 days by gas chromatography. In addition, the effect of Al on the respiration rate of soil organic matter was studied by stripping exchangeable Al from SOM by means of extensive leaching with $CaCl_2$. Subsequently, different levels of bound Al were obtained by flushing with various amounts of $AlCl_3$ (Figure 2).

3. Results and discussion

3.1 CHEMICAL PROPERTIES OF THE O HORIZON AFTER MANIPULATION

Addition of Al as indicated in Table I resulted in a significant increase in exchangeable Al and in significant decreases in exchangeable Ca and Mg and base saturation (Table II). The increase in the content of exchangeable acidity was not significant.

Table II

Selected soil chemical characteristics of the O horizon at Nordmoen after the treatments were finished in 1999. Different letters indicate significant differences between the treatments at the 0.05 level (Tukey's t-test).

Treatment	pH (H₂O)	CEC	BS	Ex. acid	Al	Ca	K	Mg	Na
		mmol/kg	%	---------mmol/kg---------					
C	4.05	303	48.2a	132	27c	94a	28	23a	1.9
A0	4.08	279	40.1b	148	72b	68b	25	17b	2.8
A1	4.02	291	31.7c	183	103a	47b	30	13c	1.6
A2	4.07	279	31.2c	179	121a	45b	29	11c	1.9

3.2 CONCENTRATION OF DOC IN O HORIZON WATER

The chemical composition of O horizon water is strongly affected by the Al treatments (Table III). pH is lowest in O horizon water of the A1 and A2 treatments due to increasing ionic strength of soil water in the order C ≈ A0 < A1 < A2 and decomplexation of H^+ caused by Al binding. Not surprisingly, the concentration of total dissolved Al is highest in A2, where the concentration of Al in irrigation water is also highest. The organically-bound fraction of Al(tot) decreases from about 90% in C and A0 to 25% (A1) and <10% (A2; data not shown). The mobility of DOC is strongly affected by the treatment with Al. DOC concentrations, although variable from year-to-year, are highest at C and A0 and considerably lower at A1 and A2 (Table III). Water extracts of forest floor samples indicate a reduction in DOC concentration in A0 of about 25% compared to C (De Wit, 2000). This supports previous findings from laboratory experiments showing a decrease in DOC concentration when the concentration of H^+ and Al in soil water increase (De Wit et al., 1999).

TABLE III

Composition of O horizon water at Nordmoen. Treatments as explained in the text. Values are averages and standard deviations of monthly observations during the growing season. The concentration of total dissolved Al (Al(tot)) in $\mu mol\ L^{-1}$ and of DOC in $mg\ L^{-1}$.

	Treatments 1998				Treatments 1999			
	C	A0	A1	A2	C	A0	A1	A2
pH	4.21/0.07	4.34/0.44	3.68/0.15	3.44/0.21	4.34/0.07	4.36/0.07	3.83/0.05	3.56/0.07
Al(tot)	10.2/5.4	15.4/10.2	60.4/36.9	211/133	6.4/1.8	9.6/2.4	59.5/46.0	191/129
DOC	40.7/20.4	29.8/8.0	13.9/6.1	13.7/9.0	22.5/7.2	22.8/8.4	12.7/4.1	9.6/4.1

3.3 DECOMPOSITION OF SOIL ORGANIC MATTER

The initial respiration rate of SOC in the litter layer of the reference was about 0.014% per hour. For the A2 the respiration rate was significantly reduced (by about 40%) to 0.0084% per hour (Figure 1). After about 2 weeks of incubation

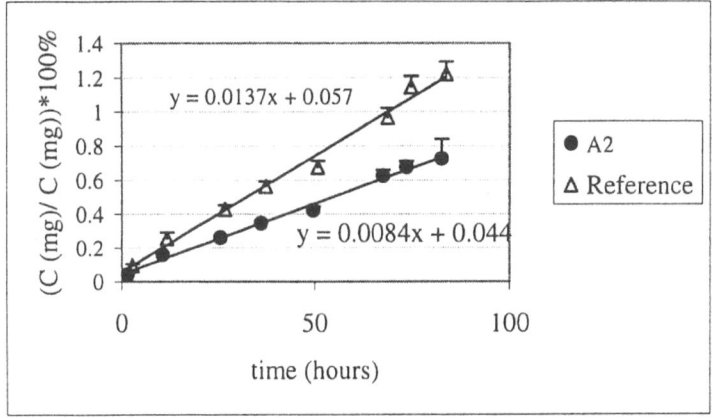

Figure 1. Respiration of SOC in the forest floor (mg C released as CO_2 per mg SOC).

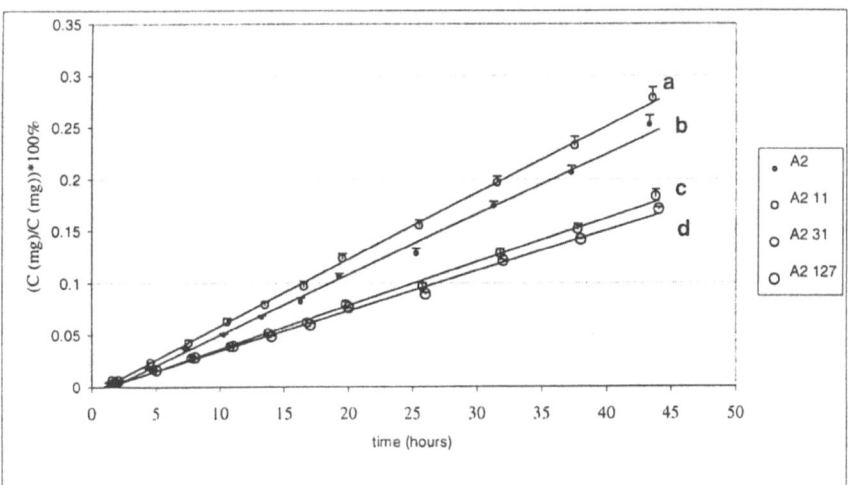

Figure 2. Respiration of SOC in the litter layer of A2. Litter samples were untreated (no number), flushed with $CaCl_2$ or flushed with $CaCl_2$ and subsequently $AlCl_3$. Numbers following the different samples indicate the content of exchangeable Al ($mmol_c$ kg^{-1}).

the respiration rate declined to 0.0090% per hour (reference) and 0.0061% per hour (A2). These results confirm previous findings concerning the effect of Al as found in soil manipulation studies in the laboratory (e.g. Boudot et al., 1989).

Excessive flushing in the laboratory of litter samples from A2 (exchangeable Al of 121 $mmol_c$ kg^{-1}) with $CaCl_2$ resulted in reduced amounts of exchangeable Al (11 $mmol_c$ kg^{-1}). Subsequent flushing with 12 mM $AlCl_3$ caused an increase in exchangeable Al (31 $mmol_c$ kg^{-1}). The highest level of exchangeable Al (127 $mmol_c$ kg^{-1}) was obtained by excessive flushing with $AlCl_3$. The considerable reduction of the content of exchangeable Al in the forest floor of A2 resulted in a small, but significant increase in the respiration rate of SOC. Subsequent increases in exchangeable Al caused a significant reduction in the respiration rate. Note that the respiration of the field-fresh A2 litter was much higher than that of the A2 litter excessively flushed with $AlCl_3$, even though the contents of exchangeable Al were similar. This illustrates that artefacts (e.g. leaching of low-molecular weight organic acids), introduced by manipulation in the laboratory may have a pronounced effects on the experimental results.

4. Conclusions

Increased levels of Al in soils and soil water in a Norway spruce stand in South-eastern Norway caused a significant decrease in the decomposition rate of SOM by 30% to 40%. Simultaneously, we found a considerable decrease in the mobility of SOC, resulting in decreased leaching of DOC from the forest floor. These findings confirm results from laboratory studies. Therefore, a regional increase in the contents of adsorbed and dissolved Al in the forest floor due to acid deposition, as was recently reported, is likely to stimulate the accumulation of organic C in forest soils.

Acknowledgement

Technical assistance of Magne Huse, Egil Kortnes and Claude Belanger are gratefully acknowledged. Thanks are due to NISK's chemical laboratories and the Dept. Soil and Water Sciences, Agric. Univ. Norway for chemical analysis.

References

Boudot, J.P., Bel-Hadj, B.A. and Chone, T.:1986, *Soil Biol. Biochem.* **18**, 457.
Boudot, J.P., Brahm, A.B.H., Steinman, R. and Seigle-Murandi, F. :1989, *Soil Biol. Biochem.* **21**, 7.
De Wit, H.A., Kotowski, M. and Mulder, J.:1999. *Soil Sci. Soc. Am. J.* **63**, 1141
De Wit, H.A., Mulder, J., Nygaard, P.H., Aamlid, D., Huse, M., Kortnes, E., Wollebæk, G. and Brean, R.: in press, *Water, Air and Soil Pollution*
De Wit, H.A. 2000. Doctoral Thesis, Agricultural University of Norway, Aas, Norway
De Wit, H.A., Mulder, J., Nygaard, P.H. and Aamlid, D.: *Water, Air Soil Pollut.* this volume
Klemmedson, J.O. and Blaser, P.: 1988, *Z. Pflanzenernähr. Bodenk.*, **151**, 261
Lawrence, G.B., David, M.B. and Shortle, W.C.: 1995, *Nature* **378**, 162
Lofts, S., Woof, C., Tipping, E., Lawlor, A.J., Clarke, N., Mulder, J. and Wollebæk, G.: 2000, *Europ. J. Soil Sci.*(in press)
Ogner, G. Wickstrøm, T., Remedios, G., Gjelsvik, S., Hensel, G. R., Jacobsen, J. E., Olsen, M., Skretting, E. and Sørlie, B.: 1999, Norwegian Forest Res. Inst., Aas, Norway, 23pp.
Tipping, E.:1994, *Computers and Geosciences* **20**, 973.

TESTING THE ALUMINIUM TOXICITY HYPOTHESIS: A FIELD MANIPULATION EXPERIMENT IN MATURE SPRUCE FOREST IN NORWAY

HELENE A. DE WIT[1*], JAN MULDER[2], PER H. NYGAARD[1] and DAN AAMLID[1]

[1]*Norwegian Forest Research Institute, Hoegskoleveien 12, 1432 Aas, Norway;* [2] *Dept. of Soil and Water Sciences, Agric. Univ. Norway, P.O. Box 5028, 1432 Aas, Norway*
(*author for correspondence, e-mail: Heleen@NISK.no)

Abstract. Aluminium (Al) has been considered to be a central element for risk evaluation of forest damage due to acidification. It has been hypothesized that Al reduces root growth, nutrient uptake and forest vitality. However, forest monitoring studies fail to show correlations between soil acidification and forest health. In general, no direct relation between Al concentration and forest health has been established. Here, Al concentrations in soil solution were manipulated by weekly additions of dilute $AlCl_3$ to levels that are believed to be unfavorable for plant growth. Four treatments (in triplicate), including a reference and three Al addition levels, were established. Effects of enhanced Al concentrations on fine root growth, nutrient uptake and crown condition in a mature Norway spruce forest in Norway were tested (1996-1999). After three years of manipulation, crown condition, tree growth and fine root growth were not affected by potentially toxic Al concentrations. However, the Mg content in current year's needles decreased at the highest Al addition treatment. The Mg/Al ratio of fine roots of the same treatment had declined too, which suggests that Al blocked Mg uptake at the root surface. The manipulation will be continued for two more years.

Keywords: Al toxicity, forest vitality, Norway spruce, needle Mg content, fine root growth, fine root chemistry

1. Introduction

Aluminium (Al) has been considered to be a central element for risk evaluation of forest damage due to acidification. Ulrich *et al.* (1980) hypothesized that enhanced concentrations of Al reduce root growth, nutrient uptake and forest vitality. In recent years, many studies have focussed on the effects of aluminium on trees. For reviews see Taylor (1991), Matzner and Murach (1995) and Cronan and Grigal (1995). Most of the experiments referred to deal with young seedlings in nutrient solution studies over short experimental periods, or to studies where polluted and non-polluted forest stands are compared. Matzner and Murach (1995) mention the lack of meaningful field experiments that are designed to test relationships between air pollution and forest health. However, they postulated that Al stress and low magnesium (Mg) supply reduce rooting depth and root biomass. So far, forest monitoring studies fail to show correlations between human-induced soil acidification and forest vitality, expressed in crown condition (Klap *et al.*, 2000). Here, we present results of a field manipulation experiment in Norway that was designed to test the Al toxicity hypothesis formulated by Ulrich *et al.* (1980). Potentially toxic Al concentrations (200 µmol L^{-1}) were established in the root-zone of a Norway spruce stand. Data on the nutrient status of trees, indicated by the chemical composition of needles and fine litterfall, are presented.

2. Materials and Methods

2.1. EXPERIMENTAL DESIGN AND LOCATION

A field manipulation experiment was set up in 1996 in a planted 45-year old Norway spruce (*Picea abies* (L.) Karst.) stand at the Nordmoen experimental forest in south-east Norway, an area with little acid deposition. The homogeneous, sandy soil is classified as a Typic Udipsamment. Base saturation in the O horizon is 65%, whereas base saturation in the mineral soil is below 8%. The dominant base cation at the exchange complex in the O horizon and the upper 5 cm of the mineral soil is calcium (Ca). The vegetation belongs to the *Eu-Piceetum myrtilletoseum* subassociation. For further details on soil and vegetation reference is made to De Wit *et al.* (in press).

The manipulation experiment was designed to establish potentially toxic Al concentrations in the soil solution of the root zone by means of weekly irrigation of 10 mm deionized groundwater to which $AlCl_3$ was added. Solution was spread evenly over 20x20 m^2 plots using 30 sprinklers. Target levels of potentially toxic inorganic Al (100 µmol L^{-1} and 200 µmol L^{-1} for treatment A-1 and A-2, respectively) have previously been shown to have negative effects on growth and nutrient uptake of Norway spruce seedlings in hydroponic studies (Göransson and Eldhuset, 1991; Godbold and Kettner, 1991). Twelve adjacent plots were established, and three plots were assigned to each of four treatments according to a randomized block design. The four treatments consisted of three Al addition treatments (A-0, A-1 and A-2) and a reference treatment (C) where only deionized water was added. Treatment A-2 received the highest Al additions. For details regarding the addition of Al, reference is made to Mulder *et al.* (this volume). The treatments will be continued in 2000 and 2001.

2.2. SAMPLING PROCEDURES

In every plot of 20x20 m^2, a central plot of 10x10 m^2 was marked. We assumed that trees within the inner plot received their nutrition from the treated area. Procedures for recording height and diameter of trees and evaluation of crown density and color are described in De Wit *et al.* (in press). Needles were sampled annually except in 1996 from five trees from the central plot and separated in current year's and previous year's needles. Litterfall was collected twice a year by means of six litterfall collectors (1.5 m height and 0.45 m diameter in the growing season, and square 0.25 m^2 collectors, placed on the ground in the winter season) that were placed alongside the central plots at 5 m intervals. The needle-fraction of the litter was analyzed chemically. Twelve root-ingrowth cores were installed in each inner plot to study fine root growth, according to the description of Vogt and Persson (1991). For further details reference on root sampling is made to De Wit (2000). Methods for sampling of soil and soil solution are described in Mulder *et al.* (this volume). Sample preparation and chemical analyses of needles, roots, soil and soil solution were done according to the procedures described in Ogner *et al.* (1999).

3. Results

3.1. SOIL SOLUTION AND SOIL CHEMISTRY

Effects of Al addition on soil and soil solution chemistry in the O horizon are described in Mulder *et al.* (this volume). For the mineral soil, effects are described in De Wit (2000) and De Wit *et al.* (in press).

The addition of $AlCl_3$ increased concentrations of total Al and inorganic Al to potentially toxic levels in 1997 in treatment A-0, A-1 and A-2, and in 1998 and 1999 in treatment A-1 and A-2. Concentrations in treatment A-2 were always higher than in treatment A-1. Compared with the reference treatment C, soil solution concentrations of Ca and Mg were elevated in treatment A-1 and A-2 whereas concentrations of potassium (K) were similar for all treatments. Considerable amounts of Ca and Mg were leached from the forest floor, causing a significant decline in base saturation in treatment A-0, A-1 and A-2. Soil chemistry in the mineral soil was not affected significantly.

3.2. FOREST VITALITY, FINE ROOT GROWTH AND LITTERFALL

Forest vitality, indicated by crown condition and tree growth, was not affected by three years of Al addition (De Wit *et al.*, in press). Crown condition indicated a normal health condition of the forest. The production and mortality of fine roots and litterfall (Table I) were not affected by treatment. Thus, no

TABLE I

Total, living and dead fine root biomass at 0-40 cm averaged per treatment (n=3). Roots obtained by root ingrowth cores technique, installed in 1996 and sampled in 1999. Needlefall averaged per treatment for 5 sample periods from 1997 to 1999 (n=15). Standard deviation in brackets.

| Treatment | fine root mass | | | needlefall |
	total (g m^{-2})	living (g m^{-2})	dead (g m^{-2})	g m^{-2} day^{-1}
C	662 (146)	604 (178)	59 (44)	0.202 (0.048)
A-0	884 (140)	854 (138)	31 (19)	0.226 (0.083)
A-1	804 (216)	754 (214)	50 (23)	0.236 (0.089)
A-2	609 (35)	563 (57)	46 (23)	0.224 (0.063)

support was found for the hypothesized link between increased Al concentrations, decreased fine root growth and loss of needles. The lack of a response of fine root production to enhanced Al concentrations, in contrast to earlier findings in hydroponic studies (Göransson and Eldhuset, 1995), may be related to factors such as the presence of mycorrhiza and rhizosphere processes such as exudation of organic acids that may detoxify Al. Those factors are often absent or negligible in hydroponic studies (Högberg and Jensén, 1994).

3.3. NUTRIENT CONTENTS OF NEEDLES, LITTER AND FINE ROOTS

After three years of treatment, the Mg-content of current year's needles had declined significantly in treatment A-2 compared with the reference treatment C. The same trend was found for the Mg-content of previous year's needles, but here the difference was not significant (Table II). This indicates a reduction in Mg-uptake in treatment A-2 compared with reference treatment C. The Mg-content of the needles was higher than levels that are reported to indicate Mg-deficiency in Norway spruce (Brække, 1994).

Soil solution concentrations of Mg in treatment A-2 were considerable higher than in reference treatment C, which suggests that the bio-availability of Mg had increased. No change was detected in the Mg-content of the mineral soil due to Al addition (De Wit, 2000). Thus, the reduced Mg uptake in treatment A-2 was not likely to be caused by a decrease in Mg-availability in the soil. Fine root investigations in the mineral soil did not show an increase in root mortality or a decrease in fine root growth (Table II; De Wit, 2000) which might explain a reduced uptake of Mg. However, chemical analysis of fine roots showed a significant decrease in Mg/Al ratio of fine roots at 20-40 cm depth (Table III). At 0-20 cm depth, the decrease was not significant. Thus, Al adsorbed to the root surface may have reduced Mg uptake, as has been found previously in nutrient solution studies (Göransson and Eldhuset, 1995). Tveite et al. (1993) reported a similar decrease in Mg-content of current and previous year's needles of Norway spruce in a field manipulation experiment where acid deposition was simulated but contributed this to a decrease Mg supply in the soil. No data on root growth and chemical composition were available in the study by Tveite et al. (1993).

The needle contents of K, Ca and Mg in reference treatment C, which received only water, increased in the first year of treatment and remained at the higher level throughout the treatment period (Table II and data not shown). This suggests that increased water availability affected the nutrient status of the trees positively. The Mg/Al ratios in current and previous year's needles were significantly lower in the Al addition treatments than in reference treatment C (Table II). The Ca/Al ratio of the needles showed the same trend as the Mg/Al ratio but the decrease was not always significant.

Few effects of the Al addition treatment were detected in the chemical composition of fine litter (Table II). The content of Al was lowest in the reference treatment C. Compared with needle contents, Al and Ca had accumulated in fine litter whereas contents of Mg and K were depleted. This illustrates the regulating ability of the plant to re-allocate Mg and K (Marschner, 1995).

TABLE II

Selected chemical characteristics (total element content[1]) of current year's (CurrYe) and previous year's needles (PrevYe) in 1995 (before start of the field manipulation) and 1999, and of needle litterfall collected after the growing season of 1997 and 1999. Different letters indicate significant differences between treatments at the 0.05 level (Tukey's t-test).

		1995/1999 Al	Ca	Mg	K	Ca/Al	Mg/Al
		----------------------mmol kg^{-1}------------------					
CurrYe's	C	3.2a / 3.2a	63a / 71a	39a / 45a	135a / 183a	20a / 22a	12.3a / 14.2a
needles	A-0	4.4a / 4.4a	70a / 71a	35a /43ab	140a / 174a	17a / 16a	8.8a / 9.9b
	A-1	4.0a / 5.1a	65a / 69a	43a /40ab	147a / 186a	16a / 15a	10.7a / 8.3b
	A-2	4.1a / 4.7a	71a / 69a	42a / 36b	133a / 165a	18a / 16a	10.9a / 8.0b
PrevYe's	C	4.0a / 5.4b	75a / 95ab	33a / 38a	113a / 156a	19a / 18a	8.6a / 7.3a
needles	A-0	4.9a / 7.9ab	87a / 81a	34a / 31a	127a / 163a	19a / 10b	7.3a / 4.0b
	A-1	4.9a / 9.0b	87a / 91ab	38a / 32a	123a / 163a	18a / 11b	7.8a / 3.6b
	A-2	5.0a / 8.4ab	84a / 114b	34a / 31a	122a / 149a	17a/ 15ab	7.4a / 4.0b
		1997/1999					
fine litter	C	13.1b / 12.9c	172a / 180b	25a / 24a	43a / 34a	13a / 14a	1.9a / 1.9a
(needles)	A-0	14.6a / 15.8a	154a / 181b	26a / 24a	44a / 34a	11a / 11a	1.8a / 1.5a
	A-1	14.4a / 13.8b	146a / 159a	28a / 27a	47a / 36a	10a / 12a	1.9a / 1.9a
	A-2	13.8ab /15.0ba	175a / 194b	25a / 25a	44a / 34a	13a / 13a	1.8a / 1.7a

[1] Determined by acid-digestion of plant material and analysis by ICP

TABLE III

Total contents[1] of Al and Mg in living fine roots (0-2 mm diameter) obtained from root ingrowth cores in mineral soil at 0-20 cm and 20-40 cm depth. Different letters indicate significant differences between treatments at the 0.05 level (Tukey's t-test).

	Al (mmol kg^{-1})		Mg (mmol kg^{-1})		Mg/Al	
Treatment	0 - 20 cm	20-40 cm	0 - 20 cm	20-40 cm	0 - 20 cm	20-40 cm
C	309a	302a	67.6a	68.1a	0.22a	0.23b
A-0	306a	314a	74.1a	84.2a	0.24a	0.27a
A-1	341a	344a	73.8a	66.2a	0.22a	0.19bc
A-2	326a	376a	61.0a	57.4a	0.19a	0.15c

[1] Determined by acid-digestion of plant material and analysis by ICP

4. Conclusions

Three years of potentially toxic Al concentrations in the root zone of a Norway spruce stand did not affect forest vitality, needle fall and fine root growth. In the third year of treatment, Mg contents in current year's needles declined in the plots with the highest Al concentrations in the soil solution. The chemical composition of fine roots indicated that Al possibly blocked uptake of Mg. The field manipulation experiment reported here will be continued in 2000 and 2001.

Acknowledgments

Technical assistance of Magne Huse, Egil Kortnes and Claude Bélanger are gratefully acknowledged. Thanks are due to NISK's chemical laboratories for chemical analysis.

References

Brække, F.: 1994, Report nr. 15, Norwegian Forest Research Institute, Aas, Norway. in Norwegian.
Cronan, C. S. and Grigal, D. F.: 1995, *Journal of Environmental Quality* 24, 209
De Wit, H.A., Mulder, J., Nygaard, P.H., Aamlid, D., Huse, M., Kortnes, E., Wollebæk, G. and Brean, R.: *Water, Air and Soil Pollution,* **in press**
De Wit, H.A. 2000. Doctoral Thesis, Agricultural University of Norway, Aas, Norway
Godbold, D.L. and Kettner, C.: 1991, *Journal of Plant Physiology* 138, 231
Göransson, A. and Eldhuset, T. D.: 1991, *Trees* 5, 136
Göransson, A. and Eldhuset, T. D.: 1995, *Water, Air and Soil Pollution* 83, 351
Högberg, P. and Jensén, P.: 1994, *Water, Air and Soil Pollution* 75, 121
Klap, J.M., Voshaar, J.H.O., De Vries, W. and Erisman, J.W.: 2000, *Water, Air and Soil Pollution* 119, 387
Marschner, H.: 1995, *Mineral nutrition of higher plants*, Academic Press Limited, London
Matzner, H. and Murach, D.: 1995, *Water, Air and Soil Pollution* 85, 63
Mulder, J., De Wit, H.A., Boonen, L. and Bakken, L. *Water, Air and Soil Pollution,* **this volume**
Ogner, G. Wickstrøm, T., Remedios, G., Gjelsvik, S., Hensel, G. R., Jacobsen, J. E., Olsen, M., Skretting, E. and Sørlie, B.: 1999, Norwegian Forest Research Institute, Aas, Norway, p23
Taylor, G.J.: 1991, *Current Topics in Plant Biogeochemistry and Physiology* 10, 57
Tveite, B., Abrahamsen, G. and Huse, M.: 1993, 'Trees: Nutrition', in: Abrahamsen, G., A.O. Stuanes and B. Tveite, (eds.), *Long-Term experiments with Acid Rain in Norwegian Forest Ecosystems*. Ecological Studies 104, Springer-Verlag, New York, pp 140-179
Ulrich, B., Mayer, R. and Khanna, P. K.: 1980, *Soil Science* 130, 193
Vogt, K.A. and Persson H.: 1991, 'Measuring growth and development of roots', in: T. Hinckley and G.P. Laffoie (eds), *Technics and approaches in forest tree ecophysiology*, CRC Press, Florida, pp 477-501

SOIL SOLUTION CHEMISTRY IN FORESTS WITH GRANITE BEDROCK IN JAPAN

KAZUO SATO[*] and TAKASHI WAKAMATSU

Komae Research Laboratory, Central Research Institute of Electric Power Industry (CRIEPI), 2-11-1 Iwado-kita, Komae, Tokyo 201-8511, Japan
(*author for correspondence, e-mail: ks@criepi.denken.or.jp)

Abstract. Soil solutions were taken from three forest areas with granite bedrock in Japan (Abukuma, Tateyama and Hiroshima) to investigate pH values, forms of Al and the molar BC/Al ratios. In each area, 10 sites were chosen for study. At each site, a target tree was selected, and two soil solution samples were taken from 10 cm depth at points 10 cm and 100 cm from the trunk of the tree to evaluate the effects of stemflow and throughfall on soil solution chemistry. Values of pH of samples taken 10 cm from the trunks (referred to as S samples) and 100 cm from the trunks (referred to as T samples) ranged from 3.66 to 6.52 and from 4.55 to 6.48, respectively. For Japanese cedar (*Cryptomeria japonica*) and Japanese cypress (*Chamaecyparis obtusa*) trees, S samples showed lower pH than T samples, whereas the inverse relation was observed for broadleaf trees. In the Abukuma and Tateyama areas, the concentrations of monomeric Al (Al_m) were mostly below 30 µmol L^{-1}. In the Hiroshima area, however, extremely high Al_m concentrations (up to 293 µmol L^{-1}) were observed at some sites. The molar ratio of BC (= Ca + Mg + K) to inorganic monomeric Al was higher than 1 for all samples, except for an S sample from the Hiroshima area having a ratio of 0.72.

Keywords: soil solution, pH, aluminum speciation, BC/Al ratio, granite

1. Introduction

Aluminum is toxic to the roots of plants (Godbold *et al.*, 1988). This toxicity can be mitigated by base cations such as Ca^{2+} (Cronan and Grigal, 1995). Thus, it may be reasonable to use the molar ratio of BC (= Ca + Mg + K) to Al in soil solution as an indicator of the potential for forest decline (Sverdrup and Warfvinge, 1993). In Japan, however, the present status of the BC/Al ratios in forest areas is virtually unknown. Furthermore, there have been only a few studies concerning the speciation of Al in soil solutions (Funakawa *et al.*, 1993; Kato *et al.*, 1995, 1999; Tsunoda *et al.*, 1997; Sato *et al.*, 2000). Cronan and Grigal (1995) recommended the use of inorganic Al in the BC/Al ratio to consider phytotoxicity.

The objectives of this study were to investigate pH values, forms of Al and the BC/Al ratios of soil solutions in forests with granite bedrock in Japan. Granite was chosen because soils developed from it show significantly low rates of base-cation weathering (Shindo and Hakamata, 1996), which offers a long-term acid-neutralizing capacity.

2. Materials and Methods

Soil solution sampling was conducted in the three forest areas (Abukuma, Tateyama and Hiroshima) shown in Figure 1. In each area, 10 sites were chosen

Figure 1. Location of the three forest areas studied

for study. The geology at all sites is granite. At each site, a target tree was selected, and two soil solution samples were taken at points 10 cm and 100 cm from the trunk of the tree to evaluate the effects of stemflow and throughfall on soil solution chemistry (Sato et al., 2000). Hereafter, the former samples are referred to as S samples (i.e., receiving stemflow) and the latter as T samples (i.e., receiving throughfall). The sampling depth was set to 10 cm because surface soils are subjected to higher acidity loading from external (acid deposition) and internal (e.g., nitrification and base-cation uptake) sources (Wakamatsu et al., 2000). Soil solution was collected in a 30 mL polyethylene tube, simultaneously using five ceramic porous cups that had been installed one day before the sampling. Suction

TABLE I
Site description

Area	Site	Dates of sampling (1998)	Target tree *	DBH (cm)	Soil type
Abukuma	A 1	7/25	Cryptomeria japonica (C)	37.5	Brown forest soil
	A 2	7/25 - 26	Quercus acutissima (B)	23.8	Brown forest soil
	A 3	7/24 - 25	Pinus densiflora (C)	20.7	Brown forest soil (dry)
	A 4	7/24 - 25	Cryptomeria japonica (C)	20.3	Brown forest soil (dry)
	A 5	7/27 - 29	Quercus acutissima (B)	16.8	Brown forest soil
	A 6	7/27 - 28	Quercus acutissima (B)	44.4	Brown forest soil
	A 7	7/28 - 29	Quercus acutissima (B)	35.3	Brown forest soil
	A 8	7/29 - 30	Cryptomeria japonica (C)	23.5	Brown forest soil
	A 9	7/31 - 8/01	Pinus densiflora (C)	30.5	Brown forest soil
	A10	7/26 - 27	Cryptomeria japonica (C)	36.9	Brown forest soil
Tateyama	T 1	7/10 - 11	Larix leptolepis (C)	42.0	Brown forest soil (wet)
	T 2	7/09 - 10	Pterocarya rhoifolia (B)	40.0	Brown forest soil (wet)
	T 3	7/11	Cryptomeria japonica (C)	29.0	Brown forest soil (wet)
	T 4	7/08 - 09	Cryptomeria japonica (C)	38.8	Brown forest soil (wet)
	T 5	7/09 - 10	Cryptomeria japonica (C)	28.6	Brown forest soil
	T 6	7/10	Cryptomeria japonica (C)	38.2	Brown forest soil
	T 7	7/11 - 12	Quercus mongolica (B)	42.0	Brown forest soil
	T 8	7/12 - 13	Cryptomeria japonica (C)	27.0	Brown forest soil
	T 9	7/12	Quercus mongolica (B)	31.0	Brown forest soil
	T10	7/08 - 09	Pinus densiflora (C)	23.0	Brown forest soil
Hiroshima	H 1	7/14 - 16	Pinus densiflora (C)	42.0	Residual regosol (coarse)
	H 2	7/15 - 16	Quercus acutissima (B)	24.2	Residual regosol (coarse)
	H 3	7/16 - 18	Cryptomeria japonica (C)	28.0	Brown forest soil (dry)
	H 4	9/25 - 26	Cryptomeria japonica (C)	19.7	Brown forest soil (dry)
	H 5	7/16 - 21	Quercus acutissima (B)	20.4	Brown forest soil (dry)
	H 6	8/06 - 10	Cryptomeria japonica (C)	39.8	Residual regosol (coarse)
	H 7	9/22 - 23	Larix leptolepis (C)	21.0	Residual regosol (coarse)
	H 8	8/06 - 09	Cryptomeria japonica (C)	19.4	Residual regosol (coarse)
	H 9	9/20 - 22	Chamaecyparis obtusa (C)	33.1	Residual regosol (coarse)
	H10	9/23 - 24	Castanea crenata (B)	9.8	Residual regosol (coarse)

* C and B in parentheses stand for conifers and broadleaf trees, respectively.

pressure was maintained at 667 hPa. The sampling was continued until more than 100 mL of solution overflowed from the tube to a flask attached downstream of the tube to reduce the pH error due to CO_2 degassing (Wakamatsu et al., 1999). Table I shows the target tree species selected and the soil type at each site.

The samples collected were analyzed for pH, major cations (Ca, Mg, K, Na and NH_4) and anions (SO_4, NO_3, Cl and F) by ion chromatography and total Al (Al_t) by ICP-AES. The speciation of monomeric Al was conducted using HPLC with fluorescence detection of the Al-lumogallion complex in combination with equilibrium calculations (Sato et al., 2000). The equilibrium constants used were those of Lindsay (1979).

3. Results and Discussion

3.1. pH

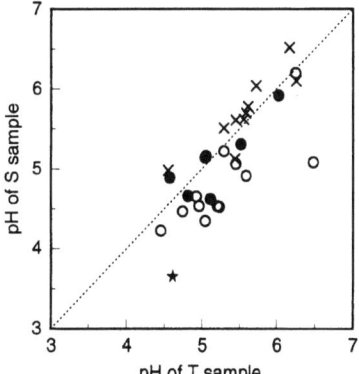

Figure 2. Correlation of pH between S and T samples. O : Japanese cedar, ★ : Japanese cypress, ● : pine, × : broadleaf trees

pH values of S samples and T samples ranged from 3.66 to 6.52 and from 4.55 to 6.48, respectively. For Japanese cedar (*Cryptomeria japonica*) and Japanese cypress (*Chamaecyparis obtusa*) trees, the pH of S samples was lower than that of T samples in all cases (Fig. 2). For pine trees (*Pinus densiflora* and *Larix leptolepis*), the difference between the two was not significant. In contrast, for broadleaf trees, the pH of S samples was slightly higher than that of T samples in most cases. These results are consistent with the following information. For Japanese cedar and Japanese cypress trees, the pH of stemflow is in the range of 3 to 4 and is always lower than that of throughfall, which causes marked acidification of soils and soil solutions around the trunk (Sassa et al., 1991; Yoneda et al., 1995; Sato and Takahashi, 1996; Baba and Okazaki, 1999; Sato et al., 2000). For broadleaf trees, however, the pH of stemflow is generally higher than that of throughfall (Sassa et al., 1991; Yoneda et al., 1995) and hence, higher values of soil pH are observed around the trunk (e.g., Yoneda et al., 1995). For pine trees, the pH of stemflow is always lower than that of throughfall (Sassa et al., 1991), but stemflow itself is not strongly acidic (pH 4.5-5.2 for *Pinus densiflora* and pH 4.2-4.8 for *Larix leptolepis*; Sassa et al., 1991). This may be the reason why pH values of S and T samples for pine trees were not significantly different.

3.2. FORMS OF AL

Figure 3 shows the results of Al speciation for soil solution samples. Values of pH of the samples are also shown. In the Abukuma and Tateyama areas, the total concentrations of monomeric Al species (Al_m) were mostly below 30 µmol L^{-1} and were composed of more than 50 % organic Al. In the Hiroshima area, however, extremely high Al_m concentrations were observed. For example, the Al_m concentration of the S sample from the H9 site was as high as 293 µmol L^{-1}. This was attributable to a considerably high Al^{3+} concentration (245 µmol L^{-1}). At the H6 site, even the T sample contained 69 µmol L^{-1} of Al_m. Regardless of these high Al levels, the target trees in the Hiroshima area were not visibly deteriorated.

David and Lawrence (1996) investigated the soil solution chemistry in the Oa horizon at red spruce (*Picea rubens*) stands in the northeastern United States where the species is undergoing regional decline. They reported that the concentrations of Al_t, Al_m and inorganic monomeric Al (Al_i) ranged from 11 to 54 µmol L^{-1}, from 6 to 29 µmol L^{-1} and from < 1.5 to 26 µmol L^{-1}, respectively. Their samples seemed to correspond to T samples in our study. The concentrations of Al_t,

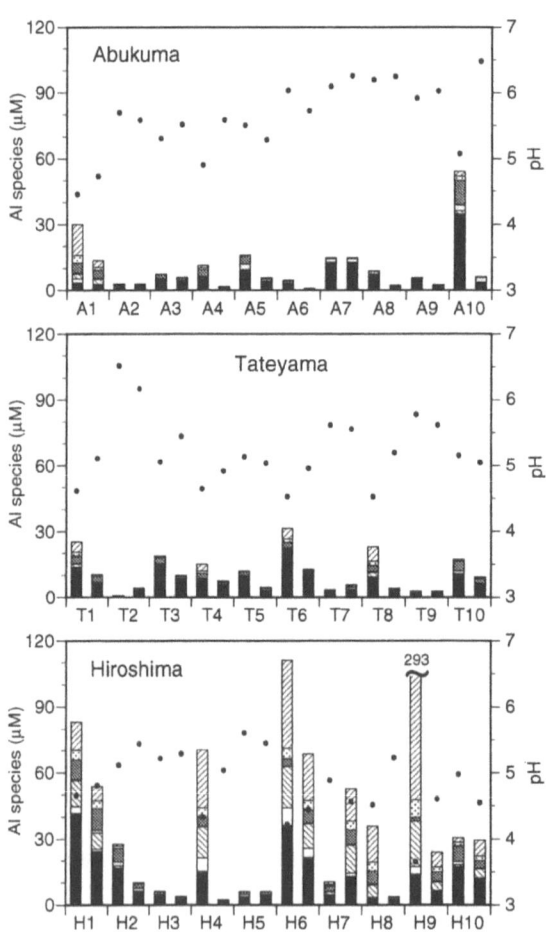

Figure 3. Concentrations of monomeric Al in S sample (left) and T sample (right) at each site.

Al_m and Al_i in T samples from the Abukuma and Tateyama areas were all within the ranges quoted above. In the Hiroshima area, however, they exceeded these ranges at three sites (H1, H6 and H7). According to the report of the Japan Environment Agency (1999), annual average pH values of precipitation during 1996-1998 at Kurahashijima in the Hiroshima area were 4.5-4.6. These values were not markedly different from those observed at Tateyama (pH 4.7-4.8) in the Tateyama area (there was no monitoring station in the Abukuma area). Thus, it is unlikely that acid deposition caused the high Al levels in the Hiroshima area. One of the possible explanations is the increase in ionic strength of soil solutions due to evaporation (Mulder and van Breemen, 1987; Sato *et al.*, 2000). The Hiroshima area is located in the Setouchi district that includes the Inland Sea. The climate in the district is characterized by little rain and a high percentage of sunshine (Nakamura *et al.*, 1990). Thus, surface soils in the area in July 1998 were extremely dry so that soil solutions were difficult to obtain at some sites (this was the reason for the postponement of sampling until August or September; see Table I). This would cause ionic strengths of the T samples to be higher (9.30×10^{-4} on average) than those in the Abukuma (3.07×10^{-4}) and Tateyama (2.87×10^{-4}) areas. Increases in ionic strength can be a cause of pH decline (Ross and Bartlett, 1992), as well as the enhancement of Al release (Sato *et al.*, 2000) due to cation exchange. It is noteworthy that Residual regosols widely distributed in the Hiroshima area possess low amounts of exchangeable base cations (Sato and Ohkishi, 1993).

3.3. BC/AL RATIOS

Figure 4 shows the molar BC/Al_i ratios of S and T samples from each site. The ratios ranged from 0.72 to 630 (geomean 22) for S samples and from 4.5 to 266 (geomean 40) for T samples. When Al_t was used instead of Al_i, the ratios ranged from 0.66 to 98 (geomean 6.1) for S samples and from 2.2 to 66 (geomean 9.4) for T samples. In Europe, the value of 1 is often used as the critical BC/Al ratio for coniferous forests (Posch *et al.*, 1995). In our study, only an S sample from the H9 site had BC/Al ratios lower than 1 ($BC/Al_i = 0.72$, $BC/Al_t = 0.66$). This is likely due to strongly acidic stemflow of the target tree (*Chamaecyparis obtusa*). Aluminum concentrations were significantly higher at some sites in the Hiroshima area (see Fig. 3), but BC concentrations were also high, resulting in BC/Al_i (and BC/Al_t) ratios not markedly different from those of other areas.

In order to investigate the BC/Al ratios in deeper soil horizons (root zone), we

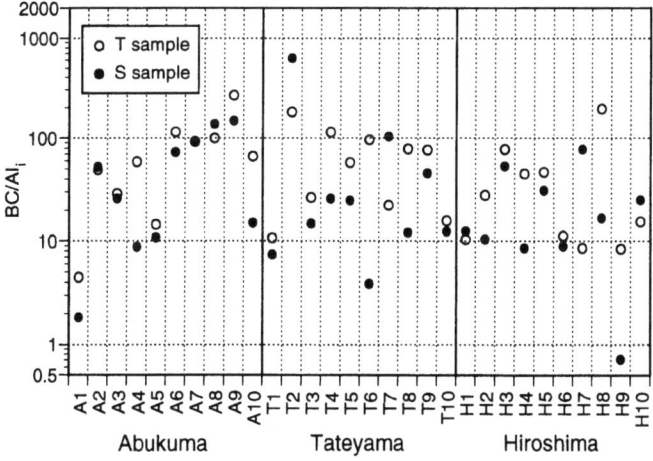

Figure 4. Molar ratios of BC to inorganic monomeric Al.

collected T samples from 25 cm and 50 cm depths at 25 sites in the Hiroshima area including the H1-H10 sites in August 1999. The BC/Al_t ratios ranged from 1.4 to 268 (geomean 12) for samples from 25 cm depth and from 1.5 to 435 (geomean 25) for samples from 50 cm depth (Al_i was not detected at some sites). These results suggest that the BC/Al_t ratios become higher in deeper soil horizons and that roots of trees are not exposed to severe Al stress in Japan. This situation is fairly different from that at red spruce sites in the northeastern U.S. where the ratios of soil solution Ca to inorganic Al in the B horizon at the 12 sites studied are all below 1 (David and Lawrence, 1996).

Acknowledgement

This study was supported by the Japanese Ministry of International Trade and Industry.

References

Baba, M. and Okazaki, M.: 1999, *Soil Sci. Plant Nutr.* **45**, 321.
Cronan, C. S. and Grigal, D. F.: 1995, *J. Environ. Qual.* **24**, 209.
David, M. B. and Lawrence, G. B.: 1996, *Soil Sci.* **161**, 314.
Funakawa, S., Hirai, H. and Kyuma, K.: 1993, *Soil Sci. Plant Nutr.* **39**, 281.
Godbold, D. L., Fritz, E. and Huttermann, A.: 1988, *Proc. Natl. Acad. Sci. U.S.A.* **85**, 3888.
Japan Environment Agency: 1999, *Acid Deposition Survey: Phase 3, Final Report*.
Kato, H., Shirai, M. and Matsukawa, S.: 1995, *Jpn. J. Soil Sci. Plant Nutr.* **66**, 39.
Kato, H., Hoshino, K., Matsukawa, S., Hirai, H. and Zhu, X.: 1999, *Jpn. J. Soil Sci. Plant Nutr.* **70**, 291.
Lindsay, W. L.: 1979, *Chemical Equilibria in Soils*. John Wiley & Sons, Inc., New York.
Mulder, J. and van Breemen, N.: 1987, 'Differences in aluminium mobilization on Spodosols in New Hampshire (USA) and The Netherlands as a result of acid deposition', in T. C. Hutchinson and K. M. Meema (eds.), *Effects of atmospheric pollutants on forests, wetlands and agricultural ecosystems*, Springer, Berlin, pp. 361-376.
Nakamura, K., Kimura, R. and Uchijima, Z.: 1990, *Nippon no Kikou*. Iwanami Shoten.
Posch, M., de Smet, P. A. M., Hettelingh, J.-P. and Downing, R. J.: 1995, *Calculation and mapping of critical thresholds in Europe*. RIVM Report No. 259101004.
Ross, D. S. and Bartlett, R. J.: 1992, *Soil Sci. Soc. Am. J.* **56**, 1796.
Sassa, T., Goto, K., Hasegawa, K. and Ikeda, S.: 1991, *Jpn. J. For. Environ.* **32**, 43.
Sato, K. and Ohkishi, H.: 1993, *Ambio* **22**, 232.
Sato, K. and Takahashi, A.: 1996, *Environ. Sci.* **9**, 221.
Sato, K., Wakamatsu, T. and Takahashi, A.: 2000, *Jpn. J. Soil Sci. Plant Nutr.* **71**, 615.
Shindo, J. and Hakamata, T.: 1996, *Application and Evaluation of the Critical Load Model in Japanese Ecosystems*. Proc. Int. Sym. Acid. Dep. Its Impacts, 10-12, Dec., Tsukuba, Japan.
Sverdrup, H. and Warfvinge, P.: 1993, *The effect of soil acidification on the growth of trees, grass and herbs as expressed by the (Ca+Mg+K)/Al ratio*. Lund Univ., Rep. 1993:2.
Tsunoda, K., Yagasaki, T., Aizawa, S., Akaiwa, H. and Satake, K.: 1997, *Anal. Sci.* **13**, 757.
Wakamatsu, T., Sato, K. and Takahashi, A.: 1999, *Jpn. J. Soil Sci. Plant Nutr.* **70**, 775.
Wakamatsu, T., Sato, K., Takahashi, A. and Shibata, H.: 2000, *Water Air Soil Pollut.* (this volume)
Yoneda, Y., Shibata, E., Sumi, T. and Waguchi, Y.: 1995, *Trans. Kansai Branch Jpn. For. Soc.* **4**, 47.

GROWTH, NET PHOTOSYNTHETIC RATE, NUTRIENT STATUS AND SECONDARY XYLEM ANATOMICAL CHARACTERISTICS OF *Fagus crenata* SEEDLINGS GROWN IN BROWN FOREST SOIL ACIDIFIED WITH H_2SO_4 SOLUTION

TAKESHI IZUTA[1]*, TAEKO YAMAOKA[1], TATSURO NAKAJI [1], TETSUSHI YONEKURA[1], MASAAKI YOKOYAMA[2], HIDEYUKI MATSUMURA[3], SACHIE ISHIDA[4], KENICHI YAZAKI [4], RYO FUNADA[4] and TAKAYOSHI KOIKE [4]

[1] *Tokyo University of Agriculture and Technology, Fuchu, Tokyo 183-8509, Japan;* [2] *Horiba Ltd., Higashi-Kanda, Chiyoda, Tokyo 101-0031, Japan;* [3] *Central Research Institute of Electric Power Industry, Abiko, Chiba 270-1194, Japan;* [4] *Hokkaido University, Sapporo 060-8589, Japan*
(* *author for correspondence, e-mail: izuta@cc.tuat.ac.jp*)

Abstract. Dry matter production, net photosynthetic rate, leaf nutrient status and trunk anatomical characteristics of *Fagus crenata* seedlings grown in brown forest soil acidified by adding H_2SO_4 solution were investigated. The soil acidification leaded to decreased (Ca+Mg+K)/Al molar ratio in the soil solution. Dry mass per plant of the seedlings grown in the soil treated with H^+ at 120 mg·L^{-1} was significantly reduced compared with the control value at 0 mg·L^{-1}. When net photosynthetic rate was reduced in the seedlings grown in the soil treated with H^+ at 120 mg·L^{-1}, the carboxylation efficiency and maximum net photosynthetic rate at saturated CO_2-concentration were lower than the control values. The addition of H^+ to the soil at 120 mg·L^{-1} induced a reduction in the concentration of Ca in the leaf. By contrast, the concentration of Al in the leaf was increased with increasing the amount of H^+ added to the soil. The annual ring formed in the seedlings grown in the soil treated with H^+ at 120 mg·L^{-1} was significantly narrower than that at 0 (control), 10, 30, 60 or 90 mg·L^{-1}. Based on the results obtained in the present study, we conclude that *Fagus crenata* is relatively sensitive to a reduction in the (Ca+Mg+K)/Al molar ratio of soil solution compared with *Picea abies*.

Keywords: soil acidification, (Ca+Mg+K)/Al ratio, *Fagus crenata* Blume

1. Introduction

During the past decade, acid deposition is considered to be one of the serious environmental problems in Japan and other Asian countries. However, very limited information is available on the effects of soil acidification and/or phytotoxic metals dissolved in acidic soil such as Al on Japanese forest tree species (Izuta *et al.*, 1996a ; Izuta *et al.*, 1997; Lee *et al.*, 1997a; Miwa *et al.*, 1998). Furthermore, the effects of soil acidification on physiological functions and anatomical characteristics of Japanese forest tree species are still not clarified.

To obtain basic data concerning the effects of soil acidification due to acid deposition on Japanese deciduous broad-leaved tree species and the evaluation of critical load of acid deposition for protecting Japanese forest ecosystems, we investigated dry matter production, leaf gas exchange rates, leaf nutrient status and trunk anatomical characteristics of *Fagus crenata* seedlings grown in brown forest soil acidified with H_2SO_4 solution. We selected *Fagus crenata* as the plant material because this tree is the most typical Japanese deciduous broad-leaved tree and forest decline of this tree can be observed in Japan.

Water, Air, and Soil Pollution **130:** 1007–1012, 2001.
© 2001 *Kluwer Academic Publishers.*

2. Materials and Methods

Brown forest soil originated from granite as a mother rock was collected from the University Forest (Kusaki, Gunma, Japan). In March 1997, five different amounts of H^+ (10, 30, 60, 90 and 120 mg per one liter of air-dried soil) were gradually added to the soil as H_2SO_4 solution. Control soil was not supplemented with H^+. To artificially change the (Ca+Mg+K)/Al molar ratios of soil solution, control or acidified soil was put in a plastic container and soaked for 3 days in deionized water to conduct leaching of cations from the soil. Then, the soil was air-dried and sieved through a 5 mm-screen.

Immediately before transplanting *Fagus crenata* seedlings to the control or acidified soil, soil solution was collected from the potted soil using a soil solution sampler (Eijkelkamp, The Netherlands). The soil solution pH, cation concentration and total sulfur content of the soil were measured with a pH meter (Model M-12, Horiba, Japan), atomic absorption spectrophotometer (AA-670/GV-6, Shimadzu, Japan) and combustion sulfur content analyzer (EMIA-120, Horiba, Japan), respectively. After the final harvest of the seedlings, elemental concentrations in the leaf were determined with the atomic absorption spectrophotometer and combustion sulfur content analyzer.

On 26 April 1997, three-year-old seedlings of *Fagus crenata* Blume were transplanted into 1.5-L pots filled with the control or acidified soil. Then, all the seedlings were grown for 153 days from 26 April to 26 September 1997 in a naturally-lit phytotron at $20.0 \pm 1.0/13.0 \pm 1.0$ °C (6:00-18:00/18:00-6:00) and $70 \pm 5\%$ R.H. On 6 June, 4 August and 26 September, the seedlings were harvested to determine leaf area and dry mass. On 24 July, measurements of leaf gas exchange rates were carried out with a portable steady-state infra-red gas analysis system (LCA-4A, ADC Ltd., UK).

At the final harvest (26 September), small blocks were cut from the trunk of the seedlings at 3-4 cm above the ground, and then fixed in a solution of formaldehyde-acetic acid-ethanol-water (1:1:9:9, v/v). Fixed samples were dehydrated through a graded ethanol series, and then embedded in celloidin. Transverse sections of 13 μm in thickness were cut with a steel knife on a sliding microtome (LS-113, Yamato-Kohki, Japan), stained with a 1% aqueous solution of safranin, and observed with a light microscope (BHS-2, Olympus, Japan).

TABLE I

The pH and elemental concentrations in soil solution, and total sulfur content of brown forest soil immediately before transplanting *Fagus crenata* seedlings into the potted soil. Each value is the mean of 4 determinations.

H^+ load (mg·L^{-1})	pH	Ca (mmol·L^{-1})	Mg (mmol·L^{-1})	K (mmol·L^{-1})	Al (mmol·L^{-1})	(Ca+Mg+K)/Al (mol·mol^{-1})	S (%)
0	4.23	6.38	1.90	0.49	0.78	10.90	0.048
10	4.31	5.08	1.73	0.30	0.76	9.31	0.051
30	4.44	2.57	0.87	0.35	0.36	10.62	0.073
60	4.25	2.00	0.52	0.40	0.56	5.34	0.097
90	4.08	2.50	0.65	0.38	2.03	1.69	0.109
120	3.84	2.30	0.58	0.52	4.06	0.84	0.137

3. Results

Table I shows the pH and elemental concentrations in soil solution of brown forest soil immediately before transplanting *Fagus crenata* seedlings into the potted soil. The pH was less in the soil treated with H^+ at 90 or 120 mg·L^{-1} than the control value (0 mg $H^+·L^{-1}$). The addition of H^+ to the soil at 90 or 120 mg·L^{-1} and leaching of cations from the soil induced an increase in the concentration of Al and a reduction in the concentrations of Ca and Mg, respectively. As a result, (Ca+Mg+K)/Al molar ratio in the soil treated with H^+ at 60, 90 or 120 mg·L^{-1} was less than the control value. Total sulfur content was increased with increasing the amount of H^+ added as H_2SO_4 solution to the soil.

Total dry mass per plant of *Fagus crenata* seedlings and photosynthetic parameters are shown in Figure 1. On 6 June, there were no significant differences among the total dry mass per plant of the seedlings grown in the soil treated with H^+ at 0-120 mg·L^{-1}. On 4 August and 26 September, total dry mass per plant of the seedlings grown in the soil treated with H^+ at 120 mg·L^{-1} was significantly less than the control value. The trend in the leaf area per plant was similar to that in the total dry mass per plant (data not shown). On 24 July, net photosynthetic rate at 350 μmol·mol^{-1} CO_2 (A_{350}) was reduced with increasing the amount of H^+ added to the soil. However, the addition of H^+ to the soil did not induce significant changes in the stomatal diffusive conductance to CO_2 and leaf dark respiration rate (data not shown). The carboxylation efficiency (CE) and maximum net photosynthetic rate at saturated CO_2-concentration (A_{max}) of the seedlings grown in the soil treated with H^+ at 90 or 120 mg·L^{-1} were significantly less than the control values. By contrast, the quantum yield (QY) was not significantly affected by the addition of H^+ to the soil.

Table II shows the elemental concentrations in the leaf of *Fagus crenata* seedlings at the final harvest. The addition of H^+ to the soil induced a significant reduction in the concentration of Ca in the leaf. By contrast, the concentrations

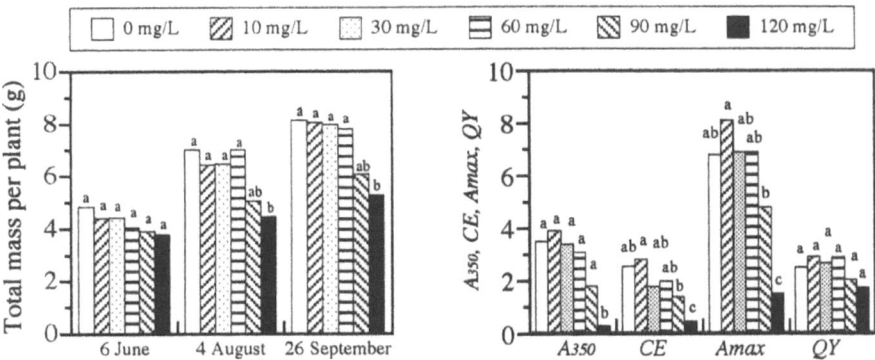

Figure 1. Total dry mass per plant, net photosynthetic rate at 350 μmol·mol^{-1} CO_2 (A_{350}, μmol·m^{-2}·s^{-1}), carboxylation efficiency (CE, 10^{-2}·μmol·m^{-2}·s^{-1}), maximum net photosynthetic rate at saturated CO_2-concentration (A_{max}, μmol·m^{-2}·s^{-1}) and quantum yield (QY, mol·mol^{-1}) of *Fagus crenata* seedlings grown in brown forest soil treated with H^+ at 0 (control), 10, 30, 60, 90 or 120 mg·L^{-1}. Each bar shows the mean of 4 determinations. The bars below different letters are significantly different according to Duncan's multiple range test ($p<0.05$).

TABLE II

The concentrations of Ca, Mg, K, Al and S in the leaf of *Fagus crenata* seedlings grown for 153 days in brown forest soil treated with H⁺ at 0 (control), 10, 30, 60, 90 or 120 mg·L⁻¹. Each value is the mean of 4 determinations. The values followed by different letters in each column are significantly different according to Duncan's multiple range test ($p<0.05$).

H⁺ load (mg·L⁻¹)	Ca ($\mu g \cdot g^{-1}$)	Mg ($\mu g \cdot g^{-1}$)	K ($\mu g \cdot g^{-1}$)	Al ($\mu g \cdot g^{-1}$)	S (%)
0	17.3 a	2.6 ab	4.6 ab	0.3 c	0.166 b
10	11.5 b	1.8 b	4.3 bc	0.7 c	0.172 b
30	10.4 b	2.0 ab	3.3 d	0.5 c	0.256 b
60	13.2 ab	2.0 ab	3.2 d	1.1 b	0.399 a
90	13.1 b	2.8 a	3.4 cd	1.3 b	0.528 a
120	10.7 b	2.6 ab	5.4 a	1.7 a	0.531 a

of Al and S in the leaf of the seedlings grown in the soil treated with H⁺ at 60, 90 or 120 mg·L⁻¹ were significantly greater than the control values.

As shown in Figure 2, the widths of annual ring formed in 1997 of *Fagus crenata* seedlings grown in the soil treated with H⁺ at 0, 30, 60 or 90 mg·L⁻¹ were almost similar to those in 1996. The annual ring formed in 1997 of the seedlings grown in the soil treated with H⁺ at 10 mg·L⁻¹ was approximately 1.8 times wider than that in 1996. By contrast, the annual ring formed in 1997 of the seedlings grown in the soil treated with H⁺ at 120 mg·L⁻¹ was approximately one-fifth of that in 1996. The annual ring formed in 1997 of the seedlings treated with H⁺ at 120 mg·L⁻¹ was significantly narrower than that at 0, 10, 30, 60 or 90 mg·L⁻¹. In addition, the seedlings grown in the soil treated with H⁺ at 120 mg·L⁻¹ tended to form small vessel elements in 1997.

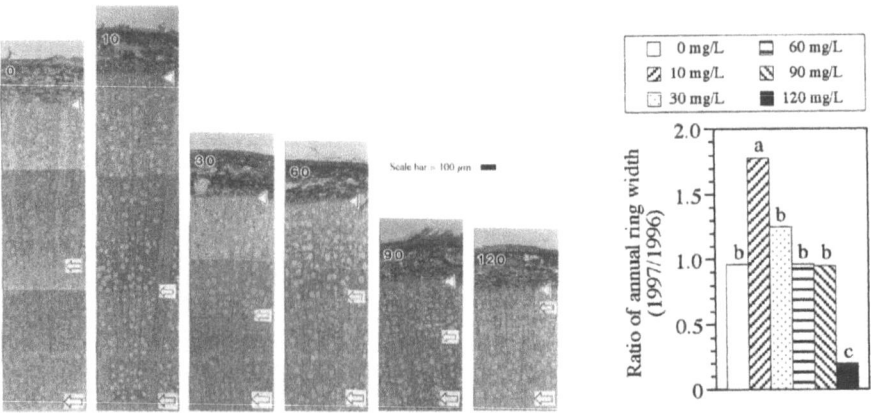

Figure 2. Light micrographs of transverse sections of trunk and change in the width of annual ring of *Fagus crenata* seedlings grown in the control or acidified soil. In the photograph, large and small arrows indicate the boundary of annual rings formed between 1995 and 1996, and that formed between 1996 and 1997, respectively. The arrowheads indicate the cambial cells. Scale bar = 100 μm. In the figure, the ratios of width of annual ring formed in 1997 to that formed in 1996 are shown, and each bar shows the mean of 6 determinations. The bars below different letter are significantly different according to Duncan's multiple range test ($p<0.05$).

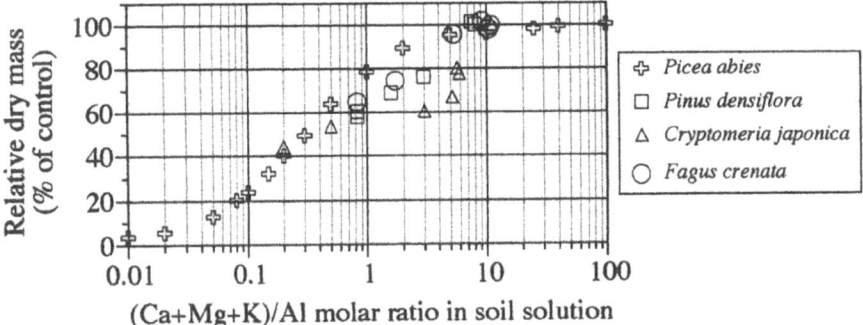

Figure 3. The relationships between (Ca+Mg+K)/Al molar ratio in soil solution and relative dry mass of *Fagus crenata* (this study), *Cryptomeria japonica* (Izuta *et al.*, 1997) and *Pinus densiflora* (Lee *et al.*, 1997a). The relationship between the molar ratio of nutrient or soil solution and biomass growth in % of control of *Picea abies* is also shown in this figure (Sverdrup *et al.*, 1994).

4. Discussion

In general, important factors related to the reduction in the dry matter production of woody plants grown in acidic soil are considered to be Al itself and molar ratios of base cations to Al in the soil solution, such as Ca/Al and (Ca+Mg+K)/Al (Cronan and Grigal, 1995; Sverdrup *et al.*, 1994). Total dry mass per plant of *Fagus crenata* seedlings was reduced with increasing the concentration of Al or with reducing the (Ca+Mg+K)/Al molar ratio in the soil solution (Figure 1, Table I and Figure 3). Therefore, the reduction in the dry matter production of the seedlings grown in the acidified soil is considered to be mainly induced by the toxicity of Al dissolved in the soil solution.

When net photosynthetic rate (A) was reduced in *Fagus crenata* seedlings grown in the soil treated with H^+ at 120 mg·L^{-1}, the CE and $Amax$ of the seedlings were significantly less than the control values (Figure 1). These results indicate that excess Al-induced reduction in the net photosynthetic rate of *Fagus crenata* seedlings grown in the acidified soil was mainly due to the reductions in the quantity and/or activity of ribulose-1,5-bisphosphate (RuBP) carboxylase/oxygenase and the regeneration rate of RuBP (von Caemmerer and Farquhar, 1981). As shown in Table II, the concentrations of Ca in the leaf of *Fagus crenata* seedlings grown in the acidified soil were significantly less than the control value. Similar results induced by artificial soil acidification or excess Al in nutrient solution were obtained in *Cryptomeria japonica* (Izuta *et al.*, 1996b), *Pinus densiflora* (Lee *et al.*, 1997b) and European tree species such as *Betula pendula* and *Picea abies* (Göransson and Eldhuset, 1987, 1991). Therefore, Al-induced inhibition of net photosynthesis and nutrient deficiency are considered to be the limiting factors related to the reduction in the dry matter production of *Fagus crenata* seedlings grown in the acidified soil.

The addition of H^+ to the soil at 120 mg·L^{-1} suppressed radial increments in the trunk of *Fagus crenata* seedlings (Figure 2). Such a change in the width of annual ring might be due to the low frequency of cambial division and/or the short period of cambial division in the trunk (Funada *et al.* 1990). Dunisch and Bauch (1994) reported that soil condition such as fertilization and soil substrate affected cambial activity and content of mineral elements in the cambial region, which resulted in changes in the width of annual ring. Therefore, the reduction in the annual ring width of *Fagus crenata* seedlings grown in the soil treated with H^+ at 120 mg·L^{-1} may be due to changes in the cambial activity caused by low content of mineral elements in the cambial cells. Furthermore, *Fagus crenata* seedlings grown in the soil treated with H^+ at 120 mg·L^{-1} tended to form small vessel elements (Figure 2). Vessel elements play an important role in the water transport of dicotyledon trees (Utsumi et al., 1998). The rate at which water flows through a vessel is proportional to the fourth power of the radius of vessel (Zimmerman, 1983; Aloni, 1991). Therefore, both of fewer vessel elements related to narrow annual ring and narrow diameters of vessel elements of trunk might decrease water conductivity of *Fagus crenata* seedlings grown in the soil treated with H^+ at 120 mg·L^{-1}.

In Europe, critical load of acid deposition for protecting forest ecosystem has been already evaluated by several models based on a (Ca+Mg+K)/Al molar ratio in nutrient or soil solution of 1.0 (Sverdrup and de Vries, 1994). As shown in Figure 3, when the (Ca+Mg+K)/Al molar ratio in soil solution of brown forest soil was 1.0, relative whole-plant dry mass of *Fagus crenata* (this study), *Cryptomeria japonica* (Izuta et al., 1997) or *Pinus densiflora* (Lee et al., 1997a) was reduced by 35~40% compared with the control values. Therefore, these typical Japanese forest tree species are considered to be relatively sensitive to a reduction in the (Ca+Mg+K)/Al molar ratio in soil solution compared with European forest tree species such as *Picea abies* (Sverdrup et al., 1994). It is necessary to investigate whether it is possible to use the (Ca+Mg+K)/Al molar ratio in soil solution of 1.0 as the criterion for assessing the critical load of acid deposition for protecting Japanese forest ecosystem.

References

Aloni, R.: 1991, 'Wood formation in deciduous hardwood trees', in: A. S. Raghavendra (ed.), *Physiology of Trees*. John Wiley & Sons, New York, pp. 175-197.
Cronan, C. S. and Grigal, D. F.: 1995, *Journal of Environmental Quality* **24**, 209.
Dunisch, O. and Bauch, J.: 1994, *Holzforschung* **48**, 447.
Funada, R., Kubo, T. and Fushitani, M.: 1990, *Inter. Assoc. Wood Anato. Bull.* **11**, 281.
Göransson, A. and Eldhuset, T. D.: 1987, *Physiologia Plantarum* **69**, 193.
Göransson, A. and Eldhuset, T. D.: 1991, *Trees* **5**, 136-142.
Izuta, T., Seki, T. and Totsuka, T.: 1996a, *Environmental Sciences* **4**, 233.
Izuta, T., Yamada, A., Miwa, M., Aoki, M. and Totsuka, T.: 1996b, *Environmental Sciences* **4**, 113.
Izuta, T., Ohtani, T. and Totsuka, T.: 1997, *Environmental Sciences* **5**, 177.
Lee, C. H., Izuta, T., Aoki, M. and Totsuka, T.: 1997a, *Journal of Japan Society for Atmospheric Environment* **32**, 46.
Lee, C. H., Izuta, T., Aoki, M. and Totsuka, T.: 1997b, *Journal of Japan Society for Atmospheric Environment* **32**, 371.
Miwa, M., Izuta, T. and Totsuka, T.: 1998, *Journal of Japan Society for Atmospheric Environment* **33**, 81.
Sverdrup, H. and de Vries, W.: 1994, *Water, Air and Soil Pollution* **72**, 143.
Sverdrup, H., Warfvinge, P. and Nihlgård, B.: 1994, *Water, Air and Soil Pollution* **78**, 1.
Utsumi, Y., Sano, Y., Fujikawa, S., Funada, R. and Ohtani, J.: 1998, *Plant Physiology* **117**, 1463.
von Caemmerer, S. and Farquhar, G. D.: 1981, *Planta* **153**, 376.
Zimmermann, M. H.: 1983, *Xylem Structure and Ascent of Saps*, Springer-Verlag, Berlin , New York.

DIFFERENTIAL ABILITY OF THE ROOT TO CHANGE RHIZOSPHERE PH BETWEEN *CHAMAECYPARIS OBTUSA SIEB.* (HINOKI) AND *QUERCUS SERRATA* THUMB.(KONARA) UNDER ALUMINIUM STRESS

RIE TOMIOKA* and CHISATO TAKENAKA

Graduate School of Bioagricultural Sciences, Nagoya University, Nagoya 464-8601, Japan
(* author for correspondence, e-mail: i991027m@mbox.media.nagoya-u.ac.jp)

Abstract. Aluminium has a strong effect on many plant species, and there are reports of differing responses to Al between coniferous and broadleaf trees. A study was conducted on the effect of various concentrations of Al in nutrient solutions on tree root systems, using *Chamaecyparis obtusa* Sieb. (hinoki) and *Quercus serrata* Thumb. (konara) seedlings. Konara showed a tendency to enhance its root growth with increasing Al concentration in nutrient solution. The ability to change the rhizosphere pH decreased for both hinoki and konara with increasing Al concentration, but konara showed a higher pH change ability. It is suggested that konara and hinoki have different strategies to adjust the pH of acid soil.

Keywords: aluminium, rhizoshpere pH, root, *Chamaecyparis obtusa* Sieb., *Quercus serrata* Thumb.

1. Introduction

Forest decline has been a major problem in Europe and North America, and one of the suggested causes is considered soil acidification induced by acid deposition (Matzner and Ulrich, 1985). There are no similar reports for Japan but declines have been reported for *Cryptomeria japonica* D. Don (Japanese cedar) (Sekiguti *et al.*, 1986), *Abies firma* (Igawa *et al.*, 1991) and *Fagus crenata* Blume (Maruta and Usui, 1997). The toxic effects of aluminium (Al) on plants is well known and is one of the main factors limiting plant growth in acidic soil (Wright, 1989). Most forest soils, especially coniferous forest soils, are acidic (pH 4.0- 6.0). Japanese cedar grows well in soil at pH 5.0 and *Chamaecyparis obtusa* Sieb. (hinoki) shows the best growth at pH 4.0 (Tutumi, 1962). In addition, some researchers have reported that growth of both below- and above-ground tissues of broadleaf trees may be enhanced at Al concentrations harmful to coniferous trees (Oda *et al.*, 1996). It is thought that some trees have an ability to adapt to naturally acidic soils, including a tolerance to the Al dissolved by soil acidification.

Most of the reports on the effects of Al on trees deal with conifers (*e.g.* Kohno *et al.*, 1998; Wissemeier *et al.*, 1998) and focus on the below- and above-ground biomass and on the nutrient contents (Lee *et al.*, 1997; Kohno *et al.*, 1995). There are a few reports about the effects of Al on broadleaf trees. Few studies which deal with both Al induced changes in tree root metabolism and rhizosphere pH have been reported. In order to understand the rhizosphere environment of trees and the adjustment strategies of tree roots to acid soil, we studied the influence of Al on changes in rhizosphere pH and root morphology.

2. Materials and Methods

One-year-old *Quercus serrata*-Thumb. (konara) and 2-year-old *Chamaecyparis obutusa* (hinoki) seedlings were grown in a phytotoron using hydroponics, at 27 °C and 70% relative humidity. The nutrient solution was 1/5 Hoagland's No.2.

After 2 weeks growth in the nutrient solution, the seedlings were exposed for 9 weeks to nutrient solution containing 0, 0.1, 0.25, 0.5, 1.0. 2.5 or 5.0 mM aluminium chloride 6 hydrate ($AlCl_3.6H_2O$). The activity of Al in nutrient solution was 0.005 at 5.0 mM concentration. The volume of nutrient solutions were 1L for hinoki and 200mL for konara per seedling. The solution pH was adjusted to 3.5 ± 0.1 to prevent precipitation of the Al and readjusted every 2 days with 1N hydrochloric acid (HCl). At this pH Al exists as Al^{3+} in the nutrient solution. Every 2 weeks the solution was changed and the solution pH was measured. Before and after the exposure to Al for 9 weeks, the fresh weight and the shoot length of the seedlings were measured. In addition, the morphology of their root tips was observed after exposure to Al. The seedlings were oven-dried at 80 °C for 48 h to determine root and shoot dry weight.

3. Results and Discussion

3.1. Dry weight ratio

There was no significant difference in growth among the treatments. The influence of Al concentration on the dry weight ratio of root to whole plant for konara and hinoki is shown in Figure 1. However, the ratio for konara showed a tendency

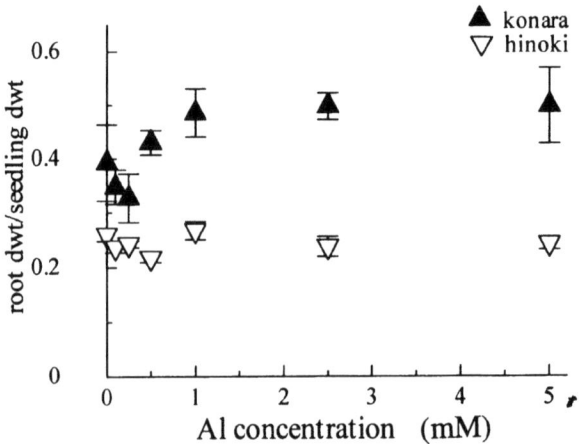

Figure 1. Effects of Al concentration in the nutrient solution on the ratio of root to whole plant dry weight. Error bars indicate standard error.

to increase at high Al concentration (above 0.5 mM). This agrees with the report that the dry weight of roots increased with Al treatment in *Quercus acutissima* Carr (Oda *et al.*, 1996). These results suggest that the growth of konara roots is enhanced by Al treatment.

Kohno *et al.* (1998) reported that the dry weight of hinoki roots decreased significantly after 14-weeks hydroponic treatment with 5.0 mM Al. Hirano *et al.* (2000) reported that there was no significant difference in the dry weight ratio between Al treatment and control, although the proportion of newly grown roots was decreased by Al treatment. In our study, there was no significant difference in the ratio after 9 weeks of Al treatment, but the root tip, which is the most vital part of the root, changed morphologically in response to Al treatment (2.5 mM). Thus if the treatment were continued, the ratio might decrease.

3.2. ABILITY TO CHANGE pH

In all nutrient solutions the pH rose during exposure to the solution (Fig. 2A and B). The pH change ability was calculated using the following equation (Wagatuma, 1982):

pH change ability = (measured pH – 3.5) / root dry weight (g)

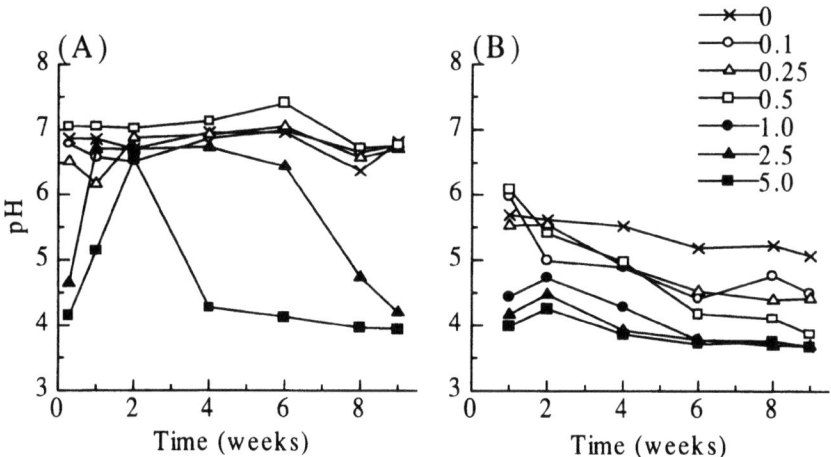

Figure 2. Change in pH of the nutrient solution containing various concentration of Al (A) konara, (B) hinoki.

The calculated abilities of konara and hinoki under various concentrations of Al treatments are shown in Figures. 3A and B. A high value indicates that the ability to neutralize the solution is high. In the controls, the pH change ability of konara was about 75 times as high as that of hinoki. The change in pH of nutrient solution was different between konara and hinoki depending on whether Al was present or not. In konara, the pH change ability decreased as the Al concentration increased (Fig. 3A), but pH of nutrient solution with Al treatment maintained as high as that of control (Fig. 2A). This suggests that the root of konara may adjust its rhizosphere pH at the individual level.

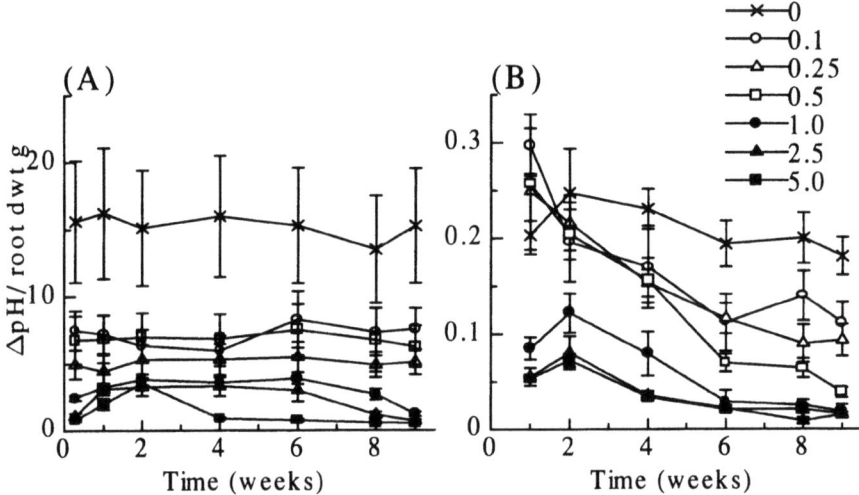

Figure 3. The pH change ability of konara (A) and hinoki (B) by treatments with various concentration of Al. Error bars indicate standard error.

In konara the pH change ability decreased as the Al concentration increased except for 0.5 mM Al treatment. The pH of nutrient solution with 0.5 mM Al rose more than that of control and kept higher pH through the treatment (Fig. 2A). In addition, the dry weight ratio was increasing at the point of 0.5 mM Al in the nutrient solution (Fig. 1). These results suggest that 0.5 mM might be a critical point of Al concentration to promote the Al tolerance mechanism.

It has been thought that Al resistance requires Al exclusion from roots or Al detoxification in the root cell (Ma *et al.*, 1998). Al exclusion mechanism includes binding Al in the cell wall, pH barrier and secretion of Al chelate substances (Delhaize *et al.*, 1993a, b; Pellet *et al*, 1995). When treated with Al, Al tolerant wheat rises its rhizosphere pH more than sensitive wheat (Taylor and Foy, 1985; Miyasaka *et al.*, 1989). And Degenhardt *et al.* (1998) demonstrated that Al tolerant Arabidopsis mutant induces net H^+ influx increase by Al treatment and causes a rise in root surface pH. A small increase in pH causes a significant alternation of Al species and less Al toxicity (Martell and Motelaitis, 1989). Although the pH change abilities of root with Al treatments were lower than those of control, the solution pHs maintained as high as those of control except 5.0 mM Al treatment. This indicates konara root might reduce Al toxicity by rising its rhizosphrer pH.

In hinoki, the pH of nutrient solution and the value of pH change ability decreased gradually with time which was by Al. With 0.1, 0.25, and 0.5 mM Al treatment, the pH values were higher than those of control at the initial stage but they were lower with time. This indicates that root cell membrane of hinoki was injured by additional Al treatment and its functions were gradually inhibited. In the case of 1.0, 2.5 and 5.0 mM Al treatment, the values were low initially, but increased once and dropped after 2 weeks as the treatment on konara with 5.0 mM Al. The morphological change was observed in the roots of hinoki with 2.5 and 5.0 mM Al treatment, for example, stunted and brown brunching roots were shown. But that was not observed in the konara root with the same treatment. This indicates that the roots of hinoki is more sensitive to Al and cell membrane function is damaged.

4. Conclusion

It is concluded that konara and hinoki have different strategies for responding to increases acidity and Al. Konara has a high ability to change rhizosphere pH and to enhance the growth of root in the presence of Al. Hinoki has a poor ability to change pH in its rhizosphere.

References

Degenhardt, J., Larsen, P. B., Howell, S. H. and Kochian, L. V.: 1998, *Plant Physiol.* **117**, 19.
Delhaize, E., Ryan, P. R. and Randall, P. J.: 1993, *Plant Physiol*, **103**, 695.
Hirano, Y., Matsumoto, C. and Takenaka, C.: 2000, *Environ. Sciences*
Igawa, M., Uraga, E., Hosono, T., Iwase, K. and Nagashima, R.: 1991, *The Chemical Society Japan* **91**, 698.
Kohno, Y., Matumura, H. and Kobayasi, T.: 1995, *J. Jpn. Soc. Atmos. Environ.* **30** (5), 316.
Kohno, Y., Umezawa, T. and Murakosi, M.: 1998, *J. Jpn. Soc. Atmos. Environ.* **33** (6), 335.
Lee, C. H., Izuta, T., Aoki, M. and Totsuka, T.: 1997, *J. Jpn. Soc. Atmos. Environ.* **32** (1), 46.
Ma, J. F., Hiradate, S. and Matsumoto, H.: 1998, *Plant Physiol*, **117**, 753.
Martell A. E. and Motekaitis, R. J.: 1989, `Coordination chemistry and specification of Al (III) in aqueous solution`, in TE Lewis (eds) Environmental Chemistry and Toxicology of Aluminum. Lewis publishers, Chelsea, MI, pp3-17.
Maruta, E. and Usui, N.: 1997, *Tanzawa Ohyama Shizen Kankyou Sougou Tyousa Houkokusho* (Kanagawa Prefecture), 78
Matzner, E. and Ulrich, B.: 1985, *Experientia41, Birkhauser Verlag. CH-4010 Basel/ Switzerland*, 578.
Miyasaka, S. C., Kochian, L. V., Shaff, J. E. and Foy, C. D.: 1989, *Plant Physiol*, **91**, 1188.
Oda A., Honma, T. and Yamamoto, F.: 1996, Effect of Aluminum on growth and physiology of woody plants,Proceeding on *Plant Growth regulation Society of America*, 23rd annual meeting, p195, (Calgary, Canada).

Pellet, D. M., Grunes, D. L. and Kochian, L. V.: 1995: *Planta*, **196**, 788.
Sekiguti, K., Hara, Y. and Ujie, A.: 1986, *Environ. Tech. Lett* **7**, 189
Taylor, G. J. and Foy, C. D.: 1985, *Amer. J. Bot.* **72** (5), 695.
Tutumi, T.: 1962, *Bull. Govern. For. Exp. Stn.* **137**, 27
Wissemeier, A. H., Hahn, G. and Marschner, H.:1998, *Plant and Soil*, **199**, 53.
Wright, RL.: 1989, *Soil Sci. Plant Anal.* **20**, 1479.
Wagatuma, T.: 1982, 'The characteristics of uptake and shift and behavior in the plant of Al`, in A. Tanaka (eds.), Plant and metal element, Japan. J. Sci. Plant Nutr: pp37-86.

SENSITIVITY OF A JAPANESE EARTHWORM (*ALLOLOBOPHORA JAPONICA*) TO SOIL ACIDITY

MASAHIKO OHNO*

Tokyo Metropolitan Research Institute for Environmental Protection, Shinsuna 1-7-5, Koto-ku, Tokyo, 136-0075 Japan
(* author for correspondence, e-mail: ohnom@kankyoken.koto.tokyo.jp)

Abstract. Earthworms commonly dominate the soil fauna and they play important roles in terrestrial ecosystems, but their numbers decrease in acidic conditions. Their sensitivity to low pH was studied in Europe and their distributions were explained by their sensitivity. These kinds of studies have not been conducted in Japan. Sensitivity of a common Japanese earthworm, *Allolobophora japonica*, to acidified soils was studied. Worms withdrew its prostomium (oral organs) abruptly from Clark-Lubs buffer solutions of pH 3.9-4.1 when they were dipped into the solutions. The species did not burrow into soils of pH (H_2O) < 3.6 and died on and in the soils of pH (H_2O) < 4.0. It seemed that the species did not tolerate the soils of pH (H_2O) < 4.0. The responses to acid soil were almost same as those of European species. Soils should be maintained above pH 4 for this species.

Keywords: earthworm, *Allolobophora japonica*, soil acidity, sensitivity, Japan

1. Introduction

Earthworms commonly dominate the soil fauna, decompose organic matter and stir up soil. They play important roles in terrestrial ecosystems. They are sparse in soils with pH < 4.0-4.5, generally absent bellow pH 3.5. There are differences in their preferred pH range between species (Lee, 1985; Curry, 1998). Their sensitivity to low pH was studied in Europe and their distributions were explained by sensitivity (Satchell, 1955,1967; Laverack, 1961). Regarding acid deposition, the abundance of earthworms decreased in plots treated with simulated acid rain (Esher *et al.*, 1992; Ammer and Makeschin, 1994), in high acidic deposition site with pH 4.5 (Kuperman, 1996) and in acidified soil samples with pH ≦ 3.9 (Hågvar and Abrahamsen, 1980). These kinds of studies have not been conducted in Japan. The sensitivity of a Japanese earthworm (*Allolobophora japonica* Michaelsen) to acidic solutions and soils was studied in order to compare its threshold pH value with those of European species.

2. Materials and Methods

2.1. EARTHWORM

A.japonica is a common earthworm in Japan. It distributes widely from mountainous areas to meadows throughout Japan (Yamaguchi, 1965; Nakamura, 1998). It lives in surface soil layer mainly.

It was collected from a secondary forest in the Yoyogi Park, a typical urban park, in Tokyo. It was brought into a laboratory (20-22°C) with soil of the forest and kept for 3 days before the experiments. Mature earthworms showing clitellum were examined and their wet body weight ranged from 0.2 to 0.4 g.

2.2. DIPPING EXPERIMENTS

Two dilute acid solutions (sulfuric acid, nitric acid) and Clark-Lubs buffer solution consisting of potassium hydrogen phthalate and hydrochloric acid or sodium hydroxide were used in the experiments. Two methods were used.

Tube method: The earthworm was inserted into polyvinyl chloride tube (Fig.1 left). It was allowed to crawl down spontaneously along the tube and submerge the anterior segments (prostomium) in the acid solution. It remained submerged at higher pH but withdrew its prostomium abruptly from solutions with low pH. Time immersed was recorded. Five to seven individuals were examined at each pH level at 20-22°C. Exposed worms were not tested again.

Suspending method: The earthworm was suspended by the mid-region with tweezers and allowed to extend into the solutions (Fig.1 right). Time immersed was recorded. Other procedures were the same as the "tube method".

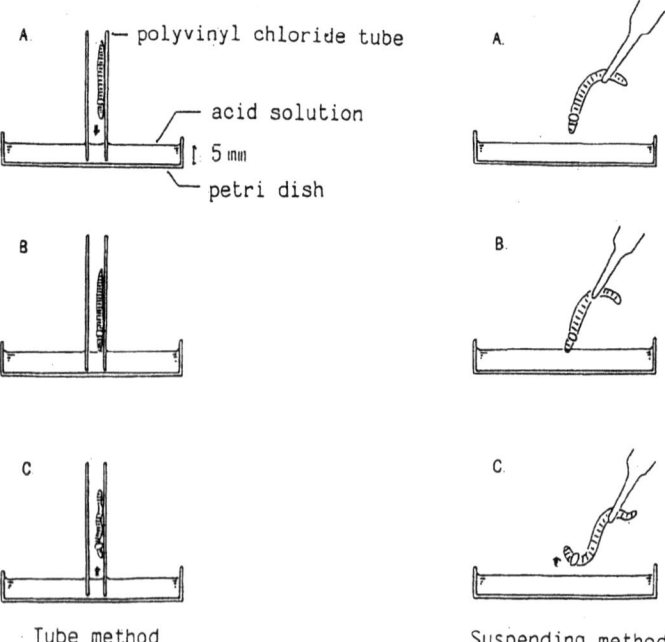

Figure 1. Schema of dipping experiments. Time for withdrawal means the time between B and C.

2.3. SOIL EXPOSURE EXPERIMENTS

Three kinds of soils, soil which was regarded as andosol from Yoyogi Park (Shibuya, an urban area of Tokyo), andosol from the Rolling Land Laboratory of Tokyo University of Agriculture and Technology (Hachioji, a suburb of Tokyo) and brown forest soil from the Aichi Experimental Plantation of Tokyo University (Seto, Aichi), were used in the experiments. These soils were passed through a 2 mm sieve and those pH were 7.28, 4.95 and 4.04 respectively. The soils were treated with two concentrations of sulfuric acid.

1N-sulfuric acid treatment: 1N-sulfuric acid solution was added to 300g of soil in a glass beaker. Adding different amount of the solution made various pH soils.

0.1N-sulfuric acid treatment: Three hundred milliliters of 0.1N solution were poured into 300g of the soil in a filtration vessel. The solution was kept in the vessel for 1 day. After draining the solution, the same amount of the new solution was applied again. This procedure was repeated daily. Changing the amount of the passing solution made various pH soils.

The acidified soils were dried in the room and the water content was adjusted to about 30%. Five earthworms were put on the surface of the soil (40g, about 3 cm depth) in a polypropylene vessel (6 cm in diameter, 7 cm in height). After observing their reactions, the mouth of the vessel was covered with cotton gauze. Five vessels (25 worms in total) were made at each pH level. The vessels were kept in a small chamber (20°C, 90%R.H., dark condition). During the exposure, the soils were watered with a small amount of distilled water every other day. The burrowing rate, ratio of individuals burying themselves perfectly to totals, was calculated after one day of exposure. Individuals moving after physical stimuli were regarded as living. The survival rate was estimated after 1-day and 2-week exposures.

3. Results and Discussion

A.japonica had pH thresholds and showed quick avoidance reactions below this value. The earthworm withdrew its prostomium abruptly from dilute acid solutions of pH 2.5 in the tube method and pH2.3 in the suspending method (Fig.2). It showed similar avoidance reactions for both acids, but the reaction against the buffer solution was different for the acids. It withdrew its oral organ from the buffer solution of pH3.9-4.1. European earthworm *Lumbricus terrestris* had threshold values at pH 2.6 of sulfuric acid solution in the tube-method experiment and at pH 4.0 of the buffer solution in the suspending-method experiment (Satchell, 1955, 1967; Laverack, 1961). The reason for the difference was not clear, but it may be due to the different buffering abilities of these solutions.

The threshold pH value in the tube method was 0.2 higher than that in the

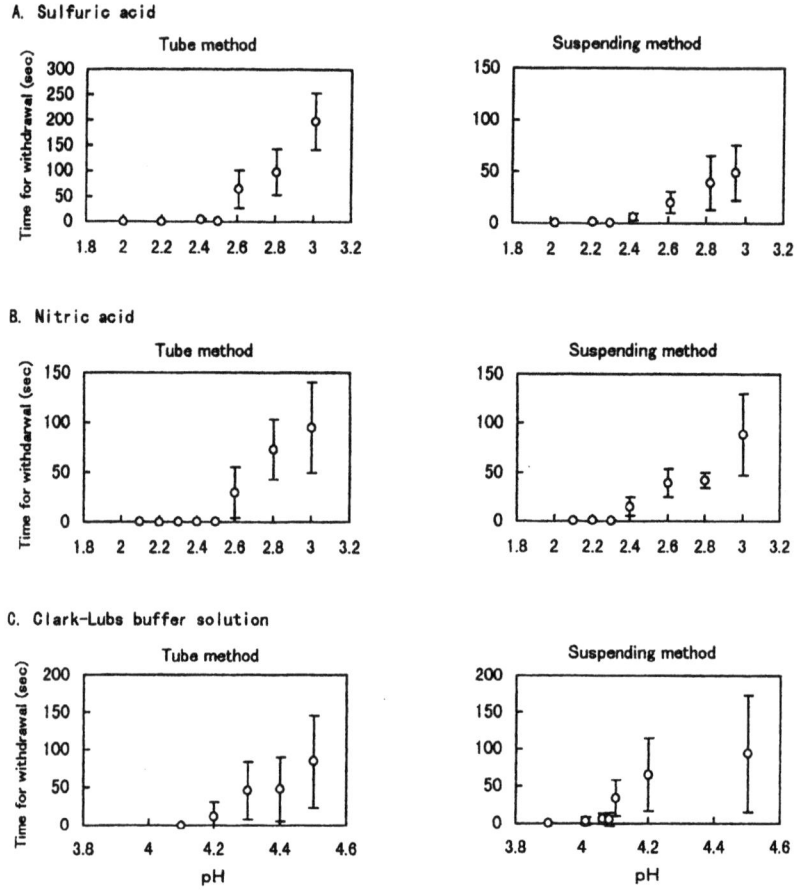

Figure 2. Time taken to withdraw the prostomium from three acid solutions in an earthworm (*Allolobophora japonica*). Circles and vertical bars mean the average values and standard deviations, respectively.

suspending method in each acid solution. The earthworms may behave more easily or spontaneously in the tube method than in the suspending one. This difference of the values between two methods seemed to be ascribed the different positions of the earthworms.

Most of the individuals did not burrow into the soils of pH<3.6 and all of them died at pH <4.0 after two-week exposure (Fig.3). Noticeable differences of survival rates were not seen among the three acidified soils. There were no differences of the avoidance reactions or survival rates between the two treatments (1N and 0.1N). The species did not tolerate soils of pH<4.0. This value, pH4.0, was the threshold of the earthworm in the soil exposure experiments, and similar to those (pH 3.9-4.1) in the buffer solution (Fig.2).

The value was also similar to the experimental thresholds of European ubiquitous earthworms, *L.rubellus* (pH 3.8) and *L.terrestris* (pH 4.0-4.3), whose limits of field distribution were pH3.7 and 4.1 respectively (Laverck, 1961).

A. Soil treated with 1N-sulfuric acid

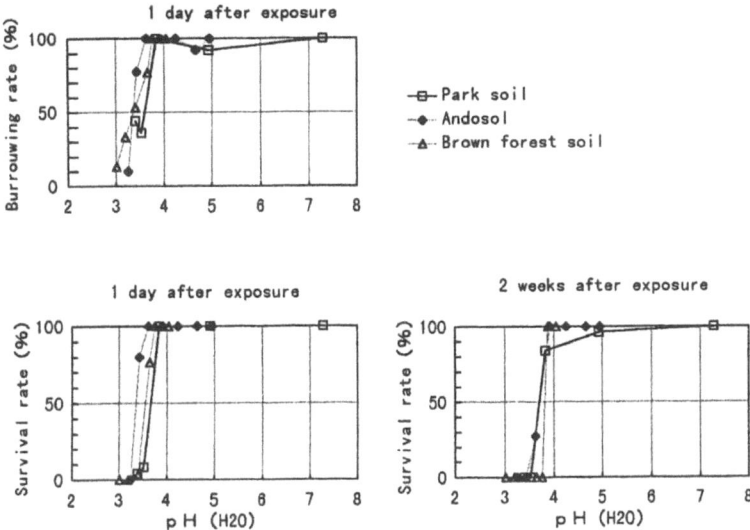

B. Soil treated with 0.1N-sulfuric acid

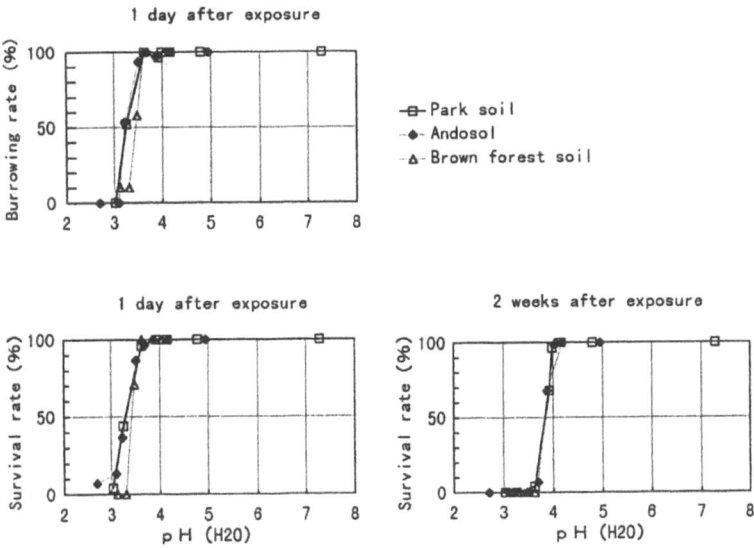

Figure 3. Relations between soil pH and burrowing, survival rates of an earthworm (*Allolobophora japonica*). Burrowing rate is ratio of individuals burrowing into the soil perfectly to total numbers. Survival rate was estimated by counting worms moving their bodies after physical stimuli.

Threshold pH values when dipped into the buffer solution or when exposed to various pH soils closely reflected the limits of field distributions of earthworms (Satchell, 1955,1967; Laverck, 1961; Madge, 1969). The distribution of *A.japonica* related to soil pH is not known in detail, but judging from the thresholds estimated by the current experiments, the species is considered to be unable to live in soils of pH<4.0 in the fields.

4. Conclusion

A.japonica showed violent quick reactions against Clark-Lubs buffer solution of pH 3.9-4.1. It died on or in soils of pH<4.0, which were treated with sulfuric acid. Noticeable differences of survival rates were not seen among three types of soils or between acidification treatments. The responses to acid soil were almost same as those of European species. From the results, *A.japonica* was considered to be unable to live in soil of pH<4.0 in the fields. Soil should be maintained above pH 4 for *A.japonica*.

Acknowledgements

The author would like to thank Dr. Yoshio Nakamura (Tohoku National Agricultural Experiment Station) for identification of the earthworm and kind suggestions.

References

Ammer, S. and Makeschin, F.: 1994, *Forstw. Cbl.* 113, 70.
Curry, J.P.: 1998, 'Factors Affecting Earthworm Abundance in Soils' in C. A. Edwards, (eds.), *Earthworm Ecology*, St. Lucie Press, Boca Raton, pp.37-64.
Esher, R.J., Marx, D.H., Baker, R.L., Brown, L.R. and Coleman, D.C.: 1992, *Water, Air, and Soil Pollution* 61, 269
Kuperman,R.G. : 1996, *Appl. Soil Ecol.*,4, 125
Hågvar, S. and Abrahamsen, G.: 1980, *Oikos* 34, 245
Laverack,M.S.: 1961, *Comp. Biochem. Physiol.* 2, 22.
Lee, K.E.: 1985, Earthworms, their Ecology and Relationships with Soils and Landuse, Academic Press, London.
Madge, D.M.: 1969, *Pedobiologia 9*, 188.
Nakamura, Y.: 1998, '*Mimizu to tsuchi to yukinoho*', Soshinsha, Tokyo (in Japanese).
Satchell, J.E.: 1955, 'Some Aspects of Earthworm Ecology', in D.K.M.Kevan (eds.), *Soil Zoology*, Butterworths Scientific Publisher, London, pp.180-201.
Satchell, J.E.: 1967, ' Lumbricidae' in A. Burges and F. Raw (eds.), *Soil Biology*, Academic Press, London, pp.259-322.
Yamaguchi, E.: 1965, 'Oligochaeta' in K. Okada, S. Uchida and T. Uchida (eds.), *New Encyclopedia of the Fauna of Japan (I)*, Hokuryukan, Tokyo. p. 547 (in Japanese).

EFFECTS OF NATURAL SOIL ACIDIFICATION ON BIODIVERSITY IN BOREAL FOREST ECOSYSTEMS

SERGUEI KOPTSIK[1*], NATALIYA BEREZINA[2] and S. LIVANTSOVA[3]

[1] *Faculty of Physics;* [2] *Biological Faculty;* [3] *Soil Science Faculty, Moscow State University, Moscow 119899, Russia*
(* *author for correspondence, e-mail: koptsik@skop.phys.msu.su*)

Abstract. Variations in soil acidity and in biodiversity were analysed in the National Natural Park "Russian North", European Russia. Improving soil quality from podzol, podzolic soil, derno-podzolic soil, brown earth to pararendzina leads to increase in diversity and changes in floristical composition, followed by changing of pine and spruce forest to mixed and birch forests. In PCA ordination species diversity, richness and evenness of trees, shrubs and vascular plants are closely connected with each other, and are represented by the first principal component. They are strongly correlated to the thickness of A1 horizon, pH_{H2O} and pH_{CaCl2} in organic, surface and subsurface mineral horizons. Only bryophyte species richness and diversity are directly related to the thickness and weight of organic horizon, soil exchangeable acidity, and inversely related to the thickness of A1 horizon and pH. Thus, the ordination of the major species diversity variables is highly related to soil pH, suggesting that pH is the best soil related predictor of species diversity parameters. Our study shows that plants notably respond to soil acidification in boreal forest ecosystems.

Keywords: soil acidification, biodiversity, forest ecosystems, ordination

1. Introduction

Soil quality has fundamental significance for biodiversity conservation (Whittaker, 1975). In boreal forest ecosystems, soil acidity plays important role in formation of plant communities, their species and structure diversity. The demanding species may have their distribution limited due to soil acidification, both anthropogenic and natural (Falkengren-Grerup *et al.*, 1995; Okland and Eilertsen, 1996; Vanmechelen *et al.*, 1997). In our study the effects of variations in soil characteristic, mainly in soil acidity, on plant diversity were examined within the National Park "Russian North", European Russia.

2. Materials and Methods

2.1. STUDY AREA

The National Park "Russian North" is situated in the northern part of European Russia, about 200 km north from Vologda (between 59°43' and 60°18'N; between 38°09' and 39°00'E). It is a hilly glaciated terrain covered mainly by loamy carbonate and noncarbonate till, fluvio- and lymnoglacial deposits. Podzolic and derno-podzolic soils are the most common soils there.

TABLE I

Forest and soil types. Numbers 1, 2, 3 correspond to denotations in Table II-III and Figure 1.

Forest type	Soil type	
	Russian classification	FAO/UNESCO
Oxalis acetocella - spruce forest	Non-deep podzolic soil	Eutric Podzoluvisol, Pde[1]
Vaccinium myrtillis - spruce forest with pine	Residually calcareous non-deep podzolic soil	Eutric Podzoluvisol, Pde[2]
Oxalis acetocella – spruce forest with rowan	Anthropomorphic residually calcareous derno-non-deep podzolic soil	Eutric Podzoluvisol, Pde[3]
Green mosses - pine forest	Feebly developed podzol	Cambic Podzol, PZb
Rubus saxatilis - spruce forest	Base saturated brown earth	Eutric Cambisol, CMe
Aegopodium podagraria – birch forest with aspen	Pararendzina	Rendzic Leptosol, LPk

Woods cover the park territory entirely: taiga is intermittent with the birch woods traditional for the Central Russia plains. The peculiarity of the flora is due to the composition of Boreal, Siberian, Arctic and European species. Northern species make the main group of the local flora. Broadleaf species play a noticeable role in the flora of «Russian North»: the northern boundary of areas of oak trees, maples, limes, elms and forest apple-trees occurs here. Ground vegetation is dominated by bilberry, Rubus saxatilis, orpine, and green mosses.

2.2. SOIL SAMPLING AND CHEMICAL METHODS

Comprehensive investigations of forest ecosystems were started at six plots representative for the area (Table I). The soil samples were collected from deep profiles by horizon. In addition 10 subplots in each of the six plots, 60 in total, were selected for the conjugate vegetation and soil study. There the soil samples were collected from the organic (O), surface (M1) and subsurface (M2) mineral horizons. The samples were air-dried at 20°C and sieved through a 2 mm sieve.

Chemical analysis of the soil included pH in H_2O and 0.01 M $CaCl_2$ suspensions with a soil to solution ratio of 1:25 and 1:2.5 for O and mineral horizons, respectively. Amount of exchangeable (EA) and total acidity (TA) was determined in 1 M NH_4NO_3 and 1 M NH_4Ac extracts, respectively, by an endpoint titration to pH 7.0. Exchangeable base cations were measured in extract of 0.1 M NH_4Cl in 70% ethanol by AAS.

2.3. DATA TREATMENT

Vegetation and soil data from the sixty subplots of six permanent plots were examined to identify what biotic and edaphic factors control vegetation structure, composition and diversity. Species richness, evenness and diversity, based on the estimated composition of tree and shrub species and on the midsummer percentage cover of vascular plant and bryophyte species, were employed for characterizing the species assemblages of communities. Species richness was expressed as logarithm of the number of observed species, diversity as the Shannon-Wiener index and evenness as the Shannon index divided by the species richness (Whittaker, 1975). Species assemblage variables were analysed by means of principal component analysis (PCA). Soil properties (thickness of O and A1 horizons, litter pool, pH_{H2O}, pH_{CaCl2}, EA and exchangeable Al in O, M1 and M2 horizons), determined in the same subplots, were used as additional variables.

3. Results and Discussion

3.1. SOIL CHEMISTRY

The data given in Table II show that appreciable differences in the soil acidity and exchangeable cations exist in the area. The type of vegetation growing on a soil is likely to have a marked influence on soil acidity due to the inherent differences in base content of the litter. Soils supporting conifers tend to be more acid than those supporting deciduous trees. This dependence is not always obvious, nor is the cause-effect connection always clear. Due to species differences in tolerance to soil acidity, soils may affect the composition of the plant community more than the community influences soil reaction (Falkengren-Grerup et al., 1995; Vanmechelen et al., 1997). However soil nutrient and moisture supply may modify effects of soil acidity on tree growth.

TABLE II
Topsoil chemical properties. Soil symbols correspond to denotations in Table I.

Soil	Hori Zon	Depth, cm	pH_{H2O}	pH_{CaCl2}	EA	TA	Exchangeable cations, $cmol_c$/kg				BS, %
					$cmol_c$/kg		Ca	Mg	K	Na	
Pde[1]	O	0-6	5.44	5.28	21.4	55.7	46.1	4.99	0.26	0.15	48.1
	AE	6-8(9)	5.35	5.14	6.45	18.1	2.18	0.37	0.02	0.02	12.5
	E	8-26	5.14	4.76	4.99	6.81	0.65	0.15	0.01	0.03	11.0
Pde[2]	O	0-6(8)	4.35	4.03	29.8	111	27.0	3.51	0.21	0.13	21.8
	E	6-25	4.76	3.96	7.98	14.8	0.26	0.11	0.03	0.02	2.8
	B1	25-38	5.68	4.67	6.92	9.18	5.12	3.42	0.02	0.03	48.3
Pde[3]	O	0-1	5.85	5.61	23.4	24.0	25.5	8.34	0.23	0.16	58.8
	A1	1-6	6.35	6.16	4.07	8.88	3.46	2.54	0.02	0.07	40.7
	AE	6-22	5.19	4.55	6.92	13.0	1.8	0.82	0.03	0.04	17.1
PZb	O	0-4	4.86	4.5	26.6	119	25.0	3.13	0.21	0.14	19.3
	Ae	4-6(7)	4.39	3.64	5.32	12.4	0.55	0.08	0.04	0.02	5.3
	Bh	6-18	4.86	4.77	5.59	6.05	0.13	0.06	0.02	0.03	3.9
CMe	O	0-4	5.86	5.64	20.2	58.0	46.1	6.54	0.21	0.20	47.8
	A1	4-10	5.6	5.39	5.32	16.3	9.11	3.56	0.06	0.47	44.8
	Bm	10-35	6.12	5.98	4.52	7.68	11.4	4.18	0.04	0.67	67.9
LPk	O	0-0.5	7.02	6.96	0	0	64.7	6.35	0.75	0.85	100
	A1	0.5-10	7.47	7.37	0	0	22.5	2.1	0.47	0.08	100
	AD	10-28	7.72	7.54	0	0	19.3	1.45	0.15	0.10	100

3.2. VEGETATION

The short conspectus of species composition of trees and ground flora in the six permanent plots is presented in Table III. Species richness differs between species assemblages, and between forest types. Ground-cover species richness and diversity are close related to tree and shrub richness and diversity. The number of vascular plant species increases, while bryophyte species number decreases from pine forest to spruce and birch forests.

TABLE III

Species composition. Trees are presented as a number per 25·25 m^2; vascular plants - as a plant cover (%) of the most spread species. Soil symbols correspond to denotations in Table I.

	Species	Pde^1	Pde^2	Pde^3	PZb	CMe	LPk
Trees:	Picea abies	38	94	110		37	3
	Pinus sylvestris		13		70	6	
	Betula pendula	6	6				21
	Populus tremula	1		2			9
	Alnus incana	4					13
	Sorbus aucuparia		6		2	6	7
Vascular plants:	Total plant cover, %	40	40-60	60-70	10-20	30-50	50-70
	Vaccinium myrtillis	20	30				
	Rubus saxatilis	<5	5			20	
	Fragaria moschata					5	
	Anthyrium filix-femina			5			
	Oxalis acetocella	10	20	50			
	Aegopodium podagraria	<1		10			20
	Convallaria majalis	10	1-5				10
	Pyrola rotundifolia	<1	5			15	
	Calamagrostis arundinacea	1-2	5		5		
	Geranium sylvaticum	<1				<1	10
	Stellaria nemorum			10			
	Actaea spicata	<1		5			
	Trollius europaeus	<1		<1		<1	5
	Linnaea borealis		10				
	Melampyrum pratense		<1		5		
Mosses:	Cover, %	20	70-90	20	100	80-100	
	Dicranum scoparium		10			15	
	Dicranum rugosum		20		15		
	Hylocomium splendens				15	15	
	Pleurozium schreberi	20	50		70	60	
	Thurdium abientinum			20			

3.3. SOIL-VEGETATION RELATIONSHIPS

Results show a close relationship among different soils and conjugate plant communities. The six plant communities, that corresponded to six soil types, are clearly separated in the plane of the first two PCs of site scores (Fig. 1a). Species

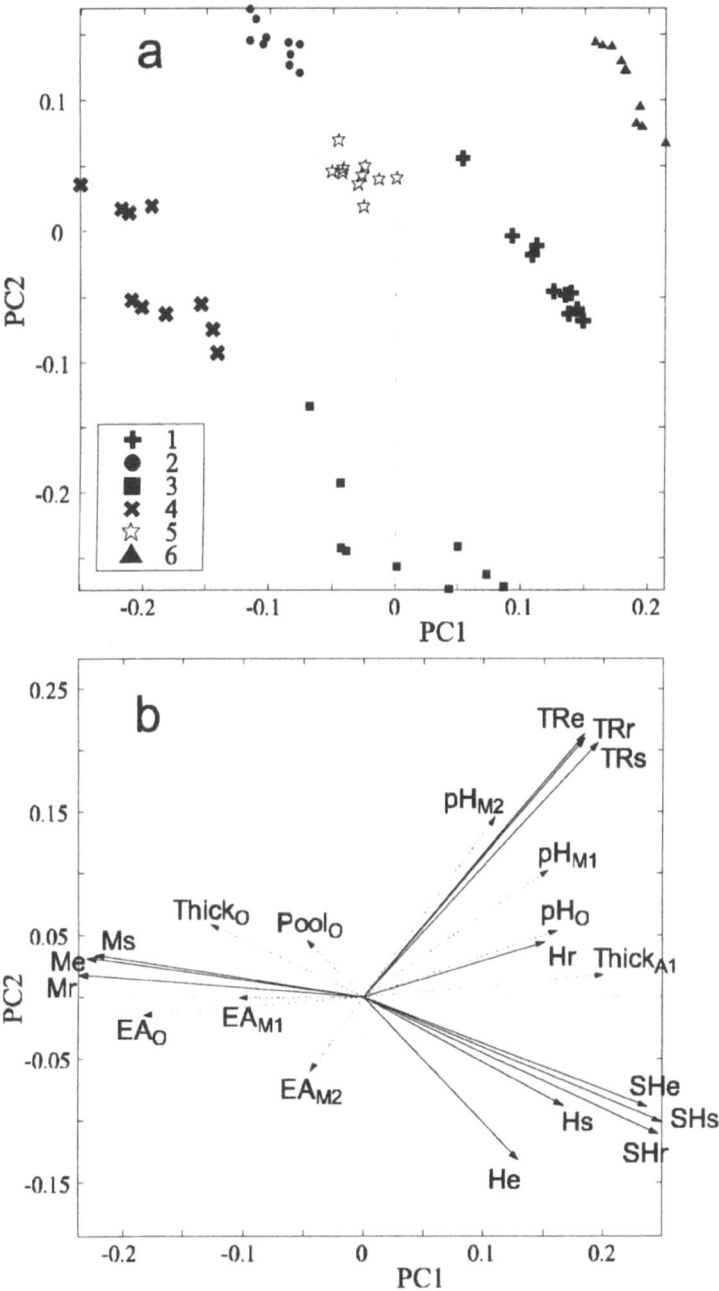

Figure 1. Ordination of sites (a) and species diversity (b, dashed lines show soil gradients). a: 1 - Oxalis acetocella - spruce forest, 2 - Vaccinium myrtillis - spruce forest with pine, 3 - Oxalis acetocella - spruce forest with rowan, 4 - Green mosses - pine forest, 5 - Rubus saxatilis - spruce forest, 6 - Aegopodium podagraria - birch forest with aspen. b: TR - trees, SH - shrubs, H - herbs and dwarf shrubs, M - mosses, s - species diversity, r - richness, e - evenness, EA - exchangeable acidity, O, M1 and M2 - organic, surface and subsurface mineral horizons.

diversity indexes of trees, vascular plants and bryophytes are almost independent from each other in PCA ordination; they are represented by different PCs (Fig. 1b). Only species diversity of bryophytes is represented mostly by the first PC. Species diversity, richness and evenness are closely connected with each other for each assemblage. Diversity indexes of tree species strongly correlate with pH_{H2O} and pH_{CaCl2} in O horizon and especially in surface and subsurface mineral horizons. Species richness and diversity of shrubs correlate weakly with soil properties. Species richness of vascular plants correlates positively with pH in O horizon and with the thickness of A1 horizon. Species richness and diversity of bryophytes relate positively to the thickness of O horizon, organic matter pool, and soil EA; they relate inversely to the thickness of A1 horizon and pH. Thus, the ordination of the major species diversity indexes is highly related to soil pH and thickness of A1 horizon, suggesting that pH and organic matter content are the best soil related predictors of species diversity. This study illustrates that trees and mosses show an obvious response to soil acidification. The results stress the complementarity of multivariate methods to direct correlation and regression analyses in interpretation of soil-vegetation relationships.

4. Conclusions

The soil acidity plays an important role in the formation of plant communities and influences their species richness and structural diversity. In typical forest ecosystems in middle the taiga of European Russia, forest soils differ greatly in the average acidity and exchangeable cations, organic carbon and nitrogen content, total elements, O and A1 layer thickness. Improving of soil quality leads to increased diversity and changes in floristic composition of phytocoenoses, followed by changing of forest types. Tree species diversity indexes correlate strongly with topsoil pH. Species richness of vascular plants correlates positively with pH in O horizon and with the thickness of A1 horizon, while the bryophyte species richness and diversity inversely relate to these variables. The complicity of trends observed is connected with high natural plant and soil variability, competition between species and effects of environmental and anthropogenic factors.

References

Falkengren-Grerup, U., Brunet, J., and Quist M.E.: 1995, *Water, Air, and Soil Pollution* **85** (3): 1233.
Okland R.H. and Eilertsen O.: 1996, *Journal of Vegetation Science* **7** (5), 747.
Vanmechelen, L., R. Groenemans, and Van Ranst, E.: 1997, Forest soil conditions in Europe. Results of a Large-Scale Soil Survey, EC-UN/ECE, Brussels, Geneva. 259 pp.
Whittaker, R.H.: 1975, Communities and Ecosystems, Macmillan Publishing Co., New York. 385pp.

IMPACTS FROM DEPOSITION ON SWEDISH FOREST ECOSYSTEMS IDENTIFIED BY INTEGRATED MONITORING

LARS LUNDIN[1*], MATS AASTRUP[2], LAGE BRINGMARK[1], SVEN BRÅKENHIELM[1], HANS HULTBERG[3], KJELL JOHANSSON, KARIN KINDBOM[3], HANS KVARNÄS[1], STEFAN LÖFGREN[1].

[1] *Department of Environmental Assessment, SLU, Box 7050, SE-750 07 Uppsala, Sweden;*
[2] *Geological Survey of Sweden, Box 670, SE-751 28 Uppsala, Sweden;* [3] *Swedish Environmental Research Institute, Box 47086, SE-402 58 Gothenburg, Sweden.*
(* Author for correspondence, e-mail:Lars.Lundin@ma.slu.se)

Abstract. Integrated monitoring of ecosystems (IM) is an international co-operative programme (ICP) to control effects of air pollution and climate change on water, soil and biological systems. It is a part of the Convention on Long-Range Transboundary Air Pollution (CLRTAP) of the United Nations Economic Commission of Europe (UN/ECE). The ICP-IM is undertaken on sites/catchments to investigate acidification, eutrophication and heavy metals with an integrated approach. In Sweden, long-term time series from forest ecosystems, with a long and stable continuity, will reveal trends and changes in processes and enable modelling to be undertaken. Investigations of acidity/alkalinity in relation to mineral and organic acids indicated the importance of atmospheric deposition. Recent results show very high inorganic nitrogen retention (99%), a net loss of sulphur originating mainly from organic horizons, and a high inorganic aluminium content in the illuvial soil horizons which could be detrimental to forests. Forest deficiency could also be caused by an observed ongoing translocation of Zn to deeper soil layers implying a movement towards increased release to surface waters.

Keywords: Acidification, chemical budgets, forest, hydrochemistry, IM, metals, nitrogen

1. Introduction

Acidifying atmospheric deposition has increased dramatically through the 20th century and, in Sweden, as well as in many other areas of Europe, peaked in the 1970s (WGE, 1999). Efforts to reduce the deposition led to the 1979 UN ECE convention on "Long Range Transboundary Air Pollution". Results from these agreements have been a decline in acid deposition during the past two decades. From 1988 to 1995 the deposition of sulphur in Europe decreased by 34% and for nitrogen by 16% (Olendrzynski, 1997). In Sweden, the sulphur decrease may have been greater, whilst the nitrogen decrease is minimal.

Related to the convention was the implementation of several monitoring programmes for Europe. In Sweden, one of these programmes was directed on forest ecosystems studying effects from acid compounds on soil, water and biota. This programme, the Integrated Monitoring of Ecosystems (IM) was altered in 1995 to include new monitoring sites. Important in the new set up was the catchment approach with well defined topographic water divides, old forests (> 100 years without major forest management), no lakes and only small wet-

Water, Air, and Soil Pollution **130:** 1031–1036, 2001.
© 2001 *Kluwer Academic Publishers.*

land areas. With this approach, effects of acid deposition could be separated from other disturbances. This resulted in four intensively monitored catchments for input-output budget studies and cross media determination to search for causal relationships and processes involved in ecosystem changes.

Main focus was initially on sulphur but later as this deposition declined, nitrogen and heavy metals became of increasing concern. Priority tasks are now directed on recovery processes, nitrogen and eutrophication, heavy metals and dynamic modelling. Conditions in the forest environment are dynamic and there are remaining effects of previous deposition at the same time that ongoing processes cause new impacts. For the Swedish IM sites indications of changes are elucidated.

2. Methods

The Swedish IM sites constitute well defined catchments where input and output of substances are controlled by measurements. The programme restarted in 1995 with new sites without lakes and only small wetland areas. All variables in the water balance equation are determined; i.e. measurements of precipitation, throughfall, soil water, groundwater and stream discharge. Water samples for chemical analysis were collected during 1997-99, mainly on bi-weekly or monthly intervals, providing calculations for substance flows. Soil water and groundwater were sampled four times annually. Soil water was collected by suction lysimeters in the 30 cm and 50 cm layers, and groundwater at two depths. Chemical analyses were performed according to the international IM manual*. Vegetation conditions were surveyed on 32 plots of 0.25 m^2.

*(http://www.vyh.fi/eng/intcoop/projects/icp_im/manual/index.htm).

3. Site Description

Swedish integrated monitoring is performed at four catchments, with three (SE 14, 15 and 16) located on a north-south central transect through the country, with site SE14 representing the south, SE15 the central and SE16 the north parts. One site (SE 04), with long time-series records is located on the northern part of the south-west coast where acid deposition is most pronounced. This site is characterised by rather thin drifts (0–0.7 m), frequently occurring gneiss-granodioritic bedrock outcrops and a central water path in connection with peat soils. In the other three catchments, rather deep Podzolic till soils cover granitic bedrock. The tills were not exposed to water sorting since all sites are located above the highest coast line. Selected sites have fairly old forests and almost no forest management has been conducted during the last 100 years with a prospect of future nature conservation. Field vegetation was characterised by bilberry types and forest stands dominated by Norwegian spruce, but pine and birch also occur. Climate varies between the sites with higher temperatures in the south and higher precipitation in SW exposed sites (SE 04 and SE 15) (Table I). Also atmospheric deposition decreases in a northerly direction (Table II).

TABLE I.
Geographical and climatic characteristics for the IM-catchments. Altitude is above mean sea level, Temp. is annual means, Temp.sum. is day-degrees above +5 °C, P is precipitation, E – evapotranspiration and R – runoff (Raab and Vedin, 1995).

Catchment	SE04	SE14	SE15	SE16
Location	N 58° 03' E 12° 01'	N 57° 05' E 14° 32'	N 59° 45' E 14° 54'	N 63° 51' E 18° 06'
Area, ha	3.7	19.6	19.1	44.8
Altitude, m	114-140	210-240	312-415	420-540
Temp. a. °C	6.7	5.8	4.2	1.2
Temp.sum °C	830	1350	1260	970
Snow cover, days	50	110	150	175
P, mm	1000	750	900	750
E, mm	480	470	450	370
R, mm	520	280	450	380

4. Results

Total bulk deposition showed pH values close to 4.6 at all IM-sites. Deposition of S was 3.7 in the North and 6.3 kg SO_4-S ha^{-1}, y^{-1} in SW and inorganic-N varied from 4.0 to 11.0 kg ha^{-1}, y^{-1} (Löfgren, 1999; Löfgren, 2000; Kindbom pers comm.) (Table II). Mean streamwater pH was at SE 04 - 4.2, SE 14–4.4, SE 15–4.5 and SE 16–5.8.

TABLE II
In- and output of SO_4-S, NO_3+NH_4-N, total-N in runoff, Ca and Mg (kg ha^{-1}, y^{-1}). Values for SE 04, 14 and 15 from 1997+1998 and for SE 16 for 1999 where runoff was estimated from average regional specific discharges on monthly basis.

Catchment	SE04	SE14	SE15	SE16	SE04	SE14	SE15	SE16
	SO_4-S				NO_3-N + NH_4-N			
bulk deposition	6.3	4.7	2.9	3.7	11.0	7.5	5.5	4.0
throughfall	10.8	5.0	3.5	2.0	9.3	2.6	2.6	2.3
runoff	15.5	10.5	12.0	3.7	0.1	0.1	0.0	0.1
Tot-N runoff					1.6	2.0	1.1	1.6
	Ca				Mg			
bulk deposition	2.5	1.4	1.4	2.1	0.8	0.9	0.5	0.3
throughfall	7.6	3.2	3.5	0.7	3.0	1.7	1.2	0.3
runoff	4.7	6.9	3.3	4.2	6.8	6.9	3.3	0.9

Outflow of SO_4-S was higher than input in three of the catchments while at SE16 in the North there was equal bulk deposition and runoff. At all catchments throughfall was also lower than runoff.

The spatial deposition pattern for inorganic nitrogen shows a higher load in the south, with over 10 kg ha^{-1}, y^{-1}, and less than half of this at the northern site. Throughfall, dominated by inorganic-N, is also lower in the North but leaching was negligible at all sites. However, organic nitrogen leaching reaches 1-2 kg ha^{-1}, y^{-1} (Table II). Litterfall would add to organic-N input to the soil but was not included, while focus was on dry and wet deposition, being external input to the catchment.

The deposition patterns for base cations (Ca and Mg) were not as well stratified as for S and N, Ca especially was spatially more evenly distributed 1.4-2.5 kg ha^{-1}, y^{-1}, but low throughfall values occurred in the north. Runoff was also fairly evenly distributed. Deposition of Mg was three times higher in the south, with sea salt influence in the SE04 site meaning high dry deposition. In the south, runoff was high whilst in the north values were only c. 15% compared to south Sweden.

Streamwater relations of sulphate and total carbon (TOC) to acidity/ alkalinity indicated the importance of the anthropogenic mineral acidity and the soil organic acidity. Rather large differences could be seen with the most acidic waters at the southern site SE 14 where relatively high SO$_4$ and TOC concentrations occurred at an acidity of 0.05–0.20 meq l^{-1}. Contrasting to this, alkalinity often occurred in the northern site SE16 with fairly low both SO$_4$ and TOC concentrations. In between, there was the site SE15 with acidic waters (0.05 – 0.10 meq l^{-1}) and SO4 concentrations in the range 0.05–0.18 meq l^{-1} but fairly low TOC concentrations at 5-12 mg l^{-1} (Fig. 1).

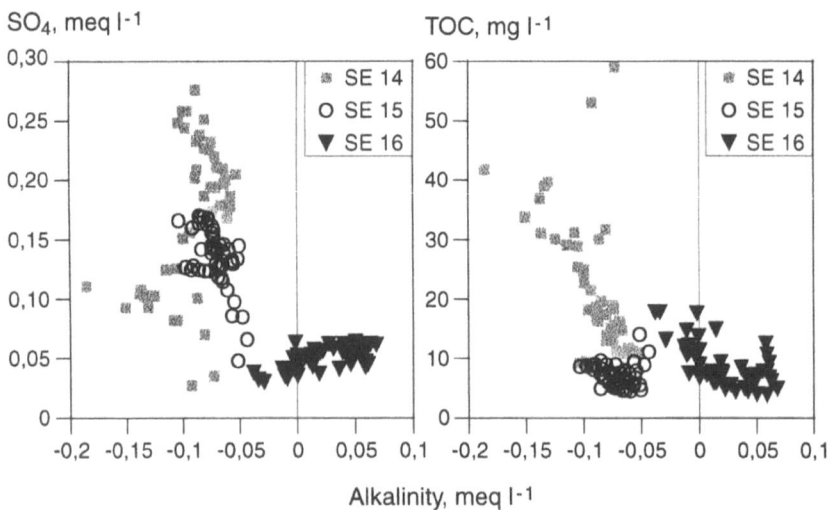

Figure 1. Streamwater SO$_4$ and TOC concentrations in the three IM-sites SE 14-16 versus acidity and alkalinity during 1998 and 1999. Negative alkalinity values are acidity.

One of the crucial acidification effects is the release of aluminium in inorganic form at c. pH 4.5. Total Al-concentrations in the catchments SE 04, 14 and 15, were highest in soil water with 1-3 mg/l, lower in groundwater with the range 0.5-1 mg l^{-1} and mostly slightly lower in streamwater, except for SE 15 where the lowest concentrations were found in the discharge area groundwater. In the less acid streamwater at SE16, Al concentrations were only 0.1-0.2 mg l^{-1}. The organic Al fraction was analysed by species separation (Driscoll, 1984) in SE 14 and this showed a decreasing concentration with soil depth but with comparably high org-Al/tot-Al ratios in the upper soil layer groundwater in discharge areas, similar to ratios in streamwater, and indicating org-Al originating from near stream organic layers (Löfgren, 2000). Also determined were inorganic-Al concentrations being 0.9 to 1.3 mg l^{-1} in soil water.

5. Discussion and Conclusions

Acid deposition has decreased during the last 20 years but with a till low pH, due also to decreased content of base cations (Löfgren, 2000). The in-output of sulphur in the IM-catchments showed releases from the soil being double the deposition. Sulphur deposition in 1998 was only 30-40% of the amount ten years earlier. This means for southern Sweden a decrease of total deposition from 20 kg ha^{-1}, y^{-1} in 1988 to 8 kg ha^{-1} y^{-1}. In central Sweden the values were 10 and 3 kg ha^{-1} y^{-1}, respectively (Westling and Lövblad, 2000). However, higher values occurred in catchment runoff, where in 1998, leaching were between 11 and 16 kg ha^{-1} y^{-1} (Table II), that would be 55-67% higher than throughfall. Sulphur concentrations in the water system of the soil-groundwater-runoff showed releases from Al- and Fe-complexes in the B-horizon and to an even larger extent, also from the organic layers close to the stream.

Nitrogen turnover in the monitoring catchments showed retention both in the tree canopy and especially in the soil with an almost negligible release of inorganic nitrogen in runoff. Instead, organic nitrogen is formed and constitute 90-95% of total nitrogen runoff.

Budgets for Ca and Mg, with separation of deposition from soil origin by use of Cl as conservative element, showed the general release from the soil for Ca constituting 63 and 78% of leaching in SE14 and SE16, respectively, but only 24% at SE 15. Contributions of Mg from the soil were 56, 45 and 37% at SE 14, 15 and 16, respectively.

Relations of SO$_4$ to Ac./Alk. and TOC to Ac./Alk. showed for the southern site negative correlations between both sulphate at concentrations above 0.14 meq l^{-1} (r–0.66) and organic content, indicating influences from organic acids. In the site SE15 the correlation to sulphate (r–0.59) was obvious while influence of organic content was negligible (r–0.06). SE16 in the north correlated (r–0.62) negatively with organic content but there was almost no change in sulphate concentration with concentration of Ac/Alk. (reg. coef. 0.2). From this could be concluded that in south Sweden, the acidity was related both to sulphur and organic acids whereas in the north, influence of organic acids dominated.

In the central part of the SE14 catchment, the soil solution was acid and inorganic Al (Al_i) dissolved. In the upper as well as lower parts of the basin, the soil profiles were less acid and the Al_i dissolution less pronounced. In the stream, pH decreased again, but the mobilisation of Al_i was lower than in the soils. Instead organic aluminium was formed in near stream zones. Critically low Ca/Al_i molar ratios (<1) occurred in the B-horizons, indicating a risk of forest decline. However, the biological observations showed no patterns of defoliation and discoloration of tree crowns deviating from those normal for the region (Löfgren, 2000).

Other metals would also be affected by low pH and Zn seems to be translocated from upper soil horizons to deeper layers with up 0.18 mg l^{-1} in soil water and 0.003-0.010 mg l^{-1} in streamwater where a negative influence on biota can be expected at 0.02 mg l^{-1}.

Legacies from high deposition would still be obvious in the Swedish forest IM ecosystems and soil recovery may yet take decades.

Acknowledgement

The Swedish IM-programme is performed in co-operation with the international programme under auspices of UN ECE Working group on Effects and financed by the Swedish Environmental Protection Agency, Environmental monitoring. Valuable help was also given from Dr. Anders Wilander.

References

Driscoll, C.T.: 1984, *International Journal of Analytical Chemistry* **16**, 267.
Löfgren, S. (ed.): 1999, *Integrated monitoring of environmental status in Swedish forest ecosystems-IM*. Annual report Swedish Environmental Protection Agency No **5031**. (In Swedish).
Löfgren, S. (ed.): 2000, *Integrated monitoring of environmental status in Swedish forest ecosystems-IM*. Annual report Swedish Environmental Protection Agency No **5071**. (In Swedish).
Olendrzynski, K. E. In: E. Berge (ed.): 1997, *EMEP/MSC-W. Report 1/97, part 1*. Emissions, dispersion and trends of acidifying and eutrophying agents, 53-63. Norwegian Meteorological Institute, Oslo.
Raab, B. and Vedin, H.: 1995, *Klimat, sjöar och vattendrag*. Sveriges Nationalatlas. Höganäs. (In Swedish).
Westling, O. and Lövblad, G.: 2000, 'Deposition trends in Sweden.', in P. Warfvinge and U. Bertills (eds.), *The environmental recovery from acidification*, Swedish Environmental Protection Agency, Report **5028**.
WGE,: 1999, *Trends in impacts of long-range transboundary air pollution*. Technical report prepared by the Bureau, the International Cooperative Programmes of the Working Group on Effects. Bull, K. Centre for Ecology and Hydrology, Monks Wood, UK.

FOLIAR NITROGEN AS AN INDICATOR OF NITROGEN DEPOSITION AND CRITICAL LOADS EXCEEDANCE ON A EUROPEAN SCALE

CAROLE E.R. PITCAIRN[1*], IAN D. LEITH[1], DAVID FOWLER[1], KEN J. HARGREAVES[1], MASOUD MOGHADDAM[1], VALERIE H. KENNEDY[2], LENNART GRANAT[3]

[1] *Centre for Ecology (Edinburgh), Bush Estate, Penicuik, Midlothian, EH26 0QB, UK;*
[2] *Centre for Ecology (Merlewood) Grange-over-Sands, Cumbria, LA11 6JU, UK;*
[3] *Department of Meteorology, Stockholm University, S-106 91 Stockholm, Sweden*
(*author for correspondence, e-mail:cerp@ceh.ac.uk)

Abstract. The foliar N content of bryophytes and *Calluna vulgaris* (L.) has been shown to be an indicator of atmospheric N deposition in the UK at a regional scale (1000km) and more recently on a smaller scale in the vicinity of intensive livestock farms. This work extends the geographical scale of the relationship between foliar N concentration of *Calluna vulgaris* and other ericaceous shrubs and N deposition with 2 measurement transects, one extending from northern Finland to southern Norway (2000 km) and the other extending from central Sweden to Stockholm, south east Sweden (330 km). Included in the second transect is a region of complex terrain in the Transtrand uplands, where the variation in N deposition with altitude and canopy cover was quantified using ^{210}Pb inventories in organic soil. The relationship between foliar N (F_N) and N deposition was shown to increase linearly with N deposition (N_D) over the range 0.8% N to 1.4% N according to $F_N = 0.040 N_D + 0.793$ ($r^2 = 0.70$). The data are entirely consistent with earlier studies which together provide a valuable indicator of critical loads exceedance, the threshold value being approximately 1.5% N, which is equivalent to a N deposition of 20 kg N ha^{-1} y^{-1}.

Keywords. foliar N, N deposition, *Calluna vulgaris*, critical load exceedance, Europe

1. Introduction

In the last 2-3 decades, changes in species composition associated with increased N deposition have been reported in widely separated parts of Europe in a range of plant communities, including lowland heathlands, forest ground flora, calcareous grasslands, coastal dunes, wetlands and upland moorland (Sutton *et al.*, 1993). Both field and experimental studies have identified increased emissions of ammonia from intensive farming as the major cause of many of the changes particularly those occurring in heathland (Bobbink & Roelofs, 1996; van der Eerden *et al.*, 1991; Pitcairn *et al.*, 1991). Remote upland areas which receive large inputs of wet and cloud deposited N are also at risk although changes in such areas are less well documented.

The foliar N content of some species has been to shown to increase with atmospheric inputs of N (Pitcairn *et al.*, 1995,1998; Poikolainen *et al.*, 1998) and the value of foliar nitrogen content of bryophytes and *Calluna vulgaris* (L.) as an indicator of N deposition has been demonstrated on a regional scale in the UK (Pitcairn *et al.*, 1995). The largest concentrations of tissue N of selected bryophytes and *C. vulgaris* were found in the Breckland, East Anglia, followed by Cumbria, both areas of high N deposition >30 kg N ha^{-1} y^{-1}. The smallest concentrations were found in north-west Scotland where N deposition was <6kg N ha^{-1} y^{-1}. More recently, foliar N content of mosses has indicated changes in species composition of woodland ground flora and critical load exceedance on a smaller scale in the vicinity of intensive livestock farms in the UK (Pitcairn et al., 1998). A critical load of 20 kg N ha^{-1} y^{-1} (species change in ground flora) was indicated by the field studies while a critical ectohydric moss foliar N concentration of 2% for woodland ground flora, was also demonstrated for the first time.

The scale of these observations has now been extended, by determining the foliar N content of samples of *C. vulgaris* and selected ericaceous shrubs along 2 transects in Northern Europe. Transect 1 extends from the Transtrand Mountains in central Sweden to Stockholm (330 km) and includes a range of altitudes, land uses and pollution climates. Transect 2 extends from northern Finland to southern Norway (2000 km) and covers a range of pollution climates. In earlier UK foliar N studies, atmospheric N deposition for each site where moss or *C. vulgaris* was sampled, was determined from UK monitoring networks and deposition models and from actual measurements of ammonia concentrations. For the Scandinavian transect studies, regional average atmospheric inputs of N were determined from EMEP monitoring and modelling data (EMEP, 1999). However for transect 1, inventories of atmospherically derived ^{210}Pb in soil were used to determine deposition of atmospheric aerosols which include wet deposited NO_3^- and NH_4^+, and cloud and aerosol dry deposition. This method does not quantify local variability in N deposition due to gaseous NO_2, HNO_3 or NH_3 deposition. However the wet deposition of NO_3^- and NH_4^+ is expected to dominate the deposited N in these transects, as ambient concentrations of HNO_3 and NH_3 are so small (EMEP 1999). The ^{210}Pb inventories have been shown to identify the local (0.5 – 1.0 km) scale variability in wet and cloud deposition due to the orographic enhancement of wet deposition in the uplands and the capture of aerosols and cloud deposition by forest canopies (Fowler et al., 1998).

The relationships between foliar N and atmospheric N deposition for the 2 Scandinavian transects are examined in this paper and the data obtained are used to test and extend the relationship for *Calluna vulgaris* already obtained for the UK. In addition, the UK relationships have been tested by including foliar N data obtained from the literature. The use of foliar N content as an indicator of atmospheric inputs of N and more importantly as an indicator of critical loads exceedance is discussed.

2. Methods

In transect 1, bulk samples of *C. vulgaris* were taken at five altitudes (480-870m) on Gammalsaters Fjallet (61°N 13°E), Transtrand Mountains, Central Sweden in October 1994. Where possible, samples were collected in the open and within the tree canopy. A further five samples were collected from open areas within the forest at approximately 50 km intervals on a transect from Transtrand to Stockholm, a distance of 330 km.

In transect 2, samples of ericaceous shrubs were collected at 6 sites (Table I) between Lapland in northern Finland and southern Norway (a distance of >2000 km), in June/July 1997.

TABLE I
Sampling sites in Transect 2

Site Name	North / Eastings	Estimated N deposition Kg N ha^{-1} y^{-1}	Species sampled
Kaamanen, Finland	69/27	1	C. vulgaris, Empetrum spp.
Haparanda, Sweden	65/24	2	Empetrum spp, Vaccinium vitis-idae
Anaset, Sweden	64/21	2	C. vulgaris, V.vitis idae
Gnarp, Sweden	62/17	2	V.vitis idae
Orebro, Sweden	59/15	10	C. vulgaris
Vestre Amoy, Norway	59/05	15	C. vulgaris, Empetrum spp.

In both transects, current year foliage of each species was collected. Samples were dried and ground and foliar N was determined by acid digestion and the indophenol blue method (Allen, 1989).

EMEP estimates were used to provide the regional mean N deposition, while the ^{210}Pb inventory studies were used to quantify the local enhancement in N deposition in the Swedish uplands (EMEP, 1999). The measurements of ^{210}Pb inventories in soil were made by γ-ray spectrometry as described by Moghaddam (1998). When using foliar N data from the literature, total N deposition for each site was estimated from EMEP maps (EMEP, 1999).

3. Results

The ^{210}Pb inventories prepared by Moghaddam (1998) showed that deposition in the Gammalsaters sites in the Transtrand mountains was enhanced both orographically and by canopy cover, due to enhanced cloud droplet and aerosol deposition over forest.

TABLE II
Mean atmospheric ^{210}Pb inventories (from Moghaddam, 1998) and foliar N content of Calluna vulgaris at Gammalsaters sites. (* D is situated on Skaftasen, an adjacent mountain)

Site	Altitude	Open fields Bq m^{-2} y^{-1}	Forest canopy Bq m^{-2} y^{-1}	% Enhancement relative to C-open:		% Foliar N content in C. vulgaris	
				altitude	canopy + altitude	Open	Forest
A	870	3636	-	31		0.74	-
B	740	3214	4376	16	58	0.88	1.09
C	540	**2708**	3668		32	0.87	0.94
D*	620	-	5381		94	-	0.88

The appropriate enhancement (Table II) was applied to the EMEP deposition value for the valley site (C-open) to quantify the local N deposition at the upland sites.

The data from both transects are shown in Figure 1. Foliar N content of the *C. vulgaris* in the Gammalsaters sites followed the pattern of canopy enhancement of the ^{210}Pb inventories and showed a good relationship with estimated N deposition throughout the transect. The very low foliar N content (0.74%) found at the highest altitude in very thin soils, and exposed conditions may be due to limiting concentrations of other nutrients (e.g. phosphorus content 0.02% compared with 0.035% at the other sites).

Foliar N concentrations from transect 2 also showed a good relationship with estimated N deposition, but values tended to be larger than in transect 1. This may be largely explained by seasonal differences in the sampling period between the 2 transects, transect 1 being sampled in October when foliar N concentrations in *C. vulgaris* are at a minimum, and transect 2 in early summer when foliar N concentrations are largest (Brunsting & Heil, 1985). Differences in species sampled, plant age (Robertson & Davies, 1965) and soil types of the sample sites will also contribute to differences between the transects.

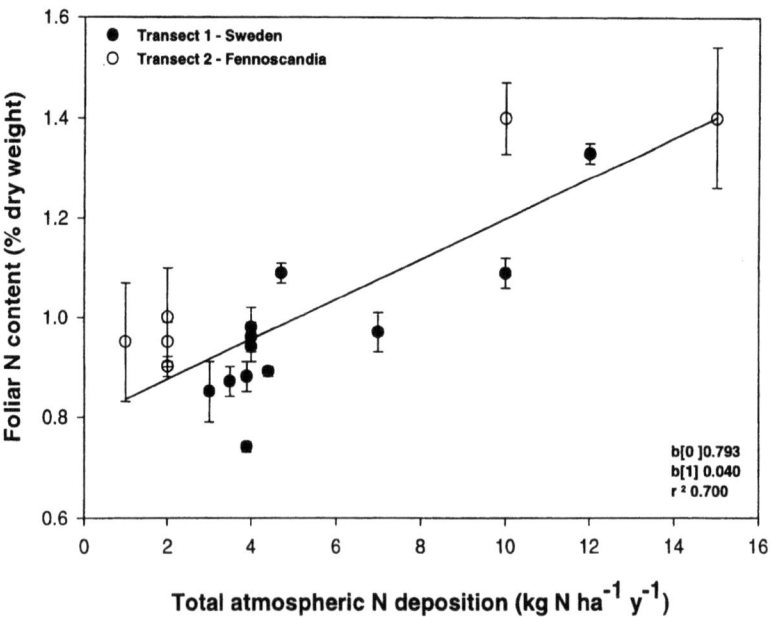

Figure 1. The relationship between foliar N content of *Calluna vulgaris* and other ericaceous species and total atmospheric N deposition, along transects from central Sweden to Stockholm, and northern Finland to southern Norway. (Error bar - standard deviation)

4. Discussion

The broad regional pattern in N deposition in Fennoscandia has been shown to be associated with a similar pattern in foliar N in *C. vulgaris* and other ericaceous species from the transect analyses in figure 1. The relationship between foliar N and N deposition in the range 0.8% to 1.4% is roughly linear with a slope of 0.04% N/kg N ha^{-1} y^{-1}. The data may be compared with more extensive data from UK and Netherlands sites (Pitcairn *et al.*, 1995), Scottish sites (Iason & Hestor, 1993; Hicks *et al.*, 2000) and other Scandinavian sites (Karlsson, 1987). These data plotted in figure 2 show a slope of 0.036 %N/ kg N ha^{-1} y^{-1}, remarkably close to that of the Scandinavian transects data, and obtained using different N deposition data sources. The robustness of the relationship between foliar N and N deposition for these species suggest the use of this simple bioassay for identifying areas of excess N deposition. The scatter in the field data is to be expected with considerable uncertainties in both variables. There is the additional problem that foliar N is influenced by other processes which regulate growth and development.

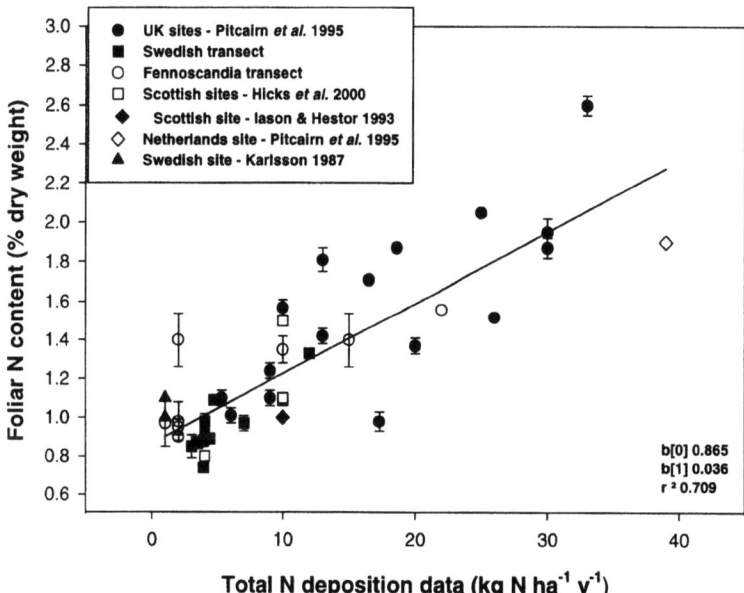

Figure 2. Relationship between foliar N content of *Calluna vulgaris* and other ericaceous shrubs and estimated total mean annual N deposition, at a range of sites in Northern Europe. (Error bar - standard deviation)

For some bryophytes which are much less dependent on soils for nutrient uptake, the relationship between foliar N and N deposition is much closer, following an equation of the form $F_N = 8.01(1-e^{0.04ND})$ where F_N is foliar N concentration and ND is N deposition (Pitcairn *et al.*, 1998).

A practical application of the foliar N assay is in detecting or monitoring critical loads exceedance in field conditions. The critical load for nutrient N effects (eutrophication) for heathland is 15-20 kg N ha^{-1} y^{-1} (Bobbink & Roelofs, 1996), which from Figure 1 and 2, is consistent with foliar N concentrations in excess of approximately 1.5%. Mapped foliar N concentrations from direct observations can be compared with deposition maps, so that areas in excess of 20 kg N ha^{-1} y^{-1} or 1.5% foliar N would be candidate locations in which to seek evidence of changes in species composition or of physiological effects. The data presented here on foliar N in ericaceous species imply that substantial areas of the UK and Netherlands exceed the critical load but only small areas of Scandinavia, largely consistent with critical loads exceedance maps (EMEP, 1999).

Acknowledgements

The authors acknowledge English Nature (formerly The Nature Conservancy Council) and the Department of the Environment, Transport and the Regions for financial support.

References

Allan, S.E. ed.: 1989, *Chemical Analysis of Ecological Materials*. Blackwell Scientific, Oxford.
Bobbink,. & Roelofs, J.: 1996, *Water, Air & Soil Pollution* **85,** 2413.
Brunsting, A.M.H. & Heil, G.W.: 1985, *Oikos* **44**, 23.
EMEP: 1999, Transboundary Acid Deposition in Europe. Summary Report (L. Tarrason & J Schaug eds.) Norwegian Meteorological Institute Report 83.
van der Eerden, L.J., de Vries, W., de Visser, P.H.B., van Dobben, H.F., Steingrover, E.G., Dueck, T.A., van Grinsen, J.J.M., Mohren, G.M.J., Boxman, A.W., Roelofs, J.G.M. & Graveland, J. ; 1997, Effects on forest ecosystems. In: .J.Heij & J.W.Erisman (eds) *Acid Deposition and its Effects on Terrestrial Ecosystems in the Netherlands*, 83-128. Elsevier Science
Fowler, D., Smith, R.I., Leith, I.D., Crossley, A., Mourne, R., Brandford,D. & Moghaddam, M.: 1998, *Water Air and Soil Pollution* **105**,459.
Hicks, W.K., Leith, I.D., Woodin, S.J. & Fowler, D.: 2000, *Environmental Pollution* **107**, 367.
Iason, G.R. & Hester, A.J.: 1993, *Journal of Ecology* **81**, 75-80.
Karlsson, P.S.: 1987, *Holartic Ecology* **10**, 114-119.
Moghaddam, M.: 1998, Study of Surface Roughness Effects on Deposition of Atmospheric Aerosol using ^{210}Pb Inventories. .PhD Thesis, University of Edinburgh.
Pitcairn, C.E.R., Fowler, D. & Grace, J.: 1991, *Changes in species composition of semi-natural vegetation associated with the increase in atmospheric inputs of nitrogen*. Institute of Terrestrial Ecology, Edinburgh, UK 80 pp
Pitcairn, C.E.R., Fowler, D. & Grace, J.: 1995, *Environmental Pollution* **88**, 193-205
Pitcairn,C.E.R., Leith,I.D., Sheppard, L.J., Sutton, M.A., Fowler,D., Munro, R.C., Tang, S. & Wilson, D.: 1998, *Environmental Pollution* **102(S1)**, 41-48.
Poikolainen,J., Lippo, H., Honigisto, Kubin, E., Mikkola, K., Lindgren, M.: 1998, *Environmental Pollution* **102(S1)**, 85.
Robertson, R.A. & Davies, G.E.: 1965, *Journal of Applied Ecology* **2**, 249.
Sutton, M.A., Pitcairn, C.E.R. & Fowler, D.: 1993, *Advances in Ecological Research* **24**, 301

COMPARISON OF THE EFFECTS OF WET N DEPOSITION (NH$_4$Cl) AND DRY N DEPOSITION (NH$_3$) ON UK MOORLAND SPECIES

IAN D. LEITH[1]*, LUCY J. SHEPPARD[1], CAROLE E.R. PITCAIRN[1], J. NEIL CAPE[1], PAUL W. HILL[1], VALERIE H. KENNEDY[2], Y. SIM TANG[1], RON I. SMITH[1] and DAVID FOWLER[1]

[1] *Centre for Ecology and Hydrology, Edinburgh Research Station, Bush Estate, Penicuik, Midlothian, EH26 0QB, UK;* [2] *Centre for Ecology and Hydrology, Merlewood Research Station, Grange-over-Sands, Cumbria, LA11 6JU, UK*

(* author for correspondence, e-mail: idl@ceh.ac.uk)

Abstract. Increases in N deposition (wet and dry) have been associated with a decline in semi-natural plant communities, adapted for growth on nutrient poor soils in the UK and Europe. The impacts of N deposition applied as either wet NH$_4^+$ or gaseous NH$_3$ on vegetation (7 species) from acid moorland in SE Scotland were compared in a dose-response study. Wet N deposition at 0, 8, 16, 32, 64, 128 kg N ha^{-1} y^{-1} was applied as NH$_4$Cl, and dry deposition as gaseous NH$_3$ (2, 6, 20, 50, 90 µg NH$_3$ m^{-3}) under controlled conditions in open-top chambers. A strong linear dose-response relationship (p<0.05) was found between foliar N content in all seven plant species and applied NH$_4$-N. However, in the NH$_3$ treatment, only *C. vulgaris* and *P. commune* showed a significant response to increasing N additions. NH$_3$ was found to increase the rate of water loss in *Calluna* in both autumn and winter by comparison with wet deposition. For *Eriophorum vaginatum*, the NH$_3$ and NH$_4^+$ treatments showed significant N dose response relationships for biomass. A significant increase in above ground biomass, proportional to the added N, was found for *Narthecium ossifragum* when N was applied as NH$_3$ compared to NH$_4^+$.

Keywords: moorland vegetation, NH$_3$, NH$_4^+$, deposition, biomass, water loss.

1. Introduction

The impacts of wet deposited N and dry deposited NH$_3$ have been widely reported for a range of communities such as lowland heaths, forest ground flora, calcareous grasslands, coastal dunes, wetlands and upland moorland (Bobbink *et al.*, 1996; Carroll *et al.*, 1999; Pitcairn *et al.*, 1995; Sutton *et al.*, 1993). Studies have generally reported effects of either wet (NH$_4^+$, NO$_3^-$) or dry N deposition (NH$_3$) with few comparing the NH$_3$ or NH$_4^+$ forms of N deposition. Enhanced wet N deposition has resulted in increased growth, increased foliar N concentration, species composition changes and increased sensitivity to abiotic and biotic stresses such as frost, drought and insect infestation in semi-natural vegetation (Caporn *et al.*, 1994; Leith *et al.*, 1999; Pitcairn *et al.*, 1995; Power *et al.*, 1998). NH$_3$ deposition to woodland plant species in close proximity to intensive livestock units has caused changes in species composition and foliar N concentration with effects diminishing with distance from NH$_3$ source (Pitcairn *et al.*, 1998). In both field and manipulation experiments, estimating the amount of wet N deposited per m^2 ground area as the ionic species NH_4^+ is relatively straightforward. By contrast estimating NH_3 deposition is more difficult, as NH_3 deposition velocities, canopy resistances and their dependence on ambient NH_3 concentrations must be known to calculate N deposition. NH_3 deposition rates have been measured over moorland in low NH_3 emission areas (Flechard and Fowler, 1998) and there have been a few studies reporting NH_3 deposition measurements at higher ambient NH_3 concentrations (Erisman et al., 1994), but these have been on lowland heath communities. Although the effects of enhanced wet N deposition have been reported for upland moorland species (Carroll et al., 1999; Leith et al., 1999) the effects of enhanced dry deposition, or a direct comparison of wet and dry deposited N on this community type has not been investigated. This open-top chamber (OTC) dose-response study exposed moorland plant species to a range of wet deposited N (0-128 kg N ha^{-1} y^{-1}) and dry N (2-90 µg NH_3 m^{-3}) concentrations, which are typical of gaseous NH_3 concentrations found in close proximity to intensive livestock units (Fowler et al., 1998). Measurements of NH_3 deposition velocities were made simultaneously using an additional OTC with a purpose built flux chamber, allowing deposition rates (kg N ha^{-1} y^{-1}) at a range of ambient NH_3 concentrations to be calculated. This enabled a direct comparison of the effects of the wet and dry N treatments on the moorland species.

2. Materials and Methods

2.1. EXPERIMENTAL DESIGN, OPEN-TOP CHAMBERS AND TREATMENTS

In each OTC, a 3m^2 area was excavated to a depth of 45cm and then back-filled with peat and plant species from an ombrotrophic bog (18 km SW of Edinburgh). Six OTCs were equipped to supply additional wet deposited N, at a range of six depositions (applied as NH_4Cl mist). The other five OTCs delivered a continuous concentration of gaseous NH_3 at five different air concentrations. The allocation of treatment chambers was by random selection. A detachable roof was fitted to each chamber in April 1999 to prevent additional precipitation reaching the 3m^2 plots. A description of the OTCs is given in Fowler et al., (1989). Six simulated mist treatments (de-ionised water, 8, 16, 32, 64, 128 kg N ha^{-1} yr^{-1}) were applied from 14 May 1999 for 25 weeks at 12 mm per week (3 x 4 mm applications wk^{-1}) from individual pressurised vessels via a spray droplet generator (Fowler et al., 1989). Continuous application of five NH_3 treatments (No addition, 6, 20, 50, 90 µg m^{-3}) for a 6-month period began on 11 May 1999. Gaseous NH_3 was generated by volatilisation of a 1% aqueous NH_3 solution in the air stream of the OTC manifold. NH_3 concentrations in each treatment chamber were monitored using passive ammonia 'Alpha' samplers (3 replicate samplers per chamber) changed every 4 weeks (Sutton et al., 1998).

2.2. PLANT MATERIAL

Ombrotrophic bog vegetation was established in each OTC in the summer of 1998 with seven individual plants per species per treatment: *Calluna vulgaris* (L.) Hull, *Narthecium ossifragum* (L.) Hudson, *Potentilla erecta* (L.) Rauschel, *Deschampsia flexuosa* (L.) Beauv., *Eriophorum vaginatum* (L.), *Polytrichum commune* Hedw. and *Molinia caerulea* (L.) Moench. In each OTC, the 3m^2 plot was divided into four sectors with an identical layout of species in each sector. All chambers were watered with de-ionised water using either an underground drip system or by overhead misting.

2.2.1. Biomass harvest, nutrients and shoot drying (C. vulgaris)
Above ground biomass of *N. ossifragum, P. erecta, D. flexuosa, E. vaginatum* and *M. caerulea* was determined by destructive harvest in October 1999. Tillers and leaves were thoroughly washed with de-ionised water, dried at 80 ^{0}C then weighed. Samples were digested and analysed for NH_4-N by the indophenol-blue method and for phosphate-P by the molybdenum-blue reaction (Grimshaw et al., 1989).

For the shoot drying curves, ten current year shoots per treatment were excised (3cm lengths) and the cut end placed in a 20 ml vial filled with de-ionised water. The shoots were kept in the dark at 4 ^{0}C overnight then blotted dry and an initial weight taken. All shoots were then placed in a temperature controlled incubator at 14^{0}C and allowed to dry. Each shoot was re-weighed at regular intervals over a 5-day period. Curves of weight loss over time were analysed using Genstat for Windows (4.1).

2.3. DRY DEPOSITION

Dry deposition (kg N ha^{-1} y^{-1}) was calculated from a series of direct NH_3 flux measurements over vegetation identical to that already reported. NH_3 flux measurements were made at a range of NH_3 concentrations (1-90 µg m^{-3}). By applying the calculated deposition velocity to the measured treatment concentrations, the NH_3 deposition can be determined, allowing a comparison of N form (NH_4^+ or NH_3) to be made. The calculated deposition of NH_3-N was 4, 9, 14, 34 and 48 kg N ha^{-1} y^{-1} respectively for the no addition, 6, 20, 50 and 90 µg NH_3 m^{-3} treatments. As the NH_3 deposition could only be determined retrospectively after NH_3 deposition velocity had been calculated the range of N deposition for the NH_3 concentrations were shown to be lower than the range for wet N deposition.

3. Results

3.1. FOLIAR INJURY

Shoot chlorosis (browning) was first observed in the highest NH_4^+ treatment (128 kg N ha^{-1}y^{-1}) after 6 weeks of treatment (wet input 30 kg N ha^{-1}) in *P. commune*. No other species showed signs of visible injury.

3.2. NUTRIENT ANALYSIS

There was a strong linear dose-response relationship (p<0.05) between foliar N content in all plant species and NH_4-N applied. However, for the NH_3 treatment, only *C. vulgaris* and *P. commune* showed a significant response to increasing N additions (Table I). Comparison of the linear regression slopes for NH_4^+ and NH_3 dose-response relationships showed a significant difference between the form of N applied and %N concentration in two species, *C. vulgaris* and *P. commune* (Table I). Both species exhibited greater N uptake when N was supplied as NH_3 rather than as NH_4^+.

3.3. BIOMASS

Biomass increased more in the dry deposited NH_3 treatment compared to the NH_4^+ in the three species measured when expressed as per unit N applied (Table I). *E. vaginatum* responded in a linear fashion to both NH_3 and NH_4^+. The regression lines for *N. ossifragum* were significantly different due to increased biomass in the NH_3 compared to the NH_4^+ treatments (Table I).

TABLE I

Linear regression analysis of foliar N and biomass with applied N as NH_4^+ or NH_3. The difference in response to NH_3 and NH_4^+ (* Prob.) was tested using a generalised linear regression model, where the slope and r^2 refer to mean values of 5 (NH_3) and 6 (NH_4^+) treatment chambers. Data in bold are statistically significant at p < 0.05.

%N concentration	NH_4^+ slope	r^2	NH_3 slope	r^2	* Prob. NH_4^+ versus NH_3
C. vulgaris	**0.004**	0.98	**0.013**	0.96	**0.003**
M. caerulea	0.006	0.79	0.090	0.70	0.572
E. vaginatum	0.006	0.91	0.001	0.01	0.314
P. erecta	0.011	0.76	0.017	0.61	0.505
D. flexuosa	0.011	0.90	0.022	0.61	0.116
P. commune	**0.010**	0.93	**0.033**	0.86	**0.007**
N. ossifragum	**0.005**	0.78	0.008	0.43	0.536
Biomass (g dwt per applied N (kg N ha^{-1} y^{-1}))					
M. caerulea	0.009	0.07	0.038	0.14	0.597
E. vaginatum	**0.018**	0.61	**0.050**	0.75	0.090
N. ossifragum	-0.010	0.51	0.041	0.35	**0.030**

3.4. DRYING CURVES FOR *C. vulgaris*

Rates of water loss were fitted to an exponential curve to give a first order decay rate. Figure 1 shows differences between NH_3 and NH_4^+ treatments (p <0.05) with the NH_3 treatments taken together having a greater rate of water loss than the NH_4^+ treatments taken together. The equations of the lines for NH_3 and NH_4^+ treatments are shown in figure1. The increased rate of water loss from the NH_3 treatments in January 2000 was also found in *C. vulgaris* drying curves measured in November 1999 (data not shown).

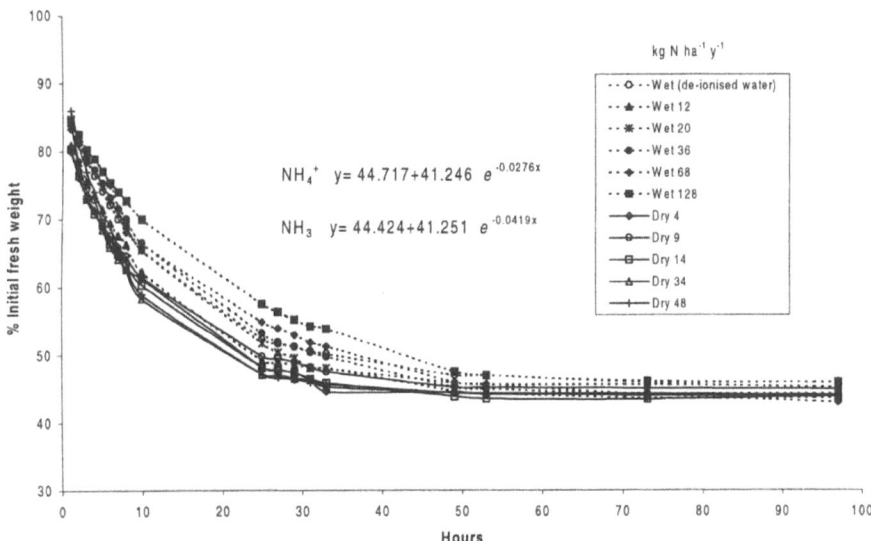

Figure 1. Effects of N form on the rates of water loss from *C. vulgaris* shoots, 30 January 2000

4. Discussion

The increase in %N found in the NH_3 treatments in this study is consistent with that reported by Van der Eerden *et al.*, (1991), who showed significant increases in foliar N in *C. vulgaris* in response to enhanced NH_3 concentrations (100 µg m^{-3}). The absence of a significant increase in biomass in response to increasing N in *M. caerulea* may be due to P deficiency in the higher N treatments with a N:P ratio of 20 and 21 respectively for the 64 and 128 kg N ha^{-1} y^{-1} compared to 15 for the 4 kg N ha^{-1} y^{-1} (Verhoeven *et al.*, 1996). Increasing foliar N concentrations in *C. vulgaris* have been associated with increased susceptibility to secondary stresses, such as drought (Van der Eerden *et al.*, 1991; Power *et al.*, 1998) and susceptibility to heather beetle damage. The differences in water loss between NH_3 and NH_4^+ treatments may reflect changes in cuticular permeability through changes in epicuticular wax morphology (Thijsse and Baas, 1990), cuticular uptake of NH_3 (Sutton *et al.*,

1995) and stomatal conductance. The highest wet NH_4^+ dose gave the slowest rate of water loss, suggesting that exposure to large NH_4^+ deposition may act to prevent drought stress in contrast to the findings of Power *et al.*, (1998). The form of N applied (NH_3 or NH_4^+) appears to affect the foliar N uptake, biomass and rate of water loss in some, but not all, moorland species.

Acknowledgements

The UK Department of the Environment, Transport and the Regions (Contract no. EPG 1/3/94) and the Natural Environment Research Council are thanked for their financial support.

References

Bobbink, R., Hornung, M. and Roelofs, J.G.M. (1996). In: *Manual of Methodologies and Criteria for Mapping Critical Levels/Loads and Geographical Areas where they are Exceeded.* UN-ECE Convention on Long-Range Transboundary air pollution, Federal Environmental Agency, Berlin.

Caporn, S.J.M., Risager, M. and Lee, J.A.: 1994, *New Phytologist* **128**, 461.

Carroll, J.A., Caporn, S.J.M., Cawley. L., Read, D.J. and Lee, J.A.: 1999, *New Phytologist* **141**, 423.

Erisman, J.W., van Elzakker, B.G., Mennen, M.G., Hogenkamp, J Zwart, E., van den Beld, L., Romer, F.G., Bobbink, R., Heil, G., Raessen, J.H., Duyzer, J.H., Verhage, H., Wyers, G.P., Otjes, R.P. and Mols, J.J.: 1994, *Atmospheric Pollution* **28**, 487.

Flechard, C.R and Fowler, D.: 1998, *Quartery Journal Royal Meteorological Society*, **124**, 733.

Fowler, D., Pitcairn, C.E.R., Sutton, M.A., Flechard, C., Loubet, B., Cape, J.N. and Munro, R.C 1998, *Environmental Pollution* **102**, S1, 41.

Fowler, D., Cape, N.J., Deans, J.D., Leith, I.D., Murray, M.B., Smith, R.I., Sheppard, L.J and Unsworth, M.H.: 1989, *New Phytologist* **113**, 321

Grimshaw, H.M., Allen, S.E. and Parkinson, J.A. 1989. Nutrient elements. In: *Chemical analysis of ecological materials (2nd ed.)*, edited by S.E. Allen, 81-159. Blackwell Scientific Publications: Oxford.

Leith, I.D., Hicks, W.K., Fowler, D. and Woodin, S.J.: 1999, *New Phytologist* **141**, 277.

Pitcairn, C.E.R., Fowler, D. and Grace, J.: 1995, *Environmental Pollution* **88**, 193.

Pitcairn, C.E.R., Leith, I.D., Sheppard, L.J., Sutton, M.A., Fowler,D., Munro, R.C., Tang, S. and Wilson, D.: 1998, *Environmental Pollution.* **102**, S1, 41.

Power, S.A., Ashmore, M.R. and Cousins, D.A.: 1998, *New Pytologist* **138**, 663.

Sutton, M.A., Pitcairn, C.E.R. and Fowler, D.: 1993, *Advances in Ecological Resources*, **24**, 301

Sutton, M.A., Fowler, D., Burkhardt, J.K. and Milford, C.: 1995, *Water, Air and Soil Pollution* **85**, 2057.

Sutton, M.A. Tang, Y.S., Miners, B.P., Coyle, M., Smith, R.I., and Fowler, D.; 1998, *Final Report to DETR. Results of National Ammonia Monitoring Network.* London, UK.

Thijsse, G and Baas, P.: 1990, *Trees* **4**, 111.

Van der Eerden, L.J., Dueck, T.A., Berdowski, J.J.M, Gevan, H. and van Dobben, H.F.: 1991, *Acta Botanica Neerlandica* **40**, 281.

Verhoeven, J.T.A, Koerselman, W. and Meuleman, A.F.M.: 1996, *Trees* **11**, 494.

ACCUMULATION OF NUTRIENTS IN ABOVE AND BELOW GROUND BIOMASS IN RESPONSE TO AMMONIUM SULPHATE ADDITION IN A NORWAY SPRUCE STAND IN SOUTHWEST SWEDEN

JOHAN BERGHOLM AND HOOSHANG MAJDI

Swedish University of Agricultural Sciences, Department of Ecology and Environmental Research, P.O. Box 7072, SE-750 07, Uppsala, Sweden

Abstract. The effects of ammonium sulphate (NS) on the accumulation of nutrients in above and below ground biomass and soil were studied in a Norway spruce stand in south-west Sweden during 1988-1993. Ammonium sulphate addition resulted in nitrogen accumulation with 326 and 16 kg ha^{-1} in above and below ground biomass, respectively. Corresponding figures for the control plots (C) were 34 and 3 kg ha^{-1}. Nitrogen accumulation in forest floor of NS was 266 kg ha^{-1} and 47 kg ha^{-1} in mineral soil. About 70% of added sulphate by fertiliser was retained in NS plots (482 kg S ha^{-1}) of which 274 kg ha^{-1} was adsorbed in the mineral soil. The sulphate addition resulted in increased leaching of nitrogen, magnesium, calcium and sulphur. It is suggested that the spruce stand at the study site has a high capacity to accumulate nitrogen with a high above ground production. The high input of ammonium sulphate may in the long run result in increased losses of cations to ground water.

Keywords: accumulation, above ground, below ground, leaching, retention, nitrogen, sulphur, Skogaby

1. Introduction

Ammonium and sulphate are major components of air pollutant deposited over widespread-forested areas of Central Europe (Schultze, 1989). Ammonium has a fertilising effect on forest stands when nitrogen availability is limited, whereas a surplus of nitrogen represents a potential stress factor that can disturb the nutrient supply balance (Aber et al., 1989; Nihlgård, 1985). Elevated nitrogen levels in soil solution can contribute to acidification and lead to ion imbalances, such as a reduction in base saturation in combination with increased concentrations of aluminium, iron and hydrogen ions (Hüttl, 1989). Increasing input of nitrogen together with the input of sulphur is supposed to cause nitrogen saturation, release of nitrate, losses of base cations and unbalanced nutrient status in the trees. However, nitrogen is considered to be a growth-limiting factor in many forest ecosystems (Boxman, 1988; Nilsson and Wiklund, 1992).

The aim of this study was to estimate the accumulation of nitrogen, phosphorous, potassium, calcium, magnesium and sulphur in above and below ground biomass and in soil and the retention of added nutrients in the Norway spruce stand during six years of treatment with ammonium sulphate.

2. Materials and methods

The experimental area is located at Skogaby, 26 km Southeast of Halmstad in south-west Sweden (56°33', 13°13'), at 95-115 m. a. s. l. The climate is maritime with an annual mean air temperature and annual mean precipitation of 7.5°C and 1100 mm, respectively. The growing season is 200 days. The soil type is a poorly developed podzol, a Haplic podzol (FAO, 1990). The soil texture is a loamy sandy till with 4% clay. The pH_{H2O}, base saturation and C/N ratio in the forest floor prior to treatments was 3.9, 30% and 26, respectively. The corresponding values for the upper B-horizon were 4.4, 7% and 22, respectively. The area was planted in 1966 with Norway spruce (*Picea abies*, (L) Karst.). The field experiment was started in 1988 in a randomised design with four blocks and plot size of 45x45m (Bergholm et al., 1995). The control (C) and the ammonium sulphate (NS) treatment have been included in this study. The NS plots received yearly 100 kg NH_4-N ha^{-1} yr^{-1} and 114 kg SO_4-S ha^{-1} yr^{-1} that was added in three portions each summer. In total 600 kg N ha^{-1} and 684 kg S ha^{-1} was added over the six years period.

Bulk deposition and throughfall were sampled using polyethene funnels at open field (n=4) and in C plots (n=6). Throughfall was also collected in NS (n=6) during two periods, 1989-1990 and 1993-1994. In each plot six samples of soil solution were collected 4-6 times a year by using suction lysimeters (ceramic cups of P80 material) at 50 cm depth. Soil samples were taken in autumn 1987, 1990 and 1993. A volume based composite sample (n=40) from forest floor and five (10 cm thick) composite samples (n=20) from mineral soil (0-50 cm) were taken on each plot. The amount of mineral soil per unit area was calculated from bulk density and stoniness indices (Viro, 1952). Data on nutrient content in above ground biomass in 1987 and 1990 was obtained from Nilsson and Wiklund (1994; 1995) and in 1993 from Nilsson (personal com.).

Bulk deposition, throughfall and soil solution were analysed for PO_4^{3-} K^+, Mg^{2+}, Ca^{2+}, NH_4^+, NO_3^-, DON, Cl^-, SO_4^{2-} and Al_{tot} (soil solution only). Base cations were analysed by atomic absorption spectrometry, SO_4^{2-} and Cl^- by ion chromatography and NH_4^+, NO_3^-, DON, PO_4^{3-} and Al_{tot} by flow injection analysis. Exchangeable base cations and acidity were determined by extractions using 1 M NH_4Cl and 1 M KCl, respectively. Exchangeable SO_4^{2-} in mineral soil was extracted by 0.01 M $CaH_4(PO_4)_2$. Soil pH was measured in distilled water. Total concentrations of nutrients in biomass and forest floor were analysed by ICP-plasma after wet digestion with nitric and perchloric acid.

Deposition of elements was calculated using measured amount of precipitation and element concentration in bulk plus dry deposition. The dry deposition of elements was estimated using the Cl-ratio in throughfall/bulk deposition. The total nitrogen deposition (NH_4^+ +NO_3^- +DON) was taken from throughfall measurements. A physically based mathematical model, SOIL (Jansson, 1998), was applied to simulate the daily water flow through the soil at 50 cm depth. Ion fluxes were calculated using concentration values multiplied by water flow. The retention of elements in the ecosystem was calculated as the difference between input (deposition and fertiliser) and output (leaching). The accumulation of nutrients in above ground biomass was calculated as the biomass increment between 1987 and 1993 times the nutrient concentrations in different biomass compartments. The biomass accumulation of coarse roots (>2mm) was calculated using an below-ground-biomass regression equation (Nihlgård, 1972). The nutrient accumulation was then calculated using the weighted mean nutrient concentration of stem, living and dead branches. Statistical analyses were carried out using the SAS GLM procedure (SAS, 1996). LSD tests were used for assessing differences between means ($p<0.05$).

3. Results

The natural deposition of total nitrogen (NH_4-N, NO_3-N and DON) was 117 kg ha^{-1} in C and 152 kg ha^{-1} in NS plots (Table 1). The deposition of P, K and Ca was similar in C and NS, while the deposition of Mg was lower and the deposition of S was higher in NS compared to that in C.

Total input of N by natural deposition in C was retained in the ecosystem except for the leaching of 4.3 kg ha^{-1} of DON (Table 1). The input of N by natural deposition and fertiliser in NS (together 752 kg ha^{-1}) was retained to 93% (700 kg ha^{-1}). The loss of N by leaching was 52 kg ha^{-1} (7.2 kg ha^{-1} was in the form of DON). About 40% (54 kg ha^{-1}) of natural deposited S (133 kg ha^{-1}) in C and 59% (482 kg ha^{-1}) of the total added amount in NS (821 kg ha^{-1}) was retained in the ecosystem. The retention of K, Ca and Mg in C was 18.4, 16.1 and 11.1 kg ha^{-1}, respectively. The application of ammonium sulphate in NS increased the leaching of Ca and Mg by two to four times compared to C. In contrast, K leaching was not affected by the ammonium sulphate application (Table 1). Phosphorus was retained completely.

The accumulation of N in above ground and below ground biomass in C was 33.8 kg ha^{-1} and 2.9 kg ha^{-1}, respectively and 37 kg ha^{-1} was accumulated in forest floor (Table 1). The NS treatment accumulated a significantly ($p<0.05$) higher amount of N in above and below ground biomass (325.7 kg ha^{-1}) compared to C. The accumulation in soil (312.7 kg ha^{-1}) was almost similar to the amount accumulated in the biomass. There was no accumulation of nutrients in fine roots in NS as the fine root biomass decreased and nutrient concentrations increased compared to C (Majdi, unpublished data). The accumulated soil nitrogen in NS consisted of 230.2 kg N_{org} and 36 kg NH_4-N in the forest floor and 30 kg NH_4-N and 16.5 kg NO_3-N in the mineral soil. The amount of nitrogen missing in C and NS was 39 kg ha^{-1} or 30% and 45.4 kg ha^{-1} or 6% of retained amounts, respectively (Table 1). The S content in above ground biomass decreased by 1 kg ha^{-1} in C, while 12.6 kg ha^{-1} was accumulated in NS plots. The main part of added S in NS was accumulated in the soil. About 93% of total S in soil in NS was in the form of adsorbed SO_4-S in the mineral soil (274 kg ha^{-1})

Table 1. Means (n=4) of input, output, the accumulation of nutrients in above and below ground biomass, forest floor and mineral soil at 0-50 cm depth and the retention of nutrients in the ecosystem during 1988-1993. All values are given in kg ha^{-1}. C = control and NS = ammonium sulphate application. Values with separate letters are significantly different ($p<0.05$)

Source	N		P		K		Ca		Mg		S	
	C	NS	C	NS	C	NS	C	NS	C	NS	C	NS
Input												
Deposition	117	152	2.7	3.0	22.0	20.8	25.9	25.0	24.1	21.3	133	137
Fertilizer	0	600	0	0	0	0	0	0	0	0	0	684
Total	117	752	2.7	3.0	22.0	20.8	25.9	25.0	24.1	21.3	133	821
Output												
Leaching	4.3^a	52^b	<0.01	<0.01	3.6^a	4.1^a	9.8^a	22.2^b	13.0^a	48.2^a	79^a	339^b
Retention	112.7	700	2.7	3.0	18.4	16.7	16.1	2.8	11.1	-26.9	54	482
Accumulation												
Above ground biomass	33.8^a	325.7^b	1.8^a	2.2^a	26.2^a	43.8^a	14.0^a	2.7^a	6.7^a	7.7^a	-1.0^a	12.6^b
Below ground Coarse roots	2.9^a	16.2^b	0.4^a	0.3^a	3.8^a	5.9^a	4.2^a	1.8^a	1.5^a	1.5^a	2.0^a	1.5^a
Forest floor	37.0^a	266.2^a	7.2^a	15.4^a	5.9^a	9.8^a	25.6^a	43.8^a	1.7^a	-4.9^a	7.1^a	29^a
Mineral soil	0	46.5	0	0	0	0	0	0	0	0	0	274
Total	73.7	654.6	9.4	17.9	35.9	59.5	43.8	48.3	9.9	4.3	8.1	317.1
Missing/surplus	-39.0	-45.4	6.7	14.9	17.5	42.8	27.7	45.5	-1.2	-31.2	-45.9	-164.9

and remaining part was in the form of S_{org} (29 kg ha^{-1}) in the forest floor. Most of the accumulated P in the ecosystem was found in the forest floor. The NS treatment accumulated more K and Mg in above and below ground biomass and in forest floor compared to C, although the differences were not significant. Less Ca was accumulated in above and below ground biomass in NS than in C, while the opposite relationship was found for the forest floor. More P, K and Ca was found in biomass and forest floor than was retained both in C and NS, indicating an uptake, probably from the mineral soil (Table 1). Magnesium was lost from the ecosystem.

4. Discussion

The nitrogen deposition was higher in NS (30%) compared to C (Table 1). The amount of other elements was almost similar in C and NS. This additional nitrogen in throughfall in NS was due to higher deposition at a few occasions and may be explained by its higher LAI-value compared to C.

The yearly average NO_3-N deposition was 7.5 kg ha^{-1} and the potential nitrification was estimated to 7 kg ha^{-1} (Persson and Wirén, 1995). Although the total yearly amount of NO_3-N could be about 14 kg ha^{-1} and at a C/N-ratio of 26, there was no leaching of nitrate from the soil in C. The high demand for nitrogen to the stand was further illustrated in NS. The first 200 kg N added was totally retained in the ecosystem. After these two years the nitrogen leaching started to increase successively. Still after six years treatment, 93% of applied nitrogen was retained in the ecosystem. The NS-treatment received 635 kg more N ha^{-1} than C (752-117 kg ha^{-1}). This resulted in an increased above ground biomass production by 44% (31% during 1987-1990 (Nilsson and Wiklund, 1992)) and an accumulation in the soil by 42% relative to control plots. The spruce stand at the study site accumulated approximately similar amounts of nitrogen in biomass and soil. In contrast to our results Alison et al. (1997) found that more nitrogen accumulated in the soil and fine roots than in the above ground biomass. However, the retention of added nitrogen was similar to our results. The different result may be explained by i.e. different stand age or climate.

Most of the retained sulphur in NS was found in the mineral soil as adsorbed sulphate in the B-horizon. The ammonium sulphate application resulted in a decrease of pH by about 0.4 pH-units in the mineral soil and an increased leaching of aluminium, which in 1993 was almost equivalent with the release of sulphate on a charge basis (Bergholm, unpublished data). The loss of Mg^{2+} from the humus layer in NS was mainly caused by ion exchange with ammonium and the transportation through the soil profile was supported by the percolation of SO_4^{2-}. The amount of Mg increased by 2.1 kg ha^{-1} in litter layer and decreased in humus layer by 7 kg ha^{-1} compared to C. The increased leaching of Ca^{2+} from NS probably come from the mineral soil because the losses from the humus layer were negligible. The increase of these elements in the forest floor in NS (Table 1) was caused by increased needle litter fall in NS.

5. Conclusions

Our results suggest that the spruce trees at the study site have a high capacity to accumulate nitrogen in above ground biomass. Losses of magnesium from humus layer and decreasing pH as effects of ammonium sulphate addition to forest floor may cause nutrient imbalances in the long run.

Acknowledgements

This work was carried out within the Skogaby project financed by the Swedish Environmental Protection Agency (NV). We would like to thank Ulf Johansson for his comments on the manuscript and the staff at the experimental station at Tönnersjöheden and the laboratory staff at the Department of Ecology and Environmental Research for the good work with sampling and analyses.

References

Aber, J., Nadelhoffer, K.J., Steudler, P. and Melillo, J.M.:1989, *Bioscience* **39**,378.
Alison, H.M., Aber J.D, Hendrickson, J.J., Browden, R. D., Melillo, J. M. and Stuedler, P. A.: 1997, *Ecological Applications* **7**, 402.
Bergholm, J., Jansson, P-E., Johansson, U., Majdi, H., Nilsson, L-O., Persson, H., Rosengren-Brink, U. and Wiklund, K.: 1995, In L-O. Nilsson, R.F. Hüttl, U. Johansson and P. Mathy (eds.). Nutrient Uptake and Cycling in Forest Ecosystems. *European Commission. Ecosystems Research* Report **21**, 69.
Boxman, D.: 1988, In Critical load for sulphur and nitrogen. J. Nilsson (ed), *UN_ECE Workshop*, Sweden.
FAO: 1990. Guidelines for Soil Profile Description, 3rd edition.
Hüttl, R.F.: 1989, 'New types of forest demages in Central Europ. *Yale University Press, New Haven, CT*. Ch. 2, 22.
Jansson, P. E.: 1998, 'Simulation Model for Soil Water and Heat Conditions, Description ofthe SOIL model', *Swedish University of Agricultural Sciences, Dept. of Soil Science*, Uppsala, 81 pp.
Nihlgård, B.: 1985, *Ambio*, **14**, 2
Nihlgård, B.: 1972, *Oikos* **23**, 69.
Nilsson, L-O. and Wiklund, K.: 1992, *Plant and Soil*, **147**, 251.
Nilsson, L-O. and Wiklund, K.: 1994, *Plant and Soil*, **164**, 221.
Nilsson, L-O. and Wiklund, K.: 1995, *Plant and Soil*, **168-169**, 437.
Persson, T. and Wrén, A. 1995: *Plant and Soil*, **168-169**, 55.
SAS Institute Inc.: 1996.The SAS system for Windows vers. 6.12
Schultze, E-D.: 1989, *Science*, **244**, 776.
Viro, P.J.: 1952, *Comm. Ins. For. Fenn.*, **40/3**, 1.

LONG-TERM MONITORING OF ATMOSPHERIC DEPOSITION AND THE IMPLICATIONS OF IONIC INPUTS FOR THE SUSTAINABILITY OF A CONIFEROUS FOREST ECOSYSTEM

E.P. FARRELL[1*], J. AHERNE[1], G.M. BOYLE[1] and N. NUNAN[2]

[1] *Forest Ecosystem Research Group, Faculty of Agriculture, University College Dublin, Belfield, Dublin 4, Ireland;*
[2] *Present Address: Unit of Soil-Plant Dynamics, Soil Biophysics, Scottish Crop Research Institute, Dundee, DD2 5DA, Scotland*
(* author for correspondence, e-mail: ted.farrell@ucd.ie)

Abstract. Ionic fluxes in a semi-mature stand of Sitka spruce (*Picea sitchensis* (Bong.) Carr.), on a spodosol in eastern Ireland, were monitored over an eight-year period, 1991–1998. The paper focuses on the long-term viability of forests in this region. Input-output balances, proton budgets and critical loads suggest that the long-term sustainability of forests in the region is threatened unless atmospheric emissions of anthropogenic substances can be controlled.

Keywords: acid deposition, atmospheric pollution, critical load, forest ecosystems, input-output budget, proton budget, sustainability

1. Introduction

The forests of Ireland are dominated by first generation coniferous plantations. The long term viability of some of these forests gives cause for concern, particularly those on sensitive soils in the east of the country which experiences significant atmospheric deposition. In this paper, input-output balances, proton budgets and critical loads are presented, based on an eight-year monitoring period (1991–1998) for a coniferous plantation forest ecosystem at Roundwood, in eastern Ireland. The objective of the paper is to test the results of these computations against a number of chemical criteria for the purpose of assessing the long-term viability of forests in the region.

2. Materials and Methods

2.1. SITE DESCRIPTION

The Roundwood monitoring site is part of an EU-wide network of forest health monitoring plots. The forest stand is a first-rotation plantation (Sitka spruce (*Picea sitchensis* (Bong.) Carr.)), established in 1955 on former extensively-managed, unenclosed hill land.

The soil is a spodosol derived directly from weathered bedrock (schist and quartzite) or thin drift. Soil pH values, in H_2O, vary between 3.5 (O horizon) and 4.6 (C horizon), which is in the range of strongly acid soils characterised by aluminium buffering. Cation exchange capacity and per cent base saturation are very low (5.6 mmol$_c$ 100g^{-1} and 16% respectively in the O/E horizon).

2.2. TOTAL DEPOSITION (INPUT)

Continuous monitoring of open-land bulk precipitation, forest bulk throughfall and stemflow have been carried out since 1991 (TABLE I). The total deposition of NO_3^-, Na^+, Cl^-, SO_4^{2-} and Al^{3+} were estimated as throughfall plus stemflow. The method of Ulrich (1983), in which Na^+ is used as a tracer, was used to estimate the total deposition of the remaining cations (Ca^{2+}, Mg^{2+}, Mn^{2+} and K^+). The canopy budget model (Draaijers and Erisman, 1995) was used to estimate total deposition of NH_4^+ and H^+.

2.3. DEEP SOIL WATER FLUX (OUTPUT)

Deep soil water (c. 75 cm, base of the solum approximately) has been collected regularly, using ceramic suction lysimeters at 75 cm depth, since 1991. Ionic fluxes for each element were calculated by multiplying soil solution concentrations by modelled soil water fluxes.

Soil water fluxes were estimated using the numerical simulation model SOIL (Nunan, 1999). The model takes meteorological data, soil water retention characteristics and fine root distribution as field inputs. Daily meteorological data (precipitation, relative humidity, mean wind speed, sunshine hours and temperature) were obtained from the nearest meteorological station. Detailed descriptions of the model are given by Jansson (1991) and Johnsson and Jansson (1991).

2.4. PROTON BUDGET

The proton budget was estimated using the method described by van Breemen *et al.* (1983, 1984) and Forsius *et al.* (1995). The base cation flux (BC_w) due to weathering (plus ion exchange reactions) was estimated as the base cation flux in deep soil water (TABLE I) plus net base cation uptake minus base cation deposition. Net uptake of base cations (39 mmol$_c$ m^{-2} yr^{-1}) was derived by combining estimates of base cation concentration in tree biomass with the average productivity over the rotation period (yield class 16) and wood density, according to de Vries (1991).

The net H^+ flux due to nitrogen transformation was determined as the difference between net output of NH_4^+ and net output of NO_3^- according to van Breemen *et al.* (1983). H^+ flux due to the dissociation of organic acids was estimated from measured DOC concentrations and pH according to the method described by Oliver *et al.* (1983).

The ratio between external and internal H^+ sources was calculated according to van Breemen *et al.* (1984). The external sources include: atmospheric deposition of H^+ and sources due to N transformations; internal sources include the dissociation of organic acids (A^-) and accumulation of base cations by biomass. It is assumed that H^+ deposition and net sources of H^+ arising from N transformations are largely anthropogenic (van Breemen *et al.*, 1984; Forsius *et al.*, 1995).

2.5. CRITICAL LOADS

According to Sverdrup and de Vries (1994) the critical load of actual acidity for mineral forest soils is estimated as the release of base cations from weathering minus the critical Acid Neutralising Capacity (ANC) leaching. This was calculated for the entire soil profile, assuming a single mixed layer. ANC leaching is determined by a critical calcium to aluminium (Ca:Al) molar ratio in soil solution. While the base cation to aluminium (BC:Al) molar ratio is the most widely used method for determining critical loads of acidity, Ca:Al is considered more appropriate in maritime regions, where the input of cations of marine origin is often significant (Hall *et al.* In Press). Sverdrup and Warfvinge (1993) have summarised limiting ratios for a variety of plant species. The value of 0.4, suggested for Sitka spruce was used in the current study.

This approach to calculating critical loads of acidity is known as the Steady State Mass Balance approach (SSMB), as steady state equilibrium is assumed. The mass balance equation for the critical load of actual acidity ($CL(Ac_{act})$) used in the current study is as follows:

$$CL(Ac_{act}) = BC_w + 1.5 \times \left(\frac{Ca_w + Ca_{dep} - Ca_u}{Ca:Al_{crit}} \right) + \left(1.5 \times \frac{Ca_w + Ca_{dep} - Ca_u}{Ca:Al_{crit} \times K_{gibb}} \right)^{1/3} \times Q^{2/3}$$

where BC_w is the base cation flux due to weathering, Ca_w is calcium weathering (estimated as a fraction of BC_w based on mineralogical ratios), Ca_{dep} is calcium deposition, Ca_u is the net uptake of calcium by biomass, $Ca:Al_{crit}$ is the critical calcium to aluminium molar ratio, Q is precipitation surplus (taken to be the deep soil water flux, m^3 ha^{-1} yr^{-1}) and K_{gibb} is the gibbsite equilibrium constant, taken to be 300 m^6 eq^{-2}.

Critical loads of nutrient nitrogen for vegetation communities are presented by Werner and Spranger (1996). For acid coniferous forests with moderate to high nitrification rates a deposition range of N of 140–210 $mmol_c$ m^{-2} yr^{-1} is recommended to prevent nutrient imbalance; 50–140 $mmol_c$ m^{-2} yr^{-1} is the limit to protect against changes in ground flora and mycorrhizae and increased leaching.

3. Results

The input-output budget is presented in Figure 1 and TABLE I. Cations and anions are balanced with no differences in the deep soil water flux (output) and only 1.4% difference in the total deposition (cation = 685 mmol$_c$ m^{-2} yr^{-1}; anions = 666 mmol$_c$ m^{-2} yr^{-1}).

Output exceeded input for NO$_3^-$, Mg^{2+}, Mn^{2+}, Al^{3+} and A$^-$, whereas retention was measured for H$^+$, NH$_4^+$, Ca^{2+}, K$^+$ and Cl$^-$. Inputs of Na$^+$ and SO$_4^{2-}$ roughly balance outputs.

The total H$^+$ load to the catchment was 123 mmol$_c$ m^{-2} yr^{-1} (pH 3.9). The difference between the sum of the H$^+$ sources and sinks was within 2% (Figure 2). The consumption of H$^+$ is dominated by weathering. Deposition and N transformations are important sources of H$^+$. The ratio between external and internal H$^+$ sources (Figure 2) was estimated to be 4.3.

Using a Ca:Al ratio of 0.4 critical loads of acidity are exceeded by approximately 73 mmol$_c$ m^{-2} yr^{-1} (critical load = 193 mmol$_c$ m^{-2} yr^{-1}). Total deposition of nitrogen (NH$_4^+$ + NO$_3^-$) is 156 mmol$_c$ m^{-2} yr^{-1}, suggesting an exceedance of critical load for nutrient nitrogen.

4. Discussion

Van Breemen *et al.* (1984) concluded that most ecosystems with external/internal H$^+$ sources ratio greater than 0.5 show significant dissolution of soil Al^{3+} and SO$_4^{2-}$ retention, resulting in export of free H$^+$ and Al^{3+} in drainage waters. While there is no evidence of sulphate retention at Roundwood, the ratio of external/internal sources of H$^+$ was estimated to be 4.3, larger than any of the values reported by van Breemen *et al.* (1984). This suggests that anthropogenic sources are making a major contribution to soil acidification at Roundwood.

Critical loads of actual acidity for forest soils are exceeded by approximately 73 mmol$_c$ m^{-2} yr^{-1}. In addition, the critical limit suggested by Sverdrup *et al.* (1990) for soil solution pH in the B horizon (pH 4.4) is exceeded.

N input fluxes are large. Output fluxes are correspondingly high, 172 mmol$_c$ m^{-2} yr^{-1} (NH$_4^+$ + NO$_3^-$), even exceeding inputs. Roundwood is nitrogen saturated following the definition of Aber *et al.* (1989). The high outputs of aluminium are clearly associated with nitrate leaching.

5. Conclusion

The impact of anthropogenic deposition is clearly seen in Roundwood. The evidence presented suggests a deterioration in the chemical condition of the soil. Nevertheless, the forest stand remains healthy. It may be unwise to conclude

TABLE I

Eight-year mean water fluxes (mm yr^{-1}) and ionic fluxes (mmol$_c$ m^{-2} yr^{-1}) in all water sources sampled at Roundwood, 1991–1998; and estimated eight-year mean total deposition (input) and deep soil water fluxes (output) at Roundwood, 1991–1998.

Parameter	H$_2$O	pH	H$^+$	NH$_4^+$	NO$_3^-$	Ca^{2+}	Mg^{2+}	K$^+$	Na$^+$	Cl$^-$	SO$_4^{2-}$	Mn^{2+}	Al^{3+}	HCO$_3^-$	A$^-$
Precipitation	1437	4.58	38	40	29	16	28	5	141	179	58	0.3	2.5	0.9	
Throughfall	913	4.23	54	63	65	47	76	51	308	372	139	13	3.9	0.3	
Stemflow	58	3.84	8	7	9	10	15	12	44	54	27	3.2	0.7	0	11
Total deposition (input)	971	3.90	123	82	74	40	70	13	352	426	166	0.7	4.6	0.1	
Deep soil water (output)	821	4.17	56	2	170	27	82	6	358	397	165	49	179	0.2	27

Al^{3+} = Total monomeric aluminium quoted as if all was present as the tripositive aluminium ion.

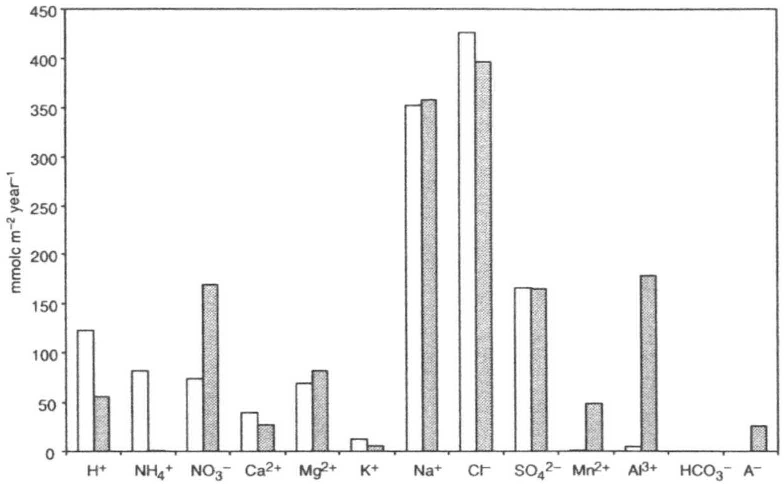

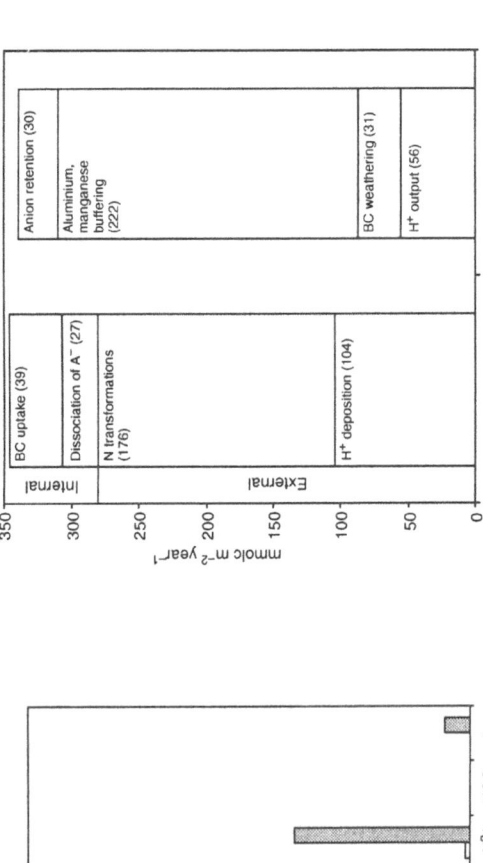

Figure 1. Eight-year mass balance (mmol$_c$ m^{-2} yr^{-1}) for Roundwood, 1991–1998 (input = white, output = grey).

Figure 2. Eight-year mean sources and sinks of hydrogen ions (mmol$_c$ m^{-2} yr^{-1}) for Roundwood, 1991–1998.

that these results are evidence of incipient deterioration of the forest condition. The apparent instability of the ecosystem may merely be a reflection of the impact of disturbance brought about by land-use change. In addition, the estimate of critical loads was preliminary and could perhaps be refined by focussing on the upper soil horizons where most of the tree roots are concentrated. Nevertheless, despite the use of a relatively generous critical limit (0.4), critical load is exceeded. It is clear that the control of emissions of anthropogenic pollutants to the atmosphere is crucial to the long-term viability of forests in the region.

Acknowledgements

Financial support for this work was provided by CEC DG VI, under EEC regulation 3528/86 and its amendments.

References

Aber, J.D., Nadelhoffer, K.J., Steudler, P. and Melillo, J.M.: 1989, *BioScience* **39:6**, 378–386.
de Vries, W.: 1991, *Methodologies for the assessment and mapping of critical loads and of the impact of abatement strategies on forest soils*, Report 46, DLO The Winand Staring Centre, Wageningen, The Netherlands.
Draaijers, G.P.J. and Erisman, J.W.: 1995, *Water, Air, and Soil Pollut.* **85**, 2253–2258.
Forsius, M., Kleemola, S., Starr, M. and Ruoho-Airola, T.: 1995, *Water, Air, and Soil Pollut.* **79**, 19–38.
Hall, J., Reynolds, B., Aherne, J. and Hornung, M.: In Press, *Water, Air, and Soil Pollut.*
Jansson, P.E.: 1991, *SOIL model, users manual*, Division of Agricultural Hydrotechnics Communications 91:7, Department of Soil Sciences, Swedish University of Agricultural Sciences, Uppsala, Sweden.
Johnsson, H. and Jansson, P.E.: 1991, *Journal of Hydrology* **129**, 149–173.
Nunan, N.: 1999, *Soil water fluxes at Brackloon and Roundwood*, Forest Ecosystem Research Group Report Number 49. Department of Environmental Resource Management, University College Dublin, Ireland.
Oliver, B.G., Thurman, E.M. and Malcolm, R.L.: 1983, *Geochimica et Cosmochimica Acta* **47**, 2031–2035.
Sverdrup, H. and de Vries, W.: 1994, *Water, Air, and Soil Pollut.* **72**, 143–162.
Sverdrup, H. and Warfvinge, P.: 1993, *The effect of soil acidification on the growth of trees, grass and herbs as expressed by the $(Ca+Mg+K)/Al$ ratio*, Reports in ecology and environmental engineering 1993:2, Department of Chemical Engineering II, Lund University.
Sverdrup, H., de Vries, W. and Henriksen, A.: 1990, *Mapping critical loads*. Miljörapport 1990:14, Nordic Council of Ministers, Copenhagen.
Ulrich, B.: 1983, 'Interaction of forest canopies with atmospheric constituents: SO2, alkali and earth alkali cations and chloride', in B. Ulrich and J. Pankrath (eds.), *Effects of Accumulation of Air Pollutants in Forest Ecosystems*,. Reidel, Dordrecht, The Netherlands, pp. 33–45.
van Breemen, N., Driscoll, C.T. and Mulder J.: 1984, *Nature* **338**, 408–410.
van Breemen, N., Mulder J. and Driscoll, C.T.: 1983, *Plant and Soil* **75**, 283–308.
Werner, B. and Spranger, T. (eds.): 1996, *Manual on methodologies and criteria for mapping critical levels/loads and geographical areas where they are exceeded*, Federal Environmental Agency (Umweltbundesamt), Berlin, Germany.

NITRIFICATION AND METHANE OXIDATION IN FOREST SOIL: ACID DEPOSITION, NITROGEN INPUT AND PLANT EFFECTS

B.K SITAULA[1*], J.I.B. SITAULA[2] Å. AAKRA[2] and L.R BAKKEN[1]

[1] *Dept. of Soil and Water Sciences;* [2] *Dept. of Chemistry and Biotechnology, Agricultural University of Norway N-1432 Aas, Norway*
(* *author for correspondence, e-mail: bishal.sitaula@ijvf.nlh.no*)

Abstract. In a laboratory incubation experiment, nitrification potential, methane oxidation, N_2O and CO_2 release were studied in the organic soil layer (0-10 cm) of field lysimeters containing re-established soil profiles from a 100-year-old Scots pine (*Pinus sylvestris*) forest of Norway. The experiment was designed as a full factorial (3 factors; N fertilisation rates, soil acidification, and plants), with three replicates. The more acidic irrigation (pH 3) significantly reduced nitrification potential and N_2O fluxes, methane oxidation and CO_2 release. We concluded that the reduction in soil N_2O release by severe acid deposition is partly due to reduction in nitrification potential. The highest N_2O fluxes were observed in the combination of fertilised planted and less acidic pH treatment. N fertilisation (90 kg N ha^{-1} y^{-1} with NH_4NO_3) increased soil N_2O release by a factor of 8 and decreased CH_4 oxidation by 60-80%. Plant effects on soil nitrification potential and methane oxidation rates are discussed.

Keywords: acidification, CH_4 oxidation, *fertilisation*, nitrification, N_2O

1. Introduction

Nitrification by ammonia oxidising bacteria (AOB) and methane oxidation by methane oxidising bacteria (MOB) are among the important soil microbial processes involved in N-cycling, soil acidification, and fluxing of N_2O and CH_4 between soil and the atmosphere. Nitrification and production of N_2O during nitrification is influenced by many environmental factors (acidity, moisture, temperature, nitrogen availability), and the main components in wet and dry deposition are both acidifying and N fertilising compounds (e.g. NO_x, NH_4, NO_3) affecting forest soils in Europe (Persson and Wiren, 1989; Galloway and Rodhe, 1991). In previous studies, the influence of various N input and acidification levels on release of N_2O and oxidation of CH_4 in forest soil has been investigated (Sitaula *et al.*, 1995a; 1995b). Acidification was found to reduce N_2O emission significantly. It was hypothesised that this was due to inhibition of nitrification by the acidity. This was now tested by measuring nitrification potential, N_2O release, and determination of viable AOB, in a factorial experiment consisting of different acidification and N fertilisation rates in planted and unplanted soil. Effects of plant growth in combination with N fertilisation and acidification were tested, since plants are an important sink for ammonium-N, and are potential competitors to AOB (Verhagen *et al.*, 1994). The presence of plants may have implications for methane oxidation as well, and it has been shown that input of ammonium N has a significantly inhibiting effect on CH_4 oxidation in forest soil (Steudler *et al.*, 1989; Sitaula and Bakken, 1993; Sitaula *et al.*, 1995a).

2. Materials and Methods

2.1. PREVIOUS TREATMENT, EXPERIMENTAL DESIGN AND SAMPLING

Earlier (1975-1990), the soil (Typic Udorthent) used for this experiment was subjected to acidification by simulated acid rain (pH 2.5 and 5.6) in the field site under the 100-year-old Scots pine (*Pinus sylvestris*) forest of Norway. In 1991, the soil profiles were established in the lysimeter and the acidification treatments were continued with pH 3 irrigation instead of pH 2.5 treatment and pH 5.5 irrigation instead of pH 5.6 treatment. Two new treatments were superimposed in the lysimeter [N fertilisation with NH_4NO_3 (0, and 90 kg N ha^{-1} y^{-1}) and plant vegetation (with and without)] in factorial design (3 replicates). The treatment with plant growth consisted of a pine sapling (*Pinus sylvestris*, 7 years old in 1997*)* and the ground vegetation (*Vaccinio-Pinetum boreale* association), and treatment without plant growth consisted of pine needle litter (1500 kg dry wt ha-1) without pine saplings or ground vegetation. More details on the previous treatment and soil properties are described by Abrahamsen *et al.* *(*1994) and Sitaula *et al.* (1995a). In 1997, soil samples were collected from the organic (0-10cm) soil layer (excluding the loose litter) from 6-8 sampling points, randomly chosen in each replicate lysimeter cylinder using a soil auger (3 cm diam.). The samples were mixed separately from each treatment to obtain bulk sample of approximately 50 g and transported to laboratory. Each sample was sieved (4 mm), and gravimetric water content was determined (Dewis and Freitas, 1970). The soil moisture content was adjusted to 38% on a volume basis.

2.2. GAS FLUX MEASUREMENT

Soil samples of 10 g dry weight were placed in 120 ml serum bottles capped with butyl rubber septa (type 20-B3P, Chromacol Ltd, London) and incubated at 15 °C for 24 hours. On 4 occasions during this incubation, the atmosphere inside the serum bottle was sampled through the septa, and analysed for N_2O, CH_4 and CO_2 by gas chromatography as described by Sitaula *et al.* (1992).

2.3. DETERMINATION OF NITRIFICATION POTENTIAL

Ten grams of soil were suspended in 100 ml of a liquid mineral medium [3.78 nM $(NH_4)_2SO_4$, MacDonald and Spokes, 1980] supplemented with a trace element solution (Donaldson and Henderson, 1989) (LM). The suspension (soil + LM) was homogenised in a Waring blender thrice for 1 min each time at maximum speed with intermittent cooling as described by Lindahl and Bakken (1995). After homogenization, the pH of the suspension was adjusted to 7, and a 5 ml sample was frozen (to be used as a control for potential nitrification measurement). The rest of the pH adjusted soil suspension was incubated in a reciprocal shaker with slow shaking at room temperature (22 °C) in darkness for two days. During this period, 5 ml of samples were collected every 24 h and stored at -20°C until potential nitrification rates were determined. Nitrate in these samples was determined using a method as described by Christensen and Tiedje (1988). The nitrification potential were estimated from the observed change in nitrate concentrations.

3. Results

3.1. NITRIFICATION POTENTIAL AND N_2O RELEASE

A statistically significant inhibiting effect of acid treatment ($p<0.001$), on both potential nitrification and N_2O release were found. The more acidic treatment (pH 3) reduced potential nitrification by 1-5 times in planted treatment and 2-10 times in soils without the plant (Fig. 1) There was a clear indication of higher nitrification potential in unplanted treatment, particularly in combination with less acidic (pH 5.5.) and unfertilised treatment. This plant pH interaction was weakly significant ($p<0.06$). N fertilisation effects on potential nitrification were somewhat variable, however, in majority of the cases the nitrification potential were higher in the fertilised treatment (Fig. 1).

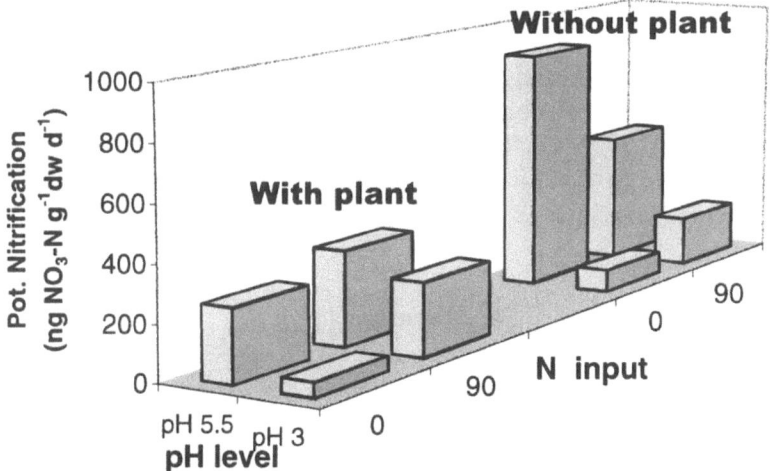

Figure 1. Nitrification potential as influenced by plant, acid irrigation (pH level 5.5 and 3) and N input (N input = 0 and 90 kg N ha^{-1}).

A significant increase in N_2O release in the fertilised soil was observed (1.5 to 9 times higher in fertilised treatment, Table I). The highest N_2O emission was found in fertilised soil with plant growth treated with the pH 5.5 irrigation. In soil without plant growth, the effect of N fertilisation on N_2O release was more pronounced in the more acidic treatment (9 times difference; Table I), whereas in the soil with plant growth, the effects of fertilisation appeared more pronounced in the less acidic (pH 5.5) treated soil (8 times difference) (Table I). This apparent plant-pH interaction was statistically significant ($p<0.05$).

The N_2O emission rates (Table I) were not closely correlated with potential nitrification, although the main effect of acidification and fertilization were similar same for both N_2O emission and nitrification rates.

TABLE I

N_2O and CO_2 release and CH_4 oxidation rates (mean±SE) as affected by plant, soil acidification and N fertilisation.

	Without plant		With plant	
	pH 3	pH 5.5	pH 3	pH 5.5
N_2O emission (ng N_2O-N g^{-1} dw d^{-1}):				
Unfertilised	2±0.3	11±0.3	4±1	15±2
Fertilised (90 kg N ha^{-1} y^{-1})	18±9	44±2	6±1	120±9
CH_4 Oxidation (ng CH_4 g^{-1} dw d^{-1}):				
Unfertilised	2±1	3±1	3±2	10±1
Fertilised (90 kg N ha^{-1} y^{-1})	<1	<1	<1	2±0.6
CO_2 emission (µg CO_2 g^{-1} dw d^{-1}):				
Unfertilised	112±13	118±22	230±35	438±37
Fertilised (90 kg N ha^{-1} y^{-1})	151±19	254±14	312±94	655±100

3.2. CH_4 OXIDATION AND CO_2 RELEASE

CH_4 oxidation was significantly decreased by N fertilisation ($p< 0.005$) and acidified soil ($p<0.008$). CH_4 oxidation rate was significantly higher in soils with plant treatment ($p<0.05$) (Table I). Planted soil had much higher CO_2 production rates than unplanted ($p<0.001$), and the planted pH 5.5 treatment much higher than the acid (pH 3) treatment ($p<0.05$) (Table I).

4. Discussion

4.1. NITRIFICATION POTENTIAL AND N_2O RELEASE

The present work demonstrates a reduction in nitrification potential and N_2O release due to severe acid deposition. Decreased N_2O release by acidification is in agreement with earlier N_2O flux measurements in the same lysimeter (Sitaula *et al.*, 1995b), where we hypothesised that the observed reduction in N_2O emission in the most acidic irrigation treatment (pH 3) was probably due to reduction in nitrification potential. This experiment supported this hypothesis. Regina *et al.* (1996) found a positive correlation between N_2O flux, nitrification potentials, soil pH and the number of nitrifiers in boreal peat soils.

The lower nitrification potentials in soil with plant growth may reflect that plant uptake of ammonium may have reduced the availability of ammonium to the ammonium oxidiser. Planted soil had Scot pine, a calcifugeous species that prefers ammonium relative to nitrate (Sogn and Abrahamsen, 1998). The pronounced plant effect on nitrification potentials with unfertilised treatment may support this explanation.

Lower N_2O release from more acidic irrigation was in contradiction to another study which reported increased N_2O release with increased acidity in acid coniferous forest soils (e.g. Martikainen and Boer, 1993). Our soil had been subjected to a long-term (15 years) application of more acidic "rain" treatment [pH 2.5 in field plots (Abrahamsen *et al.*, 1994) and continued with pH 3 irrigations in the lysimeters (Sitaula *et al.*, 1995b)]. Such a severe acidification might have reduced the extent and frequency of high pH microsites where nitrification or denitrification could take place and may also have directly affected the acid-tolerant chemolithotrophic ammonium oxidisers (Martikainen and Boer, 1993).

The N_2O-N/NO_3-N production rates ranged from 1/100 to 1/5, which is 10 to 200 times the ratio for unrestricted ammonium oxidation (Jiang and Bakken, 1999). This strongly suggested that processes other than unrestricted ammonium oxidation contribute to the N_2O release. Oxygen and acidity stressed ammonium oxidisers may be one source, but denitrification with low N_2O- efficiency may be another (Holtan-Hartwig *et al.*, 2000). The generally high N_2O release in planted verses unplanted soil suggests that other factors dominate in regulating N_2O fluxes.

The magnitude of increase in N_2O release due to fertilisation was comparable to our earlier study in the same lysimeter (Sitaula *et al.*, 1995b). This is obvious since N input increased the supply of substrate for both nitrification and denitrification (Eichner, 1990; Mosier *et al.*, 1991; Sitaula *et al.*, 1995b).

4.2. CH_4 OXIDATION AND CO_2 RELEASE

The negative effects of soil acidification on CH_4 oxidation observed in the present laboratory incubation experiment was in contrast to the *in situ* methane flux measurement 7 years ago where higher methane uptake rates were found in the pH 3 irrigation treatment (Sitaula *et al.*, 1995a). This earlier observation is likely to be a result of experimental artefacts (e.g. systematic difference in gas diffusivities) at the beginning of lysimeter establishment in 1991.

The results indicate a cumulative effect of N fertilisation on methane oxidation. There was more than 60% reduction due to N input in the present measurements compared to the 40% reduction in our previous studies of CH_4 fluxes (Sitaula *et al.*, 1995a).

The reason for observed plant stimulation of CH_4 oxidation is probably that plants are the most important sinks for ammonium in the system. Ammonium and intermediates of ammonium oxidation have been reported as competitive substrates for methane monooxygenase enzyme (Bedard and Knowles, 1989). Moreover, CH_4 can also be oxidised by ammonium oxidisers.

The increased CO_2 emission in planted versus unplanted soil is due to the input of fresh organic carbon in planted treatment (root turnover). The negative effect of acidity (pH 3) would then be due to its negative effect on plant productivity and not due to direct effect of acidity on the microbial community. The pH 3 irrigation significantly reduced the stem height and stem growth of Scot pine sapling (Sogn and Abrahamsen, 1998). The absence of acidification effect on CO_2 evolution from unplanted soil strengthens this conclusion.

References

Abrahamsen, G., Stuanes, A.O. and Tveite, B.: 1994, *Long Term Experiments with Acid Rain in Norwegian Forest Ecosystems.* Ecological Studies 104, Springer, New York..
Bedard, C. and Knowles, R.: 1989, *Microbiological Reviews* **53**, 68.
Christensen, S. and Tiedje, J.: 1988, *Applied and Environmental Microbiology* **54**, 1409.
Dewis, J. and Freitas, F.: 1970, Physical and chemical methods of soil and water analysis. *FAO Soils Bulletin* **10**, Rome.
Donaldson, J.M. and Henderson, G.S.: 1989, *Soil Science Society of America Journal* **53**, 1608.
Eichner, M.J.: 1990, *Journal of Environmental Quality* **19**, 272.
Galloway, J.N. and Rodhe, H.: 1991, 'Regional atmospheric budgets of S and N fluxes: how well can they be quantified?', *in* F. T. Last. and R. Watling (eds.), *Acidic deposition,* Royal Society of Edinburgh, Edinburgh, pp. 61-80.
Holtan-Hartwig, L, Dörsch, P. and Bakken L.R.: 2000, *Soil Biology and Biochemistry* **32**, 833.
Jiang, Q.Q. and Bakken, L.R: 1999, *Applied and Environmental Microbiology* **65**, 2679.
Lindahl, V. and Bakken, L.R.: 1995, *FEMS Microbiology Ecology* **16**, 135.
MacDonald, R. M. and Spokes, J.R.: 1980, *FEMS Microbiology Letters* **8**, 143.
Martikainen, P.J. and Boer, W.D.: 1993, *Soil Biology and Biochemistry* **25**, 343.
Mosier, A., Schimel, D., Valentine, D., Bronson, K. and Parton, W.: 1991, *Nature* **350**, 330.
Persson, T. and Wiren, A.: 1989, Microbial activity in forest soils in relation to acid/base and carbon/nitrogen status. *in* F.N. Braekke, K. Bjor and B. Halvorsen (eds.) *Air pollution as stress factor in Nordic forests,* Norwegian Institute for Forest Research, Aas pp. 83-95.
Regina, K., Nykanen, H., Silvola, J. and Martikainen, P.J.: 1996, *Biogeochemistry* **35**, 401.
Sitaula, B.K., Luo, J.F. and Bakken, L.R.: 1992, *Journal of Environmental Quality* **21**, 493-496.
Sitaula, B.K. and Bakken, L.R.: 1993, *Soil Biology and Biochemistry* **25**, 1415.
Sitaula, B.K., Bakken, L.R and Abrahamsen, G.: 1995a, *Soil Biology and Biochemistry* **27**, 871.
Sitaula, B.K., Bakken, L.R and Abrahamsen, G.: 1995b, *Soil Biology and Biochemistry* **27**, 1401.
Sogn, T. A.and Abrahamsen, G.: 1998, *Forest ecology and Management* **103**, 117.
Steudler, P.A., Bowden, R.D., Melillo, J.M. and Aber, J.D.: 1989, *Nature* **341**, 314.
Verhagen, F.J.M., Hageman, P.E.J., Woldendorp, P.J.W., Laanbroek, H.J.: 1994, *Soil Biology and Biochemistry* **26**, 89.

THE RELATIONSHIP BETWEEN TOPSOIL AND STREAM SEDIMENT HEAVY METAL CONCENTRATIONS AND ACIDIFICATION

J.R. HALL[1*], B. REYNOLDS[2], T. SPARKS[1], A. COLGAN[1], I. THORNTON[3] and S.P. McGRATH[4].

[1] *CEH Monks Wood, Abbots Ripton, Huntingdon, PE28 2LS, UK.;* [2] *CEH Bangor, University College North Wales, Bangor, LL57 2UW, UK.;* [3] *Environmental Geochemistry Research Group, Imperial College, Royal School of Mines, London, SW7 2BP, UK.;* [4] *BBSRC, Institute of Arable Crops Research, Rothamsted Experimental Station, Harpenden, AL5 2JQ, UK.*

(*author for correspondence, e-mail: jrha@ceh.ac.uk)

Abstract. Critical loads of acidity have been used by the UNECE Convention on Long Range Transboundary Air Pollution for the development of protocols to control the emissions of acidifying pollutants. Since soil acidity has an effect on the mobilisation of heavy metals in the environment, it is important to understand the relationships between acidity and heavy metal pollution. This paper examines the relationships between soil acidification and heavy metal (cadmium, copper, lead and zinc) concentrations in topsoils and in stream sediments. It makes use of published heavy metal data and two indices of acidification: soil pH and soil acidity critical loads. For cadmium and zinc, a general increase in the ratio of stream sediment to toposil metal concentrations is seen with a decrease in soil pH and soil acidity critical loads. This demonstrates that where soils are more acidic and acid sensitive the metal concentration in the stream sediments is greater relative to that in the topsoil, suggesting mobilisation of these metals under acid conditions. Results for copper are similar but the relationship weaker. However, for lead the ratios tend to decrease with a decrease in pH and critical loads suggesting that where soils are more acid, lead remains in the soil rather than being mobilised into streams and precipitating onto stream sediments. This reflects the association between soil lead concentrations and soil organic matter content, which tends to be greater in acidic, peaty soils.

Keywords: acidification, heavy metals, soils, stream sediments.

1. Introduction

The UNECE Convention on Long-Range Transboundary Air Pollution (LRTAP) is considering effects-based approaches, including critical loads, for the control and abatement of heavy metal emissions. Acidity critical loads have already been used in the LRTAP Convention Protocols. Since soil acidity has an effect on the mobilisation of heavy metals in the environment, it is important to understand the relationships between acidity and heavy metal pollution. It is also important to make use of existing national scale data to explore these relationships, since monitoring at the national scale is prohibitively expensive.

Much work has been done over the last two decades to examine the relationships between acidity and aluminium concentrations and toxicity in freshwaters, but fewer studies have focused on other key potentially toxic metals such as cadmium (Cd), zinc (Zn) and lead (Pb). Various studies have also shown that acidity is a key variable determining the mobility of metals in the environment (Gower, 1980; *Johansson et al.*, 1995a; Johansson *et al.*, 1995b; Nelson and Campbell, 1991; Nriagu, 1990; Wilson and Bell, 1996). This paper examines the relationships between soil acidification and heavy metal (Cd, Zn, Pb and copper (Cu)) concentrations in topsoils and in stream sediments. It makes use of published metal concentration data, which have not previously been compared and uses two different indices of acidification: soil pH and soil acidity critical loads. In this paper we consider the hypothesis that metals become more mobile under acid conditions, so that when acid conditions prevail, higher concentrations of metals will be expected in stream sediments relative to topsoils (assuming no input of metals from aquatic sources, eg, mine drainage directly in to the stream). This hypothesis is explored using a Geographic Information System (GIS) and statistical analysis of the data.

2. Data

The following data sets for England and Wales have been used for this study.
(i) Total concentrations of Cd, Zn, Pb and Cu in stream sediments
These data are based on a survey carried out in 1969, which aimed at sampling stream sediments at a density of one for every 2.5 km square across England and Wales. They are published in the Wolfson Geochemical Atlas of England and Wales (Webb *et al.*, 1978).
(ii) Total concentrations of Cd, Zn, Pb and Cu in topsoils
These data are published in the Soil Geochemical Atlas of England and Wales (McGrath and Loveland, 1992). The maps are based on samples taken every 5km on an orthogonal grid during a survey carried out between 1979 and 1982.
(iii) Percentage organic carbon in topsoils
These data are also published in the Atlas by McGrath and Loveland (1992). Values range from 0.1% to 65.9% with a mean value for all (5666) 5km squares of 6.66%.
(iv) Indices of acidification: (a) Topsoil pH values. These data were collected as part of the above survey of topsoils in England and Wales. *(b) Empirical critical loads of acidity for soils.* These are based on the weathering rate and mineralogy of the dominant soil series in each 1 km square of United Kingdom (Hornung *et al.*, 1995).

3. Methods

The two metal concentration data sets were examined separately in the statistical package Minitab v13.1 (Table I). Histograms of the data highlighted suspected anomalous high values, probably associated with contaminated areas, such as mine workings. There are currently no national-scale digital data sets listing all present and past mining or smelting sites to enable such areas to be easily identified. Therefore, anomalous high values were calculated according to Davies (1980): "possibly anomalous" values as those exceeding the mean plus two standard deviations, and "probably anomalous" values where they exceed the mean plus three standard deviations. The data were positively skewed, so were log transformed to normalise their distributions before calculating the anomalous values. Anti-logs provided the threshold values for identifying probable and possible anomalous data points for each metal in both the stream sediment and topsoil data sets (Table I).

TABLE I
Summary of heavy metal concentration data sets. * Based on log-transformed data. # Limits of detection.

Parameter	Minimum mg kg^{-1}	Maximum mg kg^{-1}	Mean mg kg^{-1}	N	Mean +2SD* mg kg^{-1}	Mean +3SD* mg kg^{-1}
Stream sediment						
Cd	<0.5#	310	0.81	48514	3.2	6.1
Zn	<5#	51000	183.1	48527	561	1279
Pb	<5#	2638	74.2	48603	276	824
Cu	<3#	2505	30.5	48603	108	263
Topsoils						
Cd	<0.2#	40.9	0.84	5665	2.2	3.2
Zn	5	3648	96.8	5669	257	460
Pb	3	16338	73.7	5669	217	469
Cu	1.2	1508	23.0	5669	61	111

The metal concentrations, soil organic carbon, pH and acidity critical loads data were imported into a GIS to examine the spatial patterns and to link the data sets geographically. The data were matched using their Ordnance Survey grid references. The number of sample points where data coincide varied from one metal to another but ranged from 1897 to 1919. The heavy metal and acidification data for the matching data points were extracted and imported into Minitab. The mean values for metal concentrations in this sub-set of data are similar to those of the full data sets.

To examine the relationship between the metal concentrations data and acidification, the ratios of stream sediment to topsoil heavy metal concentrations were calculated. In addition, the topsoil pH values were assigned to five classes (<4, 4-5, 5-6, 6-7, >7). The empirical soil acidity critical loads consist of 5 discreet values (0.2, 0.5, 1.0, 2.0, 4.0 keq ha^{-1} year^{-1}) (Hornung et al., 1995) and these were also assigned to a 5-band classification. Sub-sets of the ratio data that excluded the "possible" and "probable" anomalous metal concentrations were also examined. One way analysis of variance (ANOVA) was used to compare the ratios with topsoil pH classes and soil critical load classes. The ratio data were mapped to examine the spatial and geographical patterns.

4. Results and discussion

The relationship between the three sets of ratios (ie all data, excluding possibles, excluding probables) and topsoil pH classes is given in Table II. For Cd, Zn and Cu there is a general increase in the mean ratio values with decreasing pH, demonstrating that in areas of more acid soils the concentration of metal in the stream sediments becomes greater in relation to that in the topsoil. This suggests mobilisation of the metals from soils into streams where they may precipitate into the sediments. For Pb the mean ratios tend to decrease as pH decreases (Table II) suggesting that where soils are more acid, Pb remains in the soil, rather than being mobilised into the streams and stream sediments. Excluding either the probable, or the possible anomalous metal concentrations, increases the discrimination between the pH classes, ie the F ratios increase (Table II). This discrimination is greater for Cd and Zn than for Cu and Pb, the former having greater F ratios than the latter. So, although Cu behaves in a similar way to Cd and Zn, its relationship with pH is weaker. Decomposing the sums of squares for pH classes into their orthogonal polynomials suggests a linear response for Cu and Pb, but a curved, possibly asymptotic, response for Cd and Zn.

TABLE II
Mean ratios of stream sediment metal concentrations to topsoil metal concentrations by topsoil pH class. ANOVA F ratios and degree of significance also shown (* $p<0.05$, ** $p<0.01$, *** $p<0.001$, ns = not significant). Mean +3sd = probable anomalies, mean +2sd = possible anomalies.

Ratio	Topsoil pH category					F ratios
	<=4.0	4.01 – 5.0	5.01 – 6.0	6.01 – 7.0	>7.0	
All data						
Cd	3.40	2.68	1.02	0.86	0.63	7.87***
Zn	4.06	7.04	1.71	1.66	1.52	1.79^{ns}
Pb	1.18	1.20	1.31	1.36	1.36	0.25^{ns}
Cu	1.73	1.78	1.65	1.28	1.13	2.60*
Excluding mean+3sd						
Cd	2.55	1.16	0.81	0.68	0.60	22.23***
Zn	3.77	2.31	1.56	1.54	1.47	44.07***
Pb	0.98	1.10	1.10	1.22	1.34	1.18^{ns}
Cu	1.76	1.51	1.36	1.27	1.12	6.24***
Excluding mean+2sd						
Cd	2.35	0.97	0.71	0.61	0.47	37.10***
Zn	3.44	1.94	1.47	1.48	1.35	67.23***
Pb	0.60	0.96	0.97	0.96	1.15	5.58***
Cu	1.67	1.40	1.31	1.23	1.02	9.21***

Pb tends to be tightly bound to organic carbon (OC) in soil and the percentage of OC tends to be greater in the more acid soils, for example, peat soils. Where the Pb ratios are small, ie where more Pb is likely to be found in the soil compared to the stream sediment, the percentage OC is high and the soil pH low. The relationship between OC and the Pb ratios was examined using the data set excluding the possible anomalies. This demonstrated a negative relationship (possibly asymptotic) between the ratios and the percentage organic carbon. A rank correlation resulted in a value of –0.205 with a significance level <0.001 for the 1796 data points. The Cd and Zn ratios have a positive linear relationship with percentage OC, suggesting they are less likely to be bound to OC and immobilised in the soil. There was no discernible relationship between Cu and percentage OC.

The relationship between acidity critical load classes and the metal ratios gave similar results and trends to those described for pH classes, with the exception of Pb which showed no clear trend. However, for Cd and Zn the F ratios were lower, indicating a weaker relationship with critical loads classes than with pH classes. The F ratios for Cu and Pb were marginally higher, but not so great as to suggest a better relationship with acidity critical loads classes.

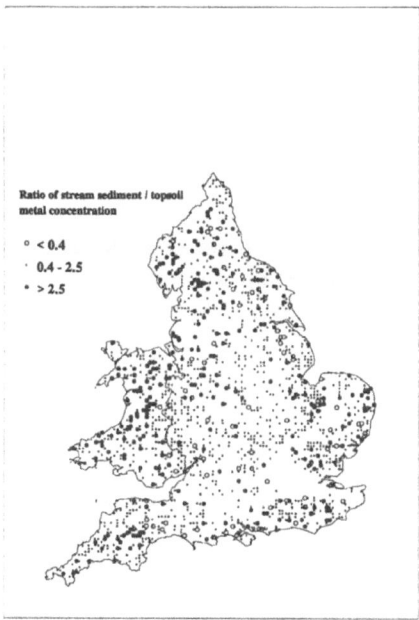

Figure 1. Map showing the ratios of stream sediment zinc concentrations to topsoil zinc concentrations.

Mapping the ratio values enables areas to be identified where mobilisation of these metals from soils into streams and stream sediments may occur. A map of the zinc ratios (Fig. 1) shows the highest ratios in the uplands of Wales and in parts of Cumbria, the Pennines, north York moors and Dartmoor. These are also the regions where soils tend to be the most acid. In addition, some high ratios are found in parts of East Anglia where there are peat or gley soils. The area of the Weald, to the south of London is also evident. However, in the south and east of England there are also sites where the topsoil metal concentrations are much greater than that in stream sediments (ie very low ratios). It must also be noted that the metal concentrations considered here are total concentrations and therefore do not necessarily represent the bioavailable toxic fraction of the metals. Such maps are useful though for identifying areas where there may be a risk of metal leaching and which require further investigation. This could include obtaining additional data on the metal speciation and activity by either (i) collecting soil and sediment samples for heavy metal analysis; (ii) using a computer model such as CHUM (Tipping, 1996) to predict metal speciation for particular sites for which sufficient data were available. These data and additional spatial data, eg on land cover, land use and topography, could be incorporated in the GIS, together with the ratio data to improve the accuracy of risk assessment of metal mobilisation in the environment.

5. Conclusions

Comparing ratios of stream sediment to topsoil metal concentrations with indices of acidification suggests that Cd, Zn and Cu are readily mobilised in areas of more acid soils, into waters where they form complexes and precipitate into steam sediments. The results clearly show that Pb behaves differently from the other metals studied, probably due to its relationship with organic carbon. The study highlights the value of using existing spatial databases to explore methods of assessing the potential risk of heavy metal pollution and the mapping the ratios identifies areas where mobilisation of heavy metals may pose an environmental threat.

References

Davies, B.E.: 1980, 'Trace element pollution', in B.E. Davies (ed.), *Applied Soil Trace Elements*, John Wiley and Sons Ltd. pp 287-351.
Gower, A.M: 1980, 'Ecological effects of changes in water quality', A.M. Gower (ed.), *Water Quality in Catchment Ecosystems*, John Wiley and Sons Ltd. pp 145-172.
Hornung, M., Bull, K.R., Cresser, M., Hall, J., Langan, S.J., Loveland, P. and Smith, C.: 1995, *Environmental Pollution*, **90**, 301.
Johansson, K., Andersson, A. and Andersson, T.: 1995a, *The Science of the Total Environment*, **160**, 373.
Johansson, K., Bringmark, E., Lindevall, L. and Wilander, A.: 1995b, *Water, Air and Soil Pollution*, **85**, 779.
McGrath, S.P. and Loveland, P.J.: 1992, *The soil geochemical atlas of England and Wales*. Chapman & Hall.
Nelson, W.O. and Campbell, P.G.C.: 1991, *Environmental Pollution*, **71**, 91.
Nriagu, J.O.: 1990, *Environment*, **32**, 7.
Tipping, E.: 1996, *Journal of Hydrology*, **174**, 305.
Webb, J.S., Thornton, I., Thompson, M., Howarth, R.J. and Lowenstein, P.L.: 1978, *The Wolfson Geochemical Atlas of England and Wales*. Clarendon Press, Oxford.
Wilson, M.J. and Bell, N.: 1996, *Applied Geochemistry*, **11**, 133.

INTEGRATED MONITORING PROGRAM ON ACIDIFICATION OF CHINESE TERRESTRIAL SYSTEMS (IMPACTS) – A CHINESE –NORWEGIAN COOPERATION PROJECT

D. TANG[1], E. LYDERSEN[2], H. M. SEIP[3]*, V. ANGELL[4], O. EILERTSEN[5], T. LARSSEN[3], X. LIU[6], G. KONG[7], J. MULDER[8], A. SEMB[9], S. SOLBERG[10], K. TORSETH[9], R. D. VOGT[3], J. XIAO[11] and D. ZHAO[12].

[1] *Atmospheric Environment Institute, Chinese Research Academy of Environmental Sciences, Beiyuan, Beijing, China 100012;* [2] *Norwegian Institute for Water Research, P.O. Box 173, Kjelsaas, N-0411 Oslo, Norway;* [3] *University of Oslo, Dept. of Chemistry, P.O. Box 1033 Blindern, N-0315 Oslo, Norway;* [4] *Norwegian Institute of International Affairs, P. O. Box. 8159 Dep., N-0033 Oslo, Norway;* [5] *Norwegian Institute of Land Inventory, P.O. Box 115, N-1430 Aas, Norway;* [6] *State Environmental Protection Agency, No. 115 Nanxiaojie, Xizhimennei, Beijing 100035, China;* [7] *South China Institute of Botany, Chinese Academy of Sciences, Guangzhou 510650, China;* [8] *Agricultural University of Norway, Dept. of Soil and Water Sciences, P.O. Box 5028, N-1432 Aas, Norway;* [9] *Norwegian Institute for Air Research, P.O. Box 100, N-2007 Kjeller, Norway;* [10] *Norwegian Forest Research Institute, Høgskolev. 12, 1432 Aas, Norway;* [11] *Guizhou Institute of Environmental Sciences, 148 Xinhua Road, 550002 Guiyang, China;* [12] *Chongqing Institute of Environmental Science and Monitoring, 37 Jia Ling VLG-1, Jiang Bei District, 630020 Chongqing, China.*
(* author for correspondence, e-mail: h.m.seip@kjemi.uio.no)

Abstract. A 5-year Chinese-Norwegian research project was launched in the autumn of 1999. Forested sites for intensive studies are or will be established in the Chongqing municipality and in Guizhou, Hunan and Guangdong provinces in southern China. Previous studies have shown that harmful effects of acid deposition are likely to be most severe in this region. The research and monitoring sites shall give information about acidification mechanisms and effects on vegetation in order to improve policy oriented acidification models and critical load estimates as well as function as interdisciplinary training centers for acid rain research. Furthermore, the project shall improve the basis for developing an efficient regional acid rain monitoring system. At one site in Guizhou and one in Chongqing, research on soil and soilwater chemistry has been ongoing for several years. The forest at these sites appears to show symptoms of reduced vitality. The sensitivity of Chinese forests to acidification is uncertain and will be focused. Decision-makers should get an improved basis for optimal mitigation measures through the project.

Keywords: Acid precipitation, China, critical loads, forest, monitoring, soil acidification, soil water, vegetation

1. Introduction

The growth of the energy consumption in China (Fig. 1), of which about 73% is produced by coal burning, has resulted in a nearly parallel increase in sulfur emissions. This has caused high sulfur concentrations in air and precipitation in large areas. Since natural alkaline dust neutralizes the precipitation in the north, the main acidification problem region is in the south. Emissions of nitrogen oxides (NO_x) are still relatively low, but are expected to increase rapidly.

Studies of acid precipitation and its effects have been given high priority in China the last two decades. However, knowledge of effects of acid rain is still mainly based on experience in Europe and North America. It is important that this research is corroborated with observations in other regions bearing in mind the large differences in soil conditions and ecosystems in general. Based on more than 10 years of cooperation between Chinese and Norwegian scientists, and in particular a pilot study carried out in 1997 (Lydersen et al., 1997; Seip et al., 1999), a five-year cooperation study was launched in the autumn of 1999. The project is a joint effort of *The State Environmental Protection Administration* (SEPA) of China and *The Norwegian Agency for Development Cooperation* (NORAD). *Chinese Research Academy of Environmental Sciences* (CRAES) and *Norwegian Institute for Water Research* (NIVA) are responsible for the implementation.

2. Some results from previous studies

Intensive catchment studies play an important role in the project. Some potentially suitable sites were identified during the pilot study (Seip et al., 1999). It was aimed at finding fairly similar sites with acid soils and Masson pine (*Pinus massoniana*) which is sensitive to acidification. The five sites finally selected for the IMPACTS project are TieShanPing (TSP) in Chongqing municipality, LiuChongGuan (LCG) and LeiGongShan (LGS) in Guizhou, CaiJaTang (CJT) in Hunan and DingHuShan (DHS) in Guangdong; see Figure 2 and Table I. There is a large difference in S-deposition between the sites. A fairly unpolluted site (LGS) has been included since there are few data on acid precipitation from remote areas in China. All five sites have acid-sensitive soils and are in the area with acid (or potentially acid) precipitation (i.e. *the acid rain zone*). In particular, increased acid deposition is likely to result in higher aluminum concentrations in soil water and higher Al/(Ca+Mg)=$R_{Al/M}$ ratios, with potentially harmful effects on vegetation. Soil pH, cation exchange capacity (CEC_E) and base saturation (BS) for the sites are shown in Table II.

At LCG and TSP, research has been going on for several years. Deposition, throughfall, soils and soil water have been studied (Larssen et al., 1998; Zhao et al., 2001). Average SO_2 concentration at LCG is about 44 μgm^{-3}. It has not been measured at TSP, but measurements in the Chongqing area indicate that the average concentration is at least as high as at LCG. The S-deposition is 8–12 g m^{-2} yr^{-1} (Table I), which is at least as high as the deposition was in "the Black Triangle" in Europe in the 1980s. Concentrations in the precipitation and S-deposition at these two sites are compared to values from one Norwegian and two Polish sites in Figure 3. The concentration of nitrogen oxides is modest at the Chinese sites, while the ammonium concentration is high at TSP. In spite of the large S-deposition, the degree of crown thinning and needle damage is not severe at LCG; at TSP the crown thinning is more severe at least partly due to insect attacks the last years. Median molar values for $R_{Al/M}$ are well below 1 at present. For most tree species one would not expect damage at this level, but the sensitivity of for example Masson pine is not well known. $R_{Al/M}$ is likely to increase if the deposition is not reduced.

Rather large soil nitrogen stores were found and the C/N ratios in soils are low compared to values from forest soils in Europe and the USA. This indicates that increased N-deposition may result in nitrate leaching. However, the C/N ratios found in our study are well within the range reported for Chinese soils in sub-tropical areas with natural vegetation (Zhu, 1990).

First order streams and smaller bodies of water seem to be acidified in some areas. During the pilot study in 1997, we also found rural mountainous areas where the surface waters appear highly sensitive to acidification due to very low acid neutralization capacity. However, based on literature studies (Larssen et al., 1999) as well as the pilot project (Seip et al., 1999), it does not seem likely that surface water acidification will become an extensive regional problem in China in the next decades. Considering larger areas, deeper soil horizons and bedrock generally neutralize the percolating water. In the acid rain zone the bedrock is generally very variable, often with intrusions of limestone in layers of sandstone.

Larssen and Carmichael (2000) and Larssen et al. (2000; 2001) demonstrated the importance of knowing the sources and deposition patterns of base cations when assessing the effects of changes in sulfur emissions.

More results from the IMPACTS sites are given in other contributions to this conference (e.g. Zhao et al. 2001; Hansen et al., 2001).

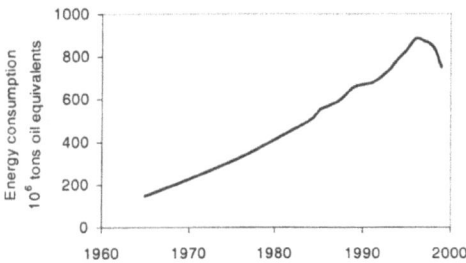

Figure 1. Energy consumption in China
(Data from UN, 1989 and BP Amoco, 2000).

Figure 2. Site locations

TABLE I

Soil sampling site locations and sulfur deposition. From Seip et al. (1999), S-deposition also from Downing et al. (1997)

Site name	Code	Soil type	Location	Sulfur deposition
DingHuShan	DHS	Acrisol	86 km W of Guangzhou	2-4 $gSm^{-2}yr^{-1}$
LiuChongGuan	LCG	Alisol	10 km NE of Guiyang	8-10 $gSm^{-2}yr^{-1}$
LeiGongShan	LGS	Acrisol	140 km E/SE of Guiyang	1-2 $gSm^{-2}yr^{-1}$
TieShanPing	TSP	Acrisol	25 km NW of Chongqing	10-12 $gSm^{-2}yr^{-1}$
CaiJiaTang	CJT	Acrisol	70 km SW of Changsha	3-5 $gSm^{-2}yr^{-1}$

TABLE II

Soil quality parameters for the investigated soils at the different sites. Median values with ranges in parenthesis. CEC and BS at soil pH.

Site	Horizon	pH(H$_2$O)	CEC meg/kg	% Base saturation
DHS	O/A	3,7 (3,4-4,4)	107 (51-206)	10 (3-21)
	B	4,3 (4,0-5,1)	35 (7-68)	4,6 (2-11)
LCG	O/A	3.8 (3.1-4.1)	108 (64-223)	20 (5-60)
	B	4.0 (3.7-4.7)	82 (25-142)	7 (3-40)
LGS	A	4.1(3.8-4.9)	9(3-13)	38(24-100)
	B	4.3(4.1-4.8)	2.1(0.9-4.5)	23(9-81)
TSP	O/A	3.8 (3.5-4.0)	63 (55-90)	12 (7-17)
	B	4.0 (3.6-4.3)	37 (25-109)	8 (5-27)
CJT	O/A	4.2 (4.1-4.3)	64 (26-92)	27 (11-38)
	B	4.4 (4.3-4.7)	37 (21-53)	11 (6-19)

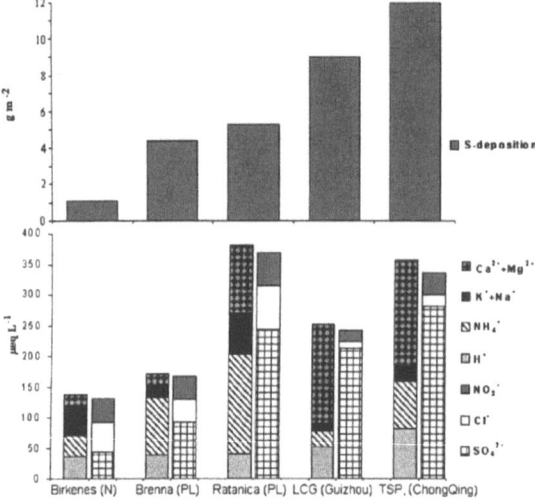

Figure 3. Comparison of S-deposition and composition of precipitation at some sites in Norway, Poland and China (from Seip et al., 1999). The bars in the upper panel show estimated total S-deposition. The bars in the lower panel show volume-weighted average ionic composition in precipitation (μeq L^{-1}). The left bar represents cations, the right bar anions. Birkenes (data for 1993-94) is in southernmost Norway which is the region of the country most affected by acid rain. Brenna and Ratanica are in southern Poland; the region in Europe with the highest S-deposition caused by long-range transported pollutants. At both Polish sites data are from 1991-94. Ratanica is about 20 km from Krakow and strongly influenced also by local pollution. Brenna is about 60 km from Katowice and 70 km from Krakow. The data for LiuChongGuan (LCG) are for 1992-95, and the data for TieShanPing (TSP) for 1995-1997 (Zhao Dawei, unpublished

3. The 5-year IMPACTS project

In China the main threat of acid deposition is to terrestrial ecosystems. This is reflected in the scientific activities in IMPACTS, which are concentrated at the 5 selected sites. The activities are organized in four work packages:
• WP-I: *Air chemistry and deposition*
At all stations, precipitation, throughfall chemistry as well as air concentrations of SO$_2$, NO$_2$ and O$_3$ will be determined on a daily or weekly basis.

- WP-II: *Soil, soil water and surface water*

Samples of soils, soil waters and surface waters in the catchments have been/will be analyzed. The aluminum chemistry will be one of the focused issues since large pools of easily available aluminum are common in these soils (Liao et al., 1998).

WP-III: *Forest and ground vegetation*

This work package focuses on vitality and nutrient status of forest ecosystems. The sensitivity of forests to $R_{Al/M}$ in soil water and other acidification parameters is uncertain and will be studied in the project.

- WP-IV: *Data base, information and integrated analysis*

This work package is responsible for dissemination of information about the project and its findings both among the participants and to third parties. The data produced in the project will be applied in integrated assessment studies, in addition, economic analyses of the impacts of acid rain will be initiated. All relevant information will be stored in a central data-base easily accessible to the researchers participating in the project. WP-IV is also responsible for publishing a biannual IMPACTS Newsletter (http://www.impacts.net.cn).

Quantitative modeling is a central part of the project, where results from WP I, II and III will be integrated. Focus will be on nutrient cycling, acidification of soils and water and on the estimation of critical loads. This, in turn, may be used to improve policy oriented acidification models, e.g. RAINS-Asia (Downing et al., 1997). In addition to provide information about acidification mechanisms and effects on vegetation, the research and monitoring sites shall function as interdisciplinary training centers for acid rain research. Furthermore, they shall improve the basis for developing an efficient regional acid rain monitoring system.

The LCG and TSP sites will first be fully equipped, then follow LGS (in Guizhou) and CJT (in Hunan). A full data collection program is expected to run from 1 January 2001 at these sites. The DHS site in Guangdong is scheduled to be running from January 2002.

4. Policy considerations

Chinese environmental strategies aim at reducing particle emissions as well as sulfur emissions. Lower particle concentrations are highly desirable for improving human health and the environmental quality in urban and industrial areas. However, removal of alkaline dust without corresponding reductions of the sulfur dioxide emissions may result in increased acidity of the precipitation, particularly in regions not to far from the emissions (Semb et al., 1995). Building of higher stacks will also improve conditions at ground level near the source but promote long-range transport of pollutants. In addition, increasing NOx emissions from both large point sources and increasing numbers of motor vehicles will also increase the acid burden. Hence, there is an obvious potential for enhanced acidification in areas today receiving high loading of sulfur as well as significant deposition of base cations which partly counteracts the effects of acid rain. In more remote, sensitive mountainous areas, which presently receive moderate acid deposition, harmful effects may occur if amount and composition of the deposition change. Careful integrated monitoring is required in such areas.

Although extensive acid rain research has been carried out in Europe and North America for about three decades, this experience cannot directly be transferred to Chinese ecosystems. Carefully planned studies in China are therefore a prerequisite for decision making. Through the IMPACTS project decision-makers will get an improved basis for designing optimal mitigation measures. The activities in IMPACTS should strengthen the Chinese influence on international cooperation on acid rain research and mitigation policies.

Reductions in emissions of air pollutants in China will have large benefits. However, it is essential that effects on local, regional and global levels are considered in an integrated way, i.e. the measures should be designed to ensure that harmful effects on humans, materials and the natural environment are minimized. Without an integrated approach unwanted negative side effects may prove considerable.

References

BP Amoco: 2000, Statistical review of world energy. The British Petroleum Company, London, <http://www.bpamoco.com/worldenergy/> [Accessed June 2000].

Downing, R.J., Ramankutty, R. and Shah, J.J.: 1997, *Rains-Asia: an assessment model for acid deposition in Asia*. The World Bank, Washington DC.

Hansen, H., Larssen, T., Seip, H.M. and Vogt, R.D.: 2001, This conference.

Larssen, T., Schnoor, J.L., Seip, H.M. and Zhao, D.: 2000, *Sci. Total Environ* **246**, 175.

Larssen, T., Seip, H.M., Carmichael, G.R. and Schnoor J.L.: 2001 This conference.

Larssen, T. and Carmichael, G.R.: 2000, *Environmental Pollution* **110**, 89.

Larssen, T., Xiong, J., Vogt, R.D., Seip, H.M., Liao, B., and Zhao, D.: 1998, *Water Air Soil Pollut.* **101**, 137.

Larssen, T., Seip, H.M., Semb, A., Mulder, J., Muniz, I., Vogt, R.D., Lydersen, E., Angell, V., Tang, D. and Eilertsen, O.: 1999, *Environmental Science & Policy* **2**, 9.

Liao, B., Larssen, T. and Seip, H.M.: 1998, *Geoderma* **86**, 295.

Lydersen, E., Angell, V., Eilertsen, O., Larssen, T., Mulder, J., Muniz, I.P., Seip, H.M., Semb, A., Vogt, R.D. and Aagaard, P.: 1997. *Planning of an integrated acidification study and survey on acid rain impacts in China*. Final Report. NIVA Report 48/97. Norwegian Institute for Water Research, P.O. Box 173 Kjelsås, 0411 Oslo, Norway.

Seip, H.M, Aagaard, P., Angell, V., Eilertsen, O., Larssen, T., Lydersen, E., Mulder, J., Muniz, I.P., Semb, A., Tang D., Vogt, R.D., Xiao J., Xiong J., Zhao, D., and Zhou, G.: 1999, *Ambio* **28**, 522.

Semb, A., Hanssen, J.E., Francois, F., Maenhaut, W., and Pacyna, J.M.: 1995. *Water Air Soil Pollut.*, **85**, 1933.

UN: 1989, *Energy Statistics Yearbook*. Dept. of International and Social Affairs, United Nations, New York.

Zhao, D., Larssen, T., Zhang, D., Gao, S., Vogt, R.D., Seip, H.M. and Lund, O.J.: 2001. This conference.

Zhu, Z.: 1990. Soil nitrogen. In: Li, C and Sun, O. (eds), *Soils of China*, Science Press, Beijing, p.554-576

CHEMICAL COMPOSITION OF PRECIPITATION, THROUGHFALL AND SOIL SOLUTIONS AT TWO FORESTED SITES IN GUANGZHOU, SOUTH CHINA

Y.G. XU [1*], G.Y. ZHOU [2], Z.M. WU [2], T.S. LUO [2] and Z.C. HE [1]

[1] *Guangzhou Institute of Geochemistry, Chinese Academy of Sciences, 510640 Guangzhou, China;*
[2] *The Research Institute of Tropical Forestry, CAF, 510520 Guangzhou, China*
(*author for correspondence, e-mail: yigangxu@gig.ac.cn)

Abstract. Rain water at two forested sites in Guangzhou (south China) show high concentrations of SO_4^{2-}, NO_3^- and Ca^{2+} and display a remarkable seasonal variation, with acid rain being more important during the spring and summer than during the autumn and winter. The amount of acid rain represents about 95% of total precipitation. The sources of pollutants from which acid rain developed includes both locally derived and long-middle distance transferred atmosphere pollutants. The seasonal variation in precipitation chemistry was largely related to the increasing neutralizing capacity of base cations in rainwater in winter. Soil acidification is highlighted by high H^+ and Al^{3+} concentrations in soil solutions. The variation in elemental concentration in soil solution was related to nitrification (H^+, NH_4^+ and NO_3^-) and cation exchange reaction (H^+, Al^{3+}) in soil. The negative effect of soil acidification is partly dampened by substantial deposition of base cations (Ca^{2+}, Mg^{2+} and K^+) in this area.

Key words: acid deposition, forest ecosystem, Guangzhou, soil solution

1. Introduction

In China, the rapid economic growth has been accompanied by the increase in pollution with acid precipitation becoming one of major environmental problems. A representative example is Guangzhou in south China, where the acid rain accounts for more than 80% of total annual precipitation during the eighties (Chen *et al.*, 1990). However, only limited reports are available in international literature on precipitation and soil solution chemistry from long term monitoring sites in this region (Larssen *et al.*, 1998; Seip *et al.*, 1999). It is therefore highly urgent to assess scientifically the current status of precipitation and the ecological effects of acid deposition. For this purpose, the precipitation, throughfall and soil solution have been collected and analyzed during the period of April 1998 to March 1999 from two monitoring forested sites in Guangzhou. These data will be used to investigate the processes/factors responsible for chemical variation in precipitation and soil solution and to assess the response of forest ecosystem at Guangzhou to acid deposition.

2. Site description and Methods

The monitoring sites with Masson pine (*Pinus massoniana* Lamb) are located at

Baiyunshan (200 m altitude) at the center of Guangzhou city (31°55'N, 108°08'E) and at Longdong about 15 km north-east of Guangzhou. The climate in this region is subtropical and monsoonal with annually averaged temperature of 22 °C. Forest soil is red lateritic soil with granite bedrock.

Precipitation was collected with a standard rain gauge after every event outside of the monitoring sites. Throughfall water was collected using self-designed collectors with 2-3 replicates at each site. Soil solution was continuously collected using ceramic plates with fixed dimension at five soil depths (0, 20, 40, 60, 80 cm). After installation, the soil section was left for two months for re-equilibration prior to collection. The throughfall and soil solutions were sampled every two weeks for chemical analyses. Water samples were analyzed using flame atomic absorption spectrometer for cations (Ca^{2+}, Na^+, K^+, Mg^{2+}, Al^{3+}), ion exchange chromatograph for major anions (SO_4^{2-}, NO_3^-, Cl^-) and ion selective electrodes for NH_4^+. Measurement of pH was conducted using a glass electrode pH meter (DF-807). Soil water fluxes were calculated on the basis of measured water volume. The annual and seasonal averages were calculated using the precipitation or water flux as weight.

3. Results

3.1. Precipitation

The total annual amount of precipitation is 1625 mm during the monitoring period. pH values vary between 3.3 and 6.3, with an average of 4.38 (Table I). The frequency of acid rain (pH<4.5) is 80%. The amount of acid rain represents about 95% of total precipitation, indicating the severe environmental status at Guangzhou. The rainwater has high concentrations of SO_4^{2-}, NO_3^- and Ca^{2+} and shows a remarkable seasonal variation, with acid rain being more important in the spring and summer than in the autumn and winter (Table I).

3.2. Throughfall

The concentrations of SO_4^{2-}, NH_4^+, Ca^{2+}, Mg^{2+} and K^+ in throughfall water are higher than in precipitation (Table I). This is due to the washing out of dry deposits on vegetation and of ions released from vegetation. However, H^+ concentration in throughfall is lower than in rainwater, probably due to the neutralization of base cations by dry alkaline dust deposited and basic leachates from plants.

3.3. Soil solutions

The element concentration is much higher in soil solution than in precipitation

TABLE I

Compositions of precipitation, throughfall and soil solution at two forested sites in Guangzhou (averaged, µeq/L)

	pH	H⁺	SO₄²⁻	NO₃⁻	NH₄⁺	Cl⁻	Al³⁺	Ca²⁺	Mg²⁺	K⁺	Na⁺
Baiyunshan											
Precipitation											
spring	4.36	43.4	75.9	13.2	65.0	35.0	15.1	55.8	8.2	7.2	37.0
summer	4.21	61.8	293.9	1.9	52.4	34.1	13.5	97.3	14.3	10.1	31.3
autumn	5.12	7.6	48.1	11.1	122.3	41.1	13.0	220.3	20.0	12.3	29.2
winter	6.37	0.4	84.2	8.4	44.8	24.2	29.0	202.4	16.7	12.6	19.2
Annual average	4.39	40.7	200.0	7.5	64.2	33.8	15.3	105.4	12.0	9.1	30.0
Throughfall	5.56	2.7	837.8	8.8	349.3	49.3	19.1	319.1	82.9	125.9	73.0
Soil solution											
0 cm	5.64	2.3	1094.3	49.6	376.3	69.5	37.2	592.7	198.8	393.1	99.5
20 cm	3.75	179.8	718.2	50.8	113.6	47.5	170.8	305.6	57.1	65.2	86.2
40 cm	4.09	80.6	603.1	58.2	47.7	42.5	183.8	234.9	63.2	23.4	94.5
60 cm	4.14	72.3	625.6	38.7	71.6	45.8	176.2	227.2	70.8	43.7	88.0
80 cm	4.57	27.1	894.5	62.8	36.0	75.6	151.1	397.9	77.3	55.7	96.8
Longdong											
Precipitation											
spring	4.54	28.7	156.8	8.6	59.6	27.8	7.3	57.4	12.5	24.8	32.1
summer	4.13	73.6	126.1	6.0	68.9	27.3	20.3	56.3	6.3	4.4	15.6
autumn	4.83	14.9	121.9	5.3	33.9	32.1	7.8	58.5	7.5	15.9	46.5
winter	6.54	0.3	214.4	5.2	33.4	16.6	26.4	115.1	10.0	6.1	7.7
Annual average	4.37	42.3	138.9	6.8	63.6	27.2	15.7	58.5	8.5	11.7	25.8
Throughfall	5.53	3.0	413.9	10.1	153.6	31.5	18.4	115.9	28.6	53.5	39.0
Soil solution											
0 cm	4.61	24.7	1047	53.6	310.8	43.6	45.9	261.6	82.8	200.8	223.0
20 cm	3.66	220.1	278.4	84.9	40.6	40.2	187.5	128.8	29.5	24.7	72.9
40 cm	3.67	214.6	279.6	123.9	35.6	40.1	182.2	155.8	36.5	18.3	91.6
60 cm	4.09	81.6	131.9	131.1	97.9	35.7	119.4	102.7	21.2	42.2	89.8
80 cm	4.11	77.9	133.8	202.1	57.6	64.6	96.7	120.1	34.0	15.4	98.1

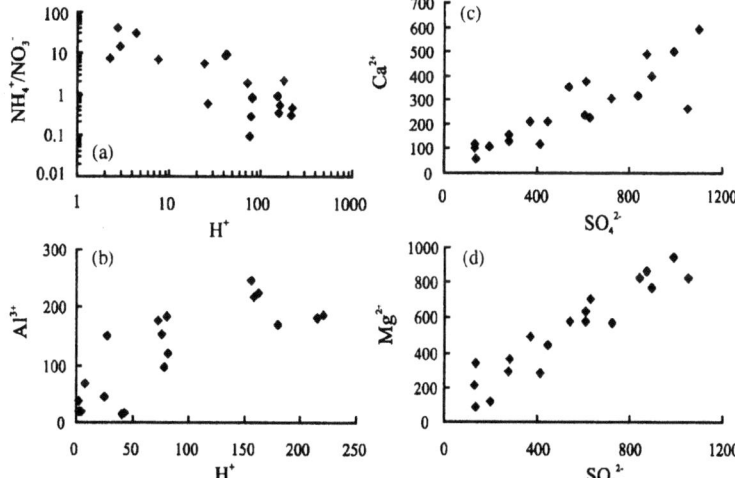

Figure 1. Diagram showing the correlation between major components (µeq/L) in precipitation, throughfall and soil solution: (a-b) H⁺ vs NH₄⁺/NO₃⁻ and Al³⁺; (a-b) SO₄²⁻ vs Ca²⁺ and Mg²⁺

(Table I). For instance, the enrichments of Al^{3+} and H^+ in soil solution relative to precipitation attain 15 and 4 times, respectively. Considerable chemical change in soil solution is observed from depth of 0 cm to 20 cm, whereas only a limited variation from 20 cm to 80 cm (except for NO_3^-). The NH_4^+/NO_3^- ratio varies between 8 and 30 in precipitation and throughfall water, suggesting ammonium nitrogen as the principle form of N in the water before penetrating the soil. This ratio becomes considerably low in soil solution (Fig. 1a), indicating nitrate as the major form of N. There is marked positive correlation between SO_4^{2-} and base cations (Ca^{2+}, Mg^{2+}, K^+) in rainwater, throughfall and soil solution (Fig. 1c, d), suggesting the geochemical affinity of these components.

4. Discussions and Conclusions

4.1. ORIGIN OF CHEMICAL VARIATION OF PRECIPITATION

The rainwater at Guangzhou shows a chemical composition similar to that in southwest China (Zhang et al., 1995b). The high SO_4^{2-} contents in precipitation from these two regions are largely related to coal combustion for the purpose of electricity generation and heating. Even if concentration of SO_4^{2-} in precipitation is high, pH of precipitation is relatively high, probably due to the neutralization of base cations (Galloway et al., 1987). Concentrations of Ca^{2+} and other base cations are indeed very high in the precipitation (Table I). The exact origin of Ca is unclear at this stage but could be derived from road dust, cement factories and long-distance transported desert dust. Marine sources cannot be ignored for the precipitation chemistry at Guangzhou, especially for Na^+ and Cl^-.

It is interesting to note that the seasonal variation in chemistry of precipitation at Guangzhou is the reverse of what observed at Chongqing (SW China), where pH in rainwater being lower in winter than in summer (Zhang et al., 1995b). This may imply the different mechanism of generation of acid rain in two areas. In the SW China, large cities are commonly situated along valleys. Limited air circulation in these regions is not favorable for the transport and diffusion of the locally emitted SO_2. Consequently, these pollutants tend to accumulate in the lower atmospheric layer and enter the rainwater when precipitation is formed. This is supported by similar $\delta^{34}S$ between atmospheric SO_2-aerosol and SO_2-particles issued during coal combustion in SW China (Zhang et al., 1995a). Therefore, the seasonal variation in precipitation chemistry in SW China may be directly related to the extent of local consumption of coal, which is greater in winter than in summer.

The source of pollutants from which acid rain develops at Guangzhou seems to be more complex. In this area, $\delta^{34}S$ of atmospheric SO_2-aerosol is significantly lower than (by about 12‰) that of SO_2-particles derived from coal combustion (Zhang et al., 1995a). Therefore, in addition to the locally produced pollutants related to coal consumption, a source with low $\delta^{34}S$ is required in order to account for the whole $\delta^{34}S$ range of precipitation at Guangzhou. In fact, the air circulation in the Pearl Delta region is strong. Long-middle distance-transported pollutants may also contribute in the formation of acid precipitation.

Ca^{2+} contents in rainwater in winter and autumn are significantly higher than in spring and summer. This is particularly true for the Baiyunshan site (Table I). It is likely that high pH in precipitation in winter is due to the increasing neutralizing capacity of base cations. Assuming that Ca^{2+} in rainwater was mainly contributed by alkaline dust, we attribute the seasonal variation in precipitation chemistry at Guangzhou to seasonal variation in weather and dry deposition. At Guangzhou, the majority of rainfall takes place in spring and summer and accounts for about 90% of annual precipitation. Soils are thus relatively wet and surface dusts are not easily incorporated into the air. Semi-continuous washing by rainfall also efficiently lowered the contents of alkaline particles in the atmosphere. This leads to the decrease in concentration of base cations in rainwater and the lowering of acid neutralizing capacity. In contrast, a high neutralizing capacity is expected for the rainwater in winter, because of high abundance of alkaline dust in atmosphere under dry weather.

4.2. EVIDENCE FOR AND MECHANISM OF SOIL ACIDIFICATION

It has been shown that soil acidification can lead to decrease of acid neutralizing capacity, losses of base cations from exchange sites and release of Al ions into soil solution as a consequence of H^+-buffer processes (Matzner and Murach 1995). Perhaps, the most direct evidence for soil acidification is the decrease in soil pH. Some of these soil acidification symptoms have been noted at the studied area, pointing to current soil acidification.

In general, the H^+ input to the soil system consists of H^+ directly deposited by precipitation and of H^+ formed within the soil. As shown in Table I, H^+ concentration in soil solution at the upper soil layer is significantly higher than in precipitation and throughfall. Even with our underestimated flux of soil water, the flux of H^+ in soil solution is considerably higher than that in precipitation and throughfall. This suggests that the H^+ production mainly occurred at the upper soil layer and the amount of internally produced H^+ was more important than the external input. The lower NH_4^+ content in soil solution than in throughfall may be related to the uptake of this component by vegetation roots. However, this cannot be the primary cause at Guangzhou because it fails to account for the negative correlation between NO_3^- and NH_4^+. Increasing H^+ along with the decreasing NH_4^+/NO_3^- ratios (Fig. 1a) is broadly consistent with the nitrification of NH_4^+ in soils. This is particularly true at Longdong where NO_3^- content in soil solution is high and increases with soil depth. However, relatively low and rather stable NO_3^- concentration is noted in the soil solutions at Baiyunshan (Table I). Plant uptake of nitrate may be another factor besides nitrification in governing nitrate chemistry in soil solution at this locality.

The H^+ concentration in soil solution decreases slightly with the soil depth (Table I), reflecting the consumption of H^+ in cation exchange and chemical weathering of primary minerals. In general, such a kind of exchange reaction would lead to the release of base cations and Al ions to soil solutions. This is widely considered as the mechanism of the soil change induced by acid deposition (Matzner and Murach, 1995). The Al concentration in soil solutions is much higher than in precipitation and in throughfall, thus supporting the hypothesis involving release of Al from soil into solution. Moreover, H^+ is positively correlated with Al^{3+} (Fig. 1b). It seems that the mobilization of Al was assisted and enhanced by the presence of H^+, consistent with experimental results with simulated acid rain (Abrahamsen et al., 1994). Because of the potential toxicity of Al ions to plant roots and soil organism (Ulrich, 1994), the increase in Al^{3+} in soil solution would leave negative impact on plants. However, the concentrations of base cations in soil solution are also high (Table I). The measured $Al^{3+}/(Ca^{2+}+Mg^{2+})$ molar ratios in soil solution are all less than the commonly accepted critical value above which forest damage occurs (Sverdrup and De Vries, 1994). It is possible that the substantial deposition of base cations at Guangzhou dampens to some degrees harmful effect of soil acidification.

Acknowledgements

This study was supported by the NSF of Guangdong province (960515). We thank C. Xie, C. Yu and L. Yang for chemical analyses, two anonymous referees and Dr. Y. Kohno for helpful reviews and the organization committee of Acid Rain 2000 for the support that allowed Y. Xu to attend the Tsukuba Conference.

References

Abrahamsen, G, Stuanes, A.O. and Tveite, B.: 1994. *Long-term experiments with acid rain in Norwegian forest ecosystem.* Springer-Verlag, New York.

Chen, Z.H., Quan, W.Z. and Li, Y.Q.: 1990. *Current status and variation trend of acid rain at Guangzhou.* National report of seven-fifth projects. (in Chinese)

Galloway, J.N., Zhao, D.W., Xiong, J.L. and Likens, G.E.: 1987. *Science*, **236**, 1559.

Larssen, T., Xiong, J., Vogt, R.D., Seip, H.M., Liao, B.H. and Zhao, D.W.: 1998. *Water, Air, and Soil Pollution.* **101**, 137.

Matzner, E. and Murach, D.: 1995. *Water, Air, and Soil Pollution.* **85**, 63.

Seip, H.M., Aagaard, P., Angell, V., Eilertsen, O., Larssen, T., Mulder, J., Muniz, I.P., Semb, A., Tang, D. Vogt, R.D., Xiao, J., Xiong, J.L., Zhao, D. and Kong, G: 1999. *Ambio*, **28**, 522.

Sverdrup, H. and de Vries, W.: 1994. *Water, Air, and Soil Pollution.* **85**, 143

Ulrich, B.: 1994. *Effects of Acid rain on Forest Processes.* D.L. Godbold and A. Huttermann (eds) Wiley-Liss, Inc. pp. 1-50.

Zhang, H.B., Chen, Y.W. and Liu, D.P.: 1995a. *Geochimica*, **24**, 126.

Zhang, F.Z., Zhang, J.Y. and Zhang, H.R.: 1995b, *Water, Air and Soil Pollution*, **90**, 407.

ACIDIFICATION OF RED PINE FOREST SOIL DUE TO ACIDIC DEPOSITION IN CHUNCHON, KOREA

YASUMI YAGASAKI[1*], TAKASHI CHISHIMA[2], MASANORI OKAZAKI[1], DU-SIK JEON[3], JEONG-HWAN YOO[4], YOUNG-KULL KIM[4]

[1] *Graduate School of Bio-Applications and Systems Engineering, Tokyo University of Agriculture and Technology, 2-24-16 Nakacho, Koganei, Tokyo, 184-8588, Japan;* [2] *Geo Green Tech co., 1-2-3 Haramachida, Machida, Tokyo, 194-0013, Japan;* [3] *Forestry Research Institute of Kangwon-Do, Woodoo-Dong Chunchon, Kangwon-Do, 132, Korea;* [4] *Korea Forestry Research Institute, Chungryangni-Dong, Tondaemun-Ku, Seoul, 130-012, Korea*
(* author for correspondence, tel. & fax: +81-(0)42-388-7276, e-mail: yasumi@cc.tuat.ac.jp)

Abstract. The effect of acidic deposition on the soil under red pine forest in Chunchon, Korea was investigated. Precipitation, stream water, and soil solution chemistry were monitored at the watershed from 1997 to 1998. Acidity of the open-bulk precipitation was often neutralized by large amounts of ammonia (NH_3) that might have originated from livestock farming and fertilization. Estimated elemental budget at the watershed showed a positive correlation between loss of base cations and proton (H^+) production due to nitrogen transformation in soil ($\triangle H^+_{NT}$: $([NH_4^+]_{in}-[NH_4^+]_{out})-([NO_3^-]_{in}-[NO_3^-]_{out})$). When $\triangle H^+_{NT}$ increased, concentrations of nitrate in soil solutions also increased. Consequently, pH values of soil solutions decreased, although ion exchange with base cations contributed to buffer reaction. Since acid buffering capacity of the red pine forest soil was small, it was concluded that the input of ammonium nitrogen enhanced nitrification in soil thus causing soil acidification represented by loss of base cations from the watershed.

Key words: acid deposition, budget, Korea, soil acidification, soil solution, nitrification

1. Introduction

A rapid increase of human population and industrialization in East Asia leads to an increase of energy use and rapid enhancement of air pollutant emission. It is estimated that emissions of nitrogen oxides and sulfur dioxide will be 3 and 3.5 times greater than the current level by the year 2020, respectively, if the emission trends continue at the present rate (van Aardenne *et al.*, 1999; Arndt *et al.*, 1997). Acidification of soil due to acidic deposition is one of the most critical environmental issues because it may result in potentially critical conditions for plants. In Korea, there is a large risk for negative effects of acidic deposition on soil acidification because:

· Dominant soils are developed from granite and granite gneiss and they are characterized by small acid buffering capacity derived mainly from small exchangeable cation content (Japan Environment Agency, 1999).
· Korea receives considerable amount of acid deposition originating from atmospheric pollutants derived from both domestic sources and long-range transport from the Asian Continent (Chung *et al.*, 1996).

In addition to these risks, possible effects of ammonia (NH_3) emission from agricultural activities must be considered. Ammonia emission in South Korea estimated by Murano et al. (1996) was relatively higher than those in other regions (Nihlgard, 1985). Because deposited atmospheric ammonium (NH_4^+) will produce H^+ through its nitrification in the soil, effect of this "potential acid" on a soil should be discussed. The objective of this study was to determine the effect of acidic deposition on the soil under red pine forest at Chunchon, Korea.

2. Materials and Methods

The research area is located in the eastern part of Chunchon, northwestern part of Kangwon Prefecture, about 80 km northeast from Seoul, Korea (Figure 1). The study sites were established in a red pine (*Pinus densiflora*) forest (Site P), 170 m above sea level, in May 1997. Mean annual precipitation in this area from 1986 to 1996 is about 1290 mm. The soil was developed from granite and was classified into Dystric Cambisols (Chishima, 1998). Chemical properties of soil at plot 1 (Chishima, 1998) are shown in Table I. Soil textures of all horizons were classified to sandy clay loam (SCL). Through all horizons, the values of pH(H_2O) and pH(KCl) were 4.79 - 5.41 and 3.99 - 4.18, respectively. Most horizons,

Figure 1. The location of research area.

excluding B2 horizon, had small cation exchange capacity (CEC) and low base saturation (BS) ranging from 9.1 - 10.4 $cmol_c\ kg^{-1}$ and 8.3 - 13.9 %, respectively. This reflects the low weathering rates of primary minerals in the granite bedrock.

Soil solutions at the depth of 10, 20, 30, and 50 cm were collected at plot 1. Ceramic porous cups were installed at each depth and soil solutions were collected in 1 L glass flasks under a vacuum of −60 cmHg at the beginning.

TABLE I
Chemical properties of soil under red pine (*Pinus densiflora*) forest at plot 1.

Hori-zon	Depth cm	pH (H_2O)	pH (KCl)	Ca	Mg	K	Na	Σ base	CEC	BS
				\multicolumn{5}{c}{$cmol_c\ kg^{-1}$}		%
A	0-10	4.79	4.02	0.4	0.1	0.2	0.1	0.8	9.1	8.3
B	10-42	5.14	4.00	0.4	0.3	0.2	0.3	1.3	10.4	12.1
B2	42-56	5.17	4.18	0.1	3.2	0.1	0.2	3.6	10.3	32.6
C	56-85+	5.41	3.99	<0.1	0.9	0.1	0.4	1.5	10.4	13.9

Throughfall samples were collected at plot 1 and plot 2. Open-bulk precipitation was collected at a gap of the forest near these plots. Precipitation and throughfall were collected by bulk samplers. Stream water in the small watershed was collected at the weir. Open-bulk precipitation, throughfall, soil solution and stream water samples were collected once or twice a month. The possible error of flow amount derived from hourly and daily variance of stream water flow rate was tentatively estimated from the flow rate variances during high and normal rainfall seasons. Preliminary input-output budget of elements at the watershed was estimated from the difference between the input through throughfall and the output through stream water. The throughfall input was calculated by averaging two throughfall inputs at plot 1 and 2.

All of the water samples were brought to Tokyo, Japan. They were filtered through a 0.45 μ m membrane filter. pH and EC were measured using a pH meter (Horiba, M-13) with a glass electrode (Horiba, 6366-10D), and an electrical conductivity meter (Toa Electric, CM-5S) with a two electrodes cell before the filtration. The water samples were stored at -25 ℃ in polyethylene bottles, thawed prior to analysis of ion concentration. Concentrations of Ca, Mg, Na and K were determined by an atomic absorption spectrophotometer (Hitachi, 170-10). The concentration of NH_4^+ was determined by a colorimetric method (indophenol blue) (Shimadzu, BioSpec-1600). Concentrations of Cl^-, SO_4^{2-}, NO_2^- and NO_3^- were determined by capillary electrophoresis (Waters, Quanta 4000E).

3. Results and Discussion

3.1. ACIDIC DEPOSITIONS AND PRECIPITATION CHEMISTRY

The weighted mean pH value through the whole sampling period was 5.50. However, pH values varied from 4.6 to 7.4. The weighted mean values of ion concentration for precipitation are shown in Table II. Characteristics of precipitation chemistry in this study such as high NH_4^+ concentrations, low H^+ concentrations, large variations in pH values, and seasonal changes in ion concentrations are approximately consistent with results of the precipitation monitoring through Korean Peninsula, including Chunchon (Lee *et al.*, 2000).

TABLE II
Weighted mean ion concentration of open-bulk precipitation in May to October of 1997 and 1998. Concentration of HCO_3^- was estimated theoretically.

H^+	NH_4^+	Na^+	Ca^{2+}	K^+	Mg^{2+}	SO_4^{2-}	nss-SO_4^{2-}	NO_3^-	Cl^-	HCO_3^-
					μ mol$_c$ L^{-1}					
3.2	152	13.7	11.3	14.1	5.1	68.4	52.7	21.6	21.5	89.9

TABLE III

Mean throughfall input (TH) and budget of ions in the watershed in May to October of 1997 and 1998. Negative (bold) and positive value of budget mean loss and possibility of accumulation in the watershed, respectively. s.d. means standard deviation.

	H^+	NH_4^+	Na^+	Ca^{2+}	K^+	Mg^{2+}	SO_4^{2-}	NO_3^-	Cl^-
				$kmol_c\ km^{-2}\ 6\ months^{-1}$					
TH	4.4	21.6	11.4	16.6	17.7	6.6	33.4	29.8	16.4
s.d.	1.4	5.9	0.7	4.8	4.1	1.9	14.4	22.7	3.6
budget	3.4	22	**-37**	**-22**	13	**-2.5**	13	23	2.5
s.d.	1.9	7	9	11	3	0.4	21	27	5.0

3.2. ESTIMATION OF INPUT-OUTPUT BUDGET AND SOIL SOLUTION CHEMISTRY

The amount of deposition to the soil under red pine forest due to throughfall (TH) is shown in Table III. The roughly estimated elemental budget based on the intermittent monitoring results (1997-1998) indicates that base cations such as Na^+ and Ca^{2+} were lost from the watershed, while SO_4^{2-}, NO_3^- and NH_4^+ were estimated to be accumulated in the watershed (Table III). The loss of base cations from the watershed resulted in acid accumulation in the watershed regardless of the origin of acid, anthropogenic or non-anthropogenic. Production of H^+ due to nitrogen transformation in soil (ΔH^+_{NT}) was quantified by a formula introduced by van Breemen et al. (1983). It is described as follows;

$$\Delta H^+_{NT} = ([NH_4^+]_{in} - [NH_4^+]_{out}) - ([NO_3^-]_{in} - [NO_3^-]_{out}) \quad (1)$$

where negative value of ΔH^+_{NT} indicates consumption of H^+ and a positive value indicates production of H^+ in the soil. Figure 2 shows the relationship between monthly base cation flux and ΔH^+_{NT} in the watershed. There was a clear positive relation between Na^+ and Ca^{2+} loss from the watershed, and ΔH^+_{NT}. Loss of cations tended to be largest in Na^+, middle in Ca^{2+}, and lowest in Mg^{2+} throughout the whole period. Since there is a positive correlation between the amount of NH_4^+ deposition and ΔH^+_{NT} ($R^2 = 0.80$) in most period, atmospheric NH_4^+ probably plays a major role for acidification of the soil in this site.

Based on the value of ΔH^+_{NT} and plots in base cation flux in Figure 2, three different levels of ΔH^+_{NT} were stated to compare soil solution chemistry with the result of elemental budget (Level I : $\Delta H^+_{NT} \leq 0$, Level II: $0 < \Delta H^+_{NT} < 5$, Level III: $5 \leq \Delta H^+_{NT}$, unit: $kmol_c\ km^{-2}\ 6\ months^{-1}$). Figure 3 shows NO_3^- concentrations and pH values of soil solutions for each level. The concentration of NO_3^- was highest in level III. The lowest pH values were also found in level III. This illustrates that nitrification affected the pH of soil solution in Level III. The observations indicate that the nitrification of deposited NH_4^+ plays a major role in production of H^+ and NO_3^- as a dominant factor for soil acidification expressed as base cation loss from the watershed in Chunchon study site.

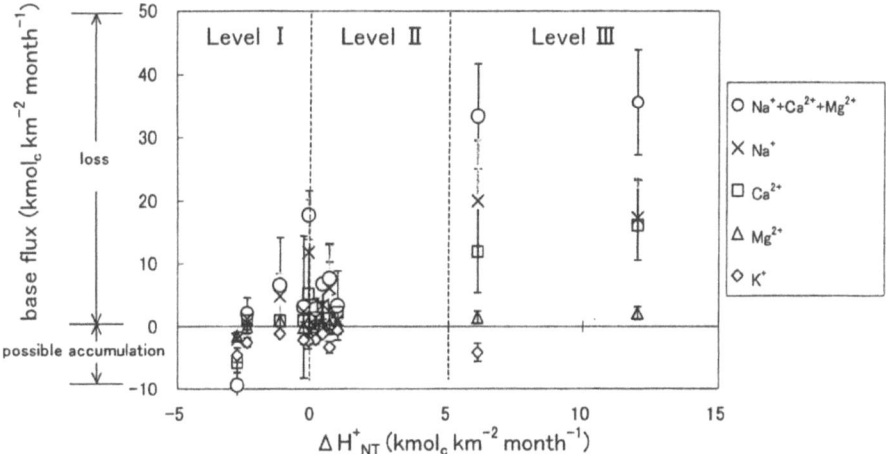

Figure 2. Relationship between monthly base cation flux and acid production due to nitrogen transformation in soil (ΔH^+_{NT}). $\Delta H^+_{NT} = ([NH_4^+]_{in} - [NH_4^+]_{out}) - ([NO_3^-]_{in} - [NO_3^-]_{out})$ (Level I: $\Delta H^+_{NT} \leq 0$, Level II: $0 < \Delta H^+_{NT} < 5$, Level III: $5 \leq \Delta H^+_{NT}$, unit: $kmol_c\ km^{-2}\ 6\ months^{-1}$).

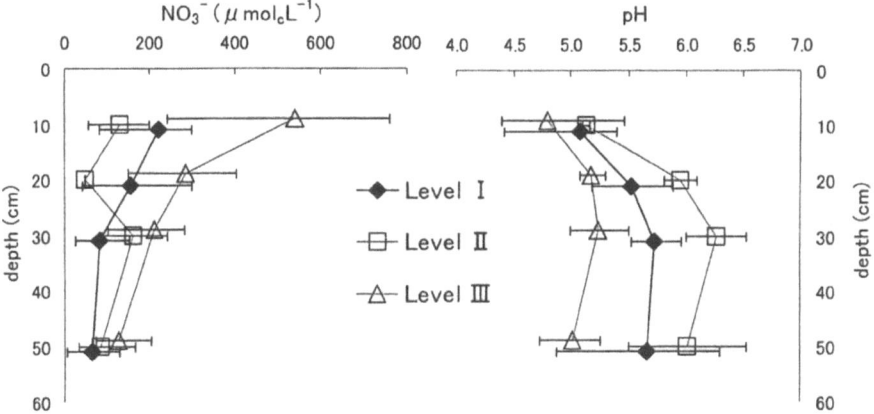

Figure 3. Relationship between ΔH^+_{NT} levels and the concentrations of NO_3^- (left) and pH values (right) of soil solutions at plot 1. Each ΔH^+_{NT} level corresponds to those of in Figure 2. Each bar shows the range between maximum and minimum value.

4. Conclusions

On a map of NH_4^+ fluxes made by Murano *et al.* (1996) based on the emission factor and statistical data of livestock and fertilizer, Chunchon was located in a grid that had the largest total NH_3 emission (3150 and 14900 tons NH_3 grid^{-1} yr^{-1}, approximately equivalent to 2.6 and 12.4 kg N ha^{-1} yr^{-1}, from fertilizer and livestock, respectively) in Japan and Korea. Despite the large acid input from the atmosphere, rain pH was often relatively high due to the high concentration of atmospheric NH_3 in Chunchon. Losses of Na^+, Ca^{2+}, and Mg^{2+} from the watershed due to H^+ production through nitrogen transformation in soil were clearly shown by the current preliminary budget study. The nitrification of deposited NH_4^+, which was probably derived from anthropogenic sources, thus resulted in acidification of soil. In the red pine forest in Chunchon, Korea, with a small acid buffering capacity of the soil (Dystric Cambisols), soil acidification will be more critical in the future assuming that the trend of NH_4^+ loading remains the current rate.

Acknowledgments

This study was funded by Global Research Fund, Japan Environment Agency. The authors would like to acknowledge Dr. Moon Gil Choi and Dr. Kazuhiro Ishizuka for organizing this research.

References

Arndt, R. L. and Carmichael, G. R.: 1997, *Atmospheric Environment* **31**, 1533.
Chishima, T.: 1998, Master's thesis, Graduate School of Tokyo University of Agriculture and Technology, Koganei, Tokyo (in Japanese).
Chung, Y. S., Kim, T.K., Kim, K. H.: 1996, *Atmospheric Environment* **30**, 2429.
Japan Environment Agency: 1999, *Report of Global Environment Research Fund (C-1)*, Japan Environment Agency (in Japanese).
Kim, J. H., Nishimura, O., Takeshita, S., Ryu, J. K., Sudo, R.: 1999, *Yousui to Haisui* **41**, 143. (in Japanese).
Lee, B. K., Hong, S. H., Lee, D. S.: 2000, *Atmospheric Environment* **34**, 563
Murano, K., Hatakeyama, S., Kubo, N. Lee, D. S., Lee, T. Y.: 1996, 'Gridded ammonia emission fluxes in Japan and Korea', *Proceedings of the International Symposium on Acidic Deposition and its Impact*, Tsukuba, Japan.
Nihlgard, B.: 1985, *AMBIO* **14**, 2.
Van Aardenne, J. A., Carmichael, G. R., Levy, II H., Streets, D. and Hordijk, L.: 1999, *Atmospheric Environment* **33**, 633.
Van Breemen, N., Mulder, J., and Driscoll, C. T.: 1983, *Plant and Soil* **75**, 283.

ELEVATIONAL PATTERNS OF ACID DEPOSITION INTO A FOREST AND NITROGEN SATURATION ON MT. OYAMA, JAPAN.

HIROSHI OKOCHI* and MANABU IGAWA

*Department of Applied Chemistry, Faculty of Engineering, Kanagawa University,
3-27-1, Rokkakubashi, Kanagawa-ku, Yokohama 221-8686 Japan
(*author for correspondence, e-mail: ookouh01@kanagawa-u.ac.jp)*

Abstract. Virgin fir trees have been dying on Mt. Oyama, which is located in the southwestern part of Kanto Plain, although the frequency of death seems to be reducing recently. We report elevational patterns of acid deposition in precipitation and throughfall under fir and cedar canopies and nitrogen saturation in the forest ecosystem on Mt. Oyama. The deposition fluxes of major inorganic ions in precipitation were nearly constant regardless of elevation except for hydrogen and ammonium ions, whereas the deposition fluxes of all major inorganic ions in throughfall among cedar increased. The 5-year average of annual nitrate deposition in precipitation from 1994 to 1998 showed 19.3 – 23.5 kg ha^{-1} yr^{-1} (annual inorganic total N deposition: 9.6 – 10.7 kgN ha^{-1} yr^{-1}) at four sites ranging in elevation from 500 to 1252 m, whereas the deposition in both cedar and fir throughfall was over 6 times greater than that in precipitation. The average soil surface nitrate concentration in 1998 was 140 µg g^{-1} (the range: 21.1 – 429 µg g^{-1}, n=80) and the 7-year average of nitrate concentration in stream water from 1992 to 1998 was 4.81 mg L^{-1} (the range: 2.38 – 20.6 mg L^{-1}, n=317). Our results indicate that nitrogen saturation is occurring in the forest ecosystem because of high N deposition, probably via acid fog, on Mt. Oyama.

Keywords: acidification, acid deposition, elevational pattern, forest ecosystem, nitrogen saturation

1. Introduction

Large numbers of fir trees have been dying on Mt. Oyama especially at high altitudes in a similar fashion to that seen in virgin forests in North America and Europe where acid deposition have found to be the cause (e.g. Schüt *et al.*, 1985; McLaughlin, 1985). In more recent years anthropogenic N deposition has received much attention because of adverse effects on plant species causing mineral imbalance, growth disturbance, and so on (Aber *et al.*, 1989; Skeffington *et al.*, 1988). We have reported that acid fog, which often has the acidity below pH 3 due to the absorption of gaseous nitric acid in the ambient air, has frequently occurred at the mountainside on Mt. Oyama (for example, Hosono *et al.*, 1994). This paper reports the composition of bulk and throughfall deposition, its variation with altitude and the effects of acid deposition on soil and stream water chemistry on Mt. Oyama.

2. Experimental methods

2.1. STUDY SITES

Our study was conducted on the southern and southeastern slopes of Mt. Oyama

(35°26'15"N, 139°14'04"E, 1252m). Virgin fir trees grow at the elevation from 400 to 1000 m (about 100 ha) on Mt. Oyama, but above 700 m large numbers of standing dead fir trees have been observed.

2.2. BULK PRECIPITATION, THROUGHFALL, AND STEMFLOW SAMPLING

Bulk precipitation and throughfall under the canopy of representative conifers on Mt. Oyama, Japanese fir (*Abies firma*) and Japanese cedar (*Cryptomeria japonica*), were measured monthly for the period from 1994 to 1998 at four sites between 500 and 1252 m. Cedar is not in decline, though fir has been dying. Collectors were fitted with filters (1.2-μm pore size membrane filter) composed of a polycarbonate filter holder (80 mm in diameter) and a 3 L polyethylene bottle. Stemflow samples (two or three per sampling point) were collected monthly with double twisted gauze (Tamaki *et al.*, 1992) at 680 m and 890 m.

2.3. SURFACE SOIL, SOIL WATER, AND STREAM WATER SAMPLING

Eighty surface soil samples (10 cm) were collected along the southeastern slope at sites ranging in elevation from 585 to 1252 m in May 1998. Pretreatment of soil and extraction for the water-soluble species was performed following standard protocols (Dojou hyoujun bunseki/sokuteihou iinkai, 1990). Soil water samples were collected daily with porous cups (Daiki Rika Kogyo Co. LTD., DIK-3900), whose inside pressure was reduced to 400 mmHg, from four depths (20, 40, 60, and 80 cm) near a fir tree at 890 m during the period 24-27 August and 1-6 September 1998. We discontinued the soil solution sampling during the period 27 August - 1 September because we had exceptionally heavy rain. Stream waters were sampled weekly from 1992 to 1998 at 700m.

2.4. CHEMICAL ANALYSIS

All water samples and soil water extracts were filtered through a 0.45-μm membrane filter, and the pH and conductivity were measured immediately. Until the analysis for the other components, the samples were stored in polypropylene bottles at 4 °C. Major inorganic ions were measured by ion chromatography (Dionex Co., 2000i/SP, QIC, and DX-100).

3. Result and Discussion

3.1. ELEVATIONAL EFFECTS ON BULK DEPOSITION AND PRECIPITATION CHEMISTRY

Figure 1 indicates that the mean precipitation amount was constant at about

2000 mm regardless of altitude though a general tendency exists for the amount of precipitation to increase with increasing altitude on mountains (Lovett and Kinsman, 1990). Despite annual variations in the patterns, the 5-year mean precipitation fluxes of H^+ and NH_4^+ were nearly constant irrespective of altitude, although other ions decreased slightly. These results imply that the concentration of chemicals in precipitation is nearly constant or decreases slightly with increasing altitude on Mt. Oyama. Lovett and Kinsman (1990) cited limited mixing of atmospheric pollutants emitted at low elevations, fall distance of hydrometers, the form of precipitation (rain or snow), and seeder-feeder mechanism as primary factors which could cause changes in precipitation chemistry with elevation. Although there is no real evidence, the limited mixing and/or fall distance might be the primary factors on Mt. Oyama because of elevational increase in the concentration and the flux by the latter two factors.

3.2. ELEVATIONAL PATTERNS OF THROUGHFALL DEPOSITION

The amount of throughfall under both species increased with altitude, but by less than that of precipitation at sites below 900 m, probably owing to transpiration by the canopy (Fig. 2). The amount of throughfall under cedar at the summit was nearly as twice that of precipitation, indicating enhancement via occult deposition, namely interception of cloud/fog water. Actually, we observed relatively high correlation with the amount of fog collected with passive sampler and the amount of throughfall. These results correspond to our observations of fog frequency on Mt. Oyama using a video camera. Fog frequency was found to increase with altitude and over 40% at the summit (Igawa *et al.*, this issue).

The deposition fluxes of all major inorganic ions in throughfall under the cedar canopy increased clearly above 700 m, but the elevational patterns of throughfall flux under the fir canopy were not clear. The ratios of the throughfall flux under the cedar or fir canopy to precipitation flux at the same altitude were shown in Table I. These data show increases in throughfall flux from both canopies. The primary leaching nutrients, K^+, Mg^{2+}, and Ca^{2+} were higher than those of other cations. The throughfall flux of NO_3^- was higher than those of other anions. These NO_3^- probably originate from nitric acid being deposited on both canopies via dry and/or fog water deposition.

TABLE I
The ratios of throughfall flux to precipitation flux of major ions

	H^+	NH_4^+	Na^+	K^+	Mg^{2+}	Ca^{2+}	Cl^-	NO_3^-	SO_4^{2-}
Fir	0.88-2.9	2.8-3.7	4.1-4.2	4.0-25	5.3-8.5	6.1-7.3	3.9-5.1	5.0-8.0	2.9-3.1
Cedar	1.1-4.8	1.4-7.4	2.9-11	1.9-36	2.8-17	3.5-38	2.8-12	4.0-20	1.6-11

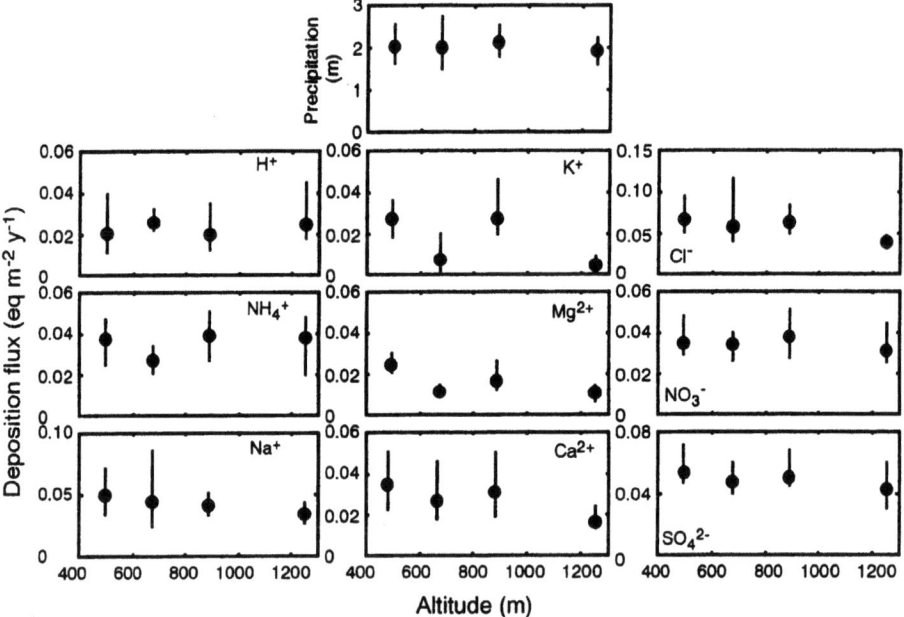

Figure 1. The 5-year annual mean deposition flux of major inorganic ions in bulk precipitation on Mt.Oyama as a function of altitude ('94-'98)

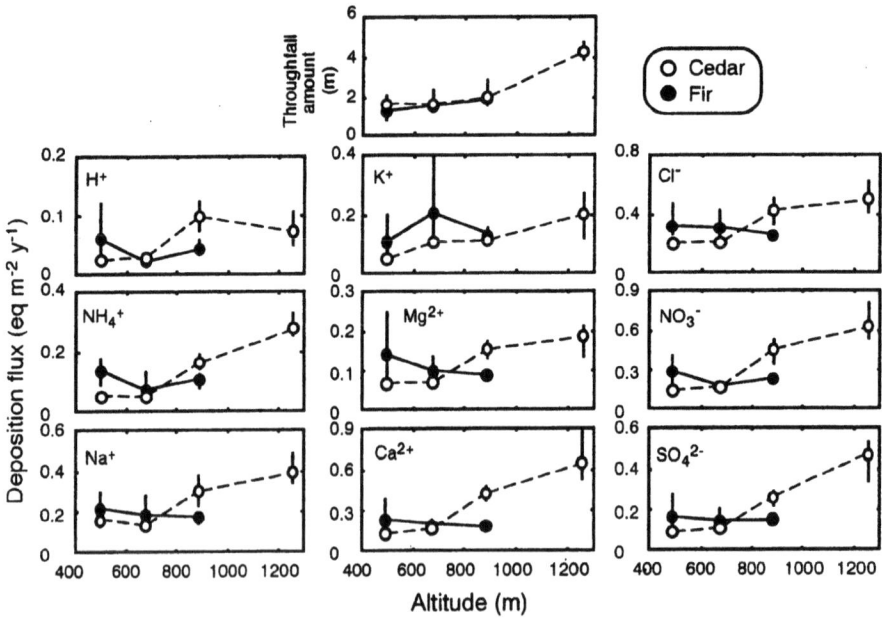

Figure 2. The 5-years annual mean deposition flux of major inorganic ions in throughfall on Mt.Oyama as a function of altitude ('94-'98)

3.3. NITROGEN SATURATION AND ACIDIFICATION ON MT. OYAMA

We have already reported that the surface soil had high concentrations of NO_3^- on Mt. Oyama, but was not acidified due to the neutralization by base cations at a point in 1989 (Okochi et al., 1993). The average NO_3^- concentration in the soil surface was 140 $\mu g\ g^{-1}$ (range: 21.1 – 429 $\mu g\ g^{-1}$, n=80) in 1998 with little change over these 9 years. However, the average $pH(H_2O)$ decreased from 5.74 to 5.14, indicating slight acidification of the surface soil.

The 7-year average of NO_3^- concentration in stream water from 1992 to 1998 was 4.81 mg L^{-1} (range: 2.38 – 20.6 mg L^{-1}, n=317). A relatively high NH_4^+ was also detected (Fig. 3). According to the definition of Williams et al. (1996), our results indicate nitrogen saturation has been occurring in the forest ecosystem. The annual mean pH of the stream seemed to decrease from 1992 through 1996 but recovered to the 1992 level in 1997. However, the minimum pH in 1998 was the lowest during this 7 year period (range: 6.91-7.79, n=317).

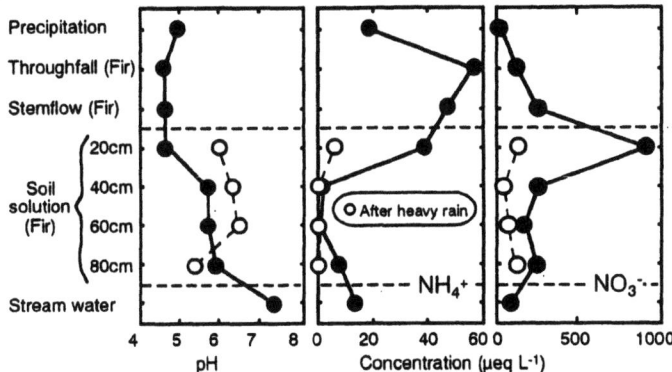

Figure 3. Comparison of the mean pH and concentration of ammonium and nitrate in precipitation, throughfall, stemflow, soil solution, and stream water. They show the 5-years average from 1994 through 1998 except stemflow and soil solution, which indicate the mean from May 1996 through 1998 for stemflow, 24-27 August (before heavy rain) and 1-6 September(after heavy rain) 1998 for soil solution, respectively. All samples except stream water were collected at 890 m.

Atmospheric N deposition and nitrification are the only source of NO_3^- in these systems. Durka et al. (1994) has reported that for the healthy and slightly declining sites, at least 16-30 % of NO_3^- in spring water originates directly from the atmosphere, whereas for more severely damaged sites almost all of the atmospheric NO_3^- leaks. The highest NO_3^- concentration in the surface soil solution indicates that nitrification is an important source of NO_3^- in stream water on Mt. Oyama (Fig. 3). However, the fraction of atmospheric NO_3^- might be higher than the estimations by Durka et al. (1994) because atmospheric inputs of NO_3^- at our sites were much higher.

The NO_3^- concentration in stream water was lower than that in the soil solution irrespective of depth, whereas NH_4^+ concentration in stream water was higher than that in soil solution except the surface (Fig. 3). In addition, the concentration of NH_4^+ and NO_3^- in soil solution decreased after heavy rain (448 mm at 1252 m), in particular in the surface. The NH_4^+ concentration in stream water increased after heavy rain, whereas there was little change of NO_3^- concentration between before and after heavy rain. This indicates that the NH_4^+ concentration in stream water on Mt. Oyama is considerably influenced by the amount of precipitation.

4. Conclusions

The elevational patterns of precipitation fluxes of major ions were dominated by limited mixing of atmospheric pollutants and/or fall distance on Mt. Oyama, whereas the patterns of throughfall fluxes are influenced by fog deposition. It is likely that nitrogen saturation and surface soil acidification has been occurring in the forest ecosystem because of the large inputs of inorganic nitrogen.

Acknowledgements

We thank many students and graduates in our laboratory for sampling and analysis, Prof. Peter Brimblecombe and Dr. Mario Massucci in UEA, U.K. for valuable comments, and two anonymous reviewers for their helpful comments. This work was supported in part by Core Research for Evolutional Science and Technology (1996-2001), Japanese Science and Technology Corporation.

References

Aber, J. D., Nadelhoffer, K. J., Steudler, P., Melillo, J. M.: 1989, *BioScience* **39**, 378.
Dojou hyoujun bunseki/sokuteihou iinkai: 1990, Dojou hyoujun bunseki/sokuteihou (Standard method for soil analysis), Hakubunsya: Tokyo, p.354.
Durka ,W., Schulze, E.-D., Gebauer, G., and Voerkelius, S.: 1994, *Nature* **372**, 765.
Hosono, T, Okochi, H., and Igawa, M.: 1994, *Bulletin of Chemical Society Japan* **67**, 368.
Lovett, G. M. and Kinsman , J. D.: 1990, *Atmospheric Environment* **24**, 2767.
McLaughlin,S.B.: 1985, *J. Air Pollut. Control Assoc.* **35**, 512.
Okochi, H., Nagashima, T., Hoka, E., and Igawa, M.: 1993, *Kankyo Kagaku Kaishi* **6**, 29.
Okochi, H., Hosono, T., Maruyama, F., and Igawa, M.: 1995, *Kankyo Kagaku Kaishi* **8**, 305.
Schüt, P. and Cowling, E. B.: 1985, *Plant Dieses* **69**, 548.
Skeffington, R. A., Wilson, E. J.: 1988, *Environ. Pollut.* **54**, 159.
Tamaki, M., Hiraki, T., Shoga, M., Nakagawa, Y., and Kobayashi, T.: 1992, *Bulletin of Hyogo Prefectural Institute of Environmental Science* **24**, 1.
Williams, M. W., Baron, J. S., Caine, N., Sommerfield, R., and Sanford, R., Jr.: 1996, *Environmental Science and Technology* **30**, 640

ACIDIFICATION OF A SMALL STREAM ON KUREHA HILL CAUSED BY NITRATE LEACHED FROM A FORESTED WATERSHED

T. KAWAKAMI[1*], H. HONOKI[2] and H. YASUDA[3]

[1] *Toyama Prefectural University, 5180 Kurokawa, Kosugi, Toyama 939-0398 Japan;* [2] *Toyama Science Museum, 1-8-31 Nishinakanomachi, Toyama-city 939-8084 Japan;* [3] *Toyama Forestry & Forest Products Research Center, 3 Yoshimine, Tateyama-cho,Toyama, 930-1362 Japan*
(* author for correspondence, e-mail: kawakami@pu-toyama.ac.jp)

Abstract. Nitrate leakage to a stream from a small forested watershed on Kureha Hill in Toyama Prefecture, Japan was investigated in order to assess its acid-base chemistry. Kureha Hill, located in Toyama City, receives high nitrogen loading from the atmosphere. In this area, there are several small streams characterized by a high concentration of nitrate and low ANC. Hyakumakidani, one of the most acidic streams on that hill, the average pH and ANC were 5.2 and -8 µeq/l, respectively. The weighted average nitrate concentration, which was 125 µeq/l, increased up to 370 µeq/l during high-discharge periods caused by heavy rainfall, while ANC decreased to -24 µeq/l at the same time. Our preliminary study using a two-stage tank model simulating the flow paths in the soil indicated that during high-discharge periods, the increased subsurface flow containing a high concentration of nitrate contributed to the increased nitrate concentration. In this watershed, the annual nitrogen budget from Aug. 1998 to Aug. 1999 showed that the loss of nitrogen exceeded the bulk nitrogen deposition, indicating that this watershed is under condition of nitrogen saturation. However, no visible attenuation has been observed.

Keywords: flow-path model, forested watershed, Kureha Hill, nitrogen budget, nitrogen saturation

1. Introduction

There has been increased attention given to the nitrogen leachate from forested watersheds to streams from the viewpoint of acidification and eutrophication. Excess atmospheric deposition of nitrogen on a forest is considered to lead its ecosystem to leach nitrate into stream water. Nitrogen leachate from a forested watershed may be caused by nitrogen saturation, under which nitrogen is no longer the limiting factor of the forest ecosystem's growth. Nitrate leachates from forested watersheds have been discussed in regard to nitrogen saturation for many watersheds in Europe (Tietema *et al.*, 1998), the U.S. (Stoddard, 1994) and Japan(Ohrui and Mitchell, 1997).

In this study, we investigated the nitrogen saturation of the small, forested Hyakumakidani watershed on Kureha Hill in Toyama Prefecture, Japan, from Aug. 1998 to Aug. 1999. The purposes of this study were (1) to estimate the nitrogen budget for the watershed in order to specify the nitrogen-saturation status proposed by Aber *et al.* (1989) and Stoddard (1994) and (2) to apply a two-stage tank model to the watershed in order to evaluate the effect of flow-path change on nitrate concentrations.

Water, Air, and Soil Pollution **130**: 1097–1102, 2001.
© 2001 *Kluwer Academic Publishers.*

2. Site description

Kureha Hill, located in Toyama City (36°41'N, 137°9'E), Japan, ranges 7 km from northwest to southeast (Fig. 1). The highest point is 145.3 m above sea level. In this area, the annual average nitrogen ($NH_4^+ + NO_3^-$) wet deposition from 1994 to 1998 was 980 eq/ha/year, sustained by a high annual precipitation of 2230 mm. There are many small streams on that hill, most of which have nitrate concentrations of more than 50 µeq/l, even during the growing season. Hyakumakidani, one of the most acidic streams, has an average nitrate concentration of 125 µeq/l, a negative ANC and an average pH of 5.2. The Hyakumakidani watershed area is 4 ha, 80% of which covered by deciduous trees, primarily 40-year-old Konara (*Quercus serrata*). The rest of the area is covered by bamboo (*Phyllostachys pubescens*) and afforested 50-year-old coniferous Sugi (*Cryptomeria japonica*). The soil, overlaid on the Quaternary lake-deposited sand and mud alternation, is classified as brown forest soils (Fujii and Yamamoto, 1979). Most of the watershed area is conserved by the government as a natural park. No agricultural areas exist within the watershed.

3. Method

3.1. SAMPLING

Precipitation was measured by a rain gauge. Within the Hyakumakidani watershed, bulk precipitation was collected weekly in a 20-cm-diameter funnel collector placed in an open area, while throughfall was collected weekly in an 18-cm-diameter funnel collector placed under a Konara canopy. At the same time, stream water was collected from Hyakumakidani. Discharge was recorded every 10 minutes at Point A as shown in Figure 1. All samples were filtered over 0.45 µm filters in the laboratory within 30 minutes after sampling. The pH was measured by the glass electrode method, ANC was measured by Gran's titration and dissolved components were analyzed by ion-chromatography.

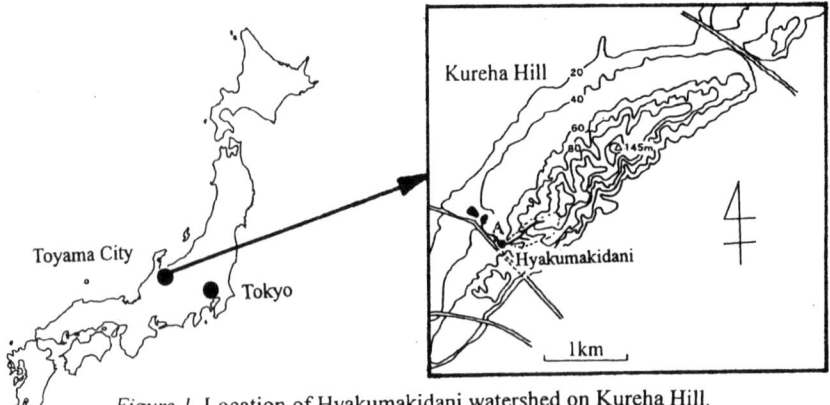

Figure 1. Location of Hyakumakidani watershed on Kureha Hill.

3.2. FLOW-PATH MODEL

We applied a simple, two-stage tank model to the Hyakumakidani watershed to investigate the effect of the flow-path change on the nitrate concentrations in the stream water. The two-stage tank model used in this study is illustrated in Figure 2. The outflow from the upper tank simulated the subsurface flow, while the outflow from the bottom tank simulated the base flow. Evapotranspiration was accounted for by abating the water level from the upper tank. First, hydrological calibration was performed by adjusting the model parameters, including nozzle sizes for both tanks and the nozzle position for the upper tank. Then the nitrate concentration (C_2 in Fig. 2) of the outflow from the bottom tank was set to be equal to the base flow concentration. The nitrate concentration of the outflow from the upper tank (C_1) was applied as a model parameter to minimize the square error between the calculated nitrate concentrations (C) and the observed nitrate concentrations of the stream water.

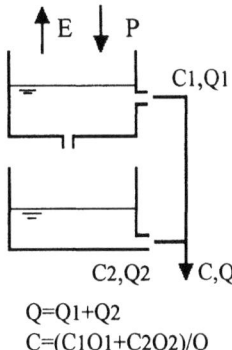

Figure 2. Two-stage tank model.
P: precipitation
E: evapotranspiration
Q: discharge
C: nitrate concentration

$Q = Q_1 + Q_2$
$C = (C_1Q_1 + C_2Q_2)/Q$

4. Results and discussion

4.1. STREAMWATER CHEMISTRY

The average water quality is shown in Table I. The average pH and ANC were 5.2 and -8 μeq/l, respectively. The nitrate, of which the average concentration was as high as 125 μeq/l, obviously affected the acid-base status of this stream water. The ammonium concentrations were under the detection limit.

Temporal patterns in the stream water pH, nitrate concentrations, sulfate concentrations and discharge from Aug. 25, 1998 to Aug. 15, 1999 are shown in Figure 3. Nitrate concentrations increased with increased discharge, while pH and sulfate concentrations decreased. These positive and negative relationships between nitrate and sulfate concentrations and discharge also have been found

TABLE I Stream water chemistry of Hyakumakidani (Aug.25, 1998-Aug.15,1999)

pH	Base cations (μeq/l)	Cl⁻ (μeq/l)	NO3⁻ (μeq/l)	SO4²⁻ (μeq/l)	Al (μg/l)	ANC (μeq/l)
5.2	504	229	125	168	83	-8

Base cations : $Na^+ + K^+ + Ca^{2+} + Mg^{2+}$

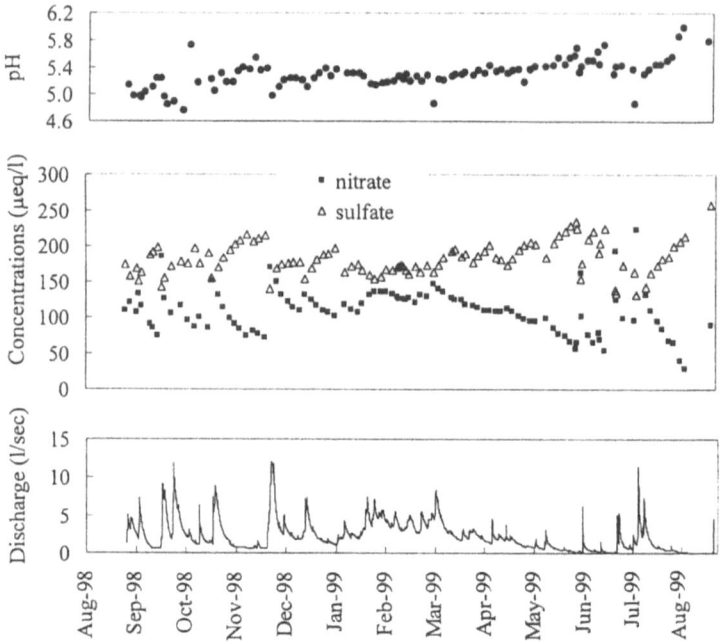

Figure 3. Temporal patterns of stream-water pH, nitrate and sulfate concentrations and discharge at Hyakumakidani.

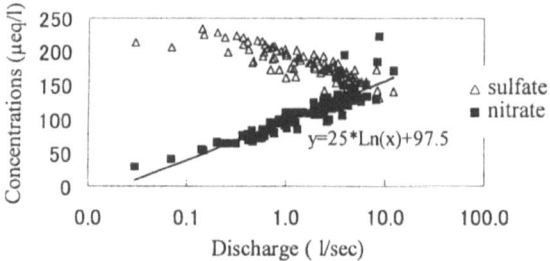

Figure 4. Relationships between the nitrate and sulfate concentrations and discharge.

in Biscuit Brook in the Catskill Mountains, U.S.A. (Murdoch and Stoddard, 1992).

In an episodic event at Hyakumakidani on September 21, 1999, a heavy rainfall caused a high discharge rate of 25 l/sec. The nitrate concentration increased to 370 µeq/l, while the pH and ANC decreased to 4.55 and -24 µeq/l, respectively.

Linear relationships between the nitrate and sulfate concentrations and discharge were found when the discharge was expressed on a logarithmic scale as illustrated in Figure 4, indicating that the nitrate and sulfate concentrations depended only on stream discharge, regardless of the season. According to Stoddard's definition, the extremely high nitrate concentrations and the lack of any seasonal pattern observed at Hyakumakidani are characteristics of the final stage of nitrogen saturation (Stoddard, 1994).

4.2. NITROGEN BUDGET

The nitrogen budget for the Hyakumakidani watershed was calculated from Aug. 25, 1998 to Aug. 24, 1999. During this period, the 2367 mm of rainfall was close to the annual average of 2230 mm. The nitrogen inputs were obtained by multiplying the nitrogen concentration ($NH_4^+ + NO_3^-$) in bulk precipitation and the amount of rainfall. The nitrogen loss from the watershed was obtained by multiplying the discharge and the nitrate concentrations estimated from the discharge by the logarithmic regression equation shown in Figure 4. The nitrogen budget revealed that the loss of nitrogen from the watershed was 2160 eq/ha/year, while nitrogen input was 1237 eq/ha/year. Even taking into account that the nitrogen flux of throughfall was 37% greater than that of bulk precipitation during the growing season (Jun. to Oct.), the loss of nitrogen far exceeded the input. The net nitrogen loss from the Hyakumakidani watershed suggested that it suffered from nitrogen saturation.

4.3. FLOW-PATH MODEL

In the Hyakumakidani watershed, discharge increases quickly in response to rainfall, reflecting the small watershed area. The quick changes in nitrate and sulfate concentrations in response to discharge suggest flow-path changes in the soil.

In order to evaluate the effects of flow-path changes in the soil on nitrate concentrations, a two-stage tank model was applied to the watershed. The

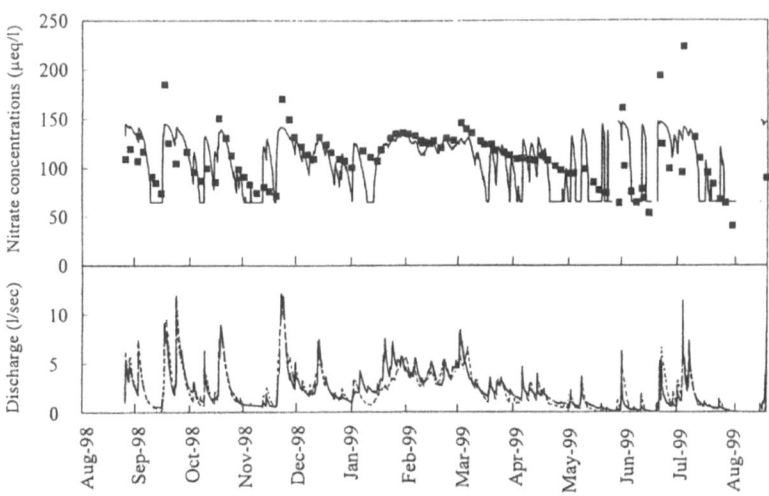

Figure 5. Upper panel: Observed nitrate concentrations and simulated nitrate concentrations (solid lines). Lower panel: Observed discharge (solid lines) and simulated discharge (dashed lines).

simulation results from Aug. 25, 1998 to Aug. 15, 1999 are shown in Figure 5. This simple, two-stage tank model simulated discharge quite well. The nitrate concentration for the outflow from the bottom tank (C2) was set to 65 µeq/l, which seemed to be the nitrate concentration when discharge is declining. The nitrate concentration of the outflow from the upper tank (C1), regarded as subsurface flow, was a variable parameter for the model. For the nitrate concentration of the outflow from the upper tank (C1), 148 µeq/l gave the best-fitting results, as shown in the upper panel of Figure 5. These results indicate that during high-discharge periods, the increased subsurface flow containing a high concentration of nitrate contributed to the increased nitrate concentrations. However, since this model with these parameters settings was only able to simulate nitrate concentrations between 65 and 148 µeq/l, further model modifications will be required to simulate episodic concentration as high as the 370 µeq/l observed in Sep. 1999.

5. Conclusions

At Hyakumakidani, a small watershed on Kureha Hill, high concentrations of nitrate throughout the year and a net loss of nitrogen from the watershed were observed. These results imply that this watershed already is at the final stage of nitrogen saturation according to Stoddard's definition. However, according to Aber's definition, the watershed is in Stage 2 because no visible attenuation has been observed.

Our preliminary study, using a two-stage tank model simulating flow-path changes in the soil, indicated that during high-discharge periods, the increased subsurface flow containing high concentrations of nitrate contributed to increased nitrate concentrations.

Acknowledgements

This research was funded in part by the Japanese Foundation of River and Watershed Environment Management.

References

Aber J. D., K. J. Nadelhoffer, P. Steudler and J. M. Melillo: 1989, *BioScience* **39**, 378.
Fujii S. and O.Yamamoto: 1979, *Bulletin of the Toyama Science Museum* **1**, 1
Murdoch, P. S. and J. L. Stoddard: 1992, *Water Resources Research*, **28**, 2707.
Ohrui, K. and M. J. Mitchell: 1997, *Ecological Applications*, **7**, 391.
Stoddard L. J.: 1994, 'Long-Term Changes in Watershed Retention of Nitrogen', in L. A. Baker (ed.), *Environmental Chemistry of Lake and Reservoirs,* Adv. Chem. Ser., 237, pp. 223-284.
Tietema A., A. W. Boxman, M. Bredemeier, B. A. Emmett, F. Molden, P. Gundersen, P. Schleppi and R. F. Wrigh: 1998, 'Nitrogen saturation experiments (NITREX) in coniferous forest ecosystem in Europe: a summary of results' in Klaas W. van der Hoek *et al.* (eds.), *Proceedings of the first international nitrogen conference*, pp. 433-437.

NITROGEN RETENTION IN JAPANESE CEDAR STANDS IN NORTHERN HONSHU, WITH HIGH NITROGEN DEPOSITION

M. BABA*, Y. SUZUKI, H. SASAKI, K. MATANO, T. SUGIURA, and H. KOBAYASHI

School of Veterinary Medicine and Animal Sciences, Kitasato Univ., Towada, Aomori 034-8628 Japan
(*author for correspondence, e-mail: baba@vmas.kitasato-u.ac.jp)

Abstract. High nitrogen, especially ammonium, input has been observed in Shichinohe, Aomori Prefecture, northeastern Japan. A monitoring study on precipitation, throughfall, and stream water has been carried out to estimate the stage of nitrogen saturation since 1996. Fifty-two to 70 % of nitrogen input in throughfall was retained in forest ecosystems. Nitrate concentration in stream water tended to decrease throughout the study. There was no symptom of nitrogen saturation at Japanese cedar stands in Shichinohe, although high nitrogen input in open bulk has been observed. Ammonium (NH_4^+) was retained in the canopy. The ratio of NH_4^+ input in throughfall to that by open bulk was 0.40 - 0.47. Total inorganic nitrogen input under the canopy amounted 0.68 - 0.72 kmolc ha^{-1} yr^{-1} (9.6 - 10.0 kg N ha^{-1} yr^{-1}). Our results suggests that atmospheric nitrogen input has benefitted the tree growth.

Keywords: Ammonium input, canopy uptake, forest, Japanese cedar, nitrogen saturation, stream water, throughfall

1. Introduction

Nitrogen saturation occurrs when nitrogen input exceeds nitrogen demands of plants and microbes in ecosystems (Aber et al., 1989). Responses of the forest ecosystem to nitrogen input would be dependent on the nitrogen saturation status. Some Japanese forest ecosystems are classified as nitrogen saturated (Ohrui and Mitchell, 1997: Baba and Okazaki, 1998). Ohrui and Mitchell (1997) revealed that 10 kg N ha^{-1} yr^{-1} was the minimum threshold of nitrogen saturation for coniferous forests throughout the world including Japanese forests. Ammonium (NH_4^+) input in Towada area, Aomori Prefecture, Japan is much higher than the Japanese average (Baba et al., 1998). Total nitrogen input in open bulk precipitation was 12.4 - 18.7 kg N ha^{-1} yr^{-1}, in 1996-1997. The objective of this study was i) to examine the stage of nitrogen saturation based on stream water chemistry and nitrogen budget and ii) to estimate the role of nitrogen retention in the canopy of Japanese cedar (*Cryptomeria japonica*).

2. Materials and Methods

2.1 STUDY SITE

The study sites are located in Minamitonai (Mt) and Haginosawa (Hs) in Shichinohe town, Aomori Prefecture (Fig. 1). The sites are 32 km west of Pacific Ocean, and 19 and 25 km from Mutsu-Bay for Mt and Hs, respectively. The

Figure 1. Location of study site.

catchment of Mt and Hs is 9.0 ha and at an elevation of 130 - 215 m a.s.l. and 10.6 ha and at an elevation of 100 - 200 m a.s.l., respectively. The forest is a Japanese cedar plantation (22 years old for Mt, and 42 years old for Hs in 1999). The mean diameter at breast height was 23 cm in Mt site and 27 cm in Hs site. The mean branch spread is 1.6 m in Mt site and 1.8 m in the Hs site. Tree densities are 1800 trunks ha^{-1} for Mt and 1000 trunks ha^{-1} for HS. The Holocene tephra had been deposited on welded tuff (Oike, 1972; Japan Economic Planning Agency, 1970), which resulted in the formation of thick Andisols, Melanudands and Hapludands. Total carbon (T-C) amounts were high in this soil because of accumulation of organic matter (Table I). Total nitrogen (T-N) amounts in O horizon and mineral soil were higher in Shichinohe sites than other Japanese cedar stands (Tsutsumi, 1989).

TABLE I
Soil nitrogen in Minamitonai and Haginosawa site.

	Minamitonai			Haginosawa		
	T-C (t ha^{-1})	T-N (t ha^{-1})	C/N	T-C (t ha^{-1})	T-N (t ha^{-1})	C/N
O	9.74	0.237	41.1	5.97	0.152	39.3
0-10 cm	87.8	6.57	13.4	90.3	6.47	14.0
10-20 cm	54.2	3.30	16.4	82.4	5.17	15.9
20-30 cm	46.0	3.15	14.6	70.8	4.90	14.4
40-50 cm	33.7	2.47	13.7	58.7	3.51	16.7

2.2. MONITORING STUDY

Rainfall (open bulk) was collected duplicately using polyethylene funnels (20 cm in diameter) which were set up 1 m above ground. In addition, polyethylene buckets were set up 1.5 m above ground to collect snowfall. Canopy throughfall was collected in triplicate close to the tree trunk (Proximal point) and under the canopy close to the edge of canopy projection (Distal point). In Hs site samplers were settled at intermediate between proximal and distal points (Middle point). Stemflow was collected through spirally rolled and glued vinyl tubes (28 mm in diameter) using 200 L vessels. Stemflow input was calculated using the mean ratio of study area : total sectional area at breast height of tree trunks in study site (Haibara and Aiba, 1982). Open bulk, throughfall, and stemflow samples were collected every second week. Area weighted mean flux was used as the throughfall flux for nitrogen budget. Stream water was collected once a week. Water level and flux was measured immediately after stream water sampling for calibration. Water level was automatically recorded at the gauge from April 1998. It was impossible to measure water level due to snow coverage from December to March. Therefore, nitrogen budget was estimated from April to November in 1997 and 1998.

Water samples were immediately analyzed after sampling for pH and electric conductivity. Subsequently all samples were filtered using membrane filters (Millipore 0.45 μm in pore size) and stored at 4°C and analyzed for major ions by ion chromatography (Shimadzu, LC 10A Ion Chromatographic analyzer). Alkalinity was determined by titration to pH 4.8. The diferences in nitrogen input between open bulk precipitation and throughfall or stemflow were examined by multiple comparison (post-hoc test using Statcel (Yanai, 1998)).

3. Results

3.1. NITROGEN INPUT

Ammonium input significantly decreased during the passage through the canopy at both sites (Fig. 2). Ammonium input in throughfall was 40 % of that in open bulk precipitation for Mt site and 47 % for Hs site. In case of nitrate (NO_3^-), throughfall input was almost same as open bulk input in Mt site. In Hs site throughfall input was 1.2 times larger than that in open bulk. Annual mean NH4+ input in throughfall amounted 0.35 kmolc ha^{-1} yr^{-1} in Mt site and 0.36 kmolc ha^{-1} yr^{-1} in Hs site. Nitrate input was comparable to NH_{4+} input (0.33 kmolc ha^{-1} yr^{-1} in Mt site and 0.36 kmolc ha^{-1} yr^{-1} in Hs site). Total inorganic nitrogen input under the canopy amounted 0.68 - 0.72 kmolc ha^{-1} yr^{-1} (9.6 - 10.0 kg N ha^{-1} yr^{-1}). Throughfall N input in Mt site increased from 0.60 kmolc ha^{-1} yr^{-1} in 1996-1997 to 0.82 kmolc ha^{-1} yr^{-1} in 1998-1999. In the case of Hs site, it increased from 0.66 kmolc ha^{-1} yr^{-1} in 1996-1997 to 0.79 kmolc ha^{-1} yr^{-1} in 1998-1999.

From April to November total nitrogen input in throughfall amounted 0.63 kmolc ha^{-1} in 1997 and 0.63 kmolc ha^{-1} in 1998 in Mt site. In Hs site it was 0.66 kmolc ha^{-1} and 0.64 kmolc ha^{-1}, respectively.

3.2. NITROGEN OUTPUT

The pH values of stream water were around pH 7 throughout the study (Result from Mt site is shown in Fig. 3). Ammonium concentration was negligible. The weighted mean NO_3^- concentration was 54 μmolc L^{-1} in 1996-1997, 54 μmolc L^{-1} in 1997-1998, and 34 μmolc L^{-1} in 1998-1999, in Mt site. In Hs site it was 47 μmolc L^{-1}, 37 μmolc L^{-1}, and 30 μmolc L^{-1}, respectively. Nitrogen output from April to November was 0.30 kmolc ha^{-1} in 1997 and 0.25 kmolc ha^{-1} in 1998 in Mt site. In Hs site it was 0.20 kmolc ha^{-1} and 0.24 kmolc ha^{-1}, respectively.

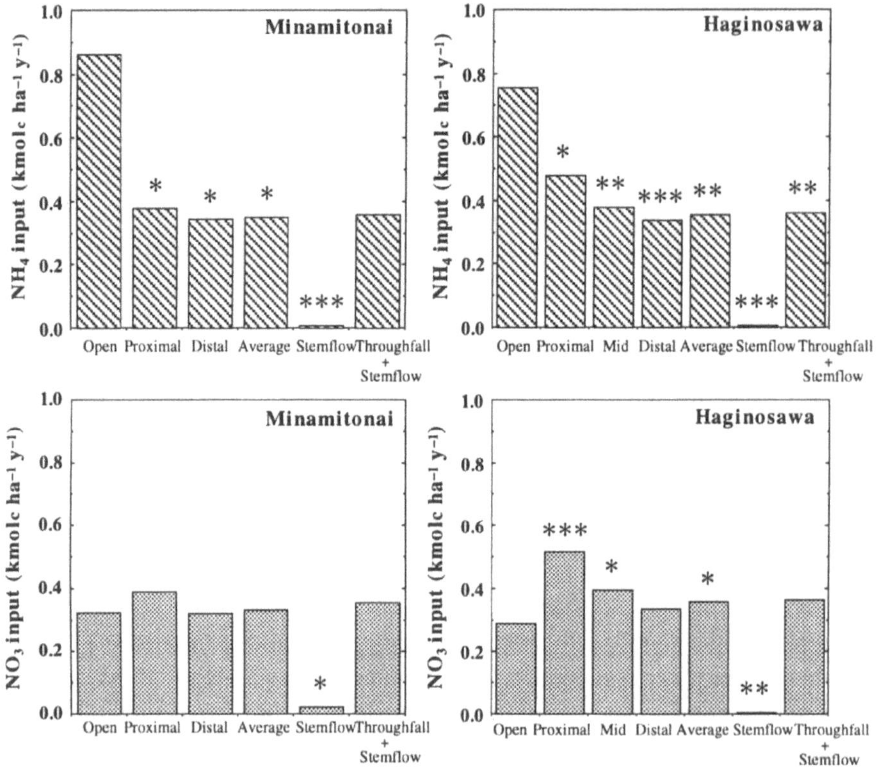

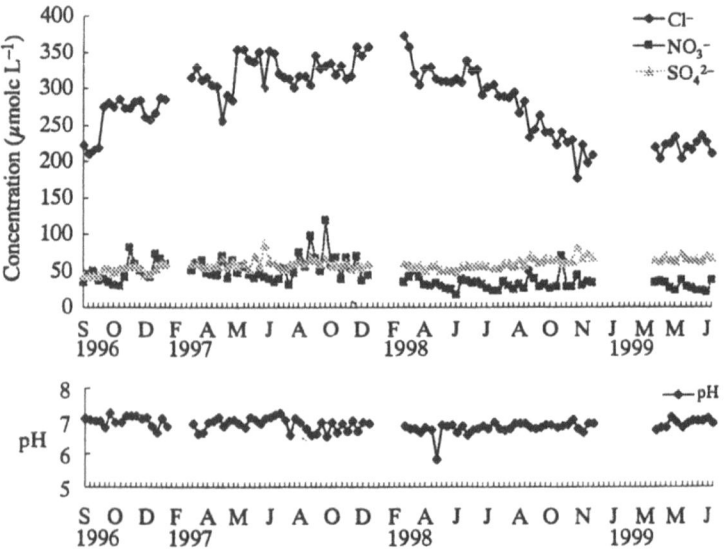

Figure 2. Nitoogen inpt in open bulk, throughfall, and stemflow. Asterisks show significance level of multiple comparison (***: $p<0.001$, **: $p<0.01$, *: $p<0.05$).

Figure 3. Changes in anion concetration and pH value of stream water.

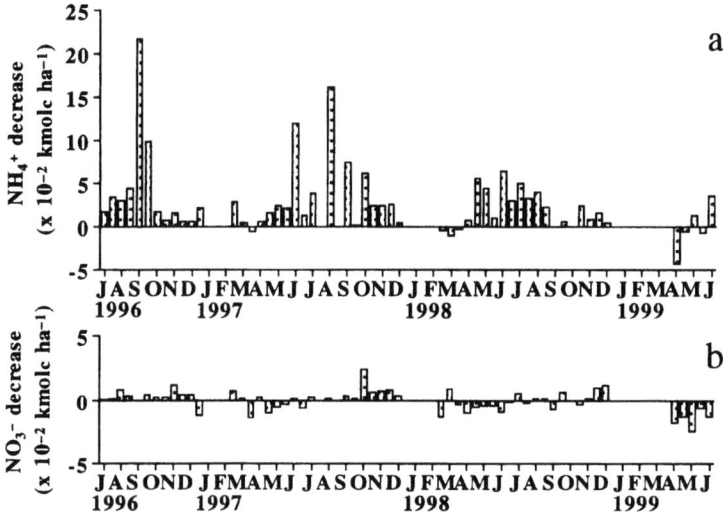

Figure 4, Decrease in nitrogen input (open bulk input - throughfall input) in the canopy.

4. Discussion

4.1. STAGE OF NITROGEN SATURATION

No obvious seasonal patterns in NO_3^- concentration in stream water was observed (Fig. 3) as also Ohrui and Mitchell (1997) reported. Although NO_3^- concentrations in stream water in both Hs and Mt sites were lower than those in nitrogen saturated stands (100 - 200 µmolc L^{-1}) (Ohrui and Mitchell, 1997; Baba and Okazaki, 1998), it was higher than the average of stream water in Japanese forests (Hirose *et al.*, 1988). Nitrate concentration at both sites was also higher than that the average for 1955-1956 in Shichinohe river, 16 µmolc L^{-1} (Kobayashi, 1961).

Although total inorganic nitrogen input in throughfall in Shichinohe was lower than in other Japanese stands (Haibara and Aiba, 1982; Kobayashi *et al.*, 1995; Sato and Takahashi, 1996; So *et al.*, 1999), it was comparable to the criterion of nitrogen saturation as indicated by Ohrui and Mitchell (1997). On the other hand, nitrogen output in stream in both Mt and Hs sites was smaller than nitrogen input. The output was 39 - 48% of input in Mt site and 30 - 38% in Hs site. That is, the older stand retained larger amounts of nitrogen than the younger stand. This corresponded NO_3^- concentration which was lower in Hs site than that in Mt site. In addition, NO_3^- concentration tended to decrease in spite of increase in throughfall nitrogen input. That is, there were no symptom of nitrogen saturation at either site.

4.2. NITROGEN RETENTION IN THE CANOPY

Decrease in NH_4^+ input (subtracting throughfall input from input in open bulk precipitation) was remarkable in summer (result in Mt site is shown in Fig. 4). This decrease was proportional to open bulk input ($R^2 = 0.82$ for Mt site and 0.84 for Hs site, $p < 0.001$). Ammonium input in open bulk significantly increased in summer and became negligible in winter, that resulted in same seasonal fluctuation of NH_4^+ uptake in the canopy. This seasonal changes would be important for our study sites to retain nitrogen in the canopy.

Decrease in NH_4^+ input in the canopy in our study sites was quite different from other Japanese cedar stands, where throughfall input exceeded input in open bulk by factors of 1.4 - 2.1. Conifer canopies can directly assimilate nitrogen (Boyce et al., 1996) and our results suggested that this is also the case for Japanese cedar. In case of NO_3^-, throughfall inputs in other study sites have been found to be 2.2 - 3.1 times higher than that in open bulk. Throughfall input in Shichinohe increased only by a factor of 1.0 - 1.2. Taking dry deposition into consideration, nitrate would also be assimilated in the canopy. Carleton and Kavanagh (1990) showed significant difference in canopy uptake among throughfall collector positions for NO_3^- in black spruce (Picea mariana) stands. This is true for NH_4^+ uptake in Hs site, Shichinohe. Throughfall input at the distal point was significantly less than that at the proximal point ($p < 0.05$) (Fig. 2).

Uptake in the canopy suggests that Japanese cedar stands are deficient in nitrogen and atmospheric nitrogen deposition is therefore likely to increase forest growth.

Acknowledgement

We are sincerely grateful to Shichinohe town government and Mr. Syoroku Kudo, owner of study sites, for their support. We thank Mr. Sei-ichi Okamura and Mr. Fumiaki Uwano for selecting study sites. This study was supported by Grant in Aid for Scientific Research (No. 09780483 and No. 11680533) from the Ministry of Education, Science and Culture of Japan and partly by the Sasakawa Scientific Research Grant from the The Japan Science Society.

References

Aber, J.D., Nadelhoffer, K.J., Steudler, P., and Melillo, J.M.: 1989, *BioScience* **39**, 378
Baba, M. and Okazaki, M.: 1998. Soil Science and Plant Nutrition 44, 513
Baba, M., Suzuki, Y., Sugiura, T., and Kobayashi H.: 1998, *Abstracts of Japanese Society of Soil Science and Plant Nutrition* **43**, 206
Boyce, R.L., Friedland, A.J., Chamberlain, C.P., and Poulson, S.R.: 1996, *Canadian Journal of Forest Research* **26**, 1539
Carleton, T.J. and Kavanagh, T.: 1990, *Canadian Journal of Forest Research* **20**, 1917
Haibara, K. and Aiba Y.: 1982, *Journal of Japanese Forestry Society* **64**, 8
Hirose, A., Iwatsubo, G., and Tsutsumi, T.: 1988, *Bulletin of the Kyoto University Forests* **60**, 162
Japan Economic Planning Agency: 1970, Geological Map of Aomori Prefecture (1:200,000). Printing Bureau, Ministry of Finance, Japan, Tokyo
Kobayashi, J.: 1961, *Nogakukenkyu* **48**, 63
Kobayashi, T., Nakagawa, Y., Tamaki, M., Hiraki, T., and Shoga, M.: 1995, *Environmental Science* **8**, 25
Ohrui, K. and Mitchell, M.: 1997, *Ecological Applications* **7**, 391
Oike, S.: 1972, *The Quaternary Research* **11**, 228
Sato, K. and Takahashi, A.: 1996, *Environmental Science* **9**, 221
So, Y., Kodaira, T., and Okazaki, M.: 1999, *Papers on Environmental Information Science* **13**, 263
Tsutsumi, T.: 1989, *Forest Ecology*, Asakura-shoten, Tokyo
Yanai, H.: 1998, Statcel. OMS Publisher, Tokyo

SOIL SOLUTION CHEMISTRY IN JAPANESE CEDAR STANDS IN NORTHERN HONSHU, WITH HIGH NITROGEN DEPOSITION

K. MATANO, M. BABA, A. SHIBUYA*, Y. SUZUKI, T. SUGIURA
and H. KOBAYASHI
*School of Veterinary Medicine and Animal Sciences, Kitasato Univ., Towada,
Aomori 034-8628 Japan*
(*author for correspondence, e-mail: ayano-s@par.odn.ne.jp)

Abstract. High nitrogen, especially ammonium, input has been observed in Shichinohe, Aomori Prefecture, northeastern Japan. The monitoring study on soil and soil solution has been carried out to determine soil acidification status since 1996. Soils and soil solutions in Minamitonai and Haginosawa are not strongly acidic. Fluctuations in nitrate concentrations coincided with sodium (Na^+) or calcium (Ca^{2+}). Produced protons due to nitrification were exchanged with Ca^{2+} or neutralized by weathering process. Exchangeable Ca^{2+} accumulated in surface layers, particular in the older Japanese cedar (*Cryptomeria japonica*) stand (42 years old). Exchangeable Ca^{2+} affected soil solution chemistry and the Ca^{2+} concentration was significantly higher in the older Japanese cedar stand than that in the younger stand (22 years old). Base cations, especially Ca^{2+}, accumulation prevented soil (solution) acidification in Shichinohe site.

Keywords: Andisols, calcium, nitrification, sodium, soil solution

1. Introduction

The Japanese cedar (*Cryptomeria japonica*) in the Kanto district might be suffering from acidic deposition (Yambe, 1978; Takahashi *et al.*, 1986). The causes of abnormal growth and defoliation in the Kanto district, however, remains controversial. For instance, water stress has been speculated as the cause (Matsumoto *et al.*, 1992). Aber *et al.* (1989) pointed out that susceptibility to water stress increase when leaf nitrogen contents elevate as the result of nitrogen saturation. Nitrogen saturation enhance nitrate leaching which contribute to soil acidification (Aber *et al.*, 1989). Nitrification via chemoautotrophic or heterotrophic processes is the source of nitrate (NO_3^-) and protons (H^+) (Gundersen and Rasmussen, 1990). Baba (1999) showed that atmospheric ammonium (NH_4^+) input contributed to soil acidification in Hachioji, Tokyo, southern Kanto district. Calcium and magnesium ions usually leached with NO_3^-. Aluminum concentration in soil solutions increased with NO_3^- concentration in summer. Japanese cedar is more susceptible to acid and aluminum than Hinoki cypress (*Chamaecyparis obtusa*) (Tsutsumi, 1962).

We carried out a monitoring study on precipitation, throughfall, soil solution and stream water in Shichinohe, Aomori Prefecture, northeastern Japan since July 1996. High NH_4^+ input (8.6 - 14.1 kg N ha^{-1} yr^{-1} in 1996-1997) with open bulk precipitation has been observed (Baba *et al.*, 1998). Although NH_4^+ input significantly decreased at the canopy, total nitrogen input was comparable to the minimum threshold of nitrogen input (10 kg N ha^{-1} yr^{-1}) for promoting nitrogen saturation, reported by Wright *et al.* (1995) (Baba *et al.*, submitted). The objective of this study was to determine the state of acidification of soil and soil solution under Japanese cedar forests.

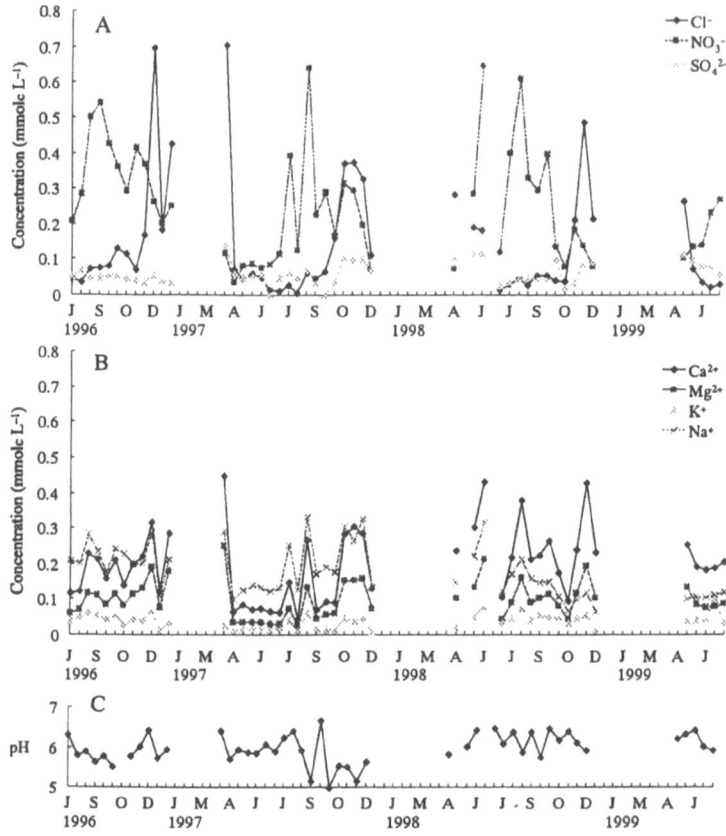

Figure 1. Changes in anion (A) and cation (B) concentration and pH values (C) of soil solution at the depth of 10 cm in Minamitonai site.

2. Materials and Methods

Site description is shown in detail by Baba *et al.* (submitted). Briefly, vegetation is Japanese cedar plantation (22 years old for Minamitonai (Mt) site and 42 years old for Haginosawa (Hs) site in 1999). The Holocene tephra, predominantly Towada-a (1,000 B.P.) and Chuseri (5,000 B.P.), was deposited on welded tuff (Oike, 1972; Shoji *et al.*, 1987; Ugolini, 1988; Japan Economic Planning Agency, 1970), which resulted in the formation of thick Andisols, Melanudands and Hapludands. The thickness of the A horizon was 25 - 35 cm in Mt site and 67 - 105 cm in Hs site.

Samplers were installed in June, 1996 and sampling started in July, 1996. Soil solutions were taken in triplicate at the different depths (10, 20, 30 and 50 cm) at least 1.5 m distant from any tree trunks at both sites. Extractors using porous ceramic cups were always connected to glass bottles at a initial vacuum (85-95 kPa). Glass bottles were evacuated using a Handy Aspirator (Nalgene) at least 1 week before the bottle exchange. Bottles were exchanged every two weeks. Samples were analyzed for pH and electric conductivity at once. All samples were stored at 4 °C and analyzed for major ions by ion chromatography

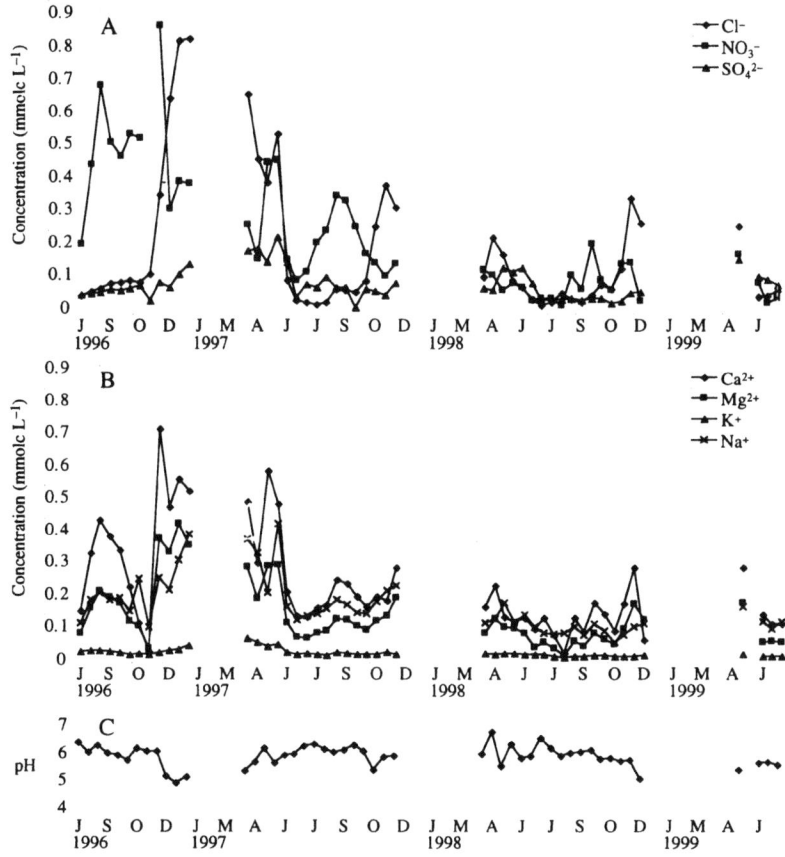

Figure 2. Changes in anion (A) and cation (B) concentration and pH values (C) of soil solution at the depth of 10cm in Haginosawa site.

(Shimadzu, LC 10A Ion Chromatographic analyzer). Alkalinity was determined by titration to pH 4.8. Silicate (SiO_2) concentration was determined by colorimetry. The differences between soil solutions in Mt site and those in Hs site were examined by t-test. It was also performed for soil depth.

Soil samples were taken at the different depths of 5, 10, 20, 30 and 50 cm. All samples were air-dried and passed through a 2 mm sieve. Then air-dried samples were used for analysis. The pH (H_2O) of a 1:5 air dried soil : water suspension was measured by glass electrode. Exchangeable cations were extracted by 1 M ammonium acetate (pH 7) after van Reeuwijk (1993) and determined by atomic absorption spectrophotometry (Hitachi, Z-6100 Polarized Zeeman Atomic Absorption Spectrophotometer). Subsequently, 200 mL of 1 M sodium acetate (pH 7) was added to replace exchangeable sites with sodium. Then, sodium ions in reagent were removed by ethyl alcohol. Retained sodium ions were extracted by 1 M ammonium acetate (pH 7) and sodium concentrations were determined to estimate cation exchange capacity (CEC).

3. Results

Soil and soil solution were not extremely acidified. Soil pH (H_2O) was 5.7 - 6.3 in Mt site and 5.5 - 6.0 in Hs site. The pH values of soil solution fluctuated between 5.0 and 6.7 in Mt site (Fig. 1) and between 5.0 and 6.8 in Hs site (Fig. 2). There was no relationship between pH values of soil solution and nitrate concentration in soil solution.

Nitrate concentration in soil solution at the depth of 10 cm in both sites increased in summer. Chloride (Cl^-) concentration increased from October to April. Sodium (Na^+) concentration was highest among base cations in soil solution in Mt site during 1996 to 1997. While calcium (Ca^{2+}) concentration became dominant ion from April 1998. Sodium concentration fluctuated with NO_3^- concentration in Mt site ($R^2 = 0.41$, $p < 0.001$). Calcium and magnesium (Mg^{2+}) concentration was proportional to Cl^- concentration ($R^2 = 0.45$, $p < 0.001$ for Ca^{2+} and $R^2 = 0.61$, $p < 0.001$ for Mg^{2+}). In case of Hs site, Ca^{2+} concentration was the highest throughout the study. Calcium concentration fluctuated with NO_3^- ($R^2 = 0.59$, $p < 0.001$) and Cl^- ($R^2 = 0.53$, $p < 0.001$) concentration in Hs site. Magnesium and Na+ concentration was proportional to Cl^- concentration regardless of seasons ($R^2 = 0.75$, $p < 0.001$ for Mg^{2+} and $R^2 = 0.62$, $p < 0.001$ for Na^+).

Anion compositions in Mt site was almost similar to those in Hs site (Fig. 3), although the percentage of NO_3^- at the depth of 10 cm in Mt site was significantly higher than that in Hs site ($p < 0.01$). The percentages of Ca^{2+} and Mg^{2+} were significantly higher in Hs site regardless of soil depth, compared to those in Mt site ($p < 0.05$). Calcium percentage at the depth of 10 cm was significantly higher than those at other soil depths ($p < 0.001$). Conversely, Na^+ percentage at the depth of 10 cm was the lowest ($p < 0.001$). Sodium percentage at the depth of 50 cm in Hs site was the highest ($p < 0.01$) and Ca^{2+} percentage was the lowest ($p < 0.01$).

Exchangeable Ca^{2+} was dominant at the all depths. Its contents in soil were also the highest in the uppermost layers (Fig. 4). Its contents were higher in Hs site than those in Mt site, although differences in contents were not statistically significant.

4. Discussion

Soils and soil solutions in Mt and Hs sites are not strongly acidic. Soil pH values in both sites were in the range which is appropriate for growth of Japanese cedar (Ohmasa, 1935). Ohrui (1997) reported that not only high nitrogen mineralization and nitrification rates but also high precipitation in growing season enhanced NO_3^- leaching from Japanese forest stand, compared European and American stands. Nitrate concentrations in soil solution at both sites were comparable other Japanese cedar stands (*e.g.* Kato *et al.*, 1993).

Fluctuations in NO_3^- concentrations coincided with Na^+ in Mt site and Ca^{2+} in Hs site. It implies that produced protons due to nitrification were exchanged with Na^+ or Ca^{2+}. In Mt site Na^+ concentration were proportional SiO_2 concentration

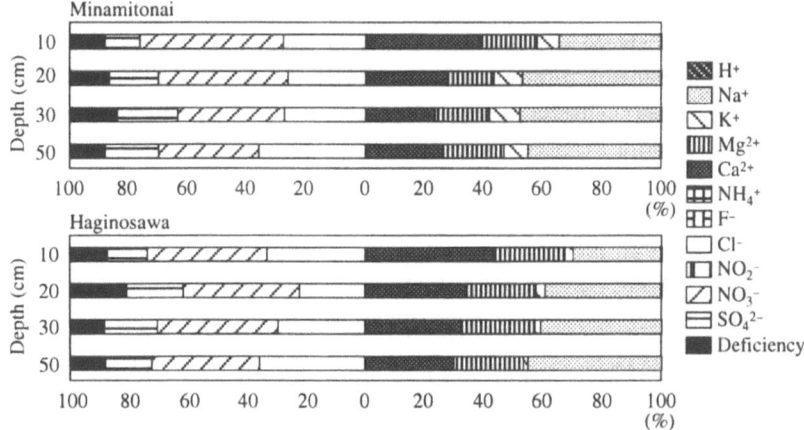

Figure 3. Composition of soil solution in Minamitonai (Upper) and Haginosawa (Bottom) site.

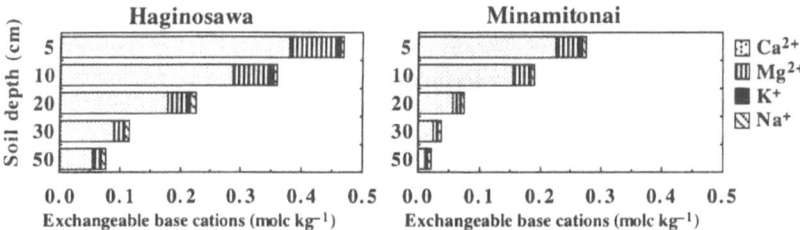

Figure 4. Exchangeable base cations in soils.

($R^2 = 0.33$, $p < 0.05$). Nitrate concentration in Mt site were also correlated with SiO_2 concentration ($R^2 = 0.43$, $p<0.01$). This suggests that mineral weathering also contributed to neutralizing protons produced by nitrification in Mt site.

Magnesium and Na^+ were also codominant cations in Mt and Hs sites. This is quite different from other study sites under Japanese cedar. High Na^+ concentration in soil solution was reported by Ugolini *et al.* (1988). They took soil solutions under Maries's fir (*Abies Mariesii*) in the southern Hakkoda region, where approximately 20 km far from Hs site and 1050 m above sea level, during mid-August and mid-September. Sodium concentration in the soil solution was also the highest among cations at their site. High Na^+ concentration would be characteristics of this area.

Exchangeable Ca^{2+} contents affected on Ca^{2+} concentration in soil solution, especially at the depth of 10 cm (Figs. 3, 4). Calcium concentration in soil solution under the Japanese cedar stands usually exceeded other cations (*e.g.* Kato *et al.*, 1993; So *et al.*, 1999). This was attributed to accumulation of Ca^{2+} due to Japanese cedar litter (Sawata and Kato, 1991, 1993). Its accumulation occurred after 35 years old and became remarkable after 45 years. Japanese cedar in Hs site is 42 years old and exchangeable Ca^{2+} contents were higher in Hs site than those in Mt site (Fig. 4). The CEC values were 0.51 molc kg^{-1} for Mt site and 0.62 molc kg^{-1} for Hs site on the average (Baba *et al.*, 1998). Base saturation was high, 7 - 37 % in Mt site and 15 - 62 % in Hs site. In particular, Ca^{2+} occupied 4 - 30 % of CEC in Mt site and 10 - 50 % in Hs site. Accumulation of Ca^{2+} would contribute to the difference in exchangeable Ca^{2+} contents and in Ca^{2+} concentration in soil solution between Hs and Mt site. Base cations accumulation prevented soil (solution) acidification in Shichinohe site.

Acknowledgements

We are sincerely grateful to Shichinohe town government and Mr. Syoroku Kudo, owner of study sites, for their support. We thank Mr. Yo-ichi Hosoda, Aomori Field Crops and Horticultural Experiment Station, for assistance to determine exchangeable cations and CEC by AAS. This study was supported by Grant in Aid for Scientific Research (No. 09780483 and No. 11680533) from the Ministry of Education, Science and Culture of Japan and partly by the Sasakawa Scientific Research Grant from the The Japan Science Society.

References

Aber, J.D., Nadelhoffer, K.J., Steudler, P., and Melillo, J.M.: 1989, *BioScience* **39**, 378.
Baba, M.: 1999. Doctoral dissertation, Tokyo Univ. of Agr. and Tech., Fuchu, Tokyo, Japan.
Baba, M., Suzuki, Y., Sasaki, H., Matano, K., Sugiura, T., and Kobayashi H.: This volume.
Baba, M., Suzuki, Y., Sugiura, T., and Kobayashi H.: 1998, *Abstracts of Japanese Society of Soil Science and Plant Nutrition* **43**, 206.
Gundersen, P. and Rasmussen, L.: 1990, *Reviews of Environmental Contamination and Toxicology* **113**, 1.
Japan Economic Planning Agency: 1970, Geological Map of Aomori Prefecture (1:200,000). Printing Bureau, Ministry of Finance, Japan, Tokyo.
Kato, H., Ishikura, T., Akama, Y., Munakata, Y., and Sawata, S.: 1993, *Japanese Journal of Soil Science and Plant Nutrition* **64**, 161.
Matsumoto, Y., Maruyama, Y. and Morikawa, Y.: 1992, *Japanese Journal Forest Environment* **34**, 2.
Ohmasa, M.: 1935, *Bulletin of the Government forest experiment station* **3**, 1 (Cited from Nashimoto, N., Takahashi, K., and Ashihara, S.: 1993, *Environmental Science* **6**, 121).
Ohrui, K.: 1997, *Japanese Journal Forest Environment* **39**, 1.
Oike, S.: 1972, *The Quaternary Research* **11**, 228.
Sawata, S. and Kato, H.: 1991, *Japanese Journal of Soil Science and Plant Nutrition* **62**, 49.
Sawata, S. and Kato, H.: 1993, *Japanese Journal of Soil Science and Plant Nutrition* **64**, 296.
Shoji, S., Takahashi, T., Saigusa, M., and Yamada, I.: 1987, *Japanese Journal of Soil Science and Plant Nutrition* **58**, 638.
So, Y., Kodaira, T., and Okazaki, M.: 1999, *Papers on Environmental Information Science* **13**, 263.
Takahashi, K., Okitsu, S. and Ueta, H.: 1986, *Japanese Journal Forest Environment* **28**, 11.
Tsutsumi, T.: 1962, *Bulletin of the Government forest experiment station* **137**, 1.
Ugolini, F.C., Dahlgren, R., Shoji, S., and Ito, T.: 1988, *Soil Science* **145**, 111.
van Reeuwijk, L.P. (Ed.) 1993, *Procedures for soil analysis (4 th Ed.)*. p. 9-1 - 9-11, ISRIC, Wageningen.
Yambe, Y.: 1978, *Bulletin of the Forestry and Forest Products Research Institute* **301**, 119.

SULFATE AND NITRATE LOADS ON A FOREST ECOSYSTEM IN KOCHI IN SOUTHWEST OF JAPAN

TSUYOSHI YAMADA[1*], SHUICHIRO YOSHINAGA[1,2], KAZUHITO MORISADA[1,2], and KEIZO HIRAI[1,2]

[1] *Shikoku Research Center, Forestry and Forest Products Research Institute (FFPRI), 2-915 Asakura-nishi, Kochi 780-8077 Japan.;* [2] *Forestry and Forest Products Research Institute, Ibaraki 305-8687 Japan.*
(* author for correspondence, e-mail: yamadan@ffpri-skk.affrc.go.jp)

Abstract. To assess the influence of acidic deposition on the forest ecosystem, it is necessary to evaluate the gross amount of acidic deposition. In this paper, we discuss the variation of sulfate (SO_4^{2-}) and nitrate (NO_3^-) loads as well as related concentration from 1991 to 1999 in the Hinoki (*Chamaecyparis obtusa*) plantation in Kochi, southwest Japan. The annual precipitation varied significantly from 1,700 to 3,900 mm during the study period. The annual sulfate concentration of rainfall was about 15 µmol L^{-1}, including about 80% non sea salt sulfate, while the annual nitrate concentration of rainfall was increased. The sulfate and nitrate concentrations of the through fall and the nitrate concentration of the stem flow were equal to or slightly higher than those of rainfall. However, the sulfate concentration of the stem flow was higher than that of rainfall, 21 to 55 µmol L^{-1}. The sulfate and nitrate loads of rainfall were measured to be 27 to 46 and 14 to 43 mmol m^{-2} y^{-1}, respectively. The sulfate and nitrate loads of the through fall were the same or slightly higher than those of rainfall. In contrast, the sulfate and nitrate loads of the stem flow were less than those of rainfall. Combined sulfate loads of the through fall and the stem flow reached about 1.5 times that of the sulfate load of rainfall.

Keywords: concentration, forest ecosystem, load, nitrate, stem flow, sulfate, through fall

1. Introduction

The influence of acidic deposition on the forest ecosystem have been reported in many forests throughout the world (Johnson and Lindberg, 1992; Likens and Bormann, 1995). In Japan, there has been no obvious damage to forest ecosystems, although acidic depositions have been observed in the last several years (Tamaki *et al.*, 1991; Tokuchi and Iwatsubo, 1992; Hara, 1997). To assess the influence of acidic depositions on the forest ecosystem, it is first necessary to evaluate the gross amount of acidic materials contained in rainfall, through fall and stem flow. In this paper, we discuss the loads of sulfate and nitrate for a Hinoki (*Chamaecyparis obtusa*) plantation in Kochi, southwest Japan, from 1991 to 1999.

2. Materials and Methods

The load of wet acid deposition was observed in a Hinoki (*Chamaecyparis obtusa*) stand planted in 1970 in Kochi city, southwest Japan. The mean height was 12.9 m; planting density, about 2,500 trees per hectare; and basal area of the stand, 265 m² in 1997. Rainfall was collected using a polyethylene funnel at neighboring open sites in the stand. Through fall was also collected using a polyethylene funnel and stem flow was collected using a polyurethane gutter wrapped on the trunk. Each sample was stored in a polyethylene tank and collected after every continuous rainfall event (average 73 times a year) from 1991 to 1999. The samples of the through fall and stem flow were collected at two sites of the stand. The sulfate and nitrate concentrations of the samples were measured by an ion-chromato-analyzer (YOKOGAWA ANALYTICAL SYSTEMS IC7000D) after the samples were passed through a 0.45µm filter.

Annual sulfate and nitrate concentrations were calculated by the summing concentrations weighted by the amount of precipitation, through fall and stem flow for each rainfall event. The sulfate and nitrate loads were calculated by multiplying the amount of precipitation, through fall and stem flow by the respective concentration.

3. Results

3.1. PRECIPITATION AND CONCENTRATION

Annual precipitation from 1991 to 1999 in Kochi ranged widely from 1,666 to 3,927 mm, average 2,555 mm (Fig. 1).

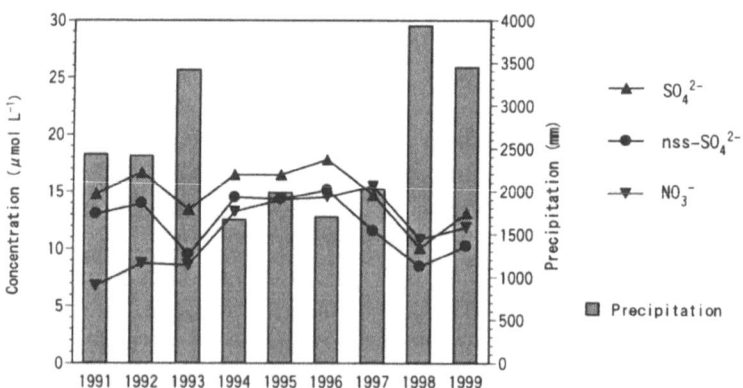

Figure 1. The sulfate and nitrate concentration of the rainfall and the yearly amount of precipitation

The annual sulfate concentration of the rainfall ranged from 10.1 to 17.8 µmol L^{-1} and was lower in heavy rain years (1993, 1998 and 1999). The non sea salt sulfate (nss-sulfate) concentration exhibited the same variation of sulfate concentration and ranged from 8.5 to 15.1 µmol L^{-1}. The proportion of nss-sulfate to total sulfate was from 71 to 88%, average 83%. The annual nitrate concentration of the rainfall ranged from 6.8 to 15.4 µmol L^{-1}.

The annual sulfate and nitrate concentrations of the through fall ranged from 13.5 to 23.5 and 5.6 to 21.8 µmol L^{-1}, respectively. The variations of the sulfate and nitrate concentrations of the through fall were similar to those of the rainfall. The stem flow sulfate concentration exhibited a maximum at 55.3 µmol L^{-1} and a minimum at 21.0 µmol L^{-1}, and the stem flow nitrate concentration, a maximum at 23.9 µmol L^{-1} and a minimum at 7.4 µmol L^{-1} (Fig. 2). In particular, the sulfate concentration of the stem flow was higher and had a larger variation than that of the rainfall and through fall.

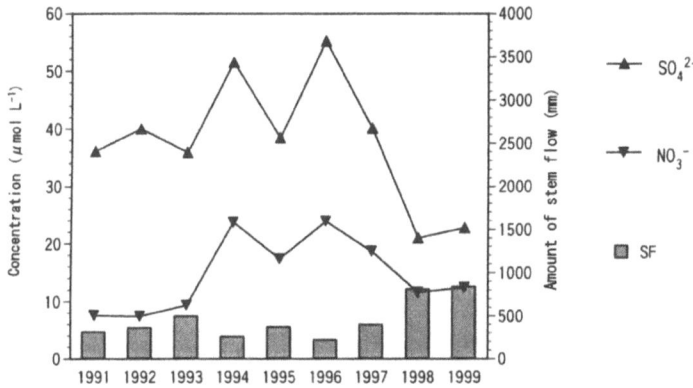

Figure 2. The sulfate and nitrate concentration and the amount of the stem flow

3.2. LOADS

The sulfate load of the rainfall in 1991, 1992, 1993, 1998 and 1999 reached about 40 mmol m^{-2} y^{-1}, in contrast to that in 1994 to 1997, which reached about 30 mmol m^{-2} y^{-1} (Fig. 3). This variation indicates that the sulfate load of the rainfall was controlled by the amount of precipitation. The nitrate load of the rainfall, however, increased gradually during the study period.

The variations of the sulfate and nitrate loads of the through fall were similar to those of the rainfall (Fig. 4). The sulfate and nitrate loads of the stem flow were about one-fifth to half those of the rainfall and through fall (Fig. 5). The sulfate load of the stem flow was stable at about 15 mmol m^{-2} y^{-1}, but the nitrate load increased like that of the rainfall and through fall.

4. Discussion

Although the annual sulfate and nitrate concentrations of the rainfall varied with precipitation, the variation observed in the sulfate concentration was insignificant (Fig. 1). The annual nitrate concentration of the rainfall, however, increased slightly during the study period. The possibility of the increasing nitrate concentration acidifying the

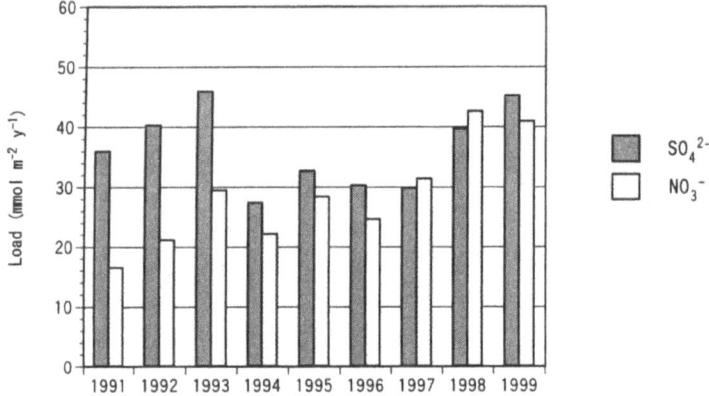

Figure 3. The sulfate and nitrate loads of the rainfall

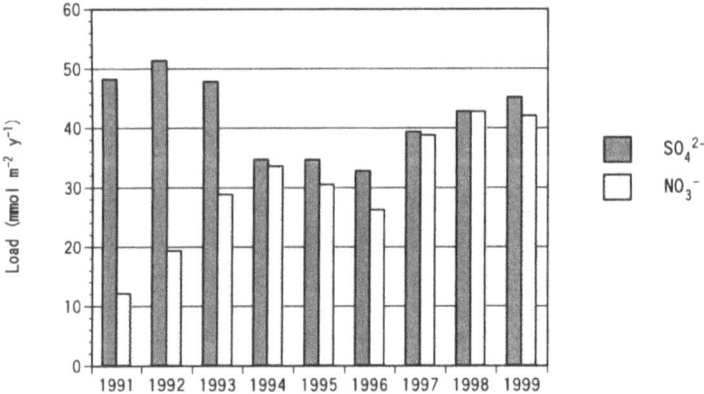

Figure 4. The sulfate and nitrate loads of the through fall

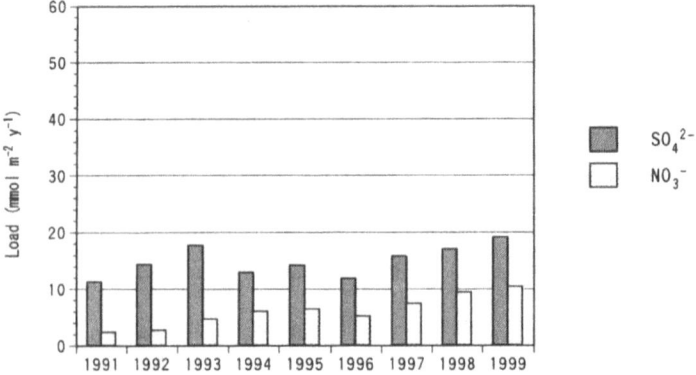

Figure 5. The sulfate and nitrate loads of the stem flow

rainfall was negligible because the mean pH of the rainfall was stable at around 4.7. The mean sulfate and nitrate concentrations of the rainfall from 29 sites monitored by the Japan Environmental Agency (Hara, 1997) were 22.2 and 14.1 µmol L^{-1}, and the means of their loads were 31.3 and 19.4 mmol m^{-2} y^{-1}. Our results agreed with the results mentioned above. The annual precipitation, however, often exceed 3,000 mm in Kochi during the study period, which is quite heavy compared with the monitored annual precipitation that ranged from 590 to 2,041 mm. The decrease in the sulfate and nitrate concentrations in heavy rain years indicated the possibility of sulfate and nitrate being diluted. This dilution might cause a slight variation in the sulfate load. In contrast, the annual nitrate load was increased in spite of the dilution. As there were a little industry and livestock in Kochi, we didn't understand the cause of the increase of the nitrate load. Nitrate increase, however, might be caused by traffic increase because the site was located near downtown and the traffic seemed to increase.

The nitrate concentration of the through fall was almost equal to that of the rainfall, whereas the sulfate concentration of the through fall was higher than that of the rainfall. This difference might indicate that sulfate was increased by wash off of dry deposited sulfate occurring in the canopy. In addition, the sulfate concentration of the stem flow was also higher than that of the rainfall and through fall (Fig. 1 and 2). Furthermore, stem flow acidity increased when the pH decreased below 4 units. These results suggest that the sulfate increase might occur in addition to the proton increase generated by leaching from the stem during the flow process.

To discuss the influence of acid deposition on the forest ecosystem, the acidic material load should be estimated not only for the rainfall but also for the through fall and stem flow. The sulfate and nitrate loads of the through fall did not differ significantly from those of the rainfall. In contrast, the loads of the stem flow were about 30 to 50 % for sulfate and 15 to 25 % for nitrate compared with those of the rainfall and through fall (Figure 3, 4 and 5). The summations of the loads of the through fall and stem flow reached 44.5 to 65.6 mmol m^{-2} y^{-1} for sulfate and 14.4 to 52.5 mmol m^{-2} y^{-1} for nitrate, 1.6 and 1.3 times as much as those of the rainfall (Fig. 6).

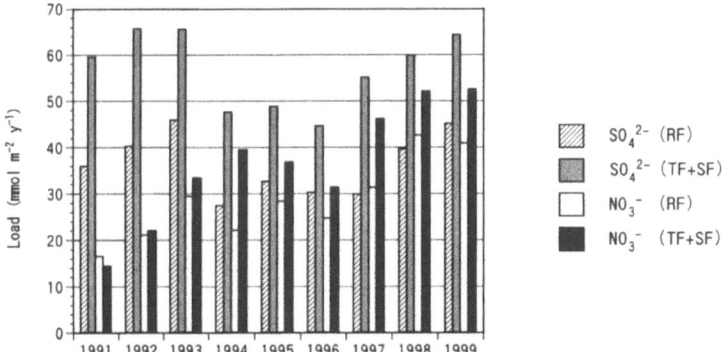

Figure 6. Comparison the sulfate and nitrate loads between the rainfall and the summation of the through fall and stem flow.

5. Conclusion

While the summation of the amounts of the through fall and the stem flow was almost equal to the precipitation in our study, sulfate and nitrate loads of the through fall and the stem flow increased 1.6 and 1.3 times as much as those of the rainfall. This fact indicates that sulfate and nitrate accumulated successively in the flow processes through the stem by leaching as well as by washing off from the canopy. The detailed mechanism for, and the precise estimation of, the sulfate and nitrate additions should be clarified in future studies.

References

Hara, H.: 1997, *The Chemical Society of Japan* **11**, 733. in Japanese with English summary.

Johnson, D. W. and Lindberg, S. E.: 1992, *Atmospheric Deposition and Nutrient Cycling in Forest Ecosystems*, Springer: Berlin.

Likens, G.E. and Bormann, F.H.: 1995, *Biogeochemistry of a Forested Ecosystem*, Springer-Verlag: New York.

Tamaki, M., Katou, T., Sekiguchi, K., Kitamura, M., Taguchi, K., Oohara, M., Mori, A., Wakamatsu, S., Murano, K., Okita, T., Yamanaka, Y. and Hara, H.: 1991, *The Chemical Society of Japan* **5**, 667. in Japanese with English summery.

Tokuchi, N. and Iwatsubo, G.: 1992, *Japanese Journal of Forest Environment* **34**, 14. in Japanese with English summery.

DEPOSITION PATTERNS OF SO_4^{2-}, NO_3^- AND H^+ IN THE BRAZILIAN TERRITORY

M. CRISTINA FORTI[1], ADILSON CARVALHO[2]*, ADOLPHO J. MELFI[3] and CELIA R. MONTES[3]

[1] *Instituto Nacional de Pesquisas Espaciais and NUPEGEL, CP 515, São Jose dos Campos, 12201-970 Brazil;* [2] *Department of Sedimentary Geology and Environment, University of São Paulo and NUPEGEL, 05508-000 - São Paulo-Brazil;* [3] *Department of Soil Science, University of São Paulo and NUPEGEL, CP 09, 13418-900-Piracicaba - Brazil.*
(* *Author for correspondence, e-mail: acarvalh@usp.br*)

Abstract. SO_4^{2-}, NO_3^- and H^+ depositions are estimated in the Brazilian territory based on the existing rainfall chemical data and on annual rainfall distribution over the whole territory. Local and regional depositions are estimated. Rainfall chemical data over the Brazilian territory shows that the average pH values are usually low (between 4.0 and 5.5). These values are observed in the tropical Amazon forest as well as in urban areas. However, the rainwater acidity in the tropical forests are due to organic acids naturally produced by the vegetation while in urban areas the acidity is mainly due to acidic anion deposition (NO_3^- and SO_4^{2-}). In some Amazonian areas, the average input values through rainfall for NO_3^- is about 0.06 keq.ha.yr^{-1} and for SO_4^{2-} is between 0.23 and 0.54 keq.ha^{-1}.yr^{-1}. On the other hand, in some urban centers, such as São Paulo, values of .072 keq.ha^{-1}.yr^{-1} for NO_3^- and 1.16 keq.ha^{-1}.yr^{-1} of SO_4^{2-} are found and in sites where sulfate sources (coal mining) are present, as for the area of Florianópolis, values as high as 5.59 keq.ha^{-1}.yr^{-1} for SO_4^{2-} are found.

Keywords: Acid rain, Brazilian territory, nitrate, sulfate

1. Introduction

The acid deposition on the surface of the earth has been subject of investigations all over the world, because of its hazardous effect to plants, animals, soils, aquatic environment, buildings and historical monuments. Up to the beginning of the 90s, several programs were established to study the acidic deposition and its effects, especially in the industrialized countries of the Northern Hemisphere. It is worthy to quote the results of the NAPAP (1993) in the USA that were undertaken by the decision makers as a parameter to impose a 50% reduction of S and N compound emission levels, up to year 2000. In spite of the reduction of the emission of the acid precursor substances, the risk of acidification of the superficial waters still remains in these countries (Likens *et al.*, 1996; Murdoch *et al.*, 1997).

In the tropical regions of the Southern Hemisphere, several works were carried out. However, they are small in number and are limited in space and time. The most comprehensive discussion, gathering papers of several authors, is the one presented by Rodhe and Herrera (1988), containing the knowledge regarding the acidification potential in five countries (Venezuela, Brazil, Nigeria, China and Australia). In Brazil, despite the fact that the contributions had appeared since the 70s, a detailed view of the acidic deposition over the entire country has not been produced. The reason for that depends on the fact that the problems of acidifying substance deposition is restricted to the more industrialized regions, situated mainly in the south and southeast coast of the country (Lisboa et al., 1992; Luca et al., 1991; Klockow et al., 1997). The knowledge about acidic deposition in Brazil has not changed in the last few years, despite the fact that a greater number of results is available now. The rainfall acidity can be divided into two groups: a) natural acidic rainwater, like the one found in the Amazon region, where the low pH is a result of the presence of organic acids coming from the vegetation (Andreae et al., 1990) and b) rainwater in which the low pH is a result of the presence of acidifying anions (NO_3^- and SO_4^{2-}) in the atmosphere. These are found mainly in the industrialized urban centers like São Paulo and Rio de Janeiro (Forti et al., 1990).

In this work, the available data for the H^+, NO_3^- and SO_4^{2-} concentrations are combined with the rainfall, to estimate their depositions for the Brazilian territory.

2. Available Data to Estimate Deposition in Brazil

Studies concerning rainfall chemistry in the Brazilian territory are scanty and the available ones are shown in Table I. It can be observed that Brazilian industrialized regions (Florianópolis and Porto Alegre) have already achieved high concentration levels of acidifying species, being comparable with the ones found in some polluted regions of the Northern Hemisphere. Even places moderately industrialized (Piracicaba/urban) already present high levels of these species. Except for these contributions, very few advances have been made, in order to achieve a satisfactory regional characterization of these depositions. However, the Cubatão region, State of São Paulo, is an exception because of the alarming environmental problems faced in recent times. Klockow et al. (1997) have shown that, due to the physiographic characteristics of the region, nitrate and sulfate deposition decreased toward the interior of the continent, while the H^+ deposition, neutralized in the polluted region, increases. A little further west, in the region of São Paulo City, recent studies (Forti, 2000a) show that the deposition level of these species increases as a consequence of the presence of large number of industrial plants.

In Brazil, rainfall is characterized by strong annual variations, with a large geographical region in its central area receiving nearly 70% of the total precipitation during the months of heavy rainfall of spring and summer (September to January). Some parts of Brazil, in the south and north region, present higher annual precipitation (> 3000 mm), (Rao et al., 1996).

To estimate the regional deposition of H^+, NO_3^- and SO_4^{2-} the existing data (Table I) were combined with the accumulated annual mean rainfall for the period from 1961 to 1990 (Leal Quadros, 1996). To build up such an estimate, the concentration values were considered as representative for different situations: areas highly/moderately industrialized, rural areas and forest areas. Considering these concentrations and using the mean values of annual precipitation, it was possible to estimate the ion deposition (keq.ha^{-1}.yr^{-1}). The calculated deposition values (Table II) as well as the rainfall (mm) distribution are shown in Figure 1.

It is clear from this figure (Fig. 1), that the combination of high content with high precipitation (>1500 mm) gives also high value of deposition. This is particularly high in the eastern part of the Amazon region, for H$^+$ where the rainfall is high. Thus it is indeed of natural origin. The presence of neutralizing pollutants in some regions induces a low H$^+$ deposition, as observed in regions of Niteroi/RJ, Cubatão/Mogi and Porto Alegre. Regarding the nitrate deposition, the general results presented (Table II) are similar to the values found for remote areas except for some urban sites such as Porto Alegre, Caraguatatuba and Niterói. Nevertheless, in the eastern part of São Paulo State, the deposition is high particularly in Cubatão/Mogi where the highest deposition is found (0.78 keq.ha^{-1}.yr^{-1}) with some regions presenting severe effects on the local ecosystem. Regarding the sulfate, again in the eastern part of São Paulo State the depositions are high particularly for Cubatão/Mogi where the highest value is found. However, differently from the nitrate deposition, São Paulo city does not present the highest deposition. Instead, it is observed that the 2nd and 3rd highest values are found in Florianópolis (Lisboa *et al.*, 1992) and Porto Alegre (Luca *et al.*, 1991). The former is a result of coal mining and the latter due to general pollution of the metropolis.

3. Deposition Pattern in Brazil

The H$^+$ deposition on the Brazilian territory shows values that correspond to low pH. This pattern is observed regionally. When the urban centers are considered it is observed that highly polluted places such as Cubatão/Mogi as well as some non-polluted places such as Natal have lower H$^+$ depositions. Differences between places like these are that in Cubatão/Mogi the H$^+$ is neutralized by the high emission of basic cations (Klockow *et al.*, 1997) combined with high rainfall (2100 mm). In sites like Natal the low H$^+$ deposition is basically due to the lower rainfall (1200 mm).

The nitrate deposition is low over the entire territory except for the two largest industrialized sites (São Paulo and Cubatão/Mogi). In these sites the deposition levels are almost ten times higher than the ones found in most parts of the Brazilian territory.

Concerning sulfate deposition, the regional pattern shows that regions with lower rainfall (< 1500 mm) have deposition values below 0.30 keq.ha^{-1}.yr^{-1}. For the regions with rainfall higher than 2100 mm the sulfate deposition is slightly higher being between 0.31 and 0.54 keq.ha^{-1}.yr^{-1}. However, heavily industrialized urban centers, such as São Paulo and Cubatão Mogi and /or with high number of combustion vehicles such as Porto Alegre, the deposition is very high. Although in Florianópolis the above mentioned anthropogenic activities are not very intense within its urban center, there is an area of coal mining exploitation nearby. It was in this area where the high sulfate deposition value was detected.

Summarizing, the deposition of the main acidic anions and H$^+$ on the Brazilian territory are low except for some restricted sites (São Paulo, Cubatão, Rio de Janeiro and Porto Alegre) that have large number of combustion vehicles or are heavily industrialized or both. As compared with the Brazilian territory dimension, these areas are rather small but presenting the highest population density. In these areas, particular attention to the acidification problems is necessary in order to preserve the water resource quality, to assure its supply to the resident population.

Table I
Concentrations of H^+, NO_3^- and SO_4^{2-} obtained by distinct authors for the Brazilian Territory in different areas.

Local	H^+	NO_3^-	SO_4^{2-}	References
		$\mu eq.L^{-1}$		
Highly/Moderately Industrialized				
Porto Alegre	3.16	7.26	124	Luca et al., 1991
Florianópolis	7.64	-	187	Lisboa et al., 1992
Fortaleza	9.12	1.77	8.33	Moreira-Nordemann et al., 1989
Niterói/Rio de Janeiro	2.88	8.71	31.3	Moreira-Nordemann et al., 1989
Cubatão/Mogi	1.0	29	240	Klockow et al., 1997
São Paulo	41.9	47.9	70.3	Forti, 2000a
Urban				
Natal	5.24	0.65	5.62	Moreira-Nordemann et al., 1989
Caraguatatuba	77.6	6.29	21.0	Moreira-Nordemann et al., 1989
Piracicaba/urbana	36.3	15.0	17.0	Lara et al, 2000
Rural				
Cubatão/Paranapiacaba	6.31	5.6	28	Klockow et al., 1997
Piracicaba/rural	39.7	13.5	12.5	Lara et al, 2000
Forest				
Serra do Navio/Nordeste Amazônia	6.94	2.99	6.42	Forti et al., 2000
Amazônia central	12.9	3.5	9.0	Lesack et al, 1991

Table II
Calculated deposition values for H^+, NO_3^- and SO_4^{2-}. Regions as indicated letters and locals indicated in Fig.1.

Regions	H^+	NO_3^-	SO_4^{2-}	Locals	H^+	NO_3^-	SO_4^{2-}
	$keq.ha^{-1}.yr^{-1}$				$keq.ha^{-1}.yr^{-1}$		
A	0.15	0.06	-	Porto Alegre	0.47	0.11	3.72
B	0.17	0.07	0.31	Florianópolis	0.12	0.08	5.59
C	0.18	0.08	0.38	São Paulo	0.63	0.72	1.05
D	0.23	0.06	0.23	Cubatão/Mogi	0.03	0.78	6.47
E	0.39	0.10	0.54	Cubatão/Paranapiacaba	0.21	0.19	0.92
F	0.35	0.10	0.48	Caraguatatuba	1.63	0.13	0.44
G	0.27	0.07	0.38	Niterói/RJ	0.04	0.13	0.47
H	0.31	0.08	0.43	Natal	0.06	0.01	0.07
I	0.19	0.05	0.27	Fortaleza	0.12	0.08	0.28
J	0.48	0.07	0.30				
K	0.48	0.16	0.25				
L	0.60	0.08	0.37				
M	0.31	0.004	0.03				

1125

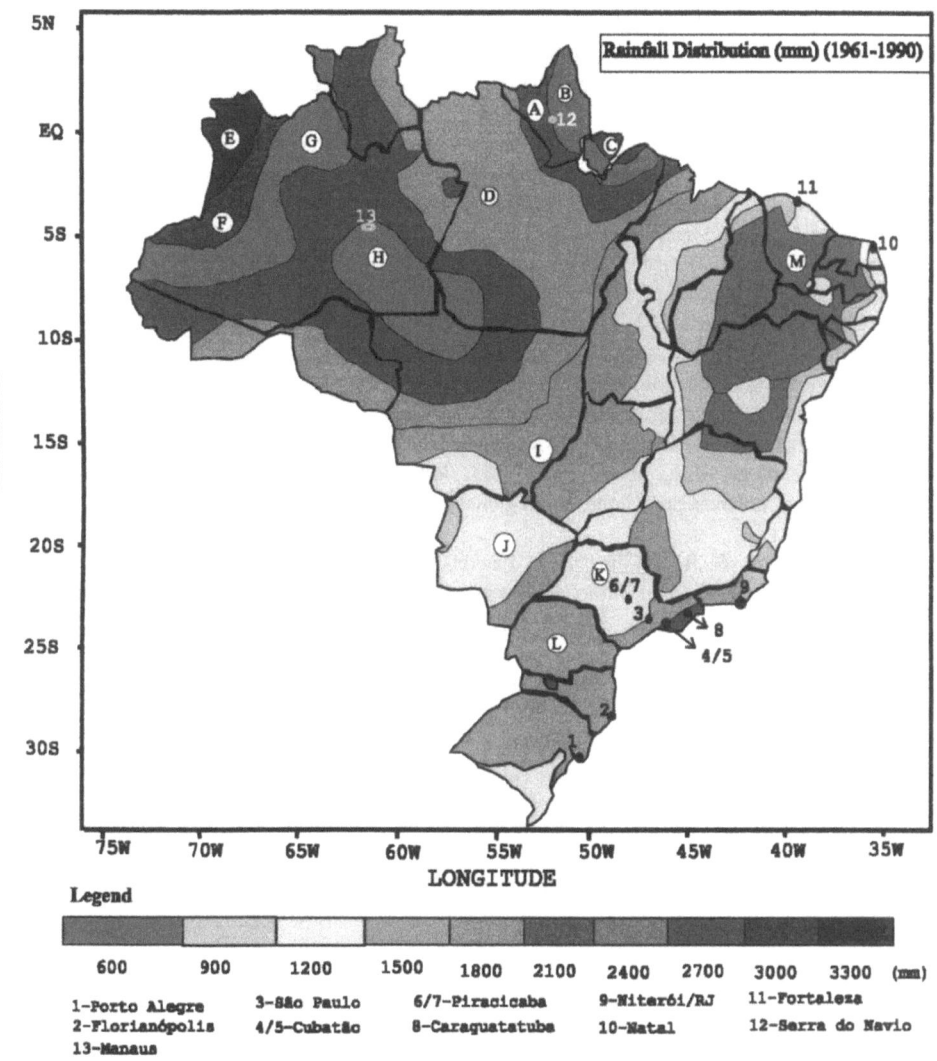

Figure 1. Rainfall distribution in Brazil (time series 1961-1990)

References

Andreae, M. O., Talbot, R. W., Berresheim, H., Beecher, K. M.: 1990, *J. Geophysic Res.*, **95**, 16987.

Forti, M.C., Moreira-Nordemann, L.M., Andrade, M.F., Orsini, C.Q.: 1990, *Atmos. Environ.* **24B**, 355.

Forti, M. C., 2000a. 'Ciclos biogeoquímicos e transferências de espécies químicas nas interfaces de ecossistemas terrestres de Mata Atlântica: estudo de duas áreas contrastantes' FAPESP No.

99/05204-4.
Forti, M. C., Boulet, R., Melfi, A. J., Neal, C.: 2000, *Water, Air and Soil Pollution*, **118**, 263.
Klockow, D., Targa, H. J., Vautz, W.: 1997, 'Air pollution and vegetation damage in the tropics - the Serra do Mar as an example' Final Report, GKSS - Forschungszentrum Geesthacht GmbH, Geesthacht, Germany.
Lara, L.B.L.S., Artaxo, P., Martinelli, L. A., Victoria, R.L., Camargo, P.B., Krusche, A., Ferraz, E.S.B.: 2000, Chemical composition of rainwater and land-use changes in Piracicaba river basin: SW Brazil. Scientific Report, CENA/USP, Brazil.
Leal de Quadro, M.F., Machado, L.H.R., Calbete, S., Batista, N.N.M., Oliveira, G.S.: 1996, Climanálise Especial - Edição Comemorativa, Out., MCT/INPE, pp.90-100.
Lesack, L.F.W. and Melack, J.M.: 1991, *Water Resour. Res.*, **27**,2953.
Likens, G. E.: 1996, *Science*, **272**: 244.
Lisboa, H.M., Costa, R.H.R. and Waltortt, L.M.B.: 1992, *Anais VII Congresso Brasileiro de Meteorologia*, São Paulo, SP, **2**,861.
Luca, S.J., Milano, L.B. and Ide, C.N., 1991: *Wat. Sci. Tech.*, **23**, 133.
Moreira-Nordemann, L. M., Ferreira, C., Magalhães, L. A., Mello, W. Z., Silva Filho, E., Panitz, C. M. N., Santiago, M. F., Souza, C. F.: 1989, *Rainwater Chemistry in the Coast of Brazil*. International Conference on Global and regional Environmental Atmospheric Chemistry. Beijing, China.
Murdoch, P.S., Wall, G. R., Phillips, P. J., Lawrence, G. B., Stoddard, J. l., Wolock, D. M., Hornbeck, J. W.: 1997, *Integrating small watershed and regional-scale data to interpret regional environments trends*. Biogeomon97-*Journal of Conference Abstracts*, **2**,255. Villanova, USA.
NAPAP, 1993. Report to Congress (June), 130pp.
Rao, V.B., Cavalcanti, I.F. A., Hada, K.,: 1996. J. Geophys. Res., **101**,26,539.
Rodhe, H. e Herrera, R. (eds.), 1988. Acidification in Tropical Countries - SCOPE 36. J. Wiley-Sons.

CARBON STOCK, AFFORESTATION AND ACIDIC DEPOSITION: AN ANALYSIS OF INTER-RELATION WITH REFERENCE TO ARID AREAS

SANJAY KUMAR[1*], R. DATTA[2], S. SINHA[1], T. KOJIMA[1],
S. KATOH[1] and M. MOHAN[3]

[1] Faculty of Engg, Seikei University, 3-3-1 Kichijoji Kitamachi, Mushashino-shi, Tokyo 180-8633 Japan; [2] 1453 Fifield Hall, Dept. of Plant Pathology, Univ of Florida, Gainesville, FL 32611-0680, USA; [3] Centre for Atmospheric Sciences, I.I.T Delhi, N. Delhi, India
(* author for correspondence, e-mail: sanjay.kumar@excite.com)

Abstract. Recent advances in desert afforestation underlines its viability and importance in combating global warming and acidification. In this paper, the inter-relation between afforestation, global warming and acid rain has been analyzed. Numerical simulations indicate that afforestation of deserts has distinct advantage as carbon sink and as an important factor for changing microclimate of the region rather than a source of energy. Acidic deposition may well be utilised as fertiliser in nutrient deficit soil of tropical arid areas. However, past trends and projections of acidic deposition in arid areas adjacent to Thar deserts indicate an early efforts are required to cap the opportunity. Delays may contribute towards more incidences of failures.

Keywords: afforestation, acidification, carbon stock, deserts

1. Introduction

The most dicey component in meeting the aspirations of humankind equitably with pro-environmental goals, is "use of energy". Most dreaded environmental problems, such as acid rain and global warming, are attributed to the use of fossil fuels. Emissions responsible for acidic deposition, also contribute towards global warming and vice versa. (Geyer, 1992). Similarly rise in the level of CO_2 also have implications on soil stabilization in ecosystems (Rillig et al., 1999). Even though the legislation to reduce acidic deposition is relatively recent and the deadline is ten years away, the chemical recovery observed (Stoddard et al., 1999) across Europe and North America has been relatively rapid. Now, it is the turn of Asian continent (Foell et al., 1995) which is experiencing population explosion and rapid economic growth.

Nevertheless, afforestation offers one of the most effective and natural remedy. They contain environmental degradation and affect directly the climate on the local, regional, and continental scales by influencing ground temperature, evapotranspiration, surface roughness, albedo or reflectivity, cloud formation, and precipitation. However, there exist serious limitations in terms of, the amount and quality of land available. The United Nations Food and Agriculture Organization estimates that an additional 200 million hectares (494.2 million acres) will be required over the next 30 years just to feed the burgeoning populations of the tropics and sub-tropics. Yet only 93 million hectares are available for farms to expand in these regions. At the same time, progressive desiccation is engulfing steadily fertile lands around arid areas and deserts. The average amount of water vapour in the earth's atmosphere i.e. water evaporated from the ocean becomes rain approximately once in a fortnight. World average rainfall is roughly 740 mm, much more than required for cultivation (500 mm). Since the rate of evaporation is relatively high from the Tropic Ocean, vegetation-induced humidity in arid and desert areas of this region has great potential to attract rainfall in the long run. Nevertheless, nutrient deficient saline soil is likely to be benefited by acidic deposition at the present state. These facts call for concerted efforts to utilize arid and semi-arid areas preferably for afforestation. In this paper, the inter-relation between desert-afforestation, carbon stock and acidic deposition are analysed with special reference to arid areas surrounding Thar desert of India, to achieve sustainable development by correcting the past misgivings.

2. Desert Afforestation – Defining Objectives

2.1. AFFORESTATION AS RENEWABLE ENERGY SOURCE

The first question arises in greening the desert is whether the species selection should be based upon their ability to transform light energy into chemical energy with a re-circulation of materials and energy flow (Kojima, 1998). An analysis of tropical rain forests indicates, less than 1% of the solar energy is stored in the form of biomass. The efficiency of the forests as a place of energy production is thus remarkably low in terms of efficiency of photovoltaic cells, which at present is easily in excess of 10%. One more important fact is that in mature natural forests, the accumulated carbon returns to the earth after the trees die because of fires, volcanic activity, changes in the earth's crust, old age, or disease etc. Consequently the overall production of organic matter (*i.e.*, the net energy accumulation) is zero. Considering the coal age, the present rate of carbon accumulation of carbon or energy by forests might be less than one millionth of the present rate of consumption of biomass based energy.

2.2. AFFORESTATION AND GLOBAL CARBON BALANCE

The role of Vegetation in the carbon balance and carbon cycle consists of the stock of ecosystems (including the living vegetation and soil) and the flow (the exchanges with the atmosphere). An analysis indicates, at present, the steady state no longer exists (Table I). Though forests account for only 34% of the total land surface area, its stock of living matter is extremely large, accounting 90% of the entire world's living stock on land and ocean. The contribution of the tropical rain forests amounts to nearly 62% of the total if dead matter is also included. In contrast to aquatic environments, the time constant for land environments is extremely large. In tropical rain forests, the amount of organic matter is small (especially in the surface layers) and, the rate of decomposition is high compared to boreal and temperate forests.

That is, in cold regions the forest ecosystems are stabilised by the soil, but in tropical forests it is quite likely that loss of the surface would lead immediately to destruction of ecosystem. The average annual rate of loss of true tropical forest has been approximately 10 million hectares, *i.e.*, the rate of carbon released would be ~1.7 gigatonnes per year (Table I). On the other hand, a comparison based on the carbon amounts between the total amount of carbon in tropical rain forests and that in deserts reveals a difference of 21 gigatonnes. This indicates that if the destruction of forests proceeds all the way to desertification, it is possible that the above figure for the rate of carbon release will increase still further.

2.3. AFFORESTATION AS NON-STEADY-STATE CARBON STOCK

There have been advances in the field of desert afforestation. The most important question is the amount of space available and the quantities of CO_2 that can be stocked. The density of the organic matter in tropical forests is over 200 tonnes per hectare (150 t/ha above ground). If deserts could be transformed into tropical forests, then to fix 200t/ha of Carbon, it would be necessary to greenify about 25 millions of hectares each year out of 4.5 billion hectares of deserts. However, if we consider only the residual amounts of carbon left in the atmosphere (i.e., 3.5 gigatonnes per year) and cap destruction of tropical forests, the annual rate of fixation is only 1.5-2.0 gigatonnes. As a result, this approach to store the carbon dioxide would be effective for 600 years. During this time, hopefully, alternative energy would be adequately developed. When considering the carbon dioxide problem, we tend to look only at the amount of carbon dioxide released from fossil fuels and the amount that accumulates in the atmosphere. However, an equal (or even greater) contribution is made by land vegetation. In the final analysis, rather than thinking of forests as a means of regularly absorbing carbon dioxide, they should be regarded as a non-steady-state storehouse of carbon, which makes a large contribution to the carbon dioxide balance.

3. Desert Afforestation And Acidic Deposition

3.1. ACIDIC DEPOSITION IN TROPICAL REGION

Theoretical neutral value of rain water pH value (5.65 at 20 deg. C) is based on the assumption that only Carbon Dioxide is present in the atmosphere and the dissolved CO_2 in rain water is in equilibrium with atmospheric CO_2. However, computations reveal that the neutral pH value of rainwater in tropics should be closer to 7.0. Solubility of CO_2 in water is quite high in tropics (~1.5 times) and the dissolution rate of CO_2 in rainwater is also higher by a factor of ~1.5 times. Experiments indicate, a trace amount of ammonia ($3 \mu g/m^3$) in the atmosphere increases the pH to ~7.0. Rapid cloud formation, meteorological conditions, alkaline soil particles, ambient temperature and intense sunlight also contributes towards higher pH value.

TABLE I
Amounts of carbon stored in ecosystems, the rate of primary production and time constant

Ecosystems	Area (10⁸ ha)	Organic matter (Gt-C)		Density (tC/ha)			Net Production (Gt-C/yr)	Production density (tC/ha/yr)	TC (Yr)	
		Living	Dead	Living	Dead	Total				
Tropical forests	18	270	126	150	70	220	13.6	7.5	20	29
Temperate Forests	12	130	153	110	130	240	7.1	5.9	19	40
Boreal forests	13	110	225	85	175	260	4.3	3.3	26	79
Woodland & Shrublands	8	40	80	50	100	150	2.4	3	17	50
Freshwater & swamps	4	4	80	10	200	210	1.25	3.1	3.2	67
Tropical grasslands	13	7	104	5	80	85	1.95	1.5	3.6	57
Temperate grasslands	9	9	135	10	150	160	2.25	2.5	4	64
Agricultural land	14	14	84	10	60	70	4.2	3	3.3	23
Tundra	8	4	160	5	200	205	0.4	0.5	10	410
Deserts	45	5	26	1	6	7	0.9	0.2	5.5	35
Abandoned land	5	15	40	30	80	110	1.25	2.5	12	44
All land area	149	608	1213	41	81	122	39.6	2.7	15.4	45
Estuaries	1.4	0.63	--	4.5	--	--	1	7.1	0.63	--
Algal beds & coral reefs	2	0.54	--	2.7	--	--	0.7	3.5	0.77	--
Region of upwelling	0.4	0.004	--	0.1	--	--	0.1	2.5	0.04	--
Continental shelf	27	0.12	--	0.045	--	--	4.3	1.6	0.03	--
Oceans	332	0.45	--	0.014	--	--	18.7	0.56	0.02	--
All aqua environment	361	1.74	900	0.048	25	25	24.8	0.69	0.07	36
All regions	510	610	2100	12	41	53	64.4	1.26	9.5	42

3.2. TRENDS OF ACIDIC DEPOSITION IN INDIA

Geographical and climatic conditions are main advantages with India. It also explains observed pH values of 6.7-6.9 off the coast as well as aircraft observations (Manju Mohan and Sanjay Kumar, 1998). Precipitation data collected from ten BAPMoN stations during 1974-1984 as per guidelines of WMO, also indicate that acidic deposition in India should largely remain well under the buffering ability of the soil. Evidently North-West exhibited high pH value due to incursion of sand-dust particles from the Thar Desert. Jodhpur (lat.–26°18', long.–73°01', elevation - 217 m, WMO Index–42339), an arid area adjacent to "Thar Desert, has considerable population density.

The Total Suspended Particle Matter (TSPM) measurement at Jodhpur reveals a significant positive correlation. Regression analysis can be represented by,

$$pH = (6.8550 \pm 0.1216) + (0.0015 \pm 0.0001)\, c$$

Where, c is total TSPM concentration. This indicates that pH value above ~ 6.8 signify contribution of TSPM concentration. However, the region has been experiencing a rapid land-use change. Prevailing meteorological conditions allows transportation and deposition of pollutants from high economic activity zones of Delhi, Bombay and Gujarat with maximum GDP growth rate (< 12%). Table II shows the SO_2 emissions and its projections for the major areas, which affect Jodhpur most. Projections are made on the basis of "Rains Asia Model", assuming application of basic co ntrol technologies to reduce emissions.

3.3. ACIDIC DEPOSITION AND AFFORESTATION IN JODHPUR

In tropics, the average life of leaf is greater than in temperate zone, which indicates that the cumulative effects of acidic deposition could be large enough to cause direct damage. Jodhpur is observed to have received rainfall with a decreasing trend in pH variation (8.70 to 6.35) over the years (Manju Mohan and Sanjay Kumar, 1998). Table III shows the monsoon period (June - September) yearly rainfall weighted mean of pH values. These data are more significant since most of the wet deposition takes place during early monsoon season. Acidic deposition is not a contin uous and equally distributed phenomenon in this region. As the precipitation amount increased, the cations and anions decreased but hydrogen ion concentration increased probably due to washing off of atmospheric constituents in earlier rain.

Ionic composition further indicates maximum neutralizing effect is by Ca^{2+} ion. The correlation between H^+ and SO_4^{2-} was not found to be significant

TABLE II
SO_2 emission from Area and Large Point Sources and projections in Kt

States/Places	1990 LPS	1990 AREA	2000 LPS	2000 AREA	2010 LPS	2010 AREA	2020 LPS	2020 AREA
Bombay	108.8	31.92	203.46	48.69	313.92	66.30	488.64	95.33
Delhi	25.32	19.25	48.04	37.10	105.56	67.53	248.16	96.15
Gujarat	288.8	10.11	404.44	150.6	660.17	294.6	1113.7	547.0

TABLE III
Mean weighted rainfall pH at Jodhpur for the monsoon season

Years	1976	1977	1978	1979	1980	1996	1997
pH value	7.96	8.28	---	7.16	---	6.1*	5.9*

* pH values correspond to the first three shower of the monsoon

indicating that some SO_4^{2-} might be present in the form of salts. Whereas, Ca^{2+} showed significant correlation indicating that it reduces H^+ ion in rainwater, rendering it alkaline. Therefore, it may be concluded that the high level of pH value in Jodhpur is not due to low sulphate and nitrate levels.

Ecosystem vulnerability depends more on the total quantity of sulphates and nitrates it receives, even if they are not acidic at the time of deposition. Projections for Jodhpur were made on the basis of mean sulphur dioxide in Delhi, for which data were collected during 1994-1995 as part of the World Bank Project "Acid Rain Potential in Asia". Average ambient concentration of SO_2 was found to be between 21.71-34.26 μ g/m^3 for Delhi. Corresponding projection ~10 μ g/m^3 for Jodhpur is well below the specified levels. However, future projections of SO_2 emissions and its likely effect on plantations are entirely different (Table II) and may cause high incidences of failures, as reported elsewhere. The ecosystem in and around Thar Desert in general and Jodhpur in particular, is likely to be affected irrecoverably by 2020 due to acidification alone even if other factors are ignored. The overall effect is likely to add to the desertification rather than acting as fertilization.

4. Conclusions

Forests play an important role in the local and global climate system. Desert afforestation, in tropics particularly, has several advantages. However, situation is likely to change in near future. Projections on ecosystem vulnerability for areas adjacent to Thar Deserts indicate acidic deposition would no longer be a positive factor by 2010 and ecosystem vulnerability would shift towards highly critical status due to added stress on account of acidic deposition. This may frustrate late afforestation efforts to contain desertification, global warming and acidification.

References

Foell, W., Amann, M., Carmichael G., Chaduich, M., Hettelingh, J., Hordjick, L. and Dianwa, Z.: 1995 : Report on the World Bank Project, Acid Rain and Emission Reduction in Asia.
Geyer, R.A.:1992, *A Global Warming Forum : Scientific, Economic and Legal Overview*, CRC Press, Inc., Florida, USA.
Kojima, T.:1998, *The Carbon Dioxide Problem.* Gordon and Breach Science Publishers, Japan.
Manju Mohan and Sanjay Kumar.: 1998, *Current Science* **75**(6), 579.
Rillig, C.M., Wright, S.F., Allen, M.F. and Field C.B.:1999 : *Nature* **400**, 628.
Stoddard J.L., et al.: 1999, *Nature* **401**, 575.

MULTI-EFFECT CRITICAL LOADS USED IN MULTI-POLLUTANT REDUCTION AGREEMENTS IN EUROPE

J.-P. HETTELINGH[*], M. POSCH and P.A.M. DE SMET

Coordination Center for Effects (CCE)
National Institute of Public Health and the Environment (RIVM)
P.O.Box 1, NL-3720 BA Bilthoven, The Netherlands
(*author for correspondence, email: jean-paul.hettelingh@rivm.nl)

Abstract. The scientific support of negotiations on emission reductions under the framework of the Convention on Long-range Transboundary Air Pollution of the UN Economic Commission for Europe has been based during the last decade on the integrated assessment of sources, including abatement costs, and risks to receptors (e.g. forests, lakes) quantified by critical loads. The shift from a single-pollutant (sulfur) protocol in 1994 to a multi-pollutant protocol in 1999 necessitated an extension of the methods by which critical loads were computed and mapped. Instead of a single critical load for acidification, methods were now developed to assess the risk of acidifying effects of both sulfur and nitrogen deposition as well as the eutrophying effects of nitrogen on sensitive elements of the environment. Collaboration with a scientific network of 24 national institutions ensured a successful implementation of the proposed methodology across countries. This paper summarizes the methodology, describes the latest input data and presents critical load maps on the basis of which about 98% and 78% of European ecosystems would be protected against acidification and eutrophication, respectively, by the year 2010 according to the multi-pollutant multi-effect protocol.

Keywords: critical loads, acidification, eutrophication, LRTAP Convention, integrated assessment.

1. Introduction

In 1999 a protocol to the 1979 Convention on Long-range Transboundary Air Pollution (LRTAP) of the United Nations Economic Commission for Europe (UN/ECE) to abate acidification, eutrophication and ground-level ozone was signed. The first such protocol, signed in 1994, addressed a single pollutant (sulfur) and a single effect (acidification). The 1999 protocol addresses emissions of sulfur, nitrogen oxides, ammonia and volatile organic compounds (VOCs) simultaneously, while considering multiple effects, i.e. acidification (by sulfur and nitrogen), eutrophication and the formation of tropospheric ozone (by nitrogen oxides and VOCs). The scientific support of this protocol required the computation and mapping of multiple critical thresholds (above which damage may occur), i.e. the critical load for acidification and eutrophication and critical levels for ozone. Results were then used in integrated assessment models, such as RAINS (Schöpp *et al.*, 1999), to identify emission reduction alternatives which limit the exceedance (see Posch *et al.*, 2001) of targeted pollutants over critical thresholds. This paper focuses on methods and data to compute and map European critical loads for the support of the 1999 Protocol.

2. Critical loads data and models

The development of methods to compute critical loads and the collection of appropriate input data are the result of close collaboration between the Coordination Center of Effects (CCE) and the scientific community currently including 24 National Focal Centers (NFCs). This section summarizes the national input data, data reliability and methods to compute critical loads for acidification and eutrophication. Detailed descriptions, including national accounts, and justifications of various simplifying assumptions can be found in Posch *et al.* (1997;1999).

2.1. INPUT DATA

National input data were collected for more than 5 million km^2 ecosystem areas comprising 1.3 million ecosystem data points in 24 participating countries, roughly 50% of the countries' area. Forest ecosystems of remaining countries (about 200,000 km^2) were assessed using European databases on forest cover and soil types. Countries provided data on base cation weathering (BC_w), (sea salt corrected) base cation deposition (BC^*_{dep}), base cation uptake (BC_u), critical leaching of Acid Neutralizing Capacity ($ANC_{le(crit)}$), nitrogen immobilization (N_i), nitrogen uptake (N_u), denitrification fraction (f_{de}), acceptable nitrogen leaching ($N_{le(acc)}$) and, based on these data, provided critical loads of acidification and eutrophication for each of these ecosystems. The mapped ecosystems can broadly be characterized in mixed, deciduous and coniferous forests, peat, grass, heath, natural vegetation and waters. In total 1,055,638 critical load data for forest soils have been submitted to the CCE by the 24 NFCs (about 80% of all critical loads submitted). The density of critical loads mapped varies greatly among countries as illustrated in Figure 1 which shows the total number of ecosystems (black bars) and their total area as a percentage of the country's area (gray bars). For example, critical loads computed for the Netherlands account for about 8% (3,196 km^2) of its land, but the number of ecosystem points (127,269) is high compared to other countries (De Smet and Posch, 1999).

2.2. DATA RELIABILITY

Ranges of national input data for critical load computations vary between countries and are based on field measurements and map information. These variations can be scientifically justified for weathering (based on parent material and soil), base cation deposition (lower in northwest in comparison to southeast Europe), and growth uptake of nutrients (based on tree species, climate and harvesting practices).

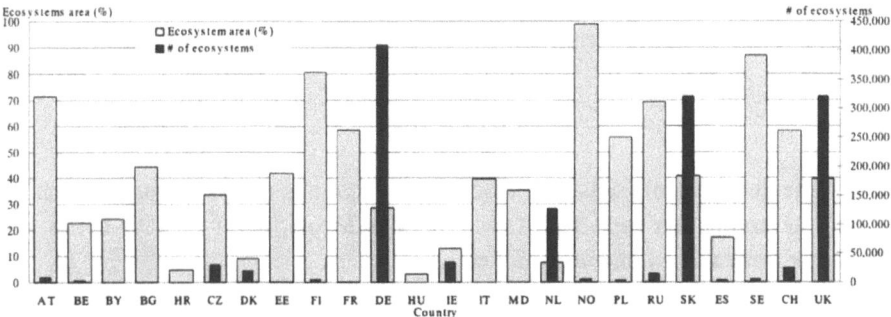

Figure 1. Histogram showing the area for which national critical loads are provided (percentage of total country area; gray shaded bars) and the number of ecosystem points (black bars) per country (indicated by their 2-letter codes).

However, for critical ANC-leaching, N immobilization, and acceptable N leaching the importance of national expert judgement increases. Some countries (CH, DK, SE) report both negative and positive critical ANC leaching values, suggesting that acid deposition should be reduced below net base cation input to allow base cation replenishment in the soil. Regarding N immobilization most countries considered higher values than those recommended in the Mapping Manual (UN/ECE, 1996) which are based on findings of the net immobilization in Swedish soils since the last glaciation (Rosén *et al.*, 1992). These higher values may affect sustainability to an extent where N saturation occurs leading to increased N leaching in the long run. A large variation between countries can be seen with respect to acceptable nitrogen leaching. Although this can be explained by the variation in net precipitation, it is likely that different criteria have been used for the acceptable nitrogen concentration in runoff.

While between-country differences in input data affect the uncertainty of critical loads, the result of a preliminary uncertainty analysis (Hettelingh and Posch, 1997) reveals that the variation of critical loads is not significantly affected by variables for which expert judgement is important. The uncertainty of critical loads in Europe is largely explained by the variation in base cation deposition and weathering. The influence of ANC leaching, including the value of the critical base cation to aluminum ratio, increases when the net base cation input is low, which is predominantly the case in northern Europe.

2.3. CRITICAL LOAD MODELS

The critical loads used in support of the multi-pollutant multi-effect protocol consists of four basic variables (a) the maximum allowable deposition of S, $CL_{max}(S)$, i.e. the highest deposition of S which does not lead to "harmful effects" in the case of zero nitrogen deposition, (b) the minimum critical load of nitrogen, (c) the maximum "harmless" acidifying deposition of N, $CL_{max}(N)$, in the case of zero sulfur deposition, and (d) the critical load of nutrient N, $CL_{nut}(N)$, preventing eutrophication. The equations are summarized as follows (Posch et al., 1999):

$$CL_{max}(S) = BC^*_{dep} - Cl^*_{dep} + BC_w - BC_u - ANC_{le(crit)} \tag{1}$$

equals the net input of (seasalt-corrected) base cations minus a critical leaching of acid neutralization capacity. As long as the deposition of N stays below the minimum critical load of nitrogen, i.e.

$$N_{dep} \leq N_i + N_u = CL_{min}(N) \tag{2}$$

all deposited N is consumed by sinks of N (immobilization and uptake), and only in this case is $CL_{max}(S)$ equivalent to a critical load of acidity. The maximum critical load for nitrogen acidity (in the case of no S deposition) is given by:

$$CL_{max}(N) = CL_{min}(N) + CL_{max}(S)/(1-f_{de}) \tag{3}$$

which not only takes into account the N sinks summarized in Equation 2, but considers also deposition-dependent denitrification. Both S and N contribute to acidification, but one equivalent of S contributes, in general, more to excess acidity than one equivalent of N. Therefore, no unique acidity critical load can be defined, but the combinations of N_{dep} and S_{dep} not causing "harmful effects" lie on the so-called *critical load function* of the ecosystem defined by the three critical loads from eqs.1-3. Examples of this function can be found elsewhere (Hettelingh, et al., 1995; Posch et al., 1999).

Excess nitrogen deposition contributes not only to acidification, but can also lead to the eutrophication of soils and surface waters. Thus a critical load of nutrient nitrogen has been defined (UN/ECE, 1996):

$$CL_{nut}(N) = CL_{min}(N) + N_{le(acc)}/(1-f_{de}) \tag{4}$$

which accounts for the N sinks and allows for an acceptable leaching of N. The four critical load variables from eqs.1-4 were asked to be submitted to the CCE by Parties to the Convention. These variables are the basis for the maps used in the effect modules of the European integrated assessment modeling effort. An important element of integrated assessment is the comparison of one deposition value for N and S to critical loads data in each of the 150×150km² EMEP grid cells. In a single grid cell, however, many (>100,000 in some cases) critical loads have been calculated. Since no unique critical load of acidity can be defined, the concept of cumulative distributions of critical loads data (Hettelingh et al., 1995) has been generalized to critical load functions, and instead of simple percentiles so-called *ecosystem protection isolines* are calculated, which – for given depositions of S and N – allow the determination of the ecosystem area protected in a grid cell (Posch *et al.* 1997; 1999).

3. Results

Figure 2 shows the 5-th percentiles of $CL_{max}(S)$ and $CL_{nut}(N)$, respectively, on the 150×150 km² EMEP-grid as used in the negotiations for the multi-pollutant multi-effect protocol. Figure 2a illustrates that the ecosystems most sensitive to sulfur based acid deposition ($CL_{max}(S)$<400 eq.ha⁻¹.yr⁻¹) are mostly located in the northern and northwestern parts of Europe. These areas are also most sensitive to acidifying nitrogen deposition (Posch *et al.*, 1999). Figure 2b shows that European ecosystems are sensitive to eutrophication ($CL_{nut}(N)$<400 eq.ha⁻¹.yr⁻¹), with some exceptions such as Germany and the United Kingdom where the critical load of nutrient nitrogen exceeds 400 eq.ha⁻¹.yr⁻¹. Maps showing the exceedance of critical loads expected in 2010 as a result of emission reductions prescribed in the multi-pollutant multi-effect protocol can be found in Posch *et al.* (2001). Results of the European integrated assessment effort analyzing the effect of emission reductions according to the 1999 multi-pollutant multi-effect protocol, are a 98% and 78% protection of European ecosystems against acidification and eutrophication, respectively. The history of the evolution of air pollution impacts in general and the excess of $CL_{max}(S)$ since 1960 in particular is also documented in WGE (1999).

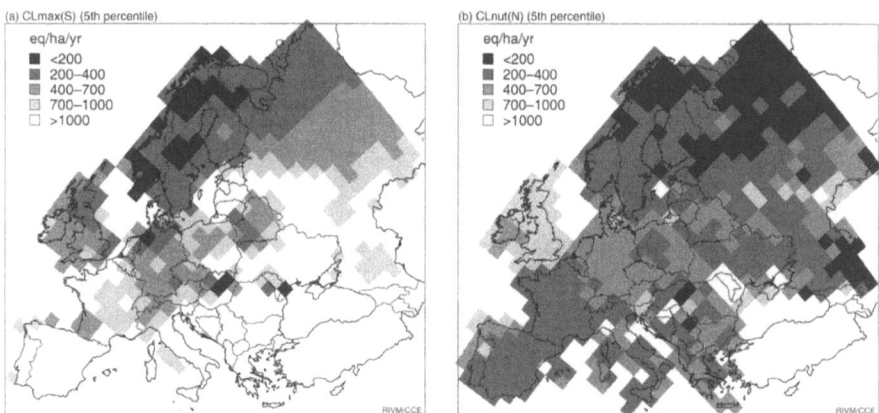

Figure 2. The 5-th percentile of (a) $CL_{max}(S)$ and (b) $CL_{nut}(N)$ on the 150×150 km² EMEP-grid covering Europe.

4. Conclusions and outlook

This paper focussed on the data, maps and modeling methodologies of critical loads as basis for the effects module of integrated assessment models used for the support of UN/ECE-LRTAP negotiations. Two effect based protocols (1994 and 1999) have been signed since the critical load approach was introduced in the nineties as part of the scientific support of effect based emission reduction negotiations. The follow-up of effects based work will target the temporal and spatial assessment of changes in ecosystem health due to protocol driven changes in emissions. For this, the further development and European-wide implementation of dynamic ecosystem models will be required. Other elements of the future effects-based UN/ECE program may include critical thresholds of other pollutants (e.g. heavy metals) and human health related endpoints

Acknowledgements

This work is supported by the Directorate for Climate Change and Industry of the Dutch Ministry of Environment. National Focal Centers and Working Groups under the LRTAP Convention of the UN/ECE and its Secretariat are acknowledged for the collaboration enabling the European assessment presented in this paper.

References

De Smet, P.A.M. and Posch, M.: 1999, Summary of national data. In: Posch *et al., op. cit.*
Hettelingh, J.-P., Posch, M., de Smet, P.A.M. and Downing, R.J.: 1995, *Water Air Soil Pollut.* **85**: 2381-2389.
Hettelingh, J.-P. and Posch, M.: 1997, An analysis of critical load and input data variability. In: Posch *et al., op. cit.*
Posch M., Hettelingh, J.-P., de Smet, P.A.M. and Downing, R.J. (eds): 1997, Calculation and Mapping of Critical Thresholds in Europe. CCE Status Report 1997, RIVM, Bilthoven, The Netherlands, 163 pp.
Posch M., de Smet, P.A.M., Hettelingh, J.-P. and Downing, R.J. (eds): 1999, Calculation and Mapping of Critical Thresholds in Europe. CCE Status Report 1999, RIVM, Bilthoven, The Netherlands, 165 pp.
Posch, M., Hettelingh, J.-P. and de Smet, P.A.M.: 2001, Characterization of critical load exceedances in Europe. *Water Air Soil Pollut.* (this volume).
Rosén, K., Gundersen, P., Tegnhammar, L., Johansson, M. and Frogner, T.: 1992, *Ambio* **21**: 364-368.
Schöpp, W., Amann, M., Cofala, J., Heyes, C., Klimont, Z.: 1999, *Environmental Modelling & Software*, **14**(1): 1-9.
UN/ECE: 1996, Manual on Methodologies and Criteria for Mapping Critical Levels/Loads, Federal Environmental Agency (Umweltbundesamt), Texte 71/96, Berlin, Germany.
WGE: 1999, Trends in Impacts of Long-range Transboundary Air Pollution, Technical report prepared by the ICPs of the Working Group on Effects (of the UN/ECE LRTAP Convention), ISBN 1 870393 52 X, Centre for Ecology and Hydrology, Monks Wood, UK.

CHARACTERIZATION OF CRITICAL LOAD EXCEEDANCES IN EUROPE

M. POSCH*, J.-P. HETTELINGH and P.A.M. DE SMET

Coordination Center for Effects (CCE)
National Institute of Public Health and the Environment (RIVM)
P.O.Box 1, NL-3720 BA Bilthoven, The Netherlands
(*author for correspondence, email: max.posch@rivm.nl)

Abstract. The excess of acidic and eutrophying depositions over critical loads (critical load exceedances) is considered a measure for the risk of harmful effects on sensitive elements of the environment. The magnitude and the geographical distribution of critical load exceedances over Europe vary with the extent to which national emissions of sulfur dioxide, nitrogen oxide and ammonia are reduced. The scientific support of negotiations on emission reductions in the framework of the Convention on Long-range Transboundary Air Pollution (LRTAP) of the UN Economic Commission for Europe has been based on the integrated assessment of sources, including abatement costs, and risks to receptors (e.g. forests, lakes) using critical load exceedances. The shift from a single-pollutant (sulfur) protocol in 1994 to a multi-pollutant protocol in 1999 necessitated an extension of the methods by which critical load exceedances are computed and mapped. The focus changed from the protection of the most sensitive ecosystem against excessive deposition of one pollutant, to an assessment of the accumulated exceedance by more pollutants of all ecosystems. This paper presents and compares the different characterisations ("gap-closure methods") used in those negotiations. It is shown that the approach finally used has several appealing features, but treats the exceedance as a linear damage function, thus going beyond the critical load definition as a simple on-off limit value.

Keywords: critical loads, acidification, eutrophication, LRTAP Convention, integrated assessment.

1. Introduction

In December 1999 the "Protocol to Abate Acidification, Eutrophication and Ground-level Ozone" to the 1979 Convention on Long-range Transboundary Air Pollution (LRTAP) of the United Nations Economic Commission for Europe (UN/ECE) was signed in Gothenburg, Sweden. This is the second protocol in which the reduction of air pollutant emissions is based on ecosystem effects, in addition to economic and technological considerations of emission reduction potentials. The first such protocol, signed in 1994, addressed only a single pollutant (sulfur) and a single effect (acidification). The 1999 protocol is more complex because it addresses emissions of sulfur, nitrogen oxides, ammonia and volatile organic compounds (VOCs) simultaneously and considers multiple effects, i.e. acidification (by sulfur and nitrogen), eutrophication (by nitrogen) and the effects of tropospheric ozone (by nitrogen oxides and VOCs). The single critical load of (sulfur) acidification was extended to include acidifying nitrogen, leading to the so-called critical load function. Other critical thresholds, such as the critical load for eutrophication and critical levels for ozone, were newly developed. Critical loads are incorporated into integrated assessment models, e.g., the RAINS model (Schöpp et al., 1999). These models seek cost-minimal emission reductions, subject to limits on the exceedance of targeted pollutants over critical thresholds. Therefore, exceedance computations became important, and several methods were designed to compute the exceedance of sulfur and nitrogen deposition over critical loads. This paper describes these methods and summarizes their advantages and disadvantages.

2. Critical Loads and their Exceedances

For the work under the LRTAP Convention critical loads have been defined at the end of the 1980s (Nilsson and Grennfelt, 1988). For the support of the 1999 multi-pollutant multi-effect Gothenburg Protocol critical loads have been computed and mapped for nutrient nitrogen, to avoid eutrophication, and for acidity, defining the maximum depositions of S and N not leading to "harmful effects" due to acidification.

While the critical load of nutrient nitrogen is a single number, $CL_{nut}(N)$, acidity critical loads for a given ecosystem are defined by a so-called critical load function, characterized by three quantities: $CL_{max}(S)$, $CL_{min}(N)$ and $CL_{max}(N)$. These quantities are computed from ecosystem parameters, e.g., with the so-called SMB-model, and the methods can be found in the Mapping Manual (UN/ECE, 1996) and are also summarized in Hettelingh et al. (1995; 2001). In Figure 1 an example of a critical load function is depicted. The gray-shaded area indicates combinations N and S depositions (N_{dep}, S_{dep}) not leading to the exceedance of critical loads.

In integrated assessment models, deposition estimates are compared to critical loads in the course of scenario analyses or optimization runs. Under the LRTAP Convention, deposition fields are provided by the "Cooperative Programme for Monitoring and Evaluation of the Long-Range Transmission of Air Pollutants in Europe" (EMEP) in 150×150 km^2 grid covering Europe. Since there can be a large number of ecosystems within one grid cell (up to 100,000; see Posch et al., 1999) critical load information has to be "summarized". This is done by constructing the cumulative distribution function (CDF) of the critical load values in each grid cell. This is used to compute a desired statistical quantity (e.g., percentile), which is then mapped and compared to deposition. Once a desired percentile critical load, CL, is computed, its *exceedance* is defined as the non-negative difference between it and the deposition: $Ex=\max\{Dep-CL, 0\}$. It was the 5-th percentile in every European grid square which was used in the integrated assessment of the 1994 Sulphur Protocol (Hettelingh et al., 1995). To compute the *protection percentage* of the ecosystems, i.e. the percent of ecosystem area within a grid cell for which the respective critical loads are not exceeded, instead of a single percentile, *all* critical load values in a grid cell have to be used.

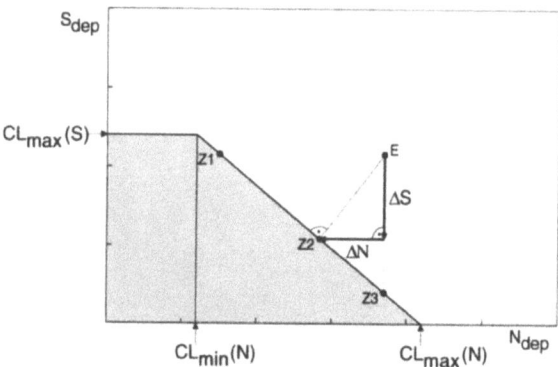

Figure 1. Example of a critical load function for S and acidifying N defined by the quantities $CL_{max}(S)$, $CL_{min}(N)$ and $CL_{max}(N)$. No unique exceedance can be defined: Let the point E denote the current deposition of N and S. By reducing N_{dep} substantially point Z1 is reached and thus non-exceedance without reducing S_{dep}; non-exceedance can also be reached by reducing S_{dep} only (by a smaller amount) until reaching Z3. For the protocol negotiations an exceedance has been *defined* as the sum of N_{dep} and S_{dep} reductions ($\Delta N+\Delta S$) needed to reach the critical load function on the shortest path (point Z2).

When comparing deposition scenarios with critical loads it appeared that non-exceedance could not be reached everywhere in Europe. Thus it was decided to use percentage reductions of the excess depositions, so-called *gap closures*, for the derivation of reduction scenarios. This is illustrated in Figure 2: the thick solid and the thick broken lines are two examples of critical load CDFs (with the same 5-th percentile critical load, indicated by 'CL'). 'D0' indicates the (present) deposition, and the difference between 'D0' and 'CL' is the exceedance. It was decided to reduce the exceedance everywhere in Europe by a fixed percentage, i.e. to "close the gap" between (present) deposition and (5-th percentile) critical load. In Figure 2a a *deposition gap closure* of 60% is shown as an example. It shows that a fixed deposition gap closure can result in very different ecosystem protection percentages (55% vs. 22%), depending on the shape of the CDF.

To account for all critical loads within a grid cell (not only the 5-th percentile), one could use an *ecosystem area gap closure* instead. This is illustrated in Figure2b: For a given deposition 'D0' the ecosystem area unprotected, i.e. with deposition above the critical loads, can be read from the vertical axis. After agreeing to a (percent) reduction of the unprotected area (e.g. 60%) it is easy to compute the required deposition reductions ('D1' and 'D2'). The area gap closure becomes problematic, however, if only a few critical load values are given for a grid cell. In such a case the CDF has large discontinuities and (small) changes in deposition may result in either no increase in the protected area at all or large jumps in the area protected.

To remedy the problem with the area gap closure caused by discontinuous CDFs, the *average accumulated exceedance* (AAE) has been introduced. It is defined as $AAE = (A_1Ex_1+...+A_nEx_n)/(A_1+...+A_n)$, where A_i is the area of the i-th

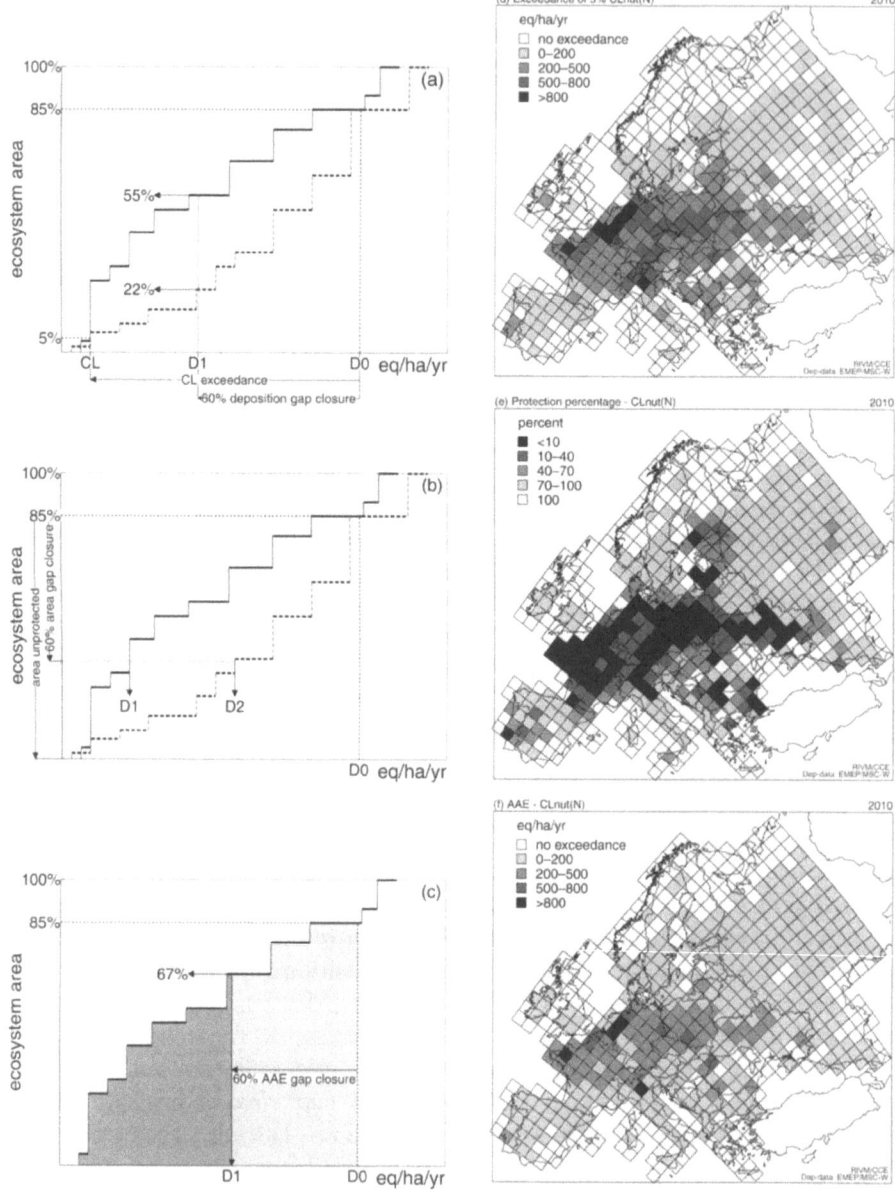

Figure 2. Left: CDFs (thick solid lines) of critical loads illustrating different gap closure methods: (a) deposition gap closure, (b) ecosystem area gap closure, and (c) average accumulated exceedance (AAE) gap closure. The thick dashed lines in (a) and (b) depict another CDF, illustrating (i) how different ecosystem protection follows from the same deposition gap closure, or (ii) how different deposition reductions are needed to reach the same protection level. Right: (d) exceedance of the 5-th percentile, (e) percent ecosystem area protected, (f) AAE of $CL_{nut}(N)$, all for the 2010 Protocol N deposition.

ecosystem in a grid cell and Ex_i its exceedance ($i=1,...,n$). The AAE is the area under the CDF of the critical loads (the entire grey-shaded area in Figure 2c). Deposition reductions have been negotiated in terms of an *AAE gap closure*: a 60% AAE gap closure is achieved by a deposition 'D1' which reduces the total grey area by 60%, resulting in the dark grey area. The biggest advantage of the AAE is that it varies smoothly with deposition, even for discontinuous CDFs, thus facilitating optimisation calculations in integrated assessments.

The three concepts introduced above (Fig.2a-c) are illustrated for the critical load of nutrient nitrogen in the corresponding maps in Figures 2d-f for the year 2010, assuming the full implementation of the 1999 Protocol. They show that in the case of nutrient N large areas will still be exceeded in 2010, and further reductions in N emissions are needed to protect the majority of European ecosystems from the eutrophying effects of nitrogen.

CDFs have been generalised for critical load functions of acidity, and instead of percentiles so-called *ecosystem protection isolines* are calculated, which allow the determination of the ecosystem area protected in a grid cell for any given depositions of S and N. As illustrated in Figure 1, no unique exceedance can be found in case of two pollutants, and thus the concept of deposition gap closure had to be abandoned. Instead, the advantages of the AAE concept was applied to the simultaneous reduction of S and N deposition by defining the exceedance as shown in Figure 1 ($Ex_i=\Delta N+\Delta S$). Details on the definitions and calculation methods can be found in (Posch *et al.*, 1999). Figure 3 shows the AAE for depositions of N and S for 1990 and the year 2010 after implementation of the 1999 Protocol, indicating that with respect to acidity large improvements will potentially be made over the next ten years.

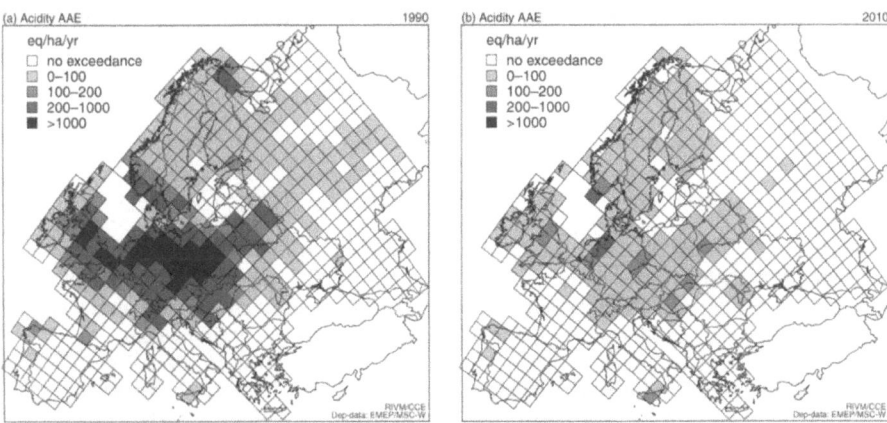

Figure 3. Average accumulated exceedance (AAE) of acidity critical loads on the 150×150 km² EMEP grid covering Europe computed with N and S deposition from (a) 1990, and (b) 2010 after full implementation of the 1999 Protocol.

3. Summary and Discussion

The advantages and disadvantages (shortcomings) of the three gap closure methods described above can be summarized as follows:

Method	Advantages	Disadvantages/Shortcomings
Deposition gap closure: (used for the 1994 Sulphur Protocol)	• Easy to use even for discontinuous CDFs (e.g. grid cells with only one CL)	• Takes only one CL value (percentile) into account • May result in no increase in protected area • Difficult to define for ≥2 pollutants
Ecosystem area gap closure:	• In line with CL definition • Easy to apply to any number of pollutants	• Difficult (or even impossible) to define a gap closure for discontinuous CDFs (e.g. grid cells with only one ecosystem)
Accumulated Exceedance (AE) gap closure: (used for the 1999 Protocol)	• AE (and AAE) is a smooth and convex function of depositions even for discontinuous CDFs	• AE interpreted as linear damage function (beyond critical load definition) • Definition not unique for ≥2 pollutants

It is the accumulated exceedance that was ultimately used for the 1999 Protocol, since it combines the advantages of accounting for all critical load values in a grid cell and has favorable mathematical properties facilitating its use in the integrated assessment of several pollutants.

Acknowledgements

This work is supported by the Directorate for Climate Change and Industry of the Dutch Ministry of Environment. EMEP/MSC-W, National Focal Centers and Working Groups under the LRTAP Convention of the UN/ECE and its Secretariat are acknowledged for their collaboration.

References

Hettelingh, J.-P., Posch, M., de Smet, P.A.M. and Downing, R.J.: 1995, *Water Air Soil Pollut.* **85**: 2381-2389.

Hettelingh, J.-P., Posch, M. and de Smet, P.A.M.: 2001, Multi-effect critical loads used in multi-pollutant reduction agreements in Europe. *Water Air Soil Pollut.* (this volume).

Nilsson, J. and Grennfelt, P. (eds): 1988, Critical Loads for Sulphur and Nitrogen. Nord 1988:97, Nordic Council of Ministers, Copenhagen, Denmark, 418 pp.

Posch M., de Smet, P.A.M., Hettelingh, J.-P. and Downing, R.J. (eds): 1999, Calculation and Mapping of Critical Thresholds in Europe. CCE Status Report 1999, RIVM, Bilthoven, The Netherlands, 165 pp.

Schöpp, W., Amann, M., Cofala, J., Heyes, C. and Klimont, Z.: 1999, *Environmental Modelling & Software*, **14**(1): 1-9.

UN/ECE: 1996, Manual on Methodologies and Criteria for Mapping Critical Levels/Loads. Federal Environmental Agency (Umweltbundesamt), Texte 71/96, Berlin, Germany.

ESTIMATION OF THE MAXIMUM CRITICAL LOAD FOR SULFUR IN SOUTH KOREA

SOON-UNG PARK* and YOUNG-HEE LEE

School of Earth and Environmental Sciences, Seoul National University, Seoul, 151-742, Korea
(*author for correspondence, e-mail: supark@snupbl.snu.ac.kr)

Abstract. The maximum critical load of sulfur and its exceedance by the sulfur deposition of 1994-1997 were mapped for South Korea with a spatial resolution of 11 × 14 km using the steady-state mass balance method. The Korean soil and geological maps were used as basis for the estimations of the critical alkalinity leaching and the weathering rate of base cations. The normalized difference vegetation index data obtained from the Advanced Very High Resolution Radiometer (AVHRR) together with the observed primary productivity of plants were used for the estimation of the critical uptake of base cations. Wet deposition of the non-sea-salt base cations was derived from measured base cation concentrations in precipitation, precipitation rate and air concentration of total suspended particulate while dry deposition of base cations was estimated using the inferential technique using scavenging ratios. The predominant ranges of base cation weathering, uptake and deposition were estimated to be of 200 - 600 eq ha^{-1} yr^{-1}, 200 - 400 eq ha^{-1} yr^{-1} and 400 - 600 eq ha^{-1} yr^{-1}, respectively. Critical alkalinity leaching was mainly in the range of 1000 - 2000 eq ha^{-1} yr^{-1} due to relatively high value of precipitation runoff. Exceedance of sulfur critical load was found at 40 % of the ecosystems considered mainly in the southeastern part of Korea, and about 60 % of Korea ecosystems were sustainable against sulfur acidity loadings.

Keywords : Exceedance, Steady-state mass balance, acidification, Asia

1. Introduction

Acidification in East Asia may be viewed as prototype of emerging environmental problems in an area of rapid economical development during the last decades. The increase of acidifying air pollutant emissions is expected to continue for the next several years due to the accelerated use of fossil fuel burning systems planned in many Asian nations. Therefore the environmental situation may dramatically exacerbate in future. Consequently, counter measures are required to keep the development sustainable which includes avoiding deterious effects on ecosystems in this region.

To assess the impact of acidifying deposition on sensitive ecosystems, we should discuss the relevance of different levels of deposition. One point of reference is the critical load concept (Nilsson and Grennfelt, 1988). The difference between the actual deposition and the critical load is called exceedance. Exceedance of regional critical loads indicates areas where damage to ecosystem structure and/or function is expected.

The purpose of this paper is to report the first application of the steady-state mass balance method to South Korea using national data in South Korea. The maximum critical load for sulfur is mapped and ecosystems under risk are identified by means of the exceedance of the critical load by current sulfur loads.

2. The steady state mass balance model

The maximum critical load for sulfur is given by (Posch et al., 1995):

$$CL_{max}(S) = BC_{dep} - Cl_{dep} + BC_w - BC_u - Alk_{le(crit)} \quad (1)$$

where BC is the sum of base cations (BC=Ca+K+Mg+Na) and $Alk_{le(crit)}$ is the critical alkalinity leaching and subscripts dep, w and u and le, respectively, stand for deposition, weathering, the net growth uptake and the leaching.

The equation is obtained by simplifying the acidity balance for an assumed homogeneous soil compartment considering the main sources and sinks of sulfur and nitrogen. For details see de Vries (1992) and Posch et al.(1995)

3. Calculation of critical load for sulfur in South Korea

Sulfur critical loads calculations basically require the regional quantification of cation fluxes from deposition, vegetation uptake and leaching.

The spatial interpolation of the wet deposition flux of non-sea-salt base cations (Ca^{2+}, Mg^{2+}, and K^+) was obtained from an optimum regression equation between the measured base cation concentration in precipitation from six wet deposition monitoring sites, precipitation rate from 63 meteorological sites and the air concentration of total suspended particulate from 25 sites scattered over South Korea. Dry deposition flux of non-sea-salt base cation was estimated using the inferential method with estimated air concentration of non-sea-salt base cations and the parameterized dry deposition velocity using micrometeorological measurements. For details see Park and Lee(2000).

Base cation weathering for various parent rock-soil type associations was estimated using the dominant 27 different parent rocks and 12 different soil types in South Korea with the available weathering rate data reported in literature and special experiments (KIGAM, 1995; UBA, 1996; Fumoto and Iwama, 1996; Shindo et al., 1995). The geological map(KIGAM, 1995) and the Korean soil map with the map scale of 1: 1,000,000 in Korea are used for the analysis of the distribution of parent rocks and soil types with a grid resolution of 11×14km.

Uptake of base cation was estimated based on the primary productivity of plant communities obtained from previous study results (Hong and Nakagoshi, 1996; Hong, 1998). The harvested biomass uptake in South Korea is less than 5% of the growth uptake. The net uptake of base cation was calculated for all forests regardless of the harvested or not harvested. The vegetation distribution in South Korea was obtained from the normalized difference Vegetation index(NDVI) data measured by the Advanced Very High Resolution Radiometer(AVHRR). The eight land-use types including mixed vegetation, agriculture and bush, coniferous, crop, evergreen, deciduous, mountain shrub and pasture shrub were obtained from the averaged AVHRR data for three years from 1991 to 1993.

The critical alkalinity leaching in equation (1) is calculated as (e.g., Posch *et al.*, 1995)

$$AlK_{le(crit)} = -Al_{le(crit)} - H_{le(crit)} = -Q([Al]_{crit} + [H]_{crit}) \quad (2)$$

where the square bracket denotes concentration (in eq m^{-3}) and Q is the precipitation runoff. Practically, the relationship between [H] and [Al] is described by a gibbsite equilibrium such that (e.g., Posch *et al.*, 1995)

$$[Al] = K_{gibb}[H]^3 \text{ or } [H] = ([Al]/K_{gibb})^{1/3} \quad (3)$$

where K_{gibb} is the gibbsite equilibrium constant. To obtain a critical alkalinity leaching we assume $[Al]_{crit}$ to be 0.2 eq m^{-3} (Hettelingh *et al.*,1992) and K_{gibb} to be given in UBA(1996). The precipitation runoff Q is calculated by the estimation of evapotranspiration using Linacre method (1977) with precipitation rates observed from 63 sites over South Korea for four years(1994-1997).

4. Results

Fig. 1 shows the spatial distribution of the weathering rate of base cations over South Korea. The weathering rates range widely from 90 eq ha^{-1} yr^{-1} to 2250 eq ha^{-1} yr^{-1}. However, 80% of the weathering rates fall between 200 - 400 eq ha^{-1} yr^{-1}.

Fig. 2 shows the spatial pattern of critical alkalinity leaching over South Korea. The lowest values of critical alkalinity leaching occur in the southeastern part of Korea where precipitation rates are low. 75% of the area analyzed has critical alkalinity leaching values between 1000-2000 eq ha^{-1} yr^{-1} while values larger than 2000 eq ha^{-1} yr^{-1} are only found at 0.1% of the analysis domain.

Fig. 3 shows the spatial pattern of mean annual total deposition of non-sea salt base cations over South Korea. The annual mean total deposition is about 420 eq ha^{-1} yr^{-1} with the predominant range of 400 to 600 eq ha^{-1} yr^{-1}.

The spatial pattern of base cations uptake is given in Fig. 4. The largest uptake of base cations is found in the southeastern and southwestern parts of the Korean peninsula. More than 85% of South Korea territory shows estimated

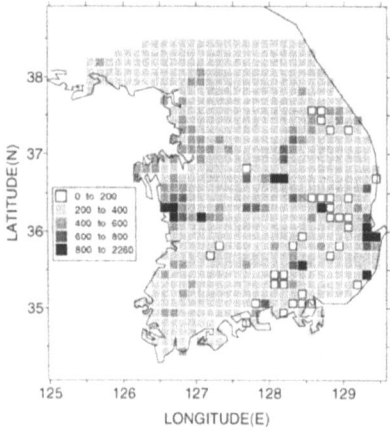

Figure 1. Geographical distribution of base cation weathering rate (eq ha^{-1} yr^{-1}) in South Korea

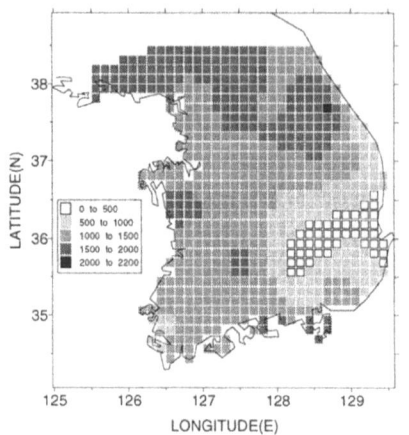

Figure 2. Geographical distribution of critical alkalinity leaching (eq ha^{-1} yr^{-1}) in South Korea

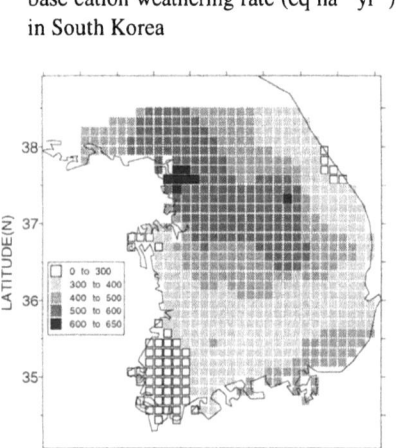

Figure 3. Geographical distribution of base cation deposition rate (eq ha^{-1} yr^{-1}) in South Korea

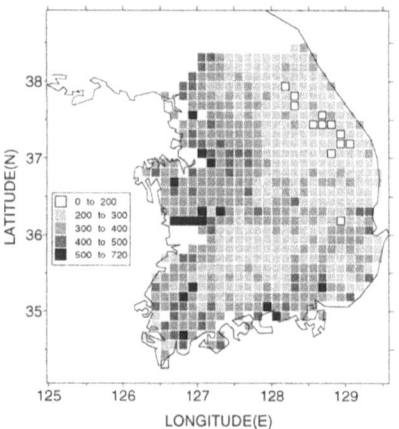

Figure 4. Geographical distribution of base cation uptake in Korean ecosystems (eq ha^{-1} yr^{-1})

base cation uptake rates between 200 and 400 eq ha^{-1} yr^{-1}.

The maximum critical load for sulfur (Fig. 5) is low (< 1000 eq ha^{-1} yr^{-1}) in the southeastern part of Korea, while high critical loads are seen in the northeastern part of Korea. More than 45% of Korean ecosystems considered have maximum critical load of sulfur between 1500 eq ha^{-1} yr^{-1} and 2000 eq ha^{-1} yr^{-1}. These values are quite different from those estimated by the RAIN-ASIA model (World Bank, 1994) covering a range of 200 to 2000 eq ha^{-1} yr^{-1} with predominant values of 200 - 500 eq ha^{-1} yr^{-1}. The differences might be attributed to more detailed and comprehensive national data sets of geology, soil, vegetation and meteorology used in this study.

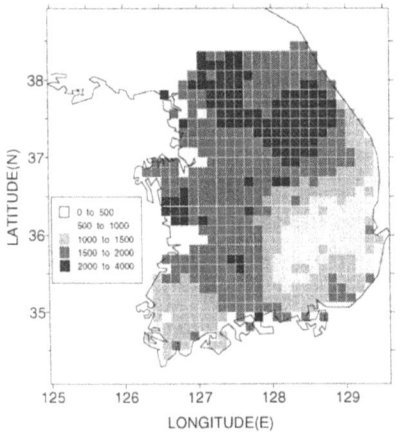

Figure 5. Geographical distribution of the maximum critical load for sulfur (eq ha^{-1} yr^{-1})

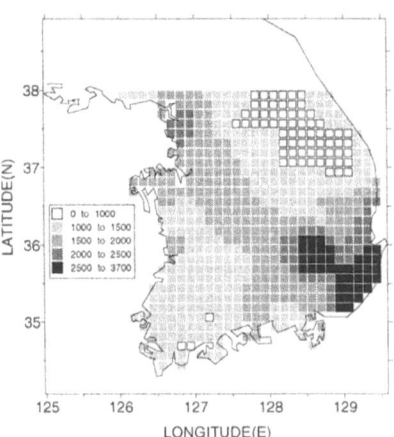

Figure 6. Mean annual total sulfur deposition (eq ha^{-1} yr^{-1}) averaged for 4 years from 1994 to 1997

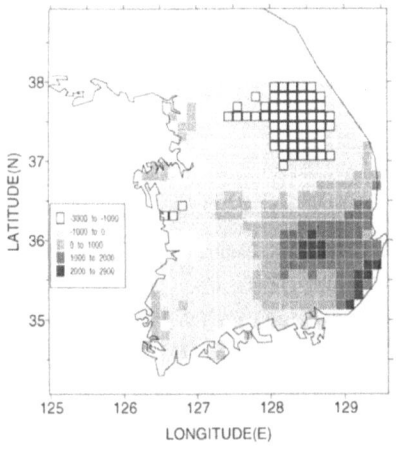

Figure 7. Geographical distribution of exceedance of sulfur (eq ha^{-1} yr^{-1})

Fig. 6 shows the distribution of the annual mean sulfur deposition averaged for the years 1994 to 1997 as estimated by Park *et al.*(2000). Exceedance of the maximum sulfur critical load is displayed in Fig. 7. Areas of high exceedance ($\geq$ 1000 eq ha^{-1} yr^{-1}) are found over southeastern part of Korea. Areas where the maximum critical load of sulfur are not exceeded are located in the northern central and southwestern parts of Korea, where the ecosystem are not adversely affected by present sulfur deposition. Ecosystems where sulfur deposition exceeds the critical load of sulfur cover an area of about 40 % of Korea, suggesting that countermeasures are required to maintain ecosystem sustainability.

5. Conclusions

The estimated base cation weathering (predominantly 200-600 eq ha^{-1} yr^{-1}), uptake (mainly 200- 400 eq ha^{-1} yr^{-1}) and deposition(mainly 400-600 ha^{-1} yr^{-1}) rates are found to be relatively low, while relatively high values of critical alkalinity leaching(predominantly 1000-2000 eq ha^{-1} yr^{-1}) due to high precipitation runoff are observed. Consequently, a large percentage of the maximum critical load values (more than 70 %) is in the range of 1000 - 2000 eq ha^{-1} yr^{-1}. Moreover, about 18 % of ecosystems mainly located in the northeastern part of Korea have maximum critical load of sulfur values larger than 2000 eq ha^{-1} yr^{-1}. High values of exceedance of the sulfur critical load (> 1,000 eq ha^{-1} yr^{-1}) occur mainly in the southeastern part of Korea where for the reference years 1994 to 1997 the highest sulfur deposition rates are found. Moderate exceedances are found in a narrow belt ranging from the northwestern to the southeastern parts and the southwestern tip of Korea. The exceedances of the maximum critical load for sulfur are found to be about 40 % of Korean ecosystems. Area with no exceedance are predominantly found in the northern and southwestern parts of Korea.

Acknowledgements

This research is partially supported by Ministry of Environment of Korea under the G-7 Project and Brain Korea 21 project supported by Ministry of Education of Korea. Special recognition goes to anonymous reviewers for their helpful comments.

References

De Vries, W.: 1992, Methodologies for the assessment and mapping of critical loads and of the impact of abatement strategies on forest soils. The Winand Staring Centre for Integrated Land, Soil and Water Research, Report 46, The Netherlands, 109pp.

Fumoto T. and Iwama H.: 1996, 'Sulfate adsorption and chemical weathering in volcanic ash soil', *International Symposium on Acid deposition and Its Impacts*, NIES, Tsukuba, pp 345-348

Hettelingh, J.-P., Posch, M. and de Smet P.A.M.: 1992, Mapping Vademecum. Report No. 259101002, RIVM, Bilthoven, The Netherlands, 33pp.

Hong, S.-K.: 1998, *Phytocoenologia* **28(1)**, 45.

Hong, S.-K. and Nakagoshi, N.: 1996, *Korean Journal of Ecology* **19(4)**, 305.

KIGAM, 1995, Geological map of Korea.

Linacre, E. T.: 1977, *Agricultural Meteorology* **18**, 409

Nilsson, J. and Grennfelt, P.: 1988, Critical loads for sulfur and nitrogen. Milj rapport 1988-15, Nordic Council of Ministers, Copenhagen.

Park, S.-U.: 1998, *Journal of Applied Meteorology* **37**, 486.

Park, S.-U., In, H.-J., Kim, S.-W. and Lee, Y.-H.: 2000, *Atmospheric Environment* **34**, 3209

Park, S.-U. and Lee, Y.-H.: 2000, *Water Air and Soil Pollution*. (In press)

Posch, M., de Smet, P.A.M., Hettelingh, J.-P. and Downing, R.J. (eds), 1995. Calculation and mapping of critical thresholds in Europe. Coordination Center for Effects, RIVM Report No. 259101004, 198pp.

Shindo, J, Bregt, A. K. and Hakamata, T.: 1995, *Water Air and Soil Pollution* **85**, 2571

World Bank, 1994. RAINS/ASIA. User's Manual, IISAA, Washington, 138pp.

REGIONAL AIR POLLUTION AND CLIMATE CHANGE IN EUROPE: AN INTEGRATED ASSESSMENT (AIR-CLIM)

PETRA MAYERHOFER[1], JOSEPH ALCAMO[1],
MAXIMILIAN POSCH[2] and JELLE G. VAN MINNEN[2]

[1] *Center for Environmental Systems Research, University of Kassel, Kurt-Wolters-Str. 3, D-34109 Kassel, Germany;* [2] *National Institute of Public Health and the Environment (RIVM), P.O. Box 1, NL-3720 BA Bilthoven, The Netherlands*

Abstract. The aim of the AIR-CLIM project is to perform an integrated analysis of the linkages between climate change and regional air pollution in Europe and to produce results that are relevant to European policy-making. Key elements of the analysis are on the impact side the exceedances of critical thresholds for air pollution *and* global change and on the cost side the estimates of costs to reduce emissions of greenhouse gases and air pollutants. The integrated modeling framework set up to meet these objectives consists of two state-of-the-art integrated models covering regional air pollution in Europe (RAINS) and global climate change (IMAGE), supplemented by new components. Based on a preliminary analysis it can be stated that climate change will make European vegetation in most regions less sensitive to acid deposition. Taking into account the emission trends the impacts of regional air pollution will decrease while the impacts of climate change increase. Different problems will be important in different regions: regional air pollution in Central and northern Europe, and climate change in southern Europe.

Keywords: air pollution, climate change, integrated assessment, Europe, critical loads, critical climate, forests, natural vegetation

1. Introduction

Despite the many overlaps of regional air pollution and climate change, these two environmental problems have been dealt with separately up to now by European policymakers. One reason for this separate approach has been that policymakers do not have the quantitative information needed to develop policies that address both regional air pollution and climate change in Europe. The aim of the AIR-CLIM project is to perform an integrated analysis of the linkage between the two problems in Europe and to produce results that are relevant to European policy-making. Specific objectives are:

1. To examine whether climate change will alter the effectiveness of agreed-upon or future policies to reduce regional air pollution-causing emissions in Europe, and vice versa.
2. To identify the relative importance and overlap of regional air pollution and climate change impacts under a consistent set of assumptions about future developments of emissions.
3. To identify and evaluate comprehensive policy strategies for controlling both regional air pollution and climate change in Europe.

It must be noted that because of the complexity of both regional air pollution and climate change the project covers only some of their overlaps.

The objective of this paper is to present intermediate results of the AIR-CLIM project. In this paper the focus is on the effects of acidification and climate change. Further results including direct air pollution effects and mitigation costs are described in the second AIR-CLIM progress report (Mayerhofer et al., 2000).

2. Compiling a Framework for Integrated Analysis

An integrated modeling framework is used to meet the objectives of the project (see Figure 1). This framework consists of parts of two state-of-the-art integrated models covering regional air pollution in Europe (RAINS) and global climate change (IMAGE), supplemented by new components. RAINS is an integrated model of regional air pollution in Europe. It describes the coupling between energy scenarios; country-scale emissions of sulfur and nitrogen; ambient concentrations and depositions of acidifying substances; and critical loads to ecosystems (Alcamo et al., 1990; Amann et al., 1995). The IMAGE 2 model is RAINS' counterpart for global climate change. It couples regional developments of energy and agriculture: emissions of greenhouse gases, and SO_2; changes in land cover and carbon fluxes between the biosphere and atmosphere; the build-up of greenhouse gases in the atmosphere; and flux of heat in the atmosphere and ocean (Alcamo et al., 1998).

2.1. AIR POLLUTION UNDER CLIMATE CHANGE

Climate change may alter the dispersion, chemical conversion and removal of pollutants in the atmosphere. This, in turn, changes the pollutant concentrations in the atmosphere and the amount of sulfur and nitrogen species deposited.

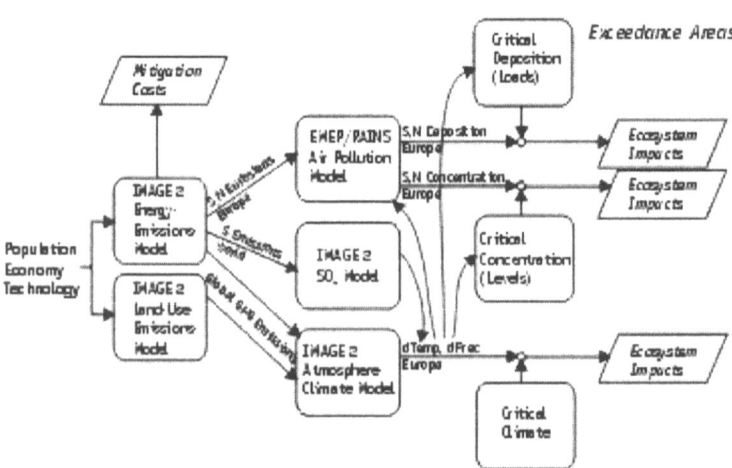

Figure 1. Integrated modeling framework for climate change and regional air pollution.

Climate-dependent source-receptor matrices (SRMs) will be used to analyze how the distribution and conversion of air pollutants in Europe will be affected by climate change. These SRMs will be derived from a long-range transport model using changed weather patterns calculated by climate models. Since these new SRMs will only be available in the later phase in the project, a provisional approach - called the *climate analogy approach* - has been used to derive preliminary matrices. Thereby, the year among the observed (1986-1995) precipitation *patterns* was identified which is closest to the precipitation pattern under climate change for a selected future year. The existing source-receptor matrix of that 'analogous' observation year is used as a surrogate to simulate long-range transport of air pollutants under climate change.

2.2. CLIMATE DEPENDENT CRITICAL LOADS OF ACIDITY

Critical loads depend among others on climate factors, which means that they could be sensitive to climate change. Critical loads of acidity for forest soils are calculated with the so-called simple mass balance (SMB) model (UBA, 1996). The SMB model is the most commonly used method for deriving acidity critical loads in the UN/ECE context (Posch *et al.*, 1999). Because weathering rates are influenced by soil temperature and acidity leaching by precipitation and evapotranspiration, the critical load of acidity is affected by temperature, precipitation and evapotranspiration. Therefore, an increase in temperature can (partially) be compensated by a decrease in precipitation surplus, i.e. precipitation minus evapotranspiration. Whether the precipitation surplus will increase or decrease under a changing climate depends on a fine balance between increasing precipitation and increasing evapotranspiration due to a higher temperature. Variation of both factors within reasonable ranges expected under global warming, indicate that critical loads will change at most about 10%.

2.3. CRITICAL CLIMATE THRESHOLDS

The critical climate approach is developed to assess negative impacts of climate change. Thereby, critical climate is defined analogous to critical loads as *a quantitative value of climate change, below which only acceptable long-term effects on ecosystem structure and functioning occur according to current knowledge*. In this study, critical climate thresholds were derived from acceptable long-term losses of net primary production (NPP) of natural ecosystems. Based on an analysis of historic NPP variations, NPP losses in the range of 10 to 20% have been identified as acceptable. Climate Isoline Diagrams (CID) depict the allowable changes in temperature and precipitation for the predefined acceptable NPP losses.

Three different types of responses can be distinguished: (1) Large parts of northern Europe are only slightly sensitive to decreased precipitation levels, even if the temperature increases; (2) Central Europe becomes less sensitive to reduced precipitation if the temperature increases. This is because higher temperatures stimulate NPP compensating negative effects of reduced precipitation; (3) southern Europe becomes even more sensitive to precipitation reductions for higher temperature.

3. Scenario analysis

An objective of the AIR-CLIM project is to derive reduction scenarios that consider reductions of both greenhouse gases and air pollutants. As an interim step the so-called *September scenarios* were analyzed that are a sub-set of the final AIR-CLIM scenarios (Mayerhofer et al., 2000). Starting point scenarios for the September (and the final) scenarios are the A1 and B1 scenarios of the SRES scenarios prepared for the Third Assessment of the IPCC as realized in the TIMER/IMAGE model (Nakicenovic et al., 2000). A1 and B1 do not assume any climate policy but rather stringent SO_2 policies. In a first step, these stringent SO_2 policies are replaced by 'AIR-CLIM' policies that keep the level of SO_2 reduction on the level of 2010. These AIR-CLIM reference scenarios are called A1-SR and B1-SR (SR for Sulfur Reference).

The September mitigation scenarios are so-called stabilization scenarios i.e. scenarios in which the CO_2 concentration is stabilized at a certain level. For A1 mitigation measures are assumed for which the CO_2 concentration stabilizes at 550 ppm; for B1 the respective level is 450 ppm. The SO_2 policies for these scenarios are as stringent as for the original A1 and B1 scenarios. The resulting mitigation scenarios are called A1-550-SA and B1-450-SA (SA for Sulfur Advanced Policy).

The analysis carried out so far within AIR-CLIM is preliminary as the methodology has still to be refined at some points. However, some preliminary conclusions can be derived from this analysis:

Emission trends. In the scenarios used here, CO_2 emissions peak around 2040 to 2060 and then decline because of a decreasing use of energy. The emissions of SO_2 and NO_x in Europe are expected to decline to the levels set in the LRTAP Protocols until 2010 and afterwards to stabilize or slowly to decline further. The global SO_2 emissions will peak around 2030 to 2040 and then (mostly because of the decreasing energy use) decline. Assuming that countries currently without international agreements on SO_2 emission reductions will react similarly as Europe and North America to high SO_2 levels the decline will be steep and the global SO_2 emissions in 2100 about the same level as in 1990 or lower.

Impact of SO_2 on climate change. Contrary to earlier studies, e.g. Posch *et al.* (1996), only a small effect of SO_2 emissions on climate change *in 2100* is calculated. The reason for this is the aforementioned decrease of the (global) SO_2 emissions while former studies assumed a continuing increase. Higher SO_2

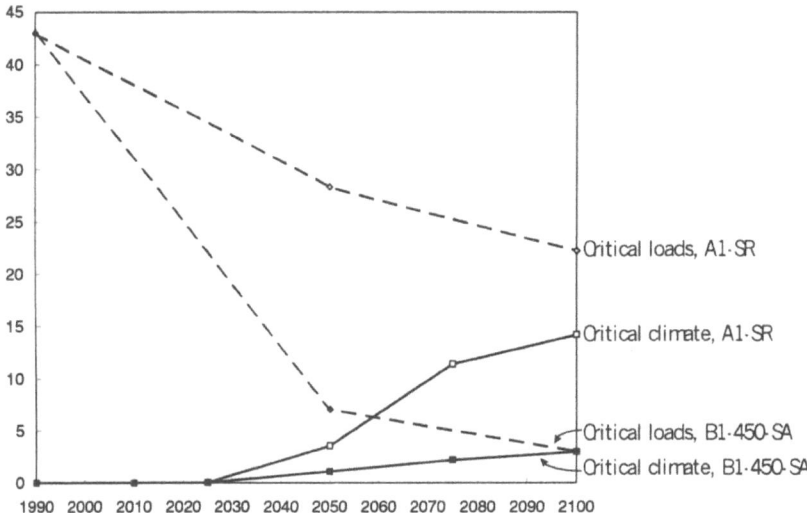

Figure 2. Exceeded areas [%] for critical thresholds for the scenarios A1-SR and B1-450-SA (critical climate in % of natural ecosystem area for 10% net primary production as acceptable effect; critical lpoad in % of forest ecosystem area).

emissions delay (but do not avoid) the exceedance of critical climate values in Europe.

Impact of climate change on critical loads of acidity and its exceedances. Under climate change the critical loads increase, i.e. the ecosystems become less sensitive with time, with the exception of southern Europe. However, even under the lowest deposition scenario, the critical loads are still exceeded in some areas (Germany, UK, East Europe) in 2100.

Critical climate and its exceedances. For the scenarios used here, critical climate values are only exceeded in few areas in southern and south-eastern Europe up to 2050. After that the exceedance area increases due to decreasing precipitation rates in combination with increasing temperature. The A1-SR scenario results in the largest area in which the critical climate is exceeded. The smallest area is computed for the B1-450-SA scenario (see Figure 2).

Development of areas for which the various critical thresholds are exceeded. While the area for which critical climate is exceeded will increase with time, the exceedance area for acid deposition will decrease. Thereby even in 2100 the exceedance areas for critical loads are still larger than those for critical climate (see Figure 2).

4. Conclusions

An integrated modeling framework has been set up that allows an integrated analysis of the linkage between climate change and regional air pollution in Europe. The results calculated with the framework obviously depend on scenario assumptions and include simplifications and uncertainties. The uncertainties will be further analyzed till the end of the project. It has to be stressed that in any case the trends identified by such an analyses are more important than the actual values.

Analyzing preliminary scenarios we have found that in some situations climate change could increase critical loads, i.e. raise the thresholds of acidification impacts. This does not exclude the possibility that vegetation could become more or less sensitive to acidification because of climate-related effects that are not covered by the critical load concept.

The analysis confirmed earlier studies that the impacts of acidification will decrease in Europe while the impacts of climate change increase. We estimated a large difference in the geographic coverage of the two problems: acidification in Central and northern Europe, and climate change in southern Europe.

Acknowledgments

AIR-CLIM is funded by the European Commission, Directorate General XII within the EC Environment and Climate Research Programme (1994-1998), Contract No. ENV4-CT97-0449.

References

Alcamo J., Kreileman, E., Krol, M., Leemans, R., Bollen, J., van Minnen, J., Schaefer, M., Toet, S., de Vries, B.: 1998, 'An instrument for building global change scenarios', in J. Alcamo, R. Leemans and E. Kreileman (eds.), *Global Change Scenarios of the 21st Century - Results from the IMAGE 2.1 Model*, Pergamon Press, pp. 3-94.

Alcamo J., Shaw, R. and Hordijk, L. (eds.): 1990, *The RAINS Model of Acidification: Science and Strategies in Europe*. Kluwer Academic Publishers, Dordrecht.

Amann M., Baldi, M., Heyes, C., Klimont, Z. and Schöpp, W.: 1995, 'Integrated assessment of emission control scenarios including the impact of tropospheric ozone', *Water Air and Soil Pollution* **85**, 2595.

Mayerhofer P., Alcamo, J., van Minnen, J. G., Posch, M., Guardans, R., Gimeno, B. S., van Harmelen, T. and Bakker, J.: 2000, *Regional Air Pollution and Climate Change in Europe: An Integrated Analysis (AIR-CLIM) - Progress Report 2*. WZ Report A0001, Center for Environmental Systems Research, University of Kassel, Kassel.

Nakicenovic N. et al.: 2000, *Special Report on Emission Scenarios: A Special Report of Working Group III of the Intergovernmental Panel on Climate Change*. Cambridge University Press, Cambridge.

Posch M., de Smet, P. A. M., Hettelingh, J.-P. and Downing, R. J. (eds.): 1999, *Calculation and Mapping of Critical Thresholds in Europe - Status Report 1999*. RIVM Report No. 259101009, RIVM, Bilthoven, 165 pp.

Posch M., Hettelingh, J.-P. and Alcamo, J.: 1996, 'Integrated scenarios of acidification and climate change in Asia and Europe', *Global Environmental Change* **6**, 375.

UBA: 1996, *Manual on methodologies and criteria for mapping critical levels/loads and geographical areas where they are exceeded*. UBA-Texte 71/96, Umweltbundesamt (UBA), Berlin.

CALCULATING CRITICAL LOADS OF SULFUR DEPOSITION FOR 100 SURFACE WATERS IN CHINA USING THE MAGIC MODEL

JIMING HAO*, XUEMEI YE, LEI DUAN, ZHONGPING ZHOU

Department of Environmental Science and Engineering, Tsinghua University, Beijing 100084, China
(* author for correspondence, e-mail: hjm-den@tsinghua.edu.cn)

Abstract. Although decades of acid deposition have apparently not resulted in surface water acidification in China, some surface waters may have the potential trend of being acidified, especially those in southern China. In this paper, a dynamic acidification model--MAGIC was applied to 100 surface waters in southern and northeastern China to evaluate the impact of acid deposition to surface waters and to determine their critical loads of S deposition, both regions having distinguishing soil, geological and acid deposition characteristics. Results indicate that most surface waters included in this paper are not sensitive to acid deposition, with critical loads of S for these waters comparatively high. On the other hand, surface waters in southern China, especially those in Fujian, Jiangxi and Guangdong provinces, are more susceptible to acidification than those in northeastern China, which coincides with their different patterns of soil, geological and acid deposition conditions. Among all the waters, a few small ponds, such as those on top of the Jinyun mountain and Emei mountain, are the most sensitive to acid deposition with critical loads of 1.84 and 3.70 keq·ha^{-1}·yr^{-1}, respectively. For the considerable ANC remaining in most 100 surface waters, it is not likely that acidification will occur in the near future for these waters.

Keywords: critical load, surface water, MAGIC, acid deposition, China

1. Introduction

Although China has suffered from serious acid deposition pollution for decades, little attention has been paid to water quality changes resulting from acid deposition, nor to critical loads for Chinese surface waters. Investigations show that although most surface waters in China are generally thought to be resistant to acid deposition, some of them may be sensitive to acid deposition, especially in southern China (Feng, 1993; Zhou, 1996). As a result, the purpose of this paper is to evaluate the effect of acid deposition on Chinese surface waters by calculating critical loads of S deposition and critical load exceedance for 100 surface waters in southern and northeastern China with a widely used dynamic acidification model MAGIC.

2. Site description

Altogether 100 surface waters located in southern and northeastern China are considered in this work, including ponds, lakes, streams and rivers with sufficient data for modeling. Except some small mountain ponds with relatively small catchment sizes less than 1 km², most other watersheds have larger areas. All the chosen watersheds span a gradient of acid deposition, climatic conditions and geological sensitivity (Table I). Compared to the north, the acid deposition load in southern China is much higher and soils are more susceptible to acidification. Although most 100 surface waters have high ANC much more than 200 µeq l^{-1}, those waters in Fujian, Jiangxi and Guangdong provinces have relatively lower ANC values. Among all the waters, a few mountain ponds in southwestern China, such as the water on top of the Jinyun mountain and Emei mountain, have the lowest ANC values, 269 and 237 µeq l^{-1} respectively.

TABLE I
Spatial distribution of the surface waters and their environmental conditions

Province (Abbreviation)	Number of waters	Rainfall (mm)	Runoff (mm)	Main soil types
Guizhou(GZ)	13	1100	600	yellow earth
Sichuan(SC)	18	900	600	yellow earth, purplish soil
Jiangxi(JX)	13	1600	900	red earth
Hubei(HB)	3	1200	700	red earth
Guangxi(GX)	4	1600	800	red earth, lateritic red earth
Guangdong(GD)	14	1600	900	lateritic red earth, latosol
Fujian(FJ)	8	1600	1100	latosol, red earth
Zhejiang(ZJ)	4	1500	1000	red earth
Jiangsu(JS)	2	900	400	yellow-brown earth, paddy soil
Anhui(AH)	1	900	400	yellow-brown earth, paddy soil
Heilongjiang (HLJ)	10	600	100	dark brown forest soil, podzolic soil, black soil
Jilin(JL)	1	600	100	paddy soil
Liaoning(LN)	9	700	100	brown forest soil, cinnamon soil, paddy soil

3. Input Data and Model Calibration

Although there were much simpler steady-state models specifically designed for calculating critical loads for surface waters (Henriksen *et al.*, 1992; Posch *et al.*, 1997), a widely used dynamic acidification model MAGIC (Cosby *et al.*, 1985) was applied in this paper. A critical load derived from a dynamic model differs from a critical load based on steady state approaches by incorporating time-dependent processes such as cation exchange and SO_4^{2-} adsorption, so that recovery of the ecosystem is inherent in the calculation.

Input data required for the MAGIC calibration procedure include those describing deposition chemistry, catchment-scale lumped soil characteristics for depth, CEC, exchangeable base cation fractions and bulk density, and also present surface water chemistry. Most of the required data were collected from the results of routine monitoring at each hydrological station and the literatures

(e. g. Chinese National Hydrological Agency, 1982; Xiong et al., 1987; Jin, 1990; Feng, 1993; Xue and Schnoor, 1994; Wang, 1995). Atmospheric deposition of base cations and strong acid anions was calculated using concentrations of each ion in precipitation and the precipitation volume. For sulfate, the total deposition was got by wet deposition multiplied a dry deposition factor set 1.7 according to Xie (1996). Parameters describing SO_4^{2-} adsorption were attained partly through laboratory experiments using the way described by Xie (1996) and partly from literature (kuang et al., 1995).

The model was calibrated for each catchment to get a set of suitable parameters for the prediction process, including base cation selectivity coefficients and weathering rate. The hindcast simulations were started in 1920, assuming the anthropogenic emissions were of little significance and the stream chemistry was in a steady state at that time. 1988 was the 'present' year, since quantities of water chemistry data were of that year. The historical trend in wet deposited non-marine SO_4^{2-} was assumed to follow the sequence of the consumption of coal in China (Xie, 1996). The calibration was accomplished following the method of Jenkin et al. (1997). The simulated present chemistry of surface waters accorded fairly well with the observed water chemistry (see Fig. 1).

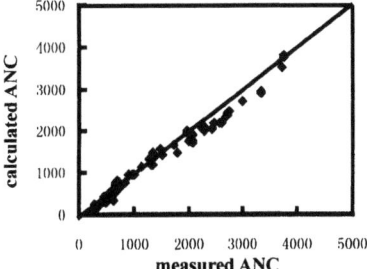 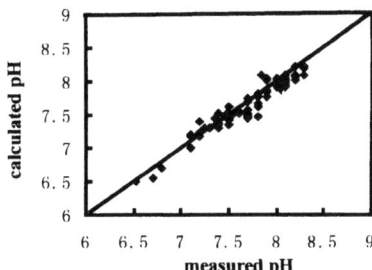

Figure 1. Comparison of the MAGIC simulated present day water chemistry with measured data

4. Results and Discussion

4.1. CRITICAL LOADS

MAGIC can be used to calculate a critical load of S by adjusting present non-marine S deposition and running the model forward in time, assuming that other ion depositions remain constant. The critical load is attained by comparing water ANC, taken as the water quality indicator, at a given time with the target value ANC_{limit} which is the critical ANC required to protect a defined biological target within the ecosystem. In this paper, an ANC_{limit} equal to 50 µeq·l^{-1} was used. This is the highest value among those used in various countries. Because Chinese domestic fishes are more readily affected by acidification than

European trout (Feng, 1993), the higher ANC_{limit} value is considered adequate to avoid negative effect on aquatic organisms in areas with higher critical loads (Henriksen et al.,1995). 2115 was taken as the limiting year for it was an enough long time for all the 100 watersheds to reach steady state.

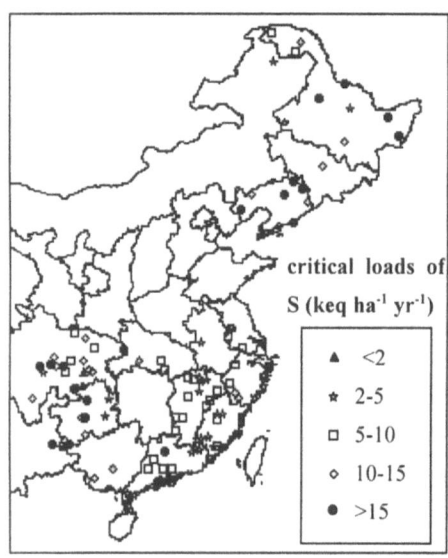

Figure 2. Spatial patterns of critical loads of S for 100 Chinese surface waters

The critical loads of S for all 100 waters are shown in Figure 2 and Figure 3. It can be seen that although critical loads of S for these waters are very high, covering a range of 2-40 keq ha^{-1} yr^{-1}, they have evident spatial patterns, i.e. the critical loads show an increasing trend from east to west and from south to north. In southern China, the waters in Fujian, Jiangxi and Guangdong province have comparatively low critical loads with average values around 5 keq·ha^{-1}·yr^{-1}. On the contrary, although catchments in southwestern China like in Sichuan and Guizhou province, suffer from heavy acidic deposition, critical loads for surface waters in this region are relatively high, excluding the two mountain ponds mentioned above. These two ponds have lower critical loads than most others, 1.84 keq·ha^{-1}·yr^{-1} and 3.70 keq·ha^{-1}·yr^{-1}, respectively. Critical loads for most surface waters in northeastern China are rather high with average values above 15 keq ha^{-1} yr^{-1}.

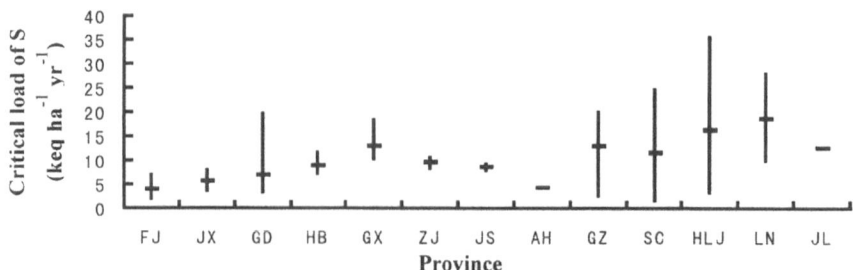

Figure 3. The range and the average value of critical loads of S for surface waters in different provinces ("-" means average values).

4.2. CRITICAL LOAD EXCEEDANCE

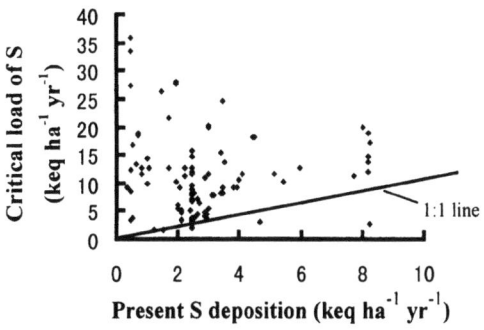

Figure 4. Comparison of critical load of S and present S deposition (Hao et al., 1996)

Critical load exceedance provides us with a quantitative estimate of ecosystem damage against which emission reduction strategies can be assessed. Figure 4 shows the status of critical load exceedance for all 100 surface waters. The results indicate that for most surface waters, critical loads of S are not exceeded by present S deposition.

4.3. DISCUSSIONS

The critical load for an ecosystem reflects its sensitivity to acid deposition. Thus, the high critical load values for the 100 Chinese surface waters indicate their high resistance to acidification. As has been mentioned above, there remains sufficient ANC in most these surface waters, even at the high elevation Emei and Jinyun pond, to neutralize acidified inputs, making them very resistant to acidification. On the other hand, the spatial pattern of critical loads for these waters can be interpreted by the patterns of soil, geological, climatic and hydrological conditions from the southeast to the southwest and northeast in China, because sensitivity of surface water to acidification greatly determined by such environmental factors. Generally, environmental factors are more adverse for surface waters in the southeast to have higher ANC than those in the southwest and northeast, i.e. the soils are more acidic and rainfall and runoff are more abundant in the southeast than in the southwest and northeast.

In spite of the relatively high critical loads for northeastern surface waters, there also exists a spatial pattern in these critical loads of S, which coincide tightly with the soil distribution pattern. Those waters flowing via podzolic soil and dark brown forest soil have lower critical loads than those flowing through black soil and paddy soil, indicating that soil characteristics may be the most important factor deciding the sensitivity of surface waters to acidification.

Compared to critical loads for soils (Duan et al., 2000), critical loads for surface waters in China are much higher. Therefore, acidification of soils may be a precondition for surface waters acidification in China.

5. Conclusions

The high critical loads and non-exceedance of S deposition calculated with MAGIC indicate that most surface waters considered in this paper are not sensitive to acid deposition. Compared to northeastern surface waters, the waters in southern China have comparatively lower critical loads of S deposition, especially surface waters on high mountain top and in Fujian, Jiangxi and Guangdong province. Critical loads for surface waters have good agreement with their present ANC, which are determined by such environmental factors as soil, geology, acid deposition and hydrology. For the considerable acid neutralizing capacity remaining in most surface waters, it is not likely that acidification will occur for most of these 100 Chinese surface waters in the near future. Nevertheless, attentions should be paid to those high mountain watersheds suffering from heavy acid deposition pollution, which may be the most important casualties due to acid deposition.

Acknowledgements

The authors would like to thank the National Natural Science Foundation of China for its financial support for this research.

References

Chinese National Hydrological Agency. Hydrological characteristics of Chinese rivers. Beijing: The Science Press, 1982. (in Chinese)
Cosby,B.J.,Wright,R.F.,Hornberger,G.M. and Galloway,J.N.: 1985, *Water Resources Research* **21**, 51.
Duan,L.,Xie,S.D.,Zhou,Z.P. and Hao,J.M.: 2000, *Water Air and Soil Pollution* **118**, 35.
Feng, Z.W.: 1993, *The Effects of Acid Rain on Ecosystem: Study on Acid Rain in Southwest China*. The China Science and Technology Press, Beijing, China. (in Chinese)
Hao,J.M.,Zhou,X.L.,Fu,L.X. and Li,Q.L.: 1996b, *China Environmental Science* **16**(5), 345. (in Chinese)
Henriksen,A.,Kamari,J. and Posch,M.: 1992, *AMBIO* **21**(5), 356.
Henriksen,A.,Posch,M.,Hultberg,H. and Lien,L.: 1995, *Water Air and Soil Pollution* **85**, 2419.
Jin,X.C.:1990, *Investigaion on the Environment of Chinese Lakes and Reservoirs*. The Chinese Environmental Press, Beijing, China. (in Chinese)
Posch,M.,Kamari,J.,Henriksen,A. and Wilander,A.: 1997, *Environmental Management* **21**(2), 291.
Jenkins A.,Renshaw,M.,Helliwell,R.,Sefton,C.,Ferrier,R. and Swingewood,P.: 1997, *Modeling surface water acidification in the UK*. IH Report No. 131, Institute of Hydrology, Wallingford, United Kingdom.
Kuang,Q.J. and Li,J.Q.: 1995, *The Journal of Envrionmental Science* **16**(4), 13. (in Chinese)
Xie,S.D.: 1996, Doctor of Philosophy Dissertation, Tsinghua University, Beijing, China.
Xiong,Y. and Li,Q.K.: 1987, *Chinese Soil*. The Science Press, Beijing, China. (in Chinese)
Xue,H.B. and Schnoor,J.L.:1994, *Water Air and Soil Pollution* **75**, 61
Wang,W.X.: 1995, *Researches on temporal and spatial distribution of acid precipitation in China*. National Key Project in the Eighth Five-year Plan, Final Report. (in Chinese)
Zhou,X.P.: 1996, *Rural Eco-Environment* **12**, 1. (in Chinese)

EFFECTS OF SITE SELECTION STRATEGY ON FRESHWATER CRITICAL LOAD EXCEEDANCES IN WALES

C.J.CURTIS[1*], R.HARRIMAN[2], M.HUGHES[1] and M.KERNAN[1]

[1] *Environmental Change Research Centre, 26 Bedford Way, London WC1H 0AP, UK.;* [2] *SOAFD Freshwater Fisheries Laboratory, Faskally, Pitlochry, Perthshire PH16 5LB, UK.*
(* author for correspondence, e-mail: ccurtis@geog.ucl.ac.uk)

Abstract. Critical loads are used in international negotiations to reduce acid deposition resulting from emissions of sulphur and nitrogen compounds within Europe. For freshwater ecosystems, the First-order Acidity Balance (FAB) model is used to generate national maps of critical loads and exceedances for both sulphur (S) and nitrogen (N). In Wales, two survey datasets have been used to calculate critical loads and exceedances; one based on water bodies selected to be "most-sensitive" to acidification within a 10 km grid and the other based on a random selection of standing waters. Both datasets indicate that critical loads were exceeded in 1990 in a significant proportion of Welsh lakes and streams; 36% of sites in the grid-based survey and 31% of sites in the random survey. However, implementation of the Gothenburg Protocol would protect all but 6% of sites in the grid-based survey and all sites in the random survey. Assessment of the relative success of the Gothenburg Protocol in protecting Welsh freshwater ecosystems therefore depends on the site selection strategy employed.

Keywords: critical loads, FAB model, acidification

1. Introduction

The multi-pollutant, multi-effect "Protocol to Abate Acidification, Eutrophication and Ground-level ozone" was signed on 1st December 1999 in Gothenburg, Sweden, under the Convention on Long Range Transboundary Air Pollution (CLRTAP) of the United Nations Economic Commission for Europe (UNECE). The Gothenburg Protocol set national emissions ceilings for three pollutants that contribute to acidifying deposition; sulphur dioxide, nitrogen oxides and ammonia. As part of the scientific input to the integrated assessment modelling programme which determined these emissions ceilings, critical loads values for sulphur (S) and nitrogen (N) deposition were submitted by the signatory nations for various ecosystems. Critical loads provide a quantitative link between deposition of a pollutant and a measurable "harmful effect" on some aspect of the ecosystem (Bull, 1991; Nilsson and Grennfelt, 1988).

National datasets submitted to the "Mapping Programme" are managed by the Coordination Centre for Effects (CCE) in the Netherlands and published in "Status Reports", the most recent of which outlines the data used in the negotiation of the Gothenburg Protocol (Posch *et al.*, 1999). Methods for the calculation of critical loads are recommended in "Mapping Manuals" published on behalf of the UNECE (UBA, 1996). For freshwaters the favoured method uses the First-order Acidity Balance (FAB) model (Posch et al., 1997). The FAB model employs a mass balance for S and N to determine which combinations of S and N deposition will result in leaching of acid anions and associated acidity leading to a depression of surface water acid neutralizing capacity (ANC) below a selected threshold value. The Mapping Programme allows for flexibility in the exact methods employed within each country and also in the ecosystems which are selected (UBA, 1996). This paper examines the potential success of the Gothenburg Protocol in protecting freshwater ecosystems in Wales in two groups of sites selected on different criteria.

2. Methods

The freshwater datasets employed here are derived from two separate surveys in Wales using very different site selection criteria. The freshwaters data submitted to the CCE from the UK are for sites selected to represent the most sensitive water bodies, both standing waters and streams, within a 10 km grid (Kreiser et al., 1993) and the distribution of these sites in Wales ("grid" survey; n=158) is shown in Figure 1a. Since these sites were selected using criteria like geology, soils and land-use to be the most sensitive to acidification within each grid square, they are not representative of the population of water bodies in Wales; instead they provide a "worst-case" picture of critical load exceedance so that if the chosen water body is protected (does not exceed its critical load), then all water bodies in that grid square should be protected. The success of this site selection strategy has been discussed previously (Curtis et al., 1995). The second dataset used here was collected for the North European Lakes Survey of 1995 (Henriksen et al., 1998) and comprises 52 randomly selected standing waters, stratified by size class ("Random" survey; Fig. 1b, Table I).

TABLE I
Site type and size

Size (ha):	0 (Streams)	0-4	4-10	10-100	>100	Total No.
Grid	55	65	20	18	0	158
Random	0	30	15	7	0	52

Linked critical loads for sulphur and nitrogen were calculated with the FAB model for all sites using the methodology described by Curtis et al. (2000). Critical load exceedances were calculated for two sets of deposition data generated with the HARM Model (Metcalfe et al., 1995, 1998); 1990 "baseline" data and modelled deposition after implementation of emissions reductions under the Gothenburg Protocol (Scenario WGS31B).

3. Results and Discussion

Summary statistics describing the water chemistry data from the two surveys are provided in Table II. It can be seen that both surveys include a wide range of chemical characteristics from very acid to very high alkalinity, with lower mean pH and ANC for the grid sites as expected. Non-marine base cation concentrations show enormous variability, while acid anion concentrations range from near zero to very high levels which are too large to be due to atmospheric inputs, and could indicate agricultural or other terrestrial sources.

TABLE II
Summary statistics of water chemistry

	Grid survey sites						Random survey sites					
	PH	Alk	BC*	SO4*	NO3	ANC	pH	Alk	BC*	SO4*	NO3	ANC
Mean	5.31	320	531	149	32	335	5.35	318	769	246	30	369
Median	6.77	119	296	92	10	140	6.21	29	191	85	11.5	41
Minimum	4.22	-78	18	1	0	-45	4.5	-35	63	35	0.16	-2
Maximum	8.77	3408	4315	1702	250	3431	7.96	2565	10654	5470	453	2588
Std.Dev.	4.93	55	660	176	49	557	5.1	631	1655	754	68	641

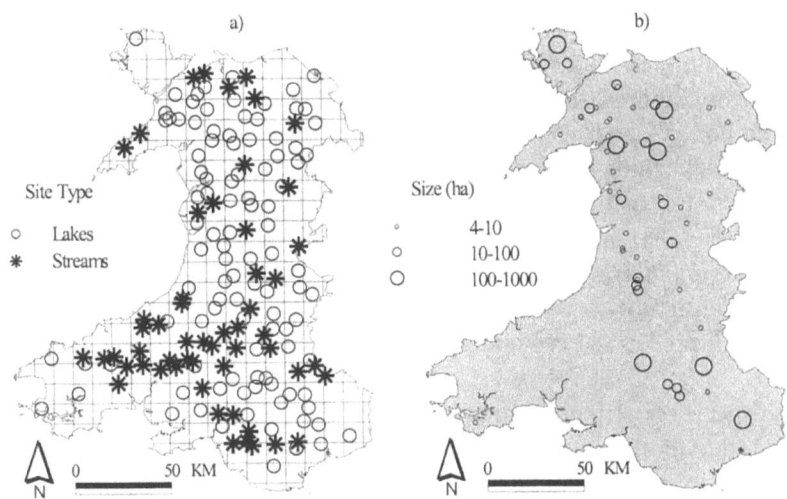

Figure 1. Distribution of Welsh survey sites for a) grid survey by type, b) random survey by size

Given the wide chemical variations within both surveys, a similar variation in critical loads and exceedance might be anticipated. Independent critical loads for S and N cannot be defined with the FAB model (Posch *et al.*, 1997) but exceedance can be calculated as the amount by which the total acidity flux exceeds the critical threshold (Fig. 2).

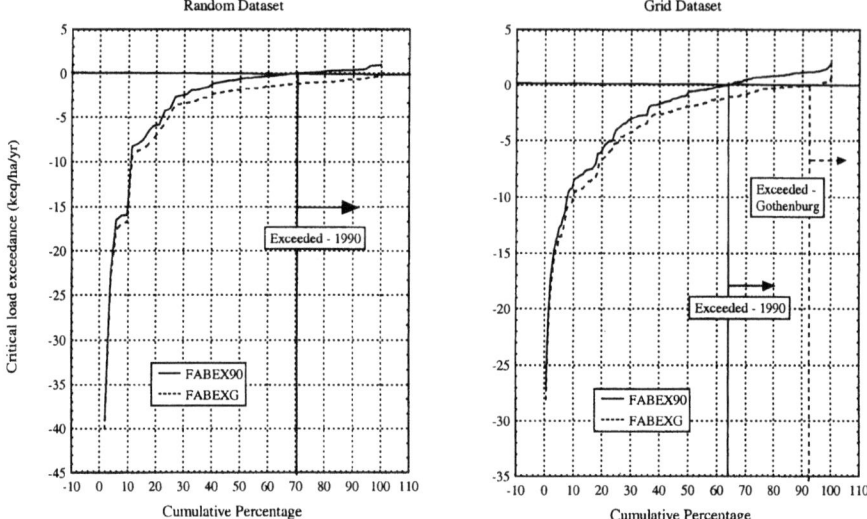

Figure 2. Cumulative frequencies of FAB model exceedance for the two Welsh surveys under 1990 ("FABEX90") and Gothenburg Protocol ("FABEXG") deposition scenarios

Exceedance is expressed in keq ha^{-1} yr^{-1} of acidity; a negative value indicates non-exceedance, while positive values indicate that the critical load is exceeded (Fig.2). Large negative values of exceedance indicate very high critical loads for paired S and N deposition. Since the maximum value of total acid deposition (S+N) for all surveyed sites in 1990 was <4 keq ha^{-1} yr^{-1} it can be seen from Figure 2 that 20-25% of sites in both surveys have exceedances of < -5 keq ha^{-1} yr^{-1} and would therefore not be exceeded even if exposed to the maximum deposition levels of both S and N. These sites are highly insensitive to acidification. In the random survey, around 30% of sites are exceeded at 1990 levels but none are exceeded under Gothenburg Protocol deposition levels. The more sensitive group of sites in the grid survey show a slightly higher proportion of exceedance under 1990 levels, but also a number of sites that are still exceeded, even under the Gothenburg Protocol levels.

The species contributing to critical load exceedances calculated for the two surveys under both deposition scenarios are summarised in Table III. According to the grid survey, 36% of sites exceeded their critical loads in 1990, with 25% requiring mandatory reductions in S deposition and 11% also requiring mandatory reductions in N deposition. The random survey indicates that of the 31% of sites exceeding their critical loads, none required a mandatory reduction in N deposition and only 6% required a mandatory reduction in S deposition (Table III). While the overall proportion of sites exceeded is similar for the two surveys, the relative importance of N deposition in causing critical load exceedance in 1990 is much greater for the grid sites.

After implementation of the Gothenburg Protocol, the proportion of sites exceeding critical loads is greatly reduced (Table III). Of the most sensitive sites represented by the grid survey, only 6% are left exceeded, with just over 1% requiring mandatory reductions in S deposition and under 1% for N deposition. According to the random survey, no sites will show critical load exceedance after full implementation of the Gothenburg Protocol.

TABLE III

Deposition reduction requirements for both grid and random (Rand.) surveys from 1990 baseline levels and after the Gothenburg Protocol

Acid species for reduction	No. sites from 1990 baseline				No. sites after Goth. Protocol			
	Grid	%	Rand.	%	Grid	%	Rand.	%
None	*101*	*64*	*36*	*69*	*148*	*93.7*	*52*	*100*
S only	0	0	0	0	0	0	0	0
S, then S or N	22	14	3	6	1	0.6	0	0
Either S or N	17	11	13	25	8	5.1	0	0
N, then S or N	0	0	0	0	0	0	0	0
Both S and N	18	11	0	0	1	0.6	0	0
Total no. sites	158	100	52	100	158	100	52	100
Total exceeded	57	36	16	31	10	6	0	0

The success of the Gothenburg Protocol in protecting Welsh freshwaters therefore varies depending upon the survey dataset used. The Welsh component of the UK grid dataset indicates that emissions reductions under the Gothenburg Protocol are insufficient to protect 6% of sampled sites, with further S reductions being the main requirement for complete protection (assessed as non-exceedance). Within the UNECE mapping programme, a five percentile critical load is employed (UBA, 1996; Posch et al., 1999) whereby exceedance of the most sensitive five percent of ecosystems is permitted. Even under this system, a further one percent of grid sites in Wales remain exceeded, so the Gothenburg Protocol just fails to prevent critical load exceedance at the five percentile level in Wales. If the random survey dataset had been submitted to the mapping programme, the Gothenburg Protocol would have been judged to be completely successful in protecting Welsh freshwaters from acidification beyond the critical chemical threshold (here ANC=0 μeql^{-1}; Curtis et al., 2000).

The difference is not unexpected, since the grid sites were selected to be the most acid-sensitive while the randomly selected sites were intended to provide a representative subsample of Welsh standing waters. The greater sensitivity of the grid dataset is not due to the inclusion of stream sites (where headwater streams might be expected to include the most acid-sensitive water bodies of all) because the proportion of stream sites exceeded is actually much lower than for all sites (15% compared with 36% overall under 1990 deposition levels). However, the difference between the surveys may be due in part to the smaller minimum size for standing waters in the grid survey of 0.5 hectares, compared with a minimum of 4 hectares in the random survey.

4. Conclusions

The success of international measures to reduce acid deposition through emissions ceilings, when judged on the proportion of ecosystems exceeding their critical loads, depends to some degree on the survey methods selected even for similar ecosystems - in this case freshwaters. The international mapping and modelling programme carried out under the auspices of the UNECE allows each nation to select its own methodologies for selecting ecosystems and modelling critical loads, within the guidelines laid down in the mapping manual (UBA, 1996). This study therefore shows the need for caution when making international comparisons of critical load exceedances, since the success of the Gothenburg Protocol in preventing acidification in freshwater ecosystems in Wales varies according to the survey dataset used.

Acknowledgements

The authors would like to thank Sarah Metcalfe, Duncan Whyatt and CEH Monks Wood for the provision of deposition data. This work was funded jointly by the NERC ED Programme and the UK DETR.

References

Bull, K.R.: 1991, *Environ. Pollut.* **69**, 105-123.
Curtis, C.J., Allott, T., Battarbee, R.W. and Harriman, R.: 1995, *Water, Air, Soil Poll.* **85**, 2467-72.
Curtis, C.J., Whyatt, J.D., Metcalfe, S.E., Allott, T.E.H. and Harriman, R.: 1999, *Energy and Environment* **10**, 571-596.
Curtis, C.J., Allott, T.E.H., Hughes, M., Hall, J., Harriman, R., Helliwell, R., Kernan, M., Reynolds, B. and Ullyett, J.: 2000, *Hydrology and Earth System Sciences* **4**, 125-140.
Henriksen, A., Skjelkvåle, B.L., Mannio, J., Wilander, A., Harriman, R., Curtis, C., Jensen, J.P., Fjeld, E. and Moiseenko, T.: 1998, *Ambio* **27**, 80-91.
Kreiser, A.M., Patrick, S.T. and Battarbee, R.W.: 1993, In M. Hornung and R.A. Skeffington (Eds.): *Critical Loads: Concept and Applications.* ITE Symposium No.28, HMSO, London, pp. 94-98.
Metcalfe, S.E., Whyatt, D.J. and Derwent, R.G.: 1995, *Quart. J. Royal Met. Soc.* **121**, 1387-1411.
Metcalfe, S.E., Derwent, R.G., Whyatt, D.J. and Dyke, H.: 1998, *Water, Air, Soil Poll.* **107**, 121-45.
Nilsson, J. and Grennfelt, P. (Eds.): 1988, *Critical loads for sulphur and nitrogen.* UNECE/Nordic Council workshop report, Skokloster, Sweden, 19-24 March, 1988. Miljørapport 1988:15. Nordic Council of Ministers, Copenhagen, 418pp.
Posch, M., Kämäri, J., Forsius, M., Henriksen, A. and Wilander, A.: 1997, *Env. Man.* **21**(2), 291-304.
Posch, M., de Smet, P.A.M., Hettelingh, J.-P. and Downing, R.J. (Eds.): 1999, *Calculation and mapping of critical thresholds in Europe: Status Report 1999.* Coordination Centre for Effects, RIVM Report No. 259101009, Bilthoven, Netherlands, 165pp.
UBA: 1996, *Manual on methodologies and criteria for mapping critical levels/loads and geographical areas where they are exceeded.* Texte 71/96, Federal Environment Agency (Umweltbundesamt), Berlin, Germany, 142 pp plus Appendices.

VARIATION IN FRESHWATER CRITICAL LOADS ACROSS TWO UPLAND CATCHMENTS IN THE UK: IMPLICATIONS FOR CATCHMENT SCALE MANAGEMENT

M. KERNAN[1*], J. HALL[2], J. ULLYET[2] and T. ALLOTT[1]

[1] *Environmental Change Research Centre, 26 Bedford Way, London WC1H OAP, UK.;*
[2] *CEH Monks Wood, Abbots Ripton, Huntingdon, PE17 2LS, UK*
(* author for correspondence, email: mkernan@geog.ucl.ac.uk)

Abstract: In the UK the "critical loads" approach has been used to derive maps based on the 10km x 10km national grid. However, this grid based approach is inappropriate for catchment scale management and these maps cannot be used for "stock at risk" assessments of the number of water bodies or lengths of streams in a given area that may be vulnerable to acidification. Critical loads are determined across two large river catchments in England (The Duddon) and Wales (The Glaslyn). High resolution, digital datasets are used to characterise the attributes of each subcatchment in terms of land cover, soil, geology, topography and topology. Empirical models used to examine the relationship between these attributes and critical loads indicate that the former can be used to account for significant variation in the latter. However, these relationships can vary from catchment to catchment. Thus, although this approach provides the potential for identifying sensitive surface waters on a catchment wide basis, it is likely that models will need to be parameterised on a catchment specific basis.

Keywords: Critical loads, surface waters, catchment characteristics, catchment management

1. Introduction

Critical loads are defined as the "quantitative estimate of an exposure to one or more pollutants below which significant harmful effects on specified sensitive elements of the environment do not occur according to present knowledge " (Nilsson and Grennfelt, 1988). Currently, the present day surface water chemistry data needed to calculate critical load are not readily available for many surface waters in the UK. An alternative means of estimating the critical loads of sites for which no (or limited) water chemistry data exist is to use empirical relationships between freshwater critical loads and variables representing the various catchment mechanisms and attributes influencing the sensitivity of surface waters to acidification. These processes are well documented (*e.g.* Reuss and Johnson, 1986; Hornung *et al.*, 1990) and include the nature of catchment soils, geology and hydrology. Previous attempts to relate critical loads to catchment characteristics at a national scale have used national sensitivity maps to predict critical load class (Hall *et al.*, 1995) or explain variation in critical loads (Kernan, 1995; Kernan *et al.*, 1998). Using nationally available digital datasets, strong relationships have been established between variables representing catchment characteristics. However, poorer relationships were observed across reduced geo-graphical and attribute space (Kernan *et al.*, in press).

Water, Air, and Soil Pollution **130**: 1169–1174, 2001.
© 2001 *Kluwer Academic Publishers.*

This paper examines whether the empirical relationships identified regionally and nationally might be used for management at the catchment level. Data from the Glaslyn (Snowdonia, north Wales) and Duddon (north west England) catchments are used in this comparative case study.

2. Study Areas and Datasets Employed

The Glaslyn and Duddon catchments were selected for this case study as they are in areas vulnerable to acidification and both exhibit a range of sensitivities as defined by the national map of freshwater sensitivity (Hornung et al., 1995). Between October 1997 and May 1999 water samples were taken from 67 sub-catchments across both river systems. Four samples were taken at each site during autumn and spring. Mean chemistry data were used to calculate Diatom Critical Load (Battarbee et al., 1996) values for each of the sub-catchments.

Sub-catchment boundaries were digitised from 1:25,000 Ordnance Survey maps. These were overlaid onto the digital datasets described below to characterise each sub-catchment according to a series of catchment attributes.

Digital geology maps at 1:250,000 scale were used to determine the percentage of each geology type in each sub-catchment. Each geological unit was also allocated to one of four sensitivity classes based on buffering capacity (Edmunds and Kinniburgh, 1986) enabling the percentage of each sub-catchment in each of the four classes to be determined. Digital versions of the 1:625,000, 1:250,000 and 1:50,000 resolution geology maps were used to enable the classification to be applied at a range of scales.

Soil type comprises the percentage of different soil series in each sub-catchment. The data are derived from 1:250,000 digital soil maps (Soil Survey of England and Wales, 1983). Soil critical loads (SCL) in each catchment were derived from a provisional map of critical loads for acidity of soils for Great Britain (Hornung et al., 1994) from which the percentage of each SCL class in each sub-catchment was calculated.

For each sub-catchment, land cover data were obtained from the 25 m resolution LANDSAT TM database held at CEH Monks Wood (Fuller and Groom, 1993).

The length of the main stream within each sub-catchment and the upstream distance from the lowest point of the main catchment were derived from a GIS. As well as the sampling point altitude and maximum catchment altitude, the percentage of the sub-catchment above 300, 400 and 600 m was generated.

3. Analysis and Results

The range of critical loads is greater in the Duddon than the Glaslyn (0.28 – 4.02 keq ha^{-1} yr^{-1} as opposed to 0.05 – 2.33 keq ha^{-1} yr^{-1}, respectively). However, the Glaslyn exhibits a greater concentration of sensitive sites with 75% of all sites exhibiting DCL values less than 1 keq ha^{-1} yr^{-1}. Duddon catchment is generally less sensitive with only 50% of DCL values less than 1 keq ha^{-1} yr^{-1}.

Redundancy analysis (RDA) was employed to examine the relationships between the catchment characteristics in more detail. RDA allows the amount of variation in a response variable explained by one or more explanatory variables to be quantified. Analysis has been undertaken using both the Glaslyn and Duddon sub-catchments together and as separate datasets. Table I shows the amount of variation significantly explained by each of the catchment datasets.

TABLE I
Results of Redundancy Analyses undertaken on catchment data and critical load for both main catchments togther (B) and separately on the Duddon (D) and Glaslyn (G). The statistically significant level of DCL variation explained by each catchment dataset is shown.

Dataset	Catchment	% DCL explained	Dataset	Catchment	% DCL explained
Geology type	B	45	Soil critical load	B	29
	D	38		G	17
	G	26		D	24
Geological sensitivity (1 50,000)	B	8	Land cover	B	36
	D	None		G	67
	G	34		D	48
Geological sensitivity (1 250,000)	B	14	Location	B	none
	D	26		D	52
	G	31		G	42
Geological sensitivity (1 625,000)	B	15	Altitude	B	16
	D	None		D	60
	G	34		G	none
Soil type	B	35			
	G	67			
	D	34			

The type of geology explains 45% of the variation in DCL across both catchments. Separate analysis of the Duddon and Glaslyn shows that 38% and 26% of DCL variance, respectively, can be explained by the type of geology underlying each sub-catchment. However, given the heterogeneity of geological type nationally, the number of potential variables would preclude applying the model to catchments spread across a wide geographical area. Using the geological sensitivity (Edmunds and Kinniburgh, 1986) dataset attempts to overcome this problem by employing a 'universal' set of variables, applicable to any catchment nationally. However, this dataset explains only 15% of the variation in DCL (using 1:625,000 scale maps) across both main systems. A separate analysis of the Duddon sub-catchments produced no significant explanation whereas 34% of DCL variation in the Glaslyn could be explained by geological sensitivity. Applying the classification scheme to higher resolution data fails to improve the level of explanation with classifications based 1:250,000 and 1:50,000 geological maps accounting for 14% and 8%, respectively of the variation in DCL, across both catchments. There was no improvement when the two main systems were analysed separately.

Soil type explains 35% of DCL variation across both catchments. Separate analysis within the Duddon produces the same level of explanation whereas, in the Glaslyn this is almost doubled (67%). As with geology, many soil types occur only in particular areas. To provide universally applicable soil variables, soil critical loads (SCL) were used, this classification enabling the sensitivity of catchment soils to acidification to be used as an explanatory variable. However, across both catchments, the level of explanation using SCL is lower (29%) than when soil type is used. This is also the case for analysis of the Duddon (24%) and Glaslyn (17%) individually.

Land cover variables explain 48% of DCL variation in the Duddon catchment, lower than in the Glaslyn (67%). When both systems are considered together, only 46% percent of DCL variance can be accounted for.

Sample point location, fails to explain significant levels of critical load variation across the two main catchments. However, location is significant when the Duddon and Glaslyn are analysed separately, accounting for a total of 52% and 42% of DCL variation, respectively. Closer examination of these analyses reveals that, whereas stream length is positively correlated with critical load within both catchments, the relationship between DCL and distance upstream is negative in the Duddon but positive in the Glaslyn. Within the Duddon altitude explains 60% of the critical load variance whereas this dataset is not significant in the Glaslyn.

4. Discussion

Although the type of geology and soil explains a significant amount of critical load variance across both catchments (and in each separately) these variables can vary widely on a regional basis. It is not practical to calibrate workable empirical models incorporating all possible geology and soil types. Both classification systems used to overcome this heterogeneity explain less of the variation than geology (regardless of the resolution of the underlying data) and soil type. The efficacy of the classification system will depend on how it has been derived and whether the theoretical concepts underpinning the classification reflect the key processes and mechanisms that determine catchment sensitivity. Additionally, the utility of soil or geological type (or indeed their associated classifications) as explanatory variables, will vary from catchment to catchment depending on how soil or geology varies within the catchment. Also important is whether the resolution of the mapped data reflects the variety of soil and geology. For example, the presence of small calcite veins can result in a considerable increase in catchment buffering capacity (Reynolds *et al.*, 1986) and these are unlikely to appear in a spatially representative way in geology maps, even at the highest resolution generally available.

The land cover dataset employed here may offer more potential for the prediction of critical loads than the soil and geology datasets as the number of classes is less and the resolution of the data are higher. However, there are errors associated with the accuracy of this dataset (Fuller and Groom, 1993). Additionally, although land cover has been shown to explain critical load where a full range of land cover types are observed, where catchments are less heterogeneous, the potential for explanation is reduced (Kernan *et al.*, 1998; in press).

The location of each sub-catchment within main catchment appears to have a substantial bearing on the sensitivity of the stream at the sampling point. The fact that the length of the main stream in each sub-catchment is positively correlated with critical load is not unexpected as, generally the larger the contributing catchment area, the greater the potential for buffering. More surprising is the positive relationship between critical load and distance upstream in the Glaslyn. Whereas the more sensitive sub-catchments are in the upper reaches of the main Duddon catchment, in the Glaslyn this relationship does not hold. In the upper reaches of the Glaslyn, some of the sub-catchments draining Mount Snowdon are relatively non sensitive due to the nature of the underlying geology on the eastern slopes of Snowdon.

Similarly, the negative relationships between altitude variables and critical load in the Duddon are not repeated when analysis is confined to the Glaslyn sub-catchments. This is also likely to be due to the geographical variation in the sensitivity of geology whereby both sensitive and non-sensitive catchments are found at high altitudes. The lack of a significant relationship between altitude and DCL in the Glaslyn catchment, together with the dipolar relationship between DCL and distance upstream in the Duddon and Glaslyn catchments illustrates the need for care when using catchment characteristics to predict critical load on a catchment basis.

5. Conclusion

There are clear relationships between variables representing catchment characteristics and critical load. However, these relationships can vary considerably between catchments because of the complex levels of process integration occurring within these systems. Regional analyses have shown that the broad relationships between these nationally available digital datasets and critical loads can be used to indicate sensitivity across broad geographical areas. If empirical models are to be used for catchment management, from the results presented here, it is apparent that a catchment specific approach, requiring catchment specific parameterisation, may be necessary for effective identification of 'stock at risk' from acidification.

Acknowledgments

This work was funded under the NERC Environmental Diagnostics Programme (GST/02/1572). We thank Ron Harriman (FFL) for water chemistry, Robin Fuller (CEH Monks Wood) for land cover data, OS for providing a digitising license, SSLRC for soils data and BGS for digital geology data.

References

Battarbee, R.W., Allott, T, Juggins, S., Kreiser, A.., Curtis, C. and Harriman, R.: 1996, *Ambio* **25**, 366.
Edmunds, W.M. and Kinniburgh, D.G.: 1986, *Journal of the Geological. Society, London* **143**, 707.
Fuller, R,M. and Groom, G.B.: 1993, *GIS Europe* **2**, 25.
Hall, J.R., Wright, S.M., Sparks, T.H., Ullyet, J., Allott, T.E.H. and Hornung, M.: 1995, *Water, Air, and Soil Pollution* **85**, 2443.
Hornung, M., Le-Grice, S., Brown, N. and Norris, D.: 1990.: 'The role of geology an soils in controlling surface water acidity in Wales', in R.W. Edwards, A.S.Gee, and J.H. Stoner (eds.), *Acid Waters in Wales*, Klewer Academic Publishers, The Netherlands, pp. 55-66.
Hornung, M., Bull, K., Cresser, M., Hall. J.R., Loveland, P.J., Langan, S., Reynolds, B. and Robertson, W.H.: 1994, 'Mapping Critical Loads for the Soils of Great Britain', in R.W. Battarbee (ed.), *Acid Rain and its Impact: the Critical Loads Debate*. Proceedings of a conference held at the Environmental Change Research Centre, University College London, pp. 43-52.
Hornung, M., Bull, K., Cresser, M., Ullyett. J., Hall. J.R., Langan, S., Loveland, P.J. and Wilson, M.J.: 1995, *Environmental Pollution,* **87**, 204.
Kernan, M.: 1995, *Water, Air and Soil Pollution.* **85**, 2479.
Kernan, M., Allott, T.E.H. and Battarbee, R.W.: 1998, *Water Air and Soil Pollution,* **185**, 31.
Kernan, M., Helliwell, R.C. and Hughes, M.: *Water Air and Soil Pollution* (in press)
Nilsson, J, and Grennfelt, P.: 1988, *UNECE/Nordic Council Workshop report, Skokloster, Sweden. March 1988*, Nordic Council of Ministers, Copenhagen.
Reuss, J.O. and Johnson, D.W.: 1986, Acid Deposition and the Acidification of Soils and Waters. Springer-Verlag Inc., New York.
Reynolds, B., Neal, C., Hornung, M. and Stevens, P.A.: 1986, *Journal of Hydrology* **87**, 167.
Soil Survey of England and Wales: 1983 *Soil Survey of England and Wales, 1:250,000 Maps.*

VARIABILITY IN MAPPING ACIDIFICATION RISK SCENARIOS FOR TERRESTRIAL ECOSYSTEMS IN ASIAN COUNTRIES

JOHAN C.I. KUYLENSTIERNA[1*], W. KEVIN HICKS[1], STEVE CINDERBY[1], HARRY W. VALLACK[1] and MAGNUZ ENGARDT[2]

[1]*Stockholm Environment Institute at York (SEI-Y), University of York, York YO10 5YW, UK.;*
[2]*Swedish Meteorological and Hydrological Institute (SMHI), Norrköping, Sweden*
(*author for correspondence, e-mail: jck1@york.ac.uk)

Abstract. Acidification has the potential to become a widespread problem in parts of Asia. Just how widespread this risk may be is discussed by comparing sulphur deposition to critical load estimates, taking into account neutralising base cation deposition from soil dust. Two scenarios for the sulphur emission in 2025 are used as inputs to the MATCH atmospheric transfer model to estimate sulphur deposition scenarios. Net acidic deposition using a low and high base cation deposition input is compared to a map of sensitivity of terrestrial ecosystems to acidic deposition. Two ranges of critical loads assigned to this sensitivity map are used. The variability in the maps showing risks of acidification using low and high estimates for critical loads and base cation deposition for two different development pathways is discussed. Certain areas are shown to be at risk in all cases whereas others are very sensitive to the values used to estimate risk.

Key words: acidification risks, Asian countries, critical load, uncertainty, base cations

1. Introduction

Methods are required to estimate the risk of ecological detriment that is likely to be caused by increasing Asian emissions of sulphur and nitrogen. In addition to local impacts in and around urban areas, currently providing much of the focus in Asia, acidification of susceptible soils and waters represents a risk over widespread areas away from urban centres (Seip *et al.,* 1999). Various methods and models have been devised to investigate these risks and the critical load approach is the most widely used. Various parameters are required to investigate the risks using the critical load approach and there are uncertainties in all of them. Therefore, one pertinent question is how sensitive the risk distributions are to the different parameters and data values used. The parameters used in the evaluation of acidification risks are: deposition values for sulphur and nitrogen, neutralising base cation deposition values and critical load values. This paper, based upon work carried out under the Sida funded, SEI co-ordinated programme on Regional Air Pollution in Developing Countries, compares the area at risk projected for the year 2025 using different values for the various parameters to assess the degree to which uncertainty in the data affects the resulting risk projections.

2. Mapping acidification risks

The risk of acidification has been assessed by calculating the excess of modelled sulphur deposition above critical loads, taking into account the neutralising influence of base cation deposition. Nitrogen deposition represents an important acidifying component of deposition but is not included here due to the complexities involved in the nitrogen cycle. Regarding the sulphur cycle, the sulphate adsorption of the soil is not included.

The deposition of sulphur in 2025 has been modelled using the MATCH model (see Engardt this vol.) based upon sulphur emission scenarios for Asian regions (see Vallack *et al.*, this vol.). The emission scenarios have been developed at a scale of 1x1 degree and these values have been used in the MATCH model to give deposition estimates at the same scale. Two variations on a 'reference' scenario have been used here to investigate risks of acidification: one assuming no further control of air pollution beyond 1995 (considered a worst case scenario), and another assuming convergence of emission control technology in developing Asian countries to 1995 average levels for OECD countries by the year 2025 (Figure 1).

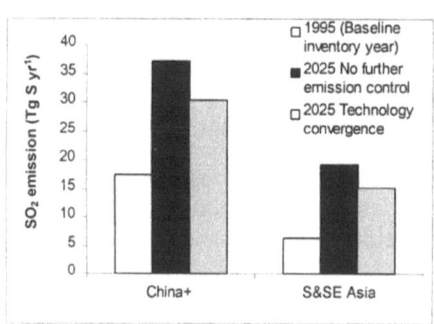

Figure 1. Sulphur emissions from two Asian regions according to two scenarios in comparison to the 1995 estimate (Note: China+: China, Vietnam, DPRK, Laos, Mongolia; S&SE Asia: South Asia and ASEAN countries, Republic of Korea (ROK), Taiwan). Source: Vallack *et al.*, (this volume).

The critical load distribution has been estimated for the globe using an assessment of five relative classes of sensitivity to which critical loads have been assigned. The sensitivity distribution is based upon soil base saturation (BS) and cation exchange capacity (CEC) for the top 50cm of soil (as described in Kuylenstierna *et al.*, 2001), mapped using the Soil Map of the World (FAO, 1995) and soil physical and chemical characteristics from the WISE database (ISRIC, 1995).

Critical load values have been assigned to the mapped classes of sensitivity based upon weathering rates (WR). Soil mineral weathering represents a long term buffering rate and has been used as a value to classify soils into critical load classes using the so-called 'Skokolster Class Approach' (Hornung, 1993; Nilsson and Grennfelt, 1988). In order to assign values to the sensitivity classes, data from sites where weathering rates were modelled and BS and CEC have been measured have been compiled, mainly from catchment studies in the UK, Scandinavia, Spain and China (Kuylenstierna *et al.*, 2001). The BS and CEC were used to classify soils at the sites into sensitivity classes and these are shown against the estimated WR in Figure 2. Most of the sites shown relate to catchments in the most sensitive areas, as most acidification work has concentrated on these sensitive sites. Critical loads have been assigned to sensitivity classes based on the comparison in Figure 2. For example, it can be seen that for sensitivity Class 1 most of the weathering rates lie between 10 and 50 meq m^{-2} yr^{-1}. A high and a low range of critical load for Class 1 (0-25 and 25-

50 meq m^{-2} yr^{-1}) were used in order to investigate the influence of critical load uncertainty on the area of excess. Critical load ranges for the other classes were also estimated (Table I). For comparison, the critical loads that were assigned to classes of weathering in the Skokloster assessment are also shown in Table I.

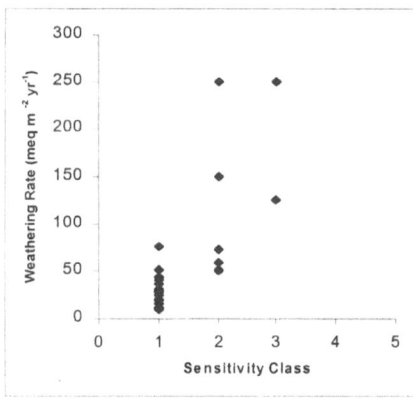

Figure 2. Site specific data from the literature for estimated weathering rate compared to sensitivity class (derived from the site BS and CEC) (source of data is described in Kuylenstierna *et al.*, 2001).

TABLE I
Preliminary critical load ranges assigned to relative sensitivity classes (the highest value has been used in exceedance calculations). The Skokloster ranges are given for comparison.

Sensitivity class	Critical (C.L.) load range (meq m^{-2} yr^{-1})		Skokloster assignment of critical loads (Nilsson and Grennfelt, 1988)
	Low range	High range	
1	<25	<50	<20
2	25-50	50-100	20-50
3	50-100	100-200	50-100
4	100-200	200-400	100-200
5	>200 (no critical load)	>400 (no C.L.)	>200 (no C.L.)

Base cation deposition from soil dust sources alone at global scale (Kuylenstierna *et al.*, 2001) has been estimated by the Stockholm Environment Institute (SEI) by taking global dust deposition estimates (from Tegen and Fung, 1995) and an estimate of the average calcium carbonate content of 10% in Asian wind-blown soil dust (based on Gomes and Gillette, 1993). The degree of neutralisation by base cation deposition is highly uncertain in Asia. The size of this uncertainty is illustrated in the comparison of the range of values suggested by different authors in Europe and China (Table II). The estimates based upon the global modelling of Tegen and Fung (1995), using the 10% calcium content, are consistently lower than the other estimates. One reason for this is that the other estimates are partly based upon measurements in wet deposition, which will include both soil dust and industrially derived base cations, whereas the Tegen and Fung dust deposition only refers to soil dust. Industrial emissions would particularly account for higher values in central Europe and the Guangxi province of China where there are large industrial emissions. To investigate the influence of uncertainty in base cation deposition distributions a low range of estimates have been used in the exceedance calculations represented by the Tegen and Fung dust deposition with 10% calcium assumed and a high range at four times that value (based upon differences between estimates in Table II).

TABLE II
Approximate base cation deposition ranges for different regions from different mapped estimates.

Region	ESTIMATED BASE CATION DEPOSITION (meq m^{-2} yr^{-1})			
	Kuylenstierna et al (2001) based on Tegen and Fung (1995)	Larssen and Carmichael (2000)	CCE (1999)	van Leeuwen et al (1996) (wet only)
S. Europe (Spain)	25-50		70-150	50-100
C. Europe	5-25		20-70	20-200
N. Europe (Nordic)	1-5		0-40	0-20
Guangxi	5-25	50-200		
Yunan	10-25	25-100		
Manchuria	25-125	50-200		

Exceedance calculations have been carried out using the two scenarios in 2025 and the different ranges of base cation deposition and critical load. Table III shows the land area exceeded in Asia when using the different combinations. Two exceedance maps are shown to illustrate where the exceedance occurs. Figure 3 shows the maximum exceedance distribution in 2025 that results from using the low range of critical loads, low base cation deposition and the high emission scenario. Figure 4 shows the lowest amount of exceedance using the lower emission scenario, high critical loads and base cation deposition values, but still there are large areas of exceedance in Asia even though the geographic extent has decreased by 54 %.

TABLE III
Area of terrestrial ecosystem critical loads exceeded by sulphur deposition in Asia in 2025 (in 000 km^2) for a high (no further control) and low (control technology convergence) emission scenario, high and low base cation deposition (BC$_{dep}$) and high and low critical load (CL) (see Table I).

		Low BC$_{dep}$		High BC$_{dep}$	
		Low CL	High CL	Low CL	High CL
High Scenario	Total area of excess	3727	2850	2589	1963
Low Scenario	Total area of excess	3333	2466	2223	1713
High Scenario	Area of high excess*	1664	1319	1237	1043
Low Scenario	Area of high excess*	1363	1079	992	720

*'high excess' refers to areas with an exceedance >100meq m^{-2} yr^{-1}

3. Conclusions

Using the range of values for parameters explained in this paper, Table III indicates that the distribution of exceedance is most influenced by quadrupling the base cation deposition. Doubling the critical load has a similar, but smaller effect on area of exceedance. The emission scenarios have least influence but this is partly due to the fact that the resulting emissions only differ by 20%. The area of high exceedance (Table III) closely follows the percentage reduction in

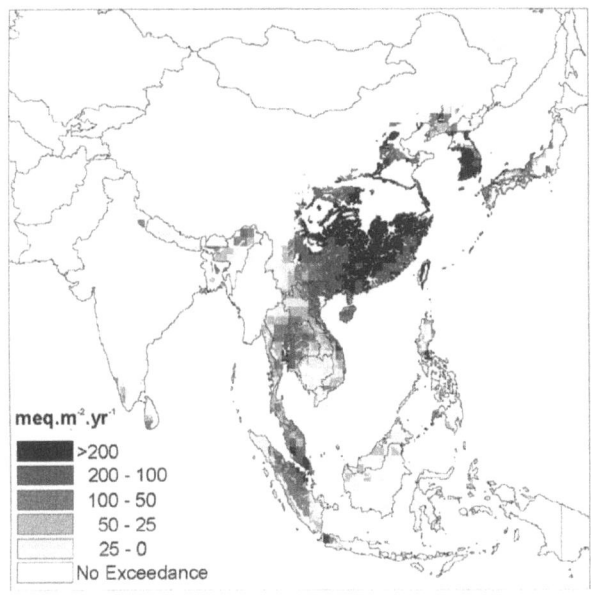

Figure 3. Map of exceedance of critical loads in 2025 using low BC_{dep} and low CL range (see Table I) and the high emission scenario (no further control of air pollution beyond 1995).

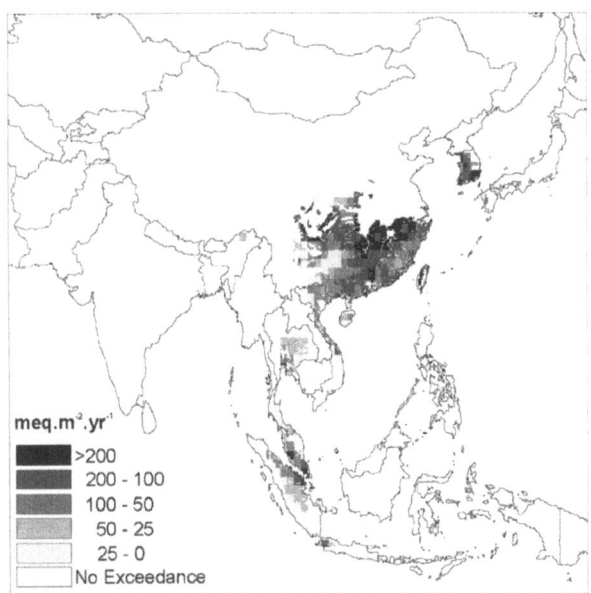

Figure 4. Map of exceedance of critical loads in 2025 using high BC_{dep} and high CL range (see Table I) and the low emission scenario (convergence of emission control technology in developing Asian countries to 1995 average levels for OECD countries by the year 2025).

emissions. The base cation deposition has a lower than expected effect on variability due to the fact that sources of dry alkaline soil dust (e.g. N. China) tend to be remote from sensitive acidic soils (e.g. in S. China and SE Asian countries) whereas the critical load relates to those soils.

It is clear that the data used to estimate areas at risk from acidic deposition have a major influence on the resulting risk distributions. It is also clear that in some regions, within the range of values used to represent parameters, there would seem to be projected risks of acidification in all cases. The area of risk is expected to decrease in line with emission reductions but in order to gain a more accurate picture of future acidification risks, further research into weathering rates and base cation deposition will be required.

Acknowledgements

We would like to acknowledge Henning Rodhe, Jeff Wilson, Philip Ineson, and Hans-Martin Seip for their support, as well as Sida for their financing of the projects.

References

CCE: 1999, Summary of national data. In Calculation and Mapping of Critical Thresholds in Europe, Status Report 1999 (ed., by M. Posch, P.A.M. de Smet, J.-P. Hettelingh and R.J. Downing). Co-ordination Centre for Effects. RIVM, Bilthoven.

Engardt, M. (this volume). Sulphur simulations for East Asia using the MATCH Model with Meteorological Data from ECMWF. *Water, Air and Soil Pollution* (this volume).

FAO: 1995, FAO Digital Soil Map of the World. Scale: 1:5,000,000. FAO, Rome, Italy.

Gomes, L. and Gillette, D.A.: 1993, A comparison of characteristics of aerosol from dust storms in central Asia with soil-derived dust from other regions. *Atmos. Env.* **27A**, 2539-2544.

Hornung, M.: 1993, The provisional map of critical loads of acidity for the soils of Great Britain. In: Critical Loads: Concept and Applications (ed. by M Hornung and R.A. Skeffington). *ITE Symposium,* **28**. HMSO, London.

ISRIC: 1995, *World Inventory of Soil Emission Potentials.* International Soil Reference and Information Centre, Wageningen, The Netherlands.

Kuylenstierna, J.C.I, Rodhe, H., Cinderby, S. and Hicks, W.K.: 2001, Acidification in developing countries: ecosystem sensitivity and the critical load approach at global scale. *Ambio*, **30** (1), 20-28.

Larsson, T. and Carmichael, G.R.: 2000, Acid rain and acidification in China: the importance of base cation deposition. *Environmental Pollution*, **110**, 89-102.

Nilsson, J. and Grennfelt, P.: 1988, *Critical Loads for Sulphur and Nitrogen*. Miljörapport **1988:15**. Nordic Council of Ministers. Copenhagen.

Seip, H.M., Aagaard, P., Angell, V., Eilertsen, O., Larssen, T., Lydersen, E., Mulder, J., Muniz, I.P., Semb, A., Tang, D.G., Vogt, R.D., Xiao, J.S., Xiong, J.L., Zhao, D.W. and Kong, G.H.: 1999, Acidification in China: Assessment based on studies at forested sites from Chongqing to Guangzhou. *Ambio*, **28**(6), 522-528

Tegen, I. and Fung, I.: 1995, Contribution to the atmospheric mineral aerosol load from land surface modification. *Journal of Geophysical Research*, **100**, 18707-18726.

Vallack, H.W., Cinderby, S., Kuylenstierna, J.C.I. and Heaps, C.: (this volume). Emissions inventories for anthropogenic sources of SO_2 and NO_x in developing country regions in 1995 with projected emissions for 2025 according to two scenarios. *Water, Air and Soil Pollution* (this volume).

Van Leeuwen,E.P., Draaijers,G.P.J. and Erisman,J.W.: 1996, Mapping wet deposition of acidifying components and base cations over Europe using measurements. *Atmos. Env.*, **30**, 2495-2511.

AIR POLLUTION EFFECTS ON TERRESTRIAL ECOSYSTEMS IN ESTONIA

LEO SAARE*, REET TALKOP and OTT ROOTS

Estonian Environment Information Centre, Mustamae tee 33, 10616 Tallinn, ESTONIA;
(* author for correspondence, Leo.Saare@ic.envir.ee)

Abstract. A number of positive changes have taken place since Estonia regained its independence in 1991. Air pollution from stationary sources has decreased over 2.5 times during 1990-1999, emissions of solid particles and SO_2 have declined 74% and 60%, respectively. The content of heavy metals in Estonian mosses has decreased in comparison with the early 1990s. Last five years occurrence of different kind of damages on decidious trees has not been frequent. Those facts indicate that air pollution with heavy metals and other pollutants has diminished during the last few years. As the pH of precipitation fluctuates in different parts of Estonia, it is very important to study the effect of precipitation on ecosystems on the basis of critical loads. Results indicate that, as for eutrophicating nitrogen, the actual nitrogen deposition in North-East Estonia and West-Estonian islands roughly coincides with the limits for pollution endurance. This pattern also applies to the total deposition of sulphur and nitrogen in South, North and North-East Estonia, although in some Northern and North-Eastern areas pollution endurance limits have been exceeded.

Keywords: air pollution, critical loads, Estonia, forests, mosses, terrestrial ecosystems

1. Introduction

Since early 1990s, certain agricultural and industrial activities have been reduced in Estonia because of the political and the economic changes. Reduced levels of S and N pollution, decreased use of fertilizers, and increasing areas of abandoned agricultural land are the main factors affecting ecosystem changes. Growth in the number of automobiles increases the potential for ozone production and changes in air pollution status.

The Baltic Sea has a particularly strong influence on the wind regime in coastal areas (Kull, 1999). Westerly and southwesterly winds are dominant with mean speed between 4 and 5 m/s. These data provide the potential to predict and characterise pollutant transfer through the air to different regions.

The sensitivity of any soil-vegetation system to atmospheric pollution can be assessed using the critical loads concept (Oja and Kull, 2000). Mapping of critical loads for sulphur (S) and nitrogen (N) in elementary ecosystems in Estonia has been undertaken (Oja *et al.*, 1998).

Currently, Estonian problems are still connected with oil-shale mining and processing (Roots *et al.*, 1998). Air pollution from stationary sources (total emission) has decreased over 2 times during 1990-1999, and emissions of solid particles and SO_2 have declined 74% and 60%, respectively. This paper estimates the environmental impact of deposition of atmospheric pollutants.

Water, Air, and Soil Pollution **130**: 1181–1186, 2001.
© 2001 *Kluwer Academic Publishers.*

2. Materials and Methods

Estonia is currently participating in 5 pan-European programs of International Cooperation under the UN/ECE Convention of Long-Range Transboundary Air Pollution. These are international programs for effects on air, mosses, forests, cultural objects and ecosystems.

Precipitation samples were taken and analysed using European Monitoring and Evaluation Program (EMEP) standardized samplers and internationally approved methods (EMEP Manual, 1996). Measured deposition of S and N loads are studied in 19 stations (Otsa and Kört, 2000).

AOT40 or ozone exposition over 40 pbb (80 µg/m^3) was calculated. AOT40 reflects accumulated O_3 exposure above a threshold of 40 ppb within the exposure period. This is the level when bioproductivity of plants is significantly impared by O_3 (reduces 10%). For cereals, the calculations are undertaken between May and July during the daytime and for forests daily from April to September (Otsa and Kört, 2000).

Concentrations of Cd, Cu, Ni, Pb and Zn were determined in mosses *Pleurozium schreberi and Hylocomium splendens*. Moss samples were taken in small gaps in pine forests of *Vaccinium* type. Each sample contained 2 litres of moss. The samples were cleaned of decaying wood and other plant remains, dried at 40°C to a constant weight. Wet ashing of the samples by HNO_3 and $HClO_4$ was used. Analyses were done using method of plasma emission spectrometry (ICP-ES) (Liiv *et al.*, 1996).

The monitoring of forests and forest soils was continued at two levels: the condition of forest trees was assessed on 91 permanent observation plots set up as a 16x16 km network (level I) and on 6 sample plots for intensive monitoring (level II) where one of the basic indicators was needle/leaf loss from the crowns of the observation trees (Terasmaa, 1999).

Critical loads for Estonia were calculated on the grounds of unified method for general acidity (S + N) and eutrophication (N), following internationally agreed methodology (Posch *et al.*, 1996). The switch to the 50x50 km grid enabled more detailed observation of different ecosystems mapped. The empirical method for raised bogs and steady-state mass balance calculations for forest soil as a receptor at forested areas were used (Oja *et al.*, 1998).

3. Results and discussion

3.1. PRECIPITATION

In 1999 the most acidic precipitation was in South, West and South-West Estonia (Table I). The influence of the Baltic Sea could be seen in North and North-East Estonia with higher concentrations of sodium and chloride. Higher calcium concentration in precipitation from the cement and oil-shale production compensates for effects of acidifying elements. Both effects have decreased during 1990s. Anyway, influence of the long-term alkaline precipitation has already caused changes in acidic state of balance of bogs ecosystem in North-East Estonia (Otsa and Kört, 2000).

TABLE I
Mean concentration of pollutants in precipitation in different parts of Estonia in 1999 (mg/l)

Sites	Bulk, mm	pH	NH_4^+	NO_3^-	Cl^-	SO_4^{2-}	Ca^{2+}	Mg^{2+}	Na^+	K^+
Kunda (N)	525	5.79	0.06	0.43	1.64	4.83	8.68	0.56	0.77	0.62
Saka (NE)	559	5.36	0.84	0.29	1.17	5.65	1.99	0.31	0.50	0.35
Jõhvi (NE)	599	6.44	0.58	0.45	2.95	8.41	3.58	0.36	1.70	1.17
Vilsandi (W)	444	4.86	0.32	0.38	1.08	0.60	0.51	0.12	0.55	0.24
Tiirikoja (E)	516	6.03	0.43	-	1.09	3.35	1.14	0.15	0.72	-
Tahkuse (SW)	421	5.38	0.37	0.37	0.44	1.54	1.89	0.34	0.34	0.28
Haanja (SE)	343	5.18	0.41	0.27	0.39	1.36	1.08	0.11	0.35	0.20
Karula (S)	377	5.24	0.98	0.31	0.48	1.91	1.28	0.21	0.32	0.51
Otepää (S)	373	5.79	0.45	0.26	0.51	1.28	1.32	0.19	0.38	0.34
Al-Pedja (E)	245	5.83	0.44	0.33	0.47	1.42	1.21	0.16	0.31	0.22
Nigula (SW)	376	5.70	0.59	0.48	0.58	1.45	1.36	0.19	0.43	0.23

N-North; W-West; S-South; E-East; SW-South –West; SE-South-East; N-E – North-East;

A comparison of pollutant concentration with modelled intensity of local pollutant emissions and wind velocity indicates that the main pollution source for South and South-West Estonia is transboundary pollution from Central Europe. In North Estonia local pollution (towns, enterprises) is dominant (Kull, 1966). West Estonia and islands are more influenced by long-range pollution brought by winds (acidic precipitation). Comparing the data with other countries (Table II) the depositions are similar to that in Norway, but much lower in comparison with the Czech Republic (Otsa and Kört, 2000).

TABLE II
Wet deposition of sulphur and nitrogen in European countries (g/m² per year)*

Country	Sulphur (S g/m² per year)	Nitrogen (N g/m² per year)
Estonia	0.5-2	0.1-0.7
Norway	0.2-1.7	0.2-2.5
Czech Republic	0.5-5.0	0.3-5.0

*Wet and dry deposition are comparatively on the same level (Otsa and Kört, 2000)

Ozone levels are quite high in Estonia in all stations. Monthly mean concentrations of O_3 in Vilsandi background station were over the daily permitted level in Estonia (65 µg/m³) on 259 days in 1999, exceeding long-term critical loads. AOT40 calculations were carried out to examine the effect of high concentrations of ozone. Table III shows that Vilsandi (and probably some other

Estonian islands) is situated in areas with poor conditions for vegetation growth.

TABLE III
Ozone expositions (AOT40) in Estonian background stations in 1999 (ppbh)

Site	Vilsandi (W)	Lahemaa (N)	Saarejärve (E)	Long-term CL
AOT40 cereals	12223	5358	983	5300
AOT40 forests	28327	2214	359	10000

A high correlation between ozone and nitrogen oxides (mainly NO_2) was observed in winter (Iher et al., 2000). Meteorological conditions determine the dispersion of NO_2 in the atmosphere. Precipitation with enhanced content of polluting substances sometimes arrives from Central Europe. For example, on 15 January, 1998, the concentration of NO_2 rose about 4 times during 3 hours at the beginning of the snowfall. The content of NO_3^-ions (nitrate) in this precipitation exceeded the annual average by 10 times.

3.2. MOSSES

Pleurozium schreberi and *Hylocomium splendens* as ectohydrous mosses accumulate heavy metals proportionally with their content in the air (Liiv, Eensaar, 1998).

TABLE IV
Heavy metals concentrations in *Pleurozium schreberi* in 1999 comparing to 1995 (µg/g)

Site	Cd 1995/1999		Cu 1995/1999		Ni 1995/1999		Pb 1995/1999		Zn 1995/1999	
Uhtna (NE)	0.20	0.18	3.86	3.39	1.1	-	6.4	2.5	27	22
Kurtna (NE)	0.12	0.18	3.26	3.61	1.8	1.2	8.5	4.5	34	25
Koorküla (S)	0.16	0.21	3.47	3.85	3.0	1.3	12.1	5.8	35	24
Kil-Nõmme (SW)	0.32	0.21	4.16	3.15	2.4	-	8.4	4.0	32	26
Aulepa (NW)	0.25	0.16	3.18	2.29	1.6	1.7	8.1	3.5	55	20
Lasila (NE)	0.33	0.13	6.43	2.87	1.0	1.1	5.8	6.1	39	29
Parika (Centr)	0.02	0.14	2.37	2.00	1.5	-	9.7	2.8	28	22

A comparison of the data 1999 with the data 1995, shows that the Cd and Zn concentration in *P. schreberi* has decreased at some sampling sites. The mean content of Ni and Pb in *P. schreberi* in North-East Estonia at Kurtna sampling plot has decreased in 1999 compared with 1995. The results reflect a constant decrease in most atmospheric heavy metal deposition in North-East Estonia. One reason for this is the reduced level heavy metals emissions from large power plants and therefore a lower accumulation into the mosses.

The ratio of the mean content of the same elements in *P. schreberi* to that in *H. splendens* was 0.9-1.1 (Liiv and Tamm, 2000). The content of heavy metals in these two mosses showed a good correlation and could be considered together without recalculation. *P.Schreberi,* more sensitive to alkaline pollution, does not grow in some sampling points in North-East Estonia and *H.splendens* could be used there. In North-East Estonia, the level of heavy metals concentration is, in most cases, higher than the Estonian average. The highest concentration of heavy metals is within 3 km from the Estonian Power Plant in north-east. The concentration of heavy metals falls to the Estonian average level 10 km to the south-west of the Power Plant.

3.3. FOREST

Today, nearly 39% of Estonian forest is pine stands, 28% birch and 23% spruce. Trees with 10% needle/leaf lost from crowns are considered to be entirely healthy. 100% defoliation is characteristic of dead trees. During the last 5 years the percentage of entirely healthy birches has fluctuated between 75% and 96%. At the same period the state of pine crowns was improving: percentage of pines with defoliation less than 20% has increased from 69.6% to 83.5% and for spruces the percentage was fluctuating between 85.4% and 90.8%. The pH of forest soil varied between slightly acidic and slightly alkaline, heavy metals (Zn, Pb and Cd) content in forest soil is on the background level (Terasmaa, 2000).

3.4. CRITICAL LOADS

Ecosystems characteristics and soil properties are the main components determining the critical load of atmospheric deposition. Between 1995-1997 critical loads were calculated on 150x150 km and 50x50 km grid cells using 5% percentiles threshold (Oja *et al.*, 1998; Oja and Kull, 2000). Tree species affect nitrogen uptake (and further removal from the site) depending on the ratio of the limiting base cation and the nitrogen in plant tissues. The proportion of birch forests will be increased in Estonia. Birch with its lower N uptake and higher leaching and nutrient leaching from foliar litter will lower critical loads. Forestation of grasslands and agricultural lands increases N uptake and thus, critical loads.

Comparing two regions with deposition equal to critical loads, North-East Estonia is characterized by substantially increased basic cation deposition buffering fully the acidic deposition while in South Estonia, basic cation depositions correspond to those of the natural backgound and deposition reaching or exceeding the critical loads in some more sensitive areas have an acidifying impact.

The areas where actual deposition is equal to critical loads (or exceeds it slightly (up to 20%) could be found in West, South and North-East Estonia. Only in Kunda area (North Estonia) the actual deposition clearly exceeds the critical loads of acidifying impact ($CL_{max}(N)$ 4808, $CL_{max}(S)$ 2150, $CL_{min}(N)$ 260 mol_c/ha yr) for the natural background by 1000 mol_c/ha yr.

4. Conclusions

During last 10 years deposition of most of the investigated substances have diminished in Estonia. The most acidic precipitation with enhanced content of polluting substances (NO_2) generally originates from south-west. An essential amount of NO_2 is transported by wind from south-west and north-east. The most acidic precipitation area is in South Estonia.

Ozone concentrations continuously show a remarkably high level all over Estonia, especially on islands, where AOT40 indicates poor conditions for plant growth.

Exceedances of critical loads found in some areas (except Kunda) remain within the limits of the calculation and measurement error and suggest just critical loads are being achieved rather than being exceeded. A further increase in nitrogen deposition will clearly bring about an exceedance of critical loads.

Acknowledgements

The authors would like to thank Siiri Liiv, Tõnu Oja, Hilja Iher, Tõnu Terasmaa and Margus Kört for their contribution on research and writing.

References

1996, *EMEP Manual for Sampling and Chemical Analysis.* EMEP/CCC Report 1/95, NILU, Kjeller, Norway.
Iher, H., Hõrrak, U. and Salm, J.: 2000, *Precipitation monitoring in Tahkuse.* Report, MoE, Envir. Inform. Centre, Tallinn.
Kull, A.: 1999, 'Estonian Wind Climate' in *Publicationes instituti Geographici Universitatis Tartuensis,* Tartu **85**, pp. 86-93.
Liiv, S., Sander, E. and Eensaar, A.:1996, *Assessment of Heavy Metal Deposition by Mosses.*Tallinn Botanical Garden, Tallinn (in Estonian).
Liiv, S. and Eensaar, A.: 1998, *Bioindication.* Report, MoE, Envir. Inform. Centre, Tallinn.
Liiv, S. and Tamm, H.: 2000, *Bioindication.* Report, MoE, Envir. Inform. Centre, Tallinn.
Oja, T. and Kull, A.: 2000, 'Sensitivity of Landscape to Atmospheric Pollution' in Ü. Mander and R.H.G. Jongman (eds), *Consequences of Land Use Changes. Advances in Ecological Sciences,* WIT Press, Southampton, Boston. pp. 147-162.
Oja, T., Kull, A. and Tamm, T.: 1998, 'Critical Loads for Air Pollution in Estonia', in O. Roots and R. Talkop (eds), *Estonian Environmental Monitoring,* MoE, Envir. Inform. Centre, Tallinn, pp.34-37.
Otsa, E. and Kört, M.: 2000, *Air Monitoring and Precipitation Chemistry.* Report, MoE, Envir. Inform. Centre, Tallinn.
Posch,M., de Smet, P.A.M., Hettelingh, J.-P. and Downing, R.J.: 1995, *Calculation and Mapping of Critical Threshoal in Europe:* Status Report, Coord. Centre for Effects, RIVM, Bilthoven.
Roots, O., Saare, L. and Talkop, R.: 1997, *Ecological Chemistry* **6**, 2, 128 (St. Petersburg, Russia).
Terasmaa, T.: 2000, *State of Estonian Forests.* Report, MoE, Envir. Inform. Centre, Tallinn (in Estonian).

CRITICAL LOADS OF ACID DEPOSITION FOR ECOSYSTEMS IN SOUTH CHINA — DERIVED BY A NEW METHOD

FULU TAO[1,2]* and ZONGWEI FENG [2]

[1] *Agrometeorology Institute, Chinese Academy of Agricultural Sciences, Beijing 100081, China;*
[2] *Research Center for Eco-Environmental Sciences, Chinese Academy of Sciences.*
Beijing 100085, China.
(* author for correspondence, e-mail: Taofl@ns.ami.ac.cn)

Abstract. Critical loads of acid deposition for ecosystems in South China are derived by synthesizing the critical loads of acid deposition for soils, the critical loads of SO_2 dry deposition for ecosystems, as well their exceedance. The results show in the southeast of Sichuan province around Chongqing municipality, the central and north of Guizhou province around Guiyang municipality, and the most areas of Jiangsu province, both the critical loads for soils and critical loads of SO_2 dry deposition are exceeded. In Guangxi Zhuang Autonomous Region and some areas among Jiangxi, Zhejiang and Anhui provinces, the critical loads of SO_2 dry deposition is the only restricting factor. There is no area where the critical load for soil is the only restricting factor in South China, so only the critical load for soil is not enough to be the basis to make sulfur abatement scheme.

Keywords: critical loads of acid deposition, SO_2 dry deposition, exceedance, South China

1. Introduction

With regard to the effects of acid deposition on vegetation it is not only the indirect impact via soil, but also the direct above-ground exposure to certain concentrations and the above-ground uptake (dry deposition) which is important. However the critical loads of acid deposition, which are usually derived by soil acidification models so far, only reflect the indirect effects of acid deposition, rather than the direct effects of dry deposition (Tao and Feng, 1999a). The concept of critical loads has consequently been criticized because of the difficulties in estimating dry deposition and deposition via fog on the required scales.

In South China, the direct effects of SO_2 and indirect effects of acid deposition (acidification) on ecosystems have both been demonstrated (Shen et al., 1995; Feng and Tao, 1998). Atmospheric concentrations of SO_2 have been shown to cause direct damage to natural ecosystems and crops, as well as having health effects on local and regional scale, and are even supposed to be the more likely cause than acid deposition for the dieback of the masson pine trees in some areas of South China (Shen et al., 1995). So both the direct effects of SO_2 and indirect effects of acid deposition on ecosystems should be simultaneously taken into account.

In this paper, the critical loads of acid deposition for ecosystems in South

China are derived by synthesizing the critical loads of acid deposition for soils, the critical loads of SO_2 dry deposition for ecosystems, as well their exceedance. The use of the critical load map for making efficient sulfur abatement is also discussed.

2. Description of the Study Area

The region influenced by acid deposition in China lies mainly in South China, including Jiangsu, Anhui, Zhejiang, Fujian, Jiangxi, Hubei, Hunan, Guangdong, Guizhou, Guangxi Zhuang Autonomous Region, the eastern part of Sichuan and Yunnan provinces, Chongqing and Shanghai municipalities (Fig. 1). See also Tao and Feng, 2000a for detailed description

Figure 1. The study area

3. Critical Loads of Acid Deposition for Soils and their Exceedance

3.1. CRITICAL LOADS OF ACID DEPOSITION FOR SOILS

Terrestrial ecosystem sensitivity to acid deposition in South China was assessed by combining the soil type, bedrock lithology, land cover and moisture profit and loss (Tao and Feng, 2000a). Then the critical loads of acid deposition for soils were mapped by combining the ecosystem sensitivity with site-specific studies conducted by MAGIC (Model of Acidification of Groundwater in Catchments) (Tao and Feng, 1999b); (Fig. 2).

The critical loads of acid deposition for soils in South China vary from 2.3~5.2 $gSm^{-2}yr^{-1}$ and increase from the southeast to the northwest on the whole. The most sensitive areas where the critical loads are less than 3.0 $gSm^{-2}yr^{-1}$ are the south of Zhejiang province, the areas between Fujian and Guangdong provinces, the southwest part of Guizhou province and the central part of Guangxi Zhuang Autonomous Region.

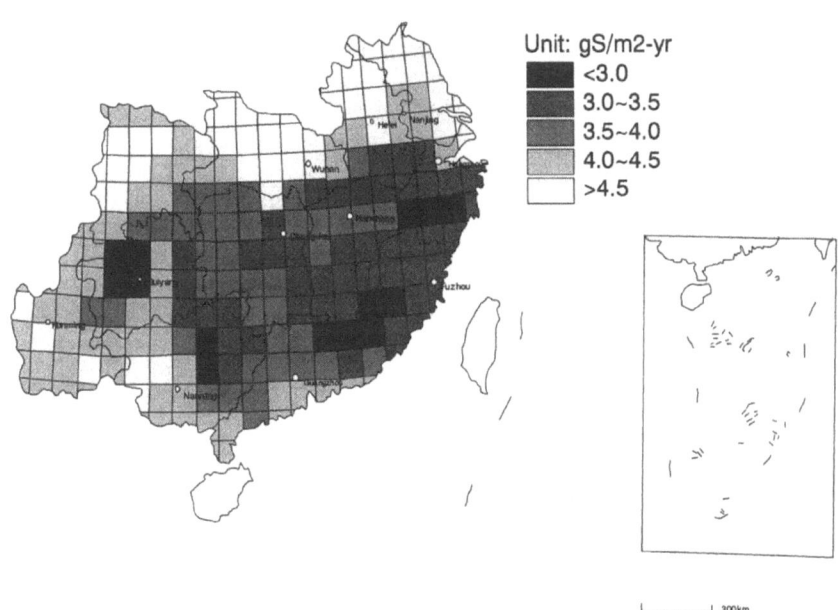

Figure 2. Critical loads of acid deposition for soils in South China

3.2. EXCEEDANCE OF CRITICAL LOADS OF ACID DEPOSITION FOR SOILS

The exceedance of critical loads for soils can be derived from the critical loads of acid deposition for soils and the total sulfur deposition (Dai *et al.*, 1998). In the Figure 1, if the values of 2.625 gSm^{-2}yr^{-1}, 3.125 gSm^{-2}yr^{-1}, 3.625 gSm^{-2}yr^{-1}, 4.125 gSm^{-2}yr^{-1} and 4.625 gSm^{-2}yr^{-1} are respectively selected to represent the ranges of <3.0 gSm^{-2}yr^{-1}, 3.0~3.5 gSm^{-2}yr^{-1}, 3.5~4.0 gSm^{-2}yr^{-1}, 4.0~4.5 gSm^{-2}yr^{-1}, >4.5 gSm^{-2}yr^{-1}, and the highest deposition is selected as the representative value in every grid cell, then the excess of critical loads for soils in South China is computed and mapped as Figure 3. The areas with excess sulfur deposition located in the southeast of Sichuan province, the central and north of Guizhou province, and the most parts of Jiangsu province. The areas with the highest excess (above 2 gSm^{-2}yr^{-1}) located in the areas around Chongqing municipality and Guiyang city.

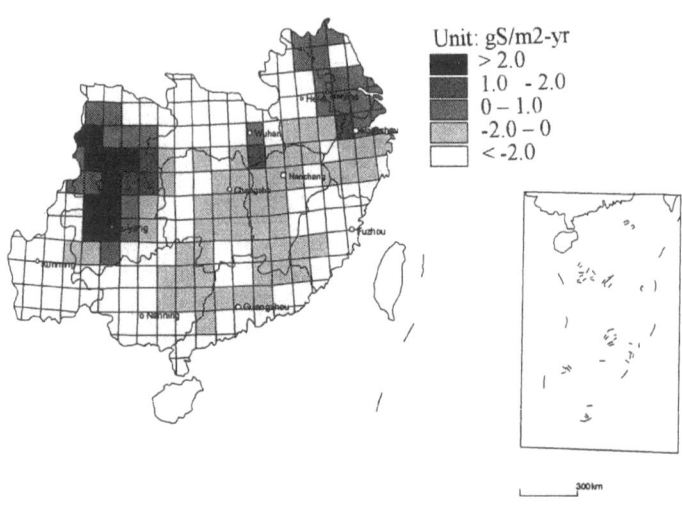

Figure 3. Excess of critical loads for soils in South China

4. Critical Loads of SO$_2$ Dry Deposition for Ecosystems and their Exceedance

Atmospheric concentrations of SO$_2$ can cause direct damage to natural ecosystems and crops, when above the SO$_2$ critical level. However the critical level, represented as air concentration, may not be conveniently used to control pollutants emission as critical load, so it is transferred to equivalent SO$_2$ dry deposition load called the critical load of SO_2 dry deposition. The critical loads of SO_2 dry deposition in South China were derived from SO_2 dry deposition velocity and critical level of SO_2 for vegetation by inferential technique. The SO_2 dry deposition velocity was from simulation and observations. The SO_2 critical levels for coniferous forest, crop and broadleaf forest are assumed to be annual average 20 g m^{-3}, 22.5 g m^{-3} and 25 g m^{-3} according to some references. The exceedance of the critical loads of SO_2 dry deposition in South China was computed and mapped (see also Tao and Feng, 2000b).

5. Synthesizing the Critical Loads for Soils and Critical Loads of SO_2 Dry Deposition

An attempt to synthesize the critical loads for soils and the critical loads of SO_2 dry deposition on one map is needed to prevent confusion among policy-makers when considering abatement schemes. Firstly the restricting factor(s) should be determined in every grid cell according to the critical loads and the degrees to which they are exceeded. The restricting factors can be classified into four

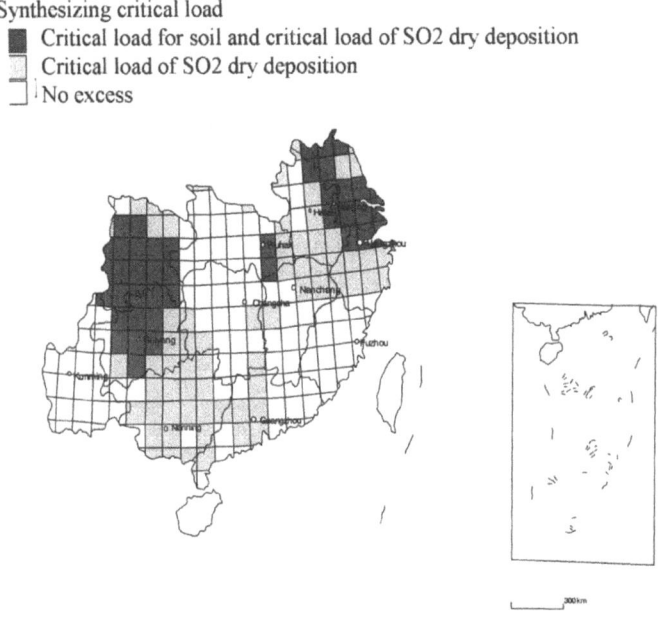

Figure 4. Synthesizing the critical loads for soils and critical loads of SO_2 dry deposition

categories: critical loads for soils, both critical loads for soils and critical loads of SO_2 dry deposition, critical loads of SO_2 dry deposition and no excess. The areas in South China are grouped into categories according to their restricting factors and mapped as Figure 4.

In the southeast of Sichuan province around Chongqing municipality, the central and north of Guizhou province around Guiyang municipality, and the most areas of Jiangsu province, both the critical loads for soils and critical loads of SO_2 dry deposition are exceeded. The both are consequently the restricting factors there, and should be taken into account while the sulfur abatement scheme is made. In Guangxi Zhuang Autonomous Region and some areas among Jiangxi, Zhejiang and Anhui provinces, the critical loads of SO_2 dry deposition is the only restricting factor. There is no area where the critical load for soil is the only restricting factor in South China, so only the critical load for soil is not enough to be the basis to make sulfur abatement scheme. In Fujian province, southeast of Yunnan province, most areas of Hubei and Hunan provinces, both the critical load for soil and critical load of SO_2 dry deposition are not exceeded.

Acknowledgements

The research was funded by the Natural Science Foundation of China (No.40001023).

References

Dai, Zh., Liu, Y., Wang, X. and Zhao, D: 1998, *Water, Air and Soil Pollution* **108**, 377.
Feng Z. and Tao, F.: 1998, *J. of Environ. Science.* **10**(4), 505-509.
Shen, J., Zhao, Q. and Tang, H.: 1995, *Water, Air and Soil Pollution* **85**, 1299.
Tao, F. and Feng, Z.: 1999a, *China Environmental Science* **19**(2), 123.
Tao, F. and Feng,.Z.: 1999b, *China Environmental Science* **19**(1), 14-.
Tao, F. and Feng, Z.: 2000a, *Water, Air and Soil Pollution* **118**, 231.
Tao, F. and Feng, Z.: 2000b, *Water, Air and Soil Pollution* **124**, 429.

CRITICAL LOADS OF ACIDITY FOR FOREST ECOSYSTEMS OF NORTH ASIA

MIKHAIL SEMENOV[1*], VLADIMIR BASHKIN[2,3], HARALD SVERDRUP[4]

[1]*Limnological Institute of Siberian Branch RAS, Ulanbatorskaya st. 3, Irkutsk 664033, Russia;*
[2]*JGSEE, KMUTT, 91 Prachauthit Rd.,Bangkok 10140 Thailand;* [3]*IBBP RAS Pushchino Moscow region 142292 Russia;* [4]*Lund University, P.O. Box 124, S-221 00 Lund, Sweden*
(*author for correspondence, e-mail: semenov@irigs.irk.ru)

Abstract. Critical loads of acidity were calculated using the PROFILE model to assess the forest ecosystem sensitivity to acid deposition in the Asian part of Russia — Siberia. The main input parameters and the output were mapped. At present atmospheric inputs of acid forming pollutants to the study territory are mainly related to transregional and transboundary pollution from Europe. It was shown that the most sensitive to acid loading are ecosystems of the Tundra zone and of the East Sayan mountains' coniferous forests with dystric cambisols and gleysoils, critical loads of actual acidity (CL(Ac)) = 0–0.3 keq/ha/yr. The most tolerant ecosystems are ecosystems of deciduous forests with podsoluvisols, luvisols and humic luvisols of South Taiga zone in West Siberia, CL(Ac) = 3.5–7.0 keq/ha/yr. Generally the values of critical loads are increasing from the North to the South and from the East to the West following the bioproductivity, annual soil temperature and alkalinity of deposition increases.

Key words: acidity, Siberia, PROFILE model, critical load, ecosystem sensitivity, mapping.

1. Introduction

In the context of worldwide agreements on atmospheric pollutants emission abatement, it is very important to assess the forest ecosystem sensitivity to acid deposition in such a vast and forested area as Asian part of Russia — Siberia. This region is situated between the Ural Mountains and the Pacific Ocean. The present significant inputs of acid forming pollutants are related mostly to transregional and transboundary pollution from Europe because of circumpolar wind directions, which are predominant in the Northern Hemisphere. In the future, ecosystem damage due to the atmospheric deposition in southern Siberian regions caused by transboundary pollution transport from China is also possible. During recent years, some research have been conducted to give quantitative estimates of Asian ecosystems sensitivity to acidity loading (Bashkin *et al.*, 1995; 1996).

These estimates have been based on semi-quantitative approaches to the assessment of soil/ecosystem buffering mechanisms like chemical weathering, base cation and nitrogen uptake by vegetation, nitrogen leaching and denitrification. As usual, the input information for the calculation of the given parameters was based predominantly on expert conclusions. The uncertainty of both input parameters and output results was very great (Bashkin *et al.*, 1996). Our latest publications were devoted to the adaptation of quantitative methods for integrated assessment of forest ecosystem stability in West Siberia (Semenov, 1999; Semenov and Bashkin, 1999). The aim of this work was to quantitatively characterize the forest ecosystem stability in the whole Siberia. As the measure of stability the critical load of acidity was chosen. The biogeochemical model PROFILE (Warfvinge and Sverdrup, 1995) and the mineralogy reconstructing model UPPSALA (Federal Environmental Agency, 1996) were used for calculations.

2. Basic information and its adaptation

The values of nutrient cycling in Siberian ecosystems (coniferous, deciduous and mixed forests on different soils) were calculated using experimental data from the literature (Arzhanova and Elpatievsky, 1990; Bazilevich, 1993; Chernyaeva et al, 1978; Targulian, 1971). The net nutrient uptake i.e. the nutrients in the biomass compartments that are expected to be removed from the system at harvest were determined to be zero, as all the ecosystems were regarded as unmanaged (virgin) ones. The information on chemical composition of solid and liquid phase of various soils and their physical parameters was also extracted from literature. In total, the information on the data required for CL calculations in PROFILE model and characterizing more than 200 sites in Siberia was collected. These sites were discussed in more details (Semenov et al., 2000).

One of the most important climatic features of Siberia is the long period of stable snow cover, about 6–7 months on average. It is well known that the snow is the most informative, integral index of element input to the ecosystem. All calculations of atmospheric deposition were made using the data on element content in snow and snow volume.

Total potential acidity input to the system, $Ac(pot)_{dep}$, was calculated according to the existing formula (Federal Environmental Agency, 1996):

$$Ac(pot)_{dep.} = SO_{xdep} + NO_{xdep} + NH_{xdep} - BC_{dep} + Cl_{dep}.$$

3. Critical loads calculation

Critical loads of acidity were calculated according to the formula: $CL(Ac) = BC_w - ANCle_{(crit)}$, where BC_w is base cation weathering and $ANCle_{(crit)}$ is critical Acid Neutralizing Capacity leaching calculated using the Bc/Al=1 ratio.

4. Mapping procedure

The maps of deposition were drawn by computing isolines. The critical loads map was derived in two steps. At the first step the basic map of territorial complexes was drawn by combining soil and vegetation maps. At the second step CL values were marked on the basic map and the contours of equal values were united. The underlying principle of mapping method was to characterize each contour by minimum 2 points (sites) taking into account the contour square and ecological conditions diversity. Finally, each contour includes 2–5 points (sites).

5. Results and discussion

During the mapping procedure in order to show the deposition of base cations as intrinsic ecosystem property, in the whole area the sea–salt correction with Na as a tracer was applied. However, we should mention that in some areas like Yamal, Central Yakutia, Far East with an excess of Na deposition due to continental sources, sea–salt correction led to underestimated non-sea–salt deposition and to distorted acid-base balance of total deposition (Table I).

TABLE I

Measured and sea-salt corrected values of atmospheric deposition and acid neutralizing capacity (keq/ha/yr). BC is non sea–salt corrected basic compounds (Ca, Mg, K, Na), Na is deposition of sodium, Bc is sea–salt corrected basic compounds (Ca, Mg, K), Ac is acidic compounds excluding nitrates, ANC = Bc − Ac excluding total mineral nitrogen, ANC is total acid neutralizing capacity

Location	Measured values				Sea-salt corrected values			
	BC − Na	Ac	ANC	ANC	Bc	Ac	ANC	ANC
Sites with marine originated sodium only								
Yakutsk	0.14–0.03	0.09	0.05	0.04	0.10	0.05	0.05	0.04
Mondy	0.24–0.02	0.78	−0.54	−0.56	0.21	0.75	−0.54	−0.56
Ilchir	0.50–0.02	0.20	0.30	0.24	0.47	0.17	0.30	0.24
Baikal	0.47–0.03	0.24	0.23	0.15	0.43	0.20	0.23	0.15
Sites with both marine and continental sodium								
Mirny	0.36–0.31	0.30	0.06	−0.03	0.017	0.15	−0.13	−0.22
Neryungri	0.49–0.26	0.10	0.39	0.35	0.16	0.00	0.16	0.12
Yamal	0.98–0.84	0.73	0.25	0.21	0.03	0.00	0.03	−0.01
Far East	1.31–0.55	0.49	0.82	0.62	0.60	0.25	0.35	0.15

The spatial distribution of Bc and sea–salt corrected Ac(pot) deposition (Figure 1, 2) are in agreement with existing data on atmospheric mass transfer, averaged for 20 year period of 1975-1995 (Kuznetsova, 1978; Semenov, 1999). Due to the predominant Atlantic air mass transfer (from West to the East) and the Ural Mountains location, both acid and base deposition values are highest in the south of West Siberia. The rates decrease with the longitude increasing (East direction) up to the Yakutia (100°E) and latitude increasing up to the Arctic Ocean. The highest deposition rates were along the 55°N belt.

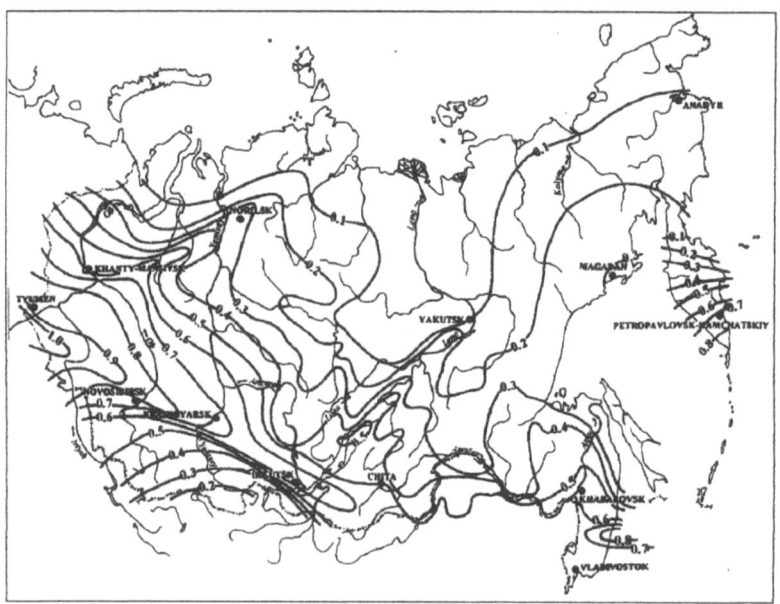

Figure 1. Bc deposition (Ca+Mg+K), keq/ha/yr

The maximum weathering rates (up to 2 keq/ha/yr) are obtained for carbonate and low-weathered soils in southern plain regions, minimum rates (down to 0.01 keq/ha/yr) are typical for the Tundra zone and highland areas.

Due to the variety of ecological conditions the critical load values vary in the range of 0–7 keq/ha/yr (Figure 3). The most sensitive ecosystems within the study area are ecosystems of Tundra zone (60°–160°E, 70°–80°N) and ecosystems of the East Sayan Mountains' coniferous forests (100°–120°E, 50°–60°N) with dystric cambisols and gleysoils, where critical loads of actual acidity (CL(Ac)) = 0–0.3 keq/ha/yr. The most tolerant ecosystems are ecosystems of deciduous forests with podsoluvisols, luvisols and humic luvisols of South Taiga zone in West Siberia (60°–90°E, 50°–60°N), CL(Ac) = 3.5–7.0 keq/ha/yr. Generally the values of critical loads are increasing from the North to the South and from the East to the West following the bioproductivity, annual soil temperature and alkalinity of deposition increases. Thus, the picture of the spatial CL distribution is similar to the intensity of biological turnover complicated by the spatial geology distribution. Preliminary evaluation of exceedances followed according to the simplest formula: $Ex = Ac_{dep} - Bc_{dep} - CL(Ac)$, showed negative values all over the Siberia. Both $(Ac_{dep} - Bc_{dep})$ and CL(Ac) values are close to each other for the territory of the East Sayan Mountains highland zone (1000 m and higher).

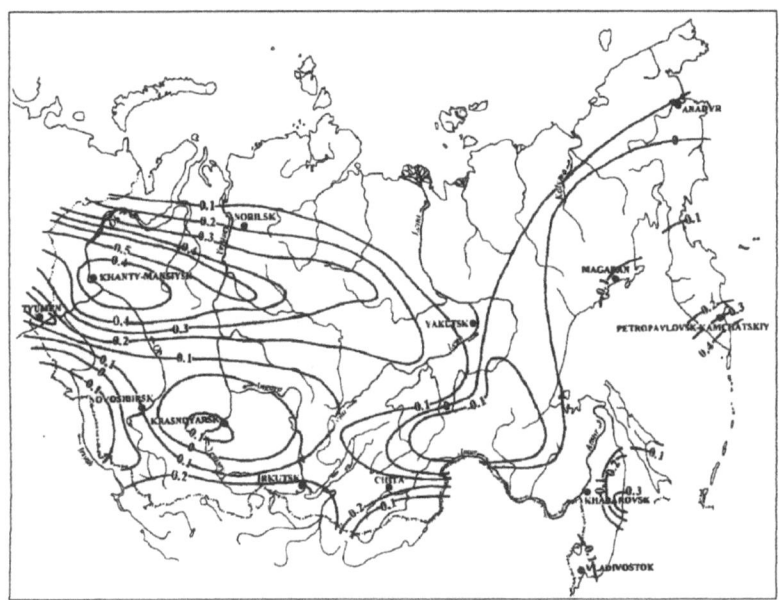

Figure 2. Ac(pot) deposition (S + N + Cl – Bc), keq/ha/yr

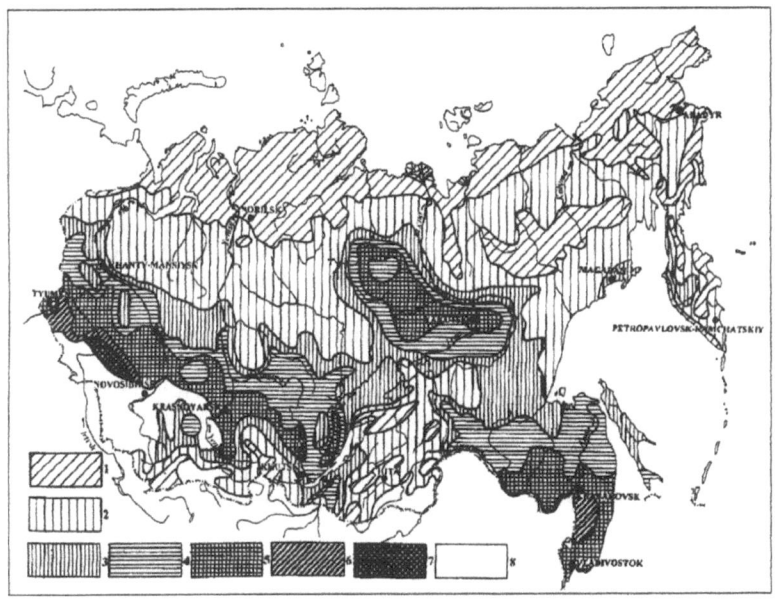

Figure 3. Critical loads of acidity, keq/ha/yr: 1: 0–1; 2: 1–2; 3: 2–3; 4: 3–4; 5: 4–5; 6: 5–6; 7: >6; 8: steppe and forest–steppe ecosystems.

Concluding the discussion of exceedances values, one should take into account that these are indeed around the large industrial complexes, e.g., Norilsk, which is one of the largest emission source for S in the world and, according to Figure 3, surrounded by the most sensitive ecosystems. However, these exceedances are out of scale of the mapping. The future more detailed mapping will be carried out to take into account these areas.

Furthermore, it should be also noted that each contour in our maps represents the several predominant ecosystem types. We understand that 200 sites are not very much for Siberia. The number of sites was restricted by the data availability. Siberia is a vast region in comparison even to the whole Europe and the population is only 17% from that in the whole Russia (less than 0,1% of European population). Until now it is still insufficiently investigated. There is no well-organized soil and precipitation monitoring in fine and meso-resolution scale with corresponding database on ecosystem changes, etc. However, all existing relevant information was collected and this CL research can be used as a basis for ongoing more detailed studies.

References

Arzhanova, V.S. and Elpatievsky P.V.: 1990, Geochemistry of landscape and technogenesis, Nauka Publ., Moscow, 196 p.

Bashkin, V.N., Kozlov, M.Ya., Priputina, I.V., Abramychev, A.Yu. and Dedkova, I.S.: 1995, *Water, Air, and Soil Pollution* **85**: 2395-2400.

Bashkin, V.N. Kozlov, M.Ya. and Abramychev, A.Yu.: 1996, *Asian–Pacific Remote Sensing and GIS Journal* **8**(2): 73–80.

Bazilevich N.I.: 1993, Biological productivity of Northern Eurasian ecosystems. Nauka Publ., Moscow, 350 p.

Chernyaeva L.E., Chernyaev A.M., Mogilenskich A.K.: 1978, Chemical composition of atmospheric deposition (Ural Mountains and around-Ural region). Gidrometeoizdat Publ., Leningrad, 177 p.

Federal Environmental Agency: 1996, Manual on Methodologies and Criteria for Mapping Critical Levels/Loads and Geographical Areas They Are Exceeded. UN ECE Convention on Long-range Transboundary Air Pollution, Texte 71/96. Federal Environmental Agency, Berlin, 142 p.

Kuznetsova, L.P.: 1978, Moisture transport in the atmosphere over the territory of USSR. Nauka Publ., Moscow, 268 p.

Semenov, M.Yu.: 1999, 'Application of Biogeochemical Model PROFILE for Critical Load Calculation', *Proceedings of the Second Training Workshop on the Calculation and Mapping of Critical Loads for Air Pollutants*, Moscow: 81–90.

Semenov, M.Yu. and Bashkin, V.N.: 1999, *Geography and Natural Recourses* **4**: 44–52

Semenov, M.Yu., Bashkin, V.N. & Sverdrup, H.: 2000, *Asian Journal of Energy & Environment*, 1(2) 143–162

Targulian V.O.: 1971, Soil formation and weathering in cold humid areas. Nauka Publ., Moscow, 268 p.

Warfvinge P. and Sverdrup H.: 1995, *Critical Loads of Acidity to Swedish Forest Soils: Methods, Data and Results.* Lund, Sweden

CALCULATION AND MAPPING OF CRITICAL LOADS FOR S, N AND ACIDITY IN CHINA*

LEI DUAN[1], SHAODONG XIE[2], ZHONGPING ZHOU[1], XUEMEI YE[1] and JIMING HAO[1*]

[1] *Department of Environmental Science and Engineering, Tsinghua University, Beijing 100084, China;* [2] *Center of Environmental Sciences, Peking University, Beijing 100871, China*
(* *author for correspondence, e-mail: hjm-den@tsinghua.edu.cn*)

Abstract. Critical loads of nutrient and acidifying nitrogen, as well as of sulphur and acidity, were derived for various ecosystems in China using the steady state mass balance (SSMB) equations. The weathering rates of major soils necessary for applying SSMB were calculated through the PROFILE model on the basis of mineralogical data from experimental analysis. The growth uptakes of nitrogen and base cations were also derived by multiplying the annual increases in biomass with the element contents of the vegetation. Using a geographical information system (GIS), 1°(latitude)×1°(longitude) critical load maps of China with different percentiles were compiled. Results indicate that low critical loads of S (< 0.5 keq·ha^{-1}·a^{-1}) occurred predominately in southwest and northeast China, and the critical loads of southeast China were intermediate and in the range of 0.5~1.0 keq·ha^{-1}·a^{-1}. In addition, the critical loads of N were very low for desert ecosystems in northwest China and high for agricultural ecosystems in east China. Among the ecosystems with intermediate critical load of N, coniferous forests may be more sensitive to N deposition than broad-leaf forests and temperate steppes.

Key words: Acid Deposition, Critical Loads, Nitrogen, SSMB, Sulphur

1. Introduction

In 1998, the Acid Rain Control Zone and Sulfur Dioxide Pollution Control Zone were designated in China for those areas that are, or could become, affected by acid deposition or ambient sulfur dioxide concentration. One of the most important principles for designating the Acid Rain Control Zone was that the critical load might be exceeded by the sulfur deposition. This was the first time critical load being applied in emission abatement in China.

As the scientific basis of the designation, critical load maps and exceedance maps of China were compiled through a modified 'level 0' method in 1996 (Duan et al., 2000). In this study, the Steady State Mass Balance (SSMB) equations were applied to update the critical load maps of China.

2. Methods and Materials

2.1. METHODS FOR CRITICAL LOAD CALCULATION

According to Posch et al. (1995), the critical load of sulfur, acidity, acidifying

---

* This project was supported by the National Natural Science Foundation of China.

Water, Air, and Soil Pollution **130**: 1199–1204, 2001.
© 2001 *Kluwer Academic Publishers.*

nitrogen, and nutrient nitrogen is given by

$$CL_{Max}(S) = BC_d + BC_w - BC_u - ALK_{le(crit)} \tag{1}$$

$$CL(AC_{act}) = BC_w - ALK_{le(crit)} \tag{2}$$

$$CL_{Max}(N) = N_i + N_u + CL_{Max}(S)/(1-f_{de}) \tag{3}$$

$$N_{nut}(N) = N_i + N_u + N_{le(crit)}/(1-f_{de}) \tag{4}$$

respectively, where the subscript d, w, u, and i refer to deposition, weathering, uptake, and immobilization, respectively, BC is the sum of base cations (BC = Ca + Mg + K + Na) for every subscript, f_{de} is the denitrification factor, $N_{le,crit}$ is the allowable (critical) leaching of N, and $ANC_{le(crit)}$ is the critical ANC leaching, which can be derived by

$$ANC_{le(crit)} = 1.5 \frac{Bc_{dep} + Bc_w - Bc_u}{(Bc/Al)_{crit}} + Q^{\frac{2}{3}} \left[1.5 \frac{Bc_{dep} + Bc_w - Bc_u}{(Bc/Al)_{crit} \cdot K_{gibb}} \right]^{\frac{1}{3}} \tag{5}$$

where Q is the precipitation surplus, K_{gibb} is the gibbsite equilibrium constant, Bc = BC – Na, and $(Bc/Al)_{crit}$ is the critical chemical value.

2.2. MAPPING MEHTODOLOGY

To compile a critical load map of China, critical load was calculated for each 0.1° (latitude) × 0.1° (longitude) gridcell. Regarding the receptors, a distinction was made between 86 vegetation types and 46 soil groups, which were distinguished on the basis of the 1:4,000,000 vegetation map and soil map of China, respectively. Parameters such as growth uptake, weathering rate, N immobilization and critical BC/Al ratio were assigned values for each gridcell based on the vegetation/soil type. Maps of precipitation and runoff were also digitized for critical load mapping. To limit the number of calculations with SSMB, if there were two or more values for a parameter in a gridcell, the average value of them weighted by area was applied to calculate the critical load of the gridcell.

For the convenience of policy-makers to formulate abatement strategies based on critical loads, the geographical representation of critical load should be consistent with the resolution at which sulfur and N deposition is modeled. Consequently, critical load map at 0.1° × 0.1° scale should be converted to 1° × 1° map so that critical load exceedances could be computed. The value chosen to present the cumulative distribution function of all critical load values within one 1° × 1° gridcell is 5th percentile.

2.3. INPUT DATA

2.3.1. *Atmospheric Deposition of Base Cation*

The sea-salt corrected wet deposition of base cation was derived from the results of the National Monitoring Network in China (Wang, 1995). The influence of dry deposition on the total deposition was accounted for by multiplying the wet deposition with a dry deposition factor

$$BC_{td} = BC_{wd}(1 + f_{dd}) \qquad (6)$$

where BC_{wd} is the wet deposition and BC_{td} the total deposition of base cation, f_{dd} is the dry deposition factor of a gridcell, which was set to the average value of the dry deposition factors of coniferous forest (1.55), broad-leaf forest (0.75) and open land (1.87) (Xie et al. 1997) weighted by the coverage fraction of each vegetation.

2.3.2. Weathering Rate

For the present study, the soil groups are particularly appropriate units for calculating weathering rates, because the soil series within a group generally, but not necessarily, have similar properties. From 1996 to 1999, 109 samples of 12 major soils, including latosol, lateritic red earth, red earth, yellow earth, yellow-brown soil, brown forest soil, dark brown forest soil, podzolic soil, cinnamon soil, black soil, albic bleached soil, and purplish soil in east China, were collected and analyzed for physicochemical property and mineralogy. Weathering rate of these noncalcareous soils were calculated through the PROFILE model (detail description of the model see Warfvinge and Sverdrup, 1992). Limited by the availability of data, weathering rates of other noncalcareous soils, such as gray-cinnamon soil, gray forest soil, paddy soil, meadow soil, bog soil, and mountain meadow soil, were derived on the basis of soil mineralogy estimated from the total analysis. In addition, the weathering rate of calcareous soils, including the pedocal soils in northwest China and the alpine soils on the Tibetan Plateau, were estimated according to their lime content (data can be found in Xiong and Li, 1987). Since the soil samples within a group may have different physicochemical and mineralogical properties, and hence different weathering rates, the smallest weathering rate was assigned to the group.

2.3.3. Growth Uptake

Assuming that whole-tree harvesting (stems and branches) is practiced for forest (including shrub), the present growth uptake of N and base cation can be derived by multiplying the annual increase in biomass with the element contents in the various compartment according to

$$X_u = K_s X_s + K_b X_b \qquad (7)$$

where X_u is the net uptake of element X, K_s and K_b are the average growth rate of stem and branch respectively, and X_s and X_b are the content of element X in stem and branch respectively. For steppe, meadow and desert, considering that excessive pasturing is very common in China, the annual uptake of N and base cation was estimated by multiplying the yield of grass with the content of element X.

In this study, literature concerning average growth rates and element contents of Chinese vegetation was critically reviewed. The growth uptake of each vegetation type was calculated using data from Liao and Jia (1999), Chen et al. (1997), Hou (1982), and so on. A detailed presentation of the critical review and a full listing of the vegetation data is beyond the scope and intent of this work and will presented elsewhere.

2.3.4. *Other Parameters*

Critical BC/Al ratio for Chinese vegetation were selected from Sverdrup and Warfvinge (1993). Al dissolution coefficient, i.e., K_{gibb}, and N transformation parameters such as immobilization, denitrification, and critical leaching, were derived on the basis of Mapping Manual (Task force on Mapping, 1996).

3. Results and Discussion

Based on the input data mentioned above, critical loads of acidity, S, acidifying N, and nutrient N were mapped using the SSMB method. As can be seen from the 5th percentile critical load map of acidity (Fig. 1), the majority of areas in southeast China were sensitive to acid deposition, while those in the northwest not. The low critical loads of acidity ($<0.5 keq \cdot ha^{-1} \cdot a^{-1}$) occurred predominately in northeast China, and the critical loads of southeast China, where acid rain occurs frequently, were almost intermediate and in the range of $0.5 \sim 1.0$ $keq \cdot ha^{-1} \cdot a^{-1}$. In comparison with the critical load derived through the modified 'lever 0' method (Duan et al. 2000), the SSMB critical load is somewhat lower because the weathering rate applied was calculated based on the most sensitive soil series in each soil group, while the classification of a soil group by 'level 0' method depended on the mineralogy of the most extensive soil series.

The critical load of S (Fig. 2) was similar to the critical load of acidity, but it was a bit higher in south China because of the high deposition of base cation. Since the present uptake of base cation was very high, even higher than the total supply of base cation by weathering and deposition, in northeast China and in southeast China (include the west part of the Tibetan Platinum), critical load of S was very low there.

In this study, critical load of N was calculated as the minimum of the critical load of nutrient N and acidifying N (shown in Figure 3). The critical load of N was lowest for desert ecosystem in northwest China, then followed by alpine and subalpine ecosystem on the Tibetan Plateau. High critical loads of N generally occurred in east China for agricultural ecosystem. Among the ecosystems with intermediate critical load of N, i.e., $10 \sim 20$ $kg \cdot ha^{-1} \cdot a^{-1}$, coniferous forest might be more sensitive to N deposition than broad-leaf forests and temperate steppes.

Acknowledgment

The authors are grateful to the National Natural Science Foundation of China for its financial support to carry out this study. Gratitude should also go to Professor Harald Sverdrup for giving us the running program of PROFILE model.

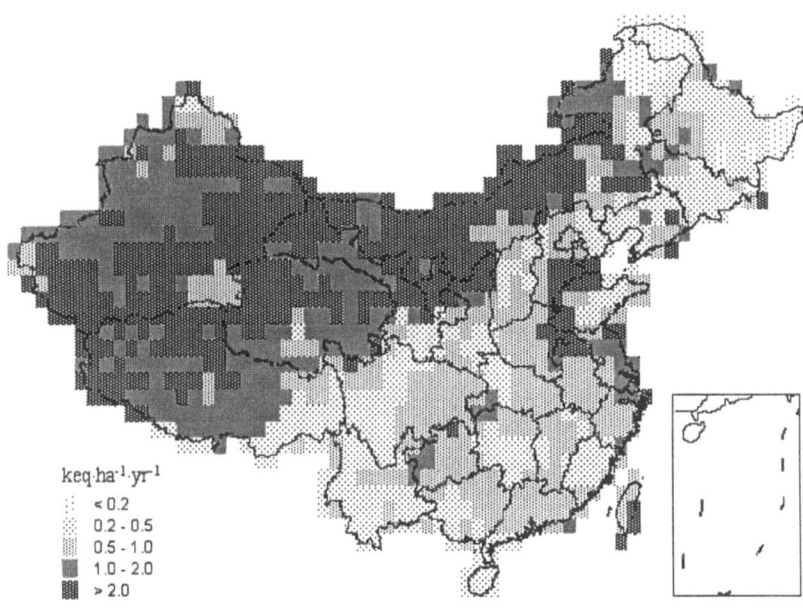

Figure 1. The 5th percentile critical load map of acidity

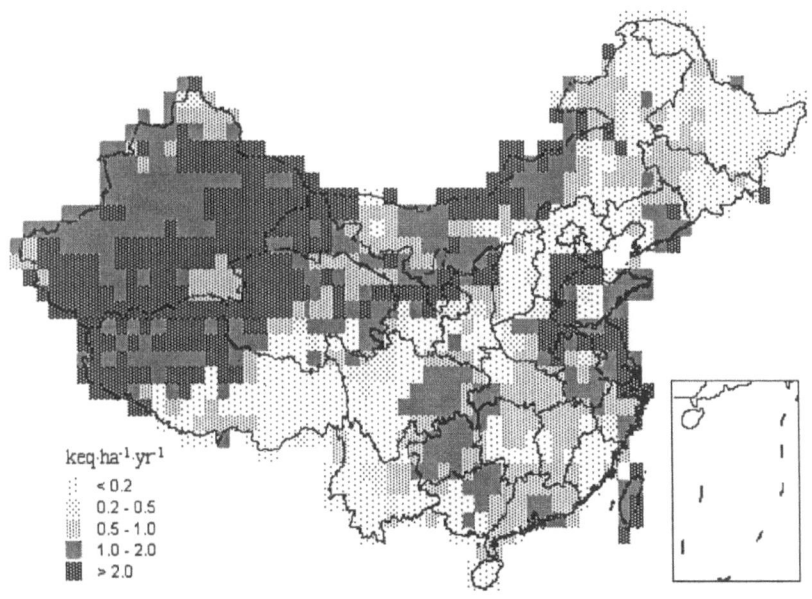

Figure 2. The 5th percentile critical load map of S

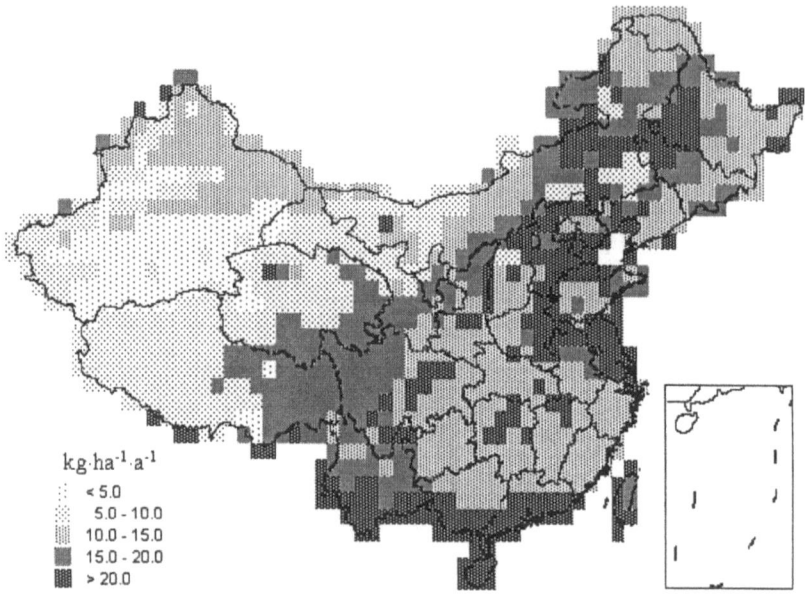

Figure 3. The 5th percentile critical load map of N

Reference

Chen, L. Z., Huang, J. H. and Yan, C. R.: 1997, *Nutrient Cycle of forest in China*, Meteorology Press, Beijing, China. (in Chinese)
Duan, L., Xie, S. D., Zhou, Z. P. and Hao, J. M.: 2000, *Water, Air and Soil Pollution* **118**(1-2), 35-51.
Hou, X. Y.: 1982, *China Vegetation Plant Geography and Chemical Constitution of Dominant Plants*, Science Press, Beijing, China. (in Chinese)
Liao, G. F. and Jia, Y. L. (eds.): 1996, *Grassland Resource of China*, China Science and Technology Press, Beijing, China. (in Chinese)
Posch, M., de Smet, P. A. M., Hettelingh, J.-P. and Downing, R. J. (eds.): 1995, *Calculation and Mapping of Critical Loads Thresholds in Europe*, CCE Technical Report, pp. 31-41.
Sverdrup, H. and Warfvinge, P.: 1993, *Effect of Soil Acidification on Growth of Trees and Plants as Expression by the (Ca+Mg+K)/Al Ratio*, Report 2:1993, Department of Chemistry Engineering II, Lund University, Lund, Sweden.
Task force on mapping: 1996, *Mapping Critical Levels/Loads*. UN/ECE Convention on Long-range Transboundary Air Pollution, Federal Environmental Agency (Umweltbundesamt), Texte 71/96, Berlin, German.
Wang, W. (eds.): 1995, *Researches on Temporal and Spatial Distribution of Acid Precipitation in China*, National Key Project in the Eighth Five-year Plan, Final Report. (in Chinese)
Warfving, P. and Sverdrup, H.: 1992, *Water, Air and Soil Pollution* **63**, 821-837.
Xie, S. D., Hao, J. M. and Zhou, Z. P.: 1997, *Environmental Science (China)* **18**(5), 6-9. (in Chinese)
Xiong, Y. and Li, Q. K.: 1987, *Chinese Soils*, the second edition, Science Press, Beijing, China. (in Chinese)

A LITERATURE REVIEW OF UNCERTAINTIES IN STUDIES OF CRITICAL LOADS FOR ACIDIC DEPOSITION

JUNLING AN[1,2*], LING ZHOU[1], MEIYUAN HUANG[1], HU LI[2], TSUNEHIKO OTOSHI[2] and KAZUHIDE MATSUDA[2]

[1]*LAPC, Institute of Atmospheric Physics, Chinese Academy of Sciences, Beijing 100029, China;* [2]*Acid Deposition and Oxidant Research Center (ADORC), 1182 Sowa, Niigata-shi, 950-2144, Japan*
(* *author for correspondence, e-mail: anjl@lycos.com*)

Abstract. Uncertainties in the assessment of critical loads for acidic deposition are caused by the choice of biological indicators (BI), critical chemical values (CCV), the current methods used to determine critical loads for an ecosystem, and deficient field data. This paper focuses on the present steady-state mass balance (SSMB) approach, dynamic models and the importance of changes in atmospheric base-cation deposition (BCD), particularly in China. It is argued that 1) for the SSMB approach much uncertainty may come from the choice of BI and the related CCV, and long-term and large-scale monitoring data on weathering rates and growth uptake are urgently needed, especially in China, 2) significant uncertainty may be caused by changes in BCD during SO_2 emission controls, particularly in China, 3) constructing a mechanistic Al submodel may be a promising direction for dynamic models, and 4) the nutrient cycle in the vegetation through biogeochemical processes should be incorporated into dynamic models but the input requirements should be moderate for broad application considerations. Generally higher BCD, different soil components and characteristics and different vegetation types in China compared to Europe and North America suggest that more field investigations on BI and their corresponding CCV be carried out before application of current approaches to specific areas, particularly in China.

Keywords: acidic deposition, critical load, biological indicator, critical chemical value, uncertainty

1. Introduction

To prevent harmful environmental effects caused by gaseous acidifying emissions, many governments are looking for a more effective and scientific means of assessing effects and planning emission controls, by increasing knowledge on emission reductions as well as environmental impacts. The critical load concept may offer the basis for an internationally acceptable solution (Bull, 1991). Mapping critical loads of acidity for soils in countries of Europe and North America began in the late 1980s (De Vries, 1988; Schulze *et al.*, 1989; Bull, 1991; Warfvinge and Sverdrup, 1992; De Vries *et al.*, 1994ab; Sverdrup and De Vries, 1994; Hettelingh *et al*, 1995a; Bashkin *et al.*, 1995; Hornung *et al.*, 1995a; Hornung *et al.*, 1995b; Party *et al.*, 1995) and in China in the beginning of 1990s (Zhao and Seip, 1991; Zhao *et al.*, 1992; Seip *et al.*, 1995; Xie *et al.*, 1995; Hettelingh *et al.*, 1995b; Xie *et al.*, 1996b; Xie *et al.*, 1998). Critical load maps provide a quantitative estimate for policy makers to implement control strategies. However, there still remain considerable uncertainties in assessment of critical loads for acidic deposition and these should be elaborated.

2. Uncertainties in Biological Indicators and Critical Chemical Values

A biological indicator, a chemical criterion, and a critical chemical value are involved in the critical load. For forest soils the tree is mainly used as the BI, but herbs and grasses are often more sensitive than trees (Andersson, 1988). Recent studies show that mycorrhizal fungi and soil animals should be considered more sensitive indicators of acidic deposition (Løkke et al., 1996). Plants /animals are generally different from one place to another (climate), so BI in different places (climate) may be different and further field investigations are required.

Up to now several chemical criteria have been proposed (Table I), but most widely used criterion is the molar BC/Al ratio of soil solutions. It is assumed that BC/Al (Ca/Al) molar ratio of less than 1.0 have a significant harmful effect on the tree growth or vitality. However, there is now an increasing evidence indicating that the Ca/Al ratio and, in particularly, the use of the critical value of 1.0 are questioned (Mulder et al., 1989; Högberg and Jensén, 1994). Recent field studies in Sweden failed to find any relationship between forest growth and soil solution BC/Al ratios (Løkke et al., 1996). Uncertainties in the CCV may lead to a decrease or an increase of the area exceeding critical loads by approximately 50% (De Vries et al., 1994ab; Xiao and Zhuo, 1995).

TABLE I
CCV used in the SMART (De Vries et al., 1989) for various parameters describing nitrogen saturation and acidification status of soils (De Vries et al., 1994b)

Criteria	Unit	Critical values	Effects
NO_3^- con.	$mol\ m^{-3}$	0.1	Harm to vegetation
C/N ratio	$g\ g^{-1}$	40[a]	NO_3^- leachate
Al^{3+} con.	$mol\ m^{-3}$	0.2[b]	Harm to roots
Al/BC ratio	$mol\ mol^{-1}$	1.0[c]	Inhibition of absorption of Ca^{2+}, Mg^{2+}

a. Taken 20 for calcareous soils;   b. Corresponding to pH=4.0
c. Corresponding to base saturation =5%;   con.= concentration

3. Uncertainty of Current Approaches

There are basically two different approaches to assess the impact of acidic deposition on soils/freshwaters. One is the steady-state approach, directly evaluates the soil's final chemical status for a given set of conditions (Warfvinge and Sverdrup, 1992; Xie et al., 1996a). This method is an attractive tool for quantifying critical loads for a large scale ecosystem due to its moderate data requirements. Under conditions of SSMB (Sverdrup and De Vries, 1994; Løkke et al., 1996) (Application of the PROFILE model is evaluated by Jonsson et al., 1995 and Hodson et al., 1997), a most widely used steady-state model in mapping critical loads, there is no net transfer of acid neutralizing capacity from the change soil matrix to the solution. This steady-state assumption cannot generally be achieved due to constantly changing climate, the dynamic nature of growth and death of organisms, and significant changes in land use or management (Løkke et al., 1996). Soil solution chemistry also shows high variability, both in space and time (Starr, 1985). For long-term effect and large-scale ecosystem considerations steady-state conditions may be approximately approached except great changes in land use or management. Much uncertainty in determination of critical loads may be caused by options of BI and their corresponding CCV. Another important problem is that long-term and large-scale monitoring data on weathering rates and net growth uptake are extremely limited, particularly in China.

Dynamic models allow soil changes to be predicted as a function of time and therefore provide guidelines as for when to apply different abatement strategies. Dynamic models are generally applied on a site-specific basis due to much more data requirements compared with the steady state approach. For dynamic models, e.g., MAGIC (Cosby et al., 1985), SMART, SAFE (Alveteg, 1998), and LTSAM (An and Huang, 1999; An et al., 1999), a key assumption is that the activity of Al^{3+} be expresses by (Cosby et al., 1985; Cronan et al., 1986; De Vries et al., 1989; An and Hunag, 1999)

$$[Al^{3+}]=K_{gibb}\times[H^+]^3$$

More studies indicate variations regarding the use of the above expression to calculate Al^{3+} activities (Mulder et al., 1987; Seip et al., 1989; Mulder and Stein, 1994). The gibbsite solubility constant (K_{gibb}) is also a highly site-specific and depth-specific constant (Alveteg et al., 1995). It is unclear whether it can be replaced by a universal rate constant. Alveteg et al. (1995) have made an effort to construct a mechanistic Al submodel. This may be an important direction.

4. Deficient Observed Data

Base cations play an integral role in the chemical processes of acidic deposition since acidification is a function of acidic and basic compounds in soils and surface waters. However, little emphasis is usually placed on the role of base cations (Hedin et al., 1994; Draaijers et al., 1995; An and Huang, 1999). The uncertainty in the total deposition of base cations can be amounted to 90 ∽ 140% (Draaijers et al., 1995). Significant uncertainty is found in dry deposition estimates. Compared with the Eastern United States and locations in Europe, precipitation in China generally has higher concentrations of sulfate, ammonium, and calcium (Galloway et al., 1987; Seip et al., 1995). The pH of precipitation is about 4 to 5 in Southern areas and about 6.5 in Northern areas of China (Galloway et al., 1987; CESS, 1989; Wang et al., 1993). Without bases in the atmosphere, the pH in Northern China would drop to about 3.5 (Galloway et al., 1987). Therefore the base cations, particularly Ca^{2+} and NH_4^+, play a greater role in mitigating precipitation acidity in China than at typical sites in North America and Europe (Galloway et al., 1987; Zhao et al., 1988; CESS, 1989; Seip et al., 1995). Application of a dynamic model, e.g., MAGIC, generally begins with a hindcast for reconstructing trends in soil and water chemistry from a steady state under background deposition to the present state. In most calculations the deposition of cations and anions is varied during this period, but after that only sulfur deposition is reduced or increased in estimates of critical loads for many areas (Zhao and Seip, 1991; Zhao et al., 1992; Feng et al., 1995). This may be applicable in Europe and North America due to low BCD but can not apply to most places in China (An et al., 1998; An and Huang, 1999; An et al., 1999). More attention to changes in base cations during SO_2 emission reductions should be and long-term monitoring of BCD needs to be strengthened, especially in China. (e.g., Sources of base cations in China, emitted locally or transported from the neighboring regions for a specific area, are scarce, so this makes it rather difficult to estimate how many changes in base-cation deposition will be caused by reducing SO_2 emissions and how many SO_2 emissions should be reduced for a specific area.) By comparison with wet deposition, dry deposition is more complex and related to many factors, such as roughness, vegetation types, and atmospheric stability. Generally dry deposition amount is estimated based on wet deposition amount (Cosby et al., 1985; Zhao et al., 1992; Xie et al., 1996ab; An and Huang, 1999). Is such estimation applicable to every specific area? Another important problem for long-term, large-scale soil acidification models is that long-term (>50 years) observations are extremely scarce, and this makes the models rather difficult to calibrate (Kros et al., 1993; Hettelingh et al., 1995ab).

5. Discussion and Conclusions

The critical load concept is gaining international recognition as a practical way of assessing effects and implementing emission controls. This concept has been applied to mapping the critical loads for countries in Europe (Hettelingh et al., 1995a; Hornung et al., 1995a) and Asia (Bashkin et al., 1995; Hettelingh et al, 1995b, Xie et al., 1998). However, substantial uncertainties still remain. For the SSMB approach much uncertainty in determination of critical loads may come from options of BI and the related CCV. Constructing a mechanistic Al (aluminum) submodel may also be a promising direction for dynamic models. The nutrient cycle in the vegetation through biochemical processes such as plant nutrient uptake, litterfall and mineralization is very important and should be incorporated into dynamic models but the input requirements should be as moderate as possible. Compared to Europe and North America, China generally has a higher BCD (Galloway et al., 1987; CESS, 1989; Seip et al., 1995) and so changes in BCD should be considered during reduction of SO_2 emissions. Different soils and vegetation types (Liu and Du, 1996; Qiu and Wu, 1997) also suggest that a choice of BI and the corresponding CCV be carefully taken and more field studies on these parameters and their critical values should also be conducted before application of current approaches to specific areas, particularly in China.

Acknowledgements

This research was supported by the Key Project A of Chinese Academy of Sciences (Grant number is KZ951-A1-403-03-03) and the Key Foundation of IAP (Grant number is 8-2210). We would like to express our gratitude to two anonymous reviewers for their key suggestions.

References

Alveteg, M., Sverdrup, H., and Warfvinge, P.:1995, *Water, Air, and Soil Pollution* **79**, 377.
Alveteg, M.:1998, *Dynamics of Forest Soil Chemistry*, Department of Chemical Engineering II, Lund University, Sweden, pp.61-81.
Andersson, M.:1988, *Water, Air, and Soil Pollution* **39**, 439.
An Junling *et al.*:1998, *Environmental Chemistry* (in Chinese) **17**(2), 136.
An Junling and Huang Meiyuan:1999, *Water, Air, and Soil Pollution* **110**, 255.
An Junling, Huang Meiyuan *et al.*:1999, *Acta Scientiae Circumstantiae* (in Chinese) **19**(3), 284.
Bashkin, V. N. *et al.*:1995, *Water, Air, and Soil Pollution* **85**, 2395.
Bull, K. R.:1991, *Environmental Pollution* **69**, 105.
China Environmental Science Society (CESS):1989, *Collections on Acid Rain* (in Chinese), China Environmental Science Press, Beijing, pp.228-230.
Cosby, B. J. *et al.*:1985, *Water Resources Research* **21**(11), 1591.
Cronan, C. S. *et al.*:1986, *Nature* **324**, 140
De Vries, W.:1988, *Water, Air, and Soil Pollution* **42**, 221.
De Vries, W. *et al.*: 1989, *Water, Air, and Soil Pollution* **48**, 349.
De Vries, W. *et al.*:1994a, *Water, Air, and Soil Pollution* **78**, 215.
De Vries, W. *et al.*:1994b, *Water, Air, and Soil Pollution* **72**, 357.
Draaijers, G. P. *et al.*:1995, *Water, Air, and Soil Pollution* **85**, 2389.
Feng Zongwei, Cao Hongfa, Zhou Xiuping, *et al.*:1995, *'Effects of acid deposition on the ecosystems and its economic losses'* (Research report in Chinese), pp.1-130.
Galloway, J. N., Zhao Dianwu, Xiong Jiling, and Likens, G. E.:1987, *Science* **236**, 1559.
Hedin, L. O. *et al.*:1994, *Nature* **367**, 351.
Hettelingh, J.-P. *et al.*:1995a, *Water, Air, and Soil Pollution* **85**, 2381.
Hettelingh, J.-P. *et al.*:1995b, *Water, Air, and Soil Pollution* **85**, 2565.
Hodson, M. E. *et al.*:1997, *Water, Air, and Soil Pollution* **98**, 79.
Högberg, P. and Jensén, P.:1994, *Water, Air, and Soil Pollution* **75**, 121.
Hornung, M. *et al.*:1995a, *'Mapping and modelling of critical loads for nitrogen a wokshop report'*, Institute of Terrestrial Ecology, UK, pp.1-192.
Hornung, M. *et al.*:1995b, *Environmental Pollution* **90**(3), 301.
Jonsson C.:1995, *Water, Air, and Soil Pollution* **81**, 1.
Kros, J. *et al.*:1993, *Water, Air, and Soil Pollution* **66**, 29.
Kros, J. *et al.*:1995, *Water, Air, and Soil Pollution* **79**, 353.
Liu Houtian and Du Xiaoming:1996, *Advances in Environmental Science* (in Chinese) **9**(5), 143.

Løkke, H. et al.:1996, *Ambio* **25**(8), 510.
Mulder, J. et al.:1987, *Soil Sci. Soc. Am. J.* **51**, 1640.
Mulder, J. et al.:1989, *Nature* **337**, 247.
Mulder, J. and Stein A.:1994, *Geochimia et Cosmochimia Acta* **58**, 85.
Party, J. P. et al.:1995, *Water, Air, and Soil Pollution* **85**, 2407.
Qiu Rongliang and Wu Qing:1997, *Advances in Environmental Science* (in Chinese) **5**(4), 8.
Schulze, E.-D., De Vries, W. et al.:1989, *Water, Air, and Soil Pollution* **48**, 451.
Seip, H. M. et al.:1989, *Journal of Hydrology* **108**,387.
Seip, H. M., Zhao Dianwu, Xiong Jiling et al.:1995, *Water, Air, and Soil Pollution* **85**, 2301.
Starr, M. R.:1985, *Soil Science* **140**, 453
Sverdrup, H. and De Vries, W.:1994, *Water, Air, and Soil Pollution* **72**, 143.
Wang Wenxing et al.:1993, *China Environmental Science* (in Chinese) **13**, 401.
Warfvinge, P. and Sverdrup, H.:1992, *Water, Air, and Soil Pollution* **63**, 119.
Xiao Huilin and Zhuo Muning:1995, *Advances in Environmental Science* (in Chinese) **3**(4), 59.
Xie Shaodong, Hao Jiming et al.:1995, *Water, Air, and Soil Pollution* **85**, 2401.
Xie Shaodong, Hao Jiming et al.:1996a, *Environmental Science* (in Chinese) **17**(1), 80.
Xie Shaodong, Hao Jiming et al.:1996b, *Environmental Science* (in Chinese) **17**(5), 1.
Xie Shaodong, Hao Jiming et al.:1998, *Environmental Science* (in Chinese) **19**(1), 13.
Zhao Dianwu, Xiong Jiling, et al.:1988, *Atmospheric Environment* **22**, 349.
Zhao Dianwu, Zhang Xiaoshan, and Xiong Jiling:1992, *China Environmental Science* (in Chinese) **12**(2), 93.
Zhao Dianwu and Seip, H. M.:1991, *Water, Air, and Soil Pollution* **60**, 83.

ACID DEPOSITION AND CRITICAL LOAD MAP OF TOKYO

KENTARO HAYASHI[1*] and MASANORI OKAZAKI[2]

[1] *Environment Dept., Pacific Consultants Co.ltd., 2-7-1 Nishi-shinjuku, Tokyo 163-0730 Japan;*
[2] *Graduate School of Bio-Applications and Systems Engineering, Tokyo University of Agriculture and Technology, Koganei, Tokyo 184-8588 Japan*
(*author for correspondence, e-mail: Kentaro.Hayashi@tk.pacific.co.jp)

Abstract. Acid deposition has been monitored in the natural vegetation of the western part of Tokyo, especially in the Okutama Mountains and surrounding areas. However, it is difficult to grasp the condition of acid deposition and the possible impacts on the vegetation in the whole area. Therefore, we attempted to make gridded acidic deposition maps and critical load maps. The grid size was 30 seconds latitude and 45 seconds longitude. Monthly wet deposition in the fiscal year of 1997 was calculated by multiplying concentration of wet deposition and precipitation. Concentration of wet deposition was estimated by averaging the data monitored at the nearest three stations with the inverse of distance as the weight. Precipitation was estimated by step-wise multiple regression with geographical factors as explanatory variables. Critical loads were estimated using the steady-state mass balance model with some modifications. As result, it was found that sulfur deposition had exceeded in most of the western part of Tokyo.

Keywords: grid map, wet deposition, critical load, exceedance, Tokyo

1. Introduction

Tokyo is the capital of Japan and its land has been industrialized. However, natural vegetation remains in the western part of Tokyo. A few decades ago, the bay area within Tokyo was highly air-polluted. Recently, the atmospheric environment within the bay area has been gradually improved and low sulfur dioxide concentrations have been detected. However, high nitrogen oxide concentrations have remained mainly due to automobile exhausts (EAJ,2000). Hence, the deposition load of acidifying substances, including transported pollutants, has not decreased (EAJ,1999). Deposition amount of acidifying substances is similar to the European condition where acidification impacts have been observed. Although in Tokyo, there are no reports of apparent damages due to acidification, there is no proof that no harmful effects will occur in future when present conditions continue. Therefore, it is important to estimate the load of acidifying substances and the quantitative tolerance of receptors, (i.e., vegetation) to acidification.

Monitoring activities of acidic deposition have been carried out in Tokyo and its surrounding areas. However, most monitoring stations are concentrated in the urban and suburb areas, resulting to insufficient data in natural mountainous areas. Furthermore, the information is even more insufficient concerning the sensitivity to acidification of the vegetation in the Tokyo area. Since acidified environment does not recover easily, it is important to establish and map information based on present knowledge concerning acidification. Thus the geographical information on acidification is a useful tool in establishing activities and policies in the prevention of acidification.

The purpose of this study is to make deposition load maps and critical load and related information maps in the western part of Tokyo, including exceedance evaluation by comparison of acidic deposition to critical load.

2. Materials and Methods

2.1. WET DEPOSITION MAP IN TOKYO

First, wet deposition concentration maps for Tokyo were determined. The dry deposition velocity was considered to be strongly dependent on the surface condition, and therefore dry deposition was excluded from mapping in this study. The monthly wet deposition concentrations of each component, H^+, Na^+, K^+, Mg^{2+}, Ca^{2+}, NH_4^+, Cl^-, NO_3^- and SO_4^{2-} were obtained from 6 stations maintained by the Tokyo Metropolitan Research Institute for Environmental Protection, and 11 stations maintained by Saitama Prefecture. The average concentration of each component in respective grid was calculated by weighing the inverse of distance from the nearest three stations to the center of the grid as the weight. The grid size was set at 30 seconds latitude and 45 seconds longitude, approximately 1 km^2 in Japan. Then, monthly wet deposition concentration maps were made for the fiscal year of 1997 (mid of Mar. 1997 - mid of Mar. 1998, strictly).

The monthly precipitation maps of Tokyo for the fiscal year of 1997 were also established. Precipitation data was obtained from 37 stations of the Automated Meteorological Data Acquisition System (AMeDAS) maintained by the Japanese Meteorological Agency (JMA) in and surrounding of Tokyo. Stepwise multi regression analysis for precipitation was carried out with monthly precipitation at each station as the objective variable and topological and geographical factors, such as elevation, undulation, distance from sea coast and other parameterized factors as explanatory variables. This method was similar to the JMA gridded map of precipitation normals (JMA,1985). Then, monthly precipitation in each respective grid was estimated with the regression model and monthly precipitation maps were made.

Finally, multiplying the concentrations of each component and the precipitation amount for each grid square, monthly wet deposition load maps were derived.

2.2. CRITICAL LOAD AND EXCEEDANCE MAP IN TOKYO

The steady-state mass balance (SMB) model was used to estimate the critical loads for acidification in Tokyo. The SMB model was developed through European studies (e.g. Posch *et al.*,1995). It contains only the long-term stable processes with some assumptions. Thus, this model is relatively easily applied in large area.

The target area where critical loads for acidification and eutrophication were estimated was the western part of Tokyo. The eastern part of Tokyo was excluded due to high industrialization.

The SMB model targets forest soils, however, managed soils are excluded. In this study, the contribution of fertilization and harvest were included for some vegetation/land-use, such as farm, paddy, orchard and pasture. The artificial effects in managed forest were neglected because their long-term contribution was considered low. The effect of fertilization was expressed by adding a fertilization term to the SMB model. Nutrient K, Mg, Ca and N were considered. And other substances brought by irrigation such as S, Cl and Na were also considered for paddy areas. The effect of harvest was included in the nutrient uptake rate of plants.

The SMB equations based on Posch *et al.* (1995) with fertilization terms are as follows (units are [eq ha^{-1} yr^{-1}], unless stated otherwise);

Critical load: S and N separated

$$CL_{max}(S) = BC_{dep} - Cl_{dep} + BC_{we} + Bc_{fert} - Bc_{gu} - Ot_{fert} + Al_{le(crit)} + Q^{2/3} \cdot \left(Al_{le(crit)} / K_{gibb} \right)^{1/3}$$

$$CL_{max}(N) = -N_{fert} + N_{gu} + N_{im} + \left\{ CL_{max}(S) / (1 - f_{de}) \right\}$$

$$CL_{min}(N) = -N_{fert} + N_{gu} + N_{im}$$

$$CL_{nut}(N) = -N_{fert} + N_{gu} + N_{im} + \left\{ N_{le(crit)} / (1 - f_{de}) \right\}$$

$$Al_{le(crit)} = \min\left(1.5 \frac{Bc_{dep} + f_{Bc} \cdot BC_{we} + Bc_{fert} - Bc_{gu}}{(Bc/Al)_{crit}} \quad , \quad r \cdot BC_{we} \right)$$

$$Bc_{gu} = \min\left\{ \max\left(Bc_{dep} + f_{Bc} \cdot BC_{we} + Bc_{fert} - Q \cdot [Bc]_{min} \quad , \quad 0 \right) \quad , \quad Bc_{gu}^{0} \right\}$$

where,
- CL : critical load (S : sulfur; N: nitrogen; suffix nut : eutrophication)
- BC, Bc : base cation; BC=Na$^+$+K$^+$+Mg^{2+}+Ca^{2+}; Bc=K$^+$+Mg^{2+}+Ca^{2+}
- BC_{dep} : base cation deposition rate
- Cl_{dep} : chloride deposition rate
- BC_{we} : base cation weathering rate of soil minerals
- Bc_{gu} : net base cation uptake rate by plant (Bcgu^0 is raw data)
- N_{gu} : net nitrogen uptake rate by plant
- Bc_{fert} : base cation fertilization rate (for cultivated land only)
- N_{fert} : nitrogen fertilization rate (for cultivated land only)
- Ot_{fert} : other supply rate by irrigation (for paddy only)
- N_{im} : nitrogen immobilization rate as stable organic compound (set to 150)
- N_{de} : denitrification rate
- Q : leachate water flux [m^3 ha^{-1} yr^{-1}]

$Al_{le(crit)}$	: critical aluminum (Al^{3+}) leachate rate
$(Bc/Al)_{crit}$	: critical Bc to Al ratio [mol mol^{-1}] (set to 1.0)
f_{Bc}	: ratio of K^++ Mg^{2+}+Ca^{2+} to BC produced by weathering (set to 0.7)
f_{de}	: denitrification factor (0-1; 1 means completely denitrified)
K_{gibb}	: gibbsite solubility constant [eq^{-2} m^6] (set as 300)
r	: stoichiometric ratio of Al to base cation weathering in primary minerals [eq eq^{-1}] (set to 2.0)
$[Bc]_{min}$	: minimum Bc concentration that plant can take up [eq m^{-3}] (set to 0.002)

Deposition loads of each substance were taken from the deposition map. Chemical weathering rate was introduced by the chemical correlation method (Sverdrup, 1990) with some modification, such as addition of some minerals (not cited in Sverdrup, 1990), and mineralogy data from the Tokyo region. Nutrient uptake rates and fertilization rates were obtained from the existing data in Japan. Leachate flux was estimated by subtracting potential evapotranspiration (PE) from precipitation, PE was obtained from the Thornthwaite method. Other factors were set mainly according to Posch et al., (1995) since there were no concrete data with respect to Japanese ecosystems.

CL(S) ranges between 0 and CL$_{max}$(S), while CL(N) ranges between CL$_{min}$(N) and the maximum of CL(N) that is determined by choosing smaller one whether CL$_{max}$(N) or CL$_{nut}$(N). Based on the estimated critical loads, a 2 dimensional graph with axis of sulfur and nitrogen load can be produced for each grid that shows the relation of CL(S) and CL(N). The exceedance of sulfur and nitrogen deposition (S$_{dep}$ and N$_{dep}$, respectively) can be possibly evaluated, when S$_{dep}$ and N$_{dep}$ are plotted on the graph of CL(S)-CL(N) relationship.

3. Results and Discussion

Estimated annual wet deposition, S$_{dep}$ = SO_4^{2-}, and N$_{dep}$ = NH_4^+ + NO_3^-, in the fiscal year of 1997 are shown in Figure 1. The east of the area is relatively developed and the northwest of the area is the Okutama Mountains. Wet deposition load, especially N, in these mountainous areas is higher compared to the hillsides. This is due to higher precipitation in these areas.

The calculation result for critical loads (CL$_{max}$(S) and CL$_{max}$(N)) is shown in Figure 2. CL$_{max}$(S) is less than 200 eq ha^{-1} yr^{-1} in almost all of the area, especially less than 100 eq ha^{-1} yr^{-1} in the middle portion of the area where managed forests are located. While, CL$_{max}$(N) is larger than CL$_{max}$(S). CL$_{max}$(N) ranged from 1000 to 1500 eq ha^{-1} yr^{-1} in almost all of the area. Particularly, CL$_{max}$(S) and CL$_{max}$(N) indicate the capacity for total loads and not for acidity loads, because the mass balance equation contains the term of base cation deposition, hence S$_{dep}$/N$_{dep}$ include sulfate/nitrate that has counter cations other than H^+.

The result for exceedance calculation is shown in Figure 3. Sulfur deposition

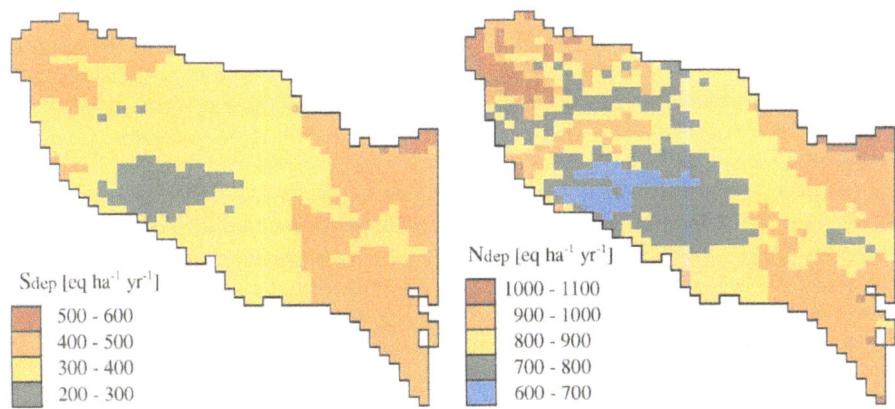

Figure 1. Annual wet Sdep (SO_4^{2-}, left) and Ndep ($NH_4^+ + NO_3^-$, right) in the fiscal year of 1997

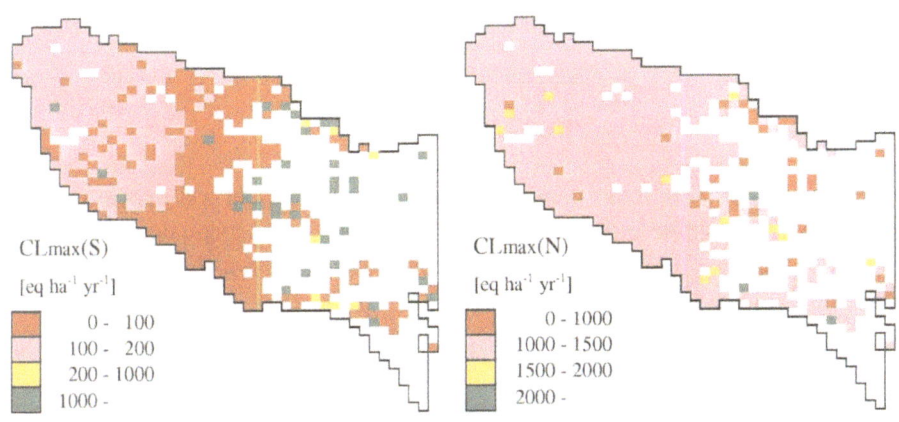

Figure 2. CLmax(S) (left) and CLmax(N) (right)

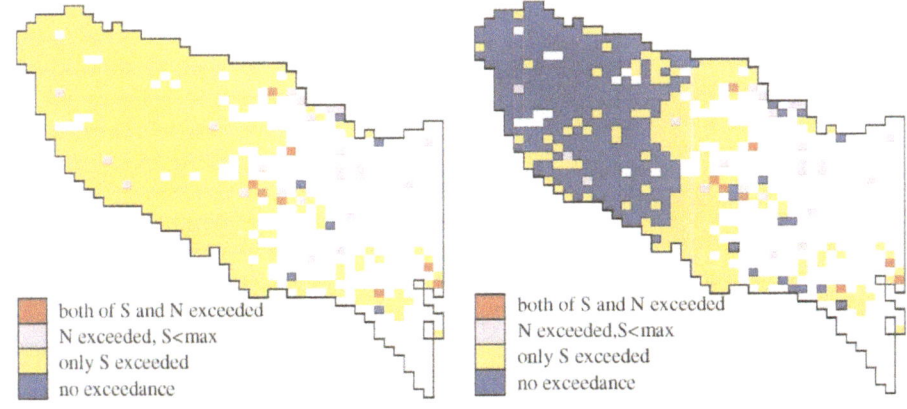

Figure 3. Exceedance under the present load

Figure 4. Exceedance under 1/4 S load

load exceeds in almost all of the western part of Tokyo. This is due to low $CL_{max}(S)$ since most of the produced or deposited base cations are consumed by plants, while large capacity of nitrogen consumption by plants results to a large $CL_{max}(N)$.

As shown in Figure 4, it is a prerequisite that the deposition load of S should be reduced to one-forth of the present to cancel the exceedance. Consequently, the capacity of S load in forest ecosystem is smaller than that of N.

We considered the effect of wet deposition only. However, the dry deposition perhaps has a comparable contribution to the total deposition. Dry deposition velocity maps should be established in the future. The wet deposition concentration at each monitoring station obtained from the sampled water was a problem, since sampled water is not entirely the same as the precipitation in volume. It affects the estimated deposition load thus some modification is needed. The data for critical load estimation is not sufficient, especially on soil mineralogy and the rate of nutrient uptake by plants. Collection of fundamental data is considered as an important task in future studies.

4. Conclusion

Although much more improvement was needed, the present wet deposition and critical loads could be estimated, and the exceedance was evaluated. At present the condition on acidification in Tokyo was assessed as not to be safe enough, especially with respect to sulfur load.

Acknowledgements

The authors would like to thank Dr. Komeiji T. and Mr. Takahashi K. for providing us the monitoring data and related useful information from the Tokyo Metropolitan Government and Saitama Prefecture for the purpose of this study.

References

Environmental Agency of Japan (EAJ): 1999, *Data Book on JEA Nationwide Surveys Phase III - Separate Book on Atmospheric Research* (in Japanese).
Environmental Agency of Japan (EAJ): 2000, *Air Pollution Condition in Japan* (in Japanese).
Japanese Meteorological Agency (JMA): 1985, *Sokko-Jiho*, **52**, 6, 357.
Posch, M., de Smet, P.A.M., Hettelingh, J-P. and Downing, R.J.: 1995, *Calculation and Mapping of Critical Thresholds in Europe*, Coordination Center for Effects (RIVM), the Netherlands.
Sverdrup, H.: 1990, *The Kinetics of Base Cation Release due to Chemical Weathering*, Lund University Press, 246p.

STEADY-STATE AND DYNAMIC ASSESSMENT OF FOREST SOIL ACIDIFICATION IN SWITZERLAND

DANIEL KURZ[1*], BEAT RIHM[2], MATTIAS ALVETEG[3] and HARALD SVERDRUP[3]

[1] EKG Geo-Science, Ralligweg 10, CH-3012 Bern, Switzerland; [2] Meteotest, Fabrikstrasse 14, CH-3012 Bern, Switzerland; [3] Lund University, Department of Chemical Engineering II, P.O. Box 124, S-221 00 Lund, Sweden
(*author for correspondence, e-mail: geoscience@bluewin.ch)

Abstract. The European steady-state Simple Mass Balance (SMB) model and the dynamic soil acidification model SAFE were used to assess the risk of future forest soil acidification in Switzerland. 2010 deposition forecasts on a 150x150 km grid resolution as well as corresponding ecosystem protection levels were obtained from RAINS model runs based on the 1999 Gothenburg Protocol obligations under the UN/ECE LRTAP Convention. Deposition values for 2010 on the national resolution were derived by scaling down present 1x1 km deposition values according to the deposition trends at the 150x150 km grid resolution. Meeting the Protocol obligations will reduce the percentage of Swiss forest ecosystems not protected against acidification between 1990 and 2010 from 41 to 4% according to the RAINS assessment and from 63 to 16% according to the assessment with the SMB at the 1x1 km resolution. The dynamic approach indicates, however, that soil conditions may not improve as much as these steady-state models suggest. By 2010, 39% of the sites considered will still have soil solution Bc/Al molar ratios below 1 at least in one soil layer. Nevertheless, deposition reductions obtained from the implementation of the new protocol will prevent the major part of Swiss forest soils from further acidification. Aiming at recovery of the more sensitive forest ecosystems would require emission reductions beyond the Protocol's obligations.

Keywords: critical loads, dynamic modeling, Gothenburg Protocol, exceedance, regional

1. Introduction

During the last decade, a series of protocols under the UN/ECE Convention on Long-Range Transboundary Air Pollution (LRTAP) have been negotiated in order to limit adverse effects of air pollution in Europe. The recently adopted Protocol to Abate Acidification, Eutrophication and Ground-Level Ozone (UN/ECE, 1999) sets mandatory national emission ceilings for sulfur dioxide, nitrogen oxides, volatile organic compounds and ammonia to be met by the year 2010. Optimum emission reductions were obtained from European scale (150x150 km grid resolution) integrated assessment modeling (Amann et al., 1999), simultaneously accounting for emission/deposition patterns, ecosystem sensitivity and abatement costs. The objective of this contribution is to closely analyze the benefits and deficits of the protocol obligations with respect to acidification effects in Swiss forest ecosystems. The sensitivity of the selected ecosystems regarding acid deposition has been mapped across Switzerland in terms of critical loads of acidity on a 1x1 km grid resolution (FOEFL, 1994), using a variant of the European Simple Mass Balance (SMB) model. Continued and temporary exceedance of critical loads of acidity affected forest ecosystems in the past or continue to do so. The dynamics of the system, however, determine if and when a violation of the critical limit used to calculate the critical loads occurs. To investigate temporal aspects of forest soil acidification, the dynamic soil chemistry model SAFE was applied to 600 evenly distributed (intersections of a 4x4 km grid) forest sites (SAEFL, 1998a). Unlike static approaches, dynamic soil chemistry models allow the quantitative assessment of the past, present and future status of the ecosystems' chemical state variables as a function of the changing environment (e.g. Sandén and Warfvinge, 1992).

2. Materials and Methods

2.1. MODEL CONCEPTS

The critical load of acidity for forest soils is derived from a mass balance equation in terms of acid neutralizing capacity (ANC) demanding all inputs of acidity to the system to be balanced by all sources of alkalinity, and permitting no net depletion of base saturation. As an intrinsic ecosystem property, it is calculated from the weathering rate (ANC_W) minus the permitted alkalinity leaching (ANC_L). In the SMB model (Posch et al., 1995), critical ANC_L is determined from a simplified expression for ANC as the maximum permitted leaching of aluminum (Al) and protons. In accordance with the Bc/Al criterion (molar $Bc/Al_{crit} = 1$), the critical Al leaching is derived from a mass balance of the relevant base cations (Bc) accounting for weathering, deposition and uptake processes as well as certain physiological limitations. Alternatively, critical Al leaching is not allowed to exceed net Al weathering from primary minerals (soil stability criterion). Al production from weathering is related to the weathering of Bc via the stoichiometry of the minerals involved. The critical proton leaching is connected to the aluminum concentration via an apparent gibbsite equilibrium (e.g. Dahlgren et al., 1990; Posch et al., 1995).

SAFE (Sandén and Warfvinge, 1992; Alveteg et al., 1998a) calculates the evolution of relevant chemical soil parameters from measured or extrapolated input. Simulations are started from 'pristine' steady-state conditions obtained from a steady-state variant of SAFE. The SAFE model includes process-oriented descriptions of cation exchange reactions, chemical weathering of minerals, leaching and accumulation of dissolved chemical components and finally solution equilibrium reactions involving carbonate, organic acid and Al-species. SAFE also includes schematic descriptions of deposition, nutrient cycling of base cations Ca^{2+}, Mg^{2+} and K^+ and nitrogen (N) as litterfall and canopy exchange, uptake of base cations and N from soil solution, net mineralization and/or immobilization of base cations and N, the contributions of these processes being specified as time-series by the model user. The temporal resolution is one year, therefore input and output parameters are expressed as annual values. The magnitude of capacity factors such as mineral abundance and cation exchange capacity is constant in time, sulfur (S) reactions serving as a net sink or source of ANC are neglected, and base cations are lumped into a divalent component Bc in the present version of the SAFE model. All deposited reduced N is assumed to be either taken up by roots or nitrified in the topmost soil layer. Dissolved organic matter is specified as input data, not generated by mineralization.

Both models were applied to the physiological relevant soil compartment (rooting zone) being one layer (thickness: 0.2 to 0.5 m) in the SMB and four layers (total 0.4 to 0.6 m thick) in the SAFE application. Each soil compartment is further assumed to be chemically isotropic and the soil solution to be perfectly mixed.

2.2. MODEL PARAMETERIZATION

By definition, the steady-state approach requires long-term average values of the input variables. This time-invariant input was either derived from national or regional survey information or from single site measurements, using transformation functions or computerized submodels. Soil properties used by the SAFE model to calculate the weathering rates were extrapolated from a series of reference soil profiles by means of a soil classification. The soil mineral composition was derived from total elemental soil analyses using a normalization procedure (for details see FOEFL, 1994; SAEFL 1998b; Alveteg et al., 1998a).

National deposition forecasts for the year 2010 were obtained from RAINS model runs of the emission ceilings established in the Gothenburg Protocol (GP99). Present deposition of each acidifying component on the national 1x1 km resolution was normalized with respect to the RAINS result on the EMEP grid scale using the grid-specific ratio of the 2010 deposition given by RAINS and the mean present deposition rate in the grid considered from the national resolution. Thus for 2010, total acidifying deposition to an EMEP grid cell can be assumed to be independent of the spatial resolution applied (national 1x1 km and EMEP 150x150 km).

In order to create the site-specific deposition time-series required by SAFE, the sites' current acidifying deposition loads were scaled with the national deposition trend. National deposition trends of S, oxidized and reduced N were inferred from historic national (FOEFL, 1995) and European (e.g. Mylona, 1993) emission trends. Emissions were transformed to deposition considering EMEP import/export budgets (Barrett and Seland, 1995). The 1990 and 2010 values were again taken from the RAINS model run (Table I), and the 2010 values were retained until 2100. Deposition is influenced by the canopy, increasing as the forest grows and decreasing as the forest is harvested. This was accounted for by linking reconstruction of deposition with forest growth in the MAKEDEP routine (Alveteg et al., 1998b), which was adapted for central European conditions. Forest growth was modeled using the stands' current biomass (Brassel and Kaufmann, 1996) and nutrient content, a n:th order Michaelis-Menten logistic growth function and information on past and future forest management. Forest exploitation in historic times and contemporary harvesting were assessed from historic sources including maps, from statistical annuals and the National Forest Inventory (NFI) 1982-86. Wood removal forecasts were derived from the Swiss base scenario to the fifth European Timber Trends Study (Pajuoja, 1995).

TABLE I
Decrease of the mean deposition of acidifying compounds in Switzerland expected from implementing the Gothenburg Protocol.

Component	Deposition			Reduction	
	peak (year)	1990	2010	1990-2010	peak-2010
	eq ha^{-1} a^{-1}	eq ha^{-1} a^{-1}	eq ha^{-1} a^{-1}	%	%
SO$_x$	1511 (1975)	731	271	-63	-82
NO$_y$	544 (1985)	518	266	-49	-51
NH$_y$	847 (1985)	837	735	-12	-13

3. Results and Discussion

According to the European scale integrated assessment modeling (Amann *et al.*, 1999), emission reductions adopted with the Gothenburg Protocol will reduce the percentage of sensitive ecosystems (forests and alpine surface waters) in Switzerland with acid deposition above their critical loads from 41% in 1990 to roughly 4% by 2010. This implies that the ecosystem protection level aimed at with this international agreement, i.e. 95%, will be accomplished with the implementation of the new Protocol. On the national 1x1 km grid resolution, however, the critical load of acidity will still be exceeded by 2010 at 19% of the ecosystems. The national detailed deposition model distinguishes several land use types which results in relatively high deposition rates for forests (the main receptor) compared to open land (e. g. grassland). The projected 2010 acidifying deposition to Swiss forests ranges from 0.47 to 4.79 keq ha^{-1} a^{-1} with a mean of 1.74 keq ha^{-1} a^{-1} compared to the national mean of 1.27 keq ha^{-1} a^{-1} obtained from EMEP grid resolution modeling. Besides its dependency on spatial resolution, exceedance of the critical load of acidity may also be influenced indirectly by the chemical criterion used to calculate the critical load. For forest soils in Switzerland, the critical load of acidity used was the minimum of the critical loads calculated from plant tolerance of Al (Bc/Al ratio = 1) and from soil stability considerations. The resulting exceedance percentage for the year 2010 is 16% using both criteria simultaneously compared to 7%, if the Bc/Al criterion alone is considered.

The Bc/Al molar ratio integrates the trends of the soil solution Bc and total inorganic Al concentrations which are both obtained from the dynamic model.

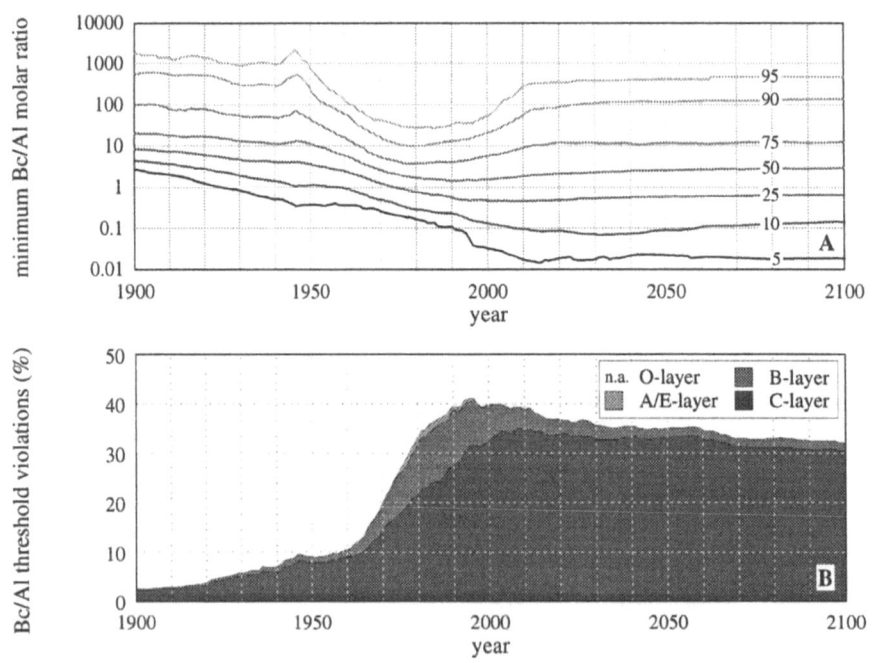

Figure 1. Time-series of the Bc/Al molar ratio minima (**A**) and percentage of sites violating the criterion (**B**). National scale, 600 sites considered, small numbers in plot A refer to percentiles.

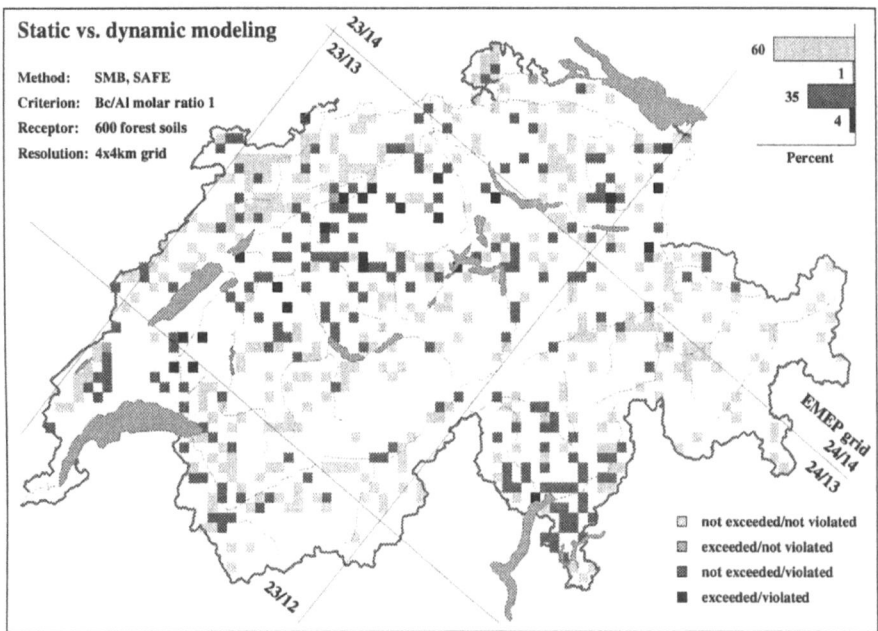

Figure 2. Spatial pattern of the combination of exceedance of critical loads of acidity (static approach) and violation of the critical Bc/Al molar ratio (dynamic approach) in 2010.

To get a single Bc/Al value for a soil profile, it is general practice to use the minimum calculated within the rooting zone independent of depth. Before 1900, up to 60% of the minima are modeled in the upper soil layers (Kurz *et al.,* 1998). By the end of the modeling period, however, more than 75% of the minima are found in the lower soil layers. 90% of the Bc/Al molar ratio minima fall between 2.7 and 1823 in 1900 and between only 0.03 and 53.2 in 2000 (Figure 1A). Deposition reductions as a result of the implementation of the Gothenburg Protocol will improve the soil chemistry in terms of Bc/Al molar ratio minima relative to the current situation at 75% of the sites considered in the long term. In the short term however, i.e. until 2010, nearly 38% of the sites will develop even lower values than today. Modeled Bc/Al molar ratio minima fall below the threshold value 1 at 5% of the sites already by 1926 and, 30 years after the peak of acidifying deposition, at a maximum of 41% of the sites in 1995, respectively (Figure 1B). The major part of the threshold violations are modeled in the lowest (C-) layer of the rooting zone. By 2010, still 39% of the sites violate the critical Bc/Al limit, although the major part of them receive acid deposition below the critical load of acidity (Figure 2). Given the adopted deposition reductions, recovery appears to be extremely sluggish with still 32% of the sites having Bc/Al molar ratio minima below 1 by 2100.

4. Conclusions

In view of the national scale modeling, ecosystem protection indicated by European scale integrated assessment modeling of emission ceilings adopted with the Gothenburg Protocol must be considered as too optimistic, although the implementation of the Protocol will reduce the ecosystem area under acidification risk in Switzerland substantially. The dynamic approach reveals the delayed response of forest soils to both increasing and decreasing acidifying input. The persistent violation of the critical Bc/Al molar ratio limit in the lower soil layers indicates that emission reductions beyond the Protocol's obligations have to be considered, if substantial recovery of more sensitive forest ecosystems is to be obtained.

Acknowledgments

We are indebted to B. Achermann and R. Volz (SAEFL), P. Blaser, P. Brassel, E. Kaufmann and S. Zimmermann (Swiss Fed. Inst. for Forest, Snow and Landscape Res.), S. Braun (Inst. for Appl. Plant biol.) and U. Eggenberger (EKG Geo-Science) for supporting the project. Funding is due to the Swiss Agency for the Environment, Forests and Landscape. The contents of this paper do not necessarily reflect the views of the SAEFL and no official endorsement should be inferred. The manuscript profited from elaborate comments of two anonymous reviewers.

References

Alveteg, M., Sverdrup, H. and Kurz, D.: 1998a, *Water, Air, and Soil Pollution*, **105**: 1.
Alveteg, M,, Walse, C. and Warfvinge, P.: 1998b, *Water, Air, and Soil Pollution*, **104**: 269.
Amann, M., Imrich, B., Cofala, J., Gyarfas, F., Heyes, C., Klimont, Z., Schöpp, W.: 1999, *Lucht & Energie 132*, Ministry of Housing, Spatial Planing and the Environment, The Hague, The Netherlands.
Barrett, K. and Seland, Ø. (eds.): 1995, *EMEP MSC-W Report 1/95*, Norwegian Meteorological Institute, Oslo, Norway.
Brassel, P. and Kaufmann, E.: 1996, NFI 1982-86 data base extract, Swiss Federal Institute for Forest, Snow and Landscape Research, Birmensdorf, Switzerland.
Dahlgren, R. A., McAvoy, D. C. and Driscoll, C. T.: 1990, *Environmental Science and Technology*, **24**: 531.
FOEFL (ed.): 1995, *Environmental Series Air, 256*, Federal Office of Environment, Forests and Landscape (FOEFL), Bern, Switzerland.
FOEFL (ed.): 1994, *Environmental Series Air 234*, Federal Office of Environment, Forests and Landscape (FOEFL), Bern, Switzerland.
Kurz, D., Alveteg, M. and Sverdrup, H.: 1998, *Water, Air, and Soil Pollution*, **105**: 11.
Mylona, S.: 1993, *EMEP MSC-W Report 2/93*, Norwegian Meteorological Institute, Oslo, Norway.
Pajuoja, H.: 1995, *UN ECE/FAO timber and forest discussion papers*, ECE/TIM/DP/4.
Posch, M., de Smet, P. A. M., Hettelingh, J-P. and Downing, R. J. (eds.): 1995, *RIVM Report 259101004*, Coordination Center for Effects, National Institute of Public Health and the Environment, Bilthoven, The Netherlands.
SAEFL (ed.): 1998a, *Environmental Documentation Air/Forest 89*, Swiss Agency for the Environment, Forest and Landscape (SAEFL), Bern, Switzerland.
SAEFL (ed.): 1998b, *Environmental Documentation Air/Forest 88*, Swiss Agency for the Environment, Forest and Landscape (SAEFL), Bern, Switzerland.
Sandén, P, and Warfvinge P. (eds.): 1992, *Reports Hydrology 5*, Swedish Meteorological and Hydrological Institute Norrköping, Sweden.
UN/ECE (ed.): 1999, Protocol to the 1979 Convention on Long-range Transboundary Air Pollution to Abate Acidification, Eutrophication and Ground-level Ozone, done at Gothenburg, Sweden, on November 30. 1999.

DEPOSITION AND CRITICAL LOADS OF NITROGEN IN SWITZERLAND

BEAT RIHM[1*] and DANIEL KURZ[2]

[1] *Meteotest, Fabrikstrasse 14, CH-3012 Bern, Switzerland;* [2] *EKG Geo-Science, Ralligweg 10, CH-3012 Bern, Switzerland*
(*author for correspondence, e-mail: rihm@meteotest.ch)

Abstract. Many ecosystems in Switzerland suffer from eutrophication due to increased atmospheric nitrogen (N) input. In order to get an overview of the problem, critical loads for nutrient N were mapped with a resolution of 1x1 km applying two methods recommended by the UN/ECE: the steady state mass balance method for productive forests, and the empirical method for semi-natural vegetation, such as natural forests, (sub-)alpine or species-rich grassland and raised bogs. The national forest inventory and a detailed atlas of vegetation types were used to identify the areas sensitive to N input. The total N input was calculated as the sum of NO_3^-, NH_4^+, NH_3, NO_2 and HNO_3 wet and dry deposition. Wet deposition was determined on the basis of a precipitation map and concentration measurements. Dry deposition was calculated with inferential methods including land-use specific deposition velocities. The concentration fields for NH_3 and NO_2 were obtained from emission inventories combined with dispersion models. Reduced N compounds account for 63% of total deposition in Switzerland. As indicated by exceeded critical loads, the highest risk for harmful effects of N deposition (decrease of ecosystem stability, species shift and losses) is expected on forests and raised bogs in the lowlands, where local emissions are intense. At high altitudes and in dry inner-alpine valleys, deposition rates are significantly lower.

Keywords: empirical sensitivities, eutrophication, exceedance, forest, mass balance, semi-natural vegetation

1. Introduction

Under the framework of the UN/ECE Convention on Long-Range Transboundary Air Pollution (LRTAP), methods were developed for calculating and mapping crititcal loads of air pollutants for various environmental receptors (UN/ ECE, 1996). Critical loads of nutrient nitrogen (CLN) reflect those N inputs to natural and semi-natural ecosystems, below which adverse effects of eutrophication should not occur.

CLN as well as critical loads of acidity (SAEFL, 1996; FOEFL, 1994; SAEFL, 1998) were determined for Switzerland on a 1x1 km grid and supplied to the bodies of the Convention as a basis for the negotiation of effect-oriented air pollution abatement scenarios. Additionally, deposition rates were modelled with the same spatial resolution in order to map the areas where critical loads are exceeded. This paper describes the methods and results of the mapping project with respect to CLN and their exceedances.

Water, Air, and Soil Pollution **130**: 1223–1228, 2001.
© 2001 *Kluwer Academic Publishers.*

2. Materials and Methods

2.1. CRITICAL LOADS FOR PRODUCTIVE FORESTS

For managed forests, CLN are calculated applying a simplified steady state mass balance (SMB) method (UN/ECE, 1996). The basic principle of this SMB method is to identify the long-term average sources and sinks of inorganic N in the system, and to determine the maximum tolerable N input that will not lead to oversaturation: $CLN_{SMB} = N_{le(acc)} + N_u + N_i + N_{de}$
The following definitions and value ranges were used:
CLN_{SMB}: critical load of nutrient N [kg N ha^{-1} yr^{-1}].
$N_{le(acc)}$: acceptable total N leaching from the rooting zone at which no damage occurs in the terrestrial or linked aquatic ecosystems (4-5 kg N ha^{-1} yr^{-1}).
N_u: net removal of N by (wood) harvesting at critical load (1-7 kg N ha^{-1} yr^{-1}). Temporal variations as a function of forest age and management are not taken into consideration.
N_i: acceptable immobilization rate of N in soil organic matter at which adverse ecosystem change will not take place (3-5 kg N ha^{-1} yr^{-1}).
N_{de}: denitrification rate (3-25 kg N ha^{-1} yr^{-1}). This is the flux to the atmosphere of N_2, N_2O and NO produced by micro-organisms in the soil. It was calculated by denitrification fractions between 0.3 and 0.8 for different soils.
All rates and fluxes of the involved processes are represented by annual means. The SMB is calculated at 10'400 sampling points from the national forest inventory (WSL, 1992), which corresponds to a forest area of 10'400 km^2 (1x1 km grid).

2.2. CRITICAL LOADS FOR (SEMI-)NATURAL ECOSYSTEMS

For natural and semi-natural ecosystems, CLN are mapped by applying the empirical method (CLN$_{emp}$) described by Bobbink et al. (1996), where empirical data on impacts of increased nitrogen deposition are summarized in several categories: e.g. nutrient imbalances in trees, increased shoot/root ratio, changes in forest ground flora and mycorrhizae, and change in diversity of grassland. Ranges of critical load values are proposed for different ecosystems. The values were derived from experiments under controlled and field conditions, from chemical analysis and comparison of vegetation and fauna composition in time and space, from ecosystem models, as well as from biological knowledge. The critical load is always given as a range reflecting (1) the real intra-ecosystem variation, (2) intervals obtained in experiments, and (3) uncertainties in deposition rates, where critical loads are based on field observations.

Table I shows the ecosystems that are mapped in Switzerland, the critical load values assigned within the proposed ranges, and the number of 1x1 km grid-cells in which the ecosystems occur. All of the selected ecosystems are of high conservation importance with respect to biodiversity (Hegg et al., 1993).

If several vegetation types occur in the same 1x1 km grid-cell then the most sensitive type was chosen for the subsequent mapping procedure, resulting in a total sensitive area of 16'400 km^2.

TABLE I

Selected vegetation types, CLN$_{emp}$ [kg N ha^{-1} yr^{-1}] and area of their occurence.

Ecosystem	CLN range	Relevant vegetation types in Switzerland	CLN	Area km^2
Acidic coniferous forests	7-20	Molinio-Pinetum	17	333
		Ononido-Pinion	12	361
		Cytiso-Pinion	12	17
		Calluno-Pinetum	12	124
Acidic deciduous forests	10-20	Quercion robori-petraeae	15	68
Calcareous forests	15-20	Quercion pubescentis	15	633
		Fraxino orno-Ostryon	15	28
		Erico-Pinion mugi (Ca)	15	592
		Erico-Pinion sylvestris	15	956
Arctic and alpine heaths	5-15	Juniperion nanae	10	1262
		Loiseleurio-Vaccinion	10	1906
Calcareous species-rich grassland	15-35	Mesobromion (erecti)	20	2686
Neutral-acid species-rich grassl.	20-30	Molinion (caeruleae)	25	612
Montane-(sub)alpine grassland	10-15	Chrysopogonetum grylli	15	6
		Seslerio-Bromion (Koelerio-Sesl.)	12	2198
		Festucetum paniculatae	12	39
		Stipo-Poion molinerii, alpine	10	523
		Elynion, alpine	10	836
		Seslerion (variae), alpine	10	5545
		Caricion ferrugineae, alpine	10	4046
Mesotrophic fens	20-35	Scheuchzerietalia	20	580
		Caricion fuscae	25	2240
		Caricion davallianae	25	2986
Ombrotrophic bogs	5-10	Sphagnion fusci	8	902
Shallow soft-water bodies	5-10	Littorellion	8	49

2.3. MODELLING AND MAPPING OF NITROGEN DEPOSITION

Wet deposition is calculated by combining the concentration field of NO$_3^-$ and NH$_4^+$ compounds in rain water with a precipitation map (FOWG, 2000). Wet concentration measurements are relatively homogenous below an altitude of 1000 m a.s.l.. At higher altitudes (alpine grassland is found up to 3000 m) concentrations decrease (Oekoscience, 1999).

Resistance analogue models are used for assessing the dry deposition of NH$_3$ and NO$_2$ gas. For these compounds, the concentration fields (annual means) are calculated from emission inventories with a resolution of 200x200 m by applying statistical dispersion models. The emission maps are a combination of precisely located emissions, such as from farm buildings or roads, and of regional statistical values that are disaggregated according to land-use patterns. For NO_2, the method is well calibrated and verifyied by monitoring results (BUWAL, 1997). For NH_3, the validation is at a preliminary stage, as only inhomogenous data sets are available from 17 monitoring sites between 1991 and 1999. Nevertheless, it was possible to reproduce the monitoring results with a RMS error of 0.75 µg m^{-3} (Fig. 1), using constant dispersion profiles given by Asman and Jaarsveld (1990). For aerosols and HNO_3, the concentration fields are calculated as a function of altitude.

The concentration fields are combined with deposition velocities, which depend on the reactivity of the pollutant, surface roughness and climatic parameters. Deposition velocity values were taken from literature (FOEFL, 1996).

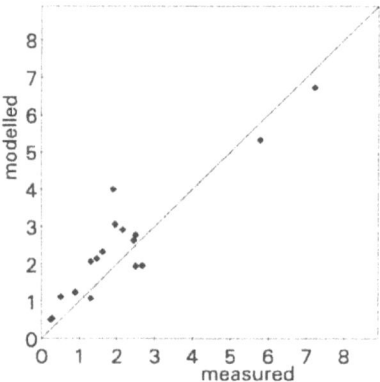

Figure 1. Comparison of measured and modelled NH_3 concentrations for 17 monitoring sites in Switzerland, 1991 to 1999 [µg m^{-3}].

3. Results and Discussion

CLN_{emp} are in the range of 8 to 25 and CLN_{SMB} in the range of 10 to 30 kg N ha^{-1} yr^{-1} and their cumulative frequency distribution is very similar (Fig. 2a).

The total calculated N deposition for Switzerland is 76.5 kt N yr^{-1}, to which reduced N compounds contribute 63% (Table II). The depositions calculated as annual means from 1993 to 1995 exceed the critical loads on 60% of the sensitive area for CLN_{emp} and on 88% for CLN_{SMB} (Fig. 2b). If values for CLN_{emp} and for CLN_{SMB} occur in the same 1x1km grid-cell then the exceedances were calculated separately.

Figure 3 shows the spatial distribution of the combined exceedances of CLN_{emp} and CLN_{SMB}. For this map, the maximum exceedance was calculated in each grid-cell, resulting in a total sensitive area of 22'800 km^2 (55% of the Swiss territory). The highest exceedances are found in the lowland and foothills of the Alps, where substantial local emissions from traffic and settlements (NO_x) and from agriculture (NH_y) occur. At (sub)alpine altitude levels, the exceedances are smaller. Areas without exceedance are mainly found in dry inner-alpine valleys with low wet deposition rates and small local emissions.

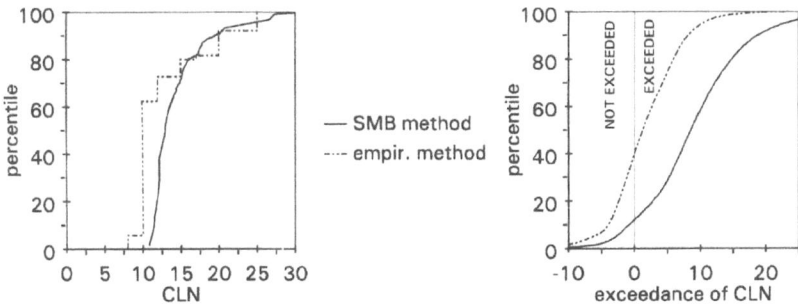

Figure 2. Cumulative frequency distributions of (a) critical loads of nutrient nitrogen (CLN) and (b) exceedances of CLN (1993-95) in Switzerland. Units: kg N ha^{-1} yr^{-1}.

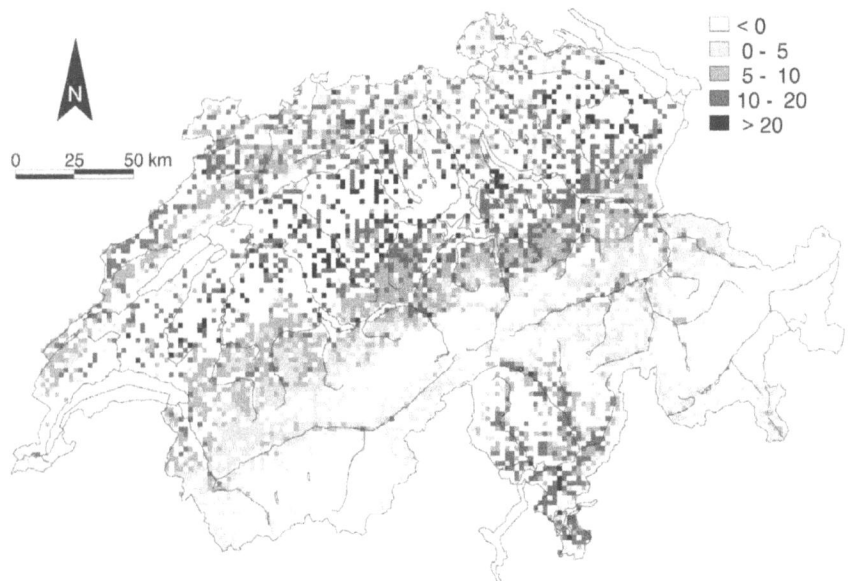

Figure 3. Exceedance of CLN 1993-95 in Switzerland calculated with 1x1 km resolution (the cartographic display is 2x2km). Units: kg N ha^{-1} yr^{-1}. Areas with negative values do not have sensitive ecosystems or deposition is less than CLN.

TABLE II

Total deposition of N compounds in Switzerland (41'200 km^2) calculated with 1x1 km resolution. Annual average 1993 to 1995. Units: kt N yr^{-1}.

reduced N:		oxidised N:		total N:	
NH_4^+ wet	23.4	NO_3^- wet	17.2	wet	40.6
NH_4^+ aerosol	2.7	NO_3^- aerosol	1.1	aerosol	3.8
NH_3 gas	21.8	NO_2 gas	8.0	gas	32.1
		HNO_3 gas	2.3		
total	**47.9**		**28.6**		**76.5**

4. Conclusions

The exceedances of CLN on large parts of forests, species-rich grassland and raised bogs indicate long-term risks for adverse ecological changes in these systems due to eutrophication effects. Such effects can be observed in the field (*e.g.* Flückiger and Braun, 1999), but it is not possible to quantify the ecological risks by using the presented mapping methods.

The application of spatially detailed emission inventories along with statistical dispersion models allows distinguishing the areas that are influenced by local/regional emissions, and remote areas, where depositions are mainly due to long-range transport. This is helpful for evaluating national emission abatement strategies.

Modelling depositions in complex terrain with patchy land use patterns are afflicted with considerable errors and remains subject to further investigations.

Acknowledgments

Funding is due to the Swiss Agency for the Environment, Forests and Landscape. The manuscript profited from elaborate comments of two anonymous reviewers.

References

Asman, W.A.H. and van Jaarsveld, H.A.: 1990, *A variable-resolution statistical transport model applied for ammonia and ammonium.* National Institute of Public Health and Environmental Protection (RIVM), Bilthoven, The Netherlands.

Flückiger, W. and Braun, S.: 1999, *Water, Air and Soil Pollution* **116**, 99.

Bobbink, R., Hornung, M. and Roelofs, J.G.M.: 1996, 'Empirical Nitrogen Critical Loads for Natural and Semi-Natural Ecosystems'. In: UN/ECE 1996.

BUWAL (ed.): 1997, NO_2 *Immissionen in der Schweiz 1990-2010.* Swiss Agency for the Environment, Forests and Landscape (SAEFL/BUWAL), Environmental Series Air 289, Berne.

FOEFL (ed.): 1994, *Critical Loads of Acidity for Forest Soils and Alpine Lakes.* Environmental Series Air 234, Federal Office of Environment, Forests and Landscape (FOEFL) Berne.

FOWG (ed.): 2000, precipitation maps 1961-1999, Hydrological Atlas of Switzerland, Federal Office for Water and Geology (FOWG), Berne, Switzerland.

Hegg, O., Béguin, C. and Zoller, H.: 1993, *Atlas schutzwuerdiger Vegetationstypen der Schweiz (Atlas of Vegetation Types Worthy of Protection in Switzerland).* Edited by Federal Office of Environment, Forests and Landscape, Berne, Switzerland.

Oekoscience: 1999, *Depositionsdatensaetze in der Schweiz.* Zuerich, Switzerland.

SAEFL (ed.): 1996, *Critical Loads of Nitrogen and their Exceedances.* Environmental Series Air 275. Swiss Agency for the Environment, Forests and Landscape (SAEFL) Berne.

SAEFL (ed.): 1998, *Critical Loads of Acidity for Forest Soils - regionalized PROFILE model.* Environmental Documentation Air/Forest 88, Swiss Agency for the Environment, Forests and Landscape (SAEFL) Berne.

UN/ECE (ed.): 1996, *Manual on Methodologies and Criteria for Mapping Critical Levels/Loads and Geographical Areas where they are Exceeded.* Convention on Long-Range Transboundary Air Pollution. Federal Environmental Agency, Texte 71/96, Berlin.

WSL: 1992, *Swiss National Forest Inventory.* Data extracts 30 May 1990 and 8 December 1992. Swiss Federal Institute for Forest, Snow and Landscape Research (WSL), Birmensdorf.

ASSESSING POTENTIAL IMPACTS ON BIODIVERSITY USING CRITICAL LOADS

K.R. BULL[1,*], J.R HALL[1], J. COOPER[1], S.E. METCALFE[2], D. MORTON[1], J. ULLYETT[1], T. L. WARR[1] and J.D. WHYATT[3]

[1] *CEH Monks Wood, Abbots Ripton, Huntingdon PE28 2LS, UK;* [2] *Department of Geography, Edinburgh University, Drummond Street, Edinburgh EH8 9XP, UK;* [3] *Department of Geography, Lancaster University, Lancaster LA1 4YB, UK*

(* author for correspondence, e-mail: keith.bull@ceh.ac.uk)

Abstract. In many countries there has been much concern over maintaining biodiversity in natural ecosystems in the face of pressures such as changing land use and pollution. The 1992 UN Convention on Biodiversity calls upon signatories to develop national strategies for the conservation and sustainable use of biodiversity. In the UK, the potential impacts of sulphur and nitrogen deposition at the national level are being assessed using national critical loads and modelled deposition maps, together with available information on the occurrence of habitats and plant species. This simple approach gives an indication of the areas where atmospheric deposition may have impacts on biodiversity. The results of the analyses are presented and the strengths and weaknesses of the methods used are discussed. This first approach to considering the effects on biodiversity shows the importance of including the effects of atmospheric deposition in any biodiversity action plan. It also highlights those areas where more or improved information is required for the national strategy. With the modelled deposition data available, it would seem that reduced impacts are to be expected by 2010. However, higher resolution deposition data, better estimates of ammonium deposition, consideration of temporal aspects and the dynamics of change, and the use of higher resolution biological data sets are likely to suggest greater impacts than current predictions.

Keywords: biodiversity, critical loads, UK, sulphur and nitrogen deposition

1. Introduction

The United Kingdom was one of 153 signatories to the Convention on Biological Diversity. Adopted at the Rio *Earth Summit* in 1992, the Convention identified action to address the loss of species and genetic resources. Countries are responsible for biodiversity within their jurisdiction. The UK published its Biodiversity Action Plan in 1994 setting out a broad strategy for conserving and enhancing species and habitats for 20 years. Responding to the plan, the UK's Biodiversity Steering Group has prepared over 200 Species Action Plans and 48 Habitat Action Plans (UK Biodiversity Secretariat, 2000). Habitats have now been aggregated into 27 Broad Habitat classes, 17 terrestrial and freshwater, 10 marine and coastal. Within each class there may be a number of Priority Habitats, which are of the highest conservation concern.

The various action plans are well-established and take account of many of the pressures that need to be addressed for their successful implementation. However, at present they take little account of acidification or nutrient nitrogen effects (eutrophication) on the biodiversity of some habitat types. While the principles of acidification and eutrophication are well understood, there is a need to consider the possible impacts of sulphur and nitrogen deposition on individual habitats or species, and evaluate the requirements of an Action Plan in relation to planned emission decreases in the UK and Europe.

The work described here makes use of available data on critical loads, modelled deposition, species distributions and land cover. It was planned to give a quick indication of the possible extent of impacts on biodiversity. The paper discusses shortfalls of the work and describes what is needed to extend the current results.

2. The Data

The 1990 Land Cover Map of Great Britain (LCM) is a digital map derived by classifying satellite images from the Landsat Thematic Mapper (Fuller *et al*, 1994). Images are classified into 25 land cover classes at 25m resolution.

The National Critical Loads (NCL) data were used for negotiation of the 1999 Gothenburg Protocol of the 1979 Convention on Long-range Transboundary Air Pollution. The Protocol defined national emission reduction strategies to abate acidification, eutrophication and ground-level ozone. Critical loads values for acidification and eutrophication are defined for areas of five terrestrial "ecosystem types" for each 1km grid of Great Britain. Areas were estimated from the LCM by grouping appropriate land cover classes (Hall, 1998).

Data for three *deposition scenarios for sulphur, and for oxidized and reduced nitrogen* were modelled, at 10km resolution, using the Hull Acid Rain Model (HARM). This provides data for UK national abatement strategy development (Metcalfe *et al*, 1998). The model results were illustrative of the deposition in 1990 and the changes that may take place in the future (Table I).

TABLE I
UK emissions (kilo-tonnes) employed in the deposition scenarios

Scenario	Sulphur (SO_2)	Oxidized nitrogen (NO_x)	Reduced nitrogen (NH_3)
1990 Base	3754	2800	329
Post Kyoto	733	1074	297
H1 2010	497	1181	264

The "Base 1990" deposition scenario was used as a baseline for assessing the results of recent and future emission controls. The "Post Kyoto" scenario was a UK and European deposition forecast for the year 2010 based on emission decreases recommended at the 1997 Kyoto Earth Summit. The third scenario used, termed "H1 2010", is a 2010 forecast based on what are believed to be the maximum feasible reductions of sulphur and nitrogen emissions in the UK and throughout Europe. Matching each of these scenarios with the NCL data provided maps of "exceedances" showing where critical loads for the five "ecosystem types" were exceeded by deposition loads of nitrogen and sulphur (for acidity) or by deposition loads of nitrogen (for eutrophication).

The National Vegetation Classification (NVC), developed at Lancaster University, UK, categorises plant communities by the relative combinations of species present (Rodwell, 1991, 1992, 1995). NVC communities are associated with, and therefore indicators of, particular habitat types.

The *Biological Records Centre (BRC) database* of the Centre for Ecology and Hydrology provided 10km resolution species distributions. BRC maps are based upon reports from networks of recorders that provide information on species occurrence.

3. Defining Critical Loads for Biodiversity Action Plan (BAP) Broad Habitat Classes and Priority Habitats

Defining critical loads for the BAP Broad Habitats made use of the descriptions compiled by the UK Biodiversity Group (UK Biodiversity Secretariat, 2000). Eleven of these were matched to 25 LCM classes (*e.g.* Fig. 1), which in turn were linked to the 5 "ecosystem types " of the NCL database. Examples of this process (Table II) show the match is sometimes simple, but it can be difficult to match some Habitats to critical loads values. For some Habitats, no match was possible (*e.g.* rivers and streams, inland rock); these were omitted from the analysis.

Figure 1. Distribution of deciduous woodland from the ITELCM.

TABLE II
Examples of matching BAP Broad Habitats and LCM classes and the NCL "ecosystem types"

BAP Broad Habitat	LCM class	NCL "ecosystem type"
Coniferous woodland	Coniferous/evergreen woodland	Coniferous woodland
Acid grassland	Moorland grass, ruderal weed felled forest	Acid grassland
Dwarf shrub heath	Dense shrub heath	Heathland
Montane habitats	Moorland grass	Acid grassland

Defining critical loads for Priority Habitats was more complex. It used the descriptions of the Priority Habitats (UK Biodiversity Secretariat, 2000) and linked these to the NVC classes and associated species compositions and hence to BRC species distribution maps. The closest related "ecosystem type" NCL critical loads values were assigned to and mapped for selected Priority Habitats.

4. Exceeded Areas of BAP Broad Habitats, Priority Habitats and Sites of Special Scientific Interest

Having linked the spatial distributions of the Broad Habitats and Priority Habitats with the NCL database critical loads exceedances were calculated using a GIS. Exceedance maps and statistics for each critical loads "ecosystem type" were produced for all scenarios, both for acidity and also eutrophication. Exceeded areas decrease with time quite markedly for acidity (Figs. 2a and 2b) and for eutrophication. However, continued high deposition in upland areas in future years still cause exceedances in these sensitive western parts of Britain. These changes are reflected in the results for individual Broad Habitats that show major improvements in both acidification and eutrophication (*e.g.* Table III). Even so, the percentage changes, and the total areas, differ from habitat to habitat. For Priority Habitats the pattern is similar though the areas are much smaller (*e.g.* Table IV).

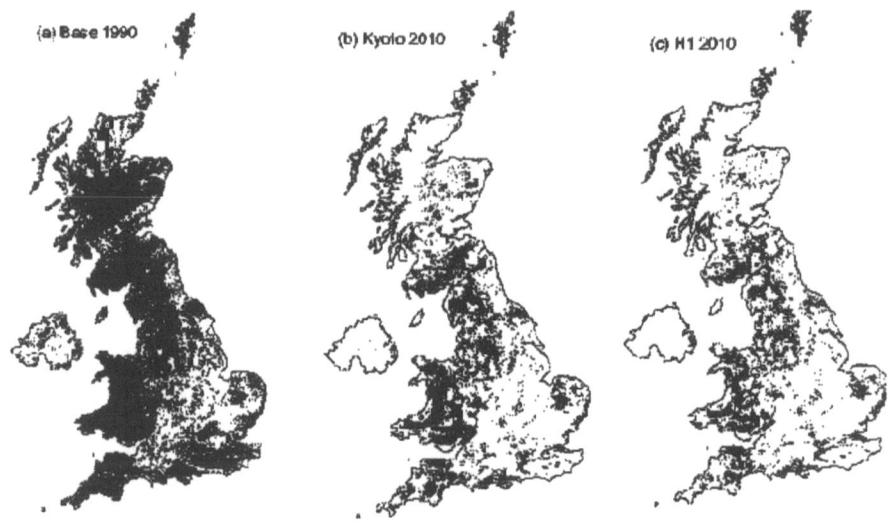

Figures 2a, 2b, 2c. Critical load exceedance maps for all "ecosystem types" for the 3 scenarios (note the position of Northern Ireland is displaced for convenience of mapping)

It is evident that the differences between the Kyoto and H1 scenarios are small. In general, as expected, the H1 scenario improves on the post-Kyoto scenario.

TABLE III
Some Broad Habitats areas and the percentage areas at risk from acidification | eutrophication

BAP Broad Habitat	Area (km^2)	Base 1990(%)	Scenario Kyoto 2010(%)	H1 2010(%)
Coniferous woodland	6830	27 | 12	5.6 | 1.4	4.6 | 0.9
Acid grassland	13570	74 | 12	28 | 6.9	20 | 7.5
Dwarf shrub heath	380	64 | 16	25 | 0.3	11 | 1.5
Montane	13330	75 | 12	28 | 7.5	20 | 5.1

TABLE IV
Some Priority Habitats areas and the percentage areas at risk from acidification | eutrophication

BAP Priority Habitat	Area (km^2)	Base 1990(%)	Scenario Kyoto 2010(%)	H1 2010(%)
Upland oak woodland	1140	68 | 77	11 | 16	9 | 9
Wet woodlands	865	75 | 71	10 | 14	7 | 8
Lowland dry acid grass	257	66 | 2	32 | 2	30 | 2
Purple moor grass and rush pastures	810	47 | 6	9 | 3	7 | 3

5. Discussion and Conclusions

The results presented show widespread effects at the national scale. They also highlight several important data issues that fall into two categories. First, the best available national data sets may not be ideal for the analysis and may be improved with further work. Second, the mapped scales of the data used are often different, which may create uncertainties in interpretation when data are overlaid. The following paragraphs identify areas where future studies of data sets may improve the analysis.

Habitats were mapped using landcover and species distributions. Both of these data sets are subject to uncertainties: satellite imagery has limited spatial and spectral resolution and there are risks of misclassification of classes; species distributions are mapped at coarse resolution. Critical loads models at present are simple, seldom directly related to biodiversity, and take no account of dynamic changes. They indicate potential damage in exceeded areas, and the potential for recovery when areas become not exceeded. Deposition data are uncertain due to current model limitations (e.g. scale), lack of knowledge of certain processes, and the inherent uncertainty of deposition. Local measurements of ammonia deposition, for example, show this may be much higher than models suggest, though the role of ammonium in acidification remains uncertain. Previous studies of uncertainties with modelled current deposition data also suggest that significant underestimates of deposition can occur (Smith et al., 1995).

Overall it is not simple to assess whether the impacts on biodiversity are under- or over-estimated in the above analysis. However, the number of factors suggesting under-estimation of effects are sufficient to merit concern over the emissions of sulphur and nitrogen pollutants and their effects on the UK environment.

In conclusion, this first approach to assessing impacts of atmospheric deposition on biodiversity shows potential widespread effects in the UK, though it also predicts marked improvements for the future. The results suggest that it is important to include atmospheric deposition of pollutants in future biodiversity action plans. Further, they demonstrate a clear need to consider the effects on biodiversity at the local scale with improved, more detailed data.

Acknowledgements

The authors acknowledge the support of the UK Department of the Environment, Transport and the Regions through contracts EPG1/3/116 (Bull, Hall, Ullyett, Cooper) and EPG 1/3/152 (Metcalfe, Whyatt)

References

Fuller, R.M., Groom, G.B., Jones, A.R.: 1994, **60**, 553.
Hall, J., Bull, K., Bradley, I., Curtis, C., Freer-Smith, P., Hornung, M., Howard, D., Langan, S., Loveland, P., Reynolds, B., Ullyett, J. & Warr, T.: 1998, Status of UK Critical Loads and Exceedances January 1998. Part 1: Critical Loads and Critical Loads Maps. Report prepared under DETR/NERC Contract EPG1/3/116, see web site
(http://www.nmw.ac.uk/ite/monk/critical_loads/nclmp.html).
Metcalfe S. E., Derwent, R. G., Whyatt, J. D. and Dyke, H.: 1998, *Water, Air, and Soil Pollution*, **107**, 121.
Rodwell, J. S. (Editor): 1991, 1992, 1995, British Plant Communities, volumes 1-4. CambridgeUniversity Press: Cambridge.
Smith, R. I., Hall, J. R. and Howard, D. C.: 1995, *Water, Air, and Soil Pollution*, **85**, 2503.
UK Biodiversity Secretariat: 2000, http://www.jncc.gov.uk/ukbg.

MAPPING THE POTENTIAL SENSITIVITY OF SURFACE WATERS TO ACIDIFICATION USING MEASURED FRESHWATER CRITICAL LOADS AS AN INDICATOR OF ACID SENSITIVE AREAS

J. M. ULLYETT[1*], J. R. HALL[1], M. HORNUNG[2] and M. KERNAN[3]

[1] *CEH, Monks Wood, Abbots Ripton, Huntingdon, PE28 2LS; UK;* [2] *CEH, Merlewood, Grange-over-Sands, Cumbria, LA11 6JU, UK;* [3] *Environmental Change Research Centre, UCL, 26 Bedford Way, London, WC1H 0AP, UK*
(* author for correspondence, e-mail: jmu@ceh.ac.uk)

Abstract. Environmental organisations in the UK have shown increasing interest in freshwater critical loads and acid sensitivity maps as a means of assessing pollution at the local and regional level. These maps can be used to identify sensitive areas when considering catchment management plans. The use of national data sets to map the sensitivity of freshwaters to acidification, highlighted the problems of relating national datasets to smaller, localised areas. The study described here investigated the use of detailed large-scale maps to predict the sensitivity of surface waters to acidification for two river catchments. Three large-scale acid sensitivity maps were produced and validated using measured freshwater critical loads. In addition, a score system relating to the buffering capacity for each soil and geology type was devised. The score value was found to have a better correlation with water chemistry and freshwater critical loads than the acid sensitivity maps. The study concluded that it was not necessary to use the largest scale data available in order to improve predictions of sensitive areas.

Keywords: critical loads, surface water acidification, freshwater sensitivity

1. Introduction

Maps showing the potential sensitivity of freshwaters to acidification have been developed in several previous studies (Hornung *et al.*, 1995; Ullyett *et al.*, 1995) and the possibility of using these maps to predict freshwater critical loads has been assessed by Hall *et al.* (1995). The latter concluded that the critical load was predicted correctly in about 50% of cases but highlighted the problems of relating small-scale national data to small stream or lake catchments.

This paper investigates improving the predictions of freshwater critical loads using large-scale data for two river catchments. The objectives of this study were:
a) to use more detailed larger scale maps to see if this improved the correlation between the freshwater sensitivity maps and the measured freshwater critical loads.
b) to use the sensitivity of the individual soil and geology types, without classification, and see if this improved the correlation with the water chemistry and freshwater critical loads.

2. Methods

2.1. THE SMALL SCALE NATIONAL FRESHWATER SENSITIVITY MAP

The small-scale national freshwater sensitivity map (Hornung *et al.*, 1995; Ullyett *et al.*, 1995) was produced at a resolution of 1km, based on national digital maps of soil and geology. The map of "The Susceptibility of UK Surface and Groundwaters to Acid Deposition" (Kinniburgh and Edmunds, 1986) provided the geological input. This was classified into four sensitivity classes (high, medium, low and non-sensitive) based on the mineralogy and geochemistry of the dominant rock types found on the 1:625 000 solid geology map of Great Britain (British Geological Survey). A digital soil map of Great Britain (Soil Survey of England and Wales, 1983; Macaulay Institute for Soil Research, 1981) at 1km resolution, but based on 1:250 000 scale data, was classified into three sensitivity classes (high, medium and low) derived from the mineralogy and geochemistry of the predominant major soil group. These two classified maps were overlaid using a geographical information system (GIS) to generate 12 classes of surface water sensitivity, which were aggregated according to Hornung *et al.*, 1995, to give a five class freshwater sensitivity map.

2.2. THE CATCHMENT BASED STUDY

For the present study two river catchments were chosen to look at in detail: the river Glaslyn in Snowdonia (North Wales) and the river Duddon in Cumbria (north west England). These catchments were chosen because both are known to be sensitive to acidification and have been the subject of previous acidification studies (Bull & Hall, 1986; Hall *et al.*, 1995; Tipping *et al.*, 1998; Wong, 1992). The Glaslyn catchment covers approximately 68km^2 and the Duddon catchment approximately 87km^2. Both catchments span a range of altitudes, from mountainous uplands down to lowland estuaries.

2.3. THE LARGE SCALE FRESHWATER SENSITIVITY MAPS

The methods used to generate the freshwater sensitivity maps for the Glaslyn and Duddon were similar to those used for the small scale national map. Digital geology data, at 50m resolution, for both catchments and at three different scales (1: 50 000, 1:250 000 and 1:625 000) were classified into the four sensitivity classes, based on the buffering capacity of the rock types. For soils, the only digital data available for the study area are at 1:250 000 scale (Soil Survey of England and Wales, 1983). This data was classified into three sensitivity classes at 50m resolution, again based on the buffering capacity of the soil types. Each of the three re-classified geology maps were overlaid with the re-classified soils data to produce three maps of freshwater sensitivity for both the Duddon and Glaslyn catchments. Despite the increasing detail on the larger scale geology maps, the subsequent classification into sensitivity classes resulted in similar maps at all three scales. This is due to a combination of reasons: (i) the classification process itself reduces the complexity and detail of the maps; (ii) using one scale of soil information throughout gives rise to similarities in all three large scale maps. An attempt to reduce the impact of the classification process is described in section 2.4 below.

2.4. DIVIDING THE CATCHMENTS FOR ANALYSIS

In order to compare the freshwater sensitivity maps with freshwater critical loads values each river catchment was divided into sub-catchments. Initially the Glaslyn and Duddon catchments were divided into 20 sub-catchments and water samples collected in the autumn of 1997. For subsequent samplings the Glaslyn was divided into 32 and the Duddon into 35 sub-catchments in order to sample the more sensitive headwater streams. In total the Glaslyn and Duddon were sampled four times (two spring and two autumn) over a two-year period (from 1997 to 1999), although the information from the first sampling was not used in the final analysis as it did not cover the same number of sub-catchments as subsequent samplings. Flow data was obtained for the river Duddon from the Environment Agency for a site on the main river at Duddon Hall, near the bottom of the catchment. These data showed that during the weeks in which water samples were taken the river was at low flow so the water chemistry represent the mean acidity critical loads for low flow conditions. The water sample information represents all upstream water chemistry from the sample point, which may include smaller sub-catchments upstream. In the context of this paper the term 'sub-catchment' refers to all upstream areas above the sample point.

For each sub-catchment the dominant freshwater sensitivity class was calculated based on the percent cover of each class. An alternative method used was the area weighted mean sensitivity class for each sub-catchment. One disadvantage of using these methods is that they degrade the sensitivity information gained from using more detailed maps.

In order to try to reduce the impact of the classification process, a system was devised for the Glaslyn to give each individual soil and geology type a score value based on its buffering capacity. This was done by:
(i) assigning a high weighting value to soils and geology types with a high buffering capacity and a low weighting value to those with a low buffering capacity;
(ii) determining the percentage of each geology and soil type within each sub-catchment;
(iii) multiplying these percentages by the weighting value to give a score for each soil and geology type;
(iv) adding these soil and geology scores together to give an "acid susceptibility score" for each sub-catchment.

Regression analysis was used to compare the acid susceptibility scores with freshwater critical loads and Alkalinity (Alk), Calcium (Ca) and Magnesium (Mg) concentrations.

2.5. CRITICAL LOADS AND WATER CHEMISTRY

The water samples collected from the Glaslyn and Duddon were analysed and the water chemistry information used to calculate acidity critical loads using two methods, the Diatom model (Battarbee *et al.*, 1996) and the Steady State Water Chemistry (SSWC) model (Henriksen and Brakke, 1988; Curtis, 2000).

The diatom model calculates the critical load for the site and sets a base line critical load for the point at which acidification first occurs. For calculations using this model measured Ca values from the water chemistry were used as an indicator of sensitivity and measured 1992-94 sulphur deposition (RGAR, 1997) as the acid loading. Acidified sites have a low Ca value in relation to sulphur loading. This model accommodates variable acid neutralising capacity (ANC) as ANC can vary from site to site depending on the natural acidity of the water.

The SSWC model calculates the critical load for a particular indicator species, or assemblage of species. In the UK this is done by setting the ANC limit value to zero, giving a probability of protecting 50% of the brown trout population. This model uses water chemistry together with rainfall and runoff data.

It is because of the different ways in which ANC is used in these two models that the diatom critical loads are usually lower than the SSWC critical loads.

3. Results

3.1. CORRELATION BETWEEN CRITICAL LOADS VALUES AND FRESHWATER SENSITIVITY CLASSES

Comparing the mean diatom critical load values and the dominant freshwater sensitivity class for each sub-catchment showed there was a large overlap between the critical loads values found within each sensitivity class. Area weighted freshwater sensitivity classes for the Glaslyn were compared with the diatom critical loads, again resulting in a large overlap between the critical loads values found within each sensitivity class. The reason for these low correlations are most likely due to the classification of the soil and geology maps, and using a single freshwater sensitivity class to represent each sub-catchment.

3.2. CORRELATION BETWEEN ACID SUSCEPTIBILITY SCORES, WATER CHEMISTRY AND CRITICAL LOADS

Comparing the acid susceptibility score for each sub-catchment of the river Glaslyn with the diatom critical loads values (Table I) gave a better correlation than the methods described above; values are almost double those of using the area weighted freshwater sensitivity classes. The 1:625 000 scale data has the best correlation, even though it might be expected that the more detailed 1:250 000 scale data would be better. This may be due to the fragmentation of the geology types on the 1:250 000 scale data, and generally higher score values for each sub-catchment using this data, which do not correlate so well with the water chemistry and critical loads values.

TABLE I
Correlation coefficients between acid susceptibility score and diatom critical loads

Sample Date	1:625000	1:250000
April 1998	0.66	0.57
October 1998	0.65	0.54
May 1999	0.62	0.53
Average	0.66	0.56

Alk and Ca were highly correlated, and comparing the Alk, Ca and Mg concentrations with the acid susceptibility scores, caused the Ca values to appear as not significant in Table II. The Alk and Mg values gave very good correlation with the acid susceptibility scores, especially at the 1:625 000 scale, reaching an R^2 value of 0.67 (67%).

TABLE II
Significance of the Regression Coefficient between acid susceptibility score and Alk, Ca and Mg

Sample Date	Alk p value	Mg p value	Ca p value	R^2 value
1:625000				
April 1998	0.009	0.121	0.266	0.58
October 1998	0.000	0.018	0.159	0.67
May 1999	0.009	0.075	0.214	0.59
Average	0.004	0.057	0.233	0.62
1:250000				
April 1998	0.009	0.083	0.111	0.52
October 1998	0.001	0.017	0.073	0.58
May 1999	0.015	0.053	0.127	0.52
Average	0.005	0.042	0.102	0.55

4. Conclusions

Although the large scale acid sensitivity maps improved the detail found on the small-scale national map, they did not improve the correlation with critical loads values. This was most likely due to the classification process used for the soil and geology data, which in effect reduced the extra information gained from using more detailed data. However, using acid susceptibility scores based on the soil and geology types within each sub-catchment of the Glaslyn improved the correlation with both the critical loads and the water chemistry of the area. The best correlation was found using the 1:625 000 scale data where the soil and geology data alone explained, on average, 0.62 (62%) of the variation found between predicted acid susceptibility and actual water chemistry values.

Acknowledgements

The authors would like to thank the Freshwater Fisheries Laboratory, Faskally, Pitlochry, Scotland, UK for analysing the water samples collected during this study.

References

Battarbee, R.W., Allott, T.E.H., Juggins, S., Kreiser, A.M., Curtis, C. and Harriman, R.: 1996, *Ambio* **25(5)**, 366.
Bull, K. R. & Hall, J. R.: 1986, *Environmental Pollution* **12**, 165.
Curtis, C.J., Allott, T.E.H., Hall, J., Harriman, R, Helliwell, R., Hughes, M., Reynolds, B. and Ullyett, J.: 2000, *Hydrology and Earth System Science*, **4(1)**, 125.
Hall, J. R., Wright, S. M., Sparks, T. H., Ullyett, J. Allott, T. E. H. and Hornung, M.: 1995, *Water, Air, and Soil Pollution*, **85**, 2443.
Henriksen, A. and Brakke, D.F.:1988, *Water, Air and Soil Pollution*, **42**, 183.
Hornung, M., Bull, K. R., Cresser, M., Ullyett, J., Hall, J.R., Langan, S., Loveland, P.J.: 1995, *Environmental Pollution* **87**, 207.
Kinniburgh, D.G., and Edmunds, W.M.: 1986, *Hydrogeological Report, 86/3*, British Geological Survey, Wallingford, UK.
Macaulay Institute for Soil Research: 1981, *Soil Survey Maps of Scotland, 1:250 000 scale*. Macaulay Institute for Soil Research, Aberdeen.
Review Group on Acid Rain (RGAR): 1997, *Acid Deposition in the United Kingdom 1992-1994*, Department of the Environment, Transport and the Regions, UK.
Soil Survey of England and Wales: 1983, *1:250 000 Soils of England and Wales*. Soil Survey of England and Wales, Harpenden, UK.
Tipping, E., Carrick, T.R., Hurley, M.A., James, J.B., Lawlor, A.J., Lofts, S., Rigg, E., Sutcliffe, D.W. and Woof, C.: 1998, *Environmental Pollution* **103**, 143.
Ullyett, J., Hall, J.R and Bull, K.R.: 1995, 'Mapping the Potential Sensitivity of Freshwaters to Acidification using Catchment Characteristics', in R.W. Battarbee (ed.) *Acid Rain and its Impact: the Critical Loads Debate*, University College London, UK. pp. 103-106.
Wong, J.L.G.: 1992, Doctor of Philosophy Dissertation, University of Wales, Bangor, UK.

CALCULATING WEATHERING RATES OF STREAM CATCHMENTS IN THE ENGLISH LAKE DISTRICT USING CRITICAL ELEMENT RATIOS, MASS-BALANCE BUDGETS AND THE MAGIC MODEL

GARETH J.P. THORNTON

Department of Earth Sciences, The Open University, Walton Hall, Milton Keynes, MK7 6AA, UK.
(Email: G.J.P.Thornton@open.ac.uk)

Abstract. The weathering rates of forty-seven stream catchments in the English Lake District were calculated using 1) critical element ratios; 2) mass-balance budgets; and 3) the MAGIC model. There was a great deal of variability in the weathering rates of the five different parent material groups (greywacke, slate, andesite, tuff and granite) found in the study area. However when individual catchments were considered, the three methods provided consistent base cation weathering rates. This suggests that any of the methods could be employed for future catchment weathering studies. This paper also explores the implications of the weathering results when considering the possibility of the area becoming acidified due to increased loads of sulphur and nitrogen in the future.

Keywords: acidification, critical element ratios, MAGIC model, mass-balance, weathering

1. Introduction

Both steady-state and dynamic models have been developed to predict the acidification of soils, ground- and surface-water systems (e.g. Cosby *et al.*, 1985; Warfvinge and Sverdrup, 1992; Van Oene and De Vries, 1994). However, the rates of release of base cations have always been difficult to measure and quantify, and different methods often mean discrepancies in the resulting estimates (Langan *et al.*, 1996). As the Skokloster classification defines critical loads of acidity on the basis of soil parent material (Nilsson and Grennfelt, 1988), it is important to be able to accurately calculate what contribution (as a buffer to catchment acidification) base-cation weathering makes.

On the whole, if a high proportion of the dissolved load is accounted for by carbonate weathering then the alkalinity will be high (and the streams less susceptible to acidification). Although the current chemical signal is not necessarily indicative of future susceptibility, for example, trace amounts of carbonate in an otherwise resistant silicate lithology can contribute disproportionately and make streams draining this bedrock less susceptible to acidification (cf. Blum *et al.*, 1998). However, it is possible that future acid precipitation may consume this trace carbonate at a faster rate than it is supplied, thus increasing the sensitivity of the stream to acidification.

This paper describes the weathering rates for five spatially extensive parent materials calculated using critical element ratios, mass-balance budgets and the MAGIC model. It also explores the implications of the results on the sensitivity of the catchments to acidification under increased loads of sulphur and nitrogen in the future.

2. Methods

2.1. STUDY AREA

The central portion of the Lake District consists of four major lithological types trending south-west to north-east, but is further characterised by significant areas of igneous intrusions (Fig. 1). Detailed geological descriptions have been published elsewhere (e.g. Firman, 1978; Jackson, 1978; Millward *et al.*, 1978).

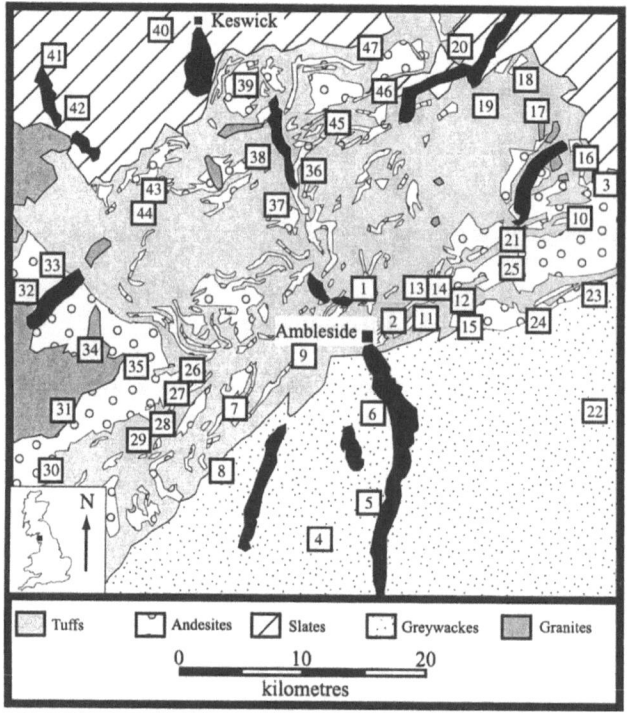

Figure 1. Location of the study streams, also showing the distribution of the five parent materials

Most of the study streams were underlain by tuff and andesite bedrock and tend to have exposed rock, thin podzolic soils and some rough sheep grazing in their catchments. Deeper brown earth soils prevail in the catchments underlain by greywacke, where rock exposures are less frequent and there are some patches of mixed deciduous woodland or conifer plantations (Table I).

Water samples were collected from the forty-seven study streams on a bimonthly basis for one year. In addition, a rock sample of the dominant catchment lithology of each stream was collected and split into two fractions (fresh and weathered). The weathering rind was removed by abrasive disk saw and was labelled as the weathered fraction. The detailed descriptions of the procedures used during sample collection, discharge measurement and analysis can be found elsewhere (Thornton and Dise, 1998; Thornton, 1998, 1999).

TABLE I

Details of the study streams divided into the five parent material groups, and showing dominant soil type, altitude and precipitation ranges. (N is the number of streams in each group)

GEOLOGY	SOIL	ALTITUDE (m)	PPTE (mm)	N
Tuff	Peaty Podzol	120-340	1693-4035	25
Andesite	Peaty Podzol	140-390	1744-3104	9
Greywacke	Brown Earth	60-280	1652-2277	7
Slate	Humic Ranker	150-240	2304-2407	3
Granite	Peaty Podzol	70-150	2430-2700	3

2.2. METHOD DESCRIPTION AND DERIVING WEATHERING RATES

2.2.1. MAGIC model

The MAGIC model was formulated to describe the long-term impacts of acid deposition, weathering and cation exchange on soil and streamwater chemistry. In the model, chemical weathering rates are determined by simulation (Cosby *et al.*, 1985) and are expressed as cation released rates. This study used an inverse method of MAGIC to calculate the release of cations due to chemical weathering. An optimisation program based on a Rosenbrock algorithm (cf. Jenkins *et al.*, 1997) used detailed streamwater and soil chemistry data to calculate the catchment weathering rate. However, a correction was required to discriminate between the sources of base cations from cation exchange and chemical weathering (cf. Thornton, 1999).

2.2.2. Mass balance method

The major source of inputs to an upland catchment is atmospheric deposition and the outputs are in streamwater solutes. The application of this technique to calculate weathering rates are widespread in the literature (e.g. Hultberg, 1985; Bain *et al.*, 1990; Langan *et al.*, 1996), and is therefore only briefly described here. For a catchment, which is assumed to be in steady state, it is possible to estimate the weathering rate from a simple input and output mass balance of cations. A water balance was carried out prior to using the mass-balance equation to correct for any leakage of groundwater. The mass-balance method can be described by the following equation:

$$C_{WR} = BC_{Stream} - (BC_{Atmos} - BC_{Biomass})$$

where C_{WR} is average catchment weathering rate, BC_{Stream} is base cations found in stream outflow, BC_{Atmos} is atmospherically derived base cations (sea-salt corrected), and $BC_{Biomass}$ is base cations retained in biomass.

2.2.3. Critical element ratio method (cf. Winchester and Floyd, 1977)

Firstly, the composition of the fresh and weathered individual rock samples are examined. Here, strontium is used as a surrogate for calcium weathering.

However, rather than comparing the direct strontium loss from the rock samples, zirconium (an immobile element) is plotted against the strontium (a mobile element) and zirconium ratio. This technique gives a good indication of calcium weathering since any change in the ratio is solely the result of a change in strontium abundance. The amount of weathering is expressed as the depletion (%) in the Sr/Zr ratio between the fresh and weathered bedrock samples, rather than as a catchment weathering rate over a one year period.

3. Results

The weathering rates of the five parent material groups for the critical element ratio, catchment budget and MAGIC methods are presented in Table II.

TABLE II

Study results for the five parent materials using proportion of strontium depletion (%) from Sr/Zr ratios and base cation release rates (keq ha^{-1} yr^{-1}) using catchment budgets and MAGIC.

GEOLOGY	Depletion	Catchment budget	MAGIC
Tuff	13.57	1.36	1.54
Andesite	22.95	1.95	2.38
Greywacke	27.38	2.38	2.20
Slate	4.90	0.95	1.25
Granite	13.75	0.66	1.00

Compared with previous studies in the UK (e.g. Whitehead *et al.*, 1988, 1997; Langan *et al.*, 1996), the catchment weathering rates are generally low for the tuff, slate and granite groups and moderately low for the andesite and greywacke groups (Table II). The weathering results are fairly similar for the catchment budget and MAGIC methods, reflecting the fact that MAGIC optimises part of its weathering rate calculation via a catchment mass balance. In addition, the results from critical element ratios follow a similar pattern (greywacke and andesite have higher rates than tuff, granite and slate).

TABLE III

Correlation coefficients of the three principal methods used for calculating weathering rates.

	Depletion	Catchment budget	MAGIC
Depletion	1.00		
Catchment budget	0.798	1.00	
MAGIC	0.796	0.884	1.00

Given the similarity of the results in Table II, the statistical relationships between the different methods were tested using a correlation matrix (Table III). A very strong positive correlation (0.80 – 0.88) was found between the three methods and the close affinity between the weathering rate results is illustrated in Figure 2. The closeness of the results is fairly surprising given that the critical element ratio method represents 'spot' data with no spatial or temporal integration, whereas the other two methods represent an integrated weathering rate over time for the whole catchment.

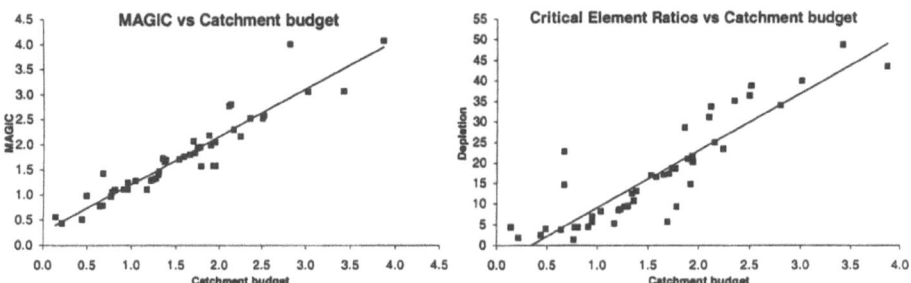

Figure 2. Comparison of the weathering rates derived using the three principal methods. Units are % for critical element depletion and keq ha^{-1} yr^{-1} for catchment budgets and MAGIC.

4. Discussion

Although public concern regarding acid rain is decreasing, the need to understand the process of catchment weathering is still important. An accurate estimate of the future weathering rate is paramount to our understanding of how the sensitivity of streams may change in the future under increased atmospheric S and N deposition. The data presented here suggest that there is a great deal of variability in the weathering rates of the five parent material groups. For instance, catchments underlain by slate, granite and tuff bedrock have lower weathering rates than the catchments underlain by greywacke and andesite. This suggests that if all the study catchments had a similar rate of acid deposition then catchments underlain by slate, granite and tuff bedrock would become more sensitive to acidification, whereas catchments underlain by greywacke and andesite would be slower to acidify.

Unfortunately the three methods used in this paper cannot predict whether an increase in acid deposition will result in an increase in the weathering rate or whether increased acid precipitation will consume base cations at a faster rate than they are supplied by weathering. However, if weathering rates stay constant, the mass-balance and MAGIC methods are capable of predicting the impact of increased sulphur and nitrogen deposition on the sensitivity of stream catchments. Theoretically a twofold increase in acid deposition should result in a concomitant increase in the catchment weathering rate, however, we cannot yet predict whether this process is linear. Therefore, to truly understand and predict the effects of increasing atmospheric S and N deposition on stream catchments a suitable method needs to be developed in the near future.

5. Conclusions

The Acid Neutralizing Capacity of streams in most upland areas depends upon mineral weathering. The results of this paper suggest that a reasonable approximation of the current weathering rate can be made despite there being many complex processes involved (e.g. heterogeneity of soil, bedrock and land use, as well as a range of hydrochemical and hydrological conditions).

In addition, assuming that weathering rates stay constant, the mass-balance and MAGIC methods are capable of making an assessment as to whether increased loads of sulphur and nitrogen will lead to the area becoming acidified in the future.

References

Bain, D.C., Mellor, A., Wilson, M.J. and Duthie, D.M.L.: 1990, 'Weathering in Scottish and Norwegian catchments', in B.J. Mason (Ed.), *The Surface Waters Acidification Programme*, Cambridge University Press.

Blum, J.D., Gazis, C.A., Jacobson, A.D. and Chamberlain, C.P.: 1998, *Geology* **26**, 411.

Cosby, B.J., Wright, R.F., Hornberger, G.M. and Galloway, J.N.: 1985, *Water Resources Research* **21**, 1591.

Firman, R.J.: 1978, 'Intrusions', in F. Moseley (Ed.), *The Geology of the Lake District*, W.S. Maney & Sons Ltd.

Hultberg, H.: 1985, *Ecological Bulletin* **37**, 133.

Jackson, D.E.: 1978, 'The Skiddaw Group', in F. Moseley (Ed.), *The Geology of the Lake District*, W.S. Maney & Sons Ltd.

Jenkins, A., Renshaw, M., Helliwell, R., Sefton, C., Ferrier, R., and Swingewood, P.: 1997, *Modelling surface water acidification in the UK* Institute of Hydrology Report No. 131.

Langan, S.J., Reynolds, B. and Bain, D.C.: 1996, *Geoderma* **69**, 275.

Millward, D., Moseley, F. and Soper, N.J.: 1978, 'The Eycott and Borrowdale Volcanics', in F. Moseley (Ed.), *The Geology of the Lake District*, W.S. Maney & Sons Ltd.

Nilsson, J. and Grennfelt, P. (Ed.): 1988, *Critical loads for sulphur and nitrogen* Nordic Council of Ministers.

Thornton, G.J.P. and Dise, N.B.: 1998, *The Science of the Total Environment* **216**, 63.

Thornton, G.J.P.: 1998, 'Chemical composition of the streams draining the English Lake District: Relationships between stream chemistry and catchment characteristics', in M.J. Haigh, J. Krecek, G.S. Rajwar and M.P. Kilmartin (eds.), *Headwaters: Water Resources and Soil Conservation*, AA Balkema: Rotterdam: Brookfield.

Thornton, G.J.P.: 1999, Doctor of Philosophy Dissertation, Open University, Milton Keynes, UK

Van Oene, H. and De Vries, W.: 1994, *Water, Air & Soil Pollution* **72**, 41.

Warfvinge, P. and Sverdrup, H.: 1992, *Water, Air & Soil Pollution* **63**, 119.

Whitehead, P.G., Reynolds, B., Hornung, M., Cosby, B.J., Neal, C. and Paricos, P.: 1988, *Hydrological Processes* **2**, 357.

Whitehead, P.G., Barlow, J., Haworth, E.Y. and Adamson, J.K.: 1997, *Hydrology and Earth Systems Science* **1**, 197.

Winchester, J.A. and Floyd, P.A.: 1977, *Chemical Geology* **20**, 325.

ADAPTING THE PROFILE MODEL TO CALCULATE THE CRITICAL LOADS FOR EAST ASIAN SOILS BY INCLUDING VOLCANIC GLASS WEATHERING AND ALTERNATIVE ALUMINUM SOLUBILITY SYSTEM

TAMON FUMOTO[1]*, JUNKO SHINDO[1], NORIKO OURA[1] and HARALD SVERDRUP[2]

[1] *National Institute of Agro-Environmental Sciences, Tsukuba, 305-8604 Japan;*
[2] *Chemistry Center, Lund University, P. O. Box 124, S-221 00, Lund, Sweden*
(*author for correspondence, e-mail: tamon@niaes.affrc.go.jp*)

Abstract. Adaptation of the steady-state soil chemistry model PROFILE was studied, on the following two parts, to calculate the critical loads for East Asian soils: (1) Dissolution rate coefficients of volcanic glass were derived from published experimental data, and calculated field weathering rate was compared with the rate estimated based on Sr isotope analysis. When BET surface area of sand fraction was regarded as mineral surface area, the calculated rates fairly agreed with the estimate, suggesting that sand fraction surface area is a reasonable estimate of weatherable mineral surface area of volcanic soils. (2) In repeated leaching experiments, Al solubility of a number of Japanese soils was explained by a model which assumed complexation of Al to soil organic matters. Such an Al solubility model is more appropriate for predicting soil chemistry than apparent gibbsite dissolution equilibrium.

Key words: Al solubility, Critical load, Soil chemistry model, Volcanic soils, Weathering

1. Introduction

PROFILE (Warfvinge and Sverdrup, 1992) is a steady-state model for predicting soil water chemistry from data on deposition, uptake and soil properties. For assessing critical acidity load, the most influential variables are mineral weathering rate and Al solubility. PROFILE calculates weathering rate of each particular mineral by a kinetic model, but the rate coefficients have been lacking for volcanic glass, the major primary mineral in volcanic ash soils common in East Asia. Also, it is uncertain whether the common approach for Al solubility, *i.e.*, apparent gibbsite dissolution equilibrium, is applicable to these soils. This study discusses the adaptation of PROFILE to East Asian soils regarding mineral weathering rate and Al solubility modeling.

2. Materials and Methods

2.1. DERIVATION OF DISSOLUTION RATE COEFFICIENTS OF VOLCANIC GLASS

In PROFILE, dissolution rate of each mineral is described in terms of base cation (BC) release rate, r (kmol$_c$ m^{-2} s^{-1}), as

$$r = k_H \frac{\left[H^+\right]^{n_H}}{f_H} + \frac{k_{H_2O}}{f_{H_2O}} + k_{CO_2} P_{CO_2}^{n_{CO_2}} + k_R \frac{\left[R^-\right]^{n_R}}{f_R}, \tag{1}$$

where k_H, k_{H2O}, k_{CO2} and k_R are the dissolution rate coefficient for reaction with H^+, water, dissolved CO_2 and organic acid anions, respectively. n_H, n_{CO2} and n_R are the reaction orders, f_H, f_{H_2O} and f_R are the rate reduction factors for product inhibition. Rate coefficients of glass were derived from perlite glass dissolution experiments (White, 1983). Si release rate (mol m^{-2} s^{-1}) was 3.9×10^{-14} (pH 1.0), 1.9×10^{-14} (pH 2.0), 1.2×10^{-14} (pH 3.0), 7.2×10^{-15} (pH 4.0), 7.5×10^{-15} (pH 5.2), 1.4×10^{-14} (pH 6.2). Using these data, k_H, k_{H2O} and n_H of Si release were determined by least square method (LSM). Reactions with CO_2 and organic acids were neglected, as the experiment excluded these species. For volcanic glasses of basaltic and dacitic tephra, k_H and k_{H2O} of BC release were calculated assuming congruent dissolution with the average chemical composition in Table I.

TABLE I
Average composition (mol kg^{-1}) of volcanic glasses from basaltic and dacitic tephras, derived from Yamada and Shoji (1983)

Rock type	SiO_2	Al_2O_3	Fe_2O_3	MgO	CaO	Na_2O_2	K_2O	TiO_2
Basaltic	9.09	1.93	0.55	0.58	1.90	0.43	0.04	0.11
Dacitic	12.61	1.30	0.14	0.17	0.36	0.54	0.24	0.04

Weathering rate, in terms of Ca^{2+} and Mg^{2+} release, was estimated on two sites (Table II) based on Sr isotope ratio and BC budget (Shindo et al., 2001). The soils at Kannondai and Yasato are classified as an Andosol and a Brown Forest soil, respectively, but the mineralogy at Yasasto suggests substantial contribution from volcanic deposits.

2.2. ALUMINIUM DISSOLUTION EXPERIMENTS

Two (Kannondai) or 4 (Yasato) grams of air-dried soil was shaken with 25 mL of 1 mM HCl in a 50 mL PE centrifuge tube for 23 hours. The suspension was then centrifuged and the supernatant was collected. Then, 25 mL of 1 mM HCl was added to repeat the same procedure 25 times. After measuring pH, the supernatant was filtered (<0.22 μ m) for further chemical analysis.

We determined Na^+, K^+, NH_4^+, Mg^{2+}, Mn^{2+}, Ca^{2+}, Cl^-, NO_3^- and SO_4^{2-} by ion chromatography, Al by ICP-AES or catechol violet method, Si and Fe by ICP-AES. Only trace amount (<0.1 ppm) Fe was detected on most samples. Aluminum ion activity was calculated by a computer program assuming that the dissolved Al consisted mostly of inorganic monomeric species, i.e., Al^{3+}, $AlOH^{2+}$, $Al(OH)_2^+$, $Al(OH)_3^0$ and $AlSO_4^+$. Stability constants of these Al complexes were taken from Stumm and Morgan (1996), and the activity coefficients were determined by Debye-Hückel approximation.

TABLE II
Properties of the soils from two sites in Ibaraki prefecture, Japan

Site	Kannondai		Yasato	
Layer	1	2	1	2
Depth (cm)	0-20	20-50	0-20	20-50
Field bulk density (kg m^{-3})	482	527	519	563
pH (H$_2$O)	5.2	5.9	4.9	5.5
Carbon (%)	6.0	1.7	3.7	3.9
Na$_4$P$_2$O$_7$-extractable element (mol kg^{-1})				
Al (Al$_p$)	0.32	0.14	0.23	0.15
(COONH$_4$)$_2$-extractable element (mol kg^{-1})				
Al (Al$_o$)	1.61	1.85	1.23	1.19
Fe (Fe$_o$)	0.36	0.53	0.32	0.43
Si (Si$_o$)	0.59	0.70	0.33	0.35
Particle size distribution (%)				
Coarse sand (2000-200 μm)	8.2	2.2	7.8	5.6
Fine sand (200-20 μm)	22.1	13.1	25.5	15.4
Silt (20-2 μm)	27.3	34.5	21.1	23.6
Clay (< 2 μm)	42.4	50.2	15.6	55.4
Specific surface area by BET method (m^2 g^{-1})				
Coarse sand	1.30	1.34	1.05	0.48
Fine sand	1.98	2.09	1.36	1.12
Silt	60.69	101.06	18.72	58.73
Clay	88.7	150.9	81.4	110.0
Mineral composition in fine sand fraction (%)				
Quartz	7.9	3.1	7.4	10.1
Feldspar	31.3	34.6	26.4	35.3
Volcanic glass	27.5	27.5	35.4	27.0
Chlorite	1.0	0.2	1.5	0.5
Plant opal	3.5	3.1	3.8	2.0
Olivine	0.3	1.3	-	0.1
Pyroxene	10.9	12.5	14.0	9.9
Hornblende	0.5	0.4	0.4	1.1
Others	17.0	18.2	10.9	14.1

2.3. ALUMINUM SOLUBILITY MODEL

The model presented in this study assumes complexation of Al^{3+} to soil organic matter (OM) in such a form:

$$nRH + Al^{3+} + (3-n)OH^- \leftrightarrow R_n Al(OH)_{3-n} + nH^+ ; \quad K, \quad (2)$$

where K is the complexation constant. Al^{3+} activity can thus be represented as

$$\begin{aligned}
\log[Al^{3+}] &= \log[Al\text{-}OM] - n\log[RH] + n\log[H^+] - (3-n)\log[OH^-] - \log K \\
&= \log[Al\text{-}OM] - n\log\{[R_T] - n[Al\text{-}OM]\} - n\text{pH} - (3-n)(\log K_w + \text{pH}) - \log K \\
&= \log\{[Al\text{-}OM]_0 - \Delta Al\} - n\log\{[R_T] - n([Al\text{-}OM]_0 - \Delta Al)\} - (3-n)\log K_w \\
&\quad - \log K - 3\text{pH}
\end{aligned} \quad (3)$$

where [R$_T$], [Al-OM], [RH], [Al-OM]$_0$ and ΔAl (mol kg^{-1}) are the contents of total complexation site, Al-OM complex, protonated complexation site, the initial

Al-OM complex and Al dissolved from Al-OM complex, respectively. K_w is the ionization constant of water.

In model application, $[Al\text{-}OM]_0$ was assumed as Al_p, $[R_T]$ was estimated as $8.21 \times C$ content (kg kg^{-1}), based on the mean C content and carboxyl group content in A-type humic acid from 10 Andisols in Japan (Yonebayashi and Hattori, 1988). ΔAl was calculated as (total Al dissolved) minus (Al dissolved from the inorganic pool) estimated from the amount of Si dissolved and Al/Si dissolution ratio of the inorganic fraction. Al/Si dissolution ratio was determined in another experiment: 4 g soil was heated with H_2O_2 and washed with 1M LiCl to digest OM and remove exchangeable cations. The soil was then leached with 0.1 mM HCl at 0.7-4 mL h^{-1} for 1150 hours. Average Al/Si dissolution ratio was 0.30, 0.077, 0.22 and 0.14 for Kannondai 1st and 2nd, Yasato 1st and 2nd layers, respectively. Using $[Al\text{-}OM]_0$, $[R_T]$, ΔAl and measured pH as the input, n and log K were optimized by LSM to fit to the experimental Al^{3+} activity.

3. Results and Discussion

3.1. CALCULATION OF FIELD WEATHERING RATE

The k_H of glass is slightly larger than feldspars, but 2 orders of magnitude smaller than pyroxene (Table III). The k_{H2O} of glass is 1-2 orders of magnitude larger than that of other minerals, leading to relatively high dissolution rate in neutral condition. These values were incorporated into PROFILE to calculate field weathering rate.

TABLE III
Dissolution rate coefficients of volcanic glass and some other minerals used in PROFILE

Mineral	-logk_H	n_H	-logk_{H2O}
Basaltic Glass	14.00	0.39	15.16
Dacitic Glass	14.49	0.39	15.65
Orthoclase	14.7	0.5	17.2
Plagioclase	14.6	0.5	16.8
Pyroxene	12.3	0.7	17.5

When BET surface area of sand to silt (2000-2 μm) fraction was entered as the mineral surface area, calculated weathering rate was 1-2 orders of magnitude higher than the field estimate. When the surface area was limited to sand (2000-20 μm) fraction, however, the difference between field estimate and calculation fell within − 53 to +36 % (Fig. 1). Some possible reasons to this are (1) highly weatherable minerals is localized in sand fraction (according to XRD analysis, dominant primary mineral in the silt fraction was quartz), (2) surface area of silt fraction is overestimated due to presence of secondary solid phase. Further study is required to attain determinative methodology, but above results suggest that BET surface area of sand fraction could be a reasonable estimate of the weatherable mineral surface area in volcanic soils for calculating weathering rate with PROFILE.

3.2. ALUMINIUM SOLUBILITY

Repeated leaching with 1 mM HCl lowered the suspension pH from 5.1 to 3.9, 5.4 to 4.0, 4.5 to 4.0 and 4.7 to 4.0. On all of the soils studied, $\log[Al^{3+}]$ showed linear relationship with pH over 4.2, but $\log[Al^{3+}]$ 'saturated' around -3.8 at pH below 4.2 (Fig. 2). When linear regression was applied to the virtually linear part of the $\log[Al^{3+}]$-pH diagram, the slope of $\log[Al^{3+}]$ against pH was -2.70, -3.01, -2.71 and -3.99. These results show that empirical linear equation like

$$\log[Al^{3+}] = \log K_O - a\text{pH} \qquad (4)$$

is insufficient to describe Al solubility when pH range below 4.2 is considered. The model proposed in this study described the Al solubility fairly well, by optimizing n and K (TableIV, Fig. 2). It predicts that (1) while Al dissolution is sufficiently small, $\log[Al^{3+}]$ is represented as

$$\log[Al^{3+}] \approx \log[Al\text{-}OM]_0 - n\log\{[R_T]-[Al\text{-}OM]_0\} - (3-n)\log K_w - \log K - 3\text{pH}, \qquad (5)$$

conforming to a line with a slope of -3 against pH, (2) as more Al is depleted from Al-OM complex, $\log[Al^{3+}]$ becomes smaller than predicted by Eq. 5.

If Al solubility is regulated by complexation to OM, empirical linear equation such as Eq. 4 is likely to give incorrect prediction, because dependence of $[Al^{3+}]$ on [Al-OM] and [RH] is not considered. When applying or evaluating similar Al-OM complex models, quite a few authors, *e.g.*, Van der Salm *et al.* (2000) and Wesselink *et al.* (1996), employed a simplification that content of protonated complexation site is virtually constant. This should be given reconsideration, however, because estimated complexation site content is not necessarily larger enough than the amount of Al-OM complex, as in TableIV.

Interpretation of the optimized parameter values requires further discussion, *e.g.*, why Kannondai 2nd layer has distinctly large K and small n. Al_p can not be concluded as the phase controlling Al solubility either, because change in Al_p

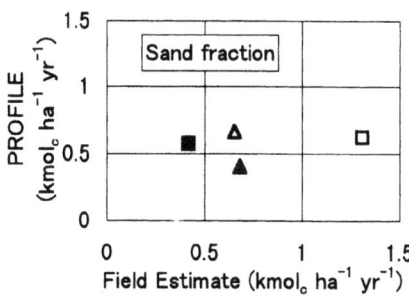
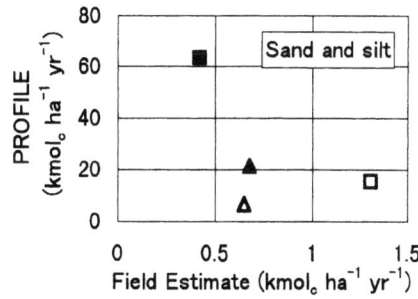

Figure 1. Comparison of weathering rate calculated by PROFILE and estimated by Sr isotope analysis. BET surface area of sand fraction or sand and silt fraction was used in PROFILE calculation. □, Kannondai 1st; ■, Kannondai 2nd; △, Yasato 1st; ▲, Yasato 2nd layers.

TABLE IV
Input and optimized parameters in the Al solubility model

Site	Layer	[R$_T$]	[Al-OM]$_0$	log$K$	n	r^2
Kannondai	1	0.487	0.317	14.9	1.42	0.956
	2	0.138	0.140	23.9	0.68	0.971
Yasato	1	0.300	0.227	17.7	1.19	0.814
	2	0.316	0.148	18.0	1.12	0.948

content was not measured during the experiment. Nevertheless, above results suggest that Al solubility is controlled by complexation to some solid phase, e.g., soil OM, and such an Al solubility system can be implemented in soil chemistry models by adding mass balance and complexation equilibrium equations of Al.

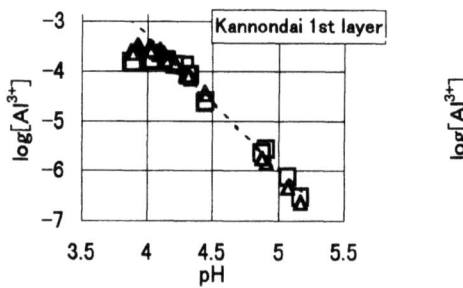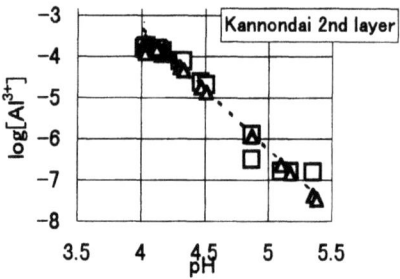

Figure 2. Aluminum solubility of soils from two sites in repeated leaching experiment. □, experiment; △, Al-OM complex model; -----, linear regression. Linear regression was applied to the virtually linear part of the log[Al^{3+}]-pH diagram.

4. Conclusions

(1) PROFILE can calculate weathering rate of volcanic soils, using the surface area of sand fraction as the estimate of surface area of weatherable minerals.
(2) Aluminum solubility of volcanic soils is likely to be controlled by complexation to some solid phase, such as organic matters.

References

Shindo, J., Fumoto, T., Oura, N., Nakano, T. and Takamatsu, T.: 2001, *Water Air Soil Pollut.* (this volume).
Stumm, W. and Morgan, J.J.: 1996, *Aquatic Chemistry*, John Wiley & Sons, N.Y.
Van der Salm, C., Westerveld, J.W. and Verstraten, J.M.:2000, *Geoderma* **96**, 173.
Wada, K. (ed.): 1986, *Ando soils in Japan*, Kyushu University Press, Fukuoka, Japan.
Warfvinge, P. and Sverdrup, H.:1992, *Water Air Soil Pollut.* **63**, 119.
Yamada, I. and Shoji, S.: 1983, *Jpn. J. Soil Sci. Plant Nutr.* **54**, 311 (in Japanese).
Wesselink, L.G., Van Breemen, N. Mulder, J. and Janssen, P.H.:1996, *Europ. J. Soil Sci.* **47**, 373.
Yonebayashi, K. and Hattori, T.: 1988, *Soil Sci. Plant Nutr.* **34**, 571.

HYDROGEOCHEMICAL CONDITIONS AFFECTING ACIDIFICATION OF STREAM WATER IN MOUNTAINOUS WATERSHEDS

HIDESHI IKEDA* and YOICHI MIYANAGA

*Central Research Institute of Electric Power Industry (CRIEPI)
1646 Abiko, Abiko - City, Chiba, 270-1194. Japan
(*author for correspondence, e-mail:ik@criepi.denken.or.jp)*

Abstract. Acid deposition in eastern Asia will increase and freshwaters in Japan are likely to become acidified in future. In order to make long-term predictions about freshwater acidification, it is necessary to evaluate acid neutralization mechanisms in Japanese watersheds. Ikeda and Miyanaga (1999) earlier proposed a method of separating acid-neutralization capacity into chemical weathering and cation exchange. By this means, we were able to assess the effect of hydrogeochemical properties on chemical weathering and stream water chemistry for three watersheds in Japan. On the basis of this assessment, acid-neutralization stream water chemistry was predicted using the ILWAS (Integrated Lake-Watershed Acidification Study) model. The main factors determining acidification are the thickness of weatherd profile and chemical weathering rates. The principal results are: (1) for non-acidified watersheds in Japan, acid deposition is neutralized by chemical weathering of primary minerals; (2) freshwaters in Japanese watersheds will not acidify even if acid deposition increases to the extent found in an acidified watershed in the U.S.A.

Keywords: acid deposition, chemical weathering, mathematical model, watershed

1. Introduction

Increased acid deposition in eastern Asia is expected to contribute to future freshwater acidification in the region. The Central Research Institute of Electric Power Industry has conducted watershed surveys in three well-defined mountainous watersheds since 1987 (Ikeda and Miyanaga, 1999). These surveys confirm that the deposited acidified water is neutralized as it moves through the watersheds. A method has been developed to evaluate acid neutralization by chemical weathering in the field, based on mass balance in the watershed and the chemical weathering reactions of primary minerals. Here we apply this method to three non-acidified watersheds in Japan and one acidified watershed in the U.S.A., and assess the effect of hydrogeochemical properties on stream water chemistry using a mathematical model.

2. Methods

2.1 EVALUATION OF ACID-NEUTRALIZATION BY CHEMICAL WEATHERING

Acid-neutralization by chemical weathering and cation exchange in watersheds is evaluated using a mass balance equation shown below (Ikeda and Miyanaga, 1999).

$$\text{Output}(x) - \text{Input}(x) = -\sum_i \left(a_{xi} \frac{dW_i}{dt} \right) \pm \Delta\text{exchange pool }(x) \quad (1)$$

where x = chemical species x; i = mineral i; Output = output flux from watershed; Input = input flux to watershed ; a_{xi} = release rate of x from i by chemical weathering. The weathering rate of primary mineral i in the watershed is given by the rate of decrease of the weight ratio of mineral i, -dwi/dt. Supply of x by weathering of mineral i is expressed as a term in parenthesis, and can be calculated by summing up the dominant primary minerals.

The value of a_{xi} is expressed as β for cations, and γ for silicic acid (H_4SiO_4) in the chemical weathering reaction shown below,

Primary mineral(i) + $\alpha \cdot H^+ \rightarrow$ Secondary mineral + $\beta \cdot$ Base Cation(x) + $\gamma \cdot H_4SiO_4$

The unknown terms, -dWi/dt and Δ exchange pool, in equation (1) are calculated from the amounts of dominant cations and silicic acid released. In this equation, Δ exchange pool = supplied by cation exchange.

2.2 EXPERIMENTAL WATERSHED SURVEY

In order to evaluate the acid neutralization mechanisms in watersheds as described in section 2.1, surveys were conducted in three non-acidified watersheds in Japan and one acidified watershed in the U.S.A. (Ikeda and Miyanaga, 1999). The survey included 1) geology and mineralogy; 2) hydrology; and 3) deposition and water chemistry. The locations, site descriptions and average annual stream water pH and alkalinity (in μ eq/L) are given in Figure 1. Based on a nationwide stream water chemistry survey (Ikeda and Miyanaga, 1994), the alkalinity of stream water is relatively high for the Chugu, moderate for the Higshidani, and low for the Jingahata. For the geological and mineralogical survey, the minerals from auger core samples were analyzed with a polarization microscope, an X-ray diffractometer. Secondary minerals were determined by concentrations of sodium (Na^+), calcium (Ca^{2+}), and silicic acid (H_4SiO_4) in soilwater and groundwater plotted on stability field diagram (Paces, 1972). The chemical composition of the primary minerals was obtained with an electron probe microanalyzer. The weight ratios of each primary mineral were calculated from its chemical composition and bulk chemistry using the Norm method (Miyashiro and Kushiro,

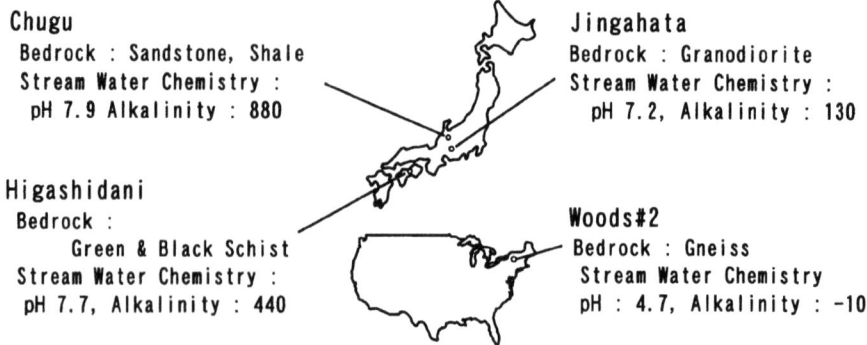

Figure 1. Locations, Site Descriptions in Experimental Watersheds

1975). For the hydrological survey, a water balance was obtained, and leakage from the watersheds was neglected. The deposition and water chemistry survey involved calculating wet and dry deposition fluxes from precipitation and atmospheric data. To this end, samples of soil water (from the vadose zone), ground water (from the aquifer), and stream water, were collected monthly, and analyzed for pH, alkalinity (end point at pH = 4.8), and chemical composition. The chemistry of soil water in the Woods#2 watershed has been described by Geary and Driscoll (1996).

2.3 SIMULATION USING THE ILWAS MODEL

The responses to acid deposition and geochemical conditions of three Japanese watersheds were simulated for 100 years using ILWAS model (Goldstein et al., 1984). Simulations were conducted in three cases; observed case in experimental watershed survey, heavy deposition case and small weathering rates case. Parameters for simulations were determined based on data in experimental watersheds. Chemical weathering rates of minerals were determined using evaluation method shown in 2.1.

3. Results and Discussion

3.1 EXPERIMENTAL WATERSHED SURVEY

The thickness of weathered profiles estimated from seismic prospecting and boring logs is 30 m for Chugu, 8 m for Higashidani, and 17 m for Jingahata. For the Woods #2, Peters and Murdoch (1985) reported a thickness of 2.3 m. The primary minerals, identified in the respective weathering profiles, are shown in Table I. Figure 1 gives the average annual values of pH and alkalinity of the stream water. For the three Japanese watersheds the water had a neutral pH because acid deposition was neutralized as it moved through the system. In the Woods#2 watershed, however, both the soil water and stream water were acidic because the extent of acid-neutralization was small relative to the acid deposition load.

3.2 ACID - NEUTRALIZATION BY CHEMICAL WEATHERING

Figure 2 shows the input–output fluxes for all four watersheds. Input fluxes include wet and dry deposition. Output fluxes are derived from the chemistry of the stream and runoff water. In the three Japanese watersheds, hydrogen ions (H^+), produced by acid deposition and dissolution of carbon dioxide (CO_2) in soil water, are neutralized and the output of H^+ ions is negligible. The supply of base cations and silicic acid (H_4SiO_4) indicates that neutralization is principally effected by the chemical weathering of primary minerals. In the Woods#2 watershed, on the other hand, only part of the deposited H^+ is neutralized and some output of H^+ ions occurs. The small supply of base cations and H_4SiO_4 in this watershed suggests that chemical weathering rates are smaller than those in the

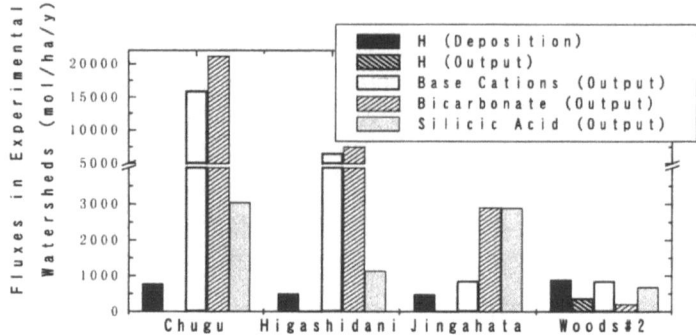

Figure 2. Input - Output Fluxes in Watersheds

non-acidified Japanese watersheds.

Weathering rates of primary minerals are calculated from input–output fluxes and chemical weathering reactions. In this calculation, the supply of silicic acid is regarded as the limiting factor because this substance does not derive from cation exchange. For example, in the Higashidani watershed, a mass balance equation can be established for each of the following species, Ca^{2+}, Mg^{2+}, Na^+, K^+, and H_4SiO_4. Unknown parameters are the chemical weathering rates of plagioclase, chlorite, muscovite, calcite, and the supply of exchangeable Na^+. In this watershed, the concentrations of exchangeable Ca^{2+}, Mg^{2+}, K^+ ions are negligible (Ikeda and Miyanaga, 1999). Then, unknown five factors, four weathering rates and a supply by ex - Na^+ can be calculated. The calculated weathering rates and annual supplies of chemical species by cation exchange are summarized in Table I. For the Chugu and Higashidani watersheds, the weathering of calcite appear to be the dominant acid-neutralization mechanisms. This conclusion is based on the following observations: the soil water, sampled at a depth (for example, 5 and 15m in Higashidani), has a high alkalinity, the concentration of exchangeable cations at these depths is very low, and calcite veins are present (Ikeda and Miyanaga, 1999). For the Jingahata watershed, on the other hand, the

Table I
Chemical Weathering Rates of Primary Minerals and Supply by Cation Exchange (mol/ha/yr)

	Chugu	Higashidani	Jingahata	Woods#2
Chemical Weathering Rate				
Plagioclase	1016	487	1562	288
K-feldspar	25		0	10
Biotite	1183		330	
Hornblende				31
Chlorite		115		
Muscovite		59		
Calcite		4485		
Supply by Cation Exchange				
Calcite and/or ex-Ca^{2+}	8902		277	239
ex-Mg^{2+}			-213	44
ex-Na^+	736	99		

H⁺ ions are apparently neutralized by the chemical weathering of plagioclase, K-feldspar and biotite. This evaluation was proved by the high concentrations of H_4SiO_4 in soilwater and groundwater. In contrast, acid deposition is not fully neutralized in the Woods #2 watershed giving rise to a significan output of H⁺ions (Fig. 2). Since the chemical weathering rates in this watershed are much smaller than those in the Japanese watersheds (Table I), stream water becomes acidified.

In these four watersheds, the mineral composition, except calcite, does not vary as much as the weathering rate. On the other hand, chemical weathering rates have a relationship with thickness of weathered profiles closely (Ikeda and Miyanaga, 1999). Therefore, it is concluded that the thickness of weathered profile is dominant factor for acid neutralization capacity.

3.3 SIMULATION FOR ACIDIFICATION OF STREAM WATER

For the three non-acidified watersheds in Japan, we have compared the simulated chemistry of soil water, ground water and stream water with the respective observed chemistry, and verified the parameters for hydrological and chemical models. These parameters are used to simulate conditions over 100 years for three distinct cases: (1) observed case; (2) heavy deposition of sulfate as in the Woods #2 watershed; (3) a small supply of base cations by chemical weathering as in the Woods #2 watershed. Simulated pH values for the Higashidani stream water are shown in Figure 3.

In the observed case (solid line), acidification of stream water is not predicted. In the heavy deposition case (dotted line), chemical weathering rates increase (with deposition), and the increased supply of base cations prevents stream water from being acidified. However, in the small weathering rate case (dashed line), stream water is predicted to acidify within 50 years. In this period, base cations in soil and bedrock are lost by cation exchange. In this case, the supply of base cations by chemical weathering (of primary minerals except calcite) is much less than that in most of Japanese watersheds (Ikeda and Miyanaga, 1994; 1998). In cases (1) and (2), base cations are supplied by the chemical weathering of pri-

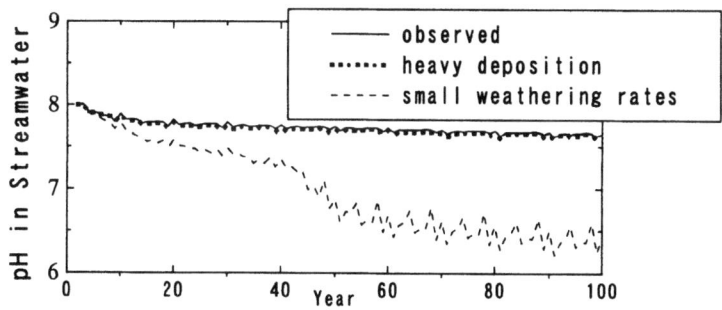

Figure 3. Simulated Stream Water Chemistry using ILWAS Model (Higashidani)

mary minerals, especially calcite. For the Chugu and Jingahata watersheds, simulations show that stream water would not acidify even if deposition were heavy and weathering rates were low. This is because the supply of base cations for the next 100 years is sufficient to compensate for an increase in deposition and a decrease in chemical weathering.

4. Conclusions

A method for assessing acid-neutralization by chemical weathering is proposed, and applied to some acidified and non-acidified watersheds. A relationship between the hydrogeochemistry of watersheds and stream water acidification is derived. For three non-acidified watersheds in Japan, acid deposition is neutralized by the chemical weathering of primary minerals. On the other hand, for the acidified Woods#2 watershed in the U.S.A., the rate of weathering is relatively small. The dominant factor for acidification is thickness of weathering profile and the magnitude of chemical weathering rates. The streamwater chemistry of the Higashidani watershed in Japan was simulated for 100 years using the ILWAS model. The result suggests that this watershed would not acidify to the same extent as the Woods#2 watershed, even if deposition loads should increase.

Acknowledgments

Field research in Japan was commissioned by the Ministry of International Trade and Industry, Japan. We wish to thank Prof. C. T. Driscoll of Syracuse University, and Prof. R. April of Colgate University for assistance in the field survey at Woods Lake. We are grateful to Dr. R.A. Goldstein of EPRI, and Dr. C.W. Chen of Systech Eng. Inc. for advice in using the ILWAS model.

References

Goldstein, R.A., Gherini, S. A. Chen C.W., Mok, L. and Hudson, R.J. : 1984, *Philosophical Transactions of Royal Society of London* **B305,** 409.
Ikeda, H. and Miyanaga, Y.: 1994, *Proceedings of 1994 Annual Meeting of the Geochemical Society of Japan,* 168 pp. (in Japanese).
Ikeda, H. and Miyanaga, Y.:1998, *CRIEPI Report,* U97098, (in Japanese).
Ikeda, H. and Miyanaga, Y.: 1999, *Journal of Japan Society on Water Environment,* **22(8),** 655-662. (Abstract in English)
Geary, R. J. and Driscoll, C.T.: 1996, *Biogeochemistry* **32(3),** 195.
Miyashiro, A. and Kushiro I.: 1975, *Ganseki-gaku* II (Petrology, in Japanese), Kyoritu -Shuppan, Tokyo, Japan pp. 162-170.
Paces, T.: 1972, *Geochimica et Cosmochimica Acta* **36,** 217.
Peters, N.E. and Murdoch, P.S.: 1985, *Water, Air, and Soil Pollution* **26,** 387.

ESTIMATION OF MINERAL WEATHERING RATES UNDER FIELD CONDITIONS BASED ON BASE CATION BUDGET AND STRONTIUM ISOTOPE RATIOS

J. SHINDO[1*], T. FUMOTO[1], N. OURA[1], T. NAKANO[2] and T. TAKAMATSU[3]

[1] *National Institute of Agro-Environmental Sciences, 3-1-1 Kannondai, Tsukuba, Ibaraki 305-8604 Japan;* [2] *University of Tsukuba, 1-1-1 Tennoudai, Tsukuba, Ibaraki 305-8572 Japan;* [3] *National Institute for Environmental Studies, 16-2 Onogawa, Tsukuba, Ibaraki 305-0053 Japan*
(* author for correspondence, email: shindo@niaes.affrc.go.jp)

Abstract. Base cation (BC) concentrations of rain, throughfall, percolation from leaf litter, and soil solution were periodically measured in two forests: Kannondai (red pine stand on volcanic soil) and Yasato (deciduous stands on granitic soil). Calculation of a BC budget gave the rate of BC release from soils; the BCs originated from mineral weathering and cation exchange. Weathering rates under field conditions were estimated from the Sr isotope ratios ($^{87}Sr/^{86}Sr$) of water and soil samples. Isotope ratios decreased in the order rain > throughfall > percolation > soil solution. Clay and silt had extremely high isotope ratios; this suggests that the sandy fraction, whose isotope ratio was smaller than that of the soil solution, was the main contributor to mineral weathering. Estimated BC weathering rates ($kmol_c \cdot ha^{-1} y^{-1}$) were 1.16 for Ca and 0.57 for Mg at Kannondai, and 0.82 for Ca and 0.51 for Mg at Yasato. The unexpected high weathering rate of granitic soil in Yasao was due to the wide coverage of the original parent material by volcanic ash. The contribution of cation exchange derived by subtraction was a little smaller than the weathering rates and was similar to the values estimated from a dynamic model that we developed.

Keywords: mineral weathering, neutralizing capacity, Sr isotope ratio, BC budget, dynamic model

1. Introduction

Chemical weathering of soil minerals is the ultimate buffering mechanism, maintaining soil neutralizing capacity by supplying the soil with base cations. Soils with low weathering rates are acidified by long-term acidic deposition. Weathering rate has been used as an important parameter to define ecosystem susceptibility in critical load estimation (Hettelingh *et al.*, 1991) and is a key parameter in the prediction of soil chemistry changes by dynamic models. The expected enhanced deposition of nitrogen in the next several decades from industrial, transport, and agricultural sectors may enhance plant growth. The estimation of mineral weathering rates is important to understanding the possibility of nutrient imbalances caused by excess uptake of nitrogen by plants. However, there is little data on Asian soils.

Methods for estimating chemical weathering rates are still under discussion. Bain and Langan (1995) reviewed them and showed that the "current rate" based on base cation budgets was sometimes more than ten times greater than the "long-term weathering rate" based on element depletion in soil layers. The current rate may include cation release from exchangeable sites (Bain and Langan, 1995). The

Sr isotope ratio ($^{87}Sr/^{86}Sr$) can give the mixing ratio of different origins of base cations: from atmosphere and soil weathering. In this paper, we estimate the weathering rates of soil minerals under field conditions based on the base cation budget in soil and on Sr isotope ratios.

2. Materials and Methods

2.1. FIELD MEASUREMENT

Field measurements were taken in a 30-year-old red pine forest growing on volcanic ash soil in Kannondai and in a deciduous mixed forest of similar age growing on granitic Brown Forest soil in Yasato, both in Ibaraki Prefecture, Kanto district, Japan. We collected rain, throughfall, litter layer percolation, and soil solution at 20 cm and 50 cm depth every one or two weeks in Kannondai and every month in Yasato. We took two replicates for rain and three for the others. After filtering the water through a 0.04 µm membrane filter, we measured the concentrations of Na^+, NH_4^+, K^+, Mg^{2+}, Ca^{2+}, Cl^-, NO_3^-, PO_4^{3-}, and SO_4^{2-} by ion chromatography, and the concentrations of Al, Sr, Si, and some heavy metals by inductively coupled plasma optical emission analyzer. We estimated the vertical water flux in the soil from the change over time in volumetric soil moisture content as measured by a time domain reflectometry (TDR) and used it to calculate the base cation discharge from 50 cm depth.

Several samples from each source were analyzed for $^{87}Sr/^{86}Sr$ ratio. The isotope ratios of the clay, silt, and sand fractions of the soil were also measured.

2.2. ESTIMATION OF THE WEATHERING RATES AND BASE CATION BUDGET

We evaluated the yearly base cation budget in soil (Fig. 1; Eq. 1) based on field measurements for one year starting October 1997. Plant uptake (BC_{UP}) consists of uptake for net growth (BC_{GU}) and refluxes as canopy reaching and litterfall (BC_{RF}). BC_{OUT}, BC_{IN}, and BC_{RF} were derived from the field measurements. BC_{GU} was calculated from tree density and DBH distributions in the field, and

$$BC_{OUT}-BC_{IN}=BC_{SOIL}-BC_{UP}=(BC_W+BC_{EX})-(BC_{GU}+BC_{RF}) \quad (1)$$

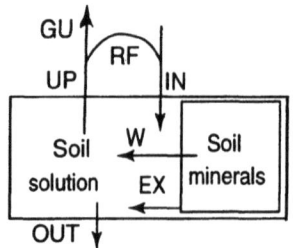

BC_{OUT}: discharge from 50cm depth of soil
BC_{IN}: infiltration into soil through litter layer
BC_{SOIL}: release from soil
BC_W: mineral weathering
BC_{EX}: exchangeable cation release
BC_{UP}: Plant uptake
BC_{GU}: uptake for net growth
BC_{RF}: refluxes as canopy leaching and litterfall

Figure 1. Base cation budget in soil layer (units are $kmol_c\ ha^{-1}y^{-1}$).

from Published data on average growth rates and cation contents in stems and branches of red pine and deciduous trees in the Kanto district. BC_W of 50 cm of soil was calculated according to Eq. (2), which is derived from the equation of Miller (1993):

$$BC_W = BC_{IN} \{(R_{SOLN} - R_{LL}) / (R_{SOIL} - R_{SOLN})\} \quad (2),$$

where $R = (^{87}Sr/^{86}Sr)/\{1+(^{87}Sr/^{86}Sr)\}$, and the subscripts $_{SOLN}$, $_{LL}$, and $_{SOIL}$ denote soil solution, litter layer percolation, and soil, respectively. We assumed that the isotope ratio of exchangeable Sr was identical to that of the soil solution, because they are in exchange equilibrium with each other. BC_{EX} was calculated based on these estimates from Eq. (1).

2.3. ESTIMATION OF SOIL CHEMICAL CHANGES WITH A DYNAMIC MODEL

We estimated temporal changes in soil chemistry, including ionic concentrations of the soil solution and composition of exchangeable cations, by applying a dynamic model that we developed (Shindo and Fumoto, 1998) to the field data. The model treated chemical processes such as cation exchange, sulfur adsorption, and dissolution of aluminum hydroxide, as equilibrium reactions. Net consumption (or production) of ions within the soil layers due to plant uptake, mineral weathering, and organic matter mineralization were treated as lumped parameters, which were determined by calibration from the observed data. The litter layer percolation was used as the ion input to the soil in the model. The predicted release rate of exchangeable base cations was compared with BC_{EX}.

3. Results and Discussion

3.1. ION BUDGET ESTIMATION FROM FIELD MEASUREMENTS

Figure 2 shows the yearly flux of Ca^{2+} and Mg^{2+} due to deposition (wet and dry), throughfall, and percolation from litter and mineral soil layers. Rain was markedly enriched in these ions after passing through the canopy and litter layer. We thought that increase of Ca^{2+} and Mg^{2+} in throughfall was due to dry deposition and canopy leaching. Assuming that the increase in Na^+ was caused only by dry

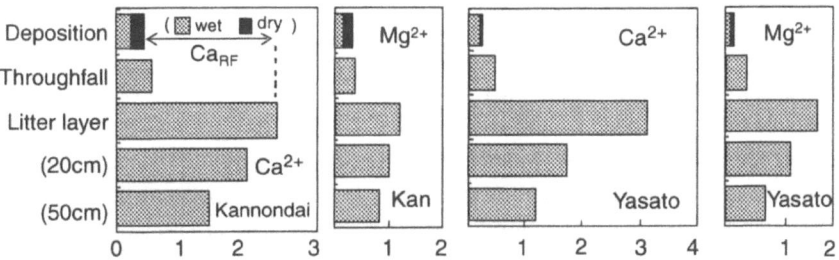

Figure 2. Yearly fluxes of calcium and magnesium ions in Kannondai and Yasato ($kmol_c ha^{-1} y^{-1}$).

deposition and chemical composition of dry deposition was similar with that of wet deposition, we estimated dry deposition flux of Ca^{2+} and Mg^{2+}. In estimation of BC_{RF}, we ignored the base cation accumulation in litter layers. Namely we assumed that the yearly average of weight and composition of the litter layer did not change, and regarded the sum of ions released from the canopy and litter layer as BC_{RF} as shown in Figure 2. BC_{RF} might be about 10 % greater than our estimate, because the amount of litter on the temperate forest floor corresponds to the amount of litterfall for three to four years (Tsutsumi, 1996). The mineral soil layer behaved as the net absorbent of cations; that is, plant uptake rates were larger than soil cation release.

3.2. Weathering Rate Estimation Basd on Sr Isotope Ratios

Sr concentrations were highly correlated with Ca and Mg concentrations (Table I). We thought that Sr isotope ratios could indicate the behavior of these two elements. Low correlation of Na^+ and K^+ with Sr indicates the different origin and different preference by plants of these cations. Figure 3 shows the means and standard deviations of the isotope ratios of the water and soil samples and typical values of some components that might control the isotopic composition of Sr in ecosystems. The values for rain in both areas were similar and were close to the value for sea water. The values decreased from rain to throughfall to litter layer percolation to soil solution. The value for exchangeable cations in one sample of topsoil from Kannondai is also shown in Figure 3. This value is very close to that of the soil solution at 20 cm depth, which supports our assumption that the isotope ratios of the exchangeable cations and the soil solution are identical. Because the ratio of the soil solution was smaller than that of the litter layer percolation, we expected the isotope ratios of the soil minerals to be smaller than that of the soil solution. In the clay and silt fractions of the soil, however, the ratios were quite high compared with the water samples from both areas. The sand fraction had the smaller isotope ratio; we considered it to be the main source of base cation weathering.

The variety of isotope ratios of soils sug-

TABLE I
Correlation coefficients between Sr and other base cations.

	Kannondai	Yasato
Na^+	0.437	0.678
K^+	0.247	0.722
Mg^{2+}	0.885	0.915
Ca^{2+}	0.938	0.965

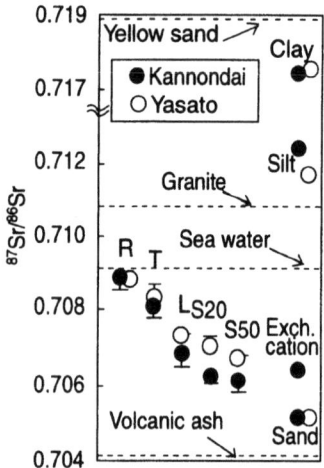

Figure 3. Means and standard deviations of Sr isotope ratios of rain (R), throughfall (T), litter layer leachate (L), soil solution at 20 and 50 cm depth (S20, S50), and soil fractions.

TABLE II
Base cation fluxes based on budget analysis and Sr isotope ratios (kmol$_c$·ha^{-1}y^{-1}).

		BC$_{IN}$	BC$_{SOIL}$	BC$_W$	BC$_{EX}$	BC$_{UP}$	BC$_{RF}$	BC$_{GU}$	BC$_{OUT}$	BC$_{EX}$(Model)
Kannondai	Ca	2.48	2.08	1.16	0.92	3.12	2.02	1.10	1.45	1.06
	Mg	1.22	0.86	0.57	0.29	1.28	0.93	0.35	0.80	0.93
Yasato	Ca	3.17	1.35	0.82	0.53	3.35	2.95	0.40	1.17	0.72
	Mg	1.73	0.89	0.51	0.38	1.90	1.59	0.31	0.72	0.36

gests a different origin of each fraction. The sand originated mainly from volcanic ash, even in Yasato, where the soil is classified as granitic. Mineral analysis showed that the soil in Yasato contains minerals typical of volcanic soils, such as pyroxene, volcanic glass, and chlorite. On the other hand, the clay fraction originated partly from yellow sand that was highly weathered, fairly stable, and poor in base cations.

Using the isotope ratio of the sand fraction, we calculated the release rates of Ca^{2+} and Mg^{2+} due to mineral weathering of soil 50 cm thick from Eq. (2). The results of the weathering rates and the other fluxes derived from the field measurements are shown in Table II. The weathering rates of these two cations were 56% to 66% of the total ion release from the soils. The balance is attributed to exchangeable cation release. Estimates of the weathering rates of soils in Europe and North America summarized by Langan *et al.* (1995) range from 0.0 to 2.07 (median 0.20) kmol$_c$·ha^{-1}y^{-1} for Ca^{2+} and from 0.01 to 1.39 (median 0.18) kmol$_c$·ha^{-1}y^{-1} for Mg^{2+}. Because many of these estimates are from catchments, which are expected to have larger rates than the soil, our results are relatively large. The weathering rate of Yasato soil was quite large compared with existing data for granitic soils because of its volcanic ash content.

3.3. ESTIMATION OF CHNGES IN EXCHANGEABLE CAIONS WITH A DYNAMIC MODEL

Figure 4 shows the changes over time in Ca^{2+} concentration in the soil solution estimated with the dynamic model and the observed values for the same period. Predicted values of pH and ionic concentration of the soil solution corresponded

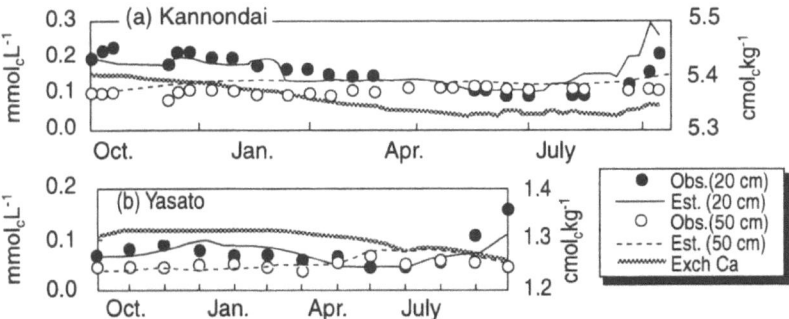

Figure 4. Changes in calcium ion concentraions in soil solution and exchangeable calcium estimated with the dynamic model.

with the measurements to some extent when we set suitable values for the net ion consumption rate. At Kannondai (the red pine stand), we assumed that the net consumption rate was constant throughout the year. At Yasato (the deciduous stand), seasonal changes in the net consumption rate had to be introduced to reproduce the measured values. Predicted changes in exchangeable Ca^{2+} concentration are also shown in Figure 4. Exchangeable Ca^{2+} and Mg^{2+} decreased slightly over the year. Cation release rates from exchangeable sites calculated from these predicted changes are shown in the last column of Table II. These values are not very different from BC_{EX}, which was derived from the budget analysis based on the weathering rate estimation.

4. Conclusions

We evaluated the release rates of base cations from chemical weathering of forest soils in Japan based on ion budgets and strontium isotope ratios. The resultant weathering rates were relatively large compared with data from Europe and North America, especially for granitic soils. Volcanic ash covering the original granitic parent material increased the weathering rate. Severe acidification of lakes and soils over a wide area has not been observed in Japan, even in areas where soils are expected to have a low buffering capacity like granitic soil. The volcanic ash might be one reason.

Acknowledgements

We thank the Kasama Forest Technology Center for allowing us to take measurements in the national forest in Yasato. This research was supported by the Global Environmental Research fund from the Environment Agency of Japan.

References

Bain, D.C. and Langan, S.J.: 1995, *Water, Air and Soil Pollution* **85**, 1051.
Hettelingh, J.P., Downing, R.J., de Smet, P.A.M.: 1991, CCE Technical report No.1, RIVM, Bilthoven, the Netherlands, 86pp.
Langan, S.J., Hodson, M.E., Bain, D.C. Skeffington, R.A. and Wilson, M.J.: 1995, *Water, Air and Soil Pollution* **85**, 1075.
Miller, E.K., Blum, J.D. and Friedland A.J.: 1993, *Nature* **362**, 438
Shindo, J. and Fumoto T.: 1998, *Global Environmental Research* **2**, 95.
Tsutsumi, T. (eds.): 1996, *Forest Ecology, Asakura*, pp. 96–111 (in Japanese).

CALCULATION AND MAPPING OF SULFUR CRITICAL LOADS FOR TERRESTRIAL ECOSYSTEMS OF THAILAND

J. MILINDALEKHA[1*], V. N. BASHKIN[2], and S. TOWPRAYOON[1,2]

[1] *Department of Environmental Technology, School of Energy and Materials;*
[2] *Joint Graduated School of Energy and Environment*
King Mongkut's University of Technology Thonburi, Bangkok 10140, Thailand.
(* author for correspondence, e-mail: milindalekha@hotmail.com)

Abstract. The essential parameters needed for the calculation of critical load of sulfur, CL(S), are base cation weathering rate, base cation uptake, acid neutralizing capacity leaching and base cation deposition. These parameters are estimated and mapped for the most area of terrestrial ecosystems of Thailand using data of national data soil survey. The values of CL(S) range from <200 to 2,225 eq.ha^{-1}yr^{-1} and about 70% of terrestrial ecosystems are characterized by low values (<200 eq.ha^{-1}yr^{-1}). These CL values reflect the sensitivity of Thai ecosystems to sulfur deposition.

Keywords: Critical load of sulfur, Terrestrial ecosystems, Thailand

1. Introduction

The main source of sulfur dioxide in Thailand is emissions from the electricity generation. In 1998, it was estimated that the sulfur dioxide emission from the energy sector was 8.84 million tones. The generation of electricity accounted for 68% of the total SO_2 emissions in 1998, mainly due to lignite, natural gas, fuel oil, diesel and biomass burning. The lignite-fired power plants are the main source of SO_2 emission. At present, the only lignite-fired power plant is in Mae Moh district, Lampang province, which is located in the northern part of the country. The lignite from Mae Moh mine is a low-grade coal with high sulfur content that varies from 1.67 to 3.35 % by weight (Khummongkol and Bowonwiwat, 1995). In order to reduce the impacts of SO_2 from this power station, the government has reduced emissions by installing a flue gas desulfurization system to each unit. However, the emission is still high due to high lignite consumption. The resulting sulfur deposition pattern is shown in Figure 1. The sulfur deposition rates must not exceed the critical loads UBA, 1996). The interest in critical loads in Thailand has increased over the past few years because the value of critical load shows the sensitivity of a particular environment for acidity deposition. By estimating critical loads for various regions, the most sensitive area at greatest risk of damage could be identified and further provision of receptor-based approach to mitigate air pollution could be consequently established. Thus, the methodology of calculating and mapping the critical load of sulfur for terrestrial ecosystems of Thailand, based on national soil survey, is described in this study. These national critical loads were compared with those from RAINS-ASIA model (World Bank, 1994).

2. Methodology

Critical loads of maximum sulfur ($CL_{max}S$) are calculated based on 4 parameters, i.e. base cation weathering rates (BC_w), base cation deposition (BC_{dep}), base cation uptake (BC_u) and critical leaching of acid neutralization capacity ($ANC_{le(crit)}$) (UBA, 1996). The formula of calculating critical load is as follows:

$$CL_{max}S = BC_w + BC_{dep} - BC_u + ANC_{le(crit)} \quad (1)$$

2.1. BASE CATION WEATHERING RATE

The base cation weathering rates (BC_w) are based on the parent material class and the texture class for each soil type. Based on actual national soil survey, Thailand has 37 types of parent material and 62 soil groups. The parent material can further be divided into 3 groups: acidic, intermediate and basic, whereas the soil groups can be divided into 6 classes depending on their clay content. The experimental weathering rates from literature (UBA, 1996) are adopted in this study with consideration of local climate, vegetation, and geology. The reference weathering rates need to be corrected based on Thai temperature condition by using Eq. (2). The values of BC_w at local mean temperature are in the range of 287 to 3,155 eq.ha^{-1}yr^{-1}.

$$BC_w(T) = BC_w(T_o)\exp(A/T_o - A/T) \quad (2)$$

where T_o is the reference temperature (281 K), T is local mean temperature (K) and A is constant, 3,600 (Sverdrup et al, 1990).

2.2. BASE CATION DEPOSITION

Base cation deposition (BC_{dep}) refers to the base cations (Ca^{2+}, Mg^{2+} and K^+) that are deposited into the ecosystem. The number of meteorological stations in Thailand is limited, therefore the monitoring data from these stations do not represent the whole country. In order to solve this problem, linear multiple regression and existing monitoring data were used to assess the relationship between BC_{dep} and temperature, precipitation and relative humidity. The relevant formula is shown in Eq. (3)

$$BC_{dep} = a + b_1.[Temperature] + b_2.[Precipitation] + b_3.[Relative\ humidity] \quad (3)$$

The data for temperature, precipitation and relative humidity were obtained from the Department of Meteorology. Since the meteorological stations do not exactly coincide with the latitude and longitude of the grids, an interpolation technique was applied to find the data for a specific grid. The meteorological data for a specific grid was estimated using the four nearest meteorological stations. The formula for interpolating is as follows:

$$X_{grid} = \left(\sum_{i=1}^{4} X_i/D_i\right) \Big/ \left(\sum_{i=1}^{4} 1/D_i\right) \qquad (4)$$

where X_{grid} is the interpolated meteorological data of specific grid, X_i is the value of meteorological data from station i, D_i is the distance from the station i to the specific grid.

2.3. BASE CATION UPTAKE

The base cation uptake (BC_u) is the net uptake by plants that is needed for long-term average growth. These values depend on forest type. The main Thai forests can be classified into 7 categories: tropical evergreen forest, scrub forest, mixed deciduous forest, dry dipterocarp forest, pine forest, rubber plantation and mangrove forest.

$$BC_u = P_{body}\left[\left(BC_{st} + f_{br/st}BC_{br}\right)\big/\left(1 + f_{br/st}\right)\right] \qquad (5)$$

where P_{body} is weight of wood part of trees and shrubs, K_{body} is the ratio of wood biomass to total biomass, BC_{st} is the concentration of base cation in stems (eq/kg), BC_{br} is the concentration of base cation in branches (eq/kg) and $f_{br/st}$ is the ratio of branch to stem biomass (Park & Bashkin, 2000).

$$P_{body} = K_{body} \times NPP \qquad (6)$$

where NPP is the net primary production of forest ecosystems (kg.ha^{-1}yr^{-1}).

$$NPP = \min\{[TP/(1+e^{(1.315-0.119t)}] \text{ or } [TP(1-e^{-0.000664P})]\} \qquad (7)$$

where TP is total productivity equal 30,000 kg·ha^{-1}yr^{-1} as the maximum NPP under the most suitable conditions, t is the local temperature (°C) and P is the precipitation (mm) (Lieth, 1975, see Kira, 1977).

2.4. CRITICAL LEACHING OF ACID NEUTRALIZATION CAPACITY

The critical leaching of acid neutralization capacity ($ANC_{le(crit)}$) can be calculated by Eq. (8)

$$ANC_{le(crit)} = -Q \cdot \left([AL]_{crit} + [H]_{crit}\right) \qquad (8)$$

where Q is the runoff (m^3/ha/yr), $[Al]_{crit}$ is 0.2 eq/m^3 and $[H]_{crit}$ can be calculated from

$$[H]_{crit} = \left([Al]_{crit}/K_{gibb}\right)^{1/3} \qquad (9)$$

Gibbsite constant (K_{gibb}) is related to the properties of soils, especially the amount of clay and organic matter in soil (Warfvinge & Sverdrup, 1995). According to USDA taxonomy, most soils in Thailand are Ultisols, Alfisols, Inceptisols and Entisols, with medium to poor organic matter and clay content. The values of runoff are calculated as precipitation minus the sum of the interception and evaporation. The values of runoff in each grid are obtained from the meteorological stations. The interpolation technique is also applied to find the value of runoff for a considered grid.

3. Results

The whole area of Thailand is divided into 73 1° x 1° LoLa grids but all calculated parameters are represented for only 56 grids, since the southern part of the country is not affected by sulfur deposition from Mae Moh power plant.

3.1. BASE CATION WEATHERING RATE

The weathering rate of base cation of the Thailand is shown in Figure 2. The weathering rates of base cation in terrestrial Thai ecosystems are from 631 eq.ha^{-1}yr^{-1} to 2,725 eq.ha^{-1}yr^{-1}. Approximately 88% of the total area were characterized by the weathering rates in the range of 1,000 to 2,000 eq.ha^{-1} yr^{-1}. The slope complex of the northern area results in a low base cation weathering rates due to lower temperatures and local parent materials. The higher BC_w rates were found in the central flood plain with variety soil types relatively enriched in clay and organic matter.

3.2. BASE CATION DEPOSITION

We have considered only wet deposition in this study. Due to insufficient data, dry deposition was not considered. With 159 samples and confidence level of 95%, the precipitation and temperature can be used to estimate base cation deposition at a significant level. In case that the relative humidity is used to assess the base cation deposition, the confidence level is reduced to 85% with R^2 of 0.40. Although the level of significance for relative humidity is generous, this parameter is the crucial importance that effects the concentration of BC_{dep} (Park and Bashkin, 2000). The linear multiple regression equation of base cation deposition and the meteorological data is shown in Eq. (10). Figure 3 shows the deposition of base cation of the country. The deposition rates range from <50 to >600 eq.ha^{-1}yr^{-1}.

Base cation deposition = −621.8977627 −0.152723049[Precipitation] +
    18.0212705[Temperature] + 168.1238323 [Relative humidity]   (10)

3.3. CRITICAL LEACHING OF ACID NEUTRALIZATION CAPACITY

The critical leaching of acid neutralization capacity is related to the Gibbsite constant and the runoff value. According to the UBA Manual (1996) and organic matter content, the Gibbsite constant values for Thailand's soils are in the range of 2,092 to 7,492 m^6/eq^2. The calculated values of critical leaching of acid neutralizing capacity are shown in Figure 4. The values of $ANC_{le(crit)}$ are from <200 to 2,393 eq.ha^{-1}yr^{-1}.

3.4. BASE CATION UPTAKE

The calculated values of BC_u vary from 37 to 1,136 $kg.ha^{-1}yr^{-1}$. The highest values of BC_u are found in the northern part of country where most area is forested mountain. The map of base cation uptake is shown in Figure 5.

3.5. CRITICAL LOAD OF SULFUR

The CL(S), estimated by using Eq.(1), range from < 200 to 2,225 $eq.ha^{-1}yr^{-1}$. About 70% of total study area can be characterized by low CL(S) values, <200 $eq.ha^{-1}yr^{-1}$ (Fig.6). Running RAINS ASIA model (World Bank, 1994), the CL(S) values were calculated mainly on a basis of the FAO soil mapping in the scale 1:5,000,000. The CL(S) values from this model range from <200 to maximum 1,000 $eq.ha^{-1}yr^{-1}$ and predominant values are <200-500 $eq.ha^{-1}yr^{-1}$. In our calculations, the values in the lowest range <200 $eq.ha^{-1}yr^{-1}$ represent the larger area than the similar range from RAINS ASIA model. This is connected with accounting more detailed national soil survey. Since clay and organic matter contents in main Thai soils are low, it directly affects the values of the weathering rate and acid neutralizing capacity leaching. The critical loads in the central area are relatively higher than in other regions. This can be related to the enrichment of predominant flood plain soils with clay and organic matter. The CL(S) values in the northern area from our study are low in accordance with low base cation weathering rates and high ANC leaching in mountain areas.

4. Conclusions

The estimated values of critical loads of sulfur for Thai ecosystems are quite low due to small fraction of clay and low organic matter in most soil types. The small critical load values reflect the sensitivity of Thai ecosystem to increasing sulfur dioxide concentration in the atmosphere. The main source of sulfur dioxide in the north of Thailand is Mae Moh coal-fired power plant, thus the feasible way to mitigate the impact of sulfur to the ecosystem is to reduce high-sulfur lignite burning for electricity generation. The recalculated national CL(S) values will be applied in running economic-ecological models for possible RAINS-ASIA energy scenarios.

Acknowledgements

The authors express their gratitude to Thailand Research Fund (TRF) under the Royal Golden Jubilee program for the financial support, Dr. P.Khummongkol and Dr. N.W.Harvey for their valuable comments.

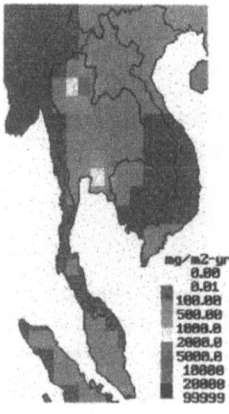

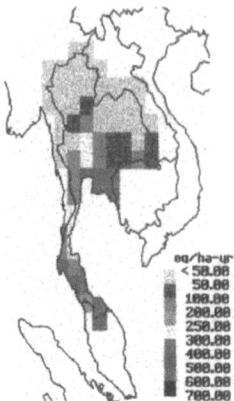

Figure 1. Sulfur deposition from RAINS ASIA model.

Figure 2. Base cation weathering rate.

Figure 3. Base cation deposition.

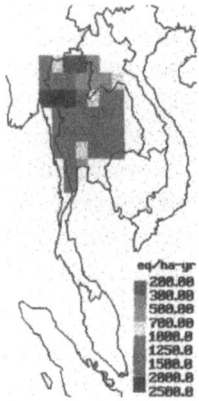
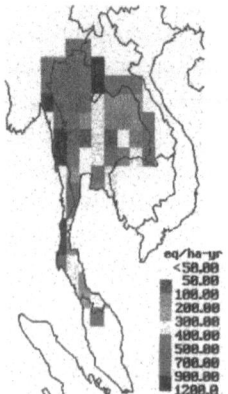
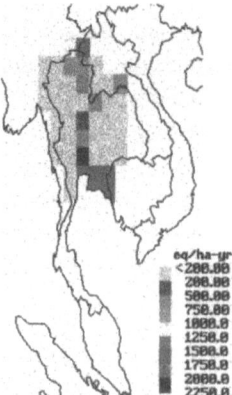

Figure 4. Critical leaching of acid neutralization capacity.

Figure 5. Base cation uptake.

Figure 6. Critical load of sulfur.

References

Khummongkol, P., Bowonwiwat, R.: 1995, *Coal and its Impact on the Environment.* Thailand Environment Institute, Bangkok.

Kira, T:1977. *Production Rates.* In: T.Shido and T.Kira, Eds. Primary Productivity of Japanese Forests. Productivity of terrestrial communities. JIBP Synthesis, vol.16, 101-162

Park, S-U. and Bashkin, V.: 2000, *Sulfur Acidity Loading at South Korean Ecosystems (Accepted for publication in Water Air and Soil Pollution Journal).*

Sverdrup, H., De Vries, W., and Henriksen, A.: 1990, *Mapping Critical Loads.* Environmental Report, 14, Copenhagen, NORD, 124 pp

UBA (Umwelt Bundes Amt): 1996, *Manual on Methodologies and Criteria for Mapping Critical Levels/Loads and Area where they are exceeded.* UNECE Convention on Long-Range Transboundary Air Pollution, Berlin, Germany.

Warfvinge, P. and Sverdrup, H.: 1995, *Critical Loads of Acidity to Swedish Forest Soils: Methods, data and results.* Reports in Ecology and Environmental Engineering Report 5, Lund University, Sweden.

World Bank: 1994, Regional Air Pollution Information and Simulation RAINS ASIA. User's Manual. International Institute for Applied Systems Analysis (IIASA), Washington, USA.

PREDICTING REVERSIBILITY OF ACIDIFICATION: THE EUROPEAN SULFUR STORY

CHRISTINE ALEWELL

Dept. of Soil Ecology, BITÖK, University of Bayreuth, 95440 Bayreuth, Germany
(author for correspondence, e-mail: christine.alewell@bitoek.uni-bayreuth.de)

Abstract. Because of the deleterious effects of acid rain and the need to predict reversibility of acidification, various scientific tools such as modeling, stable isotopes and flux/budget calculations have been used in biogeochemical sulfur (S) research. The aim of this study was to evaluate consistencies and discrepancies between these different tools. While modeling has been seemingly successful in predicting S dynamics in soil solution and stream water by considering inorganic sulfate sorption and desorption only, stable S isotopes indicate that biological S turnover plays a crucial role for the sulfate released to soil solution and stream water. A comparison of budget calculations with soil S pools reveals that inorganic sulfate sorption and desorption are the controlling processes as long as deposition is high (> 15 kg S ha^{-1}yr^{-1}) and soils have a high sulfate sorption capacity. This explains the successful model predictions of the last two decades. However, for soils with low sulfate sorption capacity and under low sulfate deposition, organic S seems to be a significant source for stream water sulfate and has to be considered in future modeling.

Keywords: forest ecosystems, modeling, reversibility of acidification, stable S isotopes

1. Introduction

From the beginning of the twentieth century emissions of sulfur (S), nitrogen and protons showed a steady increase in Europe (with setbacks during the world wars and the oil crisis, Fig 1). Since the late 1970s and early 1980s, a decreasing trend in the amount of acidifying air pollutants is continuing (EMEP, 2000). Because of the major environmental concerns connected to acid deposition, considerable effort has been placed on biogeochemical S research. Results from modeling S dynamics, however, have not always been consistent with results from mass balances as represented by S flux and pool data. Furthermore, recent results from stable S isotopes seem to tell a very different story. The aim of this study was to discuss consistencies and discrepancies between the various approaches used in biogeochemical S research.

2. Modeling sulfur dynamics in Europe: a success story?

With the major change represented by decreasing S deposition since the early 1980's, the focus of ecosystem modeling has switched from describing acidification processes to predicting the recovery of soil solution and surface waters. Ecosystem models (e.g. ILWAS, SMART, SAFE, CHESS, MAGIC) working on the basis of sulfate sorption isotherms (e.g. the Langmuir isotherm or the Freundlich isotherm) predicted a fast response of soil solution and stream water sulfate concentrations to decreased sulfate deposition for soils with low sulfate sorption capacity (Fig. 1, soils A). Low sulfate sorption capacity of soils is due to all or a combination of the following parameters: 1) shallow postglacial soil development, 2) sandy soil textures and 3) a high humus content.

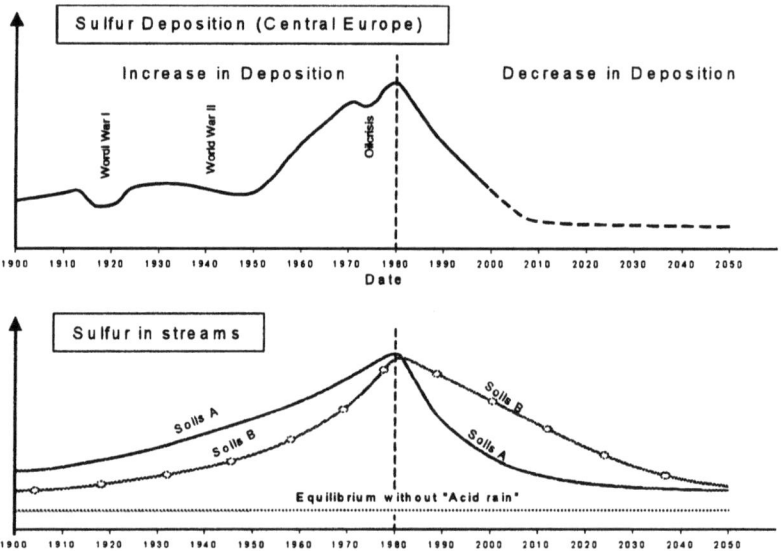

Figure 1: Long-term trend of sulfur dynamics in Central Europe. Soils A have a low, Soils B a high sulfate sorption capacity.

Examples are the geologically young, sandy soils in southern Norway and southeastern Netherlands which responded rapidly to reduced deposition with a significant decrease in H^+, sulfate and Al^{3+} concentrations and increase in alkalinity in streams within the first five to ten years (Wright et al., 1988; van Dijk et al., 1992; Skjelkvåle et al., 2000). In contrast, soils with high sulfate sorption capacity were predicted to respond slowly with the decrease in sulfate concentration to be delayed over decades (Fig. 1, soils B), due to a release of previously-stored sulfate. Examples are the deeply-weathered soils in the low mountain ranges of Germany and in south Sweden. Today, the seepage water from these soils shows no or only a slow decrease in sulfate concentrations and no recovery of pH or alkalinity (Alewell et al., 1997; Manderscheid et al., 2000; Moldan, 1999). Thus, we can conclude that modeling S dynamics in Europe seems to be indeed a success story. However, an apparently correct model prediction of element concentrations in forest ecosystems does not mean that the active processes are correctly described. Models may be right for the wrong reasons. This becomes evident, if we compare model structures with results from stable S isotope research.

3. The contrasting story: Stable sulfur isotopes

To gain insight into the importance of biological S turnover in forest ecosystems (mainly the contributions of S mineralization and sulfate reduction) relative to sulfate adsorption/ desorption processes (e.g.; Torssander and Mörth, 1997; Mitchell et al., 1999; Alewell et al., 1999), stable S isotopes have been used. The isotopic composition in forest ecosystems is controlled by isotopic composition of sources (i.e., atmospheric deposition and mineral weathering) and isotopic discrimination during S transformations. Only biological processes cause shifts in S isotope ratios under natural environmental conditions (Krouse and Grinenko, 1991). Data from Germany (Mayer et al., 1995a,b; Alewell and Gehre, 1999), the Czech Republic (Novak et al., 1996; 2000), Norway (Thorssander and Mörth, 1997) and the northeastern U.S. (Fuller et al., 1986; Zhang 1994; Alewell et al., 1999) revealed that $\delta^{34}S$ values of sulfate in soil solutions of upland sites were generally slightly depleted relative to sulfate in throughfall data. A depletion in ^{34}S of sulfate in soil solution in relation to that in throughfall indicates that S mineralization is a potential sulfate source, because the soil microflora prefers the lighter ^{32}S isotope. Mayer et al. (1995a) concluded that in addition to $\delta^{34}S$ values the oxygen isotopic composition ($\delta^{18}O$) of sulfate indicates that biological processes were contributing to the sulfate in soil solution.

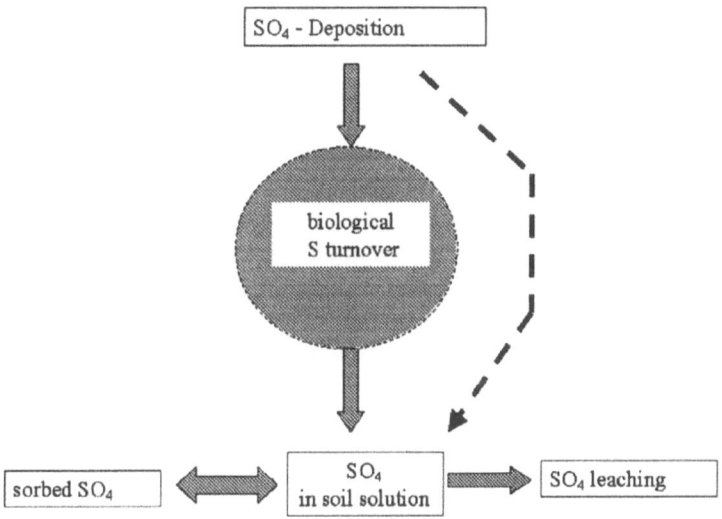

Figure 2: Sulfur cycle in forest ecosystems as indicated by results from stable sulfur isotopes.

Generally, authors agree that the consistently lower $\delta^{34}S$ values in soil solutions compared to throughfall and the differences in $\delta^{34}S$ between organic and inorganic S at upland forest sites can only be explained by S immobilization/ mineralization processes (Fuller et al., 1986; Krouse and Grinenko, 1991; Zhang, 1994; Mayer et al., 1995a; Alewell et al., 1999; Novak et al., 2000; Alewell and Novak, 2000). A sulfur isotope mixing model indicated that about 30% of sulfate in stream water was organically cycled (Novak et al., 2000). Thus, contrary to most biogeochemical models which consider only inorganic sulfate sorption and desorption, stable S isotopes indicate that a considerable proportion of the atmospherically deposited sulfate is cycled through the organic S pool before being released to soil solution and stream water (Fig. 2).

4. Closing the gap: evidence from soil sulfur pools

Soil S pools in combination with flux/ budget calculations indicate that the source for sulfate in solution might be very different for soils with low sulfate storage capacity and soils with high sulfate storage capacity. Soils with a high sulfate sorption capacity accumulated relatively large amounts of sulfate under the high deposition regime of the past. An example is the catchment *Villingen* (Black Forest, Germany). A combination of the deeply-weathered, loamy soils at *Villingen* (Cw > 80 cm) which cover a 10 to 20 m layer of sediments (Feger, 1993), cause a relatively-high sulfate sorption capacity. Pools of inorganic sulfate are large enough (113 and 376 kg S $ha^{-1} 80$ cm^{-1} of water and phosphate-extractable sulfate, respectively) to explain sulfate release within the last decade (Fig. 3, left; Prietzel, 1998; Armbruster, 1998). Another example for soils with high sulfate storage capacity is the *Lehstenbach* catchment in the Fichtelgebirge (northeastern Bavaria, Germany). The weathering depth in the catchment is ≥ 40 m; soil texture is loamy sand to loam. Sulfate storage in the catchment down to 10 m depth is > 2000 and 3000 kg S ha^{-1} for water and phosphate extractable sulfate, respectively (Manderscheid *et al.*, 2000). The average net release of sulfate (output minus input) for the years 1993-1996 at a terrestrial plot within the catchment was 11.5 kg S $ha^{-1}yr^{-1}$ (43.5 kg S $ha^{-1}yr^{-1}$ output in soil solution at 90 cm depth minus 32 kg S $ha^{-1}yr^{-1}$ input in throughfall, Matzner *et al.*, 2000). Thus, considering only inorganic soil sulfate pools the storage in the catchments is large enough to buffer decreased sulfate deposition for decades and thus delays recovery. This is consistent with modeling results.

A very different situation is found in soils with low sulfate sorption capacity (soils A). At the catchment *Schluchsee* (Black Forest, Germany), the sandy soil with a very high humus content down to 80 cm depth (Feger, 1993) has low pools of inorganic sulfate (Prietzel, 1998). A comparison of net release of sulfate over the last 9 years with soil sulfate pools demonstrates that the sulfate can not come solely from inorganic sulfate pools (Fig. 3, right; Armbruster, 1998; Prietzel, 1998). Another example of low sulfate adsorbing soil is the soil at *Risdalsheia* (southern Norway), the site of the RAIN project with experimental exclusion of acid deposition by a roof (Wright *et al.*, 1988). Soils at *Risdalsheia* are extremely shallow with an average thickness of 11 cm and have a high content of organic material. The sulfate released within the last 14 years can not be explained by losses from inorganic sulfate pools (Wright *et al.*, 1988; Wright, pers. comm.). The same mechanism has been shown for soils in North America (Hubbard Brook Experimental Forest, New Hampshire, USA), where budget calculations for the last 30 years indicate that the amount of sulfate released to the stream can not be explained solely by release from inorganic sulfate pools (Driscoll *et al.*, 1998; Alewell *et al.*, 1999; Mitchell *et al.*, 2000). The most likely sulfate source in these catchments would be the organic S pool, even though S weathering as well as dry deposition might contribute smaller amounts to the gap in the S budgets (Alewell and Gehre, 1999; Alewell et al., 1999; Driscoll et al., 1998; Prietzel, 1998; Mitchell et al., 2000). Thus, under decreasing anthropogenic deposition, budget calculations point to the contribution of organic S to sulfate in runoff for soils with low sulfate sorption capacity.

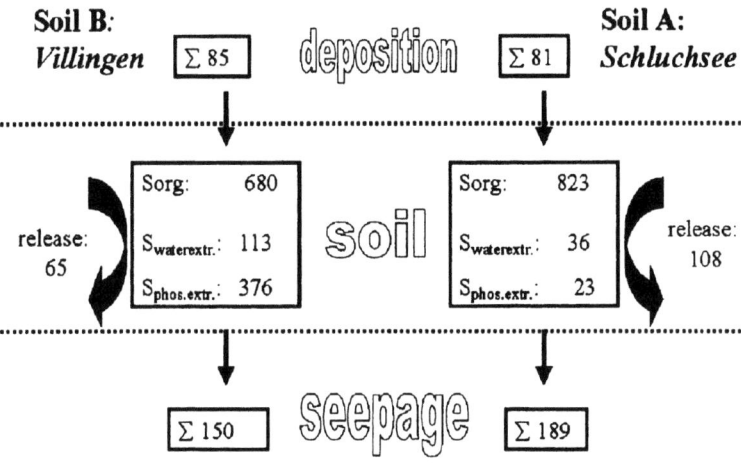

Figure 3: Comparison of cumulative sulfate fluxes (Σ 1988 - 1996) with soil sulfur pools down to 80 cm depth (kg S ha^{-1}). Flux data from Armbruster (1998). Soil pools determined in 1986 by Prietzel (1998).

5. Synthesis and Conclusions

The question arises why models of S dynamics in Europe result in successful description and prediction of sulfate fluxes without the consideration of organic S pools and biological S turnover. Stable isotope data indicates that at least part of the deposited sulfate cycles through the organic S pool. Flux /budget calculations indicate that in soils with low sulfate sorption capacity (soils A) some (or even a major part) of the released sulfate is released from organic S. However, up to the early 1980s, soils in Europe received a high deposition regime of $\geq 15 \leq 120$ kg S ha^{-1}yr^{-1}, which dominated S cycling in forest soils. Most likely inorganic sulfate dynamics controlled sulfate release under these conditions, which explains why modeling sulfate dynamics considering only inorganic sulfate sorption/ desorption was so successful. However, the significant reduction of sulfate deposition during the last 15 years, probably increases the relative contribution of biological S turnover. This explains why inorganic sulfate pools in soils with a low sulfate sorption capacity (soils A) can not completely explain sulfate release to soil solutions or stream waters over the last 15 years without the consideration of organic S pools. Thus, the contribution of organic S to sulfate in stream water will most likely become more important under the current low deposition regime and has to be considered in future modeling.

Acknowledgements:
This project was financially supported by the German Ministry of Education, Science, Research and Technology, grant no. PT BEO 51-0339476, and the Commission of European Communities RECOVER 2010 project (EVK1-CT-1999-00018). I would like to thank Richard F. Wright for his helpful comments on the manuscript.

References

Alewell, C. and Gehre, M.: 1999, *Biogeochemistry* **47**, 319.
Alewell, C. and Novak, M.: 2000, *Environmental Pollution* (in press)
Alewell, C., Bredemeier, M., Matzner, E. and Blanck, K.: 1997, *J. Environ. Qual.* **26**, 658.
Alewell, C., Mitchell, M., Likens, G.E. and Krouse, R.H.: 1999, *Biogeochemistry* **44**, 281.
Armbruster, M.: 1998, *Freiburger Bodenkundliche Berichte* **38**, 301pp.
Driscoll, C.T., Likens, G.E. & Church, M.R.: 1998, *Water Air and Soil Pollution* **105**, 319.
EMEP: 2000, http://www.emep.int/emis_tables.
Feger, K.H.: 1993, *Freiburger Bodenkundliche Abhandlungen* **31**, 237pp
Fuller R.D., Mitchell M.J., Krouse H.R., Wyskowski B.J. and Driscoll C.T.: 1986, *Water Air Soil Poll.* **28**, 163.
Krouse, H.R. and Grinenko, V.A.: 1991, *Stable Isotopes. Natural and Anthropogenic Sulphur in the Environment,* Scope **43**, John Wiley & Sons Ltd., New York, NY.
Manderscheid, B., Schweisser, T., Lischeid, G., Alewell, C. and Matzner, E.: 2000, *Soil Sci. Soc. Am. J.* **64**, 1078.
Matzner, E., Alewell, C., Bittersohl, J., Lischeid, G., Kammerer, G., Manderscheid, B., Matschonat, G., Moritz, K., Tenhunen, J.D., and Totsche, K.U.: 2000, 'Biogeochemistry of a spruce forest catchment of the Fichtelgebirge in response to changing atmospheric deposition.' in: J.D. Tenhunen, R. Lenz, and R. Hantschel, (eds.): *Ecosystem Approaches to Landscape Management in Central Europe,* Ecological Studies, Springer Verlag, Heidelberg (in press)
Mayer, B., Feger, K.H., Giesemann, A. and Jäger, H.-J.: 1995a, *Biogeochemistry* **30**, 31.
Mayer, B., Fritz, P., Prietzel, J. and Krouse, H. R.: 1995b, *Applied Geochemistry* **10**, 161.
Mitchell, M.J., Krouse, R.H., Mayer, B., Stam, A.C. and Zhang, Y.M.: 1999. 'Use of stable isotopes in evalauting biogeochemistry if forest ecosystems' in: C. Kendall, and J. McDonnell (eds.), Isotope tracers in Catchment Hydrology. Elsevier, The Netherlands, 489.
Mitchell, M.J., Mayer, B., Bailey, S.W., Hornbeck, J., Alewell, C., Driscoll, C.T., Likens, G.E.: 2000. Discrepancies in sulfur mass balance at the Hubbard Brook Experimental Forest: Use of stable isotopes for evaluating sulfur sources and sinks. *Water Air Soil Pollut.* (this volume).
Moldan, P.: 1999, Silvestria 117; Swedish University of Agricultural Sciences, Umeå.
Novák, M., Bottrell, S.H.; Fottová, D., Buzek, F., Groscheová, H. and Zák, K.: 1996, *Environmental. Science and Technology.* **30**, 3473.
Novak, M., Kirchner, J.W., Groscheova, H., Havel, M., Verny, J., Krejci, R. and Buzek, F.: 2000. *Geochimica et Cosmochimica Acta* **64**, 367.
Prietzel, J.: 1998, Habilitationsschrift der Ludwig-Maximilians-Universität München, 399pp.
Skjelkvåle, B.L., Tørseth, K., and Aas, W.: 2000. Decrease in acid deposition – recovery in Norwegian waters. *Water Air Soil Pollut.* (this volume)
Torssander, P. and Mörth, C.-M.: 1997, 'Sulfur Dynamics in the Roof Experiment at lake Gårdsjön deduced from sulfur and oxygen isotope ratios in sulfate. in: H. Hultberg and R. Skeffington (eds.), *Experimental Reversal of Acid Rain Effects: The Gårdsjön Roof Project.* John Wiley & Sons. Chichester, pp. 185-206.
van Dijk, H.F.G., Boxman, A.W. and Roelofs, J.G.M.: 1992, *Forest Ecol. Management* **51**, 207.
Wright, R.F., Lotse, E. and Semb, A.: 1988, *Nature* **334**, 670.
Zhang, Y.: 1994, Ph.D. Dissertation. SUNY, ESF, Syracuse, NY, USA, 273pp.

EFFECTS OF ACID DEPOSITION ON FOREST SOILS IN NORTHERNMOST RUSSIA: MODELLED AND FIELD DATA

SERGUEI V. KOPTSIK[1*] AND GALINA N. KOPTSIK[2]

[1]*Faculty of Physics, Moscow State University, Moscow 119899, Russia;* [2]*Soil Science Faculty, Moscow State University, Moscow 119899, Russia*
(author for correspondence, e-mail: koptsik@skop.phys.msu.su)

Abstract. In addition to strong natural stresses forest ecosystems in the Kola Subarctic, Russia, receive high loads of sulphur and heavy metals from the nickel smelter. To estimate soil response to acid deposition we compared the soil field data along a pollution gradient and simulated time effects. Multivariate technique was applied to investigate spatial distribution of soil field data. Time response of soils to acid deposition was evaluated with the SMART model. According to field observations there is no evidence for strong soil acidification effects close to the smelter. Concentrations of exchangeable Ca and base saturation increase, while acidity decrease in lower soil mineral horizons towards the pollution source. However, some features seem to reflect the early stages of the started acidification. Most soil profiles have low pH values. Despite increasing of exchangeable Ca and Mg towards the smelter in lower mineral horizons due to geological inheritance, they do not reveal the same trends in the upper ones. Concentration of exchangeable K in organic horizons decreases towards the smelter, thus confirming the starting acidification. As result, exchangeable base cations are depleted in the considerable part of shallow soil profiles. According to model simulation the present acid load does not effect considerably on forest soils in background areas, however, dramatic shift in soil chemistry near the smelter is expected within several decades. Due to low pool of exchangeable base cations and low weathering rate continued acid deposition can lead to increased soil acidification and nutrient imbalance.

Key words: forest soils, acid deposition, exchangeable cations, acidification modelling

1. Introduction

The areas surrounding the Pechenganikel smelter, the northernmost Europe, are heavily polluted from emissions of SO_2 and heavy metals (HM) for several decades. Air pollution has caused severe damage to the terrestrial ecosystems (Aamlid and Venn, 1993). The release rate of base cations due to weathering, their only natural long-term source, range from 0.05 to 0.3 $kmol_c \cdot ha^{-1} \cdot yr^{-1}$ in the 50 cm soil layer, thus demonstrating the high sensitivity of the coarse and thin podzols to acidification (Koptsik *et al.*, 1999a, b). Comparatively low sulphate adsorption in the region (Gustafsson, 1996) can not delay soil acidification under the high present sulphur loads for a long time.

We cannot directly analyse soil response to acid deposition by following a single site through time due to the absence of systematic observations. Rather, we compare sites at different distances from the smelter and simulate time effects. In so doing we implicitly assume that this substitution of space for time will provide an insight into factors that control soil changes in the vicinity of smelter. The scope of this study is to examine both field and modelled changes in soil acidity and exchangeable cations under the influence of air pollutants.

2. Materials and methods

2.1. STUDY AREA

The study area is located around the nickel smelter in the north-western part of the Kola Peninsula, Russia. It is a hilly subarctic glaciated terrain covered by tills, glaciofluvial deposits and open bedrock with coarse texture. The soil mineralogy of the study area is generally characterised by slowly weathered primary minerals, represented by mostly quartz and feldspars derived from gneisses and granites. The area is located at the polar tree line. Pine and birch forests are characterised by scarce tree stands and scarcely developed ground vegetation, dominated by crowberry, bilberry, mosses and lichens. Podzol is the most common soil type, its thickness usually does not exceed 0.3-0.5 m.

At present the nickel smelter emits about 250,000 tons of sulphur dioxide annually and is the 4th largest point SO_2 source in Europe (Barrett and Protheroe, 1995). The total annual sulphur deposition is about 5-7 $g \cdot m^{-2}$ in the vicinity of the smelter and 0.1-0.2 $g \cdot m^{-2}$ in background areas (Sivertsen *et al.*, 1992).

2.2. SOIL SAMPLING AND CHEMICAL METHODS

Soil samples were collected mainly from pine and from birch forests at different distances and directions from the smelter (28 plots in total). The samples were air dried at 20°C and ground to pass a 2 mm sieve. Chemical analysis of the soil included pH in H_2O and 0.01 M $CaCl_2$ extracts. Exchangeable acidity (EA) and exchangeable cations (Al, Ca, Mg, K, Na, Fe, Mn, Zn) were measured in the 1 M NH_4NO_3 extract. Ammonium nitrate-extractable elements in soils were determined by inductive coupled plasma emission spectroscopy.

2.3. DATA TREATMENT

Correlation, linear and nonlinear regression were used to study the current soil chemistry. Due to the large and complex set of observed data and to their high spatial variability principal component analyses (PCA, ter Braak, 1987) was used to identify most important relationships and to present soil data in the concise graphical form. Long-term response of soils to acid deposition was evaluated with the dynamic process-oriented SMART model (Simulation Model for Acidification's Regional Trends, de Vries *et al.*, 1989; de Vries, 1994). The chemical responses of the upper 25 and 50 cm layers of podzols were simulated for a period of 100 years. Three scenarios included actual (0.02-0.50 $mol_c \cdot m^{-2} \cdot yr^{-1}$ at different sites), 30% and 90% reduced actual sulphur deposition.

3. Results and Discussion

3.1. SPATIAL VARIABILITY OF FIELD DATA

The prevailing in the region thin sandy podzols are rather acidic due to the moist, cool climate, base-poor granite bedrock and mainly coniferous forest cover keeping low organic matter content. Podzols are characterised by low pH, elevated EA, low content of exchangeable base cations and low CEC (Table 1).

The first two PCs of the element scores are shown in Figure 1 with solid arrows. The first two principal components (PC) accounted for 27.6 and 23.6% of the total variation, twice that covered by the two following PCs (12 and 11%). The cosine of angle between any two arrows in the ordination diagram approximate correlations between the corresponding soil properties.

Table 1. Statistics of acidity and exchangeable cation contents in podzols. EA – exchangeable acidity; $BC = Ca^{2+}+Mg^{2+}+K^{+}+Na^{+}$ – exchangeable bases cations; CEC=(EA+BC) – cation exchange capacity; BS=(100·BC/CEC) – base saturation, std – standard deviation.

Hori zon	Para meter	pH H_2O	pH $CaCl_2$	EA	Al	Ca	Mg	K	Na	BC	CEC	BS
						$mmol \cdot kg^{-1}$				$mmol_c \cdot kg^{-1}$		%
O	min	3.6	3.0	80	3	7	3	6	1.7	28	150	18
	max	4.9	4.3	230	29	130	37	40	5.7	320	440	78
	average	4.1	3.4	140	11	65	22	19	3.2	195	330	58
	std	0.4	0.4	30	7	25	8	7	1	65	60	13
E	min	4.1	3.4	9	1.5	0.4	0.2	0.1	0.03	2.0	11	6.7
	max	4.9	3.8	30	10.5	3.0	1.5	2.6	1.9	11.5	41	40
	average	4.4	3.6	18	6	1.3	0.65	1.0	0.5	5.5	24	23
	std	0.2	0.15	7	3	0.8	0.45	0.85	0.5	3	10	8
Bf1	min	4.2	3.7	8.6	2.7	0.3	0.1	0.1	0.1	1.1	11	9.4
	max	5.4	4.7	53	18	4.5	1.4	3.3	2.4	12	59	47
	average	4.75	4.3	23	7.7	1.3	0.5	0.9	0.8	5.2	28	19
	std	0.3	0.3	15	5	1	0.4	0.9	0.7	3	17	10
Bf2	min	4.6	4.1	2	0.4	0.07	0.01	0.1	0.1	0.5	3	10
	max	5.8	5.2	53	18	5.1	1.0	3.3	2.2	11.5	59	75
	average	5.0	4.7	14	4.5	1.1	0.37	0.8	0.6	4.4	18	28
	std	0.3	0.3	14	4.8	1.2	0.3	1	0.6	3	16	16
BC	min	4.8	4.6	1.0	0.15	0.06	0.02	0.08	0.15	0.4	1.9	20
	max	6.1	5.4	12	3.4	5.0	0.4	0.7	2.7	11	16	92
	average	5.3	4.9	3.9	1.5	1.1	0.2	0.3	0.9	3.7	7.6	46
	std	0.4	0.25	3.5	1.1	1.5	0.1	0.2	0.8	3	5	25
C	min	4.9	4.7	0.2	0.1	0.08	0.02	0.08	0.15	0.4	1.8	21
	max	6.3	5.4	13	2.7	4.9	0.35	2.2	1.1	11	18	98
	average	5.4	4.9	4.0	1.2	1.3	0.18	0.5	0.6	4.0	8.0	46
	std	0.5	0.3	4	0.9	1.9	0.15	0.7	0.3	4	5	30

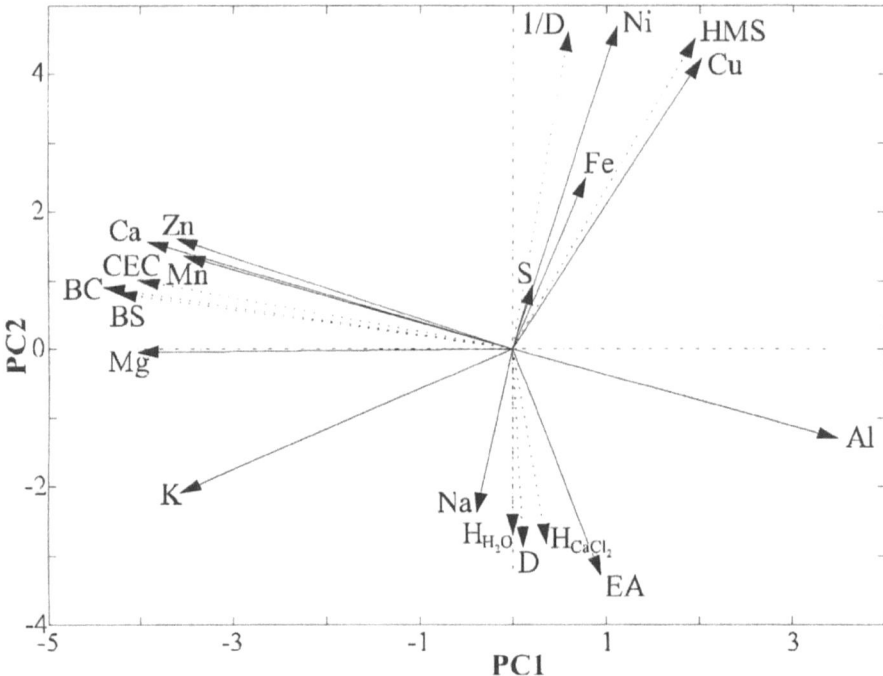

Figure 1. PCA ordination diagram of exchangeable cations for soil organic horizons. External variables are shown with dotted arrows. D - distance, 1/D - inverse distance from the smelter, HMS = (100·HM/CEC) – heavy metal saturation, other notations as in Table 1.

Natural sources of variation appeared to be the most significant; exchangeable Ca, Mg, K, Mn and Zn, the soil fertility forming nutrients that determined the first PC, formed an isolated group in the ordination plot. The first PC did not depend on the distance from the smelter. Variation of exchangeable Ni, Cu, Fe and extractable S, caused by atmospheric deposition, was very high. These variables were the main factors contributing to the second PC and formed another isolated ordination group. Like the heavy metals determining it, the second PC depended nonlinearly on the distance from the smelter. Exchangeable acidity, that is determined by pollution-induced H^+ input as well as by naturally highly variable Al, lay isolated and formed a third group in the correlation plot.

Air pollution has not resulted in topsoil acidification that is detected in the form of unusually low pH values or especially high exchangeable Al near the smelter. Contrary, PCA as well as correlation and regression analyses shows that pH in water and $CaCl_2$ suspensions increases and exchangeable acidity decreases towards the smelter. This is probably due to the high Ca and Mg deposition near the pollution source, the decreased input of H^+ as a result of organic matter transformation and plant uptake in destroyed ecosystems, and more basic lithology. However, exchangeable K, the most mobile base cation, decreases towards the pollution source. The decrease of K^+ seems to be the first sign of started acidification. This tendency may be caused by replacement of K^+ with pollutant cations, which is evidenced by negative correlations with Al^{3+}, Ni^{2+}, and Cu^{2+}. Despite increase of exchangeable Ca, Mg and base saturation towards the smelter in lower mineral horizons due to geological inheritance, they do not show the same trends in upper horizons. This fact is surprising as the anthropogenic dust input of Ca and Mg near the smelter is easily available. The deep penetration of acidity, peculiarities of spatial distribution of base cations and their losses from the topsoil reveal an early stage of ongoing soil acidification.

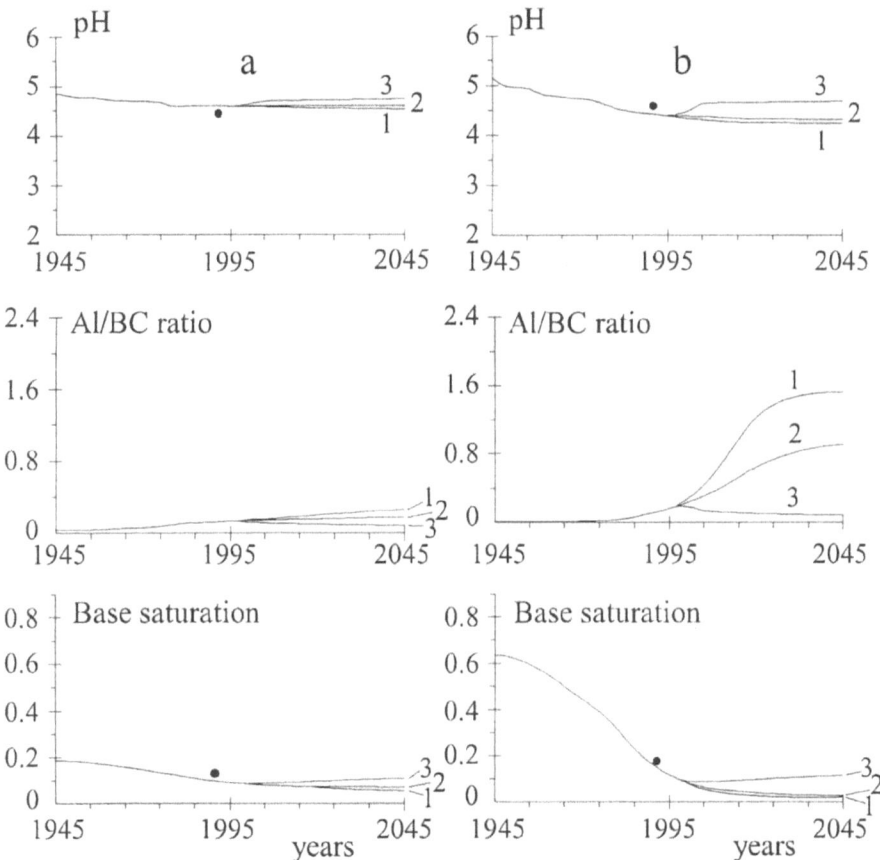

Figure 2. Simulated time dependencies of pH, Al/BC ratio and base saturation in podzols (0-25 cm) at background (a) and polluted (b) plots. Experimental data are shown with dots.

3.2. MODELLED DATA

According to calculations with the SMART model long-term effects of acid deposition on poor sandy podzols with low CEC result in a decrease in pH within the cation exchange buffer range. The changes are negligible for background soils and considerable for soils under elevated acid load close to the smelter.

Simulated time dependencies of pH, Al/BC ratio and base saturation are shown in Figure 2 for actual, 30 and 90% reduced deposition scenarios at background and polluted plots. At high acid loads BC can be exhausted in several years. An increase in CEC in podzols developed on glacial till comparing with glaciofluvial sand leads to a longer period for depleting the exchangeable BC, thus causing a longer time lag before the Al/BC ratio starts to increase. Abundance of Al hydroxide prevents the further decrease in pH. The time lag between reductions in deposition and a decrease in the Al/BC ratio is short. However, pronounced reductions, up to the background level, are needed to get Al/BC ratio below a critical value of 1.0. Results of numerical simulation agreed rather well with the experimental data. At least 90% of the total SO_2 emissions has to be reduced to avoid exceedance of critical values of soil chemical properties. In addition to the long-term chronic acidification, the powerful mass-flow of melt water in spring may release in a sharp episodic acidification of soils (Koptsik, 1997).

4. Conclusions

At present geological features of the territory and alkaline dust deposition prevent soil acidification near the pollution source. However, the deep penetration of acidity, close correlation of exchangeable acidity with base cation content and their depletion from the topsoil, decreased concentrations of exchangeable K^+ in organic horizons towards the pollution source confirm the started soil acidification in the study area. According to calculations with the SMART model the soils will acidify severely within the next 20-30 years unless there are drastic reductions of the SO_2 emissions from the "Pechenganikel" smelter.

References

Aamlid, D. and Venn, K.: 1993, *Norwegian Journal of Agricultural Sciences* **7**, 71.
Barrett, M. and Protheroe, R.: 1995, Sulphur Emission from Large Point Sources in Europe. Stockholm: Swedish NGO Secretariat on Acid Rain. 22 pp.
De Vries, W., Posch, M. and Kämäri, J.: 1989, *Water, Air and Soil Pollution* **48**, 349.
De Vries, W.: 1994, Soil Response to Acid Deposition at Different Regional Scales: Field and Laboratory Data, Critical Loads and Model Predictions. Wageningen, The Netherlands. 487 pp.
Koptsik, G., Koptsik, S., Venn, K., Aamlid, D., Strand, L. and Zhuravleva, M.: 1999a, *Eurasian Soil Science* **32** (7), 787.
Koptsik, G., Teveldal, S., Aamlid, D. and Venn, K.: 1999b, *Applied Geochemistry* **14** (2), 173.
Koptsik, S.: 1997, *Acid Snow and Rain*. Proceedings of International Congress. Niigata, Japan.
Sivertsen, B., Makarova, T., Hagen, L.O. and Baklanov, A.A.: 1992, Air pollution in the border areas of Norway and Russia. Summary report 1990-1991. Lillestrom (NILU OR 8/92).
ter Braak, C.J.F.: 1987, 'Ordination', in H.J. Jongman, C.J.F. ter Braak and O.F.R. van Tongeren (eds.), *Data analysis in community and landscape ecology*, Pudoc: Wageningen, pp. 91-173.

DYNAMIC MODELLING OF SPATIALLY VARIABLE CATCHMENT HYDROCHEMISTRY FOR CRITICAL LOADS ASSESSMENT

HELEN J. FOSTER[1], MATTHEW J. LEES[1], HOWARD S. WHEATER[1]*, COLIN NEAL[2] and BRIAN REYNOLDS[3]

[1] *Dept. of Civil and Environmental Engineering, Imperial College, London, SW7 2BU, UK;*
[2] *Centre for Ecology and Hydrology, Wallingford, Oxon, OX10 8BB, UK;* [3] *Centre for Ecology and Hydrology, Deiniol Road, Bangor, LL57 2UP, UK*
(*author for correspondence, e-mail: h.wheater@ic.ac.uk)

Abstract. Concern about acidification in upland areas has brought about the need to model the stream hydrochemical response to deposition and land-use changes and calculate critical loads. Application of dynamic models such as MAGIC are preferable to steady-state methods, since they are able to produce an estimate of the time scale required to meet some water chemistry target given a reduction in acid deposition. These models typically consider annual changes in stream chemistry at one point. However, in order to protect biota from 'acid episodes', quantification of temporal variability needs to encompass event responses; in addition spatial variability across the catchment also needs to be considered. In this paper, modelling of both spatial and temporal variability is combined in a new framework which enables quantification of catchment hydrochemical variability in time and space. Both low and high flow hydro-chemical variability are quantified in terms of statistical distributions of ANC (Acid Neutralisation Capacity). These are then input as stochastic variables to an EMMA (End-Member Mixing Analysis) model which accounts for temporal variability and ANC is hence predicted as a function of time and space across the whole catchment using Monte-Carlo simulation. The method is linked to MAGIC to predict future scenarios and may be used by iteration to calculate critical loads. The model is applied to the headwaters of the River Severn at Plynlimon, Wales, to demonstrate its capabilities.

Keywords: End-Member Mixing Analysis (EMMA), hydrochemistry, Plynlimon, critical loads

1. Introduction

Many UK upland catchments have thin acidic soils and a low buffering capacity, which makes them susceptible to chronic acidification, especially where plantation forestry is a major land-use (Harriman *et al.*, 1994). Critical load calculations are a well established method used to predict the amount of pollutant deposition that such areas may withstand without adverse ecological effects (e.g. Nilsson and Grennfelt, 1988). These often use steady state calculations and are applied on a grid square basis. However, dynamic methods for critical load calculation are preferable because they consider the time taken to reach certain chemical conditions. A catchment-scale model is also required if assessment of the effects of future land-use and pollution changes on stream chemistry are to be made and the long-term simulation model MAGIC (Cosby *et al.*, 1985) has been extensively applied to this end (Jenkins *et al.*, 1997). However, MAGIC uses mean annual chemistry at the catchment outlet which neglects consideration both of 'acid episodes', which may cause harm to biota during storm events (Ormerod and Jenkins, 1994), and the great spatial variability in stream chemistry that exists across small (<50 ha) catchments

(Foster *et al.*, 2000). These both need to be incorporated into modelling strategies when aiming to protect sensitive catchments.

Temporal variability in stream chemistry at a point may be modelled using End-Member Mixing Analysis (EMMA, Christophersen *et al.*, 1990) based on mixing of two or more chemically distinct water sources (end-members). This method may then be coupled with MAGIC to enable prediction of future temporal changes under different land-use or pollution deposition scenarios (Neal *et al.*, 1992). In this paper, this method is extended to incorporate spatial variability by considering the two end-members as stochastic variables for input into an EMMA-based Monte Carlo simulation. This can be used to quantify catchment hydrochemical variability under future scenarios, and hence to underpin improved critical loads prediction. Firstly the method is outlined and secondly its capability demonstrated by an application.

2. Temporal variability in streamwater chemistry

Rapid changes in streamwater chemistry in upland regions are predominantly the result of changing flowpaths during hydrological events (Wheater *et al.*, 1990) and streamwater may be thought of as consisting of two or more 'end-members' with different chemistries, the proportions of which change with discharge (Christophersen *et al.*, 1990; Neal *et al.*, 1992). At low flows, a stream is fed mainly by water from the lower soil horizons and deeper groundwater, which typically has a high Acid Neutralisation Capacity (ANC) due to the weathering in the lower soil horizons and bedrock (Reynolds *et al.*, 1986). At high flows, a large component of streamwater is derived from the upper, more acidic soils with a low ANC. This concept underlies two-component EMMA (Christophersen *et al.*, 1990). Using the conservative measure, ANC, the model may be written as follows;

$$\frac{Q_{g-w}}{Q_{strm-w}} = \frac{(ANC_{strm-w} - ANC_{soil-w})}{(ANC_{g-w} - ANC_{soil-w})} \tag{1}$$

where Q represents discharge, *ANC* is the Acid Neutralisation Capacity and the subscripts *strm-w*, *soil-w* and *g-w* refer to streamwater, soilwater and groundwater sources. It is assumed that the two end-members each have a distinct chemical signature which changes very slowly over time (i.e. does not change during event responses). Although this assumption has been shown to be incorrect (Chapman *et al.*, 1997), it is used to develop the technique because present knowledge is insufficient to deal with its short-comings.

To obtain estimates of end-member chemistry, available ANC data (e.g. weekly) are ranked according to flow, and the mean average of the upper and lower 5% of values chosen as being representative of soil and groundwater end-member chemistry in the model. Continuous (e.g. 15 minute) discharge data for a given representative period (e.g. one year), a time series of ANC data for the same period and the end-member estimates are used to calculate a time series of groundwater flow proportion (GWP) from Equation 1. The ANC and GWP time series data are converted into duration curves by ranking the values and plotting in terms of percentage of time that a certain value is exceeded. These duration curves summarise the variability in stream ANC and groundwater proportion of flow and are used as a basis for representing variability during the subsequent modelling stages.

3. Incorporation of Spatial variability

Although the dominant control on stream chemistry variability is discharge, different temporal responses are seen between sites on contrasting soils, bedrock mineralisation and land-use (Reynolds *et al.*, 1986; Foster *et al.*, 1997) implying the existence of a spatial dimension to the variability for both high and low flow streamwater chemistry for different sites across catchments. . Spot samples of high and low flow ANC are required from intervals along all streams in the catchment, to enable quantification of the end-member variability. Stream chemistry may not have a constant or predictable composition, for example from catchment characteristics (Foster *et al.*, 2000), however, it may be incorporated into the EMMA by a stochastic representation. Both end-members and GWP are therefore approximated by statistical distributions and are input to the EMMA model as stochastic variables using Monte-Carlo simulation. The method is applied as follows:

i. The groundwater proportion (GWP) duration curve calculated from the continuous data is approximated by a cumulative normal distribution. It is considered representative of all sites across the catchments.

ii. For given increments of duration (e.g. 1%) the GWP value is selected.

iii. 2000 end-members are randomly selected from distributions approximating each, where $ANC_{s-w} < ANC_{g-w}$.

iv. Selected end-members are input to the following equation:

$$ANC_{stm-w} = ANC_{g-w}(GWP) + ANC_{s-w}(1-GWP) \qquad (2)$$

This produces 2000 ANC values representing spatial variability in ANC which are ranked to give percentage of streamlength along which ANC is exceeded at a given time. Because values are selected from given distributions of parameters at random and then run through the model, the results represent the full range of likely scenarios given the variability of the input parameters. Application of this method results in a 'surface' of ANC consisting of a number of ANC duration curves for changing GWP which may be contoured and used to assess the current stream water quality situation, for example, the percentage of streamlength that ANC is above $0\mu eql^{-1}$ for 90% of the time.

4. Future scenarios

Future scenarios of temporal variability may be predicted by considering the chemical changes in the end-members through time under given pollutant deposition and forest scenarios. MAGIC is used to calculate changes in an upper and lower soil layer, whose chemistry is considered to represent the two end-members. A future scenario is run using methods outlined by Jenkins *et al.* (1997) assuming given future pollutant deposition and land-use scenarios. EMMA is then undertaken for the future scenario using the new end-member chemistries, assuming that hydrological conditions do not alter over time and that the standard deviation of end-members remains constant during the period of change. Normal distributions are used to characterise the 'future' end-member compositions using the new mean values and previous standard deviations. The ground-water proportion duration curve obtained previously is used to produce new ANC duration curves by selecting values from the two end-member distributions as before. A new ANC surface may then be plotted.

5. Model application to Plynlimon, Wales

The upper Severn catchments in central Wales, UK are located 24km from the Irish Sea and have an area of 1925 ha and altitude of 320 to 740m. The geology is undivided mudstones, shales and grits of upper Ordovician and Silurian age and mineralogy is dominated by iron magnesium chlorite, mica and quartz, with localised calcite veins. Soils are peats, stagnopodzols, acid brown earths and stagnogleys with locally derived drift deposits and land-use consists of plantation forestry of Sitka Spruce (*Picea sitchensis*) and Norway Spruce (*Picea abies*) and acidic moorland (Robson, 1993). Weekly stream chemistry, including high flow, has been monitored at 10 sites on first to third order streams for up to 15 years by CEH, the Centre for Ecology and Hydrology. Continuous flow and pH data are available for the Hafren (Robson, 1993) and spot-sampled low flow chemistry at 40 sites from summer 1996 (Foster *et al.*, 2000). Continuous ANC is calculated from continuous pH (Robson, 1993).

Streamwater chemistry at Plynlimon varies greatly with ANC ranging from --113.8 to 4.7μeql^{-1} at high flow and -93.2 to 217.1μeql^{-1} at low flow. The mean average ANC end-member values are -65.4μeql^{-1} and 37.8μeql^{-1} for high and low flow respectively. The low flow samples are linearly extrapolated between sites so that a value is calculated for each 25m reach, which enables these values to be considered to represent the groundwater end-member variability along the stream length of the catchment. Given the high heterogeneity of the low flow data, and other groundwater chemistry measurements from boreholes drilled at Plynlimon (Hill and Neal, 1997), there is uncertainty associated with such downscaling, but it is used here to demonstrate the potential of the technique. There are too few high flow samples to allow extrapolation. The data are approximated by normal distributions and the Monte-Carlo simulation of

EMMA run and the results are plotted (Figure 1). Stream ANC is above $0\mu eql^{-1}$ for 90% of the time along 10% of the streamlength of the catchment.

For the future scenarios, stream chemistry data are not available at the Severn catchment outlet and so the MAGIC model is calibrated to the Hafren, with an upper soil ANC of –64.6 and a lower soil of $38.7\mu eql^{-1}$. For future scenario A it is assumed that the forest trees mature and are felled at 50 years of age without replanting and for B that replanting takes place immediately. For pollution deposition, adherence to the Second Sulphur Protocol is assumed. The forest felling and decrease in sulphate deposition are found to significantly increase the ANC of both stream and soil, with an upper layer ANC of $29.1\mu eql^{-1}$(A) and $-1.8\mu eql^{-1}$(B) and a lower layer of $87.5\mu eql^{-1}$(A) and $115.2\mu eql^{-1}$(B). The Monte-Carlo EMMA simulation is then undertaken again.

The reduction in deposition improves the water quality greatly with stream ANC exceeded for 90% of the time along 55% of streamlength under scenario A and for 90% of the time along 85% of streamlength under B, compared to 90% of the time along 10% of streamlength previously (Figure 1).

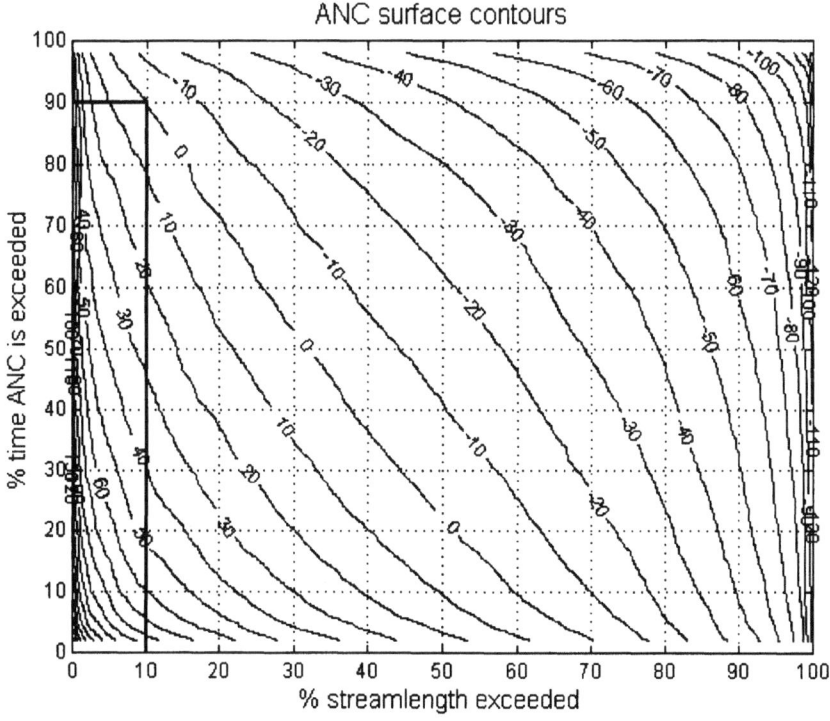

Figure 1. ANC surface contoured as a function of time and streamlength ANC is exceeded.

6. Conclusions

The modelling undertaken is simple, and yet is powerful in that it can predict the variability of stream chemistry both spatially and temporally over both the long (years) and short (daily) term. Because the method links EMMA to a process-based model that describes the relationship between acid deposition and end member chemistry, the combined methodology may be used to underpin future policy decisions. MAGIC has already been used extensively to calculate critical loads to predict the deposition reduction required to meet a mean stream chemistry ANC. The method presented here may be used to produce, by iteration, a risk-based critical load which will maintain the stream ANC above 0 for 90% of the time along 90% of the catchment streamlength (for example), rather than just calculating a mean annual value. This improvement begins to address some of the concerns about current policy decision making, which do not account for sub-catchment spatial variability.

Acknowledgements

Research was supported by UK NERC Environmental Diagnostics grant GST/02/1572, and NERC research studentship GT4/96/166/F.

References

Champan, P.J., Reynolds, B., and Wheater, H.S.: 1997, *Hydrology and Earth System Sciences* **1**(3), 671.
Christophersen, N., Neal, C., Hooper, R. P., Vogt, R. D. and Andersen, S.: 1990, *Journal of Hydrology* **116**, 307.
Cosby, B.J., Wright, R.F., Hornberger, G.M. and Galloway, J.N.: 1985, *Water Resources Research* **21**, 51.
Foster, H.J., Alexander, S., Locks, T., Wheater, H.S., Lees, M.J. and Reynolds, B.: 1997, *Hydroogy and Earth System Sciences* **1**(3), 639.
Foster, H.J., Reynolds, B., Lees, M.J., Wheater, H.S. and Locks, T.J.: 2000, *Journal of Hydrology* (submitted).
Harriman, R., Likens, G.E., Hultberg, H. and Neal, C.: 1994, 'Influence of management practices in catchments on freshwater acidification', in C.E.W. Steinberg and R.F. Wright (eds), *Acidification of freshwater ecosystems*, Wiley, pp83-101.
Jenkins, A., Renshaw, M., Helliwell, R., Sefton, C., Ferrier, R. and Swingewood, P.: 1997, *Modelling surface water acidification in the UK*. IH Report 131, Wallingford, UK.
Neal, C., Robson, A.J., Reynolds, B., and Jenkins, A.: 1992, *Journal of Hydrology* **130**, 87.
Nilsson, J. and Grennfelt, P.: 1988, *Critical loads for S and N*. Nordic Council of Ministers, Copenhagen.
Ormerod, S.J. and Jenkins, A.: 1994, 'The biological effects of acid episodes', in C.E.W. Steinberg and R.F. Wright (eds), *Acidification of freshwater ecosystems*, Wiley, pp259-272.
Reynolds, B., Neal, C., Hornung, M., and Stevens, P.A.: 1986, *Journal of Hydrology* **87**, 167.
Robson, A.J.: 1993, Doctor of Philosophy Dissertation, University of Lancaster, UK.
Wheater, H.S., Kleissen, F.M., Beck, M.B., Tuck, S., Jenkins, A. and Harriman, R.: 1990, 'Modelling short-term flow and chemical response in the Allt a' Mharcaidh catchment', in B.J. Mason (ed), *The Surface Waters Acidification Programme*, Cambridge University Press, pp477-483.

INTEGRATED MODELLING OF ACIDIFICATION EFFECTS TO FOREST ECOSYSTEMS – MODEL SONOX

WOJCIECH MILL

*Institute of Environmental Protection – Section of Integrated Modeling,
Grunwaldzka Str. 7B/2, 41-106 Siemianowice Sl., Poland
(author for correspondence, e-mail: mill@silesia.top.pl)*

Abstract. The model SONOX has been developed that replicates the sequence of events on the way from emission of sulfur and nitrogen to the potential risk to forest ecosystems as reflected by critical loads exceedance. The model produces a set of emission, deposition and critical load maps in various spatial resolutions as well as maps of critical loads exceedances dynamically generated in response to various sulfur and nitrogen emission scenarios. Optionally, also the share of forest areas protected or unprotected against acidification is spatially presented. The model provides a decision tool to develop strategies aimed at the abatement of excess sulfur and nitrogen emissions. The present version of the model has been tailored to and implemented for Polish conditions. This model has been used to analyse the trends in impacts of acid deposition on Polish forest ecosystems.

Keywords: forest ecosystems, critical loads, anthropogenic acidification, environmental modeling, Poland

1. Introduction

European research over the last 20 years into the environmental effects caused by emission of sulphur and nitrogen compounds to the atmosphere has provided quantitative estimates of the effects of acidification on forest ecosystems. This has given a basis for the development of effect based emission control policies. The dominant effect of sulfur and nitrogen emissions on forest ecosystems is acidification. If an excess of acid deposition occurs, leading to significant modification of the forest soil chemical status, forest health is potentially at risk. There are many documented examples of forest damages manifested as discoloration, defoliation and dieback especially in Central Europe caused by airborne acid pollution. Under the Convention on Long-Range Transboundary Air Pollution a number of protocols have been worked out setting SO_2, NO_x and NH_3 emission ceilings for all parties to the Convention. To establish a scientific background for negotiations of these protocols integrated models have been widely used. By use of them the most cost efficient emission abatement scenarios meeting the accepted ecological targets have been derived. These encouraging examples of using models to support negotiations on an international scale invited some countries to implement the integrated assessment modelling in the development of emission reduction programs on a national scale as well for sulfur and nitrogen compounds aimed at the fulfilment of the Protocol obligations in the most effective way. Following this idea the SONOX model has been developed for Poland. SONOX is limited just to quantification of ecological effects by means of critical loads exceedance. The emission control and economic aspects are not considered in the model.

2. Model characteristics

The model SONOX was developed to support decision-making in air quality management and forest ecosystems protection against anthropogenic acidification. In particular the model may assist the development of national emission abatement programs and their implementation. The SONOX model replicates the sequence from emission of sulfur and nitrogen through their atmospheric transport and deposition and finally to the examination of the exceedance of deposition tolerable by forest ecosystems – Figure 1. Basically SONOX has been designed to determine ecological effects of acid deposition from Polish emission sources. However, it also provides the inclusion of transboundary fluxes and simulates the acidifying effects from particular European countries to the territory of Poland. A one year time scale has been adopted in the model for calculations of the present and future acidification status.

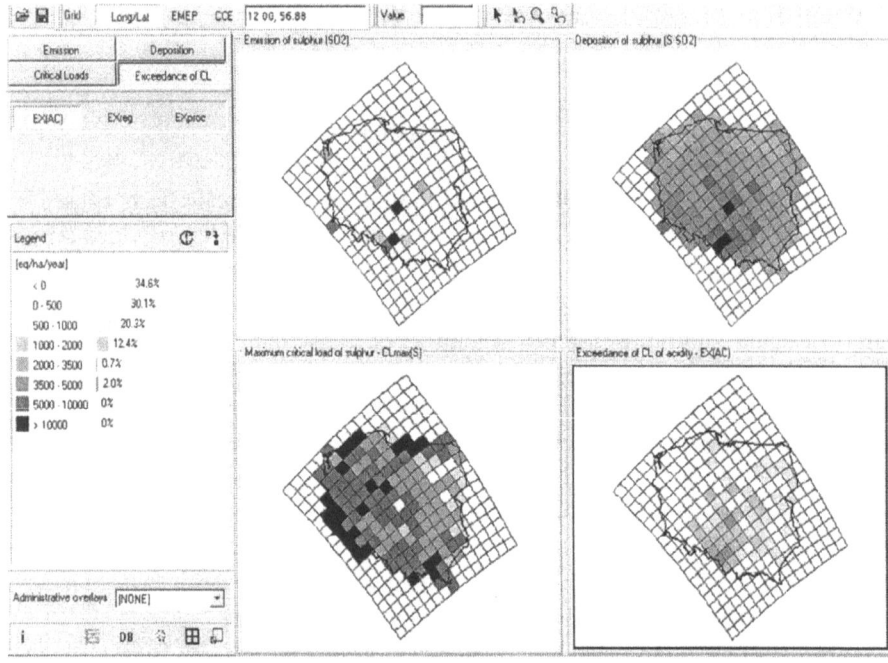

Figure 1. Illustration of the SONOX model concept

3. Model structure

Integrated models do integrate inside their structure a number of submodels simulating particular contributing processes, therefore they usually have a modular structure. Integration of these modules by the main model consists in organisation of input and output data flow among the modules according to a determined order. The elements of the model structure and the system of inter modular linkages is presented in Figure 2.

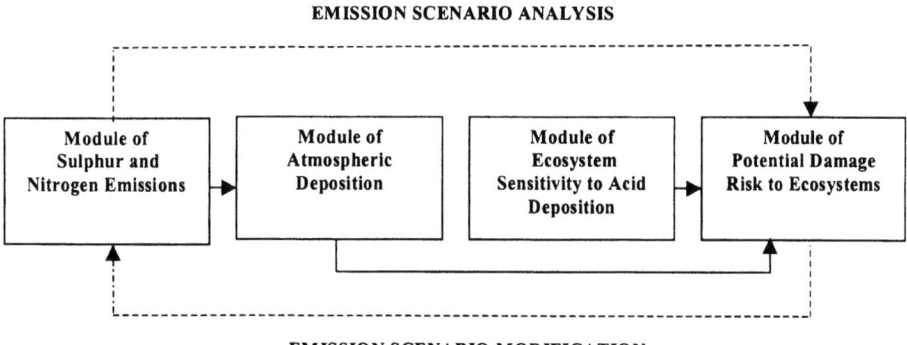

Figure 2. Diagram of the SONOX structure

3.1. MODULE OF SULFUR AND NITROGEN EMISSIONS

This module includes, in form of databases, official scenarios of sulfur and nitrogen compounds emissions, developed by the UN/ECE Co-operative program for monitoring and evaluation of the long range transmission of air pollutants in Europe (EMEP), as well as developed by the user. Emission data have been given in the 50 x 50 km spatial distribution. Each scenario includes SO_2, NO_x and NH_3 emission data for selected European countries including Poland. The emission module is equipped with the appropriate software allowing to manage the databases contents.

3.2. MODULE OF ATMOSPHERIC DEPOSITION

This module has been developed in co-operation with experts from EMEP (Olendrzynski and Bartnicki, 1998). It includes the so called "source-receptor matrices" where the source denotes emitting region and the receptors are the EMEP grid cells covering Poland's territory. Owing to these matrices a quick simulation of atmospheric transport and deposition of emitted pollutants in spatial resolution of 50 x 50 km is possible. This kind of matrices have been developed for seven main emitting regions referring to four largest power plants (4 EMEP grid cells) and three large industrial and/or urban centers (6 EMEP grid cells) of Poland. The eighth emission region was the "background" i.e. all Polish emissions except the above seven regions (154 grid cells – see Figure 3). To simulate transboundary fluxes, transfer matrixes for sulfur and nitrogen have been developed to evaluate the deposition from the following selected European countries: Germany, Czech Republic, Byelorussia, Hungary, Ukraine, Russian Federation (Kola-Karelia), France, U. K. and the rest of Europe onto Poland.

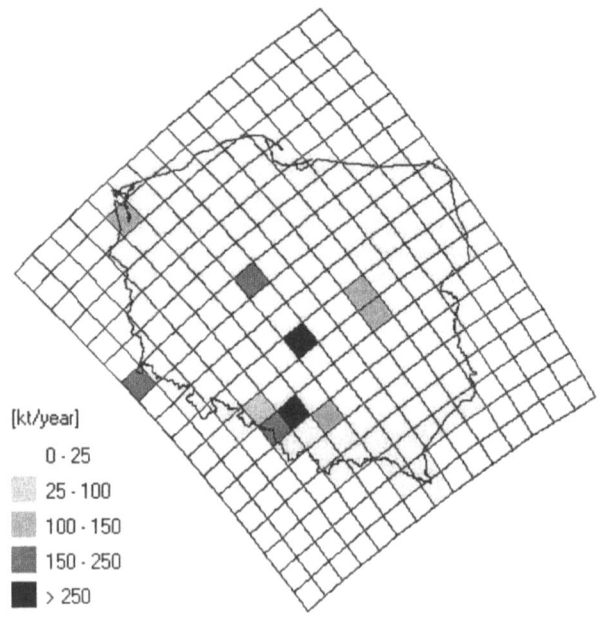

Figure 3. Distribution of the emission regions based on SO$_2$ emission in 1990

3.3. MODULE OF ECOSYSTEM SENSITIVITY TO ACID DEPOSITION

This module consists of a database of critical loads which are considered as a measure of ecosystems sensitivity to acid deposition. Critical loads determine such an amount of sulfur and nitrogen deposition which, if exceeded on a long term scale, can generate hazard to forests health (Nilsson and Grennfelt, 1988). The mathematical form of critical loads is a „critical loads function" defining an area within which every pair of sulfur and nitrogen deposition values does not induce a potential risk to functioning and structure of the considered ecosystems. Critical loads functions of acidity are of trapezoid shape and are defined by three parameters $CL_{max}(S)$, $CL_{min}(N)$ and $CL_{max}(N)$ (UBA, 1996). These critical loads parameters have been calculated and mapped for 1461 forest monitoring sites (Mill, 1997) which are installed and serviced by the Forest Research Institute within the forest monitoring program (Wawrzoniak et al., 1997). Experimental data used for calculations have been delivered from this monitoring. Basing on these point data a 5 percentile value of relevant critical load parameter has been derived for each EMEP grid cell.

3.4. MODULE OF POTENTIAL DAMAGE RISK TO ECOSYSTEMS

The amount of exceedance of tolerable sulfur and nitrogen deposition defined as critical load is a measure of damage risk to forest ecosystems. Accordingly to critical loads function, their exceedances are represented as a combination of pairs of sulfur and nitrogen deposition. This module provides quantitative information on the exceedance of critical acid deposition as well as indicates the basic directions for strategy of necessary emission reductions. The module produces maps of Poland, partitioned into grid cells of 150×150 km, 50×50 km and 10×10 km size for which values of critical loads exceedances, percentage of ecosystems protected against acidification and reduction requirements for sulfur and nitrogen deposition are given.

4. Estimation of ecological effects of sulfur and nitrogen emission reduction according to the obligations of II Sulfur Protocol – model implementation

The model has been implemented to estimate the potential damage risk of acidification on the basis of the acid deposition rate to forest ecosystems in Poland, that was predicted based on the official SO_2 emission scenarios for 1980, 1990 and 1996 as well as projections for 2000, 2005 and 2010 (Mill, 2000). The sum of exceedance values over the forested area was used as criterion of risk. The percentage of 50×50 km sized grids where the protected area lies in the each of the four ranges (0-30%, 31-60%, 61-80% and >80%) were also calculated. The results presented in Table I were obtained using SO_2 emission scenarios determined by the II Sulfur Protocol.

TABLE I

Yearly exceedance and percentages in EMEP grid cells of Poland in which protected ecosystems cover 0-30 %, 31-60 %, 61-80 % and >80 % based on the II Sulfur Protocol scenario of SO_2 emission.

Year	Total exceedance of critical loads [eq/yr]	Percentage of grid cells with ecosystem protection ranges:			
		0 – 30 %	31 – 60 %	61 – 80 %	>80 %
1980	126880	49.7	20.3	5.9	24.1
1990	69963	41.8	15.0	5.2	38.0
1996	24315	13.7	11.1	5.2	68.6
2000	31548	17.7	13.7	4.6	64.1
2005	22827	12.4	10.5	4.6	69.3
2010	14716	9.9	4.6	4.6	79.1

Table 1 shows a considerable progress in mitigation of ecological effects of emission of acidogenic pollutants in Poland in the period 1980-1996. A considerable 45% reduction of exceedance of critical load of acidity till 1990 and 80% reduction till 1996, compared to exceedance of critical load estimated for 1980 has been achieved.

The number of 50 x 50 km spatial grids, in which over 80% of ecosystems, i.e. elementary forest areas were protected against excessive acid deposition, have been growing clearly in the period of 1980-1996. It has been changing from about 40 (24% of 164 grids for whole Poland) to about 113. This growing tendency is accompanied by the opposite tendency relating to the number of these grids in which only up to 30% of the ecosystems is protected.

The exceedance of critical loads of acidity from 1996 indicates that in the period beginning from the basic year (1990) the expected improvement in forest ecosystems protection against acidification was observed. This tendency shows that there is a real possibility to achieve in the coming years the protection level required by the II Sulfur Protocol obligations for Poland.

In the development process of national strategy for sulfur emission reduction numerous emission scenarios may be considered due to technological, economic and legal targets. Each of them, should be analysed in view of ecological consequences, because even if a given scenario meets all the above mentioned targets, its execution may not remove the risk to ecosystems in areas of particular protection e.g. nature reserves, national parks etc., or even transfer this risk from other places onto these areas. Under such conditions the decision-makers will have to make choices according to objective criteria. Such criteria can be provided by this kind of modeling approach.

References

Mill, W.: 1997, National Focal Center Report No.4 in Posch M., J-P. Hettelingh , P.A.M.de Smet, R.Downing,., *Calculation and Mapping of Critical Thresholds in Europe*: CCE Status Report No.4, RIVM Report No.259101007, Bilthoven, The Netherlands.

Mill W.: 2000, *Analysis of the national program of SO_2 emission reduction in view of potential risk to forest ecosystems,* Manuscript, Institute of Environmental Protection, Warsaw.

Nilsson J., P.Grennfelt., editors: 1988, *Critical loads for sulfur and nitrogen-Report from a workshop held at Skokloster, Sweden 19-24 March,1989,page 418 pp.* UN/ECE and Nordic Council of Ministers, Copenhagen.

Olendrzynski K., Bartnicki J.: 1998, EMEP data on sulfur and nitrogen emission-deposition for Poland, Manuscript, Oslo.

UBA: 1996, Manual on methodologies for mapping critical loads/levels and geographical areas where they are exceeded, Umweltbundesamt, Berlin.

Wawrzoniak, J., Małachowska J., Wójcik J. and Liwińska A.: 1997, *Stan uszkodzenia lasów w Polsce w 1995 roku na podstawie badań monitoringowych,* (Report on *forest damages in Poland in 1996 based on monitoring research),* Biblioteka Monitoringu Środowiska, Warsaw.

PREDICTION OF SOIL ACIDIFICATION USING A DYMANIC MODEL AT A BAMBOO FOREST IN OSAKA PREFECTURE

HU LI[1]*, AKIKAZU KAGA[2] and KATSUHITO YAMAGUCHI[3]

[1] *Acid Deposition and Oxidant Research Center, 314-1 Sowa, Niigata, 950-2144 Japan;* [2] *Department of Environmental Engineering, Osaka University, 2-1 Yamadaoka, Suita, 565-0871 Japan;* [3] *Department of Global Engineering, Osaka University, 2-1 Yamadaoka, Suita, 565-0871 Japan*

(* author for correspondence, e-mail: lihu@cocoa.ocn.ne.jp)

Abstract. This study is based on research of throughfall formation and soil chemistry processes. Through the experimental verification of the cation exchange part of the simulation model; making use of results of the flux of acid deposition on the forest and the chemical weathering of soil mineral, we predicted the future soil acidification. The result indicated that chemical weathering occupies the important portion of acid neutralization capacity of the soil and that significant soil acidification will not occur in this field within 40 years, even if the present acid load continues in the future.

Keywords: prediction model, dynamic model, soil acidification, acid deposition, precipitation data

1. Introduction

In the companion paper (Kaga *et al.*, 1998) of this study, the estimation of acid deposition using a multiple regression model at a bamboo forest had been described. Another companion paper (Takano *et al.*, 1997) demonstrates the estimation of chemical weathering of soil mineral using both experimental and modeling methods. In this paper, we present results of the prediction of future soil acidification at a bamboo forest in Osaka Prefecture by a dynamic model.

This dynamic, multi-layer soil model calculates the change of the chemical characteristics of soils towards depth as well as the soil water flux in a vertical direction. Prior to carrying out the final model runs, the prediction quality of the cation exchange part in the soil model was verified by means of a 20 days percolation experiment (Li, 1997).

Using the results of these papers and the other necessary data as the input, we predicted the future soil acidification at the forested field.

2. Structure and characteristics of the model

Our simulation model is composed of the multiple-tank model that simulates the throughfall formation process of bamboo canopy, and the multiple-layer soil model that simulates the percolation process of the soil water. The throughfall formation process includes generally wet deposition, dry deposition, canopy exudation and leaching. The percolating process of soil water includes the processes of the cation exchange, the chemical weathering, the uptake of roots, the decomposition of organic matter and the evaporation of moisture (Li, 1997).

The multiple-tank sub-model described in the paper (Kaga et al., 1998), approximates the throughfall formation process, and describes the relation of the leaching parameters (e.g. particulate deposition, foliar exudation) and the chemical components in throughfall as simple algebraic expressions. Meteorological variables such as precipitation, temperature and radiation serve as input into this sub-model, and thus affect the hydraulic process in soil. The calculated results were summarized as annual input data.

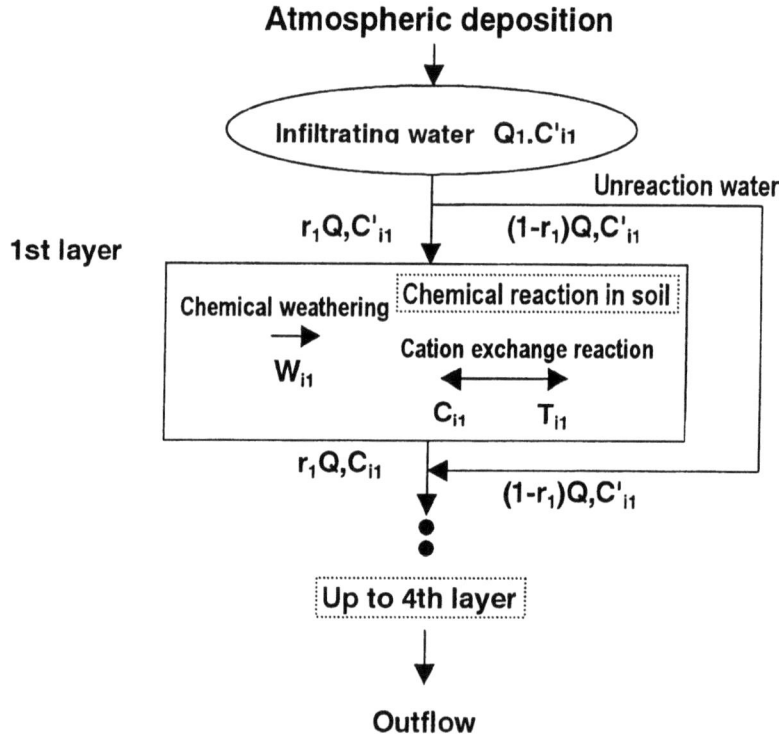

Figure 1. The structure of the multiple-layer soil model

In the multiple-layer soil model, we assumed four layers of 5 cm thickness each, because the soil characteristics change with depth and the effects of acid deposition are more important in the topsoil. The structure of the multiple-layer soil model is shown in Figure 1. Q_j indicate the flow rate of infiltrating water. C'_{ij} and C_{ij} are the concentration of ion **i** in infiltrating water to **j** layer; and that reacted in soil water in **j** layer, respectively. Only the part of the infiltrating water whose ratio is r_j was assumed to react within the soil layer **j** and remaining water percolated from the layer without any reactions. We assumed that the value of the ratio r_j was 0.6, which was determined by calibration with the model. T_{ij} is the quantity of exchangeable cation i in j layer and W_{ij} denotes the chemical weathering rate of ion i in the layer j. In the model, concentrations of nine ions: aluminum (Al^{3+}), calcium (Ca^{2+}), magnesium (Mg^{2+}), potassium (K^+), sodium (Na^+), ammonium (NH_4^+), sulfate (SO_4^{2-}), nitrate (NO_3^-), chloride (Cl^-) and the their complex ions in the aqueous solution are treated individually. The dependence of Al^{3+} concentration on pH is estimated from the observation result. The MAGIC model (Cosby et al., 1985) was referenced regarding the chemical process of soil water.

For the sake of estimating the parameters of chemical weathering rate of soil minerals, a one-layer soil sub-model was developed (Takano et al., 1997). The chemical weathering rate is described as a simple rate formula in which soil is treated as containing one kind of material. The weathering rates estimated with the one-layer soil sub-model were incorporated into the multiple-layer soil model.

3. The verification of experiment and the calculated results

3.1. COMPARISON OF THE EXPERIMENTAL RESULTS AND THE SIMULATIONS

Regarding processes involved in soil acidification model, only the ion exchange was experimentally verified. Soil samples from a selected soil profile were used for a percolation experiment with artificially acidified rainwater. The pH of the artificial rainwater was 3.0 and the concentration of SO_4^{2-}, Cl^- and NO_3^- was 0.5, 0.3 and 0.2 meq/l, respectively. We prepared the five kinds of soil columns for the experiment using soil samples that were collected from each of the successive five layers of soil profile. We filled the first layer's soil in the first column, and soils from the first to the jth layers into the jth column (j = 2 to 5) in order. These column are named as "1L", "1,2L", "1~3L", "1~4L", and "1~5L", respectively. Soil weight filled in each column was 50 grams. The artificial rainwater was continuously dropped into these soil columns at the rate of 10ml/h, separately. The percolated water were collected continuously and analyzed every 4 to 12 hours for pH and concentrations of nine ions (Al^{3+}, Ca^{2+}, Mg^{2+}, K^+, Na^+, NH_4^+, SO_4^{2-}, NO_3^- and Cl^-).

A simulation model in which only the cation exchange reaction was implemented was applied to these experiments. For the model application, soil properties such as CEC, exchangeable cation concentrations, selectivity coefficient were determined in the laboratory. The simulation results on the change of pH and ionic concentration in percolated water were compared with the measurements, as shown in Figure 2 and Figure 3.

The figures show that the changes of pH and cation concentrations in the course of time were comparable. From the convergence of experimental and simulation results we conclude that cation exchange is modeled correctly and that model results are reliable.

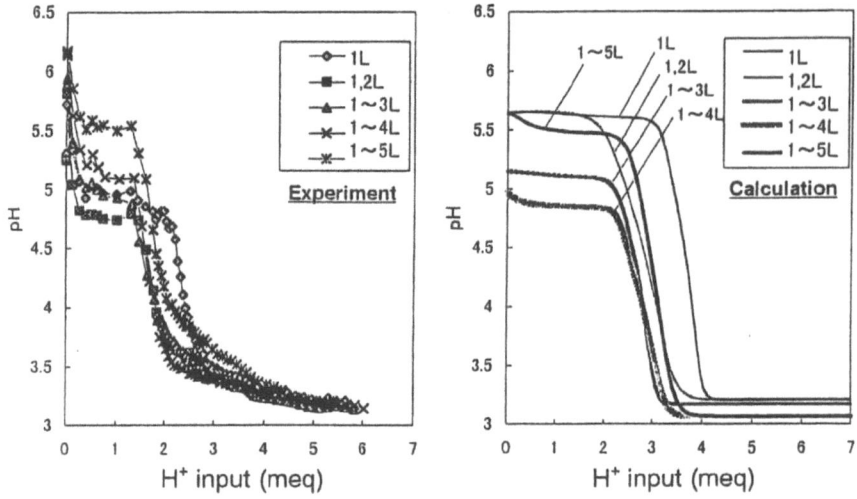

Figure 2. The comparison of the pH change in the experiment and the result of calculation in five columns

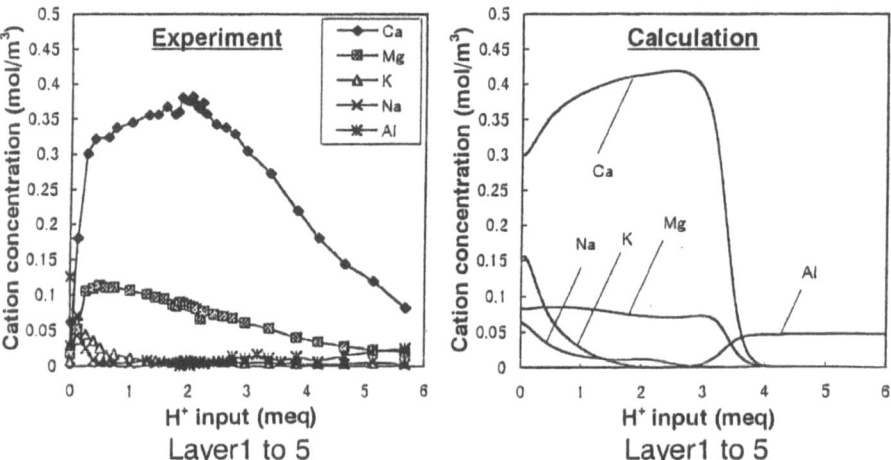

Figure 3. The comparison of the cation concentration change in the experiment and the result of calculation regarding fifth column (five kinds of 10 g soil from five soil layers)

3.2. PREDICTING SOIL ACIDIFICATION IN A BAMBOO FOREST

The dynamic model described in the section 2 was used to simulate the soil acidification of a bamboo forest in Osaka Prefecture, Japan for the period of 170 years from 1868 to 2038. The rates of acidic deposition in the past were derived based on the literature data with the methods described in the companion papers (Tomita *et al.*, 1998). As for the future, we assumed that the deposition rate would be the same as the current rate observed for the next 40 years.

Soil solution pH decreased from the beginning of the predicted period in all soil layers simultaneously, particularly sharp during times of intense industrial development in the 1950's and 1960's (Fig. 4). After recovering to pre-industrial values during the 1980's, i.e. 4 to 4.5, the soil solution pH will only slightly decrease in the future. One might speculate that a relationship may exist between the declining concentrations of alkaline fly ash in local ambient air and declining in soil solution pH in the middle of the 1980's. Base cation (BC) to aluminum molar ratio increased until 1980 particularly in the topsoil layer (Fig. 5). For the future we predict a moderate decrease of Bc/Al molar ratio, the range of values, however, still being markedly above the critical limit for bamboo forest ecosystems (Sverdrup and Warfvinge, 1993).

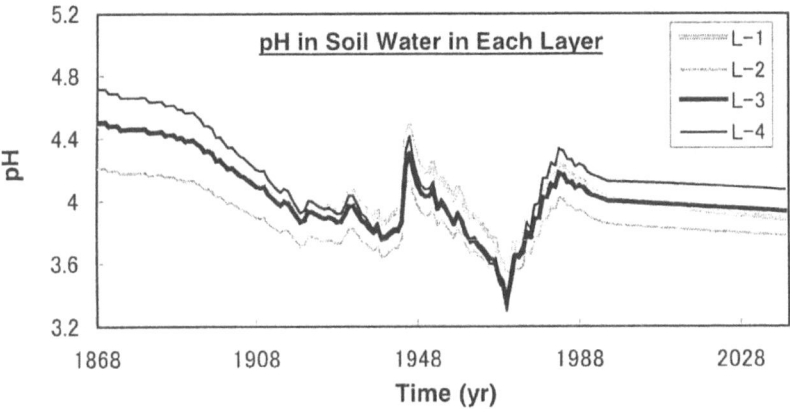

Figure 4. Predicted pH in soil water in each layer

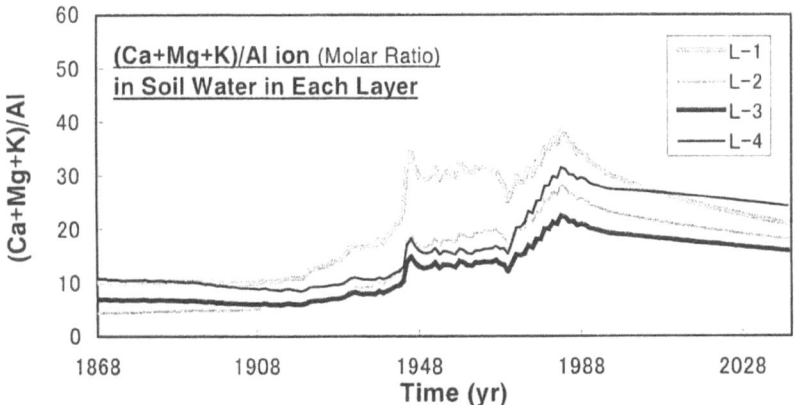

Figure 5. Predicted (Ca + Mg + K) / Al (molar ratio) in soil water in each layer

4. Conclusions

In this study, we predicted future acidification of a brown forest soil using a dynamic soil model and site-specific input data. Decomposition of organic matter due to microorganism activity and the quantification of nitrogen cycle is planned to be included in future versions of the model. We are planning to verify modeled components concentrations in the soil water in the field, and the model is intended to be applied to other soil-vegetation combinations.

References

Cosby, B. J., Hornberger, G. M., Galloway, J. N. and Wright R. F.: 1985, *Water Resources Research*, **1**, 51.

Kaga, A., Li, H., Yamaguchi, K., Takano, M. and Ogawa, K.: 1998, *J. Jpn. Soc. Atmos. Environ.* **34**(2), 65

Li, H.: 1997, Ph D Dissertation, Osaka University, Suita, Osaka. (in Japanese)

Sverdrup, H. and Warfvinge, P.: 1993, *The Effect of Soil Acidification on the Growth of Trees, Grass and Herbs as Expressed by the (Ca+Mg+K)/Al Ratio*, Dept. of Chemical Engineering II, Lund University, Lund, Sweden.

Takano, M., Kaga, A., Li, H., Yamaguchi, K., Tsuruta, T. and Muratsu, M.: 1997, *J. Jpn. Environ. Sci.* **10**(4), 287. (in Japanese)

Tomita, K., Yamaguchi, K. and Kaga, A.: 1998, 'Research concerning Prediction Model of Soil and Inland Water (7)', in *Academic Presentation Proceeding of Japan Society for Air Harmony and Sanitary Engineering Kinki Branch*, Osaka. pp. 51-54 (in Japanese)

MAGIC MODELING OF LONG-TERM LAKE WATER AND SOIL CHEMISTRY AT ABBORRTRÄSKET, NORTHERN SWEDEN

PAVEL KRÁM[*,1], HJALMAR LAUDON[1,2], KEVIN BISHOP[1], LARS RAPP[1] and JAKUB HRUŠKA[3]

[1]*Dept. of Environmental Assessment, Swedish Univ. of Agric. Sci., SE-750 07 Uppsala, Sweden;*
[2]*Dept. of Forest Ecology, Swedish Univ. of Agric. Sci., SE-901 83 Umeå, Sweden;* [3]*Dept. of Environ. Geochemistry, Czech Geol. Survey, Klárov 3, 118 21 Prague 1, Czech Republic*
[*]*(author for correspondence, e-mail: pavel.kram@ma.slu.se)*

Abstract. The geochemical model MAGIC has been applied to the Abborrträsket lake catchment in northern Sweden for the period 1843-2000. The two objectives were to 1.) simulate historical biogeochemical fluxes and pools and 2.) test whether the MAGIC model of biogeochemical cycling contradicts the published diatom record of a relatively stable pH (around 6) during the last two centuries in this weakly buffered, acid sensitive lake. Abborrträsket has received elevated sulfur and nitrogen deposition in the second half of the 20th Century and had a large part of its catchment clearcut in 1975. The MAGIC simulation of very small pH decline from 6.1 to 5.9 between 1843 and 1987 was comparable to the published diatom reconstructions of almost stable lake water pH up until the lake was limed in 1988. MAGIC also simulated the modern soil and water chemistry, including lake liming. Thus the diatom indication of stable pH cannot be dismissed as necessarily incorrect.

Keywords: acidification, catchment, diatoms, dynamic modeling, lake, MAGIC, pH, Sweden

1. Introduction

There is a concern that naturally acidic waters in northern Sweden have been limed in the mistaken belief that they were significantly acidified by acidic deposition (Bishop, 1997). Almost four thousand lakes in the two most northern Swedish counties (Norrbotten and Västerbotten) were classified as anthropogenically acidified (Bernes, 1991). This statement was undermined by paleoecological studies of more than one hundred lakes in northern Sweden which showed no regional, diatom-inferred pattern of acidification since the pre-industrial period (Korsman, 1999). We applied the MAGIC model to an acid sensitive lake (Abborrträsket) from that region to simulate biogeochemical patterns and compare our modeling to diatom indications of relatively stable pH in this lake (Korsman, 1999).

2. Materials and Methods

2.1. MODEL DESCRIPTION

MAGIC (Model of Acidification of Groundwater in Catchments) was designed to reconstruct past and predict future drainage water as well as soil chemistry (Cosby et al., 1985). The MAGIC model uses a lumped representation of biogeochemical processes because understanding the runoff chemistry may not require detailed knowledge of the spatial distribution of the parameters within a catchment. Water fluxes, atmospheric deposition, net vegetation uptake, weathering, and descriptions of organic acids are required as external inputs. This model has been used to simulate a large number of forest catchments. Examples of MAGIC model applications for European lakes are Wright *et al.* (1986), Jenkins *et al.* (1990), Whitehead *et al.* (1997) and Hinderer *et al.* (1998).

2.2. SITE AND INPUT DATA DESCRIPTION

Lake Abborrträsket (63°42'N, 19°16' E) is situated in the county of Västerbotten, 30 km inland from Sweden's eastern, Baltic Sea coast. It is a small, clear-water lake with maximum depth of 10.2 m. The area of the lake is 9 ha which is almost 20% of the area of the whole catchment (46 ha). The elevation range of the catchment is 239-275 m. The lake is used for sport-fishing and has been limed regularly since 1988 (Korsman, 1999). Lime powder (<0.2 mm) has been dispersed on the ice in winter (Karlsson, pers. com.). The catchment is underlain by migmatite, and glacial till covers most of the catchment. The site was selected to be free of sulfur-rich, unconsolidated marine sediments which are common along Sweden's eastern coast below the highest post-glacial shoreline. About 30% of the catchment is covered by an uneven-aged natural forest composed of Norway spruce (*Picea abies*), Scots pine (*Pinus sylvestris*) and European birch *(Betula verrucosa)*. Young trees of the same composition cover the remaining 70% of the catchment, which was clearcut in 1975.

Lake sediment cores at Abborrträsket were taken in 1995 (Korsman, 1999). Diatom-inferred pH decreased from 6.50 in the bottom of the core (320 cm depth, from more than 3100 years ago) to 5.85 (at a depth of 20 cm, around year 1800). The sample from 10 cm depth (ca 1955) indicated pH 5.97. In the uppermost part of the sediment, partially representing the liming, reconstructed pH increased to 6.12 (Korsman, 1999). The liming signal (and other short term changes) are "blurred" by bioturbation in the sediments, which make the diatom record approximate a 30 year running average (Korsman, pers. com.).

Forest soils were sampled in 1998. Mean depth of the O horizon was 9 cm and the E and B horizons together had a mean thickness of 24 cm. Estimated C horizon thickness was 17 cm (Table 1). Exchangeable base cations (BC: Ca^{2+}, Mg^{2+}, Na^+, K^+) were determined by NH_4Cl extraction, exchangeable acidity by KCl extraction. Cation exchange capacity (CEC) was computed as the sum of the exchangeable BC and acidity. Results were weighted to the measured thickness of soil horizons and estimated bulk density. Lumped soil base saturation (BS) at Abborrträsket was 7.9% with the following proportions (Ca^{2+}: 5.5%, Mg^{2+}: 1.2%, Na^+: 0.11%, and K^+: 1.1%). Partial pressure of CO_2 in the lake water was calculated using annual mean temperature. An empirical model of organic acid dissociation for Sweden was used to set pK values (Köhler *et al.*, 1999). Weathering rates were fitted in the model calibration.

TABLE I

Selected parameter values for the MAGIC model calibration at Abborrträsket catchment.

Parameter	Unit	Value
Soil depth	m	0.5
Soil bulk density	kg m^{-3}	1070
Soil cation exchange capacity (CEC)	meq kg^{-1}	36.27
Soil sulfate adsorption half saturation	meq m^{-3}	47.0
Soil sulfate adsorption maximum capacity	meq kg^{-1}	12.0
Partial pressure of CO_2 in soil water	atm	0.0050
Partial pressure of CO_2 in lake water	atm	0.0039
Organic acids (triprotic) in lake water	meq m^{-3}	21.0
pK_1 of organic acids	-log K_1	3.6
pK_2 of organic acids	-log K_2	4.2
pK_3 of organic acids	-log K_3	5.5
Mean soil and lake temperature	°C	2.5
Precipitation volume	m yr^{-1}	0.750
Lake fraction of the whole catchment area	unitless	0.195
Lake retention time	yr^{-1}	2.0
Whole catchment discharge	m yr^{-1}	0.400
Water bypassing soil (direct precipitation on lake)	m yr^{-1}	0.078
Water volume participating in soil reactions	m yr^{-1}	0.322
Weathering rate of Ca^{2+}	meq m^{-2} yr^{-1}	32.0
Weathering rate of Mg^{2+}	meq m^{-2} yr^{-1}	26.5
Weathering rate of Na^+	meq m^{-2} yr^{-1}	1.4
Weathering rate of K^+	meq m^{-2} yr^{-1}	1.5
Weathering rate of SO_4^{2-}	meq m^{-2} yr^{-1}	3.5
Tree net uptake of Ca^{2+} in 1843-1974	meq m^{-2} yr^{-1}	2.4
Tree net uptake of Mg^{2+} in 1843-1974	meq m^{-2} yr^{-1}	0.4
Tree net uptake of K^+ in 1843-1974	meq m^{-2} yr^{-1}	0.3

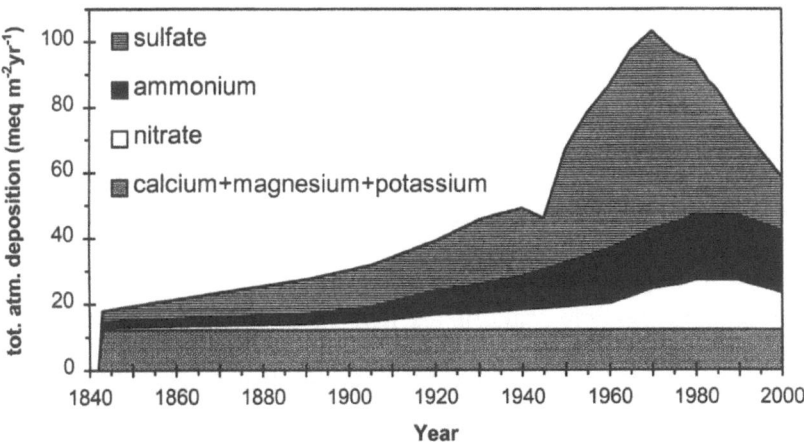

Figure 1. Scenario of historical total atmospheric inputs of SO_4^{2-}, NH_4^+, NO_3^- and the nutrient base cations (Ca^{2+}, Mg^{2+}, K^+) used in modeling Abborrträsket 1843-2000.

Prior to the large clearcut in 1975, the Abborrträsket catchment had its original forest. Net uptake of BC was set to low values due to low intensity harvesting 1834-1974. Removal of about 0.5% of timber per year was reported, and spruce was harvested more than pine or birch. An assumption of the living tree biomass of 5 kg m^{-2} in steady-state was used. Concentrations in bole wood of spruce from Lövliden (NE from Abborrträsket; Ca 0.19%, Mg 0.02%, K 0.05%; Alveteg, pers. com.) were used for an approximation of the pools in the whole tree biomass in the catchment. Empirical curves for spruce net uptake were used for the period of intense new forest growth (1979-2000).

Past sulfate, nitrate and ammonium deposition (Fig. 1) were derived from regional time series (Mylona, 1996, Mylona, pers. com., Alveteg, pers. com.) and measurements from the nearest available long-term measurements of wet deposition (Westling, pers. com.). Other major chemical constituents of precipitation (BC and Cl) were kept constant throughout the simulation at the levels observed at Docksta. The lake water chemistry for the calibration year 1983 came from several measurements of pH, alkalinity and color in the early and mid 1980's before the lake was limed (Länsstyrelsen, unpubl. data). The proportion of ions was estimated from correlation to empirical models of water chemistry in northern Sweden based on a large regional data set (Köhler *et al.*, 1999).

3. Results and Discussion

The MAGIC model was applied in a hindcast mode (1843-1983) and calibrated for the reference year 1983 at Abborrträsket. The calibration process of the lake water composition was conducted step-wise in the following order: chloride, sulfate, fluoride, nitrate, ammonium, calcium, magnesium, sodium and potassium. Preliminary calibration of soil exchangeable BC in the soil was conducted in the same order. The model was then run in the predictive mode (1984-2000) to examine the soil chemistry change between 1983 and 1998, because the soil data are available only from a sampling in 1998. The whole calibration procedure was repeated until the match between the observed and simulated lake water chemistry in 1983 and soil chemistry in 1998 was obtained. The difference between observed and simulated values for the evaluated anions and cations in the corresponding years always lay between ± 0.1 µeq L^{-1} in the lake water 1983 and ± 0.1% in the soil 1998 (Fig. 2).

After the calibration, MAGIC was run again in the predictive mode (1984-2000) to explore the time scale of the lake water liming "recovery" and the hypothetical situation without liming (Fig. 2). The lake liming simulation (using two tons of dolomitic lime annually) was compared to the observed lake water data (1998) for the model evaluation. The difference between observed and simulated concentrations of individual cations and anions in lake water for the evaluation year 1998 lay between ± 0.7 µeq L^{-1} in lake water. Only sulfate was over-predicted more (by 2.9 µeq L^{-1}), but this represents a negligible difference

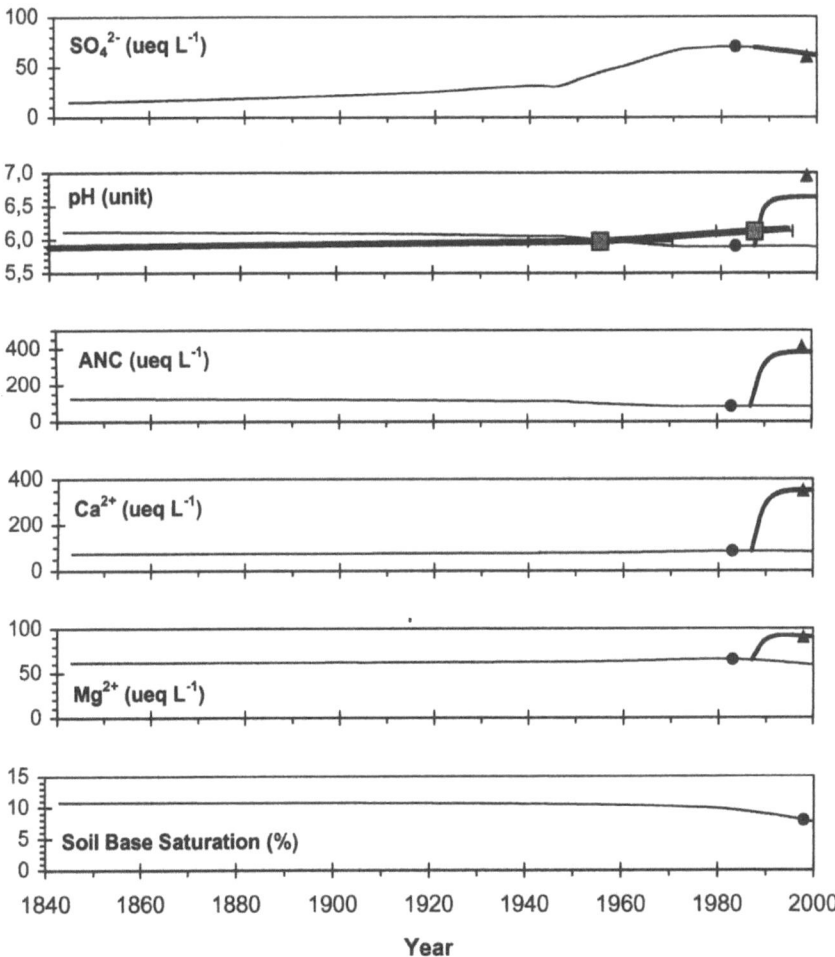

Figure 2. Results of the MAGIC simulations at Abborrträsket for the lake water sulfate, pH, acid neutralizing capacity, calcium and magnesium and the soil base saturation in 1843-2000 (lines). Comparison of the simulated values with the observed lake water and soil chemistry (circles and triangles). Lake water composition in 1988-2000 is in two versions. The first simulation (thick lines) include the actual lake liming in 1988-2000, the second one without liming (thin lines) is hypothetical. Diatom inferred pH values (squares) were obtained from Korsman (1999). Uncertainties in the age estimate of lake sediments and their bioturbation are represented by the "error bars". The very thick line connects the individual diatom-inferred estimates.

(5%). Increases of pH and acid neutralizing capacity (ANC) caused by liming were under-predicted by 0.31 pH units, and by 29.5 µeq L^{-1}, respectively. MAGIC simulated a stable lake water pH between 1843 and 1890 (pH 6.12-6.1) (Fig. 2). A slight pH decline from 6.04 to 5.9 was modeled in response to increased S deposition (1948-1970). In the hypothetical case without liming the pH stabilized at 5.9 in 1972-1998. According to our simulation, the sulfate concentration was 14.9 µeq L^{-1} in 1843. The sulfate increase was relatively slow until 1947 (31.8 µeq L^{-1}) but much faster between 1948 and 1972 when the sulfate concentration had reached 67.4 µeq L^{-1}. After 1982 MAGIC predicted a slow decrease of lake sulfate. The simulation results showed a relatively steady soil BS (10.8% in 1843-1902), slight decline (from 10.7 % in 1903 to 10.5% in 1954) and increasingly faster decline in the remaining period (10.2% in 1970, 9.9% in 1980, 9.0% in 1990, 7.7% in 2000) (Fig. 2). The simulation of soil BS suggests that most of the decline is due to the incorporation of BC into tree biomass due to fast growth of the new forest on 70% of the terrestrial catchment.

4. Conclusions

This paper represents the first MAGIC modeling results from northern Sweden. Thus it is one of the first attempts to determine whether the relatively stable lake pH's indicated by paleolimnological methods can be consistent with a biogeochemical model of how acid deposition has affected the forest ecosystem at a time when forest harvesting also applies an acidification pressure. It appears that anthropogenic atmospheric deposition and forest growth can be consistent with the slight changes in lake water pH indicated by the diatom record.

Acknowledgements

Support was provided by the Swedish EPA, MISTRA and a post-doctoral stipend from SLU Dept. of Environ. Assessment (for PK). M. Alveteg, S. Mylona, O. Westling and T. Korsman kindly provided unpublished data.

References

Bernes, C.: 1991, *Acidification and liming of Swedish freshwaters.* Monitor 12, Swedish Environmental Protection Agency, Solna.
Bishop, K.H.: 1997, *Transactions of the Institute of British Geographers* **22**, 49.
Cosby, B.J., Hornberger, G.M., Galloway, J.N. and Wright, R.F.: 1985, *Water Resources Research* **21**, 51.
Hinderer, M., Jüttner, I., Winkler, R., Steiberg, C.E.W. and Kettrup, A.: 1998, *The Science of the Total Environment* **218**, 113.
Jenkins, A., Whitehead, P.G., Cosby, B.J. and Birks, H.J.B.: 1990, *Philosophical Transactions of the Royal Society of London,* **B 327**, 435.
Köhler, S., Hruška, J. and Bishop, K.: 1999, *Canadian Journal of Fisheries and Aquatic Sciences* **56**, 1461.
Korsman, T.: 1999, *Journal of Paleolimnology* **22**, 1.
Mylona, S.: 1996, *Tellus* **48 B**, 662.
Whitehead, P.G., Barlow, J., Haworth, E.Y. and Adamson, J.K.: 1997, *Hydrology and Earth System Sciences* **1**, 197.
Wright, R.F., Cosby, B.J., Hornberger, G.M. and Galloway, J.N.: 1986, *Water, Air, and Soil Pollution* **30**, 367.

MONITORING ACID WATERS IN THE UK: 1988-1998 TRENDS

D.T. MONTEITH[1]*, C. D. EVANS[2] and S. PATRICK [1]

[1] *Environmental Change Research Centre, University College London, UK.;*
[2] *Centre for Ecology and Hydrology, Wallingford, UK*
(*author for correspondence, e-mail: dmonteit@geog.ucl.ac.uk)

Abstract. Since 1988, a network of lakes and streams has been monitored in areas of the UK sensitive to surface water acidification. Analysis of 10 years data has focused on the identification and quantification of time-trends in chemical parameters, to establish whether declines in emission of acidifying pollutants have resulted in recovery of acidified surface waters. A national decline in S deposition in the UK since 1988 has not generally been accompanied by a significant improvement in freshwater chemistry. At the three sites where xSO_4 concentrations have declined, NO_3 has increased and there has been no increase in pH or alkalinity. Upward trends in pH and alkalinity observed at several other sites are not associated with downward trends in acidic anions. Temporal variation in xSO_4, NO_3, acidity, DOC and other important meaures of surface water quality can all be linked to decadal-scale variation in climate, and this has important implications for the detection of recovery-related trends.

Keywords: acidification, UK, chemical trends, recovery, climate, NAO Index

1. Introduction

Since the mid-1980s a series of protocols, signed under the auspices of the United Nations Economic Commission for Europe (UNECE), has led to a significant reduction in emissions of S across Europe. Between 1986 and 1997, S emissions from the UK have declined by approximately 57% while those from the rest of UNECE Europe have declined by 45%. Over the same period there has also been a reduction of 14% in emissions of oxides of nitrogen (NO_x) from UK sources (Fowler and Smith, 2000).

The chemistry of atmospheric deposition has been monitored by the UK Acid Deposition Monitoring Network since 1986. Recently Fowler and Smith (2000) undertook trend analyses of data from this Network over the period 1988-1997. Large declines in the non-marine sulphate (xSO_4) concentration of wet deposition (3 µeq l^{-1} yr^{-1}) were detected in areas close to emissions sources, i.e. central and southeast England. Smaller but significant declines (1 µeq l^{-1}) were found at intermediate distances from sources. Similar rates of decline were also estimated for H$^+$ ion concentration in these regions. There were slight reductions in S deposition in the acid sensitive, high rainfall, west coast areas, but these were much smaller than the reductions in S emissions, and trends in S deposition at individual sites were not statistically significant. There was little indication of trends in NO_x deposition, although small increases in concentrations and deposition fluxes of NO_3 were found at west coast and other remote monitoring stations.

The United Kingdom Acid Waters Monitoring Network (UKAWMN) was established in 1988 in order to determine the impact of anticipated emission reductions. The UKAWMN consists of 11 lakes and 11 streams (Fig. 1), all situated on acid sensitive lithologies.

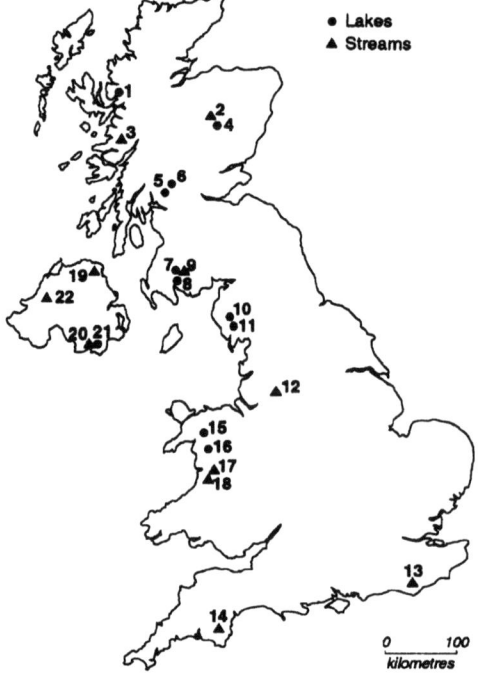

1. Loch Coire nan Arr
2. Allt a'Mharcaidh
3. Allt na Coire nan Con
4. Lochnagar
5. Loch Chon
6. Loch Tinker
7. Round Loch of Glenhead
8. Loch Grannoch
9. Dargall Lane
10. Scoat Tarn
11. Burnmoor Tarn
12. River Etherow
13. Old Lodge
14. Narrator Brook
15. Llyn Llagi
16. Llyn Cwm Mynach
17. Afon Hafren
18. Afon Gwy
19. Beagh's Burn
20. Bencrom River
21. Blue Lough
22. Coneyglen Burn

Figure 1. Location of UKAWMN sites

Diatom-based pH reconstructions from sediment cores taken at the onset of the monitoring period indicate that at least nine of the lakes have acidified over the last 150 years as a result of acid pollutant deposition (Patrick *et al.*, 1995). Most UKAWMN streams are also in an acidified state. All sites have been monitored chemically and biologically (epilithic diatoms, aquatic macrophytes, macroinvertebrates and salmonids). This paper summarises the results of a time trend testing exercise for water chemistry only, conducted on the first ten years of data (1988-1998) and reported by Monteith and Evans (2000).

2. Methodology

A suite of chemical determinands is measured in samples taken monthly from streams and quarterly from lake outflows. Changes with time in chemical determinands have been assessed using linear regression, and the Seasonal Kendall Test (SKT) as modified by Hirsch and Slack (1984). Variation in climatic conditions over the UK has been assessed with reference to the North Atlantic Oscillation Index (NAOI). The NAOI has been shown to correlate strongly with rainfall and temperature patterns over the British Isles, particularly over the winter-spring period (Hurrel, 1995). Periods of high NAOI are associated with stormy, westerly conditions, with higher than average rainfall and air temperature; negative NAOI periods are typically cold, dry and northeasterly dominated. Winter values for the NAOI were particularly high in the late 1980s and early 1990s, fell to negative values in 1996 and remained low in 1997. Over the decade of our analysis therefore, there has been a general trend of declining winter-spring storminess, rainfall and air temperature and this is particularly apparent in the western UK.

3. Results

TABLE I

Significant temporal trends ($p<0.05$) in chemical determinands and their annual slopes determined by linear regression or by the Seasonal Kendall test (denoted by *) when only the latter is significant.

UKAWMN Site no.	xSO_4 ($\mu eq\ l^{-1}\ yr$)	NO_3 ($\mu eq\ l^{-1}\ yr$)	pH (units yr^{-1})	labile Al ($\mu g\ l^{-1}\ yr^{-1}$)	alkalinity ($\mu eq\ l^{-1}\ yr$)	Cl ($\mu eq\ l^{-1}\ yr$)	DOC ($mg\ l^{-1}\ yr^{-1}$)
1							0.2
2							0.1
3						-7.7*	0.2
4	-1.0	1.4	-0.02				0.1
5		0.9	0.04	-2.7	-1.0*		0.2
6							0.2*
7		0.9		-2.6*		-9.0	0.1
8						-9.0	0.2
9				-2.2	0.9	-6.6	0.1
10			0.02	-6.9			0.1
11						-6.2	0.1
12	-8.0	2.1					0.5
13	-8.8	0.7		-6.0			0.4
14	1.0		0.05	2.9	1.8	-3.6	
15			0.05				
16						-10.1	
17						-2.9	0.1
18			0.05				0.1
19				-0.9			0.7
20		1.0					0.2
21			0.02		1.2		0.2
22							0.8

Results of trend analysis for certain key determinands are presented in Table I.

Large declines in non-marine SO_4 (xSO_4) concentrations have been recorded at UKAWMN sites in the 'close-to-source' regions, i.e. at Old Lodge (site 13) (southeast England) and the River Etherow (site 12) (Pennines). A smaller decline has occurred at Lochnagar (site 4) (northeast Scotland). However, nineteen of the twenty-two UKAWMN sites have shown no significant downward trend in xSO_4 concentration.

Significant increases in NO_3 concentration have been identified at six sites. NO_3 concentrations at Lochnagar appear to have undergone a step-change increase, which reflects a transition from a seasonal pattern to year-round leaching.

Upward trends in pH, and/or alkalinity, have been identified at seven sites. This is surprising, since there is no indication of downward trends in acid anion concentrations at most of these. Owing to the absence of this link we cannot attribute these trends to long term chemical recovery from acidification at this stage. At the three sites with declining xSO_4 concentrations, NO_3 has increased, while at Old Lodge and the River Etherow base cation concentrations have decreased. These increases in the 'secondary' acid anion and reductions in base cations appear so far to have offset any influence of xSO_4 decreases on acidity.

Time-trends have been identified for Cl, base cations, pH and alkalinity at several west coast sites. However, time-series plots indicate cyclical rather than linear temporal patterns in many cases. At these sites concentrations of Cl, largely derived from seasalt inputs, generally peak around 1990, and decline to a trough in 1996. At the same west coast sites, cyclical variation is also apparent in the concentration of non-marine base cations and Acid Neutralising Capacity (ANC, derived from ΣBase cations - ΣAcid Anions).

Significant increases in Dissolved Organic Carbon (DOC) concentrations have occurred at nineteen sites and a subset of these also exhibits increases in non-labile Al concentration.

4. Discussion and Conclusion

Evidence for the impact of large reductions in European and UK S emissions on UK acidified waters over the 1988-1998 period is mostly confined to the sites in southeast England (Old Lodge) and the Pennines (River Etherow), where deposition reductions have been strongest. It would seem that the more remote regions of the country, in which the majority of acidified freshwater systems are situated, have not benefited proportionally from the large national decline in S deposition. The precise period encompassed by the UKAWMN may also be important in explaining the wider absence of strong trends in acid deposition and surface water chemistry. Longer term records for freshwater sites in Wales and Scotland suggest there were substantial reductions in xSO_4 concentration and concomitant improvements in pH, in the early to mid-1980s (Monteith and Evans, 2000).

It is perhaps surprising that a signal of declining xSO_4 has not been detected for sites at intermediate distances (*i.e.* 100-200 km) from major sources, since small but detectable declines in xSO_4 deposition have been detected here. Catchment soils in these areas may currently act as a source of long-term stored S. However, it would also appear that climatic influences, and particularly the shift in storminess over the monitoring period, have confused interpretation of many expected chemical signals of recovery. The statistical disentangling of "recovery" and climate signals will be limited until persistently high NAOI conditions, similar to those experienced in the early part of the record, return to the UK. The findings of the 1988-1998 analysis, summarised below, should be of great value in this respect.

At stream sites with west-coast locations, Cl concentration during the winter months is positively correlated with the NAOI for the preceding month. This relationship appears to result from increased inputs of seasalts during stormy conditions. The overall decline in the winter NAOI over the decade can therefore account for the decline in Cl at many west coast sites.

Evans *et al.* (in press) have proposed that errors may have arisen in the estimate of anthropogenically derived SO_4 (i.e. xSO_4) in freshwaters, as a result of seasalt influences on catchment storage of SO_4. Although most soils in the UK are believed to be fully saturated with SO_4 in the long term (Jenkins et al., 1977), adsorption capacity is a function of concentration and also acidity, which increases the positive charge of sesqui-oxide surfaces (Johnson and Cole, 1980; Curtin and Seyers, 1990; Harriman et al. 1995). Water enriched in both SO_4 and H^+ following seasalt events (Langan, 1989) may therefore temporarily increase SO_4 adsorption capacity. Chloride on the other hand behaves conservatively, passing rapidly into the drainage network. Since a fixed SO_4:Cl ratio (based on sea water constitution) is used to estimate the proportion of total SO_4 which is derived from marine and non-marine sources, "non-marine" SO_4 is likely to be underestimated during periods of high seasalt deposition and overestimated when inputs are low. The general trend of declining seasalt inputs over the decade may, therefore, have masked small real reductions in anthropogenic SO_4 at some west coast sites. Interestingly Lochnagar, the only relatively remote lake to receive low seasalt inputs, does exhibit a significant decline in xSO_4.

Given the problems in estimating xSO4, it is possible that the amelioration in acidity detected at some sites could reflect a real response to declining, although as yet undetected, acid anion inputs. However, linear trends observed in base cations, non-marine (i.e. sea-salt corrected) base cations at several sites and, in some cases, hydrogen ion and labile aluminium, can also be linked to climate, since in many cases these determinands are correlated with Cl concentration. It has been proposed that temporal variation in these determinands is largely driven by ion displacement from catchment soil ion-exchange sites by cations deposited in seasalt (Evans et al., in press). In addition, inter-annual variation in pH and alkalinity may have been influenced by variations in rainfall, which determines the proportional contributions of relatively buffered baseflow and more acidic surface and subsurface flow, and which co-varies to some extent with seasalt deposition. Linear trends in base cations, pH and alkalinity may therefore primarily reflect a response to a decline in winter storminess, from the early 1990s to the end of the interpretative period, rather than any decline in acid deposition.

Synchronous patterns of inter-annual variation in NO_3 have been identified across most of the Network. This variation is inversely correlated with winter values for the NAOI, suggesting that low winter temperatures may lead to enhanced spring leaching (Monteith et al., 2000). The upward trends identified at six sites appear independent of this variability, providing tentative evidence for increasing N saturation of their catchment soils. At Lochnagar, the increase in NO_3 has been larger than the decline in SO_4 and this site has continued to acidify.

The remarkable spatial consistency in DOC increases suggests a regional scale trend in UK upland freshwaters. Mechanisms are unclear at this stage, although it is possible these result from an increase in microbial decomposition of soil organic matter during a recent series of relatively warm summers in the UK.

Inter-annual variation in the chemistry of acid sensitive waters, in regions of the UK where recent declines in acid deposition have at most been slight, may therefore have been dominated by climatic factors over the 1988-1998 period. In order that recovery trends may be confidentally assessed and quantified, time-series should be substantially longer than the frequency of variation of the most influential climatic factors. Since the NAOI has varied on a decadal scale in recent decades the length of the UKAWMN dataset has yet to meet, although may be approaching, this requirement. In future 'global warming' may further complicate recovery assessment as an increase in storminess for the UK, predicted by some climate models (*e.g.* Hulme and Jenkins, 1998), could for example increase the frequency and intensity of base cation dilution and seasalt episodes. However, as the dataset continues to lengthen, it should become possible to quantify variance related to climatic effects, allowing for more sensitive detection of underlying recovery responses, if, when and where they occur.

Acknowledgements

The UKAWMN is funded by the Department of Environment Transport and the Regions (DETR), and the Environment and Heritage Service Northern Ireland. The authors wish to thank the numerous organisations and individuals who have contributed to sample collation and analysis over the last decade. NAOI data were obtained from the University of East Anglia, Climatic Research Unit internet site (http://www/cru.uea.ac.uk/cru/data/nao.htm).

References

Curtin, D. and Seyers, J.K.: 1990, *Journal of Soil Science* **41**, 433.
Evans, C.D., Monteith, D.T. and Harriman, R.: In press, *Science of the Total Environment*.
Fowler, D. and Smith, R.I.: 2000, 'Spatial and temporal variability in the deposition of acidifying species in the UK between 1986 and 1997', in D.T. Monteith and C.D.Evans (eds.*), UK Acid Waters Monitoring Network 10 Ten Year Report. Analysis and Interpretation of Results April 1988-March 1998*. ENSIS, London.
Langan, S.J.: 1989, *Hydrological Processes* **3**, 25.
Harriman, R., Anderson, H. and Miller, J.D.: 1995, *Water Air and Soil Pollution* **85**, 553.
Hulme, M. and Jenkins, G.J. *Climate change scenarios for the UK*: UKCIP Technical Report No. 1, Climatic Research Unit, Norwich.
Hirsch, R.M. and Slack, J.R.: 1984, *Water Resources Research* **20**, 727.
Hurrel, J.W.: 1995, *Science* **269**, 676.
Jenkins, A., Ferrier, R. and Cosby, B.J.: 1997, *Journal of Hydrology* **197**, 111.
Johnson, D.W. and Cole, D.W.: 1980, *Environment International* **3**, 76.
Langan, S.J.: 1989, *Hydrological Processes* **3**, 393.
Monteith, D.T. and Evans, C.D.: 2000. *UK Acid Waters Monitoring Network 10 Ten Year Report. Analysis and Interpretation of Results April 1988-March 1998*. ENSIS, London.
Monteith, D.T., Evans, C.D. and Reynolds, B.: 2000, *Hydrological Processes*. **14**, 1745.
Patrick, S.T., Monteith, D.T. and Jenkins, A. 1995, *UK Acid Waters Monitoring Network: The First Five Years. Analysis and Interpretation of Results April 1988-March 1993*. ENSIS, London.

PLANKTONIC AND LITTORAL MICROCRUSTACEANS AS INDICES OF RECOVERY IN LIMED LAKES IN SE NORWAY

BJØRN WALSENG[1*] and LEIF R. KARLSEN[2]

[1] Norwegian Institute for Nature Research (NINA) P.O. Box 736 Sentrum, N-0105 Oslo, Norway;
[2] The County Governor of Østfold, P.O. Box 325, 1502 Moss, Norway.
(*author for correspondence, e-mail: bjorn.walseng@ninaosl.ninaniku.no)

Abstract. A two year study of planktonic and littoral microcrustaceans (Cladocera and Copepoda) from 15 lakes in the southeastern part of Norway, Østfold county, document the recovery of acidified lakes due to liming. Six lakes that where limed about 10 years ago, seven acid and two neutral reference lakes, were sampled twice a year (1998 and 1999). One acid lake was limed in autumn 1998. Qualitative nethaul samples from the deepest part of the lake and from the most frequent habitat in the littoral zone were used. The limed lakes had a species composition which indicates that these lakes are about to recover. Species associated with neutral lakes dominates while acid-tolerant species were rare. The acid-sensitive species, *Daphnia longispina* and *D. cristata*, were found in the limed lakes. This study shows the usefulness of a low-cost sampling program where microcrustaceans are used as bioindicators of recovery.

Keywords: acidification, liming, Cladocera, Copepoda, zooplankton, littoral zone, recovery

1. Introduction

Reduced crustacean plankton diversity caused by acidification is well documented from the southern parts of Norway (Drabløs and Tollan, 1980; Muniz, 1991). In the last decades, however, great efforts have been used to de-acidify lakes and entire river systems. Effects of liming on planktonic crustaceans are well documented from both Sweden (Degerman *et al.*, 1995) and Canada (Keller *et al.*, 1992; Yan *et al.*, 1996). In Canada both natural recovery and recovery after liming have been documented by using plankton species as indicators. In Sudbury, Ontario, where the Canadian studies have taken place, plankton communities consist of twice as many species as in lakes in Norway, and the number of sensitive species is about three times as high. The use of only plankton species as indices of recovery in Norwegian lakes would therefore rely on a few species.

In Norway 112 out of 133 species are littoral or planktonic/littoral (Schartau *et al.*, 1997). In the River Rore (Aust-Agder, Norway), a seven-year study has shown that these groups are well suited to indicate the recovery of lakes following liming. Also in Lake 302S, Canada, the composition and temporal patterns of microcrustaceans were described both during acidification and during recovery (Hann and Turner, 2000). By using littoral microcrustaceans the number of available species is 3-4 times higher and we know that for many of these species, sensitivity to pH is comparable to planktonic crustaceans.

Studies of the recovery of microcrustaceans both in Lake 302S (Hann and Turner, 2000) and in Killarney, Canada, (Schartau and Walseng, *in press*), where the community composition has been analysed by ordination techniques, have shown that pH explains most of the variance along the first ordination axis. Although changes in crustacean species composition due to liming may be caused by both abiotic and biotic factors (Svensson *et al.*, 1995), we presume that the main effect on species composition is directly and/or indirectly through pH.

The main purpose of our investigation was to evaluate recovery in six lakes, which have been limed regularly for about 10 years, using both planktonic and littoral microcrustaceans (Cladocera and Copepoda) as indicators. Our hypothesis was therefore; if we assume that pH has been relatively stable since liming started, a recovery should be reflected in a change in species composition, ending up with communities dominated by species normally associated with circum-neutral lakes.

2. Materials and methods

All lakes in this survey are situated in Østfold county. Size of the lakes varies between 0.03 km^2 and 13.6 km^2 (Table I).

Four lakes, Bredtjenn, Tvetervatn, Ravnsjøen and Store Lysern, are included in a national monitoring programme following natural recovery, which includes lakes that are not limed or will not be limed in the future. They serve as acid reference lakes in our study. Another humic acidic lake, Trollbergvatn, was limed during the study. Two more lakes, Setervatn and Damvannet, which are not limed, were included because liming of these lakes are questioned. The lakes Lundebyvannet and Rømsjøen represent the near neutral situation since they are not severely affected by acidification.

TABLE I

Characteristic data of the investigated lakes, pH and number of species. * mean 1999.

lake	Position (UTM) East	West	Elevation m a.s.l.	Surface km2	Status	pH prior to liming	pH average 1998-1999	number of species
Bredtjenn	653324	6555881	190	0,28	acid		4,8	24
Tvetervatn	628072	6569168	79	1,2	acid		5,4	24
Ravnsjøen	613546	6586658	82	0,33	acid		5,3	23
Store Lysern	618617	6618636	162	6,82	acid		5,5	33
Damvannet	649150	6615188	193	0,25	acid		5,4	33
Setertjern	662069	6629619	284	0,05	acid		5,1	29
Stensvannet	642556	6600039	189	0,44	limed	5,2	6,1	36
Kulevatn	641903	6597665	190	0,3	limed	5,0	6,0	35
Trollbergtjern	643527	6603257	188	0,03	limed	4,7	5.5*	29
Langvannet	653197	6616239	214	1,3	limed	5,5	6,7	41
Vortungen	653961	6624618	214	1,9	limed	5,4	6,4	40
Sundvatnet	660731	6630081	290	0,16	limed	4,7	6,3	39
Hølvatnet	663391	6629747	248	1,6	limed	5,0	6,9	32
Rømsjøen	659234	6621782	137	13,6	neutral	5,8	6,7	42
Lundebyvanne	640262	6603894	158	0,5	neutral		6,4	38

The latter has been limed a few times when pH was just below 6.0. Both lakes are characterised as mesotrophic lakes surrounded by marine deposits. The remaining six lakes, distributed in the central and northern part of the study area, are limed.

Liming of the investigated lakes started in 1985 when Lake Langvatnet and Lake Kulevatnet were limed. Liming of the Lakes Stensvannet, Vortungen, Hølvatnet and Sundvatnet, started in the years 1987-90. Lake Trollbergtjern was limed in autumn 1998.

The study is based on samples from 15 lakes in 1998 and 1999 (Table I). Samples were taken in summer (May/June) and in autumn (September). Altogether 120 qualitative samples have been analysed. Qualitative samples were taken by a nethaul (90 microns) and samples were taken from the deepest part of the lake (planktonic) and from the most frequent habitat in the littoral zone. Samples were preserved in diluted formaldehyde (4%).

Water samples from each visit were analysed for pH. Previous information about pH after liming indicates that pH has been quite stable except for a drop to about 4.7 in the lakes Kulevatnet and Stensvannet in 1987.

A species and site data matrix from the years 1998 and 1999 (15 lakes * 2 years) based on recording of presence/absence of species is used as input in a multivariate analysis of the dataset. Patterns in the distributions of crustaceans for the different sites were summarised by Detrended Correspondence Analysis (Hill, 1979; Hill and Gauch, 1980), using the program CANOCO with downweighting of rare species (ter Braak, 1987, 1990). The ordination results are presented in a diagram where the sites are represented by points along axes in a two-dimensional space. The diagram is a graphical summary of the data, and the axes can be interpreted as underlying environmental gradients.

3. Results

The pH in the two neutral reference lakes, Lundebyvannet and Rømsjøen, was around 6.5 during the study (Table I). pH in the acid reference lake, Lake Bredtjenn, was stable below 5.0, while in the remaining acid reference lakes, Tvetervatn, Ravnsjøen, and St Lysern, pH varied between 5.0 and 5.6. pH in the limed lakes has varied between 5.8 and 7.0. In Lake Trollbergtjern, which was limed in winter 1999, pH was 4.7 and 4.8 in May and September 1998, respectively. In 1999, after liming, the respective values were 5.6 and 5.3. Lake Damvannet (mean pH 5.4) and Lake Setertjern (mean pH 5.1) were quite acidic.

Altogether 69 species of microcrustaceans, 45 cladocerans and 24 copepods were found. The number of species in each lake varied between 23 (18 cladocerans and 5 copepods) and 42 (28 cladocerans and 14 copepods) (Table I). In the acidic reference lakes, Bredtjenn, Tvetervatn and Ravnsjøen, 24, 23 an 24 species were found, respectively. The highest number of species, 42, was found in Lake Rømsjøen.

The cladocerans *Alona rustica* and *Acantholeberis curvirostris*, species associated with acidic lakes, were found in 10 and 9 lakes, respectively. Both were lacking in the two neutral reference lakes and in the limed lakes, Vortungen, Sundvatnet and Hølvatnet. Among species associated with

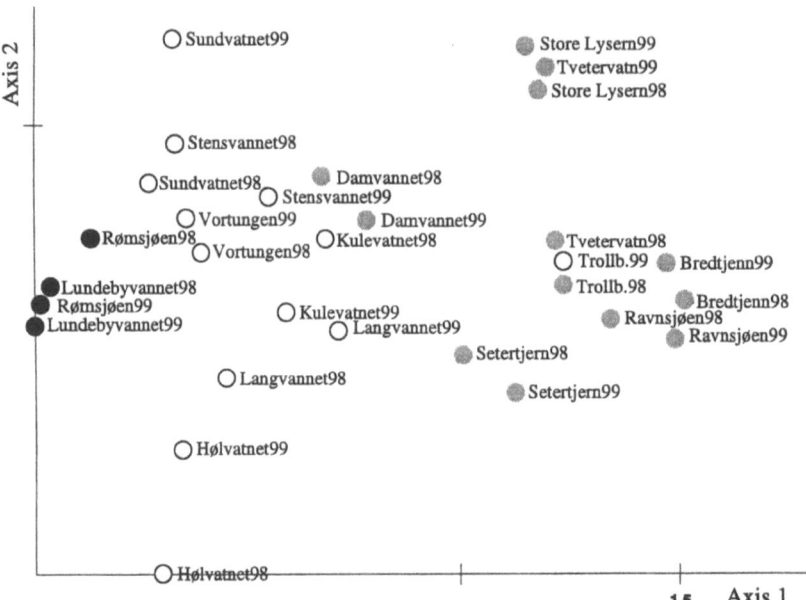

Figure 1. DCA-ordination sample plot based on qualitative (planktonic and littoral) samples. Each year is represented by one species list (combined early summer and autumn samples). Points for the acid lakes are encircled with grey colour, the near neutral reference lakes are shown as dark circles while limed lakes are shown as open circles.

neutral lakes, the cladocerans *Ceriodaphnia pulchella* and *Bosmina longirostris* occurred, in 6 and 5 of the limed/neutral lakes. Except for Setertjern, *Daphnia longispina* and *D. cristata* were only found in the limed lakes and in the neutral reference lakes.

A DCA plot based on yearly species lists showed that sample plots representing the four acid reference lakes, Bredtjenn, Ravnsjøen, Tvetervatn and St Lysern, are grouped to the right on axis 1, while plots representing the neutral reference lakes, Lundebyvannet and Rømsjøen, are found at the opposite end of axis 1 (Figure 1). Sample plots representing the six limed lakes are distributed between the reference lakes. These lakes, however, vary more in their position along axis 2. The two lakes were liming was questioned, Lake Setertjern and Lake Damvatnet, are situated between the acid reference lakes and the limed lakes. The position of Lake Trollbergtjern did not change from 1998 to 1999 though it was limed. Since the length of axis 1 was about 1.5, a principal component analysis (PCA) was also used in case this method would give a better fit. The deviation from our DCA plot was small and the PCA plot is not presented here.

Bivariate regression shows that the contribution from pH to the observed variance along axis 1 is strong, $r^2=0.72$. Using only the acidic and neutral reference lakes as input, the correlation is $r^2=0.77$. The position of the species along axis 1 is shown in Walseng and Schartau (*in press*).

Using either planktonic or littoral microcrustaceans as input in a DCA-ordination, the correlation between axis 1 and pH is respectively $r^2=0.60$ and $r^2=0.70$.

4. Discussion

Our two-year study shows that planktonic and littoral microcrustaceans have responded to changes in water quality due to liming in such a way that lake recovery can be concluded. Our hypothesis was that these lakes would reflect most of the species variation along a pH gradient. When the species lists from each lake for 1998 and 1999 were used as input in a DCA analysis, we derived a plot where the limed lakes (except for Lake Trollbergtjern 1999) were placed between the acidic and neutral reference lakes along axis 1. Unfortunately, samples were not taken before the lakes were limed, but knowing that pH in some of these lakes was 5.0 or less before liming, it seems obvious that improvement in water chemistry has resulted in a microcrustacean fauna more similar to the near neutral localities. In the limed lakes, Langvannet, Vortungen and Sundvatn, the same number of species as in the neutral reference lakes were found (about 40 species). Axis 1 was strongly correlated with pH, and when the limed lakes were excluded this barely affected the correlation. This means that the limed lakes have achieved a species composition which is in accordance to what can be expected on background of pH. Species associated with acidic water were found in the acidic reference lakes while the near-neutral reference lakes and the limed lakes were dominated by species which seem to be more or less acid-sensitive. Unpublished data including information on species composition from 2500 lakes in Norway of varying pH (Walseng, unpubl.), confirms this. *Alona costata* is a species found in our study which is only associated with neutral lakes (Flössner, 1967; Walseng, 1994). While sensitive species are known to invade lakes following liming, species which are known as acid-tolerant appear to become rarer.

Both planktonic and littoral crustaceans were included in our study because we wanted to increase the number of indicator species. In Canada both natural recovery and recovery after liming have been successfully studied by using only plankton species as indicators (Locke et al. 1994, Yan et al. 1996). An argument for not using only plankton species in Norwegian lakes was that too few acid-sensitive plankton species may be expected to respond to liming. In South Norway the use of only plankton species as indicators of recovery would have relied on 1-3 species. The easternmost parts of Norway, however, is the most species rich part of the country (Walseng and Karlsen, 1997). *Limnosida frontosa*, *Bosmina coregoni* and *Heterocope appendiculata* are examples of acid-sensitive plankton species in our study which are only found in the eastern and southeastern parts of Norway.

When only plankton species were used as input in a DCA-ordination, the correlation between axis 1 and pH was $r^2=0.60$. If only littoral samples were used the same figure was $r^2=0.70$. This tells us that by including both planktonic and littoral crustaceans we get the strongest basis to assess recovery ($r^2=0.72$).

The species composition in Lake Trollbergtjern, which was limed in winter 1999, did not respond to lime treatment the following year. Studies in Canada (Yan et al., 1996) has shown that differences in the development of recovery may be due to the degree of acidification. A fast recovery may therefore not be expected in Lake Trollbergtjern where pH was 4.6 before liming.

Setting targets for recovery is an important management question and it is obvious that knowledge about the pre-stress situation would be the best foundation for realistic assessments of baselines. When such data are not available, our approach evaluating the position to the limed lakes in a DCA-plot including acid and neutral reference lakes, may give us information of when recovery of the crustacean fauna is achieved.

Acknowledgements

We thank Erik Framstad for his comments to the manuscript. The study has recieved financial support from the Directorate for Nature management.

References

Degerman, E., Henrikson, L., Herrmann, J. and Nyberg, P.: 1995, 'The effects of liming on aquatic fauna.', in L. Henrikson and Y.W. Brodin (eds.), *Liming of acidified surface waters. A Swedish synthesis.* Springer Verlag, Berlin. pp. 221-282.
Drabløs D. and Tollan A.: 1980, Proceedings from the Internationale Conference on ecological Impact on Acid Precipitation, Sandefjord, Norway.
Flössner, D.: 1967, *Limnologica (Berlin)* **5**, 417.
Hann, B.J. and Turner, M.A.: 2000, *Freshwater Biology* **43**, 133.
Hill, M.O.: 1979, DECORANA - *A Fortran program for detrended correspondence analysis and reciprocal avaraging*. Cornell University, Ithaca, New York.
Hill, M.O. and Gauch, H.G.: 1980, *Vegetatio* **42**, 47.
Keller, W., Yan, N.D., Hovell, T., Molot, L.A. and Taylor, W.D.: 1992, *Journal of Fisheries and Aquatic Sciences* (Suppl. 1), 52.
Locke, A., Sprules, W.G., Keller, W. and Pitblado, J.R.: 1994, *Canadian Journal of Fisheries and Aquatic Sciences* **51**, 151.
Muniz, I.P.: 1991, *Proceedings of the Royal Society of Edinburgh* **97B**, 227.
Schartau, A.K., Hobæk, A., Faafeng, B., Halvorsen, G., Løvik, J.E., Nøst, T., Solheim, A.L. and Walseng, B.: 1997, *Vann og vassdrag i by- og tettstednære områder.* - NINA temahefte 14, NIVA lnr 3768-97. 58 pp.
Schartau, A.K . and Walseng, B.: in press., *Water, Air and Soil Pollution (Acid Rain 2000).*
Svensson, J.-E., Henrikson, L., Larsson, S. and Wilander, A.: 1995, 'Liming strategies and effects: The lake Gårdsjön case study.' in L. Henrikson and Y.W. Brodin (eds.), *Liming of acidified surface waters. A Swedish synthesis.,* Springer Verlag, Berlin. pp. 309-325.
ter Braak, C.J.F.: 1987, *CANOCO-a FORTRAN program for canonical community ordination (Version 2.1).* - TNO Institute of Applied Computer Sci.
ter Braak, C.J.F.: 1990, *Update notes: CANOCO version 3.10.* Agriculture Math. Group, Wageningen.
Walseng, B.: 1994, *Verh. Int. Verein. Limnol.* **25**, 2358.
Walseng, B. and Karlsen, L.F.: 1997, *NINA Oppdragsmelding 490*, Norway.
Walseng, B. and Schartau, A.K.: in press.,*Water, Air and Soil Pollution (Acid Rain 2000).*
Yan, N.D., Keller, W., Somers, K.W., Pawson, T.W. and Girard, R.E.: 1996, *Canadian Journal of Fisheries and Aquatic Science* **53**, 1301.

CRUSTACEAN COMMUNITIES IN CANADA AND NORWAY: COMPARISON OF SPECIES ALONG A pH GRADIENT

BJØRN WALSENG[1*] and ANN KRISTIN L. SCHARTAU[2]

[1] *Norwegian Institute for Nature Research (NINA) P.O. Box 736 Sentrum, N-0105 Oslo, Norway;*
[2] *Norwegian Institute for Nature Research (NINA) Tungasletta 2, N-7485 Trondheim, Norway.*
(* author for correspondence, e-mail: bjorn.walseng@ninaosl.ninaniku.no)

Abstract. Increased pH in acid lakes changes the crustacean fauna from communities dominated by acid-tolerant species to communities dominated by more acid-sensitive species. Studies from Canada (Killarney) and southeastern Norway (Østfold county) have demonstrated that planktonic and littoral crustaceans can be used as indicators of such recoveries. In both places the cladocerans *Alona rustica* and *Acantholeberis curvirostris* were found in acidic lakes, whereas *Alona costata* and the copepod *Eucyclops macrurus* were found in near neutral lakes. The calanoids *Diaptomus minutus* in North America and *Eudiaptomus gracilis* in Europe, both dominate in acidic water, and may ecologically be equivalent species. Sometimes the same species occur at different pH in the two continents. *Bosmina longirostris* and *Alonella excisa* may serve as examples, but a pertinent question is whether or not they are really the same species.

Keywords: acidification, recovery, microcrustacean, indicatorspecies, Canada, Norway

1. Introduction

Loss of crustacean species diversity because of anthropogenic acidification is well documented through studies in field and laboratory (e.g. Havens and DeCosta, 1987; Locke, 1991; Muniz, 1991; Havens and Hanazato, 1993). In attempts to restore the original fauna, acidified waters have been limed both in North America and Europe. Such remedial treatments have lead to recovery of crustacean communities (Keller *et al.*, 1992a; Henrikson and Brodin, 1995; Yan *et al.*, 1996). Similarly, reduced local acid deposition and rises in pH has resulted in crustacean recovery (Keller *et al.*, 1992b; Locke *et al.*, 1994).

Originally, only planktonic crustaceans were used as indicators of lake recovery, but in the late 1990s, also littoral crustaceans have been used (Hann and Turner, 2000; Walseng and Karlsen, *in press*). Approximately 75% of the crustacean species are either littoral or planktonic/littoral. Thus, the inclusion of this group may improve the applicability of crustaceans as indicators of recovery.

Two similar studies on planktonic and littoral microcrustaceans as indices of recovery, have been conducted, one in North America (Killarney, Sudbury, Canada) and one in North Europe (Norway, Østfold county). The main purpose of the study in Killarney was to monitor the natural recovery of freshwater crustaceans in an area that until 1972 had suffered from ecological degradation caused by acid deposition, primarily from local sources. The study from Østfold is based on limed lakes and the chemical recovery has therefore been faster. Both studies revealed that pH is a very useful parameter for assessing presence or absence of species. Some species are acid-tolerant and dominate in acidic lakes, others are acid-sensitive and are found in moderately acid or neutral lakes (Walseng, 1994;

Walseng, 1998). The studies therefore provide ample opportunity to study transcontinental differences in crustacean ecology.

Originally, North American crustacean species were identified by European keys. Sometimes this has been misleading, and recent studies using allozymic and morphological variables have revealed a number of new species (Frey, 1988; De Melo and Hebert, 1994a). Less, however, is known about differences in ecology between European and North American crustaceans.

The purpose of this study was to compare the species composition of planktonic and littoral crustaceans in lakes along a pH gradient in Canada and Norway. Based on the studies from Canada and Norway we will investigate if (i) the same species occur in lakes of similar pH in the two countries; (ii) ecologically equivalent but taxonomically different species occur in lakes with similar pH in the two countries; and (iii) the same species occur at different pH in the two countries.

2. Material and methods

In Canada the project started in 1997 when 13 Killarney lakes were sampled. In 1998 and 1999 nine additional lakes were sampled. These previously acidic lakes are now in a process of recovery due to emission controls. The current lake pH values are between 4.6 and 7.6.

In south-east Norway, 15 lakes, seven acidic (pH 4.7-5.4), six limed (5.8-7.0) and two moderately acid/neutral (pH 6.3-6.9), were monitored in 1998 and 1999 (Walseng and Karlsen, *in press*). Liming has been conducted regularly since the late 1980s.

The investigated lakes in Killarney were situated 182-325 m a.s.l., while the lakes in Østfold were situated 79-290 m a.s.l. The two areas should also be comparable with respect to the climate having relatively warm summers and dry cold winters.

In all lakes planktonic and littoral crustaceans were sampled twice a year (June and September/October). One sample was taken by a nethaul (90 microns) from approximately the deepest point in each lake. Two additional samples were taken from the littoral zone by dragging the net slowly over stony substrate and in different stands of vegetation (2-12 m). Using presence absence data, qualitative samples should ensure the majority of species present.

The samples were analyzed for presence or absence of species. Planktonic and littoral species were treated together. Patterns in the distributions of the crustaceans were summarised by Detrended Correspondence Analysis (DCA) (Hill, 1979; Hill and Gauch, 1980) using the program CANOCO with downweighting of rare species (ter Braak, 1987; ter Braak, 1990). In both Canada and Norway, axis 1 were strongly correlated with pH, $r^2=0.80$ and $r^2=0.72$, respectively. The species score from each plot (Table I) provides information about species associated with acidic and neutral lakes. We divided the species into five groups of equal size depending on their position along axis 1, starting with group I at the acidic end, ending up with group V at the opposite (near neutral) end of axis 1. The groups are acidic (I), weakly acidic (II), indifferent (III), weakly sensitive (IV) and sensitive (V) species. It is important to stress that these groups are not truly ecologically defined.

TABLE 1
Crustaceans in Norway (Østfold) and Canada (Killarney), respectively, along axis 1 in a DCA-ordination.

Østfold, Norway	axis 1	n=30 (sites)	Killarney, Ontario, Canada	axis 1	n=57 (sites)
CATEGORY I (acidic)			Chydorus gibbus Lilljeborg	6,2959	1
			Mesocyclops lauckarti Claus	5,8516	1
			Simocephalus serrulatus (Koch)	5,3911	4
Diacyclops languidus (Sars)	5.6640	2	Rhynchotalona falcata Sars	5,0455	2
Diacyclops bicuspidatus (Sars)	3.8993	1	Acantholeberis curvirostris (O.F.M.)	4,5489	16
Acanthocyclops vernalis (Fisch.)	3.6667	1	Alona quadrangularis (O.F.M.)	3,9242	5
Acantholeberis curvirostris (O.F.M.)	3.4988	14	Alona rustica Scott	3,9025	8
Ceriodaphnia quadrangula (O.F.M.)	3.2976	11	Alona intermedia Sars	3,6951	11
Alona rustica Scott	3.2268	14	Ilyocryptus sordidus (Liév)	3,5759	2
Diacyclops nanus (Sars)	2.7459	2	Ilyocryptus spinifer Herrick	3,2153	7
Macrocyclops fuscus (Jur.)	2.4965	14	Eucyclops prionophorus Kiefer	3,1742	20
Streblocerus serricaudatus (Fisch.)	2.1728	14	Orthocyclops modestus Herrick	3,0991	10
Paracyclops affinis Sars	1.9436	14	Latona setifera (O.F.M.)	2,8658	21
Alonella excisa (Fischer)	1.9377	19	Alona affinis (Leydig)	2,5713	33
Latona setifera (O.F.M.)	1.9376	3	Ophryoxus gracilis Sars	2,3249	42
Alona guttata Sars	1.6408	23	Chydorus piger (Sars)	2,1770	44
Alona affinis (Leydig)	1.2320	24	Paracyclops affinis Sars	2,1656	13
CATEGORY II (weakly acidic)			Alonella exigua (Fischer)	2,0136	44
			Macrocyclops albidus (Jur.)	1,8622	31
			Acanthocyclops vernalis Fischer	1,8621	3
Alonella nana (Baird)	1.1841	28	Bosmina tubifen (Brehm)	1,8278	9
Acroperus harpae (Baird)	1.1620	27	Chydorus brevilabris/sphaericus	1,8020	52
Rhynchotalona falcata Sars	1.1344	17	Eucyclops serrulatus (Fisch.)	1,6997	35
Scapholeberis mucronata (O.F.M.)	1.1087	28	Acroperus harpae (Baird)	1,6089	54
Macrocyclops albidus (Jur.)	1.0217	26	Eucyclops neomacruroides Dussart	1,5937	22
Chydorus piger Sars	1.0101	6	Macrocyclops fuscus (Jur.)	1,5117	9
Eucyclops serrulatus (Fisch.)	.9889	29	Alona guttata Sars	1,4488	28
Alona intermedia Sars	.9639	2	Leptodiaptomus minutus Lillj.	1,4412	56
Bosmina longispina Leydig	.9110	30	Polyphemus pediculus (Leuck.)	1,2949	56
Alonopsis elongata Sars	.9110	30	Bosmina longispina Leydig	1,2272	35
Polyphemus pediculus (Leuck.)	.9110	30	Chydorus latus (Sars)	1,1874	3
Eurycercus lamellatus (A.F.M.)	.7315	25	Alonella acutirostris Birge	1,1115	25
Acanthocyclops robustus Sars	.6706	12	Bosmina freyi De Melo	1,0115	45
Chydorus sphaericus (O.F.M.)	.6665	28	Alona bicolor Frey	1,0016	16
CATEGORY III (indifferent)			Streblocerus serricaudatus (Fisch.)	0,8925	9
			Daphnia pulex Leydig	0,8657	3
Lathonura rectirostris (O.F.M.)	.6413	1	Acanthocyclops capillatus Sars	0,7897	3
Graptoleberis testudinaria (Sars)	.6399	9	Holopedium gibberum Zaddach	0,7522	42
Bythotrephes longimanus Leydig	.5765	11	Daphnia catabwa Coker	0,6753	7
Chydorus latus Sars	.5540	2	Diaphanosoma sp	0,6138	39
Eudiaptomus gracilis Sars	.3834	26	Cyclops bicuspidatus thomasi S.A. Forbes	0,5936	39
Holopedium gibberum Zaddach	.3753	20	Alonella excisa (Fischer)	0,5810	13
Paracyclops fimbriatus (Fisch.)	.2849	1	Acanthocyclops robustus Sars	0,5663	14
Megacyclops viridis (Jur.)	.1709	7	Paracyclops fimbriatus poppei Rehberg	0,5196	8
Cyclops scutifer Sars	.1163	24	Simocephalus vetulus Schödler	0,5032	3
Mesocyclops leuckarti (Claus)	.0985	17	Scapholeberis kingi (Sars)	0,3820	27
Megacyclops gigas (Claus)	-.0514	7	Graptoleberis testudinaria (Fischer)	0,2925	3
Ophryoxus gracilis Sars	-.0510	16	Monospilus dispar Sars	0,2803	2
Pleuroxus truncatus (O.F.M.)	-.0230	22	Daphnia ambigua Scourfield	0,2425	24
Ceriodaphnia megops Sars	-.8803	3	Cyclops varicans rubellus Lillj.	0,2076	7
CATEGORY IV (weakly sensitive)			Mesocyclops edax S.A. Forbes	0,1787	34
			Chydorus bicornutus Doolittle	0,1319	9
			Eucyclops macrurus (Sars)	0,0930	2
Sida crystallina (O.F.M.)	-.1275	22	Ceriodaphnia quadrangula (O.F.M.)	0,0888	2
Leptodora kindti Focke	-.3350	13	Sida crystallina (O.F.M.)	-0,0211	32
Heterocope appendiculata Sars	-.3443	18	Pleuroxus striatus Schödler	-0,0449	8
Pseudochydorus globosus (Baird)	-.4551	5	Tropocyclops extensus Kiefer	-0,0772	25
Alona karelica Stenroos	-.4594	2	Eurycercus (Bullatifrons) sp.	-0,0977	2
Cyclops abyssorum	-.4917	2	Pleuroxus hastatus Sars	-0,1221	11
Daphnia longispina (O.F.M.)	-.5504	13	Cyclops scutifer Sars	-0,1234	9
Daphnia cristata Sars	-.5965	18	Daphnia parvula Fordyce	-0,1611	1
Eucyclops speratus (Lillj.)	-.6049	10	Daphnia dubia Herrick	-0,3440	5
Alona quadrangulasris (O.F.M.)	-.6571	1	Leptodora kindti (Focke)	-0,5185	13
Thermocyclops oithonoides (Sars)	-.7246	18	Bosmina liederi De Melo	-0,5661	21
Diaphanosoma brachyurum (Liév.)T	-.7690	11	Oxyurella tenuicaudis (Sars)	-0,5941	1
Disparalona rostrata (Koch)	-.8121	6	Epischura lacustris S.A. Forbes	-0,6270	24
CATEGORY V (sensitive)			Kurzia latissima (Kurz)	-0,6643	4
			Daphnia retrocurva S. A. Forbes	-0,6783	9
			Alona costata Sars	-0,6788	13
Acanthocyclops capillatus Sars	-.9461	1	Chydorus faviformis Birge	-0,6827	5
Eucyclops denticulatus (A.Graet.)	-.9520	4	Bythotrephes longimanus Leydig	-0,7183	2
Simocephalus vetula (O.F.M.)	-.9873	2	Daphnia galeata mendotae Birge	-0,7365	19
Alonella exigua (Fischer)	-1.0354	2	Skiptodiaptomus oregonensis Lillj.	-0,8102	13
Eucyclops macruroides (Lillj.)	-1.0820	3	Leptodiaptomus ashlandi Marsh	-0,9133	4
Ceriodaphnia pulchella Sars	-1.3357	11	Pleuroxus procurvis Birge	-0,9924	1
Bosmina longirostris (O.F.M.)	-1.4058	10	Alona circumfibriata Megard	-1,0197	3
Camptocercus rectirostris Schoedler	-1.4468	5	Daphnia longiremis (Sars)	-1,0324	10
Pleuroxus trigonellus (O.F.M.)	-1.5503	3	Lathonura rectirostris (O.F.M.)	-1,0450	1
Bosmina coregoni (Baird)	-1.9344	4	Pleuroxus truncatus (O.F.M.)	-1,0450	1
Alona costata Sars	-2.2438	4	Ectocyclops phaleratus (Koch)	-1,1430	3
Limnosida frontosa Sars	-2.2438	4	Alona setulosa Megard	-1,2403	8
Eucyclops macrurus (Sars)	-2.3601	3	Leptodiaptomus sicilis S.A. Forbes	-1,3468	6
Eudiaptomus graciloides (Lillj.)	-2.4130	1	Alonella nana (Baird)	-1,7101	2

3. Results and Discussion

The total number of species sampled in Killarney and Østfold were 83 and 69 species, respectively. Thirty-five species were found in both countries, and the majority of these species are from the acidic part of axis 1 (category I-III) (Table I).

3.1. THE SAME SPECIES THAT OCCUR IN LAKES OF SIMILAR PH IN THE TWO COUNTRIES

In acidic lakes *Acantholeberis curvirostris*, *Alona rustica* and *A. affinis* occur in both continents (category I). *Acantholeberis curvirostris*, is acid-tolerant (Potts and Fryer, 1979), and the only littoral species found to disappear after liming (Hasselrot et al., 1984). *Alona rustica* is also acid-tolerant (Flössner, 1967; Walseng, 1994), and a useful indicator of acidic conditions. Frey (1965), however, questioned whether the North American and the European *A. rustica* are identical. The third species in category I, *A. affinis*, is found in a variety of waters, and frequently occurs in acidic lakes, as do *Bosmina longispina*, *Acroperus harpae*, *Macrocyclops albidus* and *Eucyclops serrulatus* (category II).

Two of the largest cladocerans, *Leptodora kindti* and *Sida crystallina* are predators. *L. kindti*, which is rare in waters more acid than pH 5.5, has reappeared after liming. *S. crystallina*, one of the most common littoral species in Norway, is acid-tolerant and found in about 40% of lakes with pH below 4.5. Both species were in both countries classified as weakly sensitive to acid water (category IV).

Alona costata is the only catogory V species from both countries. According to Flössner (1967), *A. costata* is found in slightly acidic to alkaline, more or less eutrophic waters. This is confirmed by our data from Norwegian lakes where this species has never been reported from lakes with pH below 6.0.

3.2. ECOLOGICALLY EQUIVALENT BUT TAXONOMICALLY DIFFERENT SPECIES THAT OCCUR IN LAKES WITH SIMILAR PH IN THE TWO COUNTRIES

Related, but not taxonomically identical species found in the same category, may be ecological equivalents. This is the case for the daphnids, which were never found in acidic water. The North American *Daphnia catawba*, *D. pulex* and *D. ambigua* were indifferent with respect to pH (category III). *D. catawba* is an acid-tolerant species in North America (Locke et al. 1994), contrasting *D. ambigua*, which is one of the first "new" species found in Killarney lakes recovering after years of acidification. The remaining four North American species, *D. longiremis*, *D. retrocurva*, *D. dubia* and *D. parvula*, and the two European species, *D. longispina* and *D. galeata*, belong to category IV and V. The daphnids from Europe and North America appear to have the same feeding niche (planktonic filter feeder), and may be regarded as ecologically equivalent.

Some calanoid copepods may also be classified as ecological equivalents. For instance, the calanoids *Leptodiaptomus minutus* dominated numerically in the Killarney lakes and *Eudiaptomus gracilis* was abundant in Østfold. By numbers, these two species dominated completely plankton communities in the acidic lakes, and they were also common in quite neutral lakes. The remaining four North American and the two Norwegian calanoids belong to categories IV and V.

3.3. THE SAME SPECIES THAT OCCUR AT DIFFERENT PH IN THE TWO COUNTRIES

Bosmina longirostris occurs in near-neutral lakes (category V) in Norway. Less than 10% of all Norwegian records of this species come from lakes with pH below 6.0. When we analysed the Canadian samples from 1997, *B. longirostris* appeared in the majority of lakes, at pH values from 4.8 to 7.0. According to Locke et al. (1994), *B. longirostris* occurs at a frequency of 60.5 % in lakes with pH less than 5.0 in Killarney. By closer inspection, small morphological differences were found between Norwegian and Canadian *B. longirostris*. The proximal pecten consists of 8 long slender spines, which increase in height distally in Norwegian individuals and 1-3 wide-based strong spines in Canadian animals. According to a taxonomic reevaluation of North American Bosminanidae (De Melo and Hebert, 1994a), the Sudbury specimens are morphological similar to two of the new species, *Sinobosmina freyi* and *S. liederi*. Sinobosmidae is a new subgenus. These newly described species differ both morphologically and in allozyms from *B. longirostris* (De Melo and Hebert, 1994b). De Melo and Hebert (1994b) show a distribution map of the three species. From North America, *B. longirostris* has so far only been found in California, and *S. liederi* and *S. freyi* only in localities in the Great Lakes catchment area.

De Melo and Hebert (1994a) reported that both these species occur in lakes, but *S. freyi* may also be found in ponds. In the Killarney lakes, *S. liederi* was assosiated with more neutral lakes (category IV) than *S. freyi* (category II). It is still an open question whether *S. freyi* is more acid-tolerant than *S. liederi*. If this is true, it underpins the importance of correct identification when using crustaceans as indicators of lake restoration.

Three species from the genus *Alonella*, *A. excisa*, *A. exigua and A. nana*, were found in both areas, whereas the fourth, *A. acutirostris*, was also found in Killarney. In Norway *A. exigua* was found in circum-neutral lakes (category V), while *A. excisa* and *A. nana* were associated with the acidic lakes (category I and II). These observations are confirmed by data from 1800 lakes in Norway. In Killarney, *A. exigua* occurs in the most acidic lakes while *A. excisa* is an indifferent species. The species were morphologically different in the two continents, and detailed studies are now needed to confirm whether they are conspecific or not. According to Chengalath (1982), both *A. exigua* and *A. excisa* are common and widely distributed in Canada. The author mentions that *A. excisa* exhibits considerable morphological variation, and adresses the question if *A. excisa* is a species complex consisting of two or more separate species.

Although our discussion of crustaceans in Norway and Canada mainly have addressed the crustaceans as indicators of ecological recovery of acid-stressed lakes, differences in ecological traits between similar species may also contribute to a better understanding of their taxonomy. In the future, besides more detailed ecological studies, there is a need for genetic and morphological studies to improve our overall knowledge of the freshwater crustaceans in Europe and North America.

Acknowledgements

We thank Professor Bror Jonsson, Dr. I.P. Muniz and other colleagues at NINA for helpful discussion of earlier drafts of this paper. We also thank the staff at the Coop-Unit, Laurentian University, Ontario, Canada for carrying out the field-work in Killarney. The studies were financially supported by the Norwegian Directorate for Nature Management (DN) and Østfold county.

References

Chengalath, R.: 1982, *Canadian Journal of Zoology* **60**, 2668.
De Melo, R. and Hebert, P.D.N.: 1994a, *Canadian Journal of Zoology* **72**, 1808.
De Melo, R. and Hebert, P.D.N.: 1994b. *Canadian Journal of Fisheries and Aquatic Sciences* **51**, 873.
Flössner, D.: 1967, *Limnologica (Berlin)* **5**, 417.
Frey, D.G.: 1965, *Crustaceana (Leiden)* **8**, 159.
Frey, D.G.: 1988, *Limnology and Oceanography* **33**, 1386.
Hann, B.J. and Turner, M.A.: 2000, *Freshwater Biology* **43**, 133.
Hasselrot B., Andersson B.I. and Hultberg H.: 1984, *Report from the Institute of Freshwater Research, Drottningholm* **61**, 78.
Havens, K.E. and DeCosta, J.: 1987, *Archiv für Hydrobiologie* **111**, 37.
Havens, K.E. and Hanazato, T.: 1993, *Environmental Pollution* **82**, 277.
Henrikson, L. and Brodin, Y.W.: 1995, *Liming of acidified surface waters. A Swedish synthesis.* Berlin: Springer Verlag.
Hill, M.O.: 1979. *DECORANA - A Fortran program for detrended correspondence analysis and reciprocal avaraging.* Cornell University, Ithaca, New York.
Hill, M.O. and Gauch, H.G.: 1980, *Vegetatio* **42**, 47.
Keller, W., Yan, N.D., Hovell, T., Molot, L.A. and Taylor, W.D.: 1992a., *Journal of Fisheries and Aquatic Sciences* (Suppl. 1), 52.
Keller, W., Gunn, J.M. and Yan, N.D.: 1992b, *Environmental Pollution* **78**, 79.
Locke, A.: 1991, *Water, Air and Soil Pollution* **60**, 135.
Locke, A., Sprules, W.G., Keller, W. and Pitblado, J.R.: 1994, *Canadian Journal of Fisheries and Aquatic Sciences* **51**, 151.
Muniz, I.P.: 1991, *Proceedings of the Royal Society of Edinburgh* **97B**, 227.
Potts, W.T.W. and Fryer, G.: 1979, *Journal of Comparative Physiology* **129**, 289.
ter Braak, C.J.F.: 1987, *CANOCO-a FORTRAN program for canonical community ordination by (partial)(detrended)(canonical) correspondance analysis, principal components analysis and redundancy analysis (Version 2.1).* - TNO Institute of Applied Computer Sci.
ter Braak, C.J.F.: 1990, *Update notes: CANOCO version 3.10.* Agriculture Math. Group, Wageningen.
Walseng, B.: 1994, *Verh. Int. Verein. Limnol..* **25**, 2358.
Walseng, B.: 1998, *Verh. Int. Verein. Limnol..* **26**, 2007.
Walseng, B. and Karlsen, L.R.: in press., Will be presented on *"6th International Conference on Acidic Deposition in Japan, December 2000"*.

CORRELATION BETWEEN MICROCRUSTACEANS AND ENVIRONMENTAL VARIABLES ALONG AN ACIDIFICATION GRADIENT IN SUDBURY, CANADA

ANN KRISTIN L. SCHARTAU[1*], BJØRN WALSENG[2] and ED SNUCINS[3]

[1] *Norwegian Institute for Nature Research, Tungasletta 2, N-7485 Trondheim, Norway;*
[2] *Norwegian Institute for Nature Research, P.O. Box 736 Sentrum, N-0105 Oslo, Norway;*
[3] *Department of Biology, Laurentian University, Sudbury, Ontario P3E 2C6, Canada*
(* author for correspondence, e-mail: ann.k.schartau@ninatrd.ninaniku.no)

Abstract. Multivariate methods were used to relate microcrustacean (pelagic and littoral) richness and composition (presence/absence) to water quality and other environmental variables. All acidification variables (pH, aluminium, ANC) showed significant correlation with both species richness and composition. The variation in microcrustacean richness was best explained by the combination of dissolved organic carbon (DOC), fish species richness and lake area. Of 16 variables tested, pH showed strongest correlation with the main gradient in the crustacean composition explaining between 13 and 16% of the variance in the species data (CCA). pH, elevation, lake area, average depth, DOC, conductivity and fish species richness explained 30-54% of the total variance. Stronger correlation was obtained between species composition and environmental data in analyses which included the between-year differences than analyses based on the cumulative species records. Analyses based on the pelagic species exclusively gave similarly stronger correlation than analyses based on all crustacean species. Small changes in the species composition during the three years of study may be an indication of recovery of microcrustaceans in Killarney lakes.

Keywords: acidification, microcrustaceans, multivariate statistics, recovery, species richness

1. Introduction

The lakes in Killarney Provincial Park, located 40-60 km south-west of Sudbury, Ontario, Canada, were some of the first lakes in North America to be acidified by atmospheric pollutants (Beamish and Harvey, 1972; Neary *et al.*, 1990). Since the 1970's water quality improvements have occurred in response to reductions in atmospheric pollution and some lakes have already recovered to their pre-industrial pH levels, as inferred from microfossils preserved in lake sediments (Snucins *et al.*, 2000). Although numerous acidified lakes in Scandinavia are now responding to reduced atmospheric pollution levels with improvements in water quality (Skjelkvåle *et al.*, 2000), the lakes in Killarney Park currently provide the best evidence of natural chemical and biological recovery. In 1997 a Canadian/Norwegian cooperative project, the Northern Lakes Recovery Study (NLRS), was established in Killarney Park to develop methods for characterising chemical and biological recovery of acidified lakes.

The toxic effects of acidification on aquatic organisms can be influenced by a number of different chemical variables including pH, calcium (Ca), aluminium (Al) and humic content. In the international work on critical load modelling, acid neutralization capacity (ANC) has been used as the criterion for assessing damaging effects to fish and macro-invertebrates (Henriksen *et al.*, 1995), but this work has not yet included microcrustaceans.

A variety of univariate and multivariate metrics have been used in studies linking pelagic crustacean and other zooplankton to acidity, such as species diversity, species richness, evenness, size and species composition. Species richness has the most well documented relationship to pH and alkalinity. Reduced numbers of zooplankton species have been recorded in acidic lakes in Southern Norway (Hobæk and Raddum, 1980; Walseng *et al.*, 1995), in Ontario, Canada (Carter, 1971; Locke *et al.*, 1994), and several other parts of the northern Hemisphere. Negative relationships between species richness and pH have also been recorded for the littoral crustaceans (Roff and Kwiatkowski, 1977; Walseng *et al.*, 1995). Multivariate metrics of zooplankton community (e.g. ordination scores) have been found to show similar patterns as species richness (Siegfried *et al.*, 1984; Yan *et al.*, 1996). In addition these multivariate metrics are more sensitive to community level changes because they incorporate co-variations between species.

The objectives of this study were 1) to describe the relationship between the microcrustacean species richness and composition and environmental variables in a set of lakes spanning a wide pH gradient, and 2) to evaluate which of the acidification related variables (pH, Al, ANC) most strongly correlated with the variations in the microcrustacean communities.

2. Study Area

Killarney Park's watershed (55,980 ha) contains over 600 water bodies spanning a broad range of physical and chemical characteristics (Snucins *et al.*, 2000). The 23 lakes in our study were chosen to include the main environmental gradients. They vary relatively little in elevation (range: 182 – 325 m a.s.l), but differ substantially in lake size (3.4 – 1087.9 ha) and maximum depth (8 – 61 m). The between-lake variation in pH, total aluminium (tot-Al), ANC and DOC was 4.7 – 7.5, 4 – 462 µg l^{-1}, -74 – 442 µeq l^{-1}, and 0.1 – 5.5 mg C l^{-1}, respectively. During the three years of this study, measurable water quality improvements occurred in many of the lakes (Snucins *et al.*, 2000).

3. Material and Methods

Our analyses are based on data of pelagic and littoral microcrustaceans from 23 lakes in Killarney Provincial Park along a pH-gradient from 4.7 to 7.5. Included in the study are recovering acidified lakes (N=19) and non-acidified reference lakes (N=4).

Microcrustaceans: Sampling was conducted between 1997 and 1999. One of the lakes was studied for one year (1997), ten lakes for two years (1998-99) and twelve lakes for three years (1997-99). The sampling procedure are described in Walseng and Schartau (2000).

Environmental data: Lake water has been sampled as surface grabs or composite tube samples by the Cooperative Freshwater Ecology Unit in Sudbury (CFEU). Analyses for pH and alkalinity were performed at the CFEU. All other analyses were done at the Ministry of Environment and Energy (MOEE) laboratory in Toronto (Snucins and Gunn 1998). For this study, we used data obtained from water samples taken during October/ November, after fall turnover in these lakes. Environmental variables included in the analyses were as follows: pH, ANC, tot-Al, labile aluminium (LAl), DOC, Ca, conductivity (cond), total phosphorus (tot-P), Secchi depth (secchi-d), lake area, average lake depth (depth-a), maximum lake depth (depth-m), elevation (elev), fish species richness (fish).

Simple and multiple regression analyses were used to relate the number of microcrustacean species to environmental variables. In multiple regression, the predictive power of the model parameters are given by the adjusted coefficient of determination (r^2). The correlations between species richness and environmental data were analysed for two separate subsets including all lakes: the subset A-I representing cumulative species records, and subset A-II, representing yearly species records (see Table I).

Direct gradient analyses with forward selections (CCA: Canonical Correspondence Analysis) were used to provide an overview of the relationships between the sample sites, based on records of presence/ absence of microcrustacean species, and environmental variables. The analyses were performed with help of the program CANOCO version 4 (ter Braak and Smilauer, 1998). Downweighting was applied to all species with frequency below the median frequency (Eilertsen *et al.*, 1990) to reduce the effect of unusual samples on the ordination (ter Braak and Smilauer, 1998). All of the environmental data except pH, lake depth and fish species number were transformed (Ln $(x+1)$) to improve normality. The correlation between the microcrustacean composition and environmental data were evaluated for subsets I, III and IV (Table I). For subsets III and IV our aim was to control for autocorrelation in time and to use the ordination to focus on effects of

TABLE I
Data subsets from 23 Killarney lakes, Sudbury, Canada 1997-99 used in the analyses. The number of active samples (subsets III-IV: lakes x years) are presented in parentheses.

Data subset	Microcrustacean data	Environmental data
A-I	Pelagic + littoral, total dataset cumulative records (23)	Median, min and max
A-II	Pelagic + littoral, total dataset separate years (57)	Yearly late autumn data
A-III	Pelagic + littoral, separate years 1998-99 (22 x 2)	Yearly late autumn data
A-IV	Pelagic + littoral, separate years 1997-99 (12 x 3)	Yearly late autumn data
B-III	Pelagic, separate years 1998-99 (22 x 2)	Yearly late autumn data
B-IV	Pelagic, separate years 1997-99 (12 x 3)	Yearly late autumn data

acidification and other explanatory variables. The effect of time (sampling year) was first tested and then removed by specifying time as covariable. Each environmental variable was tested by Monte Carlo permutation significance tests with 199 unrestricted permutations (ter Braak and Smilauer, 1998) before adding it to the model. Sequential Bonferroni adjustments of the significance level were performed for all multiple tests (Rice, 1989).

4. Results and Discussion

A total number of 83 microcrustacean species; 59 Cladocera and 24 Copepoda, were recorded (Walseng and Schartau, 2000). Of these, 34 crustaceans were defined as pelagic species. Microcrustacean species richness (range 13-53 species/lake) was positively correlated with fish species richness, DOC, pH, ANC, tot-P, and lake area and negatively correlated with aluminium, secchi disk readings, and elevation (Schartau and Walseng, *unpublished*). Fish species richness and average depth explained 79% of the variance in dataset A-I whereas DOC, fish and average depth together accounted for 67% of the variance in the dataset A-II (Table II). The microcrustacean species richness was correlated with acidification related variables (ANC, pH, LAl, tot-Al), as indicated by significant linear regressions (Schartau and Walseng, *unpublished*). For subset A-I the explanatory value, given by r^2, was highest for LAl (r^2=0.54) followed by pH (0.49), ANC (0.44) and tot-Al (0.37), For subset A-II, ANC (0.30) was followed by pH (0.28), LAl (0.24) and tot-Al (0.13). The contributions made by LAl, fish species richness and average depth (subset A-I) and ANC, DOC and lake area (subset A-II) yielded r^2 of 0.79 and 0.63, respectively.

TABLE II
Relationship between microcrustacean species richness and environmental variables identified by stepwise linear regression analyses. The analyses are based on data from 23 Killarney lakes, Sudbury, Canada. Only significant regressions (p<0.05) are listed.

Subset A-I		Subset A-II	
Variables	r^2	Variables	r^2
fish	0.70	DOC	0.58
+ depth-a	0.79	+ fish	0.63
		+ depth-a	0.67

The main gradient in the microcrustacean composition (CCA) was strongest correlated with pH, which explained 13-16% of the variance in the species data (Table III). The other acidification related variables, ANC, LAl and tot-Al, explained 9-15%, 10-16%, and 12-14%, respectively. Fish species richness, conductivity and DOC was positively correlated with pH but, for three of the subsets, still significant after controlling for the effect of pH. Relative to analyses based on the total species

TABLE III

Relationship (extra fit/total inertia in %) between microcrustacean composition and environmental variables identified by CCA with forward selection. Time (sampling year) is included as covariable in analyses of subset III and IV. The analyses are based on data from 23 Killarney lakes, Sudbury, Canada. Only significant variables (p<0.05) are listed.

Subset A-I		Subset A-III		Subset A-IV		Subset B-III		Subset B-IV	
pH	16	pH	13	pH	15	pH	14	pH	16
+ fish	24	+ elev	18	+ elev	22	+ elev	22	+ elev	28
+ cond	30	+ cond	23	+ area	29	+ DOC	28	+ DOC	34
		+ area	27	+ cond	33	+ cond	34	+ area	40
		+ DOC	31	+ fish	37	+ fish	37	+ ANC	45
		+ fish	34	+ depth-a	40	+ area	40	+ cond	51
		+ tot-Al	36					+ fish	54

records (subset I), stronger correlations were obtained when between-year differences (subset III and IV) were taken into account. Furthermore, analyses of pelagic microcrustaceans gave stronger correlations between species composition and environmental data than analyses of total microcrustaceans. In total the significant variables explained 30-40% and 40-54% of the total variance based on subset A and B, respectively (Table III). Higher inter-sample variance for the littoral microcrustaceans relative to the pelagic microcrustaceans are expected due to the higher habitat heterogeneity in the littoral zone. We might have been able to explain more of the variance in the total microcrustacean composition had data been available for some of the environmental variables that are thought to be of high ecological importance to littoral species, *e.g.* biomass and diversity of macrophytes (Duigan and Kovach, 1994).

The influence of time was small (0.5-4.4%) and a significant correlation ($p<0.05$) between variation in species composition and time was found for subset A-IV only. Changes in microcrustacean species composition during the three years of study may be an indication of recovery, but may also be due to year-to-year variations in other environmental variables, *e.g.* the climate. Evidence of recovery over a longer period of time by pelagic crustaceans in Killarney lakes is presented by Locke *et al.* (1994). However, because many acid sensitive species of microcrustaceans exist in the littoral zone, we believe that littoral microcrustaceans will be important to include as indicators of biological recovery in the Killarney lakes.

Acknowledgements

This project is part of the Northern Lakes Recovery Study (NLRS). We thank Dr. Torbjørn Forseth (NINA) for valuable comments on an earlier draft of this paper. We also thank the staff at CFEU, Laurentian University, Ontario, Canada for carrying out the field work in Killarney, Jocelyne Heneberry (CFEU) for water chemistry analyses, and Svein-Erik Sloreid (NINA) for carrying out the CANOCO analyses. The study was financially supported by the Norwegian Directorate for Nature Management and the Norwegian Institute for Nature Research.

References

Beamish, R.J. and Harvey, H.H.: 1972, *Journal of the Fisheries Research Board of Canada* **29**, 1131.
Carter, J.C.H.: 1971, *Archiv für Hydrobiologie* **68**, 204.
Duigan, C.A. and Kovach, W.L.: 1994, *Marine and freshwater ecosystems* **4**, 307.
Eilertsen, O., Økland, R.H., Økland, T. and Pedersen, O.: 1990, *Journal of Vegetation Science* **1**, 261-270.
Henriksen, A., Posch, M., Hultberg, H. and Lien, L.: 1995, *Water, Air, and Soil Pollution* **85**, 2419.
Hobæk, A. and Raddum, G.G.: 1980, *Zooplankton communities in acidified lakes in South Norway*. Rapport IR 75/80, SNSF-prosjektet, Norway.
Locke, A., Sprules, W.G., Keller, W., and Pitbaldo, J.R.: 1994, *Canadian Journal of Fisheries and Aquatic Sciences* **51**, 151.
Neary, B.P., Dillon, P.J., Munro, J.R. & Clark, B.J.: 1990, *The acidification of Ontario lakes: an assessment of their sensitivity and current status with respect to biological damage*. Tech. Rep. Ontario Ministry of Environment, Dorset, Ontario, Canada..
Rice, W.R.: 1989, *Evolution* **43**, 223.
Roff, J.C. and Kwiatkowski, R.W.: 1977, *Canadian Journal of Zoology* **55**, 899.
Siegfried, C. A., Sutherland, J.W., Quinn, S.O. and Bloomfield, J.A.: 1984, *Analysis of plankton community structure in Adirondack lakes in relation to acidification*. Biol. Survey N.Y.S. Sci. Serv., N.Y.S. Dept. Environ. Conserv., Albany, NY.
Skjelkvåle, B.L., Torseth, K., Aas, W. and Andersen, T.: 2000, *Water, Air, and Soil Pollution* (This volume).
Snucins, E. and Gunn, J.: 1998, *Chemical and biological status of Killarney Park lakes (1995-1997). A study of lakes in the early stages of recovery from acidification*. Ontario Ministry of Natural Resources, CFEU, Sudbury, Ontario.
Snucins, E., Gunn, J., Keller, B., Dixit, S., Hindar, A. and Henriksen, A.: 2000, *Journal of Environmental Monitoring and Assessment* (in press).
ter Braak, C.J.F. and Smilauer, P.: 1998, *CANOCO reference manual and User's guide to Canoco for Windows. Software for canonical community ordination (version 4)*. Microcomputer Power (Ithaca, NY, USA).
Walseng, B., Raddum, G. and Kroglund, F.: 1995, *Liming in Norway. Invertebrates*. Utredning for DN 1995-6, 65 pp. (In Norwegian with English summary).
Walseng, B. and Schartau, A.K.L.: 2000, *Water, Air, and Soil Pollution* (This volume).
Yan, N.D., Keller, W., Somers, D.M., Pawson, T.W. and Girard, R.E.: 1996, *Canadian Journal of Fisheries and Aquatic Sciences* **53**, 1301.

THE RIVER BJERKREIM IN SW NORWAY - SUCCESSFUL CHEMICAL AND BIOLOGICAL RECOVERY AFTER LIMING

BJØRN WALSENG[1*], ROY M. LANGÅKER[2], TOR E. BRANDRUD[1],
PÅL BRETTUM[3], ARNE FJELLHEIM[4], TRYGVE HESTHAGEN[5],
ØYVIND KASTE[6], BJØRN M. LARSEN[5] and ELI -A. LINDSTRØM[3]

[1] NINA, P.O. Box 736 Sentrum, N-0105 Oslo, Norway; 2 DN, N-7485 Trondheim, Norway;
[3] NIVA, P.O. Box 173 Kjelsås, N-0411 Oslo, Norway; [4] LFI, Stavanger museum, Musegt 16,
N-4010 Stavanger, Norway; [5] NINA, N-7485 Trondheim, Norway; [6] NIVA Sørlandsavd,
Televeien 3, N-4879 Grimstad, Norway
(* author for correspondence, e-mail: bjorn.walseng@ninaosl.ninaniku.no)

Abstract: On a large scale, the acidified River Bjerkreim, southwestern Norway, has been treated with lime since the autumn 1996. During the Atlantic salmon (Salmon salar) smolting period pH has been above 6.2 and LAl concentrations below 10 µg L-1. Before 1996, only the western part of the watercourse harboured acid-sensitive species, such as the Atlantic salmon, snails, mayflies, daphnids and Gammarus lacustris. Prior to liming in 1996, Atlantic salmon fry (0+) and parr (≥1+) were found in 4 of 20 sampling sites, contrary to 17 (fry) and 12 (parr) in 1999. Atlantic salmon catches have increased from about 0.8 tons in 1994 to about 10 tons in 1998 and 1999. Acid-sensitive invertebrates have invaded the limed parts of the river.

Keywords: acidification, liming, recovery, diversity, Atlantic salmon, invertebrates

1. Introduction

Southern Norway has been subject to acid deposition for more than 100 years (Mylona, 1993). During the last decades there has been a focus on acid rain and its effects on biotic communities in Norway and other affected countries (Drabløs and Tollan, 1980; Schindler, 1988; Muniz, 1991; Havens, 1993).

Since the beginning of the 1920s hatchery water has been treated with lime to avoid fish kills, but not until the 1980s were efforts made to treat whole river systems (Hindar and Kleiven, 1987). Reestablishment of the lost fauna has been studied in limed systems (Henrikson and Brodin, 1995). Studies from the acidified Lake Hovvatn indicate that recovery is slow, probably due to lack of nearby faunal refuges (Fjellheim and Raddum, 1993). However, in limed lakes with short distances to refuges, the recovery may be fast (Raddum et al., 1998).

In the River Bjerkreim, a large (706 km^2), acidified river system in southwestern Norway, liming started in 1996. The river consists of three main tributaries from the northeast and one small tributary from the northwest (Fig. 1). Before liming, the three main tributaries were acidic with pH values frequently below 5.2 and with high labile aluminium (LAl) concentrations (Henriksen et al., 1997; Kaste et al., 1997). Geology, fluvial deposits and anthropogenic effects are responsible for the higher base cation concentrations in the western arm of the river. Since autumn 1996, the salmon-rearing part of the main river has been neutralised to protect acid-sensitive organisms, and especially the Atlantic salmon which disappeared from the most acid parts of the River Bjerkreim about 100 years ago.

Throughout the acidic period, the northwestern arm has supported a number of acid-sensitive species such as Atlantic salmon, snails, mayflies (*Baetis* sp.), daphnids and *Gammarus lacustris*. Because of this refuge we hypothized that acid-sensitive species should rapidly recolonize the limed parts when the watercourse was limed. We therefore started to monitor fish, invertebrates, benthic algal and aquatic vegetation in 1996.

2. Methods

Liming started in the autumn 1996 with the lakes Ørsdalsvatn (12.3 km^2) and Austrumdalsvatn (2.8 km^2). The whole river, however, was not lime treated before the autumn 1997, when the lime doser at Malmei was implemented (Figure 1). Since then, both the two large lakes have been limed annually. Limestone powder has been added continuously from the Malmei doser. In 1998 and 1999, a total of 2200-2700 tonnes of limestone powder has been added annually to the system. Since 1998, target pH values have been set to 6.2 during the salmon pre-smoltification period (February 15 – March 31), 6.4 during the smoltification period (April 1 – May 31) and 6.0 during rest of the year.

Since 1970 monthly and sometimes biweekly dip water samples have been taken at Tengs (Figure 1), near the outlet of the main river. During 1970-1979, the water was analysed for pH. Since 1980, all water samples were analysed for pH, Ca^{2+}, Mg^{2+}, Na^+, K^+, Cl^-, SO_4^{2-}, NO^{3-}, reactive Al (RAl), organic-Al (ILAl), tot-N, tot-P, and tot organic C (TOC). The Al species are measured colometrically after Røgeberg & Henriksen (1985), and labile Al (LAl) is defined as RAl minus ILAl.

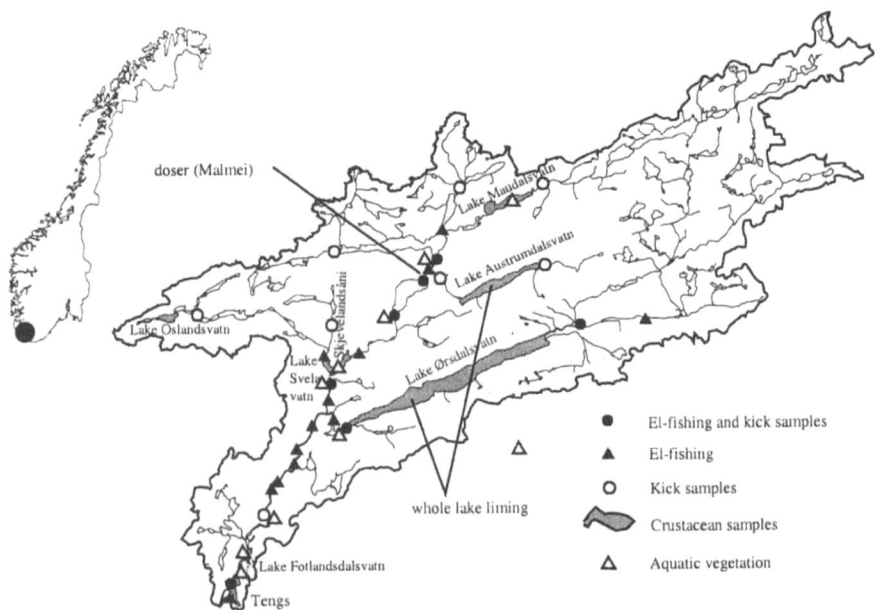

Figure 1. The catchment area of River Bjerkreim water system.

The population of Atlantic salmon (*Salmo salar*) and Brown trout (*Salmo trutta*) from 1996 to 1999 was studied. The fish were sampled by an electrofishing apparatus (1600 V, DC, unloaded) at 20 sites spread throughout salmon-rearing stretches of the river. At each location the electrofishing was carried out thrice in the upstream direction. Fish densities (number per 100m^2) were estimated by the removal method (Bohlin *et al.*, 1989), and separately for fry (0+) and parr ($\geq$1+ age groups).

Microcrustaceans were sampled twice a year (summer and fall) in six lakes (Figure 1). Qualitative plankton samples were taken by a net haul, while littoral crustaceans were sampled from the most frequent habitat in the littoral zone, also by a nethaul. At 15 river stations and 6 littoral lake sites benthic kick samples (Frost *et al.*, 1971) were collected (spring and fall). Benthic algal vegetation was sampled after methods given in Lindström (1991). The distribution, composition and structure of aquatic vegetation were surveyed in six lakes where abundance of the species was scored according to a five-step, subjective scale.

3. Results

The River Bjerkreim was poorly buffered with mean calcium concentrations of 1.1 mg l^{-1}, mean pH-values of 5.75 and mean ANC of 12 μeq l^{-1} (1993-95). Concentrations of LAl reached levels of 70 μg l^{-1}. Long-term pH data from the river outlet were between 4.9 and 6.8 (Figure 2). Until the early 1980s pH drops to about 5.0 were measured during acidic episodes, but since 1985 no pH drops below 5.2 have been observed. After liming, mean calcium concentrations and pH have increased to 1.6 mg l^{-1} and 6.4, respectively, and LAl concentrations have never exceeded 10 μg l^{-1}. pH values have been remarkably stable and since 1998 no pH depressions below 6.0 have been detected. During April 1–May 31 pH was below the target value (6.4) 70% of this period, but never below 6.2.

Prior to liming in late 1996, Atlantic salmon fry (0+) and parr ($\geq$1+ age groups), were only found at four out of 20 selected sampling sites. In 1999

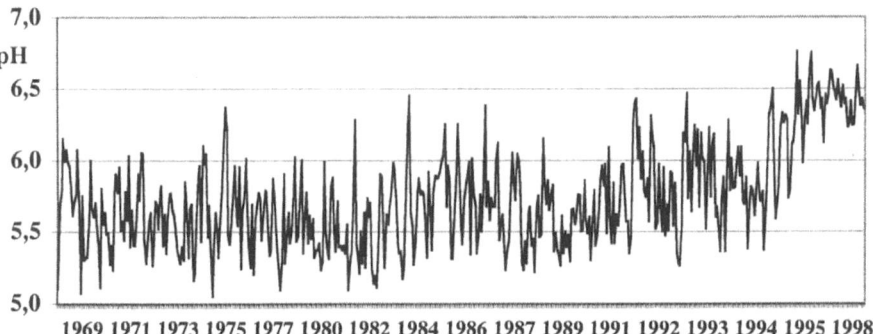

Figure 2. pH at Tengs during 1969-1999.

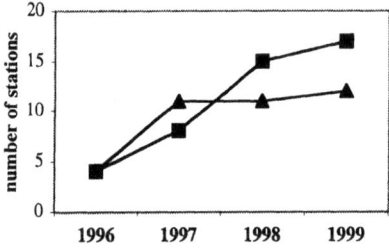
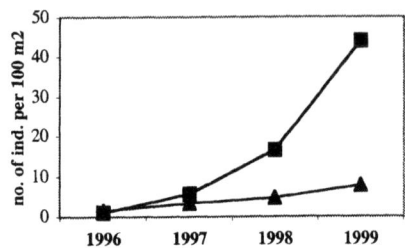

Figure 3. Fry (■) and parr (▲) caught at 20 stations during 1996-1999.

Figure 4. Density (per 100 m^2) of fry (■) and parr (▲) during 1996-1999.

the figures were 17 stations for fry and 12 for parr (Figure 3). Upstream Lake Ørsdalsvatn, neither fry nor parr were caught. Concurrently, the density of juvenile Atlantic salmon has increased (Figure 4). In 1996, a mean density was 1 fry per 100 m^2, in 1999 it was 44 (per 100 m^2). The densities of parr in 1996 and 1999 was 1.4 and 7.6, respectively. The density of fry varied between 1 and 265 induviduals per 100 m^2, while parr varied between 1 and 84 induviduals per 100 m^2. The highest number of parr was found at the site close to the outlet of river Bjerkreim, whereas the highest densities of fry were found in the main river ca 10 km upstream of the river mouth.

The number of stream-living acid sensitive-invertebrate species at 15 sites has increased from 11 before liming to 19 in 1999. Mayflies such as *Caenis horaria, C. luteolum, Baetis fuscatus* and the snails *Lymnaea peregra, L. truncatula* and *Gyraulus* sp. are new species found in the main river after liming. Acid-sensitive species found before liming have expanded to new sites. *L. peregra* is found in Lake Fotlandsvatn after liming while it has increased in number in Lake Svelavatn. Two species of mayflies, *Centroptilum luteolum* and *Cloëon dipterum*, are new to Lake Svelavatn after liming. The latter was also found in Lake Fotlandsvatn after liming.

Using planktonic and littoral crustaceans as indicis on recovery (DCA-ordination), there is a weak tendency that the species composition of the limed lakes have moved towards the neutral end of axis 1. The calanoid *Mixodiaptomus laciniatus* was found in Lake Austrumdalsvatn for the first time in 1999. It is common in the neutral reference lake Oslandsvatn and also found in the lakes Svelavatn and Fotlandsvatn. In the latter, both *D. longispina* and *D. galeata* have occurred after liming.

The aquatic macrophyte vegetation in the River Bjerkreim is species rich, including acid-sensitive species, especially in western areas. In 1998, two years after liming, an increase in acid sensitive species such as *Myriophyllum alterniflorum* was observed. Growth of the grass-like, macrophyte, *Juncus supinus bulbosus,* did not change much after liming, while vitality in the acid tolerant moss, *Nardia compressa,* was reduced just downstream the doser.

The benthic algal vegetation in River Bjerkereim is characterised by high species diversity, and there was a substantial annual variation in species composition and abundance. Occasional mass development of filamentous green algae may occur in some of the river reaches. An increase of the moderately acid-sensitive *Stigonema mamillosum* was observed in lake Ørsdalsvatn after liming.

4. Discussion

After liming the water chemistry in River Bjerkreim has been quite stable, and no pH drop below 6.0 has been detected. However, the present liming goal of pH equal to 6.4 during April 1 – May 31 has been difficult to achieve, but drops below pH 6.2 do not occur. The stable pH may be due to the presence of two large lake basins (Lake Ørsdalsvatn and Lake Austrumdalsvatn). They seem to have a stabilising effect on the hydrology and water chemistry of the lakes and the lower parts of the main river.

Before liming (1993-1995), concentrations of inorganic monomeric aluminium (LAl) reached levels of 70 µg l^{-1}, which can be harmful for many aquatic organisms including Atlantic salmon (Staurnes et al., 1995). After liming, LAl concentrations were below 10 µg l^{-1}, and in combination with a stable pH above 6.0, this provides a suitable water quality for most acid-sensitive aquatic organisms. Water quality at the main inlet of lake Ørsdalsvatn (pH < 5.0 and LAl>140 µg l^{-1} during snowmelt), however, is still poor and does not sustain Atlantic salmon.

Contrary to most acidified Norwegian rivers, the River Bjerkreim includes areas where the damage from acidification has been small. A small population of Atlantic salmon has always reproduced in the tributary, River Skjevelandsåni (Figure 1).

The catch statistics for salmon indicated that a minimum population level in River Bjerkreim was reached in the 1970s. An increase in catches was documented in the late 1980s, which may be related to the restoration of the salmon ladder at Tengs (1979) and less severe drops in pH. In more recent years, however, there has been a considerable increase in the Atlantic salmon catch; from about 0.8 tons in 1994 to about 10 tons in both 1998 and 1999. Although the catch statistics, kept since the 1880s,, were less reliable prior to 1994, the increase in catches after liming have been very substantial. Since 1993 there has been a release of unfed salmon fry in the spring (20 000 and 280 000 fish annually, mean 159 000 fry). The increased abundance of adult Atlantic salmon in the River Bjerkreim may also be due to these releases. But the main effect results appear to be from an improvement in water quality. It seems evident that the recent increase in the density of young Atlantic salmon will overrule the number of stocked individuals.

Acid-sensitive invertebrates like the mayfly *Baetis rhodani* and snails (*Lymnaea* sp and *Gyraulus* sp.) are now invading the limed parts of the river system. *Baetis* sp. are known as a group that rapidly reponds to improved water quality. It may also survive in small nearby refuges. Snails may also recolonised after liming, though at a slower pace than insects (Raddum et al. 1998).

Colonising of acid sensitive invertebrates seems to occur faster in some lakes (Svelavatn and Fotlandsvatn) than in others (Austrumdalsvatnet and Ørsdalsvatn). Fast colonisation in the former lakes was expected, because both lakes are situated downstream lake Oslandsvatn, which has never been seriously acidified. In both lakes acid-sensitive organisms such as *Daphnia* and *Lymnaea* may have survived in low numbers or they may have followed the river from intact populations upstream. The lakes Austrumdalsvatnet and Ørsdalsvatnet are more isolated and their tributaries contain acidic water. In spite of this, *Mixodiaptomus laciniatus*, which is common in the lakes Oslandsvatn, Svelavatn and Fotlandsvatn, was new to lake Austrumdalsvatn in

1999. This species is also found as a new species in other limed lakes in South Norway.

Rapid expansion of *Juncus bulbosus* has been observed in acidified softwater lakes in Western Europe (Brandrud & Roelofs 1995). In South Norwegian lakes, this species expended rapidly the first years after liming. The expansion of acid sensitive macrophytes after liming have been more rapid in River Bjerkreim than in other rivers.

Over all our study shows that acid-sensitive species, which have survived in the non-acid refuges, have rapidly invaded the areas downstream due to improved water quality. This process seems to be slower in the lakes, receiving acidified water.

Acknowledgements

We thank Professor Bror Jonsson and Dr. Nina Jonsson for their comments to the manuscript. Special thanks to all people who assisted with field sampling over several years. The study has recieved financial support from the Directorate for Nature Management, Norway.

Literature

Bohlin, T., Hamrin, S., Heggberget, T.G., Rasmussen, G. and Saltveit, S.: 1989, *Hydrobiologia* **173**, 9.
Brandrud, T.E. and Roelofs, J.G.M.: 1995, *Water, Air and Soil Pollution* **85**, 913.
Drabløs D. and Tollan A.: 1980, Proceedings from the Internationale Conference on ecological Impact on Acid Precipitation, Sandefjord, Norway.
Fjellheim, A. and Raddum,: G.G.: 1993, 'Changes in the mayfly community of Lake Hovvatn during the 12 years of Liming` in G. Guissani and C. Callieri (eds.), *Strategies for lake ecosystems beyond 2000*. Proceedings, Stresa, pp. 444-447.
Frost, S., Huni, A. and Kershaw, W.E.: 1971, *Canadian Journal of Zoology* **49**, 167.
Havens, K.E. and Hanazato, T.: 1993, *Environmental Pollution* **82**, 277.
Henriksen, A., Hindar, A., Hessen, D.O. and Kaste, Ø.: 1997, *Ambio* **26**, 304.
Henrikson, L. and Brodin, Y.W.: 1995, Liming of acidified surface waters. A Swedish synthesis. Berlin: Springer Verlag.
Hindar, A. and Kleiven, E.: 1987, Kalkingsvirksomheten i perioden 1984-1986. DN-rapport 2-1987. Direktoratet for naturforvaltning.
Kaste, Ø., Henriksen, A. and Hindar, A.: 1997, *Ambio* **26**, 296.
Lindström, E.-A.: 1991, 'Use of periphyton for monitoring rivers in Norway. in B.A. Whitton *et al.* (ed.) *Use of algae for monitoring rivers*. Universität Innsbruck. pp. 139-144.
Muniz, I.P.: 1991, *Proceedings of the Royal Society of Edinburgh* 97B, 227.
Mylona, S.: 1993. Trends of sulphur dioxide emissions, air consentrations and deposistions of sulphur in Europe since 19880. EMEP/MSC-W Report 2/93. Norway.
Raddum, G.G., Hansen, H. and Walseng, B.: 1998, *Verhandlungen der internationale Vereinigung für theoretische und angewandte Limnologie* **26**, 760.
Røgeberg, E.J.S. and Henriksen, A.: 1985, *Vatten* **41**, 48.
Schindler, D.W.: 1988, *Science* **239**, 149.
Staurnes, M., Kroglund, F. and Rosseland, B.O.: 1995, *Water, Air and Soil Pollution* **85**, 347.

CAN PHOSPHATE HELP ACIDIFIED LAKES TO RECOVER?

A. LYCHE-SOLHEIM[1*], Ø. KASTE[1] and E. DONALI[2]

[1] *Norwegian Institute for Water Research, P. Box 173 Kjelsaas, 0411 Oslo, Norway*
[2] *Norsk Hydro ASA Research Center, P. Box 2560, 3907 Porsgrunn, Norway*
(* author for correspondence, e-mail: anne.lyche@niva.no)

Abstract. Experimental addition of phosphate to enclosures in an acidified lake in Southern Norway was performed to study the effect on nitrate, pH and labile aluminium along a gradient of phosphate from 4-19 µg P L^{-1}. Nitrate decreased from 180 µg L^{-1} to below detection limit after three weeks at P-concentrations > 17 µg L^{-1}, due to phytoplankton uptake. pH increased from 4.9 to 5.2, corresponding to a 50% decrease of H$^+$-equivalents from 12 to 6 µeq L^{-1} due to algal uptake of H$^+$-ions when assimilating NO$_3^-$-ions. Due to the increased pH and probably also precipitation with phosphate, concentrations of labile aluminium decreased from 150 to 100 µg L^{-1} within the P-interval 4-19 µg L^{-1}. Algal biomass increased from 0.5 to 6 µg chlorophyll a L^{-1} along the same P-gradient. The results suggest that moderate P-addition (< 15 µg P L^{-1} to avoid eutrophication problems) can improve water quality in moderately acidified lakes, and also increase nitrate retention in strongly acidified lakes. In humic lakes, the treatment will be less efficient due to light limitation of primary production and the presence of organic acids.

Keywords: acidification, phosphate addition, nitrate retention, alkalinity generation, eutrophication

1. Introduction

In Northern Europe the relative importance of nitrate to acidification is increasing due to reduced emissions of sulphate in recent years (Henriksen *et al.*, 1997, Kaste *et al.*, 1998). In acidified lakes and rivers in Southern Norway nitrate concentrations are often very high due to atmospheric deposition, whereas total phoshorus concentrations are very low (< 4 µg L^{-1}, Skjelkvåle *et al.*, 1997). This leads to low nitrate retention in many catchments (Kaste *et al.*, 1997), and possible eutrophication in estuaries and near-shore coastal waters during summer periods. When the nutrient balance is shifted towards a high N:P-ratio, P may become the limiting nutrient for algal growth. Empirical data show increased probability for dominance of toxic dinoflagellates in coastal waters when P is the limiting nutrient, and silica is depleted (Skjoldal, 1993).

Today, liming is the most widely used method to restore acidified lakes. As an alternative method, Davison *et al.* (1995) demonstrated that phosphate addition to a small lake reduced the acidity by increasing the nitrate uptake by algae or plants. This prosess of nitrate assimilation can be written as:

$$106\ CO_2 + 16\ NO_3^- + HPO_4^{2-} + 18\ H^+ + 122 H_2O = (CH_2O)_{106}(NH_3)_{16}(H_3PO_4) + 138\ O_2 \quad (eq.\ 1)$$

Thus, an increase of aquatic N-retention could have beneficial effects both in acidified limnic ecosystems, through increased pH-levels, and in near-shore marine waters, through reduced probability for development of nuisance algae.

In the whole-lake experiment reported by Davison *et al.* (1995), P-addition was found to increase NO_3-retention and increase pH from 5.2-5.9 in a small, oligohumic British lake. However, the amounts of P added during the last year of this experiment (28 µg L^{-1}), produced a mean chlorophyll-a concentration of 13 µg L^{-1} with peaks above 30 µg L^{-1}, which is above the limit where nuisance algae may develop.

The objective of our study has been to investigate whether lower levels of P-addition can be effective to increase NO_3-retention and increase pH in acidified lakes and yet keep algal biomass below critical limits.

2. Materials and Methods

During a three-week experiment in July 1998, phosphate (NaH_2PO_4) was added to 8 pelagic mesocosms (1.5 m wide and 3 m deep) in a small, mesohumic acidified lake in Grimstad, Southern Norway (58°22'N 8°26'E). The P-additions produced an initial gradient of total phosphorus (tot-P) in the 8 mesocosms from 4 µg L^{-1} (control mesocosm), which was equal to the surrounding lake, - to 8, 12, 16, 20, 24, 28 and 32 µg L^{-1}.

The mesocosms were sampled with a tube sampler every third day throughout the experimental period. Parallel samples were also taken from the lake epilimnion (0-3 m integrated). The samples were preserved and analyzed according to standard analytical methods for total phosphorus (tot-P), nitrate (NO_3^-), ammonium (NH_4^+), pH, alkalinity (alk), chlorophyll a (chla), sulphate (SO_4^{2-}), total organic carbon (TOC), colour, and total-, reactive and organic aluminium (tot-Al, R-Al and IL-Al). Labile Al (L-Al) is defined as the difference between R-Al and IL-Al.

3. Results and Discussion

Except for TOC and colour, there were no significant differences between the analyzed parameters in the lake epilimnion and the control mesocosm throughout the experimental period. TOC and colour were clearly reduced in the control mesocosm compared to the surrounding lake. This decrease was probably caused by continuous supply of humic material in the lake, but not in the closed mesocosms. Photo-oxidation and bacterial respiration in the mesocosms during the experimental period probably broke down parts of the humic material present in the mesocosms at the start of the experiment.

Figure 1a shows that tot-P concentrations at the end of the experiment were 5, 8, 12, 12, 16, 14 and 17µg L^{-1} in the 7 P-added mesocosms. This 45-70% decrease from the initial concentrations is most likely caused by sedimentation.

As shown in Figure 1b the alkalinity changed markedly during the experiment in all the mesocosms as well as in the lake, and the change was greatest in the mesocosms with the highest tot-P-concentrations. This change is mainly due to biological

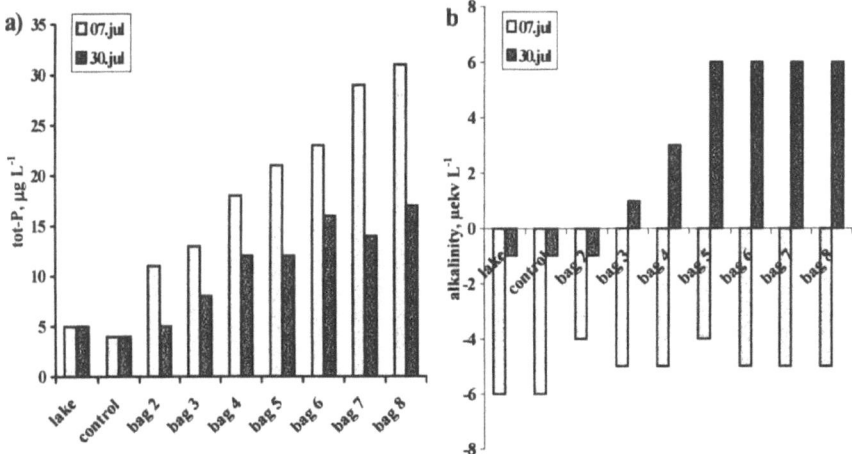

Figure 1. a) Initial and final concentrations of total-P in the bags and the lake. b) Initial and final concentrations of alkalinity in the bags and the lake.

uptake of H^+ accompanying the uptake of nitrate and phosphate according to eq.1. The alkalinity changed from negative to positive values in all the mesocosms with a final tot-P >8 µg L^{-1}. In all mesocosms with >12 µg P/L, the final alkalinity was similar at 6 µeqv L^{-1}. This was probably due to nitrate limitation in these mesocosms during the last part of the experiment (see below), which prevented further alkalinity production.

The response to the P-treatment was highly dynamic during the first half of the experiment, especially in bags receiving the highest P-additions. During the last 10 days of the experiment, however, the response parameters stabilized. In Figure 2 the response in water quality parameters is shown as means of the samples which were taken during these last 10 days (four samplings).

The nitrate concentration decreased linearly with increasing tot-P-concentration from 180 µg L^{-1} to 3 µg L^{-1} with a slope of 12.1 µg NO_3-N/ µg tot-P (R^2=0.992). This shows that a P-addition up to roughly 20 µg tot-P L^{-1} was sufficient to cause a complete retention of all nitrate present in the system. Chlorophyll a increased as expected with increasing tot-P-concentration from 0.5 to 6 µg L^{-1}. This is below the eutrophic level where nuisance algae have a high probability to develop. At tot-P concentrations above 15 µg L^{-1} there was no further increase in chlorophyll. This was most likely caused by nitrogen-limitation, since the nitrate levels were below 10 µg L^{-1} in both the bags receiving the highest P-additions. Ammonium concentrations were low (< 20 µg NH_4^+ L^{-1}) in all the mesocosms.

The effects of P-additions on acidity is shown in Figure 2c and d. As tot-P increased from 4 to 19 µg L^{-1} pH increased from 4.9 to 5.2. This corresponded to a 50% decrease in the H^+ concentration – from 12 to 6 µeqv L^{-1} along the P-gradient. No further pH increase can be expected in this system if P was added

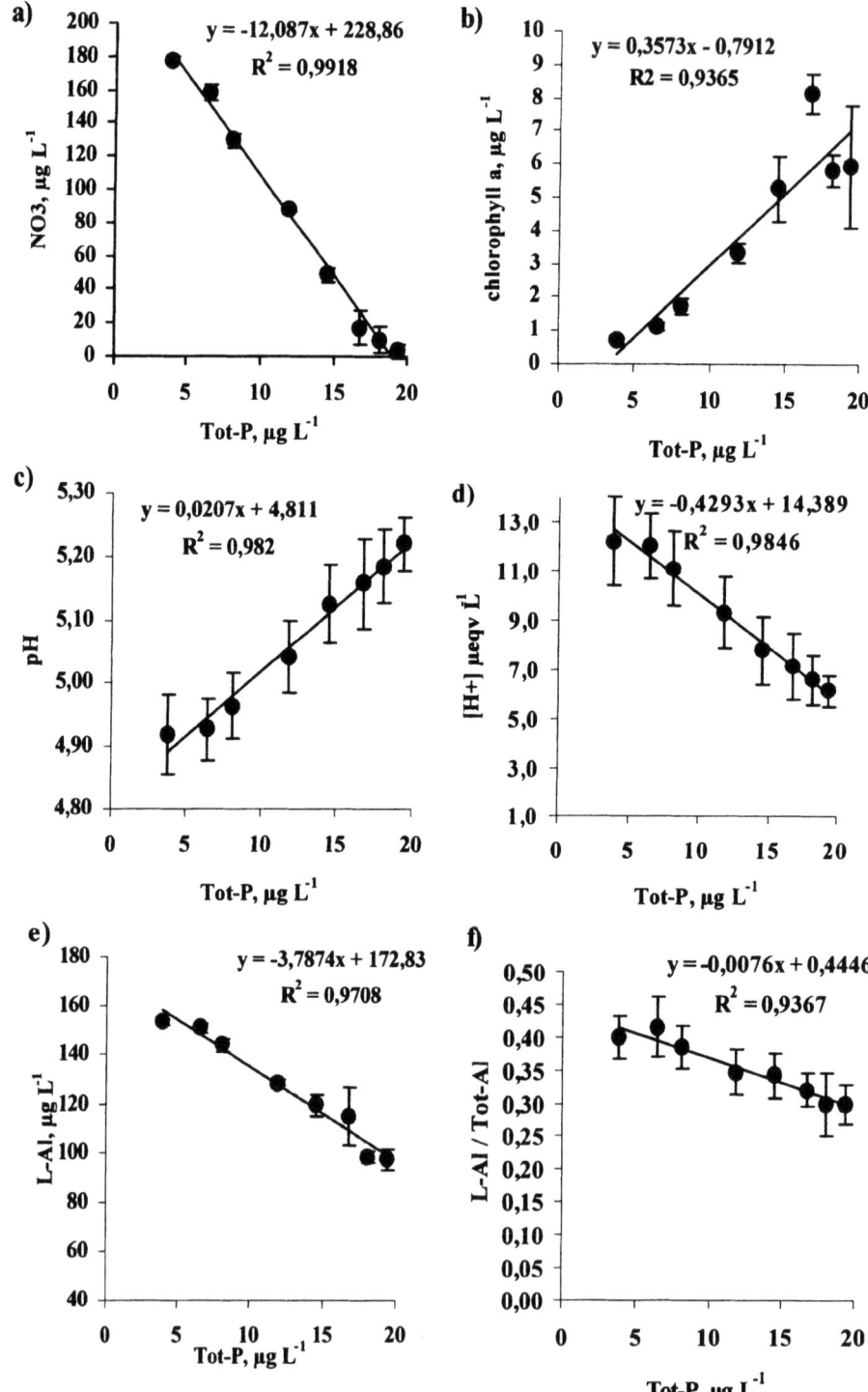

Figure 2. Change in water quality parameters versus total phosphorus-concentration. Points and error bars are means and standard error of the four last sampling dates

in higher concentrations, since there was no nitrate left during the last part of the experiment in the bag with the highest P-addition.

Along with the decrease in acidity, there was also a clear decrease in labile aluminium (L-Al) from 150 to 100 µg L^{-1} with increasing P (Fig. 2e). The decrease was -3.8 µg L-Al µg P^{-1} (R^2=0.971). Total aluminium (tot-Al) also decreased with increasing tot-P concentration with a slope of -2.9 µg tot-Al µg P^{-1} (R^2=0.77). This suggests that most of the decrease in L-Al was due to a precipitation reaction with the added PO_4^{3-}, similar to the P-removal process in sewage treatment plants where Al is added to remove PO_4^{3-}. As shown in figure 2f, the proportion of labile: tot-Al also decreased from 0.4 to 0.3 along the P-gradient, probably as a result of the decrease in acidity.

The obtained increase in pH and decrease in L-Al from the experiment was not sufficient to obtain an acceptable water quality for brown trout and other acid-sensitive species in the lake. Also, no further improvements can be expected at higher P-additions due to N-limitation of the system. In addition, the response to P-additions in the lake could be less than the response found in the mesocosms due to higher TOC and colour in the lake, which may cause light limitation of the phytoplankton and thus reduced nitrate uptake. This will be tested in an ongoing whole lake experiment.

In other lakes with even higher NO_3^- concentrations P-additions could be increased to obtain a greater reduction in H^+ and L-Al. On the other hand, higher P-additions could cause eutrophic conditions with high chlorophyll levels and a risk of persistent algal blooms. In less acidic lakes, with initial pH-values around 5.2 and similar NO_3-concentrations, however, the corresponding decrease in H^+-concentration could increase pH from 5.2 to above 6.0 at P-additions similar to what was used in this experiment.

4. Conclusions

Moderate P-additions greatly enhanced nitrate retention in mesocosms in an acidified lake during summer without causing unacceptable eutrophication problems. Transferred to whole-lake conditions, this increased nitrate retention will reduce the nitrate loading on downstream marine recipients. The increased nitrate retention caused by the added phosphorus also improved water quality in terms of decreased acidity and decreased concentrations of labile aluminium. In this highly acidic lake with initial pH of 4.9, these improvements was not sufficient to produce acceptable water quality for brown trout and other acid-sensitive species. The observed 50% decrease in H^+-concentration nevertheless suggests that a biologically acceptable water quality could be obtained with moderate P-additions (tot-P < 20 µg L^{-1}) in less acidic oligohumic lakes where intial pH is above 5.2 and TOC is sufficiently low to prevent substantial light limitation of phytoplankton production.

Acknowledgements

The authors would like to thank Jarle Haavardstun for valuable field assistance in the preparations for the mesocosm experiment. The study was funded by internal grants from the Norwegian Institute for Water Research.

References

Davison, W., George, D. G. and Edwards, N. J. A.: 1995, *Nature* **377**, 504.
Henriksen, A., Hindar, A., Hessen, D. O. and Kaste, Ø.: 1997, *Ambio* **26**, 304.
Kaste, Ø., Henriksen, A. and Hindar, A.: 1997, *Ambio* **26**, 296.
Kaste, Ø., Lyche-Solheim, A., Wright, R. F., Bakke R., Brandrud T. E., Kommedal R., Lindström, E. A. and Skiple, A.: 1998, Norwegian Institute for Water Research Report **3817-98** pp. 15-43.
Skjelkvaale, B. L., Henriksen, A., Faafeng, B. A., Fjeld, E., Traaen, T., Lien, L., Lydersen, E. and Buan, A. K.: 1997, The Norwegian State Pollution Control Authority (SFT) Report **677**
Skjoldal, H. R.: 1993, "Eutrophication and algal growth in the North Sea", in N. F. R. Della Croce (ed.), *Symposium Mediterranean Seas 2000*. Ist. Scienze Ambient. Marine, Sta. Margherita Ligure pp. 445-478.

ATLANTIC SALMON AND ACIDIFICATION IN SOUTHERN NORWAY: A DISASTER IN THE 20TH CENTURY, BUT A HOPE FOR THE FUTURE?

S. SANDØY* and R.M. LANGÅKER

Directorate for Nature Management, N-7485 Trondheim, Norway
(* author for correspondance, e-mail: steinar.sandoy@dirnat.no)

Abstract. Due to acidification, 18 Norwegian stocks of Atlantic salmon are extinct and an additional 8 are threatened. In the two southernmost counties, salmon is eradicated. Due to the high acid sensitivity, production of salmon was greatly reduced as early as 1920, several decades before acid rain was recognized as an environmental problem. International agreements on reduced atmospheric emissions will reduce acidification effects in Norway substantially during the coming 20 to 50 years. However, the extreme acid sensitivity of salmon makes the destiny of this species in Southern Norway uncertain. Liming is an effective measure to protect and restore fish populations in acidified waters. Liming of acidified salmon rivers has become important in Norway in recent years which in combination with reduced emissions will be an important contribution to protection of the Atlantic salmon species. In this paper we give an overview of the effects of acidification on Norwegian salmon and discuss different aspects of mitigation measures; the expected effect of international agreements on reduced atmospheric emissions, the expected effect of liming on salmon production and the possibilities of re-establishing self sustaining salmon stocks in limed rivers.

Keywords: Atlantic salmon, Norway, acidification, recovery, liming

1. Introduction

The causes of acidification of surface water in Scandinavia were clarified during the 1960´s and 70´s, almost one century after the first negative effect on fish populations. The first indications of acidification affecting fish are from episodic killings of Atlantic salmon (Salmo salar) in some southern rivers in Norway around 1910. Official Norwegian salmon catch statistics shows a large decline in catches around 1900. In the two southern counties, Aust-Agder and Vest-Agder, catches declined about 80% from 1885 to 1920. Sporadic catches of salmon were reported up to the late 1960´s, but the natural salmon stocks in this region were virtually extinct around 1960 (Fig. 1). There have been occasional catches of salmon also during the last twenty years, but we have no indications of reproduction of salmon. In 1995 18 Norwegian salmon stocks were extinct due to acidification, 11 in Agder counties and 7 in the western counties Rogaland and Hordaland. Twelve stocks are threatened (Anon., 1995).

The physiology, life cycle and attractiveness of Atlantic salmon as a sports fish make this species vulnerable to several other threats; overexploitation, river regulations and escaped farmed salmon. Due to a low degree of urbanisation, Norway has many intact salmon stocks in the middle and northern part.

In 1983 the Norwegian government initiated a liming programme. The funding increased year by year, until 1996, and after yearly budgets have been around 110 mill. NOK (≈13 mill. USD). Several salmon rivers were limed from

Water, Air, and Soil Pollution **130**: 1343–1348, 2001.
© 2001 *Kluwer Academic Publishers.*

the middle of 1990's, in 2000 20 rivers are limed.

The purpose of this paper is to give an overview of the status of Atlantic salmon in Norway with an emphasis on acidification, and with a management point of view. We also present the future expectations of liming of acidified salmon rivers and the planned reductions of emissions in Europe.

2. Effects of Acidification on Atlantic Salmon

During the 1800's salmon was an important resource in southern Norway, as food and for export, and wealthy people from Europe arrived during summer for salmon fishing, bringing resources to many local communities. According to official salmon catch statistic, more than 70 tons were caught in 1885 in the 7 southernmost rivers. In Mandal River, official catches were 36 tons in 1885, among the 5 best in Norway.

There are no measurements of water chemistry from that period, and the exact relationship between water quality and the development of salmon populations cannot be ascertained. Mylona (1993) has estimated of sulphur dioxide in Europe from 1880 (Fig. 1).

Sulphur deposition in southern Norway in 1890 was 400mg S/m^2. Critical load for this area is calculated to be around 300mg S/m^2 (Henriksen *et al.*, 1992), which

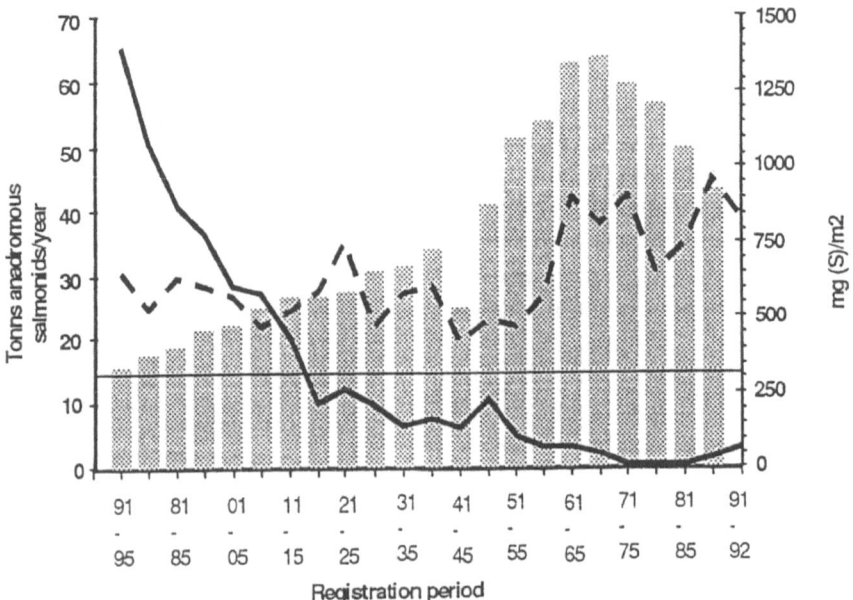

Figure 1. Deposition of sulphur and catches of salmon. Bars: Estimated historical deposition of oxidised sulphur in southern Norway. Continuous line: Catches of salmon in 7 rivers along the south coast. Dotted line: Catches of salmon in 29 rivers on western coast. (From : Mylona 1993 and Kroglund *et al.* 1994.)

indicates an exceeded critical load for acidification as early as around 1890. This may explain the declining salmon catches around 1900 and indicates that even a small exceedance of critical load of sulphur may be harmful for salmon. A further increase of the exceedance from 1900 and onwards was followed by a rapid decline in salmon catches up to 1920 (Fig. 1) and a further decline and extinction of several salmon stocks in the next 40 years. The total annual loss of salmon production of Norwegian stocks due to acidification is estimated to range between 345 and 1150 tons (Hesthagen and Hansen, 1991). Acidification is therefore the single factor causing the most substantial negative effects on salmon stocks in Norway.

3. The Norwegian Liming Program

The national liming program is an intermittent mitigating measure against the extensive damages of freshwater ecosystems by acid rain. Operational liming followed the research program on liming run in Norway from 1979 to 1983 (Baalsrud et al., 1985; Henrikson et al., 1995). The two first limed salmon rivers were River Audna (1985) and Vikedal River (1987). The salmon stock became extinct around 1970 in River Audna and was strongly reduced during the 1970´s and early 1980´s in Vikedal River (Hesthagen, 1989) (Fig. 2).

From 1994 the liming budgets reached a level that made liming of the large salmon rivers possible. A main goal is to develop an economical optimal liming program giving acceptable biological results. The main cost of most liming projects are purchase of powdered limestone. Size of lime doses is therefore an

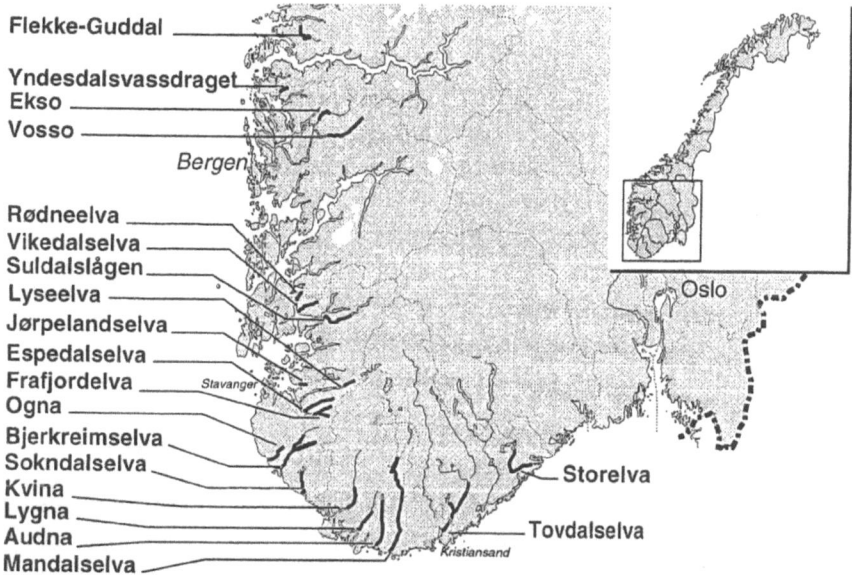

Figure 2. Limed salmon rivers in Southern Norway.

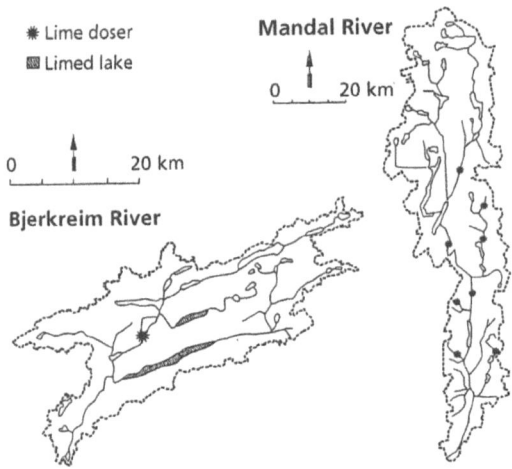

Figure 3. Liming strategies in Bjerkreim River and Mandal River.

important issue, directly related to both costs and ecological effects.

The liming strategies are under continuous development. Two main liming strategies are used for liming salmon rivers depending on the characteristics of the river and watershed (Fig. 3): Lime dosers for continuous liming of running water and liming directly on lake surface (Henrikson et al., 1995). In the lime dosers, the amount of lime is usually controlled by pH and water flow to stabilise the water quality downstream (Sandøy and Romundstad, 1995). Most projects have a consumption between 500 and 3000 tons lime a year to produce an acceptable water quality. For all 20 salmon rivers about 40 000 tons limestone/year would be required.

The biological effects of acid water and liming are studied by extensive monitoring- and research projects (Anon. 1999). Exposure studies of salmon parr and smolt have showed different water quality requirements for the different freshwater stages (Rosseland and Staurnes, 1994). Studies indicate an increasing sensitivity towards the smolt run, usually occurring in May (Staurnes et al., 1993) and even moderate acidification seems to affect physiological adaptation to see water (Staurnes et al., 1995; Kroglund and Staurnes, 1999). Therefore different pH-targets are set for liming in different seasons. During the period 1 June to 14 February salmon rivers should be limed to pH 6.0. A pH target of 6.2 is from 15 February to 31 March, and pH 6,4 from 1 April to 31 May. The elevated pH levels in late winter and spring will also contribute to an increased protection of the fish against acidic episodes during snowmelt.

The rapid increases in densities of juvenile Atlantic salmon in the limed rivers are very promising (Anon., 1999). Increased catches of adult salmon have been recorded in rivers limed for several years. Estimates of future river catches by rod indicate an increase of 80 to 100 tons a year as a result of liming.

4. Liming and Re-Establishing Extinct Salmon Stocks

In 9 rivers the main goal is to re-establish a self-reproducing salmon population. Two strategies have been used: liming with or without a stocking program. So far both strategies seem to be successful, but we do not know the genetic effect or the long-term result of either strategy. Liming without stocking gives a surprisingly rapid re-colonisation of salmon. Sokndal River in Rogaland county was limed in 1989. The first yearlings of salmon were recorded in 1990 and from 1994 density of yearlings and parr have been between 30 and 70 per 100 m2 (Anon., 1999). Catches of adult salmon in Sokndal River have the latest years been between 1 and 1.5 tons/yr. The salmon spawning after liming must have been strayers from other rivers or escaped farmed salmon.

A research project started in 1996 with the aim to study the re-colonisation process of salmon, evaluate the genetic effects of stocking strategies, compare stocking and natural re-colonisation and study population dynamics of re-colonising salmon. Mandal River started in spring 2000 and Tovdal River will start in 2002 are the main study sites. In both rivers it is possible to artificially prevent passage of salmon to the upper part, and combine the two re-colonising strategies, stocking and natural re-colonisation.

An important challenge was to choose appropriate parent populations for stocking. For both historic populations life history parameters are known, such as smolt age, age and size of spawners (Huitfeldt-Kaas, 1946). Abiotic parameters of the river, such as water flow and temperature, are believed to affect the phenotypic and genotypic characteristics of a salmon stock. Genetic mapping has shown that salmon stocks in the same region are genetically more related than geographically separated stocks. Geographic and watershed characteristics were important criteria when selecting parent stocks for re-establishing salmon in Mandal and Tovdal Rivers. River Storelva has the only remaining natural salmon stock in the Agder region, and was chosen as parental stock for Tovdal River (Fig 2). Bjerkreim River is the geographically closest remaining stock to Mandal River, and was selected for Mandal River. The life history parameters of salmon in the two rivers were also quite similar (Huitfeldt-Kaas, 1946).

5. Critical Load and Recovery

During the last two decades the European nations have made agreements to reduce atmospheric emissions of acidifying compounds. The latest and most extensive was signed in Gothenburg in December 1999. Based on the steady state critical load models, the agreed reductions are estimated to reduce the area where critical load is exceeded in Norway by about 80% compared with the 1985 level, 15-30 years after the reductions are fulfilled in 2010 (Wright, 2000). That means the water quality will make possible a recovery of e.g. brown trout (Salmo trutta) in a large number of lakes during the first half of this century.

The dose-response function, which the calculations of critical load exceedances are based on, is that of brown trout to ANC (Acid Neutralising Capacity) (Henriksen *et al.*, 1995). A dose-response relationship between ANC and Atlantic salmon population status is much more difficult to establish, due to the complicated life cycle dynamics and the several environmental threats affecting salmon. Salmon is much more sensitive to acidic water than brown trout. That means that the critical load of salmon is more exceeded than that of brown trout, and the recovery of salmon will take more time than usually presented as the general effect of the Gothenburg protocol. The present knowledge does not, however, allow us to conclude more exactly. Therefore, we cannot say when the recovery process may allow us to stop liming of the salmon rivers, or if the agreed reductions ever will give satisfactory conditions in the most acidified rivers. The moderately acidified rivers will most likely achieve non-acidified conditions within the next 10 to 20 years (Hindar et al., 1998; Wright, 2000).

6. Conclusion

In southern Norway, 18 natural Atlantic salmon stocks are extinct because of acidification and salmon has been nearly absent from the entire southern region the last 30 years. The national liming program includes 20 salmon rivers, and the aim is to re-establish self-reproducing salmon populations in 9 of the mostly acidified rivers. The liming will bring the salmon back to the southern region of Norway after 50 years of absence, and allow yearly river catches by rod of salmon to increase by 80-100 tons in the limed rivers. Reduced emissions will lead to reduced acidification. The future status of the salmon rivers is, however, uncertain due to the high acid sensitivity of salmon. Some of the populations will certainly recover, but we do not know if natural recovery from acidification will allow stopping the liming of all the salmon rivers.

References

Anonymous 1995: Directorate for Nature Management, Norway: *DN-notat*-1.
Anonymous 1999: Directorate for Nature Management, Norway: *DN-notat*-4.
Baalsrud, K., Hindar, A., Johannessen, M., and Matzow, D.: 1985, "Liming of acid water. Liming project, final report", Department of the Environment, Directorate for Nature Management (Norway).
Henrikson, L., Hindar, A. and Thörnelöf, E.: 1995, *Water, Air, and Soil Pollution* **85**, 131.
Henriksen, A., Kämäri, J., Posch, M. and Wilander, A.: 1992, *Ambio* **21**, 356.
Henriksen, A., Posch, M., Hultberg, H. and Lien, L.: 1995, *Water, Air, and Soil Pollution* **85**, 2419.
Hesthagen T.: 1989, *Fisheries* **14**, 10.
Hesthagen, T. and Hansen, L.P.: 1991, *Aquaculture and Fisheries Management* **22**, 85.
Hindar, A., Henriksen, A., Sandøy, S. and Romundstad, A.J.: 1998, *Restoration Ecology* **6**, 353.
Huitfeldt-Kaas, H.: 1946, *Nytt Mag. Naturvitensk.* **B85**, 115.
Kroglund, F. Hesthagen, T. Hindar, A. Raddum, G.G., Staurnes, M., Gausen, D. and Sandøy, S.: 1994. Directorate for Nature Management, Norway, *Utredning for DN*, **10**.
Kroglund, F. and Staurnes, M.: 1999, *Canadian Journal of Fisheries and Aquatic Sciences* **56**, 2078.
Rosseland, B.O. and Staurnes, M.: 1994. In Steinberg, C.E.W. and Wright, R.R. (eds). *Acidifica-tion of Freshwater Ecosystems: Implications for the future.* John Wiley and sons Ltd: 228-246.
Sandøy, S. and Romundstad, A.J.: 1995, *Water, Air, and Soil Pollution* **85**, 997.
Staurnes, M., Blix, P. and Reite O.B.: 1993, *Canadian Journal of Fisheries and Aquatic Sciences* **50**, 1816.
Staurnes, M., Kroglund, F. and Rosseland, B.O.: 1995, *Water, Air, and Soil Pollution* **85**, 347.
Wright, R.F.: 2000, *Water, Air, and Soil Pollution*, (in press).

THE RETURN OF THE SALMON

FRODE KROGLUND[1]*, ØYVIND KASTE[1], BJØRN O. ROSSELAND[1] and TRYGVE POPPE[2]

[1]*Norwegian Institute for Water Research, PO Box 173, Kjelsås, N-0411 Oslo, Norway.*
[2]*Norwegian School of Veterinary Science. PO Box 8146, N-0033 Oslo, Norway.*
(* author for correspondence, e-mail: frode.kroglund@niva.no)

Abstract. The Atlantic salmon population in the River Otra, southern Norway was lost during the 1960's due to acid rain and industrial and municipal pollution. The industrial and municipal pollution sources were sanitized by 1995. A concurrent reduction in acid deposition has during the last 10 years raised pH from 5.2 to 5.7 and reduced inorganic monomeric Al from 71 to 28µg Al L^{-1} above the industrial area. The water quality improvement resulted in salmon fry again being caught from 1995. Physiological measurements (blood parameters and seawater tolerance) performed on smolts of Atlantic salmon exposed within the river during the spring of 1999 suggests that the smolts were fully smoltified and seawater tolerant, despite having moderate gill morphological changes and having moderate high gill Al concentrations (70-80µg Al g^{-1} dw). The smolt quality measured suggests that the river again can support a native salmon population, provided no negative change in water quality. Winter episodes and acid tributaries within the watershed can, however, offset the recovery process.

Keywords: Atlantic salmon, recovery, acidification.

1. Introduction

The river catches of Atlantic salmon (*Salmo salar*) started to decline because of acidification already at the turn of the century in southern Norway (Leivestad *et al.*, 1976). In Otra, the catches from 1920 to 1960 were variable, but lower than the catches prior to 1920. After 1960 catches quickly fell to very low levels. The anadromous stretch of the Otra watershed, West Agder, Norway (Fig. 1a) is located within a region seriously affected by acid precipitation, but is less acidified than the surrounding rivers due to differences in geology and total acid load. The lower part of the watershed was heavily affected by industrial and municipal pollution from Vennesla, but these local sources of environmental stressors were sanitized by 1995 (Kaste *et al.*, 1999). River Otra is not limed, and any water quality change upstream the industry (at Evje) must be attributed to changes in acid deposition. The station Skråstad is located in the anadromous part of the river, downstream from industrial sites. Evje is located 40 km upstream Vennesla.

After 30 years with "zero" salmon catches, salmon have been caught in the watershed from 1995 (Fig. 1b). Atlantic salmon fry were also seen for the first time in 1995, after more than three decades with no sign of salmon reproduction. This has given a rise for an optimism regarding future catches, raising the question "is the water quality at present sufficient to support a self-reproducing salmon population", or "is the watershed still suffering from acid precipitation producing water qualities that can offset the "reestablishment of Atlantic salmon". We evaluated smolt quality on native smolts caught within the watershed in the spring of 1999 and on native smolts exposed for 12 days within River Otra Seawater tolerance, physiological and histological status and gill aluminum concentration were used to evaluate smolt quality.

2. Methods

Atlantic salmon smolts were caught in the tributaries Høyebekken, Lonanebekken and Straisbekken. Only two smolts were caught in the main river. Smolts from Høyebekken and Straisbekken were placed into cages (30×50×30) at two sites in Otra and exposed there for 12 days. Smolts retained in Lonanebekken served as a handling stress control. Blood and gill samples (n=6) of wild fish were collected within 2 minutes of capture, representing fish status at time of capture (1st May). After exposure in Otra, 6 smolts were samples representing freshwater status, while15 smolts were transferred to a 24 h seawater challenge test performed in 34 ‰ at water temperatures varying between 5 and 9°C.

Blood samples were collected from the caudal artery after killing the fish by a blow to the head. Blood glucose was measured on whole blood using Medisence glucose electrodes. Hematocrit was determined in the field. Plasma chloride was measured on a Radiometer CMT-10 Chloride Titrator. We use 120-140 mM for plasma chloride, 35-45 for hematocrit and 3-6 mM for glucose as criteria of "no effect". The second gill arch on the right side of each fish was dissected out and frozen before being freeze-dried and weighed at the laboratory. The gills were digested in HNO_3 and analyzed for Al by Inductive Coupled Plasma Emission Spectroscopy (ICP). The result states the amount of gill Al as µg Al g^{-1} gill dry weight (dw). The second gill filament on the left side of each fish was dissected out and put directly into pH 7 buffered formaldehyde. Thin sections (3-5 µm) were mounted on slides using standard methods employed at the Norwegian School of Veterinary Science. One slide was stained with solochrome azurin (Denton et al., 1984) for localization of Al on and within the gill. The other slide was stained with hematoxylin and eosin (Culling et al., 1985) for evaluation of gill structure based on light microscopy (LM). Morphological changes were ranked into four categories (A-D), where "A" represents normal gill structure and "D" represents gills showing severe signs of hypertrophy, hyperplasi, fusion of lamella together with signs of necrosis (Kroglund et al., 1998). "B" and "C" represent intermediate stages.

Water samples have been collected and analyzed at NIVA for all major ions monthly since 1980 at several sites within River Otra as a part of the Norwegian water quality monitoring program (Kaste et al., 1999). Additional samples were taken during the experimental period.

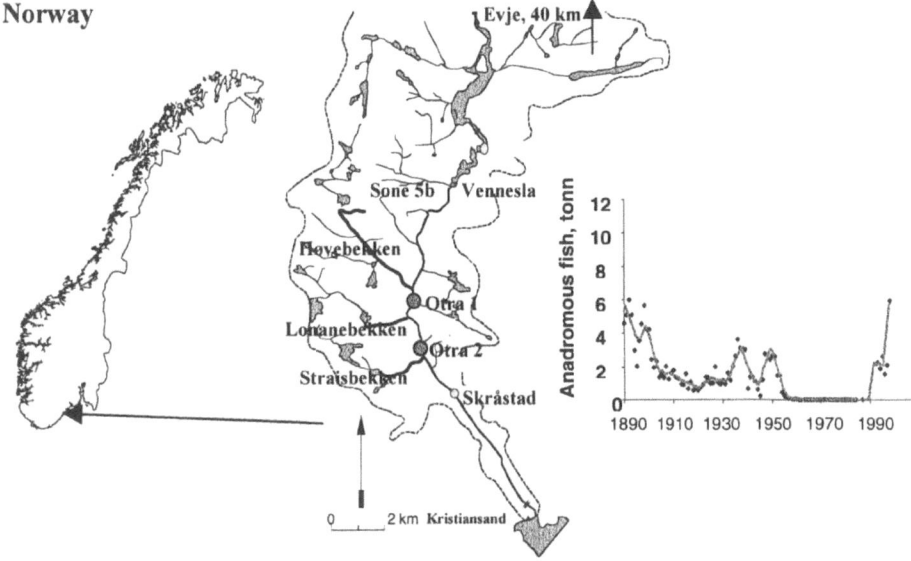

Figure 1. Map of the anadromous stretch of Otra. Atlantic salmon cannot pass Vennesla. Water samples are collected monthly at Skråstad (representing anadromous section of the river) and Evje (represented water not affected by industry). River catches of salmonids from 1880 to 1998 are inserted.

3. Results and discussion

From 1987 to 1999, pH increased from below 5.5 to levels normally in the range of 5.5 to 6.0 at Skråstad and Evje, (Fig. 2). This increase in pH was mirrored by a reduction in inorganic monomeric Al (Ali). By 1996, concentrations of Ali were generally lower than 40µg L^{-1}, as opposed to concentrations in excess of 80 µg Al L^{-1} 10 years earlier. While the changes measured at Skråstad could be related both to industrial and municipal pollution and acid rain up to 1995, changes at Evje can only be explained by reduced acid loading.

During the winter of 1998/99, pH measured at Skråstad was normally in the range of 5.7 to 6.0. In Feb. a low pH of pH 5.3 was measured. Low pH was associated with increased concentrations of Ali exceeding 40 µg Al L^{-1} from December to February. In March to May, concentrations were lower than 10 µg Al L^{-1} (Fig. 2). During the exposure period, pH in the main river never fell below 6.0 and Ali was lower than 20 µg Al L^{-1} (Table I). In the tributaries pH was higher than 6.0 and the concentration of Ali was 16 µg Al L^{-1} or lower. TOC concentrations ranged around 2 to 4 mg C L^{-1}. Calcium concentrations varied between (1 to 5 mg Ca L^{-1}) (Table I).

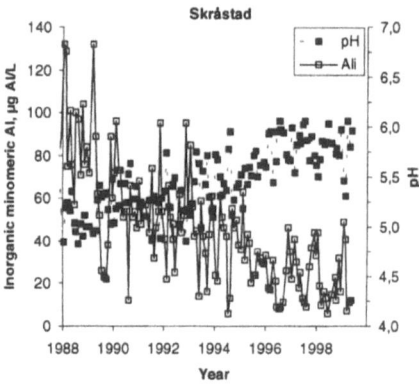 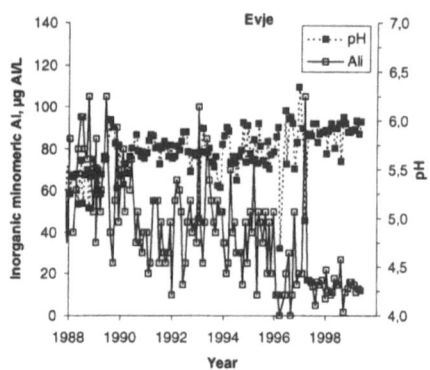

Figure 2. Changes in pH and inorganic monomeric Al (Ali) at Skråstad and Evje from 1987 to 1999.

TABLE I

pH, TOC (total organic carbon), calcium (Ca), acid reactive or total Al, reactive Al and inorganic monomeric Al (Ali) measured in Otra and in the three tributaries Høyebekken, Straisbekken and Lonanebekken prior to and at the end of the exposure period.

	Sample date	pH	TOC mg L^{-1}	Total Al μg L^{-1}	Alc μg L^{-1}	RAl μg L^{-1}	ILAl μg L^{-1}	LAl μg L^{-1}	ANC μeq L^{-1}	Ca mg L^{-1}
Otra	1st May	6.09	1.8	107	58	49	33	16	23	0.97
	12th May	6.42	3.9	202	125	77	61	16	25	3.76
Straisbkn.	1st May	6.55	2.8	143	94	49	43	6		5.26
	12th May	6.14	1.9	80	32	48	37	11		0.96
Høyebk.	1st May	6.26	3.1	171	83	88	77	11		2.1
	12th May	5.97	1.9	89	38	51	37	14		0.95
Lonanebk.	1st May	6.57	3	142	77	65	58	7		4.25
	12th May	6.46	3.6	154	74	80	69	11		2.97

The smolts were 12.7±0.9 cm long, weighted 16.0±4.1g and had a K-factor of 0.81±0.07. There was little size variation between the different localities. Fish smaller than 11 cm did not have the morphological traits of smoltifying salmon and were rejected from further analysis. The smolts exposed to River Otra waters responded as a non-stressed fish with plasma chloride concentrations between 120 and 140mM, hematocrit values between 35 and 40 and glucose concentrations between 3.5 and 6.0 mM (Table II). After 24 h exposure in a seawater challenge test, smolt should have plasma chloride levels lower than 160 mM (Hansen, 1998). A size related smolt quality/status might explain the difference in plasma chloride between Otra 1 and Otra 2. Apart from fish in Lonanebekken, having around 30 μg Al g^{-1} gill dw, all fish had gill Al concentrations ranging from 70 to 100 μg Al g^{-1} dw on both sampling dates (Table II). No aluminum was detected on the cell surface using LM, but intraepithelial Al was commonly observed. Gill Al in excess of 10 μg Al g^{-1} dw is regarded as an accumulation (B. Rosseland, unpublished data), where intraepithelial Al can damage intracellular physiological processes (Exley *et al.*, 1992). Gill histology demonstrated negative responses on the cellular level (morphological category B and C).

TABLE II

Gill Al concentration, localization of gill Al (staining), morphological and physiological status of smolts of Atlantic salmon after electrofishing and after 12 days exposure at two sites in Otra and in Lonanebekken. Seawater challenge tests (sea) were performed on fish exposed in Otra. Gill Al concentrations is given as µg Al g^{-1} gill tissue dw. Intraepithelial Al Cat A= not detected, B=, low abundance and C=very common. Histological categories B and C, see method.

		Gill-Al	Intraepithelial Al categories			Histology categories		Blood physiology			
1st May	n=		A	B	C	B	C	Plasma Cl mM	Hct %	Glucose mM	Plasma Cl sea
Høyebk.	5	81±11		20	80	60	40	137±3	34±1	3.7±0.1	
Lonanebk.	6	29±23	33	50	17	67	33	129±8	36±3	3.8±0.9	
Otra 1	2	99±12		100	0	50	50	134±4	33±4	4.3±1.1	
12th May											
Otra 1	6	76±13	33	67	0	33	67	135±4	37±4	3.1±0.2	149±11
Otra 2	6	70±14		80	20	80	20	134±1	38±3	3.4±0.6	170±20
Lonanebk.	6	29±7	83	17	0	67	33	130±10	42±8	5.0±2.9	

After many years of absence, the salmon catches in River Otra have increased since 1995, indicating a positive development towards establishment of a new self-sustainable salmon population. The detection of spawning salmon and salmon fry and smolts within the watershed is in itself not a sufficient indicator that the water quality has returned to non-affected levels. Smolts of Atlantic salmon must acquire properties that permit normal seawater migration and survival if a self-recruiting population is to be established. Seawater tolerance can be affected without freshwater life stages showing water quality related stress (Rosseland and Staurnes, 1994). Smolts can loose their capacity to ion regulate in seawater, even at low Al concentrations (Staurnes et al., 1993; 1996; Kroglund and Staurnes, 1999; Kroglund et al., 2000).

The physiological samples did not reveal any toxic response to the water quality in Otra, but both gill Al concentration and Al detected by staining suggest that Al is being accumulated. The gill morphological changes can be a response to counteract this water quality pressure, implying compensatory mechanisms. The smolts were seawater tolerant despite gill morphological changes and the accumulation of gill-Al in excess of levels that have been associated with reduced seawater tolerance in experiments (Kroglund and Staurnes, 1999; Kroglund et al., 2000). Inorganic monomeric Al was present in the watershed, with high concentrations in excess of 40µg Ali L^{-1} during the winter, but with concentrations lower than 20µg Ali L^{-1} from March to May. In experiments, smolts acquired full seawater tolerance within two weeks when they were transferred from a water quality containing 25 µg Ali L^{-1} to a water quality containing less than 10 µg Ali L^{-1} and gill bound Al was eliminated (Kroglund et al., 2000). We postulate that Al is accumulated on and in the gill structure during the winter months, but that low concentrations of Ali and high water pH during spring allow the fish to recover from prior Al exposures and develop normal smolt properties.

As far as we know, this is the first documented example that decreased acidification permits the reestablishment of a viable Atlantic salmon population. Future acidification episodes can still delay the reestablishment of the Atlantic salmon population in River Otra, but based on these data, the water quality seems to develop in a direction that could provide a healthy self-recruiting Atlantic salmon population in Otra within years.

Acknowledgement

The authors would like to thank Otra laxefiskelag (Otra fishery owners) for the financial support and two anonymous referees for valuable comments improving the final manuscript.

References

Culling, C. F. A., Allison, R.T and Barr, W.T.: 1985, *Cellular Pathology Technique*. 4th. Butterworths.
Denton, J., Freemont, A. J. and Ball, J. :1984, *J. Clin. Path.* **37**, 136.
Exley, C. and Birchall, J. D.; 1992, *J. Theor. Biol.* **159**, 1992, 83.
Hansen, T.: 1998, Oppdrett av laksesmolt. Landbruksforlaget, ISBN: 82-529-1722-4; 232 p.
Kaste, Ø., Lande, A., Larsen, B.M., Aanes, K.J. and Åsen, P.A.: 1999, *NIVA-report*, 4057-99, 58 p.
Kroglund, F., Teien, H.C., Rosseland, B.O., Salbu, B. and. Lucassen, E.C.H.E.T.: 2000, Water quality dependent recovery from aluminum stress in Atlantic salmon smolts. Submitted "Acid rain, 2000"; *Water Air and Soil Pollution*.
Kroglund, F. and Staurnes, M.: 1999, *Can. J. Fish. Aquat. Sci.* **56/11**, 2078.
Kroglund, F., Teien, H.C., Rosseland, B.O., Lucassen, E., Salbu, B. and Åtland, Å.: 1998, *NIVA-report* 3970-98. In English. 102 p.
Leivestad, H. Hendrey, G., Muniz, I.P. and Snekvik, E.: 1976, Effects of acid precipitation on freshwater organisms. In Brække, F.H. (red.), *Impact of acid precipitation on forest and freshwater ecosystems in Norway*, p. 87-111. SNSF –project, FR6/76.
Rosseland, B.O. and Staurnes, M.: 1994, Physiological mechanisms for toxic effects and resistance to acidic water: an ecophysiological and ecotoxicological approach. In: Steinberg, C.E.W and Wright, R.F. (eds). *Acidification of Freshwater Ecosystems: Implications for the Future*. John Wiley and Sons Ltd: 228-246.
Staurnes, M., Hansen, L.P., Fugelli, K. and Haraldstad, Ø.: 1996, Can. *J. Fish. Aquat. Sci.* **53**, 1695.
Staurnes, M., Blix, P. and Reite, O.B.: 1993, *Can. J. Fish. Aquat. Sci*: 1816.

RECOVERY OF YOUNG BROWN TROUT IN SOME ACIDIFIED STREAMS IN SOUTHWESTERN AND WESTERN NORWAY

TRYGVE HESTHAGEN*, TORBJØRN FORSETH, RANDI SAKSGÅRD, HANS M. BERGER and BJØRN M. LARSEN

Norwegian Institute for Nature Research, Tungasletta 2, N-7485 Trondheim, Norway.
(* author for correspondence, e-mail:trygve.hesthagen@ninatrd.ninaniku.no)

Abstract. Temporal changes in densities of young brown trout (*Salmo trutta*), mainly of age 0+, and in water quality (pH and alkalinity) were assessed by means of electrofishing in lake tributaries in three acidic, softwater watercourses in western and southwestern Norway; Gaular and Vikedal (1987-1999) and Bjerkreim (1988-1999). Approximately 74 sites were sampled each year. Most of the streams were acidic with mean annual pH levels between 5.1-5.9. Alkalinity and pH increased significantly in all three areas during the study period. Brown trout fry densities increased significantly during the period in Vikedal and Bjerkreim. Also in Gaular, the density of young brown trout has exhibited a positive trend in recent years. We suggest that the increase in the density of young brown trout is because the study areas have became less acidified during recent years due to reduction in sulphate deposition.

Keywords: Brown trout, streams, water quality, acidification, recovery

1. Introduction

A recovery phase for fish in acid-stressed waters has been recognised in recent years in several countries in western Europe and in North America as a result of reductions in acid inputs (*e.g.* Harriman *et al.*, 1995; Rask *et al.*, 1995; Gunn and Keller, 1998). Since 1980, the content of sulphate in precipitation in various areas of Norway has fallen by 51-72% (Skjelkvåle *et al.*, 2000). There has also been a widespread improvement in water quality as evidenced by a rise in pH and acid neutralizing capacity (ANC) and a reduction in sulphate and toxic inorganic aluminium (Al_i), while calcium has remained unchanged (Skjelkvåle *et al.*, 2000).

The objective of this study was to test whether a reduction in sulphate deposition and subsequent water quality improvement, leads to an increase in the density of young brown trout. This species is highly vulnerable to acid water, with embryos and alevins as their most sensitive stages (Reader *et al.*, 1991). The study was carried out in tributary streams to lakes in three catchments in southwestern and western Norway, which are less acidified than areas in southernmost Norway (Skjelkvåle *et al.*, 2000).

2. Study Area

The study streams were located in the Gaular (630 km^2), Vikedal (119 km^2) and Bjerkreim (693 km^2) catchments, at altitudes of between 145 and 715 m above sea level (Fig. 1). The streams mainly drain slowly weathering rocks such as granite and gneiss. The precipitation, at about 2000-2800 mm annually, is acidic with an

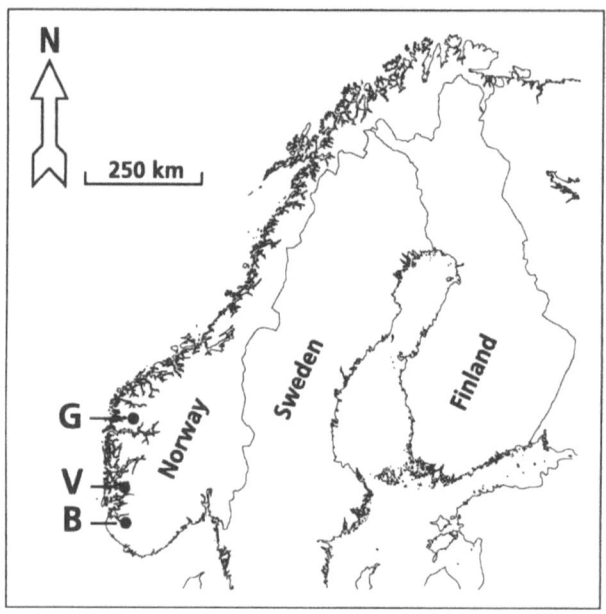

Figure 1. Location of the study areas: Gaular (G), Vikedal (V) and Bjerkreim (B).

annual range in mean pH of about 4.5-4.8 (cf. Hesthagen *et al.*, 1999). None of the study streams were affected by local water pollution, liming or habitat destruction, and they are located outside farmland, except for a few sites in Bjerkreim and Gaular. The tributary streams were typically 2-4 m, wide whereas the main inlets and outlets range between 15 and 20 m in width. Brown trout was the only fish species in the study streams, except for some juvenile specimens of brook trout (*Salvelinus fontinalis*) and Atlantic salmon (*S. salar*) in a few streams in Bjerkreim.

3. Methods

Brown trout were sampled by means of a portable back-pack electrofishing apparatus (1600 V, DC) from mid-August to early September each year between 1987-1999 in two areas (Gaular and Vikedal) and between 1988-1999 in Bjerkreim. Exceptions were that electrofishing was carried out in October in two years in Bjerkreim (1997 and 1998), and that no sampling was carried out in Gaular and Bjerkreim in 1994. We electrofished the main inlet and outlet, and secondary streams to various lakes, 26 sites in Gaular draining nine lakes, 25 sites in Vikedal draining eight lakes and 23 sites in Bjerkreim draining 12 lakes. Streams which had physical obstacles that prevented brown trout from entering them were excluded. We electrofished in an upstream direction, and both lake inlets and outlets were sampled in near proximity to the shore-line of the lake. In each stream, we established a fixed sampling area which was repeatedly electrofished each year. The lengths of all captured brown trout were measured to the nearest mm, and fish could usually be classified as belonging to either 0+ or ≥1+ age groups on the basis of their length-frequency distribution.

Each stream was sampled in a single run during the first years of the study (1987-1992), and later in three successive runs (1993-1999). Variations in water flow and water temperature may affect catches of salmonids made by electrofishing in running waters (Jensen and Johnsen, 1988). Thus, these environmental variables were included when densities were regressed against time (cf. Hesthagen et al., 1999). The mean water flow during the sampling period each year correlated significantly with fry catches (100 m^{-2} in the first fishing run) in Gaular (r^2=0.47, P<0.01) and Vikedal (r^2=0.69, P<0.001), and with parr catches in Bjerkreim (r^2=0.48, P<0.01). Then, the relationships between densities and water flow were used to standardise densities to an average flow. Finally, the probabilities (p_c) of capture were used to estimate separate densities of fry and older parr in each stream in different catchments, and hence density 100 m^{-2} (Bohlin et al., 1989). For the period between 1987-1992, we used mean values for p_c after three successive runs from 1993 to 1999 to estimate density 100 m^{-2}. Water was sampled in each stream during electrofishing and analysed for pH and alkalinity (cf. Hesthagen et al., 1999).

4. Results

Most of the streams were acidic, with a mean annual pH between 5.1-5.9, the streams in Bjerkreim being most acidic (Figure 2). Mean values for alkalinity for the entire study period in these three areas were 8.7, 10.2 and 10.6 µeq L^{-1}, respectively. Streams in all three catchments exhibited a significant increase in pH and alkalinity during the study period (linear regression, all P<0.0001 except for P<0.001 for alkalinity in Bjerkreim). In all three catchments, mean pH levels have generally remained above about 5.5 in recent years, except that in Bjerkreim in 1997, while mean alkalinity has generally risen from below 5.0 µeq L^{-1} to 10-20 µeq L^{-1}.

A total of 15,924 brown trout were caught, of which the largest fraction consisted of fry (n=11,114 or 70 %). Older parr were mainly one- and two-year-old fish. The density of fry and parr typically ranged between 10-30 and 5-10 specimens 100 m^{-2} in all three catchments, respectively (Figure 3). In Gaular, the density of neither fry or parr correlated with year (Table I), although there was a positive trend in fry density from 1993 to 1998. In Vikedal, fry density correlated positively with both water flow and year number, explaining 86-% of the variability in fry density ($F_{2,10}=37.8$, $P<0.0001$). The annual variability in parr density in this area correlated only with year number ($F_{1,11}=10.26$). In Bjerkreim, year number was the only variable that correlated with fish density, but only for fry ($F_{1,9}=5.58$).

TABLE I

Significance levels (P) and coefficient of determination (r^2) from simple linear regressions of year number (Time) and water flow and fry and parr densities in Gaular, Vikedal and Bjerkreim river systems. ns=not significantly different ($P>0.05$)

	Gaular		Vikedal		Bjerkreim	
	Fry	Parr	Fry	Parr	Fry	Parr
P for Time	ns	Ns	<0.005	<0.05	<0.05	ns
P for Water flow	ns	Ns	<0.001	ns	ns	ns
r^2 for Time			0.17	0.31	0.31	
r^2 for Water flow			0.69			

5. Discussion

Alkalinity and pH increased significantly in all three areas during the study period. pH has generally increased by about 0.4-0.5 units, and has remained above about 5.5 in recent years. Thus, it is likely that the initiation of the recovery of young brown trout in the study araes is related to a general decline in acidication. The rise in pH is of biological significance because it occurred at levels of vital importance for the survival of young brown trout. Bioassays in acid water have shown that a rise in calcium concentration from about 0.5 to 1.0 mg L^{-1}, and in pH from about 4.5 to 5.4, significantly reduces mortality in yolk-sac fry of brown trout (Brown and Lynam, 1981; Brown, 1983). The concentrations of calcium in the study streams are within this range, with values between 0.3 and 0.7 mg L^{-1} (unpubl. data). It has earlier been shown that calcium explained a relatively large fraction of the variability in brown trout density in the study streams (Hesthagen, 1997; Hesthagen et al., 1999). Although these sites had low concentrations of inorganic aluminium, the combined effect of acid water and low calcium levels may have affected their stocks of young brown trout (cf. Hesthagen et al., 1999).

Although water quality conditions in streams in Gaular has also improved, no significant increase in the density of young brown trout was found here. Streams in this area had relatively high fish densities during the late 1980s, probabaly because this area is less acid than Vikedal and Bjerkreim. However, Gaular exhibited a pronounced fall in young brown trout density in the early 1990s, which may to some extent have been an effect of sea salt episodes (cf. Hindar et al., 1994; Barlaup and Åtland, 1996).

In conclusion, the recovery of young brown trout in three study areas in western and southwestern Norway has commenced, and this process can be related to recent improvements in water quality. However, these watercourses are regarded as only moderately acidified, at least compared with areas in southernmost Norway. Thus, the initiation and duration of water chemical and biological recovery in acidified areas will be related to the critical load exceedance (cf. Henriksen et al., 1999). Although there has been a pronounced reduction in acid deposition in recent years, full chemical and biological recovery may take a long time to achieve.

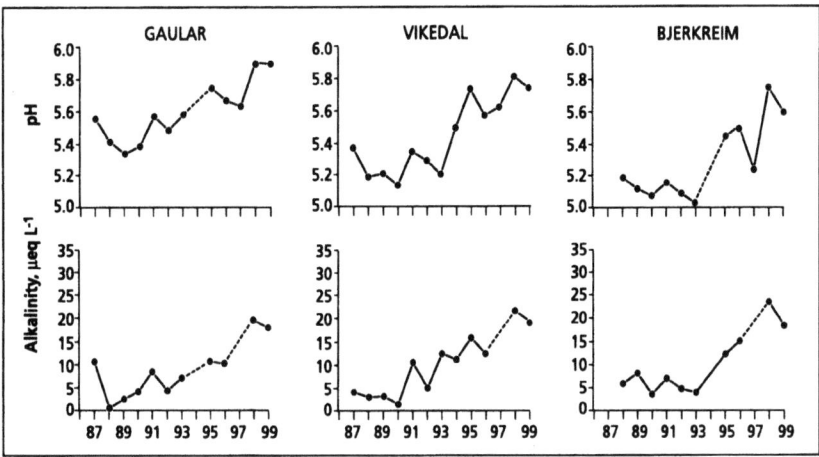

Figure 2. Mean pH and alkalinity values in streams sampled in the Gaular and Vikedal (1987-1999) and Bjerkreim (1988-1999) catchments. Water was not sampled in Gaular and Bjerkreim in 1994, and alkalinity was not measured in any of the areas in 1997 (indicated by broken lines).

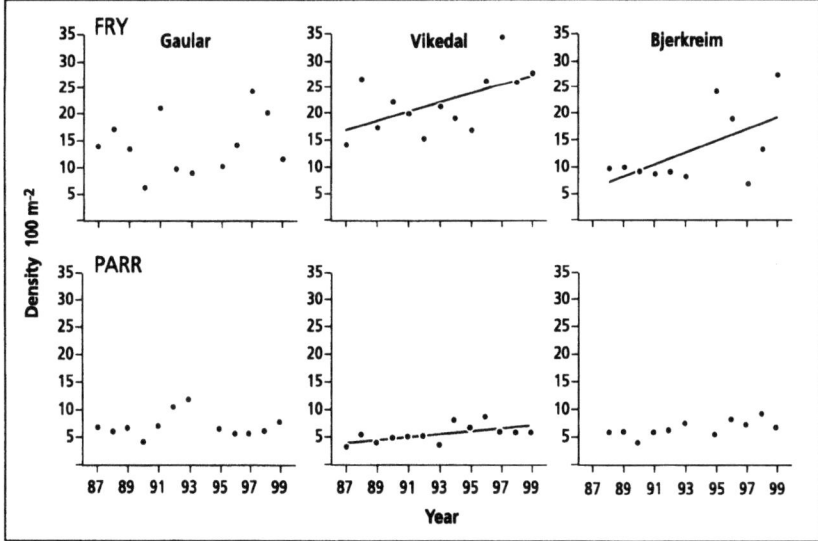

Figure 3. Estimated mean density of fry (age 0+) and parr (age ≥1+) in streams sampled in Gaular and Vikedal (1987-1999) and Bjerkreim (1988-1999).

Acknowledgements

The field work was financed by the Directorate for Nature Management, and conducted in connection with the Norwegian Monitoring Programme for Long-range Transported Air Pollutants. The authors would like to thank Dr. Hugh M. Allen for checking the English.

References

Aoyama, K., Katoh, K., Murano, T., Paces, T. and Taguchi, Y. (eds.), *Acid Snow and Rain*, Proceedings of an International Congress on Acid Snow and Rain 1997, Niigata University, Niigata, Japan. pp. 728-733..

Barlaup, B.T and Åtland, Å.: 1996, *Canadian Journal of Fisheries and Aquatic Sciences* **53**, 1835.

Bohlin, T., Hamrin, S., Heggberget, T.G., Rasmussen, G. and Saltveit, S.J.: 1989, *Hydrobiologia* **173**, 9.

Brown, D.J.A.: 1983, *Bulletin of Environmental Contamination and Toxicology* **30**, 582.

Brown, D.J.A. and Lynam, S.: 1981, *Journal of Fish Biology* **19**, 205.

Gunn, J.M. and Keller, W.: 1998, *Restoration Ecology* **6**, 316.

Harriman; R., Morrison, B.R.S., Birks, H.J.B., Christie, A.E.G., Collen, P. and Watt, A.W.: 1995, *Water, Air and Soil Pollution* **85**, 701.

Henriksen, A., Fjeld, E. and Hesthagen, T.: 1999, *Ambio* **28**, 583.

Hesthagen, T.: 1997, 'The density of young brown trout (*Salmo trutta*) in streams of different acid-neutralizing capacity in three acidic softwater river systems in south Norway' ,in

Hesthagen, T., Heggenes, J., Larsen, B.M., Berger, H.M. and Forseth, T.: 1999, *Water, Air, and Soil Pollution* **112**, 85.

Hindar, A., Henriksen, A., Tørseth, K. and Semb, A.: 1994, *Nature* **272**, 327.

Jensen, A. J. and Johnsen, B.O.: 1988, *Verhandlungen Internationaler Vereinigung für Theoretische und Angewandte Limnologie* **23**, 1724.

Reader, J.P., Dalziel, T.R.K., Morris, R., Sayer, M.D.J. and Dempsey, C.H.: 1991, *Journal of Fish Biology* **39**, 181.

Skjelkvåle, B.L., Tørseth, K., Aas, W. and Andersen, T.: 2000, *Water, Air, and Soil Pollution* (This volume).

Rask, M., Raitaniemi, J., Mannio, J., Vuorenmaa, J. and Nyberg, K: 1995, *Water, Air, and Soil Pollution* **85**, 315.

LOW SUCCESS RATE IN RE-ESTABLISHING EUROPEAN PERCH IN SOME HIGHLY ACIDIFIED LAKES IN SOUTHERNMOST NORWAY

TRYGVE HESTHAGEN*, HANS M. BERGER, ANN KRISTIN LIEN SCHARTAU, TERJE NØST, RANDI SAKSGÅRD and LEIDULF FLØYSTAD

Norwegian Institute for Nature Research, Tungasletta 2, N-7485 Trondheim, Norway
(* author for correspondence, e-mail:trygve.hesthagen@ninatrd.ninaniku.no)

Abstract. In order to test whether major reductions in acid inputs had improved water quality sufficiently for fish populations to recover, we stocked wild European perch (*Perca fluviatilis*) in three highly acidified lakes that had previously supported this species, and in one limed lake. The fish, which were introduced from a local lake (donor lake), generally ranged from 12 to 16 cm in total length, and were stocked at densities of 117-177 fish ha^{-1}. The untreated lakes were highly acid, with minimum pH values and maximum inorganic aluminium concentrations (Al$_i$) during the spring of 4.6-4.7 and 118-151 µg L^{-1} respectively. In the limed lake, the corresponding values for pH and Al$_i$ ranged between 5.8 and 6.6 and 5 and 19 µg L^{-1} respectively. Gill-netting in two subsequent years after the introduction yielded only a few recruits (0+) and one adult in one of the three acidified lakes in one year only. However, stocked perch reproduced successfully in both years in the limed lake. There was a significant linear relationship between the catches (CPUE) of juvenile perch (age 0+) in the different lakes in the autumn and the water quality in May (time of hatching), both in terms of Al$_i$ (r^2=0.934, P<0.05) and pH (r^2=0.939, P<0.05). Our data suggest unsuccessful recruitment in waters of pH <5.1 and Al$_i$ > 60 µg L^{-1}.

Keywords: European perch, acidified and limed lakes, re-establishment, aluminium

1. Introduction

In Norway, the deposition of acidifying compounds with subsequent deterioration of water quality have caused serious damage to freshwater fish resources, especially brown trout (*Salmo trutta*) and European perch (*Perca fluviatilis*) with around 8,200 and 1,000 lost populations, respectively (Hesthagen *et al.*, 1999). Perch are regarded as being relatively acid-tolerant compared with brown trout, which is the other dominant fish species in acidified areas (Sevaldrud and Skogheim, 1986).

In order to mitigate this damage, liming to improve water quality has been carried out in a large number of waters, mostly lakes, over the past 10-15 years (Sandøy and Romundstad, 1995). This large-scale restoration programme appears to have been fairly successful in improving lenthic fish stocks (Forseth *et al.*, 1997). However, in recent years there have been pronounced reductions in acid depositions, as well as a widespread raise in pH and reduction in the concentration of toxic inorganic aluminium (Skjelkvåle *et al.*, 2000). Thus, the question arises of to what extent this improvement in water quality will lead to the recovery and re-establishment of fish populations without liming. However, a critical factor in the restoration of fish in acid-damaged lakes after an improvement in water quality is re-colonisation (Gunn and Mills, 1998). For locations with lost fish populations, physical barriers may stop colonisation from other lakes in the catchment, and active rehabilitation by means of stockings is therefore necessary (Bergquist, 1991; Harvey and Jackson, 1995). In Norway, natural re-colonisation of fish into limed water systems rarely happens, except in a few cases in large watercourses as a result of downstream migration (Kleiven, 1995, 1997). For perch, the poor ability to swim in running water seems to impose another limitation on the process of natural re-colonisation (cf. Pavlov, 1987).

In this paper, we look at whether reductions in acid inputs have improved water quality sufficiently for perch to spawn and survive successfully in acidified lakes in southernmost Norway. Wild specimens were stocked in three lakes that formerly supported this species, and in one lake that was limed prior to the introduction of perch.

2. Study lakes

The study lakes were located in the Tovdal catchment in southern Norway, at altitudes between 210-380 m above sea level. The surface area and maximum depth ranged from between 2.1 to 3.5 ha and from 10.7 to 11.8 m, respectively. All four lakes formerly supported perch, and according to the land owners their populations were probably lost between the 1940s and 1960s.

3. Methods

Fish that were introduced in the experimental lakes were caught in a nearby lake (Gauslåtjern) by gill nets (12.5 and 16.6 mm mesh size) in October 1997 and in early May 1998, except that Lake Fiskevatn was stocked in May 1998 only. The stocked fish, both mature and immature, generally ranged from 12 to 16 cm in total length, and were introduced at densities of 117-177 individuals ha^{-1}. All the perch were fin-clipped in order to be able to identify recaptured individuals as having been introduced. Water samples were taken 11-15 times between September 1997 to September 1999. Lake Fiskevatn was limed with 800 kg of powdered limestone in late October in both 1997 and 1998.

In September 1998 and 1999, test-fishing was carried out by means of multi-mesh survey benthic gill nets (30.0 m long and 1.5 m deep) made up of 12 different bar-mesh sizes from 5 to 55 mm (Appelberg et al., 1995). Single nets were set at depths of between 0-3, 3-6 and 6-12 m, and left overnight for about 12 h. The catch per unit effort (CPUE) was calculated as the number of fish caught 100 m^{-2} net area 12^{-h} of fishing. Total length (mm), weight (g) and stage of maturity were recorded for each fish, and otoliths and operculum bones were removed for later age analysis.

4. Results

The three untreated lakes were highly acid, with minimum and maximum pH and inorganic aluminium (Al_i) values of between 4.6-4.7 and 118-151 µg L^{-1} respectively (Figure 1). They were also generally low in calcium, as mean concentrations ranged from 0.35 to 1.19 mg L^{-1} (Table I). The lakes, including the limed one, are relatively humic, with TOC values between 4.85-5.78 mg L^{-1} and Secchi depths of 2.5-3.5 m, except for Lake Knutetjern (1.71 mg L^{-1} and 5.0 m respectively). The limed lake (Lake Fiskevatn) had mean values of pH, calcium and Al_i of 4.83, 0.72 mg L^{-1} and 67 µg L^{-1} respectively before liming. The water quality improved considerably after liming, and pH remained between 5.4-6.6, with the lowest values being recorded before the second liming in October 1998. During the post-liming period, mean concentrations of calcium and Al_i were 2.04 mg L^{-1} and 9.38 µg L^{-1} respectively. The water quality in the donor lake (Gauslåtjern) was generally good with respect to both pH (5.6-6.2) and Al_i (5-20 µg L^{-1}).

Test-fishing showed that the introduced perch had survived and reproduced in only one of the three acidified lakes. In Lake Håndbekktjern, one adult specimen was caught in 1998 and 18 juveniles in 1999 (CPUE=6.7). However, in the limed lake (Lake Fiskevatn), the stocked perch reproduced successfully in both 1998 and 1999; the catches of juveniles were 87 (CPUE=32.2) and 119 individuals (CPUE=52.9), respectively. A relatively large number of stocked fish were also taken in Lake Fiskevatn; 49 specimens in 1998 and 21 specimens in 1999. In the donor lake (Gauslåtjern), CPUE was 37.4 in 1998 (n=101) and 31.5 in 1999 (n=85), and fish of ages 0+ and 1+ dominated. There was a significant linear relationship between the catches (CPUE) of juvenile perch (age 0+) in the different lakes and the water quality in May (time of hatching), both in terms of Al_i ($F_{1,2}$=43.38, r^2=0.934, P<0.05) and pH ($F_{1,2}$=47.42, r^2=0.939, P<0.05) (Figure 2). From these relationships, recruitment is assumed to be unsuccessful in lakes with pH < 5.1 and Al_i > 60 µg L^{-1}.

TABLE I
Means values ± standard deviation for some water chemistry variables in the study lakes, 1997-1999. The results are based on 11-15 samples for each variable in each lake, except for that of TOC, with 8-9 samples. The data for Fiskevatn are after liming. Lake 1=Håndbekktjern, 2=Lølandstjern, 3=Knutetjern, 4=Fiskevatn and 5=Gauslåtjern.

Lake	pH	Alkalinity µeq L^{-1}	TOC mg L^{-1}	Ca mg L^{-1}	Sulphate mg L^{-1}	Al_i µg L^{-1}	ANC µeq L^{-1}
1	5.25±0.32	15.26±8.80	5.34±1.10	1.08±0.18	3.36±0.38	57.88±30.79	8.22±12.51
2	5.21±0.30	17.72±12.03	5.78±1.39	1.19±0.22	3.58±0.37	60.91±28.52	9.78±17.38
3	4.81±0.08	0.00±0.00	1.71±0.50	0.35±0.15	2.85±0.40	138.34±20.79	-28.50±8.55
4	6.13±0.42	72.92±44.46	4.99±0.91	2.04±0.83	2.87±0.41	9.38±8.60	60.74±43.56
5	5.98±0.22	39.02±6.79	4.85±0.72	1.58±0.11	3.78±0.39	13.66±9.73	35.99±10.19

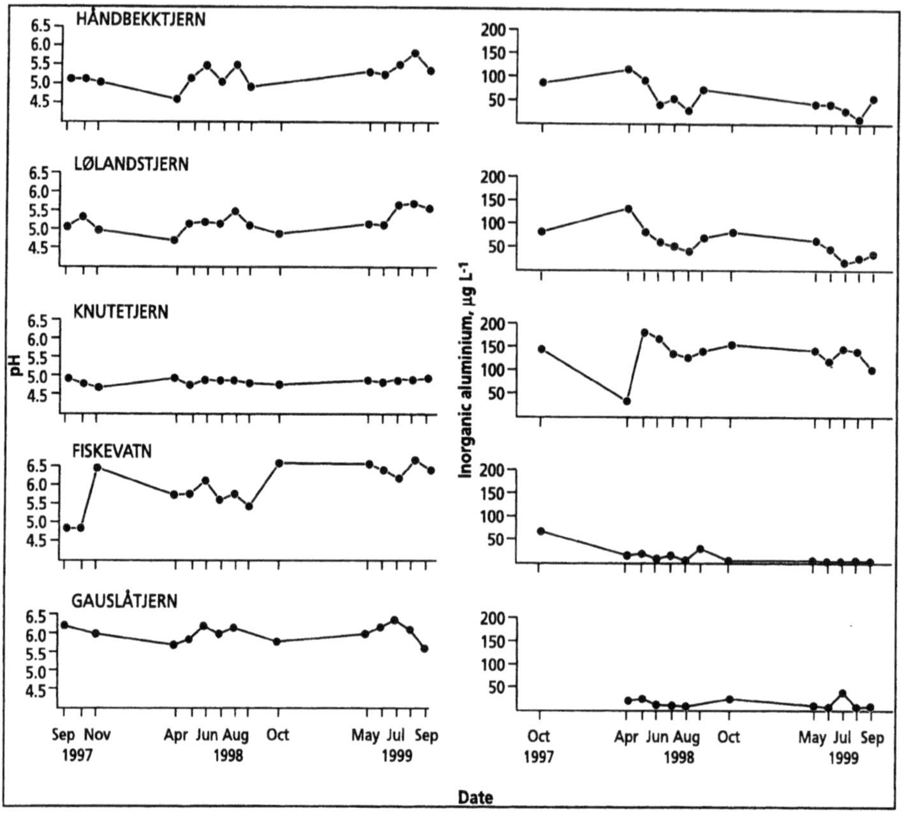

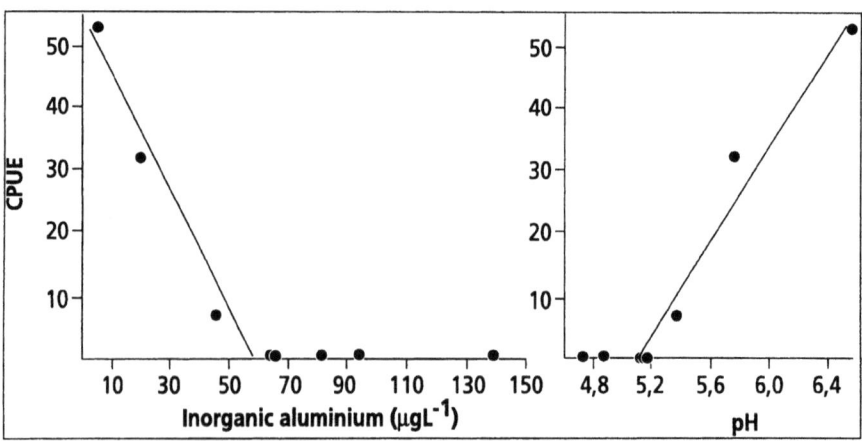

Figure 1. pH and concentrations of inorganic aluminium in the study lakes, 1997-1999.

Figure 2. Relationships between recruitment strength (CPUE) for perch from test-fishing in the study lakes in autumn 1998 and 1999, and pH and concentrations of inorganic aluminium measured in May 1998 and 1999. The relationship can be described by these two equations: CPUE=52.98 - 0.90 Al_i, and CPUE= - 192.92 + 37.82 pH. Only one point for CPUE=0 was included in the analysis (that from the lake with best water quality). Data from Lake Gauslåtjern were excluded because large number of mature fish were removed.

5. Discussion

The study suggests that the water quality in this highly acidified region of southernmost Norway (Tovdal) is still not satisfactory for the re-establishment of perch. On the other hand, stocked perch survived and reproduced successfully in two subsequent years in a limed lake. The three untreated lakes were highly acid, with minimum pH and maximum inorganic aluminium concentrations (Al_i) of 4.6-4.7 and 118-151 µg L^{-1} respectively. Perch reproduced to some extent in one of our experimental lakes in 1999 but in none in 1998 (Lake Håndbekktjern). This probably reflects differences in water quality in the springs of 1998 and 1999, with minimum pH and maximum Al_i values of 5.15 and 94 µg L^{-1} and 5.36 and 46 µg L^{-1} respectively. Corresponding values for pH and Al_i in the limed lake (Lake Fiskevatn) in May 1998 and 1999 were 5.76 and 6.57 and 19 and 5 µg L^{-1} respectively. The relationship between recruitment strength and water quality in the spring in the study lakes suggest that waters of pH < 5.1 and Al_i > 60 µg L^{-1} impaired reproduction success. This is in accordance with experimental studies that have demonstrated high mortality of perch eggs and fry immediately after hatching at pH values below 5.0-5.5 (Milbrink and Johansson, 1975; Runn et al., 1977). It is also in agreement with critical values for damage to and losses of perch in acidified lakes (Rask, 1983,1984; Lappalainen et al., 1988; Raitaniemi et al., 1988; Hesthagen et al., 1992). Thus, the sustainability of fish populations in this area will be dependent on liming for many years to come. A model predicts that with the fully implemented reduction of deposition of S and N, it will take almost 50 years to reduce the present-day liming requirement by 65% (Wright, 2000).

In addition to the direct death of introduced perch in the untreated experimental lakes due to unfavourable water quality, the losses may also to some extent be due to emigration. During test-fishing in a small lake about 1 km downstream of Lake Håndbekktjern in 1998 (Lake Ristjern), seven fin-clipped individuals were caught, indicating that they had been stocked in the lake upstream (unpubl. data). Avoidance of unfavourable water quality has been observed in several species of fish, including that related to acidic water (Gunn, 1986; Barlaup et al., 1989; Gloss et al., 1989; Gray, 1990; Gagen et al., 1994).

The large losses of fish populations in southernmost Norway, and the fact that physical barriers stop colonisation from other lakes in the catchment, suggests the necessity of active rehabilitation of individual lakes. Thus, when water chemistry conditions become satisfactory in the future, large stocking programmes for different species of fish will be needed. Moreover, the re-establishment of various fish species is important when structuring fish assemblages (Appelberg, 1998).

Acknowledgements

This study was financed by the Directorate for Nature Management, and was conducted in connection with the Norwegian Monitoring Programme for Long-range Transported Air Pollutants. The authors would like to thank Dr. Hugh M. Allen for checking the English.

References

Appelberg, M.: 1998, *Restoration Ecology* **6**, 343.
Appelberg, M., Berger, H.M., Hesthagen, T., Kleiven, E., Kurkilahti, M., Raitaniemi, J. and Rask, M.: 1995, *Water, Air, and Soil Pollution* **85**, 401.
Barlaup, B. T., Åtland, Å., Raddum, G.G. and Kleiven, E.: 1989, *Water, Air, and Soil Pollution* **47**, 139.
Bergquist, B.: 1991, *Nordic Journal of Freshwater Research* **66**, 7.
Forseth, T., Halvorsen, G.A., Ugedal, O, Fleming, I., Schartau, A.K.L, Nøst, T., Hartvigsen, R., Raddum, G., Mooij, W. and Kleiven, E.: 1997, *'Biological status in limed lakes'* ,NINA Oppdragsmelding, **508**. Trondheim. (In Norwegian with English summary).
Gagen, C.J., Sharpe, W.E. and Carline, R.F.: 1994, *Canadian Journal of Fisheries and Aquatic Sciences* **51**, 1620.
Gunn, J. M.: 1986, *Environmental Biology of Fishes* **17**, 241.
Gunn, J.M. and Mills, K.H.: 1998, *Restoration Ecology* **6**, 390.
Gloss, S.P., Schofield, C.L., Spateholts, R.L. and Plonski, B.A.: 1989, *Canadian Journal of Fishereis and Aquatic Sciences* **46**, 277.
Gray, R.H.: 1990, *Environmental Toxicology and Chemistry* **9**, 53.
Harvey, H.H., and Jackson, D.A.: 1995, *Water, Air and Soil Pollution* **85**, 383.
Hesthagen, T., Berger, H.M., Larsen, B.M., Nøst, T. and Sevaldrud, I.H.: 1992, *Environmental Pollution* **78**, 97.
Hesthagen, T., Sevaldrud, I.H. and Berger, H.M.: 1999, *Ambio* **28**, 112.
Kleiven, E.:1995, Fisk' , in A.J. Romundstad (ed.),*'Kalking i vann og vassdrag. Overvåking av større prosjekter. Årsrapport 1993,'* 'DN-Notat **1995-2**. Trondheim. pp. 108-117.
Kleiven, E.: 1997, *'Loss and recolonization of different fish species in the lake Herefossfjorden, the Tovdal watercourse, in the period 1970-1996'* ,NIVA- Rep. **3724-97**. (In Norwegian with English summary).
Lappalainen, A., Rask, M. and Vuorinen, P.J.: 1988, *Environmental Biology of Fishes* **21**, 231.
Milbrink, G. and Johansson, N.: 1975, *Report of the Institute of Freshwater Research Drottningholm* **54**, 203.
Rask, M.:1983, *Annales Zoologici Fennici* **20**, 73.
Rask, M.: 1984, *Annales Zoologici Fennici* **21**, 15.
Raitaniemi, J., Rask, M. and Vuorinen, P.J.: 1988, *Annales Zoologici Fennici* **25**, 209.
Runn, P., Johansson, N. and Milbrink, G.: 1977, *Zoon* **5**, 115.
Sandøy, S. and Romundstad, A.J.: 1995, *Water, Air and Soil Pollution* **85**, 997.
Sevaldrud, I.H. and Skogheim, O.K.: 1986, *Water, Air and Soil Pollution* **30**, 381.
Skjelkvåle, B.L., Tørseth, K., Aas, W. and Andersen, T.: 2000, *Water, Air, and Soil Pollution* (This volume)
Wright, R.F.: 2000, *Water, Air, and Soil Pollution.* (In press).

RECOVERY OF THE PERCH (PERCA FLUVIATILIS) IN AN ACIDIFIED LAKE AND SUBSEQUENT RESPONSES IN MACROINVERTEBRATES AND THE GOLDENEYE (BUCEPHALA CLANGULA)

M. RASK[1], H. PÖYSÄ[2], P. NUMMI[3] and C. KARPPINEN[1]

[1]*Finnish Game and Fisheries Research Institute, Evo Fisheries Research Station, FIN-16970 Evo, Finland;* [2]*Finnish Game and Fisheries Research Institute, Evo Game Research Station, FIN-16970 Evo, Finland;* [3]*Department of Applied Zoology, P. O. Box 27, FIN-00014 University of Helsinki, Finland*

Abstract. The perch population of Lake Vähä Valkjärvi, a two hectare clear-water lake in southern Finland, decreased due to acid precipitation during the 1980s. During the early 1990s a decrease in acidic deposition resulted in slight improvement of water quality of the lake. This was followed by recovery of the reproduction of perch starting in 1991. A mark and recapture experiment in spring 1995 indicated a hundred fold increase in the population size of perch in a four year period. A decrease in the abundance of aquatic invertebrates was recorded during 1989-1996. This decrease well coincided with the recovery of the perch population, suggesting that increased predation by fish was responsible for the decrease. The occurrence of goldeneye young also dropped in L. Vähä Valkjärvi since 1993. This was thought to be due to increased food competition with perch.

Key words: acidification, European perch, goldeneye, macroinvertebrates, recovery

1. Introduction

Starting in the early 1980s, sulphate deposition has decreased in many areas of Europe and North America affected by acidification (Kulmala *et al.*, 1988; Driscoll *et al.*, 1995). This was followed, after an interval of varying number of years, by improvements in the quality of surface waters (Mannio and Vuorenmaa, 1995; Driscoll *et al.*, 1995), and later, by recovery of the biological systems (Keller *et al.*, 1992). In Finland, the first records on the recovery of fish populations were made in early 1990s when strong year-classes of perch (*Perca fluviatilis*) appeared in lakes that were formerly almost devoid of fish due to acidification (Nyberg *et al.*, 1995; Rask *et al.*, 1995). As concerns the waterfowl, it has been predicted that the common goldeneye (*Bucephala clangula*) is a species that will suffer from the recovery of acidified lakes and their fish populations (McNicol *et al.*, 1995). This would be due to changes in competetive relations between fish and waterfowl (Eriksson, 1979; Eadie and Keast, 1982; Pöysä *et al.*, 1994; McNicol *et al.*, 1995).

In this paper, we describe the fall and rise of the perch population in Vähä Valkjärvi, an acidified forest lake in southern Finland, and the responses of aquatic macroinvertebrates and goldeneye young to the recovery of the perch.

2. Material and methods

Lake Vähä Valkjärvi is a small (area 2.3 ha, maximum depth 3 m) clear water seepage lake in the Evo forest area, southern Finland. The catchment of the lake is composed of sand and gravel with scots pine forest as the dominant vegetation. Perch is the dominant fish species but there are also some pike (*Esox lucius*) in the lake. The lake experienced a rapid acidification from pH levels of 5 to 4.5 during the 1980s. Since the early 1990s, there has been an increasing trend in lake pH (from 4.4-4.5 to 4.6-4.7) and alkalinity (from -40 to -20 µeq L^{-1}) indicating slight improvement in water quality.

The size and structure of the perch population in L. Vähä Valkjärvi has been examined since 1985 (Rask 1992). Mark-recapture methods were applied and abundance estimates were obtained in 1986, 1989, 1993, 1995, 1997, and 1999 (modified Schnabel estimate, Krebs, 1989). The fish were caught by using wire traps with 1 cm square mesh, measured for total length, fin-clipped, and released. Opercular bones were used for age determination and back-calculations of growth were conducted (Monastyrsky procedure, b=0.88, Tesch, 1971)

Aquatic invertebrates were sampled in Vähä Valkjärvi each year between 1989-1996 using methods described in detail in Nummi and Pöysä (1993). Free-swimming invertebrates were trapped with activity traps, identified, and grouped following the taxon list and length categories of Nudds and Bowlby (1984) with small modifications (Nummi and Pöysä, 1993). The abundance index of free swimming invertebrates is given per 100 trap days.

Goldeneye broods were censused each year between 1989-1996, five censuses between early June and early August using the standard point count and round count methods (Nummi and Pöysä, 1993). We used the total number of young observed during the censuses per lake as an index of goldeneye abundance.

4. Results

During 1985-1991, the perch population of Vähä Valkjärvi was dominated by 15-25 cm long individuals of the last strong year classes that had born in 1978 and 1979. During the 1980s there was almost no succesful reproduction of perch in the lake, but some recruits born in 1984 and 1986 still appeared in the trap catches in 1985 and 1988, respectively (Table I). The perch born in 1986 were longer at age three (*t-test*, $P < 0.001$) than those born in 1978 (Figure 1). In 1991 only eight individuals were caught during the sampling period of two weeks and fifty trap nights and the population was considered to be close to extinction. However, first signals of recovery of the perch population were recorded in spring 1993 when young individuals born in 1991 appeared in the catches (Table I). Strong year-classes, born in 1992 and in 1993, were found in

TABLE I

The length frequency distribution of perch caught during the marking and recapturing in L. Vähä Valkjärvi in the period 1985-1999.

cm	1985	1986	1987	1988	1989	1991	1993	1995	1997	1999
8							4			
9				1			42	1		
10	2			5			13	88		
11	1			4			5	236	2	
12		1		6				167	9	10
13		2		1				242	137	213
14		4						199	423	433
15	6	10			6			59	243	180
16	17	21	3	3	4			14	76	45
17	16	25	8	7	5			1	13	6
18	7	23	9	2	1	1			3	3
19	3	21	12	8	2	0				1
20		9	14	9		1			1	
21		8	8	9	2	0				
22			1	7		1				
23			0	4	2	0	3			
24			1	3	1	0	2			
25			0	0		1	1			
26						1	3			
27						0	4			
28						1	2			
29						2	2			
30							3			
31							3			
32							5			

1995 resulting in sharp increases both in the mean trap catches (from <1 to 25 perch) and in the population estimates (from < 100 to 10200, Table II). The growth of the perch of the new strong year-classes was similar to that of the previous strong year-classes 1978 and 1979 (Figure 1).

At the same time as the perch population recovered, the abundance of invertebrates significantly decreased. L. Vähä Valkjärvi was the only lake among the monitoring lakes in Evo area showing such a clearly decreasing pattern in invertebrate abundance between 1989-1996 (Figure 2). The number of goldeneye young in L. Vähä Valkjärvi also showed a decreasing trend during 1989-1996 (Figure 2). However, decreases of goldeneye young were recorded also in other monitoring lakes of Evo area although not as clear as in L. Vähä Valkjärvi.

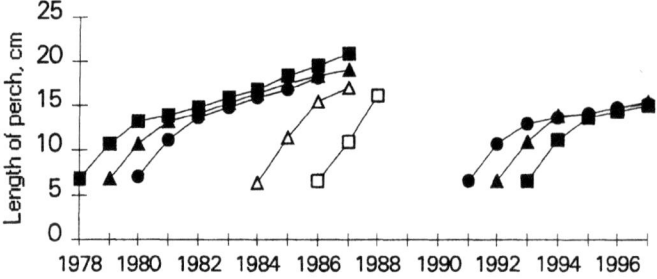

Figure 1. Back-calculated growth curves of perch for year-classes that occurred in Lake Vähä Valkjärvi during 1985-1997.

TABLE II

CPUE values (number of perch/trap/night) and population estimates with 95 % confidence intervals for perch in Lake Vähä Valkjärvi during 1986-1999.

Year	CPUE	Population est.	95%min	95%max
1986	2,1	372	258	598
1987	1,1			
1988	1,4			
1989	0,1	54	22	157
1990				
1991	0,2			
1992				
1993	2,3	58	38	95
1994				
1995	26,6	10267	8084	13397
1996				
1997	26,8	6146	4300	10764
1998				
1999	18	2167	1788	2747

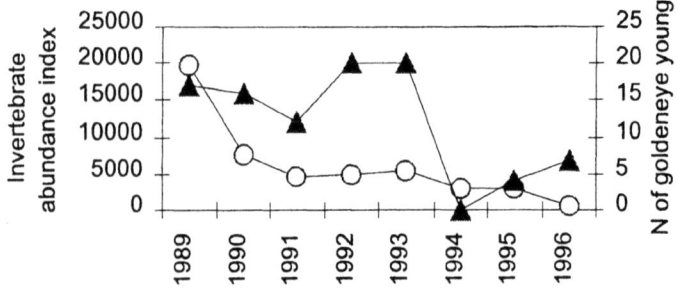

Figure 2. The invertebrate abundance index (circles) and the number of goldeneye young (triangles) in Lake Vähä Valkjärvi during 1989-1996.

5. Discussion

Acidification strongly affected the perch population of Vähä Valkjärvi in the mid 1980s. The extremely high mortality of developing embryos of perch as shown by Lappalainen et al., (1988) obviously was the primary cause of the crash of the population whereas in another less acidic lake in the same area, the perch population remained almost unchanged through the early 1990s (Rask et al., 1998).

Responses of adult perch to acidity were also recorded: a fish kill in spring 1986 (Nikinmaa et al., 1990) and delayed spawning of perch in 1987 (Rask et al., 1990). The growth of the few recruits in the mid 1980s was faster compared to the fish of strong year-classes, which is related to a decreased intraspecific food competition in a low density population. However, despite the lack of the food competition, the increase in growth was not as clear as shown in some other acidified lakes (Raitaniemi et al., 1988), suggesting that the perch in Vähä Valkjärvi suffered from a continuous physiological stress due to acidification.

The recovery of the perch population in early 1990s obviously took place after the slight improvement in water quality, which, in turn, is related to the decreased sulphate deposition in the area (Kulmala et al. 1998). Similar changes of perch populations have been recorded also in other acidified lakes in southern Finland (Nyberg et al., 1995). In the shift of 1980s and 1990s there were some exceptionally mild winters. The lack of snow cover resulted in low spring runoff and, apparently, no spring acidity peaks took place in the lakes in 1989 and 1990. This maybe affected the onset of recovery of perch populations in some acidified lakes (Nyberg et al., this volume). However, in Vähä Valkjärvi strong year-classes of perch were born later, in 1992 and 1993, in years with "normal" hydrological conditions.

As in L. Vähä Valkjärvi, decreases of fish populations due to acidification are often followed by increased abundances of macroinvertebrates that were previously regulated by fish predation (Eriksson et al., 1980, Bendell & McNicol, 1987). The clear decrease in the amount of macroinvertebrates in L. Vähä Valkjärvi coincided with the time when the perch of the year classes 1991-1993 grew from a mean length of 10 cm to 13 cm, to a size when they usually switch from zooplankton to macroinvertebrate food in small lakes of the Evo area (Rask et al., 1998).

A positive relationship between the invertebrate abundance and the goldeneye density was shown earlier in another study in the lakes of the Evo area (Nummi & Pöysä, 1993; Pöysä et al., 1994). A crash of perch population in a neighbouring lake was followed by increases in macroinvertebrate abundance and goldeneye brood density (Pöysä et al., 1994; Rask et al., 1996). Despite the general slightly decreasing trend in the goldeneye brood abundance in the lakes of Evo area, we believe that the decrease in invertebrate abundance due to recovery of the perch population was the major cause of the decrease of goldeneye in Vähä Valkjärvi. Our observations suggest adverse effects of biological recovery from acidification on goldeneyes, supporting the recent predictions of McNicol et al., (1995). Apparently high interspecific food competition between perch and goldeneye took place since 1993 when the perch of the new strong year-classes grew big enough to switch to macroinvertebrate diet.

References

Bendell, B.E. and McNicol, D.K.: 1987, *Hydrobiologia* **150**, 193.
Driscoll, C T., Postek, K.M., Kretser, W. and Raynal, D.J.: 1995, *Water, Air and Soil Pollution* **85**, 583.
Eadie, J.M. and Keast, A.: 1982, *Oecologia* **55**, 225.
Eriksson, M.O.G.: 1979, *Oecologia* **41**, 99.
Eriksson, M.O.G., Henrikson, L., Nilsson, B.-I., Nyman, G., Oscarson, H.G., Stenson, A.E. and Larsson, K.: 1980, *Ambio* **9**, 249.
Keller, W., Gunn, J.M. and Yan, N.D.: 1992, *Environmental Pollution* **78**, 79.
Krebs, C.J.: 1989, *Ecological Methodology*. Harper and Row, New York, USA.
Kulmala, A., Leinonen, L., Ruoho-Airola, T., Salmi, T. and Waldén, J.: 1998, *Finnish Meteorological Institute, Air Quality Measurements*.
Lappalainen, A., Rask, M. and Vuorinen, P.: 1988, *Environmental Biology of Fishes* **21**, 231.
Mannio, J. and Vuorenmaa, J.: 1995, *Water, Air and Soil Pollution* **85**, 571.
McNicol, D.K., Mallory, M.L. and Wedeles, C.H.R.: 1995, *Water, Air and Soil Pollution* **85**, 457.
Nikinmaa, M., Salama, A., Tuurala, H., 1990. 'Respiratory effects of environmental acidification in perch (*Perca fluviatilis*) and rainbow trout (*Salmo gairdneri*)'. In: P. Kauppi, P. Anttila and K. Kenttämies (eds.), *Acidification in Finland*. Springer-Verlag: Berlin, Germany. pp. 929-940.
Nudds, T.D. and Bowlby, J.N.: 1984, *Can. J. Zool.* **62**, 2002.
Nummi, P. and Pöysä, H.: 1993, *Ecography* **16**, 319.
Nyberg, K., Raitaniemi, J., Rask, M., Mannio, J. and Vuorenmaa, J.: 1995, *Water, Air and Soil Pollution* **85**, 395.
Pöysä, H., Rask, M. and Nummi, P.: 1994, *Annales Zoologici Fennici* **31**, 397.
Raitaniemi, J., Rask, M. and Vuorinen, P.J.: 1988, *Annales Zoologici Fennici* **25**, 209.
Rask, M.: 1992, *Environmental Pollution* **78**, 121.
Rask, M., Raitaniemi, J., Mannio, J., Vuorenmaa, J. and Nyberg, K.: 1995, *Water, Air and Soil Pollution* **85**, 315.
Rask, M., Järvinen, M., Kuoppamäki, K. and Pöysä, H.: 1996, *Annales Zoologici Fennici* **33**, 517.
Rask, M., Holopainen, A.-L., Karusalmi, A., Niinioja, R., Tammi, J., Arvola, L., Keskitalo, J., Blomqvist, I., Heinimaa, S., Karppinen, C., Salonen, K. and Sarvala, J.: 1998, *Boreal Environment Research* **3**, 263.
Tesch, F.W., 1971. 'Age and growth.' In: W.E. Ricker (ed.), *Methods for assessment of fish production in fresh waters*. IBP Handbook 3. Blackwell Scientific Publications: Oxford, UK. pp. 98-130.

PATTERNS IN WATER QUALITY AND FISH STATUS OF SOME ACIDIFIED LAKES IN SOUTHERN FINLAND DURING A DECADE: RECOVERY PROCEEDING

K. NYBERG[1], J. VUORENMAA[2], M. RASK[3], J. MANNIO[2], J. RAITANIEMI[3]

[1]*University of Helsinki, Department of Limnology, P.O.Box 27, FIN-00014 Helsinki, Finland;*
[2]*Finnish Environment Institute, P.O.Box 140, FIN-00251 Helsinki, Finland;* [3]*Finnish Game and Fisheries Research Institute, Evo Fisheries Research Station, FIN-16970 Evo, Finland*

Abstract. Since the early 1980s, the acidic deposition in the northern Europe has decreased substantially. This has resulted in corresponding improvements of the water quality in some acid sensitive small lakes of southern Finland. Among the fish of these lakes, the first signs of recovery were recorded in the early 1990s, when the European perch (*Perca fluviatilis* L.) started to reproduce in some sparse populations. Since then, the reproduction of perch has been successful in several years. The appearance of strong year-classes in lakes earlier almost empty of fish indicates recovery. This development has resulted in increased population densities, decreased mean sizes of fish and decreased growth rates. In a more acid sensitive species, roach (*Rutilus rutilus* (L.)), no clear indications of recovery have been recorded this far. However, schools of small roach (age 1+) were observed in the summer of 1998 in two acidic lakes that were inhabited by sparse roach populations during 1985-1995.

Keywords: acidification, fish status, lakes, *Perca fluviatilis*, recovery, runoff, *Rutilus rutilus*, water quality

1. Introduction

Due to the successful reduction of the emissions, sulphate deposition has decreased in southern Finland since the late 1970s (Kulmala *et al.*, 1998). The first signs of chemical recovery of acidic lake waters, including decreases in sulphate concentrations and increases in pH and alkalinity, were recorded in the early 1990s (Mannio and Vuorenmaa, 1995). At the same time, first records of abundant year-classes of perch were made in acidified lakes that had been almost empty of fish during the 1980s. In some lakes this resulted in sharp increases in population density and, consequently, slower growth rates. These changes were attributed to the chemical recovery of the lake waters (Rask *et al.*, 1995a), although the effect of favourable weather conditions during the shift of decade could not be ruled out (Nyberg *et al.*, 1995).

For this paper, we chose five lakes studied in the acidification monitoring programmes of the Finnish Environment Institute (FEI) and the Finnish Game and Fisheries Research Institute (FGFRI), in order to show examples of the recent patterns in acidity related water quality parameters and fish population parameters. It was hypothesized that in the conditions of improved water quality, the recovery of perch populations would continue. Further, when the improvement of the water quality in less acidified lakes achieved the critical level for roach reproduction, the recovery of roach populations would also begin.

Water, Air, and Soil Pollution **130**: 1373–1378, 2001.
© 2001 *Kluwer Academic Publishers.*

2. Material and methods

The five study lakes are located in southernmost Finland, three of them (Orajärvi, Saaren Musta and Kattilajärvi) ca. 20 km and two (Munajärvi, Vitsjö) ca. 100 km to the west of Helsinki. The surface area of the lakes is 7–34 ha and maximum depth 4–12 m. The catchment area is mainly coniferous forest, and its size is 29–200 ha, of which 13–37% is bedrock and 3–15% peatland. All the lakes have low ionic concentrations (conductivity 2.4–4.1 mS m^{-1}) and are oligotrophic (total phosphorus 5–12 $\mu g\ L^{-1}$). Lakes Vitsjö and Kattilajärvi have higher base cation levels than the other lakes, and correspondingly higher pH (>6) and positive Gran alkalinity. Of the highly acidified lakes, Munajärvi is characterized by a short retention time (1 yr) and a high amount of humic material (total organic carbon 7-12 mg l^{-1}), derived from the surrounding peatlands. All the others are clear water lakes (TOC <5 mg l^{-1}) with relatively long retention times.

According to water quality model calculations and paleolimnological analyses (Huttunen et al., 1990), and information from local residents, all five lakes experienced precipitation induced acidification. The responses of fish to acidification varied from slight changes in the population structures of roach in lakes Vitsjö and Kattilajärvi to an almost complete extinction of perch in Munajärvi (Raitaniemi et al., 1988).

Water samples were collected yearly in October-November during autumnal circulation, which was a part of the national monitoring of lake acidification (Mannio and Vuorenmaa 1995, Mannio, this volume). Altogether 25 chemical variables were analysed in the laboratories of the Uusimaa Regional Environment Centre and the Finnish Environment Institute according to standardized methods (Forsius et al. 1990). For trend significance analysis, we used nonparametric Kendall tau test and for the slope estimation simple linear regression (SPSS 9.0 software). In addition to the selected water quality parameters of the lakes, annual and winter runoff values are presented from a catchment monitoring site of FEI in the same region.

Fish samples from the lakes were taken every third year by means of gill net test fishing with a Finnish standard gillnet series (Raitaniemi et al., 1988) and NORDIC multimesh surveynets (Kurkilahti and Rask 1996). To obtain comparable catch per unit effort values over the study period, the difference in the net panel area between the two net types was corrected and the catches of only comparable mesh sizes (12–60 mm) were included in counting the CPUEn values (number of fish in one NORDIC net after one night). Mean weight of perch and roach in the total catch of a lake was used to characterise their population structure. In the age determination and back-calculation of growth, opercular bones were used for perch (Monastyrsky procedure, b=0.88) and scales (Fraser-Lee procedure, c=2.0 cm) for roach (Tesch, 1971).

3. Results and discussion

In general, the chemical trends in these headwater lakes followed the changes observed in deposition (Kulmala et al., 1998). Sulphate has declined significantly (-4.1-5.7 μeq L^{-1} a^{-1}) except in Munajärvi (-0.5 μeq L^{-1}) (Figure 1). The yearly rate of decline in base cations is also considerable; from 0 to -5.0 μeq L^{-1} a^{-1}. Accordingly, the rate of recovery as an increase in alkalinity is from 1.3 to 4.0 μeqL^{-1} a^{-1} (Figure 1), corresponding to the median increase of 1.74 μeq L^{-1} in 62 lakes in South Finland (Mannio, this volume). The alkalinity was lowest in 1990, and has in general increased throughout the decade. The strikingly low alkalinities in 1990 may have occurred due to a cold autumn: the lakes got frozen early and the samples had to be taken from the ice and close to the shore. The change in pH is significant only in Vitsjö (0.2-0.3 units) and Oraj rvi (0.5-0.6 units), where the decline in hydrogen ion is considerable (10 μeq L^{-1}). Lakes Saaren Musta and Munajärvi still show relatively high labile aluminum levels; between 75 and 100 μeq L^{-1}, levels considered harmful for reproduction of perch and lethal for roach (Hultberg, 1988; Rask et al., 1995b).

The data of the monitoring catchment of the FEI show that during the shift of the 1980s and the 1990s, a considerable part of the annual runoff took place in the winter, in 1990 even more than 50 % (Figure 2). It can be concluded that in those years the spring runoff was smaller than usually. Consequently, the acidic meltwater episodes in those years may have been smaller, which maybe affected the onset of the recovery of perch.

In the 1990s, successful reproduction of perch occurred in several years even in the most acidified lakes. For example in Orajärvi, new year classes of perch appeared annually during 1992–1998. Hydrological conditions in this period varied considerably from year to year (Figure 2), suggesting that general improvement in the water quality in the 1990s, rather than exceptional runoff conditions, was the reason for the recovery of perch. The CPUEn of perch in the most acidified lakes was 7–30 times higher and mean weight of fish essentially lower in the late 1990s than during the 1980s (Rask, 1989; Nyberg et al., 1995; Rask et al., 1995a) (Figure 3). In the less acidified lakes, the CPUEn of perch showed a slightly increasing trend in Vitsjö, whereas a peak occurred in Kattilajärvi in 1988. The mean weight of perch in these two lakes remained almost unchanged throughout the study period (Figure 3). The increased CPUEn values in the most acidified lakes indicate increases in the population density of perch, which was followed by a sharp decrease in their growth rate. This was attributed to increased food competition (Raitaniemi et al., 1988; Rask, 1989; Nyberg et al., 1995; Rask et al., 1995a).

In the 1980s, the levels of pH and labile aluminium in lakes Vitsjö and Kattilajärvi were close to the critical values for reproduction of roach, pH 5.8 (Hultberg, 1988), or pH 5.8 and Al$_{lab}$ < 20 μg L^{-1} (Rask et al., 1995b). Because no new year classes of roach appeared in the late 1980s, it was suggested that

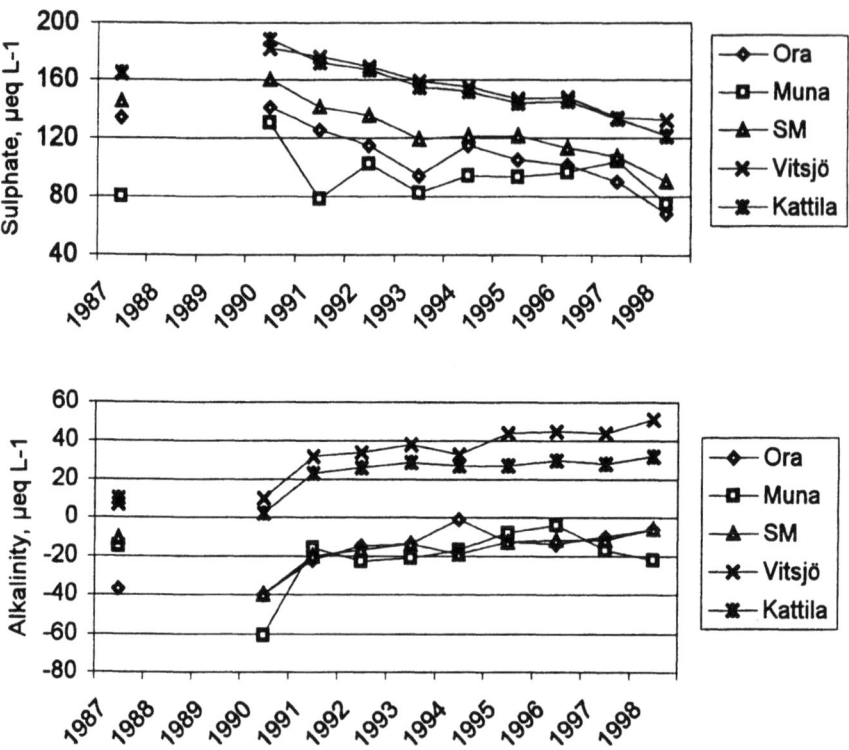

Fig. 1. Sulphate concentrations and alkalinities in the surface water of the study lakes in 1987 and 1990-1997 during the autumn turnover.

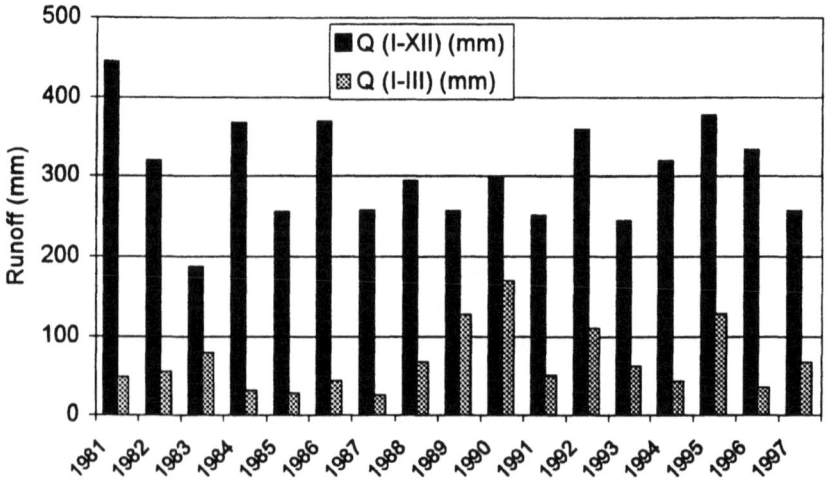

Fig. 2. Annual (Q I-XII) and winter (Q I-III, January-March) runoff in the Teeresuonoja catchment monitoring site of the Finnish Environment Institute during 1981-1997.

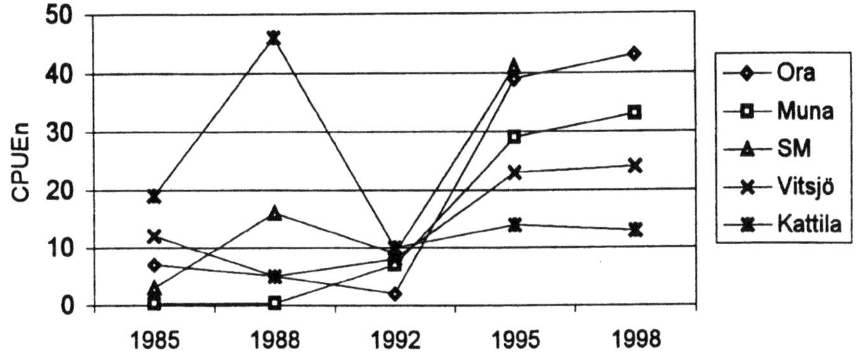

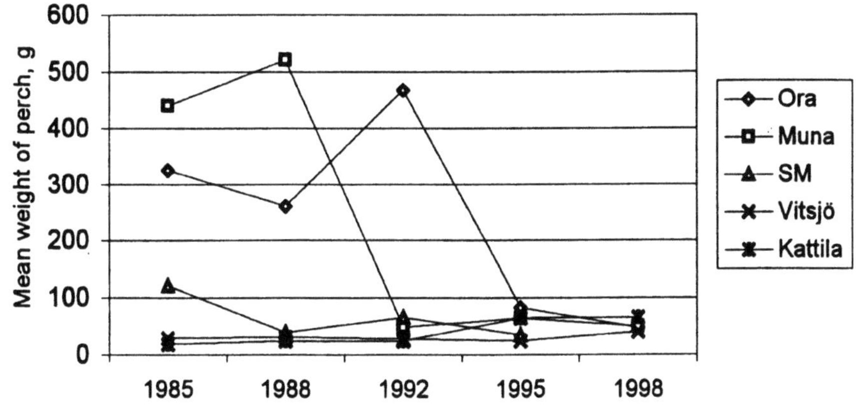

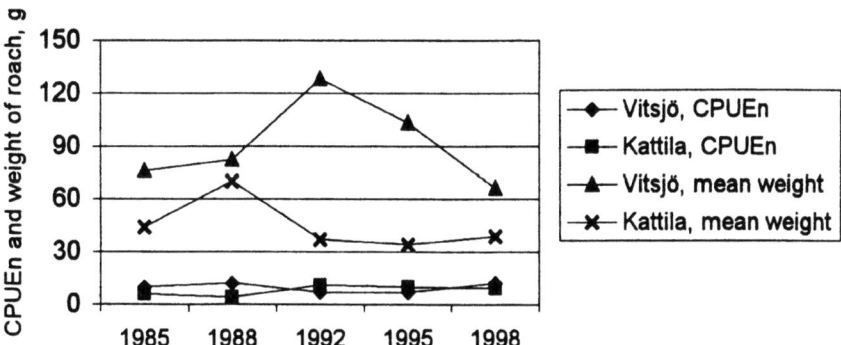

Fig. 3. The number of perch per NORDIC surveynet (CPUEn) and the mean weight of perch in the total catch from the study lakes in the test fishings during 1985-1998.

Fig. 4. The number of roach per NORDIC surveynet (CPUEn) and their mean weight in the total catch in lakes Vitsjö and Kattilajärvi in test fishings in 1985-1998.

the populations suffered from acidification induced reproduction failures (Rask, 1989), which resulted in increased mean weight of the fish (Figure 4). During the 1990s, the mean weight of roach has decreased because some successful reproduction has taken place. In Vitsjö, the first successful reproduction after the 1980s took place in 1994. However, the CPUEn remained at a quite constant low level (Figure 4). The growth rates of roach have been unchanged over the study period, which also suggests that no essential changes have taken place in the sizes or structures of the populations. In 1998, roach schools, apparently of age 1+, were visually observed in both lakes, which may be an indication of an abundant year class and a signal of accelerating recovery of the populations.

4. Conclusions

Our results indicate that improvements in water quality of the study lakes are in line with the general decreasing trends observed in the deposition. The increased relative abundance and decreases in mean weight and growth of perch in these chemically recovering lakes indicate clearly higher population densities than in the 1980s. The exceptional hydrological conditions at the beginning of the 1990s maybe affected the appearance of the first new year classes of perch, but the successful reproduction throughout the 1990s clearly indicate that the perch populations in the study lakes are recovering.

References

Forsius, M., Malin, V., Mäkinen, I., Mannio, J., Kämäri, J., Kortelainen, P. and Verta, M.:1990, *Environmetrics* **1**, 73.
Hultberg, H.: 1988, Nord 15/1988, 185-200.
Huttunen, P., Kenttämies, K., Liehu, A., Liukkonen, M., Nuotio, T., Sandman, O. and Turkia, J.: 1990, 'Paleolimnological evaluation of the recent acidification of susceptible lakes in Finland'. In: P. Kauppi, P. Anttila and K. Kenttämies (Eds.), *Acidification in Finland.* Springer-Verlag: Berlin, Germany. pp. 1071-1090.
Kulmala, A., Leinonen, L., Ruoho-Airola, T., Salmi, T. and Waldén, J.: 1998, *Finnish Meteorological Institute, Air Quality Measurements*.
Kurkilahti, M. and Rask, M.: 1996, *Fisheries Research* **27**, 243-260.
Mannio, J. and Vuorenmaa, J.: 1995, *Water, Air and Soil Pollution* **85**, 571-576.
Nyberg, K., Raitaniemi, J., Rask, M., Mannio, J. and Vuorenmaa, J.: 1995, *Water, Air and Soil Pollution* **85**, 395-400.
Raitaniemi, J., Rask, M. and Vuorinen, P.J.: 1988, *Annales Zoologici Fennici* **25**, 209-219.
Rask, M.: 1989, *Verh. Internat. Verein. Limnol.* **24**, 2425-2427.
Rask, M., Raitaniemi, J., Mannio, J., Vuorenmaa, J. and Nyberg, K.: 1995a, *Water, Air and Soil Pollution* **85**, 315-320.
Rask, M., Mannio, J., Forsius, M., Posch, M. and Vuorinen, P.J.: 1995b, *Environmental Biology of Fishes* **42**, 51-63.
Tesch, F.W., 1971. 'Age and growth.' In: W.E. Ricker (Ed.), *Methods for assessment of fish production in fresh waters.* IBP Handbook 3. Blackwell Scientific Publications: Oxford, UK. pp. 98-130.

ACIDIFICATION AND LIMING OF RIVER VIKEDAL, WESTERN NORWAY. A 20 YEAR STUDY OF RESPONSES IN THE BENTHIC INVERTEBRATE FAUNA

ARNE FJELLHEIM and GUNNAR G. RADDUM

University of Bergen, Institute of Zoology, Allegt. 41, N-5007 Bergen, Norway;
(author for correspondence, e-mail: arne.fjellheim@zoo.uib.no)

Abstract. This paper deals with benthic invertebrate population responses to acidification and liming in the lower part of River Vikedal. In 1979 the river showed signs of increasing acidification. Highly sensitive invertebrates like the mayfly *Baetis rhodani* were present in the river in low abundance, but disappeared in the subsequent years. In order to re-establish a non-toxic water quality for fish, liming of the spring snowmelt to a minimum of pH 5.7 was started in 1987. During the later years liming has been successively increased. The invertebrate fauna showed a slow, but positive, response during the first years after liming, especially during autumn. *B. rhodani* recolonized the river in low density, but the spring cohort was still weak. Since 1994 the lime dosage was increased to secure a minimum pH of 6.3 during spring snowmelt. This has resulted in an overall increased biodiversity in the limed section of the river. Several acid-sensitive species, like both cohorts of *B. rhodani* and freshwater snails have colonized this part of the river. Simultaneously biodiversity in the unlimed reference sites has slightly improved during the last years. This is correlated with decreased sulphur deposition and improved surface water chemical conditions.

Keywords: acidification, liming, benthic invertebrate response, long term trends

1. Introduction

The damage of Norwegian salmonid stocks due to acidification started more than one hundred years ago and has resulted in dramatic effects on fish populations in a large number of rivers. The most pronounced effects were found in Southern Norway, with an estimated loss between 345 and 1150 tonnes per year of Atlantic salmon (*Salmo salar* L.) (Hesthagen and Hansen, 1991).

River Vikedal was one of the first rivers in the western part of Norway to be affected. During the spring 1981 a severe fish kill of young salmonids was observed in the river (Rosseland and Skogheim, 1984). During the subsequent years, eposodic fish kills of juvenile Atlantic salmon and brown trout (*Salmo trutta* L.) were observed each spring (Hesthagen 1986). Subsequently, catches of adult salmon declined markedly and the population of Atlantic salmon was characterised as strongly reduced (Lien *et al.* 1996). The acidification also affected the benthic invertebrate community in the watershed and the most sensitive invertebrate species disappeared from the main lower part of the river (Fjellheim *et al.* 1987). In order to mitigate the negative effects of acidification, a lime doser was installed in the main river in 1987.

The first benthic invertebrate sampling was performed in River Vikedal in 1979. Since 1982, as part of the Norwegian Monitoring Programme for Long

Range Transported Air Pollutants, River Vikedal has been regularly monitored with respect to invertebrates. The main purpose of this paper is to give an overview of the development of the benthic invertebrate community, with special focus on acid-sensitive species, during the last two decades.

2. Study Area

River Vikedal is located in SW Norway at 59°33'N, 5°53'E (Figure 1). The catchment area (119 km2) mainly consists of slowly weathering rocks and thin soils giving the watershed a poor capacity of buffering acid rain. Large parts of the catchment is not inhabited by man. Farmlands are located in the lower part of the watershed, from Lake Fjellgardsvatnet to the sea. Atlantic salmon and anadromous brown trout may ascend 10 km of the main river, where migrating fish are stopped by a large waterfall, Laakafoss.

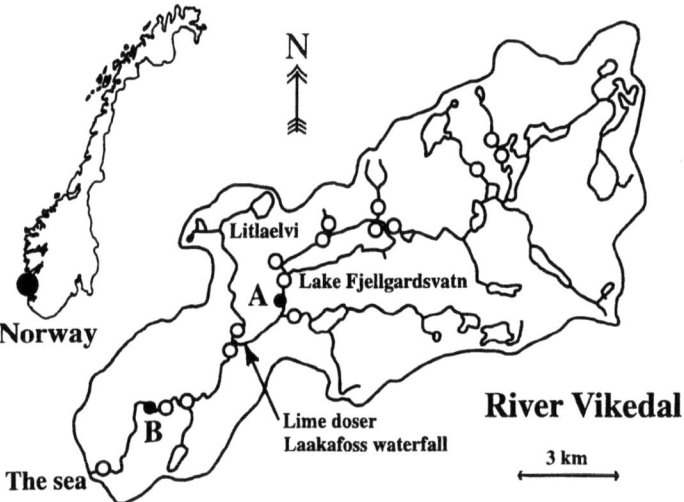

Figure 1. Map of River Vikedal showing the benthic invertebrate sampling stations (circles). Main sampling stations are indicated by filled circles.

The water chemistry of the unlimed part of River Vikedal during 1982 – 1994 shows that the river was acidified, episodic pH dropping below 5.0 (Figure 2). Concentration of labile aluminium normally ranged between 20 and 70 µg l^{-1}. After 1995 the natural chemical conditions of the unlimed river has improved with respect to acidity. In 1987 a lime doser was installed in the river (Figure 1). During the first three years liming was performed only during snowmelt, later the dosing has been intensified according to Table I. In 1998 a second lime doser was installed in the acidic tributary Litlaelvi in order to secure a more stable pH in the main river. As indicated by chemical data (Figure 3), the doser has mainly functioned according to the intentions during the period of operation (Table I).

TABLE I

Liming goals of the anadromous part of River Vikedal during the period spring 1987 – 1999.

Year(s)	Period	pH
1987-1989	Spring snow melt	5.7
1990-1993	15.02. - 31.05	6.0
	01.06 - 14.02	5.7
1994-1997	15.02. - 31.05	6.3
	01.06 - 14.02	5.7
1998	15.02. - 31.05	6.4
	01.06 - 14.02	5.7
1999	15.02. - 31.03	6.2
	01.04 – 31.05	6.4
	01.06 – 14.02	6.0

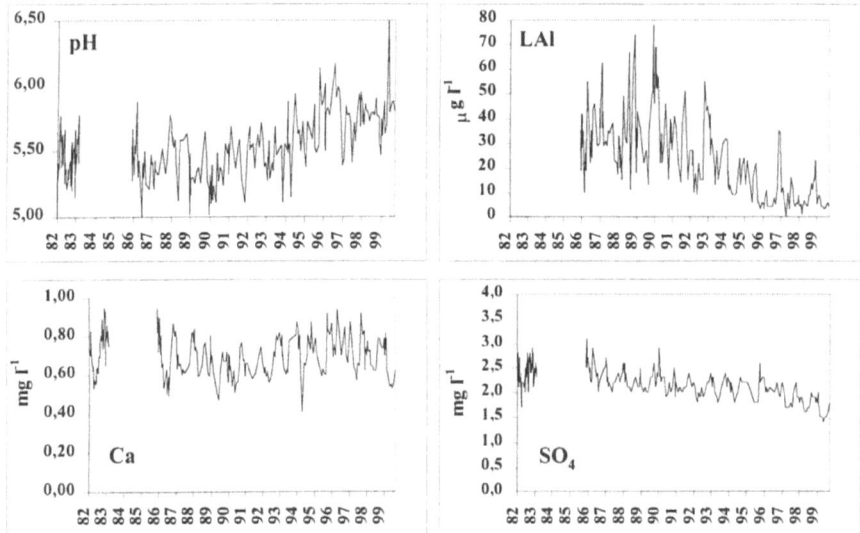

Figure 2. Water chemical data from the unlimed reference station of River Vikedal. Data provided by the Norwegian Institute for Water Research.

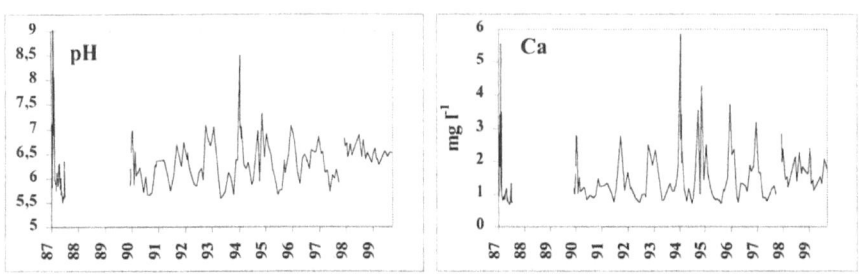

Figure 3. Water chemical data from the limed reference station of River Vikedal. Data provided by the Norwegian Institute for Water Research.

3. Methods

All qualitative benthic kick samples (Frost *et al.* 1971) were collected in spring (May) and in autumn (October-November). The two main stations, A (unlimed) and B (limed since 1987) were sampled during the period 1979 – 1999 (Figure 1). The acid rain monitoring network comprises 12 reference stations in the unlimed section of the river regularly sampled since 1982. Additional four stations in the limed main river were monitored during 1987 – 1998. All samples were sieved through a net of 250 m and later sorted and identified. Acidification scores were calculated according to Raddum *et al.* (1988) and Fjellheim and Raddum (1990).

4. Results

During the invertebrate survey in 1979, no individuals of the highly sensitive mayfly *Baetis rhodani* were found at station A, while the species was present at station B both during spring and autumn (Figure 4). During the first half of the 1980's, *B. rhodani* disappeared from both stations. After liming in 1987 a rapid population build-up of the autumn cohort was followed by a more slow response of the spring cohort. At the unlimed station, the population of *B. rhodani* seems to have stabilised at low numbers since 1995.

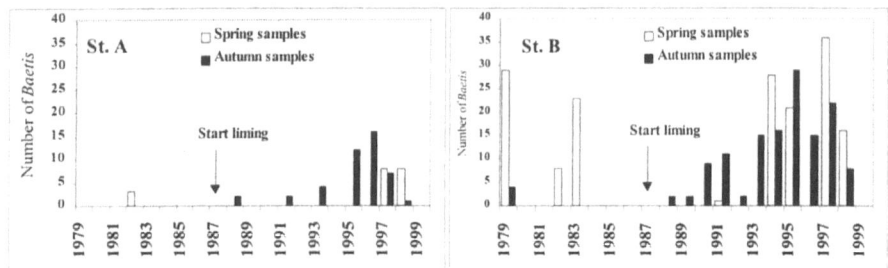

Figure 4. Number of *Baetis rhodani* found in benthic samples at station A and B in River Vikedal during 1979 – 1999.

The number of species within the groups Ephemeroptera, Plecoptera, Trichoptera hve increased from 12 to 25 (St. A) and 14 to 29 (St. B) during 1982 to 1999 (Figure 5). The recorded numbers of species from the reference station A were low during the first 12 years. After 1994, a marked increase in species richness was observed. On the contrary, species numberss in the samples from the limed St. B increased markedly during the first years, but later more slowly.

The acidification score of the unlimed section of the river mostly varied between 0.2 and 0.4 during 1982 – 1991, characterising the benthic invertebrate communities as strongly affected by acidification. During the last ten years the score has improved, and the benthic communities are now regarded as moderately affected. The limed part of the river showed marked differences between spring and autumn samples during the first 8 years of liming (Figure 6). Later, as lime dosing was intensified, also spring samples responded in a more positive direction, and the benthic invertebrate community of the main lower part of the river is now only slightly affected by acidification.

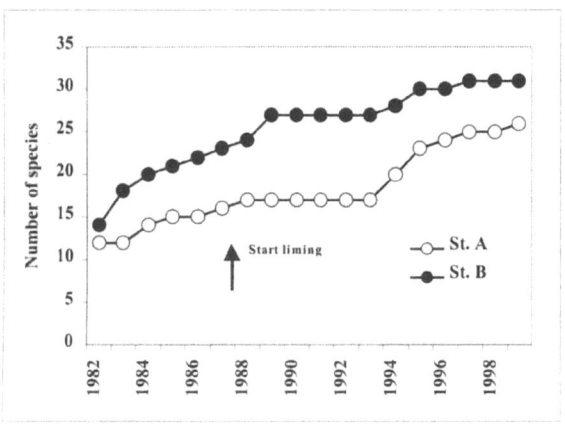

Figure 5. Cumulative number of species within Ephemeroptera, Plecoptera and Trichoptera from St. A and St. B.

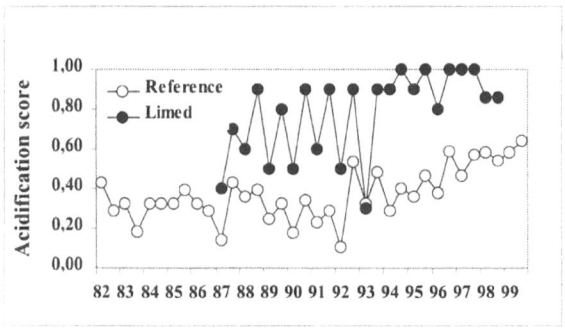

Figure 6. Acidification scores (after Raddum *et al*. 1988, Fjellheim and Raddum, 1990) of the limed (since 1987) and unlimed section of River Vikedal. Sampling stations are indicated on Figure 1.

5. Discussion

Data from the survey in 1979 showed that River Vikedal was strongly affected by acidification in the late 1970's. Besides small tributaries of good water quality (Fjellheim and Raddum 1995), only the lowermost part of the main river hosted populations of strongly sensitive species. During the beginning of the 1980's the acidification also stroke the anadromous section of the river, eliminating the most sensitive invertebrates and reducing the salmon stock. Data obtained during the first period after liming indicate that the lime dosage was incomplete, and that acidic episodes was harmful to parts of the invertebrate populations. The lime dose during spring had to be increased to match a pH goal raised from 6.0 to 6.3 before the benthic communities returned towards a more normal composition. This includes a population build-up of both spring and fall cohorts of *B. rhodani*. The genus *Baetis* contains key organisms in many rivers, which are frequently reported to respond positively to liming (Fjellheim and Raddum 1992, Lingdell and Engblom 1995, Eggleton *et al.* 1996). Also the freshwater gastropod *Lymnaea peregra* responded to the increased lime dose. The first specimen was recorded in 1995 and later *L. peregra* has been recorded regularly. This snail is normally not found below pH 6.0 and Ca < 1 mg l^{-1} in Norway (Økland 1990). The monitoring of the liming programme show that a high dosage is necessary to restore populations of highly sensitive species. The natural recovery observed in the unlimed section of the river is most probably a consequence of reduced sulphur deposition and is also supported by a positive development of brown trout populations in the upper part of the catchment (Hesthagen and Forseth, 1998, Hesthagen, pers comm.).

Acknowledgements

This study is part of the Norwegian Liming Programme, the Norwegian Monitoring Programme for Long Range Transported Air Pollutants and the Programme Research and Reference Rivers (FORSKREF). We want to thank the Norwegian Directorate of Nature Management who has financed all these programmes. We will also thank the Norwegian Institute for Water Research for providing chemical data.

References

Eggleton, M.A., Morgan, and Pennington, W.L.: 1996, *Restoration Ecology*, **4**, 247.
Fjellheim, A., Raddum, G.G.: 1990, *The Science of the Total Environment*, **96**, 57-66.
Fjellheim, A., Raddum, G.G.: 1992. *Environmental Pollution*, 78, 173.
Fjellheim, A., Raddum, G.G.: 1995, *Water, Air and Soil Pollution*, **85**, 931.
Fjellheim, A., Hesthagen, T., Raddum, G.G. and Mejdell Larsen, B.: 1987, 'Production, growth and food of young Atlantic salmon in two rivers with different acidification', in R. Perry, R.M. Harrison, J.N.B. Bell, and J.N. Lester (eds.) *Acid Rain: Scientific and Technical Advances*. Selper Publications Ltd., London. pp 500-507.
Frost, S., Huni, A., Kershaw, W.E.: 1971, *Can. J. Zool.*, **49**, 167-173.
Hesthagen, T.: 1986, *Water, Air and Soil Pollution*, **30**, 619.
Hesthagen, T. and Hansen, L. P.: 1991, *Aquaculture and Fisheries Management* **22**, 85.
Hesthagen, T. and Forseth, T..: 1998, *Verh. Internat. Verein Limnol.*, **26**, 2255.
Lien, L., Raddum, G. G. Fjellheim, A. and Henriksen, A.: 1996, *The science of the total environment* **177**, 173 .
Lingdell, P.E. and Engblom, E.: 1995, *Water, Air and Soil Pollution*, **85**, 955.
Økland, J.: 1990, *Lakes and snails*, Universal Book Services/Dr. W Backhuys, Oegstgeest.
Rosseland, B. O. and Skogheim, O. K. 1984. *Rep. Inst. Freshwater Res. Drottningholm*, **63**, 185.
Raddum, G.G., Fjellheim, A. and Hesthagen, T. 1988, *Verh. Internat. Verein. Limnol.*, **23**, 2291.

SUBLITTORAL CHIRONOMIDS AS INDICATORS OF ACIDITY (DIPTERA: CHIRONOMIDAE)

GODTFRED A. HALVORSEN[1*], JOCELYNE H. HENEBERRY[2] and ED SNUCINS[2]

[1] *Laboratory for Freshwater Ecology and Inland Fisheries (LFI), Department of Zoology, University of Bergen, Allégt. 41, N-5007 Bergen, Norway;* [2] *Ontario Ministry of Natural Resources, Cooperative Freshwater Ecology Unit, Laurentian University, Sudbury, Ontario, Canada, P3E 2C6.*
(* *author for correspondence, e-mail: godtfred.halvorsen@zoo.uib.no*)

Abstract. The sublittoral chironomid fauna of 22 lakes in Killarney Park, Ontario, Canada were examined for their response to different levels of acidification. Included in the analysis were naturally acidic lakes, lakes acidified by atmospheric deposition but now recovering, and unacidified circumneutral lakes. pH in the study lakes ranged from 4.6 to 7.7. No correlation was found between species richness and pH, nor between abundance and pH. Acid neutralizing capacity (ANC), the temperature at the sampling depth, pH, and dissolved organic carbon were the variables contributing significantly in a canonical correspondence analysis of the abundance data. ANC was the most important variable in describing the chironomid community, accounting for about 9 % of the variance in the species data. This study is the first step in an effort to model the changes in the chironomid community of recovering acidified lakes and the results suggest that ANC may be an important predictor variable.

Keywords: acidification, ANC, chironomids, recovery

1. Introduction

Chemical recovery of lakes and streams acidified by atmospheric pollutants has been occurring in parts of North America and Europe since the 1980's (e.g. Stoddard *et al.*, 1999). One of the areas with lakes recovering in response to reduced atmospheric pollution levels is Killarney Provincial Park, located 40-60 km southwest of Sudbury, Ontario, Canada (Snucins *et al.*, 2000). Many lakes in the Sudbury area were acidified by acid deposition originating from the local metal smelters, but also from an array of other sources in both Canada and the United States. Since the 1970's emission reductions at both local and long-range sources have led to water quality improvements in many of those lakes. Although many lakes remain acidic, some have recovered to approximately their pre-industrial pH levels. The history of the affected area and the beginnings of recovery have been described in several articles in Gunn (1995), Keller *et al.* (1998) and Snucins *et al.* (2000).

The Northern Lakes Recovery Study (NLRS) was initiated in 1997 to study the biological recovery of acidified lakes in Killarney Park. The study is a joint project between Canada and Norway involving the Cooperative Freshwater Ecology Unit (CFEU), Department of Biology, Laurentian University, Sudbury, the Norwegian Institute for Nature Research (NINA), and the Laboratory for Freshwater Ecology and Inland Fisheries (LFI), Department of Zoology, University of Bergen.

The LFI is examining the response of the profundal/sublittoral chironomid community to natural water quality recovery. This paper is a first step in an attempt to elucidate how well chironomids reflect the different levels of acidification and it will serve as a basis for future evaluations of the rate and nature of biological recovery.

2. Materials and Methods

Killarney Provincial Park (latitude 46°5' longitude 81°24') covers an area of 485 square kilometers. About half of the park is underlain by slow-weathering bedrock which provides little buffering against acid precipitation. The remainder of the park contains more easily weathered bedrock and glacial deposits. As a result, the many lakes within this relatively small area span a wide pH gradient and include naturally acidic lakes, anthropogenically acidified lakes, and well-buffered unacidified lakes (Snucins and Gunn, 1998). Human impacts other than acidification are limited. Logging and mining do not occur in the park and only a small number of leisure cottages exist on some lakes.

The 22 NLRS lakes were selected to span wide ranges in pH and dissolved organic carbon (DOC). A pilot study (Halvorsen, 1999) done during the spring and fall of 1997 found that some lakes exhibited oxygen depletion in the profundal zone and that the spring sampling was confounded by early chironomid emergence. As a result, the 1998 sampling was restricted to late fall and to sublittoral depths that had high end-of-summer oxygen levels.

The sampling depths were 9-15 m in most lakes, but were 3-6m in three relatively shallow lakes that experienced oxygen depletion in the profundal zone. The oxygen saturation at the sampling depths in the 22 lakes varied between 52 % and 152 %, with an average of 109 % saturation. All lakes are classified as oligotrophic (Total phosphorus 2-12 μg/l).

Five lakes had populations of the gammarid species *Diporeia hoyi* (Smith). Gammarids have been reported to interact with soft-bottom chironomids (Wiederholm, 1988, Johnson and Goedkoop, 1992) and may influence the chironomid community. The abundances of the species in the Kajak samples were accordingly included in the analyses. The ranges and averages of the selected environmental variables are shown in Table I.

TABLE I
Ranges and averages of some environmental variables from the 22 Killarney study lakes

Environmental variable	Min	Max	Avg
PH	4.6	7.7	5.8
Acid neutralising capacity – ANC (μekv/l)	- 82	400	25
Temperature at sampling depth – Temp (°C)	6.8	22.4	13.2
Dissolved organic carbon – DOC (mg/l)	0	4.6	1.9
Sampling depth – Depth (m)	3	15	11
Calcium – Ca (mg/l)	0.95	7.60	1.85
Labile aluminum – LAl (μg/l)	0	434	65
Abundance of *D. hoyi* - Gam (ind/m^2)	0	3527	262
Oxygen content at sampling depth – O (mg/l)	6.21	14.54	11.60

Ten benthic samples in each lake, taken from October 14th to 29th, 1998 using a modified Kajak sampler, were sieved through a 250 μm mesh and stored in alcohol. They were sorted using binocular microscopes. Water samples for chemical analysis were taken on November the 8th as tube composite samples from 0 – 5 m (Snucins and Gunn, 1998: Appendix A). The water chemistry analyses were done at the CFEU (pH) and at the Ontario Ministry of the Environment laboratory in Toronto (remaining variables). The ANC values were calculated by A. Schartau, NINA, Trondheim.

Statistical analyses of the data were performed with ordination analyses by means of Canoco 4.0 (ter Braak and Smilauer, 1998). The abundance data with the exception of the pH, and the environmental data were log transformed to reduce skewness and heteroscedasticity in the data. Rare species were down-weighted. The detrended correspondence analyses (DCA) were run with detrending by segments. In the canonical correspondence analyses (CCA) scaling were focused on inter-species distances and the scaling type were biplot scaling. Both the first canonical axis and all axes together were tested with a Monte-Carlo permutation test with 199 permutations under a reduced model. The included environmental variables were tested by the forward selection option, with Monte-Carlo permutation tests with 199 permutations as above.

3. Results

A total of 65 chironomid species / taxa were found in the Killarney lakes. They were identified to the lowest possible level using existing keys.

No correlations were found between the number of species and pH, and between the total abundance of chironomids and pH (Spearman rank correlations, Spearman's rho = 0.068 and – 0.109 respectively).

An initial detrended correspondence analysis (DCA) of the abundance data showed that the length of the gradient along the first ordination axis was 3.965 standard deviation units. This indicates a strong unimodal response in the abundance data, and canonical correspondense analysis (CCA) were used in the following analyses of the species abundances and the environmental data (Birks, 1995).

A CCA on the abundance data with all of the variables in Table I included as environmental variables were run. Both the first and all of the canonical axes together were significant (p = 0.03 and p = 0.005 respectively). Foreward selection showed that the environmental variables ANC, Temp, pH and DOC contributed significantly to the ordination, listed in order of decreasing amount of abundance variation explained. Of the remaining variables Gam was the one explaining most of the remaining variation when the other variables were added to the model, but this variable did not contribute significantly to the ordination.

A partial CCA was run with the variables Temp and Gam as covariables and with ANC, pH and DOC as environmental variables. This was done in order to elucidate the importance of the variables more directly connected with acidification. Both the first canonical axis and all of the axes together were significant (p = 0.025 and 0.005 respectively). The total variance in the species data (total inertia) was 3.584, the sum of all unconstrained eigenvalues 3.032 and the sum of all canonical eigenvalues 0.909. Decomposing the variance following ter Braak and Smilauer (1998: 124) shows that the covariables Temp and Gam combined explained 15 % of the total variance in the species data, calculated as follows: (100*(3.584 – 3.032) / 3.584). ANC, pH and DOC explains an additional 25 % of the variance (100 * 0.909 / 3.584). The remaining 60 % of the variation is unaccounted for in our analysis.

We wished to decompose the variance further, that is, to find the contribution made by each of the variables ANC, pH and DOC, and the shared contribution of these variables. Partial analyses were done according to the method described in ter Braak and Smilauer (1998: 258). This showed that the variance explained by ANC was 9 %, of pH 8 % and of DOC 8 %. The shared amount explained by the three variables was 2 %.

4. Discussion

The number of species /taxa found in the Killarney lakes was greater in 1998 (N = 65) than in the previous year's pilot study (N = 41) (Halvorsen, 1999). This is probably because we avoided sampling in the anoxic profundal zone of some of the lakes. In addition, three of the lakes were sampled at relatively shallow depths (3-6 m) in 1998, and some of the species found may belong to the littoral rather than the sublittoral fauna.

The lack of correlation between total species richness and abundance of chironomids and pH is consistent with earlier studies of the total benthic fauna in Laurentian Shield lakes (Dermott, 1985, Dermott et al., 1986, Carbone et al., 1998), as well as the 1997 Killarney pilot study (Halvorsen, 1999). Earlier Scandinavian studies (Wiederholm and Eriksson, 1977, Mossberg and Nyberg, 1979, Raddum and Sæther, 1981) reported a decrease in benthic species richness with reduced pH. This may be because the Scandinavian studies involved relatively few lakes (1, 7, and 5 lakes respectively) and included much more acidic waters, as low as pH 3.9, where species richness changes may be more pronounced.

The identification of ANC as the most important variable in describing the sublittoral chironomid community is of fundamental importance. The ANC can be modelled under different emmision scenarios as has recently been done for the Killarney lakes (Hindar and Henriksen, 1998). This means that it may be possible to develop models for recovering lakes that predict the chironomid community species composition and their relative abundances. The development of such models is our ultimate goal.

The amount of variation in the species data explained by the three variables

ANC, pH and DOC may seem low, especially when taken separately. However, chironomid communities are known to show large annual variations. The sampling in the NLRS lakes will continue for a further three years to provide variance estimates that will be used to improve the species response curves.

An important assumption underlying this study is that chironomid recolonization following chemical recovery has a short or negligible time-lag. We do not have data to test this assumption at the moment. However, in Killarney Park the unacidified lakes often occur in close proximity to lakes that have experienced strong acidification. Adult chironomids are flying insects and we presume this mobility allows them to be rapid colonisers. Ongoing research in Southern Norway on the full-scale liming of a long-time acidified river (Brandrud *et al.*, 2000) have shown that acid-sensitive mayflies, stoneflies and also putatively sensitive chironomids colonized the river at a locality that was close to a refuge only half a year after the liming started. Therefore, we believe that the assumption of a short time-lag in chironomid recolonisation is valid for the Killarney lakes where the recovery process has already been going on for several years.

Acknowledgements

This project is part of the Northern Lakes Recovery Study (NLRS), a cooperative study between Canada and Norway, designed to assess the factors that affect the recovery of acid stressed lakes. The samples were collected by field technicians from the CFEU, Laurentian University, Sudbury. Although too numerous to list individually, their assistance is nevertheless much appreciated. The Norwegian part of this study was financed by the Norwegian Directorate for Nature Management, with Steinar Sandøy as coordinator.

References

Birks, H.J.B.: 1995, 'Quantitative palaeoenvironmental reconstructions', in D. Maddy and J.S. Brew (eds.), *Statistical modelling of quaternary science data*, Technical Guide **5**, Quaternary Research Association, Cambridge, pp. 161-254

Brandrud, T.E., Brettum, P., Dolmen, D., Halvorsen, G., Halvorsen, G.A., Lindstrøm, E-A., Raddum, G.G., Romstad, R., and Schnell, Ø.A.: 2000, *Effects of liming on biodiversity. Results from 1997-98, the two first years after liming was started.*Utredning for DN, Direktoratet for naturforvaltning, Trondheim. (In Norwegian).

Carbone, J., Keller, W. and Griffiths, R.W.: 1998, *Restoration Ecology* **6**, 376.

Dermott, R.M.: 1985, *Hydrobiologia* **128**, 31.

Dermott, R.M., Kelso, J.R.M. and Douglas A.: 1986, *Water, Air and Soil Pollution* **28**, 283.

Gunn, J.M. (ed.): 1995, *Restoration and Recovery of an Industrial Region. Progress in Restoring the Smelter-Damaged Landscape near Sudbury, Canada.* Springer-Verlag, New York.

Halvorsen, G.A.: 1999, 'Chironomids as indicators of acidification. Report from a pilot project in the Northern Lakes Recovery Study', in G.G. Raddum, B.O. Rosseland and J. Bowman (eds.), *Workshop on biological assessment and monitoring; evaluations and models*, ICP-Waters report **50/99**, NIVA, Oslo, pp. 73-79.

Hindar, A. and Henriksen, A.: 1998, *Mapping of Critical Load and Critical Load Exceedances in the Killarney Provincial Park, Ontario, Canada*. NIVA Report **3889-98**, NIVA, Oslo.

Johnson, R.K. and Goedkoop, W.: 1992, *Neterlands Journal of Aquatic Ecoogy* **26**, 491.

Keller, W., Heneberry, J.H. and Gunn, J.M.: 1998, *Journal of Aquatic Ecosysystems Stress and Recovery* **6**, 189.

Mossberg, P. and Nyberg, P.: 1979, *Report Institute of Freshwater Research, Drottningholm* **58**, 77.

Raddum, G.G. and Sæther, O.A.: 1981, *Verhandlungen Internationale Vereinigung für Theoretischen und Angewandte Limnologie* **29**, 399.

Snucins E. and Gunn, J.M.: 1998, *Chemical and Biological Status of Killarney Park Lakes (1995-1997). A study of lakes in the early stage of recovery from acidification.* Ontario Ministry of Natural Resources, Cooperative Freshwater Ecology Unit., Sudbury, Ontario.

Snucins, E., Gunn, J.M., Keller, B., Dixit, S., Hindar, A. and Henriksen, A.: 2000, *Journal of Environmantal Monitoring and Assesment* (in press)

Stoddard, J.L., Jeffries, D.S., Lükewille, A., Clair, T.A., Dillon, P.J., Driscoll, C.T., Forsius, M., Johannesen, M., Kahl, J.S., Kellogg, J.H., Kemp, A., Mannio, J., Monteith, D.T., Murdoch, P.S., Patrick, S., Rebsdorf, A., Skjelkvåle, B.L., Stainton, M.P., Traaen, T.S., van Dam, H., Webster, K.E., Wleting, J. and Wilander, A.: 1999, *Nature* **401**, 575.

ter Braak, C.J.F. and Smilauer, P.: 1998, *CANOCO Reference Manual and User's Guide to Canoco 4.0 for Windows: Software for Canonical Community Ordination (version 4)*, Microcomputer Power, Ithaca, New York.

Wiederholm, T.: 1988, *Spixiana (Supplement)* **14**, 7.

Wiederholm, T. and Eriksson, L.: 1977, *Oikos*, **29**, 261.

REAPPEARANCE OF HIGHLY ACID-SENSITIVE INVERTEBRATES AFTER LIMING OF AN ALPINE LAKE ECOSYSTEM

ARNE FJELLHEIM[1], ÅSMUND TYSSE[2] and VILHELM BJERKNES[3]

[1] *University of Bergen, Institute of Zoology, Allegt. 41, N-5007 Bergen, Norway;*
[2] *County Governor of Buskerud, Dept. of Environment, Box 1604, N-3007, Drammen, Norway;*
[3] *Norwegian Institute of Water Research, Nordnesboder 5, N-5005 Bergen, Norway;*
(author for correspondence, e-mail: arne.fjellheim@zoo.uib.no)

Abstract. The amphipod *Gammarus lacustris* was earlier a main food item of brown trout in Lake Svartavatnet at the Hardangervidda mountain plateau in South Norway. In the middle of the 1980's, *G. lacustris* disappeared from the trout diet due to increased acidification. In order to preserve a unique genetic variant of brown trout living in the area, a liming programme was initiated in 1994. During the first years after liming, *G. lacustris* was absent both in fish stomachs and in lake littoral samples. In 1999, it reappeared in brown trout stomach samples together with two other strongly sensitive species, the tadpole shrimp *Lepidurus arcticus* and the freshwater gastropod *Lymnaea peregra*. Data from monitoring indicate that the water chemical conditions of L. Svartavatnet are still close to the critical limits of these animals. They have probably survived in small refuges of acceptable water quality, either in areas of inflowing groundwater or in the littoral, below the more acidic surface layer. The fact that these sensitive animals have not yet been found in benthic samples emphasise fish diet as an important tool in early registration of the presence/absence of invertebrates with low abundance or patchy distribution.

Keywords: monitoring, mountain lake, liming, brown trout diet, *Gammarus*, *Lepidurus*, Lymnaea

1. Introduction

During the last decades, the increasing problems of acidification due to acid precipitation have extended to parts of the South Norwegian mountain range, causing extinction of fish populations in many lakes (Hesthagen *et al.*, 1999). Less attention have been paid to the fate of acid-sensitive invertebrates in alpine lake ecosystems (Fjellheim *et al.*, 2000), but it is evident that problems of acidification may be severe in these areas due to exceedance of critical limits (Borgstrøm and Hendrey, 1976, Raddum and Fjellheim, 1984).

In Norway, most liming programmes are performed below the timber line (Sandøy and Romundstad, 1995). Monitoring shows that biodiversity increases after liming, especially through increased densities of acid-sensitive invertebrates. The rate of response depends upon factors such as the geographical situation of the limed catchment, the location of refuges within the catchment, the mobility of potential colonizers and the intensity of the liming programme (Fjellheim and Raddum, 1993). This paper gives results from the highest (with respect to altitude) situated liming project, in Norway (1460 – 1213 m a.s.l.). The project is a part of an ameliorating programme for the conservation of an endangered trout variety. Besides protecting the trout population, a main goal of the liming programme was to secure an acceptable water quality for the most important food organisms of the trout.

2. Materials and methods

Lake Svartavatnet is situated 1213 m a.s.l. at the Hardangervidda Mountain plateau, Central Norway (Figure 1). The lake is 21 m deep, has a surface area of 1.1 km^2 and a total catchment area of 11 km^2. The surroundings are dominated by bare rock (mostly gneisses), shrubs and small bogs. Soil cover is generally scarce. The lake littoral is composed of a patchy substrate consisting of sand, gravel, boulders of different size and bare rock. Brown trout is the only fish species in the area, of which a genetic variant, "the fine spotted trout" (Skaala and Jørstad, 1988) reproduces naturally in L. Svartavatnet and L. Svartavasstjørni (Figure 1).

Figure 1. Map of the research area.

L. Svartavatnet was regularly test-fished during 1970 – 1991. Around 1970 the lake hosted a viable trout population whose food mainly consisted of *Gammarus lacustris* (J. P. Madsen, summary reports, pers. comm.). In the middle of the 1980s, the trout population showed clear signs of recruitment failure. Since 1985, *G. lacustris* has been absent from the gut samples. During the late 1980s, sporadic water samples showed pH values down to 5.0. Skaala and Jørstad (1988) demonstrated large genetic differences between the fine spotted trout and other brown trout populations. It was concluded that the genetical distinctness and demographic situation strongly motivated preservation of the population and its habitat (Skaala et al., 1991).

During the 1990s the populations of fine spotted trout were strengthened by stocking fry reared in hatcheries (1991 and 1995) and by liming the catchment area and the actual lakes to improve water quality (Table 1). Additionally, during the period 1997 – 1999, specimens of *G. lacustris* and *L. arcticus* were transferred from L. Skiftesjøen to L. Svartavasstjørni (Figure 1).

TABLE 1.
Liming of the catchment of L. Svartavatnet

Year	Location	Product	Amount
1993	Inlet	Coral gravel	1 ton
1994-1996	Catchment	Crushed limestone	15 tons
	Upstream water bodies	Powdered limestone	17 tons
1997	Catchment	Crushed limestone	15 tons
	Upstream water bodies	Powdered limestone	22 tons
	L. Svartavatnet	Powdered limestone	10 tons
1998-1999	Catchment	Crushed limestone	20 tons
	Upstream water bodies	Powdered limestone	22 tons
	L. Svartavatnet	Powdered limestone	13 tons

Testfishing in L. Svartavatnet were performed in 1997, 1998 and 1999, using gillnet series identical to those used before liming. Qualitative littoral kick samples (Table II) were simultaneously taken from two different sites using a sweep-net, 25x30 cm, of mesh size 250 μm (approximately 5 min sampling time). All stomach and benthic samples were preserved in 70 % alcohol and sorted in the laboratory. The relative composition of the trout diet was estimated using the "points method" (Hynes, 1950).

3. Results

The water chemistry in 1993 (Figure 2) shows the situation before liming. Both in- and outlet had a low pH (minimum 5.17 and 5.18), low Ca values and relatively high concentrations of inorganic labile aluminium. After liming, pH and Ca concentration increased and the concentration of labile aluminium simultaneously dropped to almost zero.

Benthic sampling during 1997 –99 (Table II) show that the lake littoral of L. Svartavatnet was dominated by chironomids. Only a few individuals of moderately sensitive species were found: the flatworm *Otomesostoma*, the stoneflies *Capnia* and *Diura* and the caddisfly *Apatania*. In South Norway these taxa are all known to tolerate pH down to 5.0 (Fjellheim and Raddum 1990). In 1997 and 1998, the trout food mainly consisted of caddisflies and Cladocera. No acid-sensitive animals were detected in the diet of the trout. In 1999, the highly sensitive *G. lacustris*, *L. peregra* and *L. arcticus* were found in brown trout stomachs (Table II), but were still absent from the benthic samples.

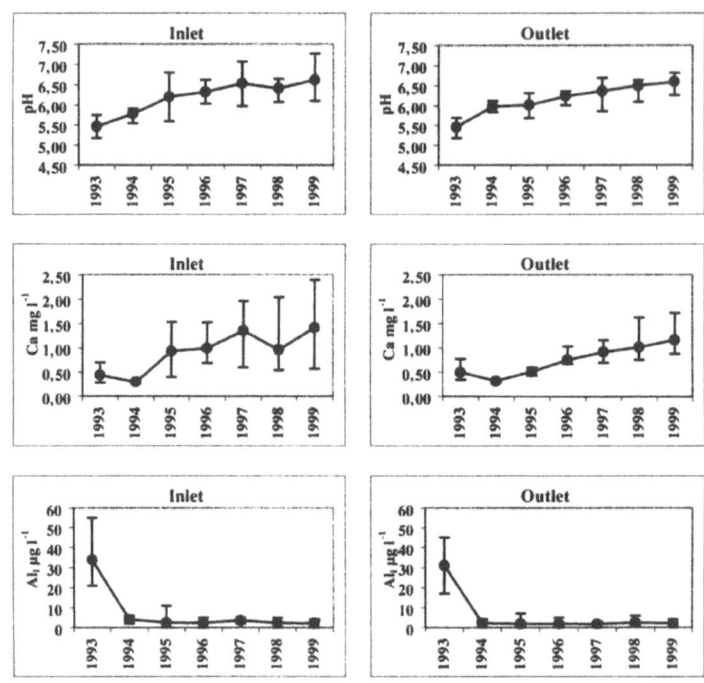

Figure 2. Surface water chemistry (mean and range) at the inlet and outlet of Lake Svartavatnet during 1993 (May–July, N=5), 1994 (Aug.–Sep., N=6) and 1995–1999 (May–Sep., N=11).

TABLE II.
Diet (volume %) of brown trout and benthic lake littoral samples (two stations pooled) sampled in Lake Svartavatnet during 1997 – 1999. *: stomach samples provided by a local fisherman.

	Brown trout stomach samples				Lake littoral samples		
Year	1997	1998	1999	1999	1997	1998	1999
Date	28.08	05.08	08.08	15.09	13.08	08.09	09.08
Sample size	7	10	21	19	2	2	2
Size range of fish (cm)	20-47	17-42	19-40	*			
Gammarus lacustris			6.2	0.9			
Lepidurus arcticus				26.0			
Lymnaea peregra			1.2				
Cladocera	36.9	40.8	33.6				
Turbellaria					2		
Nematoda					7	12	25
Acari					15	10	7
Oligochaeta			0.4		24	32	37
Plecoptera					14	29	23
Trichoptera	54.8	37.3	20.0	1.0	9	7	6
Megaloptera			1.4				
Chironomidae		11.5	5.6		101	96	124
Other Diptera	0.8	3.5	31.0	70.1	4	10	4
Coleoptera	2.6	6.8	0.6	2.0			8
Other	4.9						

4. Discussion

The littoral crustaceans *Gammarus lacustris* and *Lepidurus arcticus* are considered to be very important food items for brown trout in Norwegian alpine lakes (Aass, 1969, Økland and Økland, 1980). Both species are highly sensitive to acidification. According to Økland and Økland (1985), no population of *G. lacustris* have been recorded in waters with pH < 6.0 and Ca < 1 mg l^{-1} in Norway. *G. lacustris* has a life cycle of two years in lakes at Hardangervidda (Bjerknes, 1974). The pH-tolerance of *L. arcticus* is less well known. According to Borgstrøm *et al.* (1976), the species is not found below pH 6.1. An experimental study by Borgstrøm and Hendrey (1976) showed that development of the first larval stages was considerably slower below pH 5.5.

Lymnaea peregra normally occurs in lakes with pH>6.0, but are occasionally found in lakes with pH down to 5.2 and Ca less than 1.5 mg l^{-1} (Økland, 1990). This species is widely distributed in the Hardangervidda area, normally appearing together with *G. lacustris* and *L. arcticus*.

During 1997 – 1999, 350 individuals of *L. arcticus* were stocked in L. Svartavasstjørni, giving a total stocked density of 0.1 ind. per 1000 m^2 lake surface, or 1.5 ind. per 1000 m^2 lake littoral (0-5m). In the same period, a total of 11300 *G. lacustris* were stocked in the same lake (49 ind. per 1000 m^2 lake littoral). In spite of large sampling efforts, no individuals of *G. lacustris* or *L. arcticus* were found in benthic samples or trout stomachs from L. Svartavasstjørni during 1997 – 1999. On the basis of the short time since stocking, low fecundity of stocked species and low number of trout migrating between the lakes, we do not think that the introduction could have affected the fauna in the downstream L. Svartavatnet. This view is also strengthened by the fact that the population of *Lepidurus* in Lake Svartavatnet seems to have recovered considerably (more than 50 individuals were found in trout stomachs in September 1999). Survival in refuges with better water quality during the acidic period is considered as a more probable hypothesis. The presence of refuges is made more likely by the appearance of *L. peregra*, which has not been reintroduced in the catchment.

G. lacustris has the capability of pH-avoidance by moving into deeper waters (Costa, 1967, Borgstrøm and Hendrey, 1976) or by clinging to buffering materials (Engblom and Lingdell, 1983).

The hatching of the eggs of *L. arcticus* coincides with the spring snowmelt. Tolerance tests by Borgstrøm and Hendrey (1976) indicate that moulting of the first instars may have been delayed prior to liming of L. Svartavatnet.

Even if the surface water of L. Svartavatnet was acidic before liming, the monitoring records demonstrates that highly acid-sensitive species have been able to survive, probably in areas of inflowing groundwater or in the littoral below the more acidic surface layer. Since none of these species were registered in trout stomachs between 1985 and 1999, the rest-populations must have been small; probably close to extinction. The closest locality containing *Gammarus*,

Lepidurus and *Lymnaea* is Lake Dragøyfjord, (1180 m a.s.l.) two km further down in the watershed (Figure 1).

Assuming a stable and acceptable water quality in coming years, we expect populations of highly sensitive invertebrates to increase in L. Svartavatnet. The invertebrate community will probably be more similar to that of L. Skiftesjøen (Figure 1), situated at the same altitude. This lake is unaffected by acidification and hosts dense populations of many acid-sensitive animals including *Gammarus*, *Lepidurus* and *Lymnaea*.

This study demonstrates that ameliorating efforts are extremely important when populations of sensitive animals are in a sublethal state. It also shows that fish diet is an important tool in demonstration of the presence/absence of invertebrates with low abundance or patchy distribution.

Acknowledgements

The authors gratefully acknowledge the Norwegian Directorate for Nature Management for financing this study. A special thank is given to Herman Stakseng for help during field work and for collecting additional fish stomach samples. We also thank Bjørn Haugen of the Hardangervidda national park warden service for water sampling.

References

Aass, P.: 1969, *Rep. Inst. Freshw. Res. Drottningholm* **49**, 183-201.
Bjerknes, V.: 1974, *Norw. J. Zool*, **22**, 39-43.
Borgstrøm, R. and Hendrey, G. R.: 1976, SNSF Project IR 22/76, Oslo-Ås.
Borgstrøm, R., Brittain, J and Lillehammer, A.: 1976, SNSF Project IR 21/76, Oslo-Ås.
Costa, H. H.: 1967, *Crustaceana 13*, 1-10.
Engblom, E. and Lingdell, P. E. 1983. Swedish Environment protection Agency, snvpm 1741, 181 pp. (in Swedish).
Fjellheim, A. and Raddum, G. G.: 1990, *The Science of the Total Environment*, **96**, 57-66.
Fjellheim, A. and Raddum G. G.: 1993, Changes in the mayfly community of Lake Hovvatn During the first 12 years of liming. - In: G.Giussani and C. Callieri (eds), *Strategies for Lake Ecosystems Beyond 2000*, Proceedings, Stresa, pp. 407-410.
Fjellheim, A., Boggero, A., Halvorsen, G. A., Nocentini, A. M., Rieradevall, M., Raddum G. G. and Schnell, Ø. A.: 2000, *Verh. Internat. Verein. Limnol.* **27**: 484-488.
Hesthagen, T., Sevaldrud, I. H. and Berger, H. M.: 1999, *Ambio* **28**, 112-117.
Hynes, H. B. N.: 1950, *J. Anim. Ecol.* **19**, 36-58.
Økland, J.: 1990, *Lakes and snails*. Universal Book Services. Oegstgeest.
Økland, J. and Økland, K. A.: 1980, pH level and food organisms for fish: Studies of 1000 lakes in Norway. In: Drabløs, D. And Tollan,A. (eds.) *Ecological impact of Acid precipitation*, pp 326-327, The SNSF project, Oslo – Ås.
Økland, J. and Økland, K. A.: 1985, *Oecologia (Berl.)* **66**, 364-367.
Raddum, G.G. and Fjellheim, A.: 1984, *Verh. Internat. Verein. Limnol.* **22**, 1973-1980.
Sandøy, S. and Romundstad, A.J.: 1995, *Water Air and Soil Pollution* **85**, 997-1002.
Skaala, Ø. and Jørstad, K. E.: 1988, *Pol. Arch. Hydrobiol.* **35**, 295-304.
Skaala, Ø., Jørstad, K. E. and Borgstrøm, R.: 1991, *J. Fish Biol.* **39** (Suppl. A), 123-130.

A STRATEGIC APPRAISAL OF OPTIONS TO AMELIORATE REGIONAL ACIDIFICATION

ALUN S GEE
Environment Agency, St Mellons Business Park, Cardiff, Wales, U.K
(e-mail: alun.gee@environment-agency.gov.uk)

Abstract. Studies in the 1980's showed that there had been extensive loss of fish and invertebrates in lakes and streams in upland Wales due to acid deposition. A regional survey of these waters in 1995 showed that, despite large reductions in sulphur emissions, acidification was still widespread and no biological improvement was detectable. Policy options are needed which should include reductions in nitrogen as well as sulphur, the contribution of land use (especially conifer afforestation), liming and the introduction of species. This paper describes a screening study comprising the construction of an evaluation matrix, the use of decision workshops of expert stakeholders and trade-off analysis to identify possible combinations of options. The resulting output is being used to construct a regional strategy for achieving measurable ecosystem recovery within the next 10 years.

Keywords: acidification, appraisal, ecosystem recovery, regional strategy, options.

1. Introduction

Studies in the early 1980's showed that acidification was responsible for the widespread loss of fish in streams and lakes in upland Britain. Large areas of Wales have little or no acid neutralising capacity, where the process of acidification has been exacerbated by land use changes, particularly conifer afforestation (Edwards *et al.*, 1990).

In 1984 the Welsh Water Authority undertook the first regional survey of the acidity of rainfall and of the chemistry and biology of lakes and streams. The survey was repeated in 1995 in order to assess if there had been any measurable improvement (Stevens *et al.*, 1997).

The results of the 1995 survey showed that the annual mean rainfall pH had increased from 4.72 to 4.88 compared with 1984. Total sulphur deposition declined from 25.9 to 15.5 kgS ha^{-1} yr^{-1}. This is commensurate with the overall decline in sulphur emissions due, in part, to the implementation of the Large Combustion Plant Directive. Nevertheless, total non-sea salt sulphur deposition over the Welsh mountains (22 kgS ha^{-1} yr^{-1}) was still the highest in Great Britain, and the total deposition of nitrogen was up to 25 kgN ha^{-1} yr^{-1}. Ammonia deposition in areas of livestock farming was up to 14 kgN ha^{-1} yr^{-1}.

Annual mean sulphate concentrations decreased by about 20% to 5.0 mgl^{-1} over the decade, but mean aluminium concentrations increased from 0.039 mgl^{-1} to 0.047 mgl^{-1}.

There was no indication of any change in the biological status of the streams or lakes between 1984 and 1995. As in 1984, trout populations were reduced significantly with increasing acidity and aluminium concentrations.

The report on the 1995 survey concluded that:
- there is evidence that some chemical components of freshwaters (eg. sulphate and pH) may be responding to reductions in sulphur deposition, but others (like aluminium) show no improvement;
- there has been no evidence that biological indicators of acidification have responded because of some combination of the following factors
  (i) suitable chemical conditions for recovery have not been sufficient, sustained or widespread enough;
  (ii) because biological recovery may be intrinsically slow;
  (iii) because the effects of reduced sulphur may have been offset by atmospheric deposition of nitrogen.

A conference to consider the implications of these findings concluded that satisfactory improvements to the acidification and biological status of Welsh lakes and streams could not be expected by reliance on emissions reduction strategies alone. It was decided to examine options for improvement of acidified waters by combinations of emission reduction and remediation. This paper describes the application of a strategic options appraisal technique to this problem.

2. Options Appraisal

2.1. MAIN STEPS

The method chosen involved the construction of an evaluation matrix and the use of decision workshops to elicit the views and preferences of technical specialists, managers and political advisors. The main steps are shown in Figure 1.

Define Aim of the Options Appraisal
⇩
Identify and Screen Options
⇩
Assess the Environmental, Economic, Social and Technical Implications
⇩
Participation and Dialogue
⇩
Trade-off Analysis of Options
⇩
Recommended Options
⇩
Decision to Pursue Options
⇩
Implementation
⇩
Monitoring and Audit

Figure 1. Flow diagram illustrating the main steps in carrying out the options appraisal
Only those steps depicted in bold type are considered in this paper.

2.2. EVALUATION MATRIX

Thirteen strategic options were initially selected, which were grouped into five broad categories:

Category	Alternative	
Sulphur reduction	A1	SO_2 reduction
Nitrogen reduction	A2	reduction in NO_x from stationary sources
	A3	reduction in NO_x from mobile sources
	A4	reduction in ammonia
Land Use	A5	restrictions on new forestry
	A6	adoption of new UK standards to existing forestry
	A7	agri-environment (environmentally sensitive or enhancing practices)
Liming	A8	Direct dosing of rivers
	A9	Source area liming (addition of lime to the main source of water in a catchment)
	A10	Agricultural or catchment liming
Bio-intervention	A11	Re-introduction of species at selected sites
	A12	Biomanipulation*
	A13	Genetic manipulation of fish*

The last two options (*) were quickly considered to present unacceptable risks to warrant further consideration.

These options were assessed according to the following categories within the matrix: Sensitivity; Potential risks and costs of options (environmental, social, economic); Benefit opportunities.

For each cell in the matrix, eleven issues were considered and an assessment made about the level of impact (high, medium or low). For example, for the Sulphur Reduction option, the following assessment was made of the impact on two of the eleven issues, Water Quality (WQ) and Fish (and Aquatic Ecology):

Option	Sensitivity		Potential Risks		Benefit Opportunities	
	WQ	Fish	WQ	Fish	WQ	Fish
A1 SO_2 reduction	High	Medium	High	Medium	High	Medium

3. Applying the appraisal process

3.1. DECISION ANALYSIS

Experts in the fields of air and water quality, fisheries, ecology, forestry, agriculture and human health were represented.

Through a process of evaluation and discussion a consensus was reached at the first workshop on the scores (high, medium, low) that should be applied to each cell in the appraisal matrix in this screening. The results of that exercise are described in detail in Brookes (1999).

3.2. TRADE-OFF ANALYSIS

At the second workshop, the following four **decision factors** were used in a trade-off analysis:

F1 maximising the opportunity for public health (reducing air pollution)
F2 minimising the adverse environmental risk
F3 best value for money
F4 benefit of the opportunity to the aquatic environment.

Using the output of the first workshop in relation to each of these decision factors enabled the participants to derive the following overall summary of option desirability.

TABLE I

Overall summary of option desirability

Alternative	Decision Factor F1	Decision Factor F2	Decision Factor F3	Decision Factor F4
A1	●	○	♦	●
A2	●	○	♦	●
A3	●	♦	♦	●
A4	○	○	♦	●
A5	♦	○	○	♦
A6	♦	●	●	♦
A7	♦	●	●	♦
A8	♦	○	○	●
A9	♦	♦	○	●
A10	♦	♦	♦	♦
A11	♦	♦	●	♦

● high or large;   ○ medium or moderate;   ♦ low or minor

From this analysis there emerged an order of priority. For any one decision factor the most desirable alternatives are shown in the following table.

TABLE II
The most desirable alternative options for each of four decision factors.

	Decision Factor		Most desirable alternatives
F1	Maximising opportunity for public health	A1	SO_2 reduction
		A2	Reduction of NO_x stationary
		A3	Reduction of NO_x mobile
F2	Minimising adverse environmental impact	A6	Adopting UK standards for existing forestry
		A7	Agri-environment
F3	Best value for money	A6	Adopting UK standards for existing forestry
		A7	Agri-environment
		A11	Re-introductions at selected sites
F4	Benefit of opportunity to the aquatic environment	A1	SO_2 reduction
		A2	Reduction of NO_x stationary
		A3	Reduction of NO_x mobile
		A4	NH_3 ammonia reduction
		A8	Direct dosing
		A9	Source area liming

It is clear that most of the desirable alternatives are those which become effective in the long-term (50 years +) but some targeted intervention may be worth considering in the short term.

3.3. OVERALL RANKING OF ALTERNATIVES

During the third workshop, participants were asked to use their judgement to assess the output of the previous two workshops and to determine the overall ranking of alternatives (Table III).

TABLE III
Overall ranking of alternative options

Rank		Alternative
1	A1	SO_2 reduction
2	A2	NO_x stationary reduction
3	A6	Adoption new UK standard for existing forestry
4	A4	Ammonia reduction
5	A8	Direct lime dosing
6	A7	Agri-environment
7	A3	NO_x mobile reduction
8	A9	Source-area liming
9	A11	Re-introduction of species
10	A5	Restrictions on new forestry
11	A10	Agriculture/catchment liming

4. Discussion

Throughout most of Europe, reductions in sulphur deposition are resulting in corresponding reductions in freshwater sulphate concentrations and a partial recovery of the invertebrate fauna (European Environment Agency, 1998). Exceptionally, this is not evident in the UK nor is a reduction in sulphate concentrations in Wales sufficient to allow even a partial recovery of invertebrates and fish (Stevens *et al.*, 1997).

The appraisal process described here has allowed a consensus to be achieved on the need to examine further reductions in emissions of sulphur and nitrogen, combined with remediation options in selected sites to promote recovery in the shorter term.

The next phase of the study is to appraise the technical and economic factors in detail resulting from the various United Nations Economic Commission for Europe (UNECE) protocols using the regional MAGIC model developed by Jenkins *et al* (1990). The output of this model will be applied to a biological model to predict the invertebrate and fish communities. The likely impact of land use practices, species introductions and selective liming will also be assessed.

Acknowledgements

The author would like to thank Andrew Brookes, Clare Twigger-Ross, Jonathan Fisher and Wendy Merrett of the Environment Agency, Havard Prosser of the National Assembly for Wales, and all those who participated in the appraisal workshops.

The views expressed here are the author's and not necessarily those of the Environment Agency.

References

Brookes, A. : 1999, *Acid Waters in Wales : Appraisal of Strategic Options Phase 1*. Final Report, July 1999, Report No.11, National Centre for Risk Analysis and Options Appraisal, Environment Agency, Steel House, 11 Tothill Street, London SW1H 9NF.

Edwards, R. W., Gee A. S. and Stoner, J. H. (eds) : 1990, *Acid Waters in Wales, Monographiae Biologicae*, 66. Kluwer Academic Publishers, Dortrecht.

European Environment Agency : 1988, *Europe's Environment : The Second Assessment*, Elsevier Science Ltd.

Jenkins, A., Whitehead, P. G., Musgrove, J. T. and Cosby, B. J. : 1990, *A Regional Model of Acidification in Wales. Journal of Hydrology*, 116, 403-416.

Stevens, P. A., Ormerod, S. J. and Reynolds, B. : 1997, *Final Report on the Acid Waters Survey for Wales*. Institute of Terrestrial Ecology Project T07072R5.

EFFECTS OF IN-LAKE RETENTION OF NITROGEN ON CRITICAL LOAD CALCULATIONS

ATLE HINDAR[1*], MAXIMILIAN POSCH[2] and ARNE HENRIKSEN[3]

[1]*Norwegian Institute for Water Research, Televeien 3, N-4879 Grimstad, Norway;*
[2]*National Institute of Public Health and the Environment, NL-3720 BA Bilthoven, Netherlands;*
[3]*Norwegian Institute for Water Research, P.O. Box 173 Kjelsås, N-0411 Oslo, Norway*
(*author for correspondence, e-mail:atle.hindar@niva.no)

Abstract. Critical loads (CLs) for soils and surface waters and their exceedances have been the basis for negotiations of emission reductions in Europe and elsewhere. In Sudbury, Canada, large reductions in sulphur emissions have resulted in reduced critical load exceedances of many lakes in the Killarney Provincial Park. To achieve a more complete chemical recovery even larger reductions of acid deposition are necessary. We extended the FAB (First-order Acidity Balance) model to include in-lake retention of nitrogen in upstream lakes and applied it to calculate CLs for Killarney lakes. Three different approaches were compared; one-lake, big-lake and lake-system. Use of "lake-system" resulted in the highest N retention and thus highest CLs, indicating that lakes at the end of chains are less sensitive to nitrogen deposition than calculated by the previous version of the model. Proper description of in-lake retention in such lake systems, as well as good data on catchment properties like land use and land cover, are important for CL-calculations used for the evaluation of future emission reduction policies.

Key words: Critical loads, in-lake retention, nitrogen, model, Killarney Provincial Park, FAB

1. Introduction

Calculations of critical loads (CLs) and their exceedances have been the basis for negotiations of emission reductions in Europe and Northern America. The 1994 Sulphur Protocol is based on critical loads for soils for most of Europe, and on critical loads for surface waters for Scandinavia. The latest protocol, signed in Gothenburg, Sweden, 1 December 1999, is also based on CLs and prescribes emission targets for sulphur, oxidized and reduced nitrogen and volatile organic compounds. The First-order Acidity Balance (FAB) model (Posch *et al.*, 1997; Henriksen and Posch, 2000) was used for the CL calculations for surface waters in the 1999 Protocol.

In Sudbury (Ontario, Canada) large reductions in sulphur emissions (about 90% since 1960) have resulted in reduced CL exceedances and a partial recovery of many lakes in the Killarney Provincial Park. To achieve a more complete chemical recovery even larger reductions are necessary, and the deposition of nitrogen has to be taken into account.

The FAB-model takes into account the retention of nitrogen in both the terrestrial and aquatic part of the catchment, and requires detailed knowledge of each catchment. An improved and extended version of the FAB-model has been derived (Hindar *et al.*, 2000) and has been applied for the first time to calculate CLs for systems (chains) of lakes, i.e. sets of lakes draining into each other.

Water, Air, and Soil Pollution **130:** 1403–1408, 2001.
© 2001 *Kluwer Academic Publishers.*

2. Materials and Methods

Killarney Provincial Park (46°3'N, 81°21'W; 48,500 ha) is situated about 60 km south-west of Sudbury, Ontario, close to the Georgian Bay in the north-eastern part of Lake Huron. The over 500 lakes and ponds in the park are widely different in size, elevation, catchment characteristics and water chemistry, and they have therefore reacted differently to acid deposition (Snucins and Gunn, 1998). Most of the park is underlain by ridges of quartzite, a material that is very weathering resistant. In these areas many of the lakes are acidified with water quality levels below critical limits for many fish species.

A synoptic survey of 151 lakes in the Park was carried out during winter 1996 (Snucins and Gunn, 1998). Water chemistry data and a mean annual runoff of 0.35 m yr^{-1} (Government of Ontario, 1984) were used to calculate the acidity critical load, CL(A), for each lake with the SSWC-model (Hindar and Henriksen, 1998).

While the SSWC model is solely based on water chemistry and incorporates the influence of the terrestrial catchment in an empirical way (via the F-factor; Brakke et al., 1990), the FAB model takes into account the sinks of N in the terrestrial catchment as well as the in-lake retention of S and N in a deterministic way. Every pair of N- and S-deposition satisfying the following equation is called a critical load:

(1) $\quad (1-\rho_S) \cdot S_{dep} + (1-\rho_N) \cdot b_N \cdot N_{dep} = (1-\rho_N) \cdot M_N + CL(A)$

where M_N and b_N depend on the denitrification fraction f_{de}, the N net uptake, N_u, and the net immobilization of N, N_i, and are a function of the magnitude of N_{dep}. The in-lake retention coefficients ρ_S and ρ_N is modeled by kinetic equations (Kelly et al., 1987), making it a function of the runoff, the lake:catchment area ratio and net mass transfer coefficients s_S and s_N. Equ.1 defines the so-called critical load function of a lake, and an example is depicted in Figure 1.

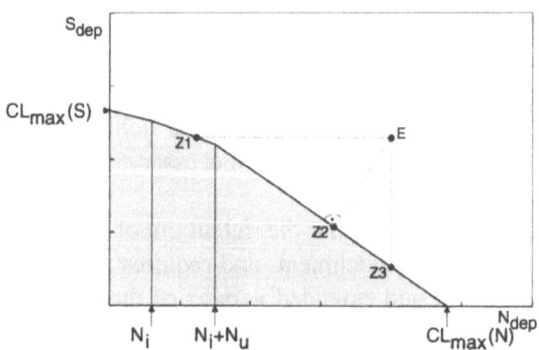

Figure 1. Example of a critical load function as defined by the FAB model. The grey-shaded area below the critical load function defines deposition pairs (N_{dep}, S_{dep}) for which there is no exceedance. Starting from E, Z1-Z3 indicate deposition pairs of non-exceedance reached by reducing N and S deposition in different proportions.

The catchment and lake sizes as well as the fraction of forest and open land have been derived from digitized maps of the park. Since there is no harvesting of forests we set $N_u=0$ for all catchments. For the long-term sustainable net immobilization of N in the soils we assumed a value of $N_i=1$ kgN ha^{-1} yr^{-1}. The denitrification fraction is modeled as a linear function of the fraction of peat lands in the catchment: $f_{de}=0.1+0.7$ f_{peat}. For the net mass transfer coefficients we used $s_S=0.5$ m yr^{-1} and $s_N=5$ m yr^{-1} for all lakes. For data sources and further information on the input data see Hindar et al. (2000).

Since many lakes in the Killarney Park are connected by streams, the FAB model has been modified to deal with such *systems of lakes*. In addition to treating a catchment as draining directly into the last lake (ignoring all upstream lakes: the "one-lake" approach), two methods have been derived for computing CLs for a system of lakes: The "big-lake" approach, which treats all lakes in the system as a single lake situated in the combined catchment, and the "lake-system" approach, which treats each sub-catchment separately and routes the water through the system, thus exposing it to sequential in-lake retention (see Appendix B in Hindar et al., 2000)

For the application of the FAB model to non-headwater lakes one has to have all input data from their upstream lakes. For each of the 151 monitored lakes in the Killarney Park a tree diagram was constructed (Hindar et al., 2000), showing which lakes drain directly into the end lake and which lakes are further upstream. For 43 of the lakes complete data sets for modelling were available. This included data for upstream lake(s) for 8 of the lakes. The remaining 35 were headwater lakes.

3. Results

Computing the CL functions with either one of the three approaches yields, of course, identical results for the 35 headwater lakes. For the other 8 lakes the values of $CL_{max}(S)$ and $CL_{max}(N)$ are listed in Table I for the three approaches.

TABLE I

$CL_{max}(S)$ and $CL_{max}(N)$ computed for the 8 lakes with sampled upstream catchments using three approaches: (i) ignoring upstream lakes ("one-lake" approach); (ii) using the "big-lake" approach and (iii) using the "lake-system" approach. Also given is CL(A), the critical load of acidity computed with the SSWC model (Hindar and Henriksen, 1998)

			One-lake		Big-lake		Lake-system	
Lake no.	Upstream lake (s) no.	CL(A)	$CL_{max}(S)$	$CL_{max}(N)$	$CL_{max}(S)$	$CL_{max}(N)$	$CL_{max}(S)$	$CL_{max}(N)$
				meq m^{-2} yr^{-1}				
2	47,102	47	51.9	109.7	55.8	152.4	58.2	157.4
43	5,100	124	137.0	288.1	165.2	608.8	202.6	1335.6
47	102	34	36.5	67.4	46.2	171.1	52.5	278.5
49	46	73	79.7	155.9	89.7	264.2	108.6	620.0
61	46	185	194.0	303.8	225.4	646.9	305.0	2472.4
112	111	17	18.5	41.6	19.8	55.3	21.6	83.6
113	114	29	32.4	75.5	32.7	79.1	33.0	81.5
124	121	28	29.7	50.5	31.8	72.5	34.6	114.1

The CLs computed with the "big-lake" approach are always smaller than the corresponding values from the "lake-system" approach. This is due to the fact that the latter allows the S and N draining from an upstream catchment to be retained again in all downstream lakes. The results show that the differences, especially in $CL_{max}(N)$, can be substantial. This depends on several factors, such as the lake:catchment ratio and land cover (fraction of forests) in the terrestrial catchment.

In Figure 2 the critical load functions for the two lake systems with three lakes each are shown. It shows, that upstream lakes are in general more sensitive, but this is less clear-cut for the chain 102→47→2 than for the system 5→43←100. Present (D1) and future (D2) depositions lead to exceedance in some lakes and to non-exceedance in others, and in-lake retention (esp. of N) has a strong influence.

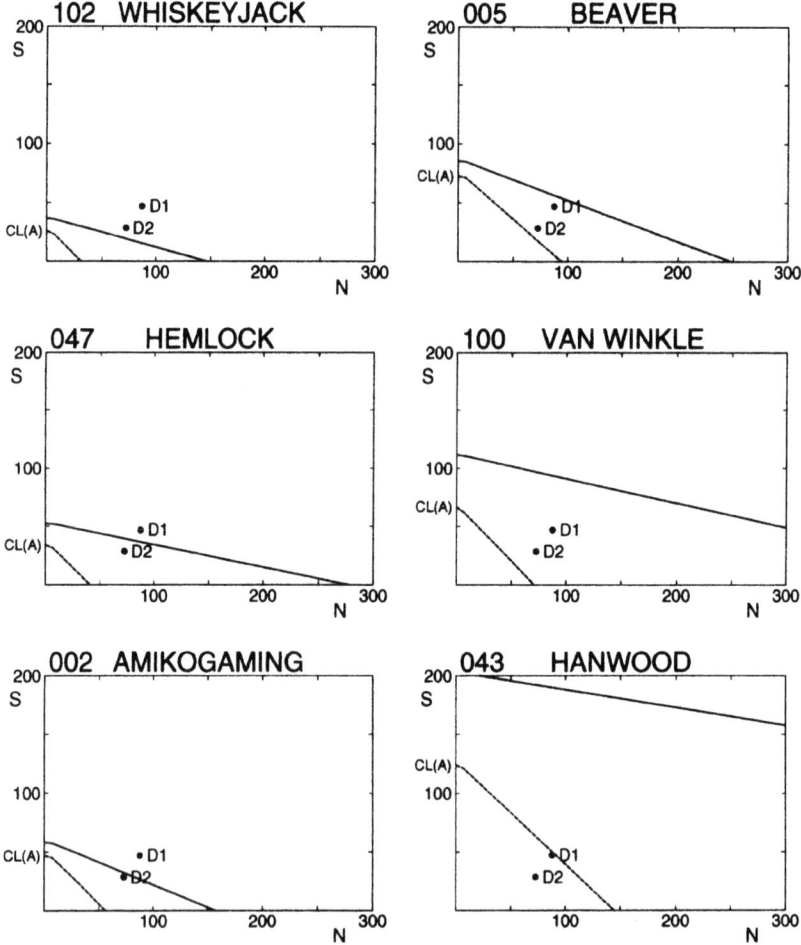

Figure 2. Critical load functions for six catchments from the Killarney Provincial Park (solid line). The dashed line shows the critical load function when neglecting in-lake retention processes (in this case $CL_{max}(S)=CL(A)$ holds). D1 and D2 are the present and a 2010 deposition scenario, resp. The 2010-scenario is based on Canadian and US sulphur emission controls by 2010 with an additional 50 % reduction of the emissions (Hindar *et al.*, 2000).

Table II summarises the effect on N retention by use of the three different methods for calculating CLs. It is worth noting that the big-lake method (including upstream lakes in a simple manner) increases the N_{lake} estimate by 47% on average and that the lake-system method increases the N_{lake} estimate by 73% on average. From this we tentatively conclude that it is more important to include upstream lakes at all ("big-lake") than to put too much emphasis on refining that inclusion ("lake-system"). Table II also shows that the relative amounts of N retained in the lakes are, in general, substantial.

TABLE II

The fraction of N (as % of deposition) retained in the terrestrial catchment, N_{terr}, and retained in the lake, N_{lake}, of the 8 lakes with sampled upstream catchments using three approaches: (i) ignoring upstream lakes ("one-lake" approach); (ii) using the "big-lake" approach" and (iii) using the "lake-system" approach. The total N deposited in 1993 (41.3 meq m^{-2} yr^{-1}) has been used (100–N_{terr}–N_{lake} is the %N_{dep} leaving the catchment).

		One-lake	Big-lake	Lake-system
Lake no.	N_{terr} %	N_{lake} %	N_{lake} %	N_{lake} %
2	13.4	44.3	56.5	57.6
43	14.7	43.7	64.9	75.3
47	10.9	37.8	69.7	77.4
49	12.3	42.0	60.9	76.3
61	13.2	28.3	59.4	79.6
112	15.4	40.0	52.5	64.2
113	15.7	45.4	47.3	48.5
124	8.6	34.3	52.5	67.1
Average		39.5	58.0	68.3

4. Discussion

Lakes along a chain are influenced by processes in lakes up-stream, which is well known, but until now not included in critical load models. CL calculations for surface waters are traditionally based on data from headwater lakes. This work has extended the methodology and computations to lake systems (lakes linked by streams), and in doing so a new module of the FAB model has been developed.

This work, which used data from lakes in the Killarney Provincial Park (Ontario, Canada), has shown that in-lake retention of nitrogen may be substantial. By comparing different computation approaches we found that lakes in chains (downstream other lakes) are favoured by the in-lake retention of a larger fraction of deposited N than headwater lakes. This increases their critical loads by making them less susceptible to acidic deposition. Since lakes in chains (lake systems) are common on shield bedrock conditions, this difference in tolerance should be taken into account.

Also, since N-retention is dependent on water retention time, with higher retention in lakes with longer retention time, both specific runoff and lake/catchment characteristics are probably important for the N-retention effect.

The quantification of the processes leading to N-retention both in the terrestrial part of the catchment and in the lake/sediment system deserves further attention. Good input-output budgets for S and N from lakes with different retention times (lake:catchment ratios) would be helpful.

Acknowledgements

Christine Brereton digitized the required land use and land cover data. We wish to thank Dr. Michael D. Moran of the Meteorological Service of Canada (Toronto, Ontario) for providing deposition data. The work has been financed by the Ontario Ministry of Environment and the Ontario Ministry of Natural Resources. Our contact has been Dr. John Gunn.

References

Brakke, D.F., Henriksen, A. and Norton, S.A.: 1990, *Verh. Internat. Verein. Limnol.* **24**, 146.

Government of Ontario: 1984, *Water quantity resources of Ontario.* Ont. Gov. Interminist. Rep.

Henriksen, A. and Posch, M.: 2001, Steady-state models for calculating critical loads of acidity for surface waters - Where do we stand today? *Water Air Soil Pollut.* (in press).

Hindar, A. and Henriksen, A.: 1998, *Mapping of critical load and critical load exceedances in the Killarney Provincial Park, Ontario, Canada.* Acid Rain Research, Report 49/98. SNO 3889-98, NIVA, Oslo, Norway.

Hindar, A., Posch, M., Henriksen, A., Gunn, J. and Snucins, E.: 2000, *Development and application of the FAB model to calculate critical loads for S and N for lakes in Killarney Provincial Park (Ontario, Canada).* Report SNO 4202-2000. NIVA, Oslo, Norway.

Kelly, C.A., Rudd, J.W.M., Hesslin, R.H., Schindler, D.W., Dillon, P.J., Driscoll, C.T., Gherini, S.A. and Heskey, R.H.: 1987, *Biogeochemistry* **3**, 129.

Posch, M., Kämäri, J., Forsius, M., Henriksen, A. and Wilander, A.: 1997, *Environmental Management* **21**, 291.

Snucins, E. and Gunn, J.: 1998, *Chemical and biological status of Killarney Park Lakes (1995-1997). A study of lakes in the early stages of recovery from acidification.* Ontario Ministry of Natural Resources. Cooperative Freshwater Ecology Unit. Sudbury, Ontario.

FLOW AND pH MODELLING TO STUDY THE EFFECTS OF LIMING IN REGULATED, ACID SALMON RIVERS

VILHELM BJERKNES[1*] and TORULV TJOMSLAND[1]

[1]*Norwegian Institute for Water Research, Regional Office West,
Nordnesboder 5, N-5005 Bergen, Norway*
(author for correspondence, e-mail: vilhelm.bjerknes@niva.no)

Abstract. In the regulated river Ekso, Western Norway, liming of the headwater has been introduced as a mitigating action to improve the water quality for Atlantic salmon (*Salmo salar* L.). Supply of lime from a dosing plant situated 5 km above the salmon producing part of the river, aims to raise pH from 5,0 to 6,5 during the smolt period for Atlantic salmon, and to 6,2 for the rest of the year. Hydrological modelling based on the relationship between $CaCO_3$ and pH is applied for the evaluation of the liming strategy, based on monitoring data from the spring 2000. The water quality demand was satisfied 80% of the time in the upper part of the salmon area, and 40% of the time in the lower part, influenced by power plant discharge. Flood forecasting and overdosing of lime ahead of floods will reduce the effects of acidified and unlimed tributaries. An additional lime doser is recommended to supply the power plant discharge.

Keywords: acidification, hydropower regulation, liming, hydrological modelling, Atlantic salmon

1. Introduction

The wild populations of Atlantic salmon (*Salmo salar* L.) are threatened by acidification in a number of rivers in Western Norway. The smolt stage is the most sensitive (Rosseland and Skogheim, 1984), and liming measures have therefore been introduced in a number of rivers to satisfy the water quality demands of Atlantic salmon smolts.

Acid rain trigs aluminium (Al) mobilisation from the bedrock (Scofield, 1977; Dickson, 1979), and water from Al-rich, acidic tributaries cause problems when mixing with limed headwater. Rosseland *et al.* (1992), demonstrated that water in the mixing zones might become more toxic than the original acid water.

Hydropower regulations may introduce additional complexity to the water quality in acidified and Al-rich river systems due to redistribution and mixture of flows of different chemical quality.

In the acidified and regulated river Ekso, Western Norway severe fish kills were frequently observed during the last decade (Hindar *et al.*, 1993; Raddum and Fjellheim, 1995), and high deposition of Al at the gills and reduced general condition of surviving smolts were demonstrated (Bjerknes *et al.*, 1998). To improve the water quality, liming was started in 1997.

Monitoring has revealed that more or less serious deviations from the target water quality occur frequently, especially during high flood episodes. Hydrological modelling based on downstream water chemistry monitoring and simulation, is presented here for the evaluation of the present liming strategy in river Ekso.

2. Materials and methods

River Ekso (Figure 1) is situated 50 km north-east of Bergen, Norway. The natural catchment area is 414 km^2, with the highest points 1300 m above sea level. The Raudfoss waterfall, 4 km above the river mouth constitutes the natural migration barrier for Atlantic salmon.

The water from the upper 160 km^2 of the catchment, including the most favourable geology and water quality (5,7 < pH< 6,2), was transferred to a power station outside the catchment area in 1986. In the lower part of Ekso, Myster power station was set in operation in 1987, utilising the 249 m altitude difference from lake Nesevatn, including water with pH values down to 5,1-5,3. The 42 km^2 unregulated rest catchment contributes with 3.2 m^3 s^{-1} water with pH values between 4,9 and 5,3. Thus, the power regulations have resulted in a less favourable water quality in the lower parts of the river, including the areas relevant for Atlantic salmon.

The licensed minimum flow in Ekso is 2 m^3 s^{-1} at Langehølen (5 km below lake Nesevatn, Figure 1) in the summer (May 15 - October 15) and 1 m^3 s^{-1} for the rest of the year.

Flow measurements presented in this paper were obtained from a runoff recording station situated at the lime doser at Langehølen, and from Myster power station. Hydrometric data from an adjacent watercourse (NVE recorder no. 55.4.0) were applied to calculate the inflow from the natural catchment below the doser.

The water quality has been monitored at the river mouth (St. 5, Figure 1) by monthly water samples and analyses since 1980. The station net for water quality monitoring was extended in 1993, and covers representative parts of the main river and subcatchments. An automatic recording facility was established close to Myster power station in 1996, registrating pH and water temperature in the river upstream of the power station and in the hydropower discharge. The water sampling program was intensified for the present study, from 1 January to 23 May 2000. Input water quality data for modelling include acidity (pH) and calcium (Ca) analysed from water samples, in addition to pH values from the recording facility.

The liming strategy (Kaste et al., 1996) involves automatic supply of lime slurry to the main river from a doser situated at Langehølen, 5 km above the salmon barrier at Raudfoss (Figure 1). Headwater- and tributary flows and upstream pH constitute the input data for lime dosing. The measure aims to raise pH to 6.5 in the salmon producing part of the river during the smoltification- and smolt migration period (February 1 - June 15), and to 6.2 for the rest of the year. The pH increases are considered to reduce the concentrations of labile Al from maximal values up to 120 µg L^{-1} to safe levels at 10-15 µg L^{-1} (Jenssen and Leivestad, 1989).

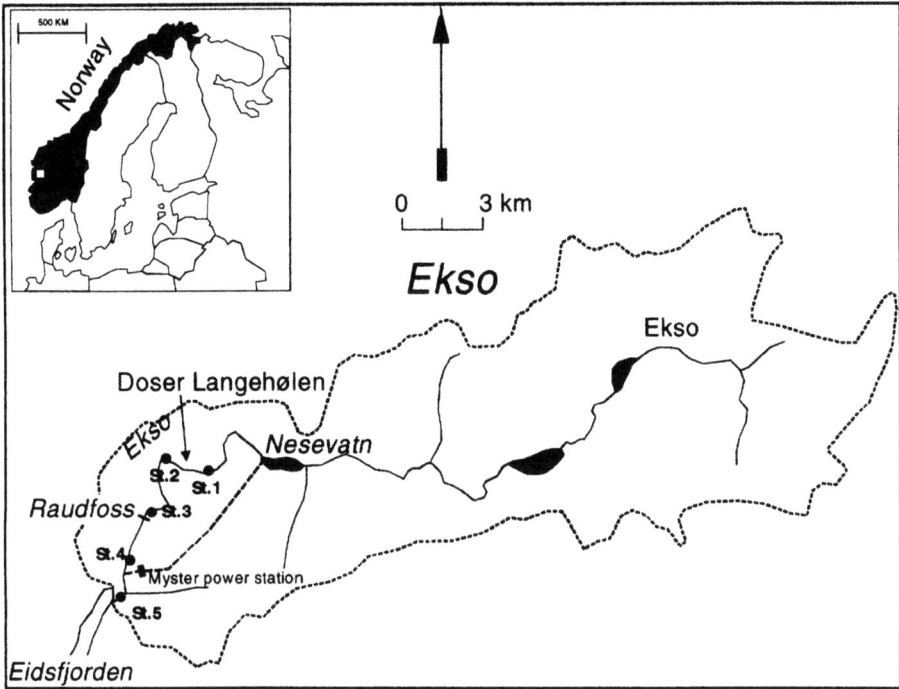

Figure 1. The present catchment of river Ekso. Station numbers 1-5 refers to locations for water sampling and pH-monitoring.

The water quality model QUAL2E (EPA, 1987) was used in this study to simulate dilution of lime as a function of flow and solid supply. The model calculated hourly Ca-concentrations in the river as a function of natural background concentrations and dosed lime slurry that consists of 86% CaCO$_3$ or 34% Ca. The relation between Ca and pH was applied to transform automatic monitored pH-values to Ca-concentrations, and thus compare the lime supply to Ca-concentrations and pH values in the river.

Comparing monitored Ca-concentrations and pH values in Ekso at Eikemo and Eide (Stations 3 and 4, Figure 1) from 1995 to 2000, gave the following relationship of Ca (mg L^{-1}) and pH:

pH=0.76 ln Ca + 6.27, R^2=0.64 (Figure 2)

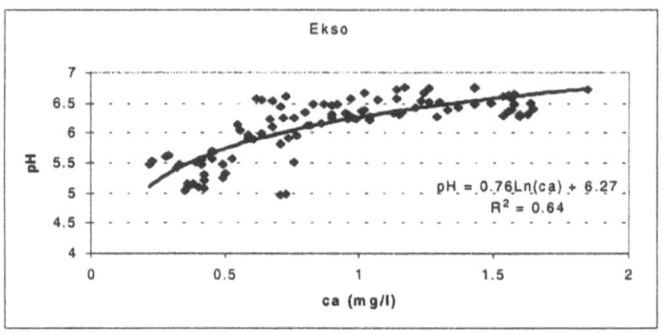

Figure 2. Relation of Ca concentration (mg L^{-1}) and pH in river Ekso 1995-2000.

3. Results

The verification of the model against observed data is shown in Figure 3 and simulations to obtain the target water quality in Figure 4 and Figure 5. The real lime supply for the study period (January 1 – May 23) was 403 tons (Table I), satisfying the water quality demand for 80% of the time at Station 4 (salmon producing area above Myster power plant) and only 40% of the time at Station 5 (river mouth), due to the power plant discharge.

TABLE I
Water flows and real, theoretical and simulated amounts of lime added during the study period from 1 January to 23 May 2000.

Catchment	Water flow average m^3 s^{-1}	Real lime dose tons	Theoretical optimal dose tons	Simulated optimal dose tons
Ekso above present doser	8.4	265	265	265
Ekso below present doser	4.2	138	103	174
Power station discharge	17.8	0	558	558
Sum	30.3	403	926	997

The theoretical optimal dose ensures that the target values are precisely obtained all the time. This would demand for a total supply of 369 tons of lime at Station 4 and 926 tons at Station 5. According to the simulations, a surplus of 71 tons of lime will be necessary in addition to the theoretical optimal dose to obtain the target quality at Station 4. This amount is calculated by calibrating the model against observed results at Station 4. The surplus will compensate for the water inflow downstream the doser, if this inflow increases more rapidly than the transport lag. 558 tons will have to be added to the presently unlimed power plant discharge to obtain the quality target at Station 5.

The transport time of lime from the existing doser to the outlet may commonly fluctuate from two days to a week depending on the water flow.

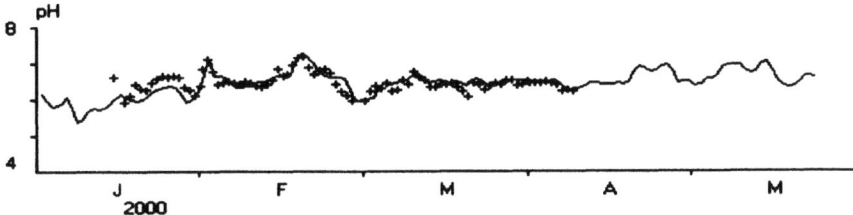

Figure 3. Verification of the model. Simulated (—) and observed (+) pH-values at St.4 Eide

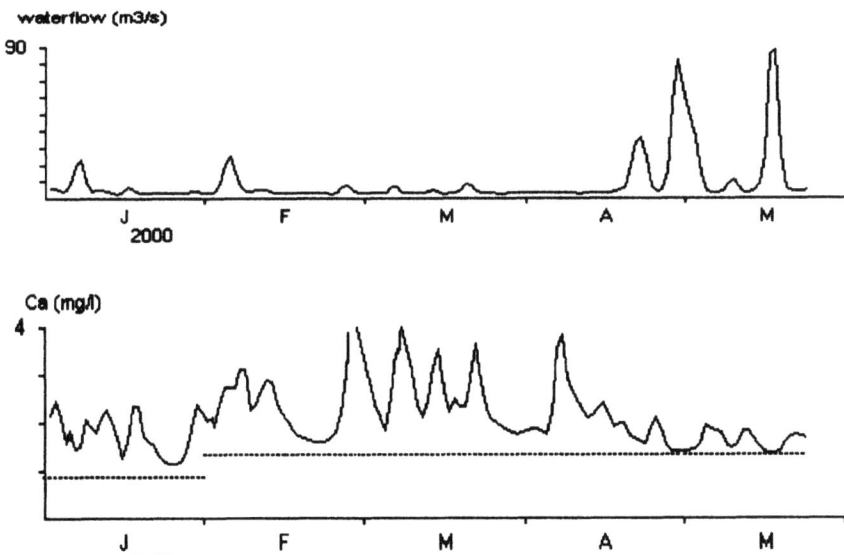

Figure 4. Waterflow and simulated Ca concentration at the doser point (—) to achieve the target pH values (.....) downstream to St.4 Eide.

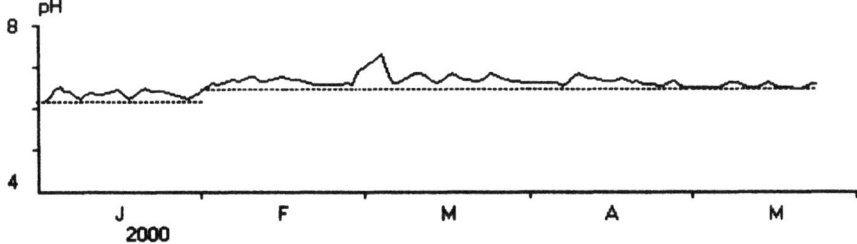

Figure 5. The simulated pH values (—) satisfied the target pH (.....) at the St.4 Eide.

4. Discussion

The water quality demand was satisfied 80% of the time at St. 4, representing the upper part of the goal area. The missing 20% are most probably due to time lag for lime transport from the doser to the points of acid tributary inflow. The present dosing plant is situated 5 km above the goal area of the liming measure, and will not respond fast enough during tributary flooding. By establishing a flow forecasting system, a lime surplus may be added to the theoretical lime demand ahead of flow increase to ensure the downstream water quality during floods.

An optimal liming strategy demands for an additional doser unit to deacidify the power station discharge, and thus ensure the water quality at the lower 1 km river stretch, below Myster power station.

Hydrological modelling should be applied, not only for the subsequent evaluation and improvement of proceeding liming projects, but as a planning tool to obtain more optimal liming strategies.

Acknowledgements

The Directorate for Nature Conservancy (DN), Bergen Municipal Electricity Supply Company (BKK) and Norwegian Institute for Water Research supported this survey.

References

Bjerknes, V., Pettersen, M.N., Salbu, B., Skiple, A., and Sælthun, N.R: 1998, 'Toxic water in mixing zones of limed salmon rivers and acid tributaries with special reference to regulated watercourses', in H. Wheather and C. Kirby (eds.), *Hydrology in a changing environment. Vol. 1.* John Wiley & Sons: Chichester. pp. 313-328.

Dickson, W.: 1979, *Aquaannalen* 1, 2.

EPA: 1987, *The Enhanced Stream Water Quality Models QUAL2E and QUAL2E-UNCAS: Dokumentation and User Model.* U.S. Environmental Protection Agency, Athens, Georgia, USA.

Hindar, A., Henriksen, A., Tørseth, K. & Lien, L.: 1993, *Deposition of seasalts and their effects in rivers and small catchment areas. Acidification and fish kills after the seasalt episode in January 1993. (In Norwegian).* Rep. No. 2917-93. Norwegian Institute for Water Research, Oslo, Norway.

Jenssen, E. A. and Leivestad, H.: 1989, *Acid water and smolt production.* Final report from the water treatment project of Salar/BP. J. Grieg, Bergen, Norway.

Kaste, Ø., Hindar, A., Skiple, A. and Henriksen, A.: 1996. *Mitigating measures against acidification of Ekso.* Rep. No. 3462-96. Norwegian Institute for Water Research, Oslo, Norway.

Raddum, G. G. & Fjellheim, A.: 1995, *Regulated rivers: Research and management* 10, 169.

Rosseland, B. O. & Skogheim, O.: 1984, *Rep. Inst. Freshw. Res. Drottningholm* 61, 186.

Rosseland, B.O., Blakar, I.A., Bulger, A., Kroglund, F., Kvellestad, A., Lydersen, E., Oughton, D., Salbu, B., Staurnes, M., and Vogt, R.: 1992, *Environ. Pollut.* 78, 3.

Schofield, C.L.: 1977, *Acid snow-melt effects on water quality and fish survival in the Adirondack Mountains of New York State.* Res. Proj. Tech. Com. Rep. Project A072-NY. Cornell University, Ithca, New York

DOES ACIDIFICATION POLICY FOLLOW RESEARCH IN NORTHERN SWEDEN? THE CASE OF NATURAL ACIDITY DURING THE 1990'S

K. BISHOP[1*], H. LAUDON[1,3], J. HRUSKA[1,2], P. KRAM[1,2],
S. KÖHLER[3] and S. LÖFGREN[1]

[1]*Department of Environmental Assessment, SLU, P.O. Box 7050, S-750 07 Uppsala, Sweden;*
[2]*Czech Geological Survey, Klarov 3, 118 21, Praha 1, Czech Republic;*
[3]*Department of Forest Ecology, SLU, S-901 86 Umeå, Sweden*
(* author for correspondence, e-mail: Kevin.Bishop@ma.slu.se)

Abstract. The situation in northern Sweden did not figure prominently in the intense period of research during the 1980's that laid the basis for many acidification-related policies now in effect in Europe and Sweden. Northern Sweden has not only relatively low acid deposition levels and significant sources of natural acidity, but also intense episodes of pH decline during spring flood that are a major focus of liming activity. Controversy over that liming and natural acidity has led to scientific advances. These include discovery of a correlation between sulfur in snow and the anthropogenic contribution to the subsequent spring flood ANC decline, but also that natural organic acidity is responsible for most of the spring pH decline in the region. This paper compares the developments in liming policy with the scientific developments of relevance to the region during the last decade. Considerable discrepancies are noted which create opportunities for revising remediation policies to better reflect the state of knowledge in 2000.

Key words: public policy, northern Sweden, episodic acidification, natural organic acidity

1. Introduction

Great strides during the 1980's resolved many scientific disagreements about the problems created by acid rain. The resulting consensus was an important basis for important policy decisions that contributed to a halving of acid deposition in Sweden during the last 15 years. Recovery of chronically acidified ecosystems, however, is slow. Thus, Sweden still invests considerable resources each year to lime surface waters, and plans to do so for decades to come (SEPA, 1999b). Liming in Northern Sweden was established on a large scale in 1991. Since then two billion kr (ca. 250 million US$) have been spent on liming in Sweden, about half of which has gone to the northern half of the country, largely to keep spring flood above pH 6. Little scientific attention, however, was paid to N. Sweden during the 1980's when the policy guidelines for liming were established.

This region has relatively low sulfur deposition levels (2-6 kg/ha in 1990, 1-3 kg/ha in 1998), but also intense episodes of pH decline during spring flood that natural factors contribute to. Research during the 1980's identified such natural factors (e.g. Oliver et al., 1983), but had relegated them to a subordinate role in Scandinavia (Mason, 1990). That research, though, had focussed on clear water systems in southern Scandinavia with much higher S deposition than found in N.

Sweden (Bishop, 1997). Expansion of liming in N. Sweden alarmed some researchers who feared that naturally acid waters would be limed to unnaturally high pH's, damaging biota adapted to acidity and diverting resources from more important drivers of fish loss (Jansson et al., 1992). That controversy spawned a number of studies of relevance for liming policy in N. Sweden. This paper reviews developments in science and policy since 1990.

2. Scientific Developments

At the start of the decade, the report making the case for liming in N. Sweden (Ahlström and Isaksson, 1990) was based on the acidity of water, reports of fish loss, and studies of the zoobenthos that indicated acid conditions. That report made no mention of the natural sources of acidity in the region. The initial opposition to liming pointed out the possibility of natural acidity (organic acids or sulfur deposits in the soils), but could not quantify the importance of these natural factors relative to acid deposition.

An important advance from this starting point was paleoecological evidence of no regional acidification since the pre-industrial period in N. Sweden's lakes (Korsman, 1999). Soil studies also failed to find acidification in all but the southeastern corner of N. Sweden (Eriksson et al., 1992). This, together with more comprehensive studies of organic acidity (Ivarsson and Jansson, 1995) gave new impetus to the concern about inappropriate liming of naturally acid waters, but also brought the issue of spring flood into focus.

Paleoecological reconstructions of surface water pH are not believed to reflect the transient pH decline during spring flood when the aquatic biota are most vulnerable. But since half of the annual acid deposition is released from the snow during a few weeks, and the contact of spring runoff with the soils is short, it is likely that episodic acidification would not be reflected in soils or the paleoecological record. On the other hand, organic acidity and dilution of acid neutralizing capacity (ANC) by low ionic strength meltwater are two natural factors that contribute to pH decline during spring (Laudon et al., 1999).

In 1995, the Swedish Environmental Protection Agency (SEPA) organized a workshop on the controversy over liming in N. Sweden. This workshop concluded that there was an urgent need for a method to distinguish natural acidity from acid deposition's contribution to spring pH decline, as well as data on spring flood chemistry and biology (Warfvinge et al., 1995). The workshop was the starting point for a comprehensive research initiative financed by SEPA to address these issues. That initiative is responsible for most of the scientific output on the situation in N. Sweden since 1995, over 20 publications. Key findings are:

- Natural spring pH decline could be modeled by accounting for organic acids and dilution of ANC by neutral snowmelt (Bishop et al, 2000).
- The anthropogenic contribution to the 1 to 2 pH unit decline in over 100 spring floods from across N. Sweden during the late 1990's almost never exceeded the range of 0.1 to 0.3 pH units (Laudon et al., in press).
- While relatively small, this anthropogenic pH decline could be significant for fish survival in the most weakly buffered streams (Laudon, 2000).
- The anthropogenic component of spring ANC decline correlates to snowpack S, thus responding rapidly to changes in S deposition (Bishop et al, 2000).
- Sulfur deposition is much less than 10 years ago (Kindbom et al., in press)

These findings are a major scientific advance on the knowledge available in 1990, and should provide a better basis for regional policy with respect to acidification and protection/restoration of biodiversity.

3. Policy Developments

Throughout the 1990's, liming policy has followed the 1988 General Guidelines (SEPA, 1988). The guidelines mention natural organic acidity, but in an ambiguous, if not contradictory manner. While acknowledging that acid deposition is not the only potential cause of a pH of less than 6.0, the guidelines go on to state that most such naturally acid waters are also acidified, and that these should be limed back to *at least* their natural level of alkalinity. This could be interpreted as an authorization of liming to alkalinities above natural levels. In fact, the final sentence of the guidelines specifies that the pH in Swedish water should always be above pH 6.0 with an alkalinity of more than 50 $\mu eq\ L^{-1}$. That is tantamount to saying that pH and alkalinity should be raised to levels above what is natural for naturally acid water. The threshold alkalinity for liming is also 30 ueq L^{-1} higher than the critical threshold for biological damage (20 ueq L^{-1}) used by Sweden in its international negotiations about the UNECE Convention on Long-Range Transport of Air Pollutants.

Liming started in southern Sweden during the 1970's, but regional authorities in N. Sweden called attention to the fact that many waters in N. Sweden that had lost fish and other species in recent decades met the chemical criteria for liming, especially during spring flood (Ahlström and Isaksson, 1990). Parliament responded by earmarking new funds for liming in N. Sweden. This put liming expenditures in N. Sweden on a par with those in S. Sweden. The next major policy development came in 1999 when "Only Natural Acidification" was made one of Sweden's 15 national environmental goals (SEPA, 1999a). That official recognition of natural acidity, however, did not change the funding and strategy for liming in N. Sweden. The relatively stable liming budget since 1991, however, meant that liming has not continued expanding into new water systems, but instead re-limes areas where liming has already been initiated and increases the objects limed within those systems.

It was not until the end of the 1990's that there was a major reconsideration of liming policy, in the form of a new national plan for the coming ten years (SEPA, 1999b). The major policy change in this document of relevance for N. Sweden is greater focus on liming to restore and maintain natural biodiversity. Another issue addressed is the effect of decreases in acid deposition. The ten-year plan states that the need for liming will not diminish until after 2010.

4. Discussion

Liming in areas with natural acidity raises policy issues that have stimulated scientific advances in the understanding of spring flood. One result is an operational method for quantitatively separating acid deposition's contribution to pH decline from natural acidity that has been published in refereed journals (Bishop et al., 2000). Have those scientific advances been used to develop policy?

Practical means for identifying naturally acid waters and treating them differently were lacking from the ten-year plan (SLU, 1999). Thus it is likely that liming practices will not change to reflect the new knowledge. It may be that not enough time has passed since the publication of the scientific advances for their inclusion in policy documents. The SEPA, however, was involved in these scientific developments and informed before the results appeared in print.

While the new ten-year liming plan has not availed itself of the new assessment methodology, there is a greater focus on biodiversity. This could provide a better basis for preserving naturally acid ecosystems at the same time as combating the effects of acidification. There is a risk, though, that the biodiversity focus in the new plan may actually accelerate the inappropriate liming of naturally acid ecosystems. In the new plan, the presence of high biodiversity value is needed to give a liming object the highest priority. This means that many of the most acidified waters in southwestern Sweden will not be given highest priority, since they have lost much of their biodiversity. Thus these waters could lose their liming subsidies, while new liming is given priority in N. Sweden where more biodiversity value is now found (SLU, 1999).

That unfortunate possibility stems from a key area in which the new liming plan misses a recent scientific finding, namely the rapid response to changes in S deposition of the anthropogenic contribution to spring flood ANC decline (Bishop et al., 2000). This means that chemical improvements are immediate, not something that will first appear after 2010. It is also hard to imagine that existing biodiversity value in N. Sweden, which survived decades of higher acid deposition levels, is currently threatened by spring flood pH acidification after a decade of sustained reductions in acid deposition levels.

The discovery of a rapid improvement after S deposition decreases is not only of regional importance for liming, but also for international policy. The timing and extent of recovery from acidification is a key issue in the evaluation of international efforts to reduce acid deposition (Lökke et al., 2000). So far, though, the response of spring flood pH decline in N. Sweden has not been considered in that context.

Ecosystem recovery (as opposed to the chemical recovery that liming can contribute to when used appropriately) is a matter of biology as well as chemistry. Much remains to be done to quantify the coupling between acidification/recovery and the biota. Local knowledge, though, can be a valuable source of information. The National Fisheries Authority surveyed local fishing officials in 1995 about the cause of fish loss in their area (SEPA, 1999b). Acidification was the single most common explanation given as to why fish status had declined in lakes. But it was twice as likely that a cause besides acidification (e.g. forestry, overfishing, habitat loss) was cited (Fig. 1).

Expenditure on liming, however, has been ca five times greater than all other efforts to improve the biological status of Swedish lakes and streams during the 1990's (Fig. 1). In light of recent scientific developments, a reconsideration of the balance between liming and other measures would seem to be in order.

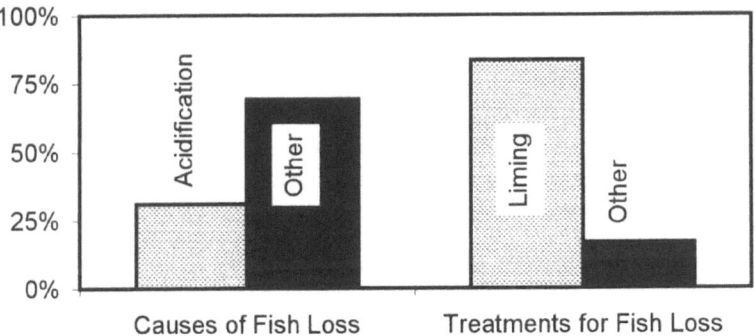

Figure 1. Causes of fish loss given by fishing officials in the 1995 national survey. "Other" causes include forestry, overfishing, disturbed spawning areas, changed stocking practices (adapted from SEPA, 1999b). Treatments for fish loss are presented as a percentage of the national expenditure (180 Mkr total) on restoration of freshwater ecosystems in 1997.

5. Conclusion

Important scientific questions concerning episodic acidification in N. Sweden have been addressed during the last decade. The anthropogenic contribution to spring flood pH decline has been quantified and a direct proportionality to winter acid deposition established. The new ten-year plan for liming, however, does not incorporate many of the key scientific advances. The plan does not foresee a decrease in acidification even in N. Sweden until after 2010 despite sustained deposition declines since 1990. There is also a possibility that liming of naturally acid waters in N. Sweden could increase as a result of this plan.

It could simply be that the scientific advances have not been in the literature long enough to have found their way into policies. Another possible obstacle to the assimilation of new knowledge into liming policy, though, is the current funding structure which separates the liming budget from the budget for other types of remedial measures, such as habitat restoration. As it stands now, a reduced need for liming will not free resources for other measures at the local level, but only reduce the resources available. This disincentive could be overcome by removing the budgetary distinction between liming and other forms of management to protect aquatic biodiversity. This would also increase the freedom of local authorities to take advantage of local knowledge when deciding what remedial measures are most appropriate in their area (cf. Fig. 1).

References

Ahlström, J., and Isaksson, K.E.: 1990, Försurningsläget i norrlands inland och fjälltrakter — Kalkning. Swedish Environmental Protection Agency Report 3781, Stockholm, Sweden.
Bishop, K. H.: 1997, *Transactions of the Institute of British Geographers* **22**, 49.
Bishop, K., Laudon, H. and Köhler, S.: 2000, *Water Resources Research* **30**, 1873.
Eriksson, E., Karltun, E. and Lundmark, J. E.: 1992, *Ambio* **21**, 150.
Ivarsson, H., and Jansson, M.: 1995, *Water, Air and Soil Pollution* **84**, 233.
Jansson, M., H. Ivarsson, H. Grip, and Bishop, K.: 1992, *Miljöaktuellt*: 29 April, p. 18.
Kindbom, K., Sjöberg, K. & Pihl-Karlsson, G.: (in press) *Trends in air concentrations and deposition in background areas in Sweden*. IVL Report, Gothenburg, Sweden.
Korsman, T.: 1999, *Journal of Paleolimnology* **22**, 1.
Laudon, H., Köhler, S. and Bishop, K.: 1999, *Science of the Total Environment* **234**, 63.
Laudon, H.: 2000, Doctor of Philosophy Dissertation, SLU, Uppsala, Sweden.
Laudon, H., Westling, O., S. Löfgren & Bishop, K.: in press, *Canadian Journal of Fisheries and Aquatic Science*.
Løkke, H. et al. Critical Loads Copenhagen 1999, Conference Report, National Environmental Research Institute, DMU Report 121, Silkeborg, Denmark.
Mason, B. J. (ed.): 1990, *The surface waters acidification programme*. Cambridge University Press, Cambridge, UK.
Oliver, B.G., E.M. Thurman, and R.L. Malcolm: 1983, *Geochimica Cosmochimica Acta,* **47**, 2031.
SLU: 1999, Evaluation of the Swedish EPA's 'Nationell plan för kalkning 2000-2009'.
SEPA: 1988, *Allmänna råd för kalkning av sjöar och vattendrag* **88:3**, Stockholm, Sweden.
SEPA: 1999a, *Fifteen : Sweden's objectives for environmental quality*, Stockholm, Sweden.
SEPA: 1999b, *Nationell plan för kalkning av sjöar och vattendrag 2000-2009 : Redovisning av ett regeringsuppdrag*, Stockholm, Sweden.
Warfvinge, P., Löfgren, S. and Lundström, U.: 1995, *Water Air and Soil Pollution* **85**, 499.

EFFECTS OF REDUCED S DEPOSITION ON LARGE-SCALE TRANSPORT OF SULPHUR IN SWEDISH RIVERS

A. WILANDER

Dept. of Environmental Assessment, Swedish University of Agricultural Sciences,
P.O. Box 7050, S-750 07 Uppsala, Sweden.
e-mail: Anders.Wilander@ma.slu.se

Abstract. Sulphur deposition has diminished by about half during the last decade. For Sweden consistent estimates of total deposition are available for 1991, and 1994–97. Based on these estimates and using GIS the deposition for large drainage areas during one decade are calculated. These values are compared with the measured S transport in rivers covering about 85% of the Swedish territory, thus enabling the construction of a S budget for Sweden. The majority of the drainage areas have a net loss of S, which can be attributed to desorption of S in the soil. During the period of high deposition in the 1980:s (>60 meq m^{-2} yr^{-1} in southern Sweden) S was adsorbed, and retarded acidification. There still seems to be some S-adsorption in the northern parts of the country, where the deposition is less (now <20 meq m^{-2} yr^{-1}).

Key words: deposition, GIS, rivers, sulphur, S budget, transport of S, Sweden

1. Introduction

Many sulphur budgets are published: on a global level (e.g. Eriksson, 1960; Brimblecombe *et al.* 1989) as well as for small catchments (e.g. Rochelle *et al.* 1987; Likens and Borman, 1995; Löfgren *et al.* 2000). All give important information about the biogeochemical behaviour of sulphur. Sweden has a range of drainage areas from those heavily affected by S deposition to those receiving a low deposition. In addition S contributions from agriculture and urban areas varies from basin to basin. The purpose of this study is to determine the extent to which reduction in S deposition has lead to changes in S losses from about 130 drainage areas.

2. Materials and methods

Data for total deposition of non-marine S calculated by the MATCH model (Persson *et al.*, 1996), with a pixel of about 10x10 km, were obtained from the Swedish Meteorological and Hydrological Institute (www.smhi.se/sgn0102/n0202/index.htm). Using GIS, the deposition was calculated for each drainage area, and for the years 1991 and 1994-97. Deposition data for 1992-93 were obtained by linear interpolation.

Sulphur export data for 128 drainage areas came from monitoring programmes initiated in 1965 run by the Department of Environmental Assessment. The drainage areas vary from very large (max 48263 km^2) to 1 km^2, with a median area of 1673 km^2. Sampling was usually monthly, while flow data were daily observations or modelled. The river transport was calculated using daily interpolations of concentrations and the corresponding flow. The chemical analyses were performed at the Department using ion

Water, Air, and Soil Pollution **130:** 1421–1426, 2001.
© 2001 *Kluwer Academic Publishers.*

chromatography for anions and plasma emission for major cations. The laboratory is accredited and participates in national and international quality control.

Calculation of non-marine S was based on Cl and the sea-salt quotient. Weathering rates for S were based on non-marine base cations and a ratio found for groundwater in granite areas (Wilander, 1994).

3. Results

Sulphur deposition in central Sweden has changed drastically with an increase at the end of the 1950:s (Fig. 1). The change in S deposition since 1985 was 0.74 meq m^{-2} year^{-1} year^{-1} and the changes in area losses were 0.67 and 1.00 for River Klarälven and River W Dalälven respectively. A pattern for the transport similar to deposition is indicated for the latest 12-year period. It is evident that for a large drainage area, it will take some time between the deposition and the arrival of this water to the surface water sampling point. The correlation coefficients between deposition and area specific loss for the two rivers were 0.50 and 0.62. When a retention of 1 year was applied the correlation coefficients fell to 0.30 and 0.33, but increased to 0.71 and 0.75 respectively when it was set to 2 years (p-value for Kendall Tau 0.004 and 0.03 respectively). This indicates a general residence time in the overburden of these drainage areas of about 2 years; of course there must be large variations. For a 2 km^2 catchment Lindström and Rodhe (1986) found a hydraulic transit time of 12 months, thus the above estimate is reasonable.

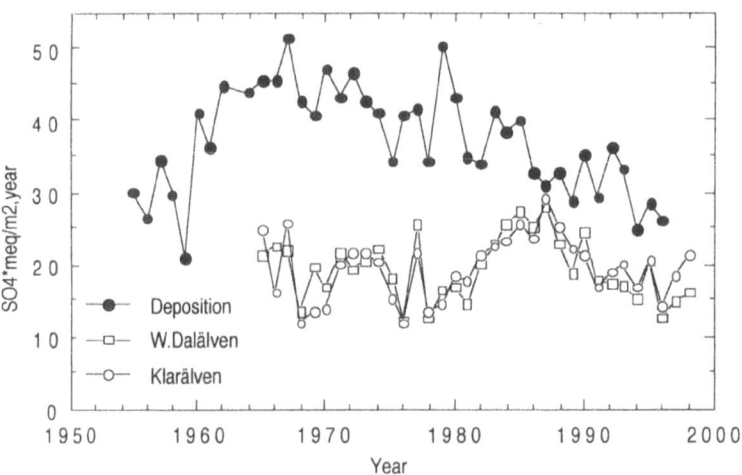

Figure 1. Annual deposition of non-marine SO4 (mean bulk deposition for 7 sampling locations in central Sweden) and annual area specific loss for two neighbouring rivers.

These observations initiated a more detailed study of S budgets in Swedish drainage areas.

3.1. SULPHUR BUDGETS FOR INDIVIDUAL CATCHMENTS

The median deposition during 1991–97 for 128 Swedish drainage areas, covering about 85% of the Swedish land, was 32.1 meq m^{-2} year^{-1} (Table I). At the same time the median specific loss was 56.9 meq m^{-2} year^{-1}. In the majority of the drainage areas there was a loss of S, which exceeded the deposition. It is evident that S-sources other than deposition must be taken into account.

The first attempt included weathering as a source (Wilander, 1994), which after deduction from the area specific loss results in an 'anthropogenic' area loss. But the median 'anthropogenic' value is still high: the median quotient anthropogenic area specific loss/deposition is 1.6 (Table I) indicating that still other sources have to be found. But the number of areas with a calculated net loss diminished. Areas with more than 3 times higher losses than present deposition (n=20) are nearly all located below the highest interglacial coastline and have a high percentage of agricultural land and dense population, both indicating S sources of fossil marine as well as anthropogenic origins.

TABLE I

Deposition, area specific losses and weathering of non-marine S. Anthropogenic area specific loss is the difference between area specific loss and weathering. (1991–97, n=128).

Parameter	Percentile		
	10	50	90
Deposition, meq m^{-2}year^{-1}	15.3	32.1	56.1
Area specific loss, meq m^{-2} year^{-1}	19.5	56.9	212
Area loss/deposition	0.97	1.9	5.2
Area specific loss – deposition, meq m^{-2} year^{-1}	-1.68	22.1	159
Weathering, meq m^{-2} year^{-1}	3.3	6.6	31.2
Anthropogenic area specific loss, meq m^{-2} year^{-1}	15.8	54.5	192
Quotient Anthropogenic areal loss/deposition	0.78	1.6	4.5
Anthropogenic areal loss – deposition, meq m^{-2}year^{-1}	-6.31	14.4	122

Accumulation (area specific loss - deposition is negative) was found for 15 areas. If the contribution from weathering is included as a natural source (anthropogenic loss–deposition) the number of drainage areas with a net accumulation increased to 24. These are located in the northern part of Sweden and have a significantly lower deposition than all areas (median 20.5 versus 32 meq m^{-2} year^{-1}). Many are located where the present land was under the sea surface during the interglacial period. Here net accumulation was not expected since there is a possible contribution of S from marine deposits. This group includes River Klarälven and River Dalälven presented in Figure 1,

3.2. TEMPORAL CHANGES IN DEPOSITION AND TRANSPORT OF S

Data for assessing temporal changes in deposition and transport of S were available for 125 areas (summarised in Table II). The calculation of changes in deposition was limited to the period 1991-97. Calculations of the rate of change of area specific losses covered the period 1990-1998 in order to minimise the influence of outliers in the linear regression analysis (Table II).

All areas had a negative trend in S deposition. A few with extreme values above 100 meq m^{-2} year^{-1} year^{-1} are located in the southern part of Sweden with a large S deposition. The smallest changes in S deposition are found in the northern areas, where the deposition is low. The correlation between the rate of change and the mean deposition is high (0.87).

While the deposition decreased in all areas, the rates of transport increased in as many as 38% of the rivers. When the S contribution from weathering is deducted to give "anthropogenic" area specific loss, the percentage of rivers with increasing losses diminishes to 30% (38 rivers). Many of these drainage areas are affected by agriculture and are relatively densely populated.

TABLE II

Rates of changes in deposition (1991-97) and area specific losses (1990-98). (n=125).

Parameter	Percentile		
	10	50	90
Deposition meq m^{-2} year^{-1} year^{-1}	-4.46	-1.84	-1.05
Area specific loss meq m^{-2} year^{-1} year^{-1}	-6.77	-0.32	5.05
Anthropogenic area specific loss meq m^{-2} year^{-1} year^{-1}	-6.56	-0.43	4.47

Rates of annual change in deposition were on a mean about five times larger in comparison with the area specific loss. This finding underlines the importance of soil as a S source suppressing the change in S export.

4. Discussion

Under steady state conditions sulphur input should equal output. The decrease in S deposition has led to a relatively short-term situation where many drainage areas had very large areal loss/deposition quotients (Table I) indicating additional S sources. Alewell *et al.* (1999) list five possible S sources: dry deposition (here included in the deposition), weathering of minerals, reoxidation of reduced S, desorption of previously adsorbed SO$_4$ and mineralisation of organic S. Weathering of S from minerals varies substantially depending of the type of rock. For granite Wilander (1994) estimated it be 0.05 times the base cations (on an equivalent basis). This value is quite similar to the value 0.037 calculated from data for Hubbard Brook (Likens and Borman, 1995). Using the first ratio the weathering for soils formed from primary rocks was estimated and deducted from the areal loss to give the 'anthropogenic' loss (Table I). The quotient diminishes of course, but high values are still found. Oxidation of reduced S is not likely to be a large source, with the exception of areas lying below the highest interglacial coastline, where pyrite may be available. Desorption remains as a possible source. Karltun (1996) found adsorbed S in amounts ranging from 420 meq m^{-2} in the northern parts of Sweden up to a maximum of 5200 meq m^{-2} in the south. Harrison *et al.* (1989) indicate that 'inorganic SO$_4$ adsorption appeared to be far more important than organic S incorporation or mineralization'. They also outline the effects of S adsorption leading to retention during the initial stages of increased S deposition and desorption with relatively high sulphate concentrations when the S deposition diminishes. The present results support these early studies of forest soils. However, Alewell *et al.* (1999) stress the high amounts of organic S in soil and the importance of mineralisation as a source. At present it is not possible to completely separate the effects of these two sources. Furthermore their relative importance is likely to vary depending on soil type and climate.

Figure 1 indicates a substantial accumulation in the drainage area during the period 1965-1985 as compared to later period. Even though the deposition data are from a wider area than the two drainage areas it is of interest to calculate the amounts stored. Between 1965 and 1998 about 600 meq m^{-2} was stored. Karltun (1996) gives a mean value of 1000 meq m^{-2} for adsorbed S for this region, thus a reasonable similarity.

It is obvious that the national scene changed dramatically due to reduced S deposition between the middle 1970:s and the 1990:s (Table III).

TABLE III

Sulphur budget for Sweden (amount S t year^{-1})

Source	Odén (1979)	1991-1997
Deposition non-marine	500000	186200
Marine S	37800	38700
Weathering	4400	1428
Fertiliser S	44300	21000
Total sources S	586500	247328
Discharge, total	648000	447555
Difference	-61500	-200227

In his discussion of the S budget Odén (1979) declared that there is an apparent imbalance. He indicated several factors; too low deposition, oxidation of S bound in soil organic matter and release of S from marine sediments and acid sulphate soils (caused by land lift after the ice age and ditching). Odén calculated the weathering rate for S from a quotient with Si. The present, lower rate was based on non-marine base cations (Wilander, 1994) which does not include the higher ratios relevant for clays and limestone. Both calculations neither include oxidation of organic matter nor release from pyrite containing soils. For one coastal area the total loss due to pyrite from the soil can be 2 to 10 times higher than that estimated here (Ivarsson and Jansson, 1995). However, only about 20% of the areas studied lie below the highest coastline and thus could be affected by marine deposits. Marine-related processes could raise the total figure for weathering to about the level given by Odén (1979), but on a national level it still is a minor part in the budget. Both estimations of marine S are similar; even though Odén used Na for the calculation, while the present one applied the quotient with Cl. The contribution from fertiliser S diminished since the 1970:s, but increased during the last decade. The difference between total sources and the calculated discharge for Sweden can now be attributed to losses from the soil: by desorption and oxidation of organic S.

5. Conclusions

On a national level there seems to be a general change from a deposition larger than the discharge in the 1970:s to a net loss of S during the 1990:s.

This change in time is mimicked by the variation in space. Budget calculations for the southern drainage areas indicate a net loss (desorption), while the situation in the northern parts still seem to be influenced by retention. Even though the deposition here is low and decreasing there are indications that many drainage areas accumulate S. Whether this accumulation is in the form of organic S or, as adsorption is not yet known. However, Karltun (1996) states that the sulphate adsorption capacity is higher in northern Sweden than in the south, which could imply a remaining S adsorption capacity.

The conclusion is thus that the diminution in S transport due to the reduction in acid S deposition is restrained by desorption of large amounts of soil sulphate as suggested by Harrison *et al.* (1989).

Acknowledgement

This work is based on data from monitoring programmes financed by the Swedish Environmental Protection Agency. The author thanks two anonymous reviewers for improving suggestions.

References

Alewell, C., Mitchell, M.J., Likens, G.E. and Krouse, H.R.: 1999, *Biogeochem.* **44**, 281.
Brimblecombe, P., Hammer, C., Rodhe, H., Ryaboshapko, A. and Boutron, C.F.: 1989. 'Human influence on the sulphur cycle', in Brimblecombe, P and Lein A.Yu. (eds.) *Evolution of the global biogeochemical sulphur cycle.* SCOPE 39, J.Wiley pp. 77–121.
Eriksson, E.: 1960, *Tellus* **12**, 63.
Harrison, R.B., Johnson, D.W., Todd, D.E.: 1989. *J. Environ. Qual.* **18**, 419.
Ivarsson, H. and Jansson, M.: 1995, *Water Air and Soil Pollut.* **84**, 233.
Karltun, E.: 1995, 'Sulphate adsorption on variable-charge minerals in podzolized soils in relation to sulphur deposition and soil acidity.' PhD thesis. Department of Soil Sciences, SLU, Sweden.
Likens, G.E. and Borman, F.H.: 1995, *Biogeochemistry of a forested ecosystem.* Springer Verlag.
Löfgren, S., Bringmark L., Aastrup M., Hultberg H., Kindbom K. and Kvarnäs, H.: 2000, Sulphur balances and dynamics in three .... *Water Air Soil Pollut.* This volume
Lindström, G. and Rodhe, A.: 1986, Nordic Hydrology **17(4/5)**, 325.
Persson, C., Langner, J., and Robertson, L.: 1996, 'Air pollution assessment studies for Sweden based on the MATCH model and air pollution measurements', in *Air Pollution Modelling and its Application.*, Vol. XI. Plenum Press, pp. 127–134.
Odén, S.: 1979, *Nordic Hydrology* **10**, 155.
Rochelle, B.P., Church, M.R. and David, M.B.: 1987, *Water Air and Soil Pollut.* **33**, 73.
Wilander, A.: 1994, *Water Air and Soil Pollut.* **75**, 371.

RECOVERY PATTERN FROM ACIDIFICATION OF HEADWATER LAKES IN FINLAND

JAAKKO MANNIO
Finnish Environment Institute, P.O.Box 140, FIN-00251 Helsinki, Finland
(email: jaakko.mannio@vyh.fi)

Abstract. Acid sensitive headwater lakes (n=163) throughout Finland have been monitored during autumn overturn between 1987-1998. Statistically significant decline in sulphate concentration is detected in 60 to 80 percent of the lakes, depending on the region. Median slope estimates are from -1.1 µeq L^{-1} in North Finland to -3.3 µeq L^{-1} in South Finland. The base cation (BC) concentrations are still declining especially in southern Finland (slope -2.5 µeq L^{-1}), where every second lake exhibits a significant downward trend. The BC slope is steeper for lakes with less peatlands, more exposed bedrock, longer retention time and southerly location, but these factors are inter correlated. Gran alkalinity slope medians for the three regions range from 1.4 to 1.8 µeq L^{-1} yr^{-1}. No significant negative alkalinity trends were detected. The similarity in the slopes of SO_4, BC and alkalinity in this data compared to seasonal sampling data from Nordic Countries can be regarded as indirect evidence that autumnal sampling is representative for long term monitoring for these ions. There are no indications of increased organic carbon in lakes, as found in some recent trend analyses of similar regional data sets. Although the processes behind the positive development in these lakes have to be revealed with site- specific intensive studies, this data suggests, that the initial recovery from lake acidification in Finland is a regional phenomenon.

Keywords: acidification, recovery, surface water quality, trends

1. Introduction

The deposition of acidifying compounds has declined in Finland significantly, sulphate with 40-60 percent in most parts of the country during the period 1987-1996, to the present level of ca. 0.5 g S m^{-2} in the southern part and 0.1 g S m^{-2} in the northern part of the country (Kulmala *et al.*, 1998). Nitrogen (NO_3+NH_4) deposition has declined less, and is presently at a similar level as S deposition; from 0.6 g N m^{-2} in southern to 0.15 g N m^{-2} in northern Finland. The acidity of deposition has not declined as consistently, however, largely due to variable declines in base cation deposition.

The regional monitoring network of small headwater lakes in Finland is aimed at detecting long-term changes caused by air pollution. These types of lakes in Finland have been estimated to have lost up to 2000 fish populations due to acidification (Rask *et al.*, 1995). The network provides data for critical load calculations and the modelling of acidification scenarios as well as a background for air pollution dose/response studies. This paper presents the development of the ionic composition in the monitoring network lakes between 1987 and 1998.

2. Material and Methods

Lakes (n=163) were sampled from mid-lake (1 m) or the outlet once in a year during autumn overturn (from early September in north to mid November in south) since 1990 and in 1987. Most lakes (103) have full set of 10 observations and only two lakes the minimum of seven observations. Catchment characteristics were determined from topographic maps (1:20 000) (Table I). Approximate water retention time was calculated based on a known bathymetry (44 cases) or maximum depths at sampling time, and regional runoff data from small hydrological catchments network (260-400 mm yr^{-1}; Kortelainen and Saukkonen, 1998).

Lakes are usually small (median area=10ha) headwater or seepage lakes (152 lakes less than 3% upstream lakes in the catchment) with no agricultural fields (only seven lakes > 3% fields). Chemically they are acid sensitive; low base cation concentrations, low alkalinity and sometimes elevated labile aluminium concentrations (Table II). Twenty-five chemical variables were analysed in the laboratories of Environmental Administration according to standardized methods (Forsius *et al.*, 1990). For trend significance analysis for each lake and each relevant variable, a nonparametric Kendall tau test was used and for the slope estimation simple linear regression (e.g.Mallory *et al.* 1998). For statistical analyses (SPSS 9.0 software), the lakes were divided in three geographical groups (Mannio and Vuorenmaa, 1995) reflecting differences in environmental conditions and northwards decreasing deposition level.

TABLE I

Morphometric characteristics of the regional lake acidification monitoring network (n=163).

Variable	Unit	South Finland (n = 62) Median (Range)		Central Finland (n = 57) Median (Range)		North Finland (n = 44) Median (Range)	
Lake area	ha	14.5	(0.7 - 211)	8	(2 - 450)	9	(1.5 - 548)
Catchment area	ha	91	(17 - 2230)	111	(16 - 2100)	112	(18 - 2810)
Max depth	m	8.0	(1-50)	6.8	(1.6 - 22)	2	(1.3 - 18.6)
Retention time	yr	2.4	(0.02 - 16)	0.9	(0.05 - 8.0)	0.3	(0.03 - 4.4)
Peatland area	%	7.4	(0 - 46)	24	(0 - 57)	23	(0 - 64)
Exposed bedrock	%	7	(0 - 51)	0	(0 - 15)	0	(0 - 78)

The group in southern Finland is characterised by lakes in upland areas with steeper topography and consequently more exposed bedrock and fewer peatlands in the catchment than in the two other groups. Water retention time is also longer in southern Finland.

TABLE II

Water quality (based on site-specific means 1987-1998) and significant trends (Kendall-t, $P < 0.05$) in 1987-1998 of the acidification monitoring network.

Variable	Unit	Median	Range	Number of lakes with a significant trend		
				negat.	n. s.	posit.
South Finland						
Conductivity	mS m^{-1}	3.0	(1.4 - 9.4)	45	17	0
pH		5.7	(4.6 - 6.8)	0	44	18
Alkalinity [1]	µeq L^{-1}	17	(-35 - 104)	0	28	34
[Ca+Mg+K+Na]*	µeq L^{-1}	176	(45 - 332)	30	31	1
SO$_4$*	µeq L^{-1}	112	(40 - 211)	50	12	0
Cl	µeq L^{-1}	39	(11 - 121)	6	56	0
NO$_3$	µeq L^{-1}	1.2	(0.2 - 15)	19	43	0
TOC	mg L^{-1}	6	(1.0 - 22)	2	54	6
tot P	µg L^{-1}	9	(4 - 27)	3	56	3
Al$_{labile}$	µg L^{-1}	19	(0 - 207)	10	52	0
Central Finland						
Conductivity	mS m^{-1}	2.1	(0.7 - 3.8)	18	38	1
pH		5.6	(4.2 - 6.8)	1	51	5
Alkalinity [1]	µeq L^{-1}	26	(-72 - 109)	0	42	15
[Ca+Mg+K+Na]*	µeq L^{-1}	140	(26 - 272)	9	44	4
SO$_4$*	µeq L^{-1}	51	(21 - 128)	44	13	0
Cl	µeq L^{-1}	18	(6 - 83)	12	44	1
NO$_3$	µeq L^{-1}	0.8	(0.1 - 5.3)	8	49	0
TOC	mg L^{-1}	11	(1.2 - 32)	2	53	2
tot P	µg L^{-1}	15	(3 - 78)	6	46	5
Al$_{labile}$	µg L^{-1}	15	(0 - 64)	7	50	0
North Finland						
Conductivity	mS m^{-1}	1.3	(0.7 - 2.6)	15	29	0
pH		5.9	(4.9 - 7.1)	0	39	5
Alkalinity [1]	µeq L^{-1}	16	(-16 - 141)	0	34	10
[Ca+Mg+K+Na]*	µeq L^{-1}	84	(27 - 212)	6	38	0
SO$_4$*	µeq L^{-1}	27	(12 - 111)	27	17	0
Cl	µeq L^{-1}	14	(8 - 46)	1	43	0
NO$_3$	µeq L^{-1}	0.5	(0.3 - 9.5)	1	42	0
TOC	mg L^{-1}	6	(0.9 - 25)	0	44	0
tot P	µg L^{-1}	9	(2 - 51)	5	39	0
Al$_{labile}$	µg L^{-1}	6	(0 - 56)	0	44	0

* denotes non-marine concentration

[1] Gran method

3. Results and Discussion

3.1. REGIONAL LAKE CHEMISTRY

The dominant feature in the regional lake chemistry is the overall decline of the constituents from south to the northern parts of the country (Table II). Only organic carbon (TOC), total phosphorus and, to a lesser extent, alkalinity level is highest in the Central Finland. All areas have still a rather similar distribution of acidic lakes, and more than half of these lakes have a pH (< 5.8) and alkalinity (<20 µeq L^{-1}) level that may induce harmful effects on the biota.

3.2. LAKE RESPONSES

The trends in the ionic composition are strongest in the southern Finland, which could be due to combined effect of largest declines in S deposition and highest initial concentration level of constituents. Up to 80 percent of the lakes in South and Central Finland show statistically significant decline in sulphate. The corresponding value, 60 percent (27 lakes), even in the northern group indicates effective diminishing of the long-range transported fraction of sulphur (Table II). No positive significant trends were detected for sulphate. The slopes of the trends, -3.3 µeq L^{-1} yr^{-1} in South and -1.9 µeq L^{-1} yr^{-1} in Central Finland (Table III) are in line with the aggregated trends from seasonally monitored sites across Nordic Countries (Stoddard et al., 1999).

The base cation (BC) concentrations are still declining especially in southern Finland, where every second lake exhibits a significant downward trend. There are at least two possible reasons for this; one is the declining deposition of base cations, and the other continuous depletion of BC pools of the watersheds. In both cases, the supply of BC to the lake is reduced due to the smaller amount of mobile anion SO_4^{2-}. Both trend significance and slope data suggests, that the BC decline is smaller than that of sulphate, which is a key condition for alkalinity increase. This applies for all regions and groupings (Tables II and III). The BC slope is steeper for lakes with less peatlands, more exposed bedrock, longer retention time and southerly location. These factors are inter correlated (Table I), but could imply better supply of BC from catchments with more organic soils, or inherently finer soil texture. One example of this is South vs. Central Finland, where smaller SO_4 but even smaller BC decline is resulting in similar alkalinity trends, 1.75 µeq L^{-1} yr^{-1} as in South Finland. This trend level is comparable to regional estimates in Nordic countries (1.3 µeq L^{-1} yr^{-1} in 1990-95, Stoddard et al. 1999). The similarity in the slopes of SO_4, BC and alkalinity in this data compared to seasonal sampling can be regarded as indirect evidence that autumnal sampling is representative for long term monitoring for these ions.

It is noticeable that no significant negative alkalinity trends were detected (Table II). The strengthening of the buffer capacity is, however, in the early stage and not very clearly reflected in the pH-level. The general increases of ca. 0.2 pH units per decade in autumn observations are probably not representative for the changes within the year. When the acidic deposition is lowered, it is expected that the acid surges caused by snow melt waters are less severe and may result in more rapid improvements in springtime pH. There are clear indications of better environmental conditions for fish in some of the lakes in this network (Nyberg et al., this volume).

TABLE III

Comparison of median change (units in Table II, per year) of the key chemical variables in different environmental classification groups. Bold numbers and same letters indicate significantly differing medians (Kruskall-Wallis or Mann-Whitney tests, $P < 0.05$)

Variable		n	pH	Alk	SO_4*	BC*	TOC
Geography							
South Finland		62	0.019	**1.74**[a]	**-3.31**[a]	**-2.47**[a,b]	**0.03**[a,b]
Central Finland		57	0.024	1.75	**-1.85**[a]	**-0.70**[a]	**-0.06**[a]
North Finland		44	0.015	**1.37**[a]	**-1.14**[a]	**-0.69**[b]	**-0.04**[b]
Soil cover							
Peatlands	< 10 %	71	0.017	1.72	**-2.52**	**-1.31**	-0.003
	> 10 %	92	0.019	1.57	**-1.69**	**-0.97**	-0.02
Exp. bedrock	< 4 %	121	0.019	**1.56**	-1.74	**-0.94**	**-0.04**
	> 4 %	42	0.014	1.78	-3.33	**-2.59**	**0.03**
Hydrology /morphometry							
Retention time	< 1 yr	88	0.016	1.60	-1.76	**-0.79**	-0.01
	> 1yr	75	0.023	1.57	-2.36	**-1.48**	-0.01
Bedrock sensitivity							
Base cations	<100 µeq L^{-1}	50	0.017	**1.30**[b]	**-1.54**[a]	-1.03	**-0.06**[a,b]
	100 – 170 µeq L^{-1}	54	0.019	**1.58**[a]	**-2.27**[a]	-1.46	**-0.01**[a]
	>170 µeq L^{-1}	59	0.017	**2.32**[a,b]	**-2.72**[a]	-1.15	**0.03**[b]

The nitrate concentrations in upland lakes in Finland are low in comparison to other monitored sites in Nordic countries or Central Europe (Traaen and Stoddard, 1995). The concentrations are negligible as long as there is biological activity, and may be elevated when the dormant season starts. The trend analysis using slope is too sensitive for outliers to be performed for NO_3 in these lakes. There is no sign of increased TOC, as found in some recent trend analyses of similar regional data sets (e.g. Bouchard, 1997; Skjelkvåle et al., 2000). Instead, lack of TOC trends suggests, that there is presently no strong unidirectional, year-to-year hydrologic or climatic factor affecting to these lakes.

4. Conclusions

Sulphate concentrations have declined in all types of small lakes throughout Finland in 1990's, indicating a straight effect of the S emission reductions. Base cation concentrations are still declining especially in southern Finland, but to a lesser extent than SO_4. This allows the buffering capacity to increase, which is significant in every third lake of this monitoring network.

Although the processes behind the positive development have to be revealed with site-specific intensive studies, this data suggests, that the initial recovery from lake acidification in Finland is a regional phenomenon.

Acknowledgements

The author thanks the laboratory and field staffs of the Environment Administration for carrying out the sampling and chemical analyses and Jussi Vuorenmaa for compiling the earlier version of the data set. The Academy of Finland is acknowledged for financial support for this work.

References

Bouchard, A.: 1997, *Water, Air and Soil Pollution*, **94**, 225.
Forsius, M., Malin, V., Mäkinen, I., Mannio, J., Kämäri, J., Kortelainen, P. and Verta, M.:1990, *Environmetrics*, **1**, 73.
Kortelainen, P. and Saukkonen, S: 1998, *Water, Air and Soil Pollution*, **105**, 239.
Kulmala, A., Leinonen, L., Ruoho-Airola, T., Salmi, T. and Waldén, J.: 1998, *Air Quality Measurements*, Finnish Meteorological Institute, 91pp.
Mallory, M.L., McNicol, D.K., Cluis, D.A. and Laberge, C.: 1998, *Canadian Journal of Fisheries and Aquatic Sciences*, **55**, 63.
Mannio, J. and Vuorenmaa, J.: 1995, *Water, Air and Soil Pollution*, **85**, 571.
Nyberg, K., Vuorenmaa, J., Rask, M., Mannio, J., Raitaniemi J.:, this volume.
Skjelkvåle, B.L., Andersen, T., Halvorsen, G.A., Raddum, G.G., Heegaard, E., Stoddard, J. and Wright, R.F.: 2000, The 12-year report: Acidification of Surface Water in Europe and North America; Trends, biological recovery and heavy metals, ICP waters report **52**, NIVA, Oslo, 115 pp.
Stoddard, J.L., Jeffries, D.S., Lükewille, A., et al.:1999, *Nature*, **401**, 575.
Rask, M., Mannio, J., Forsius, M. Posch, M. and Vuorinen, P. J.:1995, *Environmental Biology of Fishes*, **42**, 51-63.
Traaen, T. and Stoddard, J.:1995, Convention on LRTAP, International cooperative programme on assessment and monitoring of acidification of rivers and lakes. NIVA, **3201**, 39 pp.

DECREASE IN ACID DEPOSITION - RECOVERY IN NORWEGIAN WATERS

BRIT LISA SKJELKVÅLE[1*] KJETIL TØRSETH[2], WENCHE AAS[2] and TOM ANDERSEN[1]

[1]*Norwegian Institute for Water Research, P.O. Box 173 Kjelsås, N-0411 Oslo, Norway*
[2]*Norwegian Institute for Air Research, P.O. Box 100, N-2027 Kjeller, Norway*
(*author for correspondence, e-mail: brit.skjelkvaale@niva.no)

Abstract Concentrations of sulphate in precipitation in southern Norway have decreased by 50-60% from 1980-1999. This has caused a decrease in sulphate concentrations in lakes of 30-40% from 1986-1999. Nitrogen in precipitation has decreased slightly over the last 10-years. In lakewater there has also been a significant but slight decrease. Concentrations of non-marine base cations in precipitation have decreased by 40% from 1980-1999. In lakewater, non-marine base cation concentrations have been at about the same level the last 10 years. This indicates that acid deposition has decreased sufficiently such that the pool of exchangeable base cations in the soil is now being replenished. The acidification situation in lakes in Norway has thus shown a clear improvement over the last 8-10 years. pH, alkalinity and ANC (acid neutralising capacity) have all increased. Concentrations of inorganic (toxic) aluminium species have decreased. The trends in H^+ and Al^{n+} do not follow the relation expected if Al^{n+} concentrations were governed solely by a single solid phase of $Al(OH)_3$.

Key words: Recovery, deposition, water chemistry, trends

1. Introduction

The Norwegian national monitoring programme for long-range transported air pollutants was set up in 1980 with the aim to measure changes in deposition and effects of acidifying compounds (Johannessen 1995). International agreements obtained under the UN-ECE Convention on Long-range Transboundary Pollution have resulted in substantial decreases in emissions of sulphur and to a lesser extent nitrogen since 1980 (EMEP 1998).

Sulphur and nitrogen from long-range transported air pollution are deposited at the highest rates in southern and south-western Norway and decrease northwards (Figure 1). The eastern part of Finnmark in northern Norway is also affected by sulphur pollution from industry on the Kola peninsula in Russia. Due to the deposition pattern, acidified, low pH lakes with damaged fish populations are found in southern Norway (Henriksen *et al.* 1988, Henriksen *et al.* 1989).

Sulphate is the "driving force" in the acidification process and it is therefore of major interest to study the effect of the reduced deposition of sulphur on water quality. Here we report the changes in precipitation chemistry over the period 1980-1999 at 3 stations and lakewater chemistry over the period 1986-1999 for 52 lakes in southern Norway. Trends in precipitation and water chemistry have been previously treated for earlier periods by Tørseth and Semb 1995, Tørseth *et al.* 1999, Skjelkvåle and Henriksen 1995, Skjelkvåle *et al.* 1998 as well as presented in the annual reports from the Norwegian monitoring programme (SFT 1999).

Water, Air, and Soil Pollution **130**: 1433–1438, 2001.
© 2001 *Kluwer Academic Publishers.*

2. The data

Precipitation samples (bulk) are collected daily or weekly and analysed for major solutes at the Norwegian Institute for Air Research (NILU) by standard methods (Aas et al. 1999). Dry deposition of gases and aerosols are also measured, but is not included in this work. Lakewater samples are collected annually at the outlet of the lakes after the autumn circulation period. Samples are analysed at the Norwegian Institute of Water Research (NIVA) by standard methods. The lakes are a subset of the 1000-lake survey of 1986 (Henriksen et al. 1989) and are acid sensitive, headwater lakes on granitic or gneissic bedrock without any local pollution sources.

Non-marine fractions of SO_4 and base cations in lakewater are calculated under the assumptions that all Cl in lakes is of marine origin (cyclic seasalts) and is accompanied by other ions in the same proportions as in seawater. Similarly, non-marine concentrations in precipitation are estimated on the basis of Na, Mg or Cl (in order of preference). The non-marine fractions are denoted by asterisk (*). ANC (acid neutralising capacity) is defined as equivalent sum of base cations ($Ca^{2+} + Mg^{2+} + Na^+ + K^+$) minus equivalent sum of strong acid anions ($SO_4^{2-} + Cl^- + NO_3^-$). Alkalinity (alk) is determined by fixed-endpoint titration to pH 4.5.

Trends in lake water chemistry for the 11 year period 1989-1999 were analysed with site-specific trend tests using the nonparametric seasonal Kendall test (SKT) modified to account for serial dependence (Hirsch et al. 1982). To infer regional trends, we applied the meta-analytical technique of van Belle and Hughes 1984) that allows, with some restrictions, trend results from multiple sites to be combined into a single estimate of trend.

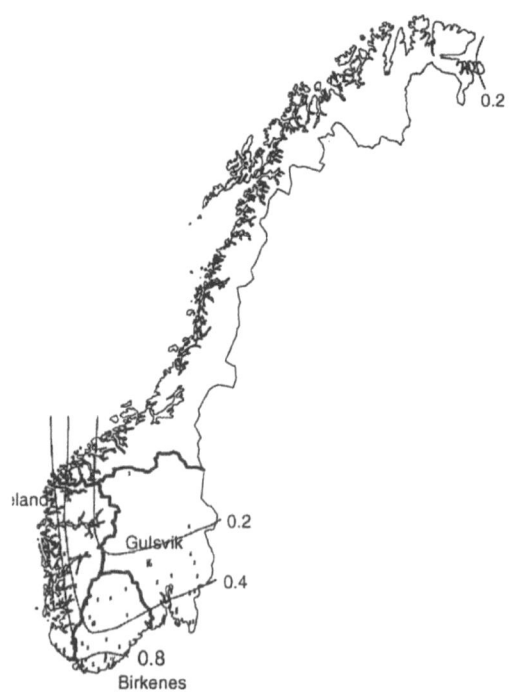

Figure 1. Location of the investigated lakes in the three regions in Southern Norway (East, West, South, divided by thick lines). Isopleths show total wet S-deposition in 1999 (unit mg S m^2 yr^{-1}). Also shown are the locations of the 3 precipitation stations (solid squares).

3. Results and discussion

3.1 PRECIPITATION

In precipitation significant reductions in concentrations of SO_4^*, nitrogen compounds, strong acid (H^+) and Ca^* have occurred over the last 20 years at the 3 representative monitoring stations in southern Norway (Figure 2). The reductions in SO_4 has been 50 – 60% from 1980-1999.

The decrease in nitrogen concentrations is smaller than that of SO_4^* and H^+, and the decrease is observed only since about 1990. Reductions of oxidised nitrogen species are somewhat larger compared to reduced compounds due to larger reductions in NO_x emissions compared to NH_3 in Europe.

Concentrations of Ca^* have decreased by about 40% since 1980 (Tørseth et al. 1999). The reduced concentrations of Ca^* offset the reductions in acidic inputs to only a minor extent as base cations correspond to less than 5-10% of the strong acid anion in deposition (Tørseth et al. 1999).

Dry deposition is an additional source of acidifying compounds. Concentrations of sulphur and nitrogen species in gases and aerosols at the 3 Norwegian monitoring stations have also decreased to about the same degree as wet components during the 1980-1999 period (data not shown; Aas et al. 1999).

3.2 LAKE WATER

In lakewater SO_4^* concentrations have decreased steadily and significant over the period 1989-99 in all 3 regions of southern Norway, and thus correspond well to the trends in precipitation chemistry (Figure 3). The reductions in SO_4^* has been 30-40% from 1986-1999. In eastern and to a lesser extent in southern regions the decrease in lake water is not as large as the decrease in precipitation, probably due to leaching of stored sulphate from the soil. In western Norway, where the soils are thin and precipitation amount high (3000-5000 mm yr^{-1}) the percent decrease in sulphate concentrations in precipitation and lakes are quite similar.

The lakewater data give no indication that the observed increase in NO_3 concentrations observed from lake surveys conducted in 1974-75 and 1986 has continued (Henriksen and Brakke 1988). On the contrary, NO_3 levels decreased significant from 1989-99 (Table I). This coincides with the decrease in concentrations of nitrogen in precipitation over the past 10 years (Figure 2). Most of the decrease has occurred the last three years (1997-99). The change in NO_3 is not visible at Figure 3 due to the scale on the y-axis.

Base cations in lakewater show very small changes over the 10-year period. There is a slight significant increase in the south and a decrease in the east (Figure 3, Table I). In the west there is no change.

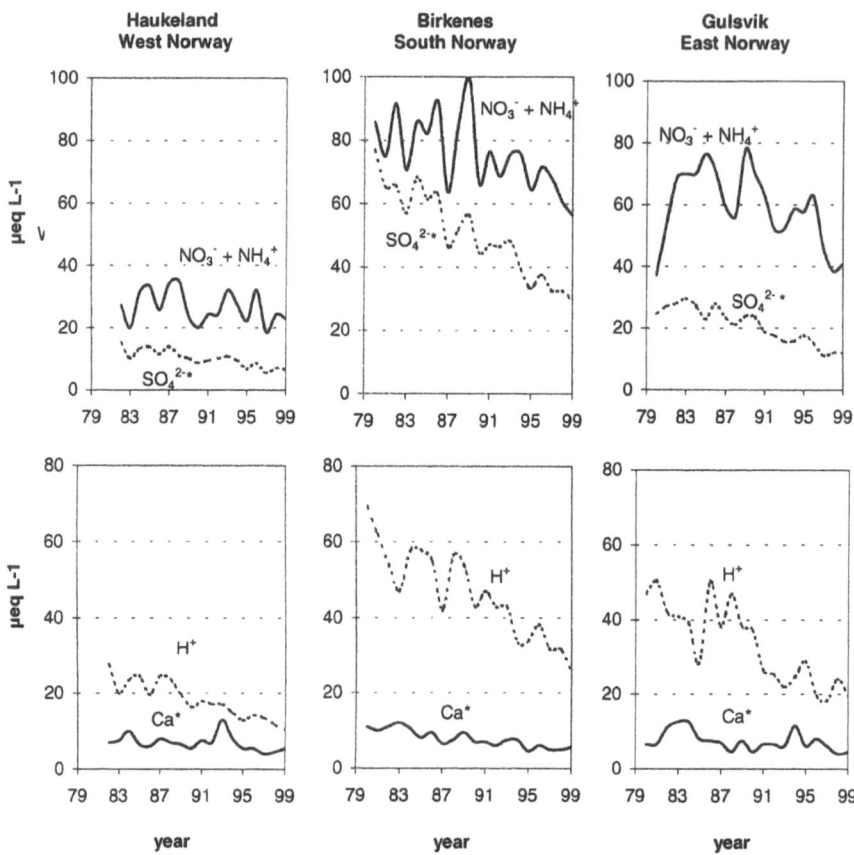

Figure 2. Trends in concentrations of $SO_4^{2-}*$, $NO_3^-+NH_4^+$ (upper panel), H^+ and Ca^* (bottom panel) in precipitation for the period 1980 to 1999 for three monitoring sites in Norway (see location in Figure 1).

The reduction in base cation deposition has been too small too influence the trend, as the soil pool and weathering input of Ca are much larger than deposition input.

As a consequence of the decreasing trends in SO_4* and the steady levels of base cations since about 1990, ANC and alkalinity have increased, and H^+ and Al^{n+} have decreased (Figure 3). For lakes with pH > 5.5 the increase in alkalinity is due to increase in HCO_3^-, but increase in TOC (Table I) and decrease in Al^{n+} also contribute to the increasing alkalinity.

Al^{n+} and H^+ are at the same concentration level and decrease at the same rate in the three regions. The decrease in Al^{n+} and H^+ do not follow the 3:1 relation expected if water was in equilibrium with a single solid phase of $Al(OH)_3$. The solubility of $Al(OH)_3$ is apparently increasing. This may indicate that the source of soluble Al is changing as pH increases.

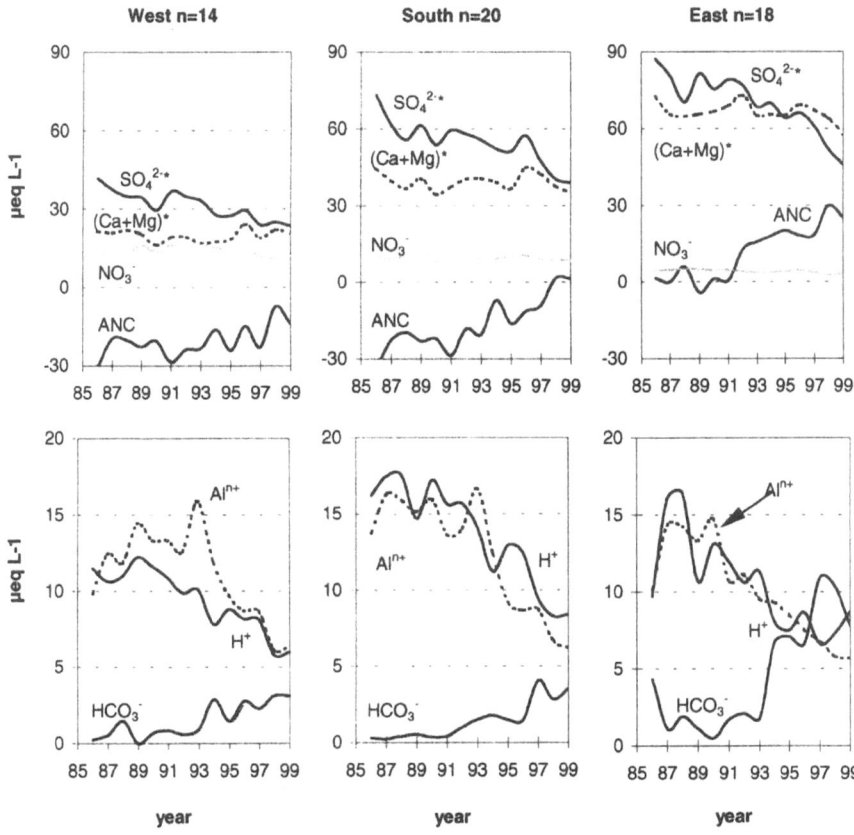

Figure 3. Trends in concentrations of SO_4^{2-*}, NO_3^-, $(Ca+Mg)^*$ and ANC (Upper panel) and Al^{n+}, H^+ and alkalinity (measured)) (Bottom panel) for lakes in three regions of Norway for the period 1986-1999 (see location in Figure 1).

4. Concluding remarks

The major and widespread improvement of water quality in acidified lakes in southern Norway due to reduced acid deposition is a success story, showing that international negotiations to reduce transboundary air pollution can work (Skjelkvåle et al. 1998). Improved water quality with regard to higher ANC and pH together with decreased concentrations of Al^{n+} has resulted in improved conditions for fish and invertebrates in lakes and rivers (Hesthagen *et al.* this volume, Raddum and Fjellheim this volume). Lakes in southern Norway are on the road to recovery, and further improvement can be expected over the next 10-20 years as the 2nd sulphur protocol of 1994 (Oslo protocol) and the multi-pollutant multi-effect protocol of 1999 (Gothenburg protocol) are implemented.

TABLE I

Yearly SKT trend slope for SO_4^*, NO_3, alkalinity, H^+, ANC, $(Ca+Mg)^*$ labile Al (unit: µeq L^{-1}) and TOC (unit: mg C L^{-1}) for the three regions for the period 1989-99. Only significant ($p < 0.005$) changes are shown. The statistical techniques (SKT and meta-analysis) are described in the text.

	unit	West n=14	South n=20	East n=18
non-marine SO_4^*	µeq L^{-1} yr^{-1}	-1.77	-1.68	-2.75
NO_3^-	µeq L^{-1} yr^{-1}	-0.16	-0.12	-0.10
Alkalinity	µeq L^{-1} yr^{-1}		0.35	0.65
TOC	mg C L^{-1} yr^{-1}	0.08	0.09	0.20
H^+	µeq L^{-1} yr^{-1}	-0.85	-0.72	-0.50
ANC	µeq L^{-1} yr^{-1}	2.10	2.35	2.69
non-marine $(Ca+Mg)^*$	µeq L^{-1} yr^{-1}		0.05	-0.29
Labile Al	µeq L^{-1} yr^{-1}	-0.87	-0.84	-0.74

References

Aas, W., Tørseth, K., Solberg, S., Berg, T., and Manø, S. 1999. Overvåking av langtransportert forurenset luft og nedbør. Atmosfærisk tilførsel, 1999. SFT 797/00, Oslo, Norway. 146 pp.

EMEP 1999. Transboundary Acid deposition in Europe. EMEP Summary Report 1999, EMEP/CCC and MSC-W Report 1/99.

Henriksen, A. and Brakke, D. F. 1988. *Water Air Soil Pollut.* **42**: 183-201.

Henriksen, A., Lien, L., Rosseland, B. O., Traaen, T. S., and Sevaldrud, I. 1989. *Ambio* **18**: 314-321.

Henriksen, A., Lien, L., Traaen, T. S., Sevaldrud, I., and Brakke, D. F. 1988. *Ambio* **17**: 259-266.

Hesthagen, T. 2000 submitted *Water Air Soil Pollut.* this volume

Hirsch, R. M., Slack, J. R., and Smith, R. A. 1982. *Water Resourc.Res.* **18**: 107-121.

Raddum, G.G. and Fjellheim, A. 2000 submitted *Water Air Soil Pollut.* this volume

Johannessen, T. 1995. *Water Air Soil Pollut.* **85**: 617-621.

SFT. 1999. Overvåking av langtransportert forurenset luft og nedbør. Årsrapport 1998. SFT 781/99, Oslo, Norway.

Skjelkvåle, B. L. and Henriksen, A. 1995. *Water Air Soil Pollut.* **85**: 629-634.

Skjelkvåle, B. L., Wright, R. F., and Henriksen, A. 1998. *Hydrol.Earth System Sci.* **2**: 555-562.

Tørseth, K., Berg, T., Hanssen, J. E., and Manø, S. 1999. Overvåking av langtransportert forurenset luft og nedbør. Atmosfærisk tilførsel, 1998. SFT 768/99, Oslo, Norway.

Tørseth, K. and Semb, A. 1995. *Water Air Soil Pollut.* **85**: 623-628.

van Belle, G. and Hughes, J. P. 1984. *Water Resourc.Res.* **20**: 127-136.

LIMING OF ACID AND METAL CONTAMINATED CATCHMENTS FOR THE IMPROVEMENT OF DRAINAGE WATER QUALITY

JOHN GUNN[1], ROD SEIN[1], BILL KELLER[1] and PETER BECKETT[2]

[1] *Cooperative Freshwater Ecology Unit, Department of Biology, Laurentian University, Sudbury, ON, Canada, P3E 2C6*
[2] *Department of Biology, Laurentian University, Sudbury, ON, Canada, P3E 2C6*

Abstract. A 38 ha near-barren experimental catchment area near an abandoned Cu and Ni smelter in Sudbury, Canada was treated with 410 tons of coarse dolomitic limestone in 1994. An additional 54 tons of pelletized fine dolomite were added to 15 wetlands within the experimental catchment in 1995. The treatments significantly increased the pH and base cation concentrations in the outlet stream. Cu and Ni concentrations initially rose after the wetland treatment, but then declined to levels below those of the reference site. Bioassay tests revealed that the toxicity of the drainage water was greatly reduced by the liming, but some localized inputs of highly toxic groundwater still posed a problem for aquatic biota. The pH of surface water in the wetlands has been maintained at >6.0 for over 4 years. The wetland liming appeared to be highly effective at neutralizing drainage water, however there may be some adverse effects on wetland plant communities as a consequence.

Keywords: liming, metal contamination, streams, catchments, toxicity tests

1. Introduction

Liming of stream catchment areas is a widely used method of reducing acid drainage from areas subject to acidic precipitation (Olem *et al.*, 1991; Henriksen and Bodin, 1995; Traaen *et al.*, 1997), but this approach has not been extensively tested in smelter-damaged lands where severe soil erosion may occur and soils are heavily contaminated with toxic metals. Sudbury, Canada, is an area where such problems exist (Gunn, 1996). Our study was designed to test whether the aerial land liming program conducted by the mining companies, mainly for land reclamation purposes, was improving drainage water quality from the treated sites. We then tested a wetland application technique as a preferred method of drainage water treatment.

2. Study Site

Daisy Lake is a 36 ha lake located 3.5 km southwest of the abandoned Coniston smelter. The Cu and Ni smelter operated from 1913 to 1972. There was also an open roast yard near the smelter that operated between 1913 and 1918. Two catchments on the northeast end of Daisy Lake, each with a well defined drainage stream, were used for this experiment - one 38 ha catchment as an experimental treatment area (J) the other 32 ha catchment as an untreated control area (I) (Figure 1). The initial water chemistry was quite similar for the two catchment streams, with low pH (approximately 4.5), low base cation levels (Ca 1.7-2.5 mg/L) and high levels of metals (Cu 56-138 µg/L; Ni 320-354 µg/L). The most striking difference in the chemistry between the two streams was the approximately 2x higher concentration of Cu in J than I at the start of the experiment. The J catchment was closer to the former smelter, but differences in loading rates cannot fully explain the higher concentration of Cu in the J stream. Ni which is also abundant in smelter emissions showed no such differences between catchments. Differences in complexation rates appears to be a more plausible explanation. Cu is usually strongly complexed by organic matter in soils and sediments (Tyler, 1978), while Ni is more soluble and mobile (Hutchinson and Whitby, 1977; Rutherford and Bray, 1978). The more extensive vegetation cover and less eroded soils in the control catchment would therefore be expected to bind more of the Cu and possibly create the observed chemical differences in the drainage water.

3. Methods

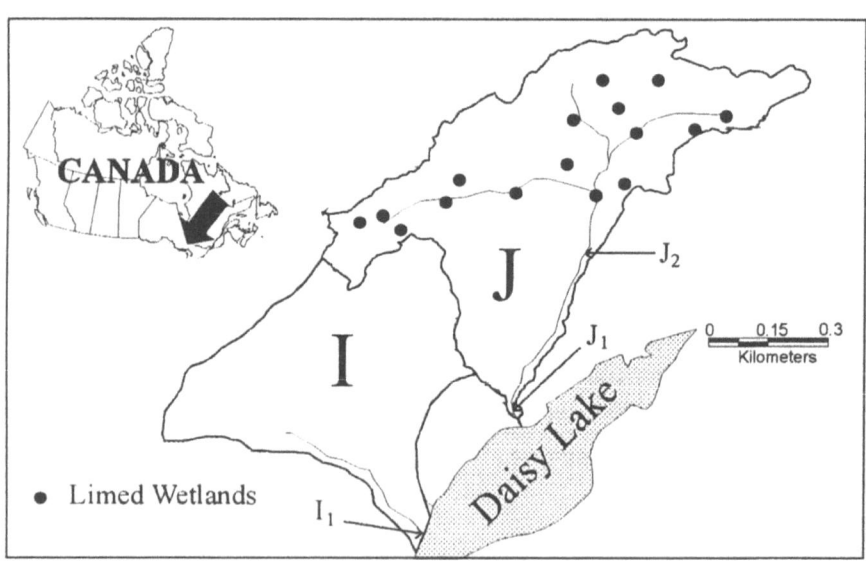

Figure 1. Daisy Lake experimental catchments, I (reference) and J (treatment) near Sudbury, Ontario. The abandoned Coniston smelter is approx. 3.5 km NE of the J catchment.

During Aug. 23 - Sept. 2, 1994 the entire J catchment area was aerially limed with 410 tons of coarse (95%>0.5 mm) dolomitic limestone (53.9% $CaCO_3$, 44.8% $MgCO_3$). During Aug. 24 -Sept. 8, 1995, an additional 54 tons of highly soluble pelletized fine (60% <0.1 mm) dolomitic limestone (54.5% $CaCO_3$, 45% $MgCO_3$) were added to 15 small wetland sites within the treated area (total surface area of wetlands 1.7 ha; application rate 31.8 tons/ha).

Sampling of the outlet stream chemistry began in 1991. Sampling was usually weekly during April-June and samples were submitted to the Ontario Ministry of the Environment for analysis. Additional sampling occurred after fall rains. A second sampling site, J2, was established in the treated catchment in 1995 as an upland site to avoid the influence of a groundwater seepage area that was discovered after the liming treatments began (Figure 1). Bulk water samples were also collected once a year for static bioassay toxicity testing using rainbow trout (*Onchorhychus mykiss*) and *Daphnia magna*. Regular water samples were taken during the ice-free season from the wetlands. Wetland vegetation was also assessed before and after treatment.

4. Results

The coarse limestone treatment produced an initial increase in the base cations in stream water but no substantial decrease in the acidity of the drainage water (pH<5.0; Figure 2). The coarse limestone treatment also appeared to have no significant effect on the toxicity of the water at the J1 site (Table I). The water remained acutely toxic to *Daphnia* and rainbow trout at both J1 and the reference stream I1. The poor performance of the coarse limestone treatment was obviously because of the size and limited solubility of the large particles (Warfvinge and Sverdrup, 1988). The fine limestone treatment of the wetlands within the J catchment greatly increased the pH and base cation concentrations, (Fig. 2) and reduced the toxicity of the stream water to the test organisms (Table I). In spite of these overall benefits there were some potentially adverse effects of the treatment, in the form of a pulse release of Cu and Ni from the catchment that lasted about 4-5 weeks after the wetlands were limed. Mean concentrations of Cu and Ni increased by 14 ug/L and 39 ug/L, respectively during this period. Once the pulse passed, metal levels continued to drop again and were well below the control stream concentrations throughout 1996 and 1997. The other surprising finding was that pH declined at the outlet sampling site (J1) at low flow periods during the post-treatment period. During these drier periods a gradient from low pH to high pH existed as one moved up the first 200 m of the stream - indicating that the lower section of J was affected by acidic groundwater. Above the groundwater input pH remained consistently above the desired criteria of pH 6.0 (a level that protects most biota) throughout the post-treatment period. At the upstream sampling site (J2) the toxicity that had previously been detected was completely eliminated by the wetland liming treatment (Table I).

There were some declines in Cu and Ni concentrations observed in the control stream at the start of the monitoring (1991-1993), but overall there was very little change in the chemistry or the toxicity results for the control stream during the study, providing good contrast with the trends in the treatment catchment.

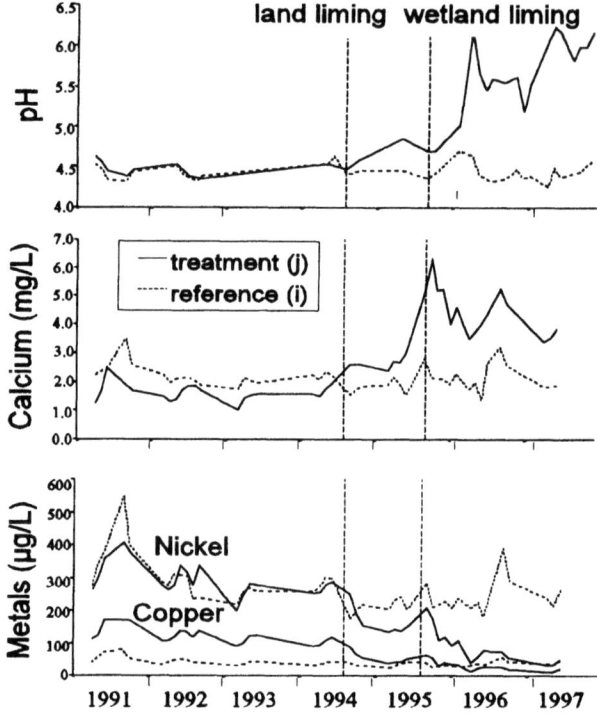

Figure 2. Effects of land liming (August 23 - September 2, 1994) and wetland liming (August 23 - September 8, 1995) on outlet stream chemistry for tributaries of Daisy Lake.

TABLE I

Lethal concentration of stream water for rainbow trout (over 96 hr) and *Daphnia magna* (over 48 hr) in static bioassays. The indicated LC_{50} values are the concentrations of water that caused 50% mortality of test animals.

Date	Rainbow Trout			*Daphnia magna*		
	J_1	J_2	I_1	J_1	J_2	I_1
May 16/91	68.4	----	77.4	15.7	----	28.3
July 13/92	70.7	70.7	70.7	19.8	20.9	31.7
			Soil Liming			
October 17/94	89.4	----	89.4	22.3	----	70.7
April 25/95	72.1	----	72.1	----	----	----
			Wetland Liming			
November 16/95	70.7	non lethal	70.7	31.5	non lethal	66.7
September 16/96	100	non lethal	70.7	82	100	46.2

The vegetation in the wetlands included *Chamaedaphne calyculata, Juncus brevicaudatus, Agrostis scabra, Scirpus cyperinus, Glyceria canadensis, Calamogrostis canadensis* and *Typha latifolia*. *Polytrichum commune* and *Pohlia nutans* mosses were abundant. Following application of the limestone there was some mortality of the *Polytrichum* and so far this has not recovered. The initial coarse-limestone treatment produced no substantial decrease in the acidity of the water in the wetlands. Following the wetland liming the pH rose to 6-6.5 and has stayed at that level for over 3 years.

5. Discussion

The decline in Cu and Ni during 1991-1994 indicated that the metal contamination of the site was declining even before the first limestone treatment. There are three potential explanations for these trends: 1) reduced atmospheric inputs, 2) increased exports from the catchment , and 3) increased complexation of metals within the catchment. We know that total emissions of metal particulates have declined by about 90% in the Sudbury area from the levels that existed in the early 1970's (Potvin and Negusanti, 1995). Declining deposition of metals may have contributed to observed patterns, but the higher initial values and more rapid decline in the J than the I catchment suggests that differences in catchment characteristics were of prime importance. For example, the Cu and Ni concentrations in surface soils have been observed to decline in recent decades near the Coniston smelter (Potvin and Negusanti, 1995), but so too has organic content (Dudka *et al.*, 1995), suggesting that the surface soils are continuing to erode by wind and rain. From visual evidence alone (from the delta formation at the mouths of the streams), it would appear that physical erosion was very important at our site and was probably greatest in the more barren J catchment area.

The observation of the brief pulse in metal concentrations immediately after the wetland liming treatments lends support to the concern (Skraba, 1989) that liming may temporarily increase metal concentrations in stream water, through displacement of metal cations at the soil exchange sites by the added Ca. This effect lasted only a few weeks. However, the potential that a pulse release of metals could accompany wetland liming needs to be the considered when planning rehabilitation efforts. In our case, the receiving lake was relatively large and already heavily contaminated so the effect was probably quite minimal.

The presence of acidic groundwater proved to be a confounding factor that reduced the effectiveness of soil and wetland treatments at our site. However, the steady improvements of the water chemistry and reduced toxicity at the outlet of the catchment in the later part of the project suggests that the acid groundwater was either a small portion of the overall flow, or was in fact influenced by the surface application during the brief period (brief in terms of groundwater recharge time) of this study. The J stream followed a steep fault line on its way to the lake. The groundwater in our study may therefore have simply been a shallow pooling of subsurface water rather than true groundwater with a long residence time.

In spite of the acid groundwater issue, the catchment treatments, particularly the wetland applications, proved to be very effective over the longer-term at improving water quality in much of the catchment stream. An integrated program of revegetation procedures to provide a plant cover, and to improve the soil conditions, as well as specialized applications to wetlands and other hydrologic sites to improve water quality, offers much promise for the rehabilitation of severely damaged ecosystems around Sudbury and other acid and metal contaminated areas.

Acknowledgements

We thank D. Bolton, M. Puro, and P. Yearwood of INCO Ltd. for organizing and supporting the experimental treatments of the catchment. T. Kaliczak of Falconbridge Ltd. supported the bioassay work.

References

Dudka S., Ponce-Hernandez, R. and Hutchinson, T. C.: 1995, *Sci. Total Environment*, 161.
Gunn, J. M.: 1995, *Restoration and Management Notes* **14**, 129.
Henriksen, L. and Bodin, Y. W.: 1995, *Liming of Acidified Surface Waters*. Springer-Verlag, Berlin.
Hutchinson, T. C. and Whitby, L. M.: 1977, *Water Air Soil Pollution* **7**, 421.
Olem, H., Schreiber, R. K., Brocksen, R. W. and Porcella D. B.: 1991, *International Lake and Watershed Liming Practices*. Terrene Institute, Washington, DC.
Potvin, R. and Negusanti, J.: 1995, 'Declining industrial emissions, improved air quality, and reduced damage to vegetation', in J. M. Gunn (ed.), *Restoration and Recovery of an Industrial Region,* Spinger-Verlag, pp. 51-62.
Rutherford, G. K. and Bray, C. R.: 1979, *Journal of Environmental Quality* **8**, 219.
Skraba, D.: 1989, Effects of surface liming of soils on stream water in a denuded, acid, metal-contaminated watershed near Sudbury, Ontario. Master of Science Dissertation, Laurentian University, Sudbury, Ontario.
Traaen, T. S., Frogen, T., Hindar, A., Kleiven, E., Lande, A. and Wright, R. F.: 1997, *Water Air Soil Pollution* **94**, 163.
Tyler, G.: 1978, *Water Air Soil Pollution* **9**, 137.
Warfvinge, P. and Sverdrup, H.: 1988, *Water Air Soil Pollution* **24**, 701.

TRENDS IN SOIL WATER COMPOSITION AT A HEAVILY POLLUTED SITE – EFFECTS OF DECREASED S-DEPOSITION AND VARIATIONS IN PRECIPITATION

R.D. VOGT[1*], H.M. SEIP[1], H. OREFELLEN[1], G. SKOTTE[1], C. IRGENS[1], and J. TYSZKA[2]

[1] *University of Oslo, Dept. of Chemistry, P.O.Box 1033 Blindern, N-0315 Oslo, Norway;*
[2] *Forest Research Institute, ul. Bitwy Warszawskiej 1920r Nr. 3, PL-00-973 Warszawa, Poland*
(* author for correspondence, e-mail: rolf.vogt@kjemi.uio.no)

Abstract. Precipitation, soils and soil water in a forested catchment in western Poland have been studied during the period 1992 – 96 (see also Vogt *et al.*, this conf.). The S-deposition in the area during the study period was 2 – 3g S m^{-2} yr^{-1}. In spite of decreasing anthropogenic emissions the S-deposition in the area did not change much during the study period mainly because the first years were exceptionally dry. However, the S-deposition was considerably higher during the previous decade. Based on soil water sulphate concentration, pH, acid neutralising capacity and the ratio of Al^{3+}/(Ca^{2+} + Mg^{2+}), there is apparently an amelioration in the conditions. A study using *inter alia* principal component analysis, indicates that this improvement is mainly due to more precipitation in the later part of the study period. Variations in precipitation amount have a pronounced effect on the soil-water chemistry, which makes it difficult to establish trends caused by changes in anthropogenic deposition. Long time series are therefore necessary to establish recovery due to reduced S-emissions.

Keywords: ANC, critical load ratio, fluctuations in climate, PCA, Trends, Poland

1. Introduction

During the last two decades emission reductions in Europe have resulted in a decrease in atmospheric sulphur deposition of up to 50%. Results from the ICP Waters programme show that sulphate concentrations are decreasing at almost all investigated sites (Stoddard, 1999). Regions with declining sulphate that fail to show recovery in acid neutralising capacity (ANC= (BC - ΣSO$_4^{2-}$,NO$_3^-$,Cl$^-$,F$^-$) µeq L^{-1}) in the 1990s are characterised by strongly declining base cation concentrations (ICP Waters, 2000). All regions showed tendencies towards increasing amounts of dissolved organic carbon (DOC) (Skjelekvåle, 2000).

The differences in runoff chemistry and temporal trends are governed by a diversity of soil/soil-water interactions within the catchment caused by changes in anthropogenic loading and climate. A striking feature of the soil-water chemistry data from a field acid addition experiment was the profound influence of fluctuating temperature and the amount of precipitation (Vogt, 1996). The increase in weak organic acidity (as inferred by an increase in DOC) plays an important role by buffering the pH at a low level, though increasing the ANC.

The purpose of this study is to improve the understanding of key factors dictating the type of recovery response. This is obtained by studying the differences in the *in situ* recovery of soil water in a set of acid soil profiles located in a heavily polluted region.

2. Site description, sampling and methods

The Czerniawka catchment (50E 48'N, 15E 35' E) in the Karkonosze National Park, in south-western Poland is a 0.93 km² large catchment situated at elevations ranging from 650 to 1050 m a.s.l. The climate is temperate/sub-alpine. The bedrock is composed of poorly weatherable gneiss. The content of feldspar and mica is low. Podzolic soil dominates in the lower alluvial deposited sandy reaches. At the intermediate altitudes poorly developed podzols have evolved from dystric cambisols (Skotte, 1995). Naturally regenerated 40 to 70 yrs. old Norway spruce stands (*Pícea ábies*) forest the catchment. The base saturation in podzol profiles is less than 20% with values about 5% in the B-horizons (Vogt *et al.*, 1994). The Czerniawka catchment is located in the region of highest sulphur and nitrogen deposition in Europe (Barrett *et al.*, 1995). The annual average estimated total deposition of sulphur and oxidised + reduced nitrogen between 1985 and 1996 was 5.4g S m^{-2}yr^{-1} and 2.3g N m^{-2}yr^{-1} (EMEP, 1998). The generally large air pollutant load in these forested mountains is found to contribute to the extensive deforestation (Mazurski, 1986; Godzik and Sienkiewicz, 1990).

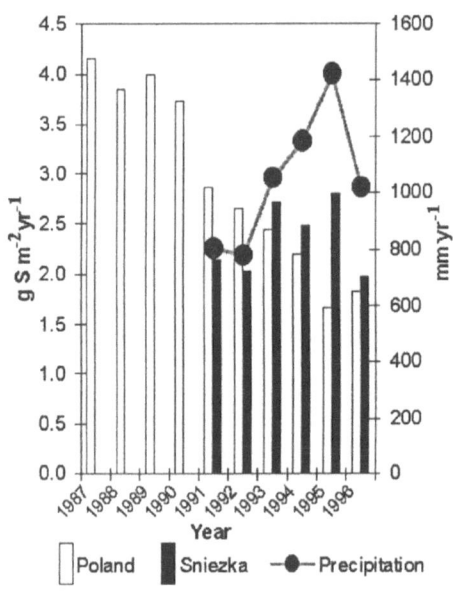

Figure 1. Precipitation amount and sulphur deposition at the EMEP station Sniezka and average S deposition in Poland (EMEP, 2000). Total deposition at Sniezka is calculated as the sum of wet and dry deposition assuming a dry deposition velocity of 8mm s^{-1}.

In Poland the average sulphur deposition has decreased from a maximum of more than 4g S m^{-2} in 1987 to less than 2g S m^{-2} in 1996 (EMEP, 2000) (Figure 1). This trend in overall anthropogenic loading is not recognised at the EMEP station Sniezka, close to the Czerniawka catchment, which was initiated in 1991 (Figure 1). The years 1991 and 1992 were exceptionally dry allowing only 1.4 and 1.3g S m^{-2}yr^{-1} to be wet deposited causing also the S deposition to fluctuate despite the overall decrease in deposition. Normal annual precipitation amount is 1289mm (Kwiatkowski and Holdys, 1985). For most components the mean volume-weighted concentrations in precipitation collected within the Czerniawka catchment are fairly similar to those observed at the Sniezka EMEP-station (Vogt *et al.*, 1994).

Soilwater suction lysimeters were installed at 8 podzolic plots in the catchment in 1991-93. Soil water samples were collected in the snow-free season 1 – 3 times a year and analysed at Dept. of Chemistry, University of Oslo using ion chromatography for anions and atomic absorption spectrophotometer for cations. DOC was estimated from UV absorbance (Edwards and Cresser, 1978).

3. Results and discussion

Although there is no clear trend in S-deposition at the nearby EMEP station during the study period, it is assumed that there must have been a decrease in S-deposition since the late 1980s as in other parts of Poland (Figure 1). It seems likely that this should have resulted in decreasing soil-water sulphate concentrations during the study period. In agreement with this hypothesis, we find a clear reduction in the median concentrations of sulphate in each soil horizon over the study period (Figure 2). Weighted least-squares regression analyses gave an average decrease in median sulphate concentration of about 45µeq L^{-1} yr^{-1} in the O-, B1- and B2 horizons. The trend was significant below the 5% level in O- and B1-, but not in the B2 horizon. The activity of aqueous aluminium ($\{Al^{3+}\}$) tends to decrease during the period, although there are important deviations from this pattern. Due also to a decline in base cation concentrations the molar critical load ratio ($R_{CL}= \{Al^{3+}\}/(\{Ca^{2+}\} + \{Mg^{2+}\})$) shows a less clear decreasing trend (Figure 2). For R_{CL} the regression analyses showed a yearly decrease of 0.14 and 0.19 in the B1- and B2 horizons respectively, though the slope was not significant in the B1 horizon. No trend was found for the O-horizon. The clear reduction in $\{SO_4^{2-}\}$ along with less clear trends for base cations (BC=$\Sigma Ca^{2+}, Mg^{2+}, Na^+, K^+$) causes significant (at the 1% level) increasing trends in the calculated ANC in the O-, B1- and B2 horizons (84, 73 and 100µeq L^{-1} yr^{-1} respectively; Figure 2). In the O-horizon the ANC increase is partly attributed to increase in weak organic acidity as inferred by an increase in DOC and pH (0.9 mg C L^{-1} and -0.15µM H^+ yr^{-1}). In the B-horizons the ANC increase is mainly caused by a reduction in the charge of inorganic labile aluminium.

Although the general trends in soil-water concentrations seem to reflect decreased deposition in the period before the investigation started, there are a number of confounding factors that need to be discussed. Conditions as temperature and wetness vary greatly for the sampling periods. Probably most important is the water flow in the soil during sampling resulting in what we might call a dilution factor. The amounts of precipitation occurring during one week, two weeks and one month before sampling were calculated. This showed that the conditions were particularly wet at the time of sampling in the fall 94 and 95 in particular if the precipitation during the previous month is considered. This is probably the main reason for the especially low sulphate concentrations

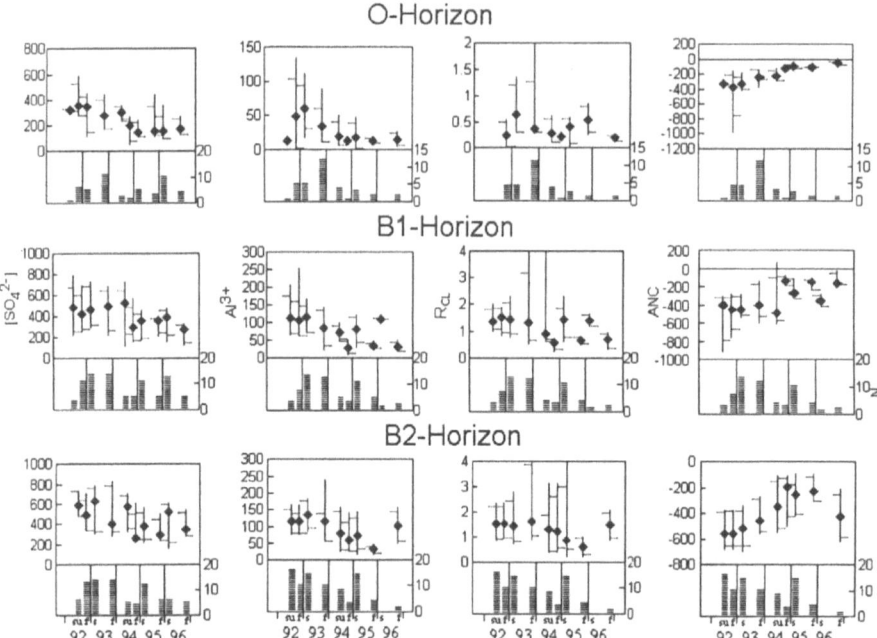

Figure 2. Concentrations (μeq L^{-1}) of sulphate and Al^{3+}, critical load ratio (R$_{CL}$) and the acid neutralising capacity (ANC) for the main soil horizons. Median (♦), 10 and 90 percentiles (horizontal lines) and range (vertical lines) are given. Lower part of each panel shows the number of samples. s, su and fa denote spring, summer and fall respectively.

and the generally favourable water quality in these samples. However, in the fall of 93 and 96 the conditions were fairly similar. Figure 2 shows that the soil water quality is considerably better in the fall of 96 than in 93, which may be an indication of beginning recovery due to reduced emissions.

It is not possible to distinguish the effect of the overall dilution relative to the previous reduction in sulphuric acid. This is mainly due to the fact that sulphur is by far the dominating anionic charge so that a reduction in sulphate gives rise to a corresponding reduction in cationic charge, causing an overall decrease in the ionic strength (i.e. apparent dilution).

To analyse the observations further, we applied the multivariate method of Principal Component Analysis (PCA) (see Minitab, 2000). In addition to the concentrations (in μeq L^{-1}) of major chemical components in the soil water samples in each horizon, we introduced *time* (i.e. Julian days) and precipitation amount during the month prior to sampling (*wetness*) as variables. Plots showing scores and loadings along the two first principle axes, PC1 and PC2, are presented in Figure 3. Note that *wetness* and *time* always are found to be very close in these plots, illustrating the problems in separating these two factors. Also sulphate and chloride show very similar loadings. If changes in chloride concentrations are mainly caused by dilution, this is likely to be the case also for sulphate. Along PC1, describing a striking 40-42% of the total data variation in the different soil horizons, all chemical parameters are negatively correlated to time and wetness. Sulphate and chloride are always among the parameters with loading furthest from *time* and *wetness* (i.e. the largest relative decrease), followed closely by inorganic aluminium equivalents in the mineral soil horizons. The strong influence on PC1 by SO_4^{2-}, Cl^-, *time* and *wetness* coupled with the fact that S-loading has decreased during the last decades and that precipitation (i.e. *wetness*) has increased over the study period (i.e. *time*), lead us to designate this component as mainly a recovery and dilution component. DOC and H^+ have low PC1 loadings in all horizons (i.e. the least relative decrease over time) due to enhanced mobilisation of DOC and, at least partly, by increased de-protonation of natural organic acids.

Along the second principal component (PC2), describing an additional 13-19% of the total variation in the soil horizons, the individual sampling plots are found to get similar scores. This component is therefore believed to express a *spatial variation* within the same genetic soil horizons. Organic related parameters, such as DOC and organically bound aluminium (Alo), on the one side and mineralogical parameters

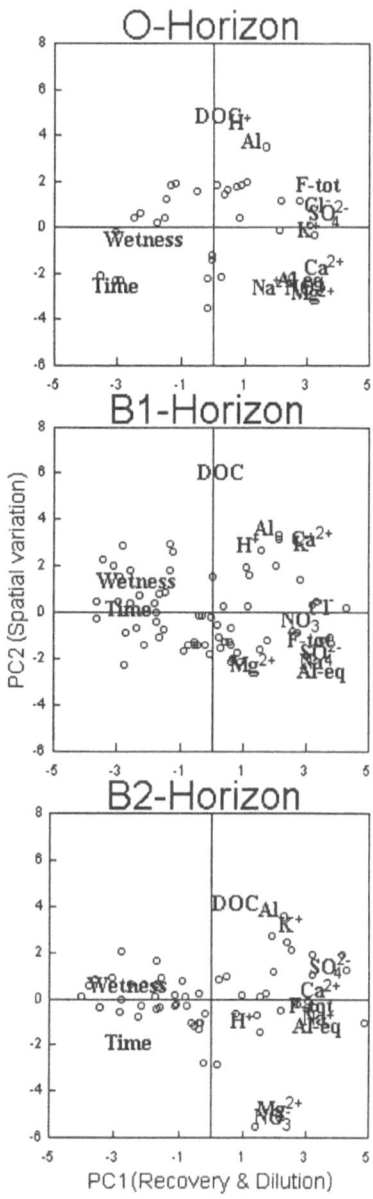

Figure 3. The 1'st and 2'nd principal components in podzol soilwater. Al-eq denotes total µeq L^{-1} aluminium specie charge

as magnesium or inorganic aluminium (Al-eq) on the other have the largest loading along the PC2. In the B-horizons the highest correlation between a measured soil chemical parameter and the PC2 scores was found for the relative amount of organic carbon (% C) (0.337 & 0.353 in the B1 & B2 horizons respectively).

4. Conclusions

There are indications of improved soil-water quality at the site Czerniawka in west Poland from 1992 - 1996. However, this is probably to a large extent due to more precipitation during the last part of the period. This should be borne in mind in studies of recovery especially if only two or a few points in time are compared. To discriminate between effects of changed meteorological conditions and effects of reduced emissions longer data series are necessary.

Acknowledgements

The Norwegian Academy of Science and Letters and the Polish Academy of Science financed the fieldwork.

References

Barrett, K., Seland, Ø., Foss, A., Mylona, S., Sandnes, H., Styve, H., and Tarrasón, L.: 1995, *EMEP, European Transboundary acidifying air pollution*, Norwegian Meterological Inst.
Edwards, A.C., and Cresser, M.S.: 1987, *Water Resour. Res.* 21/1, 49
EMEP: 1998, *Transboundery acidification air pollution in Europe. MSC-W status report 1998. Part 2: Numerical addendum.* Norwegian Meteorological Institute, Norway.
EMEP: 2000, 'Poland, Detailed reports per country, Co-operative programme for monitoring and evaluation of the long range transmission of air pollutants in Europe' <http://www.emep.int/index.html> [Accessed February 1 2000].
Godzik, S. and Sienkiewicz, J.: 1990, 'Air pollution and forest health in Central Europe, Poland, Czechoslovakia, and the German Democratic Republic', in W. Grodzinski, E.B. Cowling and A. Breymeyer (eds.),*Ecological risks*.National Academy Press, Washington, D.C.pp.155-170.
ICP Waters: 2000, 'The International Co-operative Programme on Assessment and Monitoring of Rivers and Lakes' <http://www.niva.no/icp%2Dwaters/major%5Fresults.htm> [Accessed February 1 2000].
Kwiatkowski, J. and Holdys, T.: 1985, 'Klimat, Karkonosze Polskie', in A. Jahn (ed.) *Akademia Nauk, Oddzial we Wrozlawiu*, Kakonoskie Towarzystwo Naukowe w Jeleniej Gorze.
Mazurski, K.R.: 1986, *For. Ecol. Manage.*, 17, 303.
Minitab: 2000, 'Minitab', <http://www.minitab.com/> [Accessed February 1 2000].
Skjelkvåle, B. L., Andersen, T., Halvorsen, G. A., Raddum, G.G., Heegaard, E., Stoddard, J. L., and Wright, R. F.: 2000, *The 12-year report; Acidification of Surface Water in Europe and North America; Trends, biological recovery and heavy metals.* NIVA-Report SNO 4208/2000, ICP Waters report 52/2000. ISBN 82-577-3827-1, 115 pp.
Skotte, G.: 1995, Cand. Scient Thesis, University of Oslo, Norway.
Stoddard, J. L., Jeffries, D. S., Lükewille, A., Clair, T. A., Dillon, P.J., Driscoll, C. T., Forsius, M., Johannessen, M., Kahl, J. S., Kellogg, J.H., Kemp, A., Mannio, J., Monteith, D., Murdoch, P. S., Patrick, S.,Rebsdorf, A., Skjelkvåle, B. L., Stainton, M. P., Traaen, T. S., van Dam, H., Webster, K. E., Wieting, J., and Wilander, A.: 1999, *Nature,* **401**, 575.
Vogt, R.D.: 1996, Doctor of Scient Dissertation. University of Oslo, Norway.
Vogt, R.D., Godzik, S., Kotowski, M., Niklinska, M., Pawlowski, L., Seip, H.M., Sienkiewicz, J., Skotte, G., Staszewski, T., Szarek, G., Tyszka, J. and Aagaard, P.: 1994, *J. Ecol. Chem.*, 3, 325.

SIMULATED ACID RAIN LEACHING CHARACTERISTICS OF ACID SOIL AMENDED WITH BIO-BRIQUETTE COMBUSTION ASH

KAZUHIKO SAKAMOTO*, YUGO ISOBE, XUHUI DONG
and SHIDONG GAO

Graduate School of Science and Engineering, Saitama University, 255 Shimo-ohkubo, Urawa, Saitama, 338-8570, Japan
(* author for correspondence, e-mail: sakakazu@kan.engjm.saitama-u.ac.jp)

Abstract. In Chongqing City, the rapid growth of the economy has accompanied an increase of sulfur dioxide emissions from coal combustion, bringing about an expansion of acid rain affected areas and acidification of soil. Recently, we reported that coal-biomass briquettes so-called bio-briquettes (BB), which are produced from pulverized raw coal, biomass, and a sulfur fixation agent (Ca(OH)$_2$) under 3 to 5 tons cm^{-2} pressure, have high sulfur-fixation efficiency. The BB ash contains nutritive substances such as Ca and Mg, and has a large acid-neutralizing capacity. Thus, in order to improve the acid soil in the Chongqing area, we analyzed the chemical composition of the original acid soil and the ash-amended soil, and investigated their leaching characteristics under simulated acid rain (SAR). It was found that plants and crops in Chongqing area would be injured if the present acid rain continues. We carried out a SAR leaching experiment and studeid the potential toxic effects of leachate from soil containing added ash. The results indicated that the contents of most toxic elements, with the sole exception of chromium, were below the environment standard for irrigation water. Because the BB ash was highly alkaline, the leaching aluminum (Al) species would be hydroxide rather than free Al^{3+} ion.

Keywords: coal-biomass-briquette combustion ash, simulated acid rain, acid soil, metal element

1. Introduction

With rapid development of the Chinese economy, energy consumption has been increasing. Because more than 75 % of the energy in China is obtained from coal, SO$_2$ and dust discharged from coal combustion have also been increasing, and the areas of acid rain precipitation and the acidification of soils caused by SO$_2$ has been expanding. Since 1996, a coal-biomass briquette has been used in Chongqing. This coal-biomass briquette called the bio-briquette (BB) is made by a technique that mixes coal (70 %–85 %, w/w), biomass (30 %–15 %, w/w) and a sulfur fixation agent Ca(OH)$_2$ (Ca/S = 2, equivalence ratio) under a pressure of 3 to 5 ton cm^{-2}. A high degree of sulfur fixation (70 %–90 %) is obtained by briquetting low-grade coal with biomass and slaked lime, so it seems to be an effective control technique to reduce SO$_2$ (Wang *et al.*, 1999).

On the other hand, disposal of coal ash is a serious problem. Most coal ash is used in cement and construction materials (Kikuchi, 1999). However, it is also used for remediation of acidic soil. Addition of coal ash to soil results in a change of soil properties, and it has been reported to increase the grain yield of maize (Kalra *et al.*, 1998). Therefore, utilization of coal combustion ash for amelioration of agricultural soil has prospects as a solution to the problem of ash disposal. Moreover, not only can BB ash neutralize acidic soil with its high alkalinity, but it can also supply nutrient elements such as phosphorus (P), nitrogen (N), and potassium (K).

Because coal or bio-briquettes are rich in heavy metals and aluminum, addition of coal or bio-briquette combustion ash may increase the soil contents of heavy metal elements such as copper (Cu), lead (Pb), zinc (Zn), chromium (Cr), cadmium (Cd), and aluminum (Al). Release of Al is an important acid-buffering process in acid soils: it removes H^+ but increases Al concentrations in soil water. High Al concentrations have toxic effects on both terrestrial and aquatic eco-systems. Soil acidification caused by acid rain with elevated concentrations of dissolved Al is one of several possible explanations for the vegetation damage that is reported in some places in southwestern China (Larssen *et al.*, 1999). Therefore, it is very important to evaluate the impact of these factors on both the environment and agriculture if bio-briquette combustion ash is to be used for amelioration of agricultural soil in areas affected by acid rain.

We are currently investigating construction of a zero-emission cycle (waste-minimization system) using BB ash for amelioration of acidic soil. The object of this paper is to evaluate the acid-neutralizing capacity of BB ash added to acid rain-affected soil, as well as to determine the heavy metal and Al in solution leached by simulated acid rain (SAR).

2. Material and Methods

2.1. SOIL AND BIO-BRIQUETTE COMBUSTION ASH

The acidic soil used in this study was collected in Chongqing, which is a typical acid rain-affected area, in December 1998. The sampling point was Longjingcun located about 20 km from central Chongqing. Soil was collected from 2 soil depths, 0 to 5 cm (topsoil) and 5 to 15 cm (sub-surface soil). It was dried at 60°C for 72 hours, passed through a 2-mm sieve and stored in an air-dried condition.

Raw coal (Songzao coal; total sulfur content: 2.73 %) and wheat straw as biomass was produced in Chongqing. They were air-dried, crushed, and mixed at ratio of coal:biomass (w/w) = 75:25. Bio-briquettes from this pulverized coal and biomass, plus a sulfur fixation agent $Ca(OH)_2$ (Ca/S = 2, equivalence ratio) were made applying 4 ton cm^{-2} with a high-pressure jack. The BBs were burned in an electric furnace (ISO 1171: 1981). BB ash was stored in an air-dried condition.

2.2. SIMULATED ACID RAIN

The SAR imitated the conditions of acid precipitation in Chongqing from 1991 to 1995. Table I shows the average pH, ion concentration, and amount of rainfall over

TABLE I

Chemical composition of rainfall in Chongqing in 1991~1995

pH	Rainfall [mm]	Ca^{2+}	Mg^{2+}	NH_4^+	K^+	Na^+	SO_4^{2-}	NO_3^-	Cl^-	F
					[µeq L^{-1}]					
4.53	900	7.47	1.33	3.95	1.10	0.92	23.42	4.65	1.93	0.95

these 5 years (Gao, 1998). Accordingly, the SAR was adjusted to pH4.5 with a ratio of $SO_4^{2-}:NO_3^-:Cl^- = 25:5:2$ (equivalence ratio).

2.3. ELEMENTAL DETERMINATION

To determine the total metal contents of the soil and BB ash, 0.2-g of samples were digested with a mixture of 5 ml of 46 % HF, 5 ml of 61 % HNO$_3$, and 4 ml of ca. 46 % HClO$_4$. The treated soil or BB ash was dissolved and diluted with 50 ml of 5 % HNO$_3$. The metal concentrations in the solutions were determined using on inductively coupled argon plasma emission spectrometer (ICP: JICP-1000UV, JEOL).

2.4. SAR ADDITION TO THE ASH-AMENDED SOIL

The columns for the leaching experiment were made as follows: Column I, 16 g of soil column; Column II, 16 g of soil plus BB ash (5 %, w/w); Column III, 16 g of soil plus BB ash (10 %, w/w). Each column was filled with topsoil:sub-surface soil = 1:3 (topsoil depth was 5 cm, sub-surface soil depth was 15 cm), as shown in Figure 1; the BB ash added to columns II and III was mixed into the topsoil layer only. SAR was added at 120-mm h^{-1}, the leachate was collected at the stated periods. The pH of the leachate was measured (pH meter: 716 S Titrino, Metrohm Ltd.) and metal ion concentrations were determined using ICP.

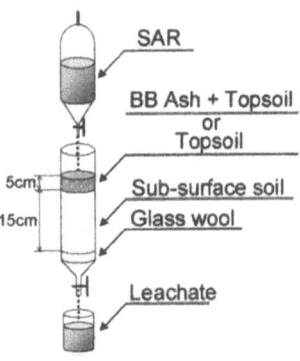

Figure 1. SAR dropping system.

3. Results and Discussion

3.1. METAL CONTENTS AND pH OF SOIL AND BIO-BRIQUETTE COMBUSTION ASH

Table II shows pH and metal contents of BB (dry base), BB ash, topsoil, and sub-surface soil. The pH was measured after sample had been mixed with ion-exchanged water (1:5, w/w) and shaken for 30-min at room temperature (*ca.* 25 °C). The soil used in this study were acidic; the topsoil pH was 4.19 and the

TABLE II
Amount of metal in test samples and its pH

	BB (Dry base)	BB Ash	Topsoil (0~5 cm)	Sub-surface soil (5~20 cm)
Ca (%)	6.30	19.04	0.02	0.02
Al (%)	2.77	8.37	5.04	5.17
Fe (%)	1.80	5.44	2.44	3.16
Mg (%)	0.15	0.45	0.27	0.29
Cr (μg)	143.50	434.09	71.47	70.46
Cu (μg)	483.30	1461.99	152.38	149.40
Zn (μg)	22.97	69.47	40.96	53.13
Ni (μg)	23.77	71.90	16.41	17.76
Cd (μg)	ND	ND	0.02	ND
Mn (μg)	68.33	206.69	99.66	113.20
pH (H$_2$O)		12.10	4.19	4.29

sub-surface soil was 4.29. On the other hand, the BB ash was highly alkaline (pH 12.10), because of the surplus sulfur fixation agent, alkali (Ca(OH)$_2$), added to the BB.

In the soil, although the contents of nutrient element (Ca, Mg) were lower than the average values of Chinese soils, Al, Fe, and Mn contents were slightly higher (Dong et al., 1999). On the other hand, Ca content in BB ash was very high, also due to the added Ca(OH)$_2$. Moreover, Mg content in BB ash was higher than that in the top and sub-surface soils. Therefore, BB ash containing a large amount of Ca will have a high buffering ability for acidic soils and may solve the problems caused by the shortage of nutrient elements in such soils.

3.2. CHARACTERISTICS OF LEACHATE OBTAINED BY DROPPING SAR

3.2.1. *Change in pH of leachates from Columns I ~ III*
The changes in pH of leachates from Columns I ~ III are shown in Figure 2. The pH of the initial leachate was almost the same (pH 3.19) in each of the 3 columns. Column I, with no BB ash added to the topsoil, the pH remained acidic. The acid neutralization capacity of this soil is low; therefore, in acid rain-affected areas, the soil used as cropland needs to be ameliorated.

In Columns II and III, the pH of the leachate rose with addition of BB ash. In the column with 5 % BB ash added to the topsoil (Column II), the pH of the leachate increased to 6.5 after addition of SAR equivalent to 5 years rainfall, increased to pH 7.6 after addition of the equivalent to 6 years rainfall, and remained at pH 7.6 until after the addition of the equivalent of 15.2 years rainfall. This result indicated that the

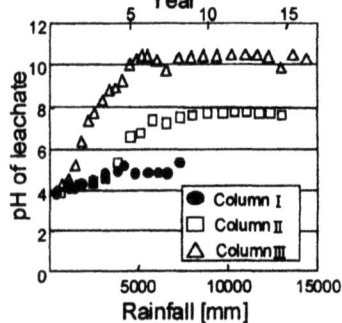

Figure 2. pH change of leachate.

acid soil was neutralized by addition of 5 % BB ash, and could maintain the neutralization ability over a long term. On the other hand, the pH of leachate from the column with 10 % added ash (Column III) became highly alkaline (pH 10.3 after addition of SAR equivalent to 5 years rainfall). A highly alkaline soil is not suitable for cultivation of crops. We think that 10 % ash addition to Chongqing acidic soil is excessive.

3.2.2. Aluminum leaching from Column I ~ III

Al ions (Al^{3+}) are toxic to plant roots and microbes in soils. Figure 3 shows the amount of Al leaching from Column I ~ III. The amount of leached Al increased after addition of BB ash. The Al^{3+} content in soil and BB ash was 0.07 % and 1.44 %, respectively (Al^{3+} was extracted by 1M KCl for 10 hours). Therefore, BB ash can account for the major source of Al in Columns II and III. However, the most toxic Al species exist in acid environments (pH< ca. 5), and cannot

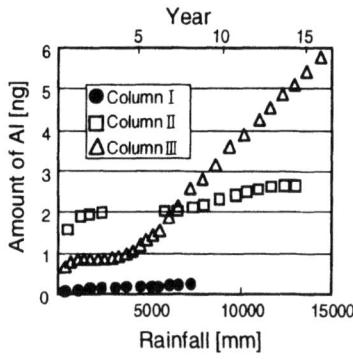

Figure 3. Integrated amounts of Al.

exist in alkaline environments. Al toxicity is reduced when the Ca concentration is very high (Takatsu *et al.*, 1998). Because of the addition of BB ash, the pH of leachate from Columns II and III ranged from neutral to alkaline and Ca concentration was high (Ca/Al >>1) due to excess sulfur fixation agent. So we consider that Al leached from BB ash does not exist as Al^{3+}, but as a hydroxide, and expect that BB ash will not diminish plant growth.

3.2.3. Toxic heavy metal

The effects and mechanism of heavy metal leaching are of significance for a comprehensive assessment of the environmental impact. To examine metals leached from BB ash, a new column was made from a mixture of topsoil and BB ash (95:5, w/w), without sub-surface soil, to avoid the effect of metal absorption by sub-surface soil (the ratio of added BB ash was equivalent to Column II). The concentrations of heavy metals (Cu, Cr, Mn, Cd, Zn) in leachates obtained by adding SAR (equivalent to 5.6 years rainfall) were compared against the Chinese environmental standards for irrigation water. Figure 4 shows that of Cu and Cr. Concentrations of metals, except that of Cr, did not exceed the standard values at an early stage (equivalent to 1.6 years rainfall). Soluble forms (*i.e.*, forms easily adsorbed by plants and accumulated in crops) exist in acidic environments (pH< ca. 5) (Evans *et al.*, 1995); however, heavy metals from the leachate would not be adsorbed by plants because its pH was alkaline. Nevertheless, attention must be paid to groundwater con-tamination risks when adding BB ash.

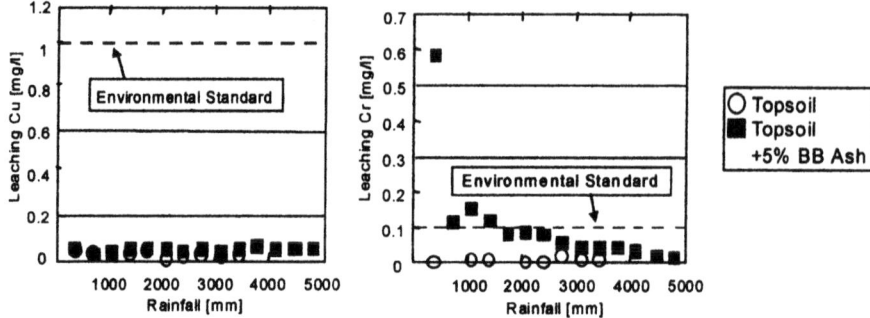

Figure 4. Comparison the concentrations of heavy metal in leachate and their environmental standard values.

4. Conclusion

The results from this study indicated that BB combustion ash has a high acid neutralization ability due to excess sulfur fixation agent (in particular, a high concentration of Ca). With the exception of chromium, aluminum and other heavy metals leached from the ash were believed to not affect the growth of plants or crops because the pH of the leachate ranged from neutral to alkaline.

To comprehensively evaluate the environmental impact, it is necessary to investigate absorption of heavy metals by plants and crops.

Acknowledgements

This study was partly supported by Steel Industry Foundation for the Advancement of Environmental Protection Technology.

References

Dong, X., Sakamoto, K., Zheng, C., Quan, H., Chen, Y. and Wang, W.: 1999, *Journal of Aerosol Research, Japan.* **14**, 171 (In Japanese).
Evans, L. J., Spiers, G. A., Zhao, G.: 1995, *International Journal of Environmental Analytical Chemistry.* **59**, 291
Gao, S.: 1998, Doctoral Dissertation, Saitama University (In Japanese).
Kalra, N., Jain, M. C., Joshi, H. C., Choudhary, R., Harti, R. C., Vatsa, B. K., Sharma, S. K and Kumar, V.: 1998, *Bioresource Technology.* **64**, 163.
Kikuchi, R.: 1999, *Resources, Conservation and Recycling.* **27**, 333
Larssen, T, Vogt, R. D., Seip, H. M., Furuberg, G., Liao, B., Xiao, J., Xiong, J.: 1999, *Geoderma* **91**, 65.
Takatsu, A., Tsunoda, K., Yoshimura, E.: 1998, *Bunseki* **10**, 772 (In Japanese).
Wang, J., Gao, S., Wang, W. and Sakamoto, K.: 1999, *Journal of Aerosol Research, Japan..* **14**, 162.

UN ECE ICP MATERIALS: DOSE-RESPONSE FUNCTIONS ON DRY AND WET ACID DEPOSITION EFFECTS AFTER 8 YEARS OF EXPOSURE

JOHAN TIDBLAD[1*], VLADIMIR KUCERA[1], ALEXANDRE A. MIKHAILOV[1,2], JAN HENRIKSEN[3], KATERINA KREISLOVA[4], TIM YATES[5], BRUNO STÖCKLE[6], and MANFRED SCHREINER[7]

[1]*Swedish Corrosion Institute, Roslagsvaegen 101, SE-104 05, Stockholm, Sweden;*
[2]*Permanent address: Institute of Physical Chemistry, Russian Academy of Sciences, Leninskij Pr. 31, 117915 Moscow, Russian Federation;*
[3]*Norwegian Institute for Air Research, P.O. Box 100, N-2007 Kjeller, Norway;*
[4]*SVUOM Praha a. s., U Mestanského Pivovaru 4, CZ-17004 PRAHA 7, Czech Republic;*
[5]*Building Research Establishment Ltd., Garston, Watford WD2 7JR, U. K.;*
[6]*Bavarian State Conservation Office, Hofgraben 4, D-80539 Münich, Germany;*
[7]*Institute of Chemistry, Academy of Fine Arts, Schillerplatz 3, A-1010 Wien, Austria;*
(*author for correspondence, e-mail: jt@corr-institute.se)

Abstract. The main results of the International Co-operative Programme on Effects on Materials, including Historic and Cultural Monuments (ICP Materials) within the United Nations Economic Commission for Europe (UN ECE) are summarised. The 8-year field exposure programme involves 39 test sites in 12 European countries and in the United States and Canada. Dose-response functions (DRF) expressing the effect of dry and wet deposition as individual terms have been obtained for a wide range of materials including bronze, copper, weathering steel, zinc, aluminium, nickel, tin, stone materials, paint coatings and glass materials. The DRF's includes parameters that are easily available on different geographical scales and can be used for mapping areas of increased corrosion rates and for calculation of costs.

Keywords: acid deposition, materials damage, NO_2, O_3, relative humidity, SO_2, temperature

1. Introduction

ICP Materials is one of several effect oriented International Co-operative Programmes (ICPs) within the United Nations Economic Commission for Europe (UN ECE). Early in the discussions on the Convention on Long-range Transboundary Air Pollution (CLRTAP) it was recognised that a good understanding of the harmful effects of air pollution was a prerequisite for reaching agreement on effective pollution control. Consequently an extensive field exposure programme was started in September 1987. It involved 39 exposure sites in 12 European countries and in the United States and Canada. A task Force is organising the programme (ICP Materials) with Sweden as lead country and the Swedish Corrosion Institute serving as the main research centre. Sub-centres in the Czech Republic, Germany, Norway, United Kingdom, Sweden and Austria have been responsible for evaluation of individual groups of materials including structural metals, stone materials, paint coatings, electric contact materials, glass and polymer materials. The aim of the programme was to perform a quantitative evaluation of the effects of sulphur pollutants in combination with NO_x and other pollutants as well as climatic parameters on the atmospheric corrosion of important materials. This was achieved by measuring gaseous pollutants, precipitation and climatic parameters at or nearby each test site and by evaluating the corrosion effects on the materials.

ICP Materials is an on-going research activity. A finalised part, however, is the extensive 8-year field exposure programme that was started in September 1987 and the results presented here are based on this 8-year programme. For further details on this programme see Tidblad *et al.* 1997.

2. Methods

2.1. SELECTION OF ENVIRONMENTAL PARAMETERS

The measured environmental data includes climatic parameters (temperature, relative humidity, time of wetness, sunshine hours and sunshine radiation), gaseous pollutants (SO_2, NO_2 and O_3) and precipitation (amount, conductivity and concentration of the ions H^+, SO_4^{2-}, NO_3^-, Cl^-, NH_4^+, Na^+, Ca^{2+}, Mg^{2+} and K^+). The main aim of the present statistical evaluation was to estimate dose-response functions suitable for mapping of areas with increased corrosion rates and for calculation of costs. This poses restriction on the parameters that can be used, as they need to be relatively easy to obtain on different geographical scales.

Table 1 shows all parameters included in the final dose-response relations. Care shall be taken when extrapolating the equations outside the intervals of environmental parameters used for their calculations. This is especially true for the [Cl^-] since this parameter is used in the equations as a substitute for the total chloride deposition, which was not measured in the programme. In the selection procedure, test sites with high deposition of chlorides were as much as possible excluded in order to facilitate the quantification of acid deposition effects.

TABLE I

Environmental parameters used in final dose-response functions including symbol, description, interval measured in the programme and unit. All parameters are expressed as annual averages.

Symbol	Description	Interval	Unit
t	Time	1-8	year
T	Temperature	2-19	°C
Rh	Relative humidity	56-86	%
[SO_2]	SO_2 concentration	1-83	µg/m³
[O_3]	O_3 concentration	14-82	µg/m³
Rain	Amount of rainfall	327-2144	mm
[H^+]	H^+ concentration	0.0006-0.13	mg/l
[Cl^-]	Cl^- concentration	0.1-12	mg/l

2.2. SEPARATION OF DRY AND WET DEPOSITION EFFECTS

For unsheltered positions the materials damage is usually discussed in terms of dry and wet deposition of pollutants. Wet deposition includes transport by means of precipitation and dry deposition transport by any other process. One important task for the programme has been to estimate the relative contribution of dry and wet deposition to the degradation of materials. Therefore, and also because it makes sense from a mechanistic point of view, the dose-response relations are all expressed on the general form

$$K = f_{dry}(T, Rh, [SO_2], [O_3]) \cdot t^k + f_{wet}(Rain, [H^+]) \cdot t^m \quad (1)$$

where K is the corrosion attack, f_{dry} is the dry deposition term, f_{wet} is the wet deposition term and k and m are estimated constants. For explanation of the other environmental parameters see table I.

The functional form used for the parameters are all simple linear, exponential or power functions with the exception of temperature, which will be explained in greater detail. The T function is more complicated and is illustrated in figure 1. A maximum is observed at about 9-11 °C. The increasing part (a) can be related to the increased time of wetness. The decreasing part (b) is attributed to a faster evaporation of moisture layers e.g. after rain or dew periods and a surface temperature above the ambient temperature due to sun radiation which result in a decrease of the surface time of wetness.

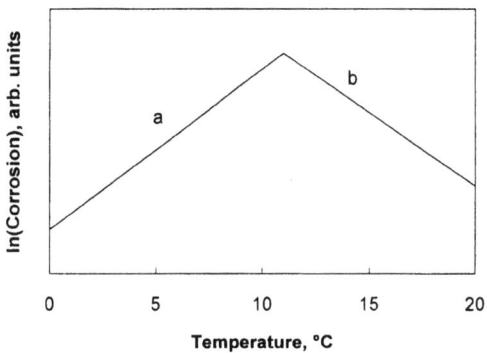

Figure 1. Schematic representation of the observed temperature dependence for many materials: a) increase of corrosion with temperature in the low temperature range and b) decrease of corrosion with temperature in the high temperature range.

3. Results and discussion

3.1. DOSE-RESPONSE FUNCTIONS FOR UNSHELTERED MATERIALS

A list of all dose-response functions, including temperature functions, for exposure of unsheltered materials is given in table II. In the following the individual groups of materials are discussed separately.

In addition functions have been obtained for weathering steel, zinc, copper, bronze, nickel, tin and glass M1 representative of medieval stained glass windows exposed in sheltered positions (Tidblad *et al.* 1997).

TABLE II

List of dose-response functions, including temperature function, for unsheltered materials. The corrosion attack is expressed as mass loss (ML in g/m^2) for metals, surface recession (R in µm) for stone materials, ASTM D 1150-55 rankings (1 to 10 where 10 means a fresh sample and 1 a completely degraded) for paint coatings or depth of leached layer (LL in nm) for glass. Abbreviations of environmental parameters were given in table I.

Material (N=number of observations, R^2 = explained variability)
Dose-response function
Temperature function

Weathering steel (N=148, R^2=0.68)
ML = 34[SO$_2$]$^{0.33}$exp{0.020Rh + f$_{Ws}$(T)}t$^{0.33}$
$f_{Ws}(T) = 0.059(T-10)$ when T≤10°C, $-0.036(T-10)$ otherwise

Zinc (N=98, R^2=0.84)
ML = 1.4[SO$_2$]$^{0.22}$exp{0.018Rh + f$_{Zn}$(T)}t$^{0.85}$ + 0.029Rain[H$^+$]t
$f_{Zn}(T) = 0.062(T-10)$ when T≤10°C, $-0.021(T-10)$ otherwise

Aluminium (N=106, R^2=0.74)
ML = 0.0021[SO$_2$]$^{0.23}$Rh·exp{f$_{Al}$(T)}t$^{1.2}$ + 0.000023Rain[Cl$^-$]t
$f_{Al}(T) = 0.031(T-10)$ when T≤10°C, $-0.061(T-10)$ otherwise

Copper (N=95, R^2=0.73)
ML = 0.0027[SO$_2$]$^{0.32}$[O$_3$]$^{0.79}$Rh·exp{f$_{Cu}$(T)}t$^{0.78}$ + 0.050Rain[H$^+$]t$^{0.89}$
$f_{Cu}(T) = 0.083(T-10)$ when T≤10°C, $-0.032(T-10)$ otherwise

Cast Bronze (N=144, R^2=0.81)
ML = 0.026[SO$_2$]$^{0.44}$Rh·exp{f$_{Br}$(T)}t$^{0.86}$ + 0.029Rain[H$^+$]t$^{0.76}$ + 0.00043Rain[Cl$^-$]t$^{0.76}$
$f_{Br}(T) = 0.060(T-11)$ when T≤11°C, $-0.067(T-11)$ otherwise

Portland limestone (N=100, R^2=0.88)
R = 2.7[SO$_2$]$^{0.48}$exp{f$_{Pl}$(T)}t$^{0.96}$ + 0.019Rain[H$^+$]t$^{0.96}$
$f_{Pl}(T) = -0.018T$

White Mansfield sandstone (N=101, R^2=0.86)
R = 2.0[SO$_2$]$^{0.52}$exp{f$_{Ms}$(T)}t$^{0.91}$ + 0.028Rain[H$^+$]t$^{0.91}$
$f_{Ms}(T) = 0$ when T≤10°C, $-0.013(T-10)$ otherwise

Coil coated galvanised steel with alkyd melamine (N=138, R^2=0.73)
(10-ASTM) = (0.0084[SO$_2$] + 0.015Rh + f$_{Cc}$(T))t$^{0.43}$ + 0.00082Rain·t$^{0.43}$
$f_{Cc}(T) = 0.040(T-10)$ when T≤10°C, $-0.064(T-10)$ otherwise

Steel panels with alkyd (N=139, R^2=0.68)
(10-ASTM) = (0.033[SO$_2$] + 0.013Rh + f$_{Sp}$(T))t$^{0.41}$ + 0.0013Rain[H$^+$]t$^{0.41}$
$f_{Sp}(T) = 0.015(T-11)$ when T≤11°C, $-0.15(T-11)$ otherwise

Glass M1 representative of medieval stained glass windows (N=46, R^2=0.56)
LL = 0.013[SO$_2$]$^{0.49}$Rh$^{2.8}$t

3.2. STRUCTURAL METALS

The equations show that for most metals dry deposition of SO$_2$ is the most important pollutant parameter except for copper where O$_3$ is also important. The exponent is generally lower than unity, i.e., the relations are non-linear with respect to SO$_2$. Wet deposition (Rain[H$^+$]) is generally important for unsheltered conditions except for weathering steel and aluminium. For aluminium there is an effect of chlorides which also applies for bronze.

The effect of wet deposition on metallic materials has traditionally been regarded as minor compared to the effect of dry deposition. However, as is illustrated in figure 2 for copper, the effect depends strongly on pH and amount of precipitation. At about pH 4.5 and lower the effect can not be neglected. Also, a conclusion from the temperature dependence of the dry deposition term (see figure 1) is that the relative importance of wet deposition is higher in warmer areas.

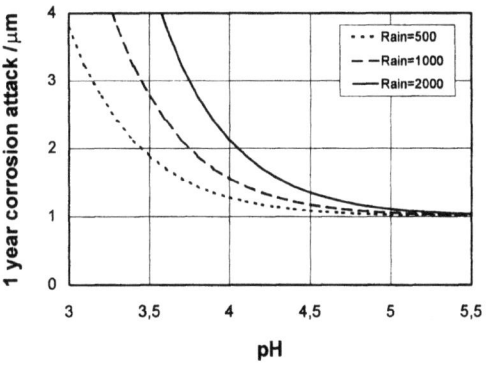

Figure 2 Corrosion attack of unsheltered copper vs. pH of precipitation calculated from the dose-response function given in table II and assuming a constant dry deposition term, $0.0027[SO_2]^{0.32}[O_3]^{0.79}Rh \cdot e^{f(T)}$, corresponding to a corrosion attack of 1.0 µm. Curves are shown for the precipitation (Rain) values of 500, 1000 and 2000 mm.

3.3. STONE MATERIALS

For both unsheltered sandstone and limestone equations separating the effect of dry and wet deposition have been obtained with a fairly high degree of explained variability. In the final equations, values from the first year of exposure have not been used due to specific behaviour at short times. The estimated temperature dependence for limestone is negative throughout the entire temperature region, in contrast to the temperature effect for metals. This may be due to the correlation between average temperature and the number of frost-thaw cycles. For sandstone the temperature effect in the low temperature region is insignificant which may be due to its higher resistance to frost-thaw cycles.

3.4. PAINT COATINGS

For both alkyd painted steel and coil coated galvanised steel with alkyd melamine dose-response functions have been obtained including SO_2 and rain parameters. ASTM rankings from 1 to 10 of the damage from cut were used where 10 means a fresh sample and 1 a completely degraded. The ASTM values represent a non-linear scale of degradation, which makes the interpretation difficult. However, the given dose-response functions can be transferred into lifetime equations using a generally accepted maintenance criterion that ASTM=5 when t=lifetime.

3.5. GLASS MATERIALS

The glass materials are representative for medieval glass windows. For unsheltered positions temperature was not significant and a time dependence could be obtained. It should be stressed, however, that the estimation of the time dependence only includes two exposure times and is thus uncertain.

4. Conclusions

Dose-response relations suitable for mapping and calculations of cost of corrosion damage have been developed and are the main result of the UN ECE exposure programme. Functions have been obtained for materials exposed in unsheltered position including weathering steel, zinc, aluminium, copper, bronze, limestone, sandstone, two paint coatings and a glass representative of medieval stained glass windows. The equations show that for most materials dry deposition of SO_2 is the most important pollutant parameter except for copper where O_3 is also important.

The effect of dry and wet deposition has been expressed as additive terms in the dose-response relations, which makes it possible to separate their individual contributions to the total corrosion attack. Time has been included in the dose-response functions, which makes it possible to calculate corrosion rates at a selected exposure time or to calculate a lifetime, given a specified critical corrosion rate.

Acknowledgements

This exposure programme is the result of co-operation between the organisations listed below. Each was responsible for gathering meteorological and pollution data, and for providing sites for the exposure of materials:
Institute of Chemistry, Academy of Fine Arts, Austria; SVUOM Praha a.s., Czech Republic; Technical Research Centre of Finland (VTT); Bavarian State Conservation Office, Germany; Agency for Energy Sources (ENEA), Italy; Institute of Environmental Sciences (TNO-MEP), the Netherlands; Norwegian Institute for Air Research (NILU); Swedish Corrosion Institute (SCI); Building Research Establishment (BRE), Department of Environment, United Kingdom; Ministerio de Fomento, Spain; Institute of Physical Chemistry, Russian Academy of Sciences, Russian Federation; Processing Centre, Ministry of Environment, Estonia; Institute of Technology, Laboratory of Mineralogy and Petrology (IST), Portugal; National Research Council of Canada and the Ministries of the Environment of Canada and of Ontario; and United States Environmental Protection Agency.

References

ASTM D 1150-55: 1987, *American Society for Testing and Materials*, Philadelphia.
Tidblad J., Kucera V. and Mikhailov A. A.: 1997, Report No 30. Statistical analysis of 8 year materials exposure and acceptable deterioration and pollution levels. *Swedish Corrosion Institute*, Stockholm, Sweden.

REGIONAL CHARACTERISTICS OF COPPER CORROSION COMPONENTS IN EAST ASIA

M. KITASE[1*], S. HATAKEYAMA[2], T. MIZOGUCHI[3] and Y. MAEDA[4]

[1] *Nagoya City Environmental Science Research Institute, 16-8 5-chome Toyoda, Minami-ku, Nagoya 457-0841 Japan;* [2] *National Institute for Environmental Studies, 16-2 ,Onogawa, Tsukuba 305-0053 Japan;* [3] *Bukkyo University, 96 Murasakino-kitanakabo-cyo, Kita-ku, Kyoto 603-8301 Japan;* [4] *Osaka Prefecture University, Gakuen-cyo, Sakai 599-8531 Japan*
(* author for correspondence, e-mail: kankaken@cjn.or.jp)

Abstract. Copper plates were exposed under shelters at 13 sampling sites in East Asia and their corrosion was analyzed. The corrosion products were first dissolved in water and then oxalic acid. Sulfate, nitrate and chloride in the solutions were measured by ion chromatography. The amounts of the three anions significantly differed depending on the atmospheric environment at the sites. Sulfate was a major part of the anions at Chongqing and Shanghai in China. Especially, at the urban sites in Japan, nitrate remarkably changed with the seasons, and often became the large anionic component in the summer. The amounts of chloride at most sites were at higher concentration levels compared with those at the rural sites in Japan. The anions in the copper corrosion must mainly reflect the impact of acid deposition.

Keywords: acid deposition , chloride, copper corrosion, East Asia, sulfate, nitrate

1. Introduction

The atmospheric corrosion of metals is an important industrial and environmental problem. Copper has been used as the material of structures and cultural properties, but they have received significant damage due to atmospheric pollution. Investigations on the corrosion rate of copper have been carried out based on interaction with the effects of air pollutants. In the meantime, the components of the copper corrosion were identified by X-ray diffraction (Tsujino *et al.*, 1999) and anions in the corrosion products were measured by ion chromatography (Katou and Akiyama, 1983). These results provided useful data when evaluating the effects of air pollutants on corrosion.

We have exposed copper plates under rain-sheltered conditions at 13 sites in China, South Korea and Japan where the extents of air pollution widely differ. The components of these samples were retained without being removed by rain, and they may clearly show the effects of air pollutants, particularly those of the acidic species. In this study, the copper corrosion products were dissolved in water and then in oxalic acid, and the components were measured by ion chromatography. We report the regional characteristics and seasonal changes in the copper corrosion components.

2. Materials and Methods

2.1. TEST PIECES

The test pieces were $0.4 \times 30 \times 40$ mm copper plates which were burnished by #400 grit paper. Before use, they were cleaned with acetone in an ultrasonic bath.

2.2. SAMPLING SITES AND PERIODS

The sampling sites were: (1) Chongqing and (2) Taiyuan as heavily polluted areas in China, (3) Shanghai as an urban site in China, (4) Taejon as the South Korea site, (5) Osaka Higashinari-ku, (6) Osaka Konohana-ku and (7) Nagoya as urban sites in Japan, (8) Fukuoka, (9) Toyama and (10) Ishikawa as the Japan Sea coastal sites, (11) Kyoto and (12) Ibaraki as rural sites in Japan, and (13) Sapporo as a cold climate site in Japan.

The test pieces, keeping exposure surface vertical, were placed under rain - sheltered conditions from June 1997 to November 1999. Sampling periods were every 3 months for the short-term exposure and from 1 to 2.5 years for the long-term exposure.

2.3. ANALYSIS OF CORROSION PRODUCTS

The test pieces were washed with 25 ml or 50 ml of water in an ultrasonic bath, and the solution was then filtered. The test pieces were then washed with 50 to 200 ml of 3×10^{-3}N oxalic acid in an ultrasonic bath. This treatment was repeated several times until the corrosion products were completely dissolved.

Sulfate (SO_4^{2-}) nitrate (NO_3^-) and chloride (Cl^-) in both the water-solutions and acid-solutions were measured by ion chromatography. In addition, sodium (Na^+), calcium (Ca^{2+}), ammonium, magnesium and potassium in the water-solutions were measured by ion chromatography or atomic absorption spectrometry.

The treated test pieces were weighed, and the corrosion losses were determined.

3. Results and Discussion

3.1. 1-YEAR EXPOSURE

Figure 1 shows the amounts of SO_4^{2-}, NO_3^- and Cl^- in the copper corrosion products during the 1-year exposure at the 13 sampling sites.

The amounts of SO_4^{2-} at Chongqing and Shanghai were significantly greater than those at the other sites, and occupied a major portion of the anions. In contrast, the amounts of SO_4^{2-} at the rural sites in Japan were quite small. These results may indicate that SO_4^{2-} in the corrosion well reflects the concentration of sulfur dioxide (SO_2) in the air.

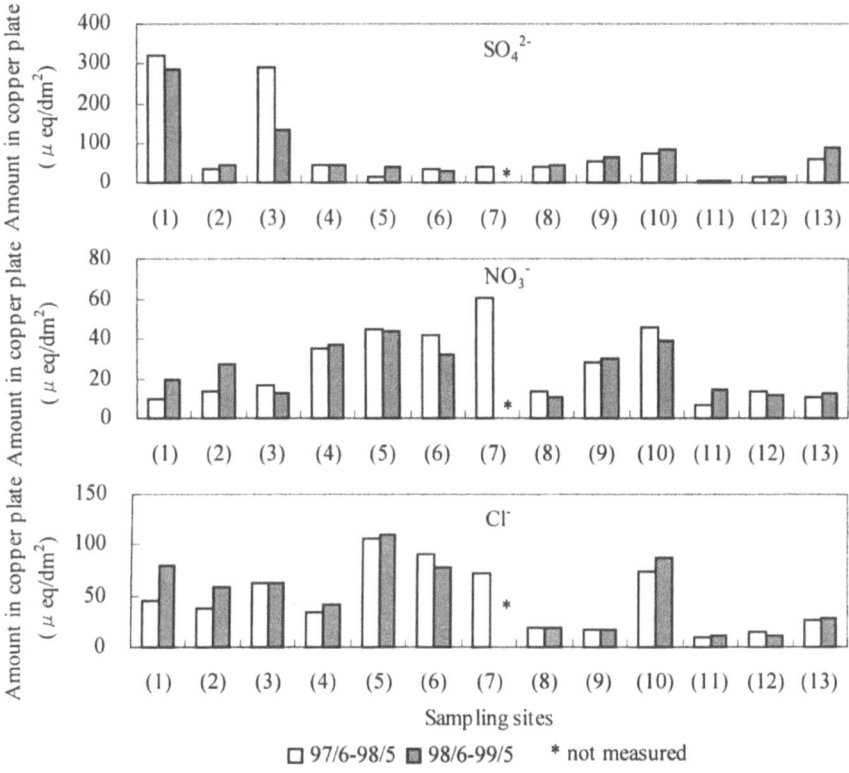

Figure 1. Amounts of anions in copper corrosion products during 1-year exposure. Sampling sites: (1) Chongqing (2) Taiyuan (3) Shanghai (4) Taejon (5) Osaka Higashinari-ku (6) Osaka Konohana-ku (7) Nagoya (8) Fukuoka (9) Ishikawa (10) Toyama (11) Kyoto (12) Ibaraki (13) Sapporo

Compared with the other ions, the amounts of NO_3^- did not differ very widely among the sites. However, they were relatively high at the urban sites in Japan, two Japan Sea coastal sites and Taejon.

The amounts of Cl^- were high at the urban sites in Japan, Toyama and the three sites in China. At any site, the amount of Cl^- cannot be fully explained by Cl^- which is from sea salt. Cl^- in the copper corrosion products will be mostly produced by the other atmospheric species, for example, hydrochloric acid (HCl) from the incineration of waste or the combustion of coal.

Except at Taiyuan and Shanghai, the ratios of total cation to total anion in the corrosions were below 0.2. This shows most anions in the corrosion products are components of the copper salts. These anions will mainly come from the deposition of acidic air pollutants on the copper plates. At Taiyuan and Shanghai the amounts of Ca^{2+} in the corrosion products were particularly high. Anions, which are regarded as calcium salts, must account for a large portion of the anions in the corrosion products at these sites.

Figure 2 shows the mass loss values at each site. The variation in the mass loss among the sites did not fit with that of each anion. These mass losses may indicate the whole effects of all the air pollutants, whereas the anions in the corrosion products may indicate the individual effect of each air pollutant.

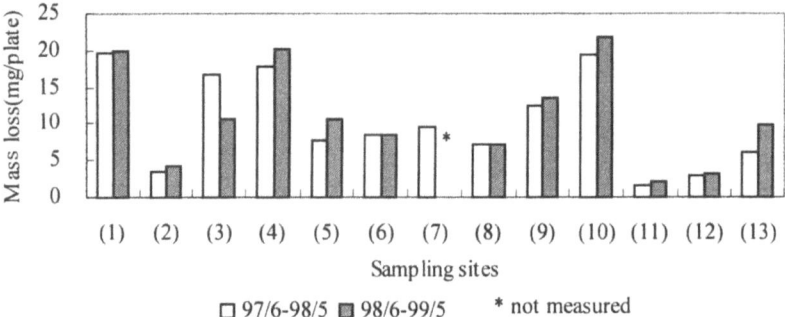

Figure 2. Mass loss of copper after 1-year exposure. Sampling site number means the same site shown in Figure1

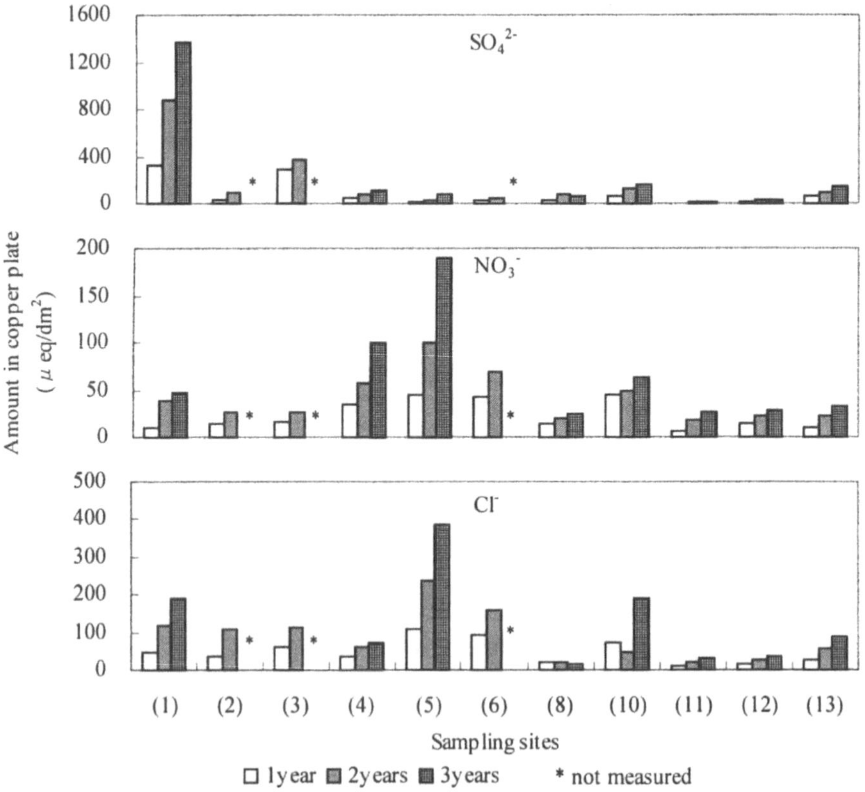

Figure 3. Amounts of anions in copper corrosion products from 1 year to 3 years. Sampling site number means the same site shown in Figure1.

3.2. LONG-TERM EXPOSURE

Figure 3 shows the amounts of SO_4^{2-}, NO_3^- and Cl^- in the copper corrosion exposed for 1 year to 2.5 years at 11 sampling sites. In the figure the value of 3- year exposure shows the converted value using each value of 2.5-year exposure.

At almost all the sites and for almost all the ions, the amounts of ions increased year by year. This indicates that the information on the corrosion components may be useful in evaluating the air pollution over the long-term.

3.3. 3-MONTH EXPOSURE

Figure 4 shows the amounts of SO_4^{2-}, NO_3^- and Cl^- in the copper corrosion products

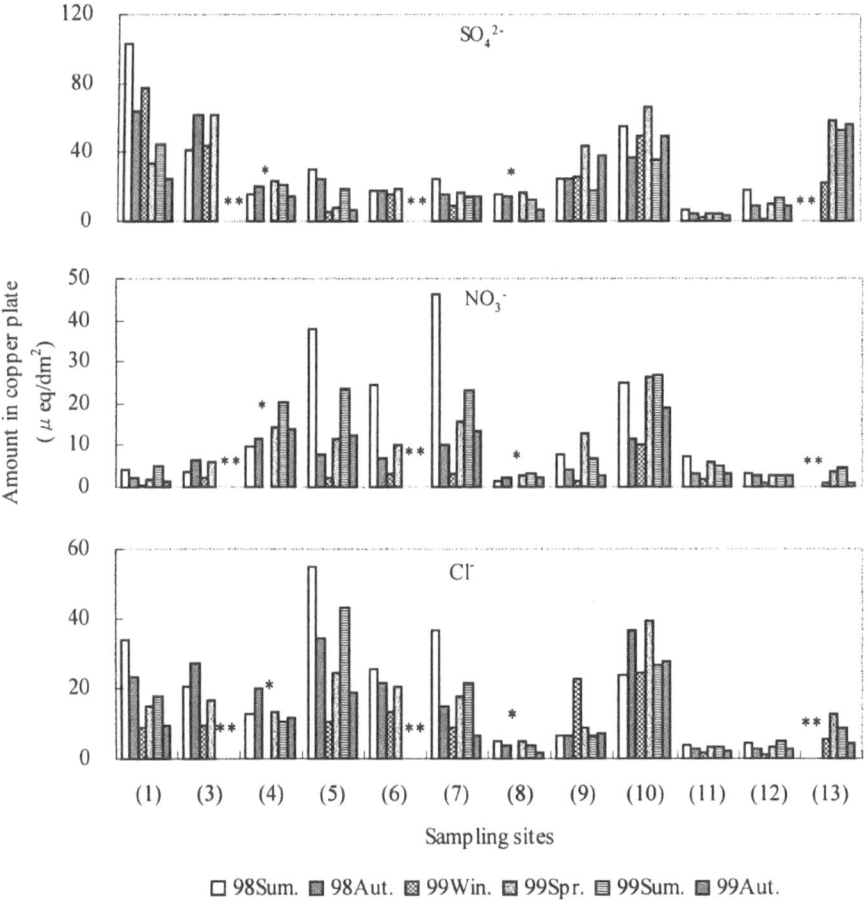

Figure 4. Amounts of anions in copper corrosion during a 3-month exposure. Sampling site number means the same site shown in Figure 1.

during the 3-month exposure.

The variations of these anions among the sites were similar to those for the 1-year exposure shown in Figure 1.

Seasonal changes of SO_4^{2-} were not very high at all the sites. On the other hand, NO_3^- at the urban sites in Japan significantly changed. The large amounts of NO_3^- in the summer probably indicates that the concentration of gaseous nitric acid (HNO_3) became especially high in the summer. The amounts of NO_3^- at Toyama and Taejon were also relatively high, but seasonal changes were not remarkable.

The amounts of Cl^- at all the sites except for two sites in Japan were high in the warm seasons and low in the cold seasons. This result may indicate the change in the gaseous HCl concentration in the air. At Toyama and Ishikawa the seasonal changes in Cl^- were not apparent. At these sites, Cl^-, which originated from sea salt, must not be ignored in the cold seasons.

4. Conclusions

Copper corrosion components showed regional variations and seasonal changes. The amounts of the anions must mainly reflect the impact of acid deposition. Therefore, anions in the copper corrosion products will be helpful in estimating the circumstances in which corrosion was formed during the short-term exposure or long-term exposure.

To evaluate the definite relation between anions in the corrosion products and acidic pollutants in the air, the influence of meteorological factors, for example, temperature and humidity, must be considered.

Acknowledgements

This work was supported by the Global Environment Research Fund of the Japan Environmental Agency and the Japan Fund for Global Environment of Japan Environment Corporation.

References

Katou T. and Akiyama K.: 1983, *Bulletin of the Institute of Environmental Science and Technology Yokohama National University* **9**, 25 (in Japanese)

Tsujino Y., Zhang D.N., Chen S.L., Leung D.Y.C., Lin. S.L., Kim S.T., Yoo Y.E., Hatakeyama S., Mizoguchi T. and Maeda Y.: 1999, *Proceedings of the 2nd International Conference on the Effects of Acid Deposition on Cultural Properties and Materials in East Asia*, pp.55-58

MAPPING OF ACID DEPOSITION EFFECTS AND CALCULATION OF CORROSION COSTS ON ZINC IN CHINA

JOHAN TIDBLAD[1*], VLADIMIR KUCERA[1]
and ALEXANDRE A. MIKHAILOV[1,2]

[1]*Swedish Corrosion Institute, Roslagsvaegen 101, SE-104 05, Stockholm, Sweden;*
[2]*Permanent address: Institute of Physical Chemistry, Russian Academy of Sciences, Leninskij Pr. 31, 117915 Moscow, Russian Federation*
(*author for correspondence, e-mail: jt@corr-institute.se*)

Abstract. Corrosion damage to materials including objects of cultural heritage due to acid deposition has been shown to cause large costs in several studies in Europe and in the United States. So far no similar extensive studies have been performed in developing countries. The World Bank has therefore initiated and financed a study of the corrosion costs in China based on available data in the literature and obtained through contacts and visits to several institutes and organisations in China. An initial assessment of the corrosion costs in China due to acidifying pollutants has been performed using a model originally developed and applied in Europe, which has been adapted to conditions in China. Here, the model is described using zinc as an example. In the calculation of corrosion costs it is assumed that the stock of materials at risk can be allocated to census data, which enables a separate calculation of the cost for each province in China. The significant differences in corrosion attack is illustrated for zinc with a corrosion map of China based on environmental data and a dose-response function adapted for Chinese conditions including the dry and wet acid deposition effects as separate terms.

Keywords: atmospheric corrosion, China, corrosion costs, mapping, zinc

1. Introduction

An initial assessment of the corrosion costs in China due to acidifying pollutants has been performed using a model originally developed and applied in Europe, which has been adapted to conditions in China. The World Bank has initiated and financed the study of the corrosion costs in China, which is based on available data in the literature and obtained through contacts and visits to several institutes and organisations in China. The present work, which deals with zinc and galvanised steel, is only part of this larger study (Tidblad and Kucera 1999).

2. Methods

For assessment of direct costs of corrosion damage caused by air pollutants a model has been developed and used first in three cities in Europe: Stockholm, Sarpsborg and Prague (Kucera et al. 1993), and subsequently also for a rough estimation for whole Europe (Cowell and ApSimon 1996). The model is shown in figure 1 and has been used here, however, with adaptation to specific conditions.

The estimated economic damage can be calculated according to the equation

$$K_a = K \cdot S \cdot (L_p^{-1} - L_c^{-1}) \qquad (1)$$

where K_a is the additional cost for maintenance/replacement, K is the cost per surface area of material, S is the surface area of material, L_p is the maintenance interval (life time) in polluted areas and L_c is the maintenance interval in clean areas.

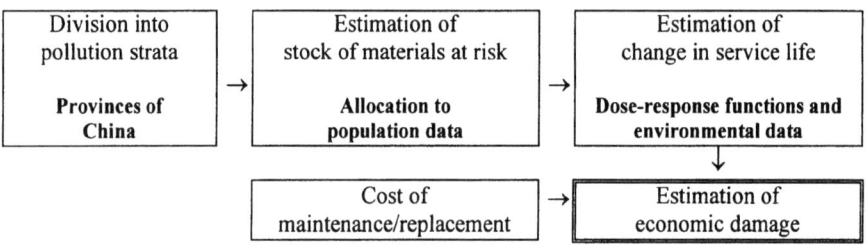

Figure 1. General approach for assessing cost of corrosion damage (adapted from Kucera *et al.* 1993).

2.1. DIVISION INTO POLLUTION STRATA

In order to perform a calculation of corrosion costs several sets of data are necessary and include environmental data (sulfur dioxide, relative humidity temperature, precipitation and acidity of precipitation), data on stock of materials at risk, i.e. the amount of materials exposed, data on prices and data on maintenance practices. Considering the available data the geographical basis of the calculations was chosen to be the provinces of China. In each province mean values are used to represent the whole province with the exception that rural and urban areas are treated separately. This is of course a simplification, which could be justified considering the amount of data available and especially that some of the data only are available as a mean value for the whole territory of China. Figure 2 shows a map of China including all the provinces.

2.2. ESTIMATION OF STOCK OF MATERIALS AT RISK

The stock of materials at risk has been assumed to correlate with census data, i.e., materials are exposed and used where people live. Each material shares a proportion of the total amount. For each province China census data, categorised into urban and rural areas, for the year 1990 have been calculated from Skinner et al. 1997.

The total amount of exposed surfaces for housing is roughly 15 billion square meters for urban areas and 36 billion square meters for rural areas (Wang 1998). Using census data this is equivalent to 66 m²/capita in urban areas and 40 in rural

Figure 2. Map of China and its provinces.

areas. This is reasonable compared to a study of three cities in Europe, Prague, Sarpsborg and Stockholm where the total amounts were 83, 165 and 132 m²/capita, respectively.

The zinc stock was divided into two categories, galvanised steel as sheet and other constructions made of zinc like fences, light poles, street signs etc. The sheet stock was estimated to 1.6 m²/capita using the total value 66 from above and a percentage value from the city of Prague (Kucera *et al.* 1993) while the stock of other constructions was estimated to 1.5 m²/capita using Stockholm data (Tolstoy *et al.* 1989). This is also reasonable compared to a recent study on the mapping of air pollution effects on materials including stock at risk in Guangzhou, China (Henriksen *et al.* 2000).

2.3. ESTIMATION OF CHANGE IN SERVICE LIFE

2.3.1. *Environmental data*

The environmental parameters consist of sulfur dioxide (SO_2), temperature (T), relative humidity (Rh), amount of precipitation (Rain) and acidity of precipitation (pH). All these were originally obtained, or could easily be converted to, a 1°x1° grid. In the NS direction 1° corresponds to 111 km while in the EW direction it varies between 66 and 105 km. Data were obtained from Carmichael 1998 and Cesar 1998 (SO_2), Legates and Willmott 1989 (T, Rain), Shiyan et al. (Rh) and Carmichael 1998 (pH).

2.3.2. *Dose-response functions*

A statistical analysis of the collected data for China from three exposure programs (1981-85, 1986-90 and 1986-94) has been performed (Tidblad and Kucera 1999) and resulted in a dose-response function for zinc giving the corrosion attack in μm, K_{Zn}, as a function of environmental parameters

$$K_{Zn} = (0.13[SO_2]^{0.34}\exp\{0.011RH - 0.028(T-10)\} + 0.21Rain[H^+]) \cdot t^{0.92} \quad (2)$$

where $[SO_2]$ is the SO_2 concentration in μg/m³, RH is the relative humidity in %, T is the temperature in °C, Rain is the amount of precipitation in mm, $[H^+]$ is the acidity of precipitation in mg/l and t is the exposure time in years.

Equation 2 is based on results from sites that are all located in parts of China with annual temperatures above 10 °C and therefore the temperature factor - 0.028(T-10) is used only when T > 10°C. For colder temperatures, the factor +0.062(T-10), taken from Tidblad *et al.* 1998, is used since it has been shown that corrosion decreases with decreasing temperatures for colder temperatures and China covers a territory having annual temperatures down to -5 °C.

2.4. MAINTENANCE CRITERIA AND COST OF MAINTENANCE/REPLACEMENT

For sheet the maintenance criterion used is after a reduction of the thickness of the zinc layer from 30 µm to 10 µm, i.e. a corrosion attack of 20 µm, according to Kucera *et al.* 1993. Using this value for K_{Zn} in equation 2 and the given environmental parameters it is possible to calculate the time, t, which in this case is equivalent to the life time used in equation 1. For other constructions surface treated by hot-dip galvanising a value of 60 µm for K_{Zn} is used.

For sheets of galvanised steel a value of 9 $/m^2 was used whereas for other constructions a value of 3 $/m^2 was used. These prices were estimated by comparison of general price differences between China and Europe combined with detailed costs in Europe for different zinc constructions from Kucera et al. 1993. It should be stressed that the Chinese prices are much lower and constitute about 10% of corresponding prices in Western Europe. The prices are taken as constants for the whole territory of China. Therefore, adjusting the final values for overall price differences is simple, should later investigations show that other prices are more reasonable.

3. Results and discussion

3.1. CORROSION MAP OF CHINA

Figure 3 shows a corrosion map of zinc for the rural territory of China

Figure 3. Calculated zinc corrosion values after one year of exposure for rural China based on collected environmental data for the year 1990.

calculated as described in section 2.3.2 and the assembled environmental data. The most corrosive areas are in the Sichuan and Guizhou provinces in the Southern part of China where a combination of climate and pollution causes the high values. The value can be as high as 2.5 µm/year in rural areas (Chongqing, 105°E/28°N). This should be compared with corrosion values in *urban* areas of Europe, which are generally lower.

3.2. CALCULATED TOTAL COSTS OF GALVANISED STEEL FOR DIFFERENT SCENARIOS

In equation 1, the maintenance interval in clean areas is always a point of reference which must be used since corrosion is a process that will take place even in the absence of pollutants. This background 'scenario' is taken as the ideal situation that would occur if the SO_2 concentration is reduced to 1 $\mu g/m^3$ and the pH is increased to 7 in all areas. In addition to the 'base case' i.e. the year of 1990 for which the main part of the data is given a few additional scenarios are investigated. In lack of advanced models for the emission and deposition of pollutants and available data for the future development three simplified scenarios have been used:

1. The SO_2 concentration is doubled in all areas
2. The SO_2 concentration is reduced to 50% in all areas
3. The pH is decreased by 0.3 units in all areas.

All these scenarios are referred to the mentioned background scenario. In other words, the cost associated with the base case 1990 is the benefit obtained from reducing 1990 levels to background levels and the cost associated with scenario 1. is the benefit obtained from reducing levels twice that of 1990 to background levels, etc. The likelihood of these events is not a subject of the discussion in the present work, they are only used for illustration purposes.

TABLE I
Calculated total costs of galvanised steel corrosion in urban areas of China due to acidifying pollutants for different scenarios compared to the situation in 1990.

Scenario	Sheet	Other constructions
Base case (1990)	$142 \cdot 10^6$ $/year	$14 \cdot 10^6$ $/year
1) SO_2 doubled	$177 \cdot 10^6$ $/year	$18 \cdot 10^6$ $/year
2) SO_2 reduced to 50%	$115 \cdot 10^6$ $/year	$12 \cdot 10^6$ $/year
3) pH decrease by 0.3 units	$196 \cdot 10^6$ $/year	$20 \cdot 10^6$ $/year

4. Conclusions

A methodology has been applied and modified taken into account conditions in

China and a database of environmental and material data has been assembled. The methodology used permits the costs to be calculated for individual provinces and mapping of areas with increased risk due to acid deposition (SO_2 and acid rain) and thereby sensitive areas in China can be identified. This will create a basis for rational decisions on cost effective measures for reduction of pollution emissions.

The model is in the present paper illustrated for zinc and galvanised steel and total corrosion costs in urban areas of China due to acidifying pollutants have been calculated for different scenarios compared to the situation in 1990. The study has shown very large differences in corrosion rates caused by acid deposition in different Chinese provinces. The most corrosive areas can be found in the Sichuan and Guizhou provinces in the Southern part of China where a combination of climate and pollution causes the high values.

Acknowledgements

The financial support from the World Bank is gratefully acknowledged.

References

Carmichael G.: 1998, University of Iowa, Personal communication
Cesar H.: 1998, The World Bank, Personal communication
Cowell and ApSimon H.: 1996, *Atmospheric Environment*, **30**, 2959.
Henriksen J., Kai T., Liangwan H. and Krigsvoll G.: 2000, UN ECE Workshop on Mapping Air Pollution Effects on Materials including Stock at Risk, Stockholm 14-16 June 2000.
Kucera V., Henriksen J., Knotkova D., and Sjöström C.: 1993, *Proceedings of the 10th European Corrosion Congress*, Barcelona.
Legates D. and Willmott C.: 1989, "Global Air Temperature and Precipitation Data Archive", Published and Disseminated by Data Support Section, Scientific Computing Division, National Center for Atmospheric Research, Colorado, at url: http://www.scd.ucar.edu/dss/datasets/ds236.0.html.
Shiyan T., Congbin F., Zhaomei Z and Qingyun Z.: "Two Long-Term Instrumental Climatic Data Bases of the People's Republic of China", CDIAC NDP-039. Available at anonymous ftp, url: ftp://CDIAC.ESD.ORNL.GOV/pub/ndp039
Skinner G. W. et al.: 1997, "China County-Level Data on Population (Census) and Agriculture, Keyed to 1:1M GIS Map", Published and Disseminated by CIESIN at url: ftp://ftpserver.ciesin.org/pub/data/China/CITAS/census_agr/, February 1997.
Tidblad J., Kucera V. and Mikhailov A. A.: 1998, "UN/ECE International co-operative programme on effects on materials including historic and cultural monuments. Report No 30: Statistical analysis of 8 years materials exposure and acceptable deterioration and pollution levels.", Swedish Corrosion Institute, Stockholm, Sweden.
Tidblad J. and Kucera V.: 1999, "Corrosion Costs in China due to Acid Deposition. An Initial Assessment", Report prepared for the World Bank.
Tolstoy N., Andersson G., Kucera V. and Sjöström C.: 1989, "Utvändiga byggnadsmaterial - mängder och nedbrytning." Statens institut för byggnadsforskning, Gävle. ISBN 91-540-9315-5.
Wang Z.: 1998, China Building Technology Development Centre, personal communication.

ENVIRONMENTAL MONITORING OF ROCK CARVINGS IN SCANDINAVIA

ELIN DAHLIN[1], PETER TORSSANDER[2], CARL-MAGNUS MÖRTH[2], HELÉNE STRANDH[2], GÖRAN ÅBERG[3], JAN F. HENRIKSEN[1], ODD ANDA[1] and RUNO LÖFVENDAHL[4],

[1] *Norwegian Institute for Air Research, P.O. Box 100, NO-2027 Kjeller, Norway;* [2] *Department of Geology and Geochemistry, Stockholm University, S-106 91 Stockholm, Sweden;* [3] *Department of Environmental Technology, Institute for Energy Technology, P.O. Box 40, NO-2027 Kjeller, Norway;* [4] *Swedish National Heritage Board, P.O. Box. 5405, S-114 84 Stockholm, Sweden*

Abstract. In the Scandinavian countries, Norway and Sweden, increased weathering rate of cultural property such as rock carvings, mostly dated between 4200 - 500 BC, has become a serious problem. Observations during the last decades have shown severe deterioration due to a combination of natural and anthropogenic influences. In order to study the rate of weathering of the rock carvings an interdisciplinary group from Norway and Sweden was formed to carry out environmental monitoring on one site in each country, one of them belonging to the UNESCO World Heritage List. The results from this investigation have elucidated the problems concerning weathering of silicate rock surfaces and the different parameters causing the deterioration. Based on the obtained results some counter measures in order to preserve the rock carvings have been tested.

Key words: weathering, environmental monitoring, rock carvings, Scandinavia

1. Introduction

In the Scandinavian countries, Norway and Sweden, there are numerous rock carvings of high preservation interest dated from 4200 to 500 BC (Hygen and Bengtsson, 2000). During the last decades deterioration of these rock carvings has increased and they begin to vanish at a high rate. Weathering and deterioration of stone is a natural process which has been accelerated due to the emission of different man-made pollutants, on local, regional and global scale. Deterioration of a stone object is caused by an interplay between different chemical, physical and biological processes, but man-made processes, especially the synergetic effects, have in many instances had a catalytic effect on the deterioration. The rate of deterioration depends on the environment, kind and amount of pollutants, and type of stone.

Many of the about 5000 rock carving sites from the Bronze Age period (1800-500 BC) in the coastal border area between Norway and Sweden have been subject to an in-depth damage inventory which showed that about 80 to 90% of the sites are damaged (Dahlin and Mandt, 1993; Löfvendahl and Bertilsson, 1996). In order to try to preserve these rock carvings, about 450 sites in the Tanum area, south western Sweden, were added to the UNESCO World Heritage List in 1994. The bedrock consists mainly of granite, which was glacially eroded during the last glaciation about 10 000 years ago. Physical, chemical and biological weathering was registered during the damage inventories.

The southern part of Scandinavia has been severely affected by acid rain during the last decades due to the prevailing south-westerly winds from Central Europe and England (Mylona, 1993). Moreover, local pollution from local industry and traffic has been an additional threat to many of the rock carving sites. The proximity to the North Sea expose the rock surfaces to high salt concentrations but also impose a relatively mild climate to the area with temperatures often undulating about the freezing point. Biological deterioration caused by lichens is another threat to the rock carvings.

An interdisciplinary group from Norway and Sweden was formed in 1997 to carry out a research project, partly financed by EU. The project, which was running during 1998 and 1999, had the emphasis on; atmospheric chemistry, petrography, geochemistry, and biology. The EU-project was a continuation of, and mainly based on, results from earlier research in Scandinavia (Löfvendahl and Bertilsson, 1996; Dahlin et al., 1998; Åberg et al., 1999).

The EU-project was carried out on one site in Norway, at Begby in Østfold county, and one site in Sweden at Litsleby, in Bohuslän county. The two chosen rock surface sites with no carvings were closely situated to rock carving areas and had similar environmental surroundings with an open coastal landscape, little soil coverage on the granitic bedrock, and few forested areas. The rock surfaces were partly covered by lichens, but the ones chosen for study were devoid of them. A difference between the areas is that the Begby site is situated close to an industrial town with the emissions from traffic and industry, while the Litsleby site is near the coast and in a rural surrounding.

The aim of the EU-project was; to study deterioration processes and compare the differences in weathering under a roof construction to atmospherically exposed surfaces; to study the influence of biological growth on the granite surfaces; to find methods to decrease the weathering rates of the rock carvings; and to compare the weathering in a rural area with an area influenced more by local industry and traffic.

2. Environmental Monitoring Methods

Weathering and deterioration of glaciated rock surfaces in granite were studied and different protective methods were evaluated. At the Litsleby site a roof-construction in metal was erected in October 1997, while at Begby an ordinary exposed bedrock surface was studied.

The following types of samples and analyses were made; drill-cores were collected to study mineralogical, chemical, biological and microbiological parameters and the alteration through surface weathering. Textural and mineralogical parameters of fresh and weathered rock cores were studied and the mineralogical composition was calculated. The dominant lichen flora was defined at both Begby and Litsleby and the relative depth of penetration of selected lichen species and micro-organisms was examined. Further identification of lichen species with chemical methods and staining techniques was made at Begby. The consequences of rock surface coverage with different natural and artificial materials were studied at Litsleby. Runoff water was collected from the exposed ponds at Litsleby and Begby while the sheltered pond at Litsleby was flushed by hand with distilled water, which was collected. These water samples were later passed through 0.45 μm filters. The particulate phase upon the study surfaces was then collected with a vacuum cleaner. All samples (water and particulates) were analysed with different standard methods. Measured temperature, pH, moisture and relative humidity were supplied to a data-logger and transferred to the laboratories at the Norwegian Institute for Air Research and the University of Stockholm. At Begby different pollution gases such as sulphur dioxide (SO_2,), nitrogen oxides (NO_x,), ozone (O_3) and ammonia (NH_3) were collected by the use of passive gas samplers together with the collection of aerosols. Micro-mapping with laser was made three times during the project period at Litsleby.

3. Results and Discussion

The granitic rock is composed mainly by K-feldspar, quartz and plagioclase. Weathering was measured in drill-cores down to 3.5 cm depth at Begby (Dahlin et al., 2000) and down to about 4-5 cm depth at Litsleby (Torssander et al., 1999). The close coastal location of Litsleby induces a marine influenced chemistry in the atmospheric input and runoff waters (Torssander et al., 1999). At Begby the marine influence is reduced due to forestation.

The analysis of uranium versus sulphur show that runoff samples contain clearly higher amounts of the weathering product uranium than the deposition samples. The throughfall samples contain dry deposition, including soil dust, to a larger extent than the bulk deposition samples. The rain collectors do not contain any dry deposition, only wet. The origin of sulphur is mainly from dry deposition, which can be seen from the high concentrations of sulphur in the throughfall samples. The rain collector and the bulk deposition had low concentrations of sulphur and practically zero concentrations of uranium. The low uranium concentration in bulk deposition in comparison to runoff water indicates the possibility to use uranium as a tracer of the weathering rate (Torssander et al., 1999).

Monitoring of atmospheric pollution at Begby show that the amount of SO_2 and NO_2 are highest during the wintertime due to the addition of emission from domestic heating. However the amount of pollution in the Fredrikstad area, has decreased by about 50% from 1980 to 1999 due to reduction of emissions from the local industry (Dahlin et. al., 2000). The pH of precipitation has increased over the last decade but is still low with values between 4-5.5 along the south-western cost, during the winter period the pH can reach below 4. The throughfall samples were even more acid than the precipitation and the content of the potassium (K), sodium (Na), magnesium (Mg) and chlorine (Cl), were about 10 times higher, which show the influence of sea salt.

In the exposed pond at Litsleby the pH was largely the same as in the bulk deposition, i.e. about 4.5 but with few exceptions the bulk deposition pH reached as high as 6.7 and the exposed pond got a pH up to 5.4 (Strandh, 1999). In contrast the sheltered pond under the roof had a rather stable pH around 6.0 due to the washing of the surface with de-ionised water. Washing of the surface was expected to reduce the total amount of elements at the surface, but the results showed that the effect in some samples were opposite when comparing to the small amount of water used for washing during 1998. So far the results from 1998 indicate that the rock surface under the roof show a lower mean maximum release rate of silicium (Si), indicating that a roof might be an alternative preservation technique against chemical weathering (Torssander *et al.*, 1999).

The strontium ($^{87}Sr/^{86}Sr$) isotope ratio in water samples from the rain collectors were very similar to that of sea-water while bulk deposition and throughfall showed a slight increase of the Sr ratio due to the varying amounts of atmospherically transported dust, salts and particles. Hence, the Sr isotope ratio together with the Sr element concentration can be used to estimate the amount of dry deposition deposited on the rock surface. The runoff samples from the exposed pond exhibit still higher $^{87}Sr/^{86}Sr$ ratios due to the input from weathering. The S-isotope investigation shows that about 25% of the sulphur come from the sea and the remaining 75% from anthropogenic activity. Strontium on the other hand comes mainly from the sea but with some addition from the rock surface. The good correlation found between calcium and strontium in the water samples concludes that the Sr isotope ratio method appears to be a promising tool for evaluation of weathering on a rock surface, as previously shown from catchment studies (Åberg et al., 1989). Especially when the Sr isotope method is combined with a quantification of silica, uranium or calcium in order to get a value of the weathering.

The results from the biological investigation at both sites have shown that more than 25 different species have been identified. Detailed studies showed that biological material have been found in micro cracks as deep as 5 mm under the rock surface. Most common are cracks with a dept of 2-3 mm filled with lichens (Dahlin et al., 2000). These cracks make the surface very porous and fragile. Investigation on effective methods in order to reduce the biological growth was carried out within the project. The results of the rock surface coverage show that a combination of 10 cm turf with an artificial fabric, covering the rock surface for at least 2 months, was one effective method to reduce the biological growth (Dahlin *et al.*, 2000).

The results from monitoring of the temperature with the WETCORR–instrument show that the number of freeze-thaw cycles on the rock surfaces are highest in spring and autumn. Figure 1 shows that during springtime the number of freeze-thaw cycles are quite frequent on the rock surface. The monitoring showed that the number of freeze-thaw cycles from under the roof was reduced with 60% compared to the exposed area outside of the roof and with almost 90% under the rock surface coverage by an artificial mat. Results of the monitoring with the WETCORR show that the best protection against freeze-thaw cycles is to cover the rock surface by an artificial mat (about 10 cm thick), the second best protection is natural soil and turf (about 10 cm thick), and the third best protection is a roof-construction.

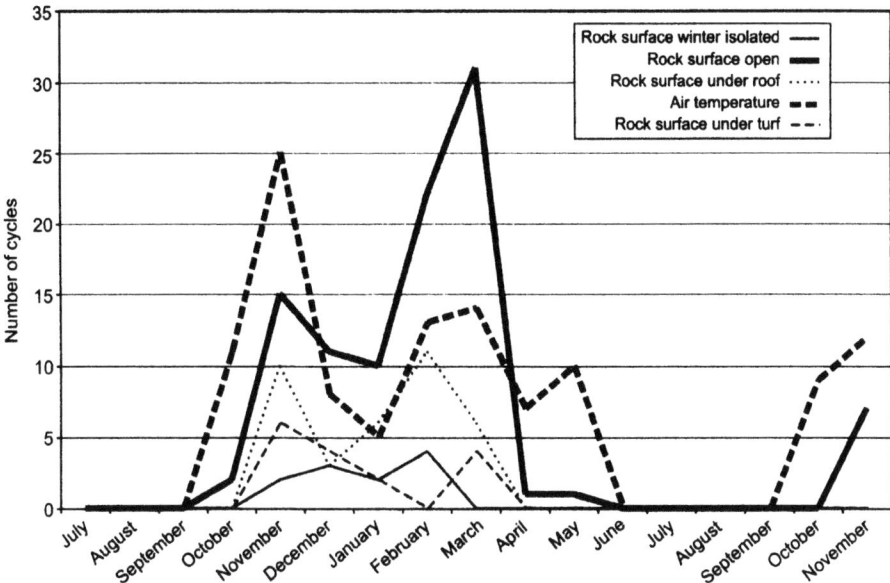

Figure 1. Results from the monitoring of freeze – thaw cycles on the rocksurfaces with the WETCORR instrument show the number of freeze-thaw cycles at both Begby and Litsleby from July 1998 – November 1999.

4. Conclusions

Using different environmental monitoring techniques have shown that there are several tools that can be used to evaluate weathering data. Especially the combination of chemical weathering studies with isotope investigations. The use of uranium and calcium as reference elements, instead of silica, seems to indicate a faster rate of weathering. The reason may be that uranium and calcium are released from more easily weathered minerals or from grain boundaries or cracks.

Monitoring of atmospheric pollution show that there is still an influence from traffic, and local industry combined with emission from domestic heating during the wintertime in the Fredrikstad area in Norway. The Bohuslän area has less pollution, but more influence of salt from the sea. In both areas the precipitation is still acid especially during wintertime.

The investigation on elimination of lichens seems promising but due to its complexity it has to be evaluated further. Different materials have been tested with varying results but have to be integrated in a maintenance program for each specific rock carving site.

The results from the investigation of the rock surface under the roof construction have shown that a roof might be an efficient way of prolonging the life-cycle of the rock-carvings. Another method which has showed promising results, but has to be investigated more, is the covering with an artificial mat during wintertime, in order to decrease the number of freeze-thaw cycles and the weathering rate.

Acknowledgements

The authors would like to thank T. Bjelland, M. Ekroth-Edebo and J. Mattsson for their research concerning the biological deterioration, I. Thorseth for her geological investigations and J. Magnusson for the co-ordination of the whole EU-project. The project was jointly funded by; the Norwegian Directorate for Cultural Heritage, the Swedish National Heritage Board, and the European Commission, as part of the Interreg II Programme.

References

Åberg, G., Jacks, G. and Hamilton, P. J.: 1989, *Journal of Hydrology* **109**, 65-78.
Åberg, G., Stray, H. and Dahlin, E.: 1999, *Journal of Archaeological Science* **26**, 1483-1488.
Dahlin, E. and Mandt, G.: 1993, *Andoranten, Scandinavian Society for Prehistoric Art*, 30-35 (in Norwegian).
Dahlin, E. M., Elvedal, U., Henriksen, J. F., Anda, O., Mattsson, J., Iden, K. and Åberg, G.: 1998, Environmental Monitoring of the Rock Art Site at Ekeberg, Oslo, Norway. Kjeller, NILU OR 22/98 (in Norwegian).
Dahlin, E., Henriksen, J. F., Anda, O., Mattsson, J., Iden, K., Åberg, G., Bjelland, T., Thorseth, I., Hamnes, G. M. and Torssander, P.: 2000, Rock Carvings in the Borderlands. Sub-project 3A: Development of knowledge on weathering processes and preservation techniques. Kjeller, NILU OR 76/99 (in Norwegian).
Hygen, A-S. and Bengtsson, L.: 2000, Rock Carvings in the Borderlands, Göteborg.
Löfvendahl, R. and Bertilsson, U.: 1996, 'Rock carvings', in E. Österlund (ed.) *Degradation of Materials and the Swedish Heritage 1992-1995*, Stockholm, pp 18-29.
Mylona, S.: 1993, Trends of sulphur dioxide emission, air concentrations and depositions of sulphur in Europe since 1880. Oslo, Norwegian Meteorological Institute (EMEP/MSC-W Report 2/93).
Strandh, H.: 1999, Mineral dissolution from molecular to field scale. Stockholm, Meddelanden från Stockholms Universitets Institution för Geologi och Geokemi, No 304,.Torssander, P., Strandh, H., Mörth, C-M. and Åberg, G.: 1999, 'Weathering of granite – A field study in SW Sweden', in H. Armannsson (ed.) *Geochemistry of the Earth Surface*, Reykjavik, pp 129-132.

EFFECT OF SIMULATED ACID RAIN ON DETERIORATION OF CONCRETE

TSUTOMU KANAZU[1*], TAKURO MATSUMURA[1], TATSUO NISHIUCHI[1] and TAKESHI YAMAMOTO[1]

[1] Structure Department, Abiko Research Laboratory, Central Research Institute of Electric Power Industry(CRIEPI), Abiko City, CHIBA 270-1194, Japan
(* author for correspondence, e-mail: kanazu@criepi.denken.or.jp)

Abstract. In the study the long-term exposure tests to simulated acid rain were performed in order to clarify the effect of acid rain on deterioration of concrete. Mortar specimens with 40 mm in width, 15 mm in thickness and 160 mm in length were used for the tests. At each time after the fixed rainfall was attained, those were tested physically and analyzed chemically. Finally total rainfall of 9000mm was given to the specimens. From the test results, it was confirmed that the eroded depth of the specimen has a good linear relation to the total rainfall under simulated acid rain with various pH. Surface erosion rates of the mortar specimens with an ordinary mix proportion under simulated acid rains with pH 3.0 and 2.5 were about 1.2 and three times larger than that under pH 5.6, respectively. It was also confirmed that flexural strength of the specimens with an ordinary mix proportion hardly changed under low pH simulated acid rain even after total rainfall of 9000mm was given.

Key words: concrete, deterioration, simulated acid rain, exposure test, mortar specimen

1. Introduction

Concrete structures as infrastructures are inclined to loose their performance through the service period according to the environmental conditions where those are exposed. Typical phenomena of deterioration are salt injury of reinforced concrete (RC) structures, freezing and thawing damage in cold climate areas, carbonation due to carbon dioxide and chemical attack by acid solution, etc. Acid rain is also one of the factors which causes deterioration of concrete structures, but it is considered that its effect is relatively small.

Concerning the studies on deterioration of concrete so far performed, the acceleration method corresponding to each phenomenon was usually used to comprehend the process and to estimate the results of deterioration. Consequently effects of salt injury, freezing and thawing and carbonation on concrete deterioration with years could be estimated quantitatively. Hereafter, the effects caused by the combined action of those phenomena should be clarified considering the natural environmental conditions, therefore the effect of acid rain should be studied independently.

In the study, long-term exposure tests to simulated acid rain were performed and the results, which were obtained from both the physical tests of the specimens and the chemical analyses of the hardened cement paste, were discussed in order to clarify the effects of acid rain on deterioration of concrete structures(T. Kanazu et al., 1996)

2. Outline of the Exposure Tests

2.1 TEST CONDITIONS

(1) Factors
Two factors were adopted in the tests in order to consider initial conditions of test specimens. One was mix proportion of mortar; water cement ratios were 60%, 80% and 100%. Another was initial defect; a half of specimens were heated at 80℃ for 48 hours to attain the defective conditions with small cracks.

(2) Measurements
The effect of acid rain on mortar specimens was usually estimated by chemical analyses of cement hydrates and total pore volume of the specimens. In the study, flexural strength and eroded depth on the surface of the specimens were also measured.

(3) Test specimen
Specimens were small mortar plates shown in Figure 1. Mix proportions were listed in Table I and were decided as cement paste volume was the same among all mixes. It is considered that water cement ratio of 65% was the maximum value in an ordinary mix proportion, therefore the qualities of the mortars with the water cement ratios of 80% and 100% were quite bad. Test conditions were shown in Table II. Three specimens were tested under the same test condition.

2.2 EXPOSURE TEST TO SIMULATED ACID RAIN

Simulated acid rain used to the tests was made by mixing sulfuric acid, hydrochloric acid and nitric acid with the fixed mixing ratio, i.e. $SO_4^{2-}: NO_3^-: Cl^-$ =5:2:3 (equivalent ratio), and those pH were arranged to be 2.5, 3.0, 4.0 and 5.6. A rainfall in a week was 48 mm through 60 mm: 16 mm through 20 mm in 6 hours at one time and three times in a week. The physical tests and the chemical analyses were done at the total rainfalls of 500 mm, 2200 mm, 3100mm (3500mm; analysis for C-S-H), 6300mm and 9000mm.

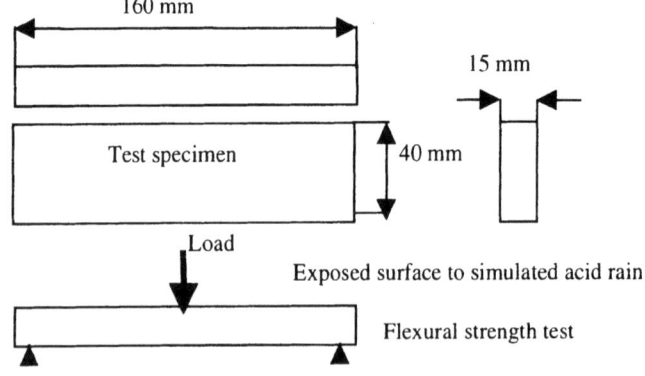

Figure 1. Mortar specimen and test method

TABLE I				
Mix proportions of mortar				
W/C (%)	Flow Value (mm)	Weight for one batch (g)		
		Cement	Water	Sand
60	240±5	170.0	102.0	510.0
80	240±5	139.5	111.6	510.0
100	240±5	118.4	118.4	510.0

W/C; water cement ratio
Paste volume of each mix is the same (156 cm^3/batch).

TABLE II	
Test conditions	
Factor	Level
Exposure condition	to simulated acid rain; pH 2.5, 3.0, 4.0, 5.6
	to air-dried condition (laboratory room)
W/C	60, 80, 100 (%)
Heating	Heated at 80°C for 48 hrs., not heated
Total rainfall	500, 2200, 3100, 6300, 9000 (mm)

3. Test Results

3.1 ERODED DEPTH ON THE SURFACE OF THE SPECIMENS AND ITS RATE WITH RAINFALL

Eroded depth on the surface of the specimen was shown in Figure 2 in relation to total rainfall. The results of the specimens heated and not heated were almost the same. Linear lines were derived from the regression' analysis. Clear linear relation between eroded depth and total rainfall was observed. Erosion rate with total rainfall was shown in Figure 3. From these figures, the following five points were derived.

(1) Influence of heating, i.e. initial defect, on the test results was not observed.
(2) Eroded depths of the specimens with W/C: 80% and 100% were slightly larger than those with W/C: 60%.
(3) In the case of W/C: 60 %, eroded depths of the specimens under the simulated acid rain with pH3.0 had little difference as compared with those under pH 5.6 (little influence of acid rain in cases that quality of mortar was good).
(4) There was no influence of W/C on eroded depth of the specimens in cases that pH of simulated acid rain was lager than 4.0 (influence of the quality of the mortar was small in cases that pH of acid rain was larger than 4.0).
(5) A clear difference of eroded depths was observed in the specimens under simulated acid rains with pH2.5 and pH3.0.

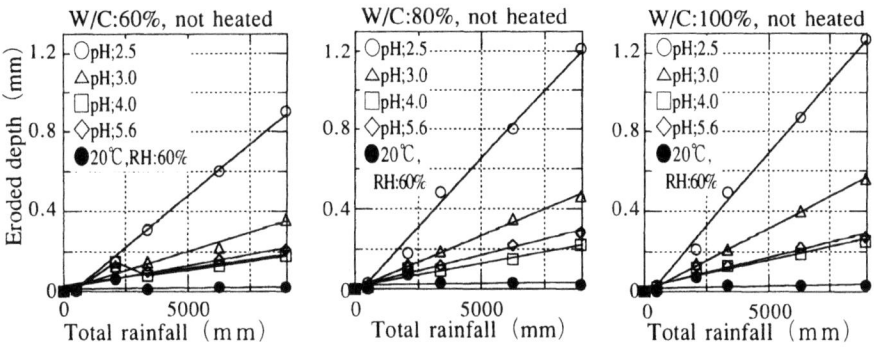

Figure 2. Relation between total rainfall and eroded depth

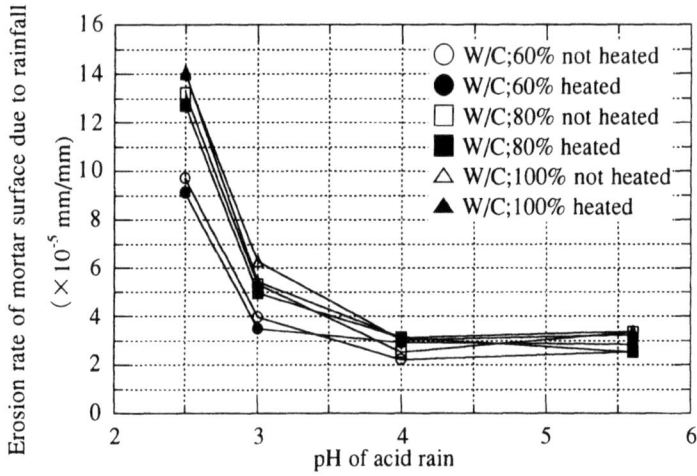

Figure 3. Relation between pH of acid rain and erosion rate of mortar

(6) Erosion rates of the specimens under the rains with pH2.5 and 3.0 were 3.0 times and 1.2 time larger than that under the rain with pH5.6, respectively.

3.2 MICROSTRUCTURE

Measured results of pore volume were shown in Figure 4. This character stands for the change of microstructure of the mortar. Measured results of this character were inclined to become large depending on the quality of the mortar i.e. the increase of W/C and to slightly decrease with total rainfall. These results showed that no influence of simulated acid rain reached the inner part of the mortar within the total rainfall of 9000mm, though it was judged from the NO_3^- observation that simulated acid rain permeated inside.

3.3 FLEXURAL STRENGTH

Figure 5 showed the change of flexural strength of the mortar specimens with total rainfall. The results of the specimens with and without initial defect were almost the same. The data under various pH conditions were standardized by the data under pH5.6 in order to eliminate the influence of the difference of mixing batches of mortar. There were slight changes of flexural strength with time at the total rainfall of 9000mm. In cases of the mortars with low quality, i.e. high W/C, flexural strength was inclined to decrease according to the increase of total rainfall and the lower the pH of simulated acid rain, the larger the decrease rates.

At the total rainfall of 9000mm, the influence of acid rain on physical property of the mortar with low quality was observed.

3.4 CHANGE OF C-S-H

Dense concentration of nitrate ion was observed at a central part of cross section of the specimen. It means that simulated acid rain might influence on cement hydrate chemically, though pore observation concluded no effect on the

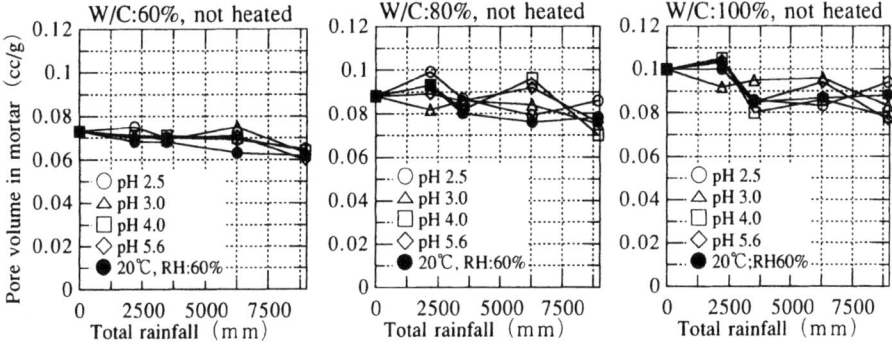

Figure 4. Relation between total rainfall and pore volume of mortar

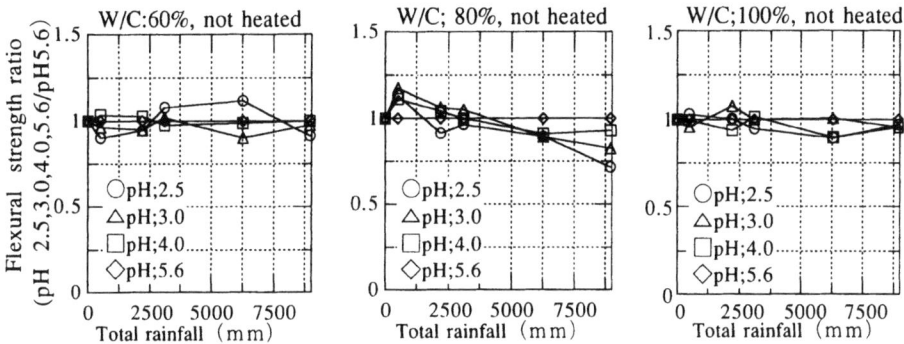

Figure 5. Relation between total rainfall and flexural strength change of the specimens

microstructure of the mortar. Figure 6 showed the change of CaO/SiO_2 molecule ratios (C/S ratio) according to the increase of total rainfall. C/S ratios at the beginning of the test were dependent on W/C of the mortars. C/S ratios of the mortars with W/C:60% scarcely changed in accordance with the increase of total rainfall, except for the case under the rain with pH2.5. In case of the mortar with poor quality, i.e. W/C: 100%, C/S ratios decreased with the increase of total rainfall irrespective of pH of simulated acid rain and it means that cement hydrate might be dissolved by acid solution. In pH range more than 3.0, dissolution of C-S-H depends on the quality of mortar, i.e. W/C.

The method of C-S-H analysis should be described. It basically followed the method suggested by K. Suzuki et al. (K. Suzuki et al., 1990) and was slightly revised. The analyzed results could be applied to the relative comparison of the change of C-S-H among the specimens.

4. Conclusions

In the study the long-term exposure tests to simulated acid rain were conducted to clarify the effect of acid rain on the deterioration of concrete structures using the mortar specimens with different qualities. Total rainfall of 9000mm was attained in the exposure tests until now. It may correspond to the rainfall for about seven years in Japan. Conclusions derived from the study are as follows.

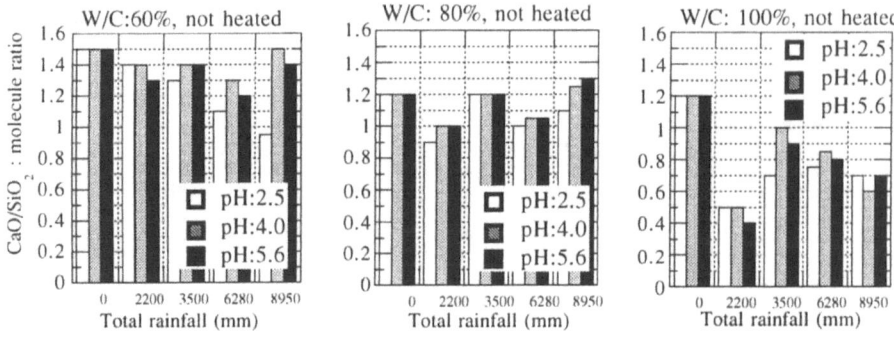

Figure 6. Change of CaO/SiO$_2$ molecule ratio with the increase of total rainfall

(1) There is a linear relation between total rainfall and eroded depth on the surface of the specimens. No effect of acid rain with pH more than 4.0 on the eroded depth is found irrespective of the quality of mortar.
(2) Erosion rates of the surface layer of the mortar under simulated acid rains with various pH are made clear experimentally. The erosion rate under the acid rain with pH3.0 is 1.2 time larger than that with pH5.6 in the case of mortar with an ordinary mix and 2.0 times larger than that with pH5.6 in the case of mortar with poor quality. The acid rain with pH2.5 increases the erosion rate of the mortar considerably; 3.0 times in the case with an ordinary mix and 4.5 times in the case with poor quality as compared with that under the rain with pH5.6.
(3) No effect of acid rain on the microstructure of the mortars is observed at the total rainfall of 9000mm, but the changes of flexural strength and CaO/SiO$_2$ molecule ratio are found only in the mortar specimens with poor quality.

Acknowledgement

The authors wish to thank the members of the Committee on Acid Rain Effects for their discussions and suggestions. This study was supported by the budget sponsored by the Agency of Natural Resources and Energy of MITI.

References

T. Kanazu et al.: 1996,*Effect of Simulated acid rain on concrete materials*. Proceedings of CRIEPI International Seminar on Transport and Effects of Acidic Substances, Tokyo, Japan.

K. Suzuki et al.; 1990,*Analysis of Hydrated Phases for Evaluating the Durability of Concrete*. Concrete Research and Technology **1**, **No.2** (in Japanese).

EFFECTS OF ACID DEPOSITION ON URUSHI LACQUER IN EAST ASIA

Y. TSUJINO[1]*, Y. SATOH[2], N. KURAMOTO[2] and Y. MAEDA[3]

[1] *Environmental Pollutant Control Center, Osaka Prefecture, Nakamichi 1-3-62, Higashinari-ku, Osaka 537-0025 Japan;* [2] *Technology Research Institute of Osaka Prefecture, Ayumino 2-7-1, Izumi city, Osaka 594-1157 Japan;* [3] *Osaka Prefecture University, Gakuen-cyo, Sakai city, Osaka 599-8231 Japan*
(*author for correspondence, e-mail: tsujino@mbox.epcc.pref.osaka.jp)

Abstract. Urushi (Japanese lacquer) plates were exposed to indoor air in 7 cities in East Asia from summer 1995 through winter 1997. The plates, collected every 3 months, were optically observed by using a gloss meter, a digital microscope and a microscopic infrared spectrophotometer for evaluation of seasonal impact by acidic air pollutants. The gloss losses were high on the urushi surface in autumn, when fog frequently appeared. Numerous fine spots were observed in 0.2-0.3 mm in diameter on the surfaces of the plates exposed at Chongqing, China, Taejon, Korea and Nara, Japan, where dense fog occurred. The spots were dark and opaque at Chongqing, where heavy air pollution was observed. A dark and opaque core was observed in each semitransparent spot at Taejon and Nara, while no spots were observed at Kyoto and Ishikawa, where fog often appeared but air pollution was at a low level. Carbonyl group, identified by microscopic infrared spectrometry, was found in the spots on the urushi surfaces. The carbonyl group may be formed by oxidation of a side-chain in urushiol (a major component of urushi sap, alkyl phenol). Urushi lacquer may be damaged by high concentration of sulfate anion, included in acid fog.

Keywords: urushi, acid fog, gloss loss, material damage, East Asia

1. Introduction

The urushi culture may arise in mountainous districts in Southeast Asia in ecological botany. The urushi culture has been developed in East Asia. Urushi material has been utilized as paints and bonding agents. Urushi material is made of the sap (kiurushi) of lacquer trees (*Rhus vernicifera*). A major component of kiurushi is urushiol in Japan, China and the Korea Peninsula. Urushiol is 3-substitiuted catechols with a side-chain of C_{15} or C_{17} (Yamauchi *et al*, 1982). Oxidation of urushiol is divided to polymerization by laccase (copper enzyme) and degradation by oxygen in air (Kumanotani, 1991). The degradation is caused by oxidation of the side chains (Kumanotani, 1991). Gloss loss and color change of urushi surface occur in the initial phase of the degradation (Toishi *et al*, 1967). The damage is accelerated by temperature, humidity and ultraviolet rays (Toishi *et al*, 1967).

However, few investigations have been carried out to assess the effects of acidic air pollution on urushi artifacts, and few approaches were discussed on protecting them from such air pollution. This paper describes the qualitative and quantitative evaluation on damage of urushi lacquer caused by acidic air pollution by using urushi plates exposed to indoor air at 7 sites in Japan, China and South Korea.

2. Methods

2.1. REAGENTS AND MATERIALS

Chemicals was of special grade from Wako (Osaka, Japan). Table I indicates the specifications of the urushi plates exposed to air.

TABLE I Specifications of urushi plates

plate size	$50 \times 50 \times 7.0$ mm
soji (wooden plate)	straightly grained hiba (*Thujopsis dolabrate*)
ji (the first-layer coating)	coating of a mixture of jinoko, kiurushi and paste
sabi (the second-layer coating)	coating of a mixture of tonoko and kiurushi
roiro shiage (the final-layer finishing)	double coating of black lacquer

2.2. APPARATUS

An instrument screen shelter was used for exposure of the urushi plates to air (for the indoor exposure). Tohsentekuno (Osaka, Japan) diffusion passive samplers were employed for monitoring the one-month averaged concentrations of sulfur dioxide (SO_2) and nitrogen dioxide (NO_2). A Nippon Test Panel Osaka (Osaka, Japan) cotton gauze collector was used for trapping the acidic pollutants such as acid fog, acidic dust and sea salt. A Suga (Tokyo, Japan) UGV-5D gloss meter (reflex angle at 60 degree), a Keyence (Osaka, Japan) VH-6200 100-power digital microscope were employed for observing the surfaces on the samples. A Nihondennshi (Tokyo, Japan) JEOL 700 with IR-MAU200 microscopic infrared spectrophotometer (with 25 micrometer scope range in diameter) was employed for the qualitative identification of products in spots on the urushi surface.

Figure 1. Location of the sampling sites (1-7). (1) the Chongqing site (Chongqing); (2) the Shanghai site (Shanghai); (3) the Taejon site (Taejon); (4) the Ishikawa site (Ishikawa); (5) the Osaka site (Osaka); (6) the Nara site (Nara) and (7) the Yawata site in Kyoto (Kyoto).

2.3. SAMPLING SITES AND SAMPLING PROCEDURE

Figure 1 shows the locations of the 7 sampling sites. Chongqing, China, is a large city where heavy air pollution is observed. The metropolitan area is located at the confluence of two big rivers (Yangtze River and Jiaring River), called as a foggy capital. Taejon, Ishikawa, Nara and Kyoto are also well known as foggy areas located at the foot of highlands. The urushi plates were suspended by nylon strings in the shelter, and exposed to acidic atmosphere. The plates were collected every 3 months (7 times) from June 1995 through January 1997 for evaluating the seasonal impacts by acidic air pollutants, and consecutively exposed for 1 year or 2 years from June 1995 trough May 1997 at the individual sites for estimating the regional impacts of air pollution on the plates.

2.4. MONITORING OF ENVIRONMENTAL FACTORS

Temperature, relative humidity (RH), wind direction (WD) and wind velocity (WV) were simultaneously monitored. The time of wetness (TOW) is defined as period (in hr or hrs) in that RH was higher than 80 % and temperature was above 0 °C. The concentrations of SO_2, NO_2, oxidant (O_X) {or ozone (O_3)} and suspended particulate matter (SPM) {or total suspended particular (TSP)} in ambient air were continuously monitored at the monitoring stations at the individual sites. At the sites in China and South Korea where air monitoring systems were not adequately established, the concentration of SO_2 and NO_2 were monitored in a one-month term by the diffusion passive samplers. About acidic pollutants, acid fog was trapped by using the cotton gauze collector and the chemical composition was determined by ion chromatography (IC). However, acidic gases, acidic dust and sea salt were partially corrected in the same time. No fog water was directly sampled.

2.5. GLOSS LOSS, MICROSCOPIC OBSERVATION AND INFRARED SPECTROMETRIC ANALYSIS

Gloss intensity on the urushi surface was determined by using the gloss meter. The gloss loss is defined as damped percentage of a gloss intensity after exposure to one before exposure. The surface state after exposure was observed by using the digital microscope. The products on the surface were identified by using the microscopic infrared spectrophotometer.

3. Results and Discussion

3.1. GLOSS LOSSES ON THE URUSHI SURFACES

The urushi samples seemed to be very dusty in the spring and the autumn at Chongqing due to yellow sand in China. All the samples at Kyoto looked like the blank one. Figure 2 shows the seasonal gloss losses at the 7 sites. The gloss losses were very large at Chongqing, Taejon and Shanghai in the autumn in 1995.

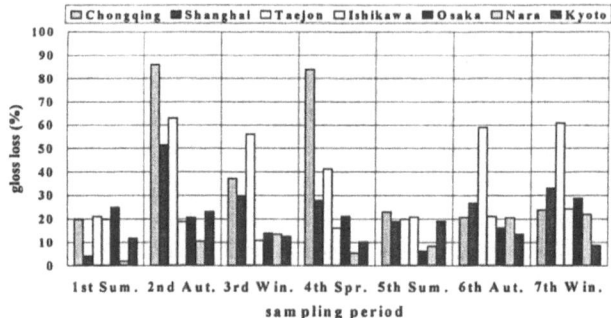

Figure 2. Seasonal gloss losses on the urushi surfaces at the 7 sites: 1st Sum: June-Aug.'95, 2nd Aut: Sept.-Nov.'95, 3rd Win: Dec.'95-Feb.'96, 4th Spr: Mar.-May'96, 5th Sum: June-Aug.'96, 6th Aut: Sept.-Nov.'96, 7th Win: Dec.'96-Feb.'97; the gross intensity of the blank samples: 66.5±2.3 % (n=3).

3.2. ENVIRONMENTAL FACTORS

The TOW was more than 1,000 hrs/90 days in the autumn in 1995 at Chongqing, Taejon, Ishikawa and Kyoto. Dense fog ought to have frequently occurred in the period at the four sites. Accordingly, we will present about the results so far obtained in the autumn in 1995. The gaseous SO_2 concentration was very high in the winter at Chongqing (195 ppb) and Shanghai (44 ppb). The NO_2 concentration was high at Osaka (36 ppb). The TSP concentrations were very high at Chongqing (340 microgram/m^3) and Shanghai (270 microgram/m^3) due to yellow sand in China.

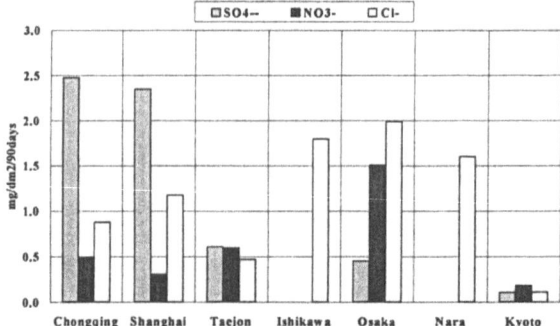

Figure 3. Amounts of SO_4^{2-}, NO_3^- and Cl^- trapped on the cotton gauze in the autumn in 1995.

The amount of sulfate anion (SO_4^{2-}) was very large at Chongqing, Shanghai and Taejon, but was at a low level at Kyoto, only 4.3 % of that at Chongqing (Figure 3). Fog water included 6,450 micromole/L of SO_4^{2-} in 1984-1990 at Chongqing (Peng et al., 1992), 25 times greater than that at Kyoto in 1993 (Takemae et al., 1993). The amount of nitrate anion (NO_3^-) was large at Osaka. The large amount of chloric anion (Cl^-) may be derived from sea salt due to existing the same equivalent of sodium cation (Na^+) at Osaka, Ishikawa and Shanghai, located on the coast. On the other hand, little sodium cation was detected at Nara, located in inland. The high concentration of gaseous hydrogen chloride may be emitted from incinerators of dumped waste at Nara.

3.3. OBSERVATION ON THE URUSHI SURFACES

Numerous fine spots were observed at Chongqing, Taejon and Nara, but no spots were observed at Shanghai, Ishikawa, Osaka and Kyoto (Table II). Figure 4 shows the typical digital microscopic images of the urushi surfaces at Chongqing, Taegu and Kyoto. The urushi layer of the blank sample was semitransparent due to finding out some straight grains on the wooden plate {see (a)}. The numerous fine spots were observed in 0.2-0.3 mm in diameter on the urushi surfaces at Chongqing, Taejon and Nara. The spots were dark and opaque at Chongqing, where the large amount of SO_4^{2-} was included in the acid fog {see (b)}. A dark and opaque core was observed in each semitransparent spot at Taejon {see (c)}. Fine droplets may gradually dry up and diminish in half size. As a result, the half size droplets, concentrated at 4-8 times, may degrade the urushi surfaces at Taejon. The spots were obscure at Nara. However, no spots were observed at Kyoto and Ishikawa, where dense fog often appeared but air pollution was at a low level {see (d)}. No fog was observed in the 7 seasons at Osaka.

TABLE II Fine spots observed on the urushi surfaces

++: dense spots; +: weak spots; 1st Sum, 2nd Aut, 3rd Win, 4th Spr, 5th Sum, 6th Aut, 7th Win: see Figure 2.

sampling site	seasonal exposure test						
	1st Sum	2nd Aut	3rd Win	4th Spr	5th Sum	6th Aut	7th Win
Chongqing	++	+	++	+		++	++
Shanghai							
Taejon		+	+			+	+
Ishikawa							
Osaka							
Nara				+		+	
Kyoto							

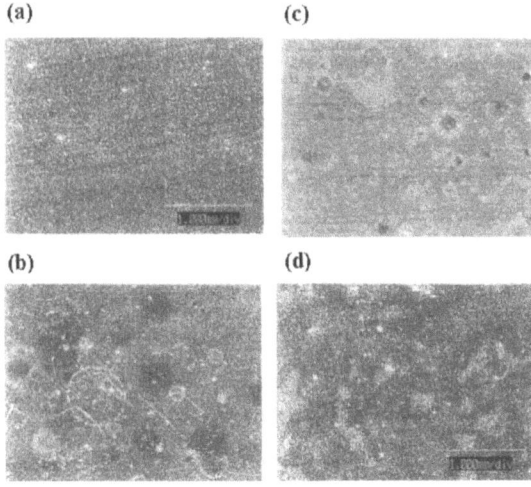

Figure 4. Digital microscopic images of the urushi surfaces: (a) in the blank test; (b) at Chongqing (No.1); (c) at Taejon (No.3); (d) at Kyoto (No.7).

3.4. PRODUCT IN THE SPOTS ON THE URUSHI SURFACES

Figure 5 shows the typical products on the urushi surfaces analyzed by the microscopic infrared spectrometry. Carbonyl group (wave number 1738 cm^{-1}) was qualitatively identified in the spots on the urushi surfaces. On the other hand, no peak of carbonyl group was found from outside the spots. The carbonyl group may be formed by oxidation of the side-chains of urushiol. The oxidation may be accelerated by high concentration of SO_4^{2-} existing in acid fog at Chongqing and Taejon, and by hydrogen chloride gas, absorbed in acid fog at Nara.

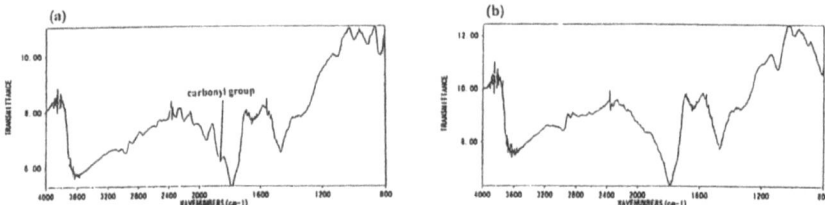

Figure 5. Infrared spectra of the urushi surface in Taejon (a) in the spot, (b) outside the spot.

4. Conclusions

The gloss losses on the urushi artifacts may occur in the first stage of degradation in that numerous fine spots appear on the surfaces. Carbonyl group is produced in the spots on the urushi surfaces exposed to indoor air at the sites where heavy air pollution and dense fog are observed.

Acknowledgements

This work was supported by the Global Environment Research Fund of the Japan Environmental Agency and the Japan Fund for Global Environment of Japan Environment Corporation.

References

Yamauchi, Y., Oshima, R. and Kumanotani, J.: 1982, *Journal of Chromatography* **243**, 71.
Kumanotani, J.: 1991, *Toso Kogaku* **26**, 251.
Toishi, K. and Kenjyo, T.: 1967, *Journal of Shikizai Society* **40**, 92.
Peng, Z., Zhang, Y and Chen, J.: 1992, *International Symposium on the Air Pollution Control Policy and Strategy, Chongqing '92*, Chongqing, China, pp.181-187.
Takemae, M. and Esaka, S.: 1993, *Annual report of Kyoto prefectural Institute of Hygienic and Environmental Sciences* **38**, 31.

If you have any concerns about our products,
you can contact us on
ProductSafety@springernature.com

In case Publisher is established outside the EU,
the EU authorized representative is:
**Springer Nature Customer Service Center GmbH
Europaplatz 3, 69115 Heidelberg, Germany**

Printed by Libri Plureos GmbH
in Hamburg, Germany